# Metals & Alloys in the UNIFIED NUMBERING SYSTEM

## 9th Edition

SAE HS-1086 OCT01
ASTM DS-56H

A Joint Publication of the
Society of Automotive Engineers, Inc.
American Society for Testing and Materials

Copublishers:

Society of Automotive Engineers, Inc.
400 Commonwealth Drive
Warrendale, PA 15096-0001, USA
Phone: (724) 776-4841
Fax: (724) 776-5760
E-mail: publications@sae.org
Website: www.sae.org
ISBN 0-7680-0879-4
SAE Order Number: HS-1086 OCT01

ASTM
100 Barr Harbor Drive
PO Box C700
West Conshohocken, PA 19428-2959, USA
Phone: (610) 832-9585
Fax: (610) 832-9555
E-mail: service@astm.org
Website: www.astm.org
ASTM Stock Number: DS-56H

# Preface to the Ninth Edition

The Ninth Edition contains more than 5000 designations in an international identification system for metals and alloys. Since the Eighth Edition was published in 1999, 150 new UNS designations have been added. This includes the initial use of the D-series for steels specified by their mechanical properties rather than chemical composition. More than 6000 changes have been made in a major updating of representative specification cross-references and specification indices.

The Index of Common Trade Names has been expanded with more than 3800 additions.

In total more than 10,000 changes have been incorporated into the Ninth Edition of *Metals & Alloys in the Unified Numbering System.*

A UNS designation is not a specification because it establishes no requirements for delivery conditions such as mechanical property requirements, heat analysis tolerances, heat treatment, packaging, marking, inspection, etc. UNS designations are identifying numbers for metal and alloys for which controlling requirements are established and published by technical societies and specifying organizations such as Aluminum Association, American Society of Mechanical Engineers (ASME), American Society for Testing and Materials (ASTM), SAE International, NACE International, American Welding Society (AWS), Copper Development Association (CDA) and ISO.

The Unified Numbering System operates though the volunteer efforts of a group of experienced, highly qualified metallurgists responsible for each of the various series of metals and alloys. Without their dedication and contributions, this book and the Unified Numbering System could not exist.

**James D. Redmond, Chairman**
**UNS Advisory Board**

# Table of Contents

# Introduction to the Unified Numbering System

The Unified Numbering System for Metals and Alloys (UNS) provides a means of correlating many internationally used metal and alloy numbering systems currently administered by societies, trade associations, and those individual users and producers of metals and alloys. This system avoids the confusion caused by the use of more than one identification number for the same metal or alloy, and the opposite situation of having the same number assigned to two or more different metals or alloys.

It provides the uniformity necessary for efficient indexing, record keeping, data storage and retrieval, and cross-referencing.

**A UNS designation is not, in itself, a specification, because it establishes no requirements for form, condition, property, or quality. It is a unified identifier of a metal or an alloy for which controlling limits have been established in specifications published elsewhere.**

The UNS establishes 18 series of designations for metals and alloys. Each UNS designation consists of a single-letter prefix followed by five digits. In most cases the letter is suggestive of the family of metals identified: for example, A for aluminum, P for precious metals, S for stainless steels.

Although some of the digits in certain UNS designation groups have special assigned meanings, each series of UNS designations is independent of the others in regard to the significance of digits, thus permitting greater flexibility and avoiding complicated and lengthy UNS designations.

Wherever feasible, and for the convenience of the user, identification "numbers" from existing systems are incorporated into the UNS designations. For example, carbon steel presently identified by the American Iron and Steel Institute as "AISI 1020" is covered by the UNS designation "G10200."

The UNS designation assignments for certain metals and alloys are established by the relevant trade associations which in the past have administered their own numbering systems; for other metals and alloys, UNS designation assignments are administered by the Society of Automotive Engineers (SAE). Each of these assigners has the responsibility for administering a specific UNS series of designations. Each considers requests for the assignment of new UNS designations, and informs the applicants of the action taken. UNS designation assigners report immediately to the office of the Unified Numbering System for Metals and Alloys the details of each new assignment for inclusion into the system.

For additional details on the UNS System, see the combined ASTM E527/SAE J1086, "Recommended Practice for Numbering Metals and Alloys," included in the Appendix.

**The listed cross-referenced specifications are representative only and are not necessarily a complete list of specifications applicable to a particular UNS designation.**

# Index to UNS Designations by Base Elements

| Element | Symbol | UNS Designation |
|---|---|---|
| Actinium | Ac | E00000 - E00999 |
| Aluminum | Al | A00001 - A99999 |
| Antimony (Stibium) | Sb | M00001 - M00999 |
| Argentum (Silver) | Ag | P07001 - P07999 |
| Arsenic | As | M01001 - M01999 |
| Aurum (Gold) | Au | P00001 - P00999 |
| Barium | Ba | M02001 - M02999 |
| Beryllium (Glucinum) | Be | R10001 - R19999 |
| Bismuth | Bi | L00001 - L00999 |
| Boron | B | R01001 - R01999 |
| Cadmium | Cd | L01001 - L01999 |
| Calcium | Ca | M03001 - M03999 |
| Cassiopeium (Lutetium) | Lu | E68000 - E68999 |
| Cerium | Ce | E01000 - E20999 |
| Cesium | Cs | L02001 - L02999 |
| Chromium | Cr | R20001 - R29999 |
| Cobalt | Co | R30001 - R39999 |
| Columbium (Niobium) | Cb | R04001 - R04999 |
| Copper (Cuprum) | Cu | C00001 - C99999 |
| Cuprum (Copper) | Cu | C00001 - C99999 |
| Dysprasium | Dy | E46000 - E47999 |
| Erbium | Er | E48000 - E49999 |
| Europium | Eu | E50000 - E51999 |
| Ferrum, Cast (Iron, Cast) | Fe | F00001 - F99999 |
| Gadolinium | Gd | E52000 - E55999 |
| Gallium | Ga | L03001 - L03999 |
| Germanium | Ge | M04001 - M04999 |
| Glucinum (Beryllium) | Be | R10001 - R19999 |
| Gold (Aurum) | Au | P00001 - P00999 |
| Hafnium | Hf | R02001 - R02999 |
| Holmium | Ho | E56000 - E57999 |
| Hydrargyrum (Mercury) | Hg | L07001 - L07999 |
| Indium | In | L04001 - L04999 |
| Iridium | Ir | P01001 - P01999 |
| Iron, Cast (Ferrum, Cast) | Fe | F00001 - F99999 |
| Kalium (Potassium) | K | L08001 - L08999 |
| Lanthanum | La | E58000 - E67999 |
| Lead (Plumbum) | Pb | L50001 - L59999 |
| Lithium | Li | L06001 - L06999 |
| Lutetium (Cassiopeium) | Lu | E68000 - E68999 |
| Magnesium | Mg | M10001 - M19999 |
| Manganese | Mn | M20001 - M29999 |
| Mercury (Hydrargyrum) | Hg | L07001 - L07999 |
| Mixed rare earths | | E21000 - E45999 |
| Molybdenum | Mo | R03001 - R03999 |
| Natrium (Sodium) | Na | L11001 - L11999 |
| Neodymium | Nd | E69000 - E73999 |
| Niobium (Columbium) | Nb | R04001 - R04999 |
| Osmium | Os | P02001 - P02999 |
| Palladium | Pd | P03001 - P03999 |
| Platinum | Pt | P04001 - P04999 |
| Plumbum (Lead) | Pb | L50001 - L59999 |
| Plutonium | Pu | M05001 - M05999 |
| Potassium (Kalium) | K | L08001 - L08999 |
| Praesodymium | Pr | E74000 - E77999 |
| Promethium | Pm | E78000 - E78999 |
| Rhenium | Re | R40001 - R49999 |
| Rhodium | Rh | P05001 - P05999 |
| Rubidium | Rb | L09001 - L09999 |
| Ruthenium | Ru | P06001 - P06999 |
| Samarium | Sm, Sa | E79000 - E82999 |
| Scandium | Sc | E83000 - E84999 |
| Selenium | Se | L10001 - L10999 |
| Silicon | Si | M30001 - M39999 |
| Silver (Argentum) | Ag | P07001 - P07999 |
| Sodium (Natrium) | Na | L11001 - L11999 |
| Stannum (Tin) | Sn | L13001 - L13999 |
| Steels - AISI H | | H00001 - H99999 |
| Steels - Cast | | J00001 - J99999 |
| Steels - Miscellaneous | | K00001 - K99999 |
| Steels - SAE/AISI Carbon and Low Alloy Steels | | G00001 - G99999 |
| Steels - Stainless | | S00001 - S59999 |
| Steels - Tool | | T00001 - T99999 |
| Steels - Valve Steels and High Temperature Alloys | | S60001 - S69999 |
| Stibium (Antimony) | Sb | M00001 - M00999 |
| Strontium | Sr | M06001 - M06999 |
| Tantalum | Ta | R05001 - R05999 |
| Tellurium | Te | M07001 - M07999 |
| Terbium | Tb | E85000 - E86999 |
| Thallium | Tl | L12001 - L12999 |
| Thorium | Th | R06001 - R06999 |
| Thulium | Tm | E87000 - E87999 |
| Tin (Stannum) | Sn | L13001 - L13999 |
| Titanium | Ti | R50001 - R59999 |
| Tungsten (Wolfram) | W | R07001 - R07999 |
| Uranium | U | M08001 - M08999 |
| Vanadium | V | R08001 - R08999 |
| Weld, Filler - Austenitic Stainless Steels | | W30000 - W39999 |
| Weld, Filler - Carbon Steels | | W00000 - W09999 |
| Weld, Filler Metal - Chromium Low Alloy Steels | | W50000 - W59999 |
| Weld, Filler - Copper Alloys | | W60000 - W69999 |
| Weld, Filler - Ferritic Stainless Steels | | W40000 - W49999 |
| Weld, Filler - Manganese-Molybdenum Alloys | | W10000 - W19999 |
| Weld, Filler - Ni Alloys | | W80000 - W89999 |
| Weld, Filler - Ni Steels | | W20000 - W29999 |
| Weld, Surfacing Alloys | | W70000 - W79999 |
| Wolfram (Tungsten) | W | R07001 - R07999 |
| Ytterbium | Yb | E88000 - E89999 |
| Yttrium | Y | E90000 - E99999 |
| Zinc | Zn | Z00001 - Z99999 |
| Zirconium | Zr | R60001 - R69999 |

# UNS NUMBERS ASSIGNED TO DATE

## With Description of Each Material Covered and References To Documents In Which The Same or Similar Materials are Described

## Axxxxx Number Series
## Aluminum and Aluminum Alloys

# ALUMINUM AND ALUMINUM ALLOYS

| UNIFIED NUMBER | DESCRIPTION | CHEMICAL COMPOSITION | CROSS REFERENCE SPECIFICATIONS |
|---|---|---|---|
| **A01001** | Aluminum Foundry Alloy, Ingot | **Al** 99.00 nom **Cu** 0.10 max **Fe** 0.6-0.8 **Si** 0.15 max **Zn** 0.05 max **Other** each 0.03 max, total 0.10 max, Mn+Cr+Ti+V 0.025 max | **AA** 100.1 **ASTM** B179 (100.1) |
| **A01301** | Aluminum Foundry Alloy, Ingot | **Al** 99.30 min **Cu** 0.10 max **Zn** 0.05 max **Other** each 0.03 max, total 0.10 max, Fe/Si ratio 2.5 min, Mn+Cr+Ti+V 0.025 max | **AA** 130.1 **ASTM** B179 (130.1) |
| **A01501** | Aluminum Foundry Alloy, Ingot | **Al** 99.50 min **Cu** 0.05 max **Zn** 0.05 max **Other** each 0.03 max, total 0.10 max, Fe/Si ratio 2.0 min, Mn+Cr+Ti+V 0.025 max | **AA** 150.1 **ASTM** B179 (150.1) |
| **A01601** | Aluminum Foundry Alloy, Ingot | **Al** 99.60 min **Fe** 0.25 max **Si** 0.10 max **Zn** 0.05 max **Other** each 0.03 max, total 0.10 max, Fe/Si ratio 2.0 min, Mn+Cr+Ti+V 0.025 max | **AA** 160.1 |
| **A01701** | Aluminum Foundry Alloy, Ingot | **Al** 99.70 min **Zn** 0.05 max **Other** each 0.03 max, total 0.10 max, Fe/Si ratio 1.5 min, Mn+Cr+Ti+V 0.025 max | **AA** 170.1 **ASTM** B179 (170.1) |
| **A02010** | Aluminum Foundry Alloy, Casting | **Ag** 0.40-1.0 **Al** rem **Cu** 4.0-5.2 **Fe** 0.15 max **Mg** 0.15-0.55 **Mn** 0.20-0.50 **Si** 0.10 max **Ti** 0.15-0.35 **Other** each 0.05 max, total 0.10 max | **AA** 201.0 **AMS** 4233; 4242 **ASTM** B26 (201.0); B618 (201.0) **SAE** J452 (382) |
| **A02012** | Aluminum Foundry Alloy, Ingot | **Ag** 0.40-1.0 **Al** rem **Cu** 4.0-5.2 **Fe** 0.10 max **Mg** 0.20-0.55 **Mn** 0.20-0.50 **Si** 0.10 max **Ti** 0.15-0.35 **Other** each 0.05 max, total 0.10 max | **AA** 201.2 **ASTM** B179 (201.2) **SAE** J452 (382) |
| * **A02020** | Aluminum Foundry Alloy, Casting | **Ag** 0.40-1.0 **Al** rem **Cr** 0.20-0.6 **Cu** 4.0-5.2 **Fe** 0.15 max **Mg** 0.15-0.55 **Mn** 0.20-0.8 **Si** 0.10 max **Ti** 0.15-0.35 **Other** each 0.05 max, total 0.10 max | **AA** 202.0 |
| * **A02022** | Aluminum Foundry Alloy, Ingot | **Ag** 0.40-1.0 **Al** rem **Cr** 0.20-0.6 **Cu** 4.0-5.2 **Fe** 0.10 max **Mg** 0.20-0.55 **Mn** 0.20-0.8 **Si** 0.10 max **Ti** 0.15-0.35 **Other** each 0.05 max, total 0.10 max | **AA** 202.2 |
| **A02030** | Aluminum Foundry Alloy, Casting | **Al** rem **Co** 0.20-0.30 **Cu** 4.5-5.5 **Fe** 0.50 max **Mg** 0.10 max **Mn** 0.20-0.30 **Ni** 1.3-1.7 **Sb** 0.20-0.30 **Si** 0.30 max **Ti** 0.15-0.25 **Zn** 0.10 max **Zr** 0.10-0.30 **Other** each 0.05 max, total 0.20 max, plus Ti+Zr 0.50 max | **AA** 203.0 **AMS** 4225 |
| **A02032** | Aluminum Foundry Alloy, Ingot | **Al** rem **Co** 0.20-0.30 **Cu** 4.8-5.2 **Fe** 0.35 max **Mg** 0.10 max **Mn** 0.20-0.30 **Ni** 1.3-1.7 **Sb** 0.20-0.30 **Si** 0.20 max **Ti** 0.15-0.25 **Zn** 0.10 max **Zr** 0.10-0.30 **Other** each 0.05 max, total 0.20 max, plus Ti+Zr 0.50 max | **AA** 203.2 |
| **A02040** | Aluminum Foundry Alloy, Casting | **Al** rem **Cu** 4.2-5.0 **Fe** 0.35 max **Mg** 0.15-0.35 **Mn** 0.10 max **Ni** 0.05 max **Si** 0.20 max **Sn** 0.05 max **Ti** 0.15-0.30 **Zn** 0.10 max **Other** each 0.05 max, total 0.15 max | **AA** 204.0 **ASTM** B26 (204.0); B108 (204.0); B618 (204.0) |
| **A02042** | Aluminum Foundry Alloy, Ingot | **Al** rem **Cu** 4.2-4.9 **Fe** 0.10-0.20 **Mg** 0.20-0.35 **Mn** 0.05 max **Ni** 0.03 max **Si** 0.15 max **Sn** 0.05 max **Ti** 0.15-0.25 **Zn** 0.05 max **Other** each 0.05 max, total 0.15 max | **AA** 204.2 **ASTM** B179 (204.2) |

*Boxed entries are no longer active and are retained for reference purposes only.

The chemical compositions listed are for identification purposes and should not be used in lieu of the cross referenced specifications.

| UNIFIED NUMBER | DESCRIPTION | CHEMICAL COMPOSITION | CROSS REFERENCE SPECIFICATIONS |
|---|---|---|---|
| **A02060** | Aluminum Foundry Alloy, Casting | **Al** rem **Cu** 4.2-5.0 **Fe** 0.15 max **Mg** 0.15-0.35 **Mn** 0.20-0.50 **Ni** 0.05 max **Si** 0.10 max **Sn** 0.05 max **Ti** 0.15-0.30 **Zn** 0.10 max **Other** each 0.05 max, total 0.15 max | **AA** 206.0 **AMS** 4237; 4244 **ASME** SFA5.10 (R206.0) **AWS** A5.10 (R206.0) |
| **A02062** | Aluminum Foundry Alloy, Ingot | **Al** rem **Cu** 4.2-5.0 **Fe** 0.10 max **Mg** 0.20-0.35 **Mn** 0.20-0.50 **Ni** 0.03 max **Si** 0.10 max **Sn** 0.05 max **Ti** 0.15-0.25 **Zn** 0.05 max **Other** each 0.05 max, total 0.15 max | **AA** 206.2 |
| * **A02080** | Aluminum Foundry Alloy, Casting | **Al** rem **Cu** 3.5-4.5 **Fe** 1.2 max **Mg** 0.10 max **Mn** 0.50 max **Ni** 0.35 max **Si** 2.5-3.5 **Ti** 0.25 max **Zn** 1.0 max **Other** total 0.50 max | **AA** 208.0 **ASTM** B26 (208.0); B108 (208.0); B618 (208.0) **MIL SPEC** MIL-F-3922; MIL-F-39000 **SAE** J452 |
| * **A02081** | Aluminum Foundry Alloy, Ingot | **Al** rem **Cu** 3.5-4.5 **Fe** 0.9 max **Mg** 0.10 max **Mn** 0.50 max **Ni** 0.35 max **Si** 2.5-3.5 **Ti** 0.25 max **Zn** 1.0 max **Other** total 0.50 max | **AA** 208.1 **ASTM** B179 (208.1) |
| * **A02082** | Aluminum Foundry Alloy, Ingot | **Al** rem **Cu** 3.5-4.5 **Fe** 0.8 max **Mg** 0.03 max **Mn** 0.30 max **Si** 2.5-3.5 **Ti** 0.20 max **Zn** 0.20 max **Other** total 0.30 max | **AA** 208.2 **ASTM** B179 (208.2) |
| * **A02130** | Aluminum Foundry Alloy, Casting | **Al** rem **Cu** 6.0-8.0 **Fe** 1.2 max **Mg** 0.10 max **Mn** 0.6 max **Ni** 0.35 max **Si** 1.0-3.0 **Ti** 0.25 max **Zn** 2.5 max **Other** total 0.50 max | **AA** 213.0 **ASTM** B108 (213.0) |
| * **A02131** | Aluminum Foundry Alloy, Ingot | **Al** rem **Cu** 6.0-8.0 **Fe** 0.9 max **Mg** 0.10 max **Mn** 0.6 max **Ni** 0.35 max **Si** 1.0-3.0 **Ti** 0.25 max **Zn** 2.5 max **Other** total 0.50 max | **AA** 213.1 |
| * **A02220** | Aluminum Foundry Alloy, Casting | **Al** rem **Cu** 9.2-10.7 **Fe** 1.5 max **Mg** 0.15-0.35 **Mn** 0.50 max **Ni** 0.50 max **Si** 2.0 max **Ti** 0.25 max **Zn** 0.8 max **Other** total 0.35 max | **AA** 222.0 **ASTM** B26 (222.0); B108 (222.0); B618 (222.0) **SAE** J452 (34) |
| * **A02221** | Aluminum Foundry Alloy, Ingot | **Al** rem **Cu** 9.2-10.7 **Fe** 1.2 max **Mg** 0.20-0.35 **Mn** 0.50 max **Ni** 0.50 max **Si** 2.0 max **Ti** 0.25 max **Zn** 0.8 max **Other** total 0.35 max | **AA** 222.1 **ASTM** B179 (222.1) **SAE** J452 (34) |
| * **A02240** | Aluminum Foundry Alloy, Casting | **Al** rem **Cu** 4.5-5.5 **Fe** 0.10 max **Mn** 0.20-0.50 **Si** 0.06 max **Ti** 0.35 max **V** 0.05-0.15 **Zr** 0.10-0.25 **Other** each 0.03 max, total 0.10 max | **AA** 224.0 **AMS** 4226 |
| * **A02242** | Aluminum Foundry Alloy, Ingot | **Al** rem **Cu** 4.5-5.5 **Fe** 0.04 max **Mn** 0.20-0.50 **Si** 0.02 max **Ti** 0.25 max **V** 0.05-0.15 **Zr** 0.10-0.25 **Other** each 0.03 max, total 0.10 max | **AA** 224.2 |

*Boxed entries are no longer active and are retained for reference purposes only.

The chemical compositions listed are for identification purposes and should not be used in lieu of the cross referenced specifications.

| UNIFIED NUMBER | DESCRIPTION | CHEMICAL COMPOSITION | CROSS REFERENCE SPECIFICATIONS |
|---|---|---|---|
| * **A02380** | Aluminum Foundry Alloy, Casting | **Al** rem **Cu** 9.0-11.0 **Fe** 1.5 max **Mg** 0.15-0.35 **Mn** 0.6 max **Ni** 1.0 max **Si** 3.5-4.5 **Ti** 0.25 max **Zn** 1.5 max **Other** total 0.50 max | **AA** 238.0 |
| * **A02381** | Aluminum Foundry Alloy, Ingot | **Al** rem **Cu** 9.0-11.0 **Fe** 1.2 max **Mg** 0.20-0.35 **Mn** 0.6 max **Ni** 1.0 max **Si** 3.5-4.5 **Ti** 0.25 max **Zn** 1.5 max **Other** total 0.50 max | **AA** 238.1 |
| * **A02382** | Aluminum Foundry Alloy, Ingot | **Al** rem **Cu** 9.5-10.5 **Fe** 1.2 max **Mg** 0.20-0.35 **Mn** 0.50 max **Ni** 0.50 max **Si** 3.5-4.5 **Ti** 0.20 max **Zn** 0.50 max **Other** total 0.50 max | **AA** 238.2 |
| **A02400** | Aluminum Foundry Alloy, Casting | **Al** rem **Cu** 7.0-9.0 **Fe** 0.50 max **Mg** 5.5-6.5 **Mn** 0.30-0.7 **Ni** 0.30-0.7 **Si** 0.50 max **Ti** 0.20 max **Zn** 0.10 max **Other** each 0.05 max, total 0.15 max | **AA** 240.0 |
| **A02401** | Aluminum Foundry Alloy, Ingot | **Al** rem **Cu** 7.0-9.0 **Fe** 0.40 max **Mg** 5.6-6.5 **Mn** 0.30-0.7 **Ni** 0.30-0.7 **Si** 0.50 max **Ti** 0.20 max **Zn** 0.10 max **Other** each 0.05 max, total 0.15 max | **AA** 240.1 |
| **A02420** | Aluminum Foundry Alloy, Casting | **Al** rem **Cr** 0.25 max **Cu** 3.5-4.5 **Fe** 1.0 max **Mg** 1.2-1.8 **Mn** 0.35 max **Ni** 1.7-2.3 **Si** 0.7 max **Ti** 0.25 max **Zn** 0.35 max **Other** each 0.05 max, total 0.15 max | **AA** 242.0 **ASTM** B26 (242.0); B108 (242.0); B618 (242.0) **SAE** J452 (39) |
| **A02421** | Aluminum Foundry Alloy, Ingot | **Al** rem **Cr** 0.25 max **Cu** 3.5-4.5 **Fe** 0.8 max **Mg** 1.3-1.8 **Mn** 0.35 max **Ni** 1.7-2.3 **Si** 0.7 max **Ti** 0.25 max **Zn** 0.35 max **Other** each 0.05 max, total 0.15 max | **AA** 242.1 **ASTM** B179 (242.1) |
| **A02422** | Aluminum Foundry Alloy, Ingot | **Al** rem **Cu** 3.5-4.5 **Fe** 0.6 max **Mg** 1.3-1.8 **Mn** 0.10 max **Ni** 1.7-2.3 **Si** 0.6 max **Ti** 0.20 max **Zn** 0.10 max **Other** each 0.05 max, total 0.15 max | **AA** 242.2 **ASTM** B179 (242.2) |
| * **A02430** | Aluminum Foundry Alloy, Casting | **Al** rem **Cr** 0.20-0.40 **Cu** 3.5-4.5 **Fe** 0.40 max **Mg** 1.8-2.3 **Mn** 0.15-0.45 **Ni** 1.9-2.3 **Si** 0.35 max **Ti** 0.06-0.20 **V** 0.06-0.20 **Zn** 0.05 max **Other** each 0.05 max, total 0.15 max | **AA** 243.0 **AMS** 4224 |
| * **A02431** | Aluminum Foundry Alloy, Ingot | **Al** rem **Cr** 0.20-0.40 **Cu** 3.5-4.5 **Fe** 0.30 max **Mg** 1.9-2.3 **Mn** 0.15-0.45 **Ni** 1.9-2.3 **Si** 0.35 max **Ti** 0.06-0.20 **Zn** 0.05 max **Other** each 0.05 max, total 0.15 max | **AA** 243.1 |
| * **A02490** | Aluminum Foundry Alloy, Casting | **Al** rem **Cu** 3.8-4.6 **Fe** 0.10 max **Mg** 0.25-0.50 **Mn** 0.25-0.50 **Si** 0.05 max **Ti** 0.02-0.35 **Zn** 2.5-3.5 **Other** each 0.03 max, total 0.10 max | **AA** 249.0 |

*Boxed entries are no longer active and are retained for reference purposes only.

The chemical compositions listed are for identification purposes and should not be used in lieu of the cross referenced specifications.

| UNIFIED NUMBER | DESCRIPTION | CHEMICAL COMPOSITION | CROSS REFERENCE SPECIFICATIONS |
|---|---|---|---|
| * **A02492** | Aluminum Foundry Alloy, Ingot | **Al** rem **Cu** 3.8-4.6 **Fe** 0.07 max **Mg** 0.30-0.50 **Mn** 0.25-0.50 **Si** 0.05 max **Ti** 0.02-0.12 **Zn** 2.5-3.5 **Other** each 0.03 max, total 0.10 max | **AA** 249.2 |
| **A02950** | Aluminum Foundry Alloy, Casting | **Al** rem **Cu** 4.0-5.0 **Fe** 1.0 max **Mg** 0.03 max **Mn** 0.35 max **Si** 0.7-1.5 **Ti** 0.25 max **Zn** 0.35 max **Other** each 0.05 max, total 0.15 max | **AA** 295.0 **ASTM** B26 (295.0); B618 (295.0) **MIL SPEC** MIL-F-3922 **SAE** J452 (38) |
| **A02951** | Aluminum Foundry Alloy, Ingot | **Al** rem **Cu** 4.0-5.0 **Fe** 0.8 max **Mg** 0.03 max **Mn** 0.35 max **Si** 0.7-1.5 **Ti** 0.25 max **Zn** 0.35 max **Other** each 0.05 max, total 0.15 max | **AA** 295.1 **ASTM** B179 (295.1) |
| **A02952** | Aluminum Foundry Alloy, Ingot | **Al** rem **Cu** 4.0-5.0 **Fe** 0.8 max **Mg** 0.03 max **Mn** 0.30 max **Si** 0.7-1.2 **Ti** 0.20 max **Zn** 0.30 max **Other** each 0.05 max, total 0.15 max | **AA** 295.2 **ASTM** B179 (295.2) |
| **A02960** | Aluminum Foundry Alloy, Casting | **Al** rem **Cu** 4.0-5.0 **Fe** 1.2 max **Mg** 0.05 max **Mn** 0.35 max **Ni** 0.35 max **Si** 2.0-3.0 **Ti** 0.25 max **Zn** 0.50 max **Other** total 0.35 max | **AA** 296.0 **ASTM** B108 (296.0) **SAE** J 452; J452 (380) |
| **A02961** | Aluminum Foundry Alloy, Ingot | **Al** rem **Cu** 4.0-5.0 **Fe** 0.9 max **Mg** 0.05 max **Mn** 0.35 max **Ni** 0.35 max **Si** 2.0-3.0 **Ti** 0.25 max **Zn** 0.50 max **Other** total 0.35 max | **AA** 296.1 **ASTM** B179 (296.1) |
| **A02962** | Aluminum Foundry Alloy, Ingot | **Al** rem **Cu** 4.0-5.0 **Fe** 0.8 max **Mg** 0.03 max **Mn** 0.30 max **Si** 2.0-3.0 **Ti** 0.20 max **Zn** 0.30 max **Other** each 0.05 max, total 0.15 max | **AA** 296.2 **ASTM** B179 (296.2) |
| **A03010** | Aluminum Foundry Alloy, Casting | **Al** rem **Cu** 3.0-3.5 **Fe** 0.8-1.5 **Mg** 0.25-0.50 **Mn** 0.50-0.8 **Ni** 1.0-1.5 **Si** 9.5-10.5 **Ti** 0.20 max **Zn** 0.05 max **Other** each 0.03 max, total 0.10 max | **AA** 301.0 |
| **A03011** | Aluminum Foundry Alloy, Ingot | **Al** rem **Cu** 3.0-3.5 **Fe** 0.8-1.2 **Mg** 0.30-0.50 **Mn** 0.50-0.8 **Ni** 1.0-1.5 **Si** 9.5-10.5 **Ti** 0.20 max **Zn** 0.05 max **Other** each 0.03 max, total 0.10 max | **AA** 301.1 |
| **A03020** | Aluminum Foundry Alloy, Casting | **Al** rem **Cu** 2.8-3.2 **Fe** 0.25 max **Mg** 0.7-1.2 **Ni** 1.0-1.5 **Si** 9.5-10.5 **Ti** 0.20 max **Zn** 0.05 max **Other** each 0.03 max, total 0.10 max | **AA** 302.0 |
| **A03021** | Aluminum Foundry Alloy, Ingot | **Al** rem **Cu** 2.8-3.2 **Fe** 0.20 max **Mg** 0.8-1.2 **Ni** 1.0-1.5 **Si** 9.5-10.5 **Ti** 0.20 max **Zn** 0.05 max **Other** each 0.03 max, total 0.10 max | **AA** 302.1 |
| **A03030** | Aluminum Foundry Alloy, Casting | **Al** rem **Cu** 0.20 max **Fe** 0.8-1.5 **Mg** 0.45-0.7 **Mn** 0.50-0.8 **Si** 9.5-10.5 **Ti** 0.20 max **Zn** 0.05 max **Other** each 0.03 max, total 0.10 max | **AA** 303.0 |
| **A03031** | Aluminum Foundry Alloy, Ingot | **Al** rem **Cu** 0.20 max **Fe** 0.8-1.2 **Mg** 0.50-0.7 **Mn** 0.50-0.8 **Si** 9.5-10.5 **Ti** 0.20 max **Zn** 0.05 max **Other** each 0.03 max, total 0.10 max | **AA** 303.1 |
| * **A03050** | Aluminum Foundry Alloy, Casting | **Al** rem **Cr** 0.25 max **Cu** 1.0-1.5 **Fe** 0.6 max **Mg** 0.10 max **Mn** 0.50 max **Si** 4.5-5.5 **Ti** 0.25 max **Zn** 0.35 max **Other** each 0.05 max, total 0.15 max | **AA** 305.0 |

*Boxed entries are no longer active and are retained for reference purposes only.

The chemical compositions listed are for identification purposes and should not be used in lieu of the cross referenced specifications.

| UNIFIED NUMBER | DESCRIPTION | CHEMICAL COMPOSITION | CROSS REFERENCE SPECIFICATIONS |
|---|---|---|---|
| * **A03052** | Aluminum Foundry Alloy, Ingot | **Al** rem **Cu** 1.0-1.5 **Fe** 0.14-0.25 **Mn** 0.05 max **Si** 4.5-5.5 **Ti** 0.20 max **Zn** 0.05 max **Other** each 0.05 max, total 0.15 max | **AA** 305.2 |
| **A03080** | Aluminum Foundry Alloy, Casting | **Al** rem **Cu** 4.0-5.0 **Fe** 1.0 max **Mg** 0.10 max **Mn** 0.50 max **Si** 5.0-6.0 **Ti** 0.25 max **Zn** 1.0 max **Other** total 0.50 max | **AA** 308.0 **ASTM** B108 (308.0) |
| **A03081** | Aluminum Foundry Alloy, Ingot | **Al** rem **Cu** 4.0-5.0 **Fe** 0.8 max **Mg** 0.10 max **Mn** 0.50 max **Si** 5.0-6.0 **Ti** 0.25 max **Zn** 1.0 max **Other** total 0.50 max | **AA** 308.1 **ASTM** B179 (308.1) |
| **A03082** | Aluminum Foundry Alloy, Ingot | **Al** rem **Cu** 4.0-5.0 **Fe** 0.8 max **Mg** 0.10 max **Mn** 0.30 max **Si** 5.0-6.0 **Ti** 0.20 max **Zn** 0.50 max **Other** total 0.50 max | **AA** 308.2 **ASTM** B179 (308.2) |
| **A03180** | Aluminum Foundry Alloy, Die Casting | **Al** rem **Cu** 3.0-4.0 **Fe** 1.0 max **Mg** 0.10-0.6 **Mn** 0.50 max **Ni** 0.35 max **Si** 5.5-6.5 **Ti** 0.25 max **Zn** 1.0 max **Other** 0.50 total | **AA** 318.0 |
| **A03181** | Aluminum Foundry Alloy, Ingot | **Al** rem **Cu** 3.0-4.0 **Fe** 0.8 max **Mg** 0.15-0.6 **Mn** 0.50 max **Ni** 0.35 max **Si** 5.5-6.5 **Ti** 0.25 max **Zn** 0.9 max **Other** 0.50 total | **AA** 318.1 |
| **A03190** | Aluminum Foundry Alloy, Casting | **Al** rem **Cu** 3.0-4.0 **Fe** 1.0 max **Mg** 0.10 max **Mn** 0.50 max **Ni** 0.35 max **Si** 5.5-6.5 **Ti** 0.25 max **Zn** 1.0 max **Other** total 0.50 max | **AA** 319.0 **ASTM** B26 (319.0); B108 (319.0); B618 (319.0) **SAE** J452 (326) |
| **A03191** | Aluminum Foundry Alloy, Ingot | **Al** rem **Cu** 3.0-4.0 **Fe** 0.8 max **Mg** 0.10 max **Mn** 0.50 max **Ni** 0.35 max **Si** 5.5-6.5 **Ti** 0.25 max **Zn** 1.0 max **Other** total 0.50 max | **AA** 319.1 **ASTM** B179 (319.1) |
| **A03192** | Aluminum Foundry Alloy, Ingot | **Al** rem **Cu** 3.0-4.0 **Fe** 0.6 max **Mg** 0.10 max **Mn** 0.10 max **Ni** 0.10 max **Si** 5.5-6.5 **Ti** 0.20 max **Zn** 0.10 max **Other** total 0.20 max | **AA** 319.2 **ASTM** B179 (319.2) |
| **A03200** | Aluminum Foundry Alloy, Casting | **Al** rem **Cu** 2.0-4.0 **Fe** 1.2 max **Mg** 0.05-0.6 **Mn** 0.8 max **Ni** 0.35 max **Si** 5.0-8.0 **Ti** 0.25 max **Zn** 3.0 max **Other** total 0.50 max | **AA** 320.0 |
| **A03201** | Aluminum Foundry Alloy, Ingot | **Al** rem **Cu** 2.0-4.0 **Fe** 0.9 max **Mg** 0.10-0.6 **Mn** 0.8 max **Ni** 0.35 max **Si** 5.0-8.0 **Ti** 0.25 max **Zn** 3.0 max **Other** total 0.50 max | **AA** 320.1 |
| * **A03240** | Aluminum Foundry Alloy, Casting | **Al** rem **Cu** 0.40-0.6 **Fe** 1.2 max **Mg** 0.40-0.7 **Mn** 0.50 max **Ni** 0.30 max **Si** 7.0-8.0 **Ti** 0.20 max **Zn** 1.0 max **Other** each 0.15 max, total 0.20 max | **AA** 324.0 |
| * **A03241** | Aluminum Foundry Alloy, Ingot | **Al** rem **Cu** 0.40-0.6 **Fe** 0.9 max **Mg** 0.45-0.7 **Mn** 0.50 max **Ni** 0.30 max **Si** 7.0-8.0 **Ti** 0.20 max **Zn** 1.0 max **Other** each 0.15 max, total 0.20 max | **AA** 324.1 |

*Boxed entries are no longer active and are retained for reference purposes only.

The chemical compositions listed are for identification purposes and should not be used in lieu of the cross referenced specifications.

| UNIFIED NUMBER | DESCRIPTION | CHEMICAL COMPOSITION | CROSS REFERENCE SPECIFICATIONS |
|---|---|---|---|
| * **A03242** | Aluminum Foundry Alloy, Ingot | **Al** rem **Cu** 0.40-0.6 **Fe** 0.6 max **Mg** 0.45-0.7 **Mn** 0.10 max **Ni** 0.10 max **Si** 7.0-8.0 **Ti** 0.20 max **Zn** 0.10 max **Other** each 0.05 max, total 0.15 max | **AA** 324.2 |
| * **A03280** | Aluminum Foundry Alloy, Casting | **Al** rem **Cr** 0.35 max **Cu** 1.0-2.0 **Fe** 1.0 max **Mg** 0.20-0.6 **Mn** 0.20-0.6 **Ni** 0.25 max **Si** 7.5-8.5 **Ti** 0.25 max **Zn** 1.5 max **Other** total 0.50 max | **AA** 328.0 **ASTM** B26 (328.0); B618 (328.0) **SAE** J452 (327) |
| * **A03281** | Aluminum Foundry Alloy, Ingot | **Al** rem **Cr** 0.35 max **Cu** 1.0-2.0 **Fe** 0.8 max **Mg** 0.25-0.6 **Mn** 0.20-0.6 **Ni** 0.25 max **Si** 7.5-8.5 **Ti** 0.25 max **Zn** 1.5 max **Other** total 0.50 max | **AA** 328.1 **ASTM** B179 (328.1) **SAE** J452 (327) |
| **A03320** | Aluminum Foundry Alloy, Casting | **Al** rem **Cu** 2.0-4.0 **Fe** 1.2 max **Mg** 0.50-1.5 **Mn** 0.50 max **Ni** 0.50 max **Si** 8.5-10.5 **Ti** 0.25 max **Zn** 1.0 max **Other** total 0.50 max | **AA** 332.0 **ASTM** B108 (332.0) **SAE** J452 (332) |
| **A03321** | Aluminum Foundry Alloy, Ingot | **Al** rem **Cu** 2.0-4.0 **Fe** 0.9 max **Mg** 0.6-1.5 **Mn** 0.50 max **Ni** 0.50 max **Si** 8.5-10.5 **Ti** 0.25 max **Zn** 1.0 max **Other** total 0.50 max | **AA** 332.1 **ASTM** B179 (332.1) |
| **A03322** | Aluminum Foundry Alloy, Ingot | **Al** rem **Cu** 2.0-4.0 **Fe** 0.6 max **Mg** 0.9-1.3 **Mn** 0.10 max **Ni** 0.10 max **Si** 8.5-10.0 **Ti** 0.20 max **Zn** 0.10 max **Other** total 0.30 max | **AA** 332.2 **ASTM** B179 (332.2) |
| **A03330** | Aluminum Foundry Alloy, Casting | **Al** rem **Cu** 3.0-4.0 **Fe** 1.0 max **Mg** 0.05-0.50 **Mn** 0.50 max **Ni** 0.50 max **Si** 8.0-10.0 **Ti** 0.25 max **Zn** 1.0 max **Other** total 0.50 max | **AA** 333.0 **ASTM** B108 (333.0) **SAE** J452 (331) |
| **A03331** | Aluminum Foundry Alloy, Ingot | **Al** rem **Cu** 3.0-4.0 **Fe** 0.8 max **Mg** 0.10-0.50 **Mn** 0.50 max **Ni** 0.50 max **Si** 8.0-10.0 **Ti** 0.25 max **Zn** 1.0 max **Other** total 0.50 max | **AA** 333.1 **ASTM** B179 (333.1) |
| **A03360** | Aluminum Foundry Alloy, Casting | **Al** rem **Cu** 0.50-1.5 **Fe** 1.2 max **Mg** 0.7-1.3 **Mn** 0.35 max **Ni** 2.0-3.0 **Si** 11.0-13.0 **Ti** 0.25 max **Zn** 0.35 max **Other** each 0.05 max | **AA** 336.0 **ASTM** B108 (336.0) **SAE** J452 (321) |
| **A03361** | Aluminum Foundry Alloy, Ingot | **Al** rem **Cu** 0.50-1.5 **Fe** 0.9 max **Mg** 0.8-1.3 **Mn** 0.35 max **Ni** 2.0-3.0 **Si** 11.0-13.0 **Ti** 0.25 max **Zn** 0.35 max **Other** each 0.05 max | **AA** 336.1 **ASTM** B179 (336.1) |
| **A03362** | Aluminum Foundry Alloy, Ingot | **Al** rem **Cu** 0.50-1.5 **Fe** 0.9 max **Mg** 0.9-1.3 **Mn** 0.10 max **N** 2.0-3.0 **Si** 11.0-13.0 **Ti** 0.20 max **Zn** 0.10 max **Other** each 0.05 max, total 0.15 max | **AA** 336.2 **ASTM** B179 (336.2) |
| * **A03370** | Aluminum Foundry Alloy, Casting | **Al** rem **Cu** 3.0-3.5 **Fe** 0.8-1.2 **Mg** 0.30-0.50 **Mn** 0.50-0.8 **Ni** 1.0-1.5 **Si** 9.5-10.5 **Ti** 0.20 max **Zn** 0.03 max **Other** each 0.03 max, total 0.10 max | **AA** 337.0 |
| * **A03371** | Aluminum Foundry Alloy, Ingot | **Al** rem **Cu** 3.0-3.5 **Fe** 0.8-0.9 **Mg** 0.35-0.50 **Mn** 0.50-0.8 **Ni** 1.0-1.5 **Si** 9.5-10.5 **Ti** 0.20 max **Zn** 0.03 max **Other** each 0.03 max, total 0.10 max | **AA** 337.1 |

*Boxed entries are no longer active and are retained for reference purposes only.

The chemical compositions listed are for identification purposes and should not be used in lieu of the cross referenced specifications.

| UNIFIED NUMBER | DESCRIPTION | CHEMICAL COMPOSITION | CROSS REFERENCE SPECIFICATIONS |
|---|---|---|---|
| *** A03380** | Aluminum Foundry Alloy, Casting | **Al** rem **Cu** 2.8-3.2 **Fe** 0.20 max **Mg** 0.8-1.2 **Ni** 1.0-1.5 **Si** 9.5-10.5 **Ti** 0.20 max **Other** each 0.03 max, total 0.10 max | **AA** 338.0 |
| *** A03381** | Aluminum Foundry Alloy, Ingot | **Al** rem **Cu** 2.8-3.2 **Fe** 0.17 max **Mg** 0.9-1.2 **Ni** 1.0-1.5 **Si** 9.5-10.5 **Ti** 0.20 max **Other** each 0.03 max, total 0.10 max | **AA** 338.1 |
| **A03390** | Aluminum Foundry Alloy, Casting | **Al** rem **Cu** 1.5-3.0 **Fe** 1.2 max **Mg** 0.50-1.5 **Mn** 0.50 max **Ni** 0.50-1.5 **Si** 11.0-13.0 **Ti** 0.25 max **Zn** 1.0 max **Other** total 0.50 max | **AA** 339.0 |
| **A03391** | Aluminum Foundry Alloy, Ingot | **Al** rem **Cu** 1.5-3.0 **Fe** 0.9 max **Mg** 0.6-1.5 **Mn** 0.50 max **Ni** 0.50-1.5 **Si** 11.0-13.0 **Ti** 0.25 max **Zn** 1.0 max **Other** total 0.50 max | **AA** 339.1 **SAE** J452 (334) |
| *** A03430** | Aluminum Foundry Alloy, Casting | **Al** rem **Cr** 0.10 max **Cu** 0.50-0.9 **Fe** 1.2 max **Mg** 0.10 max **Mn** 0.50 max **Si** 6.7-7.7 **Sn** 0.05 max **Zn** 1.2-2.0 **Other** each 0.10 max, total 0.35 max | **AA** 343.0 |
| *** A03431** | Aluminum Foundry Alloy, Ingot | **Al** rem **Cr** 0.10 max **Cu** 0.50-0.9 **Fe** 0.50-0.9 **Mg** 0.10 max **Mn** 0.50 max **Si** 6.7-7.7 **Sn** 0.05 max **Zn** 1.2-1.9 **Other** each 0.10 max, total 0.35 max | **AA** 343.1 |
| **A03540** | Aluminum Foundry Alloy, Casting | **Al** rem **Cu** 1.6-2.0 **Fe** 0.20 max **Mg** 0.40-0.6 **Mn** 0.10 max **Si** 8.6-9.4 **Ti** 0.20 max **Zn** 0.10 max **Other** each 0.05 max, total 0.15 max | **AA** 354.0 **AMS** AMS-A-21180 (354.0) **ASTM** B108 (354.0); B686 (354.0) **SAE** J452 |
| **A03541** | Aluminum Foundry Alloy, Ingot | **Al** rem **Cu** 1.6-2.0 **Fe** 0.15 max **Mg** 0.45-0.6 **Mn** 0.10 max **Si** 8.6-9.4 **Ti** 0.20 max **Zn** 0.10 max **Other** each 0.05 max, total 0.15 max | **AA** 354.1 **ASTM** B179 (354.1) |
| **A03550** | Aluminum Foundry Alloy, Casting | **Al** rem **Cr** 0.25 max **Cu** 1.0-1.5 **Fe** 0.6 max **Mg** 0.40-0.6 **Mn** 0.50 max **Si** 4.5-5.5 **Ti** 0.25 max **Zn** 0.35 max **Other** each 0.05 max, total 0.15 max. **Note:** If Fe exceeds 0.45, Mn shall not be less than 0.5\mtFe | **AA** 355.0 **AMS** 4210; 4212; 4214; 4245; 4280; 4281 **ASME** SFA5.10 (R-C355.0) **ASTM** B26 (355.0); B108 (355.0); B618 (355.0) **AWS** A5.10 (R-C355.0) **MIL SPEC** MIL-F-3922 **SAE** J452 (322) |
| **A03551** | Aluminum Foundry Alloy, Ingot | **Al** rem **Cr** 0.25 max **Cu** 1.0-1.5 **Fe** 0.50 max **Mg** 0.45-0.6 **Mn** 0.50 max **Si** 4.5-5.5 **Ti** 0.25 max **Zn** 0.35 max **Other** each 0.05 max, total 0.15 max. **Note:** If Fe exceeds 0.45, Mn shall not be less than 0.5\mtFe | **AA** 355.1 **ASTM** B179 (355.1) |
| **A03552** | Aluminum Foundry Alloy, Ingot | **Al** rem **Cu** 1.0-1.5 **Fe** 0.14-0.25 **Mg** 0.50-0.6 **Mn** 0.05 max **Si** 4.5-5.5 **Ti** 0.20 max **Zn** 0.05 max **Other** each 0.05 max, total 0.15 max | **AA** 355.2 **ASTM** B179 (355.2) |
| **A03560** | Aluminum Foundry Alloy, Casting | **Al** rem **Cu** 0.25 max **Fe** 0.6 max **Mg** 0.20-0.45 **Mn** 0.35 max **Si** 6.5-7.5 **Ti** 0.25 max **Zn** 0.35 max **Other** each 0.05 max, total 0.15 max. **Note:** If Fe exceeds 0.45, Mn shall not be less than 0.5 x Fe | **AA** 356.0 **AMS** 4217; 4260; 4261; 4284; 4285; 4286 **ASTM** B26 (356.0); B108 (356.0); B618 (356.0) **MIL SPEC** MIL-F-3922 **SAE** J452 (323) |

*Boxed entries are no longer active and are retained for reference purposes only.

The chemical compositions listed are for identification purposes and should not be used in lieu of the cross referenced specifications.

| UNIFIED NUMBER | DESCRIPTION | CHEMICAL COMPOSITION | CROSS REFERENCE SPECIFICATIONS |
|---|---|---|---|
| **A03561** | Aluminum Foundry Alloy, Ingot | **Al** rem **Cu** 0.25 max **Fe** 0.50 max **Mg** 0.25-0.45 **Mn** 0.35 max **Si** 6.5-7.5 **Ti** 0.25 max **Zn** 0.35 max **Other** each 0.05 max, total 0.15 max. **Note:** If Fe exceeds 0.45, Mn shall not be less than 0.5 x Fe | **AA** 356.1 **ASTM** B179 (356.1) |
| **A03562** | Aluminum Foundry Alloy, Ingot | **Al** rem **Cu** 0.10 max **Fe** 0.13-0.25 **Mg** 0.30-0.45 **Mn** 0.05 max **Si** 6.5-7.5 **Ti** 0.20 max **Zn** 0.05 max **Other** each 0.05 max, total 0.15 max | **AA** 356.2 **ASTM** B179 (356.2) |
| **A03570** | Aluminum Foundry Alloy, Casting | **Al** rem **Cu** 0.05 max **Fe** 0.15 max **Mg** 0.45-0.6 **Mn** 0.03 max **Si** 6.5-7.5 **Ti** 0.20 max **Zn** 0.05 max **Other** each 0.05 max, total 0.15 max | **AA** 357.0 **AMS** 4241; 4246 **ASME** SFA5.10 (RA-357.0) **ASTM** B108 (357.0) **AWS** A5.10 (RA-357.0) **SAE** J452 |
| **A03571** | Aluminum Foundry Alloy, Ingot | **Al** rem **Cu** 0.05 max **Fe** 0.12 max **Mg** 0.45-0.6 **Mn** 0.03 max **Si** 6.5-7.5 **Ti** 0.20 max **Zn** 0.05 max **Other** each 0.05 max, total 0.15 max | **AA** 357.1 **ASTM** B179 (357.1) |
| **A03580** | Aluminum Foundry Alloy | **Al** rem **Be** 0.10-0.30 **Cr** 0.20 max **Cu** 0.20 max **Fe** 0.30 max **Mg** 0.40-0.6 **Mn** 0.20 max **Si** 7.6-8.6 **Ti** 0.10-0.20 **Zn** 0.20 max **Other** each 0.05 max, total 0.15 max | **AA** 358.0 |
| **A03582** | Aluminum Foundry Alloy, Ingot | **Al** rem **Be** 0.15-0.30 **Cr** 0.05 max **Cu** 0.10 max **Fe** 0.20 max **Mg** 0.45-0.6 **Mn** 0.10 max **Si** 7.6-8.6 **Ti** 0.12-0.20 **Zn** 0.10 max **Other** each 0.05 max, total 0.15 max | **AA** 358.2 |
| **A03590** | Aluminum Foundry Alloy, Casting | **Al** rem **Cu** 0.20 max **Fe** 0.20 max **Mg** 0.50-0.7 **Mn** 0.10 max **Si** 8.5-9.5 **Ti** 0.20 max **Zn** 0.10 max **Other** each 0.05 max, total 0.15 max | **AA** 359.0 **AMS** AMS-A-21180 (359.0) **ASTM** B108 (359.0) **SAE** J452 |
| **A03592** | Aluminum Foundry Alloy, Ingot | **Al** rem **Cu** 0.10 max **Fe** 0.12 max **Mg** 0.55-0.7 **Mn** 0.10 max **Si** 8.5-9.5 **Ti** 0.20 max **Zn** 0.10 max **Other** each 0.05 max, total 0.15 max | **AA** 359.2 **ASTM** B179 (359.2) |
| **A03600** | Aluminum Foundry Alloy, Casting | **Al** rem **Cu** 0.6 max **Fe** 2.0 max **Mg** 0.40-0.6 **Mn** 0.35 max **Ni** 0.50 max **Si** 9.0-10.0 **Sn** 0.15 max **Zn** 0.50 max **Other** total 0.25 max | **AA** 360.0 **AMS** 4290 **ASTM** B85 (360.0) **SAE** J452 |
| **A03602** | Aluminum Foundry Alloy, Ingot | **Al** rem **Cu** 0.10 max **Fe** 0.7-1.1 **Mg** 0.45-0.6 **Mn** 0.10 max **Ni** 0.10 max **Si** 9.0-10.0 **Sn** 0.10 max **Zn** 0.10 max **Other** total 0.20 max | **AA** 360.2 **ASTM** B179 (360.2) |
| **A03610** | Aluminum Foundry Alloy, Casting | **Al** rem **Cr** 0.20-0.30 **Cu** 0.50 max **Fe** 1.1 max **Mg** 0.40-0.6 **Mn** 0.25 max **Ni** 0.20-0.30 **Si** 9.5-10.5 **Sn** 0.10 max **Ti** 0.20 max **Zn** 0.50 max **Other** each 0.05 max, total 0.15 max | **AA** 361.0 |
| **A03611** | Aluminum Foundry Alloy, Ingot | **Al** rem **Cr** 0.20-0.30 **Cu** 0.50 max **Fe** 0.8 max **Mg** 0.45-0.6 **Mn** 0.25 max **Ni** 0.20-0.30 **Si** 9.5-10.5 **Sn** 0.10 max **Ti** 0.20 max **Zn** 0.40 max **Other** each 0.05 max, total 0.15 max | **AA** 361.1 |
| * **A03620** | Aluminum Foundry Alloy, Casting | **Al** rem **Cu** 0.20 max **Fe** 0.8-1.2 **Mg** 0.50-0.7 **Mn** 0.50-0.8 **Si** 9.5-10.5 **Ti** 0.20 max **Zn** 0.03 max **Other** each 0.03 max, total 0.10 max | **AA** 362.0 |

*Boxed entries are no longer active and are retained for reference purposes only.

The chemical compositions listed are for identification purposes and should not be used in lieu of the cross referenced specifications.

| UNIFIED NUMBER | DESCRIPTION | CHEMICAL COMPOSITION | CROSS REFERENCE SPECIFICATIONS |
|---|---|---|---|
| * **A03621** | Aluminum Foundry Alloy, Ingot | **Al** rem **Cu** 0.20 max **Fe** 0.8-0.9 **Mg** 0.6-0.7 **Mn** 0.50-0.8 **Si** 9.5-10.5 **Ti** 0.20 max **Zn** 0.03 max **Other** each 0.03 max, total 0.10 max | **AA** 362.1 |
| **A03630** | Aluminum Foundry Alloy, Casting | **Al** rem **Cu** 2.5-3.5 **Fe** 1.1 max **Mg** 0.15-0.40 **Ni** 0.25 max **Pb** 0.25 max **Si** 4.5-6.0 **Sn** 0.25 max **Ti** 0.20 max **Zn** 3.0-4.5 **Other** total 0.30 max, plus Mn+Cr 0.8 max | **AA** 363.0 |
| **A03631** | Aluminum Foundry Alloy, Ingot | **Al** rem **Cu** 2.5-3.5 **Fe** 0.8 max **Mg** 0.20-0.40 **Ni** 0.25 max **Pb** 0.25 max **Si** 4.5-6.0 **Sn** 0.25 max **Ti** 0.20 max **Zn** 3.0-4.5 **Other** total 0.30 max, plus Mn+Cr 0.8 max | **AA** 363.1 |
| **A03640** | Aluminum Foundry Alloy, Casting | **Al** rem **Be** 0.02-0.04 **Cr** 0.25-0.50 **Cu** 0.20 max **Fe** 1.5 max **Mg** 0.20-0.40 **Mn** 0.10 max **Ni** 0.15 max **Si** 7.5-9.5 **Sn** 0.15 max **Zn** 0.15 max **Other** each 0.05 max, total 0.15 max | **AA** 364.0 |
| **A03642** | Aluminum Foundry Alloy, Ingot | **Al** rem **Be** 0.02-0.04 **Cr** 0.25-0.50 **Cu** 0.20 max **Fe** 0.7-1.1 **Mg** 0.25-0.40 **Mn** 0.10 max **Ni** 0.15 max **Si** 7.5-9.5 **Sn** 0.15 max **Zn** 0.15 max **Other** each 0.05 max, total 0.15 max | **AA** 364.2 |
| **A03650** | Aluminum Foundry Alloy, Casting | **Al** rem **Cu** 0.03 max **Fe** 0.15 max **Mg** 0.10-0.50 **Mn** 0.50-0.8 **P** 0.001 max **Si** 9.5-11.5 **Sr** 0.010-0.020 **Ti** 0.04-0.15 **Zn** 0.07 max **Other** each 0.03 max, total 0.10 max | **AA** 365.0 |
| **A03651** | Aluminum Foundry Alloy, Ingot | **Al** rem **Cu** 0.03 max **Fe** 0.12 max **Mg** 0.15-0.50 **Mn** 0.50-0.8 **P** 0.001 max **Si** 9.5-11.5 **Sr** 0.010-0.020 **Ti** 0.04-0.15 **Zn** 0.07 max **Other** each 0.03 max, total 0.10 max | **AA** 365.1 |
| **A03690** | Aluminum Foundry Alloy, Casting | **Al** rem **Cr** 0.30-0.40 **Cu** 0.50 max **Fe** 1.3 max **Mg** 0.25-0.45 **Mn** 0.35 max **Ni** 0.05 max **Si** 11.0-12.0 **Sn** 0.10 max **Zn** 1.0 max **Other** each 0.05 max, total 0.15 max | **AA** 369.0 |
| **A03691** | Aluminum Foundry Alloy, Ingot | **Al** rem **Cr** 0.30-0.40 **Cu** 0.50 max **Fe** 1.0 max **Mg** 0.30-0.45 **Mn** 0.35 max **Ni** 0.05 max **Si** 11.0-12.0 **Sn** 0.10 max **Zn** 0.9 max **Other** each 0.05 max, total 0.15 max | **AA** 369.1 |
| **A03800** | Aluminum Foundry Alloy, Casting | **Al** rem **Cu** 3.0-4.0 **Fe** 2.0 max **Mg** 0.10 max **Mn** 0.50 max **Ni** 0.50 max **Si** 7.5-9.5 **Sn** 0.35 max **Zn** 3.0 max **Other** total 0.50 max | **AA** 380.0 **ASTM** B85 (380.0) **SAE** J452 (308) |
| **A03802** | Aluminum Foundry Alloy, Ingot | **Al** rem **Cu** 3.0-4.0 **Fe** 0.7-1.1 **Mg** 0.10 max **Mn** 0.10 max **Ni** 0.10 max **Si** 7.5-9.5 **Sn** 0.10 max **Zn** 0.10 max **Other** total 0.20 max | **AA** 380.2 **ASTM** B179 (380.2) |
| **A03810** | Aluminum Foundry Alloy, Casting | **Al** rem **Cr** 0.15 max **Cu** 3.0-4.0 **Fe** 1.3 max **Mg** 0.13 max **Mn** 0.50 max **Pb** 0.15 max **Sb** 0.15 max **Si** 9.0-10.0 **Sn** 0.15 max **Ti** 0.20 max **Zn** 3.0 max **Other** total 0.50 max | **AA** 381.0 |
| **A03812** | Aluminum Foundry Alloy, Ingot | **Al** rem **Cr** 0.15 max **Cu** 3.0-4.0 **Fe** 0.7-1.0 **Mg** 0.13 max **Mn** 0.50 max **Ni** 0.50 max **Pb** 0.15 max **Sb** 0.15 max **Si** 9.0-10.0 **Sn** 0.15 max **Ti** 0.20 max **Zn** 2.9 max. **Other** total 0.50 max. | **AA** 381.2 |

*Boxed entries are no longer active and are retained for reference purposes only.

The chemical compositions listed are for identification purposes and should not be used in lieu of the cross referenced specifications.

| UNIFIED NUMBER | DESCRIPTION | CHEMICAL COMPOSITION | CROSS REFERENCE SPECIFICATIONS |
|---|---|---|---|
| **A03830** | Aluminum Foundry Alloy, Casting | **Al** rem **Cu** 2.0-3.0 **Fe** 1.3 max **Mg** 0.10 max **Mn** 0.50 max **Ni** 0.30 max **Si** 9.5-11.5 **Sn** 0.15 max **Zn** 3.0 max **Other** total 0.50 max | **AA** 383.0 **ASTM** B85 (383.0) **SAE** J452 (383) |
| **A03831** | Aluminum Foundry Alloy, Ingot | **Al** rem **Cu** 2.0-3.0 **Fe** 1.0 max **Mg** 0.10 max **Mn** 0.50 max **Ni** 0.30 max **Si** 9.5-11.5 **Sn** 0.15 max **Zn** 2.9 max **Other** total 0.50 max | **AA** 383.1 **ASTM** B179 (383.1) |
| **A03832** | Aluminum Foundry Alloy, Ingot | **Al** rem **Cu** 2.0-3.0 **Fe** 0.6-1.0 **Mg** 0.10 max **Mn** 0.10 max **Ni** 0.10 max **Si** 9.5-11.5 **Sn** 0.10 max **Zn** 0.10 max **Other** total 0.20 max | **AA** 383.2 **ASTM** B179 (383.2) |
| **A03840** | Aluminum Foundry Alloy, Casting | **Al** rem **Cu** 3.0-4.5 **Fe** 1.3 max **Mg** 0.10 max **Mn** 0.50 max **Ni** 0.50 max **Si** 10.5-12.0 **Sn** 0.35 max **Zn** 3.0 max **Other** total 0.50 max | **AA** 384.0 **ASTM** B85 (384.0) **SAE** J452 (303) |
| **A03841** | Aluminum Foundry Alloy, Ingot | **Al** rem **Cu** 3.0-4.5 **Fe** 1.0 max **Mg** 0.10 max **Mn** 0.50 max **Ni** 0.50 max **Si** 10.5-12.0 **Sn** 0.35 max **Zn** 2.9 max **Other** total 0.50 max | **AA** 384.1 **ASTM** B179 (384.1) |
| **A03842** | Aluminum Foundry Alloy, Ingot | **Al** rem **Cu** 3.0-4.5 **Fe** 0.6-1.0 **Mg** 0.10 max **Mn** 0.10 max **Ni** 0.10 max **Si** 10.5-12.0 **Sn** 0.10 max **Zn** 0.10 max **Other** total 0.20 max | **AA** 384.2 **ASTM** B179 (384.2) |
| **A03850** | Aluminum Foundry Alloy, Casting | **Al** rem **Cu** 2.0-4.0 **Fe** 2.0 max **Mg** 0.30 max **Mn** 0.50 max **Ni** 0.50 max **Si** 11.0-13.0 **Sn** 0.30 max **Zn** 3.0 max **Other** total 0.50 max | **AA** 385.0 |
| **A03851** | Aluminum Foundry Alloy, Ingot | **Al** rem **Cu** 2.0-4.0 **Fe** 1.1 max **Mg** 0.30 max **Mn** 0.50 max **Ni** 0.50 max **Si** 11.0-13.0 **Sn** 0.30 max **Zn** 2.9 max **Other** total 0.50 max | **AA** 385.1 |
| **A03900** | Aluminum Foundry Alloy, Casting | **Al** rem **Cu** 4.0-5.0 **Fe** 1.3 max **Mg** 0.45-0.65 **Mn** 0.10 max **Si** 16.0-18.0 **Ti** 0.20 max **Zn** 0.10 max **Other** each 0.10 max, total 0.20 max | **AA** 390.0 **ASTM** B85 (390.0) **SAE** J452 |
| **A03902** | Aluminum Foundry Alloy, Ingot | **Al** rem **Cu** 4.0-5.0 **Fe** 0.6-1.0 **Mg** 0.50-0.65 **Mn** 0.10 max **Si** 16.0-18.0 **Ti** 0.20 max **Zn** 0.10 max **Other** each 0.10 max, total 0.20 max | **AA** 390.2 **ASTM** B179 (390.2) |
| **A03910** | Aluminum Foundry Alloy, Casting | **Al** rem **Cu** 0.20 max **Fe** 1.2 max **Mg** 0.40-0.7 **Mn** 0.30 max **Si** 18.0-20.0 **Ti** 0.20 max **Zn** 0.10 max **Other** each 0.10 max, total 0.20 max | **AA** 391.0 |
| **A03911** | Aluminum Foundry Alloy, Ingot | **Al** rem **Cu** 0.20 max **Fe** 0.9 max **Mg** 0.45-0.7 **Mn** 0.30 max **Si** 18.0-20.0 **Ti** 0.20 max **Zn** 0.10 max **Other** each 0.10 max, total 0.20 max | **AA** 391.1 |
| **A03920** | Aluminum Foundry Alloy, Casting | **Al** rem **Cu** 0.40-0.8 **Fe** 1.5 max **Mg** 0.8-1.2 **Mn** 0.20-0.6 **Ni** 0.50 max **Si** 18.0-20.0 **Sn** 0.30 max **Ti** 0.20 max **Zn** 0.50 max **Other** each 0.15 max, total 0.50 max | **AA** 392.0 **ASTM** B85 (392.0) |
| **A03921** | Aluminum Foundry Alloy, Ingot | **Al** rem **Cu** 0.40-0.8 **Fe** 1.1 max **Mg** 0.9-1.2 **Mn** 0.20-0.6 **Ni** 0.50 max **Si** 18.0-20.0 **Sn** 0.30 max **Ti** 0.20 max **Zn** 0.40 max **Other** each 0.15 max, total 0.50 max | **AA** 392.1 **ASTM** B179 (392.1) |
| **A03930** | Aluminum Foundry Alloy, Casting | **Al** rem **Cu** 0.7-1.1 **Fe** 1.3 max **Mg** 0.7-1.3 **Mn** 0.10 max **Ni** 2.0-2.5 **Si** 21.0-23.0 **Ti** 0.10-0.20 **V** 0.08-0.15 **Zn** 0.10 max **Other** each 0.05 max, total 0.15 max | **AA** 393.0 |

The chemical compositions listed are for identification purposes and should not be used in lieu of the cross referenced specifications.

| UNIFIED NUMBER | DESCRIPTION | CHEMICAL COMPOSITION | CROSS REFERENCE SPECIFICATIONS |
|---|---|---|---|
| **A03931** | Aluminum Foundry Alloy, Ingot | **Al** rem **Cu** 0.7-1.1 **Fe** 1.0 max **Mg** 0.8-1.3 **Mn** 0.10 max **Ni** 2.0-2.5 **Si** 21.0-23.0 **Ti** 0.10-0.20 **V** 0.08-0.15 **Zn** 0.10 max **Other** each 0.05 max, total 0.15 max | **AA** 393.1 |
| **A03932** | Aluminum Foundry Alloy, Ingot | **Al** rem **Cu** 0.7-1.1 **Fe** 0.8 max **Mg** 0.8-1.3 **Mn** 0.10 max **Ni** 2.0-2.5 **Si** 21.0-23.0 **Ti** 0.10-0.20 **V** 0.08-0.15 **Zn** 0.10 max **Other** each 0.05 max, total 0.15 max | **AA** 393.2 |
| **A04082** | Aluminum Foundry Alloy, Ingot (Steel Coating Alloy) | **Al** rem **Cu** 0.10 max **Fe** 0.6-1.3 **Mn** 0.10 max **Si** 8.5-9.5 **Zn** 0.10 max **Other** each 0.10 max, total 0.20 max | **AA** 408.2 |
| **A04092** | Aluminum Foundry Alloy, Ingot (Steel Coating Alloy) | **Al** rem **Cu** 0.10 max **Fe** 0.6-1.3 **Mn** 0.10 max **Si** 9.0-10.0 **Zn** 0.10 max **Other** each 0.10 max, total 0.20 max | **AA** 409.2 |
| **A04112** | Aluminum Foundry Alloy, Ingot (Steel Coating Alloy) | **Al** rem **Cu** 0.20 max **Fe** 0.6-1.3 **Mn** 0.10 max **Si** 10.0-12.0 **Zn** 0.10 max **Other** each 0.10 max, total 0.20 max | **AA** 411.2 |
| **A04130** | Aluminum Foundry Alloy, Casting | **Al** rem **Cu** 1.0 max **Fe** 2.0 max **Mg** 0.10 max **Mn** 0.35 max **Ni** 0.50 max **Si** 11.0-13.0 **Sn** 0.15 max **Zn** 0.50 max **Other** total 0.25 max | **AA** 413.0 **ASTM** B85 (413.0) **SAE** J452 |
| **A04132** | Aluminum Foundry Alloy, Ingot | **Al** rem **Cu** 0.10 max **Fe** 0.7-1.1 **Mg** 0.07 max **Mn** 0.10 max **Ni** 0.10 max **Si** 11.0-13.0 **Sn** 0.10 max **Zn** 0.10 max **Other** total 0.20 max | **AA** 413.2 **ASTM** B179 (413.2) |
| **A04352** | Aluminum Foundry Alloy, Ingot | **Al** rem **Cu** 0.05 **Fe** 0.40 **Mg** 0.05 **Mn** 0.05 **Si** 3.3-3.9 **Zn** 0.10 **Other** each 0.05, total 0.20 | **AA** 435.2 |
| **A04430** | Aluminum Foundry Alloy, Casting | **Al** rem **Cr** 0.25 max **Cu** 0.6 max **Fe** 0.8 max **Mg** 0.05 max **Mn** 0.50 max **Si** 4.5-6.0 **Ti** 0.25 max **Zn** 0.50 max **Other** total 0.35 max | **AA** 443.0 **ASTM** B26 (443.0); B108 (443.0); B618 (443.0) **SAE** J452 (35) |
| **A04431** | Aluminum Foundry Alloy, Ingot | **Al** rem **Cr** 0.25 max **Cu** 0.6 max **Fe** 0.6 max **Mg** 0.05 max **Mn** 0.50 max **Si** 4.5-6.0 **Ti** 0.25 max **Zn** 0.50 max **Other** total 0.35 max | **AA** 443.1 **ASTM** B179 (443.1) |
| **A04432** | Aluminum Foundry Alloy, Ingot | **Al** rem **Cu** 0.10 max **Fe** 0.6 max **Mg** 0.05 max **Mn** 0.10 max **Si** 4.5-6.0 **Ti** 0.20 max **Zn** 0.10 max **Other** each 0.05 max, total 0.15 max | **AA** 443.2 **ASTM** B179 (443.2) |
| **A04440** | Aluminum Foundry Alloy, Casting | **Al** rem **Cu** 0.25 max **Fe** 0.6 max **Mg** 0.10 max **Mn** 0.35 max **Si** 6.5-7.5 **Ti** 0.25 max **Zn** 0.35 max **Other** each 0.05 max, total 0.15 max | **AA** 444.0 |
| **A04442** | Aluminum Foundry Alloy, ingot | **Al** rem **Cu** 0.10 max **Fe** 0.13-0.25 **Mg** 0.05 max **Mn** 0.05 max **Si** 6.5-7.5 **Ti** 0.20 max **Zn** 0.05 max **Other** each 0.05 max, total 0.15 max | **AA** 444.2 |
| **A04452** | Aluminum Foundry Alloy, Ingot | **Al** rem **Cu** 0.10 max **Fe** 0.6-1.3 **Mn** 0.10 max **Si** 6.5-7.5 **Zn** 0.10 max **Other** each 0.10 max, total 0.20 max | **AA** 445.2 |
| **A05110** | Aluminum Foundry Alloy, Casting | **Al** rem **Cu** 0.15 max **Fe** 0.50 max **Mg** 3.5-4.5 **Mn** 0.35 max **Si** 0.30-0.7 **Ti** 0.25 max **Zn** 0.15 max **Other** each 0.05 max, total 0.15 max | **AA** 511.0 |
| **A05111** | Aluminum Foundry Alloy, Ingot | **Al** rem **Cu** 0.15 max **Fe** 0.40 max **Mg** 3.6-4.5 **Mn** 0.35 max **Si** 0.30-0.7 **Ti** 0.25 max **Zn** 0.15 max **Other** each 0.05 max, total 0.15 | **AA** 511.1 |

The chemical compositions listed are for identification purposes and should not be used in lieu of the cross referenced specifications.

| UNIFIED NUMBER | DESCRIPTION | CHEMICAL COMPOSITION | CROSS REFERENCE SPECIFICATIONS |
|---|---|---|---|
| **A05112** | Aluminum Foundry Alloy, Ingot | **Al** rem **Cu** 0.10 max **Fe** 0.30 max **Mg** 3.6-4.5 **Mn** 0.10 max **Si** 0.30-0.7 **Ti** 0.20 max **Zn** 0.10 max **Other** each 0.05 max, total 0.15 max | **AA** 511.2 |
| **A05120** | Aluminum Foundry Alloy, Casting | **Al** rem **Cr** 0.25 max **Cu** 0.35 max **Fe** 0.6 max **Mg** 3.5-4.5 **Mn** 0.8 max **Si** 1.4-2.2 **Ti** 0.25 max **Zn** 0.35 max **Other** each 0.05 max, total 0.15 max | **AA** 512.0 |
| **A05122** | Aluminum Foundry Alloy, Ingot | **Al** rem **Cu** 0.10 max **Fe** 0.30 max **Mg** 3.6-4.5 **Mn** 0.10 max **Si** 1.4-2.2 **Ti** 0.20 max **Zn** 0.10 max **Other** each 0.05 max, total 0.15 max | **AA** 512.2 |
| **A05130** | Aluminum Foundry Alloy, Casting | **Al** rem **Cu** 0.10 max **Fe** 0.40 max **Mg** 3.5-4.5 **Mn** 0.30 max **Si** 0.30 max **Ti** 0.20 max **Zn** 1.4-2.2 **Other** each 0.05 max, total 0.15 max | **AA** 513.0 **ASTM** B108 (513.0) |
| **A05132** | Aluminum Foundry Alloy, Ingot | **Al** rem **Cu** 0.10 max **Fe** 0.30 max **Mg** 3.6-4.5 **Mn** 0.10 max **Si** 0.30 max **Ti** 0.20 max **Zn** 1.4-2.2 **Other** each 0.05 max, total 0.15 max | **AA** 513.2 **ASTM** B179 (513.2) |
| **A05140** | Aluminum Foundry Alloy, Casting | **Al** rem **Cu** 0.15 max **Fe** 0.50 max **Mg** 3.5-4.5 **Mn** 0.35 max **Si** 0.35 max **Ti** 0.25 max **Zn** 0.15 max **Other** each 0.05 max, total 0.15 max | **AA** 514.0 **ASTM** B26 (514.0); B618 (514.0) **SAE** J452 (320) |
| **A05141** | Aluminum Foundry Alloy, Ingot | **Al** rem **Cu** 0.15 max **Fe** 0.40 max **Mg** 3.6-4.5 **Mn** 0.35 max **Si** 0.35 max **Ti** 0.25 max **Zn** 0.15 max **Other** each 0.05 max, total 0.15 max | **AA** 514.1 **ASTM** B179 (514.1) |
| **A05142** | Aluminum Foundry Alloy, Ingot | **Al** rem **Cu** 0.10 max **Fe** 0.30 max **Mg** 3.6-4.5 **Mn** 0.10 max **Si** 0.30 max **Ti** 0.20 max **Zn** 0.10 max **Other** each 0.05 max, total 0.15 max | **AA** 514.2 **ASTM** B179 (514.2) |
| **A05150** | Aluminum Foundry Alloy, Casting | **Al** rem **Cu** 0.20 max **Fe** 1.3 max **Mg** 2.5-4.0 **Mn** 0.40-0.6 **Si** 0.50-1.0 **Zn** 0.10 max **Other** each 0.05 max, total 0.15 max | **AA** 515.0 |
| **A05152** | Aluminum Foundry Alloy, Ingot | **Al** rem **Cu** 0.10 max **Fe** 0.6-1.0 **Mg** 2.7-4.0 **Mn** 0.40-0.6 **Si** 0.50-1.0 **Zn** 0.05 max **Other** each 0.05 max, total 0.15 max | **AA** 515.2 |
| **A05160** | Aluminum Foundry Alloy, Casting | **Al** rem **Cu** 0.30 **Fe** 0.35-1.0 **Mg** 2.5-4.5 **Mn** 0.15-0.40 **Ni** 0.25-0.40 **Pb** 0.10 max **Si** 0.30-1.5 **Sn** 0.10 max **Ti** 0.10-0.20 **Zn** 0.20 max **Other** each 0.05 max | **AA** 516.0 |
| **A05161** | Aluminum Foundry Alloy, Ingot | **Al** rem **Cu** 0.30 max **Fe** 0.35-0.7 **Mg** 2.6-4.5 **Mn** 0.15-0.40 **Ni** 0.25-0.40 **Pb** 0.10 max **Si** 0.30-1.5 **Sn** 0.10 max **Ti** 0.10-0.20 **Zn** 0.20 max **Other** each 0.05 max | **AA** 516.1 |
| **A05180** | Aluminum Foundry Alloy, Die Casting | **Al** rem **Cu** 0.25 max **Fe** 1.8 max **Mg** 7.5-8.5 **Mn** 0.35 max **Ni** 0.15 max **Si** 0.35 max **Sn** 0.15 max **Zn** 0.15 max **Other** total 0.25 max | **AA** 518.0 **ASTM** B85 (518.0) **MIL SPEC** MIL-F-3922 |
| **A05181** | Aluminum Foundry Alloy, Ingot | **Al** rem **Cu** 0.25 max **Fe** 1.1 max **Mg** 7.6-8.5 **Mn** 0.35 max **Ni** 0.15 max **Si** 0.35 max **Sn** 0.15 max **Zn** 0.15 max **Other** total 0.25 max | **AA** 518.1 **ASTM** B179 (518.1) |
| **A05182** | Aluminum Foundry Alloy, Ingot | **Al** rem **Cu** 0.10 max **Fe** 0.7 max **Mg** 7.6-8.5 **Mn** 0.10 max **Ni** 0.05 max **Si** 0.25 max **Sn** 0.05 max **Other** total 0.10 max | **AA** 518.2 **ASTM** B179 (518.2) |

The chemical compositions listed are for identification purposes and should not be used in lieu of the cross referenced specifications.

| UNIFIED NUMBER | DESCRIPTION | CHEMICAL COMPOSITION | CROSS REFERENCE SPECIFICATIONS |
|---|---|---|---|
| **A05200** | Aluminum Foundry Alloy, Casting | **Al** rem **Cu** 0.25 max **Fe** 0.30 max **Mg** 9.5-10.6 **Mn** 0.15 max **Si** 0.25 max **Ti** 0.25 max **Zn** 0.15 max **Other** each 0.05 max, total 0.15 max | **AA** 520.0 **ASTM** B26 (520.0); B618 (520.0) **SAE** J452 (324) |
| **A05202** | Aluminum Foundry Alloy, Ingot | **Al** rem **Cu** 0.20 max **Fe** 0.20 max **Mg** 9.6-10.6 **Mn** 0.10 max **Si** 0.15 max **Ti** 0.20 max **Zn** 0.10 max **Other** each 0.05 max, total 0.15 max | **AA** 520.2 **ASTM** B179 (520.2) |
| **A05350** | Aluminum Foundry Alloy, Casting | **Al** rem **B** 0.005 max **Be** 0.003-0.007 **Cu** 0.05 max **Fe** 0.15 max **Mg** 6.2-7.5 **Mn** 0.10-0.25 **Si** 0.15 max **Ti** 0.10-0.25 **Other** each 0.05 max, total 0.15 max | **AA** 535.0 **ASTM** B26 (535.0); B108 (535.0); B618 (535.0) **SAE** J452 |
| **A05352** | Aluminum Foundry Alloy, Ingot | **Al** rem **B** 0.002 max **Be** 0.003-0.007 **Cu** 0.05 max **Fe** 0.10 max **Mg** 6.6-7.5 **Mn** 0.10-0.25 **Si** 0.10 max **Ti** 0.10-0.25 **Other** each 0.05 max, total 0.15 max | **AA** 535.2 **ASTM** B179 (535.2) |
| **A07050** | Aluminum Foundry Alloy, Casting | **Al** rem **Cr** 0.20-0.40 **Cu** 0.20 max **Fe** 0.8 max **Mg** 1.4-1.8 **Mn** 0.40-0.6 **Si** 0.20 max **Ti** 0.25 max **Zn** 2.7-3.3 **Other** each 0.05 max, total 0.15 max | **AA** 705.0 **ASTM** B26 (705.0); B108 (705.0); B618 (705.0) **SAE** J452 (311) |
| **A07051** | Aluminum Foundry Alloy, Ingot | **Al** rem **Cr** 0.20-0.40 **Cu** 0.20 max **Fe** 0.6 max **Mg** 1.5-1.8 **Mn** 0.40-0.6 **Si** 0.20 max **Ti** 0.25 max **Zn** 2.7-3.3 **Other** each 0.05 max, total 0.15 max | **AA** 705.1 **ASTM** B179 (705.1) |
| **A07070** | Aluminum Foundry Alloy, Casting | **Al** rem **Cr** 0.20-0.40 **Cu** 0.20 max **Fe** 0.8 max **Mg** 1.8-2.4 **Mn** 0.40-0.6 **Si** 0.20 max **Ti** 0.25 max **Zn** 4.0-4.5 **Other** 0.05 max, total 0.15 max | **AA** 707.0 **ASTM** B26 (707.0); B108 (707.0); B618 (707.0) **SAE** J452 (312) |
| **A07071** | Aluminum Foundry Alloy, Ingot | **Al** rem **Cr** 0.20-0.40 **Cu** 0.20 max **Fe** 0.6 max **Mg** 1.9-2.4 **Mn** 0.40-0.6 **Si** 0.20 max **Ti** 0.25 max **Zn** 4.0-4.5 **Other** each 0.05 max, total 0.15 max | **AA** 707.1 **ASTM** B179 (707.1) |
| **A07100** | Aluminum Foundry Alloy, Casting | **Al** rem **Cu** 0.35-0.6 **Fe** 0.50 max **Mg** 0.6-0.8 **Mn** 0.05 max **Si** 0.15 max **Ti** 0.25 max **Zn** 6.0-7.0 **Other** each 0.05 max, total 0.15 max | **AA** 710.0 **ASTM** B26 (710.0); B618 (710.0) **SAE** J452 (313) |
| **A07101** | Aluminum Foundry Alloy, Ingot | **Al** rem **Cu** 0.35-0.6 **Fe** 0.40 max **Mg** 0.65-0.8 **Mn** 0.05 max **Si** 0.15 max **Ti** 0.25 max **Zn** 6.0-7.0 **Other** each 0.05 max, total 0.15 max | **AA** 710.1 **ASTM** B179 (710.1) |
| **A07110** | Aluminum Foundry Alloy, Casting | **Al** rem **Cu** 0.35-0.6 **Fe** 0.7-1.4 **Mg** 0.25-0.45 **Mn** 0.05 max **Si** 0.30 max **Ti** 0.20 max **Zn** 6.0-7.0 **Other** each 0.05 max, total 0.15 max | **AA** 711.0 **ASTM** B108 (711.0) **MIL SPEC** MIl-F-3922 (C712.0) |
| **A07111** | Aluminum Foundry Alloy, Ingot | **Al** rem **Cu** 0.35-0.6 **Fe** 0.7-1.1 **Mg** 0.30-0.45 **Mn** 0.05 max **Si** 0.30 max **Ti** 0.20 max **Zn** 6.0-7.0 **Other** each 0.05 max, total 0.15 max | **AA** 711.1 **ASTM** B179 (711.1) |
| **A07120** | Aluminum Foundry Alloy, Casting | **Al** rem **Cr** 0.40-0.6 **Cu** 0.25 max **Fe** 0.50 max **Mg** 0.50-0.65 **Mn** 0.10 max **Si** 0.30 max **Ti** 0.15-0.25 **Zn** 5.0-6.5 **Other** each 0.05 max, total 0.20 max | **AA** 712.0 **ASTM** B26 (712.0); B618 (712.0) **MIL SPEC** MIL-F-3922 (D712.0) **SAE** J452 (310) |
| **A07122** | Aluminum Foundry Alloy, Ingot | **Al** rem **Cr** 0.40-0.6 **Cu** 0.25 max **Fe** 0.40 max **Mg** 0.50-0.65 **Mn** 0.10 max **Si** 0.15 max **Ti** 0.15-0.25 **Zn** 5.0-6.5 **Other** each 0.05 max, total 0.20 max | **AA** 712.2 **ASTM** B179 (712.2) |
| **A07130** | Aluminum Foundry Alloy, Casting | **Al** rem **Cr** 0.35 max **Cu** 0.40-1.0 **Fe** 1.1 max **Mg** 0.20-0.50 **Mn** 0.6 max **Ni** 0.15 max **Si** 0.25 max **Ti** 0.25 max **Zn** 7.0-8.0 **Other** each 0.10 max, total 0.25 max | **AA** 713.0 **ASTM** B26 (713.0); B108 (713.0); B618 (713.0) **SAE** J452 (315) |

The chemical compositions listed are for identification purposes and should not be used in lieu of the cross referenced specifications.

| UNIFIED NUMBER | DESCRIPTION | CHEMICAL COMPOSITION | CROSS REFERENCE SPECIFICATIONS |
|---|---|---|---|
| **A07131** | Aluminum Foundry Alloy, Ingot | **Al** rem **Cr** 0.35 max **Cu** 0.40-1.0 **Fe** 0.8 max **Mg** 0.25-0.50 **Mn** 0.6 max **Ni** 0.15 max **Si** 0.25 max **Ti** 0.25 max **Zn** 7.0-8.0 **Other** each 0.10 max, total 0.25 max | **AA** 713.1 **ASTM** B179 (713.1) |
| **A07710** | Aluminum Foundry Alloy, Casting | **Al** rem **Cr** 0.06-0.20 **Cu** 0.10 max **Fe** 0.15 max **Mg** 0.8-1.0 **Mn** 0.10 max **Si** 0.15 max **Ti** 0.10-0.20 **Zn** 6.5-7.5 **Other** each 0.05 max, total 0.15 max | **AA** 771.0 **ASTM** B26 (771.0); B618 (771.0) |
| **A07712** | Aluminum Foundry Alloy, Ingot | **Al** rem **Cr** 0.06-0.20 **Cu** 0.10 max **Fe** 0.10 max **Mg** 0.85-1.0 **Mn** 0.10 max **Si** 0.10 max **Ti** 0.10-0.20 **Zn** 6.5-7.5 **Other** each 0.05 max, total 0.15 max | **AA** 771.2 **ASTM** B179 (771.2) |
| **A07720** | Aluminum Foundry Alloy, Casting | **Al** rem **Cr** 0.06-0.20 **Cu** 0.10 max **Fe** 0.15 max **Mg** 0.6-0.8 **Mn** 0.10 max **Si** 0.15 max **Ti** 0.10-0.20 **Zn** 6.0-7.0 **Other** each 0.05 max, total 0.15 max | **AA** 772.0 |
| **A07722** | Aluminum Foundry Alloy, Ingot | **Al** rem **Cr** 0.06-0.20 **Cu** 0.10 max **Fe** 0.10 max **Mg** 0.65-0.8 **Mn** 0.10 max **Si** 0.10 max **Ti** 0.10-0.20 **Zn** 6.0-7.0 **Other** each 0.05 max, total 0.15 max | **AA** 772.2 |
| **A08500** | Aluminum Foundry Alloy, Casting | **Al** rem **Cu** 0.7-1.3 **Fe** 0.7 max **Mg** 0.10 max **Mn** 0.10 max **Ni** 0.7-1.3 **Si** 0.7 max **Sn** 5.5-7.0 **Ti** 0.20 max **Other** total 0.30 max | **AA** 850.0 **ASTM** B26 (850.0); B108 (850.0); B618 (850.0) |
| **A08501** | Aluminum Foundry Alloy, Ingot | **Al** rem **Cu** 0.7-1.3 **Fe** 0.50 max **Mg** 0.10 max **Mn** 0.10 max **Ni** 0.7-1.3 **Si** 0.7 max **Sn** 5.5-7.0 **Ti** 0.20 max **Other** total 0.30 max | **AA** 850.1 **ASTM** B179 (850.1) |
| **A08510** | Aluminum Foundry Alloy, Casting | **Al** rem **Cu** 0.7-1.3 **Fe** 0.7 max **Mg** 0.10 max **Mn** 0.10 max **Ni** 0.30-0.7 **Si** 2.0-3.0 **Sn** 5.5-7.0 **Ti** 0.20 max **Other** total 0.30 max | **AA** 851.0 **ASTM** B26 (851.0); B108 (851.0); B618 (851.0) |
| **A08511** | Aluminum Foundry Alloy, Ingot | **Al** rem **Cu** 0.7-1.3 **Fe** 0.50 max **Mg** 0.10 max **Mn** 0.10 max **Ni** 0.30-0.7 **Si** 2.0-3.0 **Sn** 5.5-7.0 **Ti** 0.20 max **Other** total 0.30 max | **AA** 851.1 **ASTM** B179 (851.1) |
| **A08520** | Aluminum Foundry Alloy, Casting | **Al** rem **Cu** 1.7-2.3 **Fe** 0.7 max **Mg** 0.6-0.9 **Mn** 0.10 max **Ni** 0.9-1.5 **Si** 0.40 max **Sn** 5.5-7.0 **Ti** 0.20 max **Other** total 0.30 max | **AA** 852.0 **ASTM** B26 (852.0); B108 (852.0); B618 (852.0) |
| **A08521** | Aluminum Foundry Alloy, Ingot | **Al** rem **Cu** 1.7-2.3 **Fe** 0.50 max **Mg** 0.7-0.9 **Mn** 0.10 max **Ni** 0.9-1.5 **Si** 0.40 max **Sn** 5.5-7.0 **Ti** 0.20 max **Other** total 0.30 max | **AA** 852.1 **ASTM** B179 (852.1) |
| **A08530** | Aluminum Foundry Alloy, Casting | **Al** rem **Cu** 3.0-4.0 **Fe** 0.7 max **Mn** 0.50 max **Si** 5.5-6.5 **Sn** 5.5-7.0 **Ti** 0.20 max **Other** total 0.30 max | **AA** 853.0 |
| **A08532** | Aluminum Foundry Alloy, Ingot | **Al** rem **Cu** 3.0-4.0 **Fe** 0.50 max **Mn** 0.10 max **Si** 5.5-6.5 **Sn** 5.5-7.0 **Ti** 0.20 max **Other** total 0.30 max | **AA** 853.2 |
| **A12010** | Aluminum Foundry Alloy, Casting | **Ag** 0.40-1.0 **Al** rem **Cu** 4.0-5.0 **Fe** 0.10 max **Mg** 0.15-0.35 **Mn** 0.20-0.40 **Si** 0.05 max **Ti** 0.15-0.35 **Other** each 0.03 max, total 0.10 max | **AA** A201.0 **AMS** 4223; 4229; AMS-A-21180 (A210.0) **ASTM** B686 (A201.0) |
| **A12011** | Aluminum Foundry Alloy, Ingot | **Ag** 0.40-1.0 **Al** rem **Cu** 4.0-5.0 **Fe** 0.07 max **Mg** 0.20-0.35 **Mn** 0.20-0.40 **Si** 0.05 max **Ti** 0.15-0.35 **Other** each 0.03 max, total 0.10 max | **AA** A201.1 |
| *** A12012** | Replaced by A12011 | | |

*Boxed entries are no longer active and are retained for reference purposes only.

The chemical compositions listed are for identification purposes and should not be used in lieu of the cross referenced specifications.

| UNIFIED NUMBER | DESCRIPTION | CHEMICAL COMPOSITION | CROSS REFERENCE SPECIFICATIONS |
|---|---|---|---|
| **A12060** | Aluminum Foundry Alloy, Casting | **Al** rem **Cu** 4.2-5.0 **Fe** 0.10 max **Mg** 0.15-0.35 **Mn** 0.20-0.50 **Ni** 0.05 max **Si** 0.05 max **Sn** 0.05 max **Ti** 0.15-0.30 **Zn** 0.10 max **Other** each 0.05 max, total 0.15 max | **AA** A206.0 **AMS** 4235; 4236 |
| **A12062** | Aluminum Foundry Alloy, Ingot | **Al** rem **Cu** 4.2-5.0 **Fe** 0.07 max **Mg** 0.20-0.35 **Mn** 0.20-0.50 **Ni** 0.03 max **Si** 0.05 max **Sn** 0.05 max **Ti** 0.15-0.25 **Zn** 0.05 max **Other** each 0.05 max, total 0.15 max | **AA** A206.2 |
| **A12400** | Aluminum Foundry Alloy, Casting replaced by A02400 | | |
| **A12401** | Aluminum Foundry Alloy, Ingot replaced by A02401 | | |
| **A12420** | Aluminum Foundry Alloy, Casting | **Al** rem **Cr** 0.15-0.25 **Cu** 3.7-4.5 **Fe** 0.8 max **Mg** 1.2-1.7 **Mn** 0.10 max **Ni** 1.8-2.3 **Si** 0.6 max **Ti** 0.07-0.20 **Zn** 0.10 max **Other** each 0.05 max, total 0.15 max | **AA** A242.0 |
| **A12421** | Aluminum Foundry Alloy, Ingot | **Al** rem **Cr** 0.15-0.25 **Cu** 3.7-4.5 **Fe** 0.6 max **Mg** 1.3-1.7 **Mn** 0.10 max **Ni** 1.8-2.3 **Si** 0.6 max **Ti** 0.07-0.20 **Zn** 0.10 max **Other** each 0.05 max, total 0.15 max | **AA** A242.1 **ASTM** B179 (A242.1) |
| **A12422** | Aluminum Foundry Alloy, Ingot | **Al** rem **Cr** 0.15-0.25 **Cu** 3.7-4.5 **Fe** 0.6 max **Mg** 1.3-1.7 **Mn** 0.10 max **Ni** 1.8-2.3 **Si** 0.35 max **Ti** 0.07-0.20 **Zn** 0.10 max **Other** each 0.05 max, total 0.15 max | **AA** A242.2 **ASTM** B179 (A242.2) |
| * **A13050** | Aluminum Foundry Alloy, Casting | **Al** rem **Cu** 1.0-1.5 **Fe** 0.20 max **Mg** 0.10 max **Mn** 0.10 max **Si** 4.5-5.5 **Ti** 0.20 max **Zn** 0.10 max **Other** each 0.05 max, total 0.15 max | **AA** A305.0 |
| * **A13051** | Aluminum Foundry Alloy, Ingot | **Ag** 4.5-5.5 **Al** rem **Cu** 1.0-1.5 **Fe** 0.15 max **Mn** 0.05 max **Ti** 0.20 max **Zn** 0.05 max **Other** each 0.05 max, total 0.15 max | **AA** A305.1 |
| * **A13052** | Aluminum Foundry Alloy, Ingot | **Al** rem **Cu** 1.0-1.5 **Fe** 0.13 max **Mn** 0.05 max **Si** 4.5-5.5 **Ti** 0.20 max **Zn** 0.05 max **Other** each 0.05 max, total 0.15 max | **AA** A305.2 |
| **A13190** | Aluminum Foundry Alloy, Casting | **Al** rem **Cu** 3.0-4.0 **Fe** 1.0 max **Mg** 0.10 max **Mn** 0.50 max **Ni** 0.35 max **Si** 5.5-6.5 **Ti** 0.25 max **Zn** 3.0 max **Other** total 0.50 max | **AA** A319.0 |
| **A13191** | Aluminum Foundry Alloy, Ingot | **Al** rem **Cu** 3.0-4.0 **Fe** 0.8 max **Mg** 0.10 max **Mn** 0.50 max **Ni** 0.35 max **Si** 5.5-6.5 **Ti** 0.25 max **Zn** 3.0 max **Other** total 0.50 max | **AA** A319.1 |
| **A13320** | Aluminum Foundry Alloy, Casting replaced by A03360 | | |
| **A13321** | Aluminum Foundry Alloy, Ingot replaced by A03361 | | |
| **A13322** | Aluminum Foundry Alloy, Ingot replaced by A03362 | | |

*Boxed entries are no longer active and are retained for reference purposes only.

The chemical compositions listed are for identification purposes and should not be used in lieu of the cross referenced specifications.

| UNIFIED NUMBER | DESCRIPTION | CHEMICAL COMPOSITION | CROSS REFERENCE SPECIFICATIONS |
|---|---|---|---|
| **A13330** | Aluminum Foundry Alloy, Casting | **Al** rem **Cu** 3.0-4.0 **Fe** 1.0 max **Mg** 0.05-0.50 **Mn** 0.50 max **Ni** 0.50 max **Si** 8.0-10.0 **Ti** 0.25 max **Zn** 3.0 max **Other** total 0.50 max | **AA** A333.0 |
| **A13331** | Aluminum Foundry Alloy, Ingot | **Al** rem **Cu** 3.0-4.0 **Fe** 0.8 max **Mg** 0.10-0.50 **Mn** 0.50 max **Ni** 0.50 max **Si** 8.0-10.0 **Ti** 0.25 max **Zn** 3.0 max **Other** total 0.50 max | **AA** A333.1 |
| **A13550** | Aluminum Foundry Alloy, Casting | **Al** rem **Cu** 1.0-1.5 **Fe** 0.09 max **Mg** 0.45-0.6 **Mn** 0.05 max **Si** 4.5-5.5 **Ti** 0.04-0.20 **Zn** 0.05 max **Other** each 0.05 max, total 0.15 max | **AA** A355.0 |
| **A13552** | Aluminum Foundry Alloy, Ingot | **Al** rem **Cu** 1.0-1.5 **Fe** 0.06 max **Mg** 0.50-0.6 **Mn** 0.03 max **Si** 4.5-5.5 **Ti** 0.04-0.20 **Zn** 0.03 max **Other** each 0.03 max, total 0.10 max | **AA** A355.2 |
| **A13560** | Aluminum Foundry Alloy, Casting | **Al** rem **Cu** 0.20 max **Fe** 0.20 max **Mg** 0.25-0.45 **Mn** 0.10 max **Si** 6.5-7.5 **Ti** 0.20 max **Zn** 0.10 max **Other** each 0.05 max, total 0.15 max | **AA** A356.0 **AMS** 4218; AMS-A-21180 (A356.0) **ASME** SFA5.10 (R-A356.0) **ASTM** B26 (A356.0); B108 (A356.0); B618 (A356.0); B686 (A356.0) **AWS** A5.10 (R-A356.0) **SAE** J452 (336) |
| **A13561** | Aluminum Foundry Alloy, Ingot | **Al** rem **Cu** 0.25 max **Fe** 0.15 max **Mg** 0.30-0.45 **Mn** 0.10 max **Si** 6.5-7.5 **Ti** 0.20 max **Zn** 0.10 max **Other** each 0.05 max, total 0.15 max | **AA** A356.1 |
| **A13562** | Aluminum Foundry Alloy, Ingot | **Al** rem **Cu** 0.10 max **Fe** 0.12 max **Mg** 0.30-0.45 **Mn** 0.05 max **Si** 6.5-7.5 **Ti** 0.20 max **Zn** 0.05 max **Other** each 0.05 max, total 0.15 max | **AA** A356.2 **ASTM** B179 (A356.2) |
| **A13570** | Aluminum Foundry Alloy, Casting | **Al** rem **Be** 0.04-0.07 **Cu** 0.20 max **Fe** 0.20 max **Mg** 0.40-0.7 **Mn** 0.10 max **Si** 6.5-7.5 **Ti** 0.04-0.20 **Zn** 0.10 max **Other** each 0.05 max, total 0.15 max | **AA** A357.0 **AMS** 4219; AMS-A-21180 (A357.0) **ASME** SFA5.10 (R-A357.0) **ASTM** B108 (A357.0); B686 (A357.0) **AWS** A5.10 (R-A357.0) **SAE** J452 |
| **A13572** | Aluminum Foundry Alloy, Ingot | **Al** rem **Be** 0.04-0.07 **Cu** 0.10 max **Fe** 0.12 max **Mg** 0.45-0.7 **Mn** 0.05 max **Si** 6.5-7.5 **Ti** 0.04-0.20 **Zn** 0.05 max **Other** each 0.03 max, total 0.10 max | **AA** A357.2 **ASTM** B179 (A357.2) |
| **A13590** | Aluminum Foundry Alloy, Casting | **Al** rem **Cu** 0.20 max **Fe** 0.25 max **Mg** 0.40-0.6 **Mn** 0.10 max **Si** 8.5-9.5 **Ti** 0.20 max **Zn** 0.05 max **Other** each 0.03 max, total 0.10 max | **AA** A359.0 |
| **A13591** | Aluminum Foundry Alloy, Ingot | **Al** rem **Cu** 0.20 max **Fe** 0.20 max **Mg** 0.45-0.6 **Mn** 0.10 max **Si** 8.5-9.5 **Ti** 0.20 max **Zn** 0.05 max **Other** each 0.03 max, total 0.10 max | **AA** A359.1 |
| **A13600** | Aluminum Foundry Alloy, Casting | **Al** rem **Cu** 0.6 max **Fe** 1.3 max **Mg** 0.40-0.6 **Mn** 0.35 max **Ni** 0.50 max **Si** 9.0-10.0 **Sn** 0.15 max **Zn** 0.50 max **Other** total 0.25 max | **AA** A360.0 **AMS** 4290 **ASTM** B85 (A360.0) **MIL SPEC** MIL-F-3922 **SAE** J452 (309) |
| **A13601** | Aluminum Foundry Alloy, Ingot | **Al** rem **Cu** 0.6 max **Fe** 1.0 max **Mg** 0.45-0.6 **Mn** 0.35 max **Ni** 0.50 max **Si** 9.0-10.0 **Sn** 0.15 max **Zn** 0.40 max **Other** total 0.25 max | **AA** A360.1 **ASTM** B179 (A360.1) |
| **A13602** | Aluminum Foundry Alloy, Ingot | **Al** rem **Cu** 0.10 max **Fe** 0.6 max **Mg** 0.45-0.6 **Mn** 0.05 max **Si** 9.0-10.0 **Zn** 0.05 max **Other** each 0.05 max, total 0.15 max | **AA** A360.2 **ASTM** B179 (A360.2) |
| **A13800** | Aluminum Foundry Alloy, Casting | **Al** rem **Cu** 3.0-4.0 **Fe** 1.3 max **Mg** 0.10 max **Mn** 0.50 max **Ni** 0.50 max **Si** 7.5-9.5 **Sn** 0.35 max **Zn** 3.0 max **Other** total 0.50 max | **AA** A380.0 **AMS** 4291 **ASTM** B85 (A380.0) **SAE** J452 (306) |

The chemical compositions listed are for identification purposes and should not be used in lieu of the cross referenced specifications.

| UNIFIED NUMBER | DESCRIPTION | CHEMICAL COMPOSITION | CROSS REFERENCE SPECIFICATIONS |
|---|---|---|---|
| **A13801** | Aluminum Foundry Alloy, Ingot | **Al** rem **Cu** 3.0-4.0 **Fe** 1.0 max **Mg** 0.10 max **Mn** 0.50 max **Ni** 0.50 max **Si** 7.5-9.5 **Sn** 0.35 max **Zn** 2.9 max **Other** total 0.50 max | **AA** A380.1 **ASTM** B179 (A380.1) |
| **A13802** | Aluminum Foundry Alloy, Ingot | **Al** rem **Cu** 3.0-4.0 **Fe** 0.6 max **Mg** 0.10 max **Mn** 0.10 max **Ni** 0.10 max **Si** 7.5-9.5 **Zn** 0.10 max **Other** each 0.05 max, total 0.15 max | **AA** A380.2 **ASTM** B179 (A380.2) |
| **A13830** | Aluminum Foundry Alloy, Die Casting | **Al** rem **Cu** 2.0-3.0 **Fe** 1.3 max **Mg** 0.10-0.30 **Mn** 0.50 max **Ni** 0.30 max **Si** 9.5-11.5 **Sn** 0.15 max **Zn** 3.0 max **Other** 0.50 total max | **AA** A383.0 |
| **A13831** | Aluminum Foundry Alloy, Ingot | **Al** rem **Cu** 2.0-3.0 **Fe** 1.0 max **Mg** 0.15-0.30 **Mn** 0.50 max **Ni** 0.30 max **Si** 9.5-11.5 **Sn** 0.15 max **Zn** 2.9 max **Other** 0.50 total max | **AA** A383.1 |
| **A13840** | Aluminum Foundry Alloy, Casting | **Al** rem **Cu** 3.0-4.5 **Fe** 1.3 max **Mg** 0.10 max **Mn** 0.50 max **Ni** 0.50 max **Si** 10.5-12.0 **Sn** 0.35 max **Zn** 1.0 max **Other** total 0.50 max | **AA** A384.0 |
| **A13841** | Aluminum Foundry Alloy, Ingot | **Al** rem **Cu** 3.0-4.5 **Fe** 1.0 max **Mg** 0.10 max **Mn** 0.50 max **Ni** 0.50 max **Si** 10.5-12.0 **Sn** 0.35 max **Zn** 0.9 max **Other** total 0.50 max | **AA** A384.1 |
| **A13900** | Aluminum Foundry Alloy, Casting | **Al** rem **Cu** 4.0-5.0 **Fe** 0.50 max **Mg** 0.45-0.65 **Mn** 0.10 max **Si** 16.0-18.0 **Ti** 0.20 max **Zn** 0.10 max **Other** each 0.10 max, total 0.20 max | **AA** A390.0 **SAE** J452 |
| **A13901** | Aluminum Foundry Alloy, Ingot | **Al** rem **Cu** 4.0-5.0 **Fe** 0.40 max **Mg** 0.50-0.65 **Mn** 0.10 max **Si** 16.0-18.0 **Ti** 0.20 max **Zn** 0.10 max **Other** each 0.10 max, total 0.20 max | **AA** A390.1 |
| **A13910** | Aluminum Foundry Alloy, Casting | **Al** rem **Cu** 0.20 max **Fe** 0.6 max **Mg** 0.40-0.7 **Mn** 0.30 max **Si** 18.0-20.0 **Ti** 0.20 max **Zn** 0.10 max **Other** each 0.10 max, total 0.20 max | **AA** A391.0 |
| **A13911** | Aluminum Foundry Alloy, Ingot | **Al** rem **Cu** 0.20 max **Fe** 0.5 max **Mg** 0.45-0.7 **Mn** 0.30 max **Si** 18.0-20.0 **Ti** 0.20 max **Zn** 0.10 max **Other** each 0.10 max, total 0.20 max | **AA** A391.1 |
| **A14130** | Aluminum Foundry Alloy, Casting | **Al** rem **Cu** 1.0 max **Fe** 1.3 max **Mg** 0.10 max **Mn** 0.35 max **Ni** 0.50 max **Si** 11.0-13.0 **Sn** 0.15 max **Zn** 0.50 max **Other** total 0.25 max | **AA** A413.0 **ASTM** B85 (A413.0) **SAE** J452 (305) |
| **A14131** | Aluminum Foundry Alloy, Ingot | **Al** rem **Cu** 1.0 max **Fe** 1.0 max **Mg** 0.10 max **Mn** 0.35 max **Ni** 0.50 max **Si** 11.0-13.0 **Sn** 0.15 max **Zn** 0.40 max **Other** total 0.25 max | **AA** A413.1 **ASTM** B179 (A413.1) |
| **A14132** | Aluminum Foundry Alloy, Ingot | **Al** rem **Cu** 0.10 max **Fe** 0.6 max **Mg** 0.05 max **Mn** 0.05 max **Ni** 0.05 max **Si** 11.0-13.0 **Sn** 0.05 max **Zn** 0.05 max **Other** total 0.10 max | **AA** A413.2 **ASTM** B179 (A413.2) |
| **A14430** | Aluminum Foundry Alloy, Casting | **Al** rem **Cr** 0.25 max **Cu** 0.30 max **Fe** 0.8 max **Mg** 0.05 max **Mn** 0.50 max **Si** 4.5-6.0 **Ti** 0.25 max **Zn** 0.50 max **Other** total 0.35 max | **AA** A443.0 |
| **A14431** | Aluminum Foundry Alloy, Ingot | **Al** rem **Cr** 0.25 max **Cu** 0.30 max **Fe** 0.6 max **Mg** 0.05 max **Mn** 0.50 max **Si** 4.5-6.0 **Ti** 0.25 max **Zn** 0.50 max **Other** total 0.35 max | **AA** A443.1 |

The chemical compositions listed are for identification purposes and should not be used in lieu of the cross referenced specifications.

| UNIFIED NUMBER | DESCRIPTION | CHEMICAL COMPOSITION | CROSS REFERENCE SPECIFICATIONS |
|---|---|---|---|
| **A14440** | Aluminum Foundry Alloy, Casting | **Al** rem **Cu** 0.10 max **Fe** 0.20 max **Mg** 0.05 max **Mn** 0.10 max **Si** 6.5-7.5 **Ti** 0.20 max **Zn** 0.10 max **Other** each 0.05 max, total 0.15 max | **AA** A444.0 **ASTM** B108 (A444.0) **SAE** J452 |
| **A14441** | Aluminum Foundry Alloy, Ingot | **Al** rem **Cu** 0.10 max **Fe** 0.15 max **Mg** 0.05 max **Mn** 0.10 max **Si** 6.5-7.5 **Ti** 0.20 max **Zn** 0.10 max **Other** each 0.05 max, total 0.15 max | **AA** A444.1 |
| **A14442** | Aluminum Foundry Alloy, Ingot | **Al** rem **Cu** 0.05 max **Fe** 0.12 max **Mg** 0.05 max **Mn** 0.05 max **Si** 6.5-7.5 **Ti** 0.20 max **Zn** 0.05 max **Other** each 0.05 max, total 0.15 max | **AA** A444.2 **ASTM** B179 (A444.2) |
| **A15140** | Aluminum Foundry Alloy, Casting replaced by A05130 | | |
| **A15142** | Aluminum Foundry Alloy, Ingot replaced by A05132 | | |
| **A15350** | Aluminum Foundry Alloy, Sand Castings | **Al** rem **Cu** 0.10 max **Fe** 0.20 max **Mg** 6.5-7.5 **Mn** 0.10-0.25 **Si** 0.20 max **Ti** 0.25 max **Other** each 0.05 max, total 0.15 max | **AA** A535.0 |
| **A15351** | Aluminum Foundry Alloy, Ingot | **Al** rem **Cu** 0.10 max **Fe** 0.15 max **Mg** 6.6-7.5 **Mn** 0.10-0.25 **Si** 0.20 max **Ti** 0.25 max **Other** each 0.05 max, total 0.15 max | **AA** A535.1 |
| **A17120** | Aluminum Foundry Alloy, Casting replaced by A07100 | | |
| **A17121** | Aluminum Foundry Alloy, Ingot replaced by A07101 | | |
| **A18500** | Aluminum Foundry Alloy, Casting replaced by A08510 | | |
| **A18501** | Aluminum Foundry Alloy, Ingot replaced by A08511 | | |
| **A22010** | Aluminum Foundry Alloy, Casting | **Ag** 0.50-1.0 **Al** rem **Cu** 4.5-5.0 **Fe** 0.05 max **Mg** 0.25-0.35 **Mn** 0.20-0.50 **Si** 0.05 max **Ti** 0.15-0.35 **Other** each 0.05 max, total 0.15 max | **AA** B201.0 |
| **A22950** | Aluminum Foundry Alloy, Casting replaced by A02960 | | |
| **A22951** | Aluminum Foundry Alloy, Ingot replaced by A02961 | | |
| **A22952** | Aluminum Foundry Alloy, Ingot replaced by A02962 | | |
| **A23190** | Aluminum Foundry Alloy, Casting | **Al** rem **Cu** 3.0-4.0 **Fe** 1.2 max **Mg** 0.10-0.50 **Mn** 0.8 max **Ni** 0.50 max **Si** 5.5-6.5 **Ti** 0.25 max **Zn** 1.0 max **Other** total 0.50 max | **AA** B319.0 **SAE** J452 (329) |
| **A23191** | Aluminum Foundry Alloy, Ingot | **Al** rem **Cu** 3.0-4.0 **Fe** 0.9 max **Mg** 0.15-0.50 **Mn** 0.8 max **Ni** 0.50 max **Si** 5.5-6.5 **Ti** 0.25 max **Zn** 1.0 max **Other** total 0.50 max | **AA** B319.1 **SAE** J452 (329) |
| **A23560** | Aluminum Foundry Alloy, Casting | **Al** rem **Cu** 0.05 max **Fe** 0.09 max **Mg** 0.25-0.45 **Mn** 0.05 max **Si** 6.5-7.5 **Ti** 0.04-0.20 **Zn** 0.05 max **Other** each 0.05, total 0.15 | **AA** B356.0 |
| **A23562** | Aluminum Foundry Alloy, Ingot | **Al** rem **Cu** 0.03 max **Fe** 0.06 max **Mg** 0.30-0.45 **Mn** 0.03 max **Si** 6.5-7.5 **Ti** 0.04-0.20 **Zn** 0.03 max **Other** each 0.03, total 0.10 | **AA** B356.2 |

The chemical compositions listed are for identification purposes and should not be used in lieu of the cross referenced specifications.

| UNIFIED NUMBER | DESCRIPTION | CHEMICAL COMPOSITION | CROSS REFERENCE SPECIFICATIONS |
|---|---|---|---|
| **A23570** | Aluminum Foundry Alloy, Casting | **Al** rem **Cu** 0.05 max **Fe** 0.09 max **Mg** 0.40-0.6 **Mn** 0.05 max **Si** 6.5-7.5 **Ti** 0.04-0.20 **Zn** 0.05 max **Other** each 0.05 max, total 0.15 max | **AA** A357.0; B357.0 |
| **A23572** | Aluminum Foundry Alloy, Ingot | **Al** rem **Cu** 0.03 max **Fe** 0.06 max **Mg** 0.45-0.6 **Mn** 0.03 max **Si** 6.5-7.5 **Ti** 0.04-0.20 **Zn** 0.03 max **Other** each 0.03 max, total 0.10 max | **AA** B357.2 |
| **A23580** | Aluminum Foundry Alloy, Casting replaced by A03580 | | |
| **A23582** | Aluminum Foundry Alloy, Ingot replaced by A03582 | | |
| **A23800** | Aluminum Foundry Alloy, Casting | **Al** rem **Cu** 3.0-4.0 **Fe** 1.3 max **Mg** 0.10 max **Mn** 0.50 max **Ni** 0.50 max **Si** 7.5-9.5 **Sn** 0.35 max **Zn** 1.0 max **Other** total 0.50 max | **AA** B380.0 |
| **A23801** | Aluminum Foundry Alloy, Ingot | **Al** rem **Cu** 3.0-4.0 **Fe** 1.0 max **Mg** 0.10 max **Mn** 0.50 max **Ni** 0.50 max **Si** 7.5-9.5 **Sn** 0.35 max **Zn** 0.9 max **Other** total 0.50 max | **AA** B380.1 |
| **A23840** | Aluminum Foundry Alloy, Casting | **Al** rem **Cu** 3.0-4.5 **Fe** 1.3 max **Mg** 0.10-0.30 **Mn** 0.50 max **Ni** 0.50 max **Si** 10.5-12.0 **Sn** 0.35 max **Zn** 1.0 max **Other** total 0.50 max | **AA** B384.0 |
| **A23841** | Aluminum Foundry Alloy, Ingot | **Al** rem **Cu** 3.0-4.5 **Fe** 1.0 max **Mg** 0.15-0.30 **Mn** 0.50 max **Ni** 0.50 max **Si** 10.5-12.0 **Sn** 0.35 max **Zn** 0.9 max **Other** 0.50 max total | **AA** B384.1 |
| **A23900** | Aluminum Foundry Alloy, Casting | **Al** rem **Cu** 4.0-5.0 **Fe** 1.3 max **Mg** 0.45-0.65 **Mn** 0.50 max **Ni** 0.10 max **Si** 16.0-18.0 **Ti** 0.20 max **Zn** 1.5 max **Other** each 0.10 max, total 0.20 max | **AA** B390.0 **ASTM** B85 (B390.0) **SAE** J452 |
| **A23901** | Aluminum Foundry Alloy, Ingot | **Al** rem **Cu** 4.0-5.0 **Fe** 1.0 max **Mg** 0.50-0.65 **Mn** 0.50 max **Ni** 0.10 max **Si** 16.0-18.0 **Ti** 0.20 max **Zn** 1.4 max **Other** each 0.10 max, total 0.20 max | **AA** B390.1 **ASTM** B179 (392.1) |
| **A23910** | Aluminum Foundry Alloy, Casting | **Al** rem **Cu** 0.20 max **Fe** 0.20 max **Mg** 0.40-0.7 **Mn** 0.30 max **Si** 18.0-20.0 **Ti** 0.20 max **Zn** 0.10 max **Other** each 0.10 max, total 0.20 max | **AA** B391.1 |
| **A23911** | Aluminum Foundry Alloy, Ingot | **Al** rem **Cu** 0.20 max **Fe** 0.15 max **Mg** 0.45-0.7 **Mn** 0.30 max **Si** 18.0-20.0 **Ti** 0.20 max **Zn** 0.10 max **Other** each 0.10 max, total 0.20 max | **AA** B391.1 |
| **A24130** | Aluminum Foundry Alloy, Casting | **Al** rem **Cu** 0.10 max **Fe** 0.50 max **Mg** 0.05 max **Mn** 0.35 max **Ni** 0.05 max **Si** 11.0-13.0 **Ti** 0.25 max **Zn** 0.10 max **Other** each 0.05 max, total 0.20 max | **AA** B413.0 |
| **A24131** | Aluminum Foundry Alloy, Ingot | **Al** rem **Cu** 0.10 max **Fe** 0.40 max **Mg** 0.05 max **Mn** 0.35 max **Ni** 0.05 max **Si** 11.0-13.0 **Ti** 0.25 max **Zn** 0.10 max **Other** each 0.05 max, total 0.20 max | **AA** B413.1 |
| **A24430** | Aluminum Foundry Alloy, Casting | **Al** rem **Cu** 0.15 max **Fe** 0.8 max **Mg** 0.05 max **Mn** 0.35 max **Si** 4.5-6.0 **Ti** 0.25 max **Zn** 0.35 max **Other** each 0.05 max, total 0.15 max | **AA** B443.0 **ASTM** B26 (B443.0); B108 (B443.0); B618 (B443.0) |
| **A24431** | Aluminum Foundry Alloy, Ingot | **Al** rem **Cu** 0.15 max **Fe** 0.6 max **Mg** 0.05 max **Mn** 0.35 max **Si** 4.5-6.0 **Ti** 0.25 max **Zn** 0.35 max **Other** each 0.05 max, total 0.15 max | **AA** B443.1 **ASTM** B179 (B443.1) |

The chemical compositions listed are for identification purposes and should not be used in lieu of the cross referenced specifications.

| UNIFIED NUMBER | DESCRIPTION | CHEMICAL COMPOSITION | CROSS REFERENCE SPECIFICATIONS |
|---|---|---|---|
| **A24442** | Aluminum Foundry Alloy, Ingot (Steel Coating Alloy) replaced by A04452 | | |
| **A25140** | Aluminum Foundry Alloy, Casting replaced by A05120 | | |
| **A25142** | Aluminum Foundry Alloy, Ingot replaced by A05122 | | |
| **A25350** | Aluminum Foundry Alloy, Casting | **Al** rem **Cu** 0.10 max **Fe** 0.15 max **Mg** 6.5-7.5 **Mn** 0.05 max **Si** 0.15 max **Ti** 0.10-0.25 **Other** each 0.05 max, total 0.15 max | **AA** B535.0 |
| **A25352** | Aluminum Foundry Alloy, Ingot | **Al** rem **Cu** 0.05 max **Fe** 0.12 max **Mg** 6.6-7.5 **Mn** 0.05 max **Si** 0.10 max **Ti** 0.10-0.25 **Other** each 0.05 max, total 0.15 max | **AA** B535.2 |
| **A27710** | Aluminum Foundry Alloy, Casting replaced by A07720 | | |
| **A27712** | Aluminum Foundry Alloy, Ingot replaced by A07722 | | |
| **A28500** | Aluminum Foundry Alloy, Casting replaced by A08520 | | |
| **A28501** | Aluminum Foundry Alloy, Ingot replaced by A08521 | | |
| **A33550** | Aluminum Foundry Alloy, Casting | **Al** rem **Cu** 1.0-1.5 **Fe** 0.20 max **Mg** 0.40-0.6 **Mn** 0.10 max **Si** 4.5-5.5 **Ti** 0.20 max **Zn** 0.10 max **Other** each 0.05 max, total 0.15 max | **AA** C355.0 **AMS** 4215; AMS-A-21180 (C355.0) **ASTM** B26 (C355.0); B108 (C355.0); B618 (C355.0); B686 (C355.0) **SAE** J452 (335) |
| **A33551** | Aluminum Foundry Alloy, Ingot | **Al** rem **Cu** 1.0-1.5 **Fe** 0.15 max **Mg** 0.45-0.6 **Mn** 0.10 max **Si** 4.5-5.5 **Ti** 0.20 max **Zr** 0.10 max **Other** each 0.05 max, total 0.15 max. | **AA** C355.1 |
| **A33552** | Aluminum Foundry Alloy, Ingot | **Al** rem **Cu** 1.0-1.5 **Fe** 0.13 max **Mg** 0.50-0.6 **Mn** 0.05 max **Si** 4.5-5.5 **Ti** 0.20 max **Zn** 0.05 max **Other** each 0.05 max, total 0.15 max | **AA** C355.2 **ASTM** B179 (C355.2) |
| **A33560** | Aluminum Foundry Alloy, Casting | **Al** rem **Cu** 0.05 max **Fe** 0.07 max **Mg** 0.25-0.45 **Mn** 0.05 max **Si** 6.5-7.5 **Ti** 0.04-0.20 **Zn** 0.05 max **Other** each 0.05 max, total 0.15 max | **AA** C356.0 |
| **A33562** | Aluminum Foundry Alloy, Ingot | **Al** rem **Cu** 0.03 max **Fe** 0.04 max **Mg** 0.30-0.45 **Mn** 0.03 max **Si** 6.5-7.5 **Ti** 0.04-0.20 **Zn** 0.03 max **Other** each 0.03 max, total 0.10 max | **AA** C356.2 |
| **A33570** | Aluminum Foundry Alloy, Casting | **Al** rem **Be** 0.04-0.07 **Cu** 0.05 max **Fe** 0.09 max **Mg** 0.45-0.7 **Mn** 0.05 max **Si** 6.5-7.5 **Ti** 0.04-0.20 **Zn** 0.05 max **Other** each 0.05 max, total 0.15 max | **AA** C357.0 |
| **A33572** | Aluminum Foundry Alloy, Ingot | **Al** rem **Be** 0.04-0.07 **Cu** 0.03 max **Fe** 0.06 max **Mg** 0.50-0.7 **Mn** 0.03 max **Si** 6.5-7.5 **Ti** 0.04-0.20 **Zn** 0.03 max **Other** each 0.03 max, total 0.10 max | **AA** C357.2 |
| **A33800** | Aluminum Foundry Alloy, Die Casting | **Al** rem **Cu** 3.0-4.0 **Fe** 1.3 max **Mg** 0.10-0.30 **Mn** 0.50 max **Ni** 0.50 max **Si** 7.5-9.5 **Sn** 0.35 max **Zn** 3.0 max **Other** 0.50 total max | **AA** C380.0 |
| **A33801** | Aluminum Foundry Alloy, Ingot | **Al** rem **Cu** 3.0-4.0 **Fe** 1.0 max **Mg** 0.15-0.30 **Mn** 0.50 max **Ni** 0.50 max **Si** 7.5-9.5 **Sn** 0.35 max **Zn** 2.9 max **Other** 0.50 total max | **AA** C380.1 |

The chemical compositions listed are for identification purposes and should not be used in lieu of the cross referenced specifications.

| UNIFIED NUMBER | DESCRIPTION | CHEMICAL COMPOSITION | CROSS REFERENCE SPECIFICATIONS |
|---|---|---|---|
| **A33840** | Aluminum Foundry Alloy, Die Casting | **Al** rem **Cu** 3.0-4.5 **Fe** 1.3 max **Mg** 0.10-0.30 **Mn** 0.50 max **Ni** 0.50 max **Si** 10.5-12.0 **Sn** 0.35 max **Zn** 3.0 max **Other** 0.50 total max | **AA** C384.0 |
| **A33841** | Aluminum Foundry Alloy, Ingot | **Al** rem **Cu** 3.0-4.5 **Fe** 1.0 max **Mg** 0.15-0.30 **Mn** 0.50 max **Ni** 0.50 max **Si** 10.5-12.0 **Sn** 0.35 max **Zn** 2.9 max **Other** 0.50 total max | **AA** C384.1 |
| **A34430** | Aluminum Foundry Alloy, Casting | **Al** rem **Cu** 0.6 max **Fe** 2.0 max **Mg** 0.10 max **Mn** 0.35 max **Ni** 0.50 max **Si** 4.5-6.0 **Sn** 0.15 max **Zn** 0.50 max **Other** total 0.25 max | **AA** C443.0 **ASTM** B85 (C443.0) **SAE** J452 (304) |
| **A34431** | Aluminum Foundry Alloy, Ingot | **Al** rem **Cu** 0.6 max **Fe** 1.1 max **Mg** 0.10 max **Mn** 0.35 max **Ni** 0.50 max **Si** 4.5-6.0 **Sn** 0.15 max **Zn** 0.40 max **Other** total 0.25 max | **AA** C443.1 **ASTM** B179 (C443.1) |
| **A34432** | Aluminum Foundry Alloy, Ingot | **Al** rem **Cu** 0.10 max **Fe** 0.7-1.1 **Mg** 0.05 max **Mn** 0.10 max **Si** 4.5-6.0 **Zn** 0.10 max **Other** each 0.05 max, total 0.15 max | **AA** C443.2 **ASTM** B179 (C443.2) |
| **A37120** | Aluminum Foundry Alloy, Casting replaced by A07110 | | |
| **A37121** | Aluminum Foundry Alloy, Ingot replaced by A07111 | | |
| * **A43560** | Aluminum Foundry Alloy, Casting | **Al** rem **Cu** 0.20 max **Fe** 0.15 max **Mg** 0.30-0.45 **Mn** 0.10 max **Si** 6.5-7.5 **Ti** 0.20 max **Zn** 0.10 max **Other** each 0.05 max, total 0.15 max | **AA** D356.0 |
| * **A43561** | Aluminum Foundry Alloy, Ingot | **Al** rem **Cu** 0.20 max **Fe** 0.12 max **Mg** 0.35-0.45 **Mn** 0.10 max **Si** 6.5-7.5 **Ti** 0.20 max **Zn** 0.10 max **Other** each 0.05 max, total 0.15 max | **AA** D356.1 |
| **A43570** | Aluminum Foundry Alloy, Casting | **Al** rem **Be** 0.04-0.07 **Fe** 0.20 max **Mg** 0.55-0.6 **Mn** 0.10 max **Si** 6.5-7.5 **Ti** 0.10-0.20 **Other** each 0.05 max, total 0.15 max | **AA** D357.0 **AMS** 4249 |
| **A43800** | Aluminum Foundry Alloy, Die Casting | **Al** rem **Cu** 3.0-4.0 **Fe** 1.3 max **Mg** 0.10-0.30 **Mn** 0.50 max **Ni** 0.50 max **Si** 7.5-9.5 **Sn** 0.35 max **Zn** 1.0 max **Other** 0.50 total max | **AA** D380.0 |
| **A43801** | Aluminum Foundry Alloy, Ingot | **Al** rem **Cu** 3.0-4.0 **Fe** 1.0 max **Mg** 0.15-0.30 **Mn** 0.50 max **Ni** 0.50 max **Si** 7.5-9.5 **Sn** 0.35 max **Zn** 0.9 max **Other** 0.50 total max | **AA** D380.1 |
| **A47120** | Aluminum Foundry Alloy, Casting replaced by A07120 | | |
| **A47122** | Aluminum Foundry Alloy, Ingot replaced by A07122 | | |
| * **A53570** | Aluminum Foundry Alloy, Casting | **Al** rem **Cu** 0.10 max **Fe** 0.10 max **Mg** 0.45-0.6 **Mn** 0.05 max **Si** 6.5-7.5 **Ti** 0.20 max **Zn** 0.05 max **Other** each 0.05 max, total 0.15 max | **AA** E357.0 |

*Boxed entries are no longer active and are retained for reference purposes only.

The chemical compositions listed are for identification purposes and should not be used in lieu of the cross referenced specifications.

| UNIFIED NUMBER | DESCRIPTION | CHEMICAL COMPOSITION | CROSS REFERENCE SPECIFICATIONS |
|---|---|---|---|
| * **A53571** | Aluminum Foundry Alloy, Ingot | **Al** rem **Cu** 0.10 max **Fe** 0.07 max **Mg** 0.50-0.6 **Mn** 0.05 max **Si** 6.5-7.5 **Ti** 0.20 max **Zn** 0.05 max **Other** each 0.05 max, total 0.15 max | **AA** E357.1 |
| **A63320** | Aluminum Foundry Alloy, Casting replaced by A03320 | | |
| **A63321** | Aluminum Foundry Alloy, Ingot replaced by A03321 | | |
| **A63322** | Aluminum Foundry Alloy, Ingot replaced by A03322 | | |
| **A63560** | Aluminum Foundry Alloy, Casting | **Al** rem **Cu** 0.20 max **Fe** 0.20 max **Mg** 0.17-0.25 **Mn** 0.10 max **Si** 6.5-7.5 **Ti** 0.04-0.20 **Zn** 0.10 max **Other** each 0.05 max, total 0.15 max | **AA** F356.0 |
| **A63562** | Aluminum Foundry Alloy, Ingot | **Al** rem **Cu** 0.10 max **Fe** 0.12 max **Mg** 0.17-0.25 **Mn** 0.05 max **Si** 6.5-7.5 **Ti** 0.04-0.20 **Zn** 0.05 max **Other** each 0.05 max, total 0.15 max | **AA** F356.2 |
| **A65140** | Aluminum Foundry Alloy, Casting replaced by A05110 | | |
| **A65141** | Aluminum Foundry Alloy, Ingot replaced by A05111 | | |
| **A65142** | Aluminum Foundry Alloy, Ingot replaced by A05112 | | |
| **A82004** | Wrought Aluminum Alloy, Heat Treatable | **Core** Alloy A92004 **Cladding** Alloy A91070 | **AA** Alclad 2004 **AMS** 4209 **MAM** 4209 |
| **A71001** | Aluminum Metal Matrix Composite, Non-heat Treatable (Thermal Spray Wire) | **Al** 88.0-92.0 (of A91060) **Other** Al2O3 particles 8.0-12.0 | **AA** 1060/Al2O3/10p **AWS** C2.25 (W-Al-Al2O3) |
| **A82014** | Wrought Aluminum Alloy Clad with Wrought Aluminum Alloy, Heat Treatable | **Core** Alloy A92014 **Cladding** Alloy A96003 | **AA** Alclad 2014 **AMS** AMS-QQ-A-250/3 **ASTM** B209 (Alclad 2014) |
| **A82024** | Wrought Aluminum Alloy Clad with Wrought Aluminum Alloy, Heat Treatable | **Core** Alloy A92024 **Cladding** Alloy A91230 | **AA** Alclad 2024; 1-1/2% Alclad 2024; Alclad One Side 2024; 1-1/2% Alclad One Side 2024 **AMS** 4036; 4040; 4041; 4077; 4194; 4195; 4279; AMS-QQ-A-250/5 **ASTM** B209 (Alclad 2024) |
| **A82219** | Wrought Aluminum Alloy Clad with Wrought Aluminum Alloy, Heat Treatable | **Core** Alloy A92219 **Cladding** Alloy A97072 | **AA** Alclad 2219 **AMS** 4094; 4095; 4096 **ASTM** B209 (Alclad 2219) |
| **A83003** | Wrought Aluminum Alloy Clad with Wrought Aluminum Alloy, Non-Heat Treatable | **Core** Alloy A93003 **Cladding** Alloy A94343 | **AA** No. 7 Brazing Sheet; No. 8 Brazing Sheet; No. 11 Brazing Sheet; No. 12 Brazing Sheet; Alclad 3003 **AMS** 4063; 4064 **ASME** SB-209; SB-210; SB-210; SB-234; SB-241 **ASTM** B209; B210; B221; B234; B361 (Alclad 3003); B241; B345; B404; B547; B548 (Alclad 3003) |
| **A86061** | Wrought Aluminum Alloy Clad with Wrought Aluminum Alloy, Heat Treatable | **Core** Alloy A96061 **Cladding** Alloy A97072 | **AA** Alclad 6061 **AMS** 4021 **ASME** SB-209 **ASTM** B209 (Alclad 6061); B548 (6061) |
| **A86951** | Wrought Aluminum Alloy Clad with Wrought Aluminum Alloy, Heat Treatable | **Core** Alloy A96951 **Cladding** Alloy A94343 | **AA** Alclad 6951; No. 13 Brazing Sheet; No. 14 Brazing Sheet; No. 21 Brazing Sheet; No. 22 Brazing Sheet; No. 23 Brazing Sheet; No. 24 Brazing Sheet; No. 33 Brazing Sheet; No. 34 Brazing Sheet; No. 44 Brazing Sheet **AMS** 4255; 4256 |
| **A87050** | Wrought Aluminum Alloy Clad with Wrought Aluminum Alloy, Heat Treatable | **Core** Alloy A97050 **Cladding** Alloy A97072 | **AA** Alclad 7050; 7050A **AMS** 4243; 4255; 4256 |
| **A87075** | Wrought Aluminum Alloy Clad with Wrought Aluminum Alloy, Heat Treatable | **Core** Alloy A97075 **Cladding** Alloy A97072 | **AA** Alclad 7075; 2-1/2% Alclad 7075; Alclad One Side 7075; 2-1/2% Alclad One Side 7075 **AMS** 4046; 4048; 4278; AMS-QQ-A-250/13; AMS-QQ-A-250/18; AMS-QQ-A-250/25; AMS-QQ-A-250/26 **ASTM** B209 (Alclad 7075) |

*Boxed entries are no longer active and are retained for reference purposes only.

The chemical compositions listed are for identification purposes and should not be used in lieu of the cross referenced specifications.

| UNIFIED NUMBER | DESCRIPTION | CHEMICAL COMPOSITION | CROSS REFERENCE SPECIFICATIONS |
|---|---|---|---|
| * **A87079** | Wrought Aluminum Alloy Clad with Wrought Aluminum Alloy, Heat Treatable | **Core** Alloy A97079 **Cladding** Alloy A97072 | **AA** Alclad 7079 |
| **A87178** | Wrought Aluminum Alloy Clad with Wrought Aluminum Alloy, Heat Treatable | **Core** Alloy A97178 **Cladding** Alloy A97072 | **AA** Alclad 7178 **AMS** AMS-QQ-A-250/15; AMS-QQ-A-250/22; AMS-QQ-A-250/28 **ASTM** B209 (Alclad 7178) |
| **A87475** | Wrought Aluminum Alloy Clad with Wrought Aluminum Alloy, Heat Treatable | **Core** Alloy A97475 **Cladding** Alloy A97072 | **AA** Alclad 7475 **AMS** 4207; 4100 |
| * **A91030** | Wrought Aluminum Alloy, Non-Heat Treatable | **Al** 99.30 min **Cu** 0.10 max **Fe** 0.6 max **Mg** 0.05 max **Mn** 0.05 max **Si** 0.35 max **Ti** 0.03 max **V** 0.05 max **Zn** 0.10 max **Other** each 0.03 max | **AA** 1030 |
| **A91035** | Wrought Aluminum Alloy, Non-Heat Treatable | **Al** 99.35 min **Cu** 0.10 max **Fe** 0.6 max **Mg** 0.05 max **Mn** 0.05 max **Si** 0.35 max **Ti** 0.03 max **V** 0.05 max **Zn** 0.10 max **Other** each 0.03 max | **AA** 1035 |
| **A91040** | Wrought Aluminum Alloy, Non-Heat Treatable | **Al** 99.40 min **Cu** 0.10 max **Fe** 0.50 max **Mg** 0.05 max **Mn** 0.05 max **Si** 0.30 max **Ti** 0.03 max **V** 0.05 max **Zn** 0.10 max **Other** each 0.03 max | **AA** 1040 |
| **A91045** | Wrought Aluminum Alloy, Non-Heat Treatable | **Al** 99.45 min **Cu** 0.10 max **Fe** 0.45 max **Mg** 0.05 max **Mn** 0.05 max **Si** 0.30 max **Ti** 0.03 max **V** 0.05 max **Zn** 0.05 max **Other** each 0.03 max | **AA** 1045 |
| **A91050** | Wrought Aluminum Alloy, Non-Heat Treatable | **Al** 99.50 min **Cu** 0.05 max **Fe** 0.40 max **Mg** 0.05 max **Mn** 0.05 max **Si** 0.25 max **Ti** 0.03 max **V** 0.05 max **Zn** 0.05 max **Other** each 0.03 max | **AA** 1050; 1050A **ASTM** B491 (1050) |
| * **A91055** | Wrought Aluminum Alloy, Non-Heat Treatable | **Al** 99.55 min **Cu** 0.05 max **Fe** 0.40 max **Mg** 0.05 max **Mn** 0.05 max **Si** 0.25 max **Ti** 0.03 max **V** 0.05 max **Zn** 0.05 max **Other** each 0.03 max | **AA** 1055 |
| **A91060** | Wrought Aluminum Alloy, Non-Heat Treatable | **Al** 99.60 min **Cu** 0.05 max **Fe** 0.35 max **Mg** 0.03 max **Mn** 0.03 max **Si** 0.25 max **Ti** 0.03 max **V** 0.05 max **Zn** 0.05 max **Other** each 0.03 max (Be 0.0008 max for welding electrode and filler metal only) | **AA** 1060 **AMS** 4000 **ASME** SB-209; SB-210; SB-221; SB-234; SB-241 **ASTM** B209 (1060); B210 (1060); B211 (1060); B221 (1060); B234 (1060); B241 (1060); B345 (1060); B361 (1060); B404 (1060); B483 (1060); B548 (1060); B736 (1060) |
| **A91065** | Wrought Aluminum Alloy, Non-Heat Treatable | **Al** 99.65 min **Cu** 0.05 max **Fe** 0.30 max **Mg** 0.03 max **Mn** 0.03 max **Si** 0.25 max **Ti** 0.03 max **V** 0.05 max **Zn** 0.05 max **Other** each 0.03 max | **AA** 1065 |
| **A91070** | Wrought Aluminum Alloy, Non-Heat Treatable | **Al** 99.70 min **Cu** 0.04 max **Fe** 0.25 max **Mg** 0.03 max **Mn** 0.03 max **Si** 0.20 max **Ti** 0.03 max **V** 0.05 max **Zn** 0.04 max **Other** each 0.03 max | **AA** 1070; 1070A |
| * **A91075** | Wrought Aluminum Alloy, Non-Heat Treatable | **Al** 99.75 min **Cu** 0.04 max **Fe** 0.20 max **Mg** 0.03 max **Mn** 0.03 max **Si** 0.20 max **Ti** 0.03 max **V** 0.05 max **Zn** 0.04 max **Other** each 0.03 max | **AA** 1075 |

*Boxed entries are no longer active and are retained for reference purposes only.

The chemical compositions listed are for identification purposes and should not be used in lieu of the cross referenced specifications.

| UNIFIED NUMBER | DESCRIPTION | CHEMICAL COMPOSITION | CROSS REFERENCE SPECIFICATIONS |
|---|---|---|---|
| **A91080** | Wrought Aluminum Alloy, Non-Heat Treatable | **Al** 99.80 min **Cu** 0.03 max **Fe** 0.15 max **Ga** 0.03 max **Mg** 0.02 max **Mn** 0.02 max **Si** 0.15 max **Ti** 0.03 max **V** 0.05 max **Zn** 0.03 max **Other** each 0.02 max | **AA** 1080; 1080A |
| **A91085** | Wrought Aluminum Alloy, Non-Heat Treatable | **Al** 99.85 min **Cu** 0.03 max **Fe** 0.12 max **Ga** 0.03 max **Mg** 0.02 max **Mn** 0.02 max **Si** 0.10 max **Ti** 0.02 max **V** 0.05 max **Zn** 0.03 max **Other** each 0.01 max | **AA** 1085 |
| **A91090** | Wrought Aluminum Alloy, Non-Heat Treatable | **Al** 99.90 min **Cu** 0.02 max **Fe** 0.07 max **Ga** 0.03 max **Mg** 0.01 max **Mn** 0.01 max **Si** 0.07 max **Ti** 0.01 max **V** 0.05 max **Zn** 0.03 max **Other** each 0.01 max | **AA** 1090 |
| * **A91095** | Wrought Aluminum Alloy, Non-Heat Treatable | **Al** 99.95 min **Cu** 0.010 max **Fe** 0.040 max **Mg** 0.010 max **Mn** 0.010 max **Si** 0.030 max **Ti** 0.005 max **Zn** 0.010 max **Other** each 0.005 max | **AA** 1095 |
| **A91098** | Wrought Aluminum Alloy, Non-Heat Treatable | **Al** 99.98 min **Cu** 0.003 max **Fe** 0.006 max **Si** 0.010 max **Ti** 0.003 max **Zn** 0.015 max **Other** 0.003 max each | **AA** 1098 |
| **A91100** | Wrought Aluminum Alloy, Non-Heat Treatable | **Al** 99.00 min **Cu** 0.05-0.20 **Mn** 0.05 max **Zn** 0.10 max **Other** each 0.05 max (Be 0.0008 max for welding electrode and filler wire only), total 0.15 max, Fe & Si 0.95 max | **AA** 1100 **AMS** 4001; 4003; 4062; 4102; 4180; AMS-QQ-A-225/1; AMS-QQ-A-250/1; AS7220 **ASME** SB-209; SB-211; SB-241; SB-241/SB-241M; SFA5.3 (E1100); SFA5.10 (ER1100) **ASTM** B209 (1100); B210 (1100); B211 (1100); B221 (1100); B241 (1100); B247 (1100); B313 (1100); B316 (1100); B361 (1100); B479 (1100); B483 (1100); B491 (1100); B547 (1100); B548 (1100); B736 (1100) **AWS** A5.3 (E1100); A5.10 (ER1100); C2.25 (W-Al-1100) **FED** QQ-A-430; QQ-A-1876; WW-T-700/1 **MIL SPEC** MIL-W-6712 **SAE** AS7220 (1100); J454 (1100) |
| **A91110** | Wrought Aluminum Alloy, Non-Heat Treatable | **Al** 99.10 min **B** 0.02 max **Cr** 0.01 max **Cu** 0.04 max **Fe** 0.8 max **Mg** 0.25 max **Mn** 0.01 max **Si** 0.30 max **Other** each 0.03 max, Ti+V 0.03 max | **AA** 1110 |
| **A91120** | Wrought Aluminum Alloy, Non-Heat Treatable | **Al** 99.20 min **B** 0.05 max **Cr** 0.01 max **Cu** 0.05-0.35 **Fe** 0.40 max **Ga** 0.03 max **Mg** 0.20 max **Mn** 0.01 max **Si** 0.10 max **Zn** 0.05 max **Other** each 0.03 max, total 0.10 max, Ti+V 0.02 max | **AA** 1120 |
| * **A91135** | Wrought Aluminum Alloy, Non-Heat Treatable | **Al** 99.35 min **Cu** 0.05-0.20 **Mg** 0.05 max **Mn** 0.04 max **Ti** 0.03 max **Zn** 0.10 max **Other** each 0.03 max, Fe+Si 0.60 max | **AA** 1135 |
| **A91145** | Wrought Aluminum Alloy, Non-Heat Treatable | **Al** 99.45 min **Cu** 0.05 max **Mg** 0.05 max **Mn** 0.05 max **Ti** 0.03 max **Zn** 0.05 max **Other** each 0.03 max, Fe+Si 0.55 max | **AA** 1145 **AMS** 4011 **ASTM** B373 (1145); B479 (1145); B737 (1145) **FED** QQ-A-1876 |
| **A91150** | Wrought Aluminum Alloy, Non-Heat Treatable | **Al** 99.50 min **Cu** 0.05-0.20 **Mg** 0.05 max **Mn** 0.05 max **Ti** 0.03 max **Zn** 0.05 max **Other** 0.03 max each, Fe+Si 0.45 max | **AA** 1150 |
| * **A91170** | Wrought Aluminum Alloy, Non-Heat Treatable | **Al** 99.70 min **Cr** 0.03 max **Cu** 0.03 max **Mg** 0.02 max **Mn** 0.03 max **Ti** 0.03 max **V** 0.05 max **Zn** 0.04 max **Other** each 0.03 max, Fe+Si 0.30 max | **AA** 1170 |

*Boxed entries are no longer active and are retained for reference purposes only.

The chemical compositions listed are for identification purposes and should not be used in lieu of the cross referenced specifications.

| UNIFIED NUMBER | DESCRIPTION | CHEMICAL COMPOSITION | CROSS REFERENCE SPECIFICATIONS |
|---|---|---|---|
| * **A91175** | Wrought Aluminum Alloy, Non-Heat Treatable | **Al** 99.75 min **Cu** 0.10 max **Ga** 0.03 max **Mg** 0.02 max **Mn** 0.02 max **Ti** 0.02 max **V** 0.05 max **Zn** 0.03 max **Other** each 0.02 max, Fe+Si 0.15 max | **AA** 1175 |
| * **A91180** | Wrought Aluminum Alloy, Non-Heat Treatable | **Al** 99.80 min **Cu** 0.01 max **Fe** 0.09 max **Ga** 0.03 max **Mg** 0.02 max **Mn** 0.02 max **Si** 0.09 max **Ti** 0.02 max **V** 0.03 max **Zn** 0.03 max **Other** each 0.02 max | **AA** 1180 |
| **A91185** | Wrought Aluminum Alloy, Non-Heat Treatable | **Al** 99.85 min **Cu** 0.01 max **Ga** 0.03 max **Mg** 0.02 max **Mn** 0.02 max **V** 0.05 max **Zn** 0.03 max **Other** each 0.01 max, Fe+Si 0.15 max | **AA** 1185 |
| **A91188** | Wrought Aluminum Alloy, Non-Heat Treatable | **Al** 99.88 min **Cu** 0.005 max **Fe** 0.06 max **Ga** 0.03 max **Mg** 0.01 max **Mn** 0.01 max **Si** 0.06 max **Ti** 0.01 max **V** 0.05 max **Zn** 0.03 max **Other** each 0.01 max (Be 0.0008 max for welding electrode and filler wire only) | **AA** 1188 **ASME** SFA5.10 (ER1188) **AWS** A5.10 (ER1188) |
| **A91190** | Wrought Aluminum Alloy, Non-Heat Treatable | **Al** 99.90 min **B** 0.01 max **Cr** 0.01 max **Cu** 0.01 max **Fe** 0.07 max **Ga** 0.02 max **Mg** 0.01 max **Mn** 0.01 max **Si** 0.05 max **Zn** 0.02 max **Other** each 0.01 max, Ti+V 0.01 max | **AA** 1190 |
| **A91193** | Wrought Aluminum Alloy, Non-Heat Treatable | **Al** 99.93 min **Cu** 0.006 max **Fe** 0.04 max **Ga** 0.03 max **Si** 0.04 max **Ti** 0.01 max **V** 0.05 max **Zn** 0.03 max **Other** each 0.01 max | **AA** 1193 |
| **A91198** | Wrought Aluminum Alloy, Non-Heat Treatable | **Al** 99.98 min **Cu** 0.006 max **Fe** 0.006 max **Ga** 0.006 max **Mn** 0.006 max **Si** 0.010 max **Ti** 0.006 max **Zn** 0.010 max **Other** 0.003 max each | **AA** 1198 |
| **A91199** | Wrought Aluminum Alloy, Non-Heat Treatable | **Al** 99.99 min **Cu** 0.006 max **Fe** 0.006 max **Ga** 0.005 max **Mg** 0.006 max **Mn** 0.002 max **Si** 0.006 max **Ti** 0.002 max **V** 0.005 max **Zn** 0.006 max **Other** each 0.002 max | **AA** 1199 |
| **A91200** | Wrought Aluminum Alloy, Non-Heat Treatable | **Al** 99.00 min **Cu** 0.05 max **Mn** 0.05 max **Ti** 0.05 max **Zn** 0.10 max **Other** each 0.05 max, total 0.15 max, Fe+Si 1.00 max | **AA** 1200; 1200A **ASTM** B491 (1200) |
| **A91230** | Wrought Aluminum Alloy, Non-Heat Treatable | **Al** 99.30 min **Cu** 0.10 max **Mg** 0.05 max **Mn** 0.05 max **Ti** 0.03 max **V** 0.05 max **Zn** 0.10 max **Other** each 0.03, Fe+Si 0.7 max | **AA** 1230 **ASTM** B209 (1230) |
| **A91235** | Wrought Aluminum Alloy, Non-Heat Treatable | **Al** 99.35 min **Cu** 0.05 max **Mg** 0.05 max **Mn** 0.05 max **Ti** 0.03 max **Zn** 0.010 max **Other** each 0.03 max, Fe+Si 0.65 max | **AA** 1235 **ASTM** B373 (1235); B479 (1235); B491 (1235); B736 (1235) **FED** QQ-A-1876 |
| * **A91250** | Wrought Aluminum Alloy, Non-Heat Treatable | **Al** 99.50 min **Cr** 0.01 max **Cu** 0.10 max **Fe** 0.40 max **Mg** 0.01 max **Mn** 0.01 max **Si** 0.20 max **Zn** 0.05 max **Other** each 0.03 max, Ti+V 0.02 max | **AA** 1250 |

*Boxed entries are no longer active and are retained for reference purposes only.

The chemical compositions listed are for identification purposes and should not be used in lieu of the cross referenced specifications.

| UNIFIED NUMBER | DESCRIPTION | CHEMICAL COMPOSITION | CROSS REFERENCE SPECIFICATIONS |
|---|---|---|---|
| **A91260** | Wrought Aluminum Alloy, Non-Heat Treatable | **Al** 99.60 min **Cu** 0.04 max **Mg** 0.03 max **Mn** 0.01 max **Ti** 0.03 max **V** 0.05 max **Zn** 0.05 max **Other** each 0.03 max (Be 0.0008 max for welding electrode and filler wire only), Fe+Si 0.40 max | **AA** 1260 |
| **A91275** | Wrought Aluminum Alloy, Non-Heat Treatable | **Al** 99.75 min **Cu** 0.05-0.10 **Fe** 0.12 max **Ga** 0.03 max **Mg** 0.02 max **Mn** 0.02 max **Si** 0.08 max **Ti** 0.02 max **V** 0.03 max **Zn** 0.03 max **Other** 0.01 max each | **AA** 1275 |
| **A91285** | Wrought Aluminum Alloy, Non-Heat Treatable | **Al** 99.85 min **Cu** 0.02 max **Fe** 0.08 max **Ga** 0.03 max **Si** 0.08 max **Ti** 0.02 max **V** 0.05 max **Zn** 0.03 max **Other** Fe+Si 0.14 max | **AA** 1285 |
| **A91300** | Wrought Aluminum Alloy, Non-Heat Treatable | **Al** 99.00 min **Cu** 0.05 max **Fe** 0.30 max **Mg** 0.03 max **Mn** 0.03 max **Si** 0.20 max **Ti** 0.03 max **Zn** 0.20-0.50 **Other** each 0.05 max, total 0.15 max | **AA** 1300 |
| **A91345** | Wrought Aluminum Alloy, Non-Heat Treatable | **Al** 99.45 min **Cu** 0.10 max **Fe** 0.40 max **Mg** 0.05 max **Mn** 0.05 max **Si** 0.30 max **Ti** 0.03 max **V** 0.05 max **Zn** 0.05 max **Other** each 0.03 max | **AA** 1345 |
| **A91350** | Wrought Aluminum Alloy, Non-Heat Treatable | **Al** 99.50 min **B** 0.05 max **Cr** 0.01 max **Cu** 0.05 max **Fe** 0.40 max **Ga** 0.03 max **Mn** 0.01 max **Si** 0.10 max **Zn** 0.05 max **Other** each 0.03 max, total 0.10, Ti+V 0.02 max | **AA** 1350; 1350A **ASTM** B230; B231; B232 (1350); B233; B236; B324; B400 (1350); B401; B524; B549; B778 (1350); B779 (1350); B901 (1350) **AWS** C2.25 (W-Al-1350) |
| **A91370** | Wrought Aluminum Alloy, Non-Heat Treatable | **Al** 99.70 min **B** 0.02 max **Cr** 0.01 max **Cu** 0.02 max **Fe** 0.25 max **Ga** 0.03 max **Mg** 0.02 max **Mn** 0.01 max **Si** 0.10 max **Zn** 0.04 max **Other** each 0.02 max, total 0.10 max, Ti+V 0.02 max | **AA** 1370 |
| **A91385** | Wrought Aluminum Alloy, Non-Heat Treatable | **Al** 99.85 min **B** 0.02 max **Cr** 0.01 max **Cu** 0.02 max **Fe** 0.12 max **Ga** 0.03 max **Mg** 0.02 max **Mn** 0.01 max **Si** 0.05 max **Zn** 0.03 max **Other** each 0.01 max, Ti+V 0.03 max | **AA** 1385 |
| **A91435** | Wrought Aluminum Alloy, Non-Heat Treatable | **Al** 99.35 min **Cu** 0.02 max **Fe** 0.30-0.50 **Mg** 0.05 max **Mn** 0.05 max **Si** 0.15 max **Ti** 0.03 max **V** 0.05 max **Zn** 0.10 max **Other** each 0.03 max | **AA** 1435 **ASTM** B483 (1435) |
| **A91445** | Wrought Aluminum Alloy, Non-Heat Treatable | **Al** 99.45 min **Cu** 0.04 max **Other** total 0.05 max, Cu+Fe+Si 0.50 max | **AA** 1445 |
| **A91450** | Wrought Aluminum Alloy, Non-Heat Treatable | **Al** 99.50 min **Cu** 0.05 max **Fe** 0.40 max **Mg** 0.05 max **Mn** 0.05 max **Si** 0.25 max **Ti** 0.10-0.20 **Zn** 0.07 max **Other** 0.03 max each | **AA** 1450 |
| **A92001** | Wrought Aluminum Alloy, Heat Treatable | **Al** rem **Cr** 0.10 max **Cu** 5.2-6.0 **Fe** 0.20 max **Mg** 0.20-0.45 **Mn** 0.15-0.50 **Ni** 0.05 max **Pb** 0.003 max **Si** 0.20 max **Ti** 0.20 max **Zn** 0.10 max **Zr** 0.05 max **Other** 0.05 max each, 0.15 max total | **AA** 2001 |
| **A92002** | Wrought Aluminum Alloy, Heat Treatable | **Al** rem **Cr** 0.20 max **Cu** 1.5-2.5 **Fe** 0.30 max **Mg** 0.50-1.0 **Mn** 0.20 max **Si** 0.35-0.8 **Ti** 0.20 max **Zn** 0.20 max **Other** 0.05 max each, 0.15 max total | **AA** 2002 |

*Boxed entries are no longer active and are retained for reference purposes only.

The chemical compositions listed are for identification purposes and should not be used in lieu of the cross referenced specifications.

| UNIFIED NUMBER | DESCRIPTION | CHEMICAL COMPOSITION | CROSS REFERENCE SPECIFICATIONS |
|---|---|---|---|
| * **A92003** | Wrought Aluminum Alloy, Heat Treatable | **Al** rem **Cd** 0.05-0.20 **Cu** 4.0-5.0 **Fe** 0.30 max **Mg** 0.02 max **Mn** 0.30-0.8 **Si** 0.30 max **Ti** 0.15 max **V** 0.05-0.20 **Zn** 0.10 max **Zr** 0.10-0.25 **Other** 0.05 max each, 0.15 max total | **AA** 2003 |
| **A92004** | Wrought Aluminum Alloy, Heat Treatable | **Al** rem **Cu** 5.5-6.5 **Fe** 0.20 max, 0.50 max **Mg** 0.50 max **Mn** 0.10 max **Si** 0.20 max **Ti** 0.05 max **Zn** 0.10 max **Zr** 0.30-0.50 **Other** each 0.05 max, total 0.15 max | **AA** 2004 **AMS** 4208 **MAM** 4208; 4209 |
| **A92005** | Wrought Aluminum Alloy, Heat Treatable | **Al** rem **Bi** 0.20 max **Cr** 0.10 max **Cu** 3.5-5.0 **Fe** 0.7 max **Mg** 0.20-1.0 **Mn** 1.0 max **Ni** 0.20 max **Pb** 1.0-2.0 **Si** 0.8 max **Ti** 0.20 max **Zn** 0.50 max **Other** 0.05 max each, 0.15 max total | **AA** 2005 |
| **A92006** | Wrought Aluminum Alloy, Heat Treatable | **Al** rem **Cu** 1.0-2.0 **Fe** 0.7 max **Mg** 0.50-1.4 **Mn** 0.6-1.0 **Ni** 0.20 max **Si** 0.8-1.3 **Ti** 0.30 max **Zn** 0.20 max **Other** 0.05 max each, 0.15 max total | **AA** 2006 |
| **A92007** | Wrought Aluminum Alloy, Heat Treatable | **Al** rem **Bi** 0.20 max **Cr** 0.10 max **Cu** 3.3-4.6 **Fe** 0.8 max **Mg** 0.40-1.8 **Mn** 0.50-1.0 **Ni** 0.20 max **Pb** 0.8-1.5 **Si** 0.8 max **Sn** 0.20 max **Ti** 0.20 max **Zn** 0.8 max **Other** 0.10 max each, 0.30 max total | **AA** 2007; 2007A |
| **A92008** | Wrought Aluminum Alloy, Heat Treatable | **Al** rem **Cr** 0.10 max **Cu** 0.7-1.1 **Fe** 0.40 max **Mg** 0.25-0.50 **Mn** 0.30 max **Si** 0.50-0.8 **Ti** 0.10 max **V** 0.05 max **Zn** 0.25 max **Other** Each 0.05 max, Total 0.15 max | **AA** 2008 |
| **A92009** | Wrought Aluminum Alloy, Heat Treatable | **Al** rem **Cu** 3.2-4.4 **Fe** 0.05 max **Mg** 1.0-1.6 **O** 0.6 max **Si** 0.25 max **Zn** 0.10 max **Other** 0.05 max each, 0.15 max total | **AA** 2009 |
| **A92010** | Wrought Aluminum Alloy, Heat Treatable | **Al** rem **Cr** 0.15 max **Cu** 0.7-1.3 **Fe** 0.50 max **Mg** 0.40-1.0 **Mn** 0.10-0.40 **Si** 0.50 max **Zn** 0.30 max **Other** 0.05 max each, 0.15 max total | **AA** 2010 |
| **A92011** | Wrought Aluminum Alloy, Heat Treatable | **Al** rem **Bi** 0.20-0.6 **Cu** 5.0-6.0 **Fe** 0.7 max **Pb** 0.20-0.6 **Si** 0.40 max **Zn** 0.30 max **Other** each 0.05 max, total 0.15 max | **AA** 2011; 2011A **AMS** AMS-QQ-A-225/3 **ASTM** B210 (2011); B211 (2011) **SAE** J454 (2011) |
| **A92012** | Wrought Aluminum Alloy, Heat Treatable | **Al** rem **Bi** 0.20-0.7 **Cu** 4.0-5.5 **Fe** 0.7 max **Si** 0.40 max **Sn** 0.20-0.6 **Zn** 0.30 max **Other** each 0.05 max, total 0.15 max | **AA** 2012 |
| **A92014** | Wrought Aluminum Alloy, Heat Treatable | **Al** rem **Cr** 0.10 max **Cu** 3.9-5.0 **Fe** 0.7 max **Mg** 0.20-0.8 **Mn** 0.40-1.2 **Si** 0.50-1.2 **Ti** 0.15 max **Zn** 0.25 max **Other** each 0.05 max, total 0.15 max, Ti+Zr 0.20 max for extruded and forged products only, but only when producer and purchaser mutually agree. | **AA** 2014; 2014A **AMS** 4028; 4029; 4121; 4133; 4134; 4153; 4314; AMS-A-22771 (2014); AMS-QQ-A-200/2; AMS-QQ-A-200/4; AMS-QQ-A-225/4 **ASME** SB211 **ASTM** B209 (2014); B210 (2014); B211 (2014); B221 (2014); B241 (2014); B247 (2014) |
| **A92017** | Wrought Aluminum Alloy, Heat Treatable | **Al** rem **Cr** 0.10 max **Cu** 3.5-4.5 **Fe** 0.7 max **Mg** 0.40-0.8 **Mn** 0.40-1.0 **Si** 0.20-0.8 **Zn** 0.25 max **Other** each 0.05 max, total 0.15 max, Ti+Zr 0.20 max for extruded and forged products only, but only when producer and purchaser mutually agree. | **AA** 2017; 2017A **AMS** 4118; AMS-QQ-A-225/5 **ASTM** B211 (2017); B316 (2017) **FED** QQ-A-430 |

*Boxed entries are no longer active and are retained for reference purposes only.

The chemical compositions listed are for identification purposes and should not be used in lieu of the cross referenced specifications.

| UNIFIED NUMBER | DESCRIPTION | CHEMICAL COMPOSITION | CROSS REFERENCE SPECIFICATIONS |
|---|---|---|---|
| **A92018** | Wrought Aluminum Alloy, Heat Treatable | **Al** rem **Cr** 0.10 max **Cu** 3.5-4.5 **Fe** 1.0 max **Mg** 0.45-0.9 **Mn** 0.20 max **Ni** 1.7-2.3 **Si** 0.9 max **Zn** 0.25 max **Other** each 0.05 max, total 0.15 max | **AA** 2018 **AMS** 4140 **ASTM** B247 (2018) |
| * **A92020** | Wrought Aluminum Alloy, Heat Treatable | **Al** rem **Cd** 0.10-0.35 **Cu** 4.0-5.0 **Fe** 0.40 max **Li** 0.9-1.7 **Mg** 0.03 max **Mn** 0.30-0.8 **Si** 0.40 max **Ti** 0.10 max **Zn** 0.25 max **Other** each 0.05 max, total 0.15 max | **AA** 2020 |
| **A92021** | Wrought Aluminum Alloy, Heat Treatable | **Al** rem **Cd** 0.05-0.20 **Cu** 5.8-6.8 **Fe** 0.30 max **Mg** 0.02 max **Mn** 0.20-0.40 **Si** 0.20 max **Sn** 0.03-0.08 **Ti** 0.02-0.10 **V** 0.05-0.15 **Zn** 0.10 max **Zr** 0.10-0.25 **Other** each 0.05 max, total 0.15 max | **AA** 2021 |
| **A92024** | Wrought Aluminum Alloy, Heat Treatable | **Al** rem **Cr** 0.10 max **Cu** 3.8-4.9 **Fe** 0.50 max **Mg** 1.2-1.8 **Mn** 0.30-0.9 **Si** 0.50 max **Ti** 0.15 max **Zn** 0.25 max **Other** each 0.05 max, total 0.15 max, Ti+Zr 0.20 max for extruded and forged products only, but only when producer and purchaser mutually agree. | **AA** 2024 **AMS** 4007; 4035; 4037; 4086; 4087; 4088; 4120; 4152; 4164; 4165; 4193; 4297; 4339; AMS-A-81596; AMS-QQ-A-200/3; AMS-QQ-A-225/6; AMS-QQ-A-250/4 **ASTM** B209 (2024); B210 (2024); B211 (2024); B221 (2024); B241 (2024); B316 (2024); F901 (2024) **FED** QQ-A-430; WW-T-700/3 **SAE** J454 (2024) |
| **A92025** | Wrought Aluminum Alloy, Heat Treatable | **Al** rem **Cr** 0.10 max **Cu** 3.9-5.0 **Fe** 1.0 max **Mg** 0.05 max **Mn** 0.40-1.2 **Si** 0.50-1.2 **Ti** 0.15 max **Zn** 0.25 max **Other** each 0.05 max, total 0.15 max | **AA** 2025 **AMS** 4130 **ASTM** B247 (2025) |
| **A92026** | Wrought Aluminum Alloy, Heat Treatable | **Al** rem **Cu** 3.6-4.3 **Fe** 0.07 max **Mg** 0.30-0.8 **Mn** 1.0-1.6 **Si** 0.05 max **Ti** 0.06 max **Zn** 0.10 max **Zr** 0.05-0.25 **Other** 0.05 max, total 0.15 max | **AA** 2026 |
| **A92030** | Wrought Aluminum Alloy, Heat Treatable | **Al** rem **Bi** 0.20 max **Cr** 0.10 max **Cu** 3.3-4.5 **Fe** 0.7 max **Mg** 0.50-1.3 **Mn** 0.20-1.0 **Pb** 0.8-1.5 **Si** 0.8 max **Ti** 0.20 max **Other** 0.10 max each, 0.30 max total | **AA** 2030 |
| **A92031** | Wrought Aluminum Alloy, Heat Treatable | **Al** rem **Cu** 1.8-2.8 **Fe** 0.6-1.2 **Mg** 0.6-1.2 **Mn** 0.50 max **Ni** 0.6-1.4 **Si** 0.50-1.3 **Ti** 0.20 max **Zn** 0.20 max **Other** 0.05 max each, 0.15 max total | **AA** 2031 |
| **A92034** | Wrought Aluminum Alloy, Heat Treatable | **Al** rem **Cr** 0.05 **Cu** 4.2-4.8 **Fe** 0.12 **Mg** 1.3-1.9 **Mn** 0.8-1.3 **Si** 0.10 **Ti** 0.15 **Zn** 0.20 **Zr** 0.08-0.15 **Other** each 0.15, total 0.15 | **AA** 2034 |
| **A92036** | Wrought Aluminum Alloy, Heat Treatable | **Al** rem **Cr** 0.10 max **Cu** 2.2-3.0 **Fe** 0.50 max **Mg** 0.30-0.6 **Mn** 0.10-0.40 **Si** 0.50 max **Ti** 0.10 max **Zn** 0.25 max **Other** each 0.05 max, total 0.15 max | **AA** 2036 **SAE** J454 (2036) |
| **A92037** | Wrought Aluminum Alloy, Heat Treatable | **Al** rem **Cr** 0.10 max **Cu** 1.4-2.2 **Fe** 0.50 max **Mg** 0.30-0.8 **Mn** 0.10-0.40 **Si** 0.50 max **Ti** 0.15 max **V** 0.05 max **Zn** 0.25 max **Other** each 0.05 max, total 0.15 max | **AA** 2037 |
| **A92038** | Wrought Aluminum Alloy, Heat Treatable | **Al** rem **Cr** 0.20 **Cu** 0.8-1.8 **Fe** 0.6 **Ga** 0.05 **Mg** 0.40-1.0 **Mn** 0.10-0.40 **Si** 0.50-1.3 **Ti** 0.15 **V** 0.05 **Zn** 0.50 **Other** each 0.05, total 0.15 | **AA** 2038 **SAE** J454 (2038) |
| **A92039** | Wrought Aluminum Alloy, Heat Treatable | **Ag** 0.05-0.50 **Al** rem **Cu** 4.5-5.5 **Fe** 0.30 max **Mg** 0.40-0.8 **Mn** 0.20-0.50 **Si** 0.20 max **Ti** 0.15 max **Zr** 0.10-0.25 **Other** each 0.05 max, total 0.15 max | **AA** 2039 |

*Boxed entries are no longer active and are retained for reference purposes only.

The chemical compositions listed are for identification purposes and should not be used in lieu of the cross referenced specifications.

| UNIFIED NUMBER | DESCRIPTION | CHEMICAL COMPOSITION | CROSS REFERENCE SPECIFICATIONS |
|---|---|---|---|
| **A92048** | Wrought Aluminum Alloy, Heat Treatable | **Al** rem **Cu** 2.8-3.8 **Fe** 0.20 max **Mg** 1.2-1.8 **Mn** 0.20-0.6 **Si** 0.15 max **Ti** 0.10 max **Zn** 0.25 max **Other** each 0.05 max, total 0.15 max | **AA** 2048 |
| **A92080** | Wrought Aluminum Alloy, Heat Treatable | **Al** rem **Be** 0.005 max **Cu** 3.3-4.1 **Fe** 0.20 max **Mg** 1.5-2.2 **Mn** 0.25 max **O** 0.05-0.50 **Si** 0.10 max **Zn** 0.10 max **Zr** 0.08-0.25 **Other** 0.05 max each, 0.15 max total | **AA** 2080 |
| **A92090** | Wrought Aluminum Alloy, Heat Treatable | **Al** rem **Cr** 0.05 max **Cu** 2.4-3.0 **Fe** 0.12 max **Li** 1.9-2.6 **Mg** 0.25 max **Mn** 0.05 max **Si** 0.10 max **Ti** 0.15 max **Zn** 0.10 max **Zr** 0.08-0.15 **Other** each 0.05 max, total 0.15 max | **AA** 2090 **AMS** 4232; 4251; 4303 |
| **A92091** | Wrought Aluminum Alloy, Heat Treatable | **Al** rem **Cr** 0.10 **Cu** 1.8-2.5 **Fe** 0.30 max **Li** 1.7-2.3 **Mg** 1.1-1.9 **Mn** 0.10 max **Si** 0.20 max **Ti** 0.10 max **Zn** 0.25 max **Zr** 0.04-0.16 **Other** each 0.05 max, total 0.15 max | **AA** 2091 |
| **A92094** | Wrought Aluminum Alloy, Heat Treatable | **Ag** 0.25-0.6 **Al** rem **Cu** 4.4-5.2 **Fe** 0.15 max **Li** 0.7-1.4 **Mg** 0.25-0.8 **Mn** 0.25 max **Si** 0.12 max **Ti** 0.10 max **Zn** 0.25 max **Zr** 0.04-0.18 **Other** 0.05 max each, 0.15 max total | **AA** 2094 |
| **A92095** | Wrought Aluminum Alloy, Heat Treatable | **Ag** 0.25-0.6 **Al** rem **Cu** 3.9-4.6 **Fe** 0.15 max **Li** 0.7-1.5 **Mg** 0.25-0.8 **Mn** 0.25 max **Si** 0.12 max **Ti** 0.10 max **Zn** 0.25 max **Zr** 0.04-0.18 **Other** 0.05 max each, 0.15 max total | **AA** 2095 |
| **A92096** | Wrought Aluminum Alloy, Heat Treatable | **Ag** 0.25-0.6 **Al** rem **Cu** 2.3-3.0 **Fe** 0.15 max **Li** 1.3-1.9 **Mg** 0.25-0.8 **Mn** 0.25 max **Si** 0.12 max **Ti** 0.10 max **Zn** 0.25 max **Zr** 0.04-0.18 **Other** each 0.05 max, total 0.15 max | **AA** X2096 |
| **A92097** | Wrought Aluminum Alloy, Heat Treatable | **Al** rem **Cu** 2.5-3.1 **Fe** 0.15 max **Li** 1.2-1.8 **Mg** 0.35 max **Mn** 0.10-0.6 **Si** 0.12 max **Ti** 0.15 max **Zn** 0.35 max **Zr** 0.08-0.16 **Other** each 0.05 max, total 0.15 max | **AA** 2097 |
| **A92098** | Wrought Aluminum Alloy, Heat Treatable | **Al** rem **Cu** 3.2-3.8 **Fe** 0.15 max **Mg** 0.25-0.8 **Mn** 0.35 max **Si** 0.12 max **Ti** 0.10 max **Zn** 035 max **Zr** 0.04-0.18 **Other** each 0.05 max, total 0.15 max | **AA** 2098 |
| **A92111** | Wrought Aluminum Alloy, Heat Treatable | **Al** rem **Bi** 0.20-0.8 **Cu** 5.0-6.0 **Fe** 0.7 max **Si** 0.40 max **Sn** 0.10-0.50 **Zn** 0.30 max **Other** each 0.05 max, total 0.15 max | **AA** 2111; 2111A |
| **A92117** | Wrought Aluminum Alloy, Heat Treatable | **Al** rem **Cr** 0.10 max **Cu** 2.2-3.0 **Fe** 0.7 max **Mg** 0.20-0.50 **Mn** 0.20 max **Si** 0.8 max **Zn** 0.25 max **Other** each 0.05 max, total 0.15 max | **AA** 2117 **AMS** AS7222 **ASTM** B316 (2117) **FED** QQ-A-430 **SAE** J454 (2117) |
| **A92124** | Wrought Aluminum Alloy, Heat Treatable | **Al** rem **Cr** 0.10 max **Cu** 3.8-4.9 **Fe** 0.30 max **Mg** 1.2-1.8 **Mn** 0.30-0.90 **Si** 0.20 max **Ti** 0.15 max **Zn** 0.25 max **Other** each 0.05 max, total 0.15 max, Ti+Zr 0.20 max for extruded and forged products only, but only when producer and purchaser mutually agree | **AA** 2124 **AMS** 4101; 4221; AMS QQ-A-250/29 **ASTM** B209 (2124) |
| **A92195** | Wrought Aluminum Alloy, Heat Treatable | **Ag** 0.25-0.6 **Al** rem **Cu** 3.7-4.3 **Fe** 0.15 max **Li** 0.8-1.2 **Mg** 0.25-0.8 **Mn** 0.25 max **Si** 0.12 max **Ti** 0.10 max **Zn** 0.25 max **Zr** 0.08-0.16 **Other** each 0.05 max, total 0.15 max | **AA** 2195 |

The chemical compositions listed are for identification purposes and should not be used in lieu of the cross referenced specifications.

| UNIFIED NUMBER | DESCRIPTION | CHEMICAL COMPOSITION | CROSS REFERENCE SPECIFICATIONS |
|---|---|---|---|
| **A92196** | Wrought Aluminum Alloy, Heat Treatable | **Ag** 0.25-0.6 **Al** rem **Cu** 2.5-3.3 **Fe** 0.15 max **Li** 1.4-2.1 **Mg** 0.25-0.8 **Mn** 0.35 max **Si** 0.12 max **Ti** 0.10 max **Zn** 0.35 max **Zr** 0.04-0.18 **Other** each 0.05 max, total 0.15 max | **AA** 2196 |
| **A92197** | Wrought Aluminum Alloy, Heat Treatable | **Al** rem **Cu** 2.5-3.1 **Fe** 0.10 max **Li** 1.3-1.7 **Mg** 0.25 max **Mn** 0.10-0.50 **Si** 0.10 max **Ti** 0.12 max **Zn** 0.05 max **Zr** 0.08-0.15 **Other** each 0.05 max, total 0.15 max | **AA** 2197 |
| **A92214** | Wrought Aluminum Alloy, Heat Treatable | **Al** rem **Cr** 0.10 max **Cu** 3.9-5.0 **Fe** 0.30 max **Mg** 0.20-0.8 **Mn** 0.40-1.2 **Si** 0.50-1.2 **Ti** 0.15 max **Zn** 0.25 max **Other** each 0.05 max, total 0.15 max, Ti+Zr 0.20 max for extruded and forged products only, but only when producer and purchaser mutually agree | **AA** 2214 |
| **A92218** | Wrought Aluminum Alloy, Heat Treatable | **Al** rem **Cr** 0.10 max **Cu** 3.5-4.5 **Fe** 1.0 max **Mg** 1.2-1.8 **Mn** 0.20 max **Ni** 1.7-2.3 **Si** 0.9 max **Zn** 0.25 max **Other** each 0.05 max, total 0.15 max | **AA** 2218 **ASTM** B247 (2218) |
| **A92219** | Wrought Aluminum Alloy, Heat Treatable | **Al** rem **Cu** 5.8-6.8 **Fe** 0.30 max **Mg** 0.02 max **Mn** 0.20-0.40 **Si** 0.20 max **Ti** 0.02-0.10 **V** 0.05-0.15 **Zn** 0.10 max **Zr** 0.10-0.25 **Other** each 0.05 max, total 0.15 max | **AA** 2219 **AMS** 4031; 4066; 4068; 4143; 4144; 4162; 4295; 4313; AMS-A-22771 (2219); AMS-QQ-A-250/30 **ASTM** B209 (2219); B210; B211 (2219); B221 (2219); B241 (2219); B247 (2219); B316 (2219) **FED** QQ-A-430 **MIL SPEC** MIL-DTL-46118 |
| **A92224** | Wrought Aluminum Alloy, Heat Treatable | **Al** rem **Cr** 0.10 max **Cu** 3.8-4.4 **Fe** 0.15 max **Mg** 1.2-1.8 **Mn** 0.30-0.9 **Si** 0.12 max **Ti** 0.15 max **Zn** 0.25 max **Other** each 0.05 max, total 0.15 max | **AA** 2224; 2224A |
| **A92297** | Wrought Aluminum Alloy, Heat Treatable | **Al** rem **Cu** 2.5-3.1 **Fe** 0.10 max **Li** 1.1-1.7 **Mg** 0.25 max **Mn** 0.10-0.50 **Si** 0.10 max **Ti** 0.12 max **Zn** 0.05 max **Zr** 0.08-0.15 **Other** each 0.05 max, total 0.15 max | **AA** 2297 |
| **A92319** | Wrought Aluminum Alloy, Heat Treatable (Filler Alloy) | **Al** rem **Be** 0.0008 max **Cu** 5.8-6.8 **Fe** 0.30 max **Mg** 0.02 max **Mn** 0.20-0.40 **Si** 0.20 max **Ti** 0.10-0.20 **V** 0.05-0.15 **Zn** 0.10 max **Zr** 0.10-0.25 **Other** each 0.05 max (Be 0.0008 max for welding electrode and filler wire only), total 0.15 max | **AA** 2319 **AMS** 4191 **ASME** SFA5.10 (ER2319) **AWS** A5.10 (ER2319) |
| **A92324** | Wrought Aluminum Alloy | **Al** rem **Cr** 0.10 max **Cu** 3.8-4.4 **Fe** 0.12 max **Mg** 1.2-1.8 **Mn** 0.30-0.9 **Si** 0.10 max **Ti** 0.15 max **Zn** 0.25 max **Other** each 0.05 max, total 0.15 max | **AA** 2324 |
| **A92419** | Wrought Aluminum Alloy, Heat Treatable | **Al** rem **Cu** 5.8-6.8 **Fe** 0.18 max **Mg** 0.02 max **Mn** 0.20-0.40 **Si** 0.15 max **Ti** 0.02-0.10 **V** 0.05-0.15 **Zn** 0.10 max **Zr** 0.10-0.25 **Other** each 0.05 max, total 0.15 max | **AA** 2419 |
| **A92424** | Wrought Aluminum Alloy, Heat Treatable | **Al** rem **Cu** 3.8-4.4 **Fe** 0.12 max **Mg** 1.2-1.6 **Mn** 0.30-0.6 **Si** 0.10 max **Ti** 0.10 max **Zn** 0.20 max **Other** each 0.05 max, total 0.15 max | **AA** 2424 **AMS** 4270; 4273 |
| **A92519** | Wrought Aluminum Alloy, Heat Treatable | **Al** rem **Cu** 5.3-6.4 **Fe** 0.30 max **Mg** 0.05-0.40 **Mn** 0.10-0.50 **Si** 0.25 max **Ti** 0.02-0.10 **V** 0.05-0.15 **Zn** 0.10 max **Zr** 0.10-0.25 **Other** each 0.05 max, total 0.15 max, Fe+Si 0.40 max | **AA** 2519 **MIL SPEC** MIL-DTL-46192 |
| **A92524** | Wrought Aluminum Alloy, Heat Treatable | **Al** rem **Cr** 0.05 max **Cu** 4.0-4.5 **Fe** 0.12 max **Mg** 1.2-1.6 **Mn** 0.45-0.7 **Si** 0.06 max **Ti** 0.10 max **Zn** 0.15 max **Other** each 0.05 max, total 0.15 max | **AA** 2524 |

The chemical compositions listed are for identification purposes and should not be used in lieu of the cross referenced specifications.

| UNIFIED NUMBER | DESCRIPTION | CHEMICAL COMPOSITION | CROSS REFERENCE SPECIFICATIONS |
|---|---|---|---|
| **A92618** | Wrought Aluminum Alloy, Heat Treatable | **Al** rem **Cu** 1.9-2.7 **Fe** 0.9-1.3 **Mg** 1.3-1.8 **Ni** 0.9-1.2 **Si** 0.10-0.25 **Ti** 0.04-0.10 **Other** each 0.05 max, total 0.15 max | **AA** 2618; 2618A **AMS** 4132; AMS-A-22771 **ASTM** B247 (2618) |
| **A93002** | Wrought Aluminum Alloy, Non-Heat Treatable | **Al** rem **Cu** 0.15 max **Fe** 0.10 max **Mg** 0.05-0.20 **Mn** 0.05-0.25 **Si** 0.08 max **Ti** 0.03 max **Zn** 0.03 max **Other** each 0.03 max, total 0.10 max | **AA** 3002 **SAE** J454 (3002) |
| **A93003** | Wrought Aluminum Alloy, Non-Heat Treatable | **Al** rem **Cu** 0.05-0.20 **Fe** 0.7 max **Mn** 1.0-1.5 **Si** 0.6 max **Zn** 0.10 max **Other** each 0.05 max, total 0.15 max | **AA** 3003 **AMS** 4006; 4008; 4065; 4067; AMS-A-22771; AMS-A-81596; AMS-QQ-A-200/1; AMS-QQ-A-225/2; AMS-QQ-A-250/2; AMS-QQ-A-250/3 **ASME** SB-209; SB-210; SB-221; SB-234; SB-241; SB-247; SFA5.3 (E3003) **ASTM** B209 (3003); B210 (3003); B211 (3003); B221 (3003); B234 (3003); B241 (3003); B247 (3003); B313 (3003); B316 (3003); B345 (3003); B361 (WP3003); B361 (3003); B404 (3003); B483 (3003); B491 (3003); B547 (3003); B548 (3003) **AWS** A5.3 (E3003) **FED** QQ-A-430; WW-T-700/2 **MIL SPEC** MIL-P-25995 **SAE** J454 (3003) |
| **A93004** | Wrought Aluminum Alloy, Non-Heat Treatable | **Al** rem **Cu** 0.25 max **Fe** 0.7 max **Mg** 0.8-1.3 **Mn** 1.0-1.5 **Si** 0.30 max **Zn** 0.25 max **Other** each 0.05 max, total 0.15 max | **AA** 3004; 3004A **ASTM** B209 (3004); B221 (3004); B313 (3004); B547 (3004); B548 (3004) **SAE** J454 (3004) |
| **A93005** | Wrought Aluminum Alloy, Non-Heat Treatable | **Al** rem **Cr** 0.10 max **Cu** 0.30 max **Fe** 0.7 max **Mg** 0.20-0.6 **Mn** 1.0-1.5 **Si** 0.6 max **Ti** 0.10 max **Zn** 0.25 max **Other** each 0.05 max, total 0.15 max | **AA** 3005; 3005A **ASTM** B209 (3005); B313 |
| **A93006** | Wrought Aluminum Alloy, Non-Heat Treatable | **Al** rem **Cr** 0.20 max **Cu** 0.10-0.30 **Fe** 0.7 max **Mg** 0.30-0.6 **Mn** 0.50-0.8 **Si** 0.50 max **Ti** 0.10 max **Zn** 0.15-0.40 **Other** each 0.05 max, total 0.15 max | **AA** 3006 |
| **A93007** | Wrought Aluminum Alloy, Non-Heat Treatable | **Al** rem **Cr** 0.20 max **Cu** 0.05-0.30 **Fe** 0.7 max **Mg** 0.6 max **Mn** 0.30-0.8 **Si** 0.50 max **Ti** 0.10 max **Zn** 0.40 max **Other** each 0.05 max, total 0.15 max | **AA** 3007 |
| * **A93008** | Wrought Aluminum Alloy, Non-Heat Treatable | **Al** rem **Cr** 0.05 max **Cu** 0.10 max **Fe** 0.7 max **Mg** 0.01 max **Mn** 1.2-1.8 **Ni** 0.05 max **Si** 0.40 max **Ti** 0.10 max **Zn** 0.05 max **Zr** 0.10-0.50 **Other** 0.05 max each, 0.15 max total | **AA** 3008 |
| **A93009** | Wrought Aluminum Alloy, Non-Heat Treatable | **Al** rem **Cr** 0.05 max **Cu** 0.10 max **Fe** 0.7 max **Mg** 0.10 max **Mn** 1.2-1.8 **Ni** 0.05 max **Si** 1.0-1.8 **Ti** 0.10 max **Zn** 0.05 max **Zr** 0.10 max **Other** each 0.05 max, total 0.15 max | **AA** 3009 |
| **A93010** | Wrought Aluminum Alloy, Non-Heat Treatable | **Al** rem **Cr** 0.05-0.40 **Cu** 0.03 max **Fe** 0.20 max **Mn** 0.20-0.9 **Si** 0.10 max **Ti** 0.05 max **V** 0.05 max **Zn** 0.05 max **Other** each 0.03 max, total 0.10 max | **AA** 3010 |
| **A93011** | Wrought Aluminum Alloy, Non Heat Treatable | **Al** rem **Cr** 0.10-0.40 **Cu** 0.05-0.20 **Fe** 0.7 max **Mn** 0.8-1.2 **Si** 0.40 max **Ti** 0.10 max **Zn** 0.10 max **Zr** 0.10-0.30 **Other** each 0.05 max, total 0.15 max | **AA** 3011 |
| **A93012** | Wrought Aluminum Alloy, Non-Heat Treatable | **Al** rem **Cr** 0.20 max **Cu** 0.10 max **Fe** 0.7 max **Mg** 0.10 max **Mn** 0.50-1.1 **Si** 0.6 max **Ti** 0.10 max **Zn** 0.10 max **Other** 0.05 max each, 0.15 max total | **AA** 3012 |
| **A93013** | Wrought Aluminum Alloy, Non-Heat Treatable | **Al** rem **Cu** 0.50 max **Fe** 1.0 max **Mg** 0.20-0.6 **Mn** 0.9-1.4 **Si** 0.6 max **Zn** 0.50-1.0 **Other** 0.05 max each, 0.15 max total | **AA** 3013 |

*Boxed entries are no longer active and are retained for reference purposes only.

The chemical compositions listed are for identification purposes and should not be used in lieu of the cross referenced specifications.

| UNIFIED NUMBER | DESCRIPTION | CHEMICAL COMPOSITION | CROSS REFERENCE SPECIFICATIONS |
|---|---|---|---|
| **A93014** | Wrought Aluminum Alloy, Non-Heat Treatable | **Al** rem **Cu** 0.50 max **Fe** 1.0 max **Mg** 0.10 max **Mn** 1.0-1.5 **Si** 0.6 max **Ti** 0.10 max **Zn** 0.50-1.0 **Other** 0.05 max each, 0.15 max total | **AA** 3014 |
| **A93015** | Wrought Aluminum Alloy, Non-Heat Treatable | **Al** rem **Cr** 0.10 max **Cu** 0.30 max **Fe** 0.8 max **Mg** 0.20-0.7 **Mn** 0.50-0.9 **Si** 0.6 max **Ti** 0.10 max **Zn** 0.25 max **Other** each 0.05 max, total 0.15 max | **AA** 3015; 3105B |
| **A93016** | Wrought Aluminum Alloy, Non-Heat Treatable | **Al** rem **Cr** 0.10 max **Cu** 0.30 max **Fe** 0.8 max **Mg** 0.50-0.8 **Mn** 0.50-0.9 **Si** 0.6 max **Ti** 0.10 **Zn** 0.25 **Other** each 0.05 max, total 0.15 max | **AA** 3016 |
| **A93017** | Wrought Aluminum Alloy, Non-Heat Treatable | **Al** rem **Cr** 0.15 max **Cu** 0.25-0.40 **Fe** 0.25-0.45 **Mg** 0.10 max **Mn** 0.8-1.2 **Si** 0.25 max **Ti** 0.05 max **Zn** 0.10 max **Other** 0.05 max each, 0.15 max total | **AA** 3017 |
| * **A93018** | Wrought Aluminum Alloy, Non-Heat Treatable | **Al** rem **Cr** 0.10 max **Cu** 0.10-0.30 **Fe** 0.15-0.25 **Mg** 0.8-1.4 **Mn** 1.1-1.4 **Pb** 0.01 max **Si** 0.30 max **Ti** 0.10 max **Zn** 0.25 max **Other** each 0.05 max, total 0.15 max | **AA** 3018 |
| **A93019** | Wrought Aluminum Alloy, Non-Heat Treatable | **Al** rem **Cr** 0.20 max **Cu** 0.9 max **Fe** 0.7 max **Mg** 0.20-0.9 **Mn** 0.30-0.9 **Ni** 0.10 max **Si** 0.6 max **Ti** 0.10 max **Zn** 0.20-0.9 **Other** each 0.05 max, total 0.15 max | **AA** 3019 |
| **A93020** | Wrought Aluminum Alloy, Non-Heat Treatable | **Al** rem **Cr** 0.20 max **Cu** 0.10 max **Fe** 0.6 max **Mg** 0.20 max **Mn** 0.6-1.2 **Si** 0.50 max **Ti** 0.05-0.25 **Zn** 0.05-0.50 **Other** each 0.05 max, total 0.15 max. | |
| **A93025** | Wrought Aluminum Alloy, Non-Heat Treatable | **Al** rem **Cr** 0.20 max **Cu** 0.30 max **Fe** 0.50-0.9 **Mg** 0.20-0.8 **Mn** 0.40-1.0 **Ni** 0.05 max **Si** 0.6 max **Ti** 0.10 max **Zn** 0.25 max **Other** each 0.05 max, total 0.15 max. | **AA** 3025 |
| **A93030** | Wrought Aluminum Alloy, Non-Heat Treatable | **Al** rem **Cr** 0.05 max **Cu** 0.10 max **Fe** 0.35 max **Mg** 0.05 max **Mn** 0.10-0.7 **Si** 0.15 max **Ti** 0.05-0.35 **Zn** 0.05-0.50 **Other** each 0.05 max, total 0.15 max | **AA** 3030 |
| **A93102** | Wrought Aluminum Alloy, Non-Heat Treatable | **Al** rem **Cu** 0.10 max **Fe** 0.7 max **Mn** 0.05-0.40 **Si** 0.40 max **Ti** 0.10 max **Zn** 0.30 max **Other** each 0.05 max, total 0.15 max | **AA** 3102 **ASME** SB-221 **ASTM** B210 (3102); B221 (3102); B483 (3102); B491 (3102) |
| **A93103** | Wrought Aluminum Alloy, Non-Heat Treatable | **Al** rem **Cr** 0.10 max **Cu** 0.10 max **Fe** 0.7 max **Mg** 0.30 max **Mn** 0.9-1.5 **Si** 0.50 max **Zn** 0.20 max **Other** each 0.05 max, 0.15 max total, Ti+Zr 0.10 max | **AA** 3103; 3103A |
| **A93104** | Wrought Aluminum Alloy, Non-Heat Treatable | **Al** rem **Cu** 0.05-0.25 **Fe** 0.8 max **Ga** 0.05 max **Mg** 0.8-1.3 **Mn** 0.8-1.4 **Si** 0.6 max **Ti** 0.10 max **V** 0.05 max **Zn** 0.25 max **Other** each 0.05 max, total 0.15 max | **AA** 3104 |
| **A93105** | Wrought Aluminum Alloy, Non-Heat Treatable | **Al** rem **Cr** 0.20 max **Cu** 0.30 max **Fe** 0.7 max **Mg** 0.20-0.8 **Mn** 0.30-0.8 **Si** 0.6 max **Ti** 0.10 max **Zn** 0.40 max **Other** each 0.05 max, total 0.15 max | **AA** 3105; 3105A **ASTM** B209 (3105) |
| **A93107** | Wrought Aluminum Alloy, Non-Heat Treatable | **Al** rem **Cu** 0.05-0.15 **Fe** 0.7 max **Mn** 0.40-0.9 **Si** 0.6 max **Ti** 0.10 max **Zn** 0.20 max **Other** each 0.05 max, total 0.15 max | **AA** 3107 |

*Boxed entries are no longer active and are retained for reference purposes only.

The chemical compositions listed are for identification purposes and should not be used in lieu of the cross referenced specifications.

| UNIFIED NUMBER | DESCRIPTION | CHEMICAL COMPOSITION | CROSS REFERENCE SPECIFICATIONS |
|---|---|---|---|
| **A93203** | Wrought Aluminum Alloy, Non-Heat Treatable | **Al** rem **Cu** 0.05 max **Fe** 0.7 max **Mn** 1.0-1.5 **Si** 0.6 max **Zn** 0.10 max **Other** 0.05 max each (Be 0.008 max for welding electrode and filler wire only), 0.15 max total | **AA** 3203 |
| **A93204** | Wrought Aluminum Alloy, Non-Heat Treatable | **Al** rem **Cu** 0.10-0.25 **Fe** 0.7 max **Mg** 0.8-1.5 **Mn** 0.8-1.5 **Si** 0.30 max **Zn** 0.25 max **Other** 0.05 max each, 0.15 max total | **AA** 3204 |
| **A93207** | Wrought Aluminum Alloy, Non-Heat Treatable | **Al** rem **Cu** 0.10 max **Fe** 0.45 max **Mg** 0.10 max **Mn** 0.40-0.8 **Si** 0.30 max **Zn** 0.10 max **Other** 0.05 max each, 0.10 max total | **AA** 3207; 3207A |
| * **A93303** | Wrought Aluminum Alloy, Non-Heat Treatable | **Al** rem **Cu** 0.05-0.20 **Fe** 0.7 max **Mn** 1.0-1.5 **Si** 0.6 max **Zn** 0.30 max **Other** each 0.05 max, total 0.15 max | **AA** 3303 **ASTM** B210 (3303) |
| **A93307** | Wrought Aluminum Alloy, Non-Heat Treatable | **Al** rem **Cr** 0.20 max **Cu** 0.30 max **Fe** 0.8 max **Mg** 0.30 max **Mn** 0.50-0.9 **Si** 0.6 max **Ti** 0.10 max **Zn** 0.40 max **Other** each 0.05 max, total 0.15 max | **AA** 3307 |
| * **A94002** | Wrought Aluminum Alloy, Non-Heat Treatable | **Al** rem **Cd** 0.8-1.4 **Cu** 0.05-0.15 **Fe** 0.35 max **Mg** 0.05-0.15 **Mn** 0.03 max **Si** 3.5-4.5 **Ti** 0.02 max **Zn** 0.15 max **Other** each 0.05 max, total 0.15 max | **AA** 4002 |
| **A94004** | Wrought Aluminum Alloy, Non-Heat Treatable (Brazing Alloy) | **Al** rem **Cu** 0.25 max **Fe** 0.8 max **Mg** 1.0-2.0 **Mn** 0.10 max **Si** 9.0-10.5 **Zn** 0.20 max **Other** each 0.05 max, total 0.15 max | **AA** 4004 **ASME** SFA5.8 (BAlSi-7) **AWS** A5.8 (BAlSi-7) |
| **A94006** | Wought Aluminum Alloy, Non-Heat Treatable | **Al** rem **Cr** 0.20 max **Cu** 0.10 max **Fe** 0.50-0.8 **Mg** 0.01 max **Mn** 0.05 max **Si** 0.8-1.2 **Zn** 0.05 max **Other** 0.05 max each, 0.15 max total | **AA** 4006 |
| **A94007** | Wrought Aluminum Alloy, Non-Heat Treatable | **Al** rem **Cr** 0.05-0.25 **Co** 0.05 max **Cu** 0.20 max **Fe** 0.40-1.0 **Mg** 0.20 max **Mn** 0.8-1.5 **Ni** 0.15-0.7 **Si** 1.0-1.7 **Ti** 0.10 max **Zn** 0.10 max **Other** 0.05 max each, 0.15 max total | **AA** 4007 |
| **A94008** | Wrought Aluminum Alloy, Non-Heat Treatable | **Al** rem **Be** 0.0008 max **Cu** 0.05 max **Fe** 0.09 max **Mg** 0.30-0.45 **Mn** 0.05 max **Si** 6.5-7.5 **Ti** 0.04-0.15 **Zn** 0.05 max **Other** each 0.05 max (Be 0.008 max for welding electrode and filler wire only), total 0.15 max | **AA** 4008 **AMS** 4181 |
| **A94009** | Wrought Aluminum Alloy, Heat Treatable | **Al** rem **Cu** 1.0-1.5 **Fe** 0.20 max **Mg** 0.45-0.6 **Mn** 0.10 max **Si** 4.5-5.5 **Ti** 0.20 max **Zn** 0.10 max **Other** each 0.05 max (Be 0.0008 max for welding electrode and filler wire only), total 0.15 max | **AA** 4009 **ASME** SFA5.10 (ER4009) **AWS** A5.10 (ER9004) |
| **A94010** | Wrought Aluminum Alloy, Non-Heat Treatable | **Al** rem **Be** 0.0008 max **Cu** 0.20 max **Fe** 0.20 max **Mg** 0.30-0.45 **Mn** 0.10 max **Si** 6.5-7.5 **Ti** 0.20 max **Zn** 0.10 max **Other** each 0.05 max (Be 0.0008 max for welding electrode and filler wire only), 0.15 max total | **AA** 4010 **ASME** SFA5.10 (ER4010) **AWS** A5.10 (ER4010) |
| **A94011** | Wrought Aluminum Alloy, Non-Heat Treatable | **Al** rem **Be** 0.04-0.07 **Cu** 0.20 max **Fe** 0.20 max **Mg** 0.45-0.7 **Mn** 0.10 max **Si** 6.5-7.5 **Ti** 0.04-0.20 max **Zn** 0.10 max **Other** each 0.05 max, total 0.15 max | **AA** 4011 **ASME** SFA5.10 (R4011) **AWS** A5.10 (R4011) |

*Boxed entries are no longer active and are retained for reference purposes only.

The chemical compositions listed are for identification purposes and should not be used in lieu of the cross referenced specifications.

| UNIFIED NUMBER | DESCRIPTION | CHEMICAL COMPOSITION | CROSS REFERENCE SPECIFICATIONS |
|---|---|---|---|
| **A94013** | Wrought Aluminum Alloy, Heat Treatable | **Al** rem **Bi** 0.6-1.5 **Cd** 0.05 max **Cu** 0.05-0.20 **Fe** 0.35 max **Mg** 0.05-0.20 **Mn** 0.03 max **Si** 3.5-4.5 **Ti** 0.02 max **Zn** 0.05 max **Other** each 0.05 max, total 0.15 max | **AA** 4013 |
| **A94014** | Wrought Aluminum Alloy, Non-Heat Treatable | **Al** rem **Cu** 0.20 max **Fe** 0.7 max **Mg** 0.30-0.8 **Mn** 0.35 max **Si** 1.4-2.2 **Zn** 0.20 max **Other** 0.05 max each, 0.15 max total | **AA** 4014 |
| **A94015** | Wrought Aluminum Alloy, Non-Heat Treatable | **Al** rem **Cu** 0.20 max **Fe** 0.7 max **Mg** 0.10-0.50 **Mn** 0.6-1.2 **Si** 1.4-2.2 **Zn** 0.20 max **Other** 0.05 max each, 0.15 max total | **AA** 4015 |
| **A94016** | Wrought Aluminum Alloy, Non-Heat Treatable | **Al** rem **Cu** 0.20 max **Fe** 0.7 max **Mg** 0.10 max **Mn** 0.6-1.2 **Si** 1.4-2.2 **Zn** 0.50-1.3 **Other** each 0.05 max, total 0.15 max | **AA** 4016 |
| **A94017** | Wrought Aluminum Alloy, Non-Heat Treatable | **Al** rem **Cu** 0.10-0.50 **Fe** 0.7 max **Mg** 0.10-0.50 **Mn** 0.6-1.2 **Si** 0.6-1.6 **Zn** 0.20 max **Other** each 0.05 max, total 0.15 max | **AA** 4017 |
| **A94018** | Wrought Aluminum Alloy, Non-Heat Treatable | **Al** rem **Cu** 0.05 max **Fe** 0.20 max **Mg** 0.50-0.8 **Mn** 0.10 max **Si** 6.5-7.5 **Ti** 0.20 max **Zn** 0.10 max **Other** each 0.05 max, total 0.10 max | **AA** 4018 |
| **A94019** | Wrought Aluminum Alloy, Non-Heat Treatable | **Al** rem **Fe** 4.6-5.4 **Ni** 1.8-2.2 **Si** 18.5-21.5 **Other** each 0.05 max, total 0.15 max | **AA** 4019 |
| **A94026** | Wrought Aluminum Alloy, Heat Treatable | **Al** rem **Bi** 1.0-2.0 **Cu** 2.5-3.5 **Fe** 0.50 max **Mg** 0.7-1.4 **Is** 9.0-11.5 **Ti** 0.05 max **Zn** 0.10 max **Other** each 0.05 max, total 0.15 max | **AA** 4026 |
| **A94032** | Wrought Aluminum Alloy, Heat Treatable | **Al** rem **Cr** 0.10 max **Cu** 0.50-1.3 **Fe** 1.0 max **Mg** 0.8-1.3 **Ni** 0.50-1.3 **Si** 11.0-13.5 **Zn** 0.25 max **Other** each 0.05 max, total 0.15 max | **AA** 4032 **AMS** 4318; 4319; AMS-A-22771 **ASTM** B247 (4032) |
| **A94043** | Wrought Aluminum Alloy, Non-Heat Treatable | **Al** rem **Cu** 0.30 max **Fe** 0.8 max **Mg** 0.05 max **Mn** 0.05 max **Si** 4.5-6.0 **Ti** 0.20 max **Zn** 0.10 max **Other** each 0.05 max (Be 0.0008 max for welding electrode and filler wire only), total 0.15 max | **AA** 4043; 4043A **AMS** 4190 **ASME** SFA5.3 (E4043); SFA5.10 (ER4043) **AWS** A5.3 (E4043); A5.10 (ER4043); C2.25 (W-Al-4043) **MIL SPEC** MIL-W-6712 |
| **A94044** | Wrought Aluminum Alloy, Non-Heat Treatable (Brazing Alloy) | **Al** rem **Cu** 0.25 max **Fe** 0.8 max **Mn** 0.10 max **Si** 7.8-9.2 **Zn** 0.20 max **Other** each 0.05 max, total 0.15 max | **AA** 4044 |
| **A94045** | Wrought Aluminum Alloy, Non-Heat Treatable | **Al** rem **Cu** 0.30 max **Fe** 0.8 max **Mg** 0.05 max **Mn** 0.05 max **Si** 9.0-11.0 **Ti** 0.20 max **Zn** 0.10 max **Other** each 0.05 max, total 0.15 max | **AA** 4045 **AMS** AMS-B-20148 **ASME** SFA5.8 (BAlSi-5) **AWS** A5.8 (BAlSi-5) |
| **A94046** | Wrought Aluminum Alloy, Non-Heat Treatable | **Al** rem **Cu** 0.03 max **Fe** 0.50 max **Mg** 0.20-0.50 **Mn** 0.40 max **Si** 9.0-11.0 **Ti** 0.15 max **Zn** 0.10 max **Other** each 0.05 max, total 0.15 max | **AA** 4046 |
| **A94047** | Wrought Aluminum Alloy, Non-Heat Treatable (Brazing Alloy) | **Al** rem **Cu** 0.30 max **Fe** 0.8 max **Mg** 0.10 max **Mn** 0.15 max **Si** 11.0-13.0 **Zn** 0.20 max **Other** each 0.05 max (Be 0.0008 max for welding electrode and filler wire only), total 0.15 max | **AA** 4047; 4047A **AMS** 4185; AMS-B-20148 **ASME** SFA5.8 (BAlSi-4); SFA5.10 (ER4047) **AWS** A5.8 (BAlSi-4); A5.10 (ER4047); C2.25 (W-Al-4047) |
| **A94048** | Wrought Aluminum Alloy, Non-Heat Treatable (Welding Alloy) | **Al** rem **Be** 0.0008 max **Cr** 0.07 max **Cu** 3.3-4.7 **Fe** 0.8 max **Mg** 0.07 max **Mn** 0.07 max **Si** 9.3-10.7 **Zn** 9.3-10.7 **Other** each 0.05 max (Be 0.0008 max for welding electrode and filler wire only), total 0.15 max | **AA** 4048 **AMS** AMS-B-20148 |

The chemical compositions listed are for identification purposes and should not be used in lieu of the cross referenced specifications.

| UNIFIED NUMBER | DESCRIPTION | CHEMICAL COMPOSITION | CROSS REFERENCE SPECIFICATIONS |
|---|---|---|---|
| **A94104** | Wrought Aluminum Alloy, Non-Heat Treatable | **Al** rem **Bi** 0.02-0.20 **Cu** 0.25 max **Fe** 0.8 max **Mg** 1.0-2.0 **Mn** 0.10 max **Si** 9.0-10.5 **Zn** 0.20 max **Other** each 0.05 max, total 0.15 max | **AA** 4104 **ASME** SFA5.8 (BAlSi-11) **AWS** A5.8 (BAlSi-11) |
| **A94145** | Wrought Aluminum Alloy, Non-Heat Treatable (brazing and Welding Alloy) | **Al** rem **Cr** 0.15 max **Cu** 3.3-4.7 **Fe** 0.8 max **Mg** 0.15 max **Mn** 0.15 max **Si** 9.3-10.7 **Zn** 0.20 max **Other** each 0.05 max (Be 0.0008 max for welding electrode and filler wire only), total 0.15 max | **AA** 4145; 4145A **AMS** 4184 **ASME** SFA5.8 (BAlSi-3); SFA5.10 (ER4145) **AWS** A5.8 (BAlSi-3); A5.10 (ER4145) |
| **A94147** | Wrought Aluminum Alloy, Non-Heat Treatable | **Al** rem **Be** 0.0008 max **Cu** 0.25 max **Fe** 0.8 max **Mg** 0.10-0.50 **Mn** 0.10 max **Si** 11.0-13.0 **Zn** 0.20 max **Other** each 0.05 max (Be 0.0008 max for welding electrode and filler wire only), total 0.15 max | **AA** 4147 **ASME** SFA5.8 (BAlSi-9) **AWS** A5.8 (BAlSi-9) |
| **A94343** | Wrought Aluminum Alloy, Non-Heat Treatable (Brazing Alloy) | **Al** rem **Cu** 0.25 max **Fe** 0.8 max **Mn** 0.10 max **Si** 6.8-8.2 **Zn** 0.20 max **Other** each 0.05 max, total 0.15 max | **AA** 4343 **AMS** AMS-B-20148 **ASME** SFA5.8 (BAlSi-2) **AWS** A5.8 (BAlSi-2) |
| * **A94543** | Wrought Aluminum Alloy, Non-Heat Treatable | **Al** rem **Cr** 0.05 max **Cu** 0.10 max **Fe** 0.50 max **Mg** 0.10-0.40 **Mn** 0.05 max **Si** 5.0-7.0 **Ti** 0.10 max **Zn** 0.10 max **Other** each 0.05 max, total 0.15 max | **AA** 4543 |
| **A94643** | Wrought Aluminum Alloy, Non-Heat Treatable (Welding Alloy) | **Al** rem **Be** 0.0008 max **Cu** 0.10 max **Fe** 0.8 max **Mg** 0.10-0.30 **Mn** 0.05 max **Si** 3.6-4.6 **Ti** 0.15 max **Zn** 0.10 max **Other** each 0.05 max (Be 0.0008 max for welding electrode and filler wire only), total 0.15 max | **AA** 4643 **AMS** 4189 **ASME** SFA5.10 (ER4643) **AWS** A5.10 (ER4643) |
| **A95005** | Wrought Aluminum Alloy, Non-Heat Treatable | **Al** rem **Cr** 0.10 max **Cu** 0.20 max **Fe** 0.7 max **Mg** 0.50-1.1 **Mn** 0.20 max **Si** 0.30 max **Zn** 0.25 max **Other** each 0.05 max, total 0.15 max | **AA** 5005; 5005A **ASTM** B209 (5005); B210 (5005); B316 (5005); B396 (5005); B397 (5005); B483 (5005); B531 (5005) **FED** QQ-A-430 (5005) **SAE** J454 (5005) |
| **A95006** | Wrought Aluminum Alloy, Non-Heat Treatable | **Al** rem **Cr** 0.10 max **Cu** 0.10 max **Fe** 0.8 max **Mg** 0.8-1.3 **Mn** 0.40-0.8 **Si** 0.40 max **Ti** 0.10 max **Zn** 0.25 max **Other** each 0.05 max, total 0.15 max | **AA** 5006 |
| **A95010** | Wrought Aluminum Alloy, Non-Heat Treatable | **Al** rem **Cr** 0.15 max **Cu** 0.25 max **Fe** 0.7 max **Mg** 0.20-0.6 **Mn** 0.10-0.30 **Si** 0.40 max **Ti** 0.10 max **Zn** 0.30 max **Other** each 0.05 max, total 0.15 max | **AA** 5010 **ASTM** B209 (5010) |
| * **A95013** | Wrought Aluminum Alloy, Non-Heat Treatable | **Al** rem **Cr** 0.03 max **Cu** 0.03 max **Fe** 0.25 max **Mg** 3.2-3.8 **Mn** 0.30-0.50 **Ni** 0.03 max **Pb** 0.003 max **Si** 0.20 max **Ti** 0.10 max **Zn** 0.10 max **Zr** 0.05 max **Other** 0.05 max each, 0.15 max total | **AA** 5013 |
| * **A95014** | Wrought Aluminum Alloy, Non-Heat Treatable | **Al** rem **Cr** 0.20 max **Cu** 0.20 max **Fe** 0.40 max **Mg** 4.0-5.5 **Mn** 0.20-0.9 **Si** 0.40 max **Ti** 0.20 max **Zn** 0.7-1.5 **Other** 0.05 max each, 0.15 max total | **AA** 5014 |
| **A95016** | Wrought Aluminum Alloy, Non-Heat Treatable | **Al** rem **Cr** 0.10 max **Cu** 0.20 max **Fe** 0.6 max **Mg** 1.4-1.9 **Mn** 0.40-0.7 **Si** 0.25 max **Ti** 0.05 max **Zn** 0.15 max **Other** each 0.05 max, total 0.15 max | **AA** 5016 |
| **A95017** | Wrought Aluminum Alloy, Non-Heat Treatable | **Al** rem **Cu** 0.18-0.28 **Fe** 0.7 max **Mg** 1.9-2.2 **Mn** 0.6-0.8 **Si** 0.40 max **Ti** 0.09 max **Other** 0.05 max each, 0.15 max total | **AA** 5017 |

*Boxed entries are no longer active and are retained for reference purposes only.

The chemical compositions listed are for identification purposes and should not be used in lieu of the cross referenced specifications.

| UNIFIED NUMBER | DESCRIPTION | CHEMICAL COMPOSITION | CROSS REFERENCE SPECIFICATIONS |
|---|---|---|---|
| **A95018** | Wrought Aluminum Alloy, Non-Heat Treatable | **Al** rem **Be** 0.0008 max (for welding electrode and filler wire only) **Cr** 0.30 max **Cu** 0.05 max **Fe** 0.40 max **Mg** 2.6-3.6 **Mn** 0.20-0.6 **Si** 0.25 max **Ti** 0.15 max **Zn** 0.20 max **Other** each 0.05 max (Be 0.0008 max for welding electrode and filler wire only), total 0.15 max, Cr+Mn 0.20-0.6 | **AA** 5018; 5018A |
| **A95019** | Wrought Aluminum Alloy, Non-Heat Treatable | **Al** rem **Cr** 0.20 max **Cu** 0.10 max **Fe** 0.50 max **Mg** 4.5-5.6 **Mn** 0.10-0.6 **Si** 0.40 max **Ti** 0.20 max **Zn** 0.20 max **Other** each 0.05 max, total 0.15 max, Cr+Mn 0.10-0.6 | **AA** 5019; 5019A |
| **A95021** | Wrought Aluminum Alloy, Non-Heat Treatable | **Al** rem **Cr** 0.15 max **Cu** 0.15 max **Fe** 0.50 max **Mg** 2.2-2.8 **Mn** 0.10-0.50 **Si** 0.40 max **Zn** 0.15 max **Other** each 0.05 max, total 0.15 max | **AA** 5021 |
| **A95022** | Wrought Aluminum Alloy, Non-Heat Treatable | **Al** rem **Cr** 0.10 max **Cu** 0.20-0.50 **Fe** 0.40 max **Mg** 3.5-4.9 **Mn** 0.20 max **Si** 0.25 max **Ti** 0.10 max **Zn** 0.25 max **Other** each 0.05 max, total 0.15 max | **AA** 5022 |
| **A95023** | Wrought Aluminum Alloy, Non-Heat Treatable | **Al** rem **Cr** 0.10 max **Cu** 0.20-0.50 **Fe** 0.40 max **Mg** 5.0-6.2 **Mn** 0.20 max **Si** 0.25 max **Ti** 0.10 max **Zn** 0.25 max **Other** each 0.05 max, total 0.15 max | **AA** 5023 |
| **A95025** | Wrought Aluminum Alloy, Non-Heat Treatable | **Al** REM **Be** 0.0008 max **Cr** 0.20 max **Cu** 0.10 max **Fe** 0.25 max **Mg** 4.5-6.0 **Mn** 0.20 max **Sc** 0.05-0.55 **Si** 0.25 max **Ti** 0.05-0.20 **Zn** 0.25 max **Zr** 0.10-0.25 **Other** each 0.05 max, total 0.15 max. | **AA** 5025 |
| * **A95034** | Wrought Aluminum Alloy | **Al** rem **Cr** 0.10 max **Cu** 0.40 max **Fe** 0.40-0.9 **Mg** 0.6-1.1 **Mn** 0.20-0.50 **Si** 0.40 max **Ti** 0.10 max **Zn** 0.40 max **Other** each 0.05 max, total 0.15 max | **AA** 5034 |
| * **A95039** | Wrought Aluminum Alloy, Non-Heat Treatable (Welding Electrode) | **Al** rem **Be** 0.0008 max **Cr** 0.10-0.20 **Cu** 0.03 max **Fe** 0.40 max **Mg** 3.3-4.3 **Mn** 0.30-0.50 **Si** 0.10 max **Ti** 0.10 max **Zn** 2.4-3.2 **Other** each 0.05 max, total 0.l0 max | **AA** 5039 |
| **A95040** | Wrought Aluminum Alloy, Non-Heat Treatable | **Al** rem **Cr** 0.10-0.30 **Cu** 0.25 max **Fe** 0.7 max **Mg** 1.0-1.5 **Mn** 0.9-1.4 **Si** 0.30 max **Zn** 0.25 max **Other** each 0.05 max, total 0.15 max | **AA** 5040 |
| **A95042** | Wrought Aluminum Alloy, Non-Heat Treatable | **Al** rem **Cr** 0.10 max **Cu** 0.15 max **Fe** 0.35 max **Mg** 3.0-4.0 **Mn** 0.20-0.50 **Si** 0.20 max **Ti** 0.10 max **Zn** 0.25 max **Other** each 0.05 max, total 0.15 max | **AA** 5042 |
| **A95043** | Wrought Aluminum Alloy, Non-Heat Treatable | **Al** rem **Cr** 0.05 max **Cu** 0.05-0.35 **Fe** 0.7 max **Ga** 0.05 max **Mg** 0.7-1.3 **Mn** 0.7-1.2 **Si** 0.40 max **Ti** 0.10 max **V** 0.05 max **Zn** 0.25 max **Other** each 0.05 max, total 0.15 max | **AA** 5043 |
| **A95049** | Wrought Aluminum Alloy, Non-Heat Treatable | **Al** rem **Cr** 0.30 max **Cu** 0.10 max **Fe** 0.50 max **Mg** 1.6-2.5 **Mn** 0.50-1.1 **Si** 0.40 max **Ti** 0.10 max **Zn** 0.20 max **Other** 0.05 max each, 0.15 max total | **AA** 5049 |
| **A95050** | Wrought Aluminum Alloy, Non-Heat Treatable | **Al** rem **Cr** 0.10 max **Cu** 0.20 max **Fe** 0.7 max **Mg** 1.1-1.8 **Mn** 0.10 max **Si** 0.40 max **Zn** 0.25 max **Other** each 0.05 max, total 0.15 max | **AA** 5050; 5050A **ASME** SB-209 **ASTM** B209 (5050); B210 (5050); B313 (5050); B483 (5050); B547 (5050); B548 (5050) |

*Boxed entries are no longer active and are retained for reference purposes only.

The chemical compositions listed are for identification purposes and should not be used in lieu of the cross referenced specifications.

| UNIFIED NUMBER | DESCRIPTION | CHEMICAL COMPOSITION | CROSS REFERENCE SPECIFICATIONS |
|---|---|---|---|
| **A95051** | Wrought Aluminum Alloy, Non-Heat Treatable | **Al** rem **Cr** 0.10 max **Cu** 0.25 max **Fe** 0.7 max **Mg** 1.7-2.2 **Mn** 0.20 max **Si** 0.40 max **Ti** 0.10 max **Zn** 0.25 max **Other** each 0.05 max, total 0.15 max | **AA** 5051; 5051A |
| **A95052** | Wrought Aluminum Alloy, Non-Heat Treatable | **Al** rem **Cr** 0.15-0.35 **Cu** 0.10 max **Fe** 0.40 max **Mg** 2.2-2.8 **Mn** 0.10 max **Si** 0.25 max **Zn** 0.10 max **Other** each 0.05 max, total 0.15 max | **AA** 5052 **AMS** 4004; 4015; 4016; 4017; 4069; 4070; 4071; 4114; AMS-QQ-A-225/7; AMS-QQ-A-250/8; AMS-A-81596 **ASME** SB-209; SB-210; SB-234; SB-241 **ASTM** B209 (5052); B210 (5052); B211 (5052); B221 (5052); B234 (5052); B241 (5052); B313 (5052); B316 (5052); B404 (5052); B483 (5052); B547 (5052); B548 (5052) **FED** QQ-A-430; WW-T-700/4 **MIL SPEC** MIL-G-18014 **SAE** J454 (5052) |
| **A95056** | Wrought Aluminum Alloy, Non-Heat Treatable | **Al** rem **Cr** 0.05-0.20 **Cu** 0.10 max **Fe** 0.40 max **Mg** 4.5-5.6 **Mn** 0.05-0.20 **Si** 0.30 max **Zn** 0.10 max **Other** each 0.05 max, total 0.15 max | **AA** 5056 **AMS** 4177; 4182; AMS-A-81596 **ASTM** B211 (5056); B316 (5056) **FED** QQ-A-430 |
| **A95058** | Wrought Aluminum Alloy, Non-Heat Treatable | **Al** rem **Cr** 0.10 max **Cu** 0.10 max **Fe** 0.50 max **Mg** 4.5-5.6 **Mn** 0.20 max **Pb** 1.2-1.8 **Si** 0.40 max **Ti** 0.20 max **Zn** 0.20 max **Other** 0.05 max each, 0.15 max total | **AA** 5058 |
| **A95059** | Wrought Aluminum Alloy, Non-Heat Treatable | **Al** rem **Cr** 0.25 max **Cu** 0.25 max **Fe** 0.50 max **Mg** 5.0-6.0 **Mn** 0.6-1.2 **Si** 0.45 max **Ti** 0.20 max **Zn** 0.40-0.9 **Zr** 0.05-0.25 **Other** each 0.05 max, total 0.15 max | **AA** 5059 |
| **A95082** | Wrought Aluminum Alloy, Non-Heat Treatable | **Al** rem **Cr** 0.15 max **Cu** 0.15 max **Fe** 0.35 max **Mg** 4.0-5.0 **Mn** 0.15 max **Si** 0.20 max **Ti** 0.10 max **Zn** 0.25 max **Other** each 0.05 max, total 0.15 max | **AA** 5082 |
| **A95083** | Wrought Aluminum Alloy, Non-Heat Treatable | **Al** rem **Cr** 0.05-0.25 **Cu** 0.10 max **Fe** 0.40 max **Mg** 4.0-4.9 **Mn** 0.40-1.0 **Si** 0.40 max **Ti** 0.15 max **Zn** 0.25 max **Other** each 0.05 max, total 0.15 max | **AA** 5083 **AMS** 4056; AMS-QQ-A-200/4; AMS-QQ-A-250/6 **ASME** SB-209; SB-221; SB-241; SB-247 **ASTM** B209 (5083); B210 (5083); B221 (5083); B241 (5083); B247 (5083); B345 (5083); B361 (5083); B547 (5083); B548 (5083) **MIL SPEC** MIL-DTL-45225; MIL-DTL-46027; MIL-DTL-46083; MIL-S-24149/2 **SAE** J454 (5083) |
| **A95086** | Wrought Aluminum Alloy, Non-Heat Treatable | **Al** rem **Cr** 0.05-0.25 **Cu** 0.10 max **Fe** 0.50 max **Mg** 3.5-4.5 **Mn** 0.20-0.7 **Si** 0.40 max **Ti** 0.15 max **Zn** 0.25 max **Other** each 0.05 max, total 0.15 max | **AA** 5086 **AMS** AMS-QQ-A-200/5; AMS-QQ-A-200/6; AMS-QQ-A-250/7; AMS-QQ-A-250/19 **ASTM** B209 (5086); B210 (5086); B221 (5086); B241 (5086); B313 (5086); B345 (5086); B361 (5086); B547 (5086); B548 (5086) **ASME** SB-209; SB-211; SB-241 **FED** WW-T-700/5 **MIL SPEC** MIL-S-24149/5 **SAE** J454 (5086) |
| **A95087** | Wrought Aluminum Alloy, Non-Heat Treatable | **Al** rem **Cr** 0.05-0.25 **Cu** 0.05 max **Fe** 0.40 max **Mg** 4.5-5.2 **Mn** 0.7-1.1 **Si** 0.25 max **Ti** 0.15 max **Zn** 0.25 max **Zr** 0.10-0.20 **Other** 0.05 max each, 0.15 max total | **AA** 5087 |
| **A95091** | Wrought Aluminum Alloy, Non-Heat Treatable | **Al** rem **C** 1.0-1.3 **Fe** 0.30 max **Li** 1.2-1.4 **Mg** 3.7-4.2 **O** 0.20-0.7 **Si** 0.20 max **Other** 0.05 max each, 0.15 max total | **AA** 5091 **AMS** 4322 |
| **A95110** | Wrought Aluminum Alloy, Non-Heat Treatable | **Al** rem **Fe** 0.08 max **Mg** 0.30-0.6 **Mn** 0.03 max **Si** 0.08 max **Ti** 0.02 max **Zn** 0.05 max **Other** 0.02 max each | **AA** 5110 |
| **A95119** | Wrought Aluminum Alloy, Non-Heat Treatable | **Al** rem **Be** 0.0008 max (for welding electrode and filler wire only) **Cr** 0.30 max **Cu** 0.05 max **Fe** 0.40 max **Mg** 4.5-5.6 **Mn** 0.20-0.6 **Si** 0.25 max **Ti** 0.15 max **Zn** 0.20 max **Other** each 0.05 max (Be 0.0008 max for welding electrode and filler wire only), total 0.15 max, Cr+Mn 0.20-0.60 | **AA** 5119 |
| **A95149** | Wrought Aluminum Alloy, Non-Heat Treatable | **Al** rem **Cr** 0.30 max **Cu** 0.05 max **Fe** 0.40 max **Mg** 1.6-2.5 **Mn** 0.50-1.1 **Si** 0.25 max **Ti** 0.15 max **Zn** 0.20 max **Other** 0.05 max each, 0.15 max total | **AA** 5149 |

The chemical compositions listed are for identification purposes and should not be used in lieu of the cross referenced specifications.

| UNIFIED NUMBER | DESCRIPTION | CHEMICAL COMPOSITION | CROSS REFERENCE SPECIFICATIONS |
|---|---|---|---|
| **A95150** | Wrought Aluminum Alloy, Non-Heat Treatable | **Al** rem **Cu** 0.10 max **Fe** 0.10 max **Mg** 1.3-1.7 **Mn** 0.03 max **Si** 0.08 max **Ti** 0.06 max **Zn** 0.10 max **Other** 0.03 max each, 0.10 max total | **AA** 5150 |
| **A95151** | Wrought Aluminum Alloy, Non-Heat Treatable | **Al** rem **Cr** 0.10 max **Cu** 0.15 max **Fe** 0.35 max **Mg** 1.5-2.1 **Mn** 0.10 max **Si** 0.20 max **Ti** 0.10 max **Zn** 0.15 max **Other** each 0.05 max, total 0.15 max | **AA** 5151 |
| **A95154** | Wrought Aluminum Alloy, Non-Heat Treatable | **Al** rem **Cr** 0.15-0.35 **Cu** 0.10 max **Fe** 0.40 max **Mg** 3.1-3.9 **Mn** 0.10 max **Si** 0.25 max **Ti** 0.20 max **Zn** 0.20 max **Other** each 0.05 max (Be 0.0008 max for welding electrode and filler wire only), total 0.15 max | **AA** 5154; 5154A; 5154B **ASME** SB-209; SB-210; SB-221 **ASTM** B209 (5154); B210 (5154); B211 (5154); B221 (5154); B313 (5154); B361 (5154); B547 (5154); B548 (5154) |
| **A95180** | Wrought Aluminum Alloy, Non-Heat Treatable | **Al** rem **Be** 0.0008 max (for welding electrode and filler wire only) **Cr** 0.10 max **Cu** 0.10 max **Mg** 3.5-4.5 **Mn** 0.20-0.7 **Ti** 0.06-0.20 **Zn** 1.7-2.8 **Zr** 0.08-0.25 **Other** each 0.05 max, total 0.15 max, Fe+Si 0.35 max | **AA** 5180 |
| **A95182** | Wrought Aluminum Alloy, Non-Heat Treatable | **Al** rem **Cr** 0.10 max **Cu** 0.15 max **Fe** 0.35 max **Mg** 4.0-5.0 **Mn** 0.20-0.50 **Si** 0.20 max **Ti** 0.10 max **Zn** 0.25 max **Other** 0.05 max, total 0.15 max | **AA** 5182 **SAE** J454 (5182) |
| **A95183** | Wrought Aluminum Alloy, Non-Heat Treatable (Welding Electrode) | **Al** rem **Be** 0.0008 max **Cr** 0.05-0.25 **Cu** 0.10 max **Fe** 0.40 max **Mg** 4.3-5.2 **Mn** 0.50-1.0 **Si** 0.40 max **Ti** 0.15 max **Zn** 0.25 max **Other** each 0.05 max, total 0.15 max | **AA** 5183 **ASME** SFA5.10 (ER5183) **AWS** A5.10 (ER5183) |
| **A95186** | Wrought Aluminum Alloy, Non-Heat Treatable | **Al** rem **Cr** 0.15 max **Cu** 0.25 max **Fe** 0.45 max **Mg** 3.8-4.8 **Mn** 0.20-0.50 **Si** 0.40 max **Ti** 0.15 max **Zn** 0.40 max **Zr** 0.05 max **Other** each 0.05 max, total 0.15 max | **AA** 5186 |
| **A95205** | Wrought Aluminum Alloy, Non-Heat Treatable | **Al** rem **Cr** 0.10 max **Cu** 0.03-0.10 **Fe** 0.7 max **Mg** 0.6-1.0 **Mn** 0.10 max **Si** 0.15 max **Zn** 0.05 max **Other** each 0.05 max, total 0.15 max | **AA** 5205 |
| **A95210** | Wrought Aluminum Alloy, Non-Heat Treatable | **Al** rem **Fe** 0.04 max **Mg** 0.35-0.6 **Mn** 0.03 max **Si** 0.06 max **Ti** 0.01 max **Zn** 0.04 max **Other** 0.01 max each | **AA** 5210 |
| **A95249** | Wrought Aluminum Alloy, Non-Heat Treatable | **Al** rem **Cr** 0.30 max **Cu** 0.05 max **Fe** 0.40 max **Mg** 1.6-2.5 **Mn** 0.50-1.1 **Si** 0.25 max **Ti** 0.15 max **Zn** 0.20 max **Zr** 0.10-0.20 **Other** 0.05 max each, 0.15 max total | **AA** 5249 |
| **A95250** | Wrought Aluminum Alloy, Non-Heat Treatable | **Al** rem **Cu** 0.10 **Fe** 0.10 **Ga** 0.03 **Mg** 1.3-1.8 **Mn** 0.04-0.15 **Si** 0.08 **V** 0.05 **Zn** 0.05 **Other** each 0.03, total 0.10 | **AA** 5250 |
| **A95251** | Wrought Aluminum Alloy, Non-Heat Treatable | **Al** rem **Cr** 0.15 max **Cu** 0.15 max **Fe** 0.50 max **Mg** 1.7-2.4 **Mn** 0.10-0.50 **Si** 0.40 max **Ti** 0.15 max **Zn** 0.15 max **Other** 0.05 max each, 0.15 max total | **AA** 5251; 5251A |
| **A95252** | Wrought Aluminum Alloy, Non-Heat Treatable | **Al** rem **Cu** 0.10 max **Fe** 0.10 max **Mg** 2.2-2.8 **Mn** 0.10 max **Si** 0.08 max **V** 0.05 max **Zn** 0.05 max **Other** each 0.03 max, total 0.10 max | **AA** 5252 **ASTM** B209 (5252) **SAE** J454 (5252) |
| **A95254** | Wrought Aluminum Alloy, Non-Heat Treatable | **Al** rem **Cr** 0.15-0.35 **Cu** 0.05 max **Fe** 0.40 max **Mg** 3.1-3.9 **Mn** 0.01 max **Si** 0.25 max **Ti** 0.05 max **Zn** 0.20 max **Other** each 0.05 max, total 0.15 max, Fe+Si 0.45 max | **AA** 5254 **ASME** SB-209 **ASTM** B209 (5254); B241 (5254); B548 (5254) |

The chemical compositions listed are for identification purposes and should not be used in lieu of the cross referenced specifications.

| UNIFIED NUMBER | DESCRIPTION | CHEMICAL COMPOSITION | CROSS REFERENCE SPECIFICATIONS |
|---|---|---|---|
| **A95257** | Wrought Aluminum Alloy, Non-Heat Treatable | **Al** rem **Cu** 0.10 max **Fe** 0.10 max **Mg** 0.20-0.6 **Mn** 0.03 max **Si** 0.08 max **Zn** 0.03 max **Other** each 0.02 max, total 0.05 max | **AA** 5257 |
| * **A95280** | Wrought Aluminum Alloy, Non-Heat Treatable | **Al** rem **Be** 0.0008 max **Cr** 0.05-0.25 **Cu** 0.10 max **Mg** 3.5-4.5 **Mn** 0.20-0.7 **Zn** 1.5-2.8 **Zr** 0.05-0.25 **Other** 0.05 max each, 0.15 max total, Fe+Si 0.35 max | **AA** 5280 |
| **A95283** | Wrought Aluminum Alloy, Non-Heat Treatable | **Al** rem **Cr** 0.05 max **Cu** 0.03 max **Fe** 0.30 max **Mg** 4.5-5.1 **Mn** 0.50-1.0 **Ni** 0.03 max **Si** 0.30 max **Ti** 0.03 max **Zn** 0.10 max **Zr** 0.05 max **Other** 0.05 max each, 0.15 max total | **AA** 5283; 5283A; 5283B |
| **A95305** | Wrought Aluminum Alloy, Non-Heat Treatable | **Al** rem **Fe** 0.08 max **Mg** 0.7-1.1 **Mn** 0.03 max **Si** 0.08 max **Ti** 0.02 max **Zn** 0.05 max **Other** 0.02 max each | **AA** 5305 |
| **A95310** | Wrought Aluminum Alloy, Non-Heat Treatable | **Al** rem **Fe** 0.008 max **Mg** 0.35-0.6 **Si** 0.01 max **Ti** 0.008 max **Zn** 0.01 max **Other** each 0.003 max, Fe+Ti 0.008 max | **AA** 5310 |
| **A95349** | Wrought Aluminum Alloy, Non-Heat Treatable | **Al** rem **Cu** 0.18-0.28 **Fe** 0.7 max **Mg** 1.7-2.6 **Mn** 0.6-1.2 **Si** 0.40 max **Ti** 0.09 max **Other** 0.05 max each, 0.15 max total | **AA** 5349 |
| **A95351** | Wrought Aluminum Alloy, Non-Heat Treatable | **Al** rem **Cu** 0.10 max **Fe** 0.10 max **Mg** 1.6-2.2 **Mn** 0.10 max **Si** 0.08 max **V** 0.05 max **Zn** 0.05 max **Other** each 0.03 max, total 0.10 max | **AA** 5351 |
| **A95352** | Wrought Aluminum Alloy, Non-Heat Treatable | **Al** rem **Cr** 0.10 max **Cu** 0.10 max **Mg** 2.2-2.8 **Mn** 0.10 max **Ti** 0.10 max **Zn** 0.10 max **Other** each 0.05 max, total 0.15, Fe+Si 0.45 max | **AA** 5352 |
| **A95354** | Wrought Aluminum Alloy, Non-Heat Treatable | **Al** rem **Cr** 0.05-0.20 **Cu** 0.05 max **Fe** 0.40 max **Mg** 2.4-3.0 **Mn** 0.50-1.0 **Si** 0.25 max **Ti** 0.15 max **Zn** 0.25 max **Zr** 0.10-0.20 **Other** each 0.05 max, total 0.15 max | **AA** 5354 |
| **A95356** | Wrought Aluminum Alloy, Non-Heat Treatable | **Al** rem **Cr** 0.05-0.20 **Cu** 0.10 max **Fe** 0.40 max **Mg** 4.5-5.5 **Mn** 0.05-0.20 **Si** 0.25 max **Ti** 0.06-0.20 **Zn** 0.10 max **Other** each 0.05 max (Be 0.0008 max for welding electrode and filler wire only), total 0.15 max | **AA** 5356 **ASME** SFA5.10 (ER5356) **AWS** A5.10 (ER5356); C2.25 (W-Al-5356) **MIL SPEC** MIL-S-24149/5 |
| **A95357** | Wrought Aluminum Alloy, Non-Heat Treatable | **Al** rem **Cu** 0.20 max **Fe** 0.17 max **Mg** 0.8-1.2 **Mn** 0.15-0.45 **Si** 0.12 max **Zn** 0.05 max **Other** each 0.05 max, total 0.15 max | **AA** 5357 |
| **A95383** | Wrought Aluminum Alloy, Non-Heat Treatable | **Al** rem **Cr** 0.25 max **Cu** 0.20 max **Fe** 0.25 max **Mg** 4.0-5.2 **Mn** 0.7-1.0 **Si** 0.25 max **Ti** 0.15 max **Zn** 0.40 max **Zr** 0.20 max **Other** each 0.05 max, total 0.15 max | **AA** 5383 |
| **A95449** | Wrought Aluminum Alloy, Non-Heat Treatable | **Al** rem **Cr** 0.30 max **Cu** 0.30 max **Fe** 0.7 max **Mg** 1.6-2.6 **Mn** 0.6-1.1 **Si** 0.40 max **Ti** 0.10 max **Zn** 0.30 max **Other** each 0.05 max, total 0.15 max | **AA** 5449 |
| **A95451** | Wrought Aluminum Alloy, Non-Heat Treatable | **Al** rem **Cr** 0.15-0.35 **Cu** 0.10 **Fe** 0.40 **Mg** 1.8-2.4 **Mn** 0.10 **Ni** 0.05 **Si** 0.25 **Ti** 0.05 **Zn** 0.10 **Other** each 0.05, total 0.15 | **AA** 5451 |

*Boxed entries are no longer active and are retained for reference purposes only.

The chemical compositions listed are for identification purposes and should not be used in lieu of the cross referenced specifications.

| UNIFIED NUMBER | DESCRIPTION | CHEMICAL COMPOSITION | CROSS REFERENCE SPECIFICATIONS |
|---|---|---|---|
| **A95454** | Wrought Aluminum Alloy, Non-Heat Treatable | **Al** rem **Cr** 0.05-0.20 **Cu** 0.10 max **Fe** 0.40 max **Mg** 2.4-3.0 **Mn** 0.50-1.0 **Si** 0.25 max **Ti** 0.20 max **Zn** 0.25 max **Other** each 0.05 max, total 0.15 max | **AA** 5454 **AMS** AMS-QQ-A-200/6; AMS-QQ-A-250/10 **ASME** SB-209; SB-211; SB-234; SB-241 **ASTM** B209 (5454); B221 (5454); B234 (5454); B241 (5454); B404 (5454); B547 (5454); B548 (5454) **MIL SPEC** MIL-G-18014 **SAE** J454 (5454) |
| **A95456** | Wrought Aluminum Alloy, Non-Heat Treatable | **Al** rem **Cr** 0.05-0.20 **Cu** 0.10 max **Fe** 0.40 max **Mg** 4.7-5.5 **Mn** 0.50-1.0 **Si** 0.25 max **Ti** 0.20 max **Zn** 0.25 max **Other** each 0.05 max, total 0.15 max | **AA** 5456; 5456A **AMS** AMS-QQ-A-200/7; AMS-QQ-A-250/9; AMS-QQ-A-250/20 **ASTM** B209 (5456); B210 (5456); B221 (5456); B241 (5456); B548 (5456) **MIL SPEC** MIL-DTL-45225; MIL-DTL-46027; MIL-DTL-46083 |
| **A95457** | Wrought Aluminum Alloy, Non-Heat Treatable | **Al** rem **Cu** 0.20 max **Fe** 0.10 max **Mg** 0.8-1.2 **Mn** 0.15-0.45 **Si** 0.08 max **Zn** 0.05 max **Other** each 0.03 max, total 0.10 max | **AA** 5457 **ASTM** B209 (5457) **SAE** J454 (5457) |
| **A95505** | Wrought Aluminum Alloy, Non-Heat Treatable | **Al** rem **Fe** 0.04 max **Mg** 0.8-1.1 **Mn** 0.03 max **Si** 0.06 max **Ti** 0.01 max **Zn** 0.04 max **Other** 0.01 max each | **AA** 5505 |
| * **A95552** | Wrought Aluminum Alloy, Non-Heat Treatable | **Al** rem **Cu** 0.10 max **Fe** 0.05 max **Mg** 2.2-2.8 **Mn** 0.10 max **Si** 0.04 max **V** 0.05 max **Zn** 0.05 max **Other** 0.03 max each, 0.10 max total | **AA** 5552 |
| **A95554** | Wrought Aluminum Alloy, Non-Heat Treatable (Welding Electrode) | **Al** rem **Be** 0.0008 max **Cr** 0.05-0.20 **Cu** 0.10 max **Fe** 0.40 max **Mg** 2.4-3.0 **Mn** 0.50-1.0 **Si** 0.25 max **Ti** 0.05-0.20 **Zn** 0.25 max **Other** each 0.05 max, total 0.15 max | **AA** 5554 **ASME** SFA5.10 (ER5554) **AWS** A5.10 (ER5554) |
| **A95556** | Wrought Aluminum Alloy, Non-Heat Treatable (Welding Electrode) | **Al** rem **Be** 0.0008 max **Cr** 0.05-0.20 **Cu** 0.10 max **Fe** 0.40 max **Mg** 4.7-5.5 **Mn** 0.50-1.0 **Si** 0.25 max **Ti** 0.05-0.20 **Zn** 0.25 max **Other** each 0.05, total 0.15 max (Be 0.0008 max for welding electrode and filler wire only) | **AA** 5556; 5556A **ASME** SFA5.10 (ER5556) **AWS** A5.10 (ER5556) |
| **A95557** | Wrought Aluminum Alloy, Non-Heat Treatable | **Al** rem **Cu** 0.15 max **Fe** 0.12 max **Mg** 0.40-0.8 **Mn** 0.10-0.40 **Si** 0.10 max **V** 0.05 max **Other** each 0.03 max, total 0.10 max | **AA** 5557 |
| **A95605** | Wrought Aluminum Alloy, Non-Heat Treatable | **Al** rem **Fe** 0.008 max **Mg** 0.8-1.1 **Si** 0.01 max **Ti** 0.008 max **Zn** 0.01 max **Other** each 0.003 max, (Fe+Ti) 0.008 max | **AA** 5605 |
| **A95652** | Wrought Aluminum Alloy, Non-Heat Treatable | **Al** rem **Cr** 0.15-0.35 **Cu** 0.04 max **Mg** 2.2-2.8 **Mn** 0.01 max **Zn** 0.10 max **Other** each 0.05 max, total 0.15 max | **AA** 5652 **ASME** SB-209 **ASTM** B209 (5652); B241 (5652); B548 (5652) |
| **A95654** | Wrought Aluminum Alloy, Non-Heat Treatable (Welding Electrode) | **Al** rem **Be** 0.0008 max **Cr** 0.15-0.35 **Cu** 0.05 max **Mg** 3.1-3.9 **Mn** 0.01 max **Ti** 0.05-0.15 **Zn** 0.20 max **Other** each 0.05 max, total 0.15 max, Fe+Si 0.45 max | **AA** 5654 **ASME** SFA5.10 (ER5654) **AWS** A5.10 (ER5654) |
| **A95657** | Wrought Aluminum Alloy, Non-Heat Treatable | **Al** rem **Cu** 0.10 max **Fe** 0.10 max **Ga** 0.03 max **Mg** 0.6-1.0 **Mn** 0.03 max **Si** 0.08 max **Zn** 0.05 max **Other** each 0.02 max, total 0.05 max | **AA** 5657 **ASTM** B209 (5657) **SAE** J454 (5657) |
| **A95754** | Wrought Aluminum Alloy, Non-Heat Treatable | **Al** rem **Cr** 0.30 max **Cu** 0.10 max **Fe** 0.40 max **Mg** 2.6-3.6 **Mn** 0.50 max **Si** 0.40 max **Ti** 0.15 max **Zn** 0.20 max **Other** each 0.05 max, total 0.15 max, Cr+Mn 0.10-0.6 | **AA** 5754 |
| * **A95854** | Wrought Aluminum Alloy, Non-Heat Treatable | **Al** rem **Cr** 0.15-0.35 **Cu** 0.10 max **Mg** 3.1-3.9 **Mn** 0.10-0.50 **Ti** 0.20 max **Zn** 0.20 max **Other** 0.05 max each, 0.15 max total, Fe+Si 0.45 max | **AA** 5854 |

*Boxed entries are no longer active and are retained for reference purposes only.

The chemical compositions listed are for identification purposes and should not be used in lieu of the cross referenced specifications.

| UNIFIED NUMBER | DESCRIPTION | CHEMICAL COMPOSITION | CROSS REFERENCE SPECIFICATIONS |
|---|---|---|---|
| **A95954** | Wrought Aluminum Alloy, Non-Heat Treatable | **Al** rem **Cr** 0.10 max **Cu** 0.10 max **Fe** 0.40 max **Mg** 3.3-4.1 **Mn** 0.10 max **Si** 0.25 max **Ti** 0.20 max **Zn** 0.20 max **Other** each 0.05 max, total 0.15 max | **AA** 5954 |
| **A96002** | Wrought Aluminum Alloy, Heat Treatable | **Al** rem **Cr** 0.05 max **Cu** 0.10-0.25 **Fe** 0.25 max **Mg** 0.45-0.7 **Mn** 0.10-0.20 **Si** 0.6-0.9 **Ti** 0.08 max **Zr** 0.09-0.14 **Other** 0.05 max each, 0.15 max total | **AA** 6002 |
| **A96003** | Wrought Aluminum Alloy, Heat Treatable (Cladding Alloy) | **Al** rem **Cr** 0.35 max **Cu** 0.10 max **Fe** 0.6 max **Mg** 0.8-1.5 **Mn** 0.8 max **Si** 0.35-1.0 **Ti** 0.10 max **Zn** 0.20 max **Other** each 0.05 max, total 0.15 max | **AA** 6003 **ASTM** B209 (6003) |
| **A96004** | Wrought Aluminum Alloy, Heat Treatable | **Al** rem **Cu** 0.10 max **Fe** 0.10-0.30 **Mg** 0.40-0.7 **Mn** 0.20-0.6 **Si** 0.30-0.6 **Zn** 0.05 max **Other** each 0.05 max, total 0.15 max | **AA** 6004 |
| **A96005** | Wrought Aluminum Alloy, Heat Treatable | **Al** rem **Cr** 0.10 max **Cu** 0.10 max **Fe** 0.35 max **Mg** 0.40-0.6 **Mn** 0.10 max **Si** 0.6-0.9 **Ti** 0.10 max **Zn** 0.10 max **Other** each 0.05 max, total 0.15 max | **AA** 6005; 6005A; 6005B **ASTM** B221 (6005, 6005A) |
| **A96006** | Wrought Aluminum Alloy, Heat Treatable | **Al** rem **Cr** 0.10 max **Cu** 0.15-0.30 **Fe** 0.35 max **Mg** 0.45-0.9 **Mn** 0.05-0.20 **Si** 0.20-0.6 **Ti** 0.10 max **Zn** 0.10 max **Other** each 0.05 max, total 0.15 max | **AA** 6006 |
| **A96007** | Wrought Aluminum Alloy, Heat Treatable | **Al** rem **Cr** 0.05-0.25 **Cu** 0.20 max **Fe** 0.7 max **Mg** 0.6-0.9 **Mn** 0.05-0.25 **Si** 0.9-1.4 **Ti** 0.15 max **Zn** 0.25 max **Zr** 0.05-0.20 **Other** each 0.05 max, total 0.15 max | **AA** 6007 |
| **A96008** | Wrought Aluminum Alloy, Heat Treatable | **Al** rem **Cr** 0.30 max **Cu** 0.30 max **Fe** 0.35 max **Mg** 0.40-0.7 **Mn** 0.30 max **Si** 0.50-0.9 **Ti** 0.10 max **V** 0.05-0.20 **Zn** 0.20 max **Other** 0.05 max each, 0.15 max total | **AA** 6008 |
| **A96009** | Wrought Aluminum Alloy, Heat Treatable | **Al** rem **Cr** 0.10 max **Cu** 0.15-0.6 **Fe** 0.50 max **Mg** 0.40-0.8 **Mn** 0.20-0.8 **Si** 0.6-1.0 **Ti** 0.10 max **Zn** 0.25 max **Other** each 0.05 max, total 0.15 max | **AA** 6009 **SAE** J454 (6009) |
| **A96010** | Wrought Aluminum Alloy, Heat Treatable | **Al** rem **Cr** 0.10 max **Cu** 0.15-0.6 **Fe** 0.50 max **Mg** 0.6-1.0 **Mn** 0.20-0.8 **Si** 0.8-1.2 **Ti** 0.10 max **Zn** 0.25 max **Other** each 0.05 max, total 0.15 max | **AA** 6010 **SAE** J454 (6010) |
| **A96011** | Wrought Aluminum Alloy, Heat Treatable | **Al** rem **Cr** 0.30 max **Cu** 0.40-0.9 **Fe** 1.0 max **Mg** 0.6-1.2 **Mn** 0.8 max **Ni** 0.20 max **Si** 0.6-1.2 **Ti** 0.20 max **Zn** 1.5 max **Other** each 0.05 max, total 0.15 max | **AA** 6011 |
| **A96012** | Wrought Aluminum Alloy, Heat Treatable | **Al** rem **Bi** 0.7 max **Cr** 0.30 max **Cu** 0.10 max **Fe** 0.50 max **Mg** 0.6-1.2 **Mn** 0.40-1.0 **Pb** 0.40-2.0 **Si** 0.6-1.4 **Ti** 0.20 max **Zn** 0.30 max **Other** 0.05 max each, 0.15 max total | **AA** 6012; 6012A |
| **A96013** | Wrought Aluminum Alloy, Heat Treatable | **Al** rem **Cr** 0.10 max **Cu** 0.6-1.1 **Fe** 0.50 max **Mg** 0.8-1.2 **Mn** 0.20-0.8 **Si** 0.6-1.0 **Ti** 0.10 max **Zn** 0.25 max **Other** each 0.05 max, total 0.15 max | **AA** 6013 **AMS** 4216; 4347 |
| **A96014** | Wrought Aluminum Alloy, Heat Treatable | **Al** rem **Cr** 0.20 max **Cu** 0.25 max **Fe** 0.35 max **Mg** 0.40-0.8 **Mn** 0.05-0.20 **Si** 0.30-0.6 **Ti** 0.10 max **V** 0.05-0.20 **Zn** 0.10 max **Other** 0.05 max each, 0.15 max total | **AA** 6014 |

The chemical compositions listed are for identification purposes and should not be used in lieu of the cross referenced specifications.

| UNIFIED NUMBER | DESCRIPTION | CHEMICAL COMPOSITION | CROSS REFERENCE SPECIFICATIONS |
|---|---|---|---|
| **A96015** | Wrought Aluminum Alloy, Heat Treatable | **Al** rem **Cr** 0.10 max **Cu** 0.10-0.25 **Fe** 0.10-0.30 **Mg** 0.8-1.1 **Mn** 0.10 max **Si** 0.20-0.40 **Ti** 0.10 max **Zn** 0.10 max **Other** 0.05 max each, 0.15 max total | **AA** 6015 |
| **A96016** | Wrought Aluminum Alloy, Heat Treatable | **Al** rem **Cr** 0.10 max **Cu** 0.20 max **Fe** 0.50 max **Mg** 0.25-0.6 **Mn** 0.20 max **Si** 1.0-1.5 **Ti** 0.15 max **Zn** 0.20 max **Other** 0.05 max each, 0.15 max total | **AA** 6016; 6016A |
| * **A96017** | Wrought Aluminum Alloy, Heat Treatable | **Al** rem **Cr** 0.10 max **Cu** 0.05-0.20 **Fe** 0.15-0.30 **Mg** 0.45-0.6 **Mn** 0.10 max **Si** 0.55-0.7 **Ti** 0.05 max **Zn** 0.05 max **Other** each 0.05 max, total 0.15 max | **AA** 6017 |
| **A96018** | Wrought Aluminum Alloy, Heat Treatable | **Al** rem **Bi** 0.40-0.7 **Cr** 0.10 max **Cu** 0.15-0.40 **Fe** 0.7 max **Mg** 0.6-1.2 **Mn** 0.30-0.8 **Pb** 0.40-1.2 **Si** 0.50-1.2 **Ti** 0.20 max **Zn** 0.30 max **Other** each 0.05 max, total 0.15 max | **AA** 6018 |
| **A96019** | Wrought Aluminum Alloy, Non-Heat Treatable | **Al** rem **Cr** 0.05-0.35 **Cu** 0.20-0.6 **Fe** 0.50 max **Mg** 0.8-1.2 **Mn** 0.10 max **Si** 0.6-1.0 **Ti** 0.15 max **Zn** 0.40-1.0 **Other** each 0.05 max, total 0.15 max. | **AA** 6019 |
| **A96020** | Wrought Aluminum Alloy, Heat Treatable | **Al** rem **Cr** 0.15 max **Cu** 0.30-0.9 **Fe** 0.50 max **Mg** 0.6-1.2 **Mn** 0.35 max **Pb** 0.05 max **Si** 0.40-0.9 **Sn** 0.9-1.5 **Ti** 0.15 max **Other** each 0.05 max, total 0.15 max | **AA** 6020 |
| **A96021** | Wrought Aluminum Alloy, Heat Treatable | **Al** rem **Cr** 0.25 max **Cu** 0.10 max **Fe** 0.40 max **Mg** 0.8-1.5 **Mn** 0.40-1.0 **Si** 0.6-1.5 **Sn** 0.6-1.5 **Ti** 0.10 max **Zn** 0.20 max **Other** each 0.05 max, total 0.15 | **AA** 6021 |
| **A96022** | Wrought Aluminum Alloy, Heat Treatable | **Al** rem **Cr** 0.10 max **Cu** 0.01-0.11 **Fe** 0.05-0.20 **Mg** 0.45-0.7 **Mn** 0.02-0.10 **Si** 0.8-1.5 **Ti** 0.15 max **Zn** 0.25 max **Other** each 0.05 max, total 0.15 max | **AA** 6022 |
| * **A96030** | Wrought Aluminum Alloy, Heat Treatable | **Al** rem **Cr** 0.04-0.35 **Cu** 0.15-0.40 **Fe** 0.7 max **In** 0.05-0.50 **Mg** 0.8-1.2 **Mn** 0.15 max **Si** 0.40-0.8 **Sn** 0.05-0.50 **Ti** 0.15 max **Zn** 0.25 max **Other** each 0.05 max, total 0.15 max | **AA** X6030 |
| **A96053** | Wrought Aluminum Alloy, Heat Treatable | **Al** rem **Cr** 0.15-0.35 **Cu** 0.10 max **Fe** 0.35 max **Mg** 1.1-1.4 **Si** 45.0-65.0 of Mg **Zn** 0.10 max **Other** each 0.05 max, total 0.15 max | **AA** 6053 **ASTM** B316 (6053) **FED** QQ-A-430 **SAE** J454 (6053) |
| **A96056** | Wrought Aluminum Alloy, Heat Treatable | **Al** rem **Cr** 0.25 max **Cu** 0.50-1.1 **Fe** 0.50 max **Mg** 0.6-1.2 **Mn** 0.40-1.0 **Si** 0.7-1.3 **Zn** 0.10-0.7 **Other** each 0.05 max, total 0.15 max, Ti+Zr 0.20 max | **AA** 6056 |
| **A96060** | Wrought Aluminum Alloy, Heat Treatable | **Al** rem **Cr** 0.05 max **Cu** 0.10 max **Fe** 0.10-0.30 **Mg** 0.35-0.6 **Mn** 0.10 max **Si** 0.30-0.6 **Ti** 0.10 max **Zn** 0.15 max **Other** each 0.05 max, total 0.15 max | **AA** 6060 **ASTM** B221 (6060) |

*Boxed entries are no longer active and are retained for reference purposes only.

The chemical compositions listed are for identification purposes and should not be used in lieu of the cross referenced specifications.

| UNIFIED NUMBER | DESCRIPTION | CHEMICAL COMPOSITION | CROSS REFERENCE SPECIFICATIONS |
|---|---|---|---|
| **A96061** | Wrought Aluminum Alloy, Heat Treatable | **Al** rem **Cr** 0.04-0.35 **Cu** 0.15-0.40 **Fe** 0.7 max **Mg** 0.8-1.2 **Mn** 0.15 max **Si** 0.40-0.8 **Ti** 0.15 max **Zn** 0.25 max **Other** each 0.05 max, total 0.15 max | **AA** 6061;6061A **AMS** 4009; 4025; 4026; 4027; 4079; 4080; 4081; 4082; 4083; 4113; 4115; 4116; 4117; 4127; 4128; 4146; 4150; 4160; 4161; 4172; 4173; 4248; 4312; AMS-A-22771; AMS-QQ-A-200/8; AMS-QQ-A-200/16; AMS-QQ-A-225/8; AMS-QQ-A-250/11; AMS-T-7081 **ASTM** B209 (6061); B210 (6061); B211 (6061); B221 (6061); B234 (6061); B241 (6061); B247 (6061); B308 (6061); B313 (6061); B316 (6061); B345 (6061); B361 (6061); B404 (6061); B429 (6061); B483 (6061); B547 (6061); B548; B632 (6061); F467; F468; F901 (6061) **ASME** SB-209; SB-210; SB-211; SB-221; SB-234; SB-241; SB-247; SB-308 **FED** QQ-A-430; WW-T-700/6 **MIL SPEC** MIL-P-25995 **SAE** J454 (6061) |
| **A96063** | Wrought Aluminum Alloy, Heat Treatable | **Al** rem **Cr** 0.10 max **Cu** 0.10 max **Fe** 0.35 max **Mg** 0.45-0.9 **Mn** 0.10 max **Si** 0.20-0.6 **Ti** 0.10 max **Zn** 0.10 max **Other** each 0.05 max, total 0.15 max | **AA** 6063; 6063A **AMS** 4156; AMS-QQ-A-200/9 **ASME** SB-221; SB-241 **ASTM** B210 (6063); B221 (6063); B241 (6063); B345 (6063); B361 (6063); B429 (6063); B483 (6063); B491 (6063) **MIL SPEC** MIL-P-25995 **SAE** J454 (6063) |
| **A96066** | Wrought Aluminum Alloy, Heat Treatable | **Al** rem **Cr** 0.40 max **Cu** 0.7-1.2 **Mg** 0.8-1.4 **Mn** 0.6-1.1 **Si** 0.9-1.8 **Ti** 0.20 max **Zn** 0.25 max **Other** each 0.05 max, total 0.15 max | **AA** 6066 **AMS** AMS-QQ-A-200/10 **ASTM** B221 (6066); B241/B241M (6066); B247 (6066) |
| **A96069** | Wrought Aluminum Alloy, Heat Treatable | **Al** rem **Cr** 0.05-0.30 **Cu** 0.55-1.0 **Fe** 0.40 max **Mg** 1.2-1.6 **Mn** 0.05 max **Si** 0.6-1.2 **Sr** 0.05 max **Ti** 0.10 max **V** 0.10-0.30 **Zn** 0.05 max **Other** each 0.05 max, total 0.15 max | **AA** 6069 |
| **A96070** | Wrought Aluminum Alloy, Heat Treatable | **Al** rem **Cr** 0.10 max **Cu** 0.15-0.40 **Fe** 0.50 max **Mg** 0.50-1.2 **Mn** 0.40-1.0 **Si** 1.0-1.7 **Ti** 0.15 max **Zn** 0.25 max **Other** each 0.05 max, total 0.15 max | **AA** 6070 **ASTM** B221 (6070); B345 (6070) |
| **A96081** | Wrought Aluminum Alloy, Heat Treatable | **Al** rem **Cr** 0.10 max **Cu** 0.10 max **Fe** 0.50 max **Mg** 0.6-1.0 **Mn** 0.10-0.45 **Si** 0.7-1.1 **Ti** 0.15 max **Zn** 0.20 max **Other** 0.05 max each, 0.15 max total | **AA** 6081 |
| **A96082** | Wrought Aluminum Alloy, Heat Treatable | **Al** rem **Cr** 0.25 max **Cu** 0.10 max **Fe** 0.50 max **Mg** 0.6-1.2 **Mn** 0.40-1.0 **Si** 0.7-1.3 **Ti** 0.10 max **Zn** 0.20 max **Other** 0.05 max each, 0.15 max total | **AA** 6082; 6082A |
| * **A96090** | Wrought Aluminum Alloy, Heat Treatable | **Al** rem **Cr** 0.15 max **Cu** 0.30-0.9 **Fe** 0.7 max **Mg** 0.8-1.2 **Mn** 0.15 max **O** 0.05-0.7 **Si** 0.40-0.8 **Ti** 0.15 max **Zn** 0.25 max **Other** 0.05 max each, 0.15 max total | **AA** 6090 |
| **A96091** | Wrought Aluminum Alloy, Heat Treatable | **Al** rem **Cr** 0.15 max **Cu** 0.15-0.40 **Fe** 0.7 max **Mg** 0.8-1.2 **Mn** 0.15 max **O** 0.05-0.50 **Si** 0.40-0.8 **Ti** 0.15 max **Zn** 0.25 max **Other** 0.05 max each, 0.15 max total | **AA** 6091 |
| **A96092** | Wrought Aluminum Alloy, Heat Treatable | **Al** rem **Cr** 0.15 max **Cu** 0.7-1.0 **Fe** 0.30 max **Mg** 0.8-1.2 **Mn** 0.15 max **O** 0.05-0.50 **Si** 0.40-0.8 **Ti** 0.15 max **Zn** 0.25 max **Other** 0.05 max each, 0.15 max total | **AA** 6092 |
| **A96101** | Wrought Aluminum Alloy, Heat Treatable | **Al** rem **B** 0.06 max **Cr** 0.03 max **Cu** 0.10 max **Fe** 0.50 max **Mg** 0.35-0.8 **Mn** 0.03 max **Si** 0.30-0.7 **Zn** 0.10 max **Other** each 0.03 max, total 0.10 max | **AA** 6101; 6101A; 6101B **ASTM** B317 (6101) |
| **A96103** | Wrought Aluminum Alloy, Heat Treatable | **Al** rem **Cr** 0.35 max **Cu** 0.20-0.30 **Fe** 0.6 max **Mg** 0.8-1.5 **Mn** 0.8 max **Si** 0.35-1.0 **Ti** 0.10 max **Zn** 0.20 max **Other** 0.05 max each, 0.15 max total | **AA** 6103 |

*Boxed entries are no longer active and are retained for reference purposes only.

The chemical compositions listed are for identification purposes and should not be used in lieu of the cross referenced specifications.

| UNIFIED NUMBER | DESCRIPTION | CHEMICAL COMPOSITION | CROSS REFERENCE SPECIFICATIONS |
|---|---|---|---|
| **A96105** | Wrought Aluminum Alloy, Heat Treatable | **Al** rem **Cr** 0.10 max **Cu** 0.10 max **Fe** 0.35 max **Mg** 0.45-0.8 **Mn** 0.15 max **Si** 0.6-1.0 **Ti** 0.10 max **Zn** 0.10 max **Other** each 0.05 max, total 0.15 max | **AA** 6105 **ASTM** B221 (6105) |
| **A96106** | Wrought Aluminum Alloy, Heat Treatable | **Al** rem **Cr** 0.20 max **Cu** 0.25 max **Fe** 0.35 max **Mg** 0.40-0.8 **Mn** 0.05-0.20 **Si** 0.30-0.6 **Zn** 0.10 max **Other** each 0.05 max, total 0.10 max | **AA** 6106 |
| **A96110** | Wrought Aluminum Alloy, Heat Treatable | **Al** rem **Cr** 0.04-0.25 **Cu** 0.20-0.7 **Fe** 0.8 max **Mg** 0.50-1.1 **Mn** 0.20-0.7 **Si** 0.7-1.5 **Ti** 0.15 max **Zn** 0.30 max **Other** each 0.05 max, total 0.15 max | **AA** 6110; 6110A **ASTM** B211 (6110) |
| **A96111** | Wrought Aluminum Alloy, Heat Treatable | **Al** rem **Cr** 0.10 **Cu** 0.50-0.9 **Fe** 0.40 **Mg** 0.50-1.0 **Mn** 0.10-0.45 **Si** 0.6-1.1 **Ti** 0.10 **Zn** 0.15 **Other** each 0.05, total 0.15 | **AA** 6111 **SAE** J454 (6111) |
| **A96113** | Wrought Aluminum Alloy, Heat Treatable | **Al** rem **Cr** 0.10 max **Cu** 0.6-1.1 **Fe** 0.30 max **Mg** 0.8-1.2 **Mn** 0.10-0.6 **O** 0.05-0.50 **Si** 0.6-1.0 **Ti** 0.10 max **Zn** 0.25 max **Other** 0.05 max each, 0.15 max total | **AA** 6113 |
| **A96116** | Wrought Aluminum Alloy, Non-Heat Treatable | **Al** rem **Cr** 0.15 max **Cu** 0.20 max **Fe** 0.25 max **Mg** 0.25-0.6 **Mn** 0.15 max **Si** 0.9-1.3 **Ti** 0.15 max **Zn** 0.20 max **Other** each 0.05 max, total 0.15 max. | **AA** 6116 |
| **A96151** | Wrought Aluminum Alloy, Heat Treatable | **Al** rem **Cr** 0.15-0.35 **Cu** 0.35 max **Fe** 1.0 max **Mg** 0.45-0.8 **Mn** 0.20 max **Si** 0.6-1.2 **Ti** 0.15 max **Zn** 0.25 max **Other** each 0.05 max, total 0.15 max | **AA** 6151 **AMS** 4125; AMS-A-22771 (6151) **ASTM** B247 (6151) |
| **A96160** | Wrought Aluminum Alloy, Heat Treatable | **Al** rem **Cr** 0.05 max **Cu** 0.20 max **Fe** 0.15 max **Mg** 0.35-0.6 **Mn** 0.05 max **Si** 0.30-0.6 **Zn** 0.05 max **Other** each 0.05 max, total 0.15 max | **AA** 6160 |
| **A96162** | Wrought Aluminum Alloy, Heat Treatable | **Al** rem **Cr** 0.10 max **Cu** 0.20 max **Fe** 0.50 max **Mg** 0.7-1.1 **Mn** 0.10 max **Si** 0.40-0.8 **Ti** 0.10 max **Zn** 0.25 max **Other** each 0.05 max, total 0.15 max | **AA** 6162 **AMS** AMS-QQ-A-200/17 **ASME** SB-241 **ASTM** B221 (6166); B241 (6162) |
| **A96181** | Wrought Aluminum Alloy, Heat Treatable | **Al** rem **Cr** 0.10 max **Cu** 0.10 max **Fe** 0.45 max **Mg** 0.6-1.0 **Mn** 0.15 max **Si** 0.8-1.2 **Ti** 0.10 max **Zn** 0.20 max **Other** 0.05 max each, 0.15 max total | **AA** 6181; 6181A |
| **A96201** | Wrought Aluminum Alloy, Heat Treatable | **Al** rem **B** 0.06 max **Cr** 0.03 max **Cu** 0.10 max **Fe** 0.50 max **Mg** 0.6-0.9 **Mn** 0.03 max **Si** 0.50-0.9 **Zn** 0.10 max **Other** each 0.03 max, total 0.10 max | **AA** 6201; 6201A **ASTM** B398 (6201); B399 (6201); B524 (6201); B711 (6201) |
| **A96205** | Wrought Aluminum Alloy, Heat Treatable | **Al** rem **Cr** 0.05-0.15 **Cu** 0.20 max **Fe** 0.7 max **Mg** 0.40-0.6 **Mn** 0.05-0.15 **Si** 0.6-0.9 **Ti** 0.15 max **Zn** 0.25 max **Zr** 0.05-0.15 **Other** each 0.05 max, total 0.15 max | **AA** 6205 |
| **A96206** | Wrought Aluminum Alloy, Heat Treatable | **Al** rem **Cr** 0.10 **Cu** 0.20-0.50 **Fe** 0.35 max **Mg** 0.45-0.8 **Mn** 0.13-0.30 **Si** 0.35-0.7 **Ti** 0.10 max **Zn** 0.20 max **Other** 0.05 max each, 0.15 max total | **AA** 6206 |
| **A96253** | Wrought Aluminum Alloy, Heat Treatable (Cladding Alloy) | **Al** rem **Cr** 0.04-0.35 **Cu** 0.10 max **Fe** 0.50 max **Mg** 1.0-1.5 **Si** 45.0-65.0 of Mg **Zn** 1.6-2.4 **Other** each 0.05 max, total 0.15 max | **AA** 6253 **ASTM** B211 (6253) |
| **A96260** | Wrought Aluminum Alloy, Non-Heat Treatable | **Al** rem **Cr** 0.03 max **Cu** 0.10 max **Fe** 0.15-0.40 **Mg** 0.45-0.7 **Mn** 0.03 max **Si** 0.40-0.6 **Ti** 0.10 max **V** 0.10-0.25 **Zn** 0.05 max **Zr** 0.05 max **Other** each 0.05 max, total 0.15 max. | **AA** 6260 |

The chemical compositions listed are for identification purposes and should not be used in lieu of the cross referenced specifications.

| UNIFIED NUMBER | DESCRIPTION | CHEMICAL COMPOSITION | CROSS REFERENCE SPECIFICATIONS |
|---|---|---|---|
| **A96261** | Wrought Aluminum Alloy, Heat Treatable | **Al** rem **Cr** 0.10 max **Cu** 0.15-0.40 **Fe** 0.40 max **Mg** 0.7-1.0 **Mn** 0.20-0.35 **Si** 0.40-0.7 **Ti** 0.10 max **Zn** 0.20 max **Other** each 0.05 max, total 0.15 max | **AA** 6261 |
| **A96262** | Wrought Aluminum Alloy, Heat Treatable, Free Machining | **Al** rem **Bi** 0.40-0.7 **Cr** 0.04-0.14 **Cu** 0.15-0.40 **Fe** 0.7 max **Mg** 0.8-1.2 **Mn** 0.15 max **Pb** 0.40-0.7 **Si** 0.40-0.8 **Ti** 0.15 max **Zn** 0.25 max **Other** each 0.05 max, total 0.15 max | **AA** 6262 **AMS** AMS-QQ-A-225/10 **ASTM** B210 (6262); B211 (6262); B221 (6262); B483 (6262); F467; F901 (6262) |
| **A96301** | Wrought Aluminum Alloy, Heat Treatable | **Al** rem **Cr** 0.10 max **Cu** 0.10 max **Fe** 0.7 max **Mg** 0.6-0.9 **Mn** 0.15 max **Si** 0.50-0.9 **Ti** 0.15 max **Zn** 0.25 max **Other** each 0.05 max, total 0.15 max | **AA** 6301 |
| **A96306** | Wrought Aluminum Alloy, Heat Treatable | **Al** rem **Cu** 0.05-0.16 **Fe** 0.10 max **Mg** 0.45-0.9 **Mn** 0.10-0.40 **Si** 0.20-0.6 **Ti** 0.05 max **Zn** 0.05 max **Other** 0.05 max each, 0.15 max total | **AA** 6306 |
| **A96351** | Wrought Aluminum Alloy, Heat Treatable | **Al** rem **Cu** 0.10 max **Fe** 0.50 max **Mg** 0.40-0.8 **Mn** 0.40-0.8 **Si** 0.7-1.3 **Ti** 0.20 max **Zn** 0.20 max **Other** each 0.05 max, total 0.15 max | **AA** 6351; 6351A **ASTM** B221 (6351); B241 (6351); B345 (6351) |
| **A96401** | Wrought Aluminum Alloy, Heat Treatable | **Al** rem **Cu** 0.05-0.20 **Fe** 0.04 max **Mg** 0.35-0.7 **Mn** 0.03 max **Si** 0.35-0.7 **Ti** 0.01 max **Zn** 0.04 max **Other** 0.01 max each | **AA** 6401 |
| **A96463** | Wrought Aluminum Alloy, Heat Treatable | **Al** rem **Cu** 0.20 max **Fe** 0.15 max **Mg** 0.45-0.9 **Mn** 0.05 max **Si** 0.20-0.6 **Zn** 0.05 max **Other** each 0.05 max, total 0.15 max | **AA** 6463; 6463A **ASTM** B221 (6463) **SAE** J454 (6463) |
| **A96763** | Wrought Aluminum Alloy, Heat Treatable | **Al** rem **Cu** 0.04-0.16 **Fe** 0.08 max **Mg** 0.45-0.9 **Mn** 0.03 max **Si** 0.20-0.6 **V** 0.05 max **Zn** 0.03 max **Other** each 0.03 max, total 0.10 max | **AA** 6763 |
| * **A96863** | Wrought Aluminum Alloy, Heat Treatable | **Al** rem **Cr** 0.05 max **Cu** 0.05-0.20 **Fe** 0.15 max **Mg** 0.50-0.8 **Mn** 0.05 max **Si** 0.40-0.6 **Ti** 0.10 max **Zn** 0.10 max **Other** 0.05 max each, 0.15 max total | **AA** 6863 |
| **A96951** | Wrought Aluminum Alloy, Heat Treatable | **Al** rem **Cu** 0.15-0.40 **Fe** 0.8 max **Mg** 0.40-0.8 **Mn** 0.10 max **Si** 0.20-0.50 **Zn** 0.20 max **Other** each 0.05 max, total 0.15 max | **AA** 6951 |
| **A96963** | Wrought Aluminum Alloy, Heat Treatable | **Al** rem **Cr** 0.10 max **Cu** 0.15-0.25 **Fe** 0.25 max **Mg** 0.35-0.7 **Mn** 0.05 max **Si** 0.40-0.6 **Ti** 0.10 max **Zn** 0.10 max **Other** each 0.05 max, total 0.15 max | **AA** 6963 |
| * **A97001** | Wrought Aluminum Alloy, Heat Treatable | **Al** rem **Cr** 0.18-0.35 **Cu** 1.6-2.6 **Fe** 0.40 max **Mg** 2.6-3.4 **Mn** 0.20 max **Si** 0.35 max **Ti** 0.20 max **Zn** 6.8-8.0 **Other** each 0.05 max, total 0.15 max | **AA** 7001 **ASTM** B209 (7011) |
| **A97003** | Wrought Aluminum Alloy, Heat Treatable | **Al** rem **Cr** 0.20 max **Cu** 0.20 max **Fe** 0.35 max **Mg** 0.50-1.0 **Mn** 0.30 max **Si** 0.30 max **Ti** 0.20 max **Zn** 5.0-6.5 **Zr** 0.05-0.25 **Other** 0.05 max each, 0.15 max total | **AA** 7003 |
| **A97004** | Wrought Aluminum Alloy, Heat Treatable | **Al** rem **Cr** 0.05 max **Cu** 0.05 max **Fe** 0.35 max **Mg** 1.0-2.0 **Mn** 0.20-0.7 **Si** 0.25 max **Ti** 0.05 max **Zn** 3.8-4.6 **Zr** 0.10-0.20 **Other** each 0.05 max, total 0.15 max | **AA** 7004 |

*Boxed entries are no longer active and are retained for reference purposes only.

The chemical compositions listed are for identification purposes and should not be used in lieu of the cross referenced specifications.

| UNIFIED NUMBER | DESCRIPTION | CHEMICAL COMPOSITION | CROSS REFERENCE SPECIFICATIONS |
|---|---|---|---|
| **A97005** | Wrought Aluminum Alloy, Heat Treatable | **Al** rem **Cr** 0.06-0.20 **Cu** 0.10 max **Fe** 0.40 max **Mg** 1.0-1.8 **Mn** 0.20-0.7 **Si** 0.35 max **Ti** 0.01-0.06 **Zn** 4.0-5.0 **Zr** 0.08-0.20 **Other** each 0.05 max, total 0.15 max | **AA** 7005 **ASTM** B221 (7005) |
| **A97008** | Wrought Aluminum Alloy, Heat Treatable (Cladding Alloy) | **Al** rem **Cr** 0.12-0.25 **Cu** 0.05 max **Fe** 0.10 max **Mg** 0.7-1.4 **Mn** 0.05 max **Si** 0.10 max **Ti** 0.05 max **Zn** 4.5-5.5 **Other** each 0.05 max, total 0.10 max | **AA** 7008 **ASTM** B209 (7008) |
| **A97009** | Wrought Aluminum Alloy, Heat Treatable | **Ag** 0.25-0.40 **Al** rem **Cr** 0.10-0.25 **Cu** 0.6-1.3 **Fe** 0.20 max **Mg** 2.1-2.9 **Mn** 0.10 max **Si** 0.20 max **Ti** 0.20 max **Zn** 5.5-6.5 **Other** 0.05 max each, 0.15 max total | **AA** 7009 |
| **A97010** | Wrought Aluminum Alloy, Heat Treatable | **Al** rem **Cr** 0.05 max **Cu** 1.5-2.0 **Fe** 0.15 max **Mg** 2.1-2.6 **Mn** 0.10 max **Ni** 0.05 max **Si** 0.12 max **Ti** 0.06 max **Zn** 5.7-6.7 **Zr** 0.10-0.16 **Other** each 0.05 max, total 0.15 max | **AA** 7010 **AMS** 4203; 4204; 4205 |
| * **A97011** | Wrought Aluminum Alloy, Heat Treatable (Cladding Alloy) | **Al** rem **Cr** 0.05-0.20 **Cu** 0.05 max **Fe** 0.20 max **Mg** 1.0-1.6 **Mn** 0.10-0.30 **Si** 0.15 max **Ti** 0.05 max **Zn** 4.0-5.5 **Other** each 0.05 max, total 0.15 max | **AA** 7011 **ASTM** B209 (7011) |
| **A97012** | Wrought Aluminum Alloy, Heat Treatable | **Al** rem **Cr** 0.04 max **Cu** 0.8-1.2 **Fe** 0.25 max **Mg** 1.8-2.2 **Mn** 0.08-0.15 **Si** 0.15 max **Ti** 0.02-0.08 **Zn** 5.8-6.5 **Zr** 0.10-0.18 **Other** 0.05 max each, 0.15 max total | **AA** 7012 |
| * **A97013** | Wrought Aluminum Alloy, Heat Treatable (Cladding Alloy) | **Al** rem **Cu** 0.10 max **Fe** 0.7 max **Mn** 1.0-1.5 **Si** 0.6 max **Zn** 1.5-2.0 **Other** each 0.05 max, total 0.15 max | **AA** 7013 |
| **A97014** | Wrought Aluminum Alloy, Heat Treatable | **Al** rem **Cu** 0.30-0.7 **Fe** 0.50 max **Mg** 2.2-3.2 **Mn** 0.30-0.7 **Ni** 0.10 max **Si** 0.50 max **Zn** 5.2-6.2 **Other** 0.05 max each, 0.15 max total, Ti+Zr 0.20 max | **AA** 7014 |
| **A97015** | Wrought Aluminum Alloy, Heat Treatable | **Al** rem **Cr** 0.15 max **Cu** 0.06-0.15 **Fe** 0.30 max **Mg** 1.3-2.1 **Mn** 0.10 max **Si** 0.20 max **Ti** 0.10 max **Zn** 4.6-5.2 **Zr** 0.10-0.20 **Other** 0.05 max each, 0.15 max total | **AA** 7015 |
| **A97016** | Wrought Aluminum Alloy, Heat Treatable | **Al** rem **Cu** 0.45-1.0 **Fe** 0.12 max **Mg** 0.8-1.4 **Mn** 0.03 max **Si** 0.10 max **Ti** 0.03 **V** 0.05 max **Zn** 4.0-5.0 **Other** each 0.03 max, total 0.10 max | **AA** 7016 |
| **A97017** | Wrought Aluminum Alloy, Heat Treatable | **Al** rem **Cr** 0.35 max **Cu** 0.20 max **Fe** 0.45 max **Mg** 2.0-3.0 **Mn** 0.05-0.50 **Ni** 0.10 max **Si** 0.35 max **Ti** 0.15 max **Zn** 4.0-5.2 **Zr** 0.10-0.25 **Other** each 0.05 max, total 0.15 max, Cr+Mn 0.15 min | **AA** 7017 |
| **A97018** | Wrought Aluminum Alloy, Heat Treatable | **Al** rem **Cr** 0.20 max **Cu** 0.20 max **Fe** 0.45 max **Mg** 0.7-1.5 **Mn** 0.15-0.50 **Ni** 0.10 max **Si** 0.35 max **Ti** 0.15 max **Zn** 4.5-5.5 **Zr** 0.10-0.25 **Other** 0.05 max each, 0.15 max total | **AA** 7018 |
| **A97019** | Wrought Aluminum Alloy, Heat Treatable | **Al** rem **Cr** 0.20 max **Cu** 0.20 max **Fe** 0.45 max **Mg** 1.5-2.5 **Mn** 0.15-0.50 **Ni** 0.10 max **Si** 0.35 max **Ti** 0.15 max **Zn** 3.5-4.5 **Zr** 0.10-0.25 **Other** 0.05 max each, 0.15 max total | **AA** 7019; 7019A |

*Boxed entries are no longer active and are retained for reference purposes only.

The chemical compositions listed are for identification purposes and should not be used in lieu of the cross referenced specifications.

| UNIFIED NUMBER | DESCRIPTION | CHEMICAL COMPOSITION | CROSS REFERENCE SPECIFICATIONS |
|---|---|---|---|
| **A97020** | Wrought Aluminum Alloy, Heat Treatable | **Al** rem **Cr** 0.10-0.35 **Cu** 0.20 max **Fe** 0.40 max **Mg** 1.0-1.4 **Mn** 0.05-0.50 **Si** 0.35 max **Zn** 4.0-5.0 **Zr** 0.08-0.20 **Other** each 0.05 max, total 0.15, Ti+Zr 0.08-0.25 | **AA** 7020 |
| **A97021** | Wrought Aluminum Alloy, Heat Treatable | **Al** rem **Cr** 0.05 max **Cu** 0.25 max **Fe** 0.40 max **Mg** 1.2-1.8 **Mn** 0.10 max **Si** 0.25 max **Ti** 0.10 max **Zn** 5.0-6.0 **Zr** 0.08-0.18 **Other** each 0.05 max, total 0.15 max | **AA** 7021 **SAE** J454 (7021) |
| **A97022** | Wrought Aluminum Alloy, Heat Treatable | **Al** rem **Cr** 0.10-0.30 **Cu** 0.50-1.0 **Fe** 0.50 max **Mg** 2.6-3.7 **Mn** 0.10-0.40 **Si** 0.50 max **Zn** 4.3-5.3 **Other** 0.05 max each, 0.15 max total, Ti+Zr 0.20 max | **AA** 7022 |
| **A97023** | Wrought Aluminum Alloy, Heat Treatable | **Al** rem **Cr** 0.05-0.35 **Cu** 0.50-1.0 **Fe** 0.50 max **Mg** 2.0-3.0 **Mn** 0.10-0.6 **Si** 0.50 max **Ti** 0.10 max **Zn** 4.0-6.0 **Other** 0.05 max each, 0.15 max total | **AA** 7023 |
| **A97024** | Wrought Aluminum Alloy, Heat Treatable | **Al** rem **Cr** 0.05-0.35 **Cu** 0.10 max **Fe** 0.40 max **Mg** 0.50-1.0 **Mn** 0.10-0.6 **Si** 0.30 max **Ti** 0.10 max **Zn** 3.0-5.0 **Other** 0.05 max each, 0.15 max total | **AA** 7024 |
| **A97025** | Wrought Aluminum Alloy, Heat Treatable | **Al** rem **Cr** 0.05-0.35 **Cu** 0.10 max **Fe** 0.40 max **Mg** 0.8-1.5 **Mn** 0.10-0.6 **Si** 0.30 max **Ti** 0.10 max **Zn** 3.0-5.0 **Other** 0.05 max each, 0.15 max total | **AA** 7025 |
| **A97026** | Wrought Aluminum Alloy, Heat Treatable | **Al** rem **Cu** 0.6-0.9 **Fe** 0.12 max **Mg** 1.5-1.9 **Mn** 0.05-0.20 **Si** 0.08 max **Ti** 0.05 max **Zn** 4.6-5.2 **Zr** 0.09-0.14 **Other** 0.03 max each, 0.10 max total | **AA** 7026 |
| * **A97027** | Wrought Aluminum Alloy, Heat Treatable | **Al** rem **Cu** 0.10-0.30 **Fe** 0.40 max **Mg** 0.7-1.1 **Mn** 0.10-0.40 **Si** 0.25 max **Ti** 0.10 max **Zn** 3.5-4.5 **Zr** 0.05-0.30 **Other** 0.05 max each, 0.15 max total | **AA** 7027 |
| **A97028** | Wrought Aluminum Alloy, Heat Treatable | **Al** rem **Cr** 0.20 max **Cu** 0.10-0.30 **Fe** 0.50 max **Mg** 1.5-2.3 **Mn** 0.15-0.6 **Si** 0.35 max **Ti** 0.05 max **Zn** 4.5-5.2 **Other** each 0.05 max, total 0.15 max, Ti+Zr 0.08-0.25 | **AA** 7028 |
| **A97029** | Wrought Aluminum Alloy, Heat Treatable | **Al** rem **Cu** 0.50-0.9 **Fe** 0.12 max **Mg** 1.3-2.0 **Mn** 0.03 max **Si** 0.10 max **Ti** 0.05 max **V** 0.05 max **Zn** 4.2-5.2 **Other** each 0.03 max, total 0.10 max | **AA** 7029 **SAE** J454 (7029) |
| **A97030** | Wrought Aluminum Alloy, Heat Treatable | **Al** rem **Cr** 0.04 max **Cu** 0.20-0.40 **Fe** 0.30 max **Ga** 0.03 max **Mg** 1.0-1.5 **Mn** 0.05 max **Si** 0.20 max **Ti** 0.03 max **Zn** 4.8-5.9 **Zr** 0.03 max **Other** 0.05 max each, 0.15 max total | **AA** 7030 |
| **A97031** | Wrought Aluminum Alloy, Heat Treatable | **Al** rem **Cu** 0.10 max **Fe** 0.8-1.4 **Mg** 0.10 max **Mn** 0.10-0.40 **Si** 0.30 max **Zn** 0.8-1.8 **Other** 0.05 max each, 0.15 max total | **AA** 7031 |
| **A97032** | Wrought Aluminum Alloy, Heat Treatable | **Al** rem **Bi** 0.01 max **Cr** 0.15-0.25 **Cu** 1.7-2.3 **Fe** 0.12 max **Mg** 1.5-2.5 **Mn** 0.05 max **Pb** 0.01 max **Si** 0.10 max **Ti** 0.10 max **Zn** 5.6-6.5 **Other** 0.05 max each, 0.15 max total | **AA** 7032 |

*Boxed entries are no longer active and are retained for reference purposes only.

The chemical compositions listed are for identification purposes and should not be used in lieu of the cross referenced specifications.

| UNIFIED NUMBER | DESCRIPTION | CHEMICAL COMPOSITION | CROSS REFERENCE SPECIFICATIONS |
|---|---|---|---|
| **A97033** | Wrought Aluminum Alloy, Non-Heat Treatable | **Al** rem **Cr** 0.20 max **Cu** 0.7-1.30 **Fe** 0.30 max **Ga** 0.05 max **Mg** 1.3-2.2 **Mn** 0.10 max **Si** 0.15 max **Ti** 0.10 max **V** 0.05 max **Zn** 4.6-5.6 **Zr** 0.08-0.15 **Other** each 0.05 max, total 0.15 max | **AA** 7033 |
| **A97034** | Wrought Aluminum Alloy, Heat Treatable | **Al** rem **Cr** 0.20 max **Cu** 0.8-1.2 **Fe** 0.12 max **Mg** 2.0-3.0 **Mn** 0.25 max **Si** 0.10 max **Zn** 11.0-12.0 **Zr** 0.08-0.30 **Other** each 0.05 max, total 0.15 max | **AA** 7034 |
| **A97039** | Wrought Aluminum Alloy, Heat Treatable | **Al** rem **Cr** 0.15-0.25 **Cu** 0.10 max **Fe** 0.40 max **Mg** 2.3-3.3 **Mn** 0.10-0.40 **Si** 0.30 max **Ti** 0.10 max **Zn** 3.5-4.5 **Other** each 0.05 max, total 0.15 max | **AA** 7039 **AMS** AMS-A-22771 (7039) **MIL SPEC** MIL-DTL-45225; MIL-DTL-46063; MIL-DTL-46083 |
| **A97040** | Wrought Aluminum Alloy, Non-Heat Treatable | **Al** rem **Cr** 0.04 max **Cu** 1.5-2.3 **Fe** 0.13 max **Mg** 1.7-2.4 **Mn** 0.04 max **Si** 0.10 max **Ti** 0.06 max **Zn** 5.7-6.7 **Zr** 0.05-0.12 **Other** each 0.05 max, total 0.15 max. | **AA** 7040 |
| **A97046** | Wrought Aluminum Alloy, Heat Treatable | **Al** rem **Cr** 0.20 max **Cu** 0.25 max **Fe** 0.40 max **Mg** 1.0-1.6 **Si** 0.20 max **Ti** 0.06 max **Zn** 6.6-7.6 **Zr** 0.10-0.18 **Other** each 0.05 max, total 0.15 max | **AA** 7046; 7046A |
| **A97049** | Wrought Aluminum Alloy, Heat Treatable | **Al** rem **Cr** 0.10-0.22 **Cu** 1.2-1.9 **Fe** 0.35 max **Mg** 2.0-2.9 **Mn** 0.20 max **Si** 0.25 max **Ti** 0.10 max **Zn** 7.2-8.2 **Other** each 0.05 max, total 0.15 max | **AA** 7049; 7049A **AMS** 4111; 4157; 4159; 4200; 4247; 4321 **ASTM** B247 (7049) **MIL SPEC** MIL-A-22771 |
| **A97050** | Wrought Aluminum Alloy, Heat Treatable | **Al** rem **Cr** 0.04 max **Cu** 2.0-2.6 **Fe** 0.15 max **Mg** 1.9-2.6 **Mn** 0.10 max **Si** 0.12 max **Ti** 0.06 max **Zn** 5.7-6.7 **Zr** 0.08-0.15 **Other** each 0.05 max, total 0.15 max | **AA** 7050; 7050A **AMS** 4050; 4107; 4108; 4201; 4340; 4341; 4342; 4333 **ASTM** B247 (7050); B316 (7050) **FED** QQ-A-430 **MIL SPEC** MIL-A-22771 |
| * **A97051** | Wrought Aluminum Alloy, Heat Treatable | **Al** rem **Cr** 0.05-0.25 **Cu** 0.15 max **Fe** 0.45 max **Mg** 1.7-2.5 **Mn** 0.10-0.45 **Si** 0.35 max **Ti** 0.15 max **Zn** 3.0-4.0 **Other** 0.05 max each, 0.15 max total | **AA** 7051 |
| **A97055** | Wrought Aluminum Alloy, Heat Treatable | **Al** rem **Cr** 0.04 max **Cu** 2.0-2.6 **Fe** 0.15 max **Mg** 1.8-2.3 **Mn** 0.05 max **Si** 0.10 max **Ti** 0.06 max **Zn** 7.6-8.4 **Zr** 0.08-0.25 **Other** 0.05 max each, 0.15 max total | **AA** 7055 **AMS** 4206; 4337 |
| **A97060** | Wrought Aluminum Alloy, Heat Treatable | **Al** rem **Cr** 0.15-0.25 **Cu** 1.8-2.6 **Fe** 0.20 max **Mg** 1.3-2.1 **Mn** 0.20 max **Pb** 0.003 max **Si** 0.15 max **Ti** 0.05 max **Zn** 6.1-7.5 **Zr** 0.05 max **Other** 0.05 max each, 0.15 max total | **AA** 7060 |
| **A97064** | Wrought Aluminum Alloy, Heat Treatable | **Al** rem **Co** 0.10-0.40 **Cr** 0.06-0.25 **Cu** 1.8-2.4 **Fe** 0.15 max **Mg** 1.9-2.9 **O** 0.05-0.30 **Si** 0.12 max **Zn** 6.8-8.0 **Zr** 0.10-0.50 **Other** 0.05 max each, 0.15 max total | **AA** 7064 |
| **A97068** | Wrought Aluminum Alloy, Non-Heat Treatable | **Al** rem **Cr** 0.05max **Cu** 1.6-2.4 **Fe** 0.15 max **Mg** 2.2-3.0 **Mn** 0.10 max **Si** 0.12 max **Ti** 0.10 max **Zn** 7.3-8.3 **Zr** 0.05-0.15 **Other** each 0.05 max, total 0.15 max | **AA** 7068 |
| * **A97070** | Wrought Aluminum Alloy, Heat Treatable | **Al** rem **Cu** 0.05 max **Fe** 0.25 max **Si** 0.15 max **Zn** 1.3-1.8 **Other** each 0.05 max, total 0.15 max | **AA** 7070 |

*Boxed entries are no longer active and are retained for reference purposes only.

The chemical compositions listed are for identification purposes and should not be used in lieu of the cross referenced specifications.

| UNIFIED NUMBER | DESCRIPTION | CHEMICAL COMPOSITION | CROSS REFERENCE SPECIFICATIONS |
|---|---|---|---|
| **A97072** | Wrought Aluminum Alloy, Non-Heat Treatable | **Al** rem **Cu** 0.10 max **Mg** 0.10 max **Mn** 0.10 max **Zn** 0.8-1.3 **Other** each 0.05 max, total 0.15 max, Fe+Si 0.7 max | **AA** 7072 **ASTM** B209 (7072); B210 (7072); B221 (7072); B234 (7072); B241 (7072); B313 (7072); B345 (7072); B404 (7072); B547 (7072) |
| **A97075** | Wrought Aluminum Alloy, Heat Treatable | **Al** rem **Cr** 0.18-0.28 **Cu** 1.2-2.0 **Fe** 0.50 max **Mg** 2.1-2.9 **Mn** 0.30 max **Si** 0.40 max **Ti** 0.20 max **Zn** 5.1-6.1 **Other** each 0.05 max, total 0.15 max, Ti+Zr 0.20 max for extruded and forged products only, but only when producer and purchaser mutually agree. | **AA** 7075 **AMS** 4044; 4045; 4078; 4122; 4123; 4124; 4126; 4131; 4141; 4147; 4154; 4166; 4167; 4168; 4169; 4174; 4186; 4277; 4310; 4311; 4323; AMS-A-22771 (7075); AMS-QQ-A-200/11; AMS-QQ-A-200/15; AMS-QQ-A-225/9; AMS-QQ-A-250/12; AMS-QQ-A-250/24 **ASTM** B209 (7075); B210 (7075); B211 (7075); B221 (7075); B241 (7075); B247 (7075); B316 (7075) **FED** QQ-A-430; WW-T-700/7 |
| **A97076** | Wrought Aluminum Alloy, Heat Treatable | **Al** rem **Cu** 0.30-1.0 **Fe** 0.6 max **Mg** 1.2-2.0 **Mn** 0.30-0.8 **Si** 0.40 max **Ti** 0.20 max **Zn** 7.0-8.0 **Other** each 0.05 max, total 0.15 max | **AA** 7076 **ASTM** B247 (7076) |
| * **A97079** | Wrought Aluminum Alloy, Heat Treatable | **Al** rem **Cr** 0.10-0.25 **Cu** 0.40-0.8 **Fe** 0.40 max **Mg** 2.9-3.7 **Mn** 0.10-0.30 **Si** 0.30 max **Ti** 0.10 max **Zn** 3.8-4.8 **Other** each 0.05 max, total 0.15 max | **AA** 7079 **AMS** 4138 |
| **A97090** | Wrought Aluminum Alloy, Heat Treatable | **Al** rem **Co** 1.0-1.9 **Cu** 0.6-1.3 **Fe** 0.15 **Mg** 2.0-3.0 **O** 0.20-0.50 **Si** 0.12 **Zn** 7.3-8.7 **Other** each 0.05 | **AA** 7090 |
| * **A97091** | Wrought Aluminum Alloy, Heat Treatable | **Al** rem **Co** 0.20-0.6 **Cu** 1.1-1.8 **Fe** 0.15 **Mg** 2.0-3.0 **O** 0.20-0.50 **Si** 0.12 **Zn** 5.8-7.1 **Other** each 0.05 max, total 0.15 max | **AA** 7091 |
| **A97093** | Wrought Aluminum Alloy, Heat Treatable | **Al** rem **Cu** 1.1-1.9 **Fe** 0.15 max **Mg** 2.0-3.0 **Ni** 0.04-0.16 **O** 0.05-0.50 **Si** 0.12 max **Zn** 8.3-9.7 **Zr** 0.08-0.20 **Other** each 0.05 max, total 0.15 max | **AA** 7093 |
| * **A97104** | Wrought Aluminum Alloy, Heat Treatable | **Al** rem **Cu** 0.03 max **Fe** 0.40 max **Mg** 0.50-0.9 **Si** 0.25 max **Ti** 0.10 max **Zn** 3.6-4.4 **Other** each 0.05 max, total 0.15 max | **AA** 7104 |
| **A97108** | Wrought Aluminum Alloy, Heat Treatable (Cladding Alloy) | **Al** rem **Cu** 0.05 max **Fe** 0.10 max **Mg** 0.7-1.4 **Mn** 0.05 max **Si** 0.10 max **Ti** 0.05 max **Zn** 4.5-5.5 **Zr** 0.12-0.25 **Other** each 0.05 max, total 0.15 max | **AA** 7108; 7108A |
| * **A97109** | Wrought Aluminum Alloy, Heat Treatable | **Ag** 0.25-0.40 **Al** rem **Cr** 0.04-0.08 **Cu** 0.8-1.3 **Fe** 0.15 max **Mg** 2.2-2.7 **Mn** 0.10 max **Si** 0.10 max **Ti** 0.10 max **Zn** 5.8-6.5 **Zr** 0.10-0.20 **Other** 0.05 max each, 0.15 max total | **AA** 7109 |
| **A97116** | Wrought Aluminum Alloy, Heat Treatable | **Al** rem **Cu** 0.50-1.1 **Fe** 0.30 max **Ga** 0.03 max **Mg** 0.8-1.4 **Mn** 0.05 max **Si** 0.15 max **Ti** 0.05 max **V** 0.05 max **Zn** 4.2-5.2 **Other** each 0.05 max, total 0.15 max | **AA** 7116 **ASTM** B221 (7116) **SAE** J454 (7116) |
| **A97122** | Wrought Aluminum Alloy, Heat Treatable | **Al** rem **Cr** 0.10 max **Cu** 0.50-1.0 **Fe** 0.35 max **Mg** 2.6-3.7 **Mn** 0.10 max **Si** 0.25 max **Ti** 0.15 max **Zn** 4.3-5.2 **Zr** 0.10-0.25 **Other** each 0.05 max, total 0.15 max | **AA** 7122 |

*Boxed entries are no longer active and are retained for reference purposes only.

The chemical compositions listed are for identification purposes and should not be used in lieu of the cross referenced specifications.

| UNIFIED NUMBER | DESCRIPTION | CHEMICAL COMPOSITION | CROSS REFERENCE SPECIFICATIONS |
|---|---|---|---|
| **A97129** | Wrought Aluminum Alloy, Heat Treatable | **Al** rem **Cr** 0.10 max **Cu** 0.50-0.9 **Fe** 0.30 max **Ga** 0.03 max **Mg** 1.3-2.0 **Mn** 0.10 max **Si** 0.15 max **Ti** 0.05 max **V** 0.05 max **Zn** 4.2-5.2 **Other** each 0.05 max, total 0.15 max | **AA** 7129 **ASTM** B221 (7129) **SAE** J454 (7129) |
| * **A97146** | Wrought Aluminum Alloy, Heat Treatable | **Al** rem **Cr** 0.10-0.22 **Cu** 1.2-1.9 **Fe** 0.40 max **Mg** 1.0-1.6 **Mn** 0.20 max **Si** 0.20 max **Ti** 0.06 max **Zn** 6.6-7.6 **Zr** 0.10-0.18 **Other** each 0.05 max, total 0.15 max | **AA** 7146 |
| **A97149** | Wrought Aluminum Alloy, Heat Treatable | **Al** rem **Cr** 0.10-0.22 **Cu** 1.2-1.9 **Fe** 0.20 max **Mg** 2.0-2.9 **Mn** 0.20 max **Si** 0.15 max **Ti** 0.10 max **Zn** 7.2-8.2 **Other** each 0.05 max, total 0.15 max | **AA** 7149 **AMS** 4320; 4343 |
| **A97150** | Wrought Aluminum Alloy, Heat Treatable | **Al** rem **Cr** 0.04 max **Cu** 1.9-2.5 **Fe** 0.15 max **Mg** 2.0-2.7 **Mn** 0.10 max **Si** 0.12 max **Ti** 0.06 max **Zn** 5.9-6.9 **Zr** 0.08-0.15 **Other** each 0.05 max, total 0.15 max | **AA** 7150 **AMS** 4252; 4306; 4307; 4345 |
| **A97175** | Wrought Aluminum Alloy, Heat Treatable | **Al** rem **Cr** 0.18-0.28 **Cu** 1.2-2.0 **Fe** 0.20 max **Mg** 2.1-2.9 **Mn** 0.10 max **Si** 0.15 max **Ti** 0.10 max **Zn** 5.1-6.1 **Other** each 0.05 max, total 0.15 max | **AA** 7175 **AMS** 4148; 4149; 4179; 4344; AMS-A-22771; AMS-QQ-A-200/15 **ASTM** B247 (7175) |
| **A97178** | Wrought Aluminum Alloy, Heat Treatable | **Al** rem **Cr** 0.18-0.35 **Cu** 1.6-2.4 **Fe** 0.50 max **Mg** 2.4-3.1 **Mn** 0.30 max **Si** 0.40 max **Ti** 0.20 max **Zn** 6.3-7.3 **Other** each 0.05 max, total 0.15 max | **AA** 7178 **AMS** AMS-QQ-A-200/13; AMS-QQ-A-200/14; AMS-QQ-A-250/14; AMS-QQ-A-250/21; AMS-QQ-A-250/22 **ASTM** B209 (7178); B221 (7178); B241 (7178); B316 (7178) **FED** QQ-A-430 |
| * **A97179** | Wrought Aluminum Alloy, Heat Treatable | **Al** rem **Cr** 0.10-0.25 **Cu** 0.40-0.8 **Fe** 0.20 max **Mg** 2.9-3.7 **Mn** 0.10-0.30 **Si** 0.15 max **Ti** 0.10 max **Zn** 3.8-4.8 **Other** each 0.05 max, total 0.15 max | **AA** 7179 |
| **A97229** | Wrought Aluminum Alloy, Heat Treatable | **Al** rem **Cu** 0.50-0.9 **Fe** 0.08 max **Mg** 1.3-2.0 **Mn** 0.03 max **Si** 0.06 **Ti** 0.05 max **V** 0.05 max **Zn** 4.2-5.2 **Other** each 0.03 max, total 0.10 max | **AA** 7229 |
| **A97249** | Wrought Aluminum Alloy, Heat Treatable | **Al** rem **Cr** 0.12-0.18 **Cu** 1.3-1.9 **Fe** 0.12 max **Mg** 2.0-2.4 **Mn** 0.10 max **Si** 0.10 max **Ti** 0.06 max **Zn** 7.5-8.2 **Other** each 0.05 max, total 0.15 max | **AA** 7249 **AMS** 4334 |
| **A97250** | Wrought Aluminum Alloy, Heat Treatable | **Al** rem **Cr** 0.04 max **Cu** 2.0-2.4 **Fe** 0.10 max **Mg** 1.9-2.3 **Mn** 0.10 max **Si** 0.08 max **Ti** 0.06 max **Zn** 5.7-6.5 **Zr** 0.08-0.13 **Other** each 0.05 max, total 0.15 max | **AA** 7250 |
| * **A97277** | Wrought Aluminum Alloy, Heat Treatable | **Al** rem **Cr** 0.18-0.35 **Cu** 0.8-1.7 **Fe** 0.7 max **Mg** 1.7-2.3 **Si** 0.50 max **Ti** 0.10 max **Zn** 3.7-4.3 **Other** each 0.05 max, total 0.15 max | **AA** 7277 **ASTM** B316 (7277) |
| **A97278** | Wrought Aluminum Alloy, Heat Treatable | **Al** rem **Cr** 0.17-0.25 **Cu** 1.6-2.2 **Fe** 0.20 max **Ga** 0.03 max **Mg** 2.5-3.2 **Mn** 0.02 max **Si** 0.15 max **Ti** 0.03 max **V** 0.05 max **Zn** 6.6-7.4 **Other** 0.03 max each, 0.10 max total | **AA** 7278; 7278A |
| **A97349** | Wrought Aluminum Alloy, Heat Treatable | **Al** rem **Cr** 0.10-0.22 **Cu** 1.4-2.1 **Fe** 0.15 max **Mg** 1.8-2.7 **Mn** 0.20 max **Si** 0.12 max **Zn** 7.5-8.7 **Other** each 0.05 max, total 0.15 max, (Ti+Zr) 0.25 max | **AA** 7349 |

*Boxed entries are no longer active and are retained for reference purposes only.

The chemical compositions listed are for identification purposes and should not be used in lieu of the cross referenced specifications.

| UNIFIED NUMBER | DESCRIPTION | CHEMICAL COMPOSITION | CROSS REFERENCE SPECIFICATIONS |
|---|---|---|---|
| **A97449** | Wrought Aluminum Alloy, Heat Treatable | **Al** rem **Cu** 1.4-2.1 **Fe** 0.15 max **Mg** 1.8-2.7 **Mn** 0.20 max **Si** 0.12 max **Zn** 7.5-8.7 **Other** each 0.05 max, total 0.15 max, (Ti+Zr) 0.25 max | **AA** 7449 |
| * **A97472** | Wrought Aluminum Alloy, Heat Treatable | **Al** rem **Cu** 0.05 max **Fe** 0.6 max **Mg** 0.9-1.5 **Mn** 0.05 max **Si** 0.25 max **Zn** 1.3-1.9 **Other** each 0.05 max, total 0.15 max | **AA** 7472 |
| **A97475** | Wrought Aluminum Alloy, Heat Treatable | **Al** rem **Cr** 0.18-0.25 **Cu** 1.2-1.9 **Fe** 0.12 max **Mg** 1.9-2.6 **Mn** 0.06 max **Si** 0.10 max **Ti** 0.06 max **Zn** 5.2-6.2 **Other** each 0.05 max, total 0.15 | **AA** 7475 **AMS** 4084; 4085; 4089; 4090; 4202 |
| * **A98001** | Wrought Aluminum Alloy, Non-Heat Treatable | **Al** rem **B** 0.001 max **Cd** 0.003 max **Co** 0.001 max **Cu** 0.15 max **Fe** 0.45-0.7 **Li** 0.008 max **Ni** 0.9-1.3 **Si** 0.17 max **Other** each 0.05 max, total 0.15 max | **AA** 8001 |
| * **A98004** | Wrought Aluminum Alloy, Non-Heat Treatable | **Al** rem **Cu** 0.03 max **Fe** 0.15 max **Mg** 0.02 max **Mn** 0.02 max **Si** 0.15 max **Ti** 0.30-0.7 **Zn** 0.03 max **Other** 0.02 max each, 0.15 max total | **AA** 8004 |
| **A98005** | Wrought Aluminum Alloy, Non-Heat Treatable | **Al** rem **Cu** 0.05 max **Fe** 0.40-0.8 **Mg** 0.05 max **Si** 0.20-0.50 **Zn** 0.05 max **Other** 0.05 max each, 0.15 max total | **AA** 8005 |
| **A98006** | Wrought Aluminum Alloy, Non-Heat Treatable | **Al** rem **Cu** 0.30 **Fe** 1.2-2.0 **Mg** 0.10 **Mn** 0.30-1.0 **Si** 0.40 **Zn** 0.10 **Other** each 0.05, total 0.15 | **AA** 8006 |
| **A98007** | Wrought Aluminum Alloy, Non-Heat Treatable | **Al** rem **Cu** 0.10 **Fe** 1.2-2.0 **Mg** 0.10 **Mn** 0.30-1.0 **Si** 0.40 **Zn** 0.8-1.8 **Other** each 0.05, total 0.15 | **AA** 8007 |
| **A98008** | Wrought Aluminum Alloy, Non-Heat Treatable | **Al** rem **Cu** 0.20 max **Fe** 0.9-1.6 **Mn** 0.50-1.0 **Si** 0.6 max **Ti** 0.10 max **Zn** 0.10 max **Other** 0.05 max each, 0.15 max total | **AA** 8008 |
| * **A98009** | Wrought Aluminum Alloy, Non-Heat Treatable | **Al** rem **Cr** 0.10 max **Fe** 8.4-8.9 **Mn** 0.10 max **O** 0.30 max **Si** 1.7-1.9 **Ti** 0.10 max **V** 1.1-1.5 **Zn** 0.25 max **Other** 0.05 max each, 0.15 max total | **AA** 8009 **AMS** 4308; 4309 |
| **A98010** | Wrought Aluminum Alloy, Non-Heat Treatable | **Al** rem **Cr** 0.20 max **Cu** 0.10-0.30 **Fe** 0.35-0.7 **Mg** 0.10-0.50 **Mn** 0.10-0.8 **Si** 0.40 max **Ti** 0.10 max **Zn** 0.40 max **Other** each 0.05 max, total 0.15 max | **AA** 8010 |
| **A98011** | Wrought Aluminum Alloy, Non-Heat Treatable | **Al** rem **Cr** 0.05 max **Cu** 0.10 max **Fe** 0.6-1.0 **Mg** 0.05 max **Mn** 0.20 max **Si** 0.50-0.9 **Ti** 0.08 max **Zn** 0.10 max **Other** each 0.05 max, total 0.15 max | **AA** 8011; 8011A |
| * **A98013** | Wrought Aluminum Alloy, Non-Heat Treatable | **Al** rem **Cr** 0.20-0.50 **Ti** 0.10 max **Other** each 0.03 max, total 0.10 max, plus Si+Fe 0.25 max | **AA** 8013 |

*Boxed entries are no longer active and are retained for reference purposes only.

The chemical compositions listed are for identification purposes and should not be used in lieu of the cross referenced specifications.

| UNIFIED NUMBER | DESCRIPTION | CHEMICAL COMPOSITION | CROSS REFERENCE SPECIFICATIONS |
|---|---|---|---|
| **A98014** | Wrought Aluminum Alloy, Non-Heat Treatable | **Al** rem **Cu** 0.20 max **Fe** 1.2-1.6 **Mg** 0.10 max **Mn** 0.20-0.6 **Si** 0.30 max **Ti** 0.10 max **Zn** 0.10 max **Other** each 0.05 max, total 0.15 max | **AA** 8014 |
| **A98015** | Wrought Aluminum Alloy, Non-Heat Treatable | **Al** rem **Cu** 0.10 max **Fe** 0.8-1.4 **Mg** 0.10 max **Mn** 0.10-0.40 **Si** 0.30 max **Zn** 0.10 max **Other** 0.05 max each, 0.15 max total | **AA** 8015 |
| **A98016** | Wrought Aluminum Alloy, Non-Heat Treatable | **Al** rem **Cu** 0.10 max **Fe** 0.7-1.1 **Mg** 0.10 max **Mn** 0.10-0.30 **Si** 0.20 max **Zn** 0.10 max **Other** 0.05 max each, 0.15 max total | **AA** 8016 |
| **A98017** | Wrought Aluminum Alloy, Non-Heat Treatable | **Al** rem **B** 0.04 max **Cu** 0.10-0.20 **Fe** 0.55-0.8 **Li** 0.003 max **Mg** 0.01-0.05 **Si** 0.10 max **Zn** 0.05 max **Other** each 0.03 max, total 0.10 max | **AA** 8017 **ASTM** B800 (8017) |
| **A98018** | Wrought Aluminum Alloy, Non-Heat Treatable | **Al** rem **Cu** 0.30-0.6 **Fe** 0.6-1.0 **Mn** 0.30 max **Si** 0.50-0.9 **Ti** 0.006-0.06 **Other** 0.05 max each, 0.15 max total | **AA** 8018 |
| **A98019** | Wrought Aluminum Alloy, Non-Heat Treatable | **Al** rem **Ce** 3.5-4.5 **Fe** 7.3-9.3 **Mn** 0.05 max **O** 0.05-0.50 **Si** 0.20 max **Ti** 0.05 max **Zn** 0.05 max **Other** each 0.05 max, total 0.15 max | **AA** 8019 |
| **A98020** | Wrought Aluminum Alloy, Non-Heat Treatable | **Al** rem **Bi** 0.10-0.50 **Cu** 0.005 max **Fe** 0.10 max **Mn** 0.005 max **Si** 0.10 max **Sn** 0.10-0.25 **V** 0.05 max **Zn** 0.005 max **Other** each 0.03 max, total 0.10 max | **AA** 8020 |
| **A98021** | Wrought Aluminum Alloy, Non-Heat Treatable | **Al** rem **Cu** 0.05 max **Fe** 1.2-1.7 **Si** 0.15 max **Other** 0.05 max each, 0.15 max total | **AA** 8021; 8021A; 8021B |
| **A98022** | Wrought Aluminum Alloy, Non-Heat Treatable | **Al** rem **Cr** 0.10 max **Fe** 6.2-6.8 **Mn** 0.10 max **O** 0.05-0.20 **Si** 1.2-1.4 **Ti** 0.10 max **V** 0.40-0.8 **Zn** 0.25 max **Other** 0.05 max each, 0.15 max total | **AA** 8022 **AMS** 4266 |
| **A98023** | Wrought Aluminum Alloy, Non-Heat Treatable | **Al** rem **B** 0.01-0.02 **Cr** 0.02 max **Cu** 0.10-0.40 **Fe** 1.3-1.6 **Mg** 0.005 max **Mn** 0.30-0.6 **Si** 0.20 max **Ti** 0.05-0.10 **Other** each 0.05 max, total 0.15 max. | **AA** 8023 |
| **A98024** | Wrought Aluminum Alloy, Heat Treatable | **Al** rem **Fe** 0.12 max **Li** 3.4-4.2 **Si** 0.10 max **Zr** 0.08-0.25 **Other** each 0.05 max, total 0.15 max | **AA** 8024 |
| **A98025** | Wrought Aluminum Alloy, Heat Treatable | **Al** rem **Cr** 0.18 max **Cu** 0.25 max **Fe** 0.06-0.25 **Mg** 0.05 max **Mn** 0.03-0.10 **Si** 0.05-0.15 **Ti** 0.005-0.02 **Zn** 0.50 max **Zr** 0.02-0.20 **Other** each 0.05 max, total 0.15 max | **AA** 8025 |
| **A98030** | Wrought Aluminum Alloy, Non-Heat Treatable | **Al** rem **B** 0.001-0.04 **Cu** 0.15-0.30 **Fe** 0.30-0.8 **Mg** 0.05 max **Si** 0.10 max **Zn** 0.05 max **Other** each 0.03 max, total 0.10 max | **AA** 8030 **ASTM** B800 (8030) |
| **A98040** | Wrought Aluminum Alloy, Non-Heat Treatable | **Al** rem **Cu** 0.20 max **Mn** 0.05 max **Zn** 0.20 max **Zr** 0.10-0.30 **Other** each 0.15 max, total 0.15, Fe+Si 1.0 max | **AA** 8040 |
| **A98050** | Wrought Aluminum Alloy, Non-Heat Treatable | **Al** rem **Cr** 0.05 max **Cu** 0.05 max **Fe** 1.1-1.2 **Mg** 0.05 max **Mn** 0.45-0.55 **Si** 0.15-0.30 **Zn** 0.10 max **Other** 0.05 max each, 0.15 max total | **AA** 8050 |
| **A98076** | Wrought Aluminum Alloy, Non-Heat Treatable | **Al** rem **B** 0.04 max **Cu** 0.04 max **Fe** 0.6-0.9 **Mg** 0.08-0.22 **Si** 0.10 max **Zn** 0.05 max **Other** each 0.03 max, total 0.10 max | **AA** 8076 **ASTM** B800 (8076) |

The chemical compositions listed are for identification purposes and should not be used in lieu of the cross referenced specifications.

| UNIFIED NUMBER | DESCRIPTION | CHEMICAL COMPOSITION | CROSS REFERENCE SPECIFICATIONS |
|---|---|---|---|
| **A98077** | Wrought Aluminum Alloy, Non-Heat Treatable | **Al** rem **B** 0.05 max **Cu** 0.05 max **Fe** 0.10-0.40 **Mg** 0.10-0.30 **Si** 0.10 max **Zn** 0.05 max **Zr** 0.02-0.08 **Other** each 0.03 max, total 0.10 max | **AA** 8077 |
| **A98079** | Wrought Aluminum Alloy, Non-Heat Treatable | **Al** rem **Cu** 0.05 max **Fe** 0.7-1.3 **Si** 0.05-0.30 **Zn** 0.10 max **Other** each 0.05 max, total 0.15 max | **AA** 8079 **ASTM** B479 (8079) |
| * **A98081** | Wrought Aluminum Alloy, Heat Treatable | **Al** rem **Cu** 0.7-1.3 **Fe** 0.7 max **Mn** 0.10 max **Si** 0.7 max **Sn** 18.0-22.0 **Ti** 0.10 max **Zn** 0.05 max **Other** each 0.05 max, total 0.15 max | **AA** 8081 |
| **A98090** | Wrought Aluminum Alloy, Heat Treatable | **Al** rem **Cr** 0.10 max **Cu** 1.0-1.6 **Fe** 0.30 max **Li** 2.2-2.7 **Mg** 0.6-1.3 **Mn** 0.10 max **Si** 0.20 max **Ti** 0.10 max **Zn** 0.25 max **Zr** 0.04-0.16 **Other** 0.05 max each, 0.15 max total | **AA** 8090 **AMS** 4259 |
| **A98091** | Wrought Aluminum Alloy, Heat Treatable | **Al** rem **Cr** 0.10 max **Cu** 1.6-2.2 **Fe** 0.50 max **Li** 2.4-2.8 **Mg** 0.50-1.2 **Mn** 0.10 max **Si** 0.30 max **Ti** 0.10 max **Zn** 0.25 max **Zr** 0.08-0.16 **Other** 0.05 max each, 0.15 max total | **AA** 8091 |
| **A98093** | Wrought Aluminum Alloy, Heat Treatable | **Al** rem **Cr** 0.10 max **Cu** 1.0-1.6 **Fe** 0.10 max **Li** 1.9-2.6 **Mg** 0.9-1.6 **Mn** 0.10 max **Si** 0.10 max **Ti** 0.10 max **Zn** 0.25 max **Zr** 0.04-0.14 **Other** each 0.05 max, total 0.15 max | **AA** 8093 |
| **A98111** | Wrought Aluminum Alloy, Non-Heat Treatable | **Al** rem **Cr** 0.05 max **Cu** 0.10 max **Fe** 0.40-1.0 **Mg** 0.05 max **Mn** 0.10 max **Si** 0.30-1.1 **Ti** 0.08 max **Zn** 0.10 max **Other** each 0.05, total 0.15 | **AA** 8111 **ASTM** B479 (8111) |
| **A98112** | Wrought Aluminum Alloy, Non-Heat Treatable | **Al** rem **Cr** 0.20 max **Cu** 0.40 max **Fe** 1.0 max **Mg** 0.7 max **Mn** 0.6 max **Si** 1.0 max **Ti** 0.20 max **Zn** 1.0 max **Other** each 0.05 max, total 0.15 max | **AA** 8112 |
| **A98130** | Wrought Aluminum Alloy, Non-Heat Treatable | **Al** rem **Cu** 0.05-0.15 **Fe** 0.40-1.0 **Si** 0.15 max **Zn** 0.10 max **Other** each 0.03 max, total 0.10 max, Fe+Si 1.0 max | **AA** 8130 **ASTM** B800 (8130) |
| **A98150** | Wrought Aluminum Alloy, Non-Heat Treatable | **Al** rem **Fe** 0.9-1.3 **Mn** 0.20-0.7 **Si** 0.30 max **Ti** 0.05 max **Other** each 0.05 max, total 0.15 max. | **AA** 8150 |
| **A98176** | Wrought Aluminum Alloy, Non-Heat Treatable | **Al** rem **Fe** 0.40-1.0 **Ga** 0.03 max **Si** 0.03-0.15 **Zn** 0.10 max **Other** each 0.05 max, total 0.15 max | **AA** 8176 **ASTM** B800 (8176) |
| * **A98177** | Wrought Aluminum Alloy, Non-Heat Treatable | **Al** rem **B** 0.04 **Cu** 0.04 **Fe** 0.25-0.45 **Mg** 0.04-0.12 **Si** 0.10 **Zn** 0.05 **Other** each 0.03 max, total 0.10 max | **AA** 8177 **ASTM** B800 (8177) |
| **A98211** | Wrought Aluminum Alloy, Non-Heat Treatable | **Al** rem **Cr** 0.15 max **Cu** 0.10 max **Fe** 0.50-1.0 **Mg** 0.10 max **Mn** 0.05-0.20 **Si** 0.40-0.8 **Ti** 0.05 max **Zn** 0.10 max **Other** 0.06 max each, 0.15 max total | **AA** 8211 |
| * **A98276** | Wrought Aluminum Alloy, Non-Heat Treatable | **Al** rem **B** 0.02 max **Cr** 0.01 max **Cu** 0.035 max **Fe** 0.50-0.8 **Ga** 0.03 max **Mg** 0.02 max **Mn** 0.01 max **Si** 0.25 max **Zn** 0.05 max **Other** 0.03 max each, 0.10 max total, V+Ti 0.03 max | **AA** 8276 |

*Boxed entries are no longer active and are retained for reference purposes only.

The chemical compositions listed are for identification purposes and should not be used in lieu of the cross referenced specifications.

| UNIFIED NUMBER | DESCRIPTION | CHEMICAL COMPOSITION | CROSS REFERENCE SPECIFICATIONS |
|---|---|---|---|
| **A98280** | Wrought Aluminum Alloy, Heat Treatable | **Al** rem **Cu** 0.7-1.3 **Fe** 0.7 max **Ni** 0.20-0.7 **Si** 1.0-2.0 **Sn** 5.5-7.0 **Ti** 0.10 max **Zn** 0.05 max **Other** each 0.05 max, total 0.15 max | **AA** 8280 |

The chemical compositions listed are for identification purposes and should not be used in lieu of the cross referenced specifications.

## UNS NUMBERS ASSIGNED TO DATE

### With Description of Each Material Covered and References To Documents In Which The Same or Similar Materials are Described

### Cxxxxx Number Series
### Copper and Copper Alloys

# COPPER AND COPPER ALLOYS

| UNIFIED NUMBER | DESCRIPTION | CHEMICAL COMPOSITION | CROSS REFERENCE SPECIFICATIONS |
|---|---|---|---|
| **C10100** | Oxygen-Free Electronic Copper (OFE) | **Ag** 0.0025 max **As** 0.0005 max **Bi** 0.0001 max **Cd** 0.0001 max **Cu** 99.99 min **Fe** 0.0010 max **Mn** 0.0005 max **Ni** 0.0010 max $O_2$ 0.0005 max **P** 0.0003 max **Pb** 0.0005 max **S** 0.0018 max **Sb** 0.0004 max **Se** 0.0010 max **Sn** 0.0002 max **Te** 0.0002 max **Zn** 0.0001 max **Other** total As, Bi, Mn, Sb, Se, Sn, and Te 0.0040 max | **AMS** 4700 **ASME** SFA5.8 (BCu-3) **ASTM** B1; B2; B3; B8; B33; B48; B49; B75; B111 (OFE); B152; B172; B173; B174; B187; B188; B189; B224 (OFE); B246; B272; B286; B298; B355; B359; B432; B447; B470; B496; B640; B698 (OFE); F68 **AWS** A5.8 (BCu-3) **FED** QQ-B-575; QQ-W-343 **MIL SPEC** MIL-W-85; MIL-B-18907; MIL-T-24107 |
| **C10200** | Oxygen-Free Copper (OF) | **Cu** (incl Ag) 99.95 min $O_2$ 0.0010 max | **AMS** 4501; 4602; 4701 **ASME** SB-42; SB-75; SB-111; SB-133; SB-152; SB-359; SB-395 **ASTM** B1; B2; B3; B8; B33; B42; B48; B49; B68; B75; B111; B152; B172; B173; B174; B187; B188; B189; B224 (OF); B226; B246; B272; B286; B298; B355; B359; B372; B395; B432; B447; B451; B470; B496; B506; B566; B638 (OF); B640; B687; B698 (OF); B738; B743 **FED** QQ-B-575; WW-T-799 **MIL SPEC** MIL-W-85; MIL-W-6712; MIL-B-18907; MIL-T-24107 **SAE** J461; J463 |
| **C10300** | Oxygen-Free Extra Low Phosphorus Copper (OFXLP) | **Cu** (incl Ag) 99.95 min **P** 0.001-0.005 | **ASTM** B42; B68; B75; B111; B152; B187; B188; B224 (OFXLP); B272; B302; B359; B372; B379 (OFXLP); B395; B432; B447; B638 (OFXLP); B640; B687; B698 (OFXLP); B743 **MIL SPEC** MIL-T-3235; MIL-T-24107 |
| **C10400** | Oxygen-Free Copper with Silver (OFS) | **Ag** 0.027 min **Cu** (incl Ag) 99.95 min $O_2$ 0.0010 max | **ASME** SB-152 **ASTM** B1; B2; B3; B49; B152; B187; B188; B189; B224 (OFS); B246; B298; B355 |
| **C10500** | Oxygen-Free Copper with Silver (OFS) | **Ag** 0.034 min **Cu** (incl Ag) 99.95 min $O_2$ 0.0010 max | **ASME** SB-152 **ASTM** B1; B2; B3; B49; B152; B187; B188; B189; B224 (OFS); B246; B272; B298; B355 |
| **C10700** | Oxygen-Free Copper with Silver (OFS) | **Ag** 0.085 min **Cu** (incl Ag) 99.95 min $O_2$ 0.0010 max | **ASME** SB-152 **ASTM** B1; B2; B3; B49; B152; B187; B188; B189; B224 (OFS); B246; B272; B298; B355 |
| **C10800** | Oxygen-Free Low Phosphorus Copper (OFLP) | **Cu** (incl Ag+P) 99.95 min **P** 0.005-0.012 | **ASTM** B42; B68; B75; B111; B152; B187; B224 (OFLP); B272; B302; B359; B360; B379; B395; B432; B447; B543; B687; B698 (OFLP); B743 **MIL SPEC** MIL-T-24107 |
| **C10910** | Copper (Controlled Oxygen) | **Cu** (incl Ag) 99.95 min **O** 0.005 max | **Other:** CDA Application Data Sheet Standard Designations for Wrought and Cast Copper and Copper Alloys as available on the Internet Web Site http://properties.copper.org/servlet/com.copper.servlet.CDAPropertiesSelectionServlet |
| **C10920** | Copper (Controlled Oxygen) | **Cu** (incl Ag) 99.90 min **O** 0.02 max | |
| **C10930** | Copper (Controlled Oxygen) | **Ag** 0.044 min **Cu** (incl Ag) 99.90 min **O** 0.02 max | **ASTM** B187 |
| **C10940** | Copper (Controlled Oxygen) | **Ag** 0.085 min **Cu** (incl Ag) 99.90 min **O** 0.02 max | **ASTM** B187 |
| **C11000** | Electrolytic Tough Pitch Copper (ETP) | **Cu** (incl Ag) 99.90 min | **AMS** 4500 **ASME** SFA5.8 (BCu-1b) **ASTM** B1; B2; B3; B8; B33; B47; B48; B49; B101; B116; B124; B152; B172; B173; B174; B187; B188; B189; B224 (ETP); B226; B228; B229; B246; B272; B283; B286; B298; B355; B370; B447; B470; B496; B566; B638; B694; B738; F467 (110); F468 (110) **AWS** A5.8 (BCu-1b) **FED** QQ-B-575 **MIL SPEC** MIL-W-6712 **SAE** J461; J463 |
| **C11010** | Remelted High Conductivity Tough Pitch Copper | **Cu** (incl Ag) 99.90 min **Other** Unspecified oxygen and trace elements | **ASTM** B224 (RHC) |
| **C11020** | Fine Refined High Conductivity Tough Pitch Copper | **Cu** (incl Ag) 99.90 min **Other** Unspecified oxygen and trace elements | **ASTM** B224 (FRHC) |
| **C11030** | Chemically Refined Tough Pitch Copper | **Cu** (incl Ag) 99.90 min **Other** Unspecified oxygen and trace elements | **ASTM** B224 (CRTP) |
| **C11040** | Electronic/Electrowon Tough Pitch Copper | **Ag** 0.0025 max **As** 0.0005 max **Bi** 0.00010 max **Cu** 99.90 min **Fe** 0.0010 max **Ni** 0.0010 max $O_2$ 0.010-0.065 **Pb** 0.0005 max **S** 0.0015 max **Sb** 0.0004 max **Se** 0.0002 max **Sn** 0.0005 max **Te** 0.0002 max **Other** Bi+Se+Te 0.0003 max total, total all (except $Cu+O_2$) 0.0065 max | **ASTM** B49 |

The chemical compositions listed are for identification purposes and should not be used in lieu of the cross referenced specifications.

| UNIFIED NUMBER | DESCRIPTION | CHEMICAL COMPOSITION | CROSS REFERENCE SPECIFICATIONS |
|---|---|---|---|
| **C11080** | Replaced by C13100 | | |
| **C11100** | Electrolytic Tough Pitch Anneal Resistant | **Cu** (incl Ag) 99.90 min **Note:** Small amounts of Cd or other elements may be added to improve the resistance to softening at elevated temperatures | **ASTM** B224 (ETP) **CDA** C11100 **SAE** J463 |
| **C11300** | Tough Pitch Copper with Silver (STP) | **Ag** 0.027 min **Cu** (incl Ag) 99.90 min **Note:** Oxygen and trace elements may vary depending on the process | **ASTM** B1; B2; B3; B5 (STP); B8; B33; B47; B48; B49; B116; B33; B48; B152; B172; B173; B174; B187; B188; B189; B224 (ETP); B226; B228; B229; B246; B286; B298; B355; B470; B496 **FED** QQ-B-575 **SAE** J461; J463 |
| **C11400** | Tough Pitch Copper with Silver (STP) | **Ag** 0.034 min **Cu** (incl Ag) 99.90 min **Note:** Oxygen and trace elements may vary depending on the process | **ASTM** B1; B2; B3; B5 (STP); B8; B47; B48; B49; B116; B152; B172; B173; B174; B187; B188; B189; B224 (ETP); B226; B229; B246; B286; B298; B355; B442 B470; B496 **FED** QQ-B-575 **SAE** J461; J463 |
| **C11500** | Tough Pitch Copper with Silver (STP) | **Ag** 0.054 min **Cu** (incl Ag) 99.90 min **Note:** Oxygen and trace elements may vary depending on the process | **ASTM** B1; B2; B3; B5 (STP); B8; B33; B47; B49; B116; B172; B173; B174; B189; B224 (ETP); B226; B228; B229; B246; B286; B298; B355; B470; B496 **FED** QQ-B-575; QQ-W-343 **SAE** J461; J463 |
| **C11600** | Tough Pitch Copper with Silver (STP) | **Ag** 0.085 min **Cu** (incl Ag) 99.90 min **Note:** Oxygen and trace elements may vary depending on the process | **ASTM** B1; B2; B3; B5 (STP); B8; B33; B47; B48; B49; B116; B152; B172; B173; B174; B187; B188; B189; B224 (ETP); B226; B229; B246; B286; B298; B355; B470; B496 **FED** QQ-B-575 **SAE** J461; J463 |
| **C11700** | Boron Deoxidized Copper | **B** 0.004-0.02 **Cu** (incl Ag+B) 99.90 min **P** 0.04 max | |
| * **C11904** | Tough Pitch Copper with Silver | **Ag** 0.027 min **Cu** (incl Ag) 99.90 min **Other** For oxygen-free and deoxidized grades, B, Li, P, etc., limits shall be as agreed upon. | |
| * **C11905** | Tough Pitch Copper with Silver | **Ag** 0.034 min **Cu** (incl Ag) 99.90 min **Other** For oxygen-free and deoxidized grades, B, Li, P, etc., limits shall be as agreed upon. | |
| * **C11906** | Tough Pitch Copper with Silver | **Ag** 0.054 min **Cu** (incl Ag) 99.90 min **Other** For oxygen-free and deoxidized grades, B, Li, P, etc., limits shall be as agreed upon. | |
| * **C11907** | Tough Pitch Copper with Silver | **Ag** 0.085 min **Cu** (incl Ag) 99.90 min **Other** For oxygen-free and deoxidized grades, B, Li, P, etc., limits shall be as agreed upon. | |
| **C12000** | Phosphorus Deoxidized, Low Residual Phosphorus Copper (DLP) | **Cu** (incl Ag) 99.90 min **P** 0.004-0.012 | **ASME** SB-42; SB-75; SB-111; SB-359; SB-395 **ASTM** B5 (DLP); B42; B68; B75; B111; B152; B187; B188; B224 (DLP); B280; B302; B359; B372; B395; B447; B640; B687; B698 (DLP); B716; B743 **FED** WW-T-799 **MIL SPEC** MIL-W-85; MIL-B-18907; MIL-T-24107 **SAE** J461; J463 |
| **C12100** | Silver-Bearing, Low Residual Phosphorus Copper (DLP) | **Ag** 0.014 min **Cu** (incl Ag) 99.90 min **P** 0.005-0.012 | **ASTM** B224 (DLPS) |
| **C12200** | Phosphorus Deoxidized, High Residual Phosphorus Copper (DHP) | **Cu** (incl Ag) 99.90 min **P** 0.015-0.040 | **ASME** B16.22; B16.29; SB-42; SB-75; SB-111; SB-152; SB-359; SB-395; SB-543 **ASTM** B5 (DHP); B42; B68; B75; B88; B101; B111; B152; B187; B224 (DHP); B272; B302; B306; B359; B360; B370; B379 (OFLP); B395; B432; B447; B543; B640; B641; B687; B698 (DHP); B716; B743; B903 **FED** WW-P-377; WW-T-799 **MIL SPEC** MIL-B-18907; MIL-T-22214; MIL-T-24107 **SAE** J461; J463 |

*Boxed entries are no longer active and are retained for reference purposes only.

The chemical compositions listed are for identification purposes and should not be used in lieu of the cross referenced specifications.

| UNIFIED NUMBER | DESCRIPTION | CHEMICAL COMPOSITION | CROSS REFERENCE SPECIFICATIONS |
|---|---|---|---|
| **C12210** | Phosphorus Deoxidized Copper | **Cu** (incl Ag) 99.90 min **P** 0.015-0.025 | |
| **C12220** | Phosphorus Deoxidized Copper | **Cu** (incl Ag) 99.90 min **P** 0.040-0.065 | |
| **C12300** | Silver-Bearing, High Residual Phosphorus Copper (DHP) | **Cu** (incl Ag) 99.90 min **P** 0.015-0.040 | **ASME** SB-152 **ASTM** B5 (DHPS); B152; B224 (DHPS) |
| **C12500** | Fire-Refined Tough Pitch Copper (FRTP) | **As** 0.012 max **Bi** 0.003 max **Cu** (incl Ag) 99.88 min **Ni** 0.050 max **Pb** 0.004 max **Sb** 0.003 max **Te** 0.025 max **Other** Se included in Te | **ASTM** B101; B216; B224 (FRTP); B370 |
| **C12510** | Fire-Refined Tough Pitch | **Bi** 0.005 max **Cu** (incl Ag) 99.9 min **Fe** 0.05 max **Ni** 0.050 max **P** 0.03 max **Pb** 0.020 max **Sb** 0.003 max **Sn** 0.05 max **(Te+Se)** 0.025 max **Zn** 0.080 max | **Other:** CDA Application Data Sheet, Standard Designations for Wrought and Cast Copper and Copper Alloys as available on the Internet Web Site, http://properties.copper.org/servlet/com.copper.servlet.CDAPropertiesSelectionServlet |
| *** C12700** | Fire-Refined Tough Pitch Copper with Silver (FRSTP) | **Ag** 0.027 min **As** 0.012 max **Bi** 0.003 max **Cu** (incl Ag) 99.88 min **Ni** 0.050 max **Pb** 0.004 max **Sb** 0.003 max **Te** 0.025 max **Other** Se included in Te | **ASTM** B224 (FRSTP) **MIL SPEC** MIL-B-20292 |
| *** C12800** | Fire-Refined Tough Pitch Copper with Silver (FRSTP) | **Ag** 0.034 min **As** 0.012 max **Bi** 0.003 max **Cu** (incl Ag) 99.88 min **Ni** 0.050 max **Pb** 0.004 max **Sb** 0.003 max **Te** 0.025 max **Other** Se included in Te | **ASTM** B224 (FRSTP) **MIL SPEC** MIL-B-20292 |
| **C12900** | Fire-Refined Tough Pitch Copper with Silver (FRSTP) | **Ag** 0.054 min **As** 0.012 max **Bi** 0.003 max **Cu** (incl Ag) 99.88 min **Ni** 0.050 max **Pb** 0.004 max **Sb** 0.003 max **Te** 0.025 max **Other** Se included in Te | **ASTM** B224 (FRSTP) |
| *** C13000** | Fire-Refined Tough Pitch Copper with Silver (FRSTP) | **Ag** 0.085 min **As** 0.012 max **Bi** 0.003 max **Cu** (incl Ag) 99.88 min **Ni** 0.050 max **Pb** 0.004 max **Sb** 0.003 max **Te** 0.025 max **Other** Se included in Te | **ASTM** B224 (FRSTP) **MIL SPEC** MIL-B-20292 |
| **C13100** | Non-Ferrous Thermal Spray Wire (W-Cu) | **Cu** (incl Ag) 99.8 min | **AWS** A5.33 (W-Cu) **Other:** CDA Application Data Sheet Standard Designations for Wrought and Cast Copper and Copper Alloys as available on the Internet Web Site, http://properties.copper.org/servlet/com.copper.servlet.CDAPropertiesSelectionServlet |
| *** C14100** | Arsenical, Tough Pitch Copper (ATP) | **As** 0.15-0.50 **Cu** 99.40 min **Other** Ag included in Cu | **ASME** SB-11; SB-12 |
| **C14180** | Copper Brazing Filler Metal | **Al** 0.01 max **Cu** (incl Ag) 99.90 min **P** 0.075 max **Pb** 0.02 max | **ASME** SFA5.8 (BCu-1) **AWS** A5.8 (BCu-1) |
| **C14181** | Copper Vacuum Grade Brazing Filler Metal | **C** 0.005 max **Cd** 0.002 max **Cu** (incl Ag) 99.90 min **P** 0.002 max **Pb** 0.002 max **Zn** 0.002 | **ASME** SFA5.8 (BVCu-1X) **AWS** A5.8 (BVCu-1X) |
| **C14200** | Phosphorus Deoxidized Copper, Arsenical (DPA) | **As** 0.15-0.50 **Cu** (incl Ag) 99.4 min **P** 0.015-0.040 | **ASME** SB-75; SB-111; SB-359; SB-395 **ASTM** B5 (DPA); B111; B152; B224 (DPA); B272; B359; B379 (DPA); B395; B447 **MIL SPEC** MIL-T-24107 |
| *** C14210** | Phosphorus Deoxidized Arsenical Copper | **As** 0.30-0.50 **Cu** (incl Ag) 99.20 min **P** 0.013-0.050 | |
| **C14300** | Cadmium Copper, Deoxidized | **Cd** 0.05-0.15 **Cu** (incl Ag+Cd) 99.90 min **Note:** Li by agreement | |

*Boxed entries are no longer active and are retained for reference purposes only.

The chemical compositions listed are for identification purposes and should not be used in lieu of the cross referenced specifications.

| UNIFIED NUMBER | DESCRIPTION | CHEMICAL COMPOSITION | CROSS REFERENCE SPECIFICATIONS |
|---|---|---|---|
| * **C14310** | Cadmium Copper, Deoxidized | **Cd** 0.10-0.30 **Cu** (incl Ag+Cd) 99.90 min **Note:** Li by agreement | |
| * **C14400** | Tin Copper Radiator Strip | **Cu** (incl Ag+Sn+P) 99.90 min **Fe** 0.03 max **Ni** 0.05 max **P** 0.013-0.025 **Sb** 0.003 max **Se+Te** 0.02 max **Sn** 0.10-0.20 **Zn** 0.05 max **Note:** Se+Te 0.02 max | **FED** QQ-B-650 |
| **C14410** | Copper-Tin-Phosphorus Alloy | **Cu** (incl Ag+Sn) 99.90 min **Fe** 0.05 max **P** 0.005-.020 **Pb** 0.05 max **Sn** 0.10-0.20 | |
| **C14415** | Copper-Tin Alloy | **Cu** (incl Ag+Sn) 99.96 min **Sn** 0.10-0.15 | |
| **C14420** | Copper-Tellurium-Tin Alloy | **Cu** (incl Ag+Sn+Te) 99.90 min **Sn** 0.05-0.15 **Te** (Te+Sn) 0.02-0.05 | **ASTM** B152; B187 |
| * **C14430** | Copper - Tin Alloy | **Cu** (incl Ag) rem **Sn** 0.25-0.35 **Other** Cu + sum of named elements 99.90 min | |
| * **C14440** | Copper Alloy | **Cu** (incl Ag+Sn) 99.96 min **Sn** 0.005-0.01 | |
| **C14500** | Phosphorus Deoxidized Copper, Tellurium Bearing (DPTE) Includes oxygen-free or deoxidized grades with deoxidizers (such as phosphorus, boron, lithium, or other) in an amount agreed upon. | **Cu** (incl Ag+Te) 99.90 min **P** 0.004-0.012 **Te** 0.40-0.70 **Note:** Other deoxidizers may be used instead of P | **ASME** B16.22 **ASTM** B5 (DPTE); B124; B224 (DPTE); B283; B301 **SAE** J461 (DPTE); J463 |
| **C14510** | Copper-Tellurium Alloy Includes oxygen-free or deoxidized grades with deoxidizers (such as phosphorus, boron, lithium, or other) in an amount agreed upon. | **Cu** (incl Ag+Te) 99.85 min **P** 0.010-0.030 **Pb** 0.05 max **Te** 0.30-0.70 | **ASTM** B301 |
| **C14520** | Copper-Tellurium | **Cu** (incl Ag+Te) 99.40 min **P** 0.004-0.020 **Te** 0.40-0.7 | **ASTM** B216; B301 |
| **C14530** | Copper - Tellurium Alloy | **Cu** (incl Ag) 99.95 min **P** 0.001-0.005 **Sn** 0.003-0.022 **Te** 0.003-0.022 | **ASTM** B152; B888 |
| **C14700** | Sulfur-Bearing Copper | **Cu** (incl Ag+P+S) 99.90 min **S** 0.20-0.50 **P** 0.002-0.005 **Note:** For oxygen-free and deoxidized grades: O, P, B, Li, etc., limits as agreed upon; Cu min includes Ag+Cu+P+S | **ASME** B16.22 **ASTM** B124; B224; B283; B301 **SAE** J461; J463 |
| * **C14710** | Sulfur Bearing Copper | **Cu** (incl AG+P+Pb+S) 99.90 min **P** 0.010-0.030 **Pb** 0.05 max **S** 0.05-0.15 **Note:** For oxygen-free and deoxidized grades O, P, B, Li, etc., limits shall be as agreed upon | **ASTM** B301 |

*Boxed entries are no longer active and are retained for reference purposes only.

The chemical compositions listed are for identification purposes and should not be used in lieu of the cross referenced specifications.

| UNIFIED NUMBER | DESCRIPTION | CHEMICAL COMPOSITION | CROSS REFERENCE SPECIFICATIONS |
|---|---|---|---|
| * **C14720** | Sulfur Copper | **Cu** (incl Ag+P+Pb+S) 99.50 min **P** 0.010-0.030 **Pb** 0.10 max **S** 0.20-0.50 **Note:** For oxygen-free and deoxidized grades O, P, B, Li, etc., limits shall be as agreed upon | **ASTM** B301 |
| * **C14730** | Sulfur Bearing Copper | **Cu** 99.80 **Other** Includes Ag, S, P, and Pb; Includes oxygen-free or deoxidized grades with deoxidizers (such as P, B, Li, or other) in an amount agreed upon. | |
| **C15000** | Zirconium Copper | **Cu** (incl Ag) 99.80 min **Zr** 0.10-0.20 **Other** Cu + sum of named elements 99.9 min | **ASTM** B224 **SAE** J461; J463 |
| **C15100** | Zirconium Copper | **Cu** (incl Ag) 99.80 min **Zr** 0.05-0.15 | **ASTM** B747; B888 |
| * **C15150** | Copper - Zirconium Alloy | **Cu** (incl Ag+Zn) 99.96 min **Zr** 0.015-0.03 | |
| **C15500** | Silver-Bearing Copper | **Ag** 0.027-0.10 **Cu** (incl Ag) 99.75 min **Mg** 0.08-0.13 **P** 0.040-0.080 | **ASTM** B888 |
| * **C15600** | Copper Cobalt | **Co** 0.20-0.30 **Cu** (incl Ag) 99.6 min **Mg** 0.02 max **P** 0.06-0.09 | |
| * **C15710** | Dispersion Strengthened Copper | **Al** 0.08-0.12 **Cu** (incl Ag) 99.71 min **Fe** 0.01 max $\mathbf{O_2}$ 0.07-0.15 **Pb** 0.01 max **Note:** 0.04% of $O_2$ as $Cu_2O$ with a negligible amount in solid solution with Cu | |
| **C15715** | Dispersion Strengthened Copper | **Al** 0.13-0.17 **Cu** (incl Ag) 99.62 min **Fe** 0.01 max $\mathbf{O_2}$ 0.12-0.19 **Pb** 0.01 max **Note:** Cu content includes Ag, and 0.04% $O_2$ present as $Cu_2O$ with a negligible amount in solid solution with Cu | |
| **C15720** | Dispersion Strengthened Copper | **Al** 0.18-0.22 **Cu** (incl Ag) 99.52 min **Fe** 0.01 max $\mathbf{O_2}$ 0.16-0.24 **Pb** 0.01 max **Note:** 0.04% $O_2$ present as $Cu_2O$ with a negligible amount in solid solution with Cu | |
| **C15725** | Dispersion Strengthened Copper | **Al** 0.23-0.27 **Cu** (incl Ag) 99.43 min **Fe** 0.01 max **O** 0.20-0.28 **Pb** 0.01 max **Note:** 0.04% $O_2$ as $Cu_2O$ with a negligible amount in solid solution with Cu | |
| * **C15735** | Dispersion Strengthened Copper | **Al** 0.33-0.37 **Cu** (incl Ag) 99.24 min **Fe** 0.01 max $\mathbf{O_2}$ 0.29-0.37 **Pb** 0.01 max **Note:** 0.04% $O_2$ present as $Cu_2O$ with a negligible amount in solid solution with Cu | |
| **C15760** | Dispersion Strengthened Copper | **Al** 0.58-0.62 **Cu** (incl Ag) 98.77 min **Fe** 0.01 max $\mathbf{O_2}$ 0.52-0.59 **Pb** 0.01 max **Note:** 0.04% $O_2$ present as $Cu_2O$ with a negligible amount in solid solution with Cu | |

*Boxed entries are no longer active and are retained for reference purposes only.

The chemical compositions listed are for identification purposes and should not be used in lieu of the cross referenced specifications.

| UNIFIED NUMBER | DESCRIPTION | CHEMICAL COMPOSITION | CROSS REFERENCE SPECIFICATIONS |
|---|---|---|---|
| **C15815** | Dispersion Strengthened Copper Alloy (CSM Metal) | **Al** 0.13-0.17 **B** 1.2-1.8 **Cu** (incl Ag) 97.82 min **Fe** 0.01 max **O** 0.19 max **Pb** 0.01 max **Other** All aluminum present as Al2O3; 0.04% oxygen present as Cu2O with a negligible amount in solid solution with copper. | **ASTM** B876 **Other:** CDA Application Data Sheet, Standard Designations for Wrought and Cast Copper and Copper Alloys as available on the Internet Web Site, http://properties.copper.org/servlet/com.copper.servlet.CDAPropertiesSelectionServlet |
| **C16200** | Cadmium Copper | **Cd** 0.7-1.2 **Cu** (incl Ag) rem **Fe** 0.02 max **Other** Cu + sum of named elements 99.5 min | **ASTM** B105; B624 **SAE** J461; J463 |
| *** C16210** | Cadmium Copper | **Cd** 0.50-1.2 **Cu** (incl Ag) rem **Other** Cu + sum of named elements 99.5 min | |
| *** C16400** | Cadmium Copper Alloy | **Cd** 0.6-0.9 **Cu** 99.8 min **Fe** 0.02 max **Sn** 0.20-0.40 **Other** Ag+Cd+Fe+Sn included in Cu | |
| **C16500** | Cadmium Copper Alloy | **Cd** 0.6-1.0 **Cu** (incl Ag) rem **Fe** 0.02 max **Sn** 0.50-0.7 **Other** Cu + sum of named elements 99.5 min | **ASTM** B9; B105 |
| **C17000** | Beryllium Copper | **Al** 0.20 max **Be** 1.60-1.79 **Cu** (incl Ag) rem **Si** 0.20 max; Ni+Co 0.20 min **Other** Ni+Fe+Co 0.6 max; Cu + sum of named elements 99.5 min | **ASTM** B194; B196; B570 **SAE** J461; J463 |
| **C17200** | Beryllium Copper | **Al** 0.20 max **Be** 1.80-2.00 **Cu** (incl Ag) rem **Si** 0.20 max; Ni+Co 0.20 min **Other** Ni+Fe+Co 0.6 max; Cu + sum of named elements 99.5 min | **AMS** 4530; 4532; 4533; 4534; 4535; 4650; 4651; 4725 **ASTM** B194; B196; B197; B570; B643 **MIL SPEC** MIL-C-21657 **SAE** J461; J463 |
| **C17300** | Beryllium Copper with Lead | **Al** 0.20 max **Be** 1.80-2.00 **Cu** (incl Ag) rem **Pb** 0.20-0.6 **Si** 0.20 max; Ni+Co 0.20 min **Other** Ni+Fe+Co 0.6 max: Cu + sum of named elements 99.5 min | **ASTM** B196; B197 |
| *** C17400** | Copper-Beryllium Alloy (Beryllium-Copper) | **Al** 0.20 max **Be** 0.15-0.50 **Co** 0.15-0.35 **Cu** (incl Ag) rem **Fe** 0.20 max **Si** 0.20 max **Other** Cu + sum of named elements 99.5 min | **ASTM** B768 |
| **C17410** | Copper-Beryllium Alloy (Beryllium-Copper) | **Al** 0.20 max **Be** 0.15-0.50 **Co** 0.35-0.6 **Cu** (incl Ag) rem **Fe** 0.20 max **Si** 0.20 max **Other** Cu + sum of named elements 99.5 min | **ASTM** B768 |
| *** C17420** | Copper-Beryllium Alloy (Beryllium-Copper) | **Al** 0.20 max **Be** 0.05-0.15 **Co** 0.05-0.6 **Cu** (incl Ag) rem **Fe** 0.20 max **Si** 0.20 max **Other** Cu + sum of named elements 99.5 min | |
| **C17450** | Copper-Beryllium Alloy | **Al** 0.20 max **Be** 0.15-0.50 **Cu** (incl Ag) rem **Fe** 0.20 max **Ni** 0.50-1.0 **Si** 0.20 max **Sn** 0.25 max **Zr** 0.50 max **Other** Cu (incl Ag) + sum of named elements 99.5 min | **Other:** CDA Application Data Sheet, Standard Designations for Wrought and Cast Copper and Copper Alloys as available on the Internet Web Site, http://properties.copper.org/servlet/com.copper.servlet.CDAPropertiesSelectionServlet |
| **C17455** | Copper Beryllium Alloy | **Al** 0.20 max **Be** 0.15-0.50 **Cu** (incl Ag) rem **Fe** 0.20 max **Ni** 0.50-1.0 **Pb** 0.20-0.6 **Si** 0.20 max **Sn** 0.25 max **Zr** 0.50 max **Other** Cu + sum of named elements 99.5 min | **Other:** CDA Application Data Sheet, Standard Designations for Wrought and Cast Copper and Copper Alloys as available on the Internet Web Site, http://properties.copper.org/servlet/com.copper.servlet.CDAPropertiesSelectionServlet |

*Boxed entries are no longer active and are retained for reference purposes only.

The chemical compositions listed are for identification purposes and should not be used in lieu of the cross referenced specifications.

| UNIFIED NUMBER | DESCRIPTION | CHEMICAL COMPOSITION | CROSS REFERENCE SPECIFICATIONS |
|---|---|---|---|
| **C17460** | Copper-Beryllium Alloy | **Al** 0.20 max **Be** 0.15-0.50 **Cu** (incl Ag) rem **Fe** 0.20 max **Ni** 1.0-1.4 **Si** 0.20 max **Sn** 0.25 max **Zr** 0.50 max **Other** Cu (incl Ag) + sum of named elements 99.5 min | **Other:** CDA Application Data Sheet, Standard Designations for Wrought and Cast Copper and Copper Alloys as available on the Internet Web Site, http://properties.copper.org/servlet/ com.copper.servlet.CDAPropertiesSelectionServlet |
| **C17465** | Copper Beryllium Alloy | **Al** 0.20 max **Be** 0.15-0.50 **Cu** (incl Ag) Rem **Fe** 0.20 max **Ni** 1.0-1.4 **Pb** 0.20-0.6 **Si** 0.20 max **Sn** 0.25 max **Zr** 0.50 max **Other** Cu + sum of named elements 99.5 min | **Other:** CDA Application Data Sheet, Standard Designations for Wrought and Cast Copper and Copper Alloys as available on the Internet Web Site, http://properties.copper.org/servlet/ com.copper.servlet.CDAPropertiesSelectionServlet |
| **C17500** | Beryllium Copper | **Be** 0.40-0.7 **Co** 2.4-2.7 **Cu** (incl Ag) rem **Fe** 0.10 max **Other** Cu + sum of named elements 99.5 min | **ASTM** B441; B534; B870 **MIL SPEC** MIL-C-81021; MIL-T-22214 **SAE** J461; J463 |
| **C17510** | Copper Beryllium Alloy | **Al** 0.20 max **Be** 0.20-0.6 **Co** 0.30 max **Cu** (incl Ag) rem **Fe** 0.10 max **Ni** 1.4-2.2 **Other** Cu + sum of named elements 99.5 min | **ASTM** B441; B534; B870 **MIL SPEC** MIL-C-81021 |
| * **C17520** | Copper Beryllium | **Be** 0.10-0.30 **Cu** (incl Ag) rem **Mg** 0.03-0.06 **Ni** 0.50-1.5 **Zr** 0.10-0.30 **Other** Cu + sum of named elements 99.5 min | |
| **C17530** | Copper-Beryllium Alloy | **Al** 0.6 max **Be** 0.20-0.40 **Cu** (incl Ag) rem **Fe** 0.20 max **Ni** (incl Co) 1.8-2.5 **Si** 0.20 max **Other** Cu (incl Ag) + sum of named elements 99.7 min | **Other:** CDA Application Data Sheet, Standard Designations for Wrought and Cast Copper and Copper Alloys as available on the Internet Web Site, http://properties.copper.org/servlet/ com.copper.servlet.CDAPropertiesSelectionServlet |
| * **C17600** | Beryllium Copper | **Ag** 0.9-1.1 **Al** 0.20 max **Be** 0.25-0.50 **Co** 1.4-1.7 **Cu** (incl Ag) rem **Fe** 0.10 max **Si** 0.20 max **Other** Cu + sum of named elements 99.5 min | **SAE** J461; J463 |
| * **C17700** | Beryllium Copper | **Al** 0.20 max **Be** 0.40-0.7 **Co** 2.4-2.7 **Cu** (incl Ag) rem **Fe** 0.10 max **Te** 0.40-0.6 **Si** 0.20 max **Other** Cu + sum of named elements 99.5 min | |
| **C18000** | Chromium Silicon Copper | **Cr** 0.10-0.6 **Cu** (incl Ag) rem **Fe** 0.15 max **Ni** (incl Co) 2.0-3.0 **Si** 0.40-0.8 **Other** Cu + sum of named elements 99.5 min | |
| **C18030** | Copper Alloy | **Cr** 0.10-0.20 **Cu** (incl Ag) 99.9 min **P** 0.005-0.15 **Sn** 0.08-0.12 | |
| **C18040** | Copper Alloy | **Cr** 0.25-0.35 **Cu** (incl Ag) 99.9 min **P** 0.005-0.015 **Sn** 0.20-0.30 **Zn** 0.05-0.15 | |
| **C18045** | Copper Alloy | **Cu** (incl Ag) 99.1 min **Cr** 0.20-0.35 **Si** 0.05 max **Sn** 0.20-0.30 **Zn** 0.15-0.30 **Other** Cu + sum of named elements 99.9% min | **Other:** CDA Application Data Sheet, Standard Designations for Wrought and Cast Copper and Copper Alloys as available on the Web Site, http://properties.copper.org/servlet/ com.copper.servlet.CDApropertiesSelectionServlet |
| **C18050** | Copper Alloy | **Cr** 0.05-0.15 **Cu** (incl Ag) 99.8 min **Te** 0.005-0.015 | |
| **C18070** | Copper Alloy | **Cr** 0.15-0.40 **Cu** (incl Ag) 99.0 min **Si** 0.02-0.07 **Ti** 0.01-0.40 **Other** Cu + sum of named elements 99.8 | |
| **C18080** | Copper Alloy | **Ag** 0.01-0.30 **Cr** 0.20-0.7 **Cu** (incl Ag) rem **Fe** 0.02-0.20 **Si** 0.01-0.10 **Ti** 0.01-0.15 **Other** Cu + sum of named elements 99.8% min | **Other:** CDA Application Data Sheet, Standard Designations for Wrought and Cast Copper and Copper Alloys as available on the Web Site, http://properties.copper.org/servlet/ com.copper.servlet.CDApropertiesSelectionServlet |

*Boxed entries are no longer active and are retained for reference purposes only.

The chemical compositions listed are for identification purposes and should not be used in lieu of the cross referenced specifications.

| UNIFIED NUMBER | DESCRIPTION | CHEMICAL COMPOSITION | CROSS REFERENCE SPECIFICATIONS |
|---|---|---|---|
| **C18090** | Copper-Tin-Nickel-Chromium-Titanium Alloy | **Cr** 0.30-1.0 **Cu** (incl Ag) 96.0 min **Ni** 0.30-1.2 **Sn** 0.50-1.2 **Ti** 0.15-0.8 **Other** Cu + sum of named elements 99.85 min | |
| **C18100** | Chromium Magnesium Copper Alloy | **Cr** 0.40-1.0 **Cu** (incl Ag) 98.7 min **Mg** 0.03-0.06 **Zr** 0.08-0.20 **Other** Cu + sum of named elements 99.5 min | |
| **C18135** | Cadmium Chromium Copper | **Cd** 0.20-0.6 **Cr** 0.20-0.6 **Cu** (incl Ag) rem **Other** Cu + sum of named elements 99.5 min | |
| **C18140** | High Copper Alloy | **Cr** 0.150-.45 **Cu** (incl Ag) rem **Si** 0.005-0.05 **Zr** 0.05-0.25 | **Other:** CDA Application Data Sheet, Standard Designations for Wrought and Cast Copper and Copper Alloys as available on the Internet Web Site, http://properties.copper.org/servlet/com.copper.servlet.CDAPropertiesSelectionServlet |
| **C18145** | Copper Alloy | **Cu** (incl Ag) rem **Cr** 0.10-0.30 **Zn** 0.10-0.30 **Zr** 0.05-0.15 **Other** Cu + sum of named elements 99.5% min | **Other:** CDA Application Data Sheet, Standard Designations for Wrought and Cast Copper and Copper Alloys as available on the Web Site, http://properties.copper.org/servlet/com.copper.servlet.CDApropertiesSelectionServlet |
| **C18150** | Copper Alloy | **Cr** 0.50-1.5 **Cu** (incl Ag) rem **Zr** 0.05-0.25 **Other** Cu + sum of named elements 99.7 min | |
| **C18200** | Chromium Copper | **Cr** 0.6-1.2 **Cu** (incl Ag) rem **Fe** 0.10 max **Pb** 0.05 max **Si** 0.10 max **Other** Cu + sum of named elements 99.5 min | |
| **C18400** | Chromium Copper | **As** 0.005 max **Ca** 0.005 max **Cr** 0.40-1.2 **Cu** (incl Ag) rem **Fe** 0.15 max **Li** 0.05 max **P** 0.05 max **Si** 0.10 max **Zn** 0.7 max **Other** Cu + sum of named elements 99.5 min | **SAE** J461; J463 |
| * **C18500** | Chromium Copper | **Ag** 0.08-0.12 **Cr** 0.40-1.0 **Cu** (incl Ag) rem **P** 0.04 max **Pb** 0.015 max **Other** Cu + sum of named elements 99.5 min | **MIL SPEC** MIL-C-19311 |
| * **C18550** | Copper Chromium Alloy | **Cr** 0.6-1.0 **Cu** (incl Ag) rem **Sn** 0.10-0.14 **Other** Cu + sum of named elements 99.8 min | |
| **C18600** | High Copper Alloy | **Co** 0.10 max **Cr** 0.10-1.0 **Cu** (incl Ag) rem **Fe** 0.25-0.8 **Ni** 0.25 max **Ti** 0.05-0.50 **Zr** 0.05-0.40 **Other** (Cu + sum of named elements) 99.5 min | **Other:** CDA Application Data Sheet, Standard Designations for Wrought and Cast Copper and Copper Alloys as available on the Internet Web Site, http://properties.copper.org/servlet/com.copper.servlet.CDAPropertiesSelectionServlet |
| **C18610** | High Copper Alloy | **Co** 0.25-0.8 **Cr** 0.10-1.0 **Cu** (incl Ag) rem **Fe** 0.10 max **Ni** 0.25 max **Ti** 0.05-0.50 **Zr** 0.05-0.40 **Other** (Cu + sum of named elements) 99.5 min | **Other:** CDA Application Data Sheet, Standard Designations for Wrought and Cast Copper and Copper Alloys as available on the Internet Web Site, http://properties.copper.org/servlet/com.copper.servlet.CDAPropertiesSelectionServlet |
| **C18665** | High Copper Alloy | **Cu** (incl Ag) 99.0 min **Mg** 0.40-0.9 **P** 0.002-0.04 | **Other:** CDA Application Data Sheet, Standard Designations for Wrought and Cast Copper and Copper Alloys as available on the Internet Web Site, http://properties.copper.org/servlet/com.copper.servlet.CDAPropertiesSelectionServlet |
| **C18700** | Leaded Copper | **Cu** (incl Ag) rem **Pb** 0.8-1.5 **Other** Cu + sum of named elements 99.5 min | **ASTM** B301 **SAE** J461; J463 |
| **C18835** | High Copper Alloy | **Cu** (incl Ag) 99.0 min **Fe** 0.10 max **Sn** 0.15-0.55 **P** 0.01 max **Pb** 0.05 max **Zn** 0.30 max **Other** (Cu + sum of named elements) 99.5 min | **Other:** CDA Application Data Sheet, Standard Designations for Wrought and Cast Copper and Copper Alloys as available on the Internet Web Site, http://properties.copper.org/servlet/com.copper.servlet.CDAPropertiesSelectionServlet |

*Boxed entries are no longer active and are retained for reference purposes only.

The chemical compositions listed are for identification purposes and should not be used in lieu of the cross referenced specifications.

| UNIFIED NUMBER | DESCRIPTION | CHEMICAL COMPOSITION | CROSS REFERENCE SPECIFICATIONS |
|---|---|---|---|
| **C18900** | High Copper Alloy | **Al** 0.01 max **Cu** (incl Ag) rem **Mn** 0.10-0.30 **P** 0.05 max **Pb** 0.02 max **Si** 0.15-0.40 **Sn** 0.6-0.9 **Zn** 0.10 max **Other** Cu + sum of named elements 99.5 min | |
| **C18980** | Copper Welding Filler Metal | **Cu** (incl Ag) 98.0 min **Mn** 0.50 max **P** 0.15 max **Pb** 0.02 max **Si** 0.50 max **Sn** 1.0 max **Other** Cu + sum of named elements 99.5 min | **ASME** SFA5.7 (ERCu); SFA5.27 (RCu) **AWS** A5.7 (ERCu); A5.27 (RCu) |
| **C18990** | Copper Tin Alloy | **Cr** 0.10-0.20 **Cu** (incl Ag) rem **P** 0.005-0.015 **Sn** 1.8-2.2 **Other** Cu + sum of named elements 99.9 min | |
| **C19000** | High Copper Alloy | **Cu** (incl Ag) rem **Fe** 0.10 max **Ni** 0.9-1.3 **P** 0.15-0.35 **Pb** 0.05 max **Zn** 0.8 max **Other** Cu + sum of named elements 99.5 min | |
| **C19010** | Copper Alloy | **Cu** (incl Ag) rem **Ni** 0.8-1.8 **P** 0.01-0.05 **Si** 0.15-0.35 **Other** Cu + sum of named elements 99.5 min | |
| **C19015** | Copper Alloy | **Cu** (incl Ag) rem **Mg** 0.02-0.15 **Ni** 0.50-2.4 **P** 0.02-0.20 **Si** 0.10-0.40 **Other** (Cu + sum of named elements) 99.8 min | **Other:** CDA Application Data Sheet, Standard Designations for Wrought and Cast Copper and Copper Alloys as available on the Internet Web Site, http://properties.copper.org/servlet/ com.copper.servlet.CDAPropertiesSelectionServlet |
| **C19020** | Copper Alloy | **Cu** (incl Ag) rem **Ni** 0.50-3.0 **P** 0.01-0.20 **Sn** 0.30-0.9 **Other** Mn+Si 0.35 max, Cu + sum of named elements 99.8 min | |
| **C19025** | High Copper Alloy (NB 109) | **Cu** rem **Fe** 0.10 max **Ni** 0.8-1.2 **P** 0.03-0.07 **Sn** 0.7-1.1 **Zn** 0.20 max **Other** Cu + sum of named elements 99.7 min | **ASTM** B422 |
| **C19030** | Copper-Nickel-Tin Alloy | **Cu** (incl Ag) rem **Fe** 0.10 max **Ni** 1.5-2.0 **P** 0.01-0.03 **Pb** 0.02 max **Sn** 1.0-1.5 **Other** Cu + sum of named elements 99.7 min | |
| **C19100** | High Copper Alloy | **Cu** (incl Ag) rem **Fe** 0.20 max **Ni** 0.9-1.3 **P** 0.15-0.35 **Pb** 0.10 max **Te** 0.35-0.6 **Zn** 0.50 max **Other** Cu + sum of all named elements 99.5 min | |
| **C19140** | Copper-Nickel-Lead-Phosphorus Alloy (Ansonia) | **Cu** (incl Ag) rem **Fe** 0.05 max **Ni** 0.8-1.2 **P** 0.15-0.35 **Pb** 0.40-0.8 **Sn** 0.05 max **Zn** 0.50 max **Other** Cu + sum of named elements 99.5 min | **Other:** CDA Application Data Sheet, Standard Designations for Wrought and Cast Copper and Copper Alloys as available on the Internet Web Site, http://properties.copper.org/servlet/ com.copper.servlet.CDAPropertiesSelectionServlet |
| **C19150** | Copper-Nickel-Lead-Phosphorus Alloy | **Cu** (incl Ag) rem **Fe** 0.05 max **Ni** 0.8-1.2 **P** 0.15-0.35 **Pb** 0.50-1.0 **Sn** 0.05 max Zn 0.50 max**Other** Cu + sum of named elements 99.5 min | |
| **C19160** | Copper-Nickel-Lead-Phosphorus Alloy (Brush Wellman) | **Cu** (incl Ag) rem **Fe** 0.05 max **Ni** 0.8-1.2 **P** 0.15-0.35 **Pb** 0.8-1.2 **Sn** 0.05 max **Zn** 0.50 max **Other** Cu + sum of named elements 99.5 min | **Other:** CDA Application Data Sheet, Standard Designations for Wrought and Cast Copper and Copper Alloys as available on the Internet Web Site, http://properties.copper.org/servlet/ com.copper.servlet.CDAPropertiesSelectionServlet |
| **C19200** | High Copper Alloy | **Cu** 98.7 min **Fe** 0.8-1.2 **P** 0.01-0.04 **Zn** 0.20 max **Other** Cu + sum of named elements 99.8 min | **ASME** SB-111; SB-359; SB-395 **ASTM** B111; B359; B395; B465; B698 **Other:** CDA Application Data Sheet, Standard Designations for Wrought and Cast Copper and Copper Alloys as available on the Internet Web Site, http://properties.copper.org/servlet/ com.copper.servlet.CDAPropertiesSelectionServlet **SAE** J461; J463 |
| **C19210** | Copper-Iron Alloy | **Cu** rem **Fe** 0.05-0.15 **P** 0.025-0.040 **Other** Cu + sum of named elements 99.8 min | |

The chemical compositions listed are for identification purposes and should not be used in lieu of the cross referenced specifications.

| UNIFIED NUMBER | DESCRIPTION | CHEMICAL COMPOSITION | CROSS REFERENCE SPECIFICATIONS |
|---|---|---|---|
| **C19215** | High Copper Alloy | **Cu** rem **Fe** 0.05-0.20 **P** 0.025-0.50 **Pb** 0.20 max **Zn** 1.1-3.5 **Other** Cu + sum of named elements 99.8 min | **Other:** CDA Application Data Sheet, Standard Designations for Wrought and Cast Copper and Copper Alloys as available on the Internet Web Site, http://properties.copper.org/servlet/com.copper.servlet.CDAPropertiesSelectionServlet |
| **C19220** | Copper-Iron Alloy | **B** 0.005-0.015 **Cu** rem **Fe** 0.10-0.30 **Ni** 0.10-0.25 **P** 0.03-0.07 **Sn** 0.05-0.10 **Other** Cu + sum of named elements 99.8 min | |
| * **C19250** | Copper-Iron Alloy | **B** 0.005-0.025 **Cu** rem **Fe** 0.05-0.40 **Ni** 0.05-0.40 **P** 0.03-0.10 **Sn** 0.05-0.20 | |
| **C19260** | Copper Iron Alloy | **Cu** 98.5 min **Fe** 0.40-0.8 **Mg** 0.02-0.15 **Ti** 0.20-0.40 **Other** Cu + sum of named elements 99.9 min | |
| **C19280** | Copper-Iron Alloy | **Cu** rem **Fe** 0.50-1.5 **P** 0.005-0.015 **Sn** 0.30-0.7 **Zn** 0.30-0.7 **Other** Cu + sum of named elements 99.8 min | |
| **C19400** | High Copper Alloy | **Cu** 97.0 min **Fe** 2.1-2.6 **P** 0.015-0.15 **Pb** 0.03 max **Zn** 0.05-0.20 **Other** Cu + sum of named elements 99.8 min | **ASME** SB-543 **ASTM** B465; B543; B698; B888 |
| **C19410** | Copper Iron Alloy | **Cu** rem **Fe** 1.8-2.3 **P** 0.015-0.050 **Sn** 0.6-0.9 **Other** Cu + sum of named elements 99.8 min | |
| **C19450** | Copper Iron Alloy | **Cu** rem **Fe** 1.5-3.0 **P** 0.005-0.05 **Sn** 0.8-2.5 **Other** Cu + sum of named elements 99.8 min | |
| **C19500** | High Copper Alloy | **Al** 0.02 max **Co** 0.30-1.3 **Cu** 96.0 min **Fe** 1.0-2.0 **P** 0.01-0.35 **Pb** 0.02 max **Sn** 0.10-1.0 **Zn** 0.20 max **Other** Cu + sum of named elements 99.8 min | **ASTM** B465; B888 |
| **C19520** | Copper-Iron-Tin Alloy | **Cu** 96.6 min **Fe** 0.50-1.5 **P** 0.01-0.35 **Sn** 0.50-1.5 **Other** Cu + sum of named elements 99.8 min | |
| * **C19600** | High Copper Alloy | **Cu** rem **Fe** 0.9-1.2 **P** 0.25-0.35 **Zn** 0.35 max **Other** Cu + sum of named elements 99.8 min | **ASTM** B105 |
| **C19700** | Copper Alloy | **Co** 0.05 max **Cu** rem **Fe** 0.30-1.2 **Mg** 0.01-0.20 **Mn** 0.05 max **Ni** 0.05 max **P** 0.10-0.40 **Pb** 0.05 max **Sn** 0.02 max **Zn** 0.20 max **Other** Cu + sum of named elements 99.8 min | **ASTM** B465; B888 |
| **C19710** | Copper Alloy | **Cu** (incl Ag) rem **Fe** 0.05-0.40 **Mg** 0.03-0.06 **Mn** 0.05 max **Ni** (incl Co) 0.10 max **P** 0.07-0.15 **Pb** 0.05 max **Sn** 0.20 max **Zn** 0.20 max **Other** Cu + sum of named elements 99.5 min | **Other:** CDA Application Data Sheet, Standard Designations for Wrought and Cast Copper and Copper Alloys as available on the Internet Web Site, http://properties.copper.org/servlet/com.copper.servlet.CDAPropertiesSelectionServlet |
| **C19720** | Copper Alloy | **Cu** rem **Fe** 0.05-0.50 **Pb** 0.05 max **Mg** 0.06-0.20 **Mn** 0.05 max **Ni** (incl Co) 0.10 max **P** 0.05-0.15 **Sn** 0.20 max **Zn** 0.20 max **Other** Cu + sum of named elements 99.5 min | **Other:** CDA Application Data Sheet, Standard Designations for Wrought and Cast Copper and Copper Alloys as available on the Internet Web Site, http://properties.copper.org/servlet/com.copper.servlet.CDAPropertiesSelectionServlet |
| **C19750** | Copper Alloy | **Co** 0.05 max **Cu** rem **Fe** 0.35-1.2 **Mg** 0.01-0.20 **Mn** 0.05 max **Ni** 0.05 max **P** 0.10-0.40 **Pb** 0.05 max **Sn** 0.05-0.40 **Zn** 0.20 max **Other** Cu + sum of named elements 99.8 min | |

*Boxed entries are no longer active and are retained for reference purposes only.

The chemical compositions listed are for identification purposes and should not be used in lieu of the cross referenced specifications.

| UNIFIED NUMBER | DESCRIPTION | CHEMICAL COMPOSITION | CROSS REFERENCE SPECIFICATIONS |
|---|---|---|---|
| **C19800** | Copper Alloy | **Cu** rem **Fe** 0.02-0.50 **Mg** 0.10-1.0 **P** 0.01-0.10 **Sn** 0.10-1.0 **Zn** 0.30-1.5 **Other** Cu + sum of named elements 99.8 min | **Other:** CDA Application Data Sheet, Standard Designations for Wrought and Cast Copper and Copper Alloys as available on the Internet Web Site, http://properties.copper.org/servlet/com.copper.servlet.CDAPropertiesSelectionServlet |
| **C19810** | Copper Alloy | **Cr** 0.10 max **Cu** rem **Fe** 1.5-3.0 **Mg** 0.10 max **P** 0.10 max **Ti** 0.10 max **Zn** 1.0-5.0 **Zr** 0.10 max **Other** Cu + sum of named elements 99.8 min | **Other:** CDA Application Data Sheet, Standard Designations for Wrought and Cast Copper and Copper Alloys as available on the Internet Web Site, http://properties.copper.org/servlet/com.copper.servlet.CDAPropertiesSelectionServlet |
| **C19900** | Copper-Titanium Alloy | **Cu+Ti** 99.5 min **Ti** 2.9-3.4 | |
| * **C20500** | Brass | **Cu** 97.0-98.0 **Fe** 0.05 max **Pb** 0.02 max **Zn** rem **Other** Cu + sum of named elements 99.8 min | |
| **C21000** | Gilding, 95% (Copper/Zinc) | **Cu** 94.0-96.0 **Fe** 0.05 max **Pb** 0.03 max **Zn** rem **Other** Cu + sum of named elements 99.8 min | **ASTM** B36; B134; B587; B642; B698 (Gilding-95%) **SAE** J461; J463 |
| **C22000** | Commercial Bronze, 90% | **Cu** 89.0-91.0 **Fe** 0.05 max **Pb** 0.05 max **Zn** rem **Other** Cu + sum of named elements 99.8 min | **ASTM** B36; B130; B131; B134; B135; B372; B587; B694 **AWS** C2.25 (W-CuZn-1) **MIL SPEC** MIL-W-85; MIL-C-3383; MIL-W-6712; MIL-B-18907 **SAE** J461; J463 |
| **C22600** | Jewelry Bronze, 87-1/2% | **Cu** 86.0-89.0 **Fe** 0.05 max **Pb** 0.05 max **Zn** rem **Other** Cu + sum of named elements 99.8 min | **ASTM** B36 |
| **C23000** | Red Brass, 85% | **Cu** 84.0-86.0 **Fe** 0.05 max **Pb** 0.05 max **Zn** rem **Other** Cu + sum of named elements 99.8 min | **ASME** B16.22; B16.29; SB-43; SB-111; SB-359; SB-395; SB-543 **ASTM** B36; B43; B111; B134; B135; B359; B395; B543; B569; B587; B687; B698 (Red Brass-85%); B888 **SAE** J461; J463 |
| **C23030** | Red Brass | **Cu** 83.5-85.5 **Fe** 0.05 max **Pb** 0.05 max **Si** 0.20-0.40 **Zn** rem **Other** Cu + sum of named elements 99.8 min | |
| **C23400** | Brass | **Cu** 81.0-84.0 **Fe** 0.05 max **Pb** 0.05 max **Zn** rem **Other** Cu + sum of named elements 99.8 min | |
| **C24000** | Low Brass, 80% | **Cu** 78.5-81.5 **Fe** 0.05 max **Pb** 0.05 max **Zn** rem **Other** Cu + sum of named elements 99.8 min | **ASTM** B36; B134 **SAE** J461; J463 |
| **C24080** | Low Brass | **Al** 0.10 max **Cu** 78.0-82.0 **Pb** 0.20 max **Zn** rem **Other** Cu + sum of named elements 99.8 min | |
| * **C25000** | Brass | **Cu** 74.0-76.0 **Fe** 0.05 max **Pb** 0.05 max **Zn** rem **Other** Cu + sum of named elements 99.7 min | |
| **C25600** | Brass | **Cu** 71.0-73.0 **Fe** 0.05 max **Pb** 0.05 max **Zn** rem **Other** (Cu + sum of named elements) 99.7 min | **Other:** CDA Application Data Sheet, Standard Designations for Wrought and Cast Copper and Copper Alloys as available on the Internet Web Site, http://properties.copper.org/servlet/com.copper.servlet.CDAPropertiesSelectionServlet |
| **C26000** | Cartridge Brass, 70% | **Cu** 68.5-71.5 **Fe** 0.05 max **Pb** 0.07 max **Zn** rem **Other** Cu + sum of named elements 99.7 min | **AMS** 4505; 4507; 4508 **ASTM** B19; B36; B129; B131; B135; B569; B587; B888 **SAE** J461; J463 |
| * **C26100** | Brass | **Cu** 68.5-71.5 **Fe** 0.05 max **P** 0.02-0.05 **Pb** 0.05 max **Zn** rem **Other** Cu + sum of named elements 99.7 min | **MIL SPEC** MIL-C-10375; MIL-S-22499 |
| **C26130** | Arsenic Inhibited Cartridge Brass | **As** 0.02-0.08 **Cu** 68.5-71.5 **Fe** 0.05 max **Pb** 0.05 max **Zn** rem **Other** Cu + sum of named elements 99.7 min | **ASTM** B569 |

*Boxed entries are no longer active and are retained for reference purposes only.

The chemical compositions listed are for identification purposes and should not be used in lieu of the cross referenced specifications.

| UNIFIED NUMBER | DESCRIPTION | CHEMICAL COMPOSITION | CROSS REFERENCE SPECIFICATIONS |
|---|---|---|---|
| **C26200** | Brass | **Cu** 67.0-70.0 **Fe** 0.05 max **Pb** 0.07 max **Zn** rem **Other** Cu + sum of named elements 99.7 min | |
| * **C26380** | Brass Brazing Rod | **Cu** 68.0-72.0 **Al** 0.10 max **Cr** 68.0-72.0 **Pb** 0.30 max **Zn** rem **Other** Cu + sum of named elements 99.7 min | **ASME** SFA5.8 (BCuZn-G) **AWS** A5.8 (BCuZn-G) |
| **C26800** | Yellow Brass, 66% | **Cu** 64.0-68.5 **Fe** 0.05 max **Pb** 0.15 max **Zn** rem **Other** Cu + sum of named elements 99.7 min | **ASTM** B36; B587 **AWS** C2.25 (W-CuZn-2) **MIL SPEC** MIL-W-6712 **SAE** J461; J463 |
| **C27000** | Yellow Brass, 65% | **Cu** 63.0-68.5 **Fe** 0.07 max **Pb** 0.10 max **Zn** rem **Other** Cu + sum of named elements 99.7 min | **AMS** 4710; 4712; 4713 **ASTM** B134; B135; B587; F467; F468 **SAE** J461; J463 |
| **C27200** | Brass | **Cu** 62.0-65.0 **Fe** 0.07 max **Pb** 0.07 max **Zn** rem **Other** Cu + sum of named elements 99.7 min | **ASTM** B36; B135; B587 |
| **C27400** | Yellow Brass, 63% | **Cu** 61.0-64.0 **Fe** 0.05 max **Pb** 0.10 max **Zn** rem **Other** Cu + sum of named elements 99.7 min | **ASTM** B134; B135 |
| **C28000** | Muntz Metal, 60% | **Cu** 59.0-63.0 **Fe** 0.07 max **Pb** 0.30 max **Zn** rem **Other** Cu + sum of named elements 99.7 min | **ASME** SB-111 **ASTM** B36; B111; B135 |
| * **C28200** | Brass | **Al** 0.005 max **Cu** 58.0-61.0 **Fe** 0.05 max **P** 0.12-0.22 **Pb** 0.03 max **Si** 0.005 max **Sn** 0.05 max **Zn** rem | |
| * **C28580** | Brass Brazing Rod | **Al** 0.10 max **Cu** 49.0-52.0 **Fe** 0.10 max **Pb** 0.50 max **Zn** rem **Other** Cu + sum of named elements 99.7 min | |
| * **C29800** | Brazing Alloy | **Al** 0.10 max **Cu** 49.0-52.0 **Fe** 0.10 max **Pb** 0.50 max **Zn** rem | |
| * **C31000** | Leaded Commercial Bronze (Low Lead) | **Cu** 89.0-91.0 **Fe** 0.10 max **Pb** 0.30-0.7 **Zn** rem | |
| **C31200** | Medium Leaded Commercial Bronze | **Cu** 87.5-90.5 **Fe** 0.10 max **Ni** 0.25 max **Pb** 0.7-1.2 **Zn** rem **Other** Cu + sum of named elements 99.6 min | |
| **C31400** | Leaded Commercial Bronze | **Cu** 87.5-90.5 **Fe** 0.10 max **Ni** 0.7 max **Pb** 1.3-2.5 **Zn** rem **Other** Cu + sum of named elements 99.6 min | **ASTM** B140 **MIL SPEC** MIL-V-18436 |
| **C31600** | Leaded Commercial Bronze (Nickel Bearing) | **Cu** 87.5-90.5 **Fe** 0.10 max **Ni** 0.7-1.2 **P** 0.04-0.10 **Pb** 1.3-2.5 **Zn** rem **Other** Cu + sum of named elements 99.6 min | **ASTM** B140 **MIL SPEC** MIL-V-18436 |
| **C32000** | Leaded Red Brass | **Cu** 83.5-86.5 **Fe** 0.10 max **Ni** 0.25 max **Pb** 1.5-2.2 **Zn** rem **Other** Cu + sum of named elements 99.6 min | **ASTM** B140 |
| * **C32500** | Leaded Brass | **Cu** 72.0-74.5 **Fe** 0.10 max **Pb** 2.5-3.0 **Zn** rem | |

*Boxed entries are no longer active and are retained for reference purposes only.

The chemical compositions listed are for identification purposes and should not be used in lieu of the cross referenced specifications.

| UNIFIED NUMBER | DESCRIPTION | CHEMICAL COMPOSITION | CROSS REFERENCE SPECIFICATIONS |
|---|---|---|---|
| * **C32510** | Leaded 70/30 Arsenical Brass | **As** 0.20-0.06 **Cu** 69.0-72.0 **Pb** 0.30-0.7 **Zn** rem **Other** Cu + sum of named elements 99.6 min | |
| **C33000** | Low Leaded Brass (Tube) | **Cu** 65.0-68.0 **Fe** 0.07 max **Pb** 0.25-0.7 **Zn** rem **Other** Cu + sum of named elements 99.6 min | **AMS** 4555 **ASTM** B135 **SAE** J461; J463 |
| * **C33100** | Leaded Brass | **Cu** 65.0-68.0 **Fe** 0.06 max **Pb** 0.8-1.5 **Zn** rem **Other** Cu + sum of named elements 99.6 min | **MIL SPEC** MIL-T-46072 **SAE** J461; J463 |
| **C33200** | High Leaded Brass (Tube) | **Cu** 65.0-68.0 **Fe** 0.07 max **Pb** 1.5-2.5 **Zn** rem **Other** Cu + sum of named elements 99.6 min | **AMS** 4558 **ASTM** B135 |
| **C33500** | Low Leaded Brass | **Cu** 62.0-65.0 **Fe** 0.15 max **Pb** 0.25-0.7 **Zn** rem **Other** Cu + sum of named elements 99.6 min **Note:** Fe 0.010 max for flat products | **ASTM** B121; B453 |
| * **C33530** | Leaded 65/35 Arsenical Brass | **As** 0.02-0.06 **Cu** 62.5-66.5 **Fe** 0.10 max **Pb** 0.30-0.8 **Zn** rem **Other** Cu + sum of named elements 99.6 min | |
| **C34000** | Medium Leaded Brass, 64-1/2% | **Cu** 62.0-65.0 **Fe** 0.15 max **Pb** 0.8-1.5 **Zn** rem **Other** Cu + sum of named elements 99.6 min **Note:** Fe 0.010 max for flat products | **ASTM** B121; B453 |
| **C34200** | High Leaded Brass, 64-1/2% | **Cu** 62.0-65.0 **Fe** 0.15 max **Pb** 1.5-2.5 **Zn** rem **Other** Cu + sum of named elements 99.6 min **Note:** Fe 0.010 max for flat products | **ASTM** B121 **SAE** J461; J463 |
| * **C34400** | Leaded Brass | **Cu** 62.0-66.0 **Fe** 0.10 max **Pb** 0.50-1.0 **Zn** rem | |
| **C34500** | Leaded Brass | **Cu** 62.0-65.0 **Fe** 0.15 max **Pb** 1.5-2.5 **Zn** rem **Other** Cu + sum of named elements 99.6 min | **ASTM** B453 **SAE** J461; J463 |
| * **C34700** | Leaded Brass | **Cu** 62.5-64.5 **Fe** 0.10 max **Pb** 1.0-1.8 **Zn** rem | |
| * **C34800** | Leaded Brass | **Cu** 61.5-63.5 **Fe** 0.10 max **Pb** 0.40-0.8 **Zn** rem | |
| * **C34900** | Leaded Brass | **Cu** 61.0-64.0 **Fe** 0.10 max **Pb** 0.10-0.50 **Zn** rem | |
| **C35000** | Medium Leaded Brass, 62% | **Cu** 60.0-63.0 **Fe** 0.15 max **Pb** 0.8-2.0 **Zn** rem **Other** Cu + sum of named elements 99.6 min **Note:** For rod, Cu 61.0 min; for flat products, Fe 0.10 max | **ASTM** B121; B453 **SAE** J461; J463 |

*Boxed entries are no longer active and are retained for reference purposes only.

The chemical compositions listed are for identification purposes and should not be used in lieu of the cross referenced specifications.

| UNIFIED NUMBER | DESCRIPTION | CHEMICAL COMPOSITION | CROSS REFERENCE SPECIFICATIONS |
|---|---|---|---|
| **C35300** | High Leaded Brass, 62% | **Cu** 60.0-63.0 **Fe** 0.15 max **Pb** 1.5-2.5 **Zn** rem **Other** Cu + sum of named elements 99.5 min **Note:** For rod, Cu 61.0 min; for flat products, Fe 0.10 max | **ASTM** B121; B453 |
| **C35330** | 63/37 Leaded Brass | **As** 0.02-0.25 **Cu** 60.5-64.0 **Pb** 1.5-3.5 **Zn** rem **Other** Lead content may be reduced to 1% by agreement between purchaser and supplier. Cu + sum of named elements 99.5 min **Note:** For rod, Cu 61.0 min; for flat products, Fe 0.10 max | |
| *** C35340** | Leaded Brass | **Cu** 60.0-63.0 **Fe** 0.10-0.30 **Pb** 1.5-2.5 **Zn** rem **Other** Cu + sum of named elements 99.5 min | **ASTM** B121; B453 |
| **C35600** | Leaded Brass Alloy | **Cu** 60.0-63.0 **Fe** 0.15 max **Pb** 2.0-3.0 **Zn** rem **Other** Cu + sum of named elements 99.5 min **Note:** For flat products, Fe 0.10 max | **ASTM** B121; B453 |
| **C36000** | Free Cutting Brass | **Cu** 60.0-63.0 **Fe** 0.35 max **Pb** 2.5-3.7 **Zn** rem **Other** Cu + sum of named elements 99.5 min | **AMS** 4610 **ASTM** B16 **SAE** J461; J463 **MIL SPEC** MIL-V-18436 |
| *** C36200** | Leaded Brass | **Cu** 60.0-63.0 **Fe** 0.15 max **Pb** 3.5-4.5 **Zn** rem **Other** Cu + sum of named elements 99.5 min | |
| **C36500** | Leaded Muntz Metal, Uninhibited (Cu-Zn-Pb) | **Cu** 58.0-61.0 **Fe** 0.15 max **Pb** 0.25-0.7 **Sn** 0.25 max **Zn** rem **Other** Cu + sum of named elements 99.6 min | **ASME** SB-171 **ASTM** B124; B171; B283; B432 |
| *** C36600** | Leaded Muntz Metal, Arsenical (Cu-Zn-Pb) | **As** 0.02-0.06 **Cu** 58.0-61.0 **Fe** 0.15 max **Pb** 0.25-0.7 **Sn** 0.25 max **Zn** rem **Other** Cu + sum of named elements 99.6 min | **ASME** SB-171 **ASTM** B171; B432 |
| *** C36700** | Leaded Muntz Metal, Antimonial (Cu-Zn-Pb) | **Cu** 58.0-61.0 **Fe** 0.15 max **Pb** 0.25-0.7 **Sb** 0.02-0.10 **Sn** 0.25 max **Zn** rem **Other** Cu + sum of named elements 99.6 min | **ASME** SB-171 **ASTM** B171; B432 |
| *** C36800** | Leaded Muntz Metal, Phosphorized (Cu-Zn-Pb) | **Cu** 58.0-61.0 **Fe** 0.15 max **P** 0.02-0.10 **Pb** 0.25-0.7 **Sn** 0.25 max **Zn** rem **Other** Cu + sum of named elements 99.6 min | **ASME** SB-171 **ASTM** B171; B432 |
| **C37000** | Free Cutting Muntz Metal (Cu-Zn-Pb) | **Cu** 59.0-62.0 **Fe** 0.15 max **Pb** 0.8-1.5 **Zn** rem **Other** Cu + sum of named elements 99.6 min | **ASTM** B135 |
| **C37100** | Leaded Brass | **Cu** 58.0-62.0 **Fe** 0.15 max **Pb** 0.6-1.2 **Zn** rem **Other** Cu + sum of named elements 99.6 min | |
| **C37700** | Forging Brass | **Cu** 58.0-61.0 **Fe** 0.30 max **Pb** 1.5-2.5 **Zn** rem **Other** Cu + sum of named elements 99.5 min | **AMS** 4614 **ASME** SB-283 **ASTM** B124; B283 **SAE** J461; J463 |

*Boxed entries are no longer active and are retained for reference purposes only.

The chemical compositions listed are for identification purposes and should not be used in lieu of the cross referenced specifications.

| UNIFIED NUMBER | DESCRIPTION | CHEMICAL COMPOSITION | CROSS REFERENCE SPECIFICATIONS |
|---|---|---|---|
| **C37710** | 2% Leaded 60/40 (Forging) Brass | **Cu** 56.5-60.0 **Fe** 0.30 max **Pb** 1.0-2.5 **Zn** rem **Other** Cu + sum of named elements 99.5 min | |
| * **C37800** | Leaded Brass | **Cu** 56.0-59.0 **Fe** 0.30 max **Pb** 1.5-2.5 **Zn** rem **Other** Cu + sum of named elements 99.5 min | |
| **C38000** | Leaded Brass | **Al** 0.50 max **Cu** 55.0-60.0 **Fe** 0.35 max **Pb** 1.5-2.5 **Sn** 0.30 max **Zn** rem **Other** Cu + sum of named elements 99.5 min | **ASTM** B455 |
| * **C38010** | Brass | **Al** 0.10-0.6 **Cu** 55.0-60.0 **Fe** 0.30 max **Pb** 1.5-3.0 **Zn** rem **Other** Cu + sum of named elements 99.5 min | |
| **C38500** | Architectural Bronze | **Cu** 55.0-59.0 **Fe** 0.35 max **Pb** 2.5-3.5 **Zn** rem **Other** Cu + sum of named elements 99.5 min | **ASTM** B455 |
| * **C38510** | Free Cutting Brass | **Cu** 56.0-60.0 **Pb** 2.5-4.5 **Zn** rem **Other** Cu + sum of named elements 99.5 min | |
| * **C38590** | Architectural Brass | **Cu** 56.5-60.0 **Fe** 0.35 max **Pb** 2.0-3.5 **Zn** rem **Other** Cu + sum of named elements 99.5 min | |
| * **C38600** | Copper | **Cu** 56.0-59.0 **Fe** 0.35 max **Pb** 2.5-3.5 **Sb** 0.02 max **Zn** rem | |
| **C40400** | Tin Brass | **Cu** rem **Sn** 0.35-0.7 **Zn** 2.0-3.0 **Other** Cu + sum of named elements 99.7 min | |
| **C40500** | Tin Brass | **Cu** 94.0-96.0 **Fe** 0.05 max **Pb** 0.05 max **Sn** 0.7-1.3 **Zn** rem **Other** Cu + sum of named elements 99.7 min | **ASTM** B591 |
| * **C40800** | Tin Brass | **Cu** 94.0-96.0 **Fe** 0.05 max **Pb** 0.05 max **Sn** 1.8-2.2 **Zn** rem **Other** Cu + sum of named elements 99.7 min | **ASTM** B591 |
| **C40810** | Tin Brass | **Cu** 94.5-96.5 **Fe** 0.08-0.12 **Ni** 0.11-0.20 **P** 0.028-0.040 **Pb** 0.05 max **Sn** 1.8-2.2 **Zn** rem **Other** Cu + sum of named elements 99.7 min | |
| **C40820** | Tin Brass | **Cu** 94.0 min **Ni** 0.10-0.50 **Pb** 0.02 max **P** 0.05 max **Sn** 1.0-2.5 **Zn** 0.20-2.5 **Other** Cu + sum of named elements 99.5 min | **Other:** CDA Application Data Sheet, Standard Designations for Wrought and Cast Copper and Copper Alloys as available on the Internet Web Site, http://properties.copper.org/servlet/com.copper.servlet.CDAPropertiesSelectionServlet |
| **C40850** | Tin Brass | **Cu** 94.5-96.5 **Fe** 0.05-0.20 **Ni** 0.05-0.20 **P** 0.02-0.04 **Pb** 0.05 max **Sn** 2.6-4.0 **Zn** rem **Other** Cu + sum of named elements 99.7 min | |
| **C40860** | Tin Brass | **Cu** 94.0-96.0 **Fe** 0.01-0.05 **Ni** 0.05-0.20 **P** 0.02-0.04 **Pb** 0.05 max **Sn** 1.7-2.3 **Zn** rem **Other** Cu + sum of named elements 99.5 min | |

*Boxed entries are no longer active and are retained for reference purposes only.

The chemical compositions listed are for identification purposes and should not be used in lieu of the cross referenced specifications.

| UNIFIED NUMBER | DESCRIPTION | CHEMICAL COMPOSITION | CROSS REFERENCE SPECIFICATIONS |
|---|---|---|---|
| * **C40900** | Tin Brass | **Cu** 92.0-94.0 **Fe** 0.05 max **Pb** 0.05 max **Sn** 0.50-0.8 **Zn** rem | |
| **C41000** | Tin Brass | **Cu** 91.0-93.0 **Fe** 0.05 max **Pb** 0.05 max **Sn** 2.0-2.8 **Zn** rem **Other** Cu + sum of named elements 99.7 min | |
| **C41100** | Tin Brass | **Cu** 89.0-92.0 **Fe** 0.05 max **Pb** 0.10 max **Sn** 0.30-0.7 **Zn** rem **Other** Cu + sum of named elements 99.7 min | **ASTM** B591 |
| **C41120** | Tin Brass | **Cu** 89.0-92.0 **Fe** 0.05- 0.20 **Ni** 0.05-0.20 **P** 0.02-0.05 **Pb** 0.05 max **Sn** 0.30-0.7 **Zn** rem **Other** (Cu + sum of named elements) 99.7 min | **Other:** CDA Application Data Sheet, Standard Designations for Wrought and Cast Copper and Copper Alloys as available on the Internet Web Site, http://properties.copper.org/servlet/com.copper.servlet.CDAPropertiesSelectionServlet |
| **C41300** | Tin Brass | **Cu** 89.0-93.0 **Fe** 0.05 max **Pb** 0.10 max **Sn** 0.7-1.3 **Zn** rem **Other** Cu + sum of named elements 99.7 min | **ASTM** B591 |
| **C41500** | Tin Brass | **Cu** 89.0-93.0 **Fe** 0.05 max **Pb** 0.10 max **Sn** 1.5-2.2 **Zn** rem **Other** Cu + sum of named elements 99.7 min | **ASTM** B591 |
| * **C41900** | Tin Brass | **Cu** 89.0-92.0 **Fe** 0.05 max **Pb** 0.10 max **Sn** 4.8-5.5 **Zn** rem | |
| **C42000** | Tin Brass | **Cu** 88.0-91.0 **P** 0.25 max **Sn** 1.5-2.0 **Zn** rem **Other** Cu + sum of named elements 99.7 min | |
| * **C42100** | Tin Brass | **Cu** 87.5-89.0 **Fe** 0.05 max **Mn** 0.15-0.35 **P** 0.35 max **Pb** 0.05 max **Sn** 2.2-3.0 **Zn** rem **Other** Cu + sum of named elements 99.7 min | |
| **C42200** | Tin Brass | **Cu** 86.0-89.0 **Fe** 0.05 max **P** 0.35 max **Pb** 0.05 max **Sn** 0.8-1.4 **Zn** rem **Other** Cu + sum of named elements 99.7 min | **ASTM** B591; B888 |
| **C42220** | Tin Brass | **Cu** 88.0-91.0 **Fe** 0.05-0.20 **Ni** 0.05-0.20 **P** 0.02-0.05 **Pb** 0.05 max **Sn** 0.7-1.4 **Zn** rem **Other** Cu + sum of named elements 99.7 min | **Other:** CDA Application Data Sheet, Standard Designations for Wrought and Cast Copper and Copper Alloys as available on the Internet Web Site, http://properties.copper.org/servlet/com.copper.servlet.CDAPropertiesSelectionServlet |
| **C42500** | Tin Brass | **Cu** 87.0-90.0 **Fe** 0.05 max **P** 0.35 max **Pb** 0.05 max **Sn** 1.5-3.0 **Zn** rem **Other** Cu + sum of named elements 99.7 min | **ASTM** B591; B888 |
| **C42510** | Replaced by C42220 | | |
| **C42520** | Tin Brass | **Cu** 88.0-91.0 **Fe** 0.05-0.20 **Ni** 0.05-0.20 **P** 0.02-0.04 **Pb** 0.05 max **Sn** 1.5-3.0 **Zn** rem **Other** Cu + sum of named elements 99.7 min | **ASTM** B888 |
| **C42600** | Tin Brass | **Cu** (incl Ag) 87.0-90.0 **Fe** 0.05-0.20 **Ni** (incl Co) 0.05-0.20 **P** 0.02-0.05 **Pb** 0.05 max **Sn** 2.5-4.0 **Zn** Rem **Other** Cu + sum of named elements 99.7 min | **Other:** CDA Application Data Sheet, Standard Designations for Wrought and Cast Copper and Copper Alloys as available on the Internet Web Site, http://properties.copper.org/servlet/com.copper.servlet.CDAPropertiesSelectionServlet |
| **C43000** | Tin Brass | **Cu** 84.0-87.0 **Fe** 0.05 max **Pb** 0.10 max **Sn** 1.7-2.7 **Zn** rem **Other** Cu + sum of named elements 99.7 min | **ASTM** B591 |

*Boxed entries are no longer active and are retained for reference purposes only.

The chemical compositions listed are for identification purposes and should not be used in lieu of the cross referenced specifications.

| UNIFIED NUMBER | DESCRIPTION | CHEMICAL COMPOSITION | CROSS REFERENCE SPECIFICATIONS |
|---|---|---|---|
| * C43200 | Tin Brass | **Cu** 85.0-88.0 **Fe** 0.05 max **P** 0.35 max **Pb** 0.05 max **Sn** 0.40-0.6 **Zn** rem **Other** Cu + sum of named elements 99.7 min | |
| C43400 | Tin Brass | **Cu** 84.0-87.0 **Fe** 0.05 max **Pb** 0.05 max **Sn** 0.40-1.0 **Zn** rem **Other** Cu + sum of named elements 99.7 min | **ASTM** B591 |
| C43500 | Tin Brass | **Cu** 79.0-83.0 **Fe** 0.05 max **Pb** 0.10 max **Sn** 0.6-1.2 **Zn** rem **Other** Cu + sum of named elements 99.7 min | |
| C43600 | Tin Brass | **Cu** 80.0-83.0 **Fe** 0.05 max **Pb** 0.05 max **Sn** 0.20-0.50 **Zn** rem **Other** Cu + sum of named elements 99.7 min | |
| * C43800 | Tin Brass | **Cu** 79.0-82.0 **Fe** 0.05 max **Pb** 0.05 max **Sn** 1.0-1.5 **Zn** rem | |
| C44250 | Copper-Zinc-Tin Alloy | **Cu** 73.0-76.0 **Fe** 0.20 max **Ni** 0.20 max **P** 0.10 max **Pb** 0.07 max **Sn** 0.50-1.5 **Zn** rem **Other** Cu + sum of named elements 99.6% min | **Other:** CDA Application Data Sheet, Standard Designations for Wrought and Cast Copper and Copper Alloys as available on the Web Site, http://properties.copper.org/servlet/ com.copper.servlet.CDApropertiesSelectionServlet |
| C44300 | Admiralty, Arsenical (Cu-Zn-Sn) | **As** 0.02-0.06 **Cu** 70.0-73.0 **Fe** 0.06 max **Pb** 0.07 max **Sn** 0.8-1.2 **Zn** rem **Other** For tubular products, Sn 0.9 min **Other** Cu + sum of named elements 99.6 min | **ASME** SB-111; SB-171; SB-359; SB-395; SB-543 **ASTM** B111; B135; B171; B359; B395; B432; B543 |
| C44400 | Admiralty, Antimonial (Cu-Zn-Sn) | **Cu** 70.0-73.0 **Fe** 0.06 max **Pb** 0.07 max **Sb** 0.02-0.10 **Sn** 0.8-1.2 **Zn** rem **Other** For tubular products, Sn 0.9 min **Other** Cu + sum of named elements 99.6 min | **ASME** SB-111; SB-171; SB-359; SB-395; SB-543 **ASTM** B111; B171; B359; B395; B432; B543 |
| C44500 | Admiralty, Phosphorized (Cu-Zn-Sn) | **Cu** 70.0-73.0 **Fe** 0.06 max **P** 0.02-0.10 **Pb** 0.07 max **Sn** 0.8-1.2 **Zn** rem **Other** For tubular products, Sn 0.9 min **Other** Cu + sum of named elements 99.6 min | **ASME** SB-111; SB-171; SB-359; SB-395; SB-543 **ASTM** B111; B171; B359; B395; B432; B543 |
| * C45450 | Copper Alloy | **Al** 0.20-0.40 **Cu** 65.0-66.0 **P** 0.10-0.30 **Sn** 0.10-0.30 **Zn** rem **Other** Cu + sum of named elements 99.8 min | |
| C46200 | Naval Brass, 63-1/2% | **Cu** 62.0-65.0 **Fe** 0.10 max **Pb** 0.20 max **Sn** 0.50-1.0 **Zn** rem **Other** Cu + sum of named elements 99.6 min | **ASTM** B21; F68; F467 **FED** QQ-B-639 |
| * C46210 | Copper Alloy | **Al** 0.03 max **Cu** 61.0-64.0 **Pb** 0.05 max **Si** 0.50 max **Sn** 1.0 max **Zn** rem **Other** Cu + sum of named elements 99.6 min | |
| C46400 | Naval Brass, Uninhibited | **Cu** 59.0-62.0 **Fe** 0.10 max **Pb** 0.20 max **Sn** 0.50-1.0 **Zn** rem **Other** Cu + sum of named elements 99.6 min | **AMS** 4611; 4612 **ASME** SB-171 **ASTM** B21; B124; B171; B283; B432; F467; F468 **FED** QQ-B-639 **MIL SPEC** MIL-W-6712 **SAE** J461; J463 |
| * C46420 | Naval Brass | **Cu** 61.0-63.5 **Fe** 0.20 max **Pb** 0.20 max **Sn** 1.0-1.4 **Zn** rem **Other** Cu + sum of named elements 99.6 min | |

*Boxed entries are no longer active and are retained for reference purposes only.

The chemical compositions listed are for identification purposes and should not be used in lieu of the cross referenced specifications.

| UNIFIED NUMBER | DESCRIPTION | CHEMICAL COMPOSITION | CROSS REFERENCE SPECIFICATIONS |
|---|---|---|---|
| **C46500** | Naval Brass, Arsenical | **As** 0.02-0.06 **Cu** 59.0-62.0 **Fe** 0.10 max **Pb** 0.20 max **Sn** 0.50-1.0 **Zn** rem **Other** Cu + sum of named elements 99.6 min | **ASME** SB-171 **ASTM** B171; B432 **SAE** J461; J463 |
| *** C46600** | Naval Brass, Antimonial | **Cu** 59.0-62.0 **Fe** 0.10 max **Pb** 0.20 max **Sb** 0.02-0.10 **Sn** 0.50-1.0 **Zn** rem **Other** Cu + sum of named elements 99.6 min | **ASME** SB-171 **ASTM** B171; B432 **SAE** J461; J463 |
| *** C46700** | Naval Brass, Phosphorized | **Cu** 59.0-62.0 **Fe** 0.10 max **P** 0.02-0.10 **Pb** 0.20 max **Sn** 0.50-1.0 **Zn** rem **Other** Cu + sum of named elements 99.6 min | **ASME** SB-171 **ASTM** B171; B432 **SAE** J461; J463 |
| **C47000** | Naval Brass, Welding and Brazing Rod | **Al** 0.01 max **Cu** 57.0-61.0 **Pb** 0.05 max **Sn** 0.25-1.0 **Zn** rem **Other** Cu + sum of named elements 99.6 min | **ASME** SFA5.8 (RBCuZn-A) **AWS** A5.8 (RBCuZn-A); C2.25 (W-CuZn-3) **MIL SPEC** MIL-B-7883 (RBCuZn-A) |
| *** C47200** | Brazing Alloy (Cu-Zn-Sn) | **Cu** 49.0-52.0 **Fe** 0.10 max **Pb** 0.50 max **Sn** 3.0-4.0 **Zn** rem | |
| *** C47600** | Tin Brass, Leaded | **Cu** 86.0-88.0 **Fe** 0.05 max **Mn** 0.05-0.15 **P** 0.03-0.07 **Pb** 1.8-2.2 **Sn** 1.8-2.2 **Zn** rem **Other** Cu + sum of named elements 99.6 min | |
| **C47940** | Brass Alloy | **Cu** 63.0-66.0 **Fe** 0.10-1.0 **Ni** (incl Co) 0.10-0.50 **Pb** 1.0-2.0 **Sn** 1.2-2.0 **Zn** rem | **ASTM** B21 |
| **C48200** | Naval Brass, Medium Leaded | **Cu** 59.0-62.0 **Fe** 0.10 max **Pb** 0.40-1.0 **Sn** 0.50-1.0 **Zn** rem **Other** Cu + sum of named elements 99.6 min | **ASTM** B21; B124; B283 **FED** QQ-B-639 |
| **C48500** | Naval Brass, High Leaded | **Cu** 59.0-62.0 **Fe** 0.10 max **Pb** 1.3-2.2 **Sn** 0.50-1.0 **Zn** rem **Other** Cu + sum of named elements 99.6 min | **ASTM** B21; B124; B283 **FED** QQ-B-639 |
| *** C48510** | Leaded Naval Brass, Arsenical | **As** 0.02-0.25 **Cu** 59.0-62.0 **Pb** 1.0-2.5 **Sn** 0.7-1.5 **Zn** rem | |
| **C48600** | Naval Brass, Arsenical, Leaded | **As** 0.02-0.25 **Cu** 59.0-62.0 **Pb** 1.0-2.5 **Sn** 0.8-1.5 **Zn** rem **Other** Cu + sum of named elements 99.6 min | |
| **C48650** | Replaced by C48600 | | |
| *** C49080** | Brass Brazing Filler Metal | **Al** 0.10 max **Cu** 49.0-52.0 **Pb** 0.50 max **Sn** 3.0-4.0 **Zn** rem **Other** Cu + sum of named elements 99.6 min | **AWS** A5.8 (BCuZn-F) |
| **C50100** | Phosphor Bronze | **Cu** rem **Fe** 0.05 max **P** 0.01-0.05 **Pb** 0.05 max **Sn** 0.50-0.8 **Other** Cu + sum of named elements 99.5 min | |
| **C50200** | Phosphor Bronze | **Cu** rem **Fe** 0.10 max **P** 0.04 max **Pb** 0.05 max **Sn** 1.0-1.5 **Other** Cu + sum of named elements 99.5 min | |

*Boxed entries are no longer active and are retained for reference purposes only.

The chemical compositions listed are for identification purposes and should not be used in lieu of the cross referenced specifications.

| UNIFIED NUMBER | DESCRIPTION | CHEMICAL COMPOSITION | CROSS REFERENCE SPECIFICATIONS |
|---|---|---|---|
| **C50500** | Phosphor Bronze, 1.25% E | **Cu** rem **Fe** 0.10 max **P** 0.03-0.35 **Pb** 0.05 max **Sn** 1.0-1.7 **Zn** 0.30 max **Other** Cu + sum of named elements 99.5 min | **ASTM** B9; B105 |
| **C50510** | Phosphor Bronze | **Cu** rem **Ni** 0.15-0.40 **P** 0.02-0.07 **Sn** 1.0-1.5 **Zn** 0.10-0.25 **Other** (Cu + sum of named elements) 99.7 min | **Other:** CDA Application Data Sheet, Standard Designations for Wrought and Cast Copper and Copper Alloys as available on the Internet Web Site, http://properties.copper.org/servlet/com.copper.servlet.CDAPropertiesSelectionServlet |
| **C50580** | Phosphor Bronze | **Cu** rem **Fe** 0.05- 0.20 **Ni** 0.05-0.20 **P** 0.02-0.10 **Pb** 0.05 max **Sn** 1.0-1.7 **Zn** 0.30 max **Other** (Cu + sum of named elements) 99.5 min | **Other:** CDA Application Data Sheet, Standard Designations for Wrought and Cast Copper and Copper Alloys as available on the Internet Web Site, http://properties.copper.org/servlet/com.copper.servlet.CDAPropertiesSelectionServlet |
| **C50590** | Phosphor Bronze | **Cu** 97.0 min **Fe** 0.05-0.40 **P** 0.02-0.15 **Pb** 0.02 max **Sn** 0.5-1.5 **Zn** 0.50 max **Other** Cu + sum of named elements 99.5 min | **Other:** CDA Application Data Sheet, Standard Designations for Wrought and Cast Copper and Copper Alloys as available on the Internet Web Site, http://properties.copper.org/servlet/com.copper.servlet.CDAPropertiesSelectionServlet |
| **C50700** | Phosphor Bronze | **Cu** rem **Fe** 0.10 max **P** 0.30 max **Pb** 0.05 max **Sn** 1.5-2.0 **Other** Cu + sum of named elements 99.5 min | |
| **C50705** | Phosphor Bronze | **Cu** 96.5 min **Fe** 0.10-0.40 **P** 0.04-0.15 **Pb** 0.02 max **Sn** 1.5-2.0 **Zn** 0.50 max **Other** Cu + sum of named elements 99.5 min | **Other:** CDA Application Data Sheet, Standard Designations for Wrought and Cast Copper and Copper Alloys as available on the Internet Web Site, http://properties.copper.org/servlet/com.copper.servlet.CDAPropertiesSelectionServlet |
| **C50710** | Phosphor Bronze Alloy (MF 202) | **Cu** rem **Ni** 0.10-0.40 **P** 0.15 max **Sn** 1.7-2.3 **Other** Cu + sum of named elements 99.5 min | |
| **C50715** | Phosphor Bronze Alloy | **Cu** rem **Fe** 0.05-0.15 **P** 0.025-0.04 **Pb** 0.02 max **Sn** 1.7-2.3 **Other** Cu + sum of named elements 99.5 min | |
| **C50725** | Copper-Zinc-Tin-Phosphorus Bronze (Kobe Steel CAC 52) | **Cu** 94.0 min **Fe** 0.05-0.20 **P** 0.02-0.06 **Pb** 0.02 max **Sn** 1.5-2.5 **Zn** 1.5-3.0 **Other** Cu + sum of named elements 99.5 min | **Other:** CDA Application Data Sheet, Standard Designations for Wrought and Cast Copper and Copper Alloys as available on the Internet Web Site, http://properties.copper.org/servlet/com.copper.servlet.CDAPropertiesSelectionServlet |
| **C50780** | Phosphor Bronze | **Cu** rem **Fe** 0.05-0.20 **Ni** 0.05-0.20 **P** 0.020-0.10 **Pb** 0.05 max **Sn** 1.7-2.3 **Zn** 0.30 max **Other** Cu + sum of named elements 99.5 min | |
| * **C50800** | Phosphor Bronze | **Cu** rem **Fe** 0.10 max **P** 0.01-0.07 **Pb** 0.05 max **Sn** 2.6-3.4 **Other** Cu + sum of named elements 99.5 min | |
| **C50900** | Phosphor Bronze | **Cu** rem **Fe** 0.10 max **P** 0.03-0.30 **Pb** 0.05 max **Sn** 2.5-3.8 **Zn** 0.30 max **Other** Cu + sum of named elements 99.5 min | |
| **C51000** | Phosphor Bronze, 5% A | **Cu** rem **Fe** 0.10 max **P** 0.03-0.35 **Pb** 0.05 max **Sn** 4.2-5.8 **Zn** 0.30 max **Other** Cu + sum of named elements 99.5 min | **AMS** 4510; 4625; 4720 **ASTM** B100; B103; B105; B139; B159; B888; F467 (510); F468 (510) **MIL SPEC** MIL-T-3595; MIL-W-6712 **SAE** J461; J463 |
| **C51000** | Phosphor Bronze | **Cu** rem **Fe** 0.05-0.20 **Ni** 0.05-0.20 **P** 0.02-0.10 **Pb** 0.05 max **Sn** 4.8-5.8 **Zn** 0.30 max **Other** Cu + sum of named elements 99.5 min | |
| **C51100** | Phosphor Bronze | **Cu** rem **Fe** 0.10 max **P** 0.03-0.35 **Pb** 0.05 max **Sn** 3.5-4.9 **Zn** 0.30 max **Other** Cu + sum of named elements 99.5 min | **ASTM** B100; B103; B888 **SAE** J461 |
| **C51180** | Phosphor Bronze | **Cu** rem **Fe** 0.05-0.20 **Ni** 0.11-0.20 **P** 0.02-0.10 **Pb** 0.05 max **Sn** 3.5-4.9 **Zn** 0.30 max **Other** Cu + sum of named elements 99.5 min | |

*Boxed entries are no longer active and are retained for reference purposes only.

The chemical compositions listed are for identification purposes and should not be used in lieu of the cross referenced specifications.

| UNIFIED NUMBER | DESCRIPTION | CHEMICAL COMPOSITION | CROSS REFERENCE SPECIFICATIONS |
|---|---|---|---|
| **C51190** | Phosphor Bronze | **Co** 0.15 max **Cu** Rem **Fe** 0.05-0.15 **P** 0.025-0.045 **Pb** 0.02 max **Sn** 3.0-6.5 **Other** Cu + sum of named elements 99.5 min | **Other:** CDA Application Data Sheet, Standard Designations for Wrought and Cast Copper and Copper Alloys as available on the Internet Web Site, http://properties.copper.org/servlet/com.copper.servlet.CDAPropertiesSelectionServlet |
| **C51800** | Phosphor Bronze | **Al** 0.01 max **Cu** rem **P** 0.10-0.35 **Pb** 0.02 max **Sn** 4.0-6.0 **Other** Cu + sum of named elements 99.5 min | **ASME** SFA5.7 (ERCuSn-A); SFA5.21 (ERCuSn-A) **AWS** A5.7 (ERCuSn-A); A5.21 (ERCuSn-A); C2.25 (W-CuSn) |
| **C51900** | Phosphor Bronze | **Cu** rem **Fe** 0.10 max **P** 0.03-0.35 **Pb** 0.05 max **Sn** 5.0-7.0 **Zn** 0.30 max **Other** Cu + sum of named elements 99.5 min | **ASTM** B103 |
| **C51980** | Phosphor Bronze | **Cu** rem **Fe** 0.05-0.20 **Ni** 0.05-0.20 **P** 0.02-0.10 **Pb** 0.05 max **Sn** 5.5-7.0 **Zn** 0.30 max **Other** Cu + sum of named elements 99.5 min | |
| **C52100** | Phosphor Bronze, 8% | **Cu** rem **Fe** 0.10 max **P** 0.03-0.35 **Pb** 0.05 max **Sn** 7.0-9.0 **Zn** 0.20 max **Other** Cu + sum of named elements 99.5 min | **ASME** SFA5.7 (ERCuSn-C) **ASTM** B103; B139; B159; B888 **AWS** A5.7 (ERCuSn-C) **MIL SPEC** MIL-E-23765/3 (MIL-CuSn-C) **SAE** J461; J463 |
| **C52180** | Phosphor Bronze | **Cu** rem **Fe** 0.05-0.20 **Ni** 0.05-0.20 **P** 0.02-0.10 **Pb** 0.05 max **Sn** 7.0-9.0 **Zn** 0.30 max **Other** Cu + sum of named elements 99.5 min | |
| **C52400** | Phosphor Bronze, 10% D | **Cu** rem **Fe** 0.10 max **P** 0.03-0.35 **Pb** 0.05 max **Sn** 9.0-11.0 **Zn** 0.20 max **Other** Cu + sum of named elements 99.5 min | **ASME** SFA5.21 (ERCuSn-D) **ASTM** B103; B139; B159 **AWS** A5.21 (ERCuSn-D) **SAE** J461; J463 |
| **C52480** | Phosphor Bronze | **Cu** rem **Fe** 0.05- 0.20 **Ni** 0.05-0.20 **P** 0.02-0.10 **Pb** 0.05 max **Sn** 9.0-11.0 **Zn** 0.30 max **Other** (Cu + sum of named elements) 99.5 min | **Other:** CDA Application Data Sheet, Standard Designations for Wrought and Cast Copper and Copper Alloys as available on the Internet Web Site, http://properties.copper.org/servlet/com.copper.servlet.CDAPropertiesSelectionServlet |
| * **C52600** | Phosphor Bronze | **Cu** 99.5 min **Fe** 0.10 max **Mn** 1.0-2.0 **P** 0.03-0.35 **Pb** 0.05 max **Sn** 2.2-3.3 **Zn** 0.20 max **Other** Mn+P+Sn included in Cu | |
| * **C52900** | Phosphor Bronze | **Cu** 99.5 min **Fe** 0.10 max **Mn** 1.0-2.0 **P** 0.03-0.35 **Pb** 0.05 max **Sn** 7.0-9.0 **Zn** 0.20 max **Other** Mn+P+Sn included in Cu | |
| * **C53200** | Phosphor Bronze B | **Cu** rem **Fe** 0.10 max **P** 0.03-0.35 **Pb** 2.5-4.0 **Sn** 4.0-5.5 **Zn** 0.20 max **Other** Cu + sum of named elements 99.5 min | |
| **C53400** | Phosphor Bronze B-1 | **Cu** rem **Fe** 0.10 max **P** 0.03-0.35 **Pb** 0.8-1.2 **Sn** 3.5-5.8 **Zn** 0.30 max **Other** Cu + sum of named elements 99.5 min | **ASTM** B103; B139 |
| **C54400** | Phosphor Bronze B-2 | **Cu** rem **Fe** 0.10 max **P** 0.01-0.50 **Pb** 3.5-4.5 **Sn** 3.5-4.5 **Zn** 1.5-4.5 **Other** Cu + sum of named elements 99.5 min | **AMS** 4520 **ASTM** B103; B139 **SAE** J461; J463 |
| * **C54600** | Phosphor Bronze B-2 (P 0.50 max) | **Cu** 99.5 min **Fe** 0.10 max **P** 0.50 max **Pb** 3.5-4.5 **Sn** 3.5-4.5 **Zn** 1.5-4.5 **Other** P+Pb+Sn+Zn included in Cu | **MIL SPEC** MIL-B-13501 |

*Boxed entries are no longer active and are retained for reference purposes only.

The chemical compositions listed are for identification purposes and should not be used in lieu of the cross referenced specifications.

| UNIFIED NUMBER | DESCRIPTION | CHEMICAL COMPOSITION | CROSS REFERENCE SPECIFICATIONS |
|---|---|---|---|
| * **C54800** | Leaded Phosphor Bronze | **Cu** 99.5 min **Fe** 0.10 max **P** 0.03-0.35 **Pb** 4.0-6.0 **Sn** 4.0-6.0 **Zn** 0.30 max **Other** P+Pb+Sn+Zn included in Cu | |
| **C55180** | Copper-Phosphorus Brazing Filler Metal | **Cu** rem **P** 4.8-5.2 **Other** Cu + P = 99.85 min | **ASME** SFA5.8 (BCuP-1) **AWS** A5.8 (BCuP-1) |
| **C55181** | Copper-Phosphorus Brazing Filler Metal | **Cu** rem **P** 7.0-7.5 **Other** Cu + P = 99.85 min | **ASME** SFA5.8 (BCuP-2) **AWS** A5.8 (BCuP-2) |
| **C55280** | Cu-Ag-P Brazing Filler Metal | **Ag** 1.8-2.2 **Cu** rem **P** 6.8-7.2 **Other** Cu + all named elements = 99.85 min | **ASME** SFA5.8 (BCuP-6) **AWS** A5.8 (BCuP-6) |
| **C55281** | Cu-Ag-P Brazing Filler Metal | **Ag** 4.8-5.2 **Cu** rem **P** 5.8-6.2 **Other** Cu + all named elements = 99.85 min | **ASME** SFA5.8 (BCuP-3) **AWS** A5.8 (BCuP-3) |
| **C55282** | Cu-Ag-P Brazing Filler Metal | **Ag** 4.8-5.2 **Cu** rem **P** 6.5-7.0 **Other** Cu + all named elements = 99.85 min | **ASME** SFA5.8 (BCuP-7) **AWS** A5.8 (BCuP-7) |
| **C55283** | Cu-Ag-P Brazing Filler Metal | **Ag** 5.8-6.2 **Cu** rem **P** 7.0-7.5 **Other** Cu + all named elements = 99.85 min | **ASME** SFA5.8 (BCuP-4) **AWS** A5.8 (BCuP-4) |
| **C55284** | Cu-Ag-P Brazing Filler Metal | **Ag** 14.5-15.5 **Cu** rem **P** 4.8-5.2 **Other** Cu + all named elements = 99.85 min | **ASME** SFA5.8 (BCuP-5) **AWS** A5.8 (BCuP-5) **FED** QQ-B-654 (BCuP-5) |
| **C55285** | Cu-Ag-P Brazing Filler Metal | **Ag** 17.2-18.0 **Cu** rem **P** 6.0-6.7 **Other** Cu + all names elements 99.5 min | **ASME** SFA5.8 (BCuP-8) **AWS** A5.8 (BCuP-8) |
| **C56000** | Non-Ferrous Thermal Spray Wire (W-CuZnAg) | **Ag** 29-31 **Cu** (incl Ag) rem **Zn** 30-34 **Other** Cu + sum of named elements = 99.5% | **AWS** C2.25 (W-CuZnAg) **Other:** CDA Application Data Sheet, Standard Designations for Wrought and Cast Copper and Copper Alloys as available on the Internet Web Site, http://properties.copper.org/servlet/com.copper.servlet.CDAPropertiesSelectionServlet |
| * **C60600** | Aluminum Bronze | **Cu** (incl Ag) rem **Al** 4.0-7.0 **Fe** 0.50 max **Other** Cu + sum of named elements 99.5 min | **ASTM** B150 **FED** QQ-C-450 |
| * **C60700** | Aluminum Bronze | **Cu** (incl Ag) rem **Al** 2.3-2.9 **Pb** 0.01 max **Sn** 1.7-2.0 **Other** Cu + sum of named elements 99.5 min | |
| **C60800** | Aluminum Bronze | **Cu** (incl Ag) rem **Al** 5.0-6.5 **As** 0.2-0.35 **Fe** 0.10 max **Pb** 0.10 max **Other** Cu + sum of named elements 99.5 min | **ASME** SB-111; SB-359; SB-395 **ASTM** B111; B359; B395 **SAE** J461, J463 |
| **C61000** | Aluminum Bronze | **Cu** (incl Ag) rem **Al** 6.0-8.5 **Fe** 0.50 max **Pb** 0.02 max **Si** 0.10 max **Zn** 0.20 max **Other** Cu + sum of named elements 99.5 min | **ASME** SB-169; SFA5.7 (ERCuAl-Al) **AWS** A5.7 (ERCuAl-Al) **FED** QQ-C-450 |
| **C61300** | Aluminum Bronze | **Cu** (incl Ag) rem **Al** 6.0-7.5 **Fe** 2.0-3.0 **Mn** 0.20 max **Ni** (incl Co) 0.15 max **P** 0.015 max **Pb** 0.01 max **Si** 0.10 max **Sn** 0.20-0.50 **Zn** 0.10 max **Other** Cu + sum of named elements 99.8 min **Note:** When the product is for subsequent welding applications and is so specified, the Cr, Cd, Zr, and Zn limits shall be 0.05 max each | **ASTM** B111; B150; B169; B171; B315; B432; B608; F467 (613); F468 (613) **FED** QQ-C-450 **SAE** J463 |
| **C61400** | Aluminum Bronze D | **Cu** (incl Ag) rem **Al** 6.0-8.0 **Fe** 1.5-3.5 **Mn** 1.0 max **P** 0.015 max **Pb** 0.01 max **Zn** 0.20 max **Other** Cu + sum of named elements 99.5 min | **ASME** SB-150; SB-169; SB-171 **ASTM** B100; B111; B150; B169; B171; B315; B432; B608; F467; F468 **FED** QQ-C-450 **SAE** J461; J463 |
| **C61470** | Aluminum Bronze Welding Electrode replaced by W60614 | | |

*Boxed entries are no longer active and are retained for reference purposes only.

The chemical compositions listed are for identification purposes and should not be used in lieu of the cross referenced specifications.

| UNIFIED NUMBER | DESCRIPTION | CHEMICAL COMPOSITION | CROSS REFERENCE SPECIFICATIONS |
|---|---|---|---|
| **C61500** | Aluminum Bronze | **Cu** (incl Ag) rem **Al** 7.7-8.3 **Ni** 1.8-2.2 **Pb** 0.015 max **Other** Cu + sum of named elements 99.5 min | |
| **C61550** | Aluminum Bronze | **Cu** (incl Ag) rem **Al** 5.5-6.5 **Cu** rem **Fe** 0.20 max **Mn** 1.0 max **Ni** (incl Co) 1.5-2.5 **Pb** 0.05 max **Sn** 0.05 max **Zn** 0.8 max **Other** Cu + all named elements = 99.5 min | |
| **C61800** | Aluminum Bronze | **Cu** (incl Ag) rem **Al** 8.5-11.0 **Fe** 0.50-1.5 **Pb** 0.02 max **Si** 0.10 max **Zn** 0.02 max **Other** Cu + sum of named elements 99.5 min | **ASME** SFA5.7 (ERCuAl-A2); SFA5.21 (ERCuAl-A2) **AWS** A5.7 (ERCuAl-A2); A5.21 (ERCuAl-A2); C2.25 (W-CuAl) **MIL SPEC** MIL-W-6712; MIL-E-23765/3 (MIL-CuAl-A2) **SAE** J461, J463 |
| *** C61810** | Aluminum Bronze (Combined with C61800) | | |
| **C61900** | Aluminum Bronze | **Cu** (incl Ag) rem **Al** 8.5-10.0 **Fe** 3.0-4.5 **Pb** 0.02 max **Sn** 0.6 max **Zn** 0.8 max **Other** Cu + sum of named elements 99.5 min | **ASTM** B124; B150; B283 |
| **C62200** | Aluminum Bronze | **Cu** (incl Ag) rem **Al** 11.0-12.0 **Fe** 3.0-4.2 **Pb** 0.02 max **Si** 0.10 max **Zn** 0.02 max **Other** Cu + sum of named elements 99.5 min | |
| **C62300** | Aluminum Bronze | **Cu** (incl Ag) rem **Al** 8.5-10.0 **Fe** 2.0-4.0 **Mn** 0.50 max **Ni** (incl Co) 1.0 max **Si** 0.25 max **Sn** 0.6 max **Other** Cu + sum of named elements 99.5 min | **AMS** 4635 **ASME** SB-150 **ASTM** B124; B150); B283 **MIL SPEC** MIL-B-16166 **SAE** J461; J463 |
| **C62400** | Aluminum Bronze | **Cu** (incl Ag) rem **Al** 10.0-11.5 **Fe** 2.0-4.5 **Mn** 0.30 max **Si** 0.25 max **Sn** 0.20 max **Other** Cu + sum of named elements 99.5 min | **ASME** SFA5.7 (ERCuAl-A3); SFA5.21 (ERCuAl-A3) **ASTM** B150 **AWS** A5.7 (ERCuAl-A3); A5.21 (ERCuAl-A3) **SAE** J461; J463 |
| **C62500** | Aluminum Bronze | **Cu** (incl Ag) rem **Al** 12.5-13.5 **Fe** 3.5-5.0 **Mn** 2.0 max **Other** Cu + sum of named elements 99.5 min | |
| **C62580** | Aluminum Bronze Welding Filler Metal | **Cu** (incl Ag) rem **Al** 12.0-13.0 **Fe** 3.0-5.0 **Pb** 0.02 max **Si** 0.04 max **Zn** 0.02 max **Other** Cu + sum of named elements 99.5 | **ASME** SFA5.21 (ERCuAl-C) **AWS** A5.21 (ERCuAl-C) |
| **C62581** | Aluminum Bronze Welding Filler Metal | **Cu** (incl Ag) rem **Al** 13.0-14.0 **Fe** 3.0-5.0 **Pb** 0.02 max **Si** 0.04 max **Zn** 0.02 max **Other** Cu + sum of named elements 99.5 | **ASME** SFA5.21 (ERCuAl-D) **AWS** A5.21 (ERCuAl-D) |
| **C62582** | Aluminum Bronze Welding Filler Metal | **Cu** (incl Ag) rem **Al** 14.0-15.0 **Fe** 3.0-5.0 **Pb** 0.02 max **Si** 0.04 max **Zn** 0.02 max **Other** Cu + sum of named elements 99.5 | **ASME** SFA5.21 (ERCuAl-E) **AWS** A5.21 (ERCuAl-E) |
| *** C62730** | Aluminum Bronze | **Cu** (incl Ag) rem **Al** 8.5-11.0 **Fe** 4.0-6.0 **Mg** 0.05 max **Mn** 0.50 max **Ni** (incl Co) 4.0-6.0 **Pb** 0.05 max **Si** 0.10 max **Sn** 0.10 max **Zn** 0.40 max **Other** Cu + sum of named elements 99.5 min | |
| **C63000** | Aluminum Bronze | **Cu** (incl Ag) rem **Al** 9.0-11.0 **Fe** 2.0-4.0 **Mn** 1.5 max **Ni** (incl Co) 4.0-5.5 **Si** 0.25 max **Sn** 0.20 max **Zn** 0.30 max **Other** Cu + sum of named elements 99.5 min | **AMS** 4640 **ASME** SB-150; SB-171 **ASTM** B124; B150; B171; B283; F467; F468 **FED** QQ-C-450 **MIL SPEC** MIL-B-16166 **SAE** J461; J463 |

*Boxed entries are no longer active and are retained for reference purposes only.

The chemical compositions listed are for identification purposes and should not be used in lieu of the cross referenced specifications.

| UNIFIED NUMBER | DESCRIPTION | CHEMICAL COMPOSITION | CROSS REFERENCE SPECIFICATIONS |
|---|---|---|---|
| **C63010** | Aluminum Bronze | **Al** 9.7-10.9 **Cu** (incl Ag) 78.0 min **Fe** 2.0-3.5 **Mn** 1.5 max **Ni** (incl Co) 4.5-5.5 **Sn** 0.20 max **Zn** 0.30 max **Other** Cu + sum of named elements 99.8 min | |
| **C63020** | Copper Alloy | **Al** 10.5-11.5 **Cr** 0.05 max **Co** 0.20 max **Cu** (incl Ag) 74.5 min **Fe** 4.0-5.5 **Mn** 1.5 max **Ni** (incl Co) 4.2-6.0 **Pb** 0.03 max **Si** 0.15 max **Sn** 0.25 max **Zn** 0.30 max **Other** Cu + sum of named elements 99.5 min | **ASTM** B124; B150; B171; B283 |
| **C63200** | Aluminum Bronze | **Cu** (incl Ag) rem **Al** 8.7-9.5 **Fe** 3.5-4.3 **Mn** 1.2-2.0 **Ni** (incl Co) 4.0-4.8 **Pb** 0.02 max **Si** 0.10 max **Other** Cu + sum of named elements 99.5 min **Note:** Fe content shall not exceed Ni content | **ASTM** B124; B150; B171; B283 |
| * **C63230** | Aluminum Bronze | **Al** 8.5-9.5 **Cu** (incl Ag) 75.9-84.5 **Fe** 3.0-5.0 **Mn** 3.5 max **Ni** (incl Co) 4.0-5.5 **Pb** 0.02 max **Si** 0.10 max **Other** Cu + sum of named elements 99.5 min **Note:** Fe content shall not exceed Ni content | |
| **C63280** | Nickel Aluminum Bronze Welding Filler Metal | **Cu** (incl Ag) rem **Al** 8.5-9.5 **Fe** 3.0-5.0 **Mn** 0.6-3.5 **Ni** 4.0-5.5 **Pb** 0.02 max **Si** 0.10 max **Other** Cu + sum of named elements 99.5 min **Note:** Fe content shall not exceed Ni content | **ASME** SFA5.7 (ERCuNiAl) **AWS** A5.7 (ERCuNiAl) **MIL SPEC** MIL-E-23765/3 (MIL-CuNiAl) |
| * **C63300** | Copper Alloy | **Al** 5.0-7.5 **Cu** (incl Ag) rem **Fe** 2.0-6.0 **Mn** 11.0-13.0 **Ni** (incl Co) 1.0-2.5 **Pb** 0.02 max **Si** 1.5 max | |
| **C63380** | Manganese Aluminum Bronze Welding Filler Metal | **Cu** (incl Ag) rem **Al** 7.0-8.5 **Fe** 2.0-4.0 **Mn** 11.0-14.0 **Ni** 1.5-3.0 **Pb** 0.02 max **Si** 0.10 max **Zn** 0.15 max **Other** Cu + sum of named elements 99.5 min **Note:** Fe content shall not exceed Ni content | **ASME** SFA5.7 (ERCuMnNiAl) **AWS** A5.7 (ERCuMnNiAl) **MIL SPEC** MIL-E-23765/3 (MIL-CuMnNiAl) |
| **C63400** | Aluminum Bronze | **Cu** (incl Ag) rem **Al** 2.6-3.2 **As** 0.15 max **Fe** 0.15 max **Ni** (incl Co) 0.15 max **Pb** 0.05 max **Si** 0.25-0.45 **Sn** 0.20 max **Zn** 0.50 max **Other** Cu + sum of named elements 99.5 min | |
| **C63600** | Aluminum Bronze | **Cu** (incl Ag) rem **Al** 3.0-4.0 **As** 0.15 max **Fe** 0.15 max **Ni** (incl Co) 0.15 max **Pb** 0.05 max **Si** 0.7-1.3 **Sn** 0.20 max **Zn** 0.50 max **Other** Cu + sum of named elements 99.5 min | |
| **C63800** | Aluminum Bronze | **Cu** (incl Ag) rem **Al** 2.5-3.1 **Co** 0.25-0.55 **Fe** 0.20 max **Mn** 0.10 max **Ni** 0.20 max **Pb** 0.05 max **Si** 1.5-2.1 **Zn** 0.8 max **Other** Cu + sum of named elements 99.5 min **Note:** Ni does not include Co | **ASTM** B422; B888 |
| * **C64110** | Aluminum Bronze, Free Machining | **Cu** (incl Ag) rem **Al** 8.0-11.0 **Mn** 0.50 max **Pb** 1.0-2.0 **Other** Cu + sum of named elements 99.5 min | |

*Boxed entries are no longer active and are retained for reference purposes only.

The chemical compositions listed are for identification purposes and should not be used in lieu of the cross referenced specifications.

| UNIFIED NUMBER | DESCRIPTION | CHEMICAL COMPOSITION | CROSS REFERENCE SPECIFICATIONS |
|---|---|---|---|
| **C64200** | Aluminum Bronze | **Cu** (incl Ag) rem **Al** 6.3-7.6 **As** 0.15 max **Fe** 0.30 max **Mn** 0.10 max **Ni** (incl Co) 0.25 max **Pb** 0.05 max **Si** 1.5-2.2 **Sn** 0.20 max **Zn** 0.50 max **Other** Cu + sum of named elements 99.5 min | **ASME** SB-150; SB-283 **ASTM** B124; B150; B283; F467; F468 **AMS** 4631 **SAE** J461; J463 |
| **C64210** | Aluminum Silicon Bronze | **Cu** (incl Ag) rem **Al** 6.3-7.0 **As** 0.15 max **Fe** 0.30 max **Mn** 0.10 max **Ni** (incl Co) 0.25 max **Pb** 0.05 max **Si** 1.5-2.0 **Sn** 0.20 max **Zn** 0.50 max **Other** Cu + sum of named elements 99.5 min | **ASTM** B124; B150; B283 |
| * **C64250** | Aluminum Silicon Bronze | **Cu** (incl Ag) rem **Al** 5.5-7.5 **Fe** 1.0 max **Mn** 0.50 max **Si** 1.5-3.0 **Other** Cu + sum of named elements 99.5 min | |
| * **C64400** | Aluminum Silicon Bronze | **Cu** (incl Ag) rem **Al** 3.5-4.5 **Cu** rem **Fe** 0.05 max **Ni** (incl Co) 4.2-5.0 **Pb** 0.03 max **Si** 0.8-1.3 **Sn** 0.10 max **Zn** 0.20 max **Other** Cu + sum of named elements 99.5 min | **ASTM** B422 |
| **C64700** | Silicon Bronze | **Cu** (incl Ag) rem **Fe** 0.10 max **Ni** (incl Co) 1.6-2.2 **Pb** 0.10 max **Si** 0.40-0.8 **Zn** 0.50 max **Other** Cu + sum of named elements 99.5 min | **ASTM** B411; B412 |
| **C64710** | Copper-Nickel-Silicon Alloy | **Cu** (incl Ag) 95.0 min **Mn** 0.10 max **Ni** (incl Co) 2.9-3.5 **Si** 0.50-0.9 **Zn** 0.20-0.50 **Other** Cu + sum of named elements 99.5 min | |
| * **C64720** | Copper Silicon Alloy | **Cu** rem **Mg** 0.01-0.03 **Ni** 1.6-2.2 **Si** 0.35-0.6 **Zn** 0.10-0.40 **Other** Cu + sum of named elements 99.5 min | |
| **C64725** | Copper-Silicon Alloys (Max 25) | **Cr** 0.20 max **Cu** (incl Ag) 95.0 min **Fe** 0.25 max **Mg** 0.20 max **Ni** (incl Co) 1.3-2.7 **Pb** 0.01 max **Si** 0.20-0.8 **Sn** 0.20-0.8 **Zn** 0.50-1.5 **Ca** 0.01 max **Other** Cu + sum of named elements 99.5 min | |
| **C64730** | Copper-Nickel-Silicon Alloy (KLF 125) | **Cu** 93.5 min **Mn** 0.10 max **Ni** 2.9-3.5 **Si** 0.50-0.9 **Sn** 1.0-1.5 **Zn** 0.20-0.50 **Other** Cu + sum of named elements 99.5 min | |
| **C64740** | Copper-Silicon Alloys (Tamac 15) | **Cu** (incl Ag) 95.0 min **Mg** 0.05 max **Ni** (incl Co) 1.0-2.0 **Pb** 0.01 max **Si** 0.05-0.50 **Sn** 1.5-2.5 **Zn** 0.20-1.0 **Ca** 0.01 max **Other** Cu + sum of named elements 99.5 min | |
| **C64750** | Copper Silicon Alloy | **Cu** (incl Ag) rem **Fe** 1.0 max **Mg** 0.10 max **Ni** (incl Co) 1.0-3.0 **P** 0.10 max **Si** 0.10-0.7 **Sn** 0.05-0.8 **Zn** 1.0 max **Zr** 0.10 max **Other** Cu + sum of named elements 99.2% min | **Other:** CDA Application Data Sheet, Standard Designations for Wrought and Cast Copper and Copper Alloys as available on the Web Site, http://properties.copper.org/servlet/com.copper.servlet.CDApropertiesSelectionServlet |
| **C64760** | Silicon Bronze | **Cu** (incl Ag) 93.5 min **Fe** 0.10-0.40 **Ni** 0.40-2.5 **Pb** 0.02 max **Si** 0.05-0.6 **Sn** 0.30 max **Zn** 0.20-2.5 **Other** Cu + sum of named elements 99.5 min | **Other:** CDA Application Data Sheet, Standard Designations for Wrought and Cast Copper and Copper Alloys as available on the Internet Web Site, http://properties.copper.org/servlet/com.copper.servlet.CDAPropertiesSelectionServlet |

*Boxed entries are no longer active and are retained for reference purposes only.

The chemical compositions listed are for identification purposes and should not be used in lieu of the cross referenced specifications.

| UNIFIED NUMBER | DESCRIPTION | CHEMICAL COMPOSITION | CROSS REFERENCE SPECIFICATIONS |
|---|---|---|---|
| C64780 | Silicon Bronze | **Cu** (incl Ag) 90.0 min **Cr** 0.01 max **Mg** 0.01 max **Mn** 0.01-1.0 **Ni** 1.0-3.5 **Pb** 0.02 max **Si** 0.20-0.9 **Ti** 0.01 max **Sn** 0.10-2.0 **Zn** 0.20-2.5 **Zr** 0.01 max **Other** Cu + sum of named elements 99.5 min | **Other:** CDA Application Data Sheet, Standard Designations for Wrought and Cast Copper and Copper Alloys as available on the Internet Web Site, http://properties.copper.org/servlet/com.copper.servlet.CDAPropertiesSelectionServlet |
| C64900 | Silicon Bronze | **Al** 0.10 max **Cu** (incl Ag) rem **Fe** 0.10 max **Ni** (incl Co) 0.10 max **Pb** 0.05 max **Si** 0.8-1.2 **Sn** 1.2-1.6 **Zn** 0.20 max **Other** Cu + sum of named elements 99.5 min | |
| C65100 | Low Silicon Bronze B | **Cu** (incl Ag) rem **Fe** 0.8 max **Mn** 0.7 max **Pb** 0.05 max **Si** 0.8-2.0 **Zn** 1.5 max **Other** Cu + sum of named elements 99.5 min | **ASME** SB-98; SB-315 **ASTM** B96; B98; B99; B105; B315; B432; F467; F468 **MIL SPEC** MIL-V-17547 |
| * C65300 | Silicon Bronze | **Cu** (incl Ag) rem **Fe** 0.8 max **Pb** 0.05 max **Si** 2.0-2.6 **Other** Cu + sum of named elements 99.5 min | |
| C65400 | Copper Silicon | **Cr** 0.01-0.12 **Cu** (incl Ag) rem **Pb** 0.05 max **Si** 2.7-3.4 **Sn** 1.2-1.9 **Zn** 0.50 max **Other** Cu + sum of named elements 99.5 min | **ASTM** B96; B888 |
| C65500 | High Silicon Bronze A | **Cu** (incl Ag) rem **Fe** 0.8 max **Mn** 0.50-1.3 **Ni** (incl Co) 0.6 max **Pb** 0.05 max **Si** 2.8-3.8 **Zn** 1.5 max **Other** Cu + sum of named elements 99.5 min | **AMS** 4615 **ASME** SB-96; SB-98; SB-315 **ASTM** B96; B98; B99; B100; B124; B283; B315; B432; F467; F468 **SAE** J461; J463 |
| C65600 | Silicon Bronze | **Al** 0.01 max **Cu** (incl Ag) rem **Fe** 0.50 max **Mn** 1.5 max **Pb** 0.02 max **Si** 2.8-4.0 **Sn** 1.5 max **Zn** 1.5 max **Other** Cu + sum of named elements 99.5 min | **ASME** SFA5.7 (ERCuSi-A); SFA5.21 (ERCuSi-A) **AWS** A5.7 (ERCuSi-A); A5.21 (ERCuSi-A) **MIL SPEC** MIL-E-23765/3 (MIL-CuSi) |
| * C65620 | Copper Silicon Alloy | **Cu** (incl Ag) 90.0 min **Fe** 1.0-2.0 **Mn** 1.0 max **P** 0.10 max **Si** 2.4-4.0 **Zn** 1.5-4.0 **Other** Cu + sum of named elements 99.5 min | |
| * C65800 | Silicon Bronze | **Cu** (incl Ag) rem **Fe** 0.25 max **Mn** 0.50-1.3 **Ni** (incl Co) 0.6 max **Pb** 0.05 max **Si** 2.8-3.8 **Other** Cu + sum of named elements 99.5 min | **ASTM** B96; B315 |
| C66100 | Silicon Bronze | **Cu** (incl Ag) rem **Fe** 0.25 max **Mn** 1.5 max **Pb** 0.20-0.8 **Si** 2.8-3.5 **Zn** 1.5 max **Other** Cu + sum of named elements 99.5 min | **ASME** SB-98 **ASTM** B98; F467; F468 |
| C66200 | Copper Zinc | **Cu** 86.6-91.0 **Fe** 0.05 max **Ni** (incl Co) 0.30-1.0 **P** 0.05-0.20 **Pb** 0.05 max **Sn** 0.20-0.7 **Zn** rem **Other** Cu + sum of named elements 99.5 min | **Other:** CDA Application Data Sheet, Standard Designations for Wrought and Cast Copper and Copper Alloys as available on the Internet Web Site, http://properties.copper.org/servlet/com.copper.servlet.CDAPropertiesSelectionServlet |
| C66300 | Misc. Copper-Zinc Alloy | **Co** 0.20 max **Cu** (incl Ag) 84.5-87.5 **P** 0.35 max **Pb** 0.05 max **Sn** 1.5-3.0 **Zn** rem **Other** (Fe+Co) 1.4-2.4, Cu + sum of named elements 99.5 | |
| C66400 | Copper Alloy | **Co** 0.30-0.7 **Cu** (incl Ag) rem **Fe** 1.3-1.7 **Mn** 0.05 max **Ni** 0.05 max **P** 0.02 max **Pb** 0.015 max **Si** 0.05 max **Sn** 0.05 max **Zn** 11.0-12.0 **Other** Fe+Co 1.8-2.3, Cu + sum of named elements 99.5 min | **ASTM** B694 |

*Boxed entries are no longer active and are retained for reference purposes only.

The chemical compositions listed are for identification purposes and should not be used in lieu of the cross referenced specifications.

| UNIFIED NUMBER | DESCRIPTION | CHEMICAL COMPOSITION | CROSS REFERENCE SPECIFICATIONS |
|---|---|---|---|
| **C66410** | Misc. Copper Zinc Alloy | **Al** 0.05 max **As** 0.05 max **Cu** (incl Ag) rem **Fe** 1.8-2.3 **Mn** 0.05 max **Ni** 0.05 max **P** 0.02 max **Pb** 0.015 max **Si** 0.05 max **Sn** 0.05 max **Zn** 11.0-12.0 **Other** Cu + sum of named elements 99.5 min | **ASTM** B694 |
| **C66420** | Copper-Zinc Alloy (Outokumpu Copper) | **Cu** (incl Ag) rem **Fe** 0.50-1.5 **Zn** 12.7-17.0 **Other** Cu + sum of named elements 99.5 min | **Other:** CDA Application Data Sheet, Standard Designations for Wrought and Cast Copper and Copper Alloys as available on the Internet Web Site, http://properties.copper.org/servlet/com.copper.servlet.CDAPropertiesSelectionServlet |
| **C66700** | Manganese Brass | **Cu** (incl Ag) 68.5-71.5 **Fe** 0.10 max **Mn** 0.8-1.5 **Pb** 0.07 max **Zn** rem **Other** Cu + sum of named elements 99.5 min | |
| **C66800** | Manganese Bronze | **Cu** (incl Ag) 60.0-63.0 **Al** 0.25 max **Fe** 0.35 max **Mn** 2.0-3.5 **Ni** (incl Co) 0.25 max **Pb** 0.50 max **Si** 0.50-1.5 **Sn** 0.30 max **Zn** rem **Other** Cu + sum of named elements 99.5 min | |
| **C66900** | Manganese Bronze | **Cu** (incl Ag) 62.5-64.5 **Fe** 0.25 max **Mn** 11.5-12.5 **Pb** 0.05 max **Zn** rem **Other** (Cu + sum of named elements) 99.8 min | |
| **C66950** | Manganese Bronze | **Al** 1.0-1.5 **Cu** (incl Ag) rem **Fe** 0.50 max **Mn** 14.0-15.0 **Pb** 0.01 max **Zn** 14.0-15.0 **Other** (Cu + sum of named elements) 99.5 min | **Other:** CDA Application Data Sheet, Standard Designations for Wrought and Cast Copper and Copper Alloys as available on the Internet Web Site, http://properties.copper.org/servlet/com.copper.servlet.CDAPropertiesSelectionServlet |
| **C67000** | Manganese Bronze B | **Al** 3.0-6.0 **Cu** (incl Ag) 63.0-68.0 **Fe** 2.0-4.0 **Mn** 2.5-5.0 **Pb** 0.20 max **Sn** 0.50 max **Zn** rem **Other** Cu + sum of named elements 99.5 min | **ASTM** B138 **SAE** J461; J463 |
| * **C67130** | Leaded Manganese Bronze | **Al** 0.10-1.0 **Cu** (incl Ag) 56.0-59.0 **Mn** 0.50-1.5 **Ni** (incl Co) 0.50-1.5 **Pb** 0.50-1.5 **Sn** 0.50-1.5 **Zn** rem **Other** Cu + sum of named elements 99.5 min | |
| **C67300** | Manganese Bronze | **Al** 0.25 max **Cu** (incl Ag) 58.0-63.0 **Fe** 0.50 max **Mn** 2.0-3.5 **Ni** (incl Co) 0.25 max **Pb** 0.40-3.0 **Si** 0.50-1.5 **Sn** 0.30 max **Zn** rem **Other** Cu + sum of named elements 99.5 min | **SAE** J461; J463 |
| **C67400** | Manganese Bronze | **Al** 0.50-2.0 **Cu** (incl Ag) 57.0-60.0 **Fe** 0.35 max **Mn** 2.0-3.5 **Ni** (incl Co) 0.25 max **Pb** 0.50 max **Si** 0.50-1.5 **Sn** 0.30 max **Zn** rem **Other** Cu + sum of named elements 99.5 min | **SAE** J461; J463 |
| * **C67410** | Manganese Bronze | **Al** 1.3-2.3 **Cu** (incl Ag) 55.5-59.0 **Fe** 1.0 max **Mn** 1.0-2.4 **Ni** (incl Co) 2.0 max **Pb** 0.8 max **Si** 0.7-1.3 **Sn** 0.50 max **Zn** rem **Other** Cu + sum of named elements 99.5 min | |
| **C67420** | Manganese Bronze | **Al** 1.0-2.0 **Cu** (incl Ag) 57.0-58.5 **Fe** 0.15-0.55 **Mn** 1.5-2.5 **Ni** (incl Co) 0.25 max **Pb** 0.25-0.8 **Si** 0.25-0.7 **Sn** 0.35 max **Zn** rem **Other** (Cu + sum of named elements) 99.5 min | **Other:** CDA Application Data Sheet, Standard Designations for Wrought and Cast Copper and Copper Alloys as available on the Internet Web Site, http://properties.copper.org/servlet/com.copper.servlet.CDAPropertiesSelectionServlet |
| **C67500** | Manganese Bronze A | **Al** 0.25 max **Cu** (incl Ag) 57.0-60.0 **Fe** 0.8-2.0 **Mn** 0.05-0.50 **Pb** 0.20 max **Sn** 0.50-1.5 **Zn** rem **Other** Cu + sum of named elements 99.5 min | **ASTM** B124; B138; B283; F467; F468 **SAE** J461; J463 |

*Boxed entries are no longer active and are retained for reference purposes only.

The chemical compositions listed are for identification purposes and should not be used in lieu of the cross referenced specifications.

| UNIFIED NUMBER | DESCRIPTION | CHEMICAL COMPOSITION | CROSS REFERENCE SPECIFICATIONS |
|---|---|---|---|
| **C67600** | Leaded Manganese Bronze | **Cu** (incl Ag) 57.0-60.0 **Fe** 0.40-1.3 **Mn** 0.05-0.50 **Pb** 0.50-1.0 **Sn** 0.50-1.5 **Zn** rem **Other** Cu + sum of named elements 99.5 min | **ASTM** B283 |
| * **C67620** | Leaded Manganese Bronze | **Cu** (incl Ag) 55.0-57.0 **Fe** 0.50-1.2 **Mn** 1.0-2.0 **Pb** 0.50-1.5 **Zn** rem **Other** Cu + sum of named elements 99.5 min | |
| * **C67700** | Copper Alloy | **As** 0.40-0.8 **Cu** (incl Ag) 55.5-58.0 **Fe** 0.7-1.5 **Mn** 0.05-0.30 **Ni** (incl Co) 1.5-2.3 **Pb** 0.50-1.0 **Zn** rem **Other** Cu + sum of named elements 99.5 min | |
| * **C67800** | Copper Alloy | **Al** 0.50-1.5 **Cu** (incl Ag) 56.0-59.0 **Fe** 0.7-1.5 **Mn** 0.20-0.6 **Pb** 0.30 max **Sn** 0.20 max **Zn** rem **Other** Cu + sum of named elements 99.5 min | |
| * **C67810** | Manganese Bronze | **Al** 0.40-1.6 **Cu** (incl Ag) 56.5-59.5 **Fe** 1.0 max **Mn** 0.40-1.8 **Ni** (incl Co) 1.5 max **Pb** 1.0 max **Si** 0.6 max **Sn** 0.50 max **Zn** rem **Other** Cu + sum of named elements 99.5 min | |
| * **C67820** | Manganese Bronze | **Al** 0.30-1.3 **Cu** (incl Ag) 56.5-59.5 **Fe** 0.50-1.2 **Mn** 0.30-2.0 **Pb** 0.10 max **Sn** 0.30-1.0 **Zn** rem **Other** Cu + sum of named elements 99.5 min | |
| **C68000** | Bronze, Low Fuming (Nickel) | **Al** 0.01 max **Cu** (incl Ag) 56.0-60.0 **Fe** 0.25-1.25 **Mn** 0.01-0.50 **Ni** (incl Co) 0.20-0.8 **Pb** 0.05 max **Si** 0.04-0.15 **Sn** 0.75-1.10 **Zn** rem **Other** Cu + sum of named elements 99.5 min | **ASME** SFA5.8 (RBCuZn-B) **AWS** A5.8 (RBCuZn-B) |
| **C68100** | Bronze, Low Fuming | **Al** 0.01 max **Cu** (incl Ag) 56.0-60.0 **Fe** 0.25-1.25 **Mn** 0.01-0.50 **Pb** 0.05 max **Si** 0.04-0.15 **Sn** 0.75-1.10 **Zn** rem **Other** Cu + sum of named elements 99.5 min | **ASME** SFA5.8 (RBCuZn-C) **AWS** A5.8 (RBCuZn-C) |
| * **C68200** | Copper Alloy | **Cu** (incl Ag) 58.0-60.0 **Fe** 0.6-1.0 **Mn** 0.6-1.0 **Si** 0.07-0.15 **Zn** rem | |
| * **C68600** | Leaded Manganese Bronze | **Al** 0.30-1.5 **Cu** (incl Ag) 56.0-60.0 **Fe** 0.50-1.2 **Mn** 0.30-2.0 **Pb** 0.50-1.5 **Sn** 0.20-1.0 **Zn** rem **Other** Cu + sum of named elements 99.5 min | |
| **C68700** | Aluminum Brass, Arsenical | **Al** 1.8-2.5 **As** 0.02-0.06 **Cu** (incl Ag) 76.0-79.0 **Fe** 0.06 max **Pb** 0.07 max **Zn** rem **Other** Cu + sum of named elements 99.5 min | **ASME** SB-111; SB-359; SB-395; SB-543 **ASTM** B111; B359; B395; B543 |
| **C68800** | Copper Alloy | **Al** 3.0-3.8 **Co** 0.25-0.55 **Cu** (incl Ag) rem **Fe** 0.20 max **Pb** 0.05 max **Zn** 21.3-24.1 **Other** Cu + sum of named elements 99.5 min | **ASTM** B592; B888 |

*Boxed entries are no longer active and are retained for reference purposes only.

The chemical compositions listed are for identification purposes and should not be used in lieu of the cross referenced specifications.

| UNIFIED NUMBER | DESCRIPTION | CHEMICAL COMPOSITION | CROSS REFERENCE SPECIFICATIONS |
|---|---|---|---|
| * **C69000** | Miscellaneous Copper-Zinc | **Al** 3.0-3.8 **Cu** (incl Ag) 72.0-74.5 **Fe** 0.05 max **Ni** (incl Co) 0.50-0.8 **Pb** 0.025 max **Zn** rem **Other** Al+Zn 25.1-27.1; Cu + sum of named elements 99.5 min | **ASTM** B592 |
| **C69050** | Copper Alloy | **Al** 3.0-4.0 **Cu** 70.0-75.0 **Ni** 0.50-1.5 **Si** 0.10-0.6 **Zn** rem **Zr** 0.01-0.20 **Other** Cu + sum of named elements 99.5 min | |
| **C69100** | Copper Alloy | **Al** 0.7-1.2 **Cu** (incl Ag) 81.0-84.0 **Fe** 0.25 max **Mn** 0.10 max **Ni** (incl Co) 0.8-1.4 **Pb** 0.05 max **Si** 0.8-1.3 **Sn** 0.10 max **Zn** rem **Other** Cu + sum of named elements 99.5 min | **ASTM** B706 |
| **C69400** | Sillcon Red Brass | **Cu** (incl Ag) 80.0-83.0 **Fe** 0.20 max **Pb** 0.30 max **Si** 3.5-4.5 **Zn** rem **Other** Cu + sum of named elements 99.5 min | **ASTM** B371 |
| **C69430** | Silicon Red Brass | **As** 0.03-0.06 **Cu** (incl Ag) 80.0-83.0 **Fe** 0.20 max **Pb** 0.30 max **Si** 3.5-4.5 **Zn** rem **Other** Cu + sum of named elements 99.5 min | **ASTM** B371 |
| * **C69440** | Silicon Red Brass | **Cu** (incl Ag) 80.0-83.0 **Fe** 0.20 max **Pb** 0.30 max **Sb** 0.03-0.06 **Si** 3.5-4.5 **Zn** rem **Other** Cu + sum of named elements 99.5 min | **ASTM** B371 |
| * **C69450** | Silicon Red Brass | **Cu** (incl Ag) 80.0-83.0 **Fe** 0.20 max **Mn** 0.40 max **P** 0.03-0.06 **Pb** 0.30 max **Si** 3.5-4.5 **Zn** rem **Other** Cu + sum of named elements 99.5 min | **ASTM** B371 |
| **C69700** | Leaded Silicon Brass | **Cu** (incl Ag) 75.0-80.0 **Fe** 0.20 max **Mn** 0.40 max **Pb** 0.50-1.5 **Si** 2.5-3.5 **Zn** rem **Other** Cu + sum of named elements 99.5 min | **ASTM** B371 |
| **C69710** | Silicon Brass | **As** 0.03-0.06 **Cu** (incl Ag) 75.0-80.0 **Fe** 0.20 max **Mn** 0.40 max **Pb** 0.50-1.5 **Si** 2.5-3.5 **Zn** Rem **Other** Cu + sum of named elements 99.5 min | **ASTM** B371 |
| * **C69720** | Silicon Brass | **Cu** (incl Ag) 75.0-80.0 **Fe** 0.20 max **Mn** 0.40 max **Pb** 0.50-1.5 **Sb** 0.03-0.06 **Si** 2.5-3.5 **Zn** rem **Other** Cu + sum of named elements 99.5 min | **ASTM** B371 |
| * **C69730** | Silicon Brass | **Cu** (incl Ag) 75.0-80.0 **Fe** 0.20 max **Mn** 0.40 max **P** 0.03-0.06 **Pb** 0.50-1.5 **Si** 2.5-3.5 **Zn** rem **Other** Cu + sum of named elements 99.5 min | **ASTM** B371 |
| * **C69800** | Copper Zinc | **Cu** (incl Ag) 66.0-70.0 **Fe** 0.4 max **Ni** (incl Co) 0.50 max **Pb** 0.8 max **Si** 0.7-1.3 **Zn** rem **Other** Cu + sum of named elements 99.5 min | |

*Boxed entries are no longer active and are retained for reference purposes only.

The chemical compositions listed are for identification purposes and should not be used in lieu of the cross referenced specifications.

| UNIFIED NUMBER | DESCRIPTION | CHEMICAL COMPOSITION | CROSS REFERENCE SPECIFICATIONS |
|---|---|---|---|
| * C69900 | Copper Manganese Alloy | **Ag** 0.05 max **Al** 1.4-2.3 **As** 0.01 max **C** 0.05 max **Cd** 0.05 max **Co** 0.20 max **Cu** (incl Ag) rem **Fe** 0.10 max **Mn** 40.0-48.0 **Ni** (incl Co) 0.10 max **Pb** 0.02 max **Zn** 0.14 max **Other** Cu + sum of named elements 99.5 min | |
| * C69910 | Brazing Alloy (Cu-Mn-Fe) | **Al** 0.25-0.8 **Cu** (incl Ag) rem **Fe** 1.0-1.4 **Mn** 28.0-32.0 **Pb** 0.01 max **Zn** 3.0-5.0 **Other** Cu + sum of named elements 99.5 min | |
| * C69950 | Copper Alloy (1600N) | **Cu** (incl Ag) 51.0-54.0 **Mn** 36.0-40.0 **Ni** (incl Co) 8.5-10.5 **Other** Cu + sum of named elements 99.5 min | **AMS** 4764 |
| C70100 | Copper-Nickel Alloy | **Cu** (incl Ag) rem **Fe** 0.05 max **Mn** 0.50 max **Ni** (incl Co) 3.0-4.0 **Zn** 0.25 max **Other** Cu + sum of named elements 99.5 min | |
| C70200 | Copper-Nickel | **Cu** (incl Ag) rem **Fe** 0.10 max **Mn** 0.40 max **Ni** (incl Co) 2.0-3.0 **Pb** 0.05 max **Other** Cu + sum of named elements 99.5 min | **MIL SPEC** MIL-B-18907 |
| C70230 | Copper Nickel | **(Ag+B)** 0.10 max **Cu** (incl Ag) rem **Ni** (incl Co) 2.2-3.2 **Si** 0.40-0.8 **Sn** 0.10-0.50 **Zn** 0.50-2.0 **Other** (Cu + sum of named elements) 99.5 min | **Other:** CDA Application Data Sheet, Standard Designations for Wrought and Cast Copper and Copper Alloys as available on the Internet Web Site, http://properties.copper.org/servlet/com.copper.servlet.CDAPropertiesSelectionServlet |
| C70250 | Copper Nickel | **Cu** (incl Ag) rem **Fe** 0.20 max **Mg** 0.05-0.30 **Mn** 0.10 max **Ni** (incl Co) 2.2-4.2 **Pb** 0.05 max **Si** 0.25-1.2 **Zn** 1.0 max **Other** Cu + sum of named elements 99.5 min | **ASTM** B422; B888 |
| C70260 | Copper Nickel | **Cu** (incl Ag) rem **Ni** (incl Co) 1.0-3.0 **P** 0.010 max **Si** 0.20-0.7 **Other** (Cu + sum of named elements) 99.5 min | **ASTM** B422; B888 **Other:** CDA Application Data Sheet, Standard Designations for Wrought and Cast Copper and Copper Alloys as available on the Internet Web Site, http://properties.copper.org/servlet/com.copper.servlet.CDAPropertiesSelectionServlet |
| C70270 | Copper Nickel | **Cu** (incl Ag) rem **Fe** 0.28-1.0 **Pb** 0.05 max **Mn** 0.15 max **Ni** (incl Co) 1.0-3.0 **Si** 0.20-1.0 **Sn** 0.10-1.0 **Zn** 1.0 max **Other** Cu + sum of named elements 99.5 min | **Other:** CDA Application Data Sheet, Standard Designations for Wrought and Cast Copper and Copper Alloys as available on the Internet Web Site, http://properties.copper.org/servlet/com.copper.servlet.CDAPropertiesSelectionServlet |
| C70280 | Copper Nickel | **Cu** (incl Ag) rem **Fe** 0.015 max **Pb** 0.02 max **Ni** (incl Co) 1.3-1.7 **P** 0.020-0.040 **Si** 0.22-0.30 **Sn** 1.0-1.5 **Zn** 0.30 max **Other** Cu + sum of named elements 99.5 min | **Other:** CDA Application Data Sheet, Standard Designations for Wrought and Cast Copper and Copper Alloys as available on the Internet Web Site, http://properties.copper.org/servlet/com.copper.servlet.CDAPropertiesSelectionServlet |
| C70290 | Copper Nickel | **Cu** (incl Ag) rem **Fe** 0.015 max **Pb** 0.02 max **Ni** (incl Co) 1.3-1.7 **P** 0.020-0.040 **Si** 0.22-0.30 **Sn** 2.1-2.7 **Zn** 0.3 max **Other** Cu + sum of named elements 99.5 min | **Other:** CDA Application Data Sheet, Standard Designations for Wrought and Cast Copper and Copper Alloys as available on the Internet Web Site, http://properties.copper.org/servlet/com.copper.servlet.CDAPropertiesSelectionServlet |
| * C70300 | Copper-Nickel Alloy | **Cu** 99.5 min **Fe** 0.05 max **Mn** 0.50 max **Ni** 4.7-5.7 **Other** Fe+Mn+Ni included in Cu | |

*Boxed entries are no longer active and are retained for reference purposes only.

The chemical compositions listed are for identification purposes and should not be used in lieu of the cross referenced specifications.

| UNIFIED NUMBER | DESCRIPTION | CHEMICAL COMPOSITION | CROSS REFERENCE SPECIFICATIONS |
|---|---|---|---|
| * **C70320** | Copper-Nickel | **Al** 0.20-1.2 **Cr** 0.18-0.50 **Cu** (incl Ag) rem **Ni** (incl Co) 2.5-5.0 **Si** 0.20-1.2 **Other** Cu + sum of named elements 99.5 min | |
| **C70400** | Copper-Nickel, 5% | **Cu** (incl Ag) rem **Fe** 1.3-1.7 **Mn** 0.30-0.8 **Ni** (incl Co) 4.8-6.2 **Pb** 0.05 max **Zn** 1.0 max **Other** Cu + sum of named elements 99.5 min | **ASME** SB-111; SB-359; SB-466; SB-543 **ASTM** B111; B359; B395; B466; B543 |
| * **C70440** | 95/5 Copper-Nickel | **C** 0.05 max **Cu** (incl Ag) rem **Fe** 1.0-1.8 **Mn** 0.50-1.5 **Ni** (incl Co) 4.5-6.0 **Pb** 0.05 max **S** 0.05 max **Si** 0.35-0.45 **Sn** 0.10 max **Zn** 1.0 max **Other** Cu + sum of named elements 99.5 min | |
| **C70500** | Copper-Nickel, 7% | **Cu** (incl Ag) rem **Fe** 0.10 max **Mn** 0.15 max **Ni** (incl Co) 5.8-7.8 **Pb** 0.05 max **Zn** 0.20 max **Other** Cu + sum of named elements 99.5 min | |
| **C70600** | Copper-Nickel, 10% | **Cu** (incl Ag) rem **Fe** 1.0-1.8 **Mn** 1.0 max **Ni** (incl Co) 9.0-11.0 **Pb** 0.05 max **Zn** 1.0 max **Other** Cu + sum of named elements 99.5 min | **ASME** SB-111; SB-171; SB-359; SB-395; SB-466; SB-467; SB-543 **ASTM** B111; B122; B151; B171; B359; B395; B432; B466; B467; B543; B552; B608 **MIL SPEC** MIL-T-15005; MIL-C-15726; MIL-C-24679; MIL-T-16420; MIL-T-22214 **SAE** J461; J463 |
| **C70610** | 90/10 Copper-Nickel | **C** 0.05 max **Cu** (incl Ag) rem **Fe** 1.0-2.0 **Mn** 0.50-1.0 **Ni** (incl Co) 10.0-11.0 **Pb** 0.01 max **S** 0.05 max **Other** Cu + sum of named elements 99.5 min | |
| **C70620** | 90/10 Copper-Nickel (Welding Rod Composition) | **C** 0.05 max **Cu** (incl Ag) 86.5 min **Fe** 1.0-1.8 **Mn1.0** max **Ni** (incl Co) 9.0-11.0 **P** 0.02max **Pb** 0.02 max **S** 0.02 max **Zn** 0.50 max **Other** Cu + sum of named elements 99.5 | **ASTM** B151 |
| **C70690** | Copper-Nickel, 90Cu-10Ni | **Al** 0.002 max **As** 0.001 max **Bi** 0.001 max **C** 0.03 max **Co** 0.02 max **Cu** (incl Ag) rem **Fe** 0.005 max **Hg** 0.0005 max **Mn** 0.001 max **Ni** (incl Co) 9.0-11.0 **P** 0.001 max **Pb** 0.001 max **S** 0.003 max **Sb** 0.001 max **Si** 0.02 max **Sn** 0.001 max **Ti** 0.001 max **Zn** 0.001 max **Other** Cu + sum of named elements 99.5 min | **ASTM** F96 |
| **C70700** | Copper-Nickel | **Cu** (incl Ag) rem **Fe** 0.05 max **Mn** 0.50 max **Ni** (incl Co) 9.5-10.5 **Other** Cu + sum of named elements 99.5 min | |
| **C70800** | Copper-Nickel, 11% | **Cu** (incl Ag) rem **Fe** 0.10 max **Mn** 0.15 max **Ni** (incl Co) 10.5-12.5 **Pb** 0.05 max **Zn** 0.20 max **Other** Cu + sum of named elements 99.5 min | |
| * **C70900** | Copper-Nickel | **Cu** (incl Ag) rem **Fe** 0.6 max **Mn** 0.6 max **Ni** (incl Co) 13.5-16.5 **Pb** 0.05 max **Zn** 1.0 max **Other** Cu + sum of named elements 99.5 min | |
| **C71000** | Copper-Nickel, 20% | **Cu** (incl Ag) rem **Fe** 1.0 max **Mn** 1.0 max **Ni** (incl Co) 19.0-23.0 **Pb** 0.05 max **Zn** 1.0 max **Other** Cu + sum of named elements 99.5 min | **ASME** SB-111; SB-359; SB-395; SB-466 **ASTM** B111; B122; B206; B359; B395; B466; B543; B694; F467; F468 **SAE** J461; J463 |

*Boxed entries are no longer active and are retained for reference purposes only.

The chemical compositions listed are for identification purposes and should not be used in lieu of the cross referenced specifications.

| UNIFIED NUMBER | DESCRIPTION | CHEMICAL COMPOSITION | CROSS REFERENCE SPECIFICATIONS |
|---|---|---|---|
| **C71100** | Copper-Nickel | **Cu** (incl Ag) rem **Fe** 0.10 max **Mn** 0.15 max **Ni** (incl Co) 22.0-24.0 **Pb** 0.05 max **Zn** 0.20 max **Other** Cu + sum of named elements 99.5 min | |
| * **C71110** | Copper-Nickel Alloy (77.5 Cu - 22.5 Ni) | **Cu** (incl Ag) rem **Mn** 0.35 max **Ni** (incl Co) 21.5-23.5 **S** 0.008 max **Ti** 0.05 max **Other** Cu + sum of named elements 99.5 min | **AMS** 4732 |
| **C71300** | Copper-Nickel | **Cu** (incl Ag) rem **Fe** 0.20 max **Mn** 1.0 max **Ni** (incl Co) 23.5-26.5 **Pb** 0.05 max **Zn** 1.0 max **Other** Cu + sum of named elements 99.5 min | |
| **C71500** | Copper-Nickel, 30% (MONEL 450) | **Cu** (incl Ag) rem **Fe** 0.40-1.0 **Mn** 1.0 max **Ni** (incl Co) 29.0-33.0 **Pb** 0.05 max **Zn** 1.0 max **Other** Cu + sum of named elements 99.5 min | **ASME** SB-111; SB-171; SB-359; SB-395; SB-466; SB-467; SB-543 **ASTM** B111; B122; B151; B171; B359; B395; B432; B466; B467; B543; B552; B608; F467; F468 **MIL SPEC** MIL-T-15005; MIL-C-15726; MIL-T-16420 **SAE** J461; J463 |
| **C71520** | 70/30 Copper-Nickel (Welding Rod Composition) | **C** 0.05 max **Cu** (incl Ag) 65.0 min **Fe** 0.40-1.0 **Mn** 1.0 max **Ni** (incl Co) 29.0-33.0 **P** 0.02 max **Pb** 0.02 max **S** 0.02 max **Zn** 0.50 max **Other** (Cu + sum of named elements) 99.5 min | **ASTM** B151 **Other:** CDA Application Data Sheet, Standard Designations for Wrought and Cast Copper and Copper Alloys as available on the Internet Web Site, http://properties.copper.org/servlet/com.copper.servlet.CDAPropertiesSelectionServlet |
| **C71580** | Copper-Nickel | **Al** 0.05 max **C** 0.07 max **Cu** (incl Ag) rem **Fe** 0.50 max **Mn** 0.30 max **Ni** (incl Co) 29.0-33.0 **P** 0.03 max **Pb** 0.05 max **S** 0.024 max **Si** 0.15 max **Zn** 0.05 max **Other** Cu + sum of named elements 99.5 min | **ASTM** F96 |
| **C71581** | Copper-Nickel Welding Filler Metal | **Cu** (incl Ag) rem **Fe** 0.40-0.7 **Mn** 1.0 max **Ni** (incl Co) 29.0-32.0 **P** 0.02 max **Pb** 0.02 max **S** 0.01 max **Si** 0.25 max **Ti** 0.20-0.50 **Other** Cu + sum of named elements 99.5 min | **ASME** SFA5.7 (ERCuNi); SFA5.27 (RCuNi); SFA5.30 (IN67) **AWS** A5.7 (ERCuNi); A5.30 (IN67) **MIL SPEC** MIL-E-21562 (RN7, EN67, RN67); MIL-I-23413 (MIL-67) |
| **C71590** | Copper-Nickel, 70Cu-30Ni | **Al** 0.002 max **As** 0.001 max **Bi** 0.001 max **C** 0.02 max **Co** 0.05 max **Cu** (incl Ag) rem **Fe** 0.15 max **Hg** 0.0005 max **Mg** 0.10 max **Mn** 0.50 max **Ni** 29.0-31.0 **P** 0.001 max **Pb** 0.001 max **S** 0.003 max **Sb** 0.001 max **Si** 0.015 max **Sn** 0.001 max **Ti** 0.001 max **Zn** 0.001 max | **ASTM** F96 |
| * **C71630** | Copper-Nickel | **C** 0.06 max **Cu** (incl Ag) rem **Fe** 0.40-1.0 **Mn** 0.50-1.5 **Ni** (incl Co) 30.0-32.0 **Pb** 0.01 max **S** 0.08 max **Other** Cu + sum of named elements 99.5 min | |
| **C71640** | Copper-Nickel | **C** 0.06 max **Cu** (incl Ag) rem **Fe** 1.7-2.3 **Mn** 1.5-2.5 **Ni** (incl Co) 29.0-32.0 **Pb** 0.01 max **S** 0.03 max **Other** Cu + sum of named elements 99.5 min | **ASTM** B111; B543; B552 |
| **C71700** | Copper-Nickel | **Be** 0.30-0.7 **Cu** (incl Ag) rem **Fe** 0.40-1.0 **Ni** (incl Co) 29.0-33.0 **Other** Cu + sum of named elements 99.5 min | |
| **C71900** | Copper-Nickel | **C** 0.04 max **Cr** 2.2-3.0 **Cu** (incl Ag) rem **Fe** 0.50 max **Mn** 0.20-1.0 **Ni** (incl Co) 28.0-33.0 **P** 0.02 max **Pb** 0.015 max **Si** 0.25 max **S** 0.015 max **Ti** 0.01-0.20 **Zn** 0.05 max **Zr** 0.02-0.35 **Other** Cu + sum of named elements 99.5 min | |

*Boxed entries are no longer active and are retained for reference purposes only.

The chemical compositions listed are for identification purposes and should not be used in lieu of the cross referenced specifications.

| UNIFIED NUMBER | DESCRIPTION | CHEMICAL COMPOSITION | CROSS REFERENCE SPECIFICATIONS |
|---|---|---|---|
| **C72150** | Copper-Nickel | **C** 0.10 max **Cu** (incl Ag) rem **Fe** 0.10 max **Mn** 0.05 max **Ni** (incl Co) 43.0-46.0 **Pb** 0.05 max **Si** 0.50 max **Zn** 0.20 max **Other** Cu + sum of named elements 99.5 min | |
| **C72200** | Copper-Nickel | **C** 0.03 max **Cr** 0.30-0.7 **Cu** (incl Ag) rem **Fe** 0.5-1.0 **Mn** 1.0 max **Ni** (incl Co) 15.0-18.0 **Pb** 0.05 max **Si** 0.03 max **Ti** 0.03 max **Zn** 1.0 max **Other** Cu + sum of named elements 99.8 min **Note:** When product is for subsequent welding applications, C shall be 0.05 max, Pb 0.02 max, P 0.02 max, S 0.02 max, and Zn 0.50 max | **ASTM** B111; B122; B171; B359; B395; B466; B543; B552 |
| * **C72400** | Copper-Nickel Alloy | **Al** 1.5-2.5 **Cu** (incl Ag) rem **Fe** 0.10 max **Mg** 0.05-0.40 **Mn** 1.0 max **Ni** (incl Co) 11.0-15.0 **Pb** 0.05 max **Sn** 0.05 max **Zn** 0.50 max **Other** Cu + sum of named elements 99.5 min | **ASTM** B422 |
| **C72420** | Copper Nickel Alloy | **Al** 1.0-2.0 **C** 0.05 max **Cr** 0.50 max **Cu** (incl Ag) rem **Fe** 0.7-1.2 **Mg** 0.05 max **Mn** 3.5-5.5 **Ni** (incl Co) 13.5-16.5 **P** 0.01 max **Pb** 0.02 max **Si** 0.15 max **S** 0.15 max **Sn** 0.10 max **Zn** 0.20 max **Other** Cu + sum of named elements 99.7 min | |
| **C72500** | Copper-Nickel | **Cu** (incl Ag) rem **Fe** 0.6 max **Mn** 0.20 max **Ni** (incl Co) 8.5-10.5 **Pb** 0.05 max **Sn** 1.8-2.8 **Zn** 0.50 max **Other** Cu + sum of named elements 99.8 min | **ASTM** B122 |
| * **C72600** | Copper-Nickel Tin Spinodal Alloy | **Cu** (incl Ag) 91.0-93.0 **Fe** 0.20 max **Mn** 0.20 max **Ni** (incl Co) 3.5-4.5 **P** 0.05 max **Pb** 0.05 max **Sn** 3.5-4.5 **Zn** 0.50 max **Other** Cu + sum of named elements 99.7 min | |
| **C72650** | Copper-Nickel-Tin Alloy | **Cu** (incl Ag) rem **Fe** 0.10 max **Mn** 0.10 max **Ni** 7.0-8.0 **Pb** 0.01 max **Sn** 4.5-5.5 **Zn** 0.10 max **Other** Cu + sum of named elements 99.7 min | **ASTM** B740 |
| **C72700** | Copper-Nickel Tin Spinodal Alloy | **Cu** (incl Ag) rem **Fe** 0.50 max **Mg** 0.15 max **Mn** 0.05-0.30 **Nb** 0.10 max **Ni** (incl Co) 8.5-9.5 **Pb** 0.02 max **Sn** 5.5-6.5 **Zn** 0.50 max **Other** Cu + sum of named elements 99.7 min | **ASTM** B740 |
| **C72800** | Copper-Nickel Tin Spinodal Alloy | **Al** 0.10 max **B** 0.001 max **Bi** 0.001 max **Cu** (incl Ag) rem **Fe** 0.50 max **Mg** 0.005-0.15 **Mn** 0.05-0.30 **Nb** 0.10-0.30 **Ni** (incl Co) 9.5-10.5 **P** 0.005 max **Pb** 0.005 max **S** 0.0025 max **Sb** 0.02 max **Si** 0.05 max **Sn** 7.5-8.5 **Ti** 0.01 max **Zn** 1.0 max **Other** Cu + sum of named elements 99.7 min | |
| **C72900** | Copper-Nickel Tin Spinodal Alloy | **Cu** (incl Ag) rem **Fe** 0.50 max **Mg** 0.15 max **Mn** 0.30 max **Nb** 0.10 max **Ni** (incl Co) 14.5-15.5 **Pb** 0.02 **Sn** 7.5-8.5 **Zn** 0.50 max **Other** Cu + sum of named elements 99.7 min;. | **ASTM** B740 |

*Boxed entries are no longer active and are retained for reference purposes only.

The chemical compositions listed are for identification purposes and should not be used in lieu of the cross referenced specifications.

| UNIFIED NUMBER | DESCRIPTION | CHEMICAL COMPOSITION | CROSS REFERENCE SPECIFICATIONS |
|---|---|---|---|
| **C72950** | Copper Nickel Tin Spinodal Alloy | **Cu** (incl Ag) rem **Fe** 0.6 max **Mn** 0.6 max **Ni** (incl Co) 20.0-22.0 **Pb** 0.05 max **Sn** 4.5-5.7 **Other** Cu + sum of named elements 99.7 min | |
| *** C73150** | Nickel Silver | **Cu** (incl Ag) rem **Fe** 0.25 max **Mn** 0.50 max **Ni** (incl Co) 4.0-7.0 **Pb** 0.10 max **Zn** 9.0-15.0 **Other** Cu + sum of named elements 99.7 min | |
| *** C73200** | Nickel Silver | **Cu** (incl Ag) rem **Fe** 0.6 max **Mn** 1.0 max **Ni** (incl Co) 19.0-23.0 **Pb** 0.05 max **Zn** 3.0-6.0 **Other** Cu + sum of named elements 99.5 min | |
| **C73500** | Nickel Silver | **Cu** (incl Ag) 70.5-73.5 **Fe** 0.25 max **Mn** 0.50 max **Ni** (incl Co) 16.5-19.5 **Pb** 0.10 max **Zn** rem **Other** Cu + sum of named elements 99.5 min | **ASTM** B122 |
| *** C73800** | Nickel Silver, 70-12 | **Cu** (incl Ag) 68.5-71.5 **Fe** 0.25 max **Mn** 0.50 max **Ni** (incl Co) 11.0-13.0 **Pb** 0.05 max **Zn** rem **Other** Cu + sum of named elements 99.5 min | |
| **C74000** | Nickel Silver | **Cu** (incl Ag) 69.0-73.5 **Fe** 0.25 max **Mn** 0.50 max **Ni** (incl Co) 9.0-11.0 **Pb** 0.10 max **Zn** rem **Other** Cu + sum of named elements 99.5 min | **ASTM** B122 |
| **C74300** | Nickel Silver | **Cu** (incl Ag) 63.0-66.0 **Fe** 0.25 max **Mn** 0.50 max **Ni** (incl Co) 7.0-9.0 **Pb** 0.10 max **Zn** rem **Other** Cu + sum of named elements 99.5 min | |
| **C74400** | Copper Nickel Zinc | **Cu** (incl Ag) 62.0-66.0 **Fe** 0.05 max **Pb** 0.05 max **Zn** rem **Other** (Cu + sum of named elements) 99.7 min | **Other:** CDA Application Data Sheet, Standard Designations for Wrought and Cast Copper and Copper Alloys as available on the Internet Web Site, http://properties.copper.org/servlet/com.copper.servlet.CDAPropertiesSelectionServlet **ASTM** B122 |
| **C74500** | Nickel Silver, 65-10 | **Cu** (incl Ag) 63.5-66.5 **Fe** 0.25 max **Mn** 0.50 max **Ni** (incl Co) 9.0-11.0 **Pb** 0.10 max **Zn** rem **Other** Cu + sum of named elements 99.5 min | **ASTM** B122; B151; B206 |
| **C75200** | Nickel Silver, 65-18 | **Cu** (incl Ag) 63.5-66.5 **Fe** 0.25 max **Mn** 0.50 max **Ni** (incl Co) 16.5-19.5 **Pb** 0.05 max **Zn** rem **Other** Cu + sum of named elements 99.5 min | **ASTM** B122; B151; B206; B458 (1); B888 **SAE** J461; J463 |
| **C75400** | Nickel Silver, 65-15 | **Cu** (incl Ag) 63.5-66.5 **Fe** 0.25 max **Mn** 0.50 max **Ni** (incl Co) 14.0-16.0 **Pb** 0.10 max **Zn** rem **Other** Cu + sum of named elements 99.5 min | |
| **C75700** | Nickel Silver, 65-12 | **Cu** (incl Ag) 63.5-66.5 **Fe** 0.25 max **Mn** 0.50 max **Ni** (incl Co) 11.0-13.0 **Pb** 0.05 max **Zn** rem **Other** Cu + sum of named elements 99.5 min | **ASTM** B151; B206 |
| *** C75720** | 12% Nickel Silver | **Cu** (incl Ag) 60.0-65.0 **Fe** 0.25 max **Mn** 0.05-0.30 **Ni** (incl Co) 11.0-13.0 **Pb** 0.04 max **Zn** rem **Other** Cu + sum of named elements 99.5 min | |

*Boxed entries are no longer active and are retained for reference purposes only.

The chemical compositions listed are for identification purposes and should not be used in lieu of the cross referenced specifications.

| UNIFIED NUMBER | DESCRIPTION | CHEMICAL COMPOSITION | CROSS REFERENCE SPECIFICATIONS |
|---|---|---|---|
| * **C75900** | Nickel Silver | **Cu** (incl Ag) 60.0-65.0 **Fe** 0.25 max **Mn** 0.50 max **Ni+Co** 17.0-19.0 **Pb** 0.10 max **Zn** rem **Other** Cu + sum of named elements 99.5 min | |
| **C76000** | Nickel Silver | **Cu** (incl Ag) 60.0-63.0 **Fe** 0.25 max **Mn** 0.50 max **Ni** (incl Co) 7.0-9.0 **Pb** 0.10 max **Zn** rem **Other** Cu + sum of named elements 99.5 min | |
| * **C76100** | Nickel Silver | **Cu** (incl Ag) 59.0-63.0 **Fe** 0.25 max **Mn** 0.50 max **Ni** (incl Co) 7.0-9.0 **Pb** 0.10 max **Zn** rem **Other** Cu + sum of named elements 99.5 min | |
| **C76200** | Nickel Silver | **Cu** (incl Ag) 57.0-61.0 **Fe** 0.25 max **Mn** 0.50 max **Ni** (incl Co) 11.0-13.5 **Pb** 0.10 max **Zn** rem **Other** Cu + sum of named elements 99.5 min | **ASTM** B122; B888 |
| * **C76300** | Nickel Silver | **Cu** (incl Ag) 60.0-64.0 **Fe** 0.50 max **Mn** 0.50 max **Ni** (incl Co) 17.0-19.0 **Pb** 0.50-2.0 **Zn** rem **Other** Cu + sum of named elements 99.5 min | |
| * **C76390** | Leaded Nickel Silver | **Cu** (incl Ag) 59.0-63.0 **Fe** 0.25 max **Mn** 0.50 max **Ni** (incl Co) 23.0-26.0 **Pb** 0.8-1.1 **Sn** 0.40-0.60 **Zn** rem **Other** Cu + sum of named elements 99.5 min | |
| **C76400** | Nickel Silver | **Cu** (incl Ag) 58.5-61.5 **Fe** 0.25 max **Mn** 0.50 max **Ni** (incl Co) 16.5-19.5 **Pb** 0.05 max **Zn** rem **Other** Cu + sum of named elements 99.5 min | **ASTM** B151; B206 |
| * **C76600** | Nickel Silver | **Cu** (incl Ag) 55.0-58.0 **Fe** 0.25 max **Mn** 0.50 max **Ni** (incl Co) 11.0-13.5 **Pb** 0.10 max **Zn** rem **Other** Cu + sum of named elements 99.5 min | |
| **C76700** | Nickel Silver, 56.5-15 | **Cu** (incl Ag) 55.0-58.0 **Mn** 0.50 max **Ni** (incl Co) 14.0-16.0 **Zn** rem **Other** Cu + sum of named elements 99.5 min | |
| **C77000** | Nickel Silver, 55-18 | **Cu** (incl Ag) 53.5-56.5 **Fe** 0.25 max **Mn** 0.50 max **Ni** (incl Co) 16.5-19.5 **Pb** 0.05 max **Zn** rem **Other** Cu + sum of named elements 99.5 min | **ASTM** B122; B151; B206 **SAE** J461; J463 |
| * **C77010** | Nickel Silver | **Cu** (incl Ag) 54.0-56.0 **Fe** 0.30 max **Mn** 0.05-0.35 **Ni** (incl Co) 17.0-19.0 **Pb** 0.03 max **Zn** rem **Other** Cu + sum of named elements 99.5 min | |
| **C77300** | Nickel Brass Welding and Brazing Filler Metal | **Al** 0.01 max **Cu** (incl Ag) 46.0-50.0 **Ni** (incl Co) 9.0-11.0 **P** 0.25 max **Pb** 0.05 max **Si** 0.04-0.25 **Zn** rem **Other** Cu + sum of named elements 99.5 min | **ASME** SFA5.8 (RBCuZn-D) **AWS** A5.8 (RBCuZn-D) |

*Boxed entries are no longer active and are retained for reference purposes only.

The chemical compositions listed are for identification purposes and should not be used in lieu of the cross referenced specifications.

| UNIFIED NUMBER | DESCRIPTION | CHEMICAL COMPOSITION | CROSS REFERENCE SPECIFICATIONS |
|---|---|---|---|
| *** C77310** | Nickel Silver | **Al** 0.01 max **Cu** (incl Ag) 46.0-56.0 **Mn** 0.50 max **Ni** (incl Co) 9.0-11.0 **P** 0.25 max **Pb** 0.05 **Si** 0.04-0.25 **Zn** rem **Other** Cu + sum of named elements 99.5 min | |
| **C77400** | Nickel Silver | **Cu** (incl Ag) 43.0-47.0 **Ni** (incl Co) 9.0-11.0 **Pb** 0.20 max **Zn** rem **Other** Cu + sum of named elements 99.5 min | **ASTM** B124; B283 |
| *** C77600** | Nickel Silver, 43.5-13 | **Cu** (incl Ag) 42.0-45.0 **Fe** 0.20 max **Mn** 0.25 max **Ni** (incl Co) 12.0-14.0 **Pb** 0.25 max **Sn** 0.15 max **Zn** rem **Other** Cu + sum of named elements 99.5 min | |
| **C78200** | Leaded Nickel Silver | **Cu** (incl Ag) 63.0-67.0 **Fe** 0.35 max **Mn** 0.50 max **Ni** (incl Co) 7.0-9.0 **Pb** 1.5-2.5 **Zn** rem **Other** Cu + sum of named elements 99.5 min | |
| *** C78800** | Leaded Nickel Silver | **Cu** (incl Ag) 63.0-67.0 **Fe** 0.25 max **Mn** 0.50 max **Ni** (incl Co) 9.0-11.0 **Pb** 1.5-2.0 **Zn** rem **Other** Cu + sum of named elements 99.5 min | |
| **C79000** | Leaded Nickel Silver | **Cu** (incl Ag) 63.0-67.0 **Fe** 0.35 max **Mn** 0.50 max **Ni** (incl Co) 11.0-13.0 **Pb** 1.5-2.2 **Zn** rem **Other** Cu + sum of named elements 99.5 min | |
| **C79200** | Leaded Nickel Silver | **Cu** (incl Ag) 59.5-66.5 **Fe** 0.25 max **Mn** 0.50 max **Ni** (incl Co) 11.0-13.0 **Pb** 0.8-1.4 **Zn** rem **Other** Cu + sum of named elements 99.5 min | **ASTM** B151; B206 |
| *** C79300** | Leaded Nickel Silver | **Cu** (incl Ag) 55.0-59.00 **Fe** 0.50 max **Mn** 0.50 max **Ni** (incl Co) 11.0-13.0 **Pb** 0.50-2.0 **Zn** rem **Other** Cu + sum of named elements 99.5 min | |
| *** C79600** | Leaded Nickel Silver, 10% | **Cu** (incl Ag) 43.5-46.5 **Mn** 1.5-2.5 **Ni** (incl Co) 9.0-11.0 **Pb** 0.8-1.2 **Zn** rem **Other** Cu + sum of named elements 99.5 min | |
| *** C79620** | Leaded Nickel Silver | **Cu** (incl Ag) 46.0-48.0 **Mn** 0.50 max **Ni** (incl Co) 8.0-11.0 **Pb** 0.50-2.0 **Zn** rem **Other** Cu + sum of named elements 99.5 min | |
| **C79800** | Leaded Nickel Silver | **Cu** (incl Ag) 45.5-48.5 **Fe** 0.25 max **Mn** 1.5-2.5 **Ni** (incl Co) 9.0-11.0 **Pb** 1.5-2.5 **Zn** rem **Other** Cu + sum of named elements 99.5 min | |

*Boxed entries are no longer active and are retained for reference purposes only.

The chemical compositions listed are for identification purposes and should not be used in lieu of the cross referenced specifications.

| UNIFIED NUMBER | DESCRIPTION | CHEMICAL COMPOSITION | CROSS REFERENCE SPECIFICATIONS |
|---|---|---|---|
| * **C79810** | Leaded Nickel Silver | **Cu** (incl Ag) 46.0-48.0 **Mn** 0.50 max **Ni** (incl Co) 8.0-11.0 **Pb** 2.0-3.5 **Zn** rem **Other** Cu + sum of named elements 99.5 min | |
| * **C79820** | Leaded Nickel Silver | **Cu** 46.0-48.0 **Mn** 0.50 max **Ni** 8.0-11.0 **Pb** 2.0-3.5 **Zn** rem **Other** Co included in Ni | |
| **C79830** | Copper Nickel Zinc | **Cu** (incl Ag) 45.5-47.0 **Fe** 0.45 max **Mn** 0.15-0.55 **Ni** 9.0-10.5 **Pb** 1.0-2.5 **Zn** rem **Other** (Cu + sum of named elements) 99.5 min | **Other:** CDA Application Data Sheet, Standard Designations for Wrought and Cast Copper and Copper Alloys as available on the Internet Web Site, http://properties.copper.org/servlet/com.copper.servlet.CDAPropertiesSelectionServlet |
| * **C79900** | Leaded Nickel Silver | **Cu** (incl Ag) 47.5-50.5 **Fe** 0.25 max **Mn** 0.50 max **Ni** (incl Co) 6.5-8.5 **Pb** 1.0-1.5 **Zn** rem **Other** Cu + sum of named elements 99.5 min | |
| **C80100** | Cast Copper | **Cu** (incl Ag) 99.95 min | |
| * **C80300** | Cast Copper, Silver Bearing | **Ag** 0.034 min **Cu** (incl Ag) 99.95 min **Other** total 0.05 max | |
| **C80410** | Cast Copper Alloy | **Cu** (incl Ag) 99.9 min **Other** total other elements 0.10 max | **Other:** CDA Application Data Sheet, Standard Designations for Wrought and Cast Copper and Copper Alloys as available on the Internet Web Site, http://properties.copper.org/servlet/com.copper.servlet.CDAPropertiesSelectionServlet |
| * **C80500** | Cast Copper, Silver Bearing | **Ag** 0.034 min **B** 0.02 max **Cu** (incl Ag) 99.75 min | |
| * **C80700** | Cast Copper | **B** 0.02 max **Cu** (incl Ag) 99.75 min | |
| * **C80900** | Cast Copper | **Ag** 0.034 min **Cu** (incl Ag) 99.70 min | |
| **C81100** | Cast Copper | **Cu** (incl Ag) 99.70 min | |
| **C81200** | Cast Phosphorus Deoxidized Copper | **Cu** (incl Ag) 99.9 min **P** 0.040-0.065 | |
| * **C81300** | Cast Beryllium Copper | **Be** 0.02-0.10 **Co** 0.6-1.0 **Cu** 98.5 min **Other** Cu + sum of named elements 99.5 min | |
| **C81400** | Cast Chromium Copper, 70C | **Be** 0.02-0.10 **Cr** 0.6-1.0 **Cu** 98.5 min **Other** Cu + sum of named elements 99.5 min | **ASTM** B30; B770 |
| **C81500** | Cast Chromium Copper | **Al** 0.10 max **Cr** 0.40-1.5 **Cu** 98.0 min **Fe** 0.10 max **Pb** 0.02 max **Si** 0.15 max **Sn** 0.10 max **Zn** 0.10 max **Other** Cu + sum of named elements 99.5 min | |

*Boxed entries are no longer active and are retained for reference purposes only.

The chemical compositions listed are for identification purposes and should not be used in lieu of the cross referenced specifications.

| UNIFIED NUMBER | DESCRIPTION | CHEMICAL COMPOSITION | CROSS REFERENCE SPECIFICATIONS |
|---|---|---|---|
| **C81540** | Cast Chromium Nickel Copper | **Al** 0.10 max **Cr** 0.10-0.6 **Cu** (incl Ag) 95.1 min **Fe** 0.15 max **Ni** (incl Co) 2.0-3.0 **Pb** 0.02 max **Si** 0.40-0.8 **Sn** 0.10 max **Zn** 0.10 max **Other** Cu + sum of named elements 99.5 min | |
| * **C81700** | Cast Beryllium Copper | **Ag** 0.8-1.2 **Be** 0.30-0.55 **Co** 0.25-1.5 **Cu** 94.2 min **Ni** 0.25-1.5 **Other** Cu + sum of named elements 99.5 min | |
| * **C81800** | Cast Chromium Copper, 50C | **Ag** 0.8-1.2 **Be** 0.30-0.55 **Co** 1.4-1.7 **Cu** 95.6 min **Other** Cu + sum of named elements 99.5 min | |
| **C82000** | Cast Beryllium Copper | **Al** 0.10 max **Be** 0.45-0.8 **Co** (incl Ni) 2.4-2.7 **Cr** 0.10 max **Cu** 95.0 min **Fe** 0.10 max **Ni** 0.20 max **Pb** 0.02 max **Si** 0.15 max **Sn** 0.10 max **Zn** 0.10 max **Other** Cu + sum of named elements 99.5 min | **ASTM** B30 (10C); B770 |
| * **C82100** | Cast Beryllium Copper | **Be** 0.35-0.8 **Co** 0.25-1.5 **Cu** 95.5 min **Ni** 0.25-1.5 **Other** Cu + sum of named elements 99.5 min | |
| **C82200** | Cast Beryllium Copper, 35B, 35C | **Be** 0.35-0.8 **Cu** 96.5 min **Ni** 1.0-2.0 **Other** Cu + sum of named elements 99.5 min | **ASTM** B30 (3C, 14C); B770 |
| **C82400** | Cast Beryllium Copper, 165C | **Al** 0.15 max **Be** 1.65-1.75 **Co** 0.20-0.40 **Cr** 0.10 max **Cu** 96.4 min **Fe** 0.20 max **Ni** 0.10 max **Pb** 0.02 max **Sn** 0.10 max **Zn** 0.10 max **Other** Cu + sum of named elements 99.5 min | **ASTM** B30 (165C, 165CT); B770 |
| **C82500** | Cast Beryllium Copper, 20C | **Al** 0.15 max **Be** 1.90-2.15 **Co** (incl Ni) 0.35-0.7 **Cr** 0.10 max **Cu** 95.5 min **Fe** 0.25 max **Ni** 0.20 max **Pb** 0.02 max **Si** 0.20-0.35 **Sn** 0.10 max **Zn** 0.10 max **Other** Cu + sum of named elements 99.5 min | **AMS** 4511; 4890 **ASTM** B30 (20C, 20CT); B770 **MIL SPEC** MIL-C-22087 (10) |
| **C82510** | Cast Beryllium Copper | **Al** 0.15 max **Be** 1.90-2.15 **Co** 1.0-2.0 **Cr** 0.10 max **Cu** 95.5 min **Fe** 0.25 max **Ni** 0.20 max **Pb** 0.02 max **Si** 0.20-0.35 **Sn** 0.10 max **Zn** 0.10 max **Other** Cu + sum of named elements 99.5 min | **ASTM** B30 (21C); B770 |
| **C82600** | Cast Beryllium Copper, 245C | **Al** 0.15 max **Be** 2.25-2.45 **Co** 0.35-0.7 **Cr** 0.10 max **Cu** 95.2 min **Fe** 0.25 max **Ni** 0.20 max **Pb** 0.02 max **Si** 0.20-0.35 **Sn** 0.10 max **Zn** 0.10 max **Other** Cu + sum of named elements 99.5 min | **ASTM** B30 (245C, 245CT); B770 **FED** WW-V-1967 |
| **C82700** | Cast Beryllium Copper | **Al** 0.15 max **Be** 2.35-2.55 **Cr** 0.10 max **Cu** 94.6 min **Fe** 0.25 max **Ni** 1.0-1.5 **Pb** 0.02 max **Si** 0.15 max **Sn** 0.10 max **Zn** 0.10 max **Other** Cu + sum of named elements 99.5 min | **FED** WW-V-1967 |
| **C82800** | Cast Beryllium Copper, 275C | **Al** 0.15 max **Be** 2.50-2.75 **Co** (incl Ni) 0.35-0.7 **Cr** 0.10 max **Cu** 94.8 min **Fe** 0.25 max **Ni** 0.20 max **Pb** 0.02 max **Si** 0.20-0.35 **Sn** 0.10 max **Zn** 0.10 max **Other** Cu + sum of named elements 99.5 min. | **ASTM** B30 (275C, 275CT); B770 **FED** WW-V-1967 |

*Boxed entries are no longer active and are retained for reference purposes only.

The chemical compositions listed are for identification purposes and should not be used in lieu of the cross referenced specifications.

| UNIFIED NUMBER | DESCRIPTION | CHEMICAL COMPOSITION | CROSS REFERENCE SPECIFICATIONS |
|---|---|---|---|
| **C83300** | Cast Leaded Red Brass | **Cu** (incl Ni) 92.0-94.0 **Pb** 1.0-2.0 **Sn** 1.0-2.0 **Zn** 2.0-6.0 **Other** Cu + sum of named elements 99.3 min | |
| **C83400** | Cast Red Brass | **Cu** (incl Ni) 88.0-92.0 **Pb** 0.50 max **Sn** 0.20 max **Zn** 8.0-12.0 **Other** Cu + sum of named elements 99.3 min | |
| * **C83410** | Red Brass | **Al** 0.005 max **Cu** (incl Ni) 88.0-91.0 **Fe** 0.05 max **Ni** (incl Co) 0.05 max **Pb** 0.10 max **Si** 0.005 max **Sn** 1.0-2.0 **Zn** rem **Other** Cu + sum of named elements 99.8 min | |
| * **C83420** | Cast Red Brass | **Cu** (incl Ni) 88.0-92.0 **Fe** 0.10 max **Pb** 0.50 max **Sn** 0.25-0.7 **Zn** rem **Other** Cu + sum of named elements 99.3 min | **SAE** J460 (795) |
| **C83450** | Cast Leaded Red Brass | **Al** 0.005 max **Cu** 87.0-89.0 **Fe** 0.30 max **Ni** (incl Co) 0.8-2.0 **P** 0.03 max **Pb** 1.5-3.0 **S** 0.08 max **Sb** 0.25 max **Si** 0.005 max **Sn** 2.0-3.5 **Zn** 5.5-7.5 **Other** Cu + sum of named elements 99.3 min **Note:** For continuous castings, P 1.5 max | **ASTM** B30; B584; B763 |
| **C83500** | Cast Leaded Red Brass | **Al** 0.008 max **Cu** 86.0-88.0 **Fe** 0.25 max **Ni** (incl Co) 0.50-1.0 **P** 0.03 max **Pb** 3.5-5.5 **S** 0.08 max **Sb** 0.25 max **Si** 0.005 max **Sn** 5.5-6.5 **Zn** 1.0-2.5 **Other** Cu + sum of named elements 99.3 min **Note:** For continuous castings, P 1.5 max | |
| * **C83520** | Cast Leaded Red Brass | **Cu** (incl Ni) rem **Fe** 0.30 max **Ni** (incl Co) 1.0 max **Pb** 3.5-4.5 **Sb** 0.25 max **Sn** 3.5-4.5 **Zn** 1.5-4.0 **Other** Cu + sum of named elements 99.8 min | **SAE** J460 (791) |
| **C83600** | Cast Leaded Red Brass | **Al** 0.005 max **Cu** (incl Ni) 84.0-86.0 **Fe** 0.30 max **Ni** (incl Co) 1.0 max **P** 0.05 max **Pb** 4.0-6.0 **S** 0.08 max **Sb** 0.25 max **Si** 0.005 max **Sn** 4.0-6.0 **Zn** 4.0-6.0 **Other** Cu + sum of named elements 99.3 min **Note:** For continuous castings, P 1.5 max | **AMS** 4855 **ASME** B16.15; B16.18; B16.23; B16.24; B16.26; B16.32; SB-62 **ASTM** B30 (4A); B62; B271; B505; B584 (formerly B145) **FED** WW-P-460; WW-T-725; WW-U-516 **MIL SPEC** MIL-V-18436 **SAE** J461; J462 **Other:** Ingot 115 |
| * **C83700** | Red Brass | **Al** 0.005 max **As** 0.05-0.20 **Cu** (incl Ni) 83.0-88.0 **Fe** 0.30 max **Ni** (incl Co) 0.30 max **Pb** 0.50 max **Si** 0.005 max **Sn** 1.0 max **Zn** rem **Other** Cu + sum of named elements 99.3 min | |
| **C83800** | Cast Leaded Red Brass | **Al** 0.005 max **Cu** (incl Ni) 82.0-83.8 **Fe** 0.30 max **Ni** (incl Co) 1.0 max **P** 0.03 max **Pb** 5.0-7.0 **S** 0.08 max **Sb** 0.25 max **Si** 0.005 max **Sn** 3.3-4.2 **Zn** 5.0-8.0 **Other** Cu + sum of named elements 99.3 min **Note:** For continuous castings, P 1.5 max | **ASME** B16.15; B16.18; B16.23 **ASTM** B30; B271; B505; B584 (formerly B145); B763 **FED** WW-T-725; WW-U-516; WW-V-1967 **SAE** J461; J462 **Other:** Ingot 120 |

*Boxed entries are no longer active and are retained for reference purposes only.

The chemical compositions listed are for identification purposes and should not be used in lieu of the cross referenced specifications.

| UNIFIED NUMBER | DESCRIPTION | CHEMICAL COMPOSITION | CROSS REFERENCE SPECIFICATIONS |
|---|---|---|---|
| **C83810** | Cast Leaded Red Brass | **Al** 0.01 max **As** 0.50 max **Bi** 0.10 max **Cu** (incl Ni) rem **Fe** 0.50 max **Ni** (incl Co) 2.0 max **Pb** 4.0-6.0 **Si** 0.005 max **Sn** 2.0-3.5 **Zn** 7.5-9.5 **Sb** 0.50 max **Other** Cu + sum of named elements 99.3 min **Note:** As+Fe+Sb 0.50 max | |
| **C84200** | Cast Semi-Red Brass | **Al** 0.005 max **Cu** (incl Ni) 78.0-82.0 **Fe** 0.40 max **Ni** (incl Co) 0.8 max **P** 0.05 max **Pb** 2.0-3.0 **S** 0.08 max **Sb** 0.25 max **Si** 0.005 max **Sn** 4.0-6.0 **Zn** 10.0-16.0 **Other** Cu + sum of named elements 99.2 min **Note:** For continuous castings, P 1.5 max | **ASTM** B30; B505 **FED** WW-P-460 |
| **C84400** | Cast Semi-Red Brass | **Al** 0.005 max **Cu** (incl Ni) 78.0-82.0 **Fe** 0.40 max **Ni** (incl Co) 1.0 max **P** 0.2 max **Pb** 6.0-8.0 **S** 0.08 max **Sb** 0.25 max **Si** 0.005 max **Sn** 2.3-3.5 **Zn** 7.0-10.0 **Other** Cu + sum of named elements 99.2 min; Ingot 123 **Note:** For continuous castings, P 1.5 max | **ASME** B16.15; B16.18; B16.23; B16.26; B16.32 **ASTM** B30; B271; B505; B584 (formerly B145); B763 **FED** WW-T-725; WW-U-516 **Other:** Ingot 123 |
| **C84410** | Cast Semi Red Brass | **Al** 0.01 max **Bi** 0.05 max **Cu** (incl Ni) rem **Ni** (incl Co) 1.0 max **Pb** 7.0-9.0 **Si** 0.20 max **Sn** 3.0-4.5 **Zn** 7.0-11.0 **Other** Cu + sum of named elements 99.2 min; Fe+Sb+As 0.8 max | |
| **C84500** | Cast Semi-Red Brass | **Al** 0.005 max **Cu** (incl Ni) 77.0-79.0 **Fe** 0.40 max **Ni** (incl Co) 1.0 max **P** 0.02 max **Pb** 6.0-7.5 **S** 0.08 max **Sb** 0.25 max **Si** 0.005 max **Sn** 2.0-4.0 **Zn** 10.0-14.0 **Other** Cu + sum of named elements 99.2 min **Note:** For continuous castings, P 1.5 max | **Other:** Ingot 125 |
| **C84800** | Cast Semi-Red Brass | **Al** 0.005 max **Cu** (incl Ni) 75.0-77.0 **Fe** 0.40 max **Ni** (incl Co) 1.0 max **P** 0.02 max **Pb** 5.5-7.0 **S** 0.08 max **Sb** 0.25 max **Si** 0.005 max **Sn** 2.0-3.0 **Zn** 13.0-17.0 **Other** Cu + sum of named elements 99.2 min **Note:** For continuous castings, P 1.5 max | **ASTM** B30; B271; B505; B584 (formerly B145); B763 **Other:** Ingot 130 |
| **C85200** | Cast Leaded Yellow Brass | **Al** 0.005 max **Cu** (incl Ni) 70.0-74.0 **Fe** 0.6 max **Ni** (incl Co) 1.0 max **P** 0.02 max **Pb** 1.5-3.8 **S** 0.05 max **Sb** 0.20 max **Si** 0.05 max **Sn** 0.7-2.0 **Zn** 20.0-27.0 **Other** Cu + sum of named elements 99.1 min | **ASTM** B30; B271; B584 (formerly B146); B763 **FED** WW-V-1967 **SAE** J461; J462 **Other:** Ingot 400 |
| * **C85210** | Yellow Brass | **Al** 0.005 max **As** 0.02-0.06 **Cu** (incl Ni) 70.0-75.0 **Fe** 0.8 max **Ni** (incl Co) 1.0 max **Pb** 2.0-5.0 **Si** 0.005 max **Sn** 1.0-3.0 **Zn** rem **Other** Cu + sum of named elements 98.9 min | |
| * **C85300** | Cast Leaded Yellow Brass | **Cu** 68.0-72.0 **Pb** 0.50 max **Sn** 0.50 max **Zn** rem **Other** total (including Pb+Sn) 1.0 max | |
| * **C85310** | Leaded Yellow Brass | **Al** 0.01 max **As** 0.02-0.06 **Cu** (incl Ni) 68.0-73.0 **Fe** 0.8 max **Ni** (incl Co) 1.0 max **Pb** 2.0-5.0 **Sn** 1.5 max **Zn** rem **Other** Cu + sum of named elements 98.9 min **Note:** Fe+Sb+As 0.8 max | |

*Boxed entries are no longer active and are retained for reference purposes only.

The chemical compositions listed are for identification purposes and should not be used in lieu of the cross referenced specifications.

| UNIFIED NUMBER | DESCRIPTION | CHEMICAL COMPOSITION | CROSS REFERENCE SPECIFICATIONS |
|---|---|---|---|
| **C85400** | Cast Leaded Yellow Brass | **Al** 0.35 max **Cu** (incl Ni) 65.0-70.0 **Fe** 0.7 max **Ni** (incl Co) 1.0 max **Pb** 1.5-3.8 **Si** 0.05 max **Sn** 0.50-1.5 **Zn** 24.0-32.0 **Other** Cu + sum of named elements 98.9 min | **ASTM** B30; B271; B584 (formerly B146); B763 **FED** WW-V-1967 **SAE** J461; J462 **Other:** Ingot 403 |
| **C85500** | Cast Leaded Yellow Brass | **Cu** (incl Ni) 59.0-63.0 **Fe** 0.20 max **Mn** 0.20 max **Ni** (incl Co) 0.20 max **Pb** 0.20 max **Sn** 0.20 max **Zn** rem **Other** Cu + sum of named elements 98.9 min | **FED** WW-V-1967 |
| * **C85600** | Cast Leaded Yellow Brass | **Cu** 59.0-63.0 **Fe** 0.20 max **Mn** 0.20 max **Ni** 0.20 max **Pb** 0.20 max **Sn** 0.20 max **Zn** rem | |
| **C85700** | Cast Leaded Yellow Brass | **Al** 0.8 max **Cu** (incl Ni) 58.0-64.0 **Fe** 0.7 max **Ni** (Incl Co) 1.0 max **Pb** 0.8-1.5 **Si** 0.05 max **Sn** 0.50-1.5 **Zn** 32.0-40.0 **Other** Cu + sum of named elements 98.7 min | **ASTM** B30; B176; B271; B584; B763 **Other:** Ingot 405.2 |
| * **C85710** | Leaded Yellow Brass | **Al** 0.20-0.8 **Cu** (incl Ni) 58.0-63.0 **Fe** 0.8 max **Mn** 0.50 max **Ni** (incl Co) 1.0 max **Pb** 1.0-2.5 **Si** 0.05 max **Sn** 1.0 max **Zn** rem **Other** Cu + sum of named elements 98.7 min | |
| **C85800** | Cast Leaded Yellow Brass | **Al** 0.50 max **As** 0.05 max **Cu** (incl Ni) 57.0 min **Fe** 0.50 max **Mn** 0.25 max **Ni** (incl Co) 0.50 max **P** 0.01 max **Pb** 1.5 max **S** 0.05 max **Sb** 0.05 max **Si** 0.25 max **Sn** 1.5 max **Zn** 31.0-41.0 **Other** Cu + sum of named elements 98.7 min | **ASTM** B30; B176 **SAE** J461; J462 **Other:** Ingot 405.1 |
| **C86100** | Cast Manganese Bronze | **Al** 4.5-5.5 **Cu** (incl Ni) 66.0-68.0 **Fe** 2.0-4.0 **Mn** 2.5-5.0 **Pb** 0.20 max **Sn** 0.20 max **Zn** rem **Other** Cu + sum of named elements 99.0 min | **ASTM** B30 **FED** WW-V-1967 **Other:** Ingot 423 |
| **C86200** | Cast Manganese Bronze | **Al** 3.0-4.9 **Cu** (incl Ni) 60.0-66.0 **Fe** 2.0-4.0 **Mn** 2.5-5.0 **Ni** (incl Co) 1.0 max **Pb** 0.20 max **Sn** 0.20 max **Zn** 22.0-28.0 **Other** Cu + sum of named elements 99.0 min | **ASTM** B30; B271; B505; B584 (formerly B147); B763 **FED** WW-V-1967 **SAE** J461; J462 **Other:** Ingot 423 |
| **C86300** | Cast Manganese Bronze | **Al** 5.0-7.5 **Cu** (incl Ni) 60.0-66.0 **Fe** 2.0-4.0 **Mn** 2.5-5.0 **Ni** (incl Co) 1.0 max **Pb** 0.20 max **Sn** 0.20 max **Zn** 22.0-28.0 **Other** Cu + sum of named elements 99.0 min | **AMS** 4862 **ASTM** B22; B30; B271; B505; B584 (formerly B147); B763 **FED** WW-V-1967 **SAE** J461; J462 **Other:** Ingot 424 |
| **C86400** | Cast Leaded Manganese Bronze | **Al** 0.50-1.5 **Cu** (incl Ni) 56.0-62.0 **Fe** 0.40-2.0 **Mn** 0.10-1.0 **Ni** (incl Co) 1.0 max **Pb** 0.50-1.5 **Sn** 0.50-1.5 **Zn** 34.0-42.0 **Other** Cu + sum of named elements 99.0 min | **ASTM** B30; B271; B505; B584 (formerly B132, B147); B763 **FED** WW-V-1967 **Other:** Ingot 420 |
| **C86500** | Cast Manganese Bronze | **Al** 0.50-1.5 **Cu** (incl Ni) 55.0-60.0 **Fe** 0.40-2.0 **Mn** 1.0-1.5 **Ni** (incl Co) 1.0 max **Pb** 0.40 max **Sn** 1.0 max **Zn** 36.0-42.0 **Other** Cu + sum of named elements 99.0 min | **AMS** 4860 **ASTM** B30; B176; B271; B505; B584 (formerly B147); B763 **FED** WW-V-1967 **SAE** J461; J462 **Other:** Ingot 421 |
| **C86550** | Cast Manganese Bronze | **Al** 0.05-2.5 **Cu** (incl Ni) 57.0 min **Fe** 0.7-2.0 **Mn** 0.10-3.0 **Ni** (incl Co) 1.0 max **Pb** 0.50 max **Si** 1.0 max **Sn** 1.0 max **Zn** rem **Other** (Cu + sum of named elements) 99.0 min | **Other:** CDA Application Data Sheet, Standard Designations for Wrought and Cast Copper and Copper Alloys as available on the Internet Web Site, http://properties.copper.org/servlet/com.copper.servlet.CDAPropertiesSelectionServlet |

*Boxed entries are no longer active and are retained for reference purposes only.

The chemical compositions listed are for identification purposes and should not be used in lieu of the cross referenced specifications.

| UNIFIED NUMBER | DESCRIPTION | CHEMICAL COMPOSITION | CROSS REFERENCE SPECIFICATIONS |
|---|---|---|---|
| **C86700** | Cast Leaded Manganese Bronze | **Al** 1.0-3.0 **Cu** (incl Ni) 53.0-60.0 **Fe** 1.0-3.0 **Mn** 1.0-3.5 **Ni** (incl Co) 1.0 max **Pb** 0.5-1.5 **Sn** 1.5 max **Zn** 30.0-38.0 **Other** Cu + sum of named elements 99.0 min | **ASTM** B30; B271; B584 (formerly B132); B763 |
| **C86800** | Cast Manganese Bronze | **Al** 2.0 max **Cu** (incl Ni) 53.5-57.0 **Fe** 1.0-2.5 **Mn** 2.5-4.0 **Ni** (incl Co) 2.5-4.0 **Pb** 0.20 max **Sn** 1.0 max **Zn** rem **Other** Cu + sum of named elements 99.0 min | **FED** WW-V-1967 |
| *** C87200** | Cast Copper Silicon Bronze | **Al** 1.5 max **Cu** 89.0 min **Fe** 2.5 max **Mn** 1.5 max **Pb** 0.50 max **Si** 1.0-5.0 **Sn** 1.0 max **Zn** 5.0 max **Other** total named elements 99.5 min | **ASTM** B271; B584; B763 **SAE** J461; J462 |
| **C87300** | Cast Copper Silicon | **Cu** 94.0 min **Fe** 0.20 max **Mn** 0.8-1.5 **Pb** 0.20 max **Si** 3.5-4.5 **Zn** 0.25 max **Other** Cu + sum of named elements 99.5 min | **ASTM** B30; B271; B584; B763 **FED** WW-V-1967 **Other:** Ingot 500A |
| **C87400** | Cast Silicon Brass | **Al** 0.8 max **Cu** 79.0 min **Pb** 1.0 max **Si** 2.5-4.0 **Zn** 12.0-16.0 **Other** Cu + sum of named elements 99.2 min | **ASTM** B30; B271; B584 (formerly B198); B763 **FED** WW-V-1967 **Other:** Ingot 500B |
| *** C87410** | Cast Silicon Brass | **Al** 0.8 max **As** 0.03-0.06 **Cu** 79.0 min **Pb** 1.0 max **Si** 2.5-4.0 **Zn** 12.0-16.0 **Other** Cu + sum of named elements 99.2 min | |
| *** C87420** | Cast Silicon Brass | **Al** 0.8 max **Cu** 79.0 min **Pb** 1.0 max **Sb** 0.03-0.06 **Si** 2.5-4.0 **Zn** 12.0-16.0 **Other** Cu + sum of named elements 99.2 min | |
| *** C87430** | Cast Silicon Brass | **Al** 0.8 max **Cu** 79.0 min **P** 0.03-0.06 **Pb** 1.0 max **Si** 2.5-4.0 **Zn** 12.0-16.0 **Other** Cu + sum of named elements 99.2 min | |
| **C87500** | Cast Silicon Brass | **Al** 0.50 max **Cu** 79.0 min **Pb** 0.50 max **Si** 3.0-5.0 **Zn** 12.0-16.0 **Other** Cu + sum of named elements 99.5 min | **ASTM** B30; B271; B584 (formerly B198); B763; B806 **SAE** J461; J462 **Other:** Ingot 500C |
| *** C87510** | Cast Silicon Brass | **Al** 0.50 max **As** 0.03-0.06 **Cu** 79.0 min **Pb** 0.50 max **Si** 3.0-5.0 **Zn** 12.0-16.0 **Other** Cu + sum of named elements 99.5 min | |
| *** C87520** | Cast Silicon Brass | **Al** 0.50 max **Cu** 79.0 min **Pb** 0.50 max **Sb** 0.03-0.06 **Si** 3.0-5.0 **Zn** 12.0-16.0 **Other** Cu + sum of named elements 99.5 min | |
| *** C87530** | Cast Silicon Brass | **Al** 0.50 max **Cu** 79.0 min **P** 0.03-0.06 **Pb** 0.50 max **Si** 3.0-5.0 **Zn** 12.0-16.0 **Other** Cu + sum of named elements 99.5 min | |

*Boxed entries are no longer active and are retained for reference purposes only.

The chemical compositions listed are for identification purposes and should not be used in lieu of the cross referenced specifications.

| UNIFIED NUMBER | DESCRIPTION | CHEMICAL COMPOSITION | CROSS REFERENCE SPECIFICATIONS |
|---|---|---|---|
| **C87600** | Cast Silicon Brass | **Cu** 88.0 min **Fe** 0.20 max **Mn** 0.25 max **Pb** 0.50 max **Si** 3.5-4.5 **Zn** 4.0-7.0 **Other** Cu + sum of named elements 99.5 min | **ASTM** B30; B584; B763 **Other:** Ingot 500D |
| **C87610** | Cast Leaded Tin Bronze | **Cu** 90.0 min **Fe** 0.20 max **Mn** 0.25 max **Pb** 0.20 max **Si** 3.0-5.0 **Zn** 3.0-5.0 **Other** Cu + sum of named elements 99.5 min | **ASTM** B30; B584; B763 **Other:** Ingot 500E |
| **C87800** | Cast Silicon Bronze | **Al** 0.15 max **As** 0.05 max **Cu** 80.0 min **Fe** 0.15 max **Mg** 0.01 max **Mn** 0.15 max **Ni** (incl Co) 0.20 max **P** 0.01 max **Pb** 0.15 max **S** 0.05 max **Sb** 0.05 max **Si** 3.8-4.2 **Sn** 0.25 max **Zn** 12.0-16.0 **Other** Cu + sum of named elements 99.5 min | **ASTM** B30; B176; B806 **SAE** J461; J462 **Other:** Ingot 500F |
| * **C87900** | Cast Silicon Bronze | **Al** 0.15 max **As** 0.05 max **Cu** 63.0 min **Fe** 0.40 max **Mn** 0.15 max **Ni** (incl Co) 0.50 max **P** 0.01 max **Pb** 0.25 max **S** 0.05 max **Sb** 0.05 max **Si** 0.8-1.2 **Sn** 0.25 max **Zn** 30.0-36.0 **Other** Cu + sum of named elements 99.5 min | **ASTM** B30 **SAE** J461; J462 **Other:** Ingot 500G |
| **C89320** | Cast Copper-Tin-Bismuth Bronze Alloy (Magnolia B-1 Lead-free Bronze) | **Al** 0.005 max **Bi** 4.0-6.0 **Cu** 87.0-91.0 **Fe** 0.20 max **Ni** (incl Co) 1.0 max **P** 0.30 max **Pb** 0.09 max **S** 0.08 max **Sb** 0.35 max **Si** 0.005 max **Sn** 5.0-7.0 **Zn** 1.0 max **Other** Cu + sum of named elements 99.5 min | **ASTM** B30 **Other:** CDA Application Data Sheet, Standard Designations for Wrought and Cast Copper and Copper Alloys as available on the Internet Web Site, http://properties.copper.org/servlet/com.copper.servlet.CDAPropertiesSelectionServlet |
| **C89325** | Cast Copper-Tin-Bismuth Bronze Alloy (Federalloy III-937) | **Al** 0.005 max **Bi** 2.7-3.7 **Cu** 84.0-88.0 **Fe** 0.15 max **Ni** (incl Co) 1.0 max **P** 0.10 max **Pb** 0.10 max **S** 0.08 max **Sb** 0.50 max **Si** 0.005 max **Sn** 9.0-11.0 **Zn** 1.0 max **Other** Misch Metal 0.10-1.0, Cu + sum of named elements (excl. Misch Metal) 99.0 min | **Other:** CDA Application Data Sheet, Standard Designations for Wrought and Cast Copper and Copper Alloys as available on the Internet Web Site, http://properties.copper.org/servlet/com.copper.servlet.CDAPropertiesSelectionServlet |
| **C89510** | Cast Copper-Tin-Zinc-Bismuth-Selenium Alloy (SeBiLOY I) | **Al** 0.005 **Bi** 0.50-1.5 **Cu** 86.0-88.0 **Fe** 0.20 max **Ni** (incl Co) 1.0 max **P** 0.05 max **Pb** 0.25 max **S** 0.08 max **Sb** 0.25 max **Se** 0.35-0.75 **Si** 0.005 max **Sn** 4.0-6.0 **Zn** 4.0-6.0 **Other** Cu + sum of named elements 99.5 min **Note:** Experience favors Bi:Se 2:1 | **Other:** CDA Application Data Sheet, Standard Designations for Wrought and Cast Copper and Copper Alloys as available on the Internet Web Site, http://properties.copper.org/servlet/com.copper.servlet.CDAPropertiesSelectionServlet |
| **C89520** | Cast Copper-Tin-Zinc-Bismuth-Selenium Alloy (SeBiLOY II) | **Al** 0.005 max **Bi** 1.6-2.2 **Cu** 85.0-87.0 **Fe** 0.20 max **Ni** (incl Co) 1.0 max **P** 0.05 max **Pb** 0.25 max **S** 0.08 max **Sb** 0.25 max **Se** 0.8-1.1 **Si** 0.005 max **Sn** 5.0-6.0 **Zn** 4.0-6.0 **Other** Cu + sum of named elements 99.5 min; Bi:Se 2:1 | **ASTM** B30; B584 **Other:** CDA Application Data Sheet, Standard Designations for Wrought and Cast Copper and Copper Alloys as available on the Internet Web Site, http://properties.copper.org/servlet/com.copper.servlet.CDAPropertiesSelectionServlet |
| **C89550** | Cast Copper-Zinc-Bismuth-Selenium Alloy (SeBiLOY III) | **Al** 0.10-0.6 **Bi** 0.6-1.2 **Cu** 58.0-64.0 **Fe** 0.50 max **Is** 0.25 max **Ni** (incl Co) 1.0 max **P** 0.01 max **Pb** 0.10 max **S** 0.05 max **Se** 0.01-0.10 **Sn** 0.50-1.5 **Zn** 32.0-40.0 **Other** Cu + sum of named elements 99.5 min | **Other:** CDA Application Data Sheet, Standard Designations for Wrought and Cast Copper and Copper Alloys as available on the Internet Web Site, http://properties.copper.org/servlet/com.copper.servlet.CDAPropertiesSelectionServlet |
| **C89831** | Cast Copper-Tin-Zinc-Bismuth Alloy (Federalloy I-844) | **Al** 0.005 max **Bi** 2.7-3.7 **Cu** 87.0-91.0 **Fe** 0.30 max **Ni** (incl Co) 1.0 max **P** 0.050 max **Pb** 0.10 max **S** 0.08 max **Sb** 0.25 max **Si** 0.005 max **Sn** 2.7-3.7 **Zn** 2.0-4.0 **Other** Misch Metal 0.10-1.0, Cu + sum of named elements (exc. Misch Metal) 99.0 min | **Other:** CDA Application Data Sheet, Standard Designations for Wrought and Cast Copper and Copper Alloys as available on the Internet Web Site, http://properties.copper.org/servlet/com.copper.servlet.CDAPropertiesSelectionServlet |

*Boxed entries are no longer active and are retained for reference purposes only.

The chemical compositions listed are for identification purposes and should not be used in lieu of the cross referenced specifications.

| UNIFIED NUMBER | DESCRIPTION | CHEMICAL COMPOSITION | CROSS REFERENCE SPECIFICATIONS |
|---|---|---|---|
| **C89833** | Cast Copper-Tin-Zinc-Bismuth Alloy (Federalloy I-836) | **Al** 0.005 max **Bi** 1.7-2.7 **Cu** 87.0-91.0 **Fe** 0.30 max **Ni** (incl Co) 1.0 max **P** 0.050 max **Pb** 0.10 max **S** 0.08 max **Sb** 0.25 max **Si** 0.005 max **Sn** 4.0-6.0 **Zn** 2.0-4.0 **Other** Misch Metal 0.10-1.0, Cu + sum of named elements (excl. Misch Metal) 99.0 min | **Other:** CDA Application Data Sheet, Standard Designations for Wrought and Cast Copper and Copper Alloys as available on the Internet Web Site, http://properties.copper.org/servlet/com.copper.servlet.CDAPropertiesSelectionServlet |
| **C89835** | Cast Copper-Tin-Zinc-Bismuth Alloy (Federalloy III-932) | **Al** 0.005 max **Bi** 1.7-2.7 **Cu** 85.0-89.0 **Fe** 0.20 max **Ni** (incl Co) 1.0 max **P** 0.10 max **Pb** 0.10 max **S** 0.08 max **Sb** 0.35 max **Si** 0.005 max **Sn** 6.0-7.5 **Zn** 2.0-4.0 **Other** Misch Metal 0.10-1.0, Cu + sum of named elements (excl. Misch Metal) 99.0 min | **Other:** CDA Application Data Sheet, Standard Designations for Wrought and Cast Copper and Copper Alloys as available on the Internet Web Site, http://properties.copper.org/servlet/com.copper.servlet.CDAPropertiesSelectionServlet |
| **C89837** | Cast Copper-Tin-Zinc-Bismuth Alloy (Federalloy I) | **Al** 0.005 max **Bi** 0.7-1.2 **Cu** 84.0-88.0 **Fe** 0.30 max **Ni** (incl Co) 1.0 max **P** 0.050 max **Pb** 0.10 max **S** 0.08 max **Sb** 0.25 max **Si** 0.005 max **Sn** 3.0-4.0 **Zn** 6.0-10.0 **Other** Misch Metal 0.10-1.0, Cu + sum of named elements (excl. Misch Metal) 99.0 min | **Other:** CDA Application Data Sheet, Standard Designations for Wrought and Cast Copper and Copper Alloys as available on the Internet Web Site, http://properties.copper.org/servlet/com.copper.servlet.CDAPropertiesSelectionServlet |
| **C89844** | Cast Copper-Zinc-Tin-Bismuth Alloy | **Al** 0.005 max **Bi** 2.0-4.0 **Cu** 83.0-86.0 **Fe** 0.30 max **Ni** (incl Co) 1.0 max **P** 0.05 max **Pb** 0.20 max **S** 0.08 max **Sb** 0.25 max **Si** 0.005 max **Sn** 3.0-5.0 **Zn** 7.0-10.0 **Other** (Cu + sum of named elements) 99.3 min | **ASTM** B30; B763 **Other:** CDA Application Data Sheet, Standard Designations for Wrought and Cast Copper and Copper Alloys as available on the Internet Web Site, http://properties.copper.org/servlet/com.copper.servlet.CDAPropertiesSelectionServlet |
| **C89940** | Cast Copper-Tin-Zinc-Nickel-Bismuth Alloy (Waukesha 119WM) | **Al** 0.005 max **Bi** 4.0-5.5 **Cu** 64.0-68.0 **Fe** 0.7-2.0 **Mn** 0.20 max **Ni** (incl Co) 20.0-23.0 **P** 0.10-0.15 **Pb** 0.01 max **S** 0.05 max **Sb** 0.10 max **Si** 0.15 max **Sn** 3.0-5.0 **Zn** 3.0-5.0 **Other** Cu + sum of named elements 99.5 min **Other:** CDA Application Data Sheet, Standard Designations for Wrought and Cast Copper and Copper Alloys as available on the Internet Web Site, http://properties.copper.org/servlet/com.copper.servlet.CDAPropertiesSelectionServlet | |
| **C90200** | Cast Tin Bronze | **Al** 0.005 **Cu** (incl Ni) 91.0-94.0 **Fe** 0.20 **Ni** (incl Co) 0.50 max **P** 0.05 max **Pb** 0.30 max **S** 0.05 **Sb** 0.20 **Si** 0.005 **Sn** 6.0-8.0 **Zn** 0.50 max **Other** Cu + sum of named elements 99.4 min **Note:** For continuous castings, P 1.5 max | **Other:** Ingot 242 |
| * **C90250** | Cast Tin Bronze | **Al** 0.005 max **Cu** (incl Ni) 89.0-91.0 **Fe** 0.25 max **Mn** 0.10 max **Ni** (incl Co) 0.50 max **P** 0.05 max **Pb** 0.30 max **S** 0.05 max **Sb** 0.20 max **Si** 0.005 max **Sn** 9.0-11.0 **Zn** 0.50 max **Other** Cu + sum of named elements 99.4 min **Note:** For continuous castings, P 1.5 max | |
| **C90300** | Cast Tin Bronze | **Al** 0.005 max **Cu** (incl Ni) 86.0-89.0 **Fe** 0.20 max **Ni** (incl Co) 1.0 max **P** 0.05 max **Pb** 0.30 max **S** 0.05 max **Sb** 0.20 max **Si** 0.005 **Sn** 7.5-9.0 **Zn** 3.0-5.0 **Other** Cu + sum of named elements 99.4 min **Note:** For continuous castings, P 1.5 max | **ASTM** B30; B271; B505; B584 (formerly B143); B763 **FED** WW-V-1967 **SAE** J461; J462 **Other:** Ingot 225 |

*Boxed entries are no longer active and are retained for reference purposes only.

The chemical compositions listed are for identification purposes and should not be used in lieu of the cross referenced specifications.

| UNIFIED NUMBER | DESCRIPTION | CHEMICAL COMPOSITION | CROSS REFERENCE SPECIFICATIONS |
|---|---|---|---|
| **C90500** | Cast Tin Bronze | **Al** 0.005 max **Cu** (incl Ni) 86.0-89.0 **Fe** 0.20 max **Ni** (incl Co) 1.0 max **P** 0.05 max **Pb** 0.30 max **S** 0.05 max **Sb** 0.20 max **Si** 0.005 max **Sn** 9.0-11.0 **Zn** 1.0-3.0 **Other** Cu + sum of named elements 99.7 min **Note:** For continuous castings, P 1.5 max | **AMS** 4845 **ASTM** B22; B30; B271; B505; B584 (formerly B143); B763 **SAE** J461; J462 **Other:** Ingot 210 |
| **C90700** | Cast Tin Bronze | **Al** 0.005 max **Cu** (incl Ni) 88.0-90.0 **Fe** 0.15 max **Ni** (incl Co) 0.50 max **P** 0.30 max **Pb** 0.50 max **S** 0.05 max **Sb** 0.20 max **Si** 0.005 max **Sn** 10.0-12.0 **Zn** 0.50 max **Other** Cu + sum of named elements 99.4 min **Note:** For continuous castings, P 1.5 max | **ASTM** B30; B427; B505 **SAE** J461; J462 **Other:** Ingot 205 |
| **C90710** | Cast Tin Bronze | **Al** 0.005 max **Cu** (incl Ni) rem **Fe** 0.10 max **Ni** (incl Co) 0.10 max **P** 0.50-1.2 **Pb** 0.25 max **Si** 0.005 max **Sn** 10.0-12.0 **Zn** 0.05 max **Other** Cu + sum of named elements 99.4 min **Note:** For continuous castings, P 1.5 max | |
| **C90800** | Cast Tin Bronze | **Al** 0.005 max **Cu** (incl Ni) 85.0-89.0 **Fe** 0.15 max **Ni** (incl Co) 0.50 max **P** 0.30 max **Pb** 0.25 max **S** 0.05 max **Sb** 0.20 max **Si** 0.005 max **Sn** 11.0-13.0 **Zn** 0.25 max **Other** Cu + sum of named elements 99.4 **Note:** For continuous castings, P 1.5 max | **ASTM** B30; B427 |
| **C90810** | Cast Tin Bronze | **Al** 0.005 max **Cu** (incl Ni) rem **Fe** 0.15 max **Ni** (incl Co) 0.50 max **P** 0.15-0.8 **Pb** 0.25 max **S** 0.05 max **Sb** 0.20 max **Si** 0.005 max **Sn** 11.0-13.0 **Zn** 0.30 max **Other** Cu + sum of named elements 99.4 min **Note:** For continuous castings, P 1.5 max | |
| **C90900** | Cast Tin Bronze | **Al** 0.005 max **Cu** (incl Ni) 86.0-89.0 **Fe** 0.15 max **Ni** (incl Co) 0.50 max **P** 0.05 max **Pb** 0.25 max **S** 0.05 max **Sb** 0.20 max **Si** 0.005 max **Sn** 12.0-14.0 **Zn** 0.25 max **Other** Cu + sum of named elements 99.4 min **Note:** For continuous castings, P 1.5 max | |
| **C91000** | Cast Tin Bronze | **Al** 0.005 max **Cu** (incl Ni) 84.0-86.0 **Fe** 0.10 max **Ni** (incl Co) 0.8 max **P** 0.05 max **Pb** 0.20 max **S** 0.5 max **Sb** 0.20 max **Si** 0.005 max **Sn** 14.0-16.0 **Zn** 1.5 max **Other** Cu + sum of named elements 99.4 min **Note:** For continuous castings, P 1.5 max | **ASTM** B30; B505 **Other:** Ingot 197 |
| **C91100** | Cast Tin Bronze | **Al** 0.005 max **Cu** (incl Ni) 82.0-85.0 **Fe** 0.25 max **Ni** (incl Co) 0.50 max **P** 1.0 max **Pb** 0.25 max **S** 0.05 max **Sb** 0.20 max **Si** 0.005 max **Sn** 15.0-17.0 **Zn** 0.25 max **Other** Cu + sum of named elements 99.4 min **Note:** For continuous castings, P 1.5 max | **ASTM** B22; B30 |
| **C91300** | Cast Tin Bronze | **Al** 0.005 max **Cu** (incl Ni) 79.0-82.0 **Fe** 0.25 max **Ni** (incl Co) 0.50 max **P** 1.0 max **Pb** 0.25 max **S** 0.05 max **Sb** 0.20 max **Si** 0.005 max **Sn** 18.0-20.0 **Zn** 0.25 max **Other** Cu + sum of named elements 99.4 min **Note:** For continuous castings, P 1.5 max | **AMS** 7322 **ASTM** B22; B30; B505 **Other:** Ingot 194 |

The chemical compositions listed are for identification purposes and should not be used in lieu of the cross referenced specifications.

| UNIFIED NUMBER | DESCRIPTION | CHEMICAL COMPOSITION | CROSS REFERENCE SPECIFICATIONS |
|---|---|---|---|
| **C91600** | Cast Tin Bronze | **Al** 0.005 max **Cu** (incl Ni) 86.0-89.0 **Fe** 0.20 max **Ni** (incl Co) 1.2-2.0 **P** 0.30 max **Pb** 0.25 max **S** 0.05 max **Sb** 0.20 max **Si** 0.005 max **Sn** 9.7-10.8 **Zn** 0.25 max **Other** Cu + sum of named elements 99.4 min **Note:** For continuous castings, P 1.5 max | **ASTM** B30; B427 **Other:** Ingot 205N |
| **C91700** | Cast Tin Bronze | **Al** 0.005 max **Cu** (incl Ni) 84.0-87.0 **Fe** 0.20 max **Ni** (incl Co) 1.2-2.0 **P** 0.30 max **Pb** 0.25 max **S** 0.05 max **Sb** 0.20 max **Si** 0.005 max **Sn** 11.3-12.5 **Zn** 0.25 max **Other** Cu + sum of named elements 99.4 min **Note:** For continuous castings, P 1.5 max | **ASTM** B30; B427 |
| **C92200** | Cast Leaded Tin Bronze | **Al** 0.005 max **Cu** (incl Ni) 86.0-90.0 **Fe** 0.25 max **Ni** (incl Co) 1.0 max **P** 0.05 max **Pb** 1.0-2.0 **S** 0.05 max **Sb** 0.25 max **Si** 0.005 max **Sn** 5.5-6.5 **Zn** 3.0-5.0 **Other** Cu + sum of named elements 99.3 min **Note:** For continuous castings, P 1.5 max | **ASME** SB-61; SB-584 **ASTM** B30; B61; B271; B505; B584 (formerly B143) **FED** WW-P-460; WW-T-725; WW-V-1967 **MIL SPEC** MIL-V-17547 **SAE** J461; J462 **Other:** Ingot 245 |
| **C92210** | Cast Leaded Tin Bronze | **Al** 0.005 max **Cu** (incl Ni) 86.0-89.0 **Fe** 0.25 max **Ni** (incl Co) 0.7-1.0 **P** 0.03 max **Pb** 1.7-2.5 **S** 0.05 max **Sb** 0.20 max **Si** 0.005 max **Sn** 4.5-5.5 **Zn** 3.0-4.5 **Other** (Cu + sum of named elements) 99.3 min **Note:** For continuous castings, P 1.5 max | **ASTM** B30; B584 **Other:** CDA Application Data Sheet, Standard Designations for Wrought and Cast Copper and Copper Alloys as available on the Internet Web Site, http://properties.copper.org/servlet/com.copper.servlet.CDAPropertiesSelectionServlet |
| **C92220** | Lead Tin Bronze Grave Markers | **Cu** 86.0-88.0 **Fe** 0.25 max **Ni** (incl Co) 0.5-1.0 **P** 0.05 max **Pb** 1.5-2.5 **Sn** 5.0-6.0 **Zn** 3.0-5.5 **Other** Cu + sum of named elements 99.3 min, In determining Cu min, copper may be calculated as Cu+Ni | |
| **C92300** | Cast Leaded Tin Bronze | **Al** 0.005 max **Cu** (incl Ni) 85.0-89.0 **Fe** 0.25 max **Ni** (incl Co) 1.0 max **P** 0.05 max **Pb** 0.30-1.0 **S** 0.05 max **Sb** 0.25 max **Si** 0.005 max **Sn** 7.5-9.0 **Zn** 2.5-5.0 **Other** Cu + sum of named elements 99.3 min **Note:** For continuous castings, P 1.5 max | **ASTM** B30; B271; B505; B584 (formerly B143); B763 **SAE** J461; J462 **Other:** Ingot 230 |
| **C92310** | Cast Leaded Tin Bronze | **Al** 0.005 max **Bi** 0.03 max **Cu** (incl Ni) rem **Ni** (Incl Co) 1.0 maxi **Pb** 0.30-1.5 **Si** 0.005 max **Sn** 7.5-8.5 **Zn** 3.5-4.5 **Other** Cu + sum of named elements 99.3 min | |
| **C92400** | Cast Leaded Tin Bronze | **Al** 0.005 max **Cu** (incl Ni) 86.0-89.0 **Fe** 0.25 max **Ni** (incl Co) 1.0 max **P** 0.05 max **Pb** 1.0-2.5 **S** 0.05 max **Sb** 0.25 max **Si** 0.005 max **Sn** 9.0-11.0 **Zn** 1.0-3.0 **Other** Cu + sum of named elements 99.3 min **Note:** For continuous castings, P 1.5 max | **Other:** Ingot 220 |
| **C92410** | Cast Leaded Tin Bronze | **Al** 0.005 max **Bi** 0.05 max **Cu** (incl Ni) rem **Fe** 0.20 max **Ni** (incl Co) 0.20 max **Pb** 2.5-3.5 **Sb** 0.25 max **Si** 0.005 max **Sn** 6.0-8.0 **Zn** 1.5-3.0 **Other** Cu + sum of named elements 99.3 min | |
| **C92500** | Cast Leaded Tin Bronze | **Al** 0.005 max **Cu** 85.0-88.0 **Fe** 0.30 max **Ni** (incl Co) 0.8-1.5 **P** 0.30 max **S** 0.05 max **Pb** 1.0-1.5 **Sb** 0.25 max **Si** 0.005 max **Sn** 10.0-12.0 **Zn** 0.50 max **Other** Cu + sum of named elements 99.3 min **Note:** For continuous castings, P 1.5 max | **ASTM** B30; B505 **FED** WW-V-1967 **SAE** J461; J462 **Other:** Ingot 200N |

The chemical compositions listed are for identification purposes and should not be used in lieu of the cross referenced specifications.

| UNIFIED NUMBER | DESCRIPTION | CHEMICAL COMPOSITION | CROSS REFERENCE SPECIFICATIONS |
|---|---|---|---|
| **C92600** | Cast Leaded Tin Bronze | **Al** 0.005 max **Cu** (incl Ni) 86.0-88.5 **Fe** 0.20 max **Ni** (incl Co) 0.7 max **P** 0.03 max **Pb** 0.8-1.5 **S** 0.05 max **Sb** 0.25 max **Si** 0.005 max **Sn** 9.3-10.5 **Zn** 1.3-2.5 **Other** Cu + sum of named elements 99.3 min **Note:** For continuous castings, P 1.5 max | **ASTM** B30; B763; B584 **Other:** Ingot 215 |
| **C92610** | Cast Leaded Tin Bronze | **Al** 0.005 max **Bi** 0.03 max **Cu** (incl Ni) rem **Fe** 0.15 max **Ni** (incl Co) 1.0 max **Pb** 0.30-1.5 **Si** 0.005 max **Sn** 9.5-10.5 **Zn** 1.7-2.8 **Other** Cu + sum of named elements 99.3 min | |
| **C92700** | Cast Leaded Tin Bronze | **Al** 0.005 max **Cu** (incl Ni) rem **Fe** 0.20 max **Ni** (incl Co) 1.0 max **P** 0.25 max **Pb** 1.0-2.5 **S** 0.05 max **Sb** 0.25 max **Si** 0.005 max **Sn** 9.0-11.0 **Zn** 0.7 max **Other** Cu + sum of named elements 99.3 min **Note:** For continuous castings, P 1.5 max | **ASTM** B30; B505 **SAE** J461; J462 **Other:** Ingot 206 |
| **C92710** | Cast Leaded Tin Bronze | **Al** 0.005 max **Cu** (incl Ni) rem **Fe** 0.20 max **Ni** (incl Co) 2.0 max **P** 0.10 max **Pb** 4.0-6.0 **S** 0.05 max **Sb** 0.25 max **Si** 0.005 max **Sn** 9.0-11.0 **Zn** 1.0 max **Other** Cu + sum of named elements 99.3 min **Note:** For continuous castings, P 1.5 max | |
| **C92800** | Cast Leaded Tin Bronze | **Al** 0.005 max **Cu** (incl Ni) 78.0-82.0 **Fe** 0.20 max **Ni** (incl Co) 0.8 max **P** 0.05 max **Pb** 4.0-6.0 **S** 0.05 max **Sb** 0.25 max **Si** 0.005 max **Sn** 15.0-17.0 **Zn** 0.8 max **Other** Cu + sum of named elements 99.3 min **Note:** For continuous castings, P 1.5 max | **AMS** 7320 **ASTM** B30; B505 |
| **C92810** | Cast Leaded Tin Bronze | **Al** 0.005 max **Cu** 78.0-82.0 **Fe** 0.50 max **Ni** (incl Co) 0.8-1.2 **P** 0.05 max **Pb** 4.0-6.0 **S** 0.05 max **Sb** 0.25 max **Si** 0.005 max **Sn** 12.0-14.0 **Zn** 0.50 max | |
| **C92900** | Cast Leaded Tin Bronze | **Al** 0.005 max **Cu** 82.0-86.0 **Fe** 0.20 max **Ni** (incl Co) 2.8-4.0 **P** 0.50 max **Pb** 2.0-3.2 **S** 0.05 max **Sb** 0.25 max **Si** 0.005 max **Sn** 9.0-11.0 **Zn** 0.25 max **Other** Cu + sum of named elements 99.3 min **Note:** For continuous castings, P 1.5 max | **ASTM** B30; B427; B505 **SAE** J461; J462 **Other:** Ingot 206N |
| **C93100** | Cast High-Leaded Tin Bronze | **Al** 0.005 max **Cu** (incl Ni) rem **Fe** 0.25 max **Ni** (incl Co) 1.0 max **P** 0.30 max **Pb** 2.0-5.0 **S** 0.05 max **Sb** 0.25 max **Si** 0.005 max **Sn** 6.5-8.5 **Zn** 2.0 max **Other** Cu + sum of named elements 99.2 min **Note:** For continuous castings, P 1.5 max | |
| **C93200** | Cast High-Leaded Tin Bronze | **Al** 0.005 max **Cu** (incl Ni) 81.0-85.0 **Fe** 0.20 max **Ni** (incl Co) 1.0 max **P** 0.15 max **Pb** 6.0-8.0 **S** 0.08 max **Sb** 0.35 max **Si** 0.005 max **Sn** 6.3-7.5 **Zn** 1.0-4.0 **Other** Cu + sum of named elements 99.2 min **Note:** For continuous castings, P 1.5 max | **ASTM** B30; B66; B271; B505; B584 (formerly B144); B763 **SAE** J461; J462 **Other:** Ingot 315 |
| **C93400** | Cast High-Leaded Tin Bronze | **Al** 0.005 max **Cu** (incl Ni) 82.0-85.0 **Fe** 0.20 max **Ni** (incl Co) 1.0 max **P** 0.50 max **Pb** 7.0-9.0 **S** 0.08 max **Sb** 0.50 max **Si** 0.005 max **Sn** 7.0-9.0 **Zn** 0.8 max **Other** Cu + sum of named elements 99.2 min **Note:** For continuous castings, P 1.5 max | **ASTM** B30; B505 **Other:** Ingot 310 |

The chemical compositions listed are for identification purposes and should not be used in lieu of the cross referenced specifications.

| UNIFIED NUMBER | DESCRIPTION | CHEMICAL COMPOSITION | CROSS REFERENCE SPECIFICATIONS |
|---|---|---|---|
| **C93500** | Cast High-Leaded Tin Bronze | **Al** 0.005 max **Cu** (incl Ni) 83.0-86.0 **Fe** 0.20 max **Ni** (incl Co) 1.0 max **P** 0.05 max **Pb** 8.0-10.0 **S** 0.08 max **Sb** 0.30 max **Si** 0.005 max **Sn** 4.3-6.0 **Zn** 2.0 max **Other** Cu + sum of named elements 99.4 min **Note:** For continuous castings, P 1.5 max | **ASTM** B30; B271; B505; B584 (formerly B144); B763 **SAE** J461; J462 **Other:** Ingot 326 |
| **C93600** | Cast High-Leaded Tin Bronze | **Al** 0.005 max **Cu** (incl Ni) 79.0-83.0 min **Fe** 0.20 max **Ni** (incl Co) 1.0 max **P** 0.15 max **Pb** 11.0-13.0 **S** 0.08 max **Sb** 0.55 max **Si** 0.005 max **Sn** 6.0-8.0 **Zn** 1.0 max **Other** Cu + sum of named elements 99.3 min **Note:** For continuous castings, P 1.5 max | **ASTM** B30; B66; B271; B505 |
| **C93700** | Cast High-Leaded Tin Bronze | **Al** 0.005 max **Cu** (incl Ni) 78.0-82.0 **Fe** 0.7 max **Ni** (incl Co) 0.50 max **P** 0.10 max **Pb** 8.0-11.0 **S** 0.08 max **Sb** 0.50 max **Si** 0.005 max **Sn** 9.0-11.0 **Zn** 0.8 max **Other** Cu may include Ni; Fe 0.35 max for steel-backed bearings; Cu + sum of named elements 99.0 min **Note:** For continuous castings, P 1.5 max | **AMS** 4842; 4827 **ASME** SB-584 **ASTM** B30; B66; B271; B505; B584 (formerly B144); B763 **SAE** J460; J461; J462 **Other:** Ingot 305 |
| **C93720** | Cast High-Leaded Tin Bronze | **Cu** (incl Ni) 83.0 min **Fe** 0.7 max **Ni** (incl Co) 0.50 max **P** 0.10 max **Pb** 7.0-9.0 **Sb** 0.50 max **Sn** 3.5-4.5 **Zn** 4.0 max **Other** Cu + sum of named elements 99.0 min **Note:** For continuous castings, P 1.5 max | **SAE** J460 |
| **C93800** | Cast High-Leaded Tin Bronze | **Al** 0.005 max **Cu** (incl Ni) 75.0-79.0 **Fe** 0.15 max **Ni** (incl Co) 1.0 max **P** 0.05 max **Pb** 13.0-16.0 **S** 0.08 max **Sb** 0.8 max **Si** 0.005 max **Sn** 6.3-7.5 **Zn** 0.8 max **Other** Cu + sum of named elements 98.9 min **Note:** For continuous castings, P 1.5 max | **ASTM** B30; B66; B271; B505; B584; B763 **SAE** J461; J462 **Other:** Ingot 319 |
| **C93900** | Cast High-Leaded Tin Bronze | **Al** 0.005 max **Cu** (incl Ni) 76.5-79.5 **Fe** 0.40 max **Ni** (incl Co) 0.8 max **P** 1.5 max **Pb** 14.0-18.0 **S** 0.08 max **Sb** 0.50 max **Si** 0.005 max **Sn** 5.0-7.0 **Zn** 1.5 max **Other** Cu + sum of named elements 98.9 min | **ASTM** B30; B505 |
| **C94000** | Cast High-Leaded Tin Bronze | **Al** 0.005 max **Cu** (incl Ni) 69.0-72.0 **Fe** 0.25 max **Ni** (incl Co) 0.50-1.0 **P** 0.05 max **Pb** 14.0-16.0 **S** 0.08 max **Sb** 0.50 max **Si** 0.005 max **Sn** 12.0-14.0 **Zn** 0.50 max **Other** Cu + sum of named elements 98.7 min **Note:** For continuous castings, P 1.5 max and S 0.25 max | **ASTM** B30; B505 |
| **C94100** | Cast High-Leaded Tin Bronze | **Al** 0.005 max **Cu** (incl Ni) 72.0-79.0 **Fe** 0.25 max **Ni** (incl Co) 1.0 max **P** 0.05 max **Pb** 18.0-22.0 **S** 0.08 max **Sb** 0.8 max **Si** 0.005 max **Sn** 4.5-6.5 **Zn** 1.0 max **Other** Cu + sum of named elements 98.7 min **Note:** For continuous castings, P 1.5 max and S 0.25 max | **ASTM** B30; B67; B505 **Other:** Ingot 325 |
| *** C94200** | Cast High-Leaded Tin Bronze | **Cu** 68.5-75.5 **Fe** 0.35 max **Ni** 0.50 max **Pb** 3.0-4.0 **Sb** 0.50 max **Sn** 3.0-4.0 **Zn** 3.0 max **Other** total 0.40 max | |

*Boxed entries are no longer active and are retained for reference purposes only.

The chemical compositions listed are for identification purposes and should not be used in lieu of the cross referenced specifications.

| UNIFIED NUMBER | DESCRIPTION | CHEMICAL COMPOSITION | CROSS REFERENCE SPECIFICATIONS |
|---|---|---|---|
| **C94300** | Cast High-Leaded Tin Bronze | **Al** 0.005 max **Cu** (incl Ni) 67.0-72.0 **Fe** 0.15 max **Ni** (incl Co) 1.0 max **P** 0.08 max **Pb** 22.0-25.0 **S** 0.08 max **Sb** 0.8 max **Si** 0.005 max **Sn** 4.5-6.0 **Zn** 0.8 max **Other** Cu + sum of named elements 99.0 **Note:** For continuous castings, P 1.5 max and S 0.25 max | **ASTM** B30; B66; B271; B505; B584 (formerly B144); B763 **SAE** J461; J462 **Other:** Ingot 322 |
| **C94310** | Cast High-Leaded Tin Bronze | **Cu** rem **Fe** 0.50 max **Ni** (incl Co) 0.25-1.0 **P** 0.05 max **Pb** 27.0-34.0 **Sb** 0.50 max **Sn** 1.5-3.0 **Zn** 0.50 max **Other** Cu + sum of named elements 99.0 min **Note:** For continuous castings, P 1.5 max | |
| **C94320** | Cast High-Leaded Tin Bronze | **Cu** (incl Ni) rem **Fe** 0.35 max **Pb** 24.0-32.0 **Sn** 4.0-7.0 **Other** Cu + sum of named elements 99.0 min | |
| **C94330** | Cast High-Leaded Tin Bronze | **Cu** (incl Ni) 68.5-75.5 **Fe** 0.7 max **Ni** (incl Co) 0.50 max **P** 0.10 max **Pb** 21.0-25.0 **Sb** 0.50 max **Sn** 3.0-4.0 **Zn** 3.0 max **Other** Cu + sum of named elements 99.0 min **Note:** For continuous castings, P 1.5 max | **SAE** J460 |
| **C94400** | Cast High-Leaded Tin Bronze | **Al** 0.005 max **Cu** (incl Ni) rem **Fe** 0.15 max **Ni** (incl Co) 1.0 max **P** 0.50 max **Pb** 9.0-12.0 **S** 0.08 max **Sb** 0.8 max **Si** 0.005 max **Sn** 7.0-9.0 **Zn** 0.8 max **Other** Cu + sum of named elements 99.0 min **Note:** For continuous castings, P 1.5 max | **ASTM** B30; B66 |
| **C94500** | Cast High-Leaded Tin Bronze | **Al** 0.005 max **Cu** (incl Ni) rem **Fe** 0.15 max **Ni** (incl Co) 1.0 max **P** 0.05 max **Pb** 16.0-22.0 **S** 0.08 max **Sb** 0.8 max **Si** 0.005 max **Sn** 6.0-8.0 **Zn** 1.2 max **Other** Cu + sum of named elements 99.0 min **Note:** For continuous castings, P 1.5 max | **ASTM** B30; B66 **Other:** Ingot 321 |
| **C94700** | Cast Nickel-Tin Bronze | **Al** 0.005 max **Cu** 85.0-89.0 **Fe** 0.25 max **Mn** 0.20 max **Ni** 4.5-6.0 **P** 0.05 max **Pb** 0.10 max **S** 0.05 max **Sb** 0.15 max **Si** 0.005 max **Sn** 4.5-6.0 **Zn** 1.0-2.5 **Other** Cu + sum of named elements 99.3 min | **ASTM** B30; B505; B584 (formerly B292); B763 **FED** WW-V-1967 **SAE** J461; J462 |
| **C94800** | Cast Nickel-Tin Bronze | **Al** 0.005 max **Cu** 84.0-89.0 **Fe** 0.25 max **Mn** 0.20 max **Ni** 4.5-6.0 **P** 0.05 max **Pb** 0.30-1.0 **S** 0.05 max **Sb** 0.15 max **Si** 0.005 max **Sn** 4.5-6.0 **Zn** 1.0-2.5 **Other** Cu + sum of named elements 99.3 min | **ASTM** B30; B505; B584 (formerly B292); B763 **FED** WW-V-1967 **SAE** J461; J462 |
| **C94900** | Cast Nickel-Tin Bronze | **Al** 0.005 max **Cu** 79.0-81.0 **Fe** 0.30 max **Mn** 0.10 max **Ni** 4.0-6.0 **P** 0.05 max **Pb** 4.0-6.0 **S** 0.08 max **Sb** 0.25 max **Si** 0.005 max **Sn** 4.0-6.0 **Zn** 4.0-6.0 **Other** Cu + sum of named elements 99.2 min | **ASTM** B30; B584 (formerly B292); B763 |
| **C95200** | Cast Aluminum Bronze | **Al** 8.5-9.5 **Cu** 86.0 min **Fe** 2.5-4.0 **Other** Cu + sum of named elements 99.0 min | **ASME** SB-148; B16.24; SB-505; SB-271 **ASTM** B30; B148; B271; B505; B763 **FED** WW-V-1967 **SAE** J461; J462 **Other:** Ingot 415A |
| **C95210** | Cast Aluminum Bronze | **Al** 8.5-9.5 **Cu** 86.0 min **Fe** 2.5-4.0 **Mg** 0.05 max **Mn** 1.0 max **Ni** (incl Co) 1.0 max **Pb** 0.05 max **Si** 0.25 max **Sn** 0.10 max **Zn** 0.50 max **Other** Cu + sum of named elements 99.0 min | |

The chemical compositions listed are for identification purposes and should not be used in lieu of the cross referenced specifications.

| UNIFIED NUMBER | DESCRIPTION | CHEMICAL COMPOSITION | CROSS REFERENCE SPECIFICATIONS |
|---|---|---|---|
| **C95220** | Cast Aluminum Bronze | **Al** 9.5-10.5 **Cu** rem **Fe** 2.5-4.0 **Mn** 0.50 max **Ni** (incl Co) 2.5 max **Other** Cu + sum of named elements 99.5 min | |
| **C95300** | Cast Aluminum Bronze | **Al** 9.0-11.0 **Cu** 86.0 min **Fe** 0.8-1.5 **Other** Cu + sum of named elements 99.0 min | **ASTM** B30; B148; B271; B505; B763; B806 **FED** WW-V-1967 **SAE** J461; J462 **Other:** Ingot 415B |
| **C95400** | Cast Aluminum Bronze | **Al** 10.0-11.5 **Cu** 83.0 min **Fe** 3.0-5.0 **Mn** 0.50 max **Ni** (incl Co) 1.5 max **Other** Cu + sum of named elements 99.5 min | **ASME** SB-148; SB-271 **ASTM** B30; B148; B271; B505; B763; B806 **FED** WW-V-1967 **SAE** J461; J462 **Other:** Ingot 415C |
| **C95410** | Cast Aluminum Bronze | **Al** 10.0-11.5 **Cu** 83.0 min **Fe** 3.0-5.0 **Mn** 0.50 max **Ni** (incl Co) 1.5-2.5 **Other** Cu + sum of named elements 99.5 min | **ASTM** B30; B148; B271; B505; B763; B806 **Other:** Ingot 415C+Ni |
| **C95420** | Cast Aluminum Bronze | **Al** 10.5-12.0 **Cu** 83.5 min **Fe** 3.0-4.3 **Mn** 0.50 max **Ni** (incl Co) 0.50 max **Other** Cu + sum of named elements 99.5 min | **AMS** 4870; 4871; 4873 |
| **C95500** | Cast Aluminum Bronze | **Al** 10.0-11.5 **Cu** 78.0 min **Fe** 3.0-5.0 **Mn** 3.5 max **Ni** (incl Co) 3.0-5.5 **Other** Cu + sum of named elements 99.5 min | **ASTM** B30; B148; B271; B505; B763; B806 **FED** WW-V-1967 **SAE** J461; J462 **Other:** Ingot 415D |
| **C95510** | Cast Aluminum Bronze | **Al** 9.7-10.9 **Cu** 78.0 min **Fe** 2.0-3.5 **Mn** 1.5 max **Ni** (incl Co) 4.5-5.5 **Sn** 0.20 max **Zn** 0.30 max **Other** Cu + sum of named elements 99.8 min **Note:** Cu includes Ag and Ni includes Co; Cu + all named elements = 99.8 min | **AMS** 4880 |
| **C95520** | Cast Aluminum Bronze | **Al** 10.5-11.5 **Co** 0.20 max **Cr** 0.05 max **Cu** 74.5 min **Fe** 4.0-5.5 **Mn** 1.5 max **Ni** (incl Co) 4.2-6.0 **Pb** 0.03 max **Si** 0.15 max **Sn** 0.25 max **Zn** 0.30 max **Other** Cu + sum of named elements 99.5 min | **AMS** 4881 **ASTM** B30; B148; B271; B305 |
| **C95600** | Cast Aluminum Bronze | **Al** 6.0-8.0 **Cu** 88.0 min **Ni** (incl Co) 0.25 max **Si** 1.8-3.3 **Other** Cu + sum of named elements 99.0 min | **ASTM** B30; B148; B763 **Other:** Ingot 415E |
| **C95700** | Cast Aluminum Bronze | **Al** 7.0-8.5 **Cu** 71.0 min **Fe** 2.0-4.0 **Mn** 11.0-14.0 **Ni** (incl Co) 1.5-3.0 **Pb** 0.03 max **Si** 0.10 max **Other** Cu + sum of named elements 99.5 min | **ASTM** B30; B148; B505 **Other:** Ingot 415F |
| **C95710** | Cast Aluminum Bronze | **Al** 7.0-8.5 **Cu** 71.0 min **Fe** 2.0-4.0 **Mn** 11.0-14.0 **Ni** (incl Co) 1.5-3.0 **P** 0.05 max **Pb** 0.05 max **Si** 0.15 max **Other** Cu + sum of named elements 99.5 min | |
| **C95720** | Cast Aluminum Bronze | **Al** 6.0-8.0 **Cu** 73.0 min **Cr** 0.20 max **Fe** 1.5-3.5 **Mn** 12.0-15.0 **Ni** (incl Co) 3.0-6.0 **Pb** 0.03 max **Si** 0.10 max **Sn** 0.10 max **Zn** 0.10 max **Other** (Cu + sum of named elements) 99.5 min | **Other:** CDA Application Data Sheet, Standard Designations for Wrought and Cast Copper and Copper Alloys as available on the Internet Web Site, http://properties.copper.org/servlet/com.copper.servlet.CDAPropertiesSelectionServlet |
| **C95800** | Cast Aluminum Bronze | **Al** 8.5-9.5 **Cu** 79.0 min **Fe** 3.5-4.5 **Mn** 0.8-1.5 **Ni** (incl Co) 4.0-5.0 **Pb** 0.03 max **Si** 0.10 max **Other** Fe less than Ni, total named elements 99.5 min **Note:** Fe content shall not exceed Ni content | **ASTM** B30; B148; B271; B505; B763; B806 **FED** WW-V-1967 **MIL SPEC** MIL-B-24480; MIL-V-17547 **SAE** J462 **Other:** Ingot 415G |
| **C95810** | Cast Aluminum Bronze | **Al** 8.5-9.5 **Cu** 79.0 min **Fe** 3.5-4.5 **Mg** 0.05 max **Mn** 0.8-1.5 **Ni** (incl Co) 4.0-5.0 **Pb** 0.10 max **Si** 0.10 max **Zn** 0.50 max **Other** Cu + sum of named elements 99.5 min | |

The chemical compositions listed are for identification purposes and should not be used in lieu of the cross referenced specifications.

| UNIFIED NUMBER | DESCRIPTION | CHEMICAL COMPOSITION | CROSS REFERENCE SPECIFICATIONS |
|---|---|---|---|
| **C95820** | Cast Nickel Aluminum Bronze | **Al** 9.0-10.0 **Cu** 77.5 min **Fe** 4.0-5.0 **Mn** 1.5 max **Ni** 4.5-5.8 **Pb** 0.02 max **Si** 0.10 max **Sn** 0.20 max **Zn** 0.20 max **Other:** (Cu + sum of named elements) 99.2 min | **ASME** SB-584 **ASTM** B30; B148; B584 |
| **C95900** | Cast Aluminum Bronze | **Al** 12.0-13.5 **Cu** rem **Fe** 3.0-5.0 **Mn** 1.5 max **Ni** (incl Co) 0.50 max **Other** Cu + sum of named elements 99.5 min | **ASTM** B30; B148; B505 |
| **C96200** | Cast Copper-Nickel | **C** 0.10 max **Cu** rem **Fe** 1.0-1.8 **Mn** 1.5 max **Nb** 1.0 max **Ni** (incl Co) 9.0-11.0 **P** 0.02 max **Pb** 0.01 max **S** 0.02 max **Si** 0.50 max **Other** Cu + sum of named elements 99.5 min | **ASTM** B30; B369 **MIL SPEC** MIL-V-18436 **SAE** J461; J462 |
| **C96300** | Cast Copper-Nickel | **C** 0.15 max **Cu** rem **Fe** 0.50-1.5 **Mn** 0.25-1.5 **Nb** 0.50-1.5 **Ni** (incl Co) 18.0-22.0 **P** 0.02 max **Pb** 0.01 max **S** 0.02 max **Si** 0.50 max **Other** Cu + sum of named elements 99.5 min | |
| **C96400** | Cast Copper-Nickel | **C** 0.15 max **Cu** rem **Fe** 0.25-1.5 **Mn** 1.5 max **Nb** 0.50-1.5 **Ni** (incl Co) 28.0-32.0 **P** 0.02 max **Pb** 0.03 max **S** 0.02 max **Si** 0.50 max **Other** Cu + sum of named elements 99.5 min **Note:** For welding grades, Pb 0.01 max | **ASTM** B30; B369; B505 **FED** WW-V-1967 |
| **C96600** | Cast Copper-Nickel | **Be** 0.40-0.7 **Cu** rem **Fe** 0.8-1.1 **Mn** 1.0 max **Ni** (incl Co) 29.0-33.0 **Pb** 0.01 max **Si** 0.15 max **Other** Cu + sum of named elements 99.5 min | |
| **C96700** | Cast Copper Nickel Beryllium Alloy | **Be** 1.1-1.2 **Cu** rem **Fe** 0.7-1.0 **Mn** 0.7 max **Ni** (incl Co) 29.0-33.0 **Pb** 0.01 max **Si** 0.15 max **Ti** 0.010-0.20 **Zr** 0.10-0.20 **Other** Cu + sum of named elements 99.5 min | **ASTM** B30 (72C); B770 |
| **C96800** | Cast Copper-Nickel Alloy | **Al** 0.10 max **B** 0.001 max **Bi** 0.001 max **Cu** rem **Fe** 0.50 max **Mg** 0.005-0.15 **Mn** 0.05-0.30 **Nb** 0.10-0.30 **Ni** (incl Co) 9.5-10.5 **P** 0.005 max **Pb** 0.005 max **S** 0.0025 max **Sb** 0.02 max **Si** 0.05 max **Sn** 7.5-8.5 **Ti** 0.01 max **Zn** 1.0 max **Other** Cu + sum of named elements 99.5 min | **ASTM** B30; B584 |
| **C96900** | Cast Copper-Nickel Alloy | **Cu** rem **Fe** 0.50 max **Mg** 0.15 max **Mn** 0.50-0.30 **Nb** 0.10 max **Ni** 14.5-15.5 **Pb** 0.02 max **Sn** 7.5-8.5 **Zn** 0.50 max **Other** Cu includes Ag, Ni includes Co | |
| **C96950** | Cast Copper Nickel | **Cu** rem **Fe** 0.50 max **Mg** 0.15 max **Mn** 0.05-0.40 **Nb** 0.10 max **Ni** (incl Co) 11.0-15.5 **Pb** 0.02 max **Si** 0.30 max **Sn** 5.8-8.5 **Other** (Cu + sum of named elements) 99.5 min | **Other:** CDA Application Data Sheet, Standard Designations for Wrought and Cast Copper and Copper Alloys as available on the Internet Web Site, http://properties.copper.org/servlet/com.copper.servlet.CDAPropertiesSelectionServlet |
| **C97300** | Cast Nickel-Silver | **Al** 0.005 max **Cu** 53.0-58.0 **Fe** 1.5 max **Mn** 0.50 max **Ni** (incl Co) 11.0-14.0 **P** 0.05 max **Pb** 8.0-11.0 **S** 0.08 max **Sb** 0.35 max **Si** 0.15 max **Sn** 1.5-3.0 **Zn** 17.0-25.0 **Other** Cu + sum of named elements 99.0 min | **ASTM** B30; B271; B505; B584 (formerly B149); B763 **Other:** Ingot 410 |
| **C97400** | Cast Nickel-Silver | **Cu** 58.0-61.0 **Fe** 1.5 max **Mn** 0.50 max **Ni** (incl Co) 15.5-17.0 **Pb** 4.5-5.5 **Sn** 2.5-3.5 **Zn** rem **Other** Cu + sum of named elements 99.0 min | **Other:** Ingot 411 |

The chemical compositions listed are for identification purposes and should not be used in lieu of the cross referenced specifications.

| UNIFIED NUMBER | DESCRIPTION | CHEMICAL COMPOSITION | CROSS REFERENCE SPECIFICATIONS |
|---|---|---|---|
| **C97600** | Cast Nickel-Silver | **Al** 0.005 max **Cu** 63.0-67.0 **Fe** 1.5 max **Mn** 1.0 max **Ni** (incl Co) 19.0-21.5 **P** 0.05 max **Pb** 3.0-5.0 **S** 0.08 max **Sb** 0.25 max **Si** 0.15 max **Sn** 3.5-4.5 **Zn** 3.0-9.0 **Other** Cu + sum of named elements 99.7 min | **ASME** SB-584 **ASTM** B30; B271; B505; B584 (formerly B149); B763 **MIL SPEC** MIL-V-18436 **SAE** J461 **Other:** Ingot 412 |
| **C97800** | Cast Nickel-Silver | **Al** 0.005 max **Cu** 64.0-67.0 **Fe** 1.5 max **Mn** 1.0 max **Ni** (incl Co) 24.0-27.0 **P** 0.05 max **Pb** 1.0-2.5 **S** 0.08 max **Sb** 0.20 max **Si** 0.15 max **Sn** 4.0-5.5 **Zn** 1.0-4.0 **Other** Cu + sum of named elements 99.6 min | **ASTM** B30; B271; B505; B584 (formerly B149); B763 **MIL SPEC** MIL-V-18436 **Other:** Ingot 413B |
| **C98200** | Cast Leaded Copper | **Cu** 73.0-79.0 **Fe** 0.7 max **Ni** 0.50 max **P** 0.10 max **Pb** 21.0-27.0 **Sb** 0.50 max **Sn** 0.6-2.0 **Zn** 0.50 max **Other** Cu + sum of named elements 99.5 min | **AMS** 4824 **SAE** J460 |
| **C98400** | Cast Leaded Copper | **Ag** 1.5 max **Cu** rem **Fe** 0.7 max **Ni** 0.50 max **P** 0.10 max **Pb** 26.0-33.0 **Sb** 0.50 max **Sn** 0.50 max **Zn** 0.50 max **Other** Cu + sum of named elements 99.5 min | **AMS** 4820; 4822 **SAE** J460 |
| **C98600** | Cast Leaded Copper | **Ag** 1.5 max **Cu** 60.0-70.0 **Fe** 0.35 max **Pb** 30.0-40.0 **Sn** 0.50 max | |
| **C98800** | Cast Leaded Copper | **Ag** 5.5 max **Cu** (incl Ag) 56.5-62.5 **Fe** 0.35 max **P** 0.02 max **Pb** 37.5-42.5 **Sn** 0.25 max **Zn** 0.10 max **Note:** Pb and Ag may be adjusted to modify the alloy hardness | |
| **C98820** | Cast Leaded Copper | **Cu** rem **Fe** 0.35 max **Pb** 40.0-44.0 **Sn** 1.0-5.0 | |
| **C98840** | Cast Leaded Copper | **Cu** rem **Fe** 0.35 max **Pb** 44.0-58.0 **Sn** 1.0-5.0 **Other** Cu + sum of named elements 99.5 min | |
| **C99300** | Incramet 800 | **Al** 10.7-11.5 **Co** 1.0-2.0 **Cu** rem **Fe** 0.40-1.0 **Ni** 13.5-16.5 **Pb** 0.02 max **Si** 0.02 max **Sn** 0.05 max **Other** Cu + sum of named elements 99.7 min | |
| **C99350** | Cast Copper Alloy | **Al** 9.5-10.5 **Cu** rem **Fe** 1.0 max **Mn** 0.25 max **Ni** (incl Co) 14.5-16.0 **Pb** 0.15 max **Zn** 7.5-9.5 **Other** Cu + sum of named elements 99.7 min | |
| **C99400** | Cast Copper Alloy | **Al** 0.50-2.0 **Cu** rem **Fe** 1.0-3.0 **Mn** 0.50 max **Ni** 1.0-3.5 **Pb** 0.25 max **Si** 0.50-2.0 **Zn** 0.50-5.0 **Other** Cu + sum of named elements 99.7 min | **ASTM** B30; B763 |
| **C99500** | Cast Copper Alloy | **Al** 0.50-2.0 **Cu** rem **Fe** 3.0-5.0 **Mn** 0.50 max **Ni** 3.5-5.5 **Pb** 0.25 max **Si** 0.50-2.0 **Zn** 0.50-2.0 **Other** Cu + sum of named elements 99.7 min | **ASTM** B30; B763 |
| **C99600** | Cast Copper (Incramute I) | **Al** 1.0-2.8 **C** 0.05 max **Co** 0.20 max **Cu** rem **Fe** 0.20 max **Mn** 39.0-45.0 **Ni** 0.20 max **Pb** 0.02 max **Si** 0.10 max **Sn** 0.10 max **Zn** 0.20 max **Other** Cu + sum of named elements 99.7 min | |
| **C99700** | Cast Copper Alloy | **Al** 0.50-3.0 **Cu** 54.0 min **Fe** 1.0 max **Mn** 11.0-15.0 **Nb** 4.0-6.0 **Ni** 4.0-6.0 **Pb** 2.0 max **Sn** 1.0 max **Zn** 19.0-25.0 **Other** Cu + sum of named elements 99.7 min | **ASTM** B30; B176 |
| **C99750** | Cast Copper Alloy | **Al** 0.25-3.0 **Cu** 55.0-61.0 **Fe** 1.0 max **Mn** 17.0-23.0 **Ni** 5.0 max **Pb** 0.50-2.5 **Sn** 0.50-2.5 **Zn** 17.0-23.0 **Other** Cu + sum of named elements 99.7 min | **ASTM** B30; B176 |

The chemical compositions listed are for identification purposes and should not be used in lieu of the cross referenced specifications.

## UNS NUMBERS ASSIGNED TO DATE

**With Description of Each Material Covered and References To Documents In Which The Same or Similar Materials are Described**

**Dxxxxx Number Series**
**Specified Mechanical Properties Steels**

# SPECIFIED MECHANICAL PROPERTIES STEELS

| UNIFIED NUMBER | DESCRIPTION | CHEMICAL COMPOSITION | CROSS REFERENCE SPECIFICATIONS |
|---|---|---|---|
| **D40450** | Carbon Steel, T.S.: 45-55 ksi | **P** 0.040 max **S** 0.050 max **Other** Cu 0.20 min (when specified), Pb 0.15-0.35 (when specified) | **ASME** SA-675 (45) **ASTM** A675 (45) |
| **D40500** | Carbon Steel, T.S.: 50-60 ksi | **P** 0.040 max **S** 0.050 max **Other** Cu 0.20 min, Pb 0.15-0.35 (when specified) | **ASME** SA-675 (50) **ASTM** A675 (50) |
| **D40550** | Carbon Steel, T.S.: 55-65 ksi | **P** 0.040 max **S** 0.050 max **Other** Cu 0.20 min, Pb 0.15-0.35 (when specified) | **ASME** SA-675 (55) **ASTM** A675 (55) |
| **D40600** | Carbon Steel, T.S.: 60-72 ksi | **P** 0.040 max **S** 0.050 max **Other** Cu 0.20 min, Pb 0.15-0.35 (when specified) | **ASME** SA-675 (60) **ASTM** A675 (60) |
| **D40650** | Carbon Steel, T.S.: 65-77 ksi | **P** 0.040 max **S** 0.050 max **Other** Cu 0.20 min, Pb 0.15-0.35 (when specified) | **ASME** SA-675 (65) **ASTM** A675 (65) |
| **D40700** | Carbon Steel, T.S.: 70-85 ksi | **P** 0.040 max **S** 0.050 max **Other** Cu 0.20 min, Pb 0.15-0.35 (when specified) | **ASME** SA-675 (70) **ASTM** A675 (70) |
| **D40750** | Carbon Steel, T.S.: 75-90 ksi | **P** 0.040 max **S** 0.050 max **Other** Cu 0.20 min, Pb 0.15-0.35 (when specified) | **ASME** SA-675 (75) **ASTM** A675 (75) |
| **D40800** | Carbon Steel, T.S.: 80 ksi min | **P** 0.040 max **S** 0.050 max **Other** Cu 0.20 min, Pb 0.15-0.35 (when specified) | **ASME** SA-675 (80) **ASTM** A675 (80) |
| **D40900** | Carbon Steel, T.S.: 90 ksi min | **P** 0.040 max **S** 0.050 max **Other** Cu 0.20 min, Pb 0.15-0.35 (when specified) | **ASME** SA-675 (90) **ASTM** A675 (90) |
| **D50400** | Alloy Steel Casting, Y.P.: 40 ksi min | **P** 0.05 max **S** 0.05 max | **ASTM** A148 (80-40) |
| **D50500** | Alloy Steel Casting, Y.P.: 50 ksi min | **P** 0.05 max **S** 0.05 max | **ASTM** A148 (80-50) |
| **D50600** | Alloy Steel Casting, Y.P.: 60 ksi min | **P** 0.05 max **S** 0.05 max | **ASTM** A148 (90-60) |
| **D50850** | Alloy Steel Casting, Y.P.: 85 ksi min | **P** 0.05 max **S** 0.05 max | **ASTM** A148 (105-85) |
| **D50950** | Alloy Steel Casting, Y.P.: 95 ksi min | **P** 0.05 max **S** 0.05 max | **ASTM** A148 (115-95) |
| **D51150** | Alloy Steel Casting, Y.P.: 115 ksi min | **P** 0.05 max **S** 0.05 max | **ASTM** A148 (130-115) |
| **D51250** | Alloy Steel Casting, Y.P.: 125 ksi min | **P** 0.05 max **S** 0.05 max | **ASTM** A148 (135-125) |
| **D51350** | Alloy Steel Casting, Y.P. 135 ksi min | **P** 0.05 max **S** 0.05 max | **ASTM** A148 (150-135) |
| **D51450** | Alloy Steel Casting, Y.P.: 145 ksi min | **P** 0.05 max **S** 0.05 max | **ASTM** A148 (160-145) |
| **D51500** | Alloy Steel Casting, Y.P.: 150 ksi min | **P** 0.020 max **S** 0.020 max | **ASTM** A148 (165-150) |
| **D51501** | Alloy Steel Casting, Y.P.: 150 ksi min, 20 ft-lb min | **P** 0.020 max **S** 0.020 max | **ASTM** A148 (165-150L) |
| **D51800** | Alloy Steel Casting, Y.P.: 180 ksi min | **P** 0.020 max **S** 0.020 max | **ASTM** A148 (210-180) |
| **D51801** | Alloy Steel Casting, Y.P.: 180 ksi min, 15 ft-lb min | **P** 0.020 max **S** 0.020 max | **ASTM** A148 (210-180L) |
| **D52100** | Alloy Steel Casting, Y.P.: 210 ksi min | **P** 0.020 max **S** 0.020 max | **ASTM** A148 (260-210) |
| **D52101** | Alloy Steel Casting, Y.P.: 210 ksi min, 6 ft-lb min | **P** 0.020 max **S** 0.020 max | **ASTM** A148 (260-210L) |

The chemical compositions listed are for identification purposes and should not be used in lieu of the cross referenced specifications.

| UNIFIED NUMBER | DESCRIPTION | CHEMICAL COMPOSITION | CROSS REFERENCE SPECIFICATIONS |
|---|---|---|---|
| **D61440** | Alloy Steel, HB: 143-352, Proof Load: 144 ksi | **Cr** 0.30-1.25 **Cu** 0.25-0.55 **Ni** 1.00 max **P** 0.01-0.15 **S** 0.050 max **Si** 0.20-0.90 | **ASME** SA-563 (C3-N) **ASTM** A563 (C3-N) |

The chemical compositions listed are for identification purposes and should not be used in lieu of the cross referenced specifications.

# UNS NUMBERS ASSIGNED TO DATE

## With Description of Each Material Covered and References To Documents In Which The Same or Similar Materials are Described

## Exxxxx Number Series
## Rare Earth and Similar Metals and Alloys

# RARE EARTH AND SIMILAR METALS AND ALLOYS

| UNIFIED NUMBER | DESCRIPTION | CHEMICAL COMPOSITION | CROSS REFERENCE SPECIFICATIONS |
|---|---|---|---|
| **E00000** | Commercial Grade Actinium | **Ac** 99.0 min | |
| **E01000** | Commercial Grade Cerium | **Ce** 99.0 min | |
| **E01401** | Cerium - Cobalt Rare Earth Alloy | **Ce** 70.0 nom **Co** 30.0 nom | |
| **E21000** | Mischmetal (Rare Earths) (Bastnasite Derived) | **Other** total mixed rare earths 99.0 min, consisting of Ce 50.0, La 38.0, Nd 12.0, Pr 4.0, other rare earth metals 1.0 | |
| **E21100** | Mischmetal (Rare Earths) (Bastnasite Derived)-2.5 Magnesium | **Mg** 2.5 nom **Other** total mixed rare earths 97.5 nom, consisting of Ce 48.8, La 32.1, Nd 11.7, Pr 3.9, other rare earth metals 1.0 | |
| **E21101** | Mischmetal (Rare Earths) (Bastnasite Derived)-5.0 Magnesium | **Mg** 5.0 nom **Other** total mixed rare earths 95.0 nom, consisting of Ce 47.5, La 31.3, Nd 11.4, Pr 3.9, other rare earth metals 1.0 | |
| **E21102** | Mischmetal (Rare Earths) (Bastnasite Derived)-10.0 Magnesium | **Mg** 10.0 nom **Other** total mixed rare earths 90.0 nom, consisting of Ce 45.0, La 29.7, Nd 10.8, Pr 3.6, other rare earth metals 0.9 | |
| **E21103** | Mischmetal (Rare Earths) (Bastnasite Derived)-20.0 Magnesium | **Mg** 20.0 nom **Other** total mixed rare earths 80.0 nom, consisting of Ce 40.0, La 26.4, Nd 9.6, Pr 3.6, other rare earth metals 0.8 | |
| **E21200** | Mischmetal (Rare Earths) (Bastnasite Derived)-1.0 Al | **Al** 1.0 nom **Other** total mixed rare earths 99.0 nom | |
| **E21201** | Mischmetal (Rare Earths) (Bastnasite Derived)-20.0 Al | **Al** 20.0 nom **Other** total mixed rare earths 80.0 nom | |
| **E21300** | Mischmetal (Rare Earths) (Bastnasite Derived)-1.0 Fe | **Fe** 1.0 nom **Other** total mixed rare earths 99.0 nom, consisting of Ce 49.4, La 32.7, Nd 11.9, Pr 4.0, other rare earth metals 1.0 | |
| **E21301** | Mischmetal (Rare Earths) (Bastnasite Derived)-5.0 Fe | **Fe** 5.0 nom **Other** total mixed rare earths 95.0 nom | |
| **E21302** | Mischmetal (Rare Earths) (Bastnasite Derived)-25.0 Fe | **Fe** 25.0 nom **Other** total mixed rare earths 75.0 nom | |
| **E21500** | Mischmetal (Rare Earths) (Bastnasite Derived)-1.0 Au | **Au** 1.0 nom **Other** total mixed rare earths 99.0 nom | |
| **E23130** | Mischmetal (Rare Earths) (Bastnasite Derived)-2.5 Mg, 1.0 Fe | **Fe** 1.0 nom **Mg** 2.5 nom **Other** total mixed rare earths 96.5 nom, consisting of Ce 48.2, La 31.8, Nd 11.6, Pr 3.9, other rare earth metals 1.0 | |
| **E23310** | Mischmetal (Rare Earths) (Bastnasite Derived)-10.0 Fe-2.5 Mg | **Fe** 10.0 nom **Mg** 2.5 nom **Other** total mixed rare earths 87.5 nom | |
| **E23311** | Mischmetal (Rare Earths) (Bastnasite Derived)-22.0 Fe-2.0 Mg | **Fe** 22.0 nom **Mg** 2.0 nom **Other** total mixed rare earths 76.0 nom | |
| **E23312** | Mischmetal (Rare Earths) (Bastnasite Derived)-23.0 Fe-6.0 Mg | **Fe** 23.0 nom **Mg** 6.0 nom **Other** total mixed rare earths 71.0 nom | |
| **E28531** | Mischmetal (Rare Earths) (Bastnasite Derived)-36.0 Si-14.0 Fe-5.0 Ca | **Ca** 5.0 nom **Fe** 14.0 nom **Si** 36.0 nom **Other** total mixed rare earths 45.0 nom, consisting of Ce 22.5, La 14.8, Nd 5.4, Pr 1.8, other rare earth metals 0.5 | |

The chemical compositions listed are for identification purposes and should not be used in lieu of the cross referenced specifications.

| UNIFIED NUMBER | DESCRIPTION | CHEMICAL COMPOSITION | CROSS REFERENCE SPECIFICATIONS |
|---|---|---|---|
| **E31000** | Mischmetal (Rare Earths) (Monazite Derived) | **Other** total mixed rare earths 99.0 min, consisting of Ce 45.0, La 20.0, Nd 19.0, Pr 6.0, Sm 4.0, Gd 2.0, Y 2.0, other rare earth metals 2.0 | |
| **E46000** | Commercial Grade Dysprosium | **Dy** 99.0 min | |
| **E48000** | Commercial Grade Erbium | **Er** 99.0 min | |
| **E50000** | Commercial Grade Europium | **Eu** 99.0 min | |
| **E52000** | Commercial Grade Gadolinium | **Gd** 99.0 min | |
| **E56000** | Commercial Grade Holmium | **Ho** 99.0 min | |
| **E58000** | Commercial Grade Lanthanum | **La** 99.0 min | |
| **E58401** | Lanthanum-Cobalt Rare Earth Alloy | **Co** 25.0 nom **La** 75.0 nom | |
| **E68000** | Commercial Grade Lutetium | **Lu** 99.0 min | |
| **E69000** | Commercial Grade Neodymium | **Nd** 99.0 min | |
| **E74000** | Commercial Grade Praseodymium | **Pr** 99.0 min | |
| **E74401** | Praseodymium-Cobalt Rare Earth Alloy | **Co** 28.0 nom **Pr** 72.0 nom | |
| **E78000** | Commercial Grade Promethium | **Pm** 99.0 min | |
| **E79000** | Commercial Grade Samarium | **Sm** 99.0 min | |
| **E79401** | Samarium-Cobalt Rare Earth Alloy | **Co** 40.0 nom **Sm** 60.0 nom | |
| **E83000** | Commercial Grade Scandium | **Sc** 99.0 min | |
| **E85000** | Commercial Grade Terbium | **Tb** 99.0 min | |
| **E87000** | Commercial Grade Thulium | **Tm** 99.0 min | |
| **E88000** | Commercial Grade Ytterbium | **Yb** 99.0 min | |
| **E90000** | Commercial Grade Yttrium | **Y** 99.0 min | |
| **E90401** | Yttrium-Cobalt Rare Earth Alloy | **Co** 36.0 nom **Y** 64.0 nom | |
| **E90501** | Yttrium-Chromium Rare Earth Alloy | **Cr** 5.0 nom **Y** 95.0 nom | |

The chemical compositions listed are for identification purposes and should not be used in lieu of the cross referenced specifications.

# UNS NUMBERS ASSIGNED TO DATE

## With Description of Each Material Covered and References To Documents In Which The Same or Similar Materials are Described

## Fxxxxx Number Series
## Cast Irons

# CAST IRONS

| UNIFIED NUMBER | DESCRIPTION | CHEMICAL COMPOSITION | CROSS REFERENCE SPECIFICATIONS |
|---|---|---|---|
| **F10001** | Cast Iron, Gray | **C** 3.50 min | **ASTM** A319 (I) |
| **F10002** | Cast Iron, Gray | **C** 3.20 min | **ASTM** A319 (II) |
| **F10003** | Cast Iron, Gray | **C** 2.80 min | **ASTM** A319 (III) |
| **F10004** | Cast Iron, Gray, HB: 187 max | **C** 3.40-3.70 **Mn** 0.50-0.80 **P** 0.25 max **S** 0.15 max **Si** 2.30-2.80 **Note:** Low carbon with high silicon and vice versa | **ASTM** A159 (G1800) **SAE** J431 (G1800) |
| **F10005** | Cast Iron, Gray, HB: 170-229 | **C** 3.20-3.50 **Mn** 0.60-0.90 **P** 0.20 max **S** 0.15 max **Si** 2.00-2.40 **Note:** Low carbon with high silicon and vice versa | **ASTM** A159 (G2500) **SAE** J431 (G2500) |
| **F10006** | Cast Iron, Gray, HB: 187-241 | **C** 3.10-3.40 **Mn** 0.60-0.90 **P** 0.15 max **S** 0.15 max **Si** 1.90-2.30 **Note:** Low carbon with high silicon and vice versa. | **ASTM** A159 (G3000) **SAE** J431 (G3000) |
| **F10007** | Cast Iron, Gray, HB: 207-255 | **C** 3.00-3.30 **Mn** 0.60-0.90 **P** 0.12 max **S** 0.15 max **Si** 1.80-2.20 **Note:** Low carbon with high silicon and vice versa | **ASTM** A159 (G3500) **SAE** J431 (G3500) |
| **F10008** | Cast Iron, Gray, HB: 217-269 | **C** 3.00-3.30 **Mn** 0.70-1.00 **P** 0.10 max **S** 0.15 max **Si** 1.80-2.10 **Note:** Low carbon with high silicon and vice versa | **ASTM** A159 (G4000) **SAE** J125 (G4000); J431 (G4000) |
| **F10009** | Cast Iron, Gray, HB: 170-229 | **C** 3.40 min **Mn** 0.60-0.90 **P** 0.15 max **S** 0.12 max **Si** 1.60-2.10 | **ASTM** A159 (G2500a) **SAE** J431 (G2500a) |
| **F10010** | Cast Iron, Gray, HB: 207-255 | **C** 3.40 min **Mn** 0.60-0.90 **P** 0.15 max **S** 0.12 max **Si** 1.30-1.80 | **ASTM** A159 (G3500b) **SAE** J431 (G3500b) |
| **F10011** | Cast Iron, Gray, HB: 207-255 | **C** 3.50 min **Mn** 0.60-0.90 **P** 0.15 max **S** 0.12 max **Si** 1.30-1.80 | **ASTM** A159 (G3500c) **SAE** J431 (G3500c) |
| **F10012** | Cast Iron, Gray, HB: 241-321 | **C** 3.10-3.60 **Cr** 0.85-1.25 **Mn** 0.60-0.90 **Mo** 0.40-0.60 **P** 0.10 max **S** 0.15 max **Si** 1.95-2.40 **Other** Ni 0.20-0.45 (optional) | **ASTM** A159 (G4000d) **SAE** J431 (G4000d) |
| **F10090** | Cast Iron Welding Filler Metal (RCI) | **C** 3.25-3.50 **Mn** 0.60-0.75 **P** 0.50-0.75 **S** 0.10 max **Si** 2.75-3.00 | **AWS** A5.15 (RCI) |
| **F10091** | Cast Iron Welding Filler Metal (RCI-A) | **C** 3.25-3.50 **Mn** 0.50-0.70 **Mo** 0.25-0.45 **Ni** 1.20-1.60 **P** 0.20-0.40 **S** 0.10 max **Si** 2.00-2.50 | **AWS** A5.15 (RCI-A) |
| **F10092** | Cast Iron Welding Filler Metal (RCI-B) | **C** 3.25-4.00 **Mg** 0.04-0.1 **Mn** 0.10-0.40 **Ni** 0.50 max **P** 0.05 max **S** 0.015 max **Si** 3.25-3.75 **Other** Ce 0.20 max | **AWS** A5.15 (RCI-B) |
| **F11401** | Cast Iron, Gray, T.S.: 138 MPa (20 ksi) Min | **Other** None specified | **ASME** SA-278 (20) **ASTM** A48 (20); A278 (20) |
| **F11501** | Cast Iron, Gray, T.S.: 145 MPa (21 ksi) Min | **Other** None specified | **ASTM** A126 (A) |
| **F11701** | Cast Iron, Gray, T.S.: 172 MPa (25 ksi) Min | **Other** None specified | **ASME** SA-278 (25) **ASTM** A48 (25); A278 (25) |
| **F12101** | Cast Iron, Gray, T.S.: 207 MPa (30 ksi) Min | **Other** None specified | **ASME** SA-278 (30) **ASTM** A48 (30); A278 (30) |
| **F12102** | Cast Iron, Gray, T.S.: 214 MPa (31 ksi) Min | **Other** None specified | **ASTM** A126 (B) |
| **F12401** | Cast Iron, Gray, T.S.: 241 MPa (35 ksi) Min | **Other** None specified | **ASME** SA-278 (35) **ASTM** A48 (35); A278 (35) |
| **F12801** | Cast Iron, Gray, T.S.: 276 MPa (40 ksi) Min | **Other** None specified | **ASTM** A48 (40) |
| **F12802** | Cast Iron, Gray, T.S.: 283 MPa (41 ksi) Min | **Other** None specified | **ASTM** A126 (C) |

The chemical compositions listed are for identification purposes and should not be used in lieu of the cross referenced specifications.

| UNIFIED NUMBER | DESCRIPTION | CHEMICAL COMPOSITION | CROSS REFERENCE SPECIFICATIONS |
|---|---|---|---|
| **F12803** | Cast Iron, Gray, T.S.: 276 MPa (40 ksi) Min | **Other** None specified | **ASME** SA-278 (40) **ASTM** A278 (40) |
| **F13101** | Cast Iron, Gray, T.S.: 310 MPa (45 ksi) Min | **Other** None specified | **ASTM** A48 (45) |
| **F13102** | Cast Iron, Gray, T.S.: 310 MPa (45 ksi) Min | **Other** None specified | **ASME** SA-278 (45) **ASTM** A278 (45) |
| **F13501** | Cast Iron, Gray, T.S.: 345 MPa (50 ksi) Min | **Other** None specified | **ASTM** A48 (50) |
| **F13502** | Cast Iron, Gray, T.S.: 345 MPa (50 ksi) Min | **Other** None specified | **ASME** SA-278 (50) **ASTM** A278 (50) |
| **F13801** | Cast Iron, Gray, T.S.: 379 MPa (55 ksi) Min | **Other** None specified | **ASTM** A48 (55) |
| **F13802** | Cast Iron, Gray, T.S.: 379 MPa (55 ksi) Min | **Other** None specified | **ASME** SA-278 (55) **ASTM** A278 (55) |
| **F14101** | Cast Iron, Gray, T.S.: 414 MPa (60 ksi) Min | **Other** None specified | **ASTM** A48 (60) |
| **F14102** | Cast Iron, Gray, T.S.: 414 MPa (60 ksi) Min | **Other** None specified | **ASME** SA-278 (60) **ASTM** A278 (60) |
| **F14801** | Cast Iron, Gray, T.S.: 483 MPa (70 ksi) Min | **Other** None specified | **ASME** SA-278 (70) **ASTM** A278 (70) |
| **F15501** | Cast Iron, Gray, T.S.: 552 MPa (80 ksi) Min | **Other** None specified | **ASME** SA-278 (80) **ASTM** A278 (80) |
| **F20000** | Cast Iron, Malleable, T.S.: 345 MPa (50 ksi) Min, Y.S.: 220.5 MPa (32 ksi) Min | **C** 2.20-2.90 **Mn** 0.15-1.25 **P** 0.02-0.15 **S** 0.02-0.20 **Si** 0.90-1.90 | **ASTM** A602 (M3210) **SAE** J158 (M3210) |
| **F20001** | Cast Iron, Malleable, T.S.: 447.9 MPa (65 ksi) Min, Y.S.: 309.7 MPa (45 ksi) Min | **C** 2.20-2.90 **Mn** 0.15-1.25 **P** 0.02-0.15 **S** 0.02-0.20 **Si** 0.90-1.90 | **SAE** J158 (M4504) |
| **F20002** | Cast Iron, Malleable, T.S.: 516.5 MPa (75 ksi) Min, Y.S.: 345 MPa (50 ksi) Min | **C** 2.20-2.90 **Mn** 0.15-1.25 **P** 0.02-0.15 **S** 0.02-0.20 **Si** 0.90-1.90 | **SAE** J158 (M5003) |
| **F20003** | Cast Iron, Malleable, T.S.: 516.5 MPa (75 ksi) Min, Y.S.: 379.3 MPa (55 ksi) Min | **C** 2.20-2.90 **Mn** 0.15-1.25 **P** 0.02-0.15 **S** 0.02-0.20 **Si** 0.90-1.90 | **SAE** J158 (M5503) |
| **F20004** | Cast Iron, Malleable, T.S.: 620.3 MPa (90 ksi) Min, Y.S.: 482.2 MPa (70 ksi) Min | **C** 2.20-2.90 **Mn** 0.15-1.25 **P** 0.02-0.15 **S** 0.02-0.20 **Si** 0.90-1.90 | **SAE** J158 (M7002) |
| **F20005** | Cast Iron, Malleable, T.S.: 723.2 MPa (105 ksi) Min, Y.S.: 586.0 MPa (85 ksi) Min | **C** 2.20-2.90 **Mn** 0.15-1.25 **P** 0.02-0.15 **S** 0.02-0.20 **Si** 0.90-1.90 | **SAE** J158 (M8501) |
| **F22000** | Cast Iron, Malleable, Cupola, T.S.: 276 MPa (40 ksi) Min, Y.S.: 207 MPa (30 ksi) Min | **Other** None specified | **ASTM** A197 |
| **F22200** | Cast Iron, Malleable, T.S.: 345 MPa (50 ksi) Min, Y.S.: 224 MPa (32 ksi) Min | **Other** None specified | **ASTM** A47 (32510) |
| **F22400** | Cast Iron, Malleable, T.S.: 365 MPa (53 ksi) Min, Y.S.: 241 MPa (35 ksi) Min | **Other** None specified | **ASTM** A47 (35018) |
| **F22830** | Cast Iron, Pearlitic Malleable, T.S.: 414 MPa (60 ksi) Min, Y.S.: 276 MPa (40 ksi) Min | **Other** None specified | **ASTM** A220 (40010) |
| **F23130** | Cast Iron, Pearlitic Malleable, T.S.: 448 MPa (65 ksi) Min, Y.S.: 310 MPa (45 ksi) Min, Elongation: 8% Min | **Other** None specified | **ASTM** A220 (45008) |

The chemical compositions listed are for identification purposes and should not be used in lieu of the cross referenced specifications.

| UNIFIED NUMBER | DESCRIPTION | CHEMICAL COMPOSITION | CROSS REFERENCE SPECIFICATIONS |
|---|---|---|---|
| **F23131** | Cast Iron, Pearlitic Malleable, T.S.: 448 MPa (65 ksi) Min, Y.S.: 310 MPa (45 ksi) Min, Elongation: 6% Min | **Other** None specified | **ASTM** A220 (45006) |
| **F23330** | Cast Iron, Pearlitic Malleable, T.S.: 483 MPa (70 ksi) Min, Y.S.: 331 MPa (48 ksi) Min | **Other** None specified | **AMS** 5310 |
| **F23530** | Cast Iron, Pearlitic Malleable, T.S.: 483 MPa (70 ksi) Min, Y.S.: 345 MPa (50 ksi) Min | **Other** None specified | **ASTM** A220 (50005) |
| **F24130** | Cast Iron, Pearlitic, T.S.: 552 MPa (80 ksi) Min, Y.S.: 414 MPa (60 ksi) Min | **Other** None specified | **ASTM** A220 (60004) |
| **F24830** | Cast Iron, Pearlitic Malleable, T.S.: 586 MPa (85 ksi) Min, Y.S.: 483 MPa (70 ksi) Min | **Other** None specified | **ASTM** A220 (70003) |
| **F25530** | Cast Iron, Pearlitic Malleable, T.S.: 655 MPa (95 ksi) Min, Y.S.: 552 MPa (80 ksi) Min | **Other** None specified | **ASTM** A220 (80002) |
| **F26230** | Cast Iron, Pearlitic Malleable, T.S.: 724 MPa (105 ksi) Min, Y.S.: 621 MPa (90 ksi) Min | **Other** None specified | **ASTM** A220 (90001) |
| **F30000** | Cast Iron, Ductile, Martensitic, HB: As specified (Q and T) | **Other** None specified | **SAE** J434 (DQ and T) |
| **F32800** | Cast Iron, Ductile, T.S.: 414 MPa (60 ksi) Min, Y.S.: 276 MPa (40 ksi) Min, HB: 170 | **Other** None specified | **ASME** SA-395 (60-40-18) **ASTM** A395 (60-40-18); A536 (60-40-18) **SAE** J434 (D4018) |
| **F32900** | Cast Iron, Ductile, T.S.: 414 MPa (60 ksi) Min, Y.S.: 290 MPa (42 ksi) Min | | **ASTM** A716 |
| **F33100** | Cast Iron, Ductile, T.S.: 448 MPa (65 ksi) Min, Y.S.: 310 MPa (45 ksi) Min, HB: 156-217 | **Other** None specified | **ASTM** A536 (65-45-12) **SAE** J434 (D4512) |
| **F33101** | Cast Iron, Ductile, T.S.: 414 MPa (60 ksi) Min, Y.S.: 310 MPa (45 ksi) Min, HB: 190 | **C** 3.0 min **Si** 2.50 max | **AMS** 5315 |
| **F33800** | Cast Iron, Ductile, T.S.: 552 MPa (80 ksi) Min, Y.S.: 379 MPa (55 ksi) Min, HB: 187-255 | **Other** None specified | **ASTM** A536 (80-55-06) **SAE** J434 (D5506) |
| **F34100** | Cast Iron, Ductile, T.S.: 552 MPa (80 ksi) Min, Y.S.: 414 MPa (60 ksi) Min | **C** 3.0 min **P** 0.08 max **S** 0.05 max **Si** 3.0 max | **AMS** 5316 **ASTM** A476 |
| **F34800** | Cast Iron, Ductile, T.S.: 689 MPa (100 ksi) Min, Y.S.: 483 MPa (70 ksi) Min, HB: 241-302 | **Other** None specified | **ASTM** A536 (100-70-03) **SAE** J434 (D7003) |
| **F36200** | Cast Iron, Ductile, T.S.: 827 MPa (120 ksi) Min, Y.S.: 621 MPa (90 ksi) Min | **Other** None specified | **ASTM** A536 (120-90-02) |
| **F41000** | Cast Iron, Gray, Austenitic, T.S.: 172 MPa (25 ksi) Min | **C** 3.00 max **Cr** 1.5-2.5 **Cu** 5.50-7.50 **Mn** 0.5-1.5 **Ni** 13.50-17.50 **S** 0.12 max **Si** 1.00-2.80 | **ASTM** A436 (1) |
| **F41001** | Cast Iron, Gray, Austenitic, T.S.: 207 MPa (30 ksi) Min | **C** 3.00 max **Cr** 2.50-3.50 **Cu** 5.50-7.50 **Mn** 0.5-1.5 **Ni** 13.50-17.50 **S** 0.12 max **Si** 1.00-2.80 | **ASTM** A436 (1b) |
| **F41002** | Cast Iron, Gray, Austenitic, Y.S.: 172 MPa (25 ksi) Min | **C** 3.00 max **Cr** 1.5-2.5 **Cu** 0.50 max **Mn** 0.50-1.50 **Ni** 18.00-22.00 **S** 0.12 max **Si** 1.00-2.80 | **ASTM** A436 (2) |

The chemical compositions listed are for identification purposes and should not be used in lieu of the cross referenced specifications.

| UNIFIED NUMBER | DESCRIPTION | CHEMICAL COMPOSITION | CROSS REFERENCE SPECIFICATIONS |
|---|---|---|---|
| **F41003** | Cast Iron, Gray, Austenitic, T.S.: 207 MPa (30 ksi) Min | **C** 3.00 max **Cr** 3.00-6.00 **Cu** 0.50 max **Mn** 0.5-1.5 **Ni** 18.00-22.00 **S** 0.12 max **Si** 1.00-2.80 | **ASTM** A436 (2b) |
| **F41004** | Cast Iron, Gray, Austenitic, T.S.: 172 MPa (25 ksi) Min | **C** 2.60 max **Cr** 2.50-3.50 **Cu** 0.50 max **Mn** 0.5-1.5 **Ni** 28.00-32.00 **S** 0.12 max **Si** 1.00-2.00 | **ASTM** A436 (3) |
| **F41005** | Cast Iron, Gray, Austenitic, T.S.: 172 MPa (25 ksi) Min | **C** 2.60 max **Cr** 4.50-5.50 **Cu** 0.50 max **Mn** 0.5-1.5 **Ni** 29.00-32.00 **S** 0.12 max **Si** 5.00-6.00 | **ASTM** A436 (4) |
| **F41006** | Cast Iron, Gray, Austenitic, T.S.: 138 MPa (20 ksi) Min | **C** 2.40 max **Cr** 0.10 max **Cu** 0.50 max **Mn** 0.5-1.5 **Ni** 34.00-36.00 **S** 0.12 max **Si** 1.00-2.00 | **ASTM** A436 (5) |
| **F41007** | Cast Iron, Gray, Austenitic, T.S.: 172 MPa (25 ksi) Min | **C** 3.00 max **Cr** 1.00-2.00 **Cu** 3.50-5.50 **Mn** 0.5-1.5 **Mo** 1.00 max **Ni** 18.00-22.00 **S** 0.12 **Si** 1.50-2.50 | **ASTM** A436 (6) |
| **F43000** | Cast Iron, Ductile, Austenitic, T.S.: 400 MPa (58 ksi) Min, Y.S.: 207 MPa (30 ksi) Min | **C** 3.00 max **Cr** 1.75-2.75 **Mn** 0.70-1.25 **Ni** 18.00-22.00 **P** 0.08 max **Si** 1.50-3.00 | **ASTM** A439 (D-2) |
| **F43001** | Cast Iron, Ductile, Austenitic, T.S.: 400 MPa (58 ksi) Min, Y.S.: 207 MPa (30 ksi) Min | **C** 3.00 max **Cr** 2.75-4.00 **Mn** 0.70-1.25 **Ni** 18.00-22.00 **P** 0.08 max **Si** 1.50-3.00 | **ASTM** A439 (D-2B) |
| **F43002** | Cast Iron, Ductile, Austenitic, T.S.: 400 MPa (58 ksi) Min, Y.S.: 193 MPa (28 ksi) Min | **C** 2.90 max **Cr** 0.50 max **Mn** 1.80-2.40 **Ni** 21.00-24.00 **P** 0.08 max **Si** 1.00-3.00 | **ASTM** A439 (D-2C) |
| **F43003** | Cast Iron, Ductile, Austenitic, T.S.: 379 MPa (55 ksi) Min, Y.S.: 207 MPa (30 ksi) Min | **C** 2.60 max **Cr** 2.50-3.50 **Mn** 1.00 max **Ni** 28.00-32.00 **P** 0.80 max **Si** 1.00-2.80 | **ASTM** A439 (D-3) |
| **F43004** | Cast Iron, Ductile, Austenitic, T.S.: 379 MPa (55 ksi) Min, Y.S.: 207 MPa (30 ksi) Min | **C** 2.60 max **Cr** 1.00-1.50 **Mn** 1.00 max **Ni** 28.00-32.00 **P** 0.08 max **Si** 1.00-2.80 | **ASTM** A439 (D-3A) |
| **F43005** | Cast Iron, Ductile, Austenitic, T.S.: 414 MPa (60 ksi) Min, HB: 202-273 | **C** 2.60 max **Cr** 4.50-5.50 **Mn** 1.00 max **Ni** 28.00-32.00 **P** 0.08 max **Si** 5.00-6.00 | **ASTM** A439 (D-4) |
| **F43006** | Cast Iron, Ductile, Austenitic, T.S.: 379 MPa (55 ksi) Min, Y.S.: 207 MPa (30 ksi) Min | **C** 2.40 max **Cr** 0.10 max **Mn** 1.00 max **Ni** 34.00-36.00 **P** 0.08 max **Si** 1.00-2.80 | **ASTM** A439 (D-5) |
| **F43007** | Cast Iron, Ductile, Austenitic, T.S.: 379 MPa (55 ksi) Min, Y.S.: 207 MPa (30 ksi) Min | **C** 2.40 max **Cr** 2.00-3.00 **Mn** 1.00 max **Ni** 34.00-36.00 **P** 0.08 max **Si** 1.00-2.80 | **ASTM** A439 (D-5B) |
| **F43010** | Cast Iron, Ductile, Austenitic, T.S.: 448 MPa (65 ksi) Min, Y.S.: 207 MPa (30 ksi) Min | **C** 2.20-2.70 **Cr** 0.20 max **Mn** 3.75-4.50 **Ni** 21.0-24.0 **P** 0.08 max **Si** 1.50-2.50 | **ASTM** A571 (D-2M) |
| **F43020** | Cast Iron, Ductile, Austenitic, T.S.: 379 MPa (50 ksi) Min, Y.S.: 207 MPa (30 ksi) Min | **C** 2.40-3.00 **Cr** 1.70-2.40 **Mn** 0.80-1.50 **Ni** 18.00-22.00 **P** 0.20 max **Si** 1.80-3.20 | |
| **F43021** | Cast Iron, Ductile, Austenitic, T.S.: 345 MPa (50 ksi) Min, Y.S.: 172 MPa (25 ksi) Min | **C** 2.70-3.10 **Cr** 0.50 max **Mn** 1.90-2.50 **Ni** 20.00-23.00 **P** 0.15 **Si** 2.00-3.00 | |
| **F43030** | Cast Iron, Ductile, T.S.: 345 MPa (50 ksi) Min, Y.S.: 173 MPa (25 ksi) Min | **C** 2.5-3.0 **Cr** 0.50 max **Mn** 1.9-2.5 **Mo** 0.30 max **Ni** 20.0-24.0 **P** 0.15 max **S** 0.05 max **Si** 2.0-3.0 | **AMS** 5395 |
| **F45000** | Cast Iron, White, HB (Sand Cast): 550 Min, HB (Chill Cast): 600 Min | **C** 3.0-3.6 **Cr** 1.4-4.0 **Mn** 1.3 max **Mo** 1.0 max **Ni** 3.3-5.0 **P** 0.30 max **S** 0.15 max **Si** 0.8 max | **ASTM** A532 (IA) |
| **F45001** | Cast Iron, White, HB (Sand Cast): 550 Min, HB (Chill Cast): 600 Min | **C** 2.5-3.0 **Cr** 1.4-4.0 **Mn** 1.3 max **Mo** 1.0 max **Ni** 3.3-5.0 **P** 0.30 max **S** 0.15 max **Si** 0.8 max | **ASTM** A532 (IB) |
| **F45002** | Cast Iron, White, HB (Sand Cast): 550 Min, HB (Chill Cast): 600 Min | **C** 2.9-3.7 **Cr** 1.1-1.5 **Mn** 1.3 max **Mo** 1.0 max **Ni** 2.7-4.0 **P** 0.30 max **S** 0.15 max **Si** 0.8 max | **ASTM** A532 (IC) |

The chemical compositions listed are for identification purposes and should not be used in lieu of the cross referenced specifications.

| UNIFIED NUMBER | DESCRIPTION | CHEMICAL COMPOSITION | CROSS REFERENCE SPECIFICATIONS |
|---|---|---|---|
| **F45003** | Cast Iron, White, HB: 400-600 | **C** 2.5-3.6 **Cr** 7.0-11.0 **Mn** 1.3 max **Mo** 1.0 max **Ni** 4.5-7.0 **P** 0.10 max **S** 0.15 max **Si** 1.0-2.2 | **ASTM** A532 (ID) |
| **F45004** | Cast Iron, White, HB: 400-600 | **C** 2.4-2.8 **Cr** 11.0-14.0 **Cu** 1.2 max **Mn** 0.5-1.5 **Mo** 0.5-1.0 **Ni** 0.5 max **P** 0.10 max **S** 0.06 max **Si** 1.0 max | **ASTM** A532 (IIA) |
| **F45005** | Cast Iron, White, HB: 400-600 | **C** 2.4-2.8 **Cr** 14.0-18.0 **Cu** 1.2 max **Mn** 0.5-1.5 **Mo** 1.0-3.0 **Ni** 0.5 max **P** 0.10 max **S** 0.06 max **Si** 1.0 max | **ASTM** A532 (IIB) |
| **F45006** | Cast Iron, White, HB: 400-600 | **C** 2.8-3.6 **Cr** 14.0-18.0 **Cu** 1.2 max **Mn** 0.5-1.5 **Mo** 2.3-3.5 **Ni** 0.5 max **P** 0.10 max **S** 0.06 max **Si** 1.0 max | |
| **F45007** | Cast Iron, White, HB: 400-600 | **C** 2.0-2.6 **Cr** 18.0-23.00 **Cu** 1.2 max **Mn** 0.5-1.5 **Mo** 1.5 max **Ni** 1.5 max **P** 0.10 max **S** 0.06 max **Si** 1.0 max | **ASTM** A532 (IID) |
| **F45008** | Cast Iron, White, HB: 400-600 | **C** 2.6-3.2 **Cr** 18.0-23.0 **Cu** 1.2 max **Mn** 0.5-1.5 **Mo** 1.0-2.0 **Ni** 1.5 max **P** 0.10 max **S** 0.06 max **Si** 1.0 max | |
| **F45009** | Cast Iron, White, HB: 400-600 | **C** 2.3-3.0 **Cr** 23.0-28.0 **Cu** 1.2 max **Mn** 0.5-1.5 **Mo** 1.5 max **Ni** 1.5 max **P** 0.10 max **S** 0.06 max **Si** 1.0 max | |
| **F45100** | Austenitic Chromium-Iron Welding Rod for Surfacing | **C** 3.7-5.0 **Cr** 27.0-35.0 **Mn** 2.0-6.0 **Si** 1.10-2.5 | |
| **F46100** | Austenitic Ni-Cr-Fe-Co Welding Rod for Surfacing (ERNiCrFeCo) | **C** 2.5-3.0 **Cr** 25-30 **Co** 10-15 **Fe** 20-25 **Mn** 1.0 max **Mo** 7-10 **Ni** 10-33 **Si** 0.6-1.5 **W** 2.0-4.0 | **ASME** SFA5.21 (ERNiCrFeCo) **AWS** A5.21 (ERNiCrFeCo) |
| **F47001** | Cast Iron, Corrosion Resistant, T.S.: 172 MPa (25 ksi) Min | **C** 2.60-3.00 **Cr** 1.80-3.50 **Cu** 5.50-7.50 **Mn** 1.00-1.50 **Ni** 13.5-17.5 **P** 0.20 max **S** 0.10 max **Si** 1.25-2.20 | |
| **F47002** | Cast Iron, Corrosion Resistant, T.S.: 172 MPa (25 ksi) Min | **C** 2.60-3.00 **Cr** 1.75-3.50 **Cu** 0.50 max **Mn** 0.80-1.30 **Ni** 18.0-22.0 **P** 0.20 max **S** 0.10 max **Si** 1.25-2.20 | |
| **F47003** | Cast Iron, Corrosion Resistant | **C** 0.70-1.10 **Cr** 0.50 max **Cu** 0.50 max **Mn** 1.50 max **Mo** 0.50 max **Si** 14.20-14.75 | |
| **F47004** | Cast Iron, Corrosion Resistant, T.S.: 207 MPa (30 ksi) Min | **C** 2.4-2.8 **Cr** 1.8-2.4 **Cu** 6.0-7.0 **Mn** 1.0-1.5 **Ni** 14.0-16.0 **P** 0.30 max **Pb** 0.003 max **S** 0.12 max **Si** 1.5-2.5 | **AMS** 5392 |
| **F47005** | Cast Iron, Corrosion and Heat Resistant, T.S.: 207 MPa (30 ksi) Min | **C** 2.4-2.8 **Cr** 1.7-2.4 **Cu** 0.5 **Mn** 0.8-1.6 **Ni** 18.0-22.0 **P** 0.30 **Pb** 0.003 **S** 0.12 **Si** 1.5-2.8 | **AMS** 5393 |
| **F47006** | Cast Iron, Ductile, Corrosion and Heat Resistant, T.S.: 379 MPa (55 ksi) Min, Y.S.: 227 MPa (32 ksi) Min | **C** 2.4-3.0 **Cr** 1.7-2.4 **Cu** 0.50 max **Mn** 0.8-1.6 **Ni** 18.0-22.0 **P** 0.25 max **Pb** 0.003 max **Si** 2.0-3.2 | **AMS** 5394 |

The chemical compositions listed are for identification purposes and should not be used in lieu of the cross referenced specifications.

## UNS NUMBERS ASSIGNED TO DATE

## With Description of Each Material Covered and References To Documents In Which The Same or Similar Materials are Described

## Gxxxxx Number Series
## AISI and SAE Carbon and Alloy Steels

# AISI AND SAE CARBON AND ALLOY STEELS

| UNIFIED NUMBER | DESCRIPTION | CHEMICAL COMPOSITION | CROSS REFERENCE SPECIFICATIONS |
|---|---|---|---|
| **G10050** | Carbon Steel | **C** 0.06 max **Mn** 0.35 max **P** 0.030 max **S** 0.050 max | **AMS** AMS-S-11310 (CS1005) **ASTM** A29 (1005); A510 (1005); A853 (1005) **SAE** J403 (1005); J412 (1005) |
| **G10060** | Carbon Steel | **C** 0.08 max **Mn** 0.25-0.40 **P** 0.030 max **S** 0.050 max **Other** Sheets and Plates, Mn 0.25-0.45 | **AMS** AMS-S-11310 (CS1006) **ASTM** A29 (1006); A510 (1006); A635 (1006); A830; A853 (1006) **SAE** J403 (1006); J412 (1006); J1397 (1006) |
| **G10080** | Carbon Steel | **C** 0.10 max **Mn** 0.30-0.50 **P** 0.030 max **S** 0.050 max **Other** Sheets, Mn 0.25-0.50; ERW Tubing, Mn 0.25-0.50 | **AMS** AMS-S-11310 (CS1008) **ASTM** A29 (1008); A29 (M1008); A108 (1008); A510 (1008); A512 (1008); A513 (1008); A519 (1008); A575 (M1008); A576 (1008); A635 (1008); A787 (1008); A830; A853 (1008) **MIL SPEC** MIL-W-6712 **SAE** J403 (1008); J403 (M1008); J412 (1008); J1397 (1008) |
| **G10090** | Carbon Steel | **C** 0.15 max **Mn** 0.60 max **P** 0.030 max **S** 0.050 max | **ASTM** A635 (1009); A827; A830 **AWS** C2.25 (W-FeC-1) |
| **G10100** | Carbon Steel | **C** 0.08-0.13 **Mn** 0.30-0.60 **P** 0.030 max **S** 0.050 max | **AMS** 5040; 5042; 5044; 5047; 5050; 5053; AMS-S-11310 (CS1010); AS7225 **ASTM** A29 (1010); A29 (M1010); A108; A108 (1010); A510 (1010); A512 (1010, MT 1010); A513 (1010, MT 1010); A519 (1010, MT 1010); A575 (M1010); A787 (1010); A787 (MT 1010); A830; A853 (1010) **SAE** J403 (1010); J403 (M1010); J412 (1010); J1397 (1010) |
| **G10110** | Carbon Steel | **C** 0.08-0.13 **Mn** 0.60-0.90 **P** 0.030 max **S** 0.050 max | **ASTM** A29 (1011); A510 (1011); A787 (1011); A853 (1011) |
| **G10120** | Carbon Steel | **C** 0.10-0.15 **Mn** 0.30-0.60 **P** 0.030 max **S** 0.050 max | **AMS** AMS-S-11310 (CS1012) **ASTM** A29 (1012); A29 (M1012); A510 (1012); A512 (1012); A513 91012); A519 (1012); A575 (M1012); A576 (1012); A611 (C, D); A635 (1012); A830 **MIL SPEC** MIL-W-5712 **SAE** J403 (1012); J403 (M1012); J412 (1012); J1397 (1012) |
| **G10130** | Carbon Steel | **C** 0.11-0.16 **Mn** 0.50-0.80 **P** 0.030 max **S** 0.050 max | **ASTM** A29 (1013) **SAE** J403; J403 (1013); J412 (1013) |
| **G10150** | Carbon Steel | **C** 0.13-0.18 **Mn** 0.30-0.60 **P** 0.030 max **S** 0.050 max **Other** Sheets, C 0.12-0.18 | **AMS** 5060 **ASTM** A29 (1015); A29 (M1015); A108 (1015); A510 (1015); A512 (1015, MT 1015); A513 (1015, MT 1015); A519 (1015, MT 1015); A575 (M1015); A576 (1015); A635 (1015); A659 (1015); A787 (1015); A787 (MT 1015); A794 (1015); A830; A853 (1015) **SAE** J403 (1015); J403 (M1015); J412 (1015); J1397 (1015) |
| **G10160** | Carbon Steel | **C** 0.13-0.18 **Mn** 0.60-0.90 **P** 0.030 max **S** 0.050 max **Other** Sheets, C 0.12-0.18; ERW Tubing, C 0.12-0.19 | **ASTM** A29 (1016); A108 (1016); A510 (1016); A512 (1016); A513 (1016); A519 (1016); A576 (1016); A635 (1016); A659 (1016); A787 (1016); A794 (1016); A830 **SAE** J403 (1016); J412 (1016); J1397 (1016) |
| **G10170** | Carbon Steel | **C** 0.15-0.20 **Mn** 0.30-0.60 **P** 0.030 max **S** 0.050 max **Other** Sheets and Plates, C 0.14-0.20; ERW Tubing, C 0.14-0.21 | **AMS** AMS-S-1310 (CS1017) **ASTM** A29 (1017); A29 (M1017); A510 (1017); A513 (1017); A519 (1017); A575 (M1017); A576 (1017); A611 (A, B, C, E); A635 (1017); A659 (1017); A787 (1017); A794 (1017); A830 **SAE** J403 (1017); J403 (M1017); J412 (1017); J1397 (1017) |
| **G10180** | Carbon Steel | **C** 0.15-0.20 **Mn** 0.60-0.90 **P** 0.030 max **S** 0.050 max **Other** Sheets and Plates, C 0.14-0.20; ERW Tubing, C 0.14-0.21 | **AMS** 5069; AMS-S-11310 (CS1018) **ASTM** A29 (1018); A108 (1018); A311 (1018); A510 (1018); A512 (1018); A513 (1018); A519 (1018); A576 (1018); A611 (D-1); A635 (1018); A659 (1018); A787 (1018); A794 (1018); A830; A853 (1018) **SAE** J403 (1018); J412 (1018); J1397 (1018) |
| **G10190** | Carbon Steel | **C** 0.15-0.20 **Mn** 0.70-1.00 **P** 0.030 max **S** 0.050 max **Other** Sheets and Plates, C 0.14-0.20; ERW Tubing, C 0.14-0.21 | **ASTM** A29 (1019); A510 (1019); A512 (1019); A513 (1019); A519 (1019); A576 (1019); A635 (1019); A787 (1019); A830 **SAE** J412 (1019); J1397 (1019) |
| **G10200** | Carbon Steel | **C** 0.18-0.23 **Mn** 0.30-0.60 **P** 0.030 max **S** 0.050 max **Other** Sheets and Plates, C 0.17-0.23 | **AMS** 5032; 5045; 5046; AMS-S-7952 (CS1020); AMS-S-11310 (CS1020) **ASTM** A29 (1020); A29 (M1020); A108 (1020); A510 (1020); A512 (1020, MT 1020); A513 (1020, MT 1020); A519 (1020, MT 1020); A575 (M1020); A576 (1020); A635 (1020); A659 (1020); A787 (MT 1020); A794 (1020); A827; A830; A853 (1020) **AWS** C2.25 (W-FeC-3) **SAE** J403 (1020); J403 (M1020); J412 (1020); J1397 (1020) |
| **G10210** | Carbon Steel | **C** 0.18-0.23 **Mn** 0.60-0.90 **P** 0.030 max **S** 0.050 max **Other** Sheets, C 0.17-0.23; ERW Tubing, C 0.17-0.24 | **ASTM** A29 (1021); A510 (1021); A512 (1021); A513 (1021); A519 (1021); A576 (1021); A635 (1021); A659 (1021); A787 (1021); A794 (1021); A830 **SAE** J403 (1021); J412 (1021); J1397 (1021) |

The chemical compositions listed are for identification purposes and should not be used in lieu of the cross referenced specifications.

| UNIFIED NUMBER | DESCRIPTION | CHEMICAL COMPOSITION | CROSS REFERENCE SPECIFICATIONS |
|---|---|---|---|
| **G10220** | Carbon Steel | **C** 0.18-0.23 **Mn** 0.70-1.00 **P** 0.030 max **S** 0.050 max **Other** Sheets, C 0.17-0.23; ERW Tubing, C 0.17-0.24 | **AMS** 5070; AMS-S-11310 (CS1022) **ASTM** A29 (1022); A108 (1022); A510 (1022); A513 (1022); A519 (1022); A576 (1022); A635 (1022; A830 **SAE** J403 (1022); J412 (1022); J1397 (1022) |
| **G10230** | Carbon Steel | **C** 0.20-0.25 **Mn** 0.30-0.60 **P** 0.030 max **S** 0.050 max **Other** Sheets, C 0.19-0.25; ERW Tubing, C 0.19-0.26 | **ASTM** A29 (1023); A29 (M1023); A510 (1023); A513 (1023); A575 (M1023); A576 (1023); A635 (1023); A659 (1023); A794 (1023); A830 **SAE** J403 (1023); J403 (1033); J403 (M1023); J412 (1023); J1397 (1023) |
| **G10250** | Carbon Steel | **C** 0.22-0.28 **Mn** 0.30-0.60 **P** 0.030 max **S** 0.050 max **Other** ERW Tubing, C 0.21-0.28 | **AMS** 5046; 5075; 5077; AMS-QQ-S-700 (C1025); AMS-S-7952 (C1025); AMS-S-11310 (CS1025); AMS-T-5066 **ASTM** A29 (1025); A29 (M1025); A108 (1025); A510 (1025); A512 (1025); A513 (1025); A519 (1025); A575 (M1025); A576 (1025); A830 **MIL SPEC** MIL-T-3520 **SAE** J403 (1025); J403 (M1025); J412 (1025); J1397 (1025) |
| **G10260** | Carbon Steel | **C** 0.22-0.28 **Mn** 0.60-0.90 **P** 0.030 max **S** 0.050 max **Other** ERW Tubing, C 0.21-0.28 | **ASTM** A29 (1026); A510 (1026); A512 (1026); A513 (1026); A519 (1026); A576 (1026); A830 **MIL SPEC** MIL-F-20670 **SAE** J403 (1026); J412 (1026); J1397 (1026) |
| **G10290** | Carbon Steel | **C** 0.25-0.31 **Mn** 0.60-0.90 **P** 0.030 max **S** 0.050 max | **ASTM** A29 (1029); A510 (1029); A576 (1029) **SAE** J403 (1029); J412 (1029) |
| **G10300** | Carbon Steel | **C** 0.28-0.34 **Mn** 0.60-0.90 **P** 0.030 max **S** 0.050 max **Other** Sheets, C 0.27-0.34, ERW Tubing, C 0.27-0.35 | **AMS** AMS-Qq-S-700 (C1030); AMS-S-11310 (CS1030) **ASTM** A29 (1030); A510 (1030); A512 (1030); A513 (1030); A519 (1030); A576 (1030); A682 (1030); A830; A866 (1030) **SAE** J403 (1030); J412 (1030); J1397 (1030) |
| **G10330** | Carbon Steel | **C** 0.29-0.36 **Mn** 0.70-1.00 **P** 0.030 max **S** 0.050 max | **ASTM** A513 (1033); A830 |
| **G10340** | Carbon Steel | **C** 0.32-0.38 **Mn** 0.50-0.80 **P** 0.030 max **S** 0.050 max | **ASTM** A29 (1034); A108 (1034); A510 (1034) |
| **G10350** | Carbon Steel | **C** 0.32-0.38 **Mn** 0.60-0.90 **P** 0.030 max **S** 0.050 max **Other** Sheets, C 0.31-0.38; ERW Tubing, C 0.31-0.39 | **AMS** 5080; 5082; AMS-QQ-S-700 (C1035) **ASTM** A29 (1035); A108 (1035); A311 (1035); A510 (1035); A512 (1035); A513 (1035); A519 (1035); A576 (1035); A682 (1035); A827; A830 **MIL SPEC** MIL-S-3289; MIL-S-19434 **SAE** J403 (1035); J412 (1035); J1397 (1035) |
| **G10370** | Carbon Steel | **C** 0.32-0.38 **Mn** 0.70-1.00 **P** 0.030 max **S** 0.050 max **Other** Sheets, C 0.31-0.38 | **ASTM** A29 (1037); A510 (1037); A576 (1037); A830 **SAE** J403 (1037); J1397 (1037) |
| **G10380** | Carbon Steel | **C** 0.35-0.42 **Mn** 0.60-0.90 **P** 0.030 max **S** 0.050 max **Other** Sheets, C 0.34-0.42 | **ASTM** A29 (1038); A510 (1038); A576 (1038); A830 **SAE** J403 (1038); J412 (1038); J1397 (1038) |
| **G10390** | Carbon Steel | **C** 0.37-0.44 **Mn** 0.70-1.00 **P** 0.030 max **S** 0.050 max **Other** Sheets, C 0.36-0.44 | **ASTM** A29 (1039); A510 (1039); A576 (1039); A830 **SAE** J403 (1039); J412 (1039); J1397 (1039) |
| **G10400** | Carbon Steel | **C** 0.37-0.44 **Mn** 0.60-0.90 **P** 0.030 max **S** 0.050 max **Other** Sheets, C 0.36-0.44 | **AMS** AMS-S-11310 (CS1040) **ASTM** A29 (1040); A108 (1040); A510 (1040); A513 (1040); A519 (1040); A576 (1040); A682 (1040); A827; A830; A866 (1040) **SAE** J403 (1040); J412 (1040); J1397 (1040) |
| **G10420** | Carbon Steel | **C** 0.40-0.47 **Mn** 0.60-0.90 **P** 0.030 max **S** 0.050 max **Other** Sheets, C 0.39-0.47 | **ASTM** A29 (1042); A183 (2-Nuts); A510 (1042); A576 (1042); A830 **SAE** J403 (1042); J412 (1042); J1397 (1042) |
| **G10430** | Carbon Steel | **C** 0.40-0.47 **Mn** 0.70-1.00 **P** 0.030 max **S** 0.050 max **Other** Sheets, C 0.39-0.47 | **ASTM** A29 (1043); A183 (2-Nuts); A510 (1043); A576 (1043); A830 **SAE** J403 (1043); J412 (1043); J1397 (1043) |
| **G10440** | Carbon Steel | **C** 0.43-0.50 **Mn** 0.30-0.60 **P** 0.030 max **S** 0.050 max | **ASTM** A29 (1044); A183 (2-Nuts); A510 (1044); A575 (M1044); A576 (1044) **SAE** J403 (1044); J403 (M1044); J412 (1044); J1397 (1044) |
| **G10450** | Carbon Steel | **C** 0.43-0.50 **Mn** 0.60-0.90 **P** 0.030 max **S** 0.050 max **Other** Sheets, C 0.42-0.50 | **AMS** AMS-QQ-S-700 (C1045) **ASTM** A29 (1045); A108 (1045); A183 (2-Nuts); A311 (1045); A510 (1045); A519 (1045); A576 (1045); A682 (1045); A827; A830 **SAE** J403 (1045); J412 (1045); J1397 (1045) |
| **G10460** | Carbon Steel | **C** 0.43-0.50 **Mn** 0.70-1.00 **P** 0.030 max **S** 0.050 max | **ASTM** A29 (1046); A510 (1046); A576 (1046); A830 **SAE** J403 (1046); J412 (1046); J1397 (1046) |
| **G10490** | Carbon Steel | **C** 0.46-0.53 **Mn** 0.60-0.90 **P** 0.030 max **S** 0.050 max | **ASTM** A29 (1049); A510 (1049); A576 (1049); A830 **SAE** J403 (1049); J412 (1049); J1397 (1049) |

The chemical compositions listed are for identification purposes and should not be used in lieu of the cross referenced specifications.

| UNIFIED NUMBER | DESCRIPTION | CHEMICAL COMPOSITION | CROSS REFERENCE SPECIFICATIONS |
|---|---|---|---|
| **G10500** | Carbon Steel | **C** 0.48-0.55 **Mn** 0.60-0.90 **P** 0.030 max **S** 0.050 max | **AMS** 5085; AMS-QQ-S-700 (C1050) **ASTM** A29 (1050); A108 (1050); A311 (1050); A510 (1050); A513 (1050); A519 (1050); A576 (1050); A682 (1050); A827; A830; A866 (1050) **SAE** J403 (1050); J412 (1050); J1397 (1050) |
| **G10530** | Carbon Steel | **C** 0.48-0.55 **Mn** 0.70-1.00 **P** 0.030 max **S** 0.050 max | **ASTM** A29 (1053); A510 (1053); A576 (1053) **SAE** J403 (1053); J412 (1053) |
| **G10550** | Carbon Steel | **C** 0.50-0.60 **Mn** 0.60-0.90 **P** 0.030 max **S** 0.050 max | **AMS** AMS-QQ-S-700 (C1055); AMS-S-700 (C1055) **ASTM** A29 (1055); A510 (1055); A576 (1055); A682 (1055); A713 (1055); A830 **SAE** J403 (1055); J412 (1055); J1397 (1055) |
| **G10590** | Carbon Steel | **C** 0.55-0.65 **Mn** 0.50-0.80 **P** 0.030 max **S** 0.050 max | **ASTM** A29 (1059); A510 (1059); A713 (1059) **SAE** J412 (1059) |
| **G10600** | Carbon Steel | **C** 0.55-0.65 **Mn** 0.60-0.90 **P** 0.030 max **S** 0.050 max | **AMS** 7240 **ASTM** A29 (1060); A510 (1060); A513 (1060); A576 (1060); A682 (1060); A713 (1060) **SAE** J403 (1060); J412 (1060); J1397 (1060) |
| **G10640** | Carbon Steel | **C** 0.60-0.70 **Mn** 0.50-0.80 **P** 0.030 max **S** 0.050 max **Other** Sheets and Plates, C 0.59-0.70 | **ASTM** A29 (1064); A510 (1064); A682 (1064); A713 (1064) **SAE** J412 (1064); J1397 (1064) |
| **G10650** | Carbon Steel | **C** 0.60-0.70 **Mn** 0.60-0.90 **P** 0.030 max **S** 0.050 max **Other** Sheets and Plates, C 0.54-0.70 | **AMS** AMS-QQ-S-700 (1065); AMS-S-700 (C1065) **ASTM** A29 (1065); A510 (1065); A682 (1065); A713 (1065) **MIL SPEC** MIL-S-46049 (1065) **SAE** J412 (1065); J1397 (1065) |
| **G10690** | Carbon Steel | **C** 0.65-0.75 **Mn** 0.40-0.70 **P** 0.030 max **S** 0.050 max | **ASTM** A29 (1069); A510 (1069); A713 (1069) **MIL SPEC** MIL-S-11713 **SAE** J412 (1069) |
| **G10700** | Carbon Steel | **C** 0.65-0.75 **Mn** 0.60-0.90 **P** 0.030 max **S** 0.050 max **Other** Plates, C 0.65-0.76 | **AMS** 5115; 7205 **ASTM** A29 (1070); A510 (1070); A576 (1070); A682 (1070); A713 (1070); A830 **MIL SPEC** MIL-S-11713 (2) **SAE** J403 (1070); J412 (1070); J1397 (1070) |
| **G10740** | Carbon Steel | **C** 0.70-0.80 **Mn** 0.50-0.80 **P** 0.030 max **S** 0.050 max **Other** Sheets and Plates, C 0.69-0.80 | **AMS** 5120; AMS-QQ-S-700 (1074) **ASTM** A29 (1074); A510 (1074); A682 (1074); A713 (1074); A830 **MIL SPEC** MIL-S-46049 (1074) **SAE** J412 (1074); J1397 (1074) |
| **G10750** | Carbon Steel | **C** 0.70-0.80 **Mn** 0.40-0.70 **P** 0.030 max **S** 0.050 max | **ASTM** A29 (1075); A510 (1075); A713 (1075) **AWS** C2.25 (W-FeC-4) **SAE** J412 (1075) |
| **G10780** | Carbon Steel | **C** 0.72-0.85 **Mn** 0.30-0.60 **P** 0.030 max **S** 0.050 max **Other** Plates, C 0.72-0.86 | **ASTM** A29 (1078); A510 (1078); A576 (1078); A713 (1078); A830 **SAE** J403 (1078); J412 (1078); J1397 (1078) |
| **G10800** | Carbon Steel | **C** 0.75-0.88 **Mn** 0.60-0.90 **P** 0.030 max **S** 0.050 max **Other** Plates, C 0.74-0.88 | **AMS** 5110; AMS-QQ-S-700 (1080) **ASTM** A29 (1080); A510 (1080); A576 (1080); A682 (1080); A713 (1080); A830 **AWS** C2.25 (W-FeC-5) **SAE** J403 (1080); J412 (1080); J1397 (1080) |
| **G10840** | Carbon Steel | **C** 0.80-0.93 **Mn** 0.60-0.90 **P** 0.030 max **S** 0.050 max **Other** Plates, C 0.80-0.94 | **AMS** AMS-QQ-S-700 (C1084) **ASTM** A29 (1084); A510 (1084); A576 (1084); A713 (1084); A830 **SAE** J1397 (1084) |
| **G10850** | Carbon Steel | **C** 0.80-0.94 **Mn** 0.70-1.00 **P** 0.030 max **S** 0.050 max | **AMS** AMS-QQ-S-700 (C1085) **ASTM** A510 (1085); A682 (1085); A830 **SAE** J403 (1085); J412 (1085); J1397 (1085) |
| **G10860** | Carbon Steel | **C** 0.80-0.93 **Mn** 0.30-0.50 **P** 0.030 max **S** 0.050 max **Other** Sheets and Plates, C 0.80-0.94 | **AMS** AMS-QQ-S-700 (C1086) **ASTM** A29 (1086); A510 (1086); A682 (1086); A713 (1086); A830 **SAE** J403 (1086); J412 (1086); J1397 (1086) |
| **G10900** | Carbon Steel | **C** 0.85-0.98 **Mn** 0.60-0.90 **P** 0.030 max **S** 0.050 max **Other** Plates, C 0.84-0.98 | **AMS** 5112 **ASTM** A29 (1090); A108 (1090); A510 (1090); A576 (1090); A713 (1090); A830 **SAE** J403 (1090); J412 (1090); J1397 (1090) |
| **G10950** | Carbon Steel | **C** 0.90-1.03 **Mn** 0.30-0.50 **P** 0.030 max **S** 0.050 max **Other** Plates, C 0.90-1.04 | **AMS** 5121; 5122; 5132; 7304; AMS-QQ-S-700 (C1095); AMS-S-7947 (C1095); AMS-S-8559 (C1095) **ASTM** A29 (1095); A108 (1095); A510 (1095); A576 (1095); A682 (1095); A713 (1095); A830 **SAE** J403 (1095); J412 (1095); J1397 (1095) |
| **G11080** | Resulfurized Carbon Steel | **C** 0.08-0.13 **Mn** 0.50-0.80 **P** 0.040 max **S** 0.08-0.13 | **ASTM** A29 (1108); A510 (1108) **SAE** J1397 (1108) |
| **G11090** | Resulfurized Carbon Steel | **C** 0.08-0.13 **Mn** 0.60-0.90 **P** 0.040 max **S** 0.08-0.13 | **ASTM** A29 (1109); A510 (1109); A576 (1109) |
| **G11100** | Resulfurized Carbon Steel | **C** 0.08-0.13 **Mn** 0.30-0.60 **P** 0.040 max **S** 0.08-0.13 | **ASTM** A29 (1110); A510 (1110); A512 (1110); A544 (1110); A576 (1110) |
| **G11160** | Resulfurized Carbon Steel | **C** 0.14-0.20 **Mn** 1.10-1.40 **P** 0.040 max **S** 0.16-0.23 | **ASTM** A29 (1116); A510 (1116); A576 (1116) |

The chemical compositions listed are for identification purposes and should not be used in lieu of the cross referenced specifications.

| UNIFIED NUMBER | DESCRIPTION | CHEMICAL COMPOSITION | CROSS REFERENCE SPECIFICATIONS |
|---|---|---|---|
| **G11170** | Resulfurized Carbon Steel | **C** 0.14-0.20 **Mn** 1.00-1.30 **P** 0.040 max **S** 0.08-0.13 | **AMS** 5022 **ASTM** A29 (1117); A108 (1117); A510 (1117); A512 (1117); A576 (1117) **SAE** J403 (1117); J412 (1117); J1397 (1117) |
| **G11180** | Resulfurized Carbon Steel | **C** 0.14-0.20 **Mn** 1.30-1.60 **P** 0.040 max **S** 0.08-0.13 | **ASTM** A29 (1118); A108 (1118); A510 (1118); A519 (1118); A576 (1118) **SAE** J403 (1118); J412 (1118) |
| **G11190** | Resulfurized Carbon Steel | **C** 0.14-0.20 **Mn** 1.00-1.30 **P** 0.040 max **S** 0.24-0.33 | **ASTM** A29 (1119); A183 (1-Bolts, 1-Nuts); A510 (1119); A576 (1119) |
| * **G11230** | Resulfurized Carbon Steel | **C** 0.20-0.27 **Mn** 1.20-1.50 **P** 0.040 max **S** 0.06-0.09 | **SAE** J1249 |
| **G11320** | Resulfurized Carbon Steel | **C** 0.27-0.34 **Mn** 1.35-1.65 **P** 0.040 max **S** 0.09-0.13 | **ASTM** A29 (1132); A510 (1132); A519 (1132); A576 (1132) **SAE** J1397 (1132) |
| **G11370** | Resulfurized Carbon Steel | **C** 0.32-0.39 **Mn** 1.35-1.65 **P** 0.030 max **S** 0.08-0.13 | **AMS** 5024 **ASTM** A29 (1137); A108 (1137); A311 (1137); A510 (1137); A519 (1137); A576 (1137) **SAE** J403 (1137); J412 (1137); J1397 (1137) |
| **G11374** | Steel, Free Machining | **C** 0.32-0.39 **Mn** 1.35-1.65 **P** 0.040 max **Pb** 0.15-0.35 **S** 0.08-0.15 | **AMS** 5020 |
| **G11390** | Resulfurized Carbon Steel | **C** 0.35-0.43 **Mn** 1.35-1.65 **P** 0.040 max **S** 0.13-0.20 | **ASTM** A29 (1139); A510 (1139); A576 (1139) |
| **G11400** | Resulfurized Carbon Steel | **C** 0.37-0.44 **Mn** 0.70-1.00 **P** 0.030 max **S** 0.08-0.13 | **ASTM** A29 (1140); A510 (1140); A576 (1140) **SAE** J403 (1140); J1397 (1140) |
| **G11410** | Resulfurized Carbon Steel | **C** 0.37-0.45 **Mn** 1.35-1.65 **P** 0.030 max **S** 0.08-0.13 | **ASTM** A29 (1141); A108 (1141); A311 (1141); A510 (1141); A519 (1141); A576 (1141) **SAE** J403 (1141); J412 (1141); J1397 (1141) |
| **G11440** | Resulfurized Carbon Steel | **C** 0.40-0.48 **Mn** 1.35-1.65 **P** 0.030 max **S** 0.24-0.33 | **ASTM** A29 (1144); A108 (1144); A311 (1144); A510 (1144); A519 (1144); A576 (1144) **SAE** J403 (1144); J412 (1144); J1397 (1144) |
| **G11450** | Resulfurized Carbon Steel | **C** 0.41-0.49 **Mn** 0.70-1.00 **P** 0.040 max **S** 0.08-0.13 | **ASTM** A29 (1145); A510 (1145); A576 (1145) **SAE** J412 (1145) |
| **G11460** | Resulfurized Carbon Steel | **C** 0.42-0.49 **Mn** 0.70-1.00 **P** 0.040 max **S** 0.08-0.13 | **ASTM** A29 (1146); A510 (1146); A576 (1146) **SAE** J412 (1146); J1397 (1146) |
| **G11510** | Resulfurized Carbon Steel | **C** 0.48-0.55 **Mn** 0.70-1.00 **P** 0.040 max **S** 0.08-0.13 | **ASTM** A29 (1151); A108 (1151); A510 (1151); A576 (1151) **SAE** J1397 (1151) |
| **G12110** | Rephosphorized and Resulfurized Carbon Steel | **C** 0.13 max **Mn** 0.60-0.90 **P** 0.07-0.12 **S** 0.10-0.15 | **ASTM** A29 (1211); A108 (1211); A510 (1211); A576 (1211) **SAE** J1397 (1211) |
| **G12120** | Rephosphorized and Resulfurized Carbon Steel | **C** 0.13 max **Mn** 0.70-1.00 **P** 0.07-0.12 **S** 0.16-0.23 | **AMS** 5010 **ASTM** A29 (1212); A108 (1212); A510 (1212); A576 (1212) **SAE** J403 (1212); J1397 (1213) |
| **G12130** | Rephosphorized and Resulfurized Carbon Steel | **C** 0.13 max **Mn** 0.70-1.00 **P** 0.07-0.12 **S** 0.24-0.33 | **ASTM** A29 (1213); A108 (1213); A510 (1213); A519 (1213); A576 (1213) **SAE** J403 (1213); J1397 (1213) |
| **G12134** | Rephosphorized, Resulphurized and Leaded Carbon Steel | **C** 0.13 max **Mn** 0.70-1.00 **P** 0.07-0.12 **Pb** 0.15-0.35 **S** 0.24-0.33 | **ASTM** A29 (12L13); A510 (12L13) |
| **G12144** | Rephosphorized and Resulfurized Carbon Steel | **C** 0.15 max **Mn** 0.85-1.15 **P** 0.04-0.09 **Pb** 0.15-0.35 **S** 0.26-0.35 | **ASTM** A29 (12L14); A108 (12L14); A510 (12L14); A519 (12L14); A576 (12L14) **SAE** J403 (12L14); J1397 (12L14) |
| **G12150** | Rephosphorized and Resulfurized Carbon Steel | **C** 0.09 max **Mn** 0.75-1.05 **P** 0.04-0.09 **S** 0.26-0.35 | **AMS** 5010 **ASTM** A29 (1215); A108 (1215); A510 (1215); A519 (1215); A576 (1215) **SAE** J403 (1215) |
| **G13300** | Mn Alloy Steel | **C** 0.28-0.33 **Mn** 1.60-1.90 **P** 0.035 max **S** 0.040 max **Si** 0.15-0.35 | **ASTM** A322 (1330); A519 (1330); A711; A752 (1330); A829 **SAE** J404 (1330); J412 (1330); J1397 (1330) |
| **G13350** | Mn Alloy Steel | **C** 0.33-0.38 **Mn** 1.60-1.90 **P** 0.035 max **S** 0.040 max **Si** 0.15-0.35 | **ASTM** A322 (1335); A519 (1335); A711 (1335); A752 (1335); A829 **SAE** J404 (1335); J412 (1335); J1397 (1335) |
| **G13400** | Mn Alloy Steel | **C** 0.38-0.43 **Mn** 1.60-1.90 **P** 0.035 max **S** 0.040 max **Si** 0.15-0.35 | **ASTM** A322 (1340); A513 (1340); A519 (1340); A752 (1340); A829 **SAE** J404 (1340); J412 (1340); J1397 (1340) |
| **G13450** | Mn Alloy Steel | **C** 0.43-0.48 **Mn** 1.60-1.90 **P** 0.035 max **S** 0.040 max **Si** 0.15-0.35 | **ASTM** A29 (1345); A322 (1345); A519 (1345); A752 (1345); A829 **SAE** J404 (1345); J1397 (1345) |

*Boxed entries are no longer active and are retained for reference purposes only.

The chemical compositions listed are for identification purposes and should not be used in lieu of the cross referenced specifications.

| UNIFIED NUMBER | DESCRIPTION | CHEMICAL COMPOSITION | CROSS REFERENCE SPECIFICATIONS |
|---|---|---|---|
| **G15116** | Alloy Steel replaced by G51986 | | |
| **G15130** | Carbon Steel | **C** 0.10-0.16 **Mn** 1.10-1.40 **P** 0.040 max **S** 0.050 max | **ASTM** A29 (1513); A510 (1513); A576 (1513) **SAE** J412 (1513) |
| **G15180** | Carbon Steel | **C** 0.15-0.21 **Mn** 1.10-1.40 **P** 0.040 max **S** 0.050 max | **ASTM** A29 (1518); A510 (1518); A519 (1518); A576 (1518) |
| **G15216** | Alloy Steel replaced by G52986 | | |
| **G15220** | Carbon Steel | **C** 0.18-0.24 **Mn** 1.10-1.40 **P** 0.030 max **S** 0.050 max | **ASTM** A29 (1522); A510 (1522); A576 (1522); A576 (1524) **SAE** J403 (1522); J412 (1522) |
| **G15240** | Carbon Steel | **C** 0.19-0.25 **Mn** 1.35-1.65 **P** 0.030 max **S** 0.050 max **Other** Sheets and Plates, C 0.18-0.25; ERW Tubing, C 0.18-0.25 Mn 1.30-1.65 | **ASTM** A29 (1524); A510 (1524); A513 (1524); A519 (1524); A576 (1524); A635 (1524); A830 **SAE** J403 (1524); J412 (1524); J1397 (1524) |
| **G15250** | Carbon Steel | **C** 0.23-0.29 **Mn** 0.80-1.10 **P** 0.040 max **S** 0.050 max | **ASTM** A29 (1525); A510 (1525); A576 (1525) |
| **G15260** | Carbon Steel | **C** 0.22-0.29 **Mn** 1.10-1.40 **P** 0.030 max **S** 0.050 max | **ASTM** A29 (1526); A510 (1526); A576 (1526) **SAE** J403 (1526); J412 (1526) |
| **G15270** | Carbon Steel | **C** 0.22-0.29 **Mn** 1.20-1.50 **P** 0.030 max **S** 0.050 max **Other** Sheets and Plates, Mn 1.20-1.55; ERW Tubing, C 0.21-0.29 Mn 1.20-1.55 | **ASTM** A29 (1527); A510 (1527); A513 (1027); A576 (1527); A830 **SAE** J403 (1527); J412 (1527) |
| * **G15330** | Carbon Steel | **C** 0.30-0.37 **Mn** 1.20-1.50 **P** 0.040 max **S** 0.050 max | **SAE** J1249 |
| * **G15340** | Carbon Steel | **C** 0.30-0.37 **Mn** 1.20-1.50 **P** 0.040 max **S** 0.050 max | **SAE** J1249 |
| **G15360** | Carbon Steel | **C** 0.30-0.37 **Mn** 1.20-1.50 **P** 0.040 max **S** 0.050 max **Other** Sheets and Plates, C 0.30-0.38 Mn 1.20-1.55 | **ASTM** A29 (1536); A510 (1536); A576 (1536); A830 **SAE** J412 (1536); J1397 (1536) |
| **G15410** | Carbon Steel | **C** 0.36-0.44 **Mn** 1.35-1.65 **P** 0.030 max **S** 0.050 max **Other** Sheets and Plates, C 0.36-0.45 Mn 1.30-1.65 | **ASTM** A29 (1541); A311 (1541); A510 (1541); A519 (1541); A576 (1541); A830; A866 (1541) **SAE** J403 (1541); J412 (1541); J1397 (1541) |
| **G15470** | Carbon Steel | **C** 0.43-0.51 **Mn** 1.35-1.65 **P** 0.040 max **S** 0.050 max | **ASTM** A29 (1547); A510 (1547); A576 (1547) **SAE** J775 (NV-2) |
| **G15480** | Carbon Steel | **C** 0.44-0.52 **Mn** 1.10-1.40 **P** 0.030 max **S** 0.050 max **Other** Sheets and Plates, C 0.43-0.52 Mn 1.05-1.40 | **ASTM** A29 (1548); A510 (1548); A576 (1548); A830 **SAE** J403 (1548); J412 (1548); J1397 (1548) |
| **G15510** | Carbon Steel | **C** 0.45-0.56 **Mn** 0.85-1.15 **P** 0.040 max **S** 0.050 max | **ASTM** A29 (1551); A510 (1551); A576 (1551) **SAE** J412 (1551) |
| **G15520** | Carbon Steel | **C** 0.47-0.55 **Mn** 1.20-1.50 **P** 0.030 max **S** 0.050 max **Other** Sheets and Plates, C 0.46-0.55 Mn 1.20-1.55 | **ASTM** A29 (1552); A510 (1552); A576 (1552); A830; A866 (1552) **SAE** J403 (1552); J412 (1552); J1397 (1552) |
| * **G15530** | Carbon Steel | **C** 0.48-0.56 **Mn** 0.80-1.10 **P** 0.040 max **S** 0.050 max | **SAE** J1249 |
| **G15610** | Carbon Steel | **C** 0.55-0.65 **Mn** 0.75-1.05 **P** 0.040 max **S** 0.050 max | **ASTM** A29 (1561); A510 (1561); A576 (1561); A713 (1561) **SAE** J412 (1561) |
| **G15660** | Carbon Steel | **C** 0.60-0.71 **Mn** 0.85-1.15 **P** 0.030 max **S** 0.050 max | **ASTM** A29 (1566); A510 (1566); A576 (1566); A713 (1566) **SAE** J403 (1566); J412 (1566) |
| * **G15700** | Carbon Steel | **C** 0.65-0.75 **Mn** 0.80-1.10 **P** 0.040 max **S** 0.050 max | **SAE** J1249 |

*Boxed entries are no longer active and are retained for reference purposes only.

The chemical compositions listed are for identification purposes and should not be used in lieu of the cross referenced specifications.

| UNIFIED NUMBER | DESCRIPTION | CHEMICAL COMPOSITION | CROSS REFERENCE SPECIFICATIONS |
|---|---|---|---|
| **G15720** | Carbon Steel | **C** 0.65-0.76 **Mn** 1.00-1.30 **P** 0.040 max **S** 0.050 max | **ASTM** A29 (1572); A510 (1572); A576 (1572); A713 (1572) **SAE** J1249 (1572) |
| * **G15800** | Carbon Steel | **C** 0.75-0.85 **Mn** 0.80-1.10 **P** 0.040 max **S** 0.050 max | **SAE** J1249 |
| * **G15900** | Carbon Steel | **C** 0.85-0.98 **Mn** 0.80-1.10 **P** 0.040 max **S** 0.050 max | **SAE** J1249 |
| **G31400** | Alloy Steel | **C** 0.38-0.43 **Cr** 0.55-0.75 **Mn** 0.70-0.90 **Ni** 1.10-1.40 **P** 0.040 max **S** 0.040 max **Si** 0.15-0.35 | **ASTM** A519 (3140) **SAE** J775 (NV 4) |
| **G33106** | Alloy Steel | **C** 0.08-0.13 **Cr** 1.40-1.75 **Mn** 0.45-0.60 **Ni** 3.25-3.75 **P** 0.025 max **S** 0.025 max **Si** 0.15-0.30 | **AMS** AMS-S-8503 **ASTM** A519 (E3310); A646 (3310-1); A834; A837 (E3310) **MIL SPEC** MIL-S-7393 |
| **G40120** | Mo Alloy Steel | **C** 0.09-0.14 **Mn** 0.75-1.00 **Mo** 0.15-0.25 **P** 0.035 max **S** 0.040 max **Si** 0.15-0.35 | **ASTM** A29 (4012); A519 (4012); A752 (4012) **SAE** J412 (4012); J1249 (4012) |
| **G40230** | Mo Alloy Steel | **C** 0.20-0.25 **Mn** 0.70-0.90 **Mo** 0.20-0.30 **P** 0.035 max **S** 0.040 max **Si** 0.15-0.35 | **ASTM** A29 (4023); A322 (4023); A519 (4023); A752 (4023) **SAE** J404 (4023); J412 (4023); J1397 (4023) |
| **G40240** | Mo Alloy Steel | **C** 0.20-0.25 **Mn** 0.70-0.90 **Mo** 0.20-0.30 **P** 0.035 max **S** 0.035-0.050 **Si** 0.15-0.35 | **ASTM** A29 (4024); A322 (4024); A519 (4024); A752 (4024) **SAE** J412 (4024); J1249 (4024); J1397 (4024) |
| **G40270** | Mo Alloy Steel | **C** 0.25-0.30 **Mn** 0.70-0.90 **Mo** 0.20-0.30 **P** 0.035 max **S** 0.040 max **Si** 0.15-0.35 | **ASTM** A29 (4027); A322 (4027); A519 (4027); A752 (4027) **SAE** J404 (4027); J412 (4027); J1397 (4027) |
| **G40280** | Mo Alloy Steel | **C** 0.25-0.30 **Mn** 0.70-0.90 **Mo** 0.20-0.30 **P** 0.035 max **S** 0.035-0.050 **Si** 0.15-0.35 | **ASTM** A29 (4028); A322 (4028); A519 (4028); A752 (4028) **SAE** J412 (4028); J1249 (4020); J1397 (4028) |
| **G40320** | Mo Alloy Steel | **C** 0.30-0.35 **Mn** 0.70-0.90 **Mo** 0.20-0.30 **P** 0.035 max **S** 0.040 max **Si** 0.15-0.35 | **SAE** J412 (4032); J1249 (4032); J1397 (4032) |
| **G40370** | Mo Alloy Steel | **C** 0.35-0.40 **Mn** 0.70-0.90 **Mo** 0.20-0.30 **P** 0.035 max **S** 0.040 max **Si** 0.15-0.35 | **AMS** 6300 **ASTM** A29 (4037); A320 (L7A, L71); A322 (4037); A519 (4037); A752 (4037) **SAE** J404 (4037); J412 (4037); J1397 (4037) |
| **G40420** | Mo Alloy Steel | **C** 0.40-0.45 **Mn** 0.70-0.90 **Mo** 0.20-0.30 **P** 0.035 max **S** 0.040 max **Si** 0.15-0.30 | **ASTM** A29 (4042); A519 (4042) **SAE** J1249 (4042); J1397 (4042) |
| **G40470** | Mo Alloy Steel | **C** 0.45-0.50 **Mn** 0.70-0.90 **Mo** 0.20-0.30 **P** 0.035 max **S** 0.040 max **Si** 0.15-0.35 | **ASTM** A29 (4047); A322 (4047); A519 (4047); A752 (4047) **SAE** J404 (4047); J412 (4047); J1397 (4047) |
| **G40630** | Mo Alloy Steel | **C** 0.60-0.67 **Mn** 0.75-1.00 **Mo** 0.20-0.30 **P** 0.035 max **S** 0.040 max **Si** 0.15-0.35 | **ASTM** A519 (4063) **SAE** J1249 (4063) |
| **G41180** | Cr-Mo Alloy Steel | **C** 0.18-0.23 **Cr** 0.40-0.60 **Mn** 0.70-0.90 **Mo** 0.08-0.15 **P** 0.035 max **S** 0.040 max **Si** 0.15-0.35 | **ASTM** A29 (4118); A322 (4118); A506 (4118); A507 (4118); A513 (4118); A519 (4118); A752 (4118); A829 **SAE** J404 (4118); J412 (4118); J1397 (4118) |
| **G41200** | Cr-Mo Alloy Steel | **C** 0.18-0.23 **Cr** 0.40-0.60 **Mn** 0.80-1.20 **Mo** 0.15-0.25 **P** 0.035 max **S** 0.040 max **Si** 0.15-0.35 | **ASTM** A29 (4120); A322 (4120) **SAE** J404 (4120) |
| **G41210** | Cr-Mo Alloy Steel | **C** 0.18-0.23 **Cr** 0.45-0.65 **Mn** 0.75-1.00 **Mo** 0.15-0.25 **P** 0.035 max **S** 0.040 max **Si** 0.15-0.35 | **ASTM** A29 (4121); A322 (4121) **SAE** J1249 (4121) |

*Boxed entries are no longer active and are retained for reference purposes only.

The chemical compositions listed are for identification purposes and should not be used in lieu of the cross referenced specifications.

| UNIFIED NUMBER | DESCRIPTION | CHEMICAL COMPOSITION | CROSS REFERENCE SPECIFICATIONS |
|---|---|---|---|
| **G41300** | Cr-Mo Alloy Steel | **C** 0.28-0.33 **Cr** 0.80-1.10 **Mn** 0.40-0.60 **Mo** 0.15-0.25 **P** 0.035 max **S** 0.040 max **Si** 0.15-0.35 | **AMS** 6346; 6348; 6350; 6351; 6356; 6360; 6361; 6362; 6370; 6371; 6373; 6374; 6528; 7496; AMS-S-6755 (4130); AMS-S-6758; AMS-S-18729 **ASTM** A29 (4130); A322 (4130); A506 (4130); A507 (4130); A513 (4130); A519 (4130); A646 (4130); A752 (4130); A829; A866 (4130) **SAE** J404 (4130); J412 (4130); J1397 (4130) |
| **G41350** | Cr-Mo Alloy Steel | **C** 0.33-0.38 **Cr** 0.80-1.10 **Mn** 0.70-0.90 **Mo** 0.15-0.25 **P** 0.035 max **S** 0.040 max **Si** 0.15-0.35 | **AMS** 6352; 6365; 6372; AMS-T-6735 **ASTM** A29 (4135); A372 (F); A519 (4135); A829 **SAE** J404 (4135); J412 (4135); J1397 (4135) |
| **G41370** | Cr-Mo Alloy Steel | **C** 0.35-0.40 **Cr** 0.80-1.10 **Mn** 0.70-0.90 **Mo** 0.15-0.25 **P** 0.035 max **S** 0.040 max **Si** 0.15-0.35 | **ASTM** A29 (4137); A320 (L7B, L72); A322 (4137); A519 (4137); A752 (4137); A829 **SAE** J404 (4137); J412 (4137); J1397 (4137) |
| **G41400** | Cr-Mo Alloy Steel | **C** 0.38-0.43 **Cr** 0.80-1.10 **Mn** 0.75-1.00 **Mo** 0.15-0.25 **P** 0.035 max **S** 0.040 max **Si** 0.15-0.35 | **AMS** 6349; 6381; 6382; 6395; 6462; 6529; AMS-S-5621 **ASTM** A29 (4140); A193 (B7, B7M); A194 (7, 7M); A320 (L7, L7M, L7C); A322 (4140); A506 (4140); A513 (4140); A519 (4140); A646 (4140); A711; A752 (4140); A829; A866 (4140) **SAE** J404 (4140); J412 (4140); J1397 (4140) |
| **G41420** | Cr-Mo Alloy Steel | **C** 0.40-0.45 **Cr** 0.80-1.10 **Mn** 0.75-1.00 **Mo** 0.15-0.25 **P** 0.035 max **S** 0.040 max **Si** 0.15-0.35 | **ASTM** A29 (4142); A322 (4142); A519 (4142); A711; A752 (4142); A829 **SAE** J404 (4142); J412 (4142); J1397 (4142) |
| **G41450** | Cr-Mo Alloy Steel | **C** 0.43-0.48 **Cr** 0.80-1.10 **Mn** 0.75-1.00 **Mo** 0.15-0.25 **P** 0.035 max **S** 0.040 max **Si** 0.15-0.35 | **ASTM** A29 (4145); A322 (4145); A519 (4145); A711; A752 (4145); A829; A866 (4150) **SAE** J404 (4145); J412 (4145); J1397 (4145) |
| **G41470** | Cr-Mo Alloy Steel | **C** 0.45-0.50 **Cr** 0.80-1.10 **Mn** 0.75-1.00 **Mo** 0.15-0.25 **P** 0.035 max **S** 0.040 max **Si** 0.15-0.35 | **ASTM** A29 (4147); A322 (4147); A519 (4147); A752 (4147) **SAE** J412 (4147); J1249 (4147); J1397 (4147) |
| **G41500** | Cr-Mo Alloy Steel | **C** 0.48-0.53 **Cr** 0.80-1.10 **Mn** 0.75-1.00 **Mo** 0.15-0.25 **P** 0.035 max **S** 0.040 max **Si** 0.15-0.35 | **ASTM** A29 (4150); A322 (4150); A519 (4150); A711; A752 (4150); A866 (4150) **SAE** J404 (4150); J1397 (4150) |
| **G41610** | Cr-Mo Alloy Steel | **C** 0.56-0.64 **Cr** 0.70-0.90 **Mn** 0.75-1.00 **Mo** 0.25-0.35 **P** 0.035 max **S** 0.40 max **Si** 0.15-0.35 | **ASTM** A29 (4161); A322 (4161); A752 (4161) **SAE** J412 (4161); J1249 (4161); J1397 (4161) |
| **G43200** | Ni-Cr-Mo Alloy Steel | **C** 0.17-0.22 **Cr** 0.40-0.60 **Mn** 0.45-0.65 **Mo** 0.20-0.30 **Ni** 1.65-2.00 **P** 0.035 max **S** 0.040 max **Si** 0.15-0.35 | **ASTM** A29 (4320); A322 (4320); A519 (4320); A752 (4320); A837 (4320) **SAE** J404 (4320); J412 (4320); J1397 (4320) |
| **G43370** | Ni-Cr-Mo Alloy Steel | **C** 0.35-0.40 **Cr** 0.70-0.90 **Mn** 0.60-0.80 **Mo** 0.20-0.30 **Ni** 1.65-2.00 **P** 0.035 max **S** 0.040 max **Si** 0.15-0.35 | **AMS** 6412; 6413; 7496 **ASTM** A519 (4337) **SAE** J1249 (4337) |
| **G43376** | Ni-Cr-Mo Alloy Steel | **C** 0.35-0.40 **Cr** 0.70-0.90 **Mn** 0.65-0.85 **Mo** 0.20-0.30 **Ni** 1.65-2.00 **P** 0.025 max **S** 0.025 max **Si** 0.15-0.35 | **ASTM** A519 (E4337) |
| **G43400** | Ni-Cr-Mo Alloy Steel | **C** 0.38-0.43 **Cr** 0.70-0.90 **Mn** 0.60-0.80 **Mo** 0.20-0.30 **Ni** 1.65-2.00 **P** 0.035 max **S** 0.040 max **Si** 0.15-0.30 | **AMS** 6359; 6409; 6414; 6415; 6454; 6484; AMS-S-5000 **ASTM** A29 (4340); A320 (L43); A322 (4340); A506 (4340); A519 (4340); A646 (4340-7); A711; A752 (4340); A829 **SAE** J404 (4340); J412 (4340); J1397 (4340) |
| **G43406** | Ni-Cr-Mo Alloy Steel | **C** 0.38-0.43 **Cr** 0.70-0.90 **Mn** 0.65-0.85 **Mo** 0.20-0.30 **Ni** 1.65-2.00 **P** 0.025 max **S** 0.025 max **Si** 0.15-0.35 | **AMS** 6415 **ASTM** A29 (43B40); A29 (E4340); A519 (E4340); A752 (E4340); A829 **SAE** J404 (E4340); J1397 (E4340) |
| **G44190** | Mo Alloy Steel | **C** 0.18-0.23 **Mn** 0.45-0.65 **Mo** 0.45-0.60 **P** 0.035 max **S** 0.040 max **Si** 0.15-0.35 | **ASTM** A29 (4419); A752 (4419) **SAE** J1249 (4419) |
| **G44220** | Mo Alloy Steel | **C** 0.20-0.25 **Mn** 0.70-0.90 **Mo** 0.35-0.45 **P** 0.035 max **S** 0.040 max **Si** 0.15-0.35 | **ASTM** A29 (4422); A519 (4422) **SAE** J412 (4422); J1249 (4422); J1397 (4422) |
| **G44270** | Mo Alloy Steel | **C** 0.24-0.29 **Mn** 0.70-0.90 **Mo** 0.35-0.45 **P** 0.035 max **S** 0.040 max **Si** 0.15-0.35 | **ASTM** A29 (4427); A519 (4427) **SAE** J412 (4427); J1249 (4427); J1397 (4427) |

The chemical compositions listed are for identification purposes and should not be used in lieu of the cross referenced specifications.

| UNIFIED NUMBER | DESCRIPTION | CHEMICAL COMPOSITION | CROSS REFERENCE SPECIFICATIONS |
|---|---|---|---|
| **G45200** | Mo Alloy Steel | **C** 0.18-0.23 **Mn** 0.45-0.65 **Mo** 0.45-0.60 **P** 0.035 max **S** 0.040 max **Si** 0.15-0.35 | **ASTM** A519 (4520) |
| **G46150** | Ni-Mo Alloy Steel | **C** 0.13-0.18 **Mn** 0.45-0.65 **Mo** 0.20-0.30 **Ni** 1.65-2.00 **P** 0.035 max **S** 0.040 max **Si** 0.15-0.35 | **AMS** 6290 **ASTM** A29 (4615); A322 (4615); A519 (4615); A752 (4615); A829 **SAE** J404 (4615); J1397 (4615) |
| **G46170** | Ni-Mo Alloy Steel | **C** 0.15-0.20 **Mn** 0.45-0.65 **Mo** 0.20-0.30 **Ni** 1.65-2.00 **P** 0.035 max **S** 0.040 max **Si** 0.15-0.35 | **AMS** 6292 **ASTM** A519 (4617); A829 **SAE** J404 (4617); J1397 (4617) |
| **G46200** | Ni-Mo Alloy Steel | **C** 0.17-0.22 **Mn** 0.45-0.65 **Mo** 0.20-0.30 **Ni** 1.65-2.00 **P** 0.035 max **S** 0.040 max **Si** 0.15-0.35 | **AMS** 6294 **ASTM** A29 (4620); A322 (4620); A519 (4620); A646 (4620-3); A752 (4620); A829; A837 (4620) **SAE** J404 (4620); J412 (4620); J1397 (4620) |
| **G46210** | Ni-Mo Alloy Steel | **C** 0.18-0.23 **Mn** 0.70-0.90 **Mo** 0.20-0.30 **Ni** 1.65-2.00 **P** 0.035 max **S** 0.040 max **Si** 0.15-0.35 | **ASTM** A29 (4621); A322 (4621); A519 (4621); A752 (4621) **SAE** J1249 (4621) |
| **G46260** | Ni-Mo Alloy Steel | **C** 0.24-0.29 **Mn** 0.45-0.65 **Mo** 0.15-0.25 **Ni** 0.70-1.00 **P** 0.035 max **S** 0.040 max **Si** 0.15-0.35 | **ASTM** A29 (4626); A322 (4626); A752 (4626) **SAE** J1249 (4626); J1397 (4626) |
| **G47150** | Ni-Cr-Mo Alloy Steel | **C** 0.13-0.18 **Cr** 0.45-0.65 **Mn** 0.70-0.90 **Mo** 0.45-0.65 **Ni** 0.70-1.00 **P** 0.035 max **S** 0.040 max **Si** 0.15-0.35 | **ASTM** A322 (4715) **SAE** J1249 (4715) |
| **G47180** | Ni-Cr-Mo Alloy Steel | **C** 0.16-0.21 **Cr** 0.35-0.55 **Mn** 0.70-0.90 **Mo** 0.30-0.40 **Ni** 0.90-1.20 **P** 0.035 max **S** 0.040 max **Si** 0.15-0.35 | **ASTM** A29 (4718); A519 (4718); A752 (4718) **SAE** J1249 (4718); J1397 (4718) |
| **G47200** | Ni-Cr-Mo Alloy Steel | **C** 0.17-0.22 **Cr** 0.35-0.55 **Mn** 0.50-0.70 **Mo** 0.15-0.25 **Ni** 0.90-1.20 **P** 0.035 max **S** 0.040 max **Si** 0.15-0.35 | **ASTM** A29 (4720); A322 (4720); A519 (4720); A711; A752 (4720) **SAE** J412 (4720); J1249 (4720); J1397 (4720) |
| **G48150** | Ni-Mo Alloy Steel | **C** 0.13-0.18 **Mn** 0.40-0.60 **Mo** 0.20-0.30 **Ni** 3.25-3.75 **P** 0.035 max **S** 0.040 max **Si** 0.15-0.35 | **ASTM** A29 (4815); A322 (4815); A519 (4815); A752 (4815); A837 (4815) **SAE** J412 (4815); J1249 (4815); J1397 (4815) |
| **G48170** | Ni-Mo Alloy Steel | **C** 0.15-0.20 **Mn** 0.40-0.60 **Mo** 0.20-0.30 **Ni** 3.25-3.75 **P** 0.035 max **S** 0.040 max **Si** 0.15-0.35 | **ASTM** A29 (4817); A322 (4817); A519 (4817); A752 (4817) **SAE** J1249 (4817); J1397 (4817) |
| **G48200** | Ni-Mo Alloy Steel | **C** 0.18-0.23 **Mn** 0.50-0.70 **Mo** 0.20-0.30 **Ni** 3.25-3.75 **P** 0.035 max **S** 0.040 max **Si** 0.15-0.35 | **ASTM** A29 (4820); A322 (4820); A519 (4820); A752 (4820) **SAE** J404 (4820); J412 (4820); J1397 (4820) |
| **G50150** | Cr Alloy Steel | **C** 0.12-0.17 **Cr** 0.30-0.50 **Mn** 0.30-0.50 **P** 0.035 max **S** 0.040 max **Si** 0.15-0.35 | **ASTM** A29 (5015); A519 (5015); A752 (5015) **SAE** J1249 (5015) |
| **G50401** | Cr-B Alloy Steel | **B** 0.0005-0.003 **C** 0.38-0.42 **Cr** 0.40-0.60 **Mn** 0.75-1.00 **P** 0.035 max **S** 0.040 max **Si** 0.15-0.35 | **ASTM** A519 (50B40) **SAE** J412 (50B40); J1249 (50B40); J1397 (50B40) |
| **G50441** | Cr-B Alloy Steel | **B** 0.0005-0.003 **C** 0.43-0.48 **Cr** 0.40-0.60 **Mn** 0.75-1.00 **P** 0.035 max **S** 0.040 max **Si** 0.15-0.35 | **ASTM** A29 (50B44); A322 (50B44); A519 (50B44); A752 (50B44) **SAE** J412 (50B44); J1249 (5046); J1397 (50B44) |
| **G50460** | Cr Alloy Steel | **C** 0.43-0.50 **Cr** 0.20-0.35 **Mn** 0.75-1.00 **P** 0.035 max **S** 0.040 max **Si** 0.15-0.35 | **AISI** 5046 **ASTM** A29 (5046); A519 (5046) **SAE** J412 (5046); J1397 (5046) |
| **G50461** | Cr-B Alloy Steel | **B** 0.0005-0.003 **C** 0.44-0.49 **Cr** 0.20-0.35 **Mn** 0.75-1.00 **P** 0.035 max **S** 0.040 max **Si** 0.15-0.35 | **ASTM** A29 (50B46); A322 (50B46); A519 (50B46); A752 (50B46) **SAE** J404 (50B46); J412 (50B46); J1397 (50B46) |
| **G50501** | Cr-B Alloy Steel | **B** 0.0005-0.003 **C** 0.48-0.53 **Cr** 0.40-0.60 **Mn** 0.75-1.00 **P** 0.035 max **S** 0.040 max **Si** 0.15-0.35 | **ASTM** A322 (50B50); A519 (50B50); A752 (50B50) **SAE** J412 (50B50); J1249 (50B50); J1397 (50B50) |
| **G50600** | Cr Alloy Steel | **C** 0.56-0.64 **Cr** 0.40-0.60 **Mn** 0.75-1.00 **P** 0.035 max **S** 0.040 max **Si** 0.15-0.35 | **SAE** J412 (5060); J1249 (5060); J1397 (5060) |

The chemical compositions listed are for identification purposes and should not be used in lieu of the cross referenced specifications.

| UNIFIED NUMBER | DESCRIPTION | CHEMICAL COMPOSITION | CROSS REFERENCE SPECIFICATIONS |
|---|---|---|---|
| **G50601** | Cr-B Alloy Steel | **B** 0.0005-0.003 **C** 0.56-0.64 **Cr** 0.40-0.60 **Mn** 0.75-1.00 **P** 0.035 max **S** 0.040 max **Si** 0.15-0.35 | **ASTM** A29 (50B60); A322 (50B60); A519 (50B60); A752 (50B60) **SAE** J412 (50B60); J1249 (50B60); J1397 (50B60) |
| **G50986** | Cr Alloy Steel | **C** 0.95-1.10 **Cr** 0.40-0.60 **Mn** 0.25-0.45 **P** 0.025 max **S** 0.025 max **Si** 0.15-0.35 | **AMS** 6442 **ASTM** A29 (E50100); A295 (50100); A519 (E50100) **SAE** J412 (50100); J1249 (50100); J1397 (50100) |
| **G51150** | Cr Alloy Steel | **C** 0.13-0.18 **Cr** 0.70-0.90 **Mn** 0.70-0.90 **P** 0.035 max **S** 0.040 max **Si** 0.15-0.35 | **ASTM** A29 (5115); A519 (5115) **SAE** J1249 (5115); J1397 (5115) |
| **G51170** | Cr Alloy Steel | **C** 0.15-0.20 **Cr** 0.70-0.90 **Mn** 0.70-0.90 **P** 0.035 max **S** 0.040 max **Si** 0.15-0.35 | **ASTM** A322 (5117) **SAE** J1249 (5117) |
| **G51190** | Alloy Steel | **C** 0.16-0.21 **Cr** 0.70-0.90 **Mn** 1.00-1.30 **P** 0.035 max **S** 0.040 max **Si** 0.15 max | **SAE** J1081 (PS 64) |
| **G51200** | Cr Alloy Steel | **C** 0.17-0.22 **Cr** 0.70-0.90 **Mn** 0.70-0.90 **P** 0.035 max **S** 0.040 max **Si** 0.15-0.35 | **ASTM** A29 (5120); A322 (5120); A519 (5120); A752 (5120) **SAE** J404 (5120); J1397 (5120) |
| **G51210** | Alloy Steel | **C** 0.18-0.23 **Cr** 0.70-0.90 **Mn** 1.00-1.30 **P** 0.035 max **S** 0.040 max **Si** 0.15-0.35 | **SAE** J1081 (PS 59) |
| **G51240** | Alloy Steel | **C** 0.21-0.26 **Cr** 0.70-0.90 **Mn** 1.00-1.30 **P** 0.035 max **S** 0.040 max **Si** 0.15-0.35 | **SAE** J1081 (PS 65) |
| **G51300** | Cr Alloy Steel | **C** 0.28-0.33 **Cr** 0.80-1.10 **Mn** 0.70-0.90 **P** 0.035 max **S** 0.040 max **Si** 0.15-0.35 | **ASTM** A29 (5130); A322 (5130); A513 (5130); A519 (5130); A752 (5130) **SAE** J404 (5130); J412 (5130); J1397 (5130) |
| **G51320** | Cr Alloy Steel | **C** 0.30-0.35 **Cr** 0.75-1.00 **Mn** 0.60-0.80 **P** 0.035 max **S** 0.040 max **Si** 0.15-0.35 | **ASTM** A29 (5132); A322 (5132); A519 (5132); A752 (5132) **SAE** J404 (5132); J412 (5132); J1397 (5132) |
| **G51350** | Cr Alloy Steel | **C** 0.33-0.38 **Cr** 0.80-1.05 **Mn** 0.60-0.80 **P** 0.035 max **S** 0.040 max **Si** 0.15-0.35 | **ASTM** A29 (5135); A322 (5135); A519 (5135); A752 (5135) **SAE** J412 (5135); J1249 (5135); J1397 (5135) |
| **G51400** | Cr Alloy Steel | **C** 0.38-0.43 **Cr** 0.70-0.90 **Mn** 0.70-0.90 **P** 0.035 max **S** 0.040 max **Si** 0.15-0.35 | **ASTM** A29 (5140); A322 (5140); A506 (5140); A519 (5140); A752 (5140); A866 (5140) **SAE** J404 (5140); J412 (5140); J1397 (5140) |
| **G51450** | Cr Alloy Steel | **C** 0.43-0.48 **Cr** 0.70-0.90 **Mn** 0.70-0.90 **P** 0.035 max **S** 0.040 max **Si** 0.15-0.35 | **ASTM** A29 (5145); A519 (5145); A752 (5145) **SAE** J1249 (5145) |
| **G51470** | Cr Alloy Steel | **C** 0.46-0.51 **Cr** 0.85-1.15 **Mn** 0.70-0.95 **P** 0.035 max **S** 0.040 max **Si** 0.15-0.35 | **ASTM** A29 (5147); A519 (5147); A752 (5147) **SAE** J412 (5147); J1249 (5147); J1397 (5147) |
| **G51500** | Cr Alloy Steel | **C** 0.48-0.53 **Cr** 0.70-0.90 **Mn** 0.70-0.90 **P** 0.035 max **S** 0.040 max **Si** 0.15-0.35 | **ASTM** A29 (5150); A322 (5150); A506 (5150); A519 (5150); A752 (5150); A866 (5150) **SAE** J404 (5150); J412 (5150); J1397 (5150) |
| **G51550** | Cr Alloy Steel | **C** 0.51-0.59 **Cr** 0.70-0.90 **Mn** 0.70-0.90 **P** 0.035 max **S** 0.040 max **Si** 0.15-0.35 | **ASTM** A29 (5155); A322 (5155); A519 (5155); A752 (5155); A866 (5155) **SAE** J412 (5155); J1249 (5155); J1397 (5155) |
| **G51600** | Cr Alloy Steel | **C** 0.56-0.64 **Cr** 0.70-0.90 **Mn** 0.75-1.00 **P** 0.035 max **S** 0.040 max **Si** 0.15-0.35 | **ASTM** A29 (5160); A295 (5160); A322 (5160); A506 (5160); A519 (5160); A752 (5160); A829 **SAE** J404 (5160); J412 (5160); J1397 (5160) |
| **G51601** | Cr-B Alloy Steel | **B** 0.0005 min **C** 0.56-0.64 **Cr** 0.70-0.90 **Mn** 0.75-1.00 **P** 0.035 max **S** 0.040 max **Si** 0.15-0.35 | **ASTM** A29 (51B60); A322 (51B60); A519 (51B60); A752 (51B60) **SAE** J404 (51B60); J412 (51B60); J1397 (51B60) |
| **G51986** | Cr Alloy Steel | **C** 0.98-1.10 **Cr** 0.90-1.15 **Mn** 0.25-0.45 **P** 0.025 max **S** 0.025 max **Si** 0.15-0.35 | **AMS** 6443; 6446; 6449 **ASTM** A29 (E51100); A295 (51100); A322 (E51100); A519 (E51100); A711; A752 (E51100) **SAE** J412 (51100); J1249 (E51100); J1397 (51100) |

The chemical compositions listed are for identification purposes and should not be used in lieu of the cross referenced specifications.

| UNIFIED NUMBER | DESCRIPTION | CHEMICAL COMPOSITION | CROSS REFERENCE SPECIFICATIONS |
|---|---|---|---|
| **G52986** | Cr Alloy Steel | **C** 0.98-1.10 **Cr** 1.30-1.60 **Mn** 0.25-0.45 **P** 0.025 max **S** 0.025 max **Si** 0.15-0.35 | **AMS** 6440; 6444; 6447; AMS-S-7420; AMS-S-22141 **ASTM** A29 (E52100); A295 (52100); A322 (E52100); A519 (E52100); A646 (E52100); A752 (E52100) **SAE** J404 (E52100); J412 (52100); J1397 (52100) |
| **G61180** | Cr-V Alloy Steel | **C** 0.16-0.21 **Cr** 0.50-0.70 **Mn** 0.50-0.70 **P** 0.035 max **S** 0.040 max **Si** 0.15-0.35 **V** 0.10-0.15 | **ASTM** A29 (6118); A322 (6118); A519 (6118); A752 (6118) **SAE** J1249 (6118); J1397 (6118) |
| **G61200** | Cr-V Alloy Steel | **C** 0.17-0.22 **Cr** 0.70-0.90 **Mn** 0.70-0.90 **P** 0.035 max **S** 0.040 max **Si** 0.15-0.35 **V** 0.10 min | **ASTM** A519 (6120); A711 **SAE** J1249 (6120) |
| **G61500** | Cr-V Alloy Steel | **C** 0.48-0.53 **Cr** 0.80-1.10 **Mn** 0.70-0.90 **P** 0.035 max **S** 0.040 max **Si** 0.15-0.35 **V** 0.15 min | **AMS** 6448; 6450; 6455; 7301; AMS-S-8503 **ASME** SA-232 **ASTM** A29 (6150); A322 (6150); A519 (6150); A646 (6150); A752 (6150); A829; A866 (6150) **SAE** J404 (6150); J412 (6150); J1397 (6150) |
| **G71406** | Replaced by K24065 | | |
| **G81150** | Ni-Cr-Mo Alloy Steel | **C** 0.13-0.18 **Cr** 0.30-0.50 **Mn** 0.70-0.90 **Mo** 0.08-0.15 **Ni** 0.20-0.40 **P** 0.035 max **S** 0.040 max **Si** 0.15-0.35 | **ASTM** A29 (8115); A519 (8115) **SAE** J1249 (8115); J1397 (8115) |
| **G81451** | Ni-Cr-Mo-B Alloy Steel | **B** 0.0005 min **C** 0.43-0.48 **Cr** 0.35-0.55 **Mn** 0.75-1.00 **Mo** 0.08-0.15 **Ni** 0.20-0.40 **P** 0.035 max **S** 0.040 max **Si** 0.15-0.35 | **ASTM** A29 (81B45); A322 (81B45); A519 (81B45); A752 (81B45) **SAE** J412 (81B45); J1249 (81B45); J1397 (81B45) |
| **G86150** | Ni-Cr-Mo Alloy Steel | **C** 0.13-0.18 **Cr** 0.40-0.60 **Mn** 0.70-0.90 **Mo** 0.15-0.25 **Ni** 0.40-0.70 **P** 0.035 max **S** 0.040 max **Si** 0.15-0.35 | **AMS** 6270 **ASTM** A29 (8615); A322 (8615); A506 (8615); A507 (8615); A519 (8615); A752 (8615); A829 **SAE** J404 (8615); J412 (8615); J1397 (8615) |
| **G86170** | Ni-Cr-Mo Alloy Steel | **C** 0.15-0.20 **Cr** 0.40-0.60 **Mn** 0.70-0.90 **Mo** 0.15-0.25 **Ni** 0.40-0.70 **P** 0.035 max **S** 0.040 max **Si** 0.15-0.35 | **AMS** 6272 **ASTM** A29 (8617); A322 (8617); A519 (8617); A752 (8617); A829 **SAE** J404 (8617); J412 (8617); J1397 (8617) |
| **G86200** | Ni-Cr-Mo Alloy Steel | **C** 0.18-0.23 **Cr** 0.40-0.60 **Mn** 0.70-0.90 **Mo** 0.15-0.25 **Ni** 0.40-0.70 **P** 0.035 max **S** 0.040 max **Si** 0.15-0.35 | **AMS** 6274; 6276; 6277; 6375; AMS-S-8690 **ASTM** A29 (8620); A322 (8620); A506 (8620); A507 (8620); A513 (8620); A519 (8620); A646 (8620-4); A752 (8620); A829; A837 (4815); A837 (8620) **SAE** J404 (8620); J412 (8620); J1397 (8620) |
| **G86220** | Ni-Cr-Mo Alloy Steel | **C** 0.20-0.25 **Cr** 0.40-0.60 **Mn** 0.70-0.90 **Mo** 0.15-0.25 **Ni** 0.40-0.70 **P** 0.035 max **S** 0.040 max **Si** 0.15-0.35 | **ASTM** A29 (8622); A322 (8622); A519 (8622); A752 (8622); A829 **SAE** J404 (8622); J412 (8622); J1397 (8622) |
| **G86250** | Ni-Cr-Mo Alloy Steel | **C** 0.23-0.28 **Cr** 0.40-0.60 **Mn** 0.70-0.90 **Mo** 0.15-0.25 **Ni** 0.40-0.70 **P** 0.035 max **S** 0.040 max **Si** 0.15-0.35 | **ASTM** A29 (8625); A322 (8625); A519 (8625); A752 (8625); A829 **SAE** J404 (8625); J412 (8625); J1397 (8625) |
| **G86270** | Ni-Cr-Mo Alloy Steel | **C** 0.25-0.30 **Cr** 0.40-0.60 **Mn** 0.70-0.90 **Mo** 0.15-0.25 **Ni** 0.40-0.70 **P** 0.035 max **S** 0.040 max **Si** 0.15-0.35 | **ASTM** A29 (8627); A322 (8627); A519 (8627); A752 (8627); A829 **SAE** J404 (8627); J412 (8627); J1397 (8627) |
| **G86300** | Ni-Cr-Mo Alloy Steel | **C** 0.28-0.33 **Cr** 0.40-0.60 **Mn** 0.70-0.90 **Mo** 0.15-0.25 **Ni** 0.40-0.70 **P** 0.035 max **S** 0.040 max **Si** 0.15-0.35 | **AMS** 6280; 6281; 6360; 6535; 7496; AMS-S-6050; AMS-S-18728; AMS-T-6735 (8630) **ASTM** A29 (8630); A322 (8630); A513 (8630); A519 (8630); A752 (8630); A829 **SAE** J404 (8630); J412 (8630); J1397 (8630) |
| **G86370** | Ni-Cr-Mo Alloy Steel | **C** 0.35-0.40 **Cr** 0.40-0.60 **Mn** 0.75-1.00 **Mo** 0.15-0.25 **Ni** 0.40-0.70 **P** 0.035 max **S** 0.040 max **Si** 0.15-0.35 | **ASTM** A29 (8637); A322 (8637); A519 (8637); A689 (8637); A752 (8637); A829 **SAE** J404 (8637); J412 (8637); J1397 (8637) |
| **G86400** | Ni-Cr-Mo Alloy Steel | **C** 0.38-0.43 **Cr** 0.40-0.60 **Mn** 0.75-1.00 **Mo** 0.15-0.25 **Ni** 0.40-0.70 **P** 0.035 max **S** 0.040 max **Si** 0.15-0.35 | **ASTM** A29 (8640); A322 (8640); A519 (8640); A752 (8640); A829 **SAE** J404 (8640); J412 (8640); J1397 (8640) |
| **G86420** | Ni-Cr-Mo Alloy Steel | **C** 0.40-0.45 **Cr** 0.40-0.60 **Mn** 0.75-1.00 **Mo** 0.15-0.25 **Ni** 0.40-0.70 **P** 0.035 max **S** 0.40 max **Si** 0.15-0.35 | **ASTM** A29 (8642); A322 (8642); A519 (8642); A752 (8642) **SAE** J412 (8642); J1249 (8642); J1397 (8642) |

The chemical compositions listed are for identification purposes and should not be used in lieu of the cross referenced specifications.

| UNIFIED NUMBER | DESCRIPTION | CHEMICAL COMPOSITION | CROSS REFERENCE SPECIFICATIONS |
|---|---|---|---|
| **G86450** | Ni-Cr-Mo Alloy Steel | **C** 0.43-0.48 **Cr** 0.40-0.60 **Mn** 0.75-1.00 **Mo** 0.15-0.25 **Ni** 0.40-0.70 **P** 0.035 max **S** 0.040 max **Si** 0.15-0.35 | **ASTM** A29 (8645); A322 (8645); A519 (8645); A752 (8645) **SAE** J404 (8645); J412 (8645); J775 (NV 5); J1397 (8645) |
| **G86451** | Ni-Cr-Mo-B Alloy Steel | **B** 0.0005-0.003 **C** 0.43-0.48 **Cr** 0.40-.060 **Mn** 0.75-1.00 **Mo** 0.15-0.25 **Ni** 0.40-0.70 **P** 0.035 max **S** 0.040 max **Si** 0.15-0.35 | **ASTM** A519 (86B45) **SAE** J412 (86B45); J1249 (86B45); J1397 (86B45) |
| **G86500** | Ni-Cr-Mo Alloy Steel | **C** 0.48-0.53 **Cr** 0.40-0.60 **Mn** 0.75-1.00 **Mo** 0.15-0.25 **Ni** 0.40-0.70 **P** 0.035 max **S** 0.040 max **Si** 0.15-0.35 | **ASTM** A29 (8650); A519 (8650); A689 **SAE** J412 (8650); J1249 (8650); J1397 (8650) |
| **G86550** | Ni-Cr-Mo Alloy Steel | **C** 0.51-0.59 **Cr** 0.40-0.60 **Mn** 0.75-1.00 **Mo** 0.15-0.25 **Ni** 0.40-0.70 **P** 0.035 max **S** 0.040 max **Si** 0.15-0.35 | **ASTM** A29 (8655); A322 (8655); A519 (8655); A689; A752 (8655); A829 **SAE** J404 (8655); J412 (8655); J1397 (8655) |
| **G86600** | Ni-Cr-Mo Alloy Steel | **C** 0.55-0.65 **Cr** 0.40-0.60 **Mn** 0.75-1.00 **Mo** 0.15-0.25 **Ni** 0.40-0.70 **P** 0.035 max **S** 0.040 max **Si** 0.15-0.35 | **ASTM** A29 (8660); A519 (8660); A689; A711 **SAE** J412 (8660); J1249 (8660); J1397 (8660) |
| **G87200** | Ni-Cr-Mo Alloy Steel | **C** 0.18-0.23 **Cr** 0.40-0.60 **Mn** 0.70-0.90 **Mo** 0.20-0.30 **Ni** 0.40-0.70 **P** 0.035 max **S** 0.040 max **Si** 0.15-0.35 | **ASTM** A29 (8720); A322 (8720); A519 (8720); A752 (8720) **SAE** J404 (8720); J412 (8720); J1397 (8720) |
| **G87350** | Ni-Cr-Mo Alloy Steel | **C** 0.33-0.38 **Cr** 0.40-0.60 **Mn** 0.75-1.00 **Mo** 0.20-0.30 **Ni** 0.40-0.70 **P** 0.035 max **S** 0.040 max **Si** 0.15-0.35 | **AMS** 6282; 6320; 6331; 6357; 7496; AMS-S-6098 **ASTM** A519 (8735) **SAE** J1249 (8735 |
| **G87400** | Ni-Cr-Mo Alloy Steel | **C** 0.38-0.43 **Cr** 0.40-0.60 **Mn** 0.75-1.00 **Mo** 0.20-0.30 **Ni** 0.40-0.70 **P** 0.035 max **S** 0.040 max **Si** 0.15-0.35 | **AMS** 6322; 6323; 6325; 6327; 6358; 7496; AMS-S-6049; AS7452; AS7456 **ASTM** A29 (8740); A320 (L7C, L73); A322 (8740); A519 (8740); A752 (8740) **SAE** J412 (8740); J1249 (8740); J1397 (8740) |
| **G87420** | Ni-Cr-Mo Alloy Steel | **C** 0.40-0.45 **Cr** 0.40-0.60 **Mn** 0.75-1.00 **Mo** 0.20-0.30 **Ni** 0.40-0.70 **P** 0.035 max **S** 0.040 max **Si** 0.15-0.35 | **ASTM** A519 (8742); A829 **SAE** J1249 (8742) |
| **G88220** | Ni-Cr-Mo Alloy Steel | **C** 0.20-0.25 **Cr** 0.40-0.60 **Mn** 0.75-1.00 **Mo** 0.30-0.40 **Ni** 0.40-0.70 **P** 0.035 max **S** 0.040 max **Si** 0.15-0.35 | **ASTM** A29 (8822); A322 (8822); A519 (8822); A752 (8822) **SAE** J404 (8822); J412 (8822); J1397 (8822) |
| **G92540** | Cr-Si Alloy Steel | **C** 0.51-0.59 **Cr** 0.60-0.80 **Mn** 0.60-0.80 **P** 0.035 max **S** 0.040 max **Si** 1.20-1.60 | **AMS** 6451 **ASTM** A29 (9254); A401; A752 (9254); A877 **SAE** J412 (9254); J1249 (9254); J1397 (9254) |
| **G92550** | Si-Mn Alloy Steel | **C** 0.51-0.59 **Mn** 0.70-0.95 **P** 0.035 max **S** 0.040 max **Si** 1.80-2.20 | **ASTM** A29 (9255); A519 (9255); A752 (9255) **SAE** J1249 (9255) |
| **G92590** | Cr-Si-Mn Alloy Steel (Grade 9259) | **C** 0.56-0.64 **Cr** 0.45-0.65 **Mn** 0.75-1.00 **P** 0.035 max **Si** 0.70-1.10 **S** 0.040 max | **ASTM** A29 (9259); A322 (9259) **SAE** J404 (9259) |
| **G92600** | Si-Mn Alloy Steel | **C** 0.56-0.64 **Mn** 0.75-1.00 **P** 0.035 max **S** 0.040 max **Si** 1.80-2.20 | **ASTM** A29 (9260); A322 (9260); A519 (9260); A689; A752 (9260) **SAE** J404 (9260); J412 (9260); J1397 (9260) |
| **G92620** | Si-Cr Alloy Steel | **C** 0.55-0.65 **Cr** 0.25-0.40 **Mn** 0.75-1.00 **P** 0.035 max **S** 0.040 max **Si** 1.80-2.20 | **ASTM** A519 (9262) **SAE** J1249 (9262) |
| **G93106** | Ni-Cr-Mo Alloy Steel | **C** 0.08-0.13 **Cr** 1.00-1.40 **Mn** 0.45-0.65 **Mo** 0.08-0.15 **Ni** 3.00-3.50 **P** 0.025 max **S** 0.025 max **Si** 0.15-0.35 | **AMS** 6260; 6265; 6267 **ASTM** A29 (E9310); A506 (E9310); A519 (E9310); A646 (9310-2) **MIL SPEC** MIL-S-7393 (9310) **SAE** J1249 (9310); J1397 (9310) |
| **G94151** | Ni-Cr-Mo-B Alloy Steel | **B** 0.0005 min **C** 0.13-0.18 **Cr** 0.30-0.50 **Mn** 0.75-1.00 **Mo** 0.08-0.15 **Ni** 0.30-0.60 **P** 0.035 max **S** 0.040 max **Si** 0.15-0.35 | **ASTM** A519 (94B15) **SAE** J412 (94B15); J1249 (94B15); J1397 (94B15) |

The chemical compositions listed are for identification purposes and should not be used in lieu of the cross referenced specifications.

| UNIFIED NUMBER | DESCRIPTION | CHEMICAL COMPOSITION | CROSS REFERENCE SPECIFICATIONS |
|---|---|---|---|
| **G94171** | Ni-Cr-Mo-B Alloy Steel | **B** 0.0005-0.003 **C** 0.15-0.20 **Cr** 0.30-0.50 **Mn** 0.75-1.00 **Mo** 0.08-0.15 **Ni** 0.30-0.60 **P** 0.035 max **S** 0.040 max **Si** 0.15-0.35 | **AMS** 6275 **ASTM** A29 (94B17); A322 (94B17); A519 (94B17); A752 (94B17) **SAE** J412 (94B17); J1249 (94B17); J1397 (94B17) |
| **G94301** | Ni-Cr-Mo-B Alloy Steel | **B** 0.0005-0.003 **C** 0.28-0.33 **Cr** 0.30-0.50 **Mn** 0.75-1.00 **Mo** 0.08-0.15 **Ni** 0.30-0.60 **P** 0.035 max **S** 0.040 max **Si** 0.15-0.35 | **ASTM** A29 (94B30); A322 (94B30); A519 (94B30); A752 (94B30) **SAE** J1249 (94B30); J1397 (94B30) |
| **G94401** | Ni-Cr-Mo-B Alloy Steel | **B** 0.0005-0.003 **C** 0.38-0.43 **Cr** 0.30-0.50 **Mn** 0.75-1.00 **Mo** 0.08-0.15 **Ni** 0.30-0.60 **P** 0.035 max **S** 0.040 max **Si** 0.15-0.35 | **ASTM** A519 (94B40) **SAE** J1249 (94B40) |
| **G98400** | Ni-Cr-Mo Alloy Steel | **C** 0.38-0.43 **Cr** 0.70-0.90 **Mn** 0.70-0.90 **Mo** 0.20-0.30 **Ni** 0.85-1.15 **P** 0.035 max **S** 0.040 max **Si** 0.15-0.35 | **AMS** 6342 **ASTM** A519 (9840); A711 **SAE** J1249 (9840) |
| **G98500** | Ni-Cr-Mo Alloy Steel | **C** 0.48-0.53 **Cr** 0.70-0.90 **Mn** 0.70-0.90 **Mo** 0.20-0.30 **Ni** 0.85-1.15 **P** 0.035 max **S** 0.040 max **Si** 0.15-0.35 | **ASTM** A519 (9850) **MIL SPEC** MIL-S-19434 **SAE** J1249 (9850) |

The chemical compositions listed are for identification purposes and should not be used in lieu of the cross referenced specifications.

# UNS NUMBERS ASSIGNED TO DATE

## With Description of Each Material Covered and References To Documents In Which The Same or Similar Materials are Described

## Hxxxxx Number Series
## AISI and SAE H-Steels

# AISI AND SAE H-STEELS

| UNIFIED NUMBER | DESCRIPTION | CHEMICAL COMPOSITION | CROSS REFERENCE SPECIFICATIONS |
|---|---|---|---|
| **H10380** | H-Carbon Steel | **C** 0.34-0.43 **Mn** 0.50-1.00 **P** 0.040 max **S** 0.050 max **Si** 0.15-0.30 | **AISI** 1038 H **ASTM** A304 (1038H) **SAE** J1268 (1038H) |
| **H10450** | H-Carbon Steel | **C** 0.42-0.51 **Mn** 0.50-1.00 **P** 0.040 max **S** 0.050 max **Si** 0.15-0.30 | **AISI** 1045 H **ASTM** A304 (1045H) **SAE** J1268 (1045H) |
| **H13300** | C-Mn H-Alloy Steel | **C** 0.27-0.33 **Mn** 1.45-2.05 **P** 0.035 max **S** 0.040 max **Si** 0.15-0.35 | **AISI** 1330 H **ASTM** A304 (1330H) **SAE** J1268 (1330H) |
| **H13350** | C-Mn H-Alloy Steel | **C** 0.32-0.38 **Mn** 1.45-2.05 **P** 0.035 max **S** 0.040 max **Si** 0.15-0.35 | **AISI** 1335 H **ASTM** A304 (1335H) **SAE** J1268 (1335H) |
| **H13400** | C-Mn H-Alloy Steel | **C** 0.37-0.44 **Mn** 1.45-2.05 **P** 0.035 max **S** 0.040 max **Si** 0.15-0.35 | **AISI** 1340 H **ASTM** A304 (1340H) **SAE** J1268 (1340H) |
| **H13450** | C-Mn H-Alloy Steel | **C** 0.42-0.49 **Mn** 1.45-2.05 **P** 0.035 max **S** 0.040 max **Si** 0.15-0.35 | **AISI** 1345 H **ASTM** A304 (1345H) **SAE** J1268 (1345H) |
| **H15211** | C-B H-Carbon Steel | **B** 0.0005-0.003 **C** 0.17-0.24 **Mn** 0.70-1.20 **P** 0.040 max **S** 0.050 max **Si** 0.15-0.30 | **AISI** 15B21 H **ASME** SA-320 (L1) **ASTM** A304 (15B21H) **SAE** J1268 (15B21H) |
| **H15220** | Mn H-Carbon Steel | **C** 0.17-0.25 **Mn** 1.00-1.50 **P** 0.040 max **S** 0.050 max **Si** 0.15-0.30 | **AISI** 1522 H **ASTM** A304 (1522H) **AWS** C2.25 (W-FeC-2) **SAE** J1268 (1522H) |
| **H15240** | Mn H-Carbon Steel | **C** 0.18-0.26 **Mn** 1.25-1.75 **P** 0.040 max **S** 0.050 max **Si** 0.15-0.30 | **AISI** 1524 H **ASTM** A304 (15B26H); A304 (1524H) **SAE** J1268 (1524H) |
| **H15260** | Mn H-Carbon Steel | **C** 0.21-0.30 **Mn** 1.00-1.50 **P** 0.040 max **S** 0.050 max **Si** 0.15-0.30 | **AISI** 1526 H **ASTM** A304 (1526H) **SAE** J1268 (1526H) |
| **H15281** | B-Mn-H Carbon Steel | **B** 0.0005-0.003 **C** 0.25-0.34 **Mn** 1.00-1.50 **P** 0.030 max **S** 0.050 max **Si** 0.15-0.35 | **SAE** J1268 (15B28H) |
| **H15301** | B-Mn-H Carbon Steel | **B** 0.0005-0.003 **C** 0.27-0.35 **Mn** 0.70-1.20 **P** 0.030 max **S** 0.050 max **Si** 0.15-0.35 | **SAE** J1268 (15B30H) |
| **H15351** | B-Mn H-Carbon Steel | **B** 0.0005-0.003 **C** 0.31-0.39 **Mn** 0.70-1.20 **P** 0.040 max **S** 0.050 max **Si** 0.15-0.30 | **AISI** 15B35 H **ASTM** A304 (15B35H) **SAE** J1268 (15B35H) |
| **H15371** | B-Mn H-Carbon Steel | **B** 0.0005-0.003 **C** 0.30-0.39 **Mn** 1.00-1.50 **P** 0.040 max **S** 0.050 max **Si** 0.15-0.30 | **AISI** 15B37 H **ASTM** A304 (15B37H) **SAE** J1268 (15B37H) |
| **H15410** | Mn H-Carbon Steel | **C** 0.35-0.45 **Mn** 1.25-1.75 **P** 0.040 max **S** 0.050 max **Si** 0.15-0.30 | **AISI** 1541 H **ASTM** A304 (1541H) **SAE** J775 (NV 1); J1268 (1541H) |
| **H15411** | C-B H-Carbon Steel | **B** 0.0005-0.003 **C** 0.35-0.45 **Mn** 1.25-1.75 **P** 0.040 max **S** 0.050 max **Si** 0.15-0.30 | **AISI** 15B41 H **ASTM** A304 (15B41H) **SAE** J1268 (15B41H) |
| **H15481** | C-B H-Carbon Steel | **B** 0.0005-0.003 **C** 0.43-0.53 **Mn** 1.00-1.50 **P** 0.040 max **S** 0.050 max **Si** 0.15-0.30 | **AISI** 15B48 H **ASTM** A304 (15B48H) **SAE** J1268 (15B48H) |
| **H15621** | C-B H-Carbon Steel | **B** 0.0005-0.003 **C** 0.54-0.67 **Mn** 1.00-1.50 **P** 0.040 max **S** 0.050 max **Si** 0.40-0.60 | **AISI** 15B62 H **ASTM** A304 (15B62H) **SAE** J1268 (15B62H) |
| **H40270** | C-Mo H-Alloy Steel | **C** 0.24-0.30 **Mn** 0.60-1.00 **Mo** 0.20-0.30 **P** 0.035 max **S** 0.040 max **Si** 0.15-0.35 | **AISI** 4027 H **ASTM** A304 (4027H) **SAE** J1268 (4027H) |
| **H40280** | C-Mo H-Alloy Steel | **C** 0.24-0.30 **Mn** 0.60-1.00 **Mo** 0.20-0.30 **P** 0.035 max **S** 0.035-0.050 **Si** 0.15-0.35 | **AISI** 4028 H **ASTM** A304 (4028H) **SAE** J1268 (4028H) |
| **H40320** | C-Mo H-Alloy Steel | **C** 0.29-0.35 **Mn** 0.60-1.00 **Mo** 0.20-0.30 **P** 0.035 max **S** 0.040 max **Si** 0.15-0.35 | **AISI** 4032H **ASTM** A304 (4032H) **SAE** J1268 (4032H) |
| **H40370** | C-Mo H-Alloy Steel | **C** 0.34-0.41 **Mn** 0.60-1.00 **Mo** 0.20-0.30 **P** 0.035 max **S** 0.040 max **Si** 0.15-0.35 | **AISI** 4037 H **ASTM** A304 (4037H) **SAE** J1268 (4037H) |
| **H40420** | C-Mo H-Alloy Steel | **C** 0.39-0.46 **Mn** 0.60-1.00 **Mo** 0.20-0.30 **Si** 0.15-0.35 | **AISI** 4042 H **ASTM** A304 (4042H) **SAE** J1268 (4042H) |

The chemical compositions listed are for identification purposes and should not be used in lieu of the cross referenced specifications.

| UNIFIED NUMBER | DESCRIPTION | CHEMICAL COMPOSITION | CROSS REFERENCE SPECIFICATIONS |
|---|---|---|---|
| **H40470** | C-Mo H-Alloy Steel | **C** 0.44-0.51 **Mn** 0.60-1.00 **Mo** 0.20-0.30 **P** 0.035 max **S** 0.040 max **Si** 0.15-0.35 | **AISI** 4047 H **ASTM** A304 (4047H) **SAE** J1268 (4047H) |
| **H41180** | Cr-Mo H-Alloy Steel | **C** 0.17-0.23 **Cr** 0.30-0.70 **Mn** 0.60-1.00 **Mo** 0.08-0.15 **P** 0.035 max **S** 0.040 max **Si** 0.15-0.35 | **AISI** 4118 H **ASTM** A304 (4118H); A534 (4118H) **SAE** J1268 (4118H) |
| **H41200** | Cr-Mo H-Alloy Steel | **C** 0.18-0.23 **Cr** 0.40-0.60 **Mn** 0.90-1.20 **Mo** 0.13-0.20 **P** 0.030 max **S** 0.040 max **Si** 0.15-0.35 | **SAE** J1268 (4120H) |
| **H41300** | Cr-Mo H-Alloy Steel | **C** 0.27-0.33 **Cr** 0.75-1.20 **Mn** 0.30-0.70 **Mo** 0.15-0.25 **P** 0.035 max **S** 0.040 max **Si** 0.15-0.35 | **AISI** 4130 H **ASTM** A304 (4130H) **SAE** J1268 (4130H) |
| **H41350** | Cr-Mo H-Alloy Steel | **C** 0.32-0.38 **Cr** 0.75-1.20 **Mn** 0.60-1.00 **Mo** 0.15-0.25 **Si** 0.15-0.35 | **AISI** 4135 H **ASTM** A304 (4135H) **SAE** J1268 (4135H) |
| **H41370** | Cr-Mo H-Alloy Steel | **C** 0.34-0.41 **Cr** 0.75-1.20 **Mn** 0.60-1.00 **Mo** 0.15-0.25 **P** 0.035 max **S** 0.040 max **Si** 0.15-0.35 | **AISI** 4137 H **ASTM** A304 (4137H) **SAE** J1268 (4137H) |
| **H41400** | Cr-Mo H-Alloy Steel | **C** 0.37-0.44 **Cr** 0.75-1.20 **Mn** 0.65-1.10 **Mo** 0.15-0.25 **P** 0.035 max **S** 0.040 max **Si** 0.15-0.35 | **AISI** 4140 H **ASTM** A304 (4140H) **SAE** J775 (NV 7); J1268 (4140H) |
| **H41420** | Cr-Mo H-Alloy Steel | **C** 0.39-0.46 **Cr** 0.75-1.20 **Mn** 0.65-1.10 **Mo** 0.15-0.25 **P** 0.035 max **S** 0.040 max **Si** 0.15-0.35 | **AISI** 4142 H **ASTM** A304 (4142H); A540 (B22) **SAE** J1268 (4142H) |
| **H41450** | Cr-Mo H-Alloy Steel | **C** 0.42-0.49 **Cr** 0.75-1.20 **Mn** 0.65-1.10 **Mo** 0.15-0.25 **P** 0.035 max **S** 0.040 max **Si** 0.15-0.35 | **AISI** 4145 H **ASTM** A304 (4145H) **SAE** J1268 (4145H) |
| **H41470** | Cr-Mo H-Alloy Steel | **C** 0.44-0.51 **Cr** 0.75-1.20 **Mn** 0.65-1.10 **Mo** 0.15-0.25 **P** 0.035 max **S** 0.040 max **Si** 0.15-0.35 | **AISI** 4147 H **ASTM** A304 (4147H) **SAE** J1268 (4147H) |
| **H41500** | Cr-Mo H-Alloy Steel | **C** 0.47-0.54 **Cr** 0.75-1.20 **Mn** 0.65-1.10 **Mo** 0.15-0.25 **P** 0.035 max **S** 0.040 max **Si** 0.15-0.35 | **AISI** 4150 H **ASTM** A304 (4150H) **SAE** J1268 (4150H) |
| **H41610** | Cr-Mo H-Alloy Steel | **C** 0.55-0.65 **Cr** 0.65-0.95 **Mn** 0.65-1.10 **Mo** 0.25-0.35 **P** 0.035 max **S** 0.040 max **Si** 0.15-0.35 | **AISI** 4161 H **ASTM** A304 (4161H) **SAE** J1268 (4161H) |
| **H43200** | Ni-Cr-Mo H-Alloy Steel | **C** 0.17-0.23 **Cr** 0.35-0.65 **Mn** 0.40-0.70 **Mo** 0.20-0.30 **Ni** 1.55-2.00 **P** 0.035 max **S** 0.040 max **Si** 0.15-0.35 | **AISI** 4320 H **AMS** 6299 **ASTM** A304 (4320H); A534 (4320H) **SAE** J1268 (4320H) |
| **H43400** | Ni-Cr-Mo H-Alloy Steel | **C** 0.37-0.44 **Cr** 0.65-0.95 **Mn** 0.55-0.90 **Mo** 0.20-0.30 **Ni** 1.55-2.00 **P** 0.035 max **S** 0.040 max **Si** 0.15-0.35 | **AISI** 4340H **ASTM** A304 (4340H); A540 (B23) **SAE** J1268 (4340H) |
| **H43406** | Ni-Cr-Mo H-Alloy Steel | **C** 0.37-0.44 **Cr** 0.65-0.95 **Mn** 0.60-0.95 **Mo** 0.20-0.30 **Ni** 1.55-2.00 **P** 0.025 max **S** 0.025 max **Si** 0.15-0.35 | **ASTM** A304 (E4340H) **SAE** J1268 (E4340H) |
| **H44190** | C-Mo H-Alloy Steel | **C** 0.17-0.23 **Mn** 0.35-0.75 **Mo** 0.45-0.60 **P** 0.035 max **S** 0.040 max **Si** 0.15-0.35 | **AISI** 4419 H **ASTM** A304 (4419H) |
| **H46200** | Ni-Mo H-Alloy Steel | **C** 0.17-0.23 **Mn** 0.35-0.75 **Mo** 0.20-0.30 **Ni** 1.55-2.00 **P** 0.035 max **S** 0.040 max **Si** 0.15-0.35 | **AISI** 4620 H **ASTM** A304 (4620H); A534 (4620H) **SAE** J1268 (4620H) |
| **H46210** | Ni-Mo H-Alloy Steel | **C** 0.17-0.23 **Mn** 0.60-1.00 **Mo** 0.20-0.30 **Ni** 1.55-2.00 **P** 0.035 max **S** 0.040 max **Si** 0.15-0.35 | **AISI** 4621 H **ASTM** A304 (4621H) |
| **H46260** | Ni-Mo H-Alloy Steel | **C** 0.23-0.29 **Mn** 0.40-0.70 **Mo** 0.15-0.25 **Ni** 0.65-1.05 **P** 0.035 max **S** 0.040 max **Si** 0.15-0.35 | **AISI** 4626 H **ASTM** A304 (4626H); A534 (4620H) |

The chemical compositions listed are for identification purposes and should not be used in lieu of the cross referenced specifications.

| UNIFIED NUMBER | DESCRIPTION | CHEMICAL COMPOSITION | CROSS REFERENCE SPECIFICATIONS |
|---|---|---|---|
| **H47180** | Ni-Cr-Mo H-Alloy Steel | **C** 0.15-0.21 **Cr** 0.30-0.60 **Mn** 0.60-0.95 **Mo** 0.30-0.40 **Ni** 0.85-1.25 **P** 0.035 max **S** 0.040 max **Si** 0.15-0.35 | **AISI** 4718 H **ASTM** A304 (4718H) **SAE** J1268 (4718H) |
| **H47200** | Ni-Cr-Mo H-Alloy Steel | **C** 0.17-0.23 **Cr** 0.30-0.60 **Mn** 0.45-0.75 **Mo** 0.15-0.25 **Ni** 0.85-1.25 **P** 0.035 max **S** 0.040 max **Si** 0.15-0.35 | **AISI** 4720 H **ASTM** A304 (4720H); A534 (4720H) **SAE** J1268 (4720H) |
| **H48150** | Ni-Mo H-Alloy Steel | **C** 0.12-0.18 **Mn** 0.30-0.70 **Mo** 0.20-0.30 **Ni** 3.20-3.80 **P** 0.035 max **S** 0.040 max **Si** 0.15-0.35 | **AISI** 4815 H **ASTM** A304 (4815H) **SAE** J1268 (4815H) |
| **H48170** | Ni-Mo H-Alloy Steel | **C** 0.14-0.20 **Mn** 0.30-0.70 **Mo** 0.20-0.30 **Ni** 3.20-3.80 **P** 0.035 max **S** 0.040 max **Si** 0.15-0.35 | **AISI** 4817 H **ASTM** A304 (4817H); A534 (4817H) **SAE** J1268 (4817H) |
| **H48200** | Ni-Mo H-Alloy Steel | **C** 0.17-0.23 **Mn** 0.40-0.80 **Mo** 0.20-0.30 **Ni** 3.20-3.80 **P** 0.035 max **S** 0.040 max **Si** 0.15-0.35 | **AISI** 4820 H **ASTM** A304 (4820H); A534 (4820H) **SAE** J1268 (4820H) |
| **H50401** | C-Cr-B H-Alloy Steel | **B** 0.0005-0.003 **C** 0.37-0.44 **Cr** 0.30-0.70 **Mn** 0.65-1.10 **Si** 0.15-0.35 | **AISI** 50B40 H **ASTM** A304 (50B40H) **SAE** J1268 (50B40H) |
| **H50441** | C-Cr-B H-Alloy Steel | **B** 0.0005-0.003 **C** 0.42-0.49 **Cr** 0.30-0.70 **Mn** 0.65-1.10 **P** 0.035 max **S** 0.040 max **Si** 0.15-0.35 | **AISI** 50B44 H **ASTM** A304 (50B44H) **SAE** J1268 (50B44H) |
| **H50460** | C-Cr H-Alloy Steel | **C** 0.43-0.50 **Cr** 0.13-0.43 **Mn** 0.65-1.10 **Si** 0.15-0.35 | **AISI** 5046 H **ASTM** A304 (5046H) **SAE** J1268 (5046H) |
| **H50461** | C-Cr-B H-Alloy Steel | **B** 0.0005-0.003 **C** 0.43-0.50 **Cr** 0.13-0.43 **Mn** 0.65-1.10 **P** 0.035 max **S** 0.040 max **Si** 0.15-0.35 | **AISI** 50B46 H **ASTM** A304 (50B46H) **SAE** J1268 (50B46H) |
| **H50501** | C-Cr-B H-Alloy Steel | **B** 0.0005-0.003 **C** 0.47-0.54 **Cr** 0.30-0.70 **Mn** 0.65-1.10 **P** 0.035 max **S** 0.040 max **Si** 0.15-0.35 | **AISI** 50B50 H **ASTM** A304 (50B50H) **SAE** J1268 (50B50H) |
| **H50601** | C-Cr-B H-Alloy Steel | **B** 0.0005-0.003 **C** 0.55-0.65 **Cr** 0.30-0.70 **Mn** 0.65-1.10 **P** 0.035 max **S** 0.040 max **Si** 0.15-0.35 | **AISI** 50B60 H **ASTM** A304 (50B60H) **SAE** J1268 (50B60H) |
| **H51200** | C-Cr H-Alloy Steel | **C** 0.17-0.23 **Cr** 0.60-1.00 **Mn** 0.60-1.00 **P** 0.035 max **S** 0.040 max **Si** 0.15-0.35 | **AISI** 5120 H **ASTM** A304 (5120H); A534 (5120H) **SAE** J1268 (5120H) |
| **H51300** | C-Cr H-Alloy Steel | **C** 0.27-0.33 **Cr** 0.75-1.20 **Mn** 0.60-1.00 **P** 0.035 max **S** 0.040 max **Si** 0.15-0.35 | **AISI** 5130 H **ASTM** A304 (5130H) **SAE** J1268 (5130H) |
| **H51320** | C-Cr H-Alloy Steel | **C** 0.29-0.35 **Cr** 0.65-1.10 **Mn** 0.50-0.90 **P** 0.035 max **S** 0.040 max **Si** 0.15-0.35 | **AISI** 5132 H **ASTM** A304 (5132H) **SAE** J1268 (5132H) |
| **H51350** | C-Cr H-Alloy Steel | **C** 0.32-0.38 **Cr** 0.70-1.15 **Mn** 0.50-0.90 **P** 0.035 max **S** 0.040 max **Si** 0.15-0.35 | **AISI** 5135 H **ASTM** A304 (5135H) **SAE** J1268 (5135H) |
| **H51400** | C-Cr H-Alloy Steel | **C** 0.37-0.44 **Cr** 0.60-1.00 **Mn** 0.60-1.00 **P** 0.035 max **S** 0.040 max **Si** 0.15-0.35 | **AISI** 5140 H **ASTM** A304 (5140H) **SAE** J1268 (5140H) |
| **H51450** | C-Cr H-Alloy Steel | **C** 0.42-0.49 **Cr** 0.60-1.00 **Mn** 0.60-1.00 **P** 0.035 max **S** 0.040 max **Si** 0.15-0.35 | **AISI** 5145 H **ASTM** A304 (5145H) |
| **H51470** | C-Cr H-Alloy Steel | **C** 0.45-0.52 **Cr** 0.80-1.25 **Mn** 0.60-1.05 **P** 0.035 max **S** 0.040 max **Si** 0.15-0.35 | **AISI** 5147 H **ASTM** A304 (5147H) **SAE** J1268 (5147H) |
| **H51500** | C-Cr H-Alloy Steel | **C** 0.47-0.54 **Cr** 0.60-1.00 **Mn** 0.60-1.00 **P** 0.035 max **S** 0.040 max **Si** 0.15-0.35 | **AISI** 5150 H **ASTM** A304 (5150H) **SAE** J775 (NV 6); J1268 (5150H) |
| **H51550** | C-Cr H-Alloy Steel | **C** 0.50-0.60 **Cr** 0.60-1.00 **Mn** 0.60-1.00 **P** 0.035 max **S** 0.040 max **Si** 0.15-0.35 | **AISI** 5155 H **ASTM** A304 (5155H) **SAE** J1268 (5155H) |

The chemical compositions listed are for identification purposes and should not be used in lieu of the cross referenced specifications.

| UNIFIED NUMBER | DESCRIPTION | CHEMICAL COMPOSITION | CROSS REFERENCE SPECIFICATIONS |
|---|---|---|---|
| **H51600** | C-Cr H-Alloy Steel | **C** 0.55-0.65 **Cr** 0.60-1.00 **Mn** 0.65-1.10 **P** 0.035 max **S** 0.040 max **Si** 0.15-0.35 | **AISI** 5160 H **ASTM** A304 (5160H) **SAE** J1268 (5160H) |
| **H51601** | C-Cr-B H-Alloy Steel | **B** 0.0005-0.003 **C** 0.55-0.65 **Cr** 0.60-1.00 **Mn** 0.65-1.10 **P** 0.035 max **S** 0.040 max **Si** 0.15-0.35 | **AISI** 51B60 H **ASTM** A304 (51B60H) **SAE** J1268 (51B60H) |
| **H61180** | Cr-V H-Alloy Steel | **C** 0.15-0.21 **Cr** 0.40-0.80 **Mn** 0.40-0.80 **P** 0.035 max **S** 0.040 max **Si** 0.15-0.35 **V** 0.10-0.15 | **AISI** 6118 H **ASTM** A304 (6118H) **SAE** J1268 (6118H) |
| **H61500** | Cr-V H-Alloy Steel | **C** 0.47-0.54 **Cr** 0.75-1.20 **Mn** 0.60-1.00 **P** 0.035 max **S** 0.040 max **Si** 0.15-0.35 **V** 0.15 min | **AISI** 6150 H **ASTM** A304 (6150H) **SAE** J1268 (6150H) |
| **H81451** | Ni-Cr-Mo-B H-Alloy Steel | **B** 0.0005-0.003 **C** 0.42-0.49 **Cr** 0.30-0.60 **Mn** 0.70-1.05 **Mo** 0.08-0.15 **Ni** 0.15-0.45 **P** 0.035 max **S** 0.040 max **Si** 0.15-0.35 | **AISI** 81B45 H **ASTM** A304 (81B45H) **SAE** J1268 (81B45H) |
| **H86170** | Ni-Cr-Mo H-Alloy Steel | **C** 0.14-0.20 **Cr** 0.35-0.65 **Mn** 0.60-0.95 **Mo** 0.15-0.25 **Ni** 0.35-0.75 **P** 0.035 max **S** 0.040 max **Si** 0.15-0.30 | **AISI** 8617 H **ASTM** A304 (8617H); A534 (8617H) **SAE** J1268 (8617H) |
| **H86200** | Ni-Cr-Mo H-Alloy Steel | **C** 0.17-0.23 **Cr** 0.35-0.65 **Mn** 0.60-0.95 **Mo** 0.15-0.25 **Ni** 0.35-0.75 **P** 0.035 max **S** 0.040 max **Si** 0.15-0.35 | **AISI** 8620 H **ASTM** A304 (8620H); A534 (8620H) **SAE** J1268 (8620H) |
| **H86220** | Ni-Cr-Mo H-Alloy Steel | **C** 0.19-0.25 **Cr** 0.35-0.65 **Mn** 0.60-0.95 **Mo** 0.15-0.25 **Ni** 0.35-0.75 **P** 0.035 max **S** 0.040 max **Si** 0.15-0.35 | **AISI** 8622 H **ASTM** A304 (8622H) **SAE** J1268 (8622H) |
| **H86250** | Ni-Cr-Mo H-Alloy Steel | **C** 0.22-0.28 **Cr** 0.35-0.65 **Mn** 0.60-0.95 **Mo** 0.15-0.25 **Ni** 0.35-0.75 **P** 0.035 max **S** 0.040 max **Si** 0.15-0.35 | **AISI** 8625 H **ASTM** A304 (8625H) **SAE** J1268 (8625H) |
| **H86270** | Ni-Cr-Mo H-Alloy Steel | **C** 0.24-0.30 **Cr** 0.35-0.65 **Mn** 0.60-0.95 **Mo** 0.15-0.25 **Ni** 0.35-0.75 **P** 0.035 max **S** 0.040 max **Si** 0.15-0.35 | **AISI** 8627 H **ASTM** A304 (8627H) **SAE** J1268 (8627H) |
| **H86300** | Ni-Cr-Mo H-Alloy Steel | **C** 0.27-0.33 **Cr** 0.35-0.65 **Mn** 0.60-0.95 **Mo** 0.15-0.25 **Ni** 0.35-0.75 **P** 0.035 max **S** 0.040 max **Si** 0.15-0.35 | **AISI** 8630 H **ASTM** A304 (8630H) **SAE** J1268 (8630H) |
| **H86301** | Ni-Cr-Mo-B H-Alloy Steel | **B** 0.0005-0.003 **C** 0.27-0.33 **Cr** 0.35-0.65 **Mn** 0.60-0.95 **Mo** 0.15-0.25 **Ni** 0.35-0.75 **Si** 0.15-0.35 | **AISI** 86B30 H **ASTM** A304 (86B30H) **SAE** J1268 (86B30H) |
| **H86370** | Ni-Cr-Mo H-Alloy Steel | **C** 0.34-0.41 **Cr** 0.35-0.65 **Mn** 0.70-1.05 **Mo** 0.15-0.25 **Ni** 0.35-0.75 **P** 0.035 max **S** 0.040 max **Si** 0.15-0.35 | **AISI** 8637 H **ASTM** A304 (8637H) **SAE** J1268 (8637H) |
| **H86400** | Ni-Cr-Mo H-Alloy Steel | **C** 0.37-0.44 **Cr** 0.35-0.65 **Mn** 0.70-1.05 **Mo** 0.15-0.25 **Ni** 0.35-0.75 **P** 0.035 max **S** 0.040 max **Si** 0.15-0.35 | **AISI** 8640 H **ASTM** A304 (8640H) **SAE** J1268 (8640H) |
| **H86420** | Ni-Cr-Mo H-Alloy Steel | **C** 0.39-0.46 **Cr** 0.35-0.65 **Mn** 0.70-1.05 **Mo** 0.15-0.25 **Ni** 0.35-0.75 **P** 0.035 max **S** 0.040 max **Si** 0.15-0.35 | **AISI** 8642 H **ASTM** A304 (8642H) **SAE** J1268 (8642H) |
| **H86450** | Ni-Cr-Mo H-Alloy Steel | **C** 0.42-0.49 **Cr** 0.35-0.65 **Mn** 0.70-1.05 **Mo** 0.15-0.25 **Ni** 0.35-0.75 **P** 0.035 max **S** 0.040 max **Si** 0.15-0.35 | **AISI** 8645 H **ASTM** A304 (8645H) **SAE** J1268 (8645H) |

The chemical compositions listed are for identification purposes and should not be used in lieu of the cross referenced specifications.

| UNIFIED NUMBER | DESCRIPTION | CHEMICAL COMPOSITION | CROSS REFERENCE SPECIFICATIONS |
|---|---|---|---|
| **H86451** | Ni-Cr-Mo-B H-Alloy Steel | **B** 0.0005-0.003 **C** 0.42-0.49 **Cr** 0.35-0.65 **Mn** 0.70-1.05 **Mo** 0.15-0.25 **Ni** 0.35-0.75 **P** 0.035 max **S** 0.040 max **Si** 0.15-0.35 | **AISI** 86B45 H **ASTM** A304 (86B45H) **SAE** J1268 (86B45H) |
| **H86500** | Ni-Cr-Mo H-Alloy Steel | **C** 0.47-0.54 **Cr** 0.35-0.65 **Mn** 0.70-1.05 **Mo** 0.15-0.25 **Ni** 0.35-0.75 **P** 0.035 max **S** 0.040 max **Si** 0.15-0.35 | **AISI** 8650 H **ASTM** A304 (8650H) **SAE** J1268 (8650H) |
| **H86550** | Ni-Cr-Mo H-Alloy Steel | **C** 0.50-0.60 **Cr** 0.35-0.65 **Mn** 0.70-1.05 **Mo** 0.15-0.25 **Ni** 0.35-0.75 **P** 0.035 max **S** 0.040 max **Si** 0.15-0.35 | **AISI** 8655 H **ASTM** A304 (8655H) **SAE** J1268 (8655H) |
| **H86600** | Ni-Cr-Mo H-Alloy Steel | **C** 0.55-0.65 **Cr** 0.35-0.65 **Mn** 0.70-1.05 **Mo** 0.15-0.25 **Ni** 0.35-0.75 **P** 0.035 max **S** 0.040 max **Si** 0.15-0.35 | **AISI** 8660 H **ASTM** A304 (8660H) **SAE** J1268 (8660H) |
| **H87200** | Ni-Cr-Mo H-Alloy Steel | **C** 0.17-0.23 **Cr** 0.35-0.65 **Mn** 0.60-0.95 **Mo** 0.20-0.30 **Ni** 0.35-0.75 **P** 0.035 max **S** 0.040 max **Si** 0.15-0.35 | **AISI** 8720 H **ASTM** A304 (8720H) **SAE** J1268 (8720H) |
| **H87400** | Ni-Cr-Mo H-Alloy Steel | **C** 0.37-0.44 **Cr** 0.35-0.65 **Mn** 0.70-1.05 **Mo** 0.20-0.30 **Ni** 0.35-0.75 **P** 0.035 max **S** 0.040 max **Si** 0.15-0.35 | **AISI** 8740 H **ASTM** A304 (8740H) **SAE** J1268 (8740H) |
| **H88220** | Ni-Cr-Mo H-Alloy Steel | **C** 0.19-0.25 **Cr** 0.35-0.65 **Mn** 0.70-1.05 **Mo** 0.30-0.40 **Ni** 0.35-0.75 **P** 0.035 max **S** 0.040 max **Si** 0.15-0.35 | **AISI** 8822 H **ASTM** A304 (8822H) **SAE** J1268 (8822H) |
| **H92590** | Cr-Si H-Alloy Steel | **C** 0.56-0.64 **Cr** 0.45-0.65 **Mn** 0.65-1.10 **P** 0.035 max **S** 0.040 max **Si** 0.70-1.20 | **ASTM** A29 (9259); A322 (9259) **SAE** J404 (9259); J1268 (9259H) |
| **H92600** | C-Si H-Alloy Steel | **C** 0.55-0.65 **Mn** 0.65-1.10 **P** 0.035 max **S** 0.040 max **Si** 1.70-2.20 | **AISI** 9260 H **ASTM** A304 (9260H) **SAE** J1268 (9260H) |
| **H93100** | Ni-Cr-Mo H-Alloy Steel | **C** 0.07-0.13 **Cr** 1.00-1.45 **Mn** 0.40-0.70 **Mo** 0.08-0.15 **Ni** 2.95-3.55 **P** 0.035 max **S** 0.040 max **Si** 0.15-0.35 | **AISI** 9310 H **ASTM** A304 (9310H); A534 (9310H) **SAE** J1268 (9310H) |
| **H93106** | Ni-Cr-Mo H-Alloy Steel | **C** 0.07-0.13 **Cr** 1.00-1.45 **Mn** 0.40-0.70 **Mo** 0.08-0.15 **Ni** 2.95-3.55 **P** 0.025 max **S** 0.025 max **Si** 0.15-0.35 | **SAE** J1268 (E9310H) |
| **H94151** | Ni-Cr-Mo-B H-Alloy Steel | **B** 0.0005-0.003 **C** 0.12-0.18 **Cr** 0.25-0.55 **Mn** 0.70-1.05 **Mo** 0.08-0.15 **Ni** 0.25-0.65 **P** 0.035 max **S** 0.040 max **Si** 0.15-0.35 | **AISI** 94B15 H **ASTM** A304 (94B15H) **SAE** J1268 (94B15H) |
| **H94171** | Ni-Cr-Mo-B H-Alloy Steel | **B** 0.0005-0.003 **C** 0.14-0.20 **Cr** 0.25-0.55 **Mn** 0.70-1.05 **Mo** 0.08-0.15 **Ni** 0.25-0.65 **P** 0.035 max **S** 0.040 max **Si** 0.15-0.35 | **AISI** 94B17 H **ASTM** A304 (94B17H) **SAE** J1268 (94B17H) |
| **H94301** | Ni-Cr-Mo-B H-Alloy Steel | **B** 0.0005-0.003 **C** 0.27-0.33 **Cr** 0.25-0.55 **Mn** 0.70-1.05 **Mo** 0.08-0.15 **Ni** 0.25-0.65 **P** 0.035 max **S** 0.040 max **Si** 0.15-0.35 | **AISI** 94B30 H **ASTM** A304 (94B30H) **SAE** J1268 (94B30H) |

The chemical compositions listed are for identification purposes and should not be used in lieu of the cross referenced specifications.

## UNS NUMBERS ASSIGNED TO DATE

### With Description of Each Material Covered and References To Documents In Which The Same or Similar Materials are Described

### Jxxxxx Number Series
### Cast Steels (Except Tool Steels)

# CAST STEELS (EXCEPT TOOL STEELS)

| UNIFIED NUMBER | DESCRIPTION | CHEMICAL COMPOSITION | CROSS REFERENCE SPECIFICATIONS |
|---|---|---|---|
| **J01700** | Carbon Steel Casting | **C** 0.12-0.22 **Mn** 0.50-0.90 **P** 0.040 max **S** 0.045 max **Si** 0.60 max | **SAE** J435 (0022) |
| * **J02000** | Carbon Steel Casting | **C** 0.15-0.25 **Mn** 0.30-0.60 **P** 0.04 **S** 0.04 **Si** 0.20-1.00 | |
| **J02001** | Carbon Steel Casting | **C** 0.15-0.25 **Mn** 0.30-0.60 **P** 0.025 max **S** 0.025 max **Si** 0.20-1.00 | **AMS** AMS-S-22141 (1020) |
| **J02002** | Carbon Steel Casting | **C** 0.15-0.25 **Cu** 0.50 max **Mn** 0.20-0.60 **P** 0.04 max **S** 0.045 max **Si** 0.20-1.00 **Other** Mo+W 0.25 max | **ASTM** A732 (1A) |
| **J02003** | Carbon Steel Casting (Grade SC1020) | **C** 0.18-0.23 **Mn** 0.40-0.80 **P** 0.040 max **Si** 0.30-0.60 **S** 0.040 max | **ASTM** A915 (SC 1020); A958 (SC 1020) |
| **J02500** | Carbon Steel Casting | **C** 0.25 max **Mn** 0.75 max **P** 0.05 max **S** 0.06 max **Si** 0.80 max | **ASTM** A27 (N1, U-60-30) |
| **J02501** | Carbon Steel Casting | **C** 0.25 max **Mn** 1.20 max **P** 0.05 max **S** 0.06 max **Si** 0.80 max | **ASTM** A27 (70-40) |
| **J02502** | Carbon Steel Casting | **C** 0.25 max **Cr** 0.40 max **Cu** 0.50 max **Mn** 0.70 max **Mo** 0.20 max **Ni** 0.50 max **P** 0.04 max **S** 0.045 max **Si** 0.60 max **V** 0.03 max **Other** Cr+Cu+Mo+Ni+V 1.00 max | **ASTM** A216 (WCA) |
| **J02503** | Carbon Steel Casting | **C** 0.25 max **Cr** 0.50 max **Cu** 0.30 max **Mn** 1.20 max **Mo** 0.20 max **Ni** 0.50 max **P** 0.04 max **S** 0.045 max **Si** 0.60 max **V** 0.03 max **Other** Cr+Cu+Mo+Ni+V 1.00 max | **ASTM** A216 (WCC); A757 (A2Q) |
| **J02504** | Carbon Steel Casting | **C** 0.25 max **Mn** 0.70 max **P** 0.04 max **S** 0.045 max **Si** 0.60 max | **ASTM** A352 (LCA); A660 (WCA) |
| **J02505** | Carbon Steel Casting | **C** 0.25 max **Mn** 1.20 max **P** 0.04 max **S** 0.045 max **Si** 0.60 max | **ASTM** A352 (LCC); A660 (WCC) |
| **J02506** | Carbon Steel Casting replaced by J22500 | | |
| **J02507** | Carbon Steel Casting | **C** 0.25 max **Mn** 0.75 max **P** 0.040 max **S** 0.045 max **Si** 0.80 max | **SAE** J435 (0025) |
| **J02508** | Carbon Steel Casting (Grade SC 1025) | **C** 0.22-0.28 **Mn** 0.40-0.80 **P** 0.040 max **Si** 0.30-0.60 **S** 0.040 max | **ASTM** A915 (SC 1025); A958 (SC 1025) |
| **J03000** | Carbon Steel Casting | **C** 0.30 max **Mn** 0.60 max **P** 0.05 max **S** 0.06 max **Si** 0.80 max | **ASTM** A27 (60-30) |
| **J03001** | Carbon Steel Casting | **C** 0.30 max **Mn** 0.70 max **P** 0.05 max **S** 0.06 max **Si** 0.80 max | **ASTM** A27 (65-35) |
| **J03002** | Carbon Steel Casting | **C** 0.30 max **Cr** 0.50 max **Cu** 0.30 max **Mn** 1.00 max **Mo** 0.20 max **Ni** 0.50 max **P** 0.04 max **S** 0.045 max **Si** 0.60 max **V** 0.03 max **Other** total Cr+Cu+Mo+Ni+V 1.00 max | **ASTM** A216 (WCB); A757 (A1Q) |
| **J03003** | Carbon Steel Casting | **C** 0.30 max **Mn** 1.00 max **P** 0.04 max **S** 0.045 max **Si** 0.60 max | **ASTM** A352 (LCB); A660 (WCB) |
| **J03004** | Carbon Steel Casting replaced by J13002 | | |
| **J03005** | Carbon Steel Casting | **C** 0.25-0.35 **Mn** 0.70-1.00 **P** 0.025 max **S** 0.025 max **Si** 0.20-1.00 | **AMS** AMS-S-22141 (IC-1030) |
| * **J03006** | Carbon Steel Casting | **C** 0.25-0.35 **Mn** 0.70-1.00 **P** 0.04 max **S** 0.04 max **Si** 0.20-1.00 | |
| **J03007** | Carbon Steel Casting | **C** 0.30 max **Mn** 0.70 max **P** 0.07 max **S** 0.06 max | |

*Boxed entries are no longer active and are retained for reference purposes only.

The chemical compositions listed are for identification purposes and should not be used in lieu of the cross referenced specifications.

| UNIFIED NUMBER | DESCRIPTION | CHEMICAL COMPOSITION | CROSS REFERENCE SPECIFICATIONS |
|---|---|---|---|
| **J03008** | Carbon Steel Casting | **C** 0.30 max **Cr** 0.20 max **Cu** 0.30 max **Mn** 0.60 max **Mo** 0.20 max **Ni** 0.50 max **P** 0.05 max **S** 0.05 max **Si** 0.20-0.60 | |
| **J03009** | Carbon Steel Casting | **C** 0.30 max **Cr** 0.35 max **Cu** 0.30 max **Mn** 0.70 max **Mo** 0.20 max **Ni** 0.50 max **P** 0.05 max **S** 0.06 max **Si** 0.80 max **Other** total Cu+Mo+Ni 1.00 max | |
| **J03010** | Carbon Steel Casting | **C** 0.30 max **Mn** 0.70 max **P** 0.040 max **S** 0.045 max **Si** 0.80 max | **SAE** J435 (0030) |
| **J03011** | Carbon Steel Casting | **C** 0.25-0.35 **Cr** 0.35 max **Cu** 0.50 max **Mn** 0.70-1.00 **Ni** 0.50 max **P** 0.04 max **S** 0.045 max **Si** 0.20-1.00 **W** 0.10 max | **ASTM** A732 (2A, 2Q) |
| **J03012** | Carbon Steel Casting (Grade SC 1030) | **C** 0.28-0.34 **Mn** 0.50-0.90 **P** 0.040 max **Si** 0.30-0.60 **S** 0.040 max | **ASTM** A915 (SC 1030); A958 (SC 1030) |
| **J03500** | Carbon Steel Casting | **C** 0.35 max **Mn** 0.60 max **P** 0.05 max **S** 0.06 max **Si** 0.80 max | **ASTM** A27 (N2) |
| **J03501** | Carbon Steel Casting | **C** 0.35 max **Mn** 0.70 max **P** 0.05 max **S** 0.06 max **Si** 0.80 max | **ASTM** A27 (70-36) |
| **J03502** | Carbon Steel Casting | **C** 0.35 max **Mn** 0.70 max **P** 0.035 max **S** 0.030 max **Si** 0.60 max | **ASTM** A356 (1) |
| **J03503** | Carbon Steel Casting | **C** 0.35 max **Mn** 0.90 max **P** 0.05 max **S** 0.06 max **Si** 0.80 max | |
| **J03504** | Carbon Steel Casting | **C** 0.35 max **P** 0.05 max **S** 0.05 max | |
| * **J04000** | Carbon Steel Casting | **C** 0.35-0.45 **Mn** 0.70-1.00 **P** 0.04 max **S** 0.04 max **Si** 0.20-1.00 | |
| **J04001** | Carbon Steel Casting | **C** 0.35-0.45 **Mn** 0.70-1.00 **P** 0.025 max **S** 0.025 max **Si** 0.20-1.00 | **AMS** AMS-S-22141 (1040) |
| **J04002** | Carbon Steel Casting | **C** 0.35-0.45 **Cr** 0.30 max **Cu** 0.50 max **Mn** 0.70-1.00 **Ni** 0.50 max **P** 0.04 max **S** 0.045 max **Si** 0.20-1.00 **W** 0.10 max | **ASTM** A732 (3A, 3Q) |
| **J04003** | Carbon Steel Casting (SC 1040) | **C** 0.37-0.44 **Mn** 0.50-0.90 **P** 0.040 max **Si** 0.30-0.60 **S** 0.040 max | **ASTM** A915 (SC 1040); A958 (SC 1040) |
| **J04500** | Carbon Steel Casting | **C** 0.40-0.50 **Cr** 0.35 max **Cu** 0.35 max **Mn** 0.50-0.90 **Ni** 0.50 max **P** 0.04 max **S** 0.045 max **Si** 0.80 max **V** 0.03 max **W** 0.10 max **Other** Mo+W 0.10 max | |
| **J04501** | Carbon Steel Casting | **C** 0.40-0.50 **Mn** 0.50-0.90 **P** 0.040 max **S** 0.045 max **Si** 0.80 max | **SAE** J435 (0050A, 0050B) |
| **J04502** | Carbon Steel Casting (Grade SC 1045) | **C** 0.43-0.50 **Mn** 0.50-0.90 **P** 0.040 max **Si** 0.30-0.60 **S** 0.040 max | **ASTM** A915 (SC 1045); A958 (SC 1045) |
| **J05000** | Carbon Steel Casting | **C** 0.45-0.55 **Mn** 0.70-1.00 **P** 0.04 max **S** 0.04 max **Si** 0.20-1.00 | |
| * **J05001** | Carbon Steel Casting | **C** 0.45-0.55 **Mn** 0.70-1.00 **P** 0.025 max **S** 0.025 max **Si** 0.20-1.00 | **AMS** AMS-S-22141 (1050) |
| **J05002** | Carbon Steel Casting | **C** 0.50 max **Mn** 0.90 max **P** 0.05 max **S** 0.06 max **Si** 0.80 max | |
| **J05003** | Carbon Steel Casting | **C** 0.30 max **Cu** 0.50 max **Mn** 0.70-1.00 **P** 0.04 max **S** 0.045 max **Si** 0.80 max **W** 0.10 max | |

*Boxed entries are no longer active and are retained for reference purposes only.

The chemical compositions listed are for identification purposes and should not be used in lieu of the cross referenced specifications.

| UNIFIED NUMBER | DESCRIPTION | CHEMICAL COMPOSITION | CROSS REFERENCE SPECIFICATIONS |
|---|---|---|---|
| **J11442** | Alloy Steel Casting (8615) | **C** 0.11-0.17 **Cr** 0.35-0.65 **Cu** 0.35 max **Mn** 0.65-1.00 **Mo** 0.15-0.35 **Ni** 0.35-0.75 **P** 0.04 max **S** 0.04 max **Si** 0.50-1.00 | **AMS** 5333 |
| **J11522** | Alloy Steel Casting | **C** 0.15 max **Cr** 0.35 max **Cu** 0.50 max **Mn** 0.30-0.60 **Mo** 0.44-0.65 **Ni** 0.50 max **P** 0.040 max **S** 0.045 max **Si** 0.15-1.65 **W** 0.10 max | **ASTM** A426 (CP15) |
| **J11547** | Alloy Steel Casting | **C** 0.10-0.20 **Cr** 0.50-0.81 **Cu** 0.50 max **Mn** 0.30-0.61 **Mo** 0.44-0.65 **Ni** 0.50 max **P** 0.040 max **S** 0.045 max **Si** 0.10-0.50 **W** 0.10 max | **ASTM** A426 (CP2) |
| **J11562** | Alloy Steel Casting | **C** 0.15 max **Cr** 0.80-1.25 **Cu** 0.50 max **Mn** 0.30-0.61 **Mo** 0.44-0.65 **Ni** 0.50 max **P** 0.040 max **S** 0.045 max **Si** 0.50 max **W** 0.10 max | **ASTM** A426 (CP12) |
| **J11697** | Alloy Steel Casting | **C** 0.13-0.20 **Cr** 1.00-1.50 **Mn** 0.50-0.90 **Mo** 0.90-1.20 **P** 0.035 max **S** 0.030 max **Si** 0.20-0.60 **V** 0.05-0.15 | **ASTM** A356 (8) |
| **J11872** | Alloy Steel Casting | **Al** 0.01 max **C** 0.15-0.21 **Cr** 1.00-1.50 **Cu** 0.35 max **Mn** 0.50-0.80 **Mo** 0.45-0.65 **Ni** 0.50 max **P** 0.020 max **S** 0.015 max **Si** 0.30-0.60 **V** 0.03 max | **ASTM** A217 (WC11) |
| **J11875** | Alloy Steel Casting | **C** 0.18 max **Cr** 1.00-1.60 max **Cu** 0.50 max **Mn** 0.40-0.70 **Mo** 0.40-0.60 **Ni** 0.60 max **P** 0.05 max **S** 0.05 max **Si** 0.60 max **V** 0.10-0.20 | |
| **J12047** | Alloy Steel Casting | **C** 0.15-0.25 **Cr** 0.40-0.60 **Mn** 0.65-0.95 **Mo** 0.15-0.25 **Ni** 0.40-0.70 **P** 0.025 max **S** 0.025 max **Si** 0.20-0.80 | **AMS** AMS-S-22141 (8620) |
| **J12048** | Alloy Steel Casting | **C** 0.15-0.25 **Cr** 0.40-0.70 **Cu** 0.50 max **Mn** 0.65-0.95 **Mo** 0.15-0.25 **Ni** 0.40-0.70 **P** 0.04 max **S** 0.045 max **Si** 0.20-0.80 **W** 0.10 max | **ASTM** A732 (13Q) |
| **J12070** | Alloy Steel Casting | **C** 0.20 max **Cr** 1.00-1.60 max **Cu** 0.50 max **Mn** 0.50-0.80 **Mo** 0.40-0.60 **Ni** 0.60 max **P** 0.05 max **S** 0.05 max **Si** 0.20-0.60 | |
| **J12072** | Alloy Steel Casting | **C** 0.20 max **Cr** 1.00-1.50 **Cu** 0.50 max **Mn** 0.50-0.80 **Mo** 0.45-0.65 **Ni** 0.50 max **P** 0.04 max **S** 0.045 max **Si** 0.60 max **W** 0.10 max | **ASTM** A217 (WC 6); A426 (CP11) |
| **J12073** | Alloy Steel Casting | **C** 0.20 max **Cr** 1.00-1.50 **Cu** 0.50 max **Mn** 0.50-0.80 **Mo** 0.45-0.65 **Ni** 0.50 max **P** 0.035 max **S** 0.030 max **Si** 0.60 max **W** 0.10 max | **ASTM** A356 (6) |
| **J12080** | Alloy Steel Casting | **C** 0.20 max **Cr** 1.00-1.50 **Mn** 0.30-0.80 **Mo** 0.45-0.65 **P** 0.04 max **S** 0.045 max **Si** 0.60 max **V** 0.15-0.25 | **ASTM** A389 (C23) |
| **J12082** | Alloy Steel Casting | **C** 0.20 max **Cr** 0.50-0.80 **Cu** 0.50 max **Mn** 0.50-0.80 **Mo** 0.45-0.65 **Ni** 0.70-1.10 **P** 0.04 max **S** 0.045 max **Si** 0.60 max **V** 0.03 max **W** 0.10 max | **ASTM** A217 (WC 4); A487 (11) |
| **J12084** | Alloy Steel Casting | **B** 0.002-0.006 **C** 0.20 max **Cr** 0.40-0.80 **Cu** 0.15-0.50 **Mn** 0.60-1.00 **Mo** 0.40-0.60 **Ni** 0.70-1.00 **P** 0.04 max **S** 0.045 max **Si** 0.80 max **V** 0.03-0.10 **W** 0.10 max | **ASTM** A487 (7) |
| **J12092** | Alloy Steel Casting | **C** 0.20 max **Cr** 0.80-1.25 **Mn** 0.30-0.80 **Mo** 0.90-1.20 **P** 0.04 max **S** 0.045 max **Si** 0.60 max **V** 0.15-0.25 | **ASTM** A389 (C24) |

The chemical compositions listed are for identification purposes and should not be used in lieu of the cross referenced specifications.

| UNIFIED NUMBER | DESCRIPTION | CHEMICAL COMPOSITION | CROSS REFERENCE SPECIFICATIONS |
|---|---|---|---|
| **J12093** | Alloy Steel Casting (4620) | **C** 0.15-0.25 **Mn** 0.40-0.70 **Mo** 0.20-0.30 **Ni** 1.65-2.00 **P** 0.025 max **S** 0.025 max **Si** 0.20-0.80 | **AMS** AMS-S-22141 (4620) |
| **J12094** | Alloy Steel Casting | **C** 0.15-0.25 **Cr** 0.35 max **Cu** 0.50 max **Mn** 0.40-0.70 **Mo** 0.20-0.30 **Ni** 1.65-2.00 **P** 0.04 max **S** 0.045 max **Si** 0.20-0.80 **W** 0.10 max | **ASTM** A732 (11Q) |
| **J12095** | Alloy Steel Casting (Grade SC 8620) | **C** 0.18-0.23 **Cr** 0.40-0.60 **Mn** 0.60-1.00 **Mo** 0.15-0.25 **Ni** 0.40-0.70 **P** 0.035 max **Si** 0.30-0.60 **S** 0.040 max | **ASTM** A915 (SC 8620); A958 (SC 8620) |
| * **J12520** | Alloy Steel Casting | **C** 0.25 max **Cr** 0.35 max **Cu** 0.50 max **Mn** 0.50-0.80 **Mo** 0.45-0.65 **Ni** 0.50 max **P** 0.04 max **S** 0.45 max **Si** 0.60 | **MIL SPEC** MIL-C-24707/2 |
| **J12521** | Alloy Steel Casting | **C** 0.25 max **Mn** 0.30-0.80 **Mo** 0.44-0.65 **P** 0.040 max **S** 0.045 max **Si** 0.10-0.50 | **ASTM** A426 (CP1) |
| **J12522** | Alloy Steel Casting | **C** 0.25 max **Mn** 0.50-0.80 **Mo** 0.45-0.65 **P** 0.04 max **S** 0.045 max **Si** 0.60 max | **ASTM** A352 (LC1) |
| **J12523** | Alloy Steel Casting | **C** 0.25 max **Mn** 0.70 max **Mo** 0.45-0.65 **P** 0.035 max **S** 0.030 max **Si** 0.60 max | **ASTM** A356 (2) |
| **J12524** | Alloy Steel Casting | **C** 0.25 max **Cu** 0.50 max **Mn** 0.50-0.80 **Mo** 0.45-0.65 **N** 0.50 max **P** 0.04 max **S** 0.045 max **Si** 0.60 max **W** 0.10 max | **ASTM** A217 (WC1) |
| **J12540** | Alloy Steel Casting | **C** 0.25 max **Cr** 0.40-0.70 **Mn** 0.70 max **Mo** 0.40-0.60 **P** 0.035 max **S** 0.030 max **Si** 0.60 max | **ASTM** A356 (5) |
| **J12545** | Alloy Steel Casting | **C** 0.25 max **Cr** 0.40 max **Cu** 0.50 max **Mn** 1.15-1.50 **Mo** 0.45-0.60 **Ni** 0.45-1.00 **P** 0.035 max **S** 0.035 max **Si** 0.60 max **V** 0.03 max | |
| **J12582** | Alloy Steel Casting | **C** 0.25 max **Cr** 0.40 max **Cu** 0.50 max **Mn** 1.20 max **Ni** 1.5-2.0 **P** 0.025 max **S** 0.025 max **Si** 0.60 max **V** 0.03 max | **ASTM** A757 (C1Q) |
| **J12595** | Alloy Steel Casting (Grade SC 8625) | **C** 0.23-0.28 **Cr** 0.40-0.60 **Mn** 0.60-1.00 **Mo** 0.15-0.25 **Ni** 0.40-0.70 **P** 0.035 max **Si** 0.30-0.60 **S** 0.040 max | **ASTM** A915 (SC 8625); A958 (SC 8625) |
| **J13002** | Alloy Steel Casting | **C** 0.30 max **Cr** 0.35 max **Cu** 0.50 max **Mn** 1.00 max **Ni** 0.50 max **P** 0.04 max **S** 0.045 max **Si** 0.80 max **V** 0.04-0.12 **Other** Mo+W 0.25 max | **ASTM** A487 (1) |
| **J13005** | Alloy Steel Casting | **C** 0.30 max **Cr** 0.35 max **Cu** 0.50 max **Mn** 1.00-1.40 **Mo** 0.10-0.30 **Ni** 0.50 max **P** 0.04 max **S** 0.045 max **Si** 0.80 max **V** 0.03 max **W** 0.10 max | **ASTM** A487 (2) |
| **J13025** | Alloy Steel Casting | **C** 0.30 max **Cr** 0.50-0.70 **Mn** 0.30-0.50 **Mo** 0.20 max **Ni** 0.70 max **Si** 0.50 max **V** 0.10 max **Other** P+S 0.07 max | **MIL SPEC** MIL-A-11356 |
| **J13042** | Alloy Steel Casting | **C** 0.25-0.35 **Cr** 0.35-0.65 **Cu** 0.35 max **Mn** 0.60-0.95 **Mo** 0.15-0.30 **Ni** 0.35-0.75 **P** 0.04 max **S** 0.04 max **Si** 1.00 max | **AMS** 5334 |

*Boxed entries are no longer active and are retained for reference purposes only.

The chemical compositions listed are for identification purposes and should not be used in lieu of the cross referenced specifications.

| UNIFIED NUMBER | DESCRIPTION | CHEMICAL COMPOSITION | CROSS REFERENCE SPECIFICATIONS |
|---|---|---|---|
| **J13045** | Alloy Steel Casting | **C** 0.25-0.35 **Cr** 0.80-1.10 **Cu** 0.50 max **Mn** 0.40-0.70 **Mo** 0.15-0.25 **P** 0.04 max **S** 0.045 max **Si** 0.20-0.80 **W** 0.10 max | **ASTM** A732 (7Q) |
| **J13046** | Alloy Steel Casting | **C** 0.25-0.35 **Cr** 0.80-1.10 **Cu** 0.35 max **Mn** 0.40-0.80 **Mo** 0.15-0.25 **Ni** 0.25 max **P** 0.04 max **S** 0.04 max **Si** 1.00 max | **AMS** 5336 |
| **J13047** | Alloy Steel Casting | **C** 0.30 max **Cr** 0.40-0.80 **Cu** 0.50 max **Mn** 1.00 max **Mo** 0.15-0.30 **Ni** 0.40-0.80 **P** 0.04 max **S** 0.045 max **Si** 0.80 max **V** 0.03 max **W** 0.10 max | **ASTM** A487 (4) |
| **J13048** | Alloy Steel Casting (4130) | **C** 0.25-0.35 **Cr** 0.80-1.10 **Mn** 0.40-0.70 **Mo** 0.15-0.25 **P** 0.025 max **S** 0.025 max **Si** 0.20-0.80 | **AMS** AMS-S-22141 (4130) |
| **J13049** | Alloy Steel Casting (8630) | **C** 0.25-0.35 **Cr** 0.40-0.60 **Mn** 0.65-0.95 **Mo** 0.15-0.25 **Ni** 0.40-0.70 **P** 0.025 max **S** 0.025 max **Si** 0.20-0.80 | |
| **J13050** | Alloy Steel Casting (8630 Modified) | **C** 0.25-0.33 **Cr** 0.40-0.90 **Cu** 0.35 max **Mn** 0.60-0.95 **Mo** 0.15-0.25 **Ni** 0.40-1.10 **P** 0.025 max **S** 0.025 max **Si** 0.50-0.90 | **AMS** 5335 |
| **J13051** | Alloy Steel Casting | **C** 0.25-0.35 **Cr** 0.40-0.70 **Cu** 0.50 max **Mn** 0.65-0.95 **Mo** 0.15-0.25 **Ni** 0.40-0.70 **P** 0.04 max **S** 0.045 max **Si** 0.20-0.80 **W** 0.10 max | **ASTM** A732 (14Q) |
| **J13052** | Alloy Steel Casting | **C** 0.30 max **Cr** 0.35 max **Cu** 0.50 max **Mn** 0.70-1.00 **Mo** 0.15-0.25 **P** 0.04 max **S** 0.045 max **Si** 0.20-0.80 **W** 0.10 max | **ASTM** A732 (5N) |
| **J13080** | Alloy Steel Casting | **C** 0.30 max **Cr** 0.40 max **Cu** 0.50 max **Mn** 0.80-1.10 **Mo** 0.20-0.30 **Ni** 1.40-1.75 **P** 0.04 max **S** 0.045 max **Si** 0.60 max **V** 0.03 max **W** 0.10 max | **ASTM** A487 (13) |
| **J13095** | Alloy Steel Casting (Grade SC 8630) | **C** 0.28-0.33 **Cr** 0.40-0.60 **Mn** 0.60-1.00 **Mo** 0.15-0.25 **Ni** 0.40-0.70 **P** 0.035 max **Si** 0.30-0.60 **S** 0.040 max | **ASTM** A915 (SC 8630); A958 (SC 8630) |
| **J13345** | Alloy Steel Casting | **C** 0.33 max **Cr** 0.75-1.10 **Cu** 0.50 max **Mn** 0.60-0.90 **Mo** 0.15-0.30 **Ni** 0.50 max **P** 0.04 max **S** 0.045 max **Si** 0.80 max **V** 0.03 max **W** 0.10 max | **ASTM** A487 (9) |
| **J13432** | Alloy Steel Casting (4335M) | **Al** 0.05 max **C** 0.30-0.38 **Mn** 0.60-1.00 **Mo** 0.65-1.00 **P** 0.04 max **S** 0.04 max **Si** 0.50-1.00 **V** 0.14 max | **AMS** AMS-S-22141 (4335M) |
| **J13442** | Alloy Steel Casting (8735) | **C** 0.30-0.38 **Cr** 0.35-0.90 **Mn** 0.30-0.70 **Mo** 0.15-0.40 **Ni** 0.35-0.75 **P** 0.04 max **S** 0.04 max **Si** 0.20-1.00 | **AMS** AMS-S-22141 (8735) |
| **J13502** | Alloy Steel Casting (Grade SC 4130) | **C** 0.28-0.33 **Cr** 0.80-1.10 **Mn** 0.40-0.80 **Mo** 0.15-0.25 **P** 0.035 max **Si** 0.30-0.60 **S** 0.040 max | **ASTM** A915 (SC 4130); A958 (SC 4130) |
| **J13512** | Alloy Steel Casting | **C** 0.35 max **Mn** 1.35-1.75 **Mo** 0.25-0.55 **P** 0.04 max **S** 0.045 max **Si** 0.20-0.80 | **ASTM** A732 (6N) |
| **J13855** | Alloy Steel Casting | **C** 0.38 max **Cr** 0.40-0.80 **Cu** 0.50 max **Mn** 1.30-1.70 **Mo** 0.30-0.40 **Ni** 0.40-0.80 **P** 0.04 max **S** 0.045 max **Si** 0.80 max **V** 0.03 max **W** 0.10 max | **ASTM** A487 (6) |
| **J14045** | Alloy Steel Casting (Grade SC 4140) | **C** 0.38-0.40 **Cr** 0.80-1.10 **Mn** 0.70-1.10 **Mo** 0.15-0.25 **P** 0.035 max **S** 0.040 max | **ASTM** A915 (SC 4140); A958 (SC 4140) |

The chemical compositions listed are for identification purposes and should not be used in lieu of the cross referenced specifications.

| UNIFIED NUMBER | DESCRIPTION | CHEMICAL COMPOSITION | CROSS REFERENCE SPECIFICATIONS |
|---|---|---|---|
| **J14046** | Alloy Steel Casting | **C** 0.35-0.45 **Cr** 0.80-1.10 **Cu** 0.35 max **Mn** 0.75-1.00 **Mo** 0.15-0.25 **Ni** 0.25 max **P** 0.04 max **S** 0.04 max **Si** 1.00 max | **AMS** 5338 |
| **J14047** | Alloy Steel Casting (4140) | **C** 0.35-0.45 **Cr** 0.80-1.10 **Mn** 0.70-1.35 **Mo** 0.15-0.25 **P** 0.025 max **S** 0.025 max **Si** 0.20-0.80 | **AMS** AMS-S-22141 (4140) |
| **J14048** | Alloy Steel Casting (8640) | **C** 0.35-0.45 **Cr** 0.40-0.60 **Mn** 0.70-1.05 **Mo** 0.15-0.25 **Ni** 0.40-0.70 **P** 0.025 max **S** 0.025 max **Si** 0.20-0.80 | **AMS** AMS-S-22141 (8640) |
| **J14049** | Alloy Steel Casting | **C** 0.35-0.45 **Cr** 0.80-1.10 **Cu** 0.50 max **Mn** 0.70-1.00 **Mo** 0.15-0.25 **Ni** 0.50 max **P** 0.04 max **S** 0.045 max **Si** 0.20-0.80 **W** 0.10 max | **ASTM** A732 (8Q) |
| **J15047** | Alloy Steel Casting (6150) | **C** 0.45-0.55 **Cr** 0.80-1.10 **Mn** 0.65-0.95 **P** 0.025 max **S** 0.025 max **Si** 0.20-0.80 **V** 0.15 min | **AMS** AMS-S-22141 (6150) |
| **J15048** | Alloy Steel Casting | **C** 0.45-0.55 **Cr** 0.80-1.10 **Cu** 0.50 max **Mn** 0.65-0.95 **Ni** 0.50 max **P** 0.04 max **S** 0.045 max **Si** 0.20-0.80 **V** 0.15 min **W** 0.10 max **Other** Mo+W 0.10 max | **ASTM** A732 (12Q) |
| **J15580** | Alloy Steel Casting | **C** 0.55 max **Cr** 0.40 max **Cu** 0.50 max **Mn** 0.80-1.10 **Mo** 0.20-0.30 **Ni** 1.40-1.75 **P** 0.04 max **S** 0.045 max **Si** 0.60 max **V** 0.03 max **W** 0.10 max | **ASTM** A487 (14) |
| **J19965** | Alloy Steel Casting (52100) | **C** 0.95-1.10 **Cr** 1.30-1.60 **Mn** 0.25-0.55 **P** 0.025 max **S** 0.025 max **Si** 0.20-0.80 | **AMS** AMS-S-22141 (52100) |
| **J19966** | Alloy Steel Casting | **C** 0.95-1.10 **Cr** 1.30-1.60 **Cu** 0.50 max **Mn** 0.25-0.55 **Ni** 0.50 max **P** 0.04 max **S** 0.045 max **Si** 0.20-0.80 **W** 0.10 max | **ASTM** A732 (15A) |
| **J21610** | Alloy Steel Casting | **C** 0.13-0.20 **Cr** 1.00-1.50 **Mn** 0.50-0.90 **Mo** 0.90-1.20 **P** 0.035 max **S** 0.030 max **Si** 0.20-0.60 **V** 0.20-0.35 | **ASTM** A356 (9) |
| **J21880** | Alloy Steel Casting | **C** 0.18 max **Cr** 2.00-2.75 **Cu** 0.50 max **Mn** 0.40-0.70 **Mo** 0.80-1.10 **Ni** 0.60 max **P** 0.05 max **S** 0.05 max **Si** 0.20-0.60 | |
| **J21890** | Alloy Steel Casting | **C** 0.18 max **Cr** 2.00-2.75 **Cu** 0.50 max **Mn** 0.40-0.70 **Mo** 0.90-1.20 **Ni** 0.50 max **P** 0.04 max **S** 0.045 max **Si** 0.60 max **W** 0.10 max | **ASTM** A217 (WC9); A426 (CP22) |
| **J22000** | Alloy Steel Casting | **C** 0.20 max **Cr** 0.50-0.90 **Cu** 0.50 max **Mn** 0.40-0.70 **Mo** 0.90-1.20 **Ni** 0.60-1.00 **P** 0.04 max **S** 0.045 max **Si** 0.60 max **V** 0.03 max **W** 0.10 max | **ASTM** A217 (WC5); A487 (12) |
| **J22090** | Alloy Steel Casting | **C** 0.20 max **Cr** 2.00-2.75 **Mn** 0.50-0.80 **Mo** 0.90-1.20 **P** 0.035 max **S** 0.030 max **Si** 0.60 max | **ASTM** A356 (10) |
| **J22091** | Alloy Steel Casting | **C** 0.20 max **Cr** 2.00-2.75 **Cu** 0.50 max **Mn** 0.50-0.90 **Mo** 0.90-1.10 **P** 0.04 max **S** 0.045 max **Si** 0.80 max **V** 0.03 max **W** 0.10 max | **ASTM** A487 (8) |
| **J22092** | Alloy Steel Casting | **C** 0.20 max **Cr** 2.00-2.75 **Cu** 0.50 max **Mn** 0.40-0.80 **Mo** 0.90-1.20 **Ni** 0.50 max **P** 0.035 max **S** 0.035 max **Si** 0.60 max **V** 0.03 max **W** 0.10 max | **ASTM** A757 (D1N, D1Q, D1N1, D1Q1, D1N2, D1Q2, D1N3, D1Q3) |
| **J22500** | Alloy Steel Casting | **C** 0.25 max **Mn** 0.50-0.80 **Ni** 2.0-3.0 **P** 0.04 max **S** 0.045 max **Si** 0.60 max | **ASTM** A352 (LC2) |

The chemical compositions listed are for identification purposes and should not be used in lieu of the cross referenced specifications.

| UNIFIED NUMBER | DESCRIPTION | CHEMICAL COMPOSITION | CROSS REFERENCE SPECIFICATIONS |
|---|---|---|---|
| **J22501** | Alloy Steel Casting | **C** 0.25 max **Cr** 0.40 max **Cu** 0.50 max **Mn** 0.50-0.80 **Mo** 0.25 max **Ni** 2.0-3.0 **P** 0.025 max **S** 0.025 max **Si** 0.60 max **V** 0.03 max | **ASTM** A757 (B2N, B2Q) |
| **J23015** | Alloy Steel Casting | **C** 0.30 max **Cr** 0.55-0.90 **Cu** 0.50 max **Mo** 0.20-0.40 **Ni** 1.40-2.00 **P** 0.04 max **S** 0.045 max **Si** 0.80 max **V** 0.03 max **W** 0.10 max | **ASTM** A487 (10) |
| **J23055** | Alloy Steel Casting | **C** 0.25-0.35 **Cr** 0.70-0.90 **Cu** 0.50 max **Mn** 0.40-0.70 **Mo** 0.20-0.30 **Ni** 1.65-2.00 **P** 0.04 max **S** 0.045 max **Si** 0.20-0.80 **W** 0.10 max | **ASTM** A732 (9Q) |
| **J23259** | Alloy Steel Casting (Grade SC 4330) | **C** 0.28-0.33 **Cr** 0.70-0.90 **Mn** 0.60-0.90 **Mo** 0.20-0.30 **Ni** 1.65-2.00 **P** 0.035 max **Si** 0.30-0.60 **S** 0.040 max | **ASTM** A915 (SC 4330); A958 (SC 4330) |
| **J23260** | Alloy Steel Casting (4330 Modified) | **C** 0.28-0.36 **Cr** 0.65-1.00 **Mn** 0.60-0.90 **Mo** 0.30-0.45 **Ni** 1.65-2.00 **P** 0.025 max **S** 0.025 max **Si** 0.50-1.00 | **AMS** 5328; 5329 |
| **J24053** | Alloy Steel Casting (Grade SC 4340) | **C** 0.38-0.43 **Cr** 0.70-0.90 **Mn** 0.60-0.90 **Mo** 0.20-0.30 **Ni** 1.65-2.00 **P** 0.035 **Si** 0.30-0.60 **S** 0.040 max | **ASTM** A915 (SC 4340); A958 (SC 4340) |
| **J24054** | Alloy Steel Casting | **C** 0.35-0.45 **Cr** 0.70-0.90 **Cu** 0.50 max **Mn** 0.70-1.00 **Mo** 0.20-0.30 **Ni** 1.65-2.00 **P** 0.04 max **S** 0.045 max **Si** 0.20-0.80 **W** 0.10 max | **ASTM** A732 (10Q) |
| **J24055** | Alloy Steel Casting (4340) | **C** 0.36-0.44 **Cr** 0.70-0.90 **Mn** 0.60-0.90 **Mo** 0.20-0.30 **Ni** 1.65-2.00 **P** 0.04 max **S** 0.04 max **Si** 0.20-0.80 | **AMS** AMS-S-22141 (4340) |
| **J24056** | Alloy Steel Casting (Nitralloy) | **Al** 0.85-1.20 **C** 0.35-0.45 **Cr** 1.40-1.80 **Mn** 0.40-0.70 **Mo** 0.30-0.45 **P** 0.025 max **S** 0.025 max **Si** 0.20-0.80 | **AMS** AMS-S-22141 (Nitralloy) |
| **J24060** | Combined with J23260 | | **AMS** 5330; 5331 |
| **J31200** | Alloy Steel Casting (Grade 16, Class A) | **C** 0.12 max **Cr** 0.20 max **Cu** 0.20 max **Mn** 2.10 max **Mo** 0.10 max **Ni** 1.00-1.40 **P** 0.02 max **Si** 0.50 max **S** 0.02 max **W** 0.10 max **V** 0.02 max | **ASTM** A487 (16) |
| **J31300** | 9% Ni Steel Casting (Grade LC9) | **C** 0.13 max **Cr** 0.50 max **Mn** 0.90 max **Mo** 0.20 max **Ni** 8.50-10.0 **P** 0.04 max **Si** 0.45 max **S** 0.045 max V 0.03 max | **ASTM** A352 (LC9) |
| **J31500** | Alloy Steel Casting | **C** 0.15 max **Cr** 0.40 max **Cu** 0.50 max **Mn** 0.50-0.80 **Mo** 0.25 max **Ni** 3.0-4.0 **P** 0.025 max **S** 0.025 max **Si** 0.60 max **V** 0.03 max | **ASTM** A757 (B3N, B3Q) |
| **J31545** | Alloy Steel Casting | **C** 0.15 max **Cr** 2.65-3.35 **Cu** 0.50 max **Mn** 0.30-0.60 **Mo** 0.80-1.06 **Ni** 0.50 max **P** 0.040 max **S** 0.045 max **Si** 0.50 max **W** 0.10 max | **ASTM** A426 (CP21) |
| **J31550** | Alloy Steel Casting | **C** 0.15 max **Mn** 0.50-0.80 **Ni** 3.00-4.00 **P** 0.04 max **S** 0.045 max **Si** 0.60 max | **ASTM** A352 (LC3) |
| *** J31575** | Replaced by D50850 | | |
| **J31670** | Cast Austenitic Cr-Ni-Mo Stainless Steel (Surgical Implant Cast 316) | **C** 0.06 max **Cr** 17.00-19.00 **Mn** 2.0 max **Mo** 2.00-3.00 **Ni** 11.00-14.00 **P** 0.045 max **S** 0.030 max **Si** 1.0 max | **ASTM** F745 |

*Boxed entries are no longer active and are retained for reference purposes only.

The chemical compositions listed are for identification purposes and should not be used in lieu of the cross referenced specifications.

| UNIFIED NUMBER | DESCRIPTION | CHEMICAL COMPOSITION | CROSS REFERENCE SPECIFICATIONS |
|---|---|---|---|
| **J32075** | Alloy Steel Casting | **C** 0.20 max **Cr** 1.15-1.65 **Cu** 0.20 max **Mn** 0.55-0.75 **Mo** 0.10-0.60 **Ni** 2.50-3.25 **P** 0.02 max **S** 0.015 max **Si** 0.50 max | |
| **J41500** | Alloy Steel Casting | **C** 0.15 max **Mn** 0.50-0.80 **Ni** 4.00-5.00 **P** 0.04 max **S** 0.045 max **Si** 0.60 max | **ASTM** A352 (LC4) |
| **J41501** | Alloy Steel Casting | **C** 0.15 max **Cr** 0.40 max **Cu** 0.50 max **Mn** 0.50-0.80 **Mo** 0.25 max **Ni** 4.0-5.0 **P** 0.025 max **S** 0.025 max **Si** 0.60 max **V** 0.03 max | **ASTM** A757 (B4N, B4Q) |
| **J42015** | Alloy Steel Casting (HY80) | **C** 0.20 max **Cr** 1.35-1.65 **Mn** 0.55-0.75 **Mo** 0.30-0.60 **Ni** 2.50-3.25 **P** 0.02 max **S** 0.015 max **Si** 0.50 max | **MIL SPEC** MIL-S-16216 (HY80); MIL-S-23008 (HY80); MIL-S-23009 (HY80) |
| **J42045** | Alloy Steel Casting | **C** 0.20 max **Cr** 4.00-6.50 **Cu** 0.50 max **Mn** 0.40-0.70 **Mo** 0.45-0.65 **Ni** 0.50 max **P** 0.04 max **S** 0.045 max **Si** 0.75 max **W** 0.10 max | **ASTM** A217 (C5); A426 (CP5) |
| **J42065** | Alloy Steel Casting | **C** 0.20 max **Cr** 1.50-2.00 **Cu** 0.50 max **Mn** 0.40-0.70 **Mo** 0.40-0.60 **Ni** 2.75-3.90 **P** 0.020 max **S** 0.020 max **Si** 0.60 max **V** 0.03 max **W** 0.10 max | **ASTM** A757 (E2N, E2Q) |
| **J42215** | Alloy Steel Casting | **C** 0.22 max **Cr** 1.35-1.85 **Mn** 0.55-0.75 **Mo** 0.30-0.60 **Ni** 2.50-3.50 **P** 0.04 max **S** 0.045 max **Si** 0.50 max | **ASTM** A352 (LC2-1) |
| **J42220** | Alloy Steel Casting | **C** 0.22 max **Cr** 1.35-1.85 **Cu** 0.50 max **Mn** 0.50-0.80 **Mo** 0.35-0.60 **Ni** 2.5-3.5 **P** 0.025 max **S** 0.025 max **Si** 0.60 max **V** 0.03 max | **ASTM** A757 (E1Q) |
| **J42240** | Alloy Steel Casting (HY100) | **C** 0.22 max **Cr** 1.35-1.85 **Mn** 0.55-0.75 **Mo** 0.30-0.60 **Ni** 2.75-3.50 **P** 0.02 max **S** 0.015 max **Si** 0.50 max | **MIL SPEC** MIL-S-16216 (HY100); MIL-S-23008 (HY100); MIL-S-23009 (HY100) |
| **J51545** | Alloy Steel Casting | **C** 0.15 max **Cr** 4.00-6.00 **Cu** 0.50 max **Mn** 0.30-0.60 **Mo** 0.45-0.65 **Ni** 0.50 max **P** 0.040 max **S** 0.045 max **Si** 1.00-2.00 **W** 0.10 max | **ASTM** A426 (CP5b) |
| **J61594** | Alloy Steel Casting | **C** 0.15 max **Cr** 6.00-8.00 **Cu** 0.50 max **Mn** 0.30-0.60 **Mo** 0.44-0.65 **Ni** 0.50 max **P** 0.040 max **S** 0.045 max **Si** 0.50-1.00 **W** 0.10 max | |
| **J80490** | Alloy Steel Casting (Grade 12) | **Al** 0.04 max **C** 0.08-0.12 **Cr** 8.0-9.5 **Cb** 0.06-0.10 **Mn** 0.30-0.60 **Mo** 0.85-1.05 **Ni** 0.40 max **N** 0.03-0.07 **P** 0.02 max **Si** 0.20-0.50 **S** 0.010 max **V** 0.18-0.25 | **ASTM** A217 (C12A); A356 (12) |
| **J82090** | Alloy Steel Casting | **C** 0.20 max **Cr** 8.00-10.00 **Cu** 0.50 max **Mn** 0.35-0.65 **Mo** 0.90-1.20 **Ni** 0.50 max **P** 0.04 max **S** 0.045 max **Si** 1.00 max **W** 0.10 max | **ASTM** A217 (C12); A426 (CP9) |
| **J84090** | Alloy Steel Casting (Grade C12A) | **Al** 0.040 max **C** 0.12 max **Cr** 8.0-9.5 **Cb** 0.060-0.10 **Mn** 0.30-0.60 **Mo** 0.85-1.05 **Ni** 0.40 max **N** 0.030-0.070 **P** 0.020 max **Si** 0.20-0.50 **S** 0.018 max **V** 0.18-0.25 | **ASTM** A217 (C12A) |
| **J91109** | Austenitic Manganese Steel Casting | **C** 1.05-1.35 **Mn** 11.0 min **P** 0.07 max **Si** 1.00 max | **ASTM** A128 (A) |
| **J91119** | Austenitic Manganese Steel Casting | **C** 0.9-1.05 **Mn** 11.5-14.0 **P** 0.07 max **Si** 1.00 max | **ASTM** A128 (B-1) |
| **J91129** | Austenitic Manganese Steel Casting | **C** 1.05-1.2 **Mn** 11.5-14.0 **P** 0.07 max **Si** 1.00 max | **ASTM** A128 (B-2) |
| **J91139** | Austenitic Manganese Steel Casting | **C** 1.12-1.28 **Mn** 11.5-14.0 **P** 0.07 max **Si** 1.00 max | **ASTM** A128 (B-3) |

The chemical compositions listed are for identification purposes and should not be used in lieu of the cross referenced specifications.

| UNIFIED NUMBER | DESCRIPTION | CHEMICAL COMPOSITION | CROSS REFERENCE SPECIFICATIONS |
|---|---|---|---|
| **J91149** | Austenitic Manganese Steel Casting | **C** 1.2-1.35 **Mn** 11.5-14.0 **P** 0.07 max **Si** 1.00 max | **ASTM** A128 (B-4) |
| **J91150** | Alloy Steel Casting (CA-15) | **C** 0.15 max **Cr** 11.5-14.0 **Mn** 1.00 max **Mo** 0.50 max **Ni** 1.00 max **P** 0.040 max **S** 0.040 max **Si** 1.50 max | **ACI** CA-15 **AMS** 5351 **ASTM** A217 (CA-15); A426 (CPCA-15); A743 (CA-15) |
| **J91151** | Alloy Steel Casting (CA-15M) | **C** 0.15 max **Cr** 11.5-14.0 **Mn** 1.00 max **Mo** 0.15-1.0 **Ni** 1.0 max **P** 0.040 max **S** 0.040 max **Si** 0.65 max | **ACI** CA-15M **ASTM** A487 (CA-15M); A743 (CA-15M) |
| **J91152** | Alloy Steel Casting (410) | **C** 0.05-0.15 **Cr** 11.5-13.5 **Cu** 0.50 max **Mn** 1.00 max **Mo** 0.50 max **Ni** 0.50 max **P** 0.04 max **S** 0.03 max **Si** 1.00 max | **AMS** 5350 |
| **J91153** | Alloy Steel Casting (CA-40) | **C** 0.20-0.40 **Cr** 11.5-14.0 **Mn** 1.00 max **Mo** 0.5 max **Ni** 1.0 max **P** 0.04 max **S** 0.04 max **Si** 1.50 max | **ACI** CA-40 **ASTM** A743 (CA-40) |
| **J91154** | Alloy Steel Casting (CA40F) | **C** 0.20-0.40 **Cr** 11.5-14.0 **Fe** rem **Mn** 1.00 max **Mo** 0.5 max **Ni** 1.00 max **P** 0.04 max **S** 0.20-0.40 **Si** 1.50 max | **ACI** CA40F **AISI** 420F **ASTM** A743 (CA40F) |
| **J91161** | Alloy Steel Casting (60416) | **C** 0.15 max **Cr** 11.5-14.0 **Cu** 0.5 max **Mn** 1.25 max **Mo** 0.5 max **Ni** 0.5 max **P** 0.05 max **S** 0.15-0.35 **Si** 1.5 max **Zr** 0.5 max | **AMS** 5349 |
| **J91171** | Alloy Steel Casting | **C** 0.15 max **Cr** 11.5-14.0 **Cu** 0.50 max **Mn** 1.00 max **Mo** 0.50 max **N** 1.00 max **P** 0.040 max **S** 0.040 max **Si** 1.50 max **V** 0.03 max **W** 0.10 max | **ASTM** A426 (CPCA-15); A487 (CA-15) |
| * **J91201** | Alloy Steel Casting (420) | **C** 0.15 min **Cr** 12.0-14.0 **Mn** 1.00 max **P** 0.04 max **S** 0.03 max **Si** 1.00 max | |
| **J91209** | Austenitic Manganese Steel Casting | **C** 1.00-1.35 **Cr** 0.75 max **Mn** 12.00-14.00 **Mo** 0.50 max **Ni** 1.00 max **P** 0.060 max **Si** 0.40-1.00 | |
| **J91249** | Austenitic Manganese Steel Casting | **C** 0.7-1.3 **Mn** 11.5-14.0 **Mo** 0.9-1.2 **P** 0.07 max **Si** 1.00 max | **ASTM** A128 (E-1) |
| **J91261** | Alloy Steel Casting | **C** 0.15 max **Cr** 11.50-14.00 **Mn** 1.00 max **Mo** 0.50-0.70 **Ni** 0.65-1.00 **P** 0.05 max **S** 0.05 max **Si** 0.50 max | |
| **J91309** | Austenitic Manganese Steel Casting | **C** 1.05-1.35 **Cr** 1.5-2.5 **Mn** 11.5-14.0 **P** 0.07 max **Si** 1.00 max | **ASTM** A128 (C) |
| **J91339** | Austenitic Manganese Steel Casting | **C** 1.05-1.45 **Mn** 11.5-14.0 **Mo** 1.8-2.1 **P** 0.07 max **Si** 1.00 max | **ASTM** A128 (E-2) |
| **J91340** | Austenitic Manganese Steel Casting (Grade F) | **C** 1.05-1.35 **Mn** 6.0-8.0 **Mo** 0.9-1.2 **P** 0.07 max **Si** 1.00max | **ASTM** A128 (F) |
| **J91422** | Alloy Steel Casting (CA28MWV) | **C** 0.20-0.28 **Cr** 11.0-12.5 **Fe** rem **Mn** 0.50-1.00 **Mo** 0.90-1.25 **Ni** 0.50-1.00 **P** 0.030 max **S** 0.030 max **Si** 1.0 max **V** 0.20-0.30 **W** 0.90-1.25 | **ACI** CA28MWV **AISI** 422 **ASTM** A743 (CA28MWV) |
| **J91459** | Austenitic Manganese Steel Casting | **C** 0.7-1.3 **Mn** 11.5-14.0 **Ni** 3.0-4.0 **P** 0.07 max **Si** 1.00 max | **ASTM** A128 (D) |
| **J91540** | Alloy Steel Casting (CA-6NM) | **C** 0.06 max **Cr** 11.5-14.0 **Mn** 1.00 max **Mo** 0.40-1.0 **Ni** 3.5-4.5 **P** 0.04 max **S** 0.03 max | **ACI** CA-6NM **ASTM** A352 (CA-6NM); A356 (CA-6NM); A487 (CA-15, CA-6NM); A743 (CA-6NM) |
| **J91550** | Alloy Steel Casting | **C** 0.06 max **Cr** 11.5-14.0 **Cu** 0.50 max **Mn** 1.00 max **Mo** 0.40-1.0 **Ni** 3.5-4.5 **P** 0.030 max **S** 0.030 max **Si** 1.00 max **W** 0.10 max | **ASTM** A757 (E3N) |

*Boxed entries are no longer active and are retained for reference purposes only.

The chemical compositions listed are for identification purposes and should not be used in lieu of the cross referenced specifications.

| UNIFIED NUMBER | DESCRIPTION | CHEMICAL COMPOSITION | CROSS REFERENCE SPECIFICATIONS |
|---|---|---|---|
| **J91601** | Alloy Steel Casting | **C** 0.12-0.20 **Cr** 14.5-17.0 **Cu** 0.50 max **Mn** 1.0 max **Mo** 0.50 max **Ni** 1.5-2.25 **P** 0.04 max **S** 0.04 max **Si** 1.0 max | **AMS** 5372 |
| **J91606** | Alloy Steel Casting (440A) | **C** 0.60-0.75 **Cr** 16.0-18.0 **Mn** 1.00 max **Mo** 0.75 max **P** 0.04 max **S** 0.03 max **Si** 1.00 max | |
| **J91631** | Alloy Steel Casting | **C** 0.12-0.20 **Cr** 12.0-14.0 **Cu** 0.50 max **Mn** 1.0 max **Mo** 0.50 max **Ni** 1.8-2.2 **P** 0.04 max **S** 0.03 max **Si** 1.0 max **W** 2.5-3.5 | **AMS** 5354 |
| **J91639** | Alloy Steel Casting (440C) | **C** 0.95-1.20 **Cr** 16.0-18.0 **Mn** 1.00 max **Mo** 0.35-0.75 **Ni** 0.75 max **P** 0.04 max **S** 0.03 max **Si** 1.00 max | **AMS** 5352 |
| **J91650** | Alloy Steel Casting | **C** 0.06 max **Cr** 10.5-12.5 **Mn** 0.50 max **Ni** 6.0-8.0 **P** 0.02 max **S** 0.02 max **Si** 1.00 max | **ACI** CA-6N **ASTM** A743 (CA6N) |
| **J91651** | Alloy Steel Casting (431) | **C** 0.08-0.15 **Cr** 15.0-17.0 **Mn** 1.0 max **N** 0.03-0.12 **Ni** 1.5-2.2 **P** 0.04 max **S** 0.04 max **Si** 1.0 max | **AMS** 5353 |
| **J91803** | Alloy Steel Casting (CB-30) | **C** 0.30 max **Cr** 18.0-21.0 **Mn** 1.00 max **Ni** 2.00 max **P** 0.04 max **S** 0.04 max **Si** 1.50 max | **ACI** CB-30 **ASTM** A743 (CB-30) |
| **J91804** | Cast Cr-Ni Stainless Steel (CB-6) | **C** 0.06 max **Cr** 15.5-17.5 **Mn** 1.00 max **Mo** 0.5 max **Ni** 3.5-5.5 **P** 0.04 max **S** 0.03 max **Si** 1.00 max | **ACI** CB-6 **ASTM** A743 |
| **J92001** | Alloy Steel Casting | **C** 0.08-0.15 **Cr** 14.5-15.5 **Mn** 0.40-1.10 **Mo** 2.0-2.6 **N** 0.05-0.13 **Ni** 3.5-4.5 **P** 0.04 max **S** 0.03 max **Si** 0.75 max | **AMS** 5359; 5368 |
| **J92110** | Alloy Steel Casting, Precipitation Hardening (15-5 PH) | **C** 0.07 max **Cb** 0.15-0.35 **Cr** 14.0-15.50 **Cu** 2.50-3.20 **Mn** 0.70 max **N** 0.05 max **Ni** 4.20-5.50 **P** 0.035 max **S** 0.03 max **Si** 1.00 max | **AMS** 5346; 5347; 5348; 5356; 5357; 5400 **ASTM** A747 (CB7Cu-2) |
| **J92130** | Alloy Steel Casting (15-5 PH) combined with J92110 | | |
| * **J92150** | Alloy Steel Casting (17-4) | **C** 0.08 max **Cr** 15.5-17.5 **Cu** 3.0-5.0 **Mn** 1.00 max **Ni** 3.0-5.0 **P** 0.04 max **S** 0.04 max **Si** 1.00 max | |
| **J92170** | Alloy Steel Casting, Precipitation Hardening | **C** 0.06 max **Cr** 15.5-16.7 **Cu** 2.5-3.2 **Mn** 0.70 max **Ni** 3.6-4.6 **P** 0.04 max **S** 0.03 max **Si** 0.50-1.0 | |
| **J92180** | Alloy Steel Casting, Precipitation Hardening | **C** 0.07 max **Cb** 0.15-0.35 **Cr** 15.50-17.70 **Cu** 2.50-3.20 **Mn** 0.70 max **N** 0.05 max **Ni** 3.60-4.60 **P** 0.035 max **S** 0.03 max **Si** 1.00 max | **ASTM** A747 (CB7Cu-1) |
| **J92200** | Alloy Steel Casting (17-4) | **C** 0.06 max **Cr** 15.5-16.7 **Cu** 2.8-3.5 **Mn** 0.70 max **N** 0.05 max **Ni** 3.6-4.6 **P** 0.04 max **S** 0.03 max **Si** 0.5-1.0 **Other** Cb+Ta 0.15-0.40 | **AMS** 5342; 5343; 5344; 5355; 5398 |
| **J92205** | Duplex Alloy Steel Casting (2205) | **C** 0.03 max **Cr** 21.0-23.5 **Cu** 1.00 max **Mn** 1.50 max **Mo** 2.5-3.5 **Ni** 4.5-6.5 **N** 0.10-0.30 **P** 0.04 max **S** 0.020 max **Si** 1.00 max | **ACI** CD-3MN **ASTM** A890 (4A); A995 (4A; CD3MN) |
| **J92240** | Alloy Steel Casting (14-4PH) | **C** 0.06 max **Cr** 13.5-14.7 **Cu** 3.0-3.5 **Mn** 0.7 max **Mo** 2.0-2.75 **N** 0.05 max **Ni** 3.75-4.75 **P** 0.02 max **S** 0.025 max **Si** 0.5-1.0 **Other** Cb+Ta 0.15-0.35 | **AMS** 5340 **SAE** J467 (14-4PH) |

*Boxed entries are no longer active and are retained for reference purposes only.

The chemical compositions listed are for identification purposes and should not be used in lieu of the cross referenced specifications.

| UNIFIED NUMBER | DESCRIPTION | CHEMICAL COMPOSITION | CROSS REFERENCE SPECIFICATIONS |
|---|---|---|---|
| **J92500** | Alloy Steel Casting (CF-3) | **C** 0.03 max **Cr** 17.0-21.0 **Mn** 1.50 max **Ni** 8.0-12.0 **P** 0.040 max **S** 0.040 max **Si** 2.00 max | **ACI** CF-3 **ASTM** A351 (CF-3, CF-3A); A451 (CPF3, CPF3A); A743 (CF-3); A744 (CF-3) |
| * **J92501** | Alloy Steel Casting (302) | **C** 0.15 max **Cr** 17.0-19.0 **Mn** 2.00 max **Ni** 8.0-10.0 **P** 0.04 max **S** 0.03 max **Si** 1.00 max | |
| **J92502** | Alloy Steel Casting | **C** 0.15-0.30 **Cr** 17.00-20.00 **Mn** 1.50 max **Mo** 0.50 max **Ni** 8.00-11.00 **P** 0.05 max **S** 0.05 max **Si** 1.50 max | |
| * **J92511** | Alloy Steel Casting (303) | **C** 0.12 max **Cr** 17.0-20.0 **Cu** 0.50 max **Mn** 2.00 max **Mo** 0.60 max **Ni** 8.0-10.0 **P** 0.17 max **S** 0.15-0.35 **Si** 1.00 max | |
| **J92512** | Alloy Steel Casting | **C** 0.25 max **Cr** 17.0-19.0 **Cu** 0.5 max **Mn** 2.0 max **Mo** 0.5 max **Ni** 8.0-10.0 **P** 0.04 max **S** 0.03 max **Si** 1.0 max | |
| **J92590** | Alloy Steel Casting (304H) | **C** 0.04-0.10 **Cr** 18.0-20.0 **Mn** 2.00 max **Ni** 8.00-11.0 **P** 0.040 max **S** 0.030 max **Si** 0.75 max | **ASTM** A351 (CF-10) |
| **J92600** | Alloy Steel Casting (CF-8) | **C** 0.08 max **Cr** 18.0-21.0 **Mn** 1.50 max **Ni** 8.0-11.0 **P** 0.04 max **S** 0.04 max **Si** 2.00 max | **ACI** CF-8 **ASTM** A351 (CF-8, CF-8A); A451 (CPF8, CPF8A); A743 (CF-8); A744 (CF-8) **MIL SPEC** MIL-C-24707/3 (CF-8) |
| **J92602** | Alloy Steel Casting (CF-20) | **C** 0.20 max **Cr** 18.0-21.0 **Mn** 1.50 max **Ni** 8.0-11.0 **P** 0.04 max **S** 0.04 max **Si** 2.00 max | **ACI** CF-20 **AMS** 5358 **ASTM** A743 (CF-20) |
| **J92603** | Alloy Steel Casting (HF) | **C** 0.20-0.40 **Cr** 18.0-23.0 **Mn** 2.00 max **Mo** 0.50 max **Ni** 8.0-12.0 **P** 0.04 max **S** 0.04 max **Si** 2.00 max | **ACI** HF **ASTM** A297 (HF) |
| **J92605** | Alloy Steel Casting (HC) | **C** 0.50 max **Cr** 26.0-30.0 **Mn** 1.00 max **Mo** 0.50 max **Ni** 4.00 max **P** 0.04 max **S** 0.04 max **Si** 2.00 max | **ACI** HC **ASTM** A297 (HC) |
| * **J92610** | Alloy Steel Casting (304) | **C** 0.08 max **Cr** 18.0-20.0 **Mn** 2.00 max **Ni** 8.0-12.0 **P** 0.04 max **S** 0.03 max **Si** 1.00 max | |
| **J92613** | Alloy Steel Casting (HC-30) | **C** 0.25-0.35 **Cr** 26.0-30.0 **Mn** 0.5-1.0 **Mo** 0.50 max **Ni** 4.0 max **P** 0.04 max **S** 0.04 max **Si** 0.50-2.00 | **ACI** HC-30 **ASTM** A608 (HC-30) |
| **J92615** | Alloy Steel Casting (CC-50) | **C** 0.50 max **Cr** 26.0-30.0 **Mn** 1.00 max **Ni** 4.00 max **P** 0.04 max **S** 0.04 max **Si** 1.50 max | **ACI** CC-50 **ASTM** A743 (CC-50) |
| **J92620** | Alloy Steel Casting (304L) | **C** 0.05 max **Cr** 18.0-21.0 **Cu** 0.50 max **Mn** 1.00-2.00 **Mo** 0.50 max **Ni** 8.0-11.0 **P** 0.04 max **S** 0.03 max **Si** 1.00 max | **AMS** 5370; 5371 |
| **J92630** | Alloy Steel Casting (321) | **C** 0.08 max **Cr** 17.0-19.0 **Mn** 2.00 max **Ni** 9.0-12.0 **P** 0.04 max **S** 0.03 max **Si** 1.00 max **Other** Ti 5\mtC min | |
| **J92640** | Alloy Steel Casting (347) | **C** 0.08 max **Cr** 17.0-19.5 **Mn** 2.00 max **Ni** 9.0-13.0 **P** 0.04 max **S** 0.03 max **Si** 1.00 max **Other** Cb+Ta 10\mtC min, 1.5 max | |

*Boxed entries are no longer active and are retained for reference purposes only.

The chemical compositions listed are for identification purposes and should not be used in lieu of the cross referenced specifications.

| UNIFIED NUMBER | DESCRIPTION | CHEMICAL COMPOSITION | CROSS REFERENCE SPECIFICATIONS |
|---|---|---|---|
| **J92641** | Alloy Steel Casting | **C** 0.10 max **Cr** 17.00-20.00 **Cu** 0.50 max **Mn** 2.00 max **Mo** 0.50 max **Ni** 9.00-12.00 **P** 0.04 max **S** 0.04 max **Si** 1.50 max **Other** Cb+Ta 10\mtC min-1.35 max | **AMS** 5363; 5364 |
| **J92650** | Alloy Steel Casting | **C** 0.08 max **Cr** 18.00-21.00 **Mn** 1.50 max **Ni** 8.00-11.00 **P** 0.05 max **S** 0.05 max **Si** 2.00 max | |
| **J92660** | Alloy Steel Casting (347H) | **C** 0.04-0.10 **Cr** 17.0-20.0 **Mn** 2.00 max **Ni** 9.00-13.0 **P** 0.040 max **S** 0.030 max **Si** 0.75 max **Other** (Cb÷Ta) 8\mtC min-1.0 max | |
| **J92700** | Cast Cr-Ni-Mo Stainless Steel (CF3M) | **C** 0.03 max **Cr** 17.0-21.0 **Mn** 1.50 max **Mo** 0.50 max **Ni** 8.0-12.0 **P** 0.04 max **S** 0.04 max **Si** 2.00 max | **ASTM** A351; A743; A744; A451 (CPF3, CPF3A) |
| **J92701** | Alloy Steel Casting (CF-16F) | **C** 0.16 max **Cr** 18.0-21.0 **Mn** 1.50 max **Ni** 9.0-12.0 **P** 0.04 max **S** 0.04 max **Si** 2.00 max | **ACI** CF-16F **ASTM** A743 (CF-16F) |
| **J92710** | Alloy Steel Casting (CF-8C) | **C** 0.08 max **Cr** 18.0-21.0 **Mn** 1.50 max **Ni** 9.0-12.0 **Nb** 8\mtC min-1.00 max, or Nb+Ta 9\mtC min-1.1 max **P** 0.04 max **S** 0.04 max **Si** 2.00 max | **ACI** CF-8C **ASTM** A351 (CF-8C); A451 (CPF8C); A743 (CF-8C); A744 (CF-8C) **MIL SPEC** MIL-C-24707/3 (CF-8C) |
| **J92711** | Alloy Steel Casting (60303) | **C** 0.16 max **Cr** 18.00-21.00 **Cu** 0.50 max **Mn** 2.00 max **Mo** 0.80 max **Ni** 9.00-12.00 **P** 0.04 max **S** 0.15-0.35 **Si** 2.00 max | **AMS** 5341 |
| **J92720** | Alloy Steel Casting | **C** 0.08 max **Cr** 17.00-20.00 **Mn** 1.50 max **Mo** 0.50 max **Ni** 10.00 min **P** 0.05 max **S** 0.05 max **Si** 1.50 max | |
| **J92730** | Alloy Steel Casting | **C** 0.08 max **Cr** 18.00-21.00 **Mn** 1.50 max **Ni** 9.00-12.00 **P** 0.05 max **S** 0.05 max **Si** 2.00 max | |
| **J92740** | Alloy Steel Casting (Durcomet 101) | **C** 0.07 max **Cr** 22.00-24.00 **Cu** 0.50 max **Mn** 0.50 max **Mo** 0.85-1.15 **Ni** 4.50-5.50 **P** 0.04 max **S** 0.04 max **Si** 1.70 max | |
| **J92800** | Alloy Steel Casting (CF-3M) | **C** 0.03 max **Cr** 17.0-21.0 **Mn** 1.50 max **Mo** 2.0-3.0 **Ni** 9.0-13.0 **P** 0.04 max **S** 0.04 max **Si** 1.50 max | **ACI** CF-3M **ASTM** A351 (CF-3M, CF-3MA); A451 (CPF-3M); A743 (CF-3M); A744 (CF-3M) |
| **J92801** | Alloy Steel Casting | **C** 0.10 max **Cr** 17.00-20.00 **Mn** 1.50 max **Mo** 0.50 max **Ni** 11.00 min **P** 0.05 max **S** 0.05 max **Si** 1.50 max | |
| **J92802** | Cast Cr-Ni Stainless Steel (CE20N) | **C** 0.20 max **Cr** 23.0-26.0 **Mn** 1.50 max **Ni** 8.0-11.0 **N** 0.08-0.20 **P** 0.040 max **S** 0.040 max **Si** 1.50 max | **ACI** CE20N **ASME** SA-351; SA-451 **ASTM** A351; A351 (CE20N); A451; A452 (CPE20N) |
| **J92803** | Alloy Steel Casting (HF-30) | **C** 0.25-0.35 **Cr** 19.0-23.0 **Mn** 1.50 max **Mo** 0.50 max **Ni** 9.0-12.0 **P** 0.04 max **S** 0.04 max **Si** 0.50-2.00 | **ASTM** A608 (HF30) |
| **J92804** | Alloy Steel Casting (CF3MN) | **C** 0.03 max **Cr** 17.0-22.0 **Fe** rem **Mn** 1.50 max **Mo** 2.0-3.0 **N** 0.10-0.20 **Ni** 9.0-13.0 **P** 0.040 max **S** 0.040 max **Si** 1.50 max | **ACI** CF3MN **AISI** 316N **ASTM** A351 (CF3MN); A743 (CF3MN) |
| **J92805** | Alloy Steel Casting (CE-8N) | **C** 0.08 max **Cr** 23.0-26.0 **Mn** 1.50 max **Mo** 0.50 max **N** 0.20-0.30 **Ni** 8.0-11.0 **P** 0.040 max **S** 0.040 max **Si** 1.50 max | **ACI** CE-8N |
| * **J92810** | Alloy Steel Casting (316) | **C** 0.08 max **Cr** 16.0-18.0 **Mn** 2.00 max **Mo** 2.0-3.0 **Ni** 10.0-14.0 **P** 0.04 max **S** 0.03 max **Si** 1.00 max | |

*Boxed entries are no longer active and are retained for reference purposes only.

The chemical compositions listed are for identification purposes and should not be used in lieu of the cross referenced specifications.

| UNIFIED NUMBER | DESCRIPTION | CHEMICAL COMPOSITION | CROSS REFERENCE SPECIFICATIONS |
|---|---|---|---|
| **J92811** | Alloy Steel Casting | **C** 0.12 max **Cr** 18.0-19.5 **Cu** 0.5 max **Mn** 2.0 max **Mo** 0.5 max **Ni** 10.0-14.0 **P** 0.04 max **S** 0.03 max **Si** 1.5 max **Other** Cb+Ta 10\mtC min-1.5 max | **AMS** 5362 |
| **J92843** | Alloy Steel Casting | **C** 0.28-0.35 **Cr** 18.00-21.00 **Cu** 0.50 max **Mn** 0.75-1.50 **Mo** 1.00-1.75 **Ni** 8.00-11.00 **P** 0.04 max **S** 0.04 max **Si** 1.00 max **Ti** 0.15-0.50 **W** 1.00-1.75 **Other** Cb+Ta 0.30-0.70 | **AMS** 5369 |
| **J92900** | Alloy Steel Casting (CF-8M) | **C** 0.08 max **Cr** 18.0-21.0 **Mn** 1.50 max **Mo** 2.0-3.0 **Ni** 9.0-12.0 **P** 0.04 max **S** 0.04 max **Si** 2.00 max | **ACI** CF-8M **ASTM** A351 (CF-8M); A451 (CPF8M); A743 (CF-8M); A744 (CF-8M) **MIL SPEC** MIL-C-24707/3 (CF-8M) |
| **J92901** | Cast Cr-Ni-Mo Stainless Steel (CF-10M) | **C** 0.04-0.10 **Cr** 18.0-21.0 **Mn** 1.50 max **Mo** 2.0-3.0 **Ni** 9.0-12.0 **P** 0.040 max **S** 0.040 max **Si** 1.50 max | **ACI** CF-10M **ASME** SA-351 **ASTM** A351 (CF-10M) |
| **J92910** | Alloy Steel Casting | **C** 0.08 max **Cr** 18.00-21.00 **Mn** 1.50 max **Mo** 2.00-3.00 **Ni** 9.00-12.00 **P** 0.05 max **S** 0.05 max **Si** 2.00 max | |
| * **J92920** | Alloy Steel Casting (316H) | **C** 0.04-0.10 **Cr** 16.0-18.0 **Mn** 2.00 max **Mo** 2.00-3.00 **Ni** 11.0-14.0 **P** 0.040 max **S** 0.030 max **Si** 0.75 max | |
| **J92951** | Alloy Steel Casting | **C** 0.15 max **Cr** 16.0-18.0 **Cu** 0.50 max **Mn** 2.0 max **Mo** 1.5-2.25 **Ni** 12.0-14.0 **P** 0.04 max **S** 0.03 max **Si** 0.75 max | **AMS** 5360 |
| **J92971** | Cast Cr-Ni-Mo Stainless Steel (CF10MC) | **C** 0.10 max **Cr** 15.0-18.0 **Mn** 1.50 max **Mo** 1.75-2.25 **Ni** 13.0-16.0 **P** 0.040 max **S** 0.040 max **Si** 1.50 max **Other** Cb 10\mtC min, 1.20 max | **ACI** CF10MC **ASME** SA-451 (CF10MC) **ASTM** A451 (CF10MC) |
| **J92972** | Cast Cr-Mn-Ni-Si-N Stainless Steel (CF10SMnN) | **C** 0.10 max **Cr** 16.0-18.0 **Fe** rem **Mn** 7.00-9.00 **N** 0.08-0.18 **Ni** 8.0-9.0 **P** 0.060 max **S** 0.030 max **Si** 3.50-4.50 | **ACI** CF10SMnN **ASME** SA-351 **ASTM** A351; A743 (CF10SMnN) |
| **J92999** | Cast Cr-Ni-Mo Stainless Steel (CG-3M) | **C** 0.03 max **Cr** 18.0-21.0 **Mn** 1.50 max **Mo** 3.0-4.0 **Ni** 9.0-13.0 **P** 0.04 max **S** 0.04 max **Si** 1.50 max | **ACI** CG-3M **ASME** SA-351 **ASTM** A351 (CG-3M); A743 (CG-3M); A744 (CG-3M) |
| **J93000** | Alloy Steel Casting (CG-8M) | **C** 0.08 max **Cr** 18.0-21.0 **Mn** 1.50 max **Mo** 3.0-4.0 **Ni** 9.0-13.0 **P** 0.04 max **S** 0.04 max **Si** 1.50 max | **ASTM** A351 (CG-8M); A743 (CG-8M); A744 (CG-8M) |
| **J93001** | Alloy Steel Casting (CG-12) | **C** 0.12 max **Cr** 20.0-23.0 **Mn** 1.50 max **Ni** 10.0-13.0 **P** 0.04 max **S** 0.04 max **Si** 2.00 max | **ACI** CG-12 **ASTM** A743 (CG-12) |
| **J93005** | Alloy Steel Casting (HD) | **C** 0.50 max **Cr** 26.0-30.0 **Mn** 1.50 max **Mo** 0.50 max **Ni** 4.0-7.0 **P** 0.04 max **S** 0.04 max **Si** 2.00 max | **ACI** HD **ASTM** A297 (HD) |
| **J93010** | Alloy Steel Casting, Maraging | **Al** 0.02-0.10 **C** 0.03 max **Co** 9.50-11.00 **Mn** 0.10 max **Mo** 4.40-4.80 **Ni** 16.00-17.50 **P** 0.01 maX **S** 0.01 max **Si** 0.10 max **Ti** 0.15-0.45 | **AMS** 5339 |
| **J93015** | Alloy Steel Casting (HD-50) | **C** 0.45-0.55 **Cr** 26.0-30.0 **Mn** 1.50 max **Mo** 0.50 max **Ni** 4.0-7.0 **P** 0.04 max **S** 0.04 max **Si** 0.50-2.00 | **ASTM** A608 (HD-50) |
| **J93072** | Alloy Steel Casting | **C** 0.15-0.25 **Cr** 17.0-20.0 **Mn** 2.0 max **Mo** 1.75-2.5 **Ni** 12.0-15.0 **P** 0.04 max **S** 0.04 max **Si** 1.0 max | **AMS** 5361 |
| **J93150** | Alloy Steel Casting, Maraging | **Al** 0.05-0.20 **C** 0.03 max **Co** 8.50-9.50 **Mn** 0.10 max **Mo** 4.50-5.50 **Ni** 18.0-19.0 **P** 0.010 max **S** 0.010 max **Si** 0.10 max **Ti** 0.55-0.85 | **AMS** 5337 |

*Boxed entries are no longer active and are retained for reference purposes only.

The chemical compositions listed are for identification purposes and should not be used in lieu of the cross referenced specifications.

| UNIFIED NUMBER | DESCRIPTION | CHEMICAL COMPOSITION | CROSS REFERENCE SPECIFICATIONS |
|---|---|---|---|
| **J93183** | Cast Ferritic-Austenitic Stainless Steel (KCR-D183) | **C** 0.03 max **Co** 0.5-1.5 **Cr** 20.0-23.0 **Cu** 1.0 max **Mn** 2.0 max **Mo** 2.0-4.0 **N** 0.08-0.25 **Ni** 4.0-6.0 **P** 0.040 max **S** 0.03 max **Si** 2.0 max | **ASTM** A872 |
| **J93254** | Austenitic Cr-Ni-Mo-Cu-N Stainless Steel Casting (CK-3MCuN, 254 SMO) | **C** 0.025 max **Cr** 19.5-20.5 **Cu** 0.50-1.0 **Mn** 1.20 max **Mo** 6.0-7.0 **N** 0.180-0.240 **Ni** 17.5-19.7 **P** 0.045 max **Si** 1.0 max **S** 0.010 max | **ACI** CK-3MCuN **ASTM** A351 (CK-3MCuN); A743 (CK-3MCuN) |
| **J93303** | Alloy Steel Casting | **C** 0.20-0.45 **Cr** 23.00-28.00 **Mn** 2.50 max **N** 0.20 max **Ni** 10.0-14.0 **P** 0.05 max **S** 0.05 max **Si** 1.75 max | **ASTM** A447 |
| **J93345** | Cast Duplex (Austenitic-Ferritic) Cr-Ni-Mo-N Stainless Steel (ESCOLOY 45 D) | **C** 0.08 max **Cr** 22.5-25.5 **Mn** 1.00 max **Mo** 3.0-4.5 **Ni** 8.0-11.0 **N** 0.10-0.30 **P** 0.04 max **S** 0.025 max **Si** 1.50 max | **ASTM** A890 (2A); A995 (2A; CE8MN) |
| **J93370** | Alloy Steel Casting | **C** 0.04 max **Cr** 24.5-26.5 **Cu** 2.75-3.25 **Mn** 1.00 max **Mo** 1.75-2.25 **Ni** 4.75-6.00 **P** 0.04 max **S** 0.04 max **Si** 1.00 max | **ACI** CD4MCu **ASTM** A351 (CD4MCu); A744 (CD4MCu); A890 (1A) |
| **J93371** | Duplex Alloy Steel Casting (3A) | **C** 0.06 max **Cr** 24.0-27.0 **Fe** rem **Mn** 1.00 max **Mo** 1.75-2.50 **N** 0.15-0.25 **Ni** 4.00-6.00 **P** 0.040 max **S** 0.040 max **Si** 1.00 max | **ACI** CD6MN **ASTM** A890 (3A); A995 (3A, CD6MN) |
| **J93372** | Cast Duplex (Austenitic-Ferritic) Cr-Ni-Cu-Mo-N Stainless Steel (CD-4MCuN) | **C** 0.04 max **Cr** 24.5-26.5 **Cu** 2.7-3.3 **Mn** 1.0 max **Mo** 1.7-2.3 **N** 0.10-0.25 **Ni** 4.7-6.0 **P** 0.04 max **S** 0.04 max **Si** 1.0 max | **ACI** CD-4MCuN **ASTM** A890 (1B); A995 (1B, CD4MCuN) |
| **J93373** | Cast Duplex (Austenitic-Ferritic) Cr-Ni-Mo-Cu-N Stainless Steel (CD3MCuN Cast Ferralium(r) 255) | **C** 0.030 max **Cr** 24.0-26.7 **Cu** 1.40-1.90 **Mn** 1.20 max **Mo** 2.9-3.8 **Ni** 5.6-6.7 **N** 0.22-0.33 **P** 0.030 max **Si** 1.10 max **S** 0.030 max | **ACI** CD3MCuN **ASTM** A890 (1C) |
| **J93380** | Ferritic - Austenitic Cr-Ni-Mo-W-N Stainless Steel Casting (Zeron 100) | **C** 0.03 max **Cr** 24.0-26.0 **Cu** 0.5-1.0 **Mn** 1.0 max **Mo** 3.0-4.0 **Ni** 6.5-8.5 **N** 0.2-0.3 **P** 0.030 max **S** 0.025 max **Si** 1.0 max **W** 0.5-1.0 **Other** %Cr+3.3 \mt %Mo+16 \mt %$N_2$ = 40 min | **ACI** CD3MVCuN **ASTM** A351 (CD3MWCuN); A743 (CD3MWCuN); A890 (6A); A995 (6A; CD3MWCuN) |
| **J93400** | Alloy Steel Casting (CH-8) | **C** 0.08 max **Cr** 22.0-26.0 **Mn** 1.50 max **Ni** 12.0-15.0 **P** 0.040 max **S** 0.040 max **Si** 1.50 max | **ACI** CH-8 **ASTM** A351 (CH-8); A451 (CPH-8) |
| **J93401** | Alloy Steel Casting (CH-10) | **C** 0.10 max **Cr** 22.0-26.0 **Mn** 1.50 max **Ni** 12.0-15.0 **P** 0.040 max **S** 0.040 max **Si** 2.00 max | **ACI** CH-10 **ASTM** A351 (CH-10); A743 (CH-10) |
| **J93402** | Alloy Steel Casting (CH-20) | **C** 0.20 max **Cr** 22.0-26.0 **Mn** 1.50 max **Ni** 12.0-15.0 **P** 0.04 max **S** 0.04 max **Si** 2.00 max | **ACI** CH-20 **ASTM** A351 (CH-20); A451 (CH-2, CPH-10, CPH-20); A743 (CH-20) |
| **J93403** | Alloy Steel Casting (HE) | **C** 0.20-0.50 **Cr** 26.0-30.0 **Mn** 2.00 max **Mo** 0.50 max **Ni** 8.0-11.0 **P** 0.04 max **S** 0.04 max **Si** 2.00 max | **ACI** HE **ASTM** A297 (HE) |
| **J93404** | Cast Ferritic-Austenitic Stainless Steel (Alloy 958) | **C** 0.03 max **Cr** 24.0-26.0 **Mn** 1.50 max **Mo** 4.0-5.0 **Ni** 6.0-8.0 **N** 0.10-0.30 **Si** 1.00 max | **ASTM** A890 (5A); A995 (5A, CE3MN) |
| **J93413** | Alloy Steel Casting (HE-35) | **C** 0.30-0.40 **Cr** 26.0-30.0 **Mn** 1.50 max **Mo** 0.50 max **Ni** 8.0-11.0 **P** 0.04 max **S** 0.04 max **Si** 0.50-2.00 | **ACI** HE-35 **ASTM** A608 (HE-35) |
| **J93423** | Alloy Steel Casting (CE-30) | **C** 0.30 max **Cr** 26.0-30.0 **Mn** 1.50 max **Ni** 8.0-11.0 **P** 0.04 max **S** 0.04 max **Si** 2.00 max | **ACI** CE-30 **ASTM** A743 (CE-30) |
| **J93503** | Alloy Steel Casting (HH) | **C** 0.20-0.50 **Cr** 24.0-28.0 **Mn** 2.00 max **Mo** 0.50 max **Ni** 11.0-14.0 **P** 0.04 max **S** 0.04 max **Si** 2.00 max | **ACI** HH **ASTM** A297 (HH); A447 (HH) |

The chemical compositions listed are for identification purposes and should not be used in lieu of the cross referenced specifications.

| UNIFIED NUMBER | DESCRIPTION | CHEMICAL COMPOSITION | CROSS REFERENCE SPECIFICATIONS |
|---|---|---|---|
| **J93513** | Alloy Steel Casting (HH-30) | **C** 0.25-0.35 **Cr** 24.0-28.0 **Mn** 1.50 max **Mo** 0.50 max **Ni** 11.0-14.0 **P** 0.04 max **S** 0.04 max **Si** 0.50-2.00 | **ACI** HH-30 **ASTM** A608 (HH30) |
| **J93550** | Cast Ferritic-Austenitic Stainless Steel (KCR-D283) | **C** 0.03 max **Co** 0.5-1.5 **Cr** 23.0-26.0 **Cu** 1.0 max **Mn** 2.0 max **Mo** 5.0-8.0 **N** 0.08-0.25 **P** 0.040 max **S** 0.03 max **Si** 2.0 max | **ASTM** A872 |
| **J93633** | Alloy Steel Casting (HH-33) | **C** 0.28-0.38 **Cr** 24.0-26.0 **Mn** 1.50 max **Mo** 0.50 max **Ni** 12.0-14.0 **P** 0.04 max **S** 0.04 max **Si** 0.50-2.00 | **ASTM** A608 (HH33) |
| **J93790** | Alloy Steel Casting (CG6MMN) | **C** 0.06 max **Cb** 0.10-0.30 **Cr** 20.50-23.50 **Mn** 4.00-6.00 **Mo** 1.50-3.00 **N** 0.20-0.40 **Ni** 11.50-13.50 **P** 0.040 max **S** 0.030 max **Si** 1.00 max **V** 0.10-0.30 | **ASTM** A351 (CG6MMN); A743 (CG6MMN) |
| **J93900** | Stainless Steel Casting (Durcomet 5) | **C** 0.025 max **Cr** 20.0-22.0 **Mn** 1.50 max **N** 0.08-0.20 **Ni** 15.0-17.0 **P** 0.040 max **S** 0.040 max **Si** 4.00-6.00 | |
| **J94003** | Alloy Steel Casting (HI) | **C** 0.20-0.50 **Cr** 26.0-30.0 **Mn** 2.00 max **Mo** 0.50 max **Ni** 14.0-18.0 **P** 0.04 max **S** 0.04 max **Si** 2.00 max | **ACI** HI **ASTM** A297 (HI) |
| **J94013** | Alloy Steel Casting (HI-35) | **C** 0.30-0.40 **Cr** 26.0-30.0 **Mn** 1.50 max **Mo** 0.50 max **Ni** 14.0-18.0 **P** 0.04 max **S** 0.04 max **Si** 0.50-2.00 | **ASTM** A608 (HI35) |
| **J94202** | Alloy Steel Casting (CK-20) | **C** 0.20 max **Cr** 23.0-27.0 **Mn** 2.00 max **Ni** 19.0-22.0 **P** 0.04 max **S** 0.04 max **Si** 2.00 max | **ACI** CK-20 **ASTM** A351 (CK-20); A451 (CPK20); A743 (CK-20) |
| **J94203** | Alloy Steel Casting (HK-30) | **C** 0.25-0.35 **Cr** 23.0-27.0 **Mn** 1.50 max **Ni** 19.0-22.0 **P** 0.040 max **S** 0.040 max **Si** 1.75 max | **ACI** HK-30 **ASTM** A351 (HK-30); A608 (HK-30) |
| **J94204** | Alloy Steel Casting (HK-40) | **C** 0.35-0.45 **Cr** 23.0-27.0 **Mn** 1.50 max **Ni** 19.0-22.0 **P** 0.040 max **S** 0.040 max **Si** 1.75 max | **ACI** HK-40 **ASTM** A351 (HK-40); A608 (HK40) |
| **J94211** | Alloy Steel Casting | **C** 0.10-0.18 **Cr** 23.0-26.0 **Cu** 0.50 max **Mn** 2.0 max **Mo** 0.50 max **Ni** 19.0-22.0 **P** 0.04 max **S** 0.04 max **Si** 0.50-1.50 | **AMS** 5365; 5366 |
| **J94213** | Alloy Steel Casting (HN) | **C** 0.20-0.50 **Cr** 19.0-23.0 **Mn** 2.00 max **Mo** 0.50 max **Ni** 23.0-27.0 **P** 0.04 max **S** 0.04 max **Si** 2.00 max | **ACI** HN **ASTM** A297 (HN) |
| **J94214** | Alloy Steel Casting (HN-40) | **C** 0.35-0.45 **Cr** 19.0-23.0 **Mn** 1.50 max **Mo** 0.50 max **Ni** 23.0-27.0 **P** 0.04 max **S** 0.04 max **Si** 0.50-2.00 | **ASTM** A608 (HN40) |
| **J94224** | Alloy Steel Casting (HK) | **C** 0.20-0.60 **Cr** 24.0-28.0 **Mn** 2.00 max **Mo** 0.50 max **Ni** 18.0-22.0 **P** 0.04 max **S** 0.04 max **Si** 2.00 max | **ACI** HK **ASTM** A297 (HK) |
| **J94302** | Alloy Steel Casting (310) | **C** 0.25 max **Cr** 24.0-26.0 **Mn** 2.00 max **Ni** 19.0-22.0 **P** 0.04 max **S** 0.03 max **Si** 1.00 max | |
| **J94603** | Replaced by N08030 | | |
| **J94604** | Replaced by N08604 | | |
| **J94605** | Replaced by N08002 | | |
| **J94613** | Replaced by N08613 | | |
| **J94614** | Replaced by N08614 | | |
| **J94650** | Alloy Steel Casting (CN-7MS) | **C** 0.07 max **Cr** 18.0-20.0 **Cu** 1.5-2.0 **Mn** 1.00 max **Mo** 2.5-3.0 **Ni** 22.0-25.0 **P** 0.04 max **S** 0.03 max **Si** 2.50-3.50 | **ASTM** A743 (CN-7MS); A744 (CN-7MS) |

The chemical compositions listed are for identification purposes and should not be used in lieu of the cross referenced specifications.

| UNIFIED NUMBER | DESCRIPTION | CHEMICAL COMPOSITION | CROSS REFERENCE SPECIFICATIONS |
|---|---|---|---|
| **J94651** | Austenitic Cr-Ni-Mo-N Stainless Steel Casting (CN-3MN, AL-6XN) | **C** 0.03 max **Cr** 20.0-22.0 **Cu** 0.75 max **Fe** rem **Mn** 2.00 max **Mo** 6.0-7.0 **N** 0.18-0.26 **Ni** 23.5-25.5 **P** 0.040 max **S** 0.010 max **Si** 1.00 max | **ACI** CN3MN **ASTM** A351 (CN3MN); A743 (CN3MN); A744 (CN3MN) |
| **J94652** | Cast Cr-Ni-Mo Stainless Steel (CN-3M) | **C** 0.03 max **Cr** 20.0-22.0 **Mn** 2.0 max **Mo** 4.5-5.5 **Ni** 23.0-27.0 **P** 0.03 max **S** 0.03 max **Si** 1.0 max | **ACI** CN-3M **ASTM** A743; A743 (CG6MMN); A743 (CN-3M) |
| **J94653** | Austenitic Cr-Ni-Mo-N Stainless Steel Casting (CK-3MN, cast SR50A) | **C** 0.035 max **Cr** 22.0-24.0 **Cu** 0.40 max **Mn** 2.00 max **Mo** 6.0-6.8 **N** 0.21-0.32 **Ni** 20.0-22.0 **P** 0.035 max **S** 0.020 max **Si** 1.00 max | **ACI** CK-3MN **ASTM** A743; A743 (CK-3MN); A744; A744 (CK-3MN) |
| **J94805** | Replaced by N08050 | | |
| **J95150** | Replaced by N08007 | | |
| **J95151** | Replaced by N08151 | | |
| **J95370** | Austenitic Cr-Ni-Mn-Mo-N Stainless Steel Casting | **Al** 0.010 max **As** 0.01 max **B** 0.003-0.007 **C** 0.030 max **Cb** 0.10 max **Co** 0.25 max **Cr** 24.0-25.0 **Cu** 0.50 max **Fe** rem **Pb** 0.01 max **Mn** 8.0-9.0 **Mo** 4.0-5.0 **N** 0.70-0.80 **Ni** 17.0-18.0 **P** 0.030 max **S** 0.010 max **Si** 0.50 max **Sn** 0.010 max **Ti** 0.10 max **W** 0.10 max **V** 0.10 max | **Other** NACE MR0175 pending |
| **J95404** | Replaced by N08005 | | |
| **J95405** | Replaced by N08004 | | |
| **J95705** | Replaced by N08705 | | |

The chemical compositions listed are for identification purposes and should not be used in lieu of the cross referenced specifications.

## UNS NUMBERS ASSIGNED TO DATE

**With Description of Each Material Covered and References To Documents In Which The Same or Similar Materials are Described**

**Kxxxxx Number Series**
**Miscellaneous Steels and Ferrous Alloys**

# MISCELLANEOUS STEELS AND FERROUS ALLOYS

| UNIFIED NUMBER | DESCRIPTION | CHEMICAL COMPOSITION | CROSS REFERENCE SPECIFICATIONS |
|---|---|---|---|
| * **K00040** | Carbon Steel, Special Killed, Deep Drawing | **C** 0.10 max **Mn** 0.50 max **P** 0.025 max **S** 0.035 max | |
| * **K00045** | Steel Welding Rod (RG45) | **Al** 0.02 max **C** 0.08 max **Cu** 0.25 max **Mn** 0.50 max **P** 0.040 max **S** 0.040 max **Si** 0.10 max | **ASME** SFA5.2 (RG45) **AWS** A5.2 (RG45) |
| * **K00060** | Steel Welding Rod (RG60) | **Al** 0.02 max **C** 0.15 max **Cu** 0.25 max **Mn** 0.90-1.40 **P** 0.040 max **S** 0.040 max **Si** 0.10-0.35 | **ASME** SFA5.2 (RG60) **AWS** A5.2 (RG60) |
| * **K00065** | Steel Welding Rod (RG65) | **Al** 0.02 max **P** 0.040 max **S** 0.040 max | **ASME** SFA5.2 (RG65) **AWS** A5.2 (RG65) |
| **K00095** | Commercially Pure Iron | **Cu** 0.15 max **P** 0.010 max **S** 0.030 max **Other** C+Mn+Si+P+S 0.10 max | **AMS** 7707 |
| * **K00100** | Enameling Steel | **C** 0.008 max **Mn** 0.60 max **S** 0.040 max | |
| * **K00400** | Enameling Steel | **C** 0.04 max **Mn** 0.12 max **P** 0.015 max **S** 0.040 max | |
| **K00600** | Carbon Steel, Special Magnetic Properties | **Al** 0.015 max **C** 0.06 max **Mn** 0.40 max **P** 0.015 max **S** 0.015 max **Si** 0.20 max | |
| **K00606** | Carbon Steel Welding Wire | **C** 0.06 max **Cu** 0.15 max **Mn** 0.25 max **P** 0.040 max **S** 0.040 max **Si** 0.08 max | **AMS** 5030 |
| **K00800** | Carbon Steel, Special Magnetic Properties | **Al** 0.02 max **C** 0.08 max **Mn** 0.40 max **P** 0.025 max **S** 0.025 max **Si** 0.20 max | |
| * **K00801** | Enameling Steel | **C** 0.08 max **Mn** 0.20 max **P** 0.015 max **S** 0.040 max | |
| **K00802** | Low Carbon Steel | **C** 0.08-0.20 **Mn** 0.40-0.80 **P** 0.040 max **S** 0.050 max | **AMS** 5061 |
| **K00912** | Steel Electrode (EL12N) | **C** 0.04-0.14 **Cu** 0.08 max **Mn** 0.25-0.60 **P** 0.012 max **S** 0.030 max **Si** 0.10 max **V** 0.05 max **Other** total other elements (except Fe) 0.50 max | **ASME** SFA5.23 (EL12N) **AWS** A5.23 (EL12N) |
| **K01000** | Carbon Steel, Special Magnetic Properties | **Al** 0.02 max **C** 0.10 max **Mn** 0.60 max **P** 0.04 max **S** 0.04 max **Si** 0.20 max | |
| **K01001** | Carbon Steel | **C** 0.05-0.15 **Mn** 0.27-0.63 **P** 0.050 max **S** 0.060 max | **ASTM** A254 |
| **K01008** | Steel Electrode (EL8) | **C** 0.10 max **Cu** 0.35 max **Mn** 0.25-0.60 **P** 0.030 max **S** 0.030 max **Si** 0.07 max **Other** total other elements (except Fe) 0.50 max | **ASME** SFA5.17 (EL8) **AWS** A5.17 (EL8) **MIL SPEC** MIL-E-23765/4 (MIL-EL8) |
| **K01009** | Steel Electrode (EL8K) | **C** 0.10 max **Cu** 0.35 max **Mn** 0.25-0.60 **P** 0.030 max **S** 0.030 max **Si** 0.10-0.25 **Other** total other elements (except Fe) 0.50 max | **ASME** SFA5.17 (EL8K) **AWS** A5.17 (EL8K) |

*Boxed entries are no longer active and are retained for reference purposes only.

The chemical compositions listed are for identification purposes and should not be used in lieu of the cross referenced specifications.

| UNIFIED NUMBER | DESCRIPTION | CHEMICAL COMPOSITION | CROSS REFERENCE SPECIFICATIONS |
|---|---|---|---|
| **K01010** | Steel Electrode (EH10K-EW) | **C** 0.06-0.14 **Mn** 1.40-2.0 **P** 0.025 max **S** 0.030 max **Si** 0.15-0.30 **Other** total other elements 0.50 max | **ASME** SFA5.25 (EH10K-EW) **AWS** A5.25 (EH10K-EW) |
| **K01012** | Steel Electrode (EL12) | **C** 0.04-0.14 **Cu** 0.35 max **Mn** 0.25-0.60 **P** 0.030 max **S** 0.030 max **Si** 0.10 max **Other** total other elements (except Fe) 0.50 max | **ASME** SFA5.17 (EL12); SFA5.23 (EL12) **AWS** A5.17 (EL12); A5.23 (EL12) **MIL SPEC** MIL-E-23765/4 (MIL-EL12) |
| **K01013** | Steel Electrode (EM12KN) | **C** 0.05-0.15 **Cu** 0.08 max **Mn** 0.80-1.25 **P** 0.012 max **S** 0.030 max **Si** 0.10-0.35 **V** 0.05 max **Other** total other elements (except Fe) 0.50 max | **ASME** SFA5.23 (EM12KN) **AWS** A5.23 (EM12KN) |
| **K01111** | Steel Electrode (EM11K) | **C** 0.07-0.15 **Cu** 0.35 max **Mn** 1.00-1.50 **P** 0.025 max **S** 0.03 max **Si** 0.65-0.85 | **ASME** SFA5.17 (EM11K) **AWS** A5.17 (EM11K) |
| **K01112** | Steel Electrode (EM12) | **C** 0.06-0.15 **Cu** 0.35 max **Mn** 0.80-1.25 **P** 0.030 max **S** 0.030 max **Si** 0.10 max **Other** total other elements (except Fe) 0.50 max | **ASME** SFA5.17 (EM12); SFA5.25 (EM12-EW) **AWS** A5.17 (EM12); A5.25 (EM12-EW) **MIL SPEC** MIL-E-23765/4 (MIL-EM12) |
| **K01113** | Steel Electrode (EM12K) (EM12K-EW) | **C** 0.05-0.15 **Cu** 0.35 max **Mn** 0.80-1.25 **P** 0.030 max **S** 0.030 max **Si** 0.10-0.35 **Other** total other elements (except Fe) 0.50 max | **ASME** SFA5.17 (EM12K); SFA5.23 (EM12K); SFA5.25 (EM12K-EW) **AWS** A5.17 (EM12K); A5.23 (EM12K); A5.25 (EM12K-EW) **MIL SPEC** MIL-E-23765/4 (MIL-EM12K) |
| **K01200** | Carbon Steel | **C** 0.06-0.18 **Mn** 0.27-0.63 **P** 0.050 max **S** 0.060 max | **ASTM** A178 (A); A179 |
| **K01201** | Carbon Steel | **C** 0.06-0.18 **Mn** 0.27-0.63 **P** 0.048 max **S** 0.058 max **Si** 0.25 max | **ASTM** A192 |
| **K01210** | Steel Electrode (EH10K) | **C** 0.07-0.15 **Cu** 0.35 max **Mn** 1.30-1.70 **P** 0.025 max **S** 0.025 max **Si** 0.05-0.25 | **ASME** SFA5.17 (EH10K) **AWS** A5.17 (EH10K) |
| **K01213** | Steel Electrode (EH12K) | **C** 0.06-0.15 **Cu** 0.35 max **Mn** 1.50-2.00 **P** 0.025 max **S** 0.025 max **Si** 0.25-0.65 | **ASME** SA5.17 (EH12K) **AWS** A5.17 (EH12K) |
| **K01313** | Steel Electrode (EM13K) | **C** 0.06-0.16 **Cu** 0.35 max **Mn** 0.90-1.40 **P** 0.030 max **S** 0.030 max **Si** 0.35-0.75 **Other** total other elements (except Fe) 0.50 max | **ASME** SFA5.17 (EM13K); SFA5.25 (EM13K-EW); SFA5.26 (EGXXS-1); SFA5.30 (INMs2) **AWS** A5.17 (EM13K); A5.25 (EM13K-EW); A5.26 (EGXXS-1); A5.30 (INMs2) |
| **K01314** | Steel Electrode (EM14K) | **C** 0.06-0.19 **Cu** 0.35 max **Mn** 0.90-1.40 **P** 0.025 max **S** 0.025 max **Si** 0.35-0.75 **Ti** 0.03-0.17 | **ASME** SFA5.17 (EM14K) **AWS** A5.17 (EM14K) |
| **K01400** | Carbon Steel | **C** 0.14 max **Mn** 0.90 max **P** 0.035 max **S** 0.04 max **Si** 0.40 max | **ASTM** A283 (A) |
| **K01500** | Carbon Steel, Special Magnetic Properties | **Al** 0.02 max **C** 0.15 max **Mn** 0.50 max **P** 0.04 max **S** 0.04 max **Si** 0.20 max | |
| **K01501** | Carbon Steel | **C** 0.15 max **Mn** 0.90 max **P** 0.035 max **S** 0.040 max **Other** Cu 0.20 min (when specified) | **ASTM** A414 (A) |
| **K01502** | Carbon Steel | **C** 0.15 max **Mn** 0.30-0.60 **P** 0.045 max **S** 0.050 max | **ASTM** A730 (A) |
| **K01503** | Carbon Steel | **C** 0.15 min | **ASTM** A194 (1) |
| * **K01504** | Carbon Steel | **C** 0.10-0.20 **Mn** 0.30-0.80 **P** 0.048 max **S** 0.058 max **Si** 0.25 max | |
| **K01505** | Carbon Steel | **C** 0.15 min **P** 0.05 max **Other** Cu 0.20 min (when specified) | **ASTM** A67 |
| **K01506** | Carbon Steel | **C** 0.15 max **Mn** 0.63 max **P** 0.050 max **S** 0.060 max | **ASTM** A539 |

*Boxed entries are no longer active and are retained for reference purposes only.

The chemical compositions listed are for identification purposes and should not be used in lieu of the cross referenced specifications.

| UNIFIED NUMBER | DESCRIPTION | CHEMICAL COMPOSITION | CROSS REFERENCE SPECIFICATIONS |
|---|---|---|---|
| **K01507** | Carbon Steel | **C** 0.15 max **Mn** 0.60 max **P** 0.035 max **S** 0.04 max **Other** Cu 0.20 min (when specified) | **ASTM** A109 (4,5) |
| **K01515** | Steel Electrode (EM15K) (EM15K-EW) | **C** 0.10-0.20 **Cu** 0.35 max **Mn** 0.80-1.25 **P** 0.030 max **S** 0.030 max **Si** 0.10-0.35 **Other** total other elements (except Fe) 0.50 max | **ASME** SFA5.17 (EM15K); SFA5.25 (EM15K-EW) **AWS** A5.17 (EM15K); A5.25 (EM15K-EW) **MIL SPEC** MIL-E-23765/1 (MIL-70S-1) |
| **K01520** | Steel Electrode | **C** 0.15 max **Mn** 0.30-0.60 **P** 0.04 max **S** 0.04 max **Si** 0.03 max | **AWS** A5.15 (ESt) |
| **K01600** | Carbon Steel | **C** 0.16 max **Mn** 0.90-1.50 **P** 0.04 max **S** 0.05 max **Si** 0.15-0.50 **Other** Cu 0.20 min (when specified) | **ASTM** A678 (A) |
| **K01601** | Carbon Steel | **C** 0.16 max **Mn** 1.00-1.35 **P** 0.04 max **S** 0.04 max **Si** 0.10-0.35 | **ASTM** A131 (CS, DS) |
| **K01602** | Carbon Steel | **C** 0.13-0.19 **Mn** 0.30-0.60 **P** 0.04 max **S** 0.05 max **Si** 0.10 max | **MIL SPEC** MIL-S-645 |
| **K01700** | Carbon Steel | **C** 0.17 max **Mn** 0.90 max **P** 0.035 max **S** 0.045 max | **ASTM** A285/A285M (A) |
| **K01701** | Carbon Steel | **C** 0.17 max **Mn** 0.84-1.46 **P** 0.035 max **S** 0.040 max **Si** 0.13-0.45 | **ASTM** A662 (A) |
| **K01702** | Carbon Steel | **C** 0.17 max **Mn** 0.90 max **P** 0.035 max **S** 0.04 max **Si** 0.40 max | **ASTM** A283 (B) |
| **K01800** | Carbon Steel | **C** 0.18 max **Mn** 0.55-0.98 **P** 0.035 max **S** 0.04 max **Si** 0.13-0.45 | **ASTM** A516 (55) |
| **K01801** | Carbon Steel | **C** 0.18 max **Mn** 0.70-1.35 **P** 0.04 max **S** 0.04 max **Si** 0.10-0.35 | **ASTM** A131 (E) |
| **K01802** | Carbon Steel | **C** 0.18 max **Cb** 0.05 max **Mn** 1.00-1.35 **P** 0.04 max **S** 0.05 max **Si** 0.15-0.50 | **ASTM** A633 (A) |
| **K01803** | Carbon Steel | **C** 0.18 max **Mn** 1.00-1.35 **P** 0.04 max **S** 0.05 max **Si** 0.15-0.50 **V** 0.10 max | **ASTM** A633 |
| * **K01804** | Carbon Steel | **C** 0.18 max **Mn** 0.90 max **P** 0.04 max **S** 0.05 max **Si** 0.10-0.30 | |
| **K01807** | Carbon Steel | **C** 0.18 max **Mn** 0.27-0.63 **P** 0.048 max **S** 0.058 max | **ASTM** A214; A822 |
| * **K01900** | Carbon Steel | **C** 0.11-0.27 **Mn** 0.27-0.93 **P** 0.048 max **S** 0.058 max **Other** Cu 0.18 min (when specified) | |
| **K01907** | Alloy Steel (C-110) | **C** 0.16-0.22 **Mn** 1.40-1.80 **P** 0.025 max **S** 0.025 max **Si** 0.25-0.45 **V** 0.07-0.18 | |
| **K02000** | Carbon Steel | **C** 0.15-0.25 **Mn** 0.30-0.60 **P** 0.045 max **S** 0.050 max | **ASTM** A730 (B) |
| **K02001** | Carbon Steel | **C** 0.20 max **Mn** 0.90 max **P** 0.04 max **S** 0.05 max **Si** 0.15-0.30 | **ASTM** A515 (55) |
| **K02002** | Carbon Steel | **C** 0.20 max **Mn** 0.70-1.35 **P** 0.04 max **S** 0.05 max **Si** 0.15-0.50 **Other** Cu 0.20 min (when specified) | **ASTM** A678 (B) |
| **K02003** | Steel, High Strength Low Alloy, replaced by K12037 | | |
| **K02004** | Carbon Steel | **C** 0.12-0.29 **Mn** 0.26-0.94 **P** 0.05 max **S** 0.06 max | **ASTM** A595 (A) |

*Boxed entries are no longer active and are retained for reference purposes only.

The chemical compositions listed are for identification purposes and should not be used in lieu of the cross referenced specifications.

| UNIFIED NUMBER | DESCRIPTION | CHEMICAL COMPOSITION | CROSS REFERENCE SPECIFICATIONS |
|---|---|---|---|
| **K02005** | Carbon Steel | **C** 0.12-0.29 **Mn** 0.35-1.40 **P** 0.05 max **S** 0.06 max | **ASTM** A595 (B) |
| **K02006** | Carbon Steel | **C** 0.20 max **Mn** 0.90-1.45 **P** 0.035 max **S** 0.040 max **Si** 0.15-0.35 | |
| **K02007** | Carbon Steel | **C** 0.20 max **Mn** 1.00-1.60 **P** 0.035 max **S** 0.040 max **Si** 0.15-0.50 | **ASTM** A662 (C) |
| **K02008** | Carbon Steel | **C** 0.20 max **Cr** 0.25 max **Cu** 0.35 max **Mn** 1.60 max **Mo** 0.09 max **Ni** 0.50 max **P** 0.025 max **S** 0.025 max **Si** 0.15-0.50 **V** 0.15 max | **ASTM** A738 (C) |
| **K02009** | Carbon Steel | **Al** 0.05 max **C** 0.20 max **Cr** 0.40 max **Cu** 0.35 max **Mn** 1.00-1.60 **Mo** 0.25 max **Ni** 0.50 max **P** 0.020 max **S** 0.020 max **Si** 0.15-0.50 **V** 0.06 max | **ASTM** A765 (IV) |
| **K02010** | Carbon Steel | **Al** 0.020 max **C** 0.20 max **Cb** 0.03 max **Cr** 0.25 max **Cu** 0.35 max **Mn** 0.70-1.35 **Mo** 0.08 max **Ni** 0.25 max **P** 0.030 max **S** 0.030 max **Si** 0.15-0.50 **V** 0.06 max | **ASTM** A841 |
| **K02011** | Carbon Steel | **Al** 0.06 max **C** 0.20 max **Cb** 0.05 max **Cr** 0.15 max **Cu** 0.15 max **Mn** 1.15-1.50 **Mo** 0.05 max **Ni** 0.25 max **P** 0.035 max **S** 0.015 max **Si** 0.10-0.35 **V** 0.12 max | **ASTM** A333 (10) |
| **K02100** | Carbon Steel | **C** 0.21 max **Mn** 0.55-0.98 **P** 0.035 max **S** 0.04 max **Si** 0.13-0.45 | **ASTM** A516 (60) |
| **K02101** | Carbon Steel | **C** 0.21 max **Mn** 0.70-1.35 **P** 0.04 max **S** 0.04 max **Si** 0.10-0.35 | **ASTM** A131 (D) |
| **K02102** | Carbon Steel | **C** 0.21 max **Mn** 0.80-1.10 **P** 0.04 max **S** 0.04 max **Si** 0.35 max | **ASTM** A131 (B) |
| **K02104** | Carbon Steel | **C** 0.21 max **Mn** 0.90-1.35 **P** 0.048 max **S** 0.058 max **Si** 0.10-0.40 | **ASTM** A524 (I, II) |
| **K02200** | Carbon Steel | **C** 0.22 max **Mn** 0.90 max **P** 0.035 max **S** 0.045 max | **ASTM** A285/A285M (B) |
| **K02201** | Carbon Steel | **C** 0.22 max **Mn** 0.90 max **P** 0.035 max **S** 0.040 max **Other** Cu 0.20 min (when specified) | **ASTM** A414 (B) |
| * **K02202** | Carbon Steel | **C** 0.22 max **Mn** 0.80-1.10 **P** 0.035 max **S** 0.04 max **Si** 0.13-0.40 | |
| **K02203** | Carbon Steel | **C** 0.22 max **Mn** 0.80-1.55 **P** 0.035 max **S** 0.040 max **Si** 0.13-0.33 | **ASTM** A662 (B) |
| **K02204** | Carbon Steel | **C** 0.22 max **Mn** 1.00-1.60 **P** 0.04 max **S** 0.05 max **Si** 0.20-0.50 **Other** Cu 0.20 min (when specified) | **ASTM** A678 (C) |
| **K02300** | Carbon Steel | **C** 0.23 max **P** 0.05 max **S** 0.05 max | **ASTM** A131 (A) |
| **K02301** | Carbon Steel | **C** 0.23 max **Mn** 0.60-0.90 **P** 0.04 max **S** 0.05 max **Si** 0.10-0.35 | **ASTM** A573 (58) |
| **K02302** | Carbon Steel | **C** 0.23 max **Mn** 1.60 max **P** 0.035 max **S** 0.040 max **Si** 0.37 max | **ASTM** A707 (L1) |
| **K02303** | Carbon Steel | **C** 0.23 max **Cb** 0.005-0.05 **Cu** 0.20 min **Mn** 1.35 max **P** 0.04 max **S** 0.05 max **Si** 0.40 max | **ASTM** A572 (50-1) |
| **K02304** | Carbon Steel | **C** 0.23 max **Cu** 0.20 min **Mn** 1.35 max **P** 0.04 max **S** 0.05 max **Si** 0.40 max **V** 0.01-0.15 | **ASTM** A572 (50-2) |

*Boxed entries are no longer active and are retained for reference purposes only.

The chemical compositions listed are for identification purposes and should not be used in lieu of the cross referenced specifications.

| UNIFIED NUMBER | DESCRIPTION | CHEMICAL COMPOSITION | CROSS REFERENCE SPECIFICATIONS |
|---|---|---|---|
| **K02305** | Carbon Steel | **C** 0.23 max **Cb** 0.05 max **Cu** 0.20 min **Mn** 1.35 max **P** 0.04 max **S** 0.05 max **Si** 0.40 max **Other** Cb+V 0.02-0.15 | **ASTM** A572 (50-3) |
| **K02306** | Carbon Steel | **C** 0.23 max **Cu** 0.20 min **Mn** 1.35 max **N** 0.015 max **P** 0.04 max **S** 0.05 max **Si** 0.40 max **V** 4xN min | **ASTM** A572 (50-4) |
| **K02400** | Carbon Steel | **C** 0.24 max **Cr** 0.25 max **Cu** 0.35 max **Mn** 0.65-1.40 **Mo** 0.08 max **Ni** 0.25 max **P** 0.035 max **S** 0.040 max **Si** 0.13-0.55 | |
| **K02401** | Carbon Steel | **C** 0.24 max **Mn** 0.90 max **P** 0.04 max **S** 0.05 max **Si** 0.15-0.30 | **ASTM** A283 (C); A515 (60) |
| * **K02402** | Carbon Steel | **C** 0.24 max **Mn** 0.80-1.10 **P** 0.35 max **S** 0.04 max **Si** 0.15-0.40 | |
| **K02403** | Carbon Steel | **C** 0.24 max **Mn** 0.79-1.30 **P** 0.035 max **S** 0.04 max **Si** 0.13-0.45 | **ASTM** A516 (65) |
| **K02404** | Carbon Steel | **C** 0.24 max **Mn** 0.85-1.20 **P** 0.04 max **S** 0.05 max **Si** 0.15-0.40 | **ASTM** A573 (65) |
| * **K02405** | Carbon Steel | **C** 0.16-0.33 **Mn** 1.14-1.71 **P** 0.048 max **S** 0.058 max **Si** 0.08-0.32 **Other** Cu 0.18 min (when specified) | |
| **K02500** | Carbon Steel | **C** 0.25 max **Mn** 0.60 max **P** 0.035 max **S** 0.04 max **Other** Cu 0.20 min (when specified) | **ASTM** A109 (1, 2, 3) |
| **K02501** | Carbon Steel | **C** 0.25 max **Mn** 0.27-0.93 **P** 0.048 max **S** 0.058 max **Si** 0.10 min | **ASTM** A369 (FPA) |
| **K02502** | Carbon Steel | **C** 0.25 max **Mn** 0.90 max **P** 0.04 max **S** 0.05 max **Other** Cu 0.20 min (when specified) | **ASTM** A570 (30, 33, 36, 40) |
| **K02503** | Carbon Steel | **C** 0.25 max **Mn** 0.90 max **P** 0.035 max **S** 0.040 max **Other** Cu 0.20 min (when specified) | **ASTM** A414 (C) |
| **K02504** | Carbon Steel | **C** 0.25 max **Mn** 0.95 max **P** 0.05 max **S** 0.06 max | **ASTM** A523 (A) |
| **K02505** | Carbon Steel | **C** 0.25 max **Mn** 1.20 max **P** 0.035 max **S** 0.040 max **Other** Cu 0.20 min (when specified) | **ASTM** A414 (D) |
| * **K02506** | Carbon Steel | **C** 0.25 max **Mn** 0.90-1.35 **P** 0.035 max **S** 0.025 max **Si** 0.15-0.30 | |
| **K02507** | Carbon Steel | **C** 0.25 max **Mn** 1.35 max **P** 0.040 max **S** 0.05 max **Other** Cu 0.20 min (when specified) | **ASTM** A570 (45, 50, 55); A857 |
| **K02508** | Carbon Steel | **C** 0.25 max **Mn** 1.00 max **P** 0.040 max **S** 0.050 max | **AMS** 5062; 7496 |
| **K02509** | Carbon Steel | **C** 0.25 max **Mn** 0.95 max **P** 0.035 max **S** 0.035 max **ASTM** A135 (A) | |
| **K02595** | Carbon Steel Structural Plate (greater than 4 in.) | **C** 0.29 max **Fe** rem **Mn** 0.85-1.20 **P** 0.04 max **S** 0.05 max | **ASTM** A36 |
| **K02596** | Carbon Steel Structural Plate (greater than 2.5 in., less than or equal to 4 in.) | **C** 0.27 max **Fe** rem **Mn** 0.85-1.20 **P** 0.04 max **S** 0.05 max | **ASTM** A36 |

*Boxed entries are no longer active and are retained for reference purposes only.

The chemical compositions listed are for identification purposes and should not be used in lieu of the cross referenced specifications.

| UNIFIED NUMBER | DESCRIPTION | CHEMICAL COMPOSITION | CROSS REFERENCE SPECIFICATIONS |
|---|---|---|---|
| **K02597** | Carbon Steel Structural Plate (greater than 1.5 in., less than or equal to 2.5 in.) | **C** 0.26 max **Fe** rem **Mn** 0.80-1.20 **P** 0.04 max **S** 0.05 max | **ASTM** A36 |
| **K02598** | Carbon Steel Structural Plate (greater than 0.75 in., less than or equal to 1.5 in.) | **C** 0.25 max **Fe** rem **Mn** 0.80-1.20 **P** 0.04 max **S** 0.05 max | **ASTM** A36 |
| **K02599** | Carbon Steel Structural Plate (to 0.75 in.) | **C** 0.25 max **Fe** rem **P** 0.04 max **S** 0.05 max | **ASTM** A36 |
| **K02600** | Carbon Steel | **C** 0.26 max **P** 0.04 max **S** 0.05 max **Other** Cu 0.20 min (when Copper Steel is specified) | **ASTM** A36 (SHAPES) |
| **K02601** | Carbon Steel | **C** 0.26 max **Mn** 1.30 max **S** 0.063 max | **ASTM** A381; A618 (Ib) |
| **K02603** | Carbon Steel | **C** 0.26 max **Mn** 0.60-0.90 **P** 0.035 max **S** 0.040 max | |
| **K02700** | Carbon Steel | **C** 0.27 max **Mn** 0.79-1.30 **P** 0.035 max **S** 0.04 max **Si** 0.13-0.45 | **ASTM** A516 (70) |
| **K02701** | Carbon Steel | **C** 0.27 max **Mn** 0.85-1.20 **P** 0.04 max **S** 0.05 max **Si** 0.15-0.40 | **ASTM** A573 (70) |
| **K02702** | Carbon Steel | **C** 0.27 max **Mn** 0.90 max **P** 0.04 max **S** 0.05 max **Si** 0.15-0.40 | **ASTM** A283 (D) |
| **K02703** | Carbon Steel | **C** 0.27 max **Mn** 1.20 max **P** 0.04 max **S** 0.05 max **Other** Cu 0.20 min (when specified) | **ASTM** A529 |
| **K02704** | Carbon Steel | **C** 0.27 max **Mn** 1.20 max **P** 0.035 max **S** 0.040 max **Other** Cu 0.20 min (when specified) | **ASTM** A414 (E) |
| **K02705** | Carbon Steel | **C** 0.27 max **Mn** 1.40 max **P** 0.05 max **S** 0.063 max **Other** Cu 0.18 min (when specified) | **ASTM** A500 (C) |
| **K02706** | Carbon Steel | **C** 0.27 min **Mn** 0.47 min **P** 0.048 max **S** 0.058 max | **ASTM** A325 (1) |
| **K02707** | Carbon Steel | **C** 0.27 max **Mn** 0.93 max **P** 0.048 max **S** 0.058 max **Si** 0.10 min | **ASTM** A210 (A-1); A556 (B2) |
| * **K02708** | Carbon Steel | **C** 0.24-0.30 **Mn** 0.60-0.90 **P** 0.04 max **S** 0.05 max **Si** 0.15-0.30 | |
| **K02709** | Carbon Steel | **C** 0.27 max **Mn** 1.00-1.50 **P** 0.030 max **S** 0.015 max **Si** 0.10 min | **ASTM** A178 (D) |
| **K02741** | Carbon Steel | **C** 0.27 max **Cr** 0.25 max **Cu** 0.35 max **Mn** 0.85-1.20 **Mo** 0.08 max **Ni** 0.25 max **P** 0.035 max **Pb** 0.05 max **S** 0.035 max **Si** 0.15-0.30 **V** 0.05 max | **ASTM** A758 |
| **K02800** | Carbon Steel | **C** 0.28 max **Mn** 0.90 max **P** 0.035 max **S** 0.04 max **Si** 0.13-0.45 | **ASTM** A515 (65) |
| **K02801** | Carbon Steel | **C** 0.28 max **Mn** 0.90 max **P** 0.035 max **S** 0.040 max | **ASTM** A285 (C) |
| **K02802** | Carbon Steel | **C** 0.28 max **Mn** 0.81-1.25 **P** 0.040 max **S** 0.050 max **Si** 0.13-0.33 | |
| **K02803** | Carbon Steel | **C** 0.28 max **Mn** 0.84-1.52 **P** 0.035 max **S** 0.040 max **Si** 0.13-0.45 | **ASTM** A299 |
| **K02900** | Carbon Steel | **C** 0.29 max **Mn** 0.92-1.46 **P** 0.035 max **S** 0.040 max **Si** 0.13-0.45 | **ASTM** A612 |
| **K02901** | Carbon Steel | **C** 0.26-0.33 **Mn** 0.60-0.90 **P** 0.040 max **S** 0.040 max **Si** 0.10 max | **MIL SPEC** MIL-S-3289 |

*Boxed entries are no longer active and are retained for reference purposes only.

The chemical compositions listed are for identification purposes and should not be used in lieu of the cross referenced specifications.

| UNIFIED NUMBER | DESCRIPTION | CHEMICAL COMPOSITION | CROSS REFERENCE SPECIFICATIONS |
|---|---|---|---|
| **K03000** | Carbon Steel | **C** 0.30 max **P** 0.05 max **S** 0.063 max **Other** Cu 0.18 min (when Copper Steel is specified) | **ASTM** A500 (A, B, D); A501 |
| **K03002** | Carbon Steel | **C** 0.30 max **Mn** 1.00 max **P** 0.04 max **S** 0.05 max **Si** 0.50 max | **ASME** SA-307 (B, C) **ASTM** A372 (A) |
| **K03003** | Carbon Steel | **C** 0.30 max **Mn** 1.00 max **P** 0.040 max **S** 0.050 max | **ASTM** A139 (B) |
| **K03004** | Carbon Steel | **C** 0.30 max **Mn** 1.20 max **P** 0.040 max **S** 0.050 max | **ASTM** A139 (C) |
| **K03005** | Carbon Steel | **C** 0.30 max **Mn** 1.20 max **P** 0.05 max **S** 0.06 max | **ASTM** A53 (E-B, S-B); A523 (B) |
| **K03006** | Carbon Steel | **C** 0.30 max **Mn** 0.29-1.06 **P** 0.048 max **S** 0.058 max **Si** 0.10 min | **ASME** SA-420 (WPL6) **ASTM** A106 (B); A234 (WPB); A333 (6); A334 (6); A369 (FPB); A420 (WPL6); A556 (C2) |
| * **K03007** | Carbon Steel | **C** 0.30 max **Mn** 0.27-0.93 **P** 0.050 max **S** 0.060 max | |
| **K03008** | Carbon Steel | **C** 0.30 max **Mn** 0.40-1.06 **P** 0.05 max **S** 0.06 max | **ASTM** A333 (1); A334 (1) |
| **K03009** | Carbon Steel | **C** 0.30 max **Mn** 1.35 max **P** 0.035 max **S** 0.040 max **Si** 0.15-0.30 | **ASTM** A350 (LF1) |
| **K03010** | Carbon Steel | **C** 0.30 max **Mn** 1.30 max **P** 0.040 max **S** 0.050 max | **ASTM** A139 (D) |
| **K03011** | Carbon Steel | **C** 0.30 max **Mn** 1.35 max **P** 0.035 max **S** 0.040 max **Si** 0.15-0.30 | **ASTM** A350 (LF2); A706 |
| **K03012** | Carbon Steel | **C** 0.30 max **Mn** 1.40 max **P** 0.040 max **S** 0.050 max | **ASTM** A139 (E) |
| * **K03013** | Carbon Steel | **C** 0.30 max **Mn** 1.50 max **P** 0.050 max **S** 0.060 max | |
| **K03014** | Carbon Steel | **C** 0.30 max **Mn** 1.50 max **P** 0.050 max **S** 0.060 max **Si** 0.13-0.37 | **ASTM** A694 |
| **K03015** | Carbon Steel | **C** 0.30 min **P** 0.04 max **S** 0.06 max | **ASTM** A183 |
| **K03016** | Carbon Steel | **C** 0.30 max **Mn** 0.60-0.90 **P** 0.04 max **S** 0.05 max **Si** 0.15-0.30 | |
| **K03017** | Carbon Steel | **C** 0.30 max **Mn** 0.80-1.35 **P** 0.035 max **S** 0.040 max **Si** 0.15-0.35 | **ASTM** A266 (4) |
| **K03018** | Carbon Steel | **C** 0.30 max **Mn** 1.20 max **P** 0.035 max **S** 0.035 max | **ASTM** A135 (B) |
| **K03019** | Carbon Steel | **C** 0.30 max **Mn** 1.00 max **S** 0.15 max **Si** 0.50 max **ASTM** A307 (A) | |
| **K03020** | Alloy Steel | **C** 0.30 max **Cr** 0.25 max **Mn** 0.70-1.35 **Mo** 0.10 max **Ni** 0.40 max **P** 0.025 max **S** 0.025 max **Si** 0.15-0.40 **V** 0.05 max **ASTM** A541 (1A) | |
| **K03046** | Carbon Steel | **Al** 0.05 max **C** 0.30 max **Cr** 0.40 max **Cu** 0.35 max **Mn** 0.60-1.05 **Mo** 0.25 max **Ni** 0.50 max **P** 0.020 max **S** 0.020 max **Si** 0.15-0.35 **V** 0.05 max | **ASTM** A765 (I) |
| **K03047** | Carbon Steel | **Al** 0.05 max **C** 0.30 max **Cr** 0.40 max **Cu** 0.35 max **Mn** 0.60-1.35 **Mo** 0.25 max **Ni** 0.50 max **P** 0.020 max **S** 0.020 max **Si** 0.15-0.35 **V** 0.05 max | **ASTM** A765 (II) |
| **K03100** | Carbon Steel | **C** 0.31 max **Mn** 0.27-0.83 **P** 0.048 max **S** 0.058 max | **ASTM** A31 (B) |

*Boxed entries are no longer active and are retained for reference purposes only.

The chemical compositions listed are for identification purposes and should not be used in lieu of the cross referenced specifications.

| UNIFIED NUMBER | DESCRIPTION | CHEMICAL COMPOSITION | CROSS REFERENCE SPECIFICATIONS |
|---|---|---|---|
| **K03101** | Carbon Steel | **C** 0.31 max **Mn** 0.90 max **P** 0.035 max **S** 0.04 max **Si** 0.13-0.33 | **ASTM** A515 (70) |
| **K03102** | Carbon Steel | **C** 0.31 max **Mn** 1.20 max **P** 0.035 max **S** 0.040 max **Other** Cu 0.20 min (when specified) | **ASTM** A414 (F) |
| **K03103** | Carbon Steel | **C** 0.31 max **Mn** 1.35 max **P** 0.035 max **S** 0.040 max **Other** Cu 0.20 min (when specified) | **ASTM** A414 (G) |
| **K03104** | Carbon Steel | **C** 0.31 min **P** 0.045 max **S** 0.040 max | **ASTM** A574 |
| **K03200** | Carbon Steel | **C** 0.32 max **Mn** 1.04 max **P** 0.035 max **S** 0.045 max **Si** 0.15-0.30 | **ASTM** A696 (B, C) |
| **K03201** | Carbon Steel | **C** 0.25-0.40 **Mn** 1.15-1.55 **P** 0.040 max **S** 0.08-0.13 **Si** 0.10-0.20 | |
| **K03300** | Carbon Steel | **C** 0.33 max **Mn** 0.81-1.25 **P** 0.040 max **S** 0.050 max **Si** 0.13 max | **ASTM** A455 (I) |
| **K03301** | Carbon Steel | **C** 0.33 max **Mn** 1.45 max **P** 0.035 max **S** 0.040 max **Si** 0.37 max | **ASTM** A707 (L2) |
| **K03500** | Carbon Steel | **C** 0.35 max **P** 0.040 max **S** 0.045 max | |
| **K03501** | Carbon Steel | **C** 0.35 max **Mn** 0.29-1.06 **P** 0.048 max **S** 0.058 max **Si** 0.10 min | **ASTM** A210 (C); A234 (WPC) |
| **K03502** | Carbon Steel | **C** 0.35 max **Mn** 1.10 max **P** 0.05 max **S** 0.05 max | **ASTM** A181 |
| **K03503** | Carbon Steel | **C** 0.35 max **Mn** 0.80 max **P** 0.050 max **S** 0.060 max | **ASTM** A178 (C) |
| * **K03504** | Carbon Steel | **C** 0.35 max **Mn** 0.60-1.05 **P** 0.040 max **S** 0.050 max **Si** 0.35 max | **ASTM** A105; A695 (B) |
| * **K03505** | Carbon Steel | **C** 0.35 max **Mn** 0.27-1.06 **P** 0.050 max **S** 0.060 max | |
| **K03506** | Carbon Steel | **C** 0.35 max **Mn** 0.40-1.05 **P** 0.040 max **S** 0.040 max **Si** 0.15-0.35 | **ASTM** A266 (1, 2); A541 (1) |
| **K03700** | Carbon Steel | **C** 0.37 max **P** 0.048 max **S** 0.058 max | **ASTM** A413 |
| **K03800** | Carbon Steel | **C** 0.18-0.58 **Mn** 0.57 min **P** 0.048 max **S** 0.058 max | **ASTM** A563 (DH) |
| * **K03810** | Carbon Steel | **B** 0.0005-0.005 **C** 0.38-0.43 **Cr** 0.30-0.55 **Cu** 0.35 max **Mn** 0.75-1.00 **Mo** 0.08-0.15 **Ni** 0.20-0.40 **P** 0.025 max **S** 0.025 max **Si** 0.15-0.35 | |
| **K03900** | Carbon Steel | **C** 0.28-0.50 **P** 0.045 max **S** 0.045 max | **ASTM** A490 (3); F385; F912 |
| **K04000** | Carbon Steel | **C** 0.35-0.45 **Cr** 0.25 max **Cu** 0.35 **Mn** 0.60-0.90 **Mo** 0.10 max **Ni** 0.30 max **P** 0.040 max **S** 0.040 max **Si** 0.15-0.30 **V** 0.06 max | **ASTM** A290 (A, B) |
| **K04001** | Carbon Steel | **C** 0.35 max **Mn** 1.35 max **P** 0.04 max **S** 0.05 max **Si** 0.15-0.35 | **ASTM** A372 (B) |
| **K04002** | Carbon Steel | **C** 0.40 min **P** 0.040 max **S** 0.050 max | **ASTM** A194 (2, 2HM, 2H) |
| **K04100** | Carbon Steel | **C** 0.28-0.55 **P** 0.040 max **S** 0.045 max | **ASTM** A354 |

*Boxed entries are no longer active and are retained for reference purposes only.

The chemical compositions listed are for identification purposes and should not be used in lieu of the cross referenced specifications.

| UNIFIED NUMBER | DESCRIPTION | CHEMICAL COMPOSITION | CROSS REFERENCE SPECIFICATIONS |
|---|---|---|---|
| **K04200** | Carbon Steel | **C** 0.25-0.58 **Mn** 0.57 min **P** 0.048 max **S** 0.058 max | **ASTM** A449 (1) |
| **K04500** | Carbon Steel | **C** 0.40-0.50 **Cr** 0.25 max **Cu** 0.35 **Mn** 0.60-0.90 **Mo** 0.10 max **Ni** 0.30 max **P** 0.040 max **S** 0.040 max **Si** 0.15-0.35 **V** 0.06 max | **ASTM** A290 (C, D) |
| **K04600** | Carbon Steel | **C** 0.43-0.50 **Mn** 0.45-0.75 **P** 0.030 max **S** 0.035 max **Si** 0.15-0.35 | **MIL SPEC** MIL-S-3039 |
| **K04700** | Carbon Steel | **C** 0.40-0.55 **Mn** 0.60-0.90 **P** 0.045 max **S** 0.050 max **Si** 0.15 min | **ASTM** A21 (U); A730 (C, D, E) |
| **K04701** | Carbon Steel | **C** 0.35-0.60 **Mn** 1.00 max **P** 0.04 max | **ASTM** A49 |
| **K04800** | Carbon Steel | **C** 0.48 max **Mn** 1.00 max **P** 0.040 max **S** 0.050 max **Si** 0.15-0.35 | **ASTM** A489 |
| **K04801** | Carbon Steel | **C** 0.48 max **Mn** 1.65 max **P** 0.04 max **S** 0.05 max **Si** 0.15-0.35 | **ASTM** A372 (C) |
| **K05000** | Carbon Steel | **C** 0.50 max **Cu** 0.35 **Mn** 0.40-0.90 **P** 0.040 max **S** 0.040 max **Si** 0.15-0.35 **V** 0.10 max | **ASTM** A291 (2) |
| **K05001** | Carbon Steel | **C** 0.50 max **Mn** 0.50-0.90 **P** 0.040 max **S** 0.040 max **Si** 0.35 max | **ASTM** A266 (3); A649 (2) |
| **K05002** | Carbon Steel | **C** 0.50 max **Mn** 0.60-1.00 **P** 0.025 max **S** 0.025 max **Si** 0.15-0.30 | **ASTM** A288 (1) |
| **K05003** | Carbon Steel | **C** 0.50 max **P** 0.05 max **S** 0.05 max | |
| **K05200** | Carbon Steel | **C** 0.45-0.59 **Mn** 0.60-0.90 **P** 0.045 max **S** 0.050 max **Si** 0.15 min | **ASTM** A21 (F, G, H); A730 (F) |
| **K05500** | Carbon Steel | **C** 0.55 max **Cr** 0.25 **Cu** 0.35 **Mn** 0.60-0.90 **Mo** 0.10 max **Ni** 0.30 max **P** 0.040 max **S** 0.040 max **Si** 0.15-0.35 **V** 0.06 | **ASTM** A291 (1) |
| **K05501** | Carbon Steel | **C** 0.55 max **Mn** 0.60-0.90 **P** 0.040 max **S** 0.050 max **Si** 0.15-0.35 **Other** Pb 0.15-0.35 (when specified) | **ASTM** A321 |
| **K05502** | Carbon Steel | **C** 0.50-0.60 **Cr** 0.30 max **Mn** 0.90-1.50 **Mo** 0.15 max **Ni** 0.60 max **P** 0.025 max **S** 0.025 max **Si** 0.15-0.35 | **ASTM** A649 (Class 5) |
| **K05700** | Carbon Steel | **C** 0.57 max **Mn** 0.60-0.85 **P** 0.05 max **S** 0.05 max **Si** 0.15 min | **ASTM** A504 (A, L) |
| **K05701** | Carbon Steel | **C** 0.50-0.65 **Mn** 0.60-0.90 **P** 0.050 max **S** 0.050 max **Si** 0.15-0.35 | **ASTM** A551 (A, AHT) |
| **K05800** | Carbon Steel | **C** 0.35-0.82 **P** 0.050 max **Other** Cu 0.20 min (when specified) | |
| **K05801** | Carbon Steel | **C** 0.58 max **Mn** 0.27 min **P** 0.048 max **S** 0.058 max | **ASTM** A563 (D) |
| **K05802** | Carbon Steel | **C** 0.58 max **P** 0.13 max **S** 0.15 max | **ASTM** A563 (O, A, B, C) |
| **K06000** | Carbon Steel | **C** 0.45-0.75 **Mn** 0.60-1.10 **P** 0.040 max **S** 0.050 max **Si** 0.10-0.35 | **AMS** AMS-W-21425 (II) |
| **K06001** | Carbon Steel | **C** 0.60 max **Mn** 1.30-1.70 **P** 0.045 max **S** 0.050 max **Si** 0.15 min | **ASTM** A729 |
| **K06002** | Carbon Steel | **C** 0.35-0.85 **P** 0.05 max **Other** Cu 0.20 min (when specified) | **ASTM** A67 (2) |
| **K06100** | Carbon Steel | **C** 0.55-0.68 **Mn** 0.60-0.90 **P** 0.04 max **S** 0.05 **Si** 0.10-0.25 | **ASTM** A1 (61-80) |
| **K06200** | Carbon Steel | **C** 0.57-0.67 **Mn** 0.60-0.85 **P** 0.05 max **S** 0.05 max **Si** 0.15 min | **ASTM** A504 (B) |

The chemical compositions listed are for identification purposes and should not be used in lieu of the cross referenced specifications.

| UNIFIED NUMBER | DESCRIPTION | CHEMICAL COMPOSITION | CROSS REFERENCE SPECIFICATIONS |
|---|---|---|---|
| **K06201** | Carbon Steel | **C** 0.50-0.75 **Mn** 0.60-1.20 **P** 0.040 max **S** 0.050 max | **ASTM** A417 |
| **K06500** | Carbon Steel | **C** 0.55-0.75 **Mn** 0.60-1.20 **P** 0.04 max **S** 0.05 max **Si** 0.10-0.30 | **AMS** AMS-W-2145 (I, II) |
| **K06501** | Carbon Steel | **C** 0.45-0.85 **Mn** 0.30-1.30 **P** 0.040 max **S** 0.050 max **Si** 0.10-0.35 | **ASTM** A227 |
| **K06700** | Carbon Steel | **C** 0.50-0.85 **Mn** 0.60-1.10 **P** 0.040 max **S** 0.050 max **Si** 0.10-0.35 | **ASTM** A648 (II); A821 |
| **K06701** | Carbon Steel | **C** 0.60-0.75 **Mn** 0.60-0.90 **P** 0.025 max **S** 0.030 max **Si** 0.15-0.35 | **ASTM** A230 |
| **K06702** | Carbon Steel | **C** 0.60-0.75 **Mn** 0.60-0.90 **P** 0.050 max **S** 0.050 max **Si** 0.15-0.35 | **ASTM** A551 (B, BHT) |
| **K06703** | Carbon Steel | **C** 0.60-0.75 **Mn** 0.60-0.90 **P** 0.04 max **Si** 0.10-0.40 | **ASTM** A2 (A) |
| **K07000** | Carbon Steel | **C** 0.64-0.77 **Mn** 0.60-0.90 **P** 0.04 max **S** 0.05 **Si** 0.10-0.25 | **ASTM** A1 (81-89) |
| **K07001** | Carbon Steel | **C** 0.55-0.85 **Mn** 0.60-1.20 **P** 0.040 max **S** 0.050 max **Si** 0.10-0.35 | **AMS** AMS-QQ-W-428 **ASTM** A229 |
| **K07100** | Carbon Steel | **C** 0.55-0.88 **Mn** 0.60-1.10 **P** 0.040 max **S** 0.050 max **Si** 0.10-0.35 | **ASTM** A648 (III) |
| **K07200** | Carbon Steel | **C** 0.65-0.80 **Mn** 0.60-0.85 **P** 0.05 max **S** 0.05 max **Si** 0.15 min | **ASTM** A504 (U) |
| **K07201** | Carbon Steel | **C** 0.67-0.77 **Mn** 0.60-0.85 **P** 0.05 max **S** 0.05 max **Si** 0.15 min | **ASTM** A504 (C) |
| **K07301** | Carbon Steel | **C** 0.67-0.80 **Mn** 0.70-1.00 **P** 0.04 max **S** 0.05 **Si** 0.10-0.25 | **ASTM** A1 (90-114) |
| **K07500** | Carbon Steel | **C** 0.69-0.82 **Mn** 0.70-1.00 **P** 0.04 max **S** 0.05 max **Si** 0.10-0.25 | **ASTM** A1 (115 and over); A759 |
| **K07700** | Carbon Steel | **C** 0.70-0.85 **Mn** 0.60-0.90 **P** 0.04 max **Si** 0.10-0.40 | **ASTM** A2 (B) |
| **K07701** | Carbon Steel | **C** 0.70-0.85 **Mn** 0.60-0.90 **P** 0.050 max **S** 0.050 max **Si** 0.15-0.35 | **ASTM** A551 (C, CHT, DHT) |
| **K08200** | Carbon Steel | **C** 0.65-1.00 **Mn** 0.20-1.30 **P** 0.040 max **S** 0.050 max **Si** 0.10-0.40 | **ASTM** A679 |
| **K08201** | Carbon Steel | **C** 0.75-0.90 **Mn** 0.60-0.90 **P** 0.04 max **Si** 0.10-0.40 | **ASTM** A2 (C) |
| **K08500** | Carbon Steel | **C** 0.70-1.00 **Mn** 0.20-0.60 **P** 0.025 max **S** 0.03 max **Si** 0.12-0.30 | **AMS** 5112 **ASTM** A228 |
| **K08501** | Carbon Steel Wire, Spring Quality Music Wire | **C** 0.70-1.00 **Mn** 0.20-0.60 **P** 0.025 max **Si** 0.10-0.30 **S** 0.030 max | **ASME** SA-905 **ASTM** A228; A905 **AMS** 5112 |
| **K08700** | Carbon Steel | **C** 0.85-0.90 **Cr** 0.04 max **Cu** 0.08 max **Mn** 0.25-0.40 **Mo** 0.04 max **Ni** 0.03 max **P** 0.025 max **S** 0.020 max **Si** 0.15-0.25 | **MIL SPEC** MIL-W-8957 |
| **K10614** | Alloy Steel | **C** 0.06 max **Cb** 0.03-0.09 **Mn** 1.20-1.90 **Mo** 0.25-0.35 **P** 0.04 max **S** 0.025 max **Si** 0.40 max | |
| **K10623** | Alloy Steel | **C** 0.06 max **Cb** 0.03-0.09 **Cu** 0.20-0.35 **Mn** 1.20-1.90 **Mo** 0.23-0.47 **P** 0.04 max **S** 0.025 max **Si** 0.40 max | **ASTM** A735 |
| **K10726** | Steel Electrode | **Al** 0.05-0.15 **C** 0.07 max **Cu** 0.50 max **Mn** 0.90-1.40 **P** 0.025 max **S** 0.035 max **Si** 0.40-0.70 **Ti** 0.05-0.15 **Zr** 0.02-0.12 | **ASME** SFA5.18 (ER70S-2); SFA5.25 (EM5K-EW); SFA5.26 (EGXXS-2); SFA5.30 (INMs1) **AWS** A5.18 (ER70S-2); A5.25 (EM5K-EW); A5.26 (EGXXS-2); A5.30 (INMs1) **MIL SPEC** MIL-I-23413 (MIL-MS-1); MIL-E-23765/1 (MIL-70S-2) |

The chemical compositions listed are for identification purposes and should not be used in lieu of the cross referenced specifications.

| UNIFIED NUMBER | DESCRIPTION | CHEMICAL COMPOSITION | CROSS REFERENCE SPECIFICATIONS |
|---|---|---|---|
| **K10882** | Alloy Steel Electrode (ER100S-1) | **Al** 0.10 max **C** 0.08 max **Cr** 0.30 max **Cu** 0.25 max **Mn** 1.25-1.80 **Mo** 0.25-0.55 **Ni** 1.40-2.10 **P** 0.010 max **S** 0.010 max **Si** 0.20-0.50 **Ti** 0.10 max **V** 0.05 max **Zr** 0.10 max **Other** total other elements (except Fe) 0.50 max | **ASME** SFA5.23 (EM2); SFA5.28 (ER100S-1) **AWS** A5.23 (EM2); A5.28 (ER100S-1) **MIL SPEC** MIL-E-23765/2 (MIL-100S-1) |
| **K10940** | Alloy Steel Electrode (ENi1N) | **C** 0.12 max **Cr** 0.15 max **Cu** 0.08 max **Mn** 0.75-1.25 **Mo** 0.30 max **Ni** 0.75-1.25 **P** 0.012 max **S** 0.010 max **Si** 0.05-0.30 **V** 0.05 max **Other** total other elements (except Fe) 0.50 max | **ASME** SFA5.23 (ENi1N) **AWS** A5.23 (ENi1N) |
| **K10943** | Alloy Steel Electrode (EB1N) | **C** 0.10 max **Cr** 0.40-0.75 **Cu** 0.08 max **Mn** 0.40-0.80 **Mo** 0.45-0.65 **P** 0.012 max **S** 0.025 max **Si** 0.05-0.30 **V** 0.05 max | **ASME** SFA5.23 (EB1N) **AWS** A5.23 (EB1N) |
| **K10945** | Alloy Steel Electrode | **C** 0.07-0.12 **Cu** 0.15 max **Mn** 1.60-2.10 **Mo** 0.40-0.60 **P** 0.03 max **S** 0.035 max **Si** 0.50-0.80 **Other** total other elements 0.50 max | **ASME** SFA5.25 (EA3K-EW); SFA5.26 (EGXXS-D2); SFA5.28 (ER80S-D2) **AWS** A5.25 (EA3K-EW); A5.26 (EGXXS-D2); A5.28 (ER80S-D2, ER90S-D2) |
| **K10958** | Alloy Steel Electrode (ENi1KN) | **C** 0.12 max **Cu** 0.08 max **Mn** 0.80-1.40 **Ni** 0.75-1.25 **P** 0.012 max **S** 0.020 max **Si** 0.40-0.80 **V** 0.05 max | **ASME** SFA5.23 (ENi1KN) **AWS** A5.23 (ENi1KN) |
| **K10982** | Alloy Steel Electrode (EM2N) | **Al** 0.10 max **C** 0.10 max **Cr** 0.30 max **Cu** 0.08 max **Mn** 1.25-1.80 **Mo** 0.25-0.55 **Ni** 1.40-2.10 **P** 0.012 max **S** 0.010 max **Si** 0.20-0.60 **Ti** 0.10 max **V** 0.05 max **Zr** 0.10 max **Other** total other elements (except Fe) 0.50 max | **ASME** SFA5.23 (EM2N) |
| **K11022** | Steel Electrode | **C** 0.06-0.15 **Cu** 0.50 max **Mn** 0.90-1.40 **P** 0.025 max **S** 0.035 max **Si** 0.45-0.70 | **ASME** SFA5.18 (ER70S-3); SFA5.26 (EGXXS-3) **AWS** A5.18 (ER70S-3); A5.26 (EGXXS-3) **MIL SPEC** MIL-E-23765/1 (MIL-70S-3) |
| **K11040** | Alloy Steel Electrode (ENil) | **C** 0.12 max **Cr** 0.15 max **Cu** 0.35 max **Mn** 0.75-1.25 **Mo** 0.30 max **Ni** 0.75-1.25 **P** 0.020 max **S** 0.020 max **Si** 0.05-0.30 **Other** total other elements (except Fe) 0.50 max | **ASME** SFA5.23 (ENil) **AWS** A5.23 (ENil) |
| **K11043** | Alloy Steel Electrode (EB1) | **C** 0.10 max **Cr** 0.40-0.75 **Cu** 0.35 max **Mn** 0.40-0.80 **Mo** 0.45-0.65 **P** 0.025 max **S** 0.025 max **Si** 0.05-0.30 | **ASME** SFA5.23 (EB1) **AWS** A5.23 (EB1) |
| **K11058** | Alloy Steel Electrode (ENi1K) | **C** 0.18 max **Cu** 0.35 max **Mn** 0.80-1.40 **Ni** 0.75-1.25 **P** 0.020 max **S** 0.020 max **Si** 0.40-0.80 | **ASME** SFA5.23 (ENi1K) **AWS** A5.23 (ENi1K) |
| **K11060** | Alloy Steel Electrode (EF1N) | **C** 0.07-0.15 **Cu** 0.08 max **Mn** 0.90-1.70 **Mo** 0.25-0.55 **Ni** 0.95-1.60 **P** 0.012 max **S** 0.025 max **Si** 0.15-0.35 **V** 0.05 max **Other** total other elements (except Fe) 0.50 max | **ASME** SFA5.23 (EF1N) **AWS** A5.23 (EF1N) |
| **K11072** | Alloy Steel Electrode (EB2N) | **C** 0.07-0.15 **Cr** 1.00-1.75 **Cu** 0.08 max **Mn** 0.45-1.00 **Mo** 0.45-0.65 **P** 0.012 max **S** 0.030 max **Si** 0.05-0.30 **V** 0.05 max **Other** total other elements (except Fe) 0.50 max | **ASME** SFA5.23 (EB2N) **AWS** A5.23 (EB2N) |
| **K11122** | Alloy Steel Electrode (EA1N) | **C** 0.07-0.17 **Cu** 0.08 max **Mn** 0.65-1.00 **Mo** 0.45-0.65 **P** 0.012 max **S** 0.030 max **Si** 0.20 max **V** 0.05 max **Other** total other elements (except Fe) 0.50 max | **ASME** SFA5.23 (EA1N) **AWS** A5.23 (EA1N) |
| **K11123** | Alloy Steel Electrode (EA2N) | **C** 0.07-0.17 **Cu** 0.08 max **Mn** 0.95-1.35 **Mo** 0.45-0.65 **P** 0.012 max **S** 0.030 max **Si** 0.20 max **V** 0.05 max **Other** total other elements (except Fe) 0.50 max | **ASME** SFA5.23 (EA2N) **AWS** A5.23 (EA2N) **MIL SPEC** MIL-E-23765/1 (MIL-70S-8) |

The chemical compositions listed are for identification purposes and should not be used in lieu of the cross referenced specifications.

| UNIFIED NUMBER | DESCRIPTION | CHEMICAL COMPOSITION | CROSS REFERENCE SPECIFICATIONS |
|---|---|---|---|
| **K11125** | Steel Electrode (ER70S-7) | **C** 0.07-0.15 **Cu** 0.50 max **Mn** 1.50-2.00 **P** 0.025 max **S** 0.035 max **Si** 0.50-0.80 | **ASME** SFA5.18 (ER70S-7) **AWS** A5.18 (ER70S-7) |
| **K11132** | Steel Electrode (ER70S-4) | **C** 0.07-0.15 **Cu** 0.50 max **Mn** 1.00-1.50 **P** 0.025 max **S** 0.035 max **Si** 0.65-0.85 | **ASME** SFA5.18 (ER70S-4) **AWS** A5.18 (ER70S-4) **MIL SPEC** MIL-E-23765/1 (MIL-70S-4) |
| **K11140** | Steel Electrode (EH 11K) | **C** 0.07-0.15 **Cu** 0.35 max **Mn** 1.40-1.85 **P** 0.030 max **S** 0.030 max **Si** 0.80-1.15 | **ASME** SFA5.17 (EH11K); SFA5.18 (ER70S-6); SFA5.25 (EH11K-EW); SFA5.26 (EGXXS-6); SFA5.30 (INMs3) **AWS** A5.17 (EH11K); A5.18 (ER70S-6); A5.25 (EH11K-EW); A5.26 (EGXXS-6); A5.30 (INMs3) **MIL SPEC** MIL-E-23765/1 (MIL-70S-6) |
| **K11160** | Alloy Steel Electrode (EF1) | **C** 0.07-0.15 **Cu** 0.35 max **Mn** 0.90-1.70 **Mo** 0.25-0.55 **Ni** 0.95-1.60 **P** 0.025 max **S** 0.025 max **Si** 0.15-0.35 **Other** total other elements (except Fe) 0.50 max | **ASME** SFA5.23 (EF1) **AWS** A5.23 (EF1) |
| **K11172** | Alloy Steel Electrode (EB2) | **C** 0.07-0.15 **Cr** 1.00-1.75 **Cu** 0.35 max **Mn** 0.45-1.00 **Mo** 0.45-0.65 **P** 0.025 max **S** 0.030 max **Si** 0.05-0.30 **Other** total other elements (except Fe) 0.50 max | **ASME** SFA5.23 (EB2) **AWS** A5.23 (EB2) |
| **K11173** | Alloy Steel Electrode (EB2RR) | **As** 0.005 max **C** 0.07-0.15 **Cr** 1.00-1.75 **Mn** 0.40-0.80 **Mo** 0.45-0.65 **P** 0.10 max **S** 0.010 max **Sb** 0.005 max **Si** 0.05-0.30 **Ti** 0.005 max | **ASME** SFA5.23 (EB2RR) **AWS** A5.23 (EB2RR) |
| **K11201** | Alloy Steel | **C** 0.12 max **Mn** 1.30 max **P** 0.035 max **S** 0.040 max **Si** 0.15-0.35 max **V** 0.020 min | |
| **K11210** | Alloy Steel | **C** 0.12 min **Cu** 0.20 min | **ASTM** A65 |
| **K11222** | Alloy Steel Electrode (EA1) | **C** 0.07-0.17 **Cu** 0.35 max **Mn** 0.65-1.00 **Mo** 0.45-0.65 **P** 0.025 max **S** 0.030 max **Si** 0.20 max **Other** total of all other elements (except Fe) 0.50 max | **ASME** SFA5.23 (EA1) **AWS** A5.23 (EA1) **MIL SPEC** MIL-E-23765/4 (MIL-EA1) |
| **K11223** | Alloy Steel Electrode (EA2) | **C** 0.07-0.17 **Cu** 0.35 max **Mn** 0.95-1.35 **Mo** 0.45-0.65 **P** 0.025 max **S** 0.030 max **Si** 0.20 max **Other** total other elements (except Fe) 0.50 max | **ASME** SFA5.23 (EA2) **AWS** A5.23 (EA2) **MIL SPEC** MIL-E-23765/1 (MIL-70S-8) |
| **K11224** | Alloy Steel | **C** 0.12 max **Cu** 0.15 max **Mn** 1.20 max **P** 0.035 max **S** 0.04 max **Si** 0.15-0.50 **Other** Ti 4\mtC min | **ASTM** A562 |
| **K11235** | Alloy Steel Electrode (ER70S-A1) | **C** 0.12 max **Cu** 0.35 max **Mn** 1.3 max **Mo** 0.4-0.65 **Ni** 0.20 max **P** 0.025 max **S** 0.025 max **Si** 0.3-0.7 | **ASME** SFA5.28 (ER70S-A1) **AWS** A5.28 (ER70S-A1) |
| **K11240** | Alloy Steel Electrode (ENi5) | **C** 0.12 max **Cu** 0.35 max **Mn** 1.20-1.60 **Mo** 0.15-0.30 **Ni** 0.75-1.25 **P** 0.020 max **S** 0.020 max **Si** 0.05-0.30 | **ASME** SFA5.23 (ENi5) **AWS** A5.23 (ENi5) |
| **K11241** | Alloy Steel Electrode (ENi5N) | **C** 0.12 max **Cu** 0.08 max **Mn** 1.20-1.60 **Mo** 0.15-0.30 **Ni** 0.75-1.25 **P** 0.012 max **S** 0.020 max **Si** 0.05-0.30 **V** 0.05 max | **ASME** SFA5.23 (ENi5N) **AWS** A5.23 (ENi5N) |
| **K11245** | Alloy Steel Electrode (EW, EWS) | **C** 0.12 max **Cr** 0.50-0.80 **Cu** 0.30-0.80 **Mn** 0.35-0.65 **Ni** 0.40-0.80 **P** 0.030 max **S** 0.040 max **Si** 0.20-0.35 **Other** total other elements (except Fe) 0.50 max | **ASME** SFA5.23 (EW); SFA5.25 (EWS-EW) **AWS** A5.23 (EW); A5.25 (EWS-EW) |

*Boxed entries are no longer active and are retained for reference purposes only.

The chemical compositions listed are for identification purposes and should not be used in lieu of the cross referenced specifications.

| UNIFIED NUMBER | DESCRIPTION | CHEMICAL COMPOSITION | CROSS REFERENCE SPECIFICATIONS |
|---|---|---|---|
| * **K11250** | Alloy Steel Electrode (ER100S-2) | **Al** 0.10 max **C** 0.12 max **Cr** 0.30 max **Cu** 0.35-0.65 **Mn** 1.25-1.80 **Mo** 0.20-0.55 **Ni** 0.80-1.25 **P** 0.010 max **S** 0.010 max **Si** 0.20-0.60 **Ti** 0.10 max **V** 0.05 max **Zr** 0.10 max **Other** total other elements (except Fe) 0.50 max | **ASME** SFA5.28 (ER100S-2) |
| **K11260** | Alloy Steel Electrode (ER80S-Nil) | **C** 0.12 max **Cr** 0.15 max **Cu** 0.35 max **Mn** 1.25 max **Mo** 0.35 max **Ni** 0.80-1.10 **P** 0.025 max **S** 0.025 max **Si** 0.40-0.80 **V** 0.05 max **Other** total other elements (except Fe) 0.50 max | **ASME** SFA5.28 (ER80S-Nil) **AWS** A5.28 (ER80S-Nil) |
| **K11267** | High Strength, Low Alloy, Precipitation Hardening Steel | **Al** 0.04-0.30 **C** 0.12 max **Cr** 0.44-1.01 **Cu** 0.40-0.75 **Mn** 0.50-1.05 **Ni** 0.47-0.98 **P** 0.04 max **S** 0.04 max **Si** 0.08-0.37 | **ASTM** A333 (4) |
| **K11268** | High Strength, Low Alloy, Precipitation Hardening Steel | **Al** 0.02-0.23 **C** 0.12 max **Cu** 0.85-1.30 **Mn** 0.50-1.00 **Mo** 0.25 max **Ni** 0.50-1.00 **P** 0.05 max **S** 0.05 max **Si** 0.15 max | |
| **K11323** | Alloy Steel Electrode Nuclear Grade (EA3N) | **C** 0.07-0.17 **Cu** 0.08 max **Mn** 1.65-2.20 **Mo** 0.45-0.65 **P** 0.012 max **S** 0.030 max **Si** 0.20 max **V** 0.05 max **Other** total other elements (except Fe) 0.50 max | **ASME** SFA5.23 (EA3N) **AWS** A5.23 (EA3N) |
| **K11324** | Alloy Steel Electrode Nuclear Grade (EA4N) | **C** 0.07-0.17 **Cu** 0.08 max **Mn** 1.20-1.70 **Mo** 0.45-0.65 **P** 0.012 max **S** 0.030 max **Si** 0.20 max **V** 0.05 max **Other** total other elements (except Fe) 0.50 max | **ASME** SFA5.23 (EA4N) **AWS** A5.23 (EA4N) |
| **K11325** | Steel Electrode | **Al** 0.50-0.90 **C** 0.07-0.19 **Cu** 0.35 max **Mn** 0.90-1.40 **P** 0.025 max **S** 0.035 max **Si** 0.30-0.60 | **ASME** SFA5.26 (EGXXS-5) **AWS** A5.26 (EGXXS-5) |
| **K11356** | Alloy Steel | **C** 0.13 max **Cr** 0.74-1.26 **Cu** 0.22-0.48 **Mn** 0.65 max **Ni** 0.17-0.53 **S** 0.06 max | **ASTM** A714 (IV) |
| **K11357** | Steel Electrode (ER70S-5) | **Al** 0.50-0.90 **C** 0.07-0.19 **Cu** 0.50 max **Mn** 1.00-1.50 **P** 0.025 **S** 0.035 **Si** 0.30-0.60 | **ASME** SFA5.18 (ER70S-5); SFA5.26 (EGXXS-5) **AWS** A5.18 (ER70S-5); A5.23 (EGXXS-5); A5.26 (EGXXS-5) **MIL SPEC** MIL-E-23765/1 (MIL-70S-5) |
| **K11365** | Alloy Steel Welding Wire | **C** 0.10-0.17 **Cr** 0.50-0.75 **Cu** 0.35 max **Mn** 0.50-0.80 **Mo** 0.15-0.25 **Ni** 0.25 max **P** 0.040 max **S** 0.040 max **Si** 0.60-0.90 **Zr** 0.05-0.15 | **AMS** 6460 |
| **K11385** | Alloy Steel Electrode (ENi4N) | **C** 0.12-0.19 **Cu** 0.08 max **Mn** 0.60-1.00 **Mo** 0.10-0.30 **Ni** 1.60-2.10 **P** 0.012 max **S** 0.020 max **Si** 0.10-0.30 **V** 0.05 max **Other** total other elements (except Fe) 0.50 max | **ASME** SFA5.23 (ENi4N) **AWS** A5.23 (ENi4N) |
| **K11422** | Alloy Steel | **C** 0.14 max **Mn** 0.30-0.80 **Mo** 0.44-0.65 **P** 0.045 max **S** 0.045 max **Si** 0.10-0.50 | **ASTM** A209 (T1b); A250 (T1b) |
| **K11423** | Alloy Steel Electrode (EA3) | **C** 0.07-0.17 **Cu** 0.35 max **Mn** 1.65-2.20 **Mo** 0.45-0.65 **P** 0.025 max **S** 0.030 max **Si** 0.20 max **Other** total other elements (except Fe) 0.50 max | **ASME** SFA5.23 (EA3) **AWS** A5.23 (EA3) **MIL SPEC** MIL-E-23765/1 (MIL-706-9); MIL-E-23765/2 (MIL-80S-2) |
| **K11424** | Alloy Steel Electrode (EA4) | **C** 0.07-0.17 **Cu** 0.35 max **Mn** 1.20-1.70 **Mo** 0.45-0.65 **P** 0.025 max **S** 0.030 max **Si** 0.20 max **Other** total other elements (except Fe) 0.50 max | **ASME** SFA5.23 (EA4) **AWS** A5.23 (EA4) |
| **K11430** | Alloy Steel | **C** 0.09-0.20 **Cr** 0.37-0.68 **Cu** 0.22-0.43 **Mn** 0.86-1.24 **P** 0.045 max **S** 0.055 max **Si** 0.13-0.32 **V** 0.01-0.11 | **ASTM** A588 (A) |

*Boxed entries are no longer active and are retained for reference purposes only.

The chemical compositions listed are for identification purposes and should not be used in lieu of the cross referenced specifications.

| UNIFIED NUMBER | DESCRIPTION | CHEMICAL COMPOSITION | CROSS REFERENCE SPECIFICATIONS |
|---|---|---|---|
| **K11485** | Alloy Steel Electrode (ENi4) | **C** 0.12-0.19 **Cu** 0.35 max **Mn** 0.60-1.00 **Mo** 0.10-0.30 **Ni** 1.60-2.10 **P** 0.015 max **S** 0.020 max **Si** 0.10-0.30 **Other** total other elements (except Fe) 0.50 max | **ASME** SFA5.23 (ENi4) **AWS** A5.23 (ENi4) |
| **K11500** | Alloy Steel | **Al** 0.02 min **C** 0.15 max **Mn** 0.27-0.63 **P** 0.048 max **S** 0.058 max | **ASTM** A587 |
| **K11501** | High-Strength Low-Alloy Steel | **C** 0.15 max **Mn** 1.65 max **P** 0.025 max **S** 0.035 max **Si** 0.10 max **Ti** 0.05 min | **ASTM** A715 |
| **K11502** | High-Strength Low-Alloy Steel | **C** 0.15 max **Mn** 1.65 max **N** 0.005 min **P** 0.025 max **S** 0.035 max **Si** 0.60 max **V** 0.02 min | |
| **K11503** | Number changed to K11509 | | |
| **K11504** | High-Strength Low-Alloy Steel | **B** 0.0025 max **C** 0.15 max **Cb** 0.005-0.06 **Cr** 0.80 max **Mn** 1.65 max **P** 0.025 max **S** 0.035 max **Si** 0.90 max **Ti** 0.10 max **Zr** 0.05 min | |
| **K11505** | High-Strength Low-Alloy Steel | **C** 0.15 max **Cb** 0.03 min **Mn** 1.65 max **Mo** 0.20 min **P** 0.025 max **S** 0.035 max **Si** 0.30 max | |
| **K11506** | High-Strength Low-Alloy Steel | **C** 0.15 max **Cb** 0.005-0.10 **Mn** 1.65 max **P** 0.025 max **S** 0.035 max **Si** 0.90 max | |
| **K11507** | High-Strength Low-Alloy Steel | **C** 0.15 max **Mn** 1.65 max **N** 0.020 max **P** 0.025 max **S** 0.035 max **Si** 0.60 max **Other** Cb or V 0.005 min; or Cb 0.005 min and V 0.005 min | |
| **K11508** | High-Strength Low-Alloy Steel | **C** 0.15 max **Cb** 0.005-0.15 **Mn** 1.65 max **P** 0.025 max **S** 0.035 max **Zr** 0.05 min | |
| **K11509** | High-Strength Low-Alloy Steel | **C** 0.18 max **Mn** 1.65 max **P** 0.025 max **S** 0.035 max **Si** 0.30 max **Ti** 0.05-0.40 | **ASTM** A656 (2) |
| **K11510** | Alloy Steel | **C** 0.15 max **Cu** 0.20 min **Mn** 1.00 max **P** 0.15 max **S** 0.05 max | **ASTM** A242 (1) |
| **K11511** | Alloy Steel | **B** 0.001-0.005 **C** 0.10-0.20 **Mn** 1.10-1.50 **Mo** 0.20-0.30 **P** 0.035 max **S** 0.04 max **Si** 0.15-0.30 | **AMS** 6386 (3) **ASTM** A514 (C); A517 (C) |
| **K11522** | Alloy Steel | **C** 0.10-0.20 **Mn** 0.30-0.80 **Mo** 0.44-0.65 **P** 0.045 max **S** 0.045 max **Si** 0.10-0.50 | **ASTM** A209 (T1); A250 (T1); A335 (P1); A369 (FP1) |
| **K11523** | Alloy Steel | **B** 0.001-0.005 **C** 0.10-0.20 **Mn** 1.10-1.50 **Mo** 0.45-0.55 **P** 0.035 max **S** 0.04 max **Si** 0.15-0.30 | |
| **K11526** | Alloy Steel | **C** 0.15 max **Cr** 0.24-1.31 **Cu** 0.22-0.58 **Mn** 0.17-0.53 **Ni** 0.68 max **P** 0.06-0.16 **S** 0.06 max **Si** 0.19-0.81 | **ASTM** A595 (C) |
| **K11535** | Alloy Steel | **C** 0.15 max **Cr** 0.24-1.31 **Cu** 0.20-0.60 **Mn** 0.55 max **Ni** 0.20-0.70 **P** 0.06-0.16 **S** 0.060 max **Si** 0.10 min | **ASTM** A423 (1) |
| **K11538** | Alloy Steel | **C** 0.15 max **Cr** 0.30-0.50 **Cu** 0.20-0.50 **Mn** 0.80-1.35 **Ni** 0.25-0.50 **P** 0.04 max **S** 0.05 max **Si** 0.15-0.30 **V** 0.01-0.10 | **ASTM** A588 (C) |
| **K11540** | Alloy Steel | **C** 0.15 max **Cu** 0.30-1.00 **Mn** 0.50-1.00 **Mo** 0.10 min **Ni** 0.40-1.10 **P** 0.04 **S** 0.05 max | **ASTM** A423 (2) |

The chemical compositions listed are for identification purposes and should not be used in lieu of the cross referenced specifications.

| UNIFIED NUMBER | DESCRIPTION | CHEMICAL COMPOSITION | CROSS REFERENCE SPECIFICATIONS |
|---|---|---|---|
| **K11541** | Alloy Steel | **C** 0.10-0.20 **Cr** 0.30 max **Cu** 0.30-1.00 **Mn** 0.50-1.00 **Mo** 0.10-0.20 **Ni** 0.40-1.10 **P** 0.04 max **S** 0.05 max **Si** 0.30 max **V** 0.01-0.10 | |
| **K11542** | Alloy Steel | **C** 0.38-0.45 **Cr** 0.80-1.10 **Cu** 0.35 max **Mn** 0.75-1.00 **Mo** 0.15-0.25 **Ni** 0.25 max **P** 0.040 max **S** 0.040 max **Si** 0.20-0.35 **Te** 0.035-0.060 | **AMS** 6378 |
| **K11546** | Alloy Steel (4140 Modified) | **C** 0.40-0.53 **Cr** 0.80-1.10 **Cu** 0.35 max **Mn** 0.75-1.00 **Mo** 0.15-0.25 **Ni** 0.25 max **P** 0.040 max **S** 0.040 max **Si** 0.20-0.35 **Te** 0.035-0.060 | **AMS** 6379 |
| **K11547** | Alloy Steel | **C** 0.10-0.20 **Cr** 0.50-0.81 **Mn** 0.30-0.61 **Mo** 0.44-0.65 **P** 0.045 max **S** 0.045 max **Si** 0.10-0.30 | **ASTM** A213 (T2); A250 (T2); A335 (P2); A369 (FP2) |
| **K11552** | Alloy Steel | **C** 0.10-0.20 **Cb** 0.04 max **Cr** 0.50-0.90 **Cu** 0.30 max **Mn** 0.75-1.25 **P** 0.04 max **S** 0.05 max **Si** 0.50-0.90 **Zr** 0.05-0.15 | |
| **K11562** | Ferritic Cr-Mo Steel (T12) | **C** 0.15 max **Cr** 0.80-1.25 **Mn** 0.30-0.61 **Mo** 0.44-0.65 **P** 0.045 max **S** 0.045 max **Si** 0.50 max | **ASTM** A182 (F12-1); A213 (T12); A250 (T12); A335 (P12); A369 (FP12) |
| **K11564** | Alloy Steel | **C** 0.10-0.20 **Cr** 0.80-1.10 **Mn** 0.30-0.80 **Mo** 0.45-0.65 **P** 0.040 max **S** 0.040 max **Si** 0.10-0.60 | **ASTM** A182 (F12-2); A336 (F12) |
| **K11567** | Alloy Steel | **C** 0.15 max **Cu** 0.50-0.80 **Mn** 1.20 max **Mo** 0.08-0.25 **Ni** 0.75-1.25 **P** 0.04 max **S** 0.05 max **Si** 0.30 max **V** 0.05 max | |
| **K11572** | Alloy Steel | **C** 0.10-0.20 **Cr** 1.00-1.50 **Mn** 0.30-0.80 **Mo** 0.44-0.65 **P** 0.040 max **S** 0.040 max **Si** 0.50-1.00 | **ASTM** A182 (F11-2); A336 (F11-2, F11-3); A336 (FN, F11A); A541 (11) |
| **K11576** | Alloy Steel | **B** 0.0005-0.006 **C** 0.10-0.20 **Cr** 0.40-0.65 **Cu** 0.15-0.50 **Mn** 0.60-1.00 **Mo** 0.40-0.60 **Ni** 0.70-1.00 **P** 0.035 max **S** 0.040 max **Si** 0.15-0.35 **V** 0.03-0.08 | **ASTM** A514 (F); A517 (F); A592 (F) |
| **K11578** | Alloy Steel | **C** 0.15 max **Mn** 0.30-0.60 **Mo** 0.44-0.65 **P** 0.030 max **S** 0.030 max **Si** 1.15-1.65 | **ASTM** A335 (P15) |
| **K11585** | Alloy Steel Electrode (EH-14) | **C** 0.10-0.20 **Cu** 0.35 max **Mn** 1.70-2.20 **P** 0.030 max **S** 0.030 max **Si** 0.10 max **Other** total other elements (except Fe) 0.50 max | **ASME** SFA5.17 (EH14); SFA5.25 (EH14-EW) **AWS** A5.17 (EH14); A5.25 (EH14-EW) **MIL SPEC** MIL-E-23765/1 (MIL-70S-7) |
| *** K11591** | Alloy Steel | **C** 0.15 max **Cr** 0.80-1.25 **Mn** 0.30-0.60 **Mo** 0.87-1.13 **P** 0.030 max **S** 0.030 max **Si** 0.10-0.35 **V** 0.15-0.25 | |
| **K11597** | Alloy Steel | **C** 0.15 max **Cr** 1.00-1.50 **Mn** 0.30-0.60 **Mo** 0.44-0.65 **P** 0.030 max **S** 0.030 max **Si** 0.50-1.00 | **ASME** SA-234 (WP 11) **ASTM** A182 (F11-1); A213 (T11); A250 (T11); A335 (P11); A336 (F11-1); A336 (F11-2); A369 (FP11) |
| **K11598** | Alloy Steel | **C** 0.10-0.20 **Cr** 1.00-1.50 **Mn** 0.30-0.80 **Mo** 0.45-0.65 **P** 0.035 max **S** 0.040 max **Si** 0.50-1.00 | |
| **K11625** | Alloy Steel | **B** 0.001-0.005 **C** 0.12-0.21 **Mn** 0.45-0.70 **Mo** 0.50-0.65 **P** 0.035 max **S** 0.04 max **Si** 0.20-0.35 | **AMS** 6386 (5) **ASTM** A514 (J); A517 (J) |
| **K11630** | Alloy Steel | **B** 0.0005-0.005 **C** 0.12-0.21 **Cr** 0.40-0.65 **Mn** 0.70-1.00 **Mo** 0.15-0.25 **P** 0.035 max **S** 0.04 max **Si** 0.20-0.35 **Ti** 0.01-0.03 **V** 0.03-0.08 | **AMS** 6386 (2) **ASTM** A514 (B); A517 (B) |

*Boxed entries are no longer active and are retained for reference purposes only.

The chemical compositions listed are for identification purposes and should not be used in lieu of the cross referenced specifications.

| UNIFIED NUMBER | DESCRIPTION | CHEMICAL COMPOSITION | CROSS REFERENCE SPECIFICATIONS |
|---|---|---|---|
| **K11640** | Alloy Steel | **C** 0.38-0.43 **Cr** 0.55-0.75 **Cu** 0.35 max **Mn** 0.75-1.00 **Mo** 0.20-0.30 **Ni** 0.55-0.85 **P** 0.025 max **S** 0.025 max **Si** 0.20-0.35 | **AMS** 6324 |
| **K11646** | Alloy Steel | **B** 0.0005-0.005 **C** 0.12-0.21 **Cr** 0.40-0.65 **Mn** 0.95-1.30 **Mo** 0.20-0.30 **Ni** 0.30-0.70 **P** 0.035 max **S** 0.04 max **Si** 0.20-0.35 **V** 0.03-0.08 | **ASTM** A514 (H); A517 (H) |
| **K11662** | Alloy Steel | **B** 0.0015-0.005 **C** 0.13-0.20 **Cr** 0.85-1.20 **Cu** 0.20-0.40 **Mn** 0.40-0.70 **Mo** 0.15-0.25 **P** 0.035 max **S** 0.04 max **Si** 0.20-0.35 **Ti** 0.04-0.10 | **AMS** 6386 (4) |
| **K11682** | Alloy Steel | **B** 0.0015-0.005 **C** 0.13-0.20 **Cr** 1.15-1.65 **Cu** 0.20-0.40 **Mn** 0.40-0.70 **Mo** 0.25-0.40 **P** 0.035 max **S** 0.04 max **Si** 0.20-0.35 **Ti** 0.04-0.10 | |
| **K11683** | Alloy Steel | **B** 0.001-0.005 **C** 0.12-0.21 **Mn** 0.45-0.70 **Mo** 0.45-0.60 **Ni** 1.20-1.50 **P** 0.035 max **S** 0.04 max **Si** 0.20-0.35 | **ASTM** A514 (M); A517 (M) |
| **K11695** | Alloy Steel | **B** 0.0015-0.005 **C** 0.10-0.22 **Cr** 1.34-2.06 **Cu** 0.17-0.43 **Mn** 0.37-0.74 **Mo** 0.36-0.64 **P** 0.035 max **S** 0.040 max **Si** 0.18-0.37 **Ti** 0.03-0.11 | **ASTM** A592 (E) |
| * **K11711** | Leaded-Resulfurized Steel | **C** 0.14-0.20 **Mn** 1.00-1.30 **P** 0.04 max **Pb** 0.15-0.35 **S** 0.08-0.13 **Si** 0.35 max | |
| **K11720** | High-Strength Low-Alloy Steel | **Al** 0.06 max **C** 0.17 max **Cb** 0.050 max **Cu** 0.35 max **Mn** 1.60 max **P** 0.035 max **S** 0.015 max **Si** 0.40 max **V** 0.11 max | **ASTM** A734 (B) |
| **K11742** | Alloy Steel | **C** 0.17 max **Cr** 0.31-0.64 **Mn** 0.97-1.52 **P** 0.035 max **S** 0.040 max **Si** 0.54-0.96 | **ASTM** A202 (A) |
| **K11757** | Alloy Steel | **C** 0.17 max **Cr** 0.74-1.21 **Mn** 0.35-0.73 **Mo** 0.40-0.65 **P** 0.035 max **S** 0.040 max **Si** 0.13-0.32 | **ASTM** A387 (12) |
| **K11789** | Alloy Steel | **C** 0.17 max **Cr** 0.94-1.56 **Mn** 0.35-0.73 **Mo** 0.40-0.70 **P** 0.035 max **S** 0.040 max **Si** 0.44-0.86 | **ASTM** A387 (11) |
| **K11797** | Alloy Steel | **C** 0.20 max **Cr** 1.00-1.50 **Mn** 0.40-0.65 **Mo** 0.45-0.65 **P** 0.035 max **S** 0.040 max **Si** 0.50-0.80 | **ASTM** A739 (B11) |
| **K11800** | Alloy Steel | **C** 0.18 max **Cr** 0.15 max **Mn** 1.30 max **Mo** 0.05 max **Ni** 0.25 max **P** 0.035 max **S** 0.040 max **Si** 0.15-0.35 **V** 0.02-0.12 | **ASTM** A541 (1C) |
| **K11801** | Alloy Steel | **C** 0.18 max **Cr** 0.15 max **Cu** 0.35 max **Mn** 1.30 max **Mo** 0.06 max **Ni** 0.25 max **P** 0.040 max **S** 0.050 max **Si** 0.15-0.35 **Ti** 0.005 min **V** 0.02 min | |
| **K11802** | Alloy Steel | **C** 0.18 max **Cr** 0.15 max **Cu** 0.35 max **Mn** 1.30 max **Mo** 0.06 max **P** 0.035 max **S** 0.040 max **Si** 0.15-0.35 **Ti** 0.005 max **V** 0.02-0.13 | |
| **K11803** | Alloy Steel | **C** 0.18 max **Mn** 1.45 max **P** 0.035 max **S** 0.040 max **Si** 0.13-0.32 **V** 0.07-0.16 | |
| **K11804** | Alloy Steel | **C** 0.18 max **Mn** 1.60 max **N** 0.005-0.030 **P** 0.025 max **S** 0.035 max **Si** 0.60 max **V** 0.030-0.20 | **ASTM** A656 (1) |

*Boxed entries are no longer active and are retained for reference purposes only.

The chemical compositions listed are for identification purposes and should not be used in lieu of the cross referenced specifications.

| UNIFIED NUMBER | DESCRIPTION | CHEMICAL COMPOSITION | CROSS REFERENCE SPECIFICATIONS |
|---|---|---|---|
| **K11805** | Alloy Steel | **C** 0.18 max **Mn** 1.30 max **P** 0.035 max **S** 0.040 max **Si** 0.15-0.35 **V** 0.02 min | |
| **K11820** | Alloy Steel | **C** 0.18 max **Mn** 0.90 max **Mo** 0.41-0.64 **P** 0.035 max **S** 0.040 max **Si** 0.13-0.32 | **ASTM** A204 (A) |
| **K11831** | High-Strength Low-Alloy Steel | **C** 0.18 max **Cr** 0.25 max **Cu** 0.35 max **Mn** 1.00-1.60 **Mo** 0.08 max **Ni** 0.25 max **P** 0.035 max **S** 0.040 max **Si** 0.55 max **V** 0.08 max | **ASTM** A724 (A) |
| **K11835** | Alloy Steel | **C** 0.18 max **Cr** 0.33 max **Cu** 0.27-1.03 **Mn** 0.45-1.05 **Mo** 0.09-0.21 **Ni** 0.35-1.15 **P** 0.045 max **S** 0.055 max | **ASTM** A714 (VI) |
| **K11846** | High-Strength Low-Alloy Steel | **Al** 0.065 max **C** 0.18 max **Cr** 0.25 max **Cu** 0.35 max **Mn** 0.90-1.60 **Mo** 0.08 max **Ni** 0.40 max **P** 0.04 max **S** 0.04 max **Si** 0.10-0.50 | **ASTM** A131 (AH32, DH32, EH32) |
| **K11847** | Alloy Steel | **B** 0.0005-0.0025 **C** 0.15-0.21 **Cr** 0.50-0.80 **Mn** 0.80-1.10 **Mo** 0.25 max **P** 0.035 max **S** 0.04 max **Si** 0.40-0.90 **Zr** 0.05-0.15 | |
| **K11852** | High-Strength Low-Alloy Steel | **C** 0.18 max **Cb** 0.05 max **Cr** 0.25 max **Cu** 0.35 max **Mn** 0.90-1.60 **Mo** 0.08 max **Ni** 0.40 max **P** 0.04 max **S** 0.04 max **Si** 0.10-0.50 **V** 0.10 max | **ASTM** A131 (AH36, DH36, EH36); A808 |
| **K11856** | Alloy Steel | **B** 0.0025 max **C** 0.15-0.21 **Cr** 0.50-0.80 **Mn** 0.80-1.10 **Mo** 0.18-0.28 **P** 0.035 max **S** 0.04 max **Si** 0.40-0.80 **Zr** 0.05-0.15 | **AMS** 6386 (1) **ASTM** A514 (A); A517 (A); A592 (A) |
| **K11872** | Alloy Steel | **B** 0.0025 max **C** 0.15-0.21 **Cr** 0.50-0.90 **Mn** 0.80-1.10 **Mo** 0.40-0.60 **P** 0.035 max **S** 0.04 max **Si** 0.50-0.90 **Zr** 0.05-0.15 | |
| * **K11900** | Alloy Steel | **B** 0.0005 min **C** 0.13-0.25 **Mn** 0.67 min **P** 0.048 max **S** 0.058 max | |
| **K11914** | Alloy Steel (NAX 9115-AC) | **C** 0.10-0.17 **Cr** 0.50-0.75 **Cu** 0.35 max **Mn** 0.50-0.80 **Mo** 0.15-0.25 **Ni** 0.25 max **P** 0.025 max **S** 0.035 max **Si** 0.60-0.90 **Zr** 0.05-0.10 | **AMS** 6354 |
| * **K11918** | Alloy Steel | **C** 0.10-0.28 **Cr** 0.30-0.40 **Mn** 0.30-0.40 **Mo** 0.07-0.15 **Ni** 0.50 max **Si** 0.20-0.40 **V** 0.10 max | **MIL SPEC** MIL-A-12560 |
| **K11940** | Alloy Steel | **B** 0.0005-0.005 **C** 0.38-0.43 **Cr** 0.70-0.90 **Cu** 0.35 max **Mn** 0.65-0.85 **Mo** 0.15-0.25 **NI** 0.70-1.00 **P** 0.025 max **S** 0.025 max **Si** 0.15-0.35 **V** 0.01-0.06 | **AMS** 6422 |
| **K11948** | Alloy Steel Electrode (EF4N) | **C** 0.16-0.23 **Cr** 0.40-0.60 **Cu** 0.08 max **Mn** 0.60-0.90 **Mo** 0.15-0.30 **Ni** 0.40-0.80 **P** 0.012 max **S** 0.035 max **Si** 0.15-0.35 **V** 0.05 max **Other** total other elements (except Fe) 0.50 max | **ASME** SFA5.23 (EF4N) **AWS** A5.23 (EF4N) |
| **K12000** | High-Strength Low-Alloy Steel | **C** 0.20 max **Cb** 0.01-0.05 **Mn** 1.15-1.50 **P** 0.04 max **S** 0.05 max **Si** 0.15-0.50 | **ASTM** A633 (C) |

*Boxed entries are no longer active and are retained for reference purposes only.

The chemical compositions listed are for identification purposes and should not be used in lieu of the cross referenced specifications.

| UNIFIED NUMBER | DESCRIPTION | CHEMICAL COMPOSITION | CROSS REFERENCE SPECIFICATIONS |
|---|---|---|---|
| **K12001** | High-Strength Low-Alloy Steel | **C** 0.20 max **Cb** 0.05 max **Mn** 1.15-1.50 **P** 0.035 max **S** 0.030 max **Si** 0.15-0.50 | **ASTM** A737 (B); A738 (B) |
| **K12003** | Alloy Steel | **C** 0.20 max **Mn** 1.45 max **P** 0.035 max **S** 0.040 max **Si** 0.13-0.32 **V** 0.07-0.16 | |
| **K12004** | Alloy Steel | **C** 0.20 max **Mn** 1.70 max **Ni** 0.40-0.70 **P** 0.035 max **S** 0.035 max **Si** 0.10-0.50 **V** 0.10-0.18 | **ASTM** A225 (D) |
| **K12010** | High-Strength Low-Alloy Steel | **C** 0.20 max **Cu** 0.20 min **Mn** 1.35 max **P** 0.04 max **S** 0.05 max | |
| **K12020** | Alloy Steel | **C** 0.20 max **Mn** 0.90 max **Mo** 0.41-0.64 **P** 0.035 max **S** 0.040 max **Si** 0.13-0.45 | **ASTM** A204 (B) |
| **K12021** | Alloy Steel | **C** 0.20 max **Mn** 0.87-1.41 **Mo** 0.41-0.64 **P** 0.035 max **S** 0.040 max **Si** 0.13-0.45 | **ASTM** A302 (A) |
| **K12022** | Alloy Steel | **C** 0.20 max **Mn** 1.07-1.62 **Mo** 0.41-0.64 **P** 0.035 max **S** 0.040 max **Si** 0.13-0.45 | **ASTM** A302 (B) |
| **K12023** | Alloy Steel | **C** 0.15-0.25 **Mn** 0.30-0.80 **Mo** 0.44-0.65 **P** 0.045 max **S** 0.045 max **Si** 0.10-0.50 | **ASTM** A209 (T1a); A250 (T1a) |
| **K12031** | High-Strength Low-Alloy Steel | **C** 0.20 max **Cr** 0.25 max **Cu** 0.35 max **Mn** 1.00-1.60 **Mo** 0.08 max **Ni** 0.25 max **P** 0.035 max **Si** 0.50 max **V** 0.08 max | **ASTM** A724 (B) |
| **K12032** | Alloy Steel | **C** 0.20 max **Cr** 0.10-0.25 **Cu** 0.20-0.35 **Mn** 1.25 max **Mo** 0.15 max **Ni** 0.30-0.60 **P** 0.035 max **S** 0.040 max **Si** 0.25-0.75 **Ti** 0.005-0.030 **V** 0.02-0.10 | |
| **K12033** | Alloy Steel | **C** 0.14-0.26 **Cr** 0.27-0.53 **Cu** 0.17-0.53 **Mn** 0.76-1.39 **Ni** 0.22-0.53 **P** 0.040 max **S** 0.045 max **Si** 0.13-0.32 **V** 0.010 min | **ASTM** A325 (C); A563 (C3C) |
| **K12037** | High-Strength Low-Alloy Steel | **C** 0.20 max **Cr** 0.25 max **Cu** 0.35 max **Mn** 0.70-1.35 **Mo** 0.08 max **Ni** 0.25 max **P** 0.04 max **S** 0.05 max **Si** 0.15-0.50 | **ASTM** A633 (D); A724 (C) |
| **K12039** | Alloy Steel | **C** 0.20 max **Mn** 1.07-1.62 **Mo** 0.41-0.64 **Ni** 0.37-0.73 **P** 0.035 max **S** 0.040 max **Si** 0.13-0.45 | **ASTM** A302 (C) |
| **K12040** | Alloy Steel | **C** 0.20 max **Cr** 0.50-1.00 **Cu** 0.30-0.50 **Mn** 1.20 max **Mo** 0.10 max **Ni** 0.80 max **P** 0.04 max **S** 0.05 max **Si** 0.25-0.70 **Ti** 0.07 max | |
| **K12042** | Alloy Steel | **C** 0.25 max **Cr** 0.25 max **Mn** 1.20-1.50 **Mo** 0.45-0.60 **Ni** 0.40-1.00 **P** 0.025 max **S** 0.025 max **Si** 0.15-0.40 **V** 0.05 max | **ASTM** A508 (3) **MIL SPEC** MIL-S-24238 (A) |
| **K12043** | Alloy Steel | **C** 0.20 max **Cr** 0.40-0.70 **Cu** 0.20-0.40 **Mn** 0.75-1.25 **Ni** 0.50 max **P** 0.04 max **S** 0.05 max **Si** 0.15-0.30 **V** 0.01-0.10 | **ASTM** A588 (B); A852 |
| **K12044** | Alloy Steel | **C** 0.20 max **Cu** 0.30 min **Mn** 0.60-1.00 **Ni** 0.50-0.70 **P** 0.04 max **S** 0.05 max **Si** 0.30-0.50 **Ti** 0.05 | |
| **K12045** | Alloy Steel | **C** 0.25 max **Mn** 1.20-1.50 **Mo** 0.45-0.60 **Ni** 0.40-1.00 **P** 0.035 max **S** 0.040 max **Si** 0.15-0.35 **V** 0.05 max | **ASTM** A541 (3) |

The chemical compositions listed are for identification purposes and should not be used in lieu of the cross referenced specifications.

| UNIFIED NUMBER | DESCRIPTION | CHEMICAL COMPOSITION | CROSS REFERENCE SPECIFICATIONS |
|---|---|---|---|
| **K12047** | Alloy Steel | **C** 0.15-0.25 **Cr** 0.80-1.25 **Mn** 0.30-0.61 **P** 0.045 max **S** 0.045 max **Si** 0.15-0.35 **V** 0.15 min | **ASTM** A213 (T17) |
| **K12048** | Alloy Steel Electrode (EF4) | **C** 0.16-0.23 **Cr** 0.40-0.60 **Cu** 0.35 max **Mn** 0.60-0.90 **Mo** 0.15-0.30 **Ni** 0.40-0.80 **P** 0.025 max **S** 0.035 max **Si** 0.15-0.35 **Other** total other elements (except Fe) 0.50 max | **ASME** SFA5.23 (EF4) **AWS** A5.23 (EF4) |
| **K12054** | Alloy Steel | **C** 0.20 max **Mn** 1.07-1.62 **Mo** 0.41-0.64 **Ni** 0.67-1.03 **P** 0.035 max **S** 0.040 max **Si** 0.13-0.45 | **ASTM** A302 (D) |
| **K12059** | Alloy Steel | **C** 0.14-0.26 **Cr** 0.45-1.05 **Cu** 0.27-0.53 **Mn** 0.36-1.24 **Mo** 0.11 max **Ni** 0.47-0.83 **P** 0.045 max **S** 0.055 max **Si** 0.20-0.55 **Ti** 0.05 max | **ASTM** A325 (D); A563 (C3D) |
| **K12062** | Alloy Steel | **C** 0.20 max **Cr** 0.80-1.25 **Mn** 0.30-0.80 **Mo** 0.44-0.65 **P** 0.045 max **S** 0.045 max **Si** 0.60 max | **ASTM** A234 (WP12) |
| **K12087** | Alloy Steel Electrode (EB5N) | **C** 0.18-0.23 **Cr** 0.45-0.65 **Cu** 0.08 max **Mn** 0.40-0.70 **Mo** 0.90-1.10 **P** 0.012 max **S** 0.025 max **Si** 0.40-0.60 **V** 0.05 max **Other** total other elements (except Fe) 0.50 max | **ASME** SFA5.23 (EB5N) **AWS** A5.23 (EB5N) |
| **K12089** | Alloy Steel | **C** 0.20 max **Mn** 0.40-0.70 **Mo** 0.19-0.33 **Ni** 1.60-2.05 **P** 0.030 max **S** 0.035 max **Si** 0.37 max | **ASTM** A707 (L4) |
| **K12103** | Alloy Steel | **C** 0.10-0.32 **Cr** 0.30-0.40 **Mn** 0.30-0.40 **Mo** 0.07-0.15 **Ni** 0.50 max **P** 0.04 max **S** 0.04 max **Si** 0.20-0.40 **V** 0.10 max | |
| * **K12121** | Alloy Steel | **C** 0.17-0.26 **Mn** 0.46-0.94 **Mo** 0.42-0.68 **P** 0.045 max **S** 0.045 max **Si** 0.18-0.37 | |
| **K12122** | Alloy Steel | **C** 0.21 max **Cr** 0.50-0.81 **Mn** 0.30-0.80 **Mo** 0.44-0.65 **P** 0.040 max **S** 0.040 max **Si** 0.10-0.60 | **ASTM** A182 (F2) |
| **K12125** | Cr-Mo Alloy Steel | **C** 0.16-0.26 **Cr** 0.40-0.60 **Mn** 1.10-1.60 **Mo** 0.10-0.25 **P** 0.025 max **S** 0.025 max **Si** 0.10-0.40 **Note:** Steel is Boron treated | |
| **K12143** | Alloy Steel | **C** 0.21 max **Cr** 0.46-0.85 **Mn** 0.50-0.88 **Mo** 0.40-0.65 **P** 0.035 max **S** 0.040 max **Si** 0.13-0.45 | **ASTM** A387 (2) |
| **K12147** | Steel Welding Rod (R100) | **Al** 0.02 max **C** 0.18-0.23 **Cr** 0.40-0.60 **Cu** 0.15 max **Mn** 0.70-0.90 **Mo** 0.15-0.25 **Ni** 0.40-0.70 **P** 0.025 max **S** 0.025 max **Si** 0.20-0.35 | **ASME** SFA5.2 (R100) **AWS** A5.2 (R100) |
| **K12187** | Alloy Steel Electrode (EB5) | **C** 0.18-0.23 **Cr** 0.45-0.65 **Cu** 0.30 max **Mn** 0.40-0.70 **Mo** 0.90-1.10 **P** 0.025 max **S** 0.025 max **Si** 0.40-0.60 **Other** total other elements (except Fe) 0.50 max | **ASME** SFA5.23 (EB5) **AWS** A5.23 (EB5) |
| **K12202** | Alloy Steel | **C** 0.22 max **Mn** 1.15-1.50 **N** 0.01-0.03 **P** 0.04 max **S** 0.05 max **Si** 0.15-0.50 **V** 0.04-0.11 | **ASTM** A350 (LF6); A633 (E); A678 (D); A737 (C) |
| * **K12211** | Alloy Steel | **C** 0.22 max **Cu** 0.20 min **Mn** 0.85-1.25 **P** 0.04 max **S** 0.05 max **Si** 0.40 max **V** 0.02 min | |

*Boxed entries are no longer active and are retained for reference purposes only.

The chemical compositions listed are for identification purposes and should not be used in lieu of the cross referenced specifications.

| UNIFIED NUMBER | DESCRIPTION | CHEMICAL COMPOSITION | CROSS REFERENCE SPECIFICATIONS |
|---|---|---|---|
| **K12220** | Alloy Steel | **C** 0.22 max **Cr** 0.25 max **Cu** 0.25 max **Mn** 0.80 max **Mo** 0.40-0.60 **Ni** 0.25 max **P** 0.04 max **S** 0.04 max **Si** 0.15-0.40 | |
| **K12238** | Alloy Steel | **C** 0.19-0.26 **Cr** 0.42-0.68 **Cu** 0.17-0.43 **Mn** 0.86-1.24 **Ni** 0.17-0.43 **P** 0.045 max **S** 0.045 max **Si** 0.13-0.32 | **ASTM** A325 (F); A563 (C3F) |
| * **K12244** | Alloy Steel | **C** 0.21 max **Cr** 0.37-0.73 **Cu** 0.17-0.43 **Mn** 0.71-1.29 **Ni** 0.22-0.53 **P** 0.045 max **S** 0.055 max **Si** 0.13-0.32 **V** 0.11 max | |
| **K12249** | Alloy Steel | **C** 0.22 max **Cu** 0.50 min **Mn** 0.60-0.90 **Ni** 0.40-0.75 **P** 0.08-0.15 **S** 0.05 max **Si** 0.10 max | **ASTM** A690 |
| **K12254** | Alloy Steel | **C** 0.18-0.27 **Cr** 0.55-0.95 **Cu** 0.27-0.63 **Mn** 0.56-1.04 **Ni** 0.27-0.63 **P** 0.045 max **S** 0.045 max **Si** 0.13-0.32 | **ASTM** A325 (E); A563 (C3E) |
| **K12320** | Alloy Steel | **C** 0.23 max **Mn** 0.90 max **Mo** 0.41-0.64 **P** 0.035 max **S** 0.040 max **Si** 0.13-0.45 | **ASTM** A204 (C) |
| **K12437** | High-Strength Low-Alloy Steel | **C** 0.24 max **Cr** 0.25 max **Cu** 0.35 max **Mn** 0.70-1.35 **Mo** 0.08 max **Ni** 0.25 max **P** 0.035 max **S** 0.040 max **Si** 0.13-0.55 | **ASTM** A537 |
| **K12447** | Carbon-Manganese-Silicon Steel | **C** 0.24 max **Cr** 0.25 max **Cu** 0.35 max **Mn** 1.50 max **Mo** 0.08 max **Ni** 0.50 max **P** 0.035 max **S** 0.040 max **Si** 0.15-0.50 | **ASTM** A738 (A) |
| **K12510** | Alloy Steel | **C** 0.25 max **Cu** 0.18 min **Mn** 1.05-1.60 **N** 0.005-0.035 **P** 0.030 max **S** 0.035 max **Si** 0.32 max **V** 0.03-0.13 | **ASTM** A707 (L3) |
| **K12520** | Alloy Steel | **C** 0.20-0.30 **Mn** 0.60-0.80 **Mo** 0.40-0.60 **P** 0.040 max **S** 0.040 max **Si** 0.20-0.35 | **ASTM** A336 (F1) |
| **K12521** | Alloy Steel | **C** 0.25 max **Mn** 1.07-1.62 **Mo** 0.41-0.64 **P** 0.035 max **S** 0.040 max **Si** 0.13-0.45 | **ASTM** A533 (A) |
| **K12524** | Alloy Steel | **C** 0.25 max **Mn** 1.60 max **Ni** 0.37-0.73 **P** 0.035 max **S** 0.040 max **Si** 0.13-0.45 **V** 0.11-0.20 | **ASTM** A225/A225M (C) |
| **K12529** | Alloy Steel | **C** 0.25 max **Mn** 1.07-1.62 **Mo** 0.41-0.64 **Ni** 0.17-0.43 **P** 0.035 max **S** 0.040 max **Si** 0.13-0.45 | **ASTM** A533 (D) |
| **K12539** | Alloy Steel | **C** 0.25 max **Mn** 1.07-1.62 **Mo** 0.41-0.64 **Ni** 0.37-0.73 **P** 0.035 max **S** 0.040 max **Si** 0.13-0.45 | **ASTM** A533 (B) |
| **K12542** | Alloy Steel | **C** 0.25 max **Cr** 0.31-0.64 **Mn** 0.97-1.52 **P** 0.035 max **S** 0.040 max **Si** 0.54-0.96 | **ASTM** A202 (B) |
| **K12554** | Alloy Steel | **C** 0.25 max **Mn** 1.07-1.62 **Mo** 0.41-0.64 **Ni** 0.67-1.03 **P** 0.035 max **S** 0.040 max **Si** 0.13-0.45 | **ASTM** A533 (C) |
| **K12608** | Alloy Steel | **C** 0.26 max **Cu** 0.18 min **Mn** 1.30 max **S** 0.063 max | **ASTM** A714 (I) |
| **K12609** | High-Strength Low-Alloy Steel | **C** 0.26 max **Cu** 0.18 min **Mn** 1.30 max **P** 0.05 max **S** 0.063 max **Si** 0.33 max **V** 0.01 min | **ASTM** A618 (II); A714 (II) |

*Boxed entries are no longer active and are retained for reference purposes only.

The chemical compositions listed are for identification purposes and should not be used in lieu of the cross referenced specifications.

| UNIFIED NUMBER | DESCRIPTION | CHEMICAL COMPOSITION | CROSS REFERENCE SPECIFICATIONS |
|---|---|---|---|
| **K12700** | High-Strength Low-Alloy Steel | **C** 0.27 max **Mn** 1.40 max **P** 0.05 max **S** 0.06 max **Si** 0.35 max **V** 0.01 min | **ASTM** A618 (III) |
| **K12709** | High-Strength Low-Alloy Steel | **C** 0.27 max **Cu** 0.18 min **Mn** 1.40 max **P** 0.05 max **S** 0.06 max **Si** 0.35 max **V** 0.01 min | **ASTM** A714 (III) |
| **K12765** | Alloy Steel | **C** 0.27 max **Cr** 0.25-0.45 **Mn** 0.50-0.90 **Mo** 0.55-0.70 **Ni** 0.50-1.00 **P** 0.035 max **S** 0.040 max **Si** 0.15-0.35 **V** 0.05 max | **ASTM** A541 (2) |
| **K12766** | Alloy Steel | **C** 0.27 max **Cr** 0.25-0.45 **Mn** 0.50-1.00 **Mo** 0.55-0.70 **Ni** 0.50-1.00 **P** 0.025 max **S** 0.025 max **Si** 0.15-0.40 **V** 0.05 max | **ASTM** A508 (2) |
| **K12810** | Alloy Steel | **C** 0.28 max **Cu** 0.20 min **Mn** 1.10-1.60 **P** 0.04 max **S** 0.05 max **Si** 0.30 max | |
| **K12821** | Alloy Steel | **C** 0.28 max **Mn** 0.30-0.90 **Mo** 0.44-0.65 **P** 0.045 max **S** 0.045 max **Si** 0.10-0.50 | **ASTM** A234 (WP1) |
| **K12822** | Alloy Steel | **C** 0.28 max **Mn** 0.60-0.90 **Mo** 0.44-0.65 **P** 0.045 max **S** 0.045 max **Si** 0.15-0.35 | **ASTM** A182 (F1) |
| **K13020** | Alloy Steel | **C** 0.30 max **Cr** 0.25 max **Cu** 0.25 max **Mn** 0.80 max **Mo** 0.40-0.60 **Ni** 0.25 max **P** 0.04 max **S** 0.04 max **Si** 0.15-0.40 | |
| **K13047** | Alloy Steel | **C** 0.26-0.34 **Cr** 0.80-1.15 **Mn** 0.40-0.70 **Mo** 0.15-0.25 **P** 0.035 max **S** 0.04 max **Si** 0.15-0.35 | **ASTM** A372 (E); A649 (3) |
| **K13048** | Alloy Steel | **C** 0.28-0.33 **Cr** 0.40-0.60 **Cu** 0.35 max **Mn** 0.70-0.90 **Mo** 0.15-0.25 **Ni** 0.40-0.70 **P** 0.025 max **S** 0.025 max **Si** 0.20-0.35 | **AMS** 6550 |
| **K13049** | Alloy Steel | **C** 0.26-0.34 **Cr** 0.40-0.65 **Mn** 0.70-1.00 **Mo** 0.15-0.25 **Ni** 0.40-0.70 **P** 0.035 max **S** 0.04 max **Si** 0.15-0.35 | |
| **K13050** | Alloy Steel | **C** 0.30 max **Mn** 1.35 max **Ni** 1.0-2.0 **P** 0.035 max **S** 0.040 max **Si** 0.20-0.35 | **ASTM** A350 (LF5) |
| **K13051** | Alloy Steel, Superstrength | **C** 0.27-0.33 **Cr** 0.80-1.10 **Mn** 0.40-0.60 **Mo** 0.15-0.25 **P** 0.025 max **S** 0.025 max **Si** 0.20-0.35 **V** 0.05-0.10 | **ASTM** A579 (13) |
| **K13147** | Alloy Steel Welding Wire | **C** 0.28-0.33 **Cr** 0.80-1.10 **Cu** 0.10 max **H** 0.0010 max **Mn** 0.40-0.60 **Mo** 0.15-0.25 **N** 0.005 max **Ni** 0.25 max **O** 0.0025 max **P** 0.008 max **S** 0.008 max **Si** 0.15-0.35 **V** 0.06 max **Other** S+P 0.012 max | **AMS** 6457 |
| **K13148** | Alloy Steel Welding Wire (6130) | **C** 0.28-0.33 **Cr** 0.80-1.10 **Cu** 0.35 max **H** 0.0025 max **Mn** 0.60-0.90 **Mo** 0.06 max **N** 0.005 max **Ni** 0.25 max **O** 0.0025 max **P** 0.008 max **S** 0.008 max **Si** 0.15-0.35 **V** 0.15-0.25 **Other** S+P 0.012 max | **AMS** 6461 |
| **K13149** | Alloy Steel Welding Wire (6130) | **C** 0.28-0.33 **Cr** 0.80-1.10 **Cu** 0.35 max **Mn** 0.70-0.90 **Mo** 0.06 max **Ni** 0.25 max **P** 0.025 max **S** 0.025 max **Si** 0.15-0.35 **V** 0.15-0.25 | **AMS** 6462 |
| **K13247** | Alloy Steel | **C** 0.30-0.35 **Cr** 0.80-1.10 **Cu** 0.35 max **Mn** 0.40-0.60 **Mo** 0.15-0.25 **Ni** 0.25 max **P** 0.025 max **S** 0.025 max **Si** 0.20-0.50 | **AMS** 6356 |

The chemical compositions listed are for identification purposes and should not be used in lieu of the cross referenced specifications.

| UNIFIED NUMBER | DESCRIPTION | CHEMICAL COMPOSITION | CROSS REFERENCE SPECIFICATIONS |
|---|---|---|---|
| **K13262** | Alloy Steel | **C** 0.32 max **Cr** 0.40 min **Mn** 1.00 max **Mo** 0.15 min **Ni** 0.70 min **P** 0.025 max **S** 0.025 max **Si** 0.75 max | |
| **K13502** | Alloy Steel | **C** 0.35 max **Cr** 0.25 max **Mn** 0.40-1.05 **Mo** 0.10 max **Ni** 0.40 max **P** 0.025 max **S** 0.025 max **Si** 0.15-0.40 **V** 0.05 max | **ASTM** A508 (1) |
| * **K13521** | Alloy Steel | **C** 0.31-0.40 **Cr** 0.22-0.43 **Mn** 0.91-1.29 **Mo** 0.04-0.11 **Ni** 0.17-0.43 **P** 0.040 max **S** 0.045 max **Si** 0.18-0.37 | |
| **K13547** | Alloy Steel | **C** 0.31-0.39 **Cr** 0.40-0.65 **Mn** 0.75-1.05 **Mo** 0.15-0.25 **P** 0.035 max **S** 0.04 max **Si** 0.15-0.35 | **ASTM** A372 (G) |
| **K13548** | Alloy Steel | **C** 0.31-0.39 **Cr** 0.80-1.15 **Mn** 0.70-1.00 **Mo** 0.15-0.25 **P** 0.035 max **S** 0.04 max **Si** 0.15-0.35 | **ASTM** A372 (J) |
| **K13550** | Alloy Steel | **C** 0.48-0.53 **Cr** 0.40-0.60 **Cu** 0.35 max **Mn** 0.75-1.00 **Mo** 0.20-0.30 **Ni** 0.40-1.00 **P** 0.025 max **S** 0.025 max **Si** 0.15-0.35 | **AMS** 6328 |
| **K13586** | Alloy Steel | **C** 0.35 max **Cr** 0.70 max **Mn** 0.60-0.90 **Mo** 0.20 min **Ni** 1.50-3.50 **P** 0.025 max **S** 0.025 max **Si** 0.15-0.35 **V** 0.03-0.12 | |
| **K13643** | Alloy Steel | **C** 0.31-0.42 **Cr** 0.42-0.68 **Cu** 0.22-0.48 **Mn** 0.86-1.24 **Ni** 0.22-0.48 **P** 0.045 max **S** 0.055 max **Si** 0.13-0.32 | **ASTM** A325 (A); A563 (C3A) |
| **K13647** | Alloy Steel | **B** 0.0005-0.003 **C** 0.33-0.40 **Cr** 0.95-1.25 **Mn** 0.30-0.50 **P** 0.035 max **S** 0.040 max **Si** 0.15-0.35 | **ASTM** A574 (51B37M) |
| **K13650** | Alloy Steel | **C** 0.20-0.53 **Cr** 0.45 min **Cu** 0.20 min **Mn** 0.40 min **Mo** 0.15 min **Ni** 0.20 min **P** 0.046 max **S** 0.050 max | **ASTM** A563 (DH3) |
| **K13745** | Alloy Steel | **C** 0.35-0.40 **Cr** 0.90-1.20 **Mn** 0.30-0.50 **P** 0.035 max **S** 0.040 max **Si** 0.15-0.35 | **ASTM** A574 (5137M) |
| * **K14044** | Alloy Steel | **C** 0.36-0.45 **Cr** 0.75-1.15 **Mn** 0.71-1.04 **Mo** 0.13-0.27 **P** 0.040 max **S** 0.045 max **Si** 0.18-0.37 | |
| **K14047** | Alloy Steel | **C** 0.38-0.43 **Cr** 0.80-1.10 **Mn** 0.75-1.00 **Mo** 0.15-0.25 **P** 0.025 max **S** 0.025 max **Si** 0.20-0.35 | |
| **K14048** | Alloy Steel | **C** 0.35-0.45 **Cr** 0.80-1.15 **Mn** 0.70-1.00 **Mo** 0.15-0.25 **P** 0.040 max **S** 0.040 max **Si** 0.20-0.35 **V** 0.06 max | **ASTM** A290 (E, F) |
| **K14072** | Alloy Steel | **C** 0.36-0.44 **Cr** 0.80-1.15 **Mn** 0.45-0.70 **Mo** 0.50-0.65 **P** 0.04 max **S** 0.04 max **Si** 0.15-0.35 **V** 0.25-0.35 | **ASTM** A193 (B16); A194 (16); A437 (B4D); A508 (1) **SAE** J775 (K14072) |
| **K14073** | Alloy Steel | **C** 0.36-0.44 **Cr** 0.80-1.15 **Mn** 0.45-0.70 **Mo** 0.50-0.65 **P** 0.025 max **S** 0.025 max **Si** 0.15-0.35 **V** 0.25-0.35 | **ASTM** A540 (B21) |
| **K14185** | Alloy Steel | **C** 0.38-0.45 **Cr** 0.95-1.35 **Mn** 0.75-1.00 **Mo** 0.55-0.70 **P** 0.025 max **S** 0.020 max **Si** 0.20-0.35 **V** 0.20-0.30 | **MIL SPEC** MIL-S-46047 |

*Boxed entries are no longer active and are retained for reference purposes only.

The chemical compositions listed are for identification purposes and should not be used in lieu of the cross referenced specifications.

| UNIFIED NUMBER | DESCRIPTION | CHEMICAL COMPOSITION | CROSS REFERENCE SPECIFICATIONS |
|---|---|---|---|
| **K14245** | Alloy Steel | **C** 0.35-0.50 **Cr** 0.75-1.50 **Mn** 0.60-1.00 **Mo** 0.15 min **P** 0.025 max **S** 0.025 max **Si** 0.15-0.35 | |
| **K14247** | Alloy Steel | **C** 0.35-0.55 **Cr** 0.80-1.15 **Mn** 0.55-1.05 **Mo** 0.15-0.50 **P** 0.035 max **S** 0.040 max **Si** 0.15-0.35 | **ASTM** A649 (1A) |
| **K14248** | Alloy Steel | **C** 0.35-0.50 **Cr** 0.80-1.15 **Mn** 0.75-1.05 **Mo** 0.15-0.25 **P** 0.035 max **S** 0.04 max **Si** 0.15-0.35 | **ASME** A372 (5E, 8) |
| **K14358** | Alloy Steel | **C** 0.36-0.50 **Cr** 0.47-0.83 **Cu** 0.17-0.43 **Mn** 0.67-0.93 **Mo** 0.07 max **Ni** 0.47-0.83 **P** 0.06-0.125 **S** 0.055 max **Si** 0.25-0.55 | **ASTM** A325 (B); A563 (C3B) |
| **K14394** | Alloy Steel | **C** 0.41-0.46 **Mn** 0.75-1.00 **Mo** 0.45-0.60 **P** 0.025 max **S** 0.025 max **Si** 1.40-1.75 **V** 0.03-0.08 | **ASTM** A579 (33) |
| **K14501** | Alloy Steel | **C** 0.45 max **Mn** 0.90 max **P** 0.035 max **S** 0.035 max **Si** 0.15-0.35 **V** 0.03-0.12 | **ASTM** A469 (1); A470 (1) |
| **K14507** | Alloy Steel | **C** 0.45 max **Cr** 1.25 max **Cu** 0.35 **Mn** 0.40-0.90 **Mo** 0.15 min **Ni** 0.50 **P** 0.040 max **S** 0.040 max **Si** 0.15-0.30 **V** 0.10 max | **ASTM** A291 (3) |
| **K14508** | Alloy Steel | **C** 0.40-0.50 **Mn** 1.40-1.80 **Mo** 0.17-0.27 **P** 0.035 max **S** 0.04 max **Si** 0.15-0.35 | **ASTM** A372 (D) |
| **K14510** | Alloy Steel | **C** 0.40-0.50 **Mn** 0.70-0.90 **Mo** 0.20-0.30 **P** 0.035 max **S** 0.040 max **Si** 0.15-0.35 | **ASTM** A194 (4) |
| **K14520** | Alloy Steel | **C** 0.45 max **Mn** 0.50-0.90 **Mo** 0.30-0.60 **P** 0.040 max **S** 0.040 max **Si** 0.15-0.45 **V** 0.10-0.25 | |
| **K14542** | Alloy Steel | **C** 0.45 max **Cr** 0.70-1.25 **Mn** 0.60-1.00 **Mo** 0.15 min **P** 0.025 max **S** 0.020 max **Si** 0.15-0.35 | **ASTM** A288 (2, 3) |
| **K14557** | Alloy Steel | **C** 0.45 max **Cr** 1.50 max **Cu** 0.35 **Mn** 0.40-0.90 **Mo** 0.15 min **Ni** 1.00-3.00 **P** 0.040 max **S** 0.040 max **Si** 0.15-0.30 **V** 0.10 max | **ASTM** A291 (3A) |
| **K14675** | Alloy Steel (17-22A) | **C** 0.42-0.50 **Cr** 0.80-1.10 **Cu** 0.35 max **Mn** 0.40-0.70 **Mo** 0.45-0.65 **Ni** 0.25 max **P** 0.025 max **S** 0.025 max **Si** 0.20-0.35 **V** 0.25-0.35 | **AMS** 6304; 6305; AS7454; AS7455; AS7458; AS7459 **MIL SPEC** MIL-S-24502 **SAE** J467 (17-22-A) |
| **K15047** | Alloy Steel | **C** 0.48-0.53 **Cr** 0.80-1.10 **Mn** 0.70-0.90 **P** 0.020 max **S** 0.035 max **Si** 0.20-0.35 **V** 0.15 min | **ASTM** A232 |
| **K15048** | Alloy Steel | **C** 0.48-0.53 **Cr** 0.80-1.10 **Mn** 0.70-0.90 **P** 0.040 max **S** 0.040 max **Si** 0.20-0.35 **V** 0.15 min | **ASTM** A231 |
| **K15590** | Combined with G92540 | | |
| **K15747** | Alloy Steel | **C** 0.52-0.62 **Cr** 0.80-1.15 **Mn** 0.75-1.05 **Mo** 0.15-0.25 **P** 0.04 max **S** 0.04 max **Si** 0.20-0.35 | |
| **K17145** | Alloy Steel | **C** 0.65-0.77 **Mn** 0.75-1.05 **Mo** 0.90-1.10 **P** 0.04 max **S** 0.04 max **Si** 0.20-0.35 | |
| **K18597** | Alloy Steel | **C** 0.80-0.90 **Cr** 0.85-1.15 **Cu** 0.15 max **Mn** 0.20-0.50 **Mo** 0.50-0.65 **Ni** 0.15 max **P** 0.015 max **S** 0.015 max **Si** 0.60-0.90 | **AMS** 6426 |

The chemical compositions listed are for identification purposes and should not be used in lieu of the cross referenced specifications.

| UNIFIED NUMBER | DESCRIPTION | CHEMICAL COMPOSITION | CROSS REFERENCE SPECIFICATIONS |
|---|---|---|---|
| **K19195** | Alloy Steel (Bearing Steel) | **C** 0.85-1.00 **Cr** 1.40-1.80 **Cu** 0.35 max **Mn** 1.40-1.70 **Mo** 0.10 max **Ni** 0.25 max **P** 0.025 max **S** 0.025 max **Si** 0.50-0.80 | **ASTM** A485 (2) |
| **K19667** | Alloy Steel (Bearing Steel) | **C** 0.90-1.05 **Cr** 0.90-1.20 **Cu** 0.35 max **Mn** 0.95-1.25 **Mo** 0.10 max **Ni** 0.25 max **P** 0.025 max **S** 0.025 max **Si** 0.45-0.75 | **ASTM** A485 (1) |
| * **K19964** | Alloy Steel (Bearing Steel) | **C** 0.95-1.10 **Cr** 1.30-1.60 **Cu** 0.25 max **Mn** 0.25-0.45 **Mo** 0.08 max **Ni** 0.35 max **P** 0.025 max **S** 0.025 max **Si** 0.20-0.35 | **MIL SPEC** MIL-S-980 |
| **K19965** | Alloy Steel (Bearing Steel) | **C** 0.95-1.0 **Cr** 1.10-1.50 **Cu** 0.35 max **Mn** 0.65-0.90 **Mo** 0.20-0.30 **Ni** 0.25 max **P** 0.025 max **S** 0.025 max **Si** 0.15-0.35 | **ASTM** A485 (3) |
| **K19990** | Alloy Steel (Bearing Steel) | **C** 0.95-1.10 **Cr** 1.10-1.50 **Cu** 0.35 max **Mn** 1.05-1.35 **Mo** 0.45-0.60 **Ni** 0.25 max **P** 0.025 max **S** 0.025 max **Si** 0.15-0.35 | **ASTM** A485 (4) |
| **K20500** | Alloy Steel Electrode (ER70S-BL2) | **C** 0.05 max **Cr** 1.20-1.50 **Cu** 0.35 max **Mn** 0.40-0.70 **Mo** 0.40-0.65 **Ni** 0.20 max **P** 0.025 max **S** 0.025 max **Si** 0.40-0.70 **Other** total other elements (except Fe) 0.50 max | **ASME** SFA5.28 (ER70S-B2L) **AWS** A5.28 (ER70S-B2L) |
| **K20622** | Alloy Steel | **C** 0.06 max **Cb** 0.02 min **Cu** 1.00-1.30 **Mn** 0.40-0.65 **Ni** 1.20-1.50 **P** 0.025 max **S** 0.025 max **Si** 0.15-0.40 | **ASTM** A710 (B) |
| **K20747** | Alloy Steel (A350, Grade LF787) | **C** 0.07 max **Cb** 0.02 min **Cr** 0.60-0.90 **Cu** 1.00-1.30 **Mn** 0.40-0.70 **Mo** 0.15-0.25 **Ni** 0.70-1.00 **P** 0.025 max **S** 0.025 max **Si** 0.40 max **V** 0.03 max | **ASTM** A350 (LF787); A710 (A); A736; A859 |
| **K20900** | Alloy Steel Electrode (ER80S-B2) | **C** 0.07-0.12 **Cr** 1.20-1.50 **Cu** 0.35 max **Mn** 0.40-0.70 **Mo** 0.40-0.65 **Ni** 0.20 max **P** 0.025 max **S** 0.025 max **Si** 0.40-0.70 **Other** total other elements (except Fe) 0.50 max | **ASME** SFA5.28 (ER80S-B2); SFA5.30 (IN515) **AWS** A5.28 (ER80S-B2); A5.30 (IN515) **MIL SPEC** MIL-I-23413 (MIL-515) |
| **K20902** | Alloy Steel | **C** 0.09 max **Cb** 0.05-0.11 **Mn** 1.75-2.30 **Mo** 0.22-0.38 **P** 0.030 max **S** 0.035 max **Si** 0.17 max | **ASTM** A707 (L6) |
| **K20910** | Alloy Steel Electrode (ENi2N) | **C** 0.12 max **Cu** 0.08 max **Mn** 0.75-1.25 **Ni** 2.10-2.90 **P** 0.012 max **S** 0.020 max **Si** 0.05-0.25 **V** 0.05 max **Other** total other elements (except Fe) 0.50 max | **ASME** SFA5.23 (ENi2N) **AWS** A5.23 (ENi2N) |
| **K20915** | Alloy Steel Electrode (EM3N) | **Al** 0.10 max **C** 0.10 max **Cr** 0.55 max **Cu** 0.08 max **Mn** 1.40-1.80 **Mo** 0.25-0.65 **Ni** 1.90-2.60 **P** 0.012 max **S** 0.010 max **Si** 0.20-0.60 **Ti** 0.10 max **V** 0.05 max **Zr** 0.10 max **Other** total other elements (except Fe) 0.50 max | **ASME** SFA5.23 (EM3N) |
| **K20930** | Alloy Steel Electrode (EM4N) | **Al** 0.10 max **C** 0.10 max **Cr** 0.60 max **Cu** 0.08 max **Mn** 1.40-1.80 **Mo** 0.30-0.65 **Ni** 2.00-2.80 **P** 0.012 max **S** 0.010 max **Si** 0.20-0.60 **Ti** 0.10 max **V** 0.05 max **Zr** 0.10 max **Other** total other elements (except Fe) 0.50 max | **ASME** SFA5.23 (EM4N) **AWS** A5.23 (EM4N) |
| **K20934** | Alloy Steel | **C** 0.09 max **Cb** 0.02 min **Cr** 0.56-0.94 **Cu** 0.95-1.35 **Mn** 0.35-0.75 **Mo** 0.14-0.28 **Ni** 0.67-1.03 **P** 0.030 max **S** 0.035 max **Si** 0.37 max | **ASTM** A707 (L5) |

*Boxed entries are no longer active and are retained for reference purposes only.

The chemical compositions listed are for identification purposes and should not be used in lieu of the cross referenced specifications.

| UNIFIED NUMBER | DESCRIPTION | CHEMICAL COMPOSITION | CROSS REFERENCE SPECIFICATIONS |
|---|---|---|---|
| **K21010** | Alloy Steel Electrode (ENi2) | **C** 0.12 max **Cu** 0.35 max **Mn** 0.75-1.25 **Ni** 2.10-2.90 **P** 0.020 max **S** 0.020 max **Si** 0.05-0.30 **Other** total other elements (except Fe) 0.50 max | **ASME** SFA5.23 (ENi2) **AWS** A5.23 (ENi2) |
| **K21015** | Alloy Steel Electrode (EM3) | **Al** 0.10 max **C** 0.10 max **Cr** 0.55 max **Cu** 0.25 max **Mn** 1.40-1.80 **Mo** 0.25-0.65 **Ni** 1.90-2.60 **P** 0.010 max **S** 0.010 max **Si** 0.20-0.60 **Ti** 0.10 max **V** 0.04 max **Zr** 0.10 max **Other** total other elements (except Fe) 0.50 max | **ASME** SFA5.23 (EM3); SFA5.28 (ER110S-1) **AWS** A5.23 (EM3); A5.28 (ER110S-1) **MIL SPEC** MIL-E-23765/2 (MIL-110S-1) |
| **K21028** | Alloy Steel | **B** 0.007 max **C** 0.08-0.13 **Cr** 0.40-0.60 **Cu** 0.35 max **Mn** 0.75-1.00 **Mo** 0.20-0.30 **Ni** 1.65-2.00 **P** 0.025 max **S** 0.025 max **Si** 0.20-0.40 **V** 0.03-0.08 | **AMS** 6266 (43BV12) |
| **K21030** | Alloy Steel Electrode (EM4) | **Al** 0.10 max **C** 0.10 max **Cr** 0.60 max **Cu** 0.25 max **Mn** 1.40-1.80 **Mo** 0.30-0.65 **Ni** 2.00-2.80 **P** 0.010 max **S** 0.010 max **Si** 0.20-0.60 **Ti** 0.10 max **V** 0.03 max **Zr** 0.10 max **Other** total other elements (except Fe) 0.50 max | **ASME** SFA5.23 (EM4); SFA5.28 (ER120S-1) **AWS** A5.23 (EM4); A5.28 (ER120S-1) **MIL SPEC** MIL-E-23765/2 (MIL-120S-1, MIL-120S-2) |
| **K21035** | Alloy Steel Electrode (EF6N) | **C** 0.07-0.15 **Cr** 0.20-0.55 **Cu** 0.08 max **Mn** 1.45-1.90 **Mo** 0.40-0.65 **Ni** 1.75-2.25 **P** 0.012 max **S** 0.015 max **Si** 0.10-0.30 **V** 0.05 max **Other** total other elements (except Fe) 0.50 max | **ASME** SFA5.23 (EF6N) **AWS** A5.23 (EF6N) |
| **K21135** | Alloy Steel Electrode (EF6) | **C** 0.07-0.15 **Cr** 0.20-0.55 **Cu** 0.35 max **Mn** 1.45-1.90 **Mo** 0.40-0.65 **Ni** 1.75-2.25 **P** 0.015 max **S** 0.015 max **Si** 0.10-0.30 **Other** total other elements (except Fe) 0.50 max | **ASME** SFA5.23 (EF6) **AWS** A5.23 (EF6) |
| **K21205** | Alloy Steel | **Al** 0.06 max **C** 0.12 max **Cr** 0.90-1.20 **Mn** 0.45-0.75 **Mo** 0.25-0.40 **Ni** 0.90-1.20 **P** 0.035 max **S** 0.015 max **Si** 0.40 max | **ASTM** A734 (A) |
| **K21240** | Alloy Steel Electrode (ER80S-Ni2) | **C** 0.12 max **Cu** 0.35 max **Mn** 1.25 max **Ni** 2.00-2.75 **P** 0.025 max **S** 0.025 max **Si** 0.40-0.80 **Other** total other elements (except Fe) 0.50 max | **ASME** SFA5.28 (ER80S-Ni2) **AWS** A5.28 (ER80S-Ni2) |
| **K21350** | Alloy Steel Electrode (EF2N) | **C** 0.10-0.18 **Cu** 0.08 max **Mn** 1.70-2.40 **Mo** 0.40-0.65 **Ni** 0.40-0.80 **P** 0.012 max **S** 0.025 max **Si** 0.20 max **V** 0.05 max **Other** total other elements (except Fe) 0.50 max | **ASME** SFA5.23 (EF2N) **AWS** A5.23 (EF2N) |
| **K21351** | Alloy Steel Electrode (EA3KN) | **C** 0.07-0.12 **Cu** 0.08 max **Mn** 1.60-2.10 **Mo** 0.40-0.60 **P** 0.012 max **S** 0.025 max **Si** 0.50-0.80 **V** 0.05 max | **ASME** SFA5.23 (EA3KN) **AWS** A5.23 (EA3KN) |
| **K21370** | Alloy Steel Electrode | **Al** 0.10 max **C** 0.13 max **Cr** 0.60 max **Mn** 1.40-2.35 **Mo** 0.30-0.60 **Ni** 1.00-2.80 **P** 0.012 max **Si** 0.25-0.65 **S** 0.008 max **Ti** 0.10 max **V** 0.03 max **Zr** 0.10 max | |
| **K21385** | Alloy Steel Electrode (EF3N) | **C** 0.10-0.18 **Cu** 0.08 max **Mn** 1.70-2.40 **Mo** 0.40-0.65 **Ni** 0.70-1.10 **P** 0.012 max **S** 0.025 max **Si** 0.30 max **V** 0.05 max **Other** total other elements (except Fe) 0.50 max | **ASME** SFA5.23 (EF3N) **AWS** A5.23 (EF3N) |
| **K21390** | Alloy Steel | **C** 0.05-0.15 **Cr** 2.00-2.50 **Mn** 0.30-0.60 **Mo** 0.90-1.10 **P** 0.025 max **S** 0.025 max **Si** 0.50 max | **ASTM** A541 (22B); A739 (B22) |
| **K21450** | Alloy Steel Electrode (EF2) | **C** 0.10-0.18 **Cu** 0.35 max **Mn** 1.70-2.40 **Mo** 0.40-0.65 **Ni** 0.40-0.80 **P** 0.025 max **S** 0.025 max **Si** 0.20 max **Other** total other elements (except Fe) 0.50 max | **ASME** SFA5.23 (EF2) **AWS** A5.23 (EF2) |

The chemical compositions listed are for identification purposes and should not be used in lieu of the cross referenced specifications.

| UNIFIED NUMBER | DESCRIPTION | CHEMICAL COMPOSITION | CROSS REFERENCE SPECIFICATIONS |
|---|---|---|---|
| **K21451** | Alloy Steel Electrode (EA3K) | **C** 0.07-0.12 **Cu** 0.35 max **Mn** 1.60-2.10 **Mo** 0.40-0.60 **P** 0.025 max **S** 0.025 max **Si** 0.50-0.80 | **ASME** SFA5.23 (EA3K) **AWS** A5.23 (EA3K) **MIL SPEC** MIL-23765/2 (MIL-80S-3) |
| **K21485** | Alloy Steel Electrode (EF3) | **C** 0.10-0.18 **Cu** 0.35 max **Mn** 1.70-2.40 **Mo** 0.40-0.65 **Ni** 0.70-1.10 **P** 0.025 max **S** 0.025 max **Si** 0.30 max **Other** total other elements (except Fe) 0.50 max | **ASME** SFA5.23 (EF3) **AWS** A5.23 (EF3) **MIL SPEC** MIL-E-23765/2 (MIL-80S-1) |
| **K21509** | Alloy Steel | **C** 0.15 max **Cr** 1.65-2.35 **Mn** 0.30-0.60 **Mo** 0.44-0.65 **P** 0.030 max **S** 0.030 max **Si** 0.50 max | **ASTM** A369 (FP3b) |
| **K21590** | Alloy Steel | **C** 0.15 max **Cr** 2.00-2.50 **Mn** 0.30-0.60 **Mo** 0.90-1.10 **P** 0.030 max **S** 0.030 max **Si** 0.50 max | **ASTM** A182 (F22-1, F22-2, F22-3); A213 (T22); A234 (WP22); A250 (T22); A335 (P22); A336 (F22, F22a); A369 (FP22); A387 (22, 22L); A508 (22); A542 (A, B); A989 |
| **K21604** | Alloy Steel | **B** 0.0015-0.005 **C** 0.12-0.20 **Cr** 1.40-2.00 **Cu** 0.20-0.40 **Mn** 0.40-0.70 **Mo** 0.40-0.60 **P** 0.035 max **S** 0.04 max **Si** 0.20-0.35 **Ti** 0.04-0.10 | **ASTM** A514 (E); A517 (E) |
| **K21650** | Alloy Steel | **B** 0.001-0.005 **C** 0.12-0.21 **Cr** 0.85-1.20 **Mn** 0.45-0.70 **Mo** 0.45-0.60 **Ni** 1.20-1.50 **P** 0.035 max **S** 0.04 max **Si** 0.20-0.35 | **ASTM** A514 (P); A517 (P) |
| **K21703** | Alloy Steel | **C** 0.17 max **Mn** 0.70 max **Ni** 2.03-2.57 **P** 0.035 max **S** 0.040 max **Si** 0.13-0.32 | **ASTM** A203 (A) |
| **K21903** | Alloy Steel | **C** 0.19 max **Mn** 0.90 max **Ni** 2.03-2.57 **P** 0.04 max **S** 0.05 max **Si** 0.13-0.32 | **ASTM** A333 (7); A334 (7) |
| **K21940** | Alloy Steel (CBS600) | **Al** 0.03-0.12 **C** 0.16-0.22 **Cr** 1.25-1.65 **Cu** 0.35 max **Mn** 0.45-0.75 **Mo** 0.90-1.10 **Ni** 0.25 max **P** 0.010 max **S** 0.010 max **Si** 0.90-1.25 | **AMS** 6255 |
| **K22033** | Alloy Steel | **C** 0.33-0.38 **Cr** 0.55-0.75 **Cu** 0.35 max **Mn** 0.60-0.80 **Mo** 0.06 max **Ni** 1.10-1.40 **P** 0.025 max **S** 0.025 max **Si** 0.20-0.35 | **AMS** 6330 |
| **K22035** | Alloy Steel | **C** 0.20 max **Cu** 0.75-1.25 **Mn** 0.40-1.06 **Ni** 1.60-2.24 **P** 0.045 max **S** 0.050 max | **ASTM** A182 (FR); A234 (WPR); A333 (9); A334 (9); A714 (V) |
| **K22036** | Alloy Steel | **C** 0.20 max **Cu** 0.75-1.25 **Mn** 0.40-1.06 **Ni** 1.60-2.24 **P** 0.035 max **S** 0.040 max | **ASTM** A350 (LF9) |
| **K22094** | Alloy Steel | **C** 0.20 max **Cr** 1.00-1.50 **Mn** 0.30-0.80 **Mo** 0.44-0.65 **P** 0.040 max **S** 0.040 max **Si** 0.50-1.00 | |
| **K22097** | Alloy Steel | **C** 0.92-1.02 **Cr** 0.90-1.15 **Cu** 0.35 max **Mn** 0.95-1.25 **Mo** 0.08 max **Ni** 0.25 max **P** 0.015 max **S** 0.015 max **Si** 0.50-0.70 | **AMS** 6445 |
| **K22103** | Alloy Steel | **C** 0.21 max **Mn** 0.70 max **Ni** 2.03-2.57 **P** 0.035 max **S** 0.040 max **Si** 0.13-0.45 | **ASTM** A203 (B) |
| * **K22375** | Alloy Steel | **C** 0.23 max **Cr** 1.50-2.00 **Mn** 0.20-0.40 **Mo** 0.40-0.60 **Ni** 2.75-3.90 **P** 0.020 max **S** 0.020 max **Si** 0.15-0.40 **V** 0.03 max | |
| **K22440** | Alloy Steel | **C** 0.38-0.43 **Cr** 0.20 max **Cu** 0.35 max **Mn** 0.60-0.80 **Mo** 0.20-0.30 **Ni** 1.65-2.00 **P** 0.025 max **S** 0.025 max **Si** 0.15-0.35 | **AMS** 6312; 6317 |

*Boxed entries are no longer active and are retained for reference purposes only.

The chemical compositions listed are for identification purposes and should not be used in lieu of the cross referenced specifications.

| UNIFIED NUMBER | DESCRIPTION | CHEMICAL COMPOSITION | CROSS REFERENCE SPECIFICATIONS |
|---|---|---|---|
| **K22573** | Alloy Steel | **C** 0.25 max **Cr** 0.50 max **Mn** 0.60 max **Mo** 0.20-0.50 **Ni** 2.50 min **P** 0.015 max **S** 0.018 max **Si** 0.15-0.30 **V** 0.03 min | **ASTM** A469 (2) |
| **K22578** | Alloy Steel | **C** 0.25 max **Cr** 0.75 max **Mn** 0.020-0.60 **Mo** 0.25 min **Ni** 2.50 min **P** 0.015 max **S** 0.018 max **Si** 0.15-0.30 **V** 0.03 min | **ASTM** A470 (2) |
| **K22720** | Alloy Steel | **C** 0.18-0.23 **Cr** 0.80-1.20 **Cu** 0.50 max **H** 0.0010 max **Mn** 0.40-0.60 **Mo** 0.80-1.20 **N** 0.005 max **O** 0.0025 max **P** 0.015 max **S** 0.008 max **Si** 0.60-0.90 **V** 0.08-0.15 | **AMS** 6459 |
| **K22770** | Alloy Steel [17-22A(V)] | **C** 0.25-0.30 **Cr** 1.00-1.50 **Cu** 0.50 max **Mn** 0.60-0.90 **Mo** 0.40-0.60 **Ni** 0.50 max **P** 0.025 max **S** 0.025 max **Si** 0.55-0.75 **V** 0.75-0.95 | **AMS** 6303; 6436; 7496 **SAE** J467 (17-22-AV) |
| **K22773** | Alloy Steel | **C** 0.27 max **Cr** 0.50 max **Mn** 0.60 max **Mo** 0.20-0.50 **Ni** 2.50 min **P** 0.015 max **S** 0.018 max **Si** 0.15-0.30 **V** 0.03 min | **ASTM** A469 (3) |
| **K22878** | Alloy Steel | **C** 0.28 max **Cr** 0.75 max **Mn** 0.20-0.60 **Mo** 0.25 min **Ni** 2.50 min **P** 0.015 max **S** 0.018 max **Si** 0.15-0.30 **V** 0.03 min | **ASTM** A470 (3, 4) |
| **K22925** | Alloy Steel Welding Wire | **C** 0.26-0.32 **Cr** 0.90-1.20 **Cu** 0.35 max **H** 0.0010 max **Mn** 0.60-0.90 **Mo** 0.90-1.10 **N** 0.0050 max **Ni** 0.40-0.70 **O** 0.0025 max **P** 0.010 max **S** 0.010 max **Si** 0.10-0.30 **V** 0.05-0.10 | **AMS** 5027 **ASTM** A646 (D6AC) |
| **K22950** | Alloy Steel | **C** 0.49-0.55 **Cr** 0.70-0.90 **Cu** 0.35 max **Mn** 0.65-0.85 **Mo** 0.20-0.30 **Ni** 1.65-2.00 **P** 0.025 max **S** 0.025 max **Si** 0.15-0.35 | **AMS** 6396; 6424 |
| **K23010** | Alloy Steel | **C** 0.25-0.35 **Cr** 0.90-1.50 **Mn** 1.00 max **Mo** 1.00-1.50 **Ni** 0.75 max **P** 0.015 max **S** 0.018 max **Si** 0.15-0.35 **V** 0.20-0.30 | **ASTM** A470 (8) |
| **K23015** | Alloy Steel [17-22A(S)] | **C** 0.27-0.33 **Cr** 1.00-1.50 **Cu** 0.35 max **Mn** 0.45-0.65 **Mo** 0.40-0.60 **Ni** 0.25 max **P** 0.025 max **S** 0.025 max **Si** 0.55-0.75 **V** 0.20-0.30 | **AMS** 6302; 6385; 6458; 7496 **SAE** J467 (17-22-AS) |
| **K23016** | Alloy Steel Electrode (EB2H) | **C** 0.28-0.33 **Cr** 1.00-1.50 **Cu** 0.30 max **Mn** 0.45-0.65 **Mo** 0.40-0.65 **P** 0.015 max **S** 0.015 max **Si** 0.55-0.75 **V** 0.20-0.30 **Other** total other elements (except Fe) 0.50 max | **ASME** SFA5.23 (EB2H) **AWS** A5.23 (EB2H) |
| **K23028** | Alloy Steel | **C** 0.30 max **Cr** 0.75 max **Mn** 0.70 max **Mo** 0.25 min **Ni** 2.00 min **P** 0.035 max **S** 0.035 max **Si** 0.15-0.35 **V** 0.03-0.12 | |
| **K23080** | Alloy Steel | **C** 0.28-0.33 **Cr** 0.75-1.00 **Cu** 0.35 max **Mn** 0.75-1.00 **Mo** 0.35 0.60 **Ni** 1.65-2.00 **P** 0.015 max **S** 0.015 max **Si** 0.20-0.35 **V** 0.05-0.10 | **AMS** 6411; 6427 **ASTM** A646 (4330 Mod.-5) |
| **K23116** | Alloy Steel Welding Wire | **C** 0.28-0.33 **Cr** 1.00-1.50 **Cu** 0.08 max **Mn** 0.45-0.65 **Mo** 0.40-0.65 **P** 0.012 max **S** 0.015 max **Si** 0.55-0.75 **V** 0.20-0.30 **Other** total other elements (except Fe) 0.50 max | **ASME** SFA5.23 (EB2HN) **AWS** A5.23 (EB2HN) |
| **K23205** | Alloy Steel | **C** 0.27-0.37 **Cr** 0.85-1.25 **Mn** 0.70-1.00 **Mo** 1.00-1.50 **Ni** 0.50 max **P** 0.015 max **S** 0.015 max **Si** 0.20 min **V** 0.20-0.30 | **ASTM** A471 (10) |

The chemical compositions listed are for identification purposes and should not be used in lieu of the cross referenced specifications.

| UNIFIED NUMBER | DESCRIPTION | CHEMICAL COMPOSITION | CROSS REFERENCE SPECIFICATIONS |
|---|---|---|---|
| **K23477** | Alloy Steel | **C** 0.31-0.38 **Cr** 0.65-0.90 **Mn** 0.60-0.90 **Mo** 0.30-0.60 **Ni** 1.65-2.00 **P** 0.025 max **S** 0.025 max **Si** 0.20-0.35 **V** 0.17-0.23 | **ASTM** A579 (21); A646 (4335 Mod.-6) **AMS** 6428 |
| **K23505** | Tool Steel, Non-Tempering | **C** 0.32-0.39 **Cr** 0.55-0.75 **Cu** 0.50-0.70 **Mn** 0.35-0.50 **Mo** 0.50-0.70 **P** 0.025 max **S** 0.025 max **Si** 0.50-0.70 | |
| **K23510** | Nitriding Steel | **Al** 0.95-1.30 **C** 0.33-0.38 **Cr** 1.00-1.30 **Mn** 0.50-0.70 **Mo** 0.15-0.25 **P** 0.035 max **S** 0.040 max **Si** 0.15-0.35 | **ASTM** A355 (D) |
| **K23545** | Alloy Steel | **C** 0.35 max **Mn** 0.50-0.90 **Mo** 0.20-0.50 **Ni** 2.25-3.00 **P** 0.040 max **S** 0.040 max **Si** 0.10-0.40 **V** 0.15 max | |
| **K23550** | Alloy Steel | **C** 0.35 max **Cr** 0.80-2.00 **Mn** 0.90 max **Mo** 0.20-0.40 **Ni** 1.5-2.25 **P** 0.015 max **S** 0.015 max **Si** 0.35 max **V** 0.20 max | **ASTM** A723 (1) |
| * **K23577** | Alloy Steel Welding Wire | **C** 0.33-0.38 **Cr** 0.65-0.90 **H** 0.0010 max **Mn** 0.60-0.90 **Mo** 0.30-0.40 **N** 0.0050 max **Ni** 1.65-2.00 **O** 0.0025 max **P** 0.008 max **S** 0.008 max **Si** 0.25 max **V** 0.17-0.23 | |
| **K23578** | Alloy Steel | **C** 0.35 max **Cr** 0.75 max **Mn** 0.70 max **Mo** 0.25 min **Ni** 2.50 min **P** 0.035 max **S** 0.035 max **Si** 0.15-0.35 **V** 0.03-0.12 | |
| **K23579** | Alloy Steel | **C** 0.35 max **Cr** 1.25 max **Mn** 0.70 max **Mo** 0.25 min **Ni** 2.50 min **P** 0.035 max **S** 0.035 max **Si** 0.15-0.35 **V** 0.03 min | |
| **K23705** | Alloy Steel | **C** 0.37 max **Cr** 0.85-1.25 **Mn** 1.00 max **Mo** 1.00-1.50 **Ni** 0.50 max **P** 0.035 max **S** 0.035 max **Si** 0.15-0.35 **V** 0.20-0.30 | |
| **K23725** | Alloy Steel Welding Wire | **C** 0.34-0.40 **Cr** 0.90-1.20 **Cu** 0.35 max **H** 0.0010 max **Mn** 0.60-0.90 **Mo** 0.90-1.10 **N** 0.0050 max **Ni** 0.40-0.70 **O** 0.0025 max **P** 0.008 max **S** 0.008 max **Si** 0.15-0.30 **V** 0.05-0.10 | **AMS** 5028 |
| **K23745** | Nitriding Steel | **Al** 0.95-1.30 **C** 0.35-0.40 **Cr** 1.20-1.50 **Mn** 0.70-0.95 **Mo** 0.15-0.25 **P** 0.035 max **S** 0.060 max **Se** 0.15-0.25 **Si** 0.15-0.35 | **ASTM** A355 (B) |
| **K24040** | Alloy Steel | **C** 0.37-0.44 **Cr** 0.65-0.95 **Mn** 0.60-0.95 **Mo** 0.20-0.30 **Ni** 1.55-2.00 **P** 0.025 max **S** 0.025 max **Si** 0.15-0.35 | **ASTM** A649 (1B) |
| **K24045** | Alloy Steel | **C** 0.35-0.45 **Cr** 0.60-0.90 **Mn** 0.60-0.90 **Mo** 0.20-0.50 **Ni** 1.65-2.00 **P** 0.040 max **S** 0.040 max **Si** 0.20-0.35 **V** 0.10 max | **ASTM** A290 (G, H, I, J, K, L) |
| **K24055** | Alloy Steel | **C** 0.38-0.43 **Cr** 0.70-0.90 **Mn** 0.60-0.80 **Mo** 0.20-0.30 **Ni** 1.65-2.00 **P** 0.035 max **S** 0.04 max **Si** 0.15-0.35 | **ASTM** A372 (L) |
| **K24056** | Alloy Steel Welding Wire | **C** 0.35-0.40 **Cr** 0.70-0.90 **Cu** 0.30 max **H** 0.0010 max **Mn** 0.60-0.90 **Mo** 0.20-0.30 **N** 0.0050 max **Ni** 1.65-2.00 **O** 0.0025 max **P** 0.008 max **S** 0.008 max **Si** 0.15-0.35 | **AMS** 6456 |

*Boxed entries are no longer active and are retained for reference purposes only.

The chemical compositions listed are for identification purposes and should not be used in lieu of the cross referenced specifications.

| UNIFIED NUMBER | DESCRIPTION | CHEMICAL COMPOSITION | CROSS REFERENCE SPECIFICATIONS |
|---|---|---|---|
| **K24064** | Alloy Steel | **C** 0.37-0.44 **Cr** 0.70-0.95 **Mn** 0.70-0.90 **Mo** 0.30-0.40 **Ni** 1.65-2.00 **P** 0.025 max **S** 0.025 max **Si** 0.20-0.35 | **ASTM** A540 (B24) |
| **K24065** | Nitriding Steel (135) | **Al** 0.95-1.30 **C** 0.38-0.43 **Cr** 1.40-1.80 **Cu** 0.35 max **Mn** 0.50-0.80 **Mo** 0.30-0.40 **Ni** 0.25 max **P** 0.025 max **S** 0.025 max **Si** 0.20-0.40 | **AMS** 6470; 6471; 6472; AMS-S-6709 **ASTM** A355 (A); A519 (E7140); A646 (Nit. 135) |
| **K24070** | Alloy Steel | **C** 0.38-0.43 **Cr** 0.70-0.90 **Mn** 0.60-0.90 **Mo** 0.30-0.60 **Ni** 1.65-2.00 **P** 0.025 max **S** 0.025 max **Si** 0.20-0.35 **V** 0.05-0.10 | **ASTM** A540 (B24V); A579 (22) |
| **K24245** | Alloy Steel | **C** 0.35-0.50 **Cr** 0.60 min **Cu** 0.35 **Mn** 0.40-0.90 **Mo** 0.20-0.60 **Ni** 1.65 min **P** 0.040 max **S** 0.040 max **Si** 0.15-0.35 **V** 0.10 max | **ASTM** A291 (4 to 7) |
| **K24336** | Alloy Steel | **B** 0.0005-0.005 **C** 0.40-0.46 **Cr** 0.80-1.05 **Cu** 0.35 max **Mn** 0.75-1.00 **Mo** 0.45-0.60 **Ni** 0.60-0.90 **P** 0.025 max **S** 0.025 max **Si** 0.50-0.80 **V** 0.01-0.06 | **AMS** 6423 **ASTM** A646 (98BV40) |
| **K24340** | Cr-O-Mo-V Martensitic Steel | **C** 0.29-0.36 **Cr** 2.80-3.30 **Cu** 0.10 max **Mn** 0.40-0.70 **Mo** 0.70-1.20 **Ni** 0.30 max **P** 0.015 max **S** 0.005 max **Si** 0.10-0.40 **V** 0.15-0.35 | **AMS** 6481 **Other:** EN 1.7765 |
| **K24535** | Alloy Steel | **C** 0.45 max **Cr** 0.50-1.25 **Mn** 0.60-1.00 **Mo** 0.20 min **Ni** 1.65-3.50 **P** 0.025 max **S** 0.025 max **Si** 0.15-0.35 | |
| **K24562** | Alloy Steel | **C** 0.45 max **Cr** 0.70-1.25 **Mn** 1.00 max **Mo** 0.20 min **Ni** 1.65-3.50 **P** 0.025 max **S** 0.020 max **Si** 0.15-0.35 **V** 0.07-0.12 | **ASTM** A288 (4, 5, 6, 7, 8) |
| **K24628** | Alloy Steel combined with K24728 | | |
| **K24728** | Alloy Steel (D-6) | **C** 0.45-0.50 **Cr** 0.90-1.20 **Mn** 0.60-0.90 **Mo** 0.90-1.10 **Ni** 0.40-0.70 **P** 0.010 max **S** 0.010 max **Si** 0.15-0.30 **V** 0.08-0.15 | **AMS** 6431; 6432; 6438; 6439 **ASTM** A579 (23) **SAE** J467 (D6A) |
| **K24729** | Alloy Steel combined with K24728 | | |
| **K30560** | Alloy Steel Electrode (ER80S-B3L) | **C** 0.05 max **Cr** 2.30-2.70 **Cu** 0.35 max **Mn** 0.40-0.70 **Mo** 0.90-1.20 **Ni** 0.20 max **P** 0.025 max **S** 0.025 max **Si** 0.40-0.70 **Other** total other elements (except Fe) 0.50 max | **ASME** SFA5.28 (ER80S-B3L) **AWS** A5.28 (ER80S-B3L) |
| **K30736** | Alloy Steel (A 213, Grade T24) | **Al** 0.02 max **B** 0.0015-0.0070 **C** 0.05-0.10 **Cr** 2.20-2.60 **Mn** 0.30-0.60 **Mo** 0.90-1.10 **N** 0.012 max **P** 0.020 max **S** 0.010 max **Si** 0.15-0.45 **V** 0.20-0.30 | **ASTM** A213 (T24) **ASME** SA-213 (T24) |
| **K30960** | Alloy Steel Electrode (ER-90S-B3) | **C** 0.07-0.12 **Cr** 2.30-2.70 **Cu** 0.35 max **Mn** 0.40-0.70 **Mo** 0.90-1.20 **Ni** 0.20 max **P** 0.025 max **S** 0.025 max **Si** 0.40-0.70 **Other** total other elements (except Fe) 0.50 max | **ASME** SFA5.28 (ER90S-B3); SFA5.30 (IN521) **AWS** A5.28 (ER90S-B3); A5.30 (IN521) **MIL SPEC** MIL-I-23413 (MIL-521) |
| **K31015** | Alloy Steel Electrode (EB3N) | **C** 0.06-0.15 **Cr** 2.25-3.00 **Cu** 0.08 max **Mn** 0.40-0.80 **Mo** 0.90-1.10 **P** 0.012 max **S** 0.025 max **Si** 0.05-0.30 **V** 0.05 max **Other** total other elements (except Fe) 0.50 max | **ASME** SFA5.23 (EB3N) **AWS** A5.23 (EB3N) |

The chemical compositions listed are for identification purposes and should not be used in lieu of the cross referenced specifications.

| UNIFIED NUMBER | DESCRIPTION | CHEMICAL COMPOSITION | CROSS REFERENCE SPECIFICATIONS |
|---|---|---|---|
| **K31115** | Alloy Steel Electrode (EB3) | **C** 0.06-0.15 **Cr** 2.25-3.00 **Cu** 0.35 max **Mn** 0.40-0.80 **Mo** 0.90-1.10 **P** 0.025 max **S** 0.025 max **Si** 0.05-0.30 **Other** total other elements (except Fe) 0.50 max | **ASME** SFA5.23 (EB3) **AWS** A5.23 (EB3) |
| * **K31116** | Alloy Steel Electrode (EB3RR) | **As** 0.005 max **C** 0.05-0.15 **Cr** 2.25-3.30 **Cu** 0.15 max **Mn** 0.40-0.80 **Mo** 0.90-1.10 **P** 0.010 max **S** 0.010 max **Sb** 0.005 max **Si** 0.05-0.30 **Sn** 0.010 max | **ASME** SFA5.23 (EB3RR) **AWS** A5.23 (EB3RR) |
| * **K31117** | Alloy Steel Electrode (EB3LRN) | **As** 0.005 max **C** 0.05-0.15 **Cr** 2.25-3.30 **Cu** 0.08 max **Mn** 0.40-0.80 **Mo** 0.90-1.10 **P** 0.010 max **S** 0.010 max **Sb** 0.005 max **Si** 0.05-0.30 **V** 0.05 max | **ASME** SFA5.23 (EB3LRN) **AWS** A5.23 (EB3LRN) |
| **K31200** | High Strength Low Alloy Steel Electrode | **Al** 0.04 max **C** 0.12 max **Cr** 0.65-1.05 **Cu** 0.15 max **Mn** 1.50-2.00 **Mo** 0.40-1.00 **Ni** 1.95-3.10 **P** 0.010 max **Si** 0.30-0.50 **S** 0.010 max **Ti** 0.04 max **V** 0.04 max **Zr** 0.04 max | **MIL SPEC** MIL-E-24355 (MIL-140S-1) |
| **K31210** | Alloy Steel Electrode (ENi-3N) | **C** 0.13 max **Cr** 0.15 max **Cu** 0.08 max **Mn** 0.60-1.20 **Ni** 3.10-3.80 **P** 0.012 max **S** 0.020 max **Si** 0.05-0.30 **V** 0.05 max **Other** total other elements (except Fe) 0.50 max | **ASME** SFA5.23 (ENi3N) **AWS** A5.23 (ENi3N) |
| **K31240** | Alloy Steel Electrode (ER80S-Ni3) | **C** 0.12 max **Cu** 0.35 max **Mn** 1.25 max **Ni** 3.00-3.75 **P** 0.025 max **S** 0.025 max **Si** 0.40-0.80 **Other** total other elements (except Fe) 0.50 max | **ASME** SFA5.28 (ER80S-Ni3) **AWS** A5.28 (ER80S-Ni3) |
| **K31286** | Alloy Steel | **C** 0.10-0.15 **Cb** 0.015-0.070 **Cr** 2.75-3.25 **Mn** 0.30-0.60 **Mo** 0.90-1.10 **P** 0.025 max **S** 0.010 max **Si** 0.10 max **V** 0.20-0.30 | **ASTM** A832 (23V) |
| **K31310** | Alloy Steel Electrode (ENi3) | **C** 0.13 max **Cr** 0.15 max **Cu** 0.35 max **Mn** 0.60-1.20 **Ni** 3.10-3.80 **P** 0.012 max **S** 0.020 max **Si** 0.05-0.30 **Other** total other elements (except Fe) 0.50 max | **ASME** SFA5.23 (ENi3) **AWS** A5.23 (ENi3) |
| **K31390** | Alloy Steel | **C** 0.10-0.15 **Cb** 0.015-0.070 **Cr** 2.7-3.3 **Cu** 0.25 max **Mn** 0.30-0.60 **Mo** 0.90-1.10 **Ni** 0.25 max **P** 0.020 max **S** 0.010 max **Si** 0.10 max **Ti** 0.015 max **V** 0.20-0.30 | **ASTM** A182 (F3VCb); A541 (3VCb); A542 (E) |
| * **K31509** | Alloy Steel | **C** 0.15 max **Cr** 2.15-2.85 **Mn** 0.30-0.60 **Mo** 0.44-0.65 **P** 0.030 max **S** 0.030 max **Si** 0.50-1.00 | |
| **K31545** | Alloy Steel | **C** 0.15 max **Cr** 2.65-3.35 **Mn** 0.30-0.60 **Mo** 0.80-1.06 **P** 0.030 max **S** 0.030 max **Si** 0.50 max | **ASTM** A182 (F21); A213 (T21); A335 (P21); A336 (F21-1, F21-3); A369 (FP21); A387 (21); A989; K31545 |
| **K31718** | Alloy Steel | **C** 0.17 max **Mn** 0.70 max **Ni** 3.18-3.82 **P** 0.035 max **S** 0.040 max **Si** 0.13-0.32 | **ASTM** A203 (D) |
| **K31820** | Alloy Steel (HY80) | **C** 0.18 max **Cr** 1.00-1.80 **Cu** 0.25 max **Mn** 0.10-0.40 **Mo** 0.20-0.60 **Ni** 2.00-3.25 **P** 0.015 max **S** 0.008 max **Si** 0.15-0.35 **Ti** 0.02 max **V** 0.03 max | **ASTM** A372 (K) **MIL SPEC** MIL-S-16216; MIL-S-21952; MIL-S-22664; MIL-S-23009; MIL-S-24451 |

*Boxed entries are no longer active and are retained for reference purposes only.

The chemical compositions listed are for identification purposes and should not be used in lieu of the cross referenced specifications.

| UNIFIED NUMBER | DESCRIPTION | CHEMICAL COMPOSITION | CROSS REFERENCE SPECIFICATIONS |
|---|---|---|---|
| **K31830** | Alloy Steel | **B** 0.001-0.003 **C** 0.18 max **Cr** 2.75-3.25 **Mn** 0.30-0.60 **Mo** 0.90-1.10 **P** 0.02 max **S** 0.02 max **Si** 0.12 max **Ti** 0.015-0.035 **V** 0.20-0.30 | **ASTM** A182 (F3V); A508 (3V); A541 (3V); A542 (C); A832 |
| **K31835** | 2.25% Chromium, 1% Molybdenum, 0.25% Vanadium | **B** 0.0020 max **C** 0.11-0.15 **Ca** 0.015 max **Cb** 0.07 max **Cr** 2.00-2.50 **Cu** 0.20 max **Mn** 0.30-0.60 **Mo** 0.90-1.10 **Ni** 0.25 max **P** 0.015 **S** 0.010 max **Si** 0.10 max **Ti** 0.030 max **V** 0.25-0.35 **Other** rare earth metals 0.020 max | **ASME** SA-336 (F22V); SA-541 (22V); SA-542 (D); SA-832 (22V) **ASTM** A182 (F22V); A541 (22V); A542 (D); A832 (22V) |
| **K31918** | Alloy Steel | **C** 0.19 max **Mn** 0.31-0.64 **Ni** 3.18-3.82 **P** 0.05 max **S** 0.05 max **Si** 0.18-0.37 | **ASME** SA-420 (WPL3) **ASTM** A333 (3); A334 (3); A420 (3); A420 (WPL3) |
| **K32018** | Alloy Steel | **C** 0.20 max **Mn** 0.70 max **Ni** 3.18-3.82 **P** 0.035 max **S** 0.040 max **Si** 0.13-0.32 | **ASTM** A203 (E) |
| **K32025** | Alloy Steel | **C** 0.20 max **Mn** 0.90 max **Ni** 3.25-3.75 **P** 0.035 max **S** 0.040 max **Si** 0.20-0.35 | **ASTM** A350 (LF3) |
| **K32026** | Alloy Steel | **Al** 0.05 max **C** 0.20 max **Cr** 0.20 max **Cu** 0.35 max **Mn** 0.90 max **Mo** 0.06 max **Ni** 3.25-3.75 **P** 0.020 max **S** 0.020 max **Si** 0.15-0.35 **V** 0.05 max | **ASTM** A765 (III) |
| **K32045** | Alloy Steel (HY100) | **C** 0.20 max **Cr** 1.00-1.80 **Cu** 0.25 max **Mn** 0.10-0.40 **Mo** 0.20-0.60 **Ni** 2.25-3.50 **P** 0.015 max **S** 0.008 max **Si** 0.15-0.35 **Ti** 0.02 max **V** 0.03 max | **MIL SPEC** MIL-S-16216; MIL-S-21952; MIL-S-22664; MIL-S-23009; MIL-S-24451 |
| **K32218** | Alloy Steel | **C** 0.22 max **Mn** 1.00 max **Ni** 3.18-3.82 **P** 0.030 max **S** 0.035 max **Si** 0.37 max | **ASTM** A707 (L7) |
| **K32550** | Alloy Steel | **C** 0.23-0.28 **Cr** 0.20-0.40 **Mn** 1.20-1.50 **Mo** 0.35-0.45 **Ni** 1.65-2.00 **P** 0.025 max **S** 0.025 max **Si** 1.30-1.70 | **AMS** 6418; 6425; 6453; 7496; AMS-S-7108 **ASTM** A579 (31) |
| **K32723** | Alloy Steel | **C** 0.27 max **Cr** 0.50 max **Mn** 0.70 max **Mo** 0.20-0.60 **Ni** 3.00 min **P** 0.015 max **S** 0.018 max **Si** 0.15-0.30 **V** 0.03 min | **ASTM** A469 (4) |
| **K32800** | Alloy Steel | **C** 0.28 max **Cr** 0.75-2.00 **Mn** 0.70 max **Mo** 0.20-0.70 **Ni** 2.00-4.00 **P** 0.015 max **S** 0.015 max **Si** 0.15-0.35 **V** 0.05 min | **ASTM** A471 (1-6) |
| **K33020** | Alloy Steel | **C** 0.27-0.33 **Cr** 1.00-1.35 **Cu** 0.35 max **Mn** 0.60-0.80 **Mo** 0.35-0.55 **Ni** 1.85-2.25 **P** 0.025 max **S** 0.025 max **Si** 0.40-0.70 | **AMS** 6407 |
| **K33125** | Alloy Steel | **C** 0.31 max **Cr** 0.50 max **Mn** 0.70 max **Mo** 0.20-0.70 **Ni** 3.00 min **P** 0.015 max **S** 0.018 max **Si** 0.15-0.30 **V** 0.05-0.15 | **ASTM** A469 (5) |
| **K33370** | Alloy Steel | **C** 0.31-0.36 **Cr** 0.65-0.90 **Mn** 0.60-0.90 **Mo** 0.30-0.50 **Ni** 2.75 min **O** 25 ppm max **P** 0.005 max **S** 0.005 max **Si** 0.20-0.30 | **MIL SPEC** MIL-A-46173 |
| **K33517** | Alloy Steel | **C** 0.32-0.38 **Cr** 0.65-0.90 **Mn** 0.60-0.90 **Mo** 0.30-0.40 **Ni** 1.65-2.00 **P** 0.040 max **S** 0.040 max **Si** 0.20-0.60 **V** 0.17-0.23 | **AMS** 5029; 6429; 6430; 6433; 6434; 6435 |
| **K33585** | Alloy Steel | **C** 0.35 max **Cr** 3.00-3.60 **Mn** 0.50-0.90 **Mo** 0.30-0.50 **Ni** 0.50-1.00 **P** 0.040 max **S** 0.040 max **Si** 0.15-0.45 **V** 0.05-0.15 | |

*Boxed entries are no longer active and are retained for reference purposes only.

The chemical compositions listed are for identification purposes and should not be used in lieu of the cross referenced specifications.

| UNIFIED NUMBER | DESCRIPTION | CHEMICAL COMPOSITION | CROSS REFERENCE SPECIFICATIONS |
|---|---|---|---|
| * **K34025** | Alloy Steel | **C** 0.40 max **Mn** 0.85 max **Ni** 3.25-3.75 **P** 0.025 max **S** 0.025 max **Si** 0.20-0.35 | |
| **K34035** | Alloy Steel | **C** 0.40 max **Cr** 0.80-2.00 **Mn** 0.90 max **Mo** 0.30-0.50 **Ni** 2.25-3.25 **P** 0.015 max **S** 0.015 max **Si** 0.35 max **V** 0.20 max | **ASTM** A723 (2) |
| **K34378** | Alloy Steel | **C** 0.41-0.46 **Cr** 1.90-2.25 **Cu** 0.35 max **Mn** 0.75-1.00 **Mo** 0.45-0.60 **Ni** 0.25 max **P** 0.015 max **S** 0.015 max **Si** 1.40-1.75 **V** 0.03-0.08 | **AMS** 6406 |
| **K40712** | Alloy Steel | **B** 0.0005-0.006 **C** 0.04-0.10 **Cb** 0.02-0.08 **Cr** 1.90-2.60 **Mn** 0.10-0.60 **Mo** 0.05-0.30 **N** 0.030 max **P** 0.030 max **S** 0.010 max **Si** 0.50 max **W** 1.45-1.75 **V** 0.20-0.30 | **ASTM** A213 (T23) **ASME** SA-213 (T23) |
| **K41245** | Alloy Steel | **C** 0.12 max **Cr** 4.00-6.00 **Mn** 0.30-0.60 **Mo** 0.45-0.65 **P** 0.03 max **S** 0.03 max **Si** 0.50 max | **ASTM** A213 (T5c); A335 (P5c) |
| **K41370** | Alloy Steel Electrode (EF5) | **C** 0.10-0.17 **Cr** 0.25-0.50 **Cu** 0.50 max **Mn** 1.70-2.20 **Mo** 0.45-0.65 **Ni** 2.30-2.80 **P** 0.010 max **S** 0.010 max **Si** 0.20 max **Other** total other elements (except Fe) 0.50 max | **ASME** SFA5.23 (EF5) **AWS** A5.23 (EF5) |
| **K41545** | Alloy Steel | **C** 0.15 max **Cr** 4.00-6.00 **Mn** 0.30-0.60 **Mo** 0.45-0.65 **P** 0.030 max **S** 0.030 max **Si** 0.50 max | **ASTM** A182 (F5); A213 (T5); A234 (WP5); A335 (P5); A336 (F5); A369 (FP5); A387 (5) |
| **K41583** | Alloy Steel | **Al** 0.01-0.16 **C** 0.15 max **Mn** 0.25-0.66 **Mo** 0.17-0.38 **N** 0.025 max **Ni** 4.65-5.35 **P** 0.035 max **S** 0.035 max **Si** 0.18-0.37 | **ASTM** A645 |
| * **K41650** | Replaced by K40712 | | |
| **K42247** | Alloy Steel | **C** 0.22 max **Cr** 1.44-2.06 **Mn** 0.15-0.45 **Mo** 0.35-0.65 **Ni** 2.68-3.97 **P** 0.025 max **S** 0.025 max **Si** 0.37 max **V** 0.05 max | **ASTM** A707 (L8) |
| **K42338** | Alloy Steel | **C** 0.23 max **Cr** 1.44-2.06 **Mn** 0.40 max **Mo** 0.41-0.64 **Ni** 2.53-3.32 **P** 0.035 max **S** 0.040 max **Si** 0.18-0.37 **V** 0.03 max | |
| **K42339** | Alloy Steel | **C** 0.23 max **Cr** 1.44-2.06 **Mn** 0.40 max **Mo** 0.41-0.64 **Ni** 2.53-3.32 **P** 0.020 max **S** 0.020 max **V** 0.03 max | **ASTM** A543/A543M(B) |
| **K42343** | Alloy Steel | **C** 0.23 max **Cr** 1.25-2.00 **Mn** 0.20-0.40 **Mo** 0.40-0.60 **Ni** 2.75-3.90 **P** 0.035 max **S** 0.040 max **Si** 0.30 max **V** 0.03 max | **ASTM** A541 (4N) |
| **K42348** | Alloy Steel | **C** 0.23 max **Cr** 1.25-2.00 **Mn** 0.20-0.40 **Mo** 0.40-0.60 **Ni** 2.75-3.90 **P** 0.035 max **S** 0.040 max **Si** 0.30 max **V** 0.08 max | **ASTM** A372 (M); A541 (5) |
| **K42365** | Alloy Steel | **C** 0.23 max **Cr** 1.50-2.00 **Mn** 0.20-0.40 **Mo** 0.40-0.60 **Ni** 2.75-3.90 **P** 0.020 max **S** 0.020 max **Si** 0.30 max **V** 0.08 max | **ASME** SA-372 (M) **ASTM** A508 (5); S372 (m) |
| **K42544** | Alloy Steel | **C** 0.25 max **Cr** 4.0-6.0 **Mn** 0.60 max **Mo** 0.44-0.65 **Ni** 0.50 max **P** 0.040 max **S** 0.030 max **Si** 0.50 max | **ASTM** A182 (F5a); A336 (F5A) |

*Boxed entries are no longer active and are retained for reference purposes only.

The chemical compositions listed are for identification purposes and should not be used in lieu of the cross referenced specifications.

| UNIFIED NUMBER | DESCRIPTION | CHEMICAL COMPOSITION | CROSS REFERENCE SPECIFICATIONS |
|---|---|---|---|
| **K42570** | Alloy Steel | **C** 0.23-0.27 **Cr** 1.25-1.75 **Mn** 0.70-1.00 **Mo** 0.20-0.30 **Ni** 3.25-3.75 **P** 0.020 max **S** 0.020 **Si** 0.20-0.35 | |
| **K42598** | Alloy Steel | **C** 0.23-0.28 **Cb** 0.03-0.07 **Cr** 1.40-1.65 **Mn** 0.20 max **Mo** 0.8-1.0 **Ni** 2.75-3.25 **P** 0.01 max **S** 0.01 max **Si** 0.10 max | **ASTM** A579 (11) |
| **K42885** | Alloy Steel | **C** 0.28 max **Cr** 1.25-2.00 **Mn** 0.60 max **Mo** 0.30-0.60 **Ni** 3.25-4.00 **P** 0.015 max **S** 0.018 max **Si** 0.15-0.30 **V** 0.05-0.15 | **ASTM** A469 (6, 7, 8); A470 (5, 6, 7) |
| **K43170** | Alloy Steel | **C** 0.29-0.33 **Mn** 0.70-1.00 **Mo** 0.20-0.30 **Ni** 3.25-3.75 **P** 0.020 max **S** 0.020 max **Si** 0.20-0.35 | |
| **K44045** | Alloy Steel | **C** 0.40 max **Cr** 0.80-2.00 **Mn** 0.90 max **Mo** 0.40-0.80 **Ni** 3.25-4.50 **P** 0.015 max **S** 0.015 max **Si** 0.35 max **V** 0.20 max | **ASTM** A723 (3) |
| **K44210** | Alloy Steel replaced by K44220 | | |
| **K44220** | Alloy Steel (300M) | **C** 0.38-0.46 **Cr** 0.70-0.95 **Mn** 0.60-0.90 **Mo** 0.30-0.65 **Ni** 1.65-2.00 **P** 0.010 max **S** 0.010 max **Si** 1.45-1.80 **V** 0.05 min | **AMS** 6257; 6417; 6419 **ASTM** A579 (32); A646 (300M-8) **SAE** J467 (300M) |
| **K44315** | Alloy Steel (300M) combined with K44220 | | |
| **K44414** | Alloy Steel | **C** 0.11-0.17 **Cr** 1.00-1.40 **Cu** 0.35 max **Mn** 0.40-0.70 **Mo** 0.08-0.15 **Ni** 3.00-3.50 **P** 0.025 max **S** 0.025 max **Si** 0.15-0.35 | **AMS** 6263; 6264 |
| **K44910** | Alloy Steel | **C** 0.07-0.13 **Cr** 1.25-1.75 **Cu** 0.35 max **Mn** 0.40-0.70 **Mo** 0.06 max **Ni** 3.25-3.75 **P** 0.025 max **S** 0.025 max **Si** 0.20-0.35 | **AMS** 6250 |
| **K51210** | Alloy Steel, (Krupp) | **C** 0.10-0.15 **Cr** 1.35-1.75 **Mn** 0.45-0.65 **Ni** 3.75-4.25 **Si** 0.15-0.35 | |
| **K51255** | Alloy Steel (HY30) | **C** 0.12 max **Cr** 0.40-0.70 **Mn** 0.60-0.90 **Mo** 0.30-0.65 **Ni** 4.75-5.25 **P** 0.010 max **S** 0.010 max **Si** 0.20-0.35 **V** 0.05-0.10 | **ASTM** A579 (12) **MIL SPEC** MIL-S-24371 |
| **K51545** | Alloy Steel | **C** 0.15 max **Cr** 4.00-6.00 **Mn** 0.30-0.60 **Mo** 0.45-0.65 **P** 0.03 max **S** 0.03 max **Si** 1.00-2.00 | **ASTM** A213 (T5b); A335 (P5b) |
| **K52355** | Alloy Steel | **Al** 1.10-1.40 **C** 0.21-0.26 **Cr** 1.00-1.25 **Cu** 0.35 max **Mn** 0.50-0.70 **Mo** 0.20-0.30 **Ni** 3.25-3.75 **P** 0.025 max **S** 0.025 max **Si** 0.20-0.40 | **AMS** 6475 |
| **K52440** | Nitriding Steel | **Al** 0.95-1.30 **C** 0.22-0.27 **Cr** 1.00-1.35 **Mn** 0.50-0.70 **Mo** 0.20-0.30 **Ni** 3.25-3.75 **P** 0.035 max **S** 0.040 max **Si** 0.15-0.35 | **ASTM** A355 (C) |
| **K61595** | Alloy Steel replaced by S50300 | | |
| **K61650** | Alloy Steel (Alloy X-2M, Vasco X-2M) | **C** 0.16 max **Cr** 4.5-5.5 **Mn** 0.5 max **Mo** 1.0-2.0 **P** 0.02 max **S** 0.015 max **Si** 1.0 max **V** 0.5 max **W** 1.0-2.0 | **Other:** BMS 7-223 |
| **K63005** | Valve Steel (CNS) replaced by S63005 | | |
| **K63007** | Valve Steel (21-55 N) replaced by S63007 | | |

The chemical compositions listed are for identification purposes and should not be used in lieu of the cross referenced specifications.

| UNIFIED NUMBER | DESCRIPTION | CHEMICAL COMPOSITION | CROSS REFERENCE SPECIFICATIONS |
|---|---|---|---|
| **K63008** | Valve Steel (21-4 N) replaced by S63008 | | |
| **K63011** | Valve Steel (746) replaced by S63011 | | |
| **K63012** | Valve Steel (21-2 N) replaced by S63012 | | |
| **K63013** | Valve Steel (Gaman H) replaced by S63013 | | |
| **K63014** | Valve Steel (10) replaced by S63014 | | |
| **K63015** | Valve Steel (10 N) replaced by S63015 | | |
| **K63016** | Valve Steel (21-12) replaced by S63016 | | |
| **K63017** | Valve Steel (21-12 N) replaced by S63017 | | |
| **K63198** | Iron Base Superalloy (19-9-DL) replaced by S63198 | | |
| **K63199** | Iron Base Superalloy (19-9-DX or 19-9-W-Mo) replaced by S63199 | | |
| **K64005** | Valve Steel (2) replaced by S64005 | | |
| **K64006** | Valve Steel (F) replaced by S64006 | | |
| **K64152** | High-Strength Alloy Steel (M152) replaced by S64152 | | |
| **K64299** | Iron Base Superalloy (29-9) replaced by S64299 | | |
| **K65006** | Valve Steel (XB) replaced by S65006 | | |
| **K65007** | Valve Steel (1) replaced by S65007 | | |
| **K65150** | Iron Base Superalloy (Pyromet X-15) replaced by S65150 | | |
| **K65770** | Iron Base Superalloy (AFC77) replaced by S65770 | | |
| **K66009** | Valve Steel (TPA) replaced by S66009 | | |
| **K66220** | Iron Base Superalloy (Discaloy) replaced by S66220 | | |
| **K66286** | Iron Base Superalloy (A286) replaced by S66286 | | |
| **K66545** | Iron Base Superalloy (W545) replaced by S66545 | | |
| **K66979** | Iron-Nickel Base Superalloy (D-979) replaced by N09979 | | |
| **K70640** | Iron, Chromium, Nickel, Titanium Alloy replaced by N09902 | | |
| **K71040** | Steel (Pyrowear Alloy 53) | **C** 0.07-0.13 **Cr** 0.75-1.25 **Cu** 1.80-2.30 **Mn** 0.25-0.50 **Mo** 3.00-3.50 **Ni** 1.60-2.40 **P** 0.015 max **S** 0.010 max **Si** 0.60-1.20 **V** 0.05-0.15 | **AMS** 6308 |

The chemical compositions listed are for identification purposes and should not be used in lieu of the cross referenced specifications.

| UNIFIED NUMBER | DESCRIPTION | CHEMICAL COMPOSITION | CROSS REFERENCE SPECIFICATIONS |
|---|---|---|---|
| **K71340** | Alloy Steel | **C** 0.13 max **Mn** 0.90 max **Ni** 7.40-8.60 **P** 0.035 max **S** 0.040 max **Si** 0.13-0.32 | **ASTM** A522 (II); A553 (II) |
| **K71350** | Cr-Ni-Mo Alloy Steel | **Al** 0.03-0.12 **C** 0.10-0.16 **Cr** 0.90-1.20 **Cu** 0.35 max **Mn** 0.40-0.70 **Mo** 4.00-5.00 **Ni** 2.75-3.25 **P** 0.010 max **S** 0.010 max **Si** 0.40-0.60 **V** 0.25-0.50 | **AMS** 6256 |
| **K74015** | Alloy Steel replaced by T20811 | | |
| **K81340** | Alloy Steel | **C** 0.13 max **Mn** 0.90 max **Ni** 8.40-9.60 **P** 0.045 max **S** 0.045 max **Si** 0.13-0.32 | **ASTM** A333 (8); A334 (8); A353M; A420 (WPL8); A522 (I); A553M (I); A844 |
| **K81590** | Alloy Steel replaced by S50400 | | |
| **K88165** | Bearing Steel replaced by T11350 | | |
| **K90901** | Ferritic Cr-Mo-V Steel (Grade 91, T91) | **Al** 0.04 max **C** 0.08-0.12 **Cr** 8.00-9.50 **Cb** 0.06-0.1 **Mn** 0.30-0.60 **Mo** 0.85-1.05 **Ni** 0.40 max **N** 0.030-0.070 **P** 0.020 max **Si** 0.20-0.50 **S** 0.010 max **V** 0.18-0.25 | **ASME** SA-188 (F91); SA-213 (T91); SA-234 (WP91); SA-335 (P91); SA-336 (F91); SA-369 (FP91); SA-387 (91) **ASTM** A213 (91); A182 (F91); A234 (WP91); A335 (P91); A336 (F91); A369 (FP91); A387 (91); A831 (T91) |
| **K90941** | Alloy Steel | **C** 0.15 max **Cr** 8.00-10.00 **Mn** 0.30-0.60 **Mo** 0.90-1.10 **P** 0.030 max **S** 0.030 max **Si** 0.50-1.00 | **ASTM** A182 (F9); A234 (WP9); A336 (F9); A369 (FP9); A989 |
| **K90987** | High-Speed Steel Welding Rod (ERFe-6) | **C** 0.5-0.9 **Cr** 3.0-5.0 **Mn** 0.50 max **Mo** 5.0-9.5 **Si** 0.50 max **V** 0.8-1.3 **W** 1.0-2.5 **Other** total other elements (except Fe) 1.0 max | |
| **K91002** | Alloy Steel (H-46, Nickelvac H-46) | **C** 0.20 max **Cr** 10.0-12.0 **Cu** 0.5 max **Mn** 1.0 max **Mo** 1.0 max **N** 0.05 max **Ni** 1.0 max **P** 0.03 max **S** 0.015 max **Si** 0.5 max **V** 0.3 max **Other** Cb+Ta 0.50 max | **Other:** PWA 1091; PDS 10705Z2 |
| **K91061** | Alloy Steel (A213, Grade T911, A335, Grade P911) | **Al** 0.04 max **B** 0.0003-0.006 **C** 0.09-0.13 **Cb** 0.06-0.10 **Cr** 8.5-9.5 **Mn** 0.30-0.60 **Mo** 0.90-1.10 **N** 0.040-0.090 **Ni** 0.40 max **P** 0.020 max **S** 0.010 max **Si** 0.10-0.50 **V** 0.18-0.25 **W** 0.90-1.10 | **ASTM** A213 (T911); A335 (P911) **ASME** SA-213 (T911) |
| **K91094** | Alloy Steel, Superstrength | **C** 0.42-0.47 **Co** 3.5-4.5 **Cr** 0.20-0.35 **Mn** 0.10-0.35 **Mo** 0.20-0.35 **Ni** 7.0-8.5 **P** 0.01 max **S** 0.01 max **Si** 0.10 max **V** 0.06-0.12 | **ASTM** A579 (83) |
| **K91122** | Alloy Steel, Superstrength (HP9-4-25) | **C** 0.24-0.30 **Co** 3.50-4.50 **Cr** 0.35-0.60 **Mn** 0.10-0.35 **Mo** 0.35-0.60 **Ni** 7.00-9.00 **P** 0.010 max **S** 0.010 max **Si** 0.10 max **V** 0.06-0.12 | **AMS** 6546 **ASTM** A579 (81) |
| **K91151** | Alloy Steel replaced by S41000 | | |
| **K91161** | Alloy Steel, Superstrength replaced by S41001 | | |
| * **K91209** | Manganese Steel, Non-magnetic | **C** 1.20-1.50 **Cr** 0.60 max **Mn** 12.00-15.00 **Mo** 0.10 max **Ni** 0.75 max **P** 0.08 max **S** 0.04 max **Si** 0.55 max | |
| **K91231** | Bearing Steel (M50 NIL) | **C** 0.11-0.15 **Co** 0.25 max **Cr** 4.00-4.25 **Cu** 0.10 max **Fe** rem **Mn** 0.15-0.35 **Mo** 4.00-4.50 **Ni** 3.20-3.60 **P** 0.015 max **S** 0.010 max **Si** 0.10-0.25 **V** 1.13-1.33 **W** 0.15 max | **AMS** 6278 |

*Boxed entries are no longer active and are retained for reference purposes only.

The chemical compositions listed are for identification purposes and should not be used in lieu of the cross referenced specifications.

| UNIFIED NUMBER | DESCRIPTION | CHEMICAL COMPOSITION | CROSS REFERENCE SPECIFICATIONS |
|---|---|---|---|
| **K91271** | Alloy Steel | **Al** 0.04 max **B** 0.0005-0.005 **C** 0.07-0.14 **Cb** 0.04-0.10 **Cr** 10.0-12.5 **Cu** 0.30-1.70 **Mn** 0.70 max **Mo** 0.25-0.60 **N** 0.040-0.100 **Ni** 0.50 max **P** 0.020 max **S** 0.010 max **Si** 0.50 max **V** 0.15-0.30 **W** 1.50-2.50 | **ASTM** A213 (T122); A335 (P122) **ASME** SA-213 (T122); SA-335 (P122) |
| **K91283** | Alloy Steel, Superstrength (HP9-4-30) | **C** 0.28-0.34 **Co** 4.0-5.0 **Cr** 0.90-1.10 **Mn** 0.10-0.35 **Mo** 0.90-1.10 **Ni** 7.0-8.5 **P** 0.01 max **S** 0.01 max **Si** 0.10 max **V** 0.06-0.12 | **AMS** 6524; 6526 **ASTM** A579 (82) |
| **K91308** | High-Speed Steel Welding Rod (RFe5-A) | **C** 0.7-1.0 **Cr** 3.0-5.0 **Fe** rem **Mn** 0.50 max **Mo** 4.0-6.0 **Si** 0.50 max **V** 1.0-2.5 **W** 5.0-7.0 **Other** total other elements (except Fe) 1.0 max | |
| * **K91313** | Alloy Steel, Superstrength | **C** 0.29-0.34 **Co** 4.25-4.75 **Cr** 0.90-1.10 **Mn** 0.10-0.35 **Mo** 0.90-1.10 **Ni** 7.00-8.00 **P** 0.010 max **S** 0.010 max **Si** 0.10 max **V** 0.06-0.12 | |
| **K91342** | Alloy Steel, Superstrength replaced by S42201 | | |
| **K91352** | Alloy Steel, Superstrength | **Al** 0.05 max **C** 0.20-0.25 **Cr** 11.00-12.50 **Mn** 0.50-1.00 **Mo** 0.90-1.25 **Ni** 0.50-1.00 **P** 0.025 max **S** 0.025 max **Si** 0.20-0.50 **Sn** 0.04 max **Ti** 0.05 max **V** 0.20-0.30 **W** 0.90-1.25 | **ASTM** A437 (B4B, B4C) |
| **K91401** | Alloy Steel | **C** 0.16-0.23 **Co** 4.15-5.10 **Cr** 0.61-0.89 **Mn** 0.20-0.40 **Mo** 0.86-1.14 **Ni** 8.40-9.60 **P** 0.010 max **S** 0.010 max **Si** 0.12 max **V** 0.04-0.14 | |
| **K91456** | Steel, High Expansion | **C** 0.55-0.65 **Mn** 5.00-6.00 **Ni** 8.50-10.50 **P** 0.040 max **S** 0.030 max **Si** 1.00 max | **AMS** 5623; 5625 |
| **K91461** | Nickel-Cobalt Steel Welding Wire | **C** 0.14-0.17 **Co** 3.50-4.00 **Cr** 0.90-1.05 **Cu** 0.10 max **H** 0.0010 max **Mn** 0.40-0.55 **Mo** 0.40-0.50 **N** 0.0080 max **Ni** 9.75-10.25 **O** 0.0050 max **P** 0.008 max **S** 0.008 max **Si** 0.15-0.25 **V** 0.06-0.10 | **AMS** 6468 |
| **K91470** | Iron, Electrical Heating Element Alloy | **Al** 2.75-3.75 **Cr** 12-15 | **ASTM** B603 (IV) |
| **K91472** | Alloy Steel (HP 9-4-20) | **C** 0.17-0.23 **Co** 4.25-4.75 **Cr** 0.65-0.85 **Cu** 0.35 max **Mn** 0.20-0.40 **Mo** 0.90-1.10 **Ni** 8.50-9.50 **P** 0.010 max **S** 0.010 max **Si** 0.20 max **V** 0.06-0.12 | **AMS** 6523; 6525 |
| **K91505** | Steel, High Expansion | **C** 0.50-0.60 **Cr** 3.00-5.00 **Cu** 0.50 max **Mn** 3.50-5.50 **Mo** 0.50 max **Ni** 11.0-14.0 **P** 0.040 max **S** 0.030 max **Si** 0.50 max | **AMS** 5624 |
| * **K91555** | Alloy Steel, Non-magnetic | **C** 0.40-0.75 **Cr** 3.50-6.00 **Mn** 6.00-10.00 **Ni** 6.00-10.00 **P** 0.05 max **S** 0.045 max **Si** 0.20-0.65 | |
| * **K91560** | Replaced by K90901 | | |
| **K91670** | Iron, Electrical Heating Element Alloy | **Al** 3.75-4.75 **Cr** 13-16 | **ASTM** B603 (III) |

*Boxed entries are no longer active and are retained for reference purposes only.

The chemical compositions listed are for identification purposes and should not be used in lieu of the cross referenced specifications.

| UNIFIED NUMBER | DESCRIPTION | CHEMICAL COMPOSITION | CROSS REFERENCE SPECIFICATIONS |
|---|---|---|---|
| **K91800** | Iron, Chromium Sealing Alloy | **C** 0.08 max **Cr** 18 nom **Fe** rem **Mn** 1.00 max **Ni** 0.50 max **P** 0.04 max **S** 0.03 max **Si** 0.75 max **Other** Ti 5\mtC-0.60 | **ASTM** F256 (I) |
| **K91870** | Iron, Electrical Heating Element Alloy | **Al** 4.75-5.75 **Cr** 14-17 | **ASTM** B603 (IIA) |
| **K91890** | Alloy Steel | **Al** 0.40 max **C** 0.03 max **Cr** 4.50-5.50 **Mn** 0.10 max **Mo** 2.75-3.25 **Ni** 11.50-12.50 **P** 0.010 max **S** 0.010 max **Si** 0.10 max **Ti** 0.20-0.35 | |
| **K91930** | Alloy Steel, Superstrength | **Al** 0.25-0.40 **C** 0.03 max **Cr** 4.75-5.25 **Mn** 0.10 max **Mo** 2.75-3.25 **Ni** 11.5-12.5 **P** 0.01 max **S** 0.01 max **Si** 0.12 max **Ti** 0.05-0.15 | **ASTM** A579 (74) |
| **K91940** | Alloy Steel, Superstrength | **Al** 0.35-0.50 **C** 0.03 max **Cr** 4.75-5.25 **Mn** 0.10 max **Mo** 2.75-3.25 **Ni** 11.5-12.5 **P** 0.01 max **S** 0.01 max **Si** 0.12 max **Ti** 0.10-0.25 | **ASTM** A579 (75) |
| **K91955** | Alloy Steel, Non-magnetic | **C** 0.40-0.60 **Cr** 3.50-6.00 **Mn** 16.00-20.00 **Ni** 2.00 max **P** 0.08 max **S** 0.025 max **Si** 0.20-0.65 | |
| **K91970** | Combined with K92571 | | |
| **K91971** | Nickel-Cobalt Steel Welding Wire | **Al** 0.01-0.03 **C** 0.10-0.14 **Co** 7.50-8.50 **Cr** 1.80-2.20 **H** 0.0003 max **Mn** 0.07-0.17 **Mo** 0.90-1.10 **N** 0.005 max **Ni** 9.50-10.50 **O** 0.0025 max **P** 0.006 max **S** 0.006 max **Si** 0.15-0.25 **Ti** 0.02 max **V** 0.04-0.09 | **AMS** 6465 |
| **K92100** | Iron, Thermostat Alloy | **Cr** 2 **Fe** 79 **Ni** 19 | **ASTM** B388 |
| **K92150** | Alloy Steel, Maraging (Alloy T-250, VascoMax T-250) | **Al** 0.05-0.20 **C** 0.03 max **Co** 0.50 max **Mn** 0.10 max **Mo** 2.5-3.5 **Ni** 18.0-19.0 **P** 0.01 max **S** 0.01 max **Si** 0.10 max **Ti** 1.00-1.50 | **AMS** 6518; 6519; 6591 **MIL SPEC** MIL-S-47319 |
| **K92350** | Iron, Thermostat Alloy | **Cr** 8.5 **Fe** 66.5 **Ni** 25 | **ASTM** B388 |
| **K92400** | Iron, Electrical Heating Element Alloy | **Al** 4.00-5.25 **Cr** 20-24 | **ASTM** B603 (IIB) |
| **K92460** | Ferritic Cr-Mo-W-V-Cb Steel (T92) | **Al** 0.04 max **B** 0.006 max **C** 0.06-0.13 **Cb** 0.03-0.10 **Cr** 8.00-9.50 **Mn** 0.30-0.60 **Mo** 0.30-0.60 **N** 0.030-0.070 **Ni** 0.40 max **P** 0.020 max **S** 0.010 max **Si** 0.50 max **V** 0.15-0.25 **W** 1.50-2.20 | **ASTM** A213 (T92); A335 (T92) |
| **K92500** | Iron, Electrical Heating Element Alloy | **Al** 5.00-6.00 **Cr** 20-24 | **ASTM** B603 (I) |
| **K92510** | Iron, Thermostat Alloy | **Cr** 3 **Fe** 75 **Ni** 22 | **ASTM** B388 |
| **K92571** | Alloy Steel (AF 1410) | **Al** 0.015 max **C** 0.13-0.17 **Co** 13.5-14.5 **Cr** 1.80-2.20 **Mn** 0.10 max **Mo** 0.90-1.10 **N** 0.0015 max **Ni** 9.50-10.50 **O** 0.0015 max **P** 0.008 max **S** 0.005 max **Si** 0.10 max **Ti** 0.015 max | **AMS** 6522; 6527; 6533; 6543; 6544 |
| **K92580** | Martensitic Age-Hardenable Alloy | **Al** 0.015 max **C** 0.21-0.25 **Co** 12.5-14.5 **Cr** 2.85-3.35 **Mn** 0.10 max **N** 0.0015 max **Ni** 10.5-12.5 **O** 0.0015 max **P** 0.008 max **S** 0.005 max **Si** 0.10 max **Ti** 0.015 max | **AMS** 6478; 6532; 7908; 7910 |
| **K92650** | Fe-Co Soft Magnetic Alloy | **C** 0.025 max **Co** 26.50-28.50 **Cr** 0.75 max **Fe** rem **Mn** 0.35 max **Ni** 0.75 max **P** 0.015 max **S** 0.015 max **Si** 0.35 max **V** 0.35 max | **ASTM** A801 (Alloy 2) |

The chemical compositions listed are for identification purposes and should not be used in lieu of the cross referenced specifications.

| UNIFIED NUMBER | DESCRIPTION | CHEMICAL COMPOSITION | CROSS REFERENCE SPECIFICATIONS |
|---|---|---|---|
| **K92710** | Alloy Steel, Maraging (Alloy C-200, VascoMax C-200) | **Al** 0.05-0.15 **C** 0.03 max **Co** 7.0-9.0 **Mn** 0.10 max **Mo** 3.0-3.6 **Ni** 17.0-19.0 **P** 0.01 max **S** 0.01 max **Si** 0.10 max **Ti** 0.10-0.30 | **MIL SPEC** MIL-S-46850 |
| **K92801** | Iron, Chromium Sealing Alloy | **C** 0.12 max **Cr** 28 nom **Fe** rem **Mn** 1.00 max **N** 0.20 max **Ni** 0.50 max **P** 0.04 max **S** 0.03 max **Si** 0.75 max | **ASTM** F256 (II) |
| * **K92810** | Alloy Steel, Maraging | **Al** 0.05-0.15 **C** 0.03 max **Co** 7.0-8.5 **Mn** 0.10 max **Mo** 4.0-4.5 **Ni** 17.0-19.0 **P** 0.010 max **S** 0.010 max **Si** 0.10 max **Ti** 0.10-0.25 | |
| **K92820** | Alloy Steel, Superstrength | **Al** 0.05-0.15 **C** 0.03 max **Co** 8.0-9.0 **Mn** 0.10 max **Mo** 3.0-3.5 **Ni** 17.0-19.0 **P** 0.01 max **S** 0.01 max **Si** 0.10 max **Ti** 0.15-0.25 | **ASTM** A579 (71); A646 (18); A646 (Marage 200) **SAE** J1099 (A538A) |
| **K92850** | Iron, Thermostat Alloy | **Al** 5 **Fe** 71.5 **Mn** 9.5 **Ni** 14 | **ASTM** B388 |
| **K92890** | Alloy Steel, Maraging | **Al** 0.05-0.15 **C** 0.03 max **Co** 7.0-8.5 **Mn** 0.10 max **Mo** 4.6-5.1 **Ni** 17.0-19.0 **P** 0.010 max **S** 0.010 max **Si** 0.10 max **Ti** 0.30-0.50 | **AMS** 6501; 6512; 6520 **ASTM** A646 (19); A646 (Marage 250) **SAE** J1099 (A538B) |
| * **K92930** | Replaced by K91271 | | |
| **K92940** | Alloy Steel, Superstrength | **Al** 0.05-0.15 **C** 0.03 max **Co** 7.5-8.5 **Mn** 0.10 max **Mo** 4.6-5.2 **Ni** 17.0-19.0 **P** 0.01 max **S** 0.01 max **Si** 0.10 max **Ti** 0.30-0.50 | **ASTM** A579 (72) |
| **K93050** | Fe-Ni Low Expansion Alloy, Free Machining (Invar FC) | **C** 0.15 max **Co** 0.50 max **Cr** 0.25 max **Fe** rem **Mn** 1.00 max **Ni** 34.0-38.0 **P** 0.020 max **S** 0.020 max **Se** 0.15-0.30 **Si** 0.35 max | **ASTM** F1684 |
| **K93120** | Alloy Steel, Maraging | **Al** 0.05-0.15 **C** 0.03 max **Co** 8.0-9.5 **Mn** 0.10 max **Mo** 4.6-5.2 **Ni** 18.0-19.0 **P** 0.010 max **S** 0.010 max **Si** 0.10 max **Ti** 0.55-0.80 | **AMS** 6514; 6521 **ASTM** A579 (73); A646 (20); A646 (Marage 300) **SAE** J1099 (A538C) |
| **K93130** | Nickel-Cobalt Steel Welding Wire | **Al** 0.05-0.15 **C** 0.010 max **Co** 8.00-9.00 **H** 0.0025 max **Mn** 0.10 max **Mo** 4.50-6.00 **N** 0.005 max **Ni** 18.00-19.00 **O** 0.0025 max **P** 0.010 max **S** 0.010 max **Si** 0.10 max **Ti** 0.65-0.80 | **AMS** 6463 |
| **K93160** | Combined with K93120 | | |
| **K93500** | Fe-Ni-Co Low Expansion Alloy (Super Invar) | **Al** 0.10 max **C** 0.05 max **Co** 4.0-6.5 **Cr** 0.25 max **Fe** rem **Mg** 0.10 max **Mn** 0.60 max **Ni** 30.5-33.5 **P** 0.015 max **S** 0.015 max **Si** 0.25 max **Ti** 0.10 max **Zr** 0.10 max | **ASTM** F1684 |
| **K93540** | Alloy Steel, Maraging (Alloy C-350, VascoMax C-350) | **Al** 0.05-0.20 **C** 0.03 max **Co** 11.5-12.5 **Mn** 0.10 max **Mo** 4.6-5.2 **Ni** 18.0-19.0 **P** 0.01 max **S** 0.01 max **Si** 0.10 max **Ti** 1.25-1.50 | **MIL SPEC** MIL-S-46850 |
| **K93600** | Iron, Thermostat Alloy (Invar) | **Fe** 64 **Ni** 36 | **ASTM** B388 |
| **K93601** | Nickel Steel, 36% Ni (A 333 Gr 11, A 334 Gr 11) | **C** 0.10 max **Co** 0.50 max **Cr** 0.50 max **Mn** 0.60 max **Mo** 0.50 max **Ni** 35.0-37.0 **P** 0.025 max **S** 0.025 max **Si** 0.35 max | **ASTM** A333 (11); A334 (11) |
| **K93602** | Nickel Steel, Free Cutting, 36% Ni | **C** 0.15 max **Mn** 1.00 max **Ni** 35.0-36.5 **Se** 0.15-0.25 **Si** 0.35 max | |

*Boxed entries are no longer active and are retained for reference purposes only.

The chemical compositions listed are for identification purposes and should not be used in lieu of the cross referenced specifications.

| UNIFIED NUMBER | DESCRIPTION | CHEMICAL COMPOSITION | CROSS REFERENCE SPECIFICATIONS |
|---|---|---|---|
| **K93603** | Fe-Ni Low Expansion Alloy (Invar) | **Al** 0.10 max **C** 0.05 max **Co** 0.50 max **Cr** 0.25 max **Fe** rem **Mg** 0.10 max **Mn** 0.60 max **Ni** 34.0-38.0 **P** 0.015 max **S** 0.015 max **Si** 0.40 max **Ti** 0.10 max **Zr** 0.10 max | **ASTM** F1684 |
| **K93800** | Iron, Thermostat Alloy | **Co** 7 **Fe** 62 **Ni** 31 | **ASTM** B388 |
| **K94000** | Iron, Thermostat Alloy | **Fe** 60 **Ni** 40 | **ASTM** B388 |
| **K94100** | Iron, Nickel Sealing Alloy | **Al** 0.10 max **C** 0.05 max **Cr** 0.25 max **Mn** 0.80 max **Ni** 41 nom **P** 0.025 max **S** 0.025 max **Si** 0.30 max | **ASTM** F30 (42) |
| **K94101** | Iron, Nickel Sealing Alloy (DUMET) | **C** 0.10 max **Fe** rem **Mn** 0.75-1.25 **Ni** 41-43 **P** 0.02 max **S** 0.02 max **Si** 0.30 max | **ASTM** F29 **AMS** 7734 |
| **K94200** | Iron, Thermostat Alloy | **Fe** 58 **Ni** 42 | **ASTM** B388 |
| **K94490** | Fe-Ni Ferromagnetic Alloy | **C** 0.05 max **Co** 0.50 max **Cr** 0.30 max **Cu** 0.30 max **Fe** rem **Mn** 0.80 max **Mo** 0.30 max **Ni** 43.5-46.5 **P** 0.03 max **S** 0.01 max **Si** 0.50 max | **ASTM** A753 (Alloy 1) |
| **K94500** | Iron, Thermostat Alloy | **Fe** 55 **Ni** 45 | **ASTM** B388 |
| **K94600** | Iron, Nickel Sealing Alloy | **Al** 0.10 max **C** 0.05 max **Cr** 0.25 max **Mn** 0.80 max **Ni** 46 nom **P** 0.025 max **S** 0.025 max **Si** 0.30 max | **ASTM** F30 (46) |
| **K94610** | Iron, Nickel-Cobalt Sealing Alloy (KOVAR) | **Al** 0.10 max **C** 0.04 max **Co** 17 nom **Cr** 0.20 max **Cu** 0.20 max **Fe** 53 nom **Mg** 0.10 max **Mn** 0.50 max **Mo** 0.20 max **Ni** 29 nom **Si** 0.20 max **Ti** 0.10 max **Zr** 0.10 max **Other** Al+Mg+Zr+Ti 0.20 max | **AMS** 7726; 7727; 7728 **ASTM** F15 |
| **K94620** | Fe-Ni-Co, Metal-to-Ceramic Sealing Alloy | **Al** 0.01 max **C** 0.02 max **Cr** 0.03 max **Co** 25 nom **Mg** 0.01 max **Mn** 0.35 max **Mo** 0.06 max **Ni** 27 nom **P** 0.006 max **S** 0.006 max **Si** 0.15 max **Ti** 0.01 max **Zr** 0.01 max | **ASTM** F1466 |
| **K94630** | Fe-Ni-Co Metal-to-Ceramic Sealing Alloy | **Al** 0.01 max **C** 0.02 max **Co** 17.0 nom **Cr** 0.03 max **Cu** 0.20 max **Mg** 0.01 max **Mn** 0.35 max **Mo** 0.06 max **Ni** 29 nom **P** 0.006 max **S** 0.006 max **Si** 0.15 max **Ti** 0.01 max **Zr** 0.01 max | **ASTM** F1466 |
| **K94760** | Iron, Nickel-Chromium Sealing Alloy | **Al** 0.20 max **C** 0.07 max **Cr** 5.6 nom **Fe** rem **Mn** 0.25 max **Ni** 42 nom **P** 0.025 max **S** 0.025 max **Si** 0.30 max | **ASTM** F31 |
| **K94800** | Iron, Nickel Sealing Alloy | **Al** 0.10 max **C** 0.05 max **Cr** 0.25 max **Fe** rem **Mn** 0.80 max **Ni** 48 nom **P** 0.025 max **S** 0.025 max **Si** 0.30 max | **ASTM** F30 (48) |
| **K94840** | Fe-Ni Ferromagnetic Alloy | **C** 0.05 max **Co** 0.50 max **Cr** 0.30 max **Cu** 0.30 max **Fe** rem **Mn** 0.80 max **Mo** 0.30 max **Ni** 47.0-49.0 **P** 0.03 max **S** 0.01 max **Si** 0.50 max | **ASTM** A753 (Alloy 2) |
| **K95000** | Iron, Nickel Thermostat Alloy | **Fe** 50 **Ni** 50 | **AMS** 7717; 7718; 7719 **ASTM** B388 |
| **K95050** | Iron, Nickel Sealing Alloy replaced by N14052 | | |
| **K95100** | Iron, Cobalt, Vanadium Ferromagnetic Alloy replaced by R30005 | | |

The chemical compositions listed are for identification purposes and should not be used in lieu of the cross referenced specifications.

# UNS NUMBERS ASSIGNED TO DATE

## With Description of Each Material Covered and References To Documents In Which The Same or Similar Materials are Described

## Lxxxxx Number Series
## Low Melting Metals and Alloys

# LOW MELTING METALS AND ALLOYS

| UNIFIED NUMBER | DESCRIPTION | CHEMICAL COMPOSITION | CROSS REFERENCE SPECIFICATIONS |
|---|---|---|---|
| **L01900** | Cadmium Metal | **Ag** 0.01 max **As** 0.003 max **Cd** 99.90 min **Cu** 0.015 max **Pb** 0.025 max **Sb** 0.001 max **Sn** 0.01 max **Zn** 0.035 max **Other** Tl 0.003 max | **ASTM** B440 |
| **L01950** | Cadmium Metal | **Cd** 99.95 min **Zn** 0.020 max | **AWS** C2.25 (W-Cd) |
| **L01951** | Cadmium Metal (99.95 Cd) | **Cd** 99.95 min **Cu** 0.0150 max **Pb** 0.0250 max **Zn** 0.0350 max **Other** Tl 0.0030 max | **ASTM** B440 |
| **L01971** | Cadmium Metal (99.99 Cd) | **Cd** 99.99 min **Cu** 0.0020 max **Fe** 0.0010 max **Ni** 0.0010 max **Pb** 0.0100 max **Zn** 0.0030 max **Other** Tl 0.0030 max **ASTM** B440 | |
| **L01981** | Cadmium Metal (99.995 Cd) | **Ag** 0.001 max **As** 0.001 max **Cd** 99.995 min **Cu** 0.005 max **Fe** 0.005 max **Hg** 0.00001 max **Ni** 0.005 max **Pb** 0.020 max **Sb** 0.00001 max **Sn** 0.001 max **Zn** 0.005 max **Other** Tl 0.005 max | **ASTM** B440 |
| **L05020** | Lead-Tin Solder replaced by L54210 | | |
| **L05021** | Lead-Tin Solder replaced by L54211 | | |
| **L05025** | Lead-Silver Solder replaced by L50150 | | |
| **L05050** | Lead-Tin Solder replaced by L54320 | | |
| **L05051** | Lead-Tin Solder replaced by L54321 | | |
| **L05055** | Lead-Silver Solder replaced by L50180 | | |
| **L05070** | Lead Alloy replaced by L53131 | | |
| **L05090** | Lead Die Casting Alloy replaced by L53340 | | |
| **L05100** | Lead-Tin Solder replaced by L54520 | | |
| **L05110** | Lead-Tin Solder replaced by L50151 | | |
| **L05120** | White Metal Bearing Alloy (Babbitt Metal) | **Al** 0.005 max **As** 0.20 max **Cu** 0.50 max **Fe** 0.10 max **Pb** 83.0-88.0 **Sb** 8.0-10.0 **Sn** 4.0-6.0 **Zn** 0.005 max **Other** total 0.75 max | **FED** QQ-T-390 |
| **L05150** | Lead-Tin Solder replaced by L54560 | | |
| **L05153** | White Metal Bearing Alloy (Babbitt Metal) replaced by L53346 | | |
| **L05155** | White Metal Bearing Alloy (Babbitt Metal) replaced by L53620 | | |
| **L05180** | Lead Die Casting Alloy replaced by L53560 | | |
| **L05188** | White Metal Bearing Alloy (Babbitt Metal) replaced by L53565 | | |
| **L05201** | Lead-Tin Solder replaced by L54711 | | |

The chemical compositions listed are for identification purposes and should not be used in lieu of the cross referenced specifications.

| UNIFIED NUMBER | DESCRIPTION | CHEMICAL COMPOSITION | CROSS REFERENCE SPECIFICATIONS |
|---|---|---|---|
| **L05202** | Lead-Tin Solder replaced by L54712 | | |
| **L05237** | White Metal Bearing Alloy (Babbitt Metal) replaced by L53585 | | |
| **L05250** | Lead-Tin Solder replaced by L54720 | | |
| **L05251** | Lead-Tin Solder replaced by L54721 | | |
| **L05252** | Lead-Tin Solder replaced by L54722 | | |
| **L05300** | Lead-Tin Solder replaced by L54820 | | |
| **L05301** | Lead-Tin Solder replaced by L54821 | | |
| **L05302** | Lead-Tin Solder replaced by L54822 | | |
| **L05350** | Lead-Tin Solder replaced by L54850 | | |
| **L05351** | Lead-Tin Solder replaced by L54851 | | |
| **L05352** | Lead Tin Solder replaced by L54852 | | |
| **L05400** | Lead-Tin Solder replaced by L54915 | | |
| **L05401** | Lead-Tin Solder replaced by L54916 | | |
| **L05402** | Lead-Tin Solder replaced by L54918 | | |
| **L05450** | Lead-Tin Solder replaced by L54950 | | |
| **L05451** | Lead-Tin Solder replaced by L54951 | | |
| **L05500** | Lead-Tin (50-50) Solder replaced by L55030 | | |
| **L05501** | Lead-Tin (50-50) Solder replaced by L55031 | | |
| **L06990** | Lithium Metal | **Fe** 0.005 max **K** 0.025 max **Li** rem **Na** 0.800 max **Other** Chlorides 0.04 max | |
| **L13002** | Pig Tin | **As** 0.0005 max **Bi** 0.001 max **Cd** 0.001 max **Cu** 0.002 max **Fe** 0.005 max **Pb** 0.010 max **S** 0.002 max **Sb** 0.008 max **Sn** 99.98 max **Zn** 0.001 max **Other** Ni+Co 0.005 max | |
| **L13004** | Pig Tin | **As** 0.01 max **Bi** 0.01 max **Cd** 0.001 max **Cu** 0.02 max **Fe** 0.01 max **Pb** 0.02 max **S** 0.01 max **Sb** 0.02 max **Sn** 99.95 max **Zn** 0.005 max **Other** Ni+Co 0.01 max | |
| **L13006** | Pig Tin | **As** 0.05 max **Sn** 99.80 min | |
| **L13007** | Pig Tin | **Ag** 0.01 max **Al** 0.001 max **As** 0.04 max **Bi** 0.015 max **Cd** 0.001 max **Cu** 0.03 max **Fe** 0.015 max **Pb** 0.05 max **S** 0.003 max **Sb** 0.04 max **Sn** 99.80 min **Zn** 0.001 max **Other** Ni+Co 0.015 max | **FED** QQ-T-371 |

The chemical compositions listed are for identification purposes and should not be used in lieu of the cross referenced specifications.

| UNIFIED NUMBER | DESCRIPTION | CHEMICAL COMPOSITION | CROSS REFERENCE SPECIFICATIONS |
|---|---|---|---|
| **L13008** | Pig Tin | **As** 0.05 max **Bi** 0.015 max **Cd** 0.001 max **Cu** 0.04 max **Fe** 0.015 max **Pb** 0.05 max **S** 0.01 max **Sb** 0.04 max **Sn** 99.80 min **Zn** 0.005 max **Other** Ni+Co 0.01 max | **ASTM** B339 (A) |
| **L13010** | Pig Tin | **Sn** 99.65 min | |
| **L13012** | Pig Tin | **Sn** 99.50 min | **AWS** C2.25 (W-Sn) |
| **L13014** | Pig Tin | **Sn** 99.00 min | |
| **L13201** | Non-Ferrous Thermal Spray Wire (W-SnZn-1) | **Sn** rem **Zn** 19-21 | **AWS** C2.25 (W-SnZn-1) |
| **L13202** | Non-Ferrous Thermal Spray Wire (W-SnZn-2) | **Cu** 0.01 max **Sn** rem **Zn** 26-34 | **AWS** C2.25 (W-SnZn-2) |
| **L13600** | Tin-Lead Solder | **Al** 0.005 max **As** 0.03 max **Bi** 0.25 max **Co** 0.08 max **Fe** 0.02 max **Pb** 40 nom **Sb** 0.12 max **Sn** 60 nom **Zn** 0.005 max | **ASTM** B32 (60A) |
| **L13601** | Tin-Lead Solder | **Al** 0.005 max **As** 0.03 max **Bi** 0.25 max **Cu** 0.08 max **Fe** 0.02 max **Pb** 40 nom **Sb** 0.20-0.50 **Sn** 60 nom **Zn** 0.005 max | **ASTM** B32 (60B) |
| **L13630** | Tin-Lead Solder | **Al** 0.005 max **As** 0.03 max **Bi** 0.25 max **Cu** 0.08 max **Fe** 0.02 max **Pb** 37 nom **Sb** 0.12 max **Sn** 63 nom **Zn** 0.005 max | **AMS** 4751 **ASTM** B32 (63A, Sn63) |
| **L13631** | Tin-Lead Solder | **Al** 0.005 max **As** 0.03 max **Bi** 0.25 max **Cu** 0.08 max **Fe** 0.02 max **Pb** 37 nom **Sb** 0.20-0.50 **Sn** 63 nom **Zn** 0.005 max | **ASTM** B32 (63B) |
| **L13650** | Tin Die Casting Alloy | **Al** 0.01 max **As** 0.15 max **Cu** 1.5-2.5 **Fe** 0.08 max **Pb** 17-19 **Sb** 14-16 **Sn** 64-66 **Zn** 0.01 max | **ASTM** B102 (PY1815A) |
| **L13700** | Tin-Lead Solder | **Al** 0.005 max **As** 0.03 max **Bi** 0.25 max **Cu** 0.08 max **Fe** 0.02 max **Pb** 30 nom **Sb** 0.20 nom **Sn** 70 nom **Zn** 0.005 max | **ASTM** B32 (70A, Sn70) |
| **L13701** | Tin-Lead Solder | **Al** 0.005 max **As** 0.03 max **Bi** 0.25 max **Cu** 0.08 max **Fe** 0.02 max **Pb** 30 nom **Sb** 0.20-0.50 **Sn** 70 nom **Zn** 0.005 max | **ASTM** B32 (70B) |
| **L13820** | Tin Die Casting Alloy | **Al** 0.01 max **As** 0.08 max **Cu** 4-6 **Fe** 0.08 max **Pb** 0.35 max **Sb** 12-14 **Sn** 80-84 **Zn** 0.01 max | **ASTM** B102 (YC135A) |
| **L13840** | White Metal Bearing Alloy (Babbitt Metal) | **Al** 0.005 max **As** 0.10 max **Bi** 0.08 max **Cd** 0.05 max **Cu** 7.5-8.5 **Fe** 0.08 max **Pb** 0.35 max **Sb** 7.5-8.5 **Sn** 83.0-85.0 **Zn** 0.005 max | **ASTM** B23 (3) |
| **L13870** | White Metal Bearing Alloy (Babbitt Metal) | **Al** 0.005 max **As** 0.10 max **Bi** 0.08 max **Cd** 0.05 max **Cu** 5.0-6.5 **Fe** 0.08 max **Pb** 0.50 max **Sb** 6.0-7.5 **Sn** 86.0-89.0 **Zn** 0.005 max | **ASTM** B23 (11) |
| **L13890** | White Metal Bearing Alloy (Babbitt Metal) | **Al** 0.005 max **As** 0.10 max **Bi** 0.08 max **Cd** 0.05 max **Cu** 3.0-4.0 **Fe** 0.08 max **Pb** 0.35 max **Sb** 7.0-8.0 **Sn** 88.0-90.0 **Zn** 0.005 max | **ASTM** B23 (2) **AWS** C2.25 (W-SnSbCu-1) |
| **L13892** | Non-Ferrous Thermal Spray Wire (W-SnSbCu-2) | **Cu** 3-5 **Pb** 0.5 max **Sb** 6-9 **Sn** rem | **AWS** C2.25 (W-SnSbCu-2) |
| **L13893** | Non-Ferrous Thermal Spray Wire (W-SnSbCu-3) | **Cu** 2-4 **Sb** 6-8 **Sn** rem | **AWS** C2.25 (W-SnSbCu-3) |

The chemical compositions listed are for identification purposes and should not be used in lieu of the cross referenced specifications.

| UNIFIED NUMBER | DESCRIPTION | CHEMICAL COMPOSITION | CROSS REFERENCE SPECIFICATIONS |
|---|---|---|---|
| **L13910** | White Metal Bearing Alloy (Babbitt Metal) | **Al** 0.005 max **As** 0.10 max **Bi** 0.08 max **Cd** 0.05 max **Cu** 4.0-5.0 **Fe** 0.08 max **Pb** 0.35 max **Sb** 4.0-5.0 **Sn** 90.0-92.0 **Zn** 0.005 max | **AMS** 4800 **ASTM** B23 (1) **AWS** C2.25 (W-SnCuSb) |
| **L13911** | Modern Pewter Casting Alloys | **As** 0.05 max **Cu** 0.25-2.0 **Fe** 0.015 max **Pb** 0.05 max **Sb** 6-8 **Sn** 90-93 **Zn** 0.005 max | **ASTM** B560 (1) |
| **L13912** | Modern Pewter Sheet Alloy | **As** 0.05 max **Cu** 1.5-3.0 **Fe** 0.015 max **Pb** 0.05 max **Sb** 5-7.5 **Sn** 90-93 **Zn** 0.005 max | **ASTM** B560 (2) |
| **L13913** | Tin Die-Casting Alloy | **Al** 0.01 max **As** 0.08 max **Cu** 4-5 **Fe** 0.08 max **Pb** 0.35 max **Sb** 4-5 **Sn** 90-92 **Zn** 0.01 max | **ASTM** B102 (CY44A) |
| **L13940** | Tin-Antimony Solder | **Al** 0.03 max **As** 0.06 max **Cd** 0.03 max **Cu** 0.08 max **Fe** 0.08 max **Pb** 0.20 max **Sb** 4.0-6.0 **Sn** 94 min **Zn** 0.03 max **Other** total 0.03 max | |
| **L13950** | Tin-Antimony Solder | **Al** 0.005 max **As** 0.05 max **Bi** 0.15 max **Cu** 0.08 max **Fe** 0.04 max **Pb** 0.20 max **Sb** 4.5-5.5 **Sn** 95 nom **Zn** 0.005 max | **ASTM** B32 (95TA, Sb5) |
| **L13960** | Tin-Silver Solder | **Ag** 3.6-4.4 **Al** 0.005 max **As** 0.05 max **Bi** 0.15 max **Cu** 0.08 max **Fe** 0.02 max **Pb** 0.20 max **Sb** 0.20-0.50 **Sn** 96 nom **Zn** 0.005 max | |
| **L13961** | Tin-Silver Solder | **Ag** 3.6-4.4 **As** 0.05 max **Cd** 0.005 max **Cu** 0.20 max **Pb** 0.10 maX **Sn** rem **Zn** 0.005 max | |
| **L13963** | Modern Pewter | **As** 0.05 max **Cu** 1.0-2.0 **Fe** 0.015 max **Pb** 0.05 max **Sb** 1.0-3.0 **Sn** 95-98 **Zn** 0.005 max | **ASTM** B560 (3) |
| **L13965** | Tin-Silver Solder | **Ag** 3.3-3.7 **Al** 0.005 max **As** 0.05 max **Bi** 0.15 max **Cu** 0.08 max **Fe** 0.02 max **Pb** 0.20 max **Sb** 0.20-0.50 **Sn** 96.5 nom **Zn** 0.005 max | **ASTM** B32 (96.5 TS) |
| **L50001** | Zone Refined Lead | **Pb** 99.9999 min | |
| **L50005** | Refined Soft Lead | **Pb** 99.999 min | |
| **L50006** | Low Bismuth, Low Silver Pure Lead | **Ag** 0.0010 max **As** 0.0005 max **Bi** 0.0015 max **Cu** 0.0010 max **Fe** 0.0002 max **Ni** 0.0002 max **Pb** 99.995 min **Sb** 0.0005 max **Sn** 0.0005 max **Te** 0.0001 max **Zn** 0.0005 max | **Other:** BS 334 |
| **L50010** | Refined Soft Lead | **Pb** 99.99 min | |
| **L50020** | Refined Soft Lead | **Pb** 99.985 | |
| **L50021** | Refined Pure Lead | **Ag** 0.0025 max **As** 0.0005 max **Bi** 0.025 max **Cu** 0.0010 max **Fe** 0.001 max **Ni** 0.0002 max **Pb** 99.97 min **Sb** 0.0005 max **Sn** 0.0005 max **Te** 0.0001 max **Zn** 0.0005 max | **Other:** BS 334 |
| **L50025** | LME Grade Pure Lead | **Pb** 99.97 | |
| **L50035** | Refined Soft Lead | **Pb** 99.95 min | |
| **L50040** | Refined Soft Lead | **Pb** 99.94 min | |
| **L50042** | Corroding Lead | **Ag** 0.0015 max **Bi** 0.050 max **Cu** 0.0015 max **Fe** 0.002 max **Pb** 99.94 min **Zn** 0.001 max **Other** Cu+Ag 0.0025 max, As+Sb+Sn 0.002 max | **ASTM** B29 (Corroding Lead); B749 |

The chemical compositions listed are for identification purposes and should not be used in lieu of the cross referenced specifications.

| UNIFIED NUMBER | DESCRIPTION | CHEMICAL COMPOSITION | CROSS REFERENCE SPECIFICATIONS |
|---|---|---|---|
| **L50045** | Common Lead | **Ag** 0.005 max **Bi** 0.050 max **Cu** 0.0015 max **Fe** 0.002 max **Pb** 99.94 min **Zn** 0.001 max **Other** As+Sb+Sn 0.002 max | **ASTM** B29 (Common Lead); B749 (Common Lead) **AWS** C2.25 (W-Pb) |
| **L50049** | Pure Lead | **Ag** 0.005 max **As** 0.001 max **Bi** 0.05 max **Cu** 0.0015 max **Fe** 0.001 max **Ni** 0.001 max **Pb** 99.94 min **Sb** 0.001 max **Sn** 0.001 max **Zn** 0.001 max **Other** (Sb+As+Sn) 0.002 max | **Other:** BS 334 |
| **L50050** | Grade A Lead | **Pb** 99.90 min **Other** total other elements 0.10 max | **FED** QQ-L-171 (A) |
| **L50060** | Type III Lead | **Pb** 99.85 min | |
| **L50065** | Grade AA Lead, Grade C Lead | **Pb** 99.7 min **Sb** 0.02 max | **FED** QQ-C-40 (AA, C) |
| **L50070** | Grade B Remelted Lead | **Bi** 0.025 max (Optional) **Pb** 39.5 min | **FED** QQ-L-201 |
| **L50080** | Grade B Lead | **Pb** 95.0 min **Other** total other elements 5.0 max | **FED** QQ-L-171 (B) |
| **L50090** | Chemical B Grade Lead | **Pb** 99.99 min | **Other:** BS 334 |
| **L50101** | Cable Sheathing Alloy | **Ag** 0.2 nom **Pb** 99.8 nom | |
| **L50110** | Electrowinning Anode Alloy | **Ag** 0.5 nom **Pb** 99.5 nom | |
| **L50113** | Solder Alloy | **Ag** 0.5 nom **Pb** 97.0 nom **Sn** 2.5 nom | |
| **L50115** | Electrowinning Anode Alloy | **Ag** 0.75 nom **Pb** 99.25 nom | |
| **L50120** | Electrowinning Anode Alloy | **Ag** 1.0 nom **Pb** 99.0 nom | |
| **L50121** | Solder Alloy | **Ag** 1.0 nom **Pb** 98. nom **Sn** 1.0 nom | |
| **L50122** | Electrowinning Anode Alloy | **Ag** 1.0 nom **As** 1.0 nom **Pb** 98. nom | |
| **L50131** | Solder Alloy-Grade 1.5S | **Ag** 1.3-1.7 **Al** 0.005 max **As** 0.02 max **Bi** 0.25 max **Cu** 0.08 max **Fe** 0.02 max **Pb** 97.5 nom **Sn** 0.75-1.25 **Zn** 0.005 max | **ASTM** B32 (1.5S) |
| **L50132** | Solder Alloy-Grade Ag1.5 | **Ag** 1.3-1.7 **Al** 0.005 max **As** 0.02 max **Bi** 0.25 max **Cd** 0.001 max **Cu** 0.30 max **Fe** 0.02 max **Pb** rem **Sb** 0.40 max **Sn** 0.75-1.25 **Zn** 0.005 max **Other** total of all others 0.08 max | **ASTM** B32 (Ag1.5) |
| **L50134** | Solder Alloy-Grade 5S | **Ag** 1.5 nom **Pb** 93.5 nom **Sn** 5.0 nom | |
| **L50140** | Cathodic Protection Anode Alloy | **Ag** 2.0 nom **Pb** 98.0 nom | |
| **L50150** | Solder Alloy-Grade 2.5S | **Ag** 2.3-2.7 **Al** 0.005 max **As** 0.02 max **Bi** 0.25 max **Cu** 0.08 max **Fe** 0.02 max **Pb** 97.5 nom **Sn** 0.25 max **Zn** 0.005 max | **ASTM** B32 (2.5S) |
| **L50151** | Solder Alloy-Grade Ag2.5 | **Ag** 2.3-2.7 **Al** 0.005 max **As** 0.02 max **Bi** 0.25 max **Cd** 0.001 max **Cu** 0.30 max **Fe** 0.02 max **Pb** rem **Sb** 0.40 max **Sn** 0.25 max **Zn** 0.005 max **Other** total of all others 0.03 max | **ASTM** B32 (Ag2.5) |
| **L50152** | Solder Alloy | **Ag** 2.5 nom **Pb** 95.5 nom **Sn** 2.0 nom | |
| **L50170** | Solder Alloy | **Ag** 5.0 nom **Pb** 95.0 nom | |
| **L50172** | Solder Alloy | **Ag** 5.0 nom **In** 5.0 nom **Pb** 90.0 nom | |
| **L50180** | Solder Alloy-Grade Ag5.5 | **Ag** 5.0-6.0 **Al** 0.005 max **As** 0.02 max **Bi** 0.25 max **Cd** 0.001 max **Cu** 0.30 max **Fe** 0.002 max **Pb** rem **Sb** 0.40 max **Sn** 0.25 max **Zn** 0.005 max **Other** total of all others 0.03 max | **ASTM** B32 (Ag5.5) |
| **L50310** | Arsenical Lead Cable Sheathing Alloy | **As** 0.15 nom **Bi** 0.10 nom **Pb** 99.6 nom **Sn** 0.10 nom | |

The chemical compositions listed are for identification purposes and should not be used in lieu of the cross referenced specifications.

| UNIFIED NUMBER | DESCRIPTION | CHEMICAL COMPOSITION | CROSS REFERENCE SPECIFICATIONS |
|---|---|---|---|
| **L50510** | Lead-Barium Alloy | **Pb** 99.9 nom **Other** Ba 0.05 nom | |
| **L50520** | Lead-Tin-Barium Alloy | **Pb** 99.0 nom **Sn** 1.0 nom **Other** Ba 0.05 nom | |
| **L50521** | Lead-Tin-Barium Alloy | **Pb** 98.5 nom **Sn** 1.5 nom **Other** Ba 0.05 nom | |
| **L50522** | Lead-Tin-Barium Alloy | **Pb** 98.0 nom **Sn** 2.0 nom **Other** Ba 0.05 nom | |
| **L50530** | Lead-Tin-Barium Alloy | **Pb** 98.9 nom **Sn** 1.0 nom **Other** Ba 0.10 nom | |
| **L50535** | Lead-Tin-Barium Alloy | **Pb** 97.9 nom **Sn** 2.0 nom **Other** Ba 0.10 nom | |
| **L50540** | Frary Metal | **Pb** 98.8 nom **Other** Ba 0.4 nom, Ca 0.8 nom | |
| **L50541** | Frary Metal | **Pb** 98.5 nom **Other** Ba 1.0 nom, Ca 0.5 nom | |
| **L50542** | Frary Metal | **Pb** 98.0 nom **Other** Ba 1.2 nom, Ca 0.8 nom | |
| **L50543** | Frary Metal | **Pb** 97.2 nom **Other** Ba 2.0 nom, Ca 0.8 nom | |
| **L50710** | Lead-Calcium Alloy | **Pb** 99.9 nom **Other** Ca 0.008 nom | |
| **L50712** | Cable Sheathing Alloy | **Pb** 99.9 nom **Other** Ca 0.025 nom | |
| **L50713** | Cable Sheathing Alloy | **Pb** 99.7 nom **Sn** 0.3 nom **Other** Ca 0.025 nom | |
| **L50720** | Lead-Calcium Alloy | **Pb** 99.9 nom **Other** Ca 0.03 nom | |
| **L50722** | Lead-Copper-Calcium Alloy | **Cu** 0.06 nom **Pb** 99.9 nom **Other** Ca 0.03 nom | |
| **L50725** | Cable Sheathing Alloy | **Cu** 0.06 nom **Pb** 99.9 nom **Other** Ca 0.035 nom | |
| **L50728** | Battery Grid Alloy, Lead-Calcium-Tin | **Pb** 99.5 nom **Sn** 0.5 nom **Other** Ca 0.04 nom | |
| **L50730** | Electrowinning Anode Alloy | **Ag** 0.5 nom **Pb** 99.4 nom **Other** Ca 0.06 nom | |
| **L50735** | Battery Grid Alloy, Lead-Calcium | **Pb** 99.9 nom **Other** Ca 0.06 nom | |
| **L50736** | Battery Grid Alloy, Lead-Calcium-Tin | **Pb** 99.7 nom **Sn** 0.2 nom **Other** Ca 0.065 nom | |
| **L50737** | Battery Grid Alloy, Lead-Calcium-Tin | **Pb** 99.4 nom **Sn** 0.5 nom **Other** Ca 0.065 nom | |
| **L50740** | Battery Grid Alloy, Lead-Calcium-Tin | **Pb** 99.2 nom **Sn** 0.7 nom **Other** Ca 0.065 nom | |
| **L50745** | Battery Grid Alloy, Lead-Calcium-Tin | **Pb** 98.9 nom **Sn** 1.0 nom **Other** Ca 0.065 nom | |
| **L50750** | Battery Grid Alloy, Lead-Calcium-Tin | **Pb** 98.6 nom **Sn** 1.3 nom **Other** Ca 0.065 nom | |
| **L50755** | Battery Grid Alloy, Lead-Calcium-Tin | **Pb** 98.4 nom **Sn** 1.5 nom **Other** Ca 0.065 nom | |
| **L50760** | Battery Grid Alloy, Lead-Calcium | **Pb** 99.9 nom **Other** Ca 0.07 nom | |
| **L50765** | Battery Grid Alloy, Lead-Calcium-Tin | **Pb** 99.2 nom **Sn** 0.7 nom **Other** Ca 0.07 nom | |
| **L50770** | Battery Grid Alloy, Lead-Calcium | **Pb** 99.9 nom **Other** Ca 0.10 nom | |

The chemical compositions listed are for identification purposes and should not be used in lieu of the cross referenced specifications.

| UNIFIED NUMBER | DESCRIPTION | CHEMICAL COMPOSITION | CROSS REFERENCE SPECIFICATIONS |
|---|---|---|---|
| **L50775** | Battery Grid Alloy, Lead-Calcium-Tin | **Pb** 99.6 nom **Sn** 0.3 nom **Other** Ca 0.10 nom | |
| **L50780** | Battery Grid Alloy, Lead-Calcium-Tin | **Pb** 99.4 nom **Sn** 0.5 nom **Other** Ca 0.10 nom | |
| **L50790** | Battery Grid Alloy, Lead-Calcium-Tin | **Pb** 98.9 nom **Sn** 1.0 nom **Other** Ca 0.10 nom | |
| **L50795** | Battery Grid Alloy, Lead-Calcium-Tin | **Pb** 99.6 nom **Sn** 0.3 nom **Other** Ca 0.12 nom | |
| **L50800** | Battery Grid Alloy, Lead-Calcium-Tin | **Pb** 99.5 nom **Sn** 0.3 nom **Other** Ca 0.21 nom | |
| **L50810** | Bahnmetal, Bearing Metal | **Al** 0.02 nom **Li** 0.04 nom **Pb** 98.7 nom **Other** Ca 0.7 nom, Na 0.6 nom | |
| **L50820** | Bahnmetal, Bearing Metal | **Al** 0.02 nom **Li** 0.04 nom **Pb** 98.7 nom **Other** Ca 0.7 nom, Na 0.2 nom, Ba 0.4 nom | |
| **L50840** | Lead-Calcium Alloy | **Pb** 99.0 nom **Other** Ca 1.0 nom | |
| **L50850** | Lead-Calcium Master Alloy | **Pb** 98 nom **Other** Ca 2.0 nom | |
| **L50880** | Lead-Calcium Alloy | **Pb** 94.0 nom **Other** Ca 6.0 nom | |
| **L50940** | Lead-Cadmium Alloy-Eutectic | **Cd** 17.0 nom **Pb** 83.0 nom | |
| **L51110** | Copperized Lead | **Cu** 0.05 nom **Pb** 99.9 nom | |
| **L51120** | Chemical Lead | **Ag** 0.002-0.02 **Bi** 0.005 max **Cu** 0.04-0.08 **Fe** 0.002 max **Pb** 99.90 min **Zn** 0.001 max **Other** As+Sb+Sn 0.002 max | **ASTM** B29 (Chemical Lead); B749 |
| **L51121** | Copper Bearing Lead | **Ag** 0.020 max **AS** 0.001 max **Bi** 0.025 max **Cu** 0.040-0.080 **Fe** 0.002 max **Ni** 0.002 max **Pb** 99.90 min **Sb** 0.001 max **Sn** 0.001 max **Zn** 0.001 max **Other** As+Sb+Sn 0.002 max | **ASTM** B29 (Copper Bearing Lead); B749 **FED** QQ-L-171 (C); QQ-L-201 (C) **Other:** BS 334 |
| **L51123** | Lead Alloy Grade D | **Ag** 0.020 max **Bi** 0.025 max **Cu** 0.04-0.08 **Fe** 0.002 max **Pb** 99.85 min **Te** 0.035-0.060 **Zn** 0.001 max **Other** As+Sb+Sn 0.002 max | **ASTM** B749 **FED** QQ-L-201 (D) |
| **L51124** | Lead Alloy Grade D | **Ag** 0.020 max **Bi** 0.025 max **Cu** 0.04-0.08 **Fe** 0.002 max **Pb** 99.82 min **Sn** 0.016 max **Te** 0.035-0.055 **Zn** 0.001 max **Other** Sb+As 0.002 max | **FED** QQ-C-40 (D) |
| **L51125** | Pure Copper Lead, Copperized Soft Lead | **Cu** 0.06 **Pb** 99.9 min | |
| **L51180** | Copper-Lead Bearing Alloy | **Cu** rem **Fe** 0.35 max **Pb** 44.0-58.0 **Sn** 1.0-5.0 **Other** others, total 0.45 max, others, each 0.15 max | |
| **L51510** | Lead-Indium-Silver Solder Alloy | **Ag** 2.38 nom **In** 4.76 nom **Pb** 92.8 nom | |
| **L51511** | Lead-Indium Solder Alloy | **In** 5.0 nom **Pb** 95.0 nom | |
| **L51512** | Lead-Indium-Silver Solder Alloy | **Ag** 2.5 nom **In** 5.0 nom **Pb** 92.5 nom | |
| **L51530** | Lead-Indium Alloy | **In** 19.0 nom **Pb** 81.0 nom | |
| **L51532** | Lead-Indium Alloy | **In** 20.0 nom **Pb** 80.0 nom | |
| **L51535** | Lead-Indium Alloy | **In** 25.0 nom **Pb** 75.0 nom | |
| **L51540** | Lead-Indium Alloy | **In** 40.0 nom **Pb** 60.0 nom | |
| **L51550** | Lead-Indium Alloy | **In** 50.0 nom **Pb** 50.0 nom | |
| **L51705** | Lead-Lithium Alloy | **Li** 0.01 nom **Pb** 99.9 nom | |
| **L51708** | Lead-Lithium Alloy | **Li** 0.02 nom **Pb** 99.9 nom | |

The chemical compositions listed are for identification purposes and should not be used in lieu of the cross referenced specifications.

| UNIFIED NUMBER | DESCRIPTION | CHEMICAL COMPOSITION | CROSS REFERENCE SPECIFICATIONS |
|---|---|---|---|
| **L51710** | Lead-Lithium Alloy | **Li** 0.03 nom **Pb** 99.9 nom | |
| **L51720** | Lead-Lithium Alloy | **Li** 0.06 nom **Pb** 99.9 nom | |
| **L51730** | Lead-Lithium Alloy | **Li** 0.7 nom **Pb** 99.3 nom | |
| **L51740** | Lead-Tin-Lithium Alloy | **Li** 0.02 nom **Pb** 99.6 nom **Sn** 0.35 nom | |
| **L51748** | Lead-Tin-Lithium Alloy | **Li** 0.04 nom **Pb** 99.2 nom **Sn** 0.7 nom | |
| **L51770** | Lead-Tin-Lithium-Calcium Alloy | **Li** 0.08 nom **Pb** 98.8 **Sn** 1.0 nom **Other** Ca 0.03 nom | |
| **L51775** | Lead-Tin-Lithium-Calcium Alloy | **Li** 0.12 nom **Pb** 98.8 nom **Sn** 1.0 nom **Other** Ca 0.03 nom | |
| **L51778** | Lead-Tin-Lithium-Calcium Alloy | **Li** 0.15 nom **Pb** 97.8 nom **Sn** 2.0 nom **Other** Ca 0.04 nom | |
| **L51780** | Lead-Tin-Lithium-Calcium Alloy | **Li** 0.15 nom **Pb** 99.6 nom **Sn** 0.1 nom **Other** Ca 0.15 nom | |
| **L51790** | Lead-Tin-Lithium-Calcium Alloy | **Li** 0.65 nom **Pb** 98.3 nom **Sn** 1.0 nom **Other** Ca 0.02 nom | |
| **L52505** | Lead-Antimony Alloy | **Pb** 99.9 nom **Sb** 0.1 nom | |
| **L52510** | 99.8 Pb | **Cu** 0.06 **Pb** 99.8 nom **Sb** 0.1 **Sn** 0.1 | |
| **L52515** | Cable Sheathing Alloy | **As** 0.015 nom **Pb** 99.8 nom **Sb** 0.2 nom | |
| **L52520** | Cable Sheathing Alloy | **Pb** 99.4 **Sb** 0.2 **Sn** 0.4 | |
| **L52525** | Cable Sheathing Alloy | **As** 0.035 nom **Pb** 99.7 nom **Sb** 0.3 nom **Te** 0.035 nom | |
| **L52530** | Lead-Antimony Alloy | **Pb** 99.6 nom **Sb** 0.35 nom | |
| **L52535** | Cable Sheathing Alloy | **As** 0.03 nom **Pb** 99.6 nom **Sb** 0.4 nom | |
| **L52540** | Cable Sheathing Alloy | **Cd** 0.25 nom **Pb** 99.2 nom **Sb** 0.5 nom | |
| **L52545** | Lead-Antimony Alloy | **Pb** 99.4 nom **Sb** 0.6 nom | |
| **L52550** | Cable Sheathing Alloy | **Cu** 0.04 nom **Pb** 99.3 nom **Sb** 0.6 nom **Te** 0.04 nom | |
| **L52555** | Alloy B per DIN 17640 | **Pb** 99.3 **Sb** 0.7 | |
| **L52560** | Bullet Alloy | **Pb** 99.2 nom **Sb** 0.75 nom | |
| **L52565** | Overhead Cable Alloy | **Pb** 99.2 nom **Sb** 0.75 nom | |
| **L52570** | Cable Sheathing Alloy, Alloy B | **Cu** 0.06 max **Pb** 99.1 **Sb** 0.85 | |
| **L52595** | Cable Sheathing Alloy | **Pb** 99.0 nom **Sb** 0.95 nom | |
| **L52605** | 1% Antimonial Lead | **Pb** 99.0 nom **Sb** 1.0 nom | |
| **L52615** | Lead-Base Die Casting Alloy | **As** 0.1 nom **Pb** 98.6 nom **S** 0.003 nom **Sb** 1.0 nom **Sn** 0.3 nom | |
| **L52618** | Lead-Antimony-Gallium Alloy | **Pb** 98.0 nom **Sb** 1.2 nom **Other** Ga 0.8 nom | |
| **L52620** | Battery Alloy | **Cd** 1.45 nom **Pb** 97.0 nom **Sb** 1.5 nom | |
| **L52625** | Shot Alloy | **As** 0.45 nom **Pb** 98.0 nom **Sb** 1.55 nom **Sn** 0.0005 max | |
| **L52630** | Battery Alloy | **As** 0.3 **Pb** 98 nom **S** 0.005 **Sb** 1.6 **Se** 0.02 **Sn** 0.1 | |
| **L52705** | 2% Antimonial Lead | **Pb** 98.0 nom **Sb** 2.0 nom | |

The chemical compositions listed are for identification purposes and should not be used in lieu of the cross referenced specifications.

| UNIFIED NUMBER | DESCRIPTION | CHEMICAL COMPOSITION | CROSS REFERENCE SPECIFICATIONS |
|---|---|---|---|
| **L52710** | Battery Alloy | **As** 0.15 **Pb** 97.5 nom **S** 0.003 **Sb** 2.0 **Sn** 0.3 | |
| **L52715** | Battery Alloy | **As** 0.25 **Pb** 97.4 nom **Sb** 2.25 **Se** 0.02 **Sn** 0.1 | |
| **L52720** | Battery Alloy | **As** 0.3 **Pb** 97.2 nom **S** 0.008 **Sb** 2.25 **Se** 0.02 **Sn** 0.2 | |
| **L52725** | Bullet Alloy | **Pb** 97.5 nom **Sb** 2.5 nom | |
| **L52730** | Electrotype-General | **Pb** 95 nom **Sb** 2.5 nom **Sn** 2.5 nom | |
| **L52750** | Battery Alloy | **As** 0.4 **Pb** 96.5 nom **S** 0.005 **Sb** 2.75 **Sn** 0.3 **Other** Ca 0.075 | |
| **L52755** | Battery Alloy | **As** 0.5 **Pb** 96.4 nom **S** 0.007 **Sb** 2.75 **Sn** 0.3 **Other** Ca 0.075 | |
| **L52760** | Battery Alloy | **As** 0.18 **Cu** 0.075 **Pb** 96.8 nom **S** 0.008 **Sb** 2.75 **Sn** 0.2 | |
| **L52765** | Battery Alloy | **As** 0.3 **Cu** 0.075 **Pb** 96.6 nom **S** 0.008 **Sb** 2.75 **Sn** 0.3 | |
| **L52770** | Battery Alloy | **As** 0.15 **Cu** 0.04 max **Pb** 96.6 nom **S** 0.001 **Sb** 2.9 **Sn** 0.3 | |
| **L52775** | Battery Alloy | **As** 0.15 **Cu** 0.05 **Pb** 96.6 nom **S** 0.004 **Sb** 2.9 **Sn** 0.3 | |
| **L52805** | 3% Antimonial Lead | **Pb** 97 nom **Sb** 3.0 nom | |
| **L52810** | Battery Alloy | **As** 0.15 **Pb** 96.5 nom **S** 0.003 **Sb** 3.0 **Sn** 0.3 | |
| **L52815** | Shot Alloy | **As** 0.6 **Pb** 96.4 nom **Sb** 3.0 **Sn** 0.0005 max | |
| **L52830** | Electrotype-General | **Pb** 94 nom **Sb** 3 nom **Sn** 3 nom | |
| **L52840** | Battery Alloy | **As** 0.5 **Cu** 0.12 **Pb** 95.7 nom **Sb** 3.25 **Sn** 0.4 **Other** Ca 0.06 | |
| **L52860** | Lead-Base Bearing Alloy No. 16 (Overlay) | **Al** 0.005 max **As** 0.05 max **Bi** 0.10 max **Cd** 0.005 max **Cu** 0.10 max **Pb** rem **Sb** 3.0-4.0 **Sn** 3.5-4.7 **Zn** 0.005 max **Other** total of others 0.40 max | |
| **L52901** | 4% Antimonial Lead | **Pb** 96 nom **Sb** 4 nom | |
| **L52905** | Battery Alloy | **As** 0.15 **Pb** 95.5 **S** 0.003 **Sb** 4.0 **Sn** 0.3 | |
| **L52910** | Lead Antimony Tin Alloy | **Pb** 95 nom **Sb** 4 nom **Sn** 1 nom | |
| **L52915** | Type Metal Alloy | **Pb** 93 nom **Sb** 4 nom **Sn** 3 nom | |
| **L52920** | 4.5% Antimonial Lead | **Pb** 95.5 nom **Sb** 4.5 nom | |
| **L52922** | Anode Alloy | **Pb** 95 nom **S** 0.5 nom **Sb** 4.5 nom | |
| **L52930** | Battery Alloy | **As** 0.3 **Cu** 0.05 **Pb** 94.6 nom **S** 0.007 **Sb** 4.75 **Sn** 0.3 | |
| **L52940** | Battery Alloy | **As** 0.15 **Cu** 0.05 **Pb** 94.8 nom **S** 0.004 **Sb** 4.75 **Sn** 0.3 | |
| **L53020** | Lyman"s 2 Alloy-Bullet Alloy | **Pb** 90.0 nom **Sb** 5.0 nom **Sn** 5.0 nom | |
| **L53105** | 6% Antimonial Lead | **Pb** 94 nom **Sb** 6.0 nom | **AWS** C2.25 (W-Pb-Sb) |
| **L53110** | Rolled Sheet Alloy | **Pb** 94 nom **Sb** 6 nom | |
| **L53115** | Lead Alloy | **Pb** 93.7 nom **Sb** 6.0 nom **Sn** 0.3 nom | |
| **L53120** | Electrowinning Anode Alloy | **As** 0.4 nom **Pb** 93.6 nom **Sb** 6.0 nom | |
| **L53122** | High-Strength Sheet Lead | **As** 0.4 nom **Pb** 93.6 nom **Sb** 6.0 nom | |
| **L53125** | Creep Resistant Pipe and Sheet Alloy | **As** 0.65 nom **Pb** 93.3 nom **Sb** 6.0 nom | |

The chemical compositions listed are for identification purposes and should not be used in lieu of the cross referenced specifications.

| UNIFIED NUMBER | DESCRIPTION | CHEMICAL COMPOSITION | CROSS REFERENCE SPECIFICATIONS |
|---|---|---|---|
| **L53130** | Lead Alloy | **As** 0.6 nom **Pb** 92.8 nom **Sb** 6.0 nom **Sn** 0.6 nom | |
| **L53131** | Lead Alloy | **Pb** rem **Sb** 6.0-7.0 **Sn** 0.25-0.75 | **AMS** 7721 |
| **L53135** | Battery Alloy | **As** 0.3 **Cu** 0.07 **Pb** 93.4 nom **S** 0.006 **Sb** 6.0 **Sn** 0.3 | |
| **L53140** | Hard Shot Alloy | **As** 1.2 nom **Pb** 92.6 nom **Sb** 6.2 nom **Sn** 0.0005 max | |
| **L53210** | Type Metal Alloy | **Pb** 89 nom **Sb** 7 nom **Sn** 4 nom | |
| **L53220** | Lead Alloy | **As** 0.6 **Pb** 91.8 nom **Sb** 7.0 **Sn** 0.6 | |
| **L53230** | 8% Antimonial Lead | **Pb** 92 nom **Sb** 8 nom | |
| **L53235** | Hard Shot Alloy | **As** 1.25 nom **Pb** 90.7 nom **Sb** 8.0 nom **Sn** 0.0005 max | |
| **L53238** | Hard Shot Alloy | **As** 2 nom **Pb** 90.0 nom **Sb** 8 nom **Sn** 0.0005 max | |
| **L53260** | Spin Casting Alloy | **Pb** 88.9 nom **Sb** 8.0 nom **Sn** 3.1 nom | |
| **L53265** | Type Metal Alloy | **Pb** 88 nom **Sb** 8.0 nom **Sn** 4.0 nom | |
| **L53305** | 9% Antimonial Lead | **Pb** 91 nom **Sb** 9.0 nom | |
| **L53310** | Lead Alloy | **Pb** 90 nom **Sb** 9 nom **Sn** 1 nom | |
| **L53320** | Whitemetal Bearing Alloy | **Pb** 86 nom **Sb** 9 nom **Sn** 5 nom | |
| **L53340** | Lead-Base Die Casting Alloy | **As** 0.15 max **Cu** 0.50 max **Pb** 89-91 **Sb** 9.25-10.75 **Zn** 0.01 max | **ASTM** B102 (Y10A) |
| **L53343** | Lead-Base Bearing Alloy | **Al** 0.005 max **As** 0.25 max **Bi** 0.10 max **Cd** 0.05 max **Cu** 0.50 max **Fe** 0.1 max **Pb** rem **Sb** 9.5-10.5 **Sn** 5.5-6.5 **Zn** 0.005 max | |
| **L53345** | Lead-Base Bearing Alloy | **Al** 0.005 max **As** 0.25 max **Bi** 0.10 max **Cd** 0.05 max **Cu** 0.50 max **Pb** rem **Sb** 9.0-11.0 **Sn** 5.0-7.0 **Zn** 0.005 max **Other** total of others 0.20 max | **SAE** J460 (13) |
| **L53346** | Lead-Base Bearing Alloy | **Al** 0.005 max **As** 0.25 max **Bi** 0.10 max **Cd** 0.05 max **Cu** 0.50 max **Fe** 0.10 max **Pb** rem **Sb** 9.5-10.5 **Sn** 5.5-6.5 **Zn** 0.005 max | **ASTM** B23 (13) |
| **L53405** | 11% Antimonial Lead | **Pb** 89 nom **Sb** 11 nom | |
| **L53420** | Linotype Alloy | **Pb** 86 nom **Sb** 11 nom **Sn** 3 nom | |
| **L53425** | Linotype-Special Alloy | **Pb** 84 nom **Sb** 11 nom **Sn** 5 nom | |
| **L53454** | Type Metal Alloy | **Pb** 85 nom **Sb** 12 nom **Sn** 3 nom | |
| **L53455** | Linotype B (Eutectic) Alloy | **Pb** 84 nom **Sb** 12 nom **Sn** 4 nom | |
| **L53456** | Type Metal Alloy | **Pb** 83 nom **Sb** 12 nom **Sn** 5 nom | |
| **L53460** | Type Metal Alloy | **Pb** 82 nom **Sb** 12 nom **Sn** 6 nom | |
| **L53465** | Lead Alloy | **Cu** 0.05 **Pb** 77.5 nom **Sb** 12.5 nom **Sn** 10.0 nom | **AWS** C2.25 (W-PbSbSn) |
| **L53470** | CT Metal | **As** 0.4 nom **Pb** 86 nom **Sb** 12.75 nom **Sn** 0.75 nom | |
| **L53480** | Arsenical Lead-G Babbitt | **As** 3.0 nom **Pb** 83.5 nom **Sb** 12.75 nom **Sn** 0.75 nom | |
| **L53505** | CT Metal | **As** 1 nom **Pb** 85 nom **Sb** 13 nom **Sn** 1 nom | |
| **L53510** | Stereotype-General Alloy | **Pb** 80.5 nom **Sb** 13 nom **Sn** 6.5 nom | |
| **L53530** | Stereotype-Flat Alloy | **Pb** 80 nom **Sb** 14 nom **Sn** 6 nom | |
| **L53550** | 15% Antimonial Lead | **Pb** 85 nom **Sb** 15 nom | |

The chemical compositions listed are for identification purposes and should not be used in lieu of the cross referenced specifications.

| UNIFIED NUMBER | DESCRIPTION | CHEMICAL COMPOSITION | CROSS REFERENCE SPECIFICATIONS |
|---|---|---|---|
| **L53555** | Lead Alloy | **Pb** 83 nom **Sb** 15 nom **Sn** 2 nom | |
| **L53558** | Type Metal Alloy | **Pb** 81 nom **Sb** 15 nom **Sn** 4 nom | |
| **L53560** | Lead-Base Die Casting Alloy | **Al** 0.01 max **As** 0.15 max **Cu** 0.50 max **Pb** 79-81 **Sb** 14-16 **Sn** 4-6 **Zn** 0.01 max | **ASTM** B102 (YT155A) |
| **L53565** | Lead-Base White Metal Bearing Alloy | **Al** 0.005 max **As** 0.30-0.60 **Bi** 0.10 max **Cd** 0.05 max **Cu** 0.50 max **Fe** 0.10 max **Pb** rem **Sb** 14.0-16.0 **Sn** 4.5-5.5 **Zn** 0.005 max | **ASTM** B23 (8) |
| **L53570** | Monotype-Ordinary Alloy | **Pb** 78 nom **Sb** 15 nom **Sn** 7 nom | |
| **L53575** | Stereotype-Curved Alloy | **Pb** 77 nom **Sb** 15 nom **Sn** 8 nom | |
| **L53580** | Rules Monotype Alloy | **Pb** 75 nom **Sb** 15 nom **Sn** 10 nom | |
| **L53585** | Lead-Base Whitemetal Bearing Alloy | **Al** 0.005 max **As** 0.30-0.60 **Bi** 0.10 max **Cd** 0.05 max **Cu** 0.50 max **Fe** 0.10 max **Pb** rem **Sb** 14.0-16.0 **Sn** 9.3-10.7 **Zn** 0.005 max | **ASTM** B23 (7) |
| **L53620** | Lead-Base Bearing Alloy | **Al** 0.005 max **As** 0.8-1.4 **Bi** 0.10 max **Cd** 0.05 max **Cu** 0.6 max **Fe** 0.10 max **Pb** rem **Sb** 14.5-17.5 **Sn** 0.8-1.2 **Zn** 0.005 max | **ASTM** B23 (15) |
| **L53650** | Display Monotype Alloy | **Pb** 75 nom **Sb** 17 nom **Sn** 8 nom | |
| **L53655** | Type Metal Alloy | **Pb** 74 nom **Sb** 17 nom **Sn** 9 nom | |
| **L53685** | Lanston Standard Case Type Monotype Alloy | **Pb** 72 nom **Sb** 19 nom **Sn** 9 nom | |
| **L53710** | Hard Foundry Type Alloy | **Cu** 1.5 nom **Pb** 58.5 nom **Sb** 20 nom **Sn** 20 nom | |
| **L53740** | Monotype Case Type Alloy | **Pb** 67 nom **Sb** 24 nom **Sn** 9 nom | |
| **L53750** | Monotype Case Type Alloy | **Pb** 64 nom **Sb** 24 nom **Sn** 12 nom | |
| **L53780** | Hard Foundry Type Alloy | **Cu** 1.5 nom **Pb** 60.5 nom **Sb** 25 nom **Sn** 13 nom | |
| **L53790** | Type Metal Alloy | **Pb** 67 nom **Sb** 28 nom **Sn** 5 nom | |
| **L53795** | Type Metal Alloy | **Pb** 65.5 nom **Sb** 29 nom **Sn** 5.5 nom | |
| **L54030** | Alloy 1/2C-Cable Sheathing Alloy | **Cd** 0.075 nom **Pb** 99.7 nom **Sn** 0.2 nom | |
| **L54050** | Alloy C-Cable Sheathing Alloy | **Cd** 0.15 nom **Pb** 99.4 nom **Sn** 0.4 nom | |
| **L54210** | 2% Tin Solder | **Al** 0.005 max **As** 0.02 max **Bi** 0.25 max **Cu** 0.08 max **Fe** 0.02 max **Pb** 98 nom **Sb** 0.12 max **Sn** 1.5-2.5 **Zn** 0.005 max | **ASTM** B32 (2A, Sn2) |
| **L54211** | 2% Tin Antimonial Solder | **Al** 0.005 max **As** 0.02 max **Bi** 0.25 max **Cu** 0.08 max **Fe** 0.02 max **Pb** 98 nom **Sb** 0.20-0.50 **Sn** 1.5-2.5 **Zn** 0.005 max | **ASTM** B32 (2B) |
| **L54250** | Solder Alloy | **Al** 0.005 max **As** 0.40-0.60 **Bi** 0.25 max **Cu** 0.08 max **Fe** 0.02 max **Pb** rem **Sb** 4.90-5.40 **Sn** 2.50-2.75 **Zn** 0.005 max **Other** other elements, total 0.08 max | **SAE** J473 (No. 9B) |
| **L54280** | Solder Alloy | **Pb** 91.9 nom **Sb** 5.1 nom **Sn** 3.0 nom | |
| **L54310** | Electrotype Curved Plate Alloy | **Pb** 93 nom **Sb** 3 nom **Sn** 4 nom | |
| **L54320** | 5/95 Solder | **Al** 0.005 max **As** 0.02 max **Bi** 0.25 max **Cu** 0.08 max **Fe** 0.02 max **Pb** 95 nom **Sb** 0.12 max **Sn** 4.5-5.5 **Zn** 0.005 max | **ASTM** B32 (5A) |

The chemical compositions listed are for identification purposes and should not be used in lieu of the cross referenced specifications.

| UNIFIED NUMBER | DESCRIPTION | CHEMICAL COMPOSITION | CROSS REFERENCE SPECIFICATIONS |
|---|---|---|---|
| **L54321** | 5% Tin Antimonial Solder | **Al** 0.005 max **As** 0.02 max **Bi** 0.25 max **Cu** 0.08 max **Fe** 0.02 max **Pb** 95 nom **Sb** 0.20-0.50 **Sn** 4.5-5.5 **Zn** 0.005 max | **ASTM** B32 (5B) |
| **L54322** | SN5 Solder | **Ag** 0.015 max **Al** 0.005 max **As** 0.02 max **Bi** 0.25 max **Cd** 0.001 max **Cu** 0.08 max **Fe** 0.02 max **Pb** rem **Sb** 0.50 max **Sn** 4.5-5.5 **Zn** 0.005 max **Other** total of all others 0.08 max | **ASTM** B32 (Sn5) |
| **L54360** | Solder Alloy | **As** 0.5 nom **Pb** 90.5 nom **Sb** 4.0 nom **Sn** 5.0 nom | |
| **L54370** | Plated Overlay for Bearings | **Pb** rem **Sn** 5.0-9.0 **Other** other elements, total 3.5 max | |
| **L54410** | 8% Tin Solder | **Pb** 92 nom **Sb** 0.3 nom **Sn** 8 nom | |
| **L54510** | Plated Overlay for Bearings | **Pb** rem **Sn** 8.0-12.0 **Other** other elements, total 3.5 max | |
| **L54520** | 10/90 Solder | **Al** 0.005 max **As** 0.02 max **Bi** 0.25 max **Cu** 0.08 max **Fe** 0.02 max **Pb** 90 nom **Sb** 0.20-0.50 **Sn** 10 nom **Zn** 0.005 max | **ASTM** B32 (10B, Sn10A) |
| **L54525** | 88-10-2 Solder | **Ag** 1.7-2.4 **Al** 0.005 max **As** 0.02 max **Bi** 0.03 max **Cd** 0.001 max **Cu** 0.08 max **Pb** rem **Sb** 0.20 max **Sn** 9.0-11.0 **Zn** 0.005 max **Other** total of all others 0.10 max | **ASTM** B32 (Sn10B) |
| **L54530** | Solder Alloy | **As** 0.5 nom **Pb** 85.5 nom **Sb** 4 nom **Sn** 10 nom | |
| **L54540** | Solder Alloy | **Pb** 87.5 nom **Sb** 0.45 nom **Sn** 12 nom | |
| **L54555** | Solder Alloy | **Al** 0.005 max **Bi** 0.25 max **Cu** 0.08 max **Fe** 0.02 max **Pb** rem **Sb** 2.75 max **Sn** 14.0-15.0 **Zn** 0.005 max **Other** other elements, total 0.08 max | **SAE** J473 (No. 6B) |
| **L54560** | 15/85 Solder | **Al** 0.005 max **As** 0.02 max **Bi** 0.25 max **Cu** 0.08 max **Fe** 0.02 max **Pb** 85 nom **Sb** 0.20-0.50 **Sn** 15 desired **Zn** 0.005 max | **ASTM** B32 (15B, Sn15) |
| **L54570** | Solder Alloy | **Pb** 82.5 nom **Sb** 2.5 nom **Sn** 15 nom | |
| **L54580** | Type Metal Alloy | **Pb** 80.5 nom **Sb** 4.5 nom **Sn** 15 nom | |
| **L54610** | Solder Alloy | **Al** 0.005 max **Bi** 0.25 max **Cu** 0.08 max **Fe** 0.02 max **Pb** rem **Sb** 1.25-1.75 **Sn** 19.0-20.0 **Zn** 0.005 max **Other** other elements, total 0.08 max | **SAE** J473 (No. 5B) |
| **L54710** | 20% Tin Solder | **Pb** 80 nom **Sb** 0.2 max **Sn** 20 nom | |
| **L54711** | Solder Alloy 20B | **Al** 0.005 max **As** 0.02 max **Bi** 0.25 max **Cu** 0.08 max **Fe** 0.02 max **Pb** 80 nom **Sb** 0.20-0.50 **Sn** 20 desired **Zn** 0.005 max | **ASTM** B32 (20B, Sn20A) |
| **L54712** | Solder Alloy 20C | **Al** 0.005 max **As** 0.02 max **Bi** 0.25 max **Cu** 0.08 max **Fe** 0.02 max **Pb** 79 nom **Sb** 0.8-1.2 **Sn** 20 desired **Zn** 0.005 max | **ASTM** B32 (20C, Sn20B) |
| **L54713** | Solder Alloy | **Pb** 77 nom **Sb** 1.2 nom **Sn** 22 nom | |
| **L54715** | Solder Alloy | **Pb** 74 nom **Sb** 3 nom **Sn** 23 nom | |
| **L54720** | 25/75 Solder | **Al** 0.005 max **As** 0.02 max **Bi** 0.25 max **Cu** 0.08 max **Fe** 0.02 max **Pb** 75 nom **Sb** 0.25 max **Sn** 25 nom **Zn** 0.005 max | **ASTM** B32 (25A) |

The chemical compositions listed are for identification purposes and should not be used in lieu of the cross referenced specifications.

| UNIFIED NUMBER | DESCRIPTION | CHEMICAL COMPOSITION | CROSS REFERENCE SPECIFICATIONS |
|---|---|---|---|
| **L54721** | Solder Alloy 25B | **Al** 0.005 max **As** 0.02 max **Bi** 0.25 max **Cu** 0.08 max **Fe** 0.02 max **Pb** 75 nom **Sb** 0.20-0.50 **Sn** 25 desired **Zn** 0.005 max | **ASTM** B32 (25B, Sn25A) |
| **L54722** | Solder Alloy 25C | **Al** 0.005 max **As** 0.02 max **Bi** 0.25 max **Cu** 0.08 max **Fe** 0.02 max **Pb** 73.7 nom **Sb** 1.1-1.5 **Sn** 25 desired **Zn** 0.005 max | **ASTM** B32 (25C, Sn25B) |
| **L54727** | Lead-Base Bearing Alloy | **Cu** 3 nom **Pb** 59 nom **Sb** 13 nom **Sn** 25 nom | |
| **L54750** | Silver-Loaded Solder | **Ag** 3 nom **Pb** 70 nom **Sn** 27 nom | |
| **L54755** | Fusible Alloy | **Bi** 21.5 nom **Pb** 51.5 nom **Sn** 27 nom | |
| **L54805** | Solder Alloy | **Pb** 70.5 nom **Sb** 1.5 nom **Sn** 28 nom | |
| **L54810** | Fusible Alloy | **Bi** 28.5 nom **Pb** 43 nom **Sn** 28.5 nom | |
| **L54815** | Solder Alloy | **Al** 0.005 max **Bi** 0.25 max **Cu** 0.08 max **Fe** 0.02 max **Pb** rem **Sb** 0.75-1.25 **Sn** 29.0-30.0 **Zn** 0.005 max **Other** other elements, total 0.08 max | **SAE** J473 (No. 3B) |
| * **L54820** | 30/70 Solder | **Al** 0.005 max **As** 0.02 max **Bi** 0.25 max **Cu** 0.08 max **Fe** 0.02 max **Pb** 70 nom **Sb** 0.25 max **Sn** 30 desired **Zn** 0.005 max | **ASTM** B32 (30A) |
| **L54821** | 30/70 Solder Alloy (Sn30A) | **Al** 0.005 max **As** 0.02 max **Bi** 0.25 max **Cu** 0.08 max **Fe** 0.02 max **Pb** rem **Sb** 0.50 max **Sn** 29.5-31.5 **Zn** 0.005 max | **ASTM** B32 (30B, Sn30A) |
| **L54822** | 30/70 Solder Alloy (Sn30B) | **Al** 0.005 max **As** 0.02 max **Bi** 0.25 max **Cu** 0.08 max **Fe** 0.02 max **Pb** rem **Sb** 1.4-1.8 **Sn** 29.5-31.5 **Zn** 0.005 max | **ASTM** B32 (30C, Sn30B) |
| **L54827** | Type Metal Alloy | **Pb** 64 nom **Sb** 6 nom **Sn** 30 nom | |
| **L54829** | Fusible Alloy | **Bi** 20 nom **Pb** 50 nom **Sn** 30 nom | |
| **L54830** | Fusible Alloy | **Bi** 30.8 nom **Pb** 38.4 nom **Sn** 30.8 nom | |
| **L54832** | Solder Alloy | **Pb** 66.7 nom **Sb** 1.8 nom **Sn** 31.5 nom | |
| **L54833** | Solder Alloy | **Pb** 65 nom **Sb** 3 nom **Sn** 32 nom | |
| **L54835** | Fusible Alloy | **Bi** 33.3 nom **Pb** 33.4 nom **Sn** 33.3 nom | |
| **L54840** | Fusible Alloy | **Bi** 32 nom **Pb** 34 nom **Sn** 34 nom | |
| **L54850** | 35/65 Solder | **Al** 0.005 max **As** 0.02 max **Bi** 0.25 max **Cu** 0.08 max **Fe** 0.02 max **Pb** 65 nom **Sb** 0.25 max **Sn** 35 nom **Zn** 0.005 max | **ASTM** B32 (35A) |
| **L54851** | Solder Alloy | **Al** 0.005 max **As** 0.02 max **Bi** 0.25 max **Cu** 0.08 max **Fe** 0.02 max **Pb** 65 nom **Sb** 0.20-0.50 **Sn** 35 nom **Zn** 0.005 max | **ASTM** B32 (35B, Sn35A) |
| **L54852** | Solder Alloy | **Al** 0.005 max **As** 0.02 max **Bi** 0.25 max **Cu** 0.08 max **Fe** 0.02 max **Pb** 63.2 nom **Sb** 1.6-2.0 **Sn** 35 nom **Zn** 0.005 max | **ASTM** B32 (35C, Sn35B) |
| **L54855** | Silver-Loaded Solder Alloy | **Ag** 3.0 nom **Pb** 61.5 nom **Sn** 35.5 nom | |
| **L54860** | Fusible Alloy | **Bi** 21 nom **Pb** 43 nom **Sn** 36 nom | |

*Boxed entries are no longer active and are retained for reference purposes only.

The chemical compositions listed are for identification purposes and should not be used in lieu of the cross referenced specifications.

| UNIFIED NUMBER | DESCRIPTION | CHEMICAL COMPOSITION | CROSS REFERENCE SPECIFICATIONS |
|---|---|---|---|
| **L54865** | Fusible Alloy | **Bi** 21 nom **Pb** 42 nom **Sn** 37 nom | |
| **L54905** | Antimonial Solder Alloy | **Al** 0.005 max **Bi** 0.25 max **Cu** 0.08 max **Fe** 0.02 max **Pb** rem **Sb** 1.5-2.00 **Sn** 38.0-38.5 **Zn** 0.005 max **Other** other elements, total 0.08 max | **SAE** J473 (No. 2B) |
| **L54910** | Fusible Alloy | **Bi** 12.6 nom **Pb** 47.5 nom **Sn** 39.9 nom | |
| **L54915** | 40/60 Solder | **Al** 0.005 max **As** 0.02 max **Bi** 0.25 max **Cu** 0.08 max **Fe** 0.02 max **Pb** 60 nom **Sb** 0.12 max **Sn** 40 nom **Zn** 0.005 max | **ASTM** B32 (40A, Sn40A) |
| **L54916** | Solder Alloy | **Al** 0.005 max **As** 0.02 max **Bi** 0.25 max **Cu** 0.08 max **Fe** 0.02 max **Pb** 60 nom **Sb** 0.20-0.50 **Sn** 40 nom **Zn** 0.005 max | **ASTM** B32 (40B) |
| **L54918** | Solder Alloy | **Al** 0.005 max **As** 0.02 max **Bi** 0.25 max **Cu** 0.08 max **Fe** 0.02 max **Pb** 58 nom **Sb** 1.8-2.4 **Sn** 40 nom **Zn** 0.005 max | **ASTM** B32 (40C, Sn40B) |
| **L54925** | Lead-Base Bearing Alloy | **Cu** 2 nom **Pb** 46 nom **Sb** 12 nom **Sn** 40 nom | |
| **L54930** | Fusible Alloy | **Bi** 4 nom **Pb** 55.5 nom **Sn** 40.5 nom | |
| **L54935** | Fusible Alloy | **Bi** 14 nom **Pb** 43 nom **Sn** 43 nom | **ASTM** B774 (291-325) |
| **L54940** | Solder Alloy | **Al** 0.005 max **Bi** 0.25 max **Cu** 0.08 max **Fe** 0.02 max **Pb** rem **Sb** 1.5-2.00 **Sn** 43.0-43.5 **Zn** 0.005 max **Other** other elements, total 0.08 max | **SAE** J473 (No. 1B) |
| **L54945** | Silver-Loaded Solder | **Ag** 1 nom **Pb** 55 nom **Sn** 44 nom | |
| **L54950** | 45/55 Solder | **Al** 0.005 max **As** 0.03 max **Bi** 0.25 max **Cu** 0.08 max **Fe** 0.02 max **Pb** 55 nom **Sb** 0.12 max **Sn** 45 nom **Zn** 0.005 max | **AMS** 4750 **ASTM** B32 (45A) |
| **L54951** | Solder Alloy | **Al** 0.005 max **As** 0.03 max **Bi** 0.25 max **Cu** 0.08 max **Fe** 0.02 max **Pb** 55 nom **Sb** 0.20-0.50 **Sn** 45 nom **Zn** 0.005 max | **ASTM** B32 (45B, Sn45) |
| **L54955** | Solder Alloy | **Pb** 52.5 nom **Sb** 2.5 nom **Sn** 45 nom | |
| **L55005** | Solder Alloy | **Pb** 52 nom **Sn** 48 nom | |
| **L55030** | 50/50 Solder | **Al** 0.005 max **As** 0.03 max **Bi** 0.25 max **Cu** 0.08 max **Fe** 0.02 max **Pb** 50 nom **Sb** 0.12 max **Sn** 50 nom **Zn** 0.005 max | **ASTM** B32 (50A) |
| **L55031** | Solder Alloy | **Al** 0.005 max **As** 0.03 max **Bi** 0.25 max **Cu** 0.08 max **Fe** 0.02 max **Pb** 50 nom **Sb** 0.20-0.50 **Sn** 50 nom **Zn** 0.005 max | **ASTM** B32 (50B, Sn50) |
| **L55210** | Battery Alloy | **Pb** 99.6 nom **Sn** 0.3 nom **Other** Ca 0.06 nom; Sr 0.06 nom | |
| **L55230** | Battery Alloy | **Al** 0.03 nom **Pb** 99 nom **Sn** 0.8 nom **Other** Sr 0.16 nom | |
| **L55260** | Battery Alloy | **Al** 0.03 nom **Pb** 99.8 nom **Other** Sr 0.2 nom | |
| **L55290** | Lead-Strontium Alloy | **Pb** 98. nom **Other** Sr 2. nom | |
| **L91090** | Aluminum Solder (KAPALLOY) | **Ag** 0.002 max **As** 0.030 max **Bi** 0.030 max **Cd** 0.005 max **Cu** 0.005 max **Fe** 0.004 max **Ni** 0.006 max **Pb** 0.030 max **Sb** 0.025 max **Sn** 90.00-92.00 **Zn** 8.00-10.00 | |

The chemical compositions listed are for identification purposes and should not be used in lieu of the cross referenced specifications.

# UNS NUMBERS ASSIGNED TO DATE

## With Description of Each Material Covered and References To Documents In Which The Same or Similar Materials are Described

## Mxxxxx Number Series
## Miscellaneous Nonferrous Metals and Alloys

# MISCELLANEOUS NONFERROUS METALS AND ALLOYS

| UNIFIED NUMBER | DESCRIPTION | CHEMICAL COMPOSITION | CROSS REFERENCE SPECIFICATIONS |
|---|---|---|---|
| **M00995** | Metallic Antimony | **Ag** 0.10 max **As** 0.10 max **Cu** 0.10 max **Fe** 0.10 max **Ni** 0.10 max **Pb** 0.20 max **S** 0.10 max **Sb** 99.50 min **Sn** 0.10 max | **MIL SPEC** MIL-A-10841 |
| **M00998** | Metallic Antimony | **Ag** 0.05 max **As** 0.05 max **Cu** 0.05 max **Fe** 0.05 max **Ni** 0.05 max **Pb** 0.15 max **S** 0.10 max **Sb** 99.80 min **Sn** 0.05 max | **MIL SPEC** MIL-A-10841 |
| * **M05995** | Plutonium Nuclear Grade | **Fe** 500 ppm max **Other** plutonium Pu 99.5 min, F+Cl 200 ppm max; Mg+Ca 500 ppm max | |
| **M08990** | Depleted Uranium | **C** 0.07 max **U** 99.00 min | **AMS** 7730 |
| **M10101** | Magnesium Alloy Casting (AM100A) | **Al** 9.4-10.6 **Cu** 0.08 max **Mg** rem **Mn** 0.13-0.35 **Ni** 0.010 max **Si** 0.20 max **Zn** 0.2 max **Other** total 0.30 max | **ASTM** B93 (AM100A) |
| **M10030** | Magnesium Alloy-Casting (A3A) | **Ag** 0.001 max **Al** 2.5-3.5 **Cd** 0.001 max **Cu** 0.005 max **Fe** 0.005 max **Mg** rem **Mn** 0.005 max **Ni** 0.001 max **Zn** 0.10 max **Other** total 0.30 max, Rare Earths 0.001 max | **ASTM** B275 (A3A) |
| **M10100** | Magnesium Alloy-Casting (AM100A) | **Al** 9.3-10.7 **Cu** 0.10 max **Mg** rem **Mn** 0.10 min **Ni** 0.01 max **Si** 0.30 max **Zn** 0.30 max **Other** total 0.30 max | **AMS** 4455; 4483 **ASTM** B80 (AM100A); B93 (AM100A); B199 (AM100A); B275 (AM100A); B403 (AM100A) **SAE** J465 (AM100-A) |
| **M10102** | Magnesium-Die Alloy Casting (AM100B) | **Al** 9.4-10.6 **Cu** 0.08 max **Mg** rem **Mn** 0.13 min **Ni** 0.01 max **Si** 1.0 max **Other** total 0.30 max | **ASTM** B275 (AM100B) |
| **M10410** | Magnesium-Die Alloy Casting (AS41A) | **Al** 3.5-5.0 **Cu** 0.06 max **Mg** rem **Mn** 0.20-0.50 **Ni** 0.03 max **Si** 0.50-1.5 **Zn** 0.12 max | **ASTM** B93; B94 (AS41A); B275 (AS41A) **SAE** J465 (AS41-A) |
| **M10411** | Magnesium Alloy-Casting (AS41A) | **Al** 3.7-4.8 **Cu** 0.04 max **Mg** rem **Mn** 0.22-0.48 **Ni** 0.01 max **Si** 0.60-1.4 **Zn** 0.10 max **Other** impurities 0.30 max | **ASTM** B93 (AS41A) |
| **M10412** | Magnesium-Die Alloy Casting (AS41B) | **Al** 3.5-5.0 **Cu** 0.02 **Fe** 0.0035 max **Mg** rem **Mn** 0.35-0.4 **Ni** 0.002 **Si** 0.50-1.5 **Zn** 0.12 **Other** metallic impurities 0.02 max each | **ASTM** B94 (AS41B) |
| **M10413** | Magnesium Alloy-Casting (AS41B) | **Al** 3.7-4.8 **Cu** 0.015 max **Fe** 0.0035 max **Mg** rem **Mn** 0.35-0.6 **Ni** 0.001 max **Si** 0.60-1.4 **Zn** 0.10 max **Other** metallic impurities 0.01 max each | **ASTM** B93 (AS41B) |
| **M10500** | Magnesium Alloy (AM50A) | **Al** 4.4-5.4 **Cu** 0.010 max **Fe** 0.004 max **Mg** rem **Mn** 0.26-0.6 **Ni** 0.002 max **Si** 0.10 max **Zn** 0.22 max **Other** metallic each 0.02 max. If either the minimum Mn limit or maximum Fe limit is not met, then Fe/Mn ratio shall not exceed 0.015 | **ASTM** B94 (AM50A) |
| **M10501** | Magnesium Alloy (AM50A) | **Al** 4.5-5.3 **Cu** 0.008 max **Fe** 0.004 max **Mg** rem **Mn** 0.28-0.50 **Ni** 0.001 max **Si** 0.05 max **Zn** 0.20 max **Other** metallic each 0.02 max | **ASTM** B93/M93M |
| **M10600** | Magnesium-Die-Mold Alloy Casting (AM60A) | **Al** 5.5-6.5 **Cu** 0.35 max **Mg** rem **Mn** 0.13 min **Ni** 0.03 max **Si** 0.50 max **Zn** 0.22 max **Other** total 0.30 max | **ASTM** B93; B94 (AM60A); B275 (AM60A) **SAE** J465 (AM60-A) |
| **M10601** | Magnesium Alloy-Casting (AM60A) | **Al** 5.7-6.3 **Cu** 0.25 max **Mg** rem **Mn** 0.15 min **Ni** 0.01 max **Si** 0.20 max **Zn** 0.20 max **Other** impurities 0.30 max | **ASTM** B93 (AM60A) |

*Boxed entries are no longer active and are retained for reference purposes only.

The chemical compositions listed are for identification purposes and should not be used in lieu of the cross referenced specifications.

| UNIFIED NUMBER | DESCRIPTION | CHEMICAL COMPOSITION | CROSS REFERENCE SPECIFICATIONS |
|---|---|---|---|
| **M10602** | Magnesium Alloy-Casting (AM60B) | **Al** 5.5-6.5 **Cu** 0.010 max **Fe** 0.005 max **Mg** rem **Mn** 0.25 min **Ni** 0.002 max **Si** 0.10 max **Zn** 0.22 max **Other** metallic impurities, each, 0.02 max. If Iron exceeds 0.005%, then the permissable Fe to Mn ratio shall not exceed 0.021. | **ASTM** B94 (AM60B) |
| **M10603** | Magnesium Alloy-Casting (AM60B) | **Al** 5.7-6.3 **Cu** 0.008 max **Fe** 0.004 max **Mg** rem **Mn** 0.27 min **Ni** 0.001 max **Si** 0.05 max **Zn** 0.20 max **Other** metallic impurities 0.01 max each | **ASTM** B93 (AM60B) |
| **M10800** | Magnesium Alloy-Casting (AM80A) | **Al** 8.0-9.0 **Cu** 0.08 max **Mg** rem **Mn** 0.18 min **Ni** 0.01 max **Si** 0.20 max **Zn** 0.20 max **Other** total 0.30 max | **ASTM** B275 (AM80A) |
| **M10900** | Magnesium Alloy-Casting (AM90A) | **Ag** 0.008 max **Al** 8.5-9.5 **Cu** 0.10 max **Fe** 0.008 max **Mg** rem **Mn** 0.15 min **Ni** 0.005 max **Si** 0.15 max **Zn** 0.20 max **Other** total 0.10 max; each 0.02 max | **ASTM** B275 (AM90A) |
| **M11100** | Magnesium Alloy-Casting (AZ10A) | **Al** 1.0-1.5 **Cu** 0.10 max **Fe** 0.005 max **Mg** rem **Mn** 0.20 min **N** 0.005 max **Si** 0.10 max **Zn** 0.20-0.6 | **ASTM** B275 (AZ10A) |
| **M11101** | Magnesium Alloy-Casting (AZ101A) | **Al** 9.5-10.5 **Be** 0.0002-0.0008 **Cu** 0.05 max **Fe** 0.005 max **Mg** rem **Mn** 0.13 min **Ni** 0.005 max **Si** 0.05 max **Zn** 0.75-1.25 **Other** total 0.30 max | **ASME** SFA5.19 (ER AZ101A) **ASTM** B275 (AZ101A) **AWS** A5.19 (ER AZ101A) |
| **M11125** | Magnesium Alloy-Casting (AZ125A) | **Al** 11.0-13.0 **Be** 0.0002-0.0008 **Mg** rem **Zn** 4.5-5.5 **Other** total 0.30 max | **ASTM** B275 (AZ125A) |
| **M11210** | Magnesium Alloy-Casting (AZ21A) | **Al** 1.6-2.5 **Cu** 0.05 max **Fe** 0.005 max **Mg** rem **Mn** 0.15 max **Ni** 0.002 max **Si** 0.05 max **Zn** 0.8-1.6 **Other** total 0.30 max, Ca 0.10-0.25 | **ASTM** B275 (AZ21A) |
| **M11310** | Magnesium Alloy-Sheet and Plate | **Al** 2.5-3.5 **Cu** 0.05 max **Fe** 0.005 max **Mg** rem **Mn** 0.20 min **Ni** 0.005 max **Si** 0.30 max **Zn** 0.6-1.4 **Other** total 0.30 max, Ca 0.30 max | **ASTM** B275 (AZ31A) |
| **M11311** | Magnesium Alloy-Wrought-Extruded Shapes-Forgings (AZ31B) | **Al** 2.5-3.5 **Cu** 0.05 max **Fe** 0.005 max **Mg** rem **Mn** 0.20 min **Ni** 0.005 max **Si** 0.10 max **Zn** 0.6-1.4 **Other** total 0.30 max, Ca 0.04 max | **AMS** 4375; 4376; 4377; 4382 **ASTM** B90 (AZ31B); B91 (AZ31B); B107 (AZ31B); B275 (AZ31B); B843 (AZ31B) **SAE** J466 (AZ31B) |
| **M11312** | Magnesium Alloy-Wrought-Extruded Shapes (AZ31C) | **Al** 2.4-3.6 **Cu** 0.10 max **Mg** rem **Mn** 0.15 min **Ni** 0.03 max **Si** 0.10 max **Zn** 0.50-1.5 **Other** total 0.30 max | **ASTM** B90; B107 (AZ31C); B275 (AZ31C) |
| **M11313** | Magnesium Alloy, High Purity, for Anodes (AZ 31D) | **Al** 2.5-3.5 **Cu** 0.04 max **Fe** 0.002 max **Mg** rem **Mn** 0.2-1.0 **Ni** 0.0010 max **Si** 0.05 max **Zn** 0.6-1.4 | **ASTM** B275 (AZ 31D); B843 (AZ 31D) |
| **M11610** | Magnesium Alloy-Forgings-Extruded Shapes (AZ61A) | **Al** 5.8-7.2 **Cu** 0.05 max **Fe** 0.005 max **Mg** rem **Mn** 0.15 min **Ni** 0.005 max **Si** 0.10 max **Zn** 0.40-1.5 **Other** total 0.30 max | **AMS** 4350 **ASTM** B91 (AZ61A); B107 (AZ61A); B275 (AZ61A) **SAE** J466 (AZ61A) |
| **M11611** | Magnesium Welding Wire | **Al** 5.8-7.2 **Be** 0.0002-0.0008 **Cu** 0.05 max **Fe** 0.005 max **Mg** rem **Mn** 0.15 min **Ni** 0.005 max **Si** 0.05 max **Zn** 0.40-1.5 **Other** total 0.30 max | **ASME** SFA5.19 (ER AZ61A) **AWS** A5.19 (ER AZ61A) |
| **M11630** | Magnesium Alloy-Casting (AZ63A) | **Al** 5.3-6.7 **Cu** 0.25 nom **Mg** rem **Mn** 0.15 min **Ni** 0.01 max **Si** 0.30 nom **Zn** 2.5-3.5 **Other** total 0.30 max | **AMS** 4420; 4422; 4424 **ASTM** B80 (AZ63A); B93 (AZ63A); B93 (2E63A); B275 (AZ63A) **SAE** J465 (AZ63A) |
| **M11631** | Magnesium Alloy-Casting (AZ63A) | **Al** 5.5-6.5 **Cu** 0.20 max **Mg** rem **Mn** 0.15-0.35 **Ni** 0.010 max **Si** 0.20 max **Zn** 2.7-3.3 **Other** impurities 0.30 max | **ASTM** B93 (AZ63A) |

The chemical compositions listed are for identification purposes and should not be used in lieu of the cross referenced specifications.

| UNIFIED NUMBER | DESCRIPTION | CHEMICAL COMPOSITION | CROSS REFERENCE SPECIFICATIONS |
|---|---|---|---|
| **M11632** | Magnesium Alloy, High Purity, for Anodes (AZ 63B) | **Al** 5.3-6.7 **Cu** 0.025 max **Fe** 0.003 max **Mg** rem **Mn** 0.15-0.35 **Ni** 0.002 max **Si** 0.10 max **Zn** 2.5-3.5 | **ASTM** B843 (AZ63B) |
| **M11634** | Magnesium Alloy, Moderate Purity, for Anodes (AZ 63C) | **Al** 5.3-6.7 **Cu** 0.05 max **Fe** 0.003 max **Mg** rem **Mn** 0.15-0.7 **Ni** 0.005 max **Si** 0.30 max **Zn** 2.5-3.5 **Other** total 0.30 max | **ASTM** B843 (AZ63C) |
| **M11636** | Magnesium Alloy, Low Purity, for Anodes (AZ 63D) | **Al** 5.0-7.0 **Cu** 0.10 max **Fe** 0.003 max **Mg** rem **Mn** 0.15-0.7 **Ni** 0.003 max **Si** 0.30 max **Zn** 2.0-4.0 **Other** total 0.30 max | **ASTM** B843; B843 (AZ63D) |
| **M11800** | Magnesium Alloy-Forgings-Extrusions (AZ80A) | **Al** 7.8-9.2 **Cu** 0.05 max **Fe** 0.005 max **Mg** rem **Mn** 0.12 min **Ni** 0.005 max **Si** 0.10 max **Zn** 0.20-0.8 **Other** total 0.30 max | **AMS** 4360 **ASTM** B80; B91 (AZ80A); B93; B107 (AZ80A); B199; B275 (AZ80A); B403 **SAE** J466 (AZ80A) |
| **M11810** | Magnesium Alloy-Casting (AZ81A) | **Al** 7.0-8.1 **Cu** 0.10 max **Mg** rem **Mn** 0.13 min **Ni** 0.01 max **Si** 0.30 max **Zn** 0.40-1.0 **Other** total 0.30 max | **ASTM** B80 (AZ81A); B199 (AZ81A); B275 (AZ81A); B403 (AZ81A) **SAE** J465 (AZ81A) |
| **M11811** | Magnesium Alloy-Casting (AZ81A) | **Al** 7.2-8.0 **Cu** 0.08 max **Mg** rem **Mn** 0.15-0.35 **Ni** 0.010 max **Si** 0.20 max **Zn** 0.5-0.9 **Other** impurities 0.30 max | **ASTM** B93 (AZ81A) |
| **M11900** | Magnesium Alloy-Casting (AZ90A) | **Al** 8.5-9.5 **Cu** 0.02 max **Fe** 0.015 max **Mg** rem **Mn** 0.15 min **Ni** 0.005 max **Si** 0.20 max **Zn** 0.20 max **Other** each 0.07 max; total 0.30 max | **ASTM** B275 (AZ90A) |
| **M11910** | Magnesium Alloy-Casting (AZ91A) | **Al** 8.3-9.7 **Cu** 0.10 max **Mg** rem **Mn** 0.13 min **Si** 0.50 max **Zn** 0.35-1.0 **Other** total 0.30 max | **AMS** 4490 **ASTM** B93 (AZ91A); B94 (AZ91A); B275 (AZ91A) **SAE** J465 (AZ91A) |
| **M11911** | Magnesium Alloy-Casting (AZ91A) | **Al** 8.5-9.5 **Cu** 0.08 max **Mg** rem **Mn** 0.15 min **Ni** 0.01 max **Si** 0.20 max **Zn** 0.45-0.9 **Other** impurities 0.30 max | **ASTM** B93 (AZ91A) |
| **M11912** | Magnesium Alloy-Casting | **Al** 8.5-9.5 **Cu** 0.25 max **Mg** rem **Mn** 0.15 min **Ni** 0.01 min **Si** 0.20 max **Zn** 0.45-0.9 **Other** total 0.30 max | **ASTM** B93; B94 (AZ91B); B275 (AZ91B) **SAE** J465 (AZ91B) |
| **M11913** | Magnesium Alloy-Casting (AZ91B) | **Al** 8.5-9.5 **Cu** 0.25 max **Mg** rem **Mn** 0.15 min **Ni** 0.01 max **Si** 0.20 max **Zn** 0.45-0.9 **Other** impurities 0.30 max | **ASTM** B93 (AZ91B) |
| **M11914** | Magnesium Alloy-Casting (AZ91C) | **Al** 8.1-9.3 **Cu** 0.10 max **Mg** rem **Mn** 0.13 min **Ni** 0.01 max **Si** 0.30 max **Zn** 0.40-1.0 **Other** total 0.30 max | **AMS** 4437; 4452 **ASTM** B80 (AZ91C); B93 (AZ91C); B199 (AZ91C); B275 (AZ91C); B403 (AZ91C) **SAE** J465 (AZ91C) |
| **M11915** | Magnesium Alloy-Casting (AZ91C) | **Al** 8.3-9.2 **Cu** 0.08 max **Mg** rem **Mn** 0.15-0.35 **Ni** 0.010 max **Si** 0.20 max **Zn** 0.45-0.9 **Other** impurities 0.30 max | **ASTM** B93 (AZ91C) |
| **M11916** | Magnesium Alloy-Casting (AZ91D) | **Al** 8.3-9.7 **Cu** 0.030 max **Fe** 0.005 max **Mg** rem **Mn** 0.15 min **Ni** 0.002 max **Si** 0.10 max **Zn** 0.35-1.0 **Other** Metallic impurities, each 0.20 max. If either the minimum manganese or the maximum iron is not met, then the iron/manganese ratio shall not exceed 0.032. | **ASTM** B275 (AZ91D) |
| **M11917** | Magnesium Alloy-Casting (Ingot) (AZ91D) | **Al** 8.5-9.5 **Cu** 0.015 max **Fe** 0.004 max **Mg** rem **Mn** 0.17 min **Ni** 0.001 max **Si** 0.05 max **Zn** 0.45-0.9 **Other** metallic impurities, each, 0.01 max. | **ASTM** B93 (AZ91D) |
| **M11918** | Magnesium Alloy-Casting (AZ91E) | **Al** 8.3-9.2 **Cu** 0.015 max **Fe** 0.005 max **Mg** rem **Mn** 0.17-0.50 **Ni** 0.0010 max **Si** 0.20 max **Zn** 0.45-0.9 **Other** 0.01 max each, impurities 0.30 max | **ASTM** B93 (AZ91E) |

The chemical compositions listed are for identification purposes and should not be used in lieu of the cross referenced specifications.

| UNIFIED NUMBER | DESCRIPTION | CHEMICAL COMPOSITION | CROSS REFERENCE SPECIFICATIONS |
|---|---|---|---|
| **M11919** | Magnesium Alloy High Purity Low Fe, Ni, and Cu, Sand Casting (AZ91E) | **Al** 8.1-9.3 **Cu** 0.015 max **Fe** 0.005 max **Mg** rem **Mn** 0.17-0.35 **Ni** 0.001 max **Si** 0.20 max **Zn** 0.40-1.0 | **AMS** 4446 **ASTM** B80 (AZ91E); B199 (AZ91E); B275 (AZ91E); B403 (AZ91E) |
| **M11920** | Magnesium Alloy-Casting (AZ92A) | **Al** 8.3-9.7 **Cu** 0.25 max **Mg** rem **Mn** 0.10 min **Ni** 0.01 max **Si** 0.30 max **Zn** 1.6-2.4 **Other** total 0.30 max | **AMS** 4434; 4453; 4484 **ASTM** B80 (AZ92A); B93 (AZ92A); B199 (AZ92A); B275 (AZ92A); B403 (AZ92A) **SAE** J465 (AZ92A) |
| **M11921** | Magnesium Alloy-Casting (AZ92A) | **Al** 8.5-9.5 **Cu** 0.20 max **Mg** rem **Mn** 0.13-0.35 **Ni** 0.010 max **Si** 0.20 max **Zn** 1.7-2.3 **Other** impurities 0.30 max | **ASTM** B93 (AZ92A) |
| **M11922** | Magnesium Alloy Welding Wire (AZ92A) | **Al** 8.3-9.7 **Be** 0.0002-0.0008 **Cu** 0.05 max **Fe** 0.005 max **Mg** rem **Mn** 0.15 min **Ni** 0.005 max **Si** 0.05 max **Zn** 1.7-2.3 **Other** total 0.30 max | **AMS** 4395 **ASME** SFA5.19 (ER AZ92A) **AWS** A5.19 (ER AZ92A) |
| **M12300** | Magnesium Alloy-Casting (EK30A) | **Cu** 0.10 max **Mg** rem **Ni** 0.01 max **Zn** 0.30 max **Zr** 0.20 min **Other** total 0.30 max, Rare Earths 2.5-4.0 | **ASTM** B275 (EK30A) |
| **M12330** | Magnesium Alloy-Sand-Permanent Mold-Investment Casting (EZ33A) | **Cu** 0.10 max **Mg** rem **Ni** 0.01 max **Zn** 2.0-3.1 **Zr** 0.50-1.0 **Other** total 0.30 max, Rare Earths 2.5-4.0 | **AMS** 4442 **ASTM** B80 (EZ33A); B93 (EZ33A); B199 (EZ33A); B275 (EZ33A); B403 (EZ33A) **SAE** J465 (EZ33A) |
| **M12331** | Magnesium Alloy (EZ33A) | **Mg** rem **Re** 2.5-4.0 **Zn** 2.0-3.1 **Zr** 0.45-1.0 **Other** total 0.30 max | **AMS** 4396 **ASME** SFA5.19 (ER EZ33A) **ASTM** B93 (EZ33A) **AWS** A5.19 (ER EZ33A) |
| **M12350** | Combined with M18210 | | |
| **M12410** | Magnesium Alloy-Sand Casting (EK41A) | **Cu** 0.10 max **Mg** rem **Ni** 0.01 max **Zn** 0.30 max **Zr** 0.40-1.0 **Other** total 0.30 max, Rare Earths 3.0-5.0 | **AMS** 4441 **ASTM** B275 (EK41A) |
| **M13210** | Magnesium Alloy-Wrought Products-Forgings (HM21A) | **Mg** rem **Mn** 0.45-1.1 **Th** 1.5-2.5 **Other** total 0.30 max | **AMS** 4363; 4383; 4390 **ASTM** B90 (HM21A); B275 (HM21A) **SAE** J466 (HM21A) |
| **M13310** | Magnesium Alloy-Casting (HK31A) | **Cu** 0.10 max **Mg** rem **Ni** 0.01 max **Th** 2.5-4.0 **Zn** 0.30 max **Zr** 0.40-1.0 **Other** total 0.30 max | **AMS** 4384; 4385; 4445 **ASTM** B80 (HK31A); B90 (HK31A); B275 (HK31A); B403 (HK31A) **SAE** J465 (HK31A); J466 (HK31A) |
| **M13311** | Magnesium Alloy-Casting (HK31B) | **Cu** 0.10 max **Mg** rem **Ni** 0.010 max **Si** 0.01 max **Th** 2.6-3.8 **Zn** 0.25 max **Zr** 0.3-1.0 **Other** impurities 0.30 max | **ASTM** B93 (HK31B) |
| **M13312** | Magnesium Alloy-Casting (HM31A) | **Mg** rem **Mn** 1.2 min **Th** 2.5-3.5 **Other** 0.30 max total | **AMS** 4388 **ASTM** B275 (HM31A) **SAE** J466 (HM31A) |
| **M13320** | Magnesium Alloy-Casting (HZ32A) | **Cu** 0.10 max **Mg** rem **Ni** 0.01 max **Th** 2.5-4.0 **Zn** 1.7-2.5 **Zr** 0.50-1.0 **Other** total 0.30 max, Rare Earths 0.10 max | **AMS** 4447 **ASTM** B80 (HZ32A); B93 (HZ32A); B275 (HZ32A) **SAE** J465 (HZ32A) |
| **M13321** | Magnesium Alloy-Casting (HZ32A) | **Cu** 0.03 max **Mg** rem **Ni** 0.010 max **Re** 0.10 max **Si** 0.01 max **Th** 2.6-3.8 **Zn** 1.8-2.4 **Zr** 0.3-1.0 **Other** impurities 0.30 max | **ASTM** B93 (HZ32A) |
| **M14141** | Magnesium Alloy-Wrought Products (LA141A) | **Al** 1.0-1.5 **Cu** 0.005 max **Fe** 0.005 max **Li** 13.0-15.0 **Mg** rem **Mn** 0.15 min **Ni** 0.005 max **Si** 0.004 max **Other** total 0.30 max, Na 0.005 max | **AMS** 4386 **ASTM** B90 (LA141A); B275 (LA141A) **SAE** J466 (LA141A) |
| **M14142** | Magnesium Alloy-Wrought Products (LS141A) | **Al** 0.05 max **Cu** 0.05 max **Fe** 0.005 max **Li** 12.0-15.0 **Mg** rem **Mn** 0.15 max **Ni** 0.005 max **Si** 0.50-0.6 **Other** Na 0.005 max | **ASTM** B275 (LS141A) **MIL SPEC** MIL-M-46143 |
| **M14145** | Magnesium Alloy-Wrought Products (LZ145A) | **Ag** 2.0-3.0 **Al** 0.05 max **Cu** 0.05 max **Fe** 0.005 max **Li** 12.0-15.0 **Mg** rem **Mn** 0.15 max **Ni** 0.005 max **Si** 1.5-2.0 **Zn** 4.5-5.0 **Other** total 0.20 max, Na 0.005 max | **ASTM** B275 (LZ145A) **MIL SPEC** MIL-M-46143 |
| **M15100** | Magnesium Alloy-Extruded Shapes-Wire (M1A) | **Cu** 0.05 max **Mg** rem **Mn** 1.2-2.0 **Ni** 0.01 max **Si** 0.10 nom **Other** total 0.30 max, Ca 0.30 max | **ASTM** B107 (M1A); B275 (M1A) **SAE** J466 (M1A) |

The chemical compositions listed are for identification purposes and should not be used in lieu of the cross referenced specifications.

| UNIFIED NUMBER | DESCRIPTION | CHEMICAL COMPOSITION | CROSS REFERENCE SPECIFICATIONS |
|---|---|---|---|
| **M15101** | Magnesium Alloy- Casting (M1B) | **Cu** 0.08 **Mg** rem **Mn** 1.3 min **Ni** 0.01 max **Si** 0.10 max **Other** total 0.20 max | **ASTM** B275 (M1B) |
| **M15102** | Magnesium Alloy- Forgings (M1C) | **Al** 0.01 max **Cu** 0.02 max **Fe** 0.03 max **Mg** rem **Mn** 0.9-1.2 **Ni** 0.001 max **Other** each 0.05 max; total 0.30 max | **ASTM** B275 (M1C); B483; B843 (MIC) |
| **M16100** | Magnesium Alloy- Wrought Products (ZE10A) | **Mg** rem **Zn** 1.0-1.5 **Other** total 0.30 max, Rare Earths 0.12-0.22 | **ASTM** B90 (ZE10A); B275 (ZE10A) **SAE** J466 (ZE10A) |
| **M16210** | Magnesium Alloy (ZK21A) | **Cu** 0.10 **Mg** rem **Ni** 0.01 max **Zn** 2.0-2.6 **Zr** 0.45-0.8 **Other** total 0.30 max | **AMS** 4387 **ASTM** B275 (ZK21A) |
| **M16331** | Magnesium Alloy-Casting (ZC63A) | **Cu** 2.4-3.00 **Mg** rem **Mn** 0.25-0.75 **Ni** 0.001 max **Si** 0.20 max **Zn** 5.5-6.5 **Other** impurities 0.30 max | **ASTM** B80 (ZC63A); B93 (ZC63A) |
| **M16400** | Magnesium Alloy- Extruded Shapes (ZK40A) | **Mg** rem **Zn** 3.5-4.5 **Zr** 0.45 min **Other** total 0.30 max | **ASTM** B107 (ZK40A); B275 (ZK40A) **SAE** J466 (ZK40A) |
| **M16410** | Magnesium Alloy- Casting (ZE41A) | **Cu** 0.10 max **Mg** rem **Mn** 0.15 max **Ni** 0.01 max **Zn** 3.5-5.0 **Zr** 0.40-1.0 **Other** total 0.30 max, Rare Earths 0.75-1.75 | **AMS** 4439 **ASTM** B80 (ZE41A); B93 (ZE41A); B275 (ZE41A) **SAE** J465 (ZE41A) |
| **M16411** | Magnesium Alloy-Casting (ZE41A) | **Cu** 0.03 max **Mg** rem **Mn** 0.15 max **Ni** 0.010 max **Re** 1.0-1.75 **Si** 0.01 max **Zn** 3.7-4.8 **Zr** 0.3-1.0 **Other** impurities 0.30 max | **ASTM** B93 (ZE41A) |
| **M16510** | Magnesium Alloy- Casting (ZK51A) | **Cu** 0.10 max **Mg** rem **Ni** 0.01 max **Zn** 3.6-5.5 **Zr** 0.50-1.0 **Other** total 0.30 max | **AMS** 4443 **ASTM** B80 (ZK51A); B93 (ZK51A); B275 (ZK51A) **SAE** J465 (ZK51A) |
| **M16511** | Magnesium Alloy-Casting (ZK51A) | **Cu** 0.03 max **Mg** rem **Ni** 0.010 max **Si** 0.01 max **Zn** 3.8-5.3 **Zr** 0.3-1.0 **Other** impurities 0.30 max | **ASTM** B93 (ZK51A) |
| **M16600** | Magnesium Alloy-Wrought Products-Forgings (ZK60A) | **Mg** rem **Zn** 4.8-6.2 **Zr** 0.45 min **Other** total 0.30 max | **AMS** 4352; 4362 **ASTM** B91 (ZK60A); B107 (ZK60A); B275 (ZK60A) **SAE** J466 (ZK60A) |
| **M16601** | Magnesium Alloy (ZK60B) | **Mg** rem **Zn** 4.8-6.8 **Zr** 0.45 min **Other** total 0.30 max | **ASTM** B275 (ZK60B) |
| **M16610** | Magnesium Alloy- Casting (ZK61A) | **Cu** 0.10 **Mg** rem **Ni** 0.01 **Zn** 5.5-6.5 **Zr** 0.6-1.0 **Other** total 0.30 max | **AMS** 4444 **ASTM** B80 (ZK61A); B93 (ZK61A); B275 (ZK61A); B403 (ZK61A) **SAE** J465 (ZK61A) |
| **M16611** | Magnesium Alloy-Casting (ZK61A) | **Cu** 0.03 max **Mg** rem **Ni** 0.010 max **Si** 0.01 max **Zn** 5.7-6.3 **Zr** 0.3-1.0 **Other** impurities 0.30 max | **ASTM** B93 (ZK61A) |
| **M16620** | Magnesium Alloy- Casting (ZH62A) | **Cu** 0.10 max **Mg** rem **Ni** 0.01 max **Th** 1.4-2.2 **Zn** 5.2-6.2 **Zr** 0.50-1.0 **Other** total 0.30 max | **AMS** 4438 **ASTM** B80 (ZH62A); B93 (ZH62A); B275 (ZH62A) **SAE** J465 (ZH62A) |
| **M16621** | Magnesium Alloy-Casting (ZH62A) | **Cu** 0.03 max **Mg** rem **Ni** 0.010 max **Si** 0.01 max **Th** 1.5-2.1 **Zn** 5.3-6.0 **Zr** 0.3-1.0 **Other** impurities 0.30 max | **ASTM** B93 (ZH62A) |
| **M16630** | Magnesium Alloy- Casting (ZE63A) | **Cu** 0.10 **Mg** rem **Ni** 0.01 **Zn** 5.5-6.0 **Zr** 0.40-1.0 **Other** total 0.30 max, Rare Earths 2.1-3.0 | **AMS** 4425 **ASTM** B80 (ZE63A); B275 (ZE63A) **SAE** J465 (ZE63A) |
| **M16631** | Magnesium Alloy-Casting (ZC63) | **Cu** 2.4-3.0 **Mg** rem **Mn** 0.25-0.75 **Ni** 0.010 max **Si** 0.20 max **Zn** 5.5-6.5 **Other** metallic impurities 0.30 max | **ASTM** B93/B93M (ZC63) |
| **M16710** | Magnesium Alloy (Wrought) | **Cu** 0.03 max **Mg** rem **Mn** 0.5-1.0 **Ni** 0.010 max **Si** 0.20 max **Zn** 6.0-7.0 **Other** 0.30 max total | **ASTM** B107; B661 |
| **M18010** | Magnesium Alloy- Casting (K1A) | **Mg** rem **Zr** 0.40-1.0 **Other** total 0.30 max | **ASTM** B80 (K1A); B93 (K1A); B275 (K1A); B403 (K1A) **SAE** J465 (K1A) |

The chemical compositions listed are for identification purposes and should not be used in lieu of the cross referenced specifications.

| UNIFIED NUMBER | DESCRIPTION | CHEMICAL COMPOSITION | CROSS REFERENCE SPECIFICATIONS |
|---|---|---|---|
| **M18011** | Magnesium Alloy-Casting (K1A) | **Cu** 0.03 max **Mg** rem **Ni** 0.010 max **Si** 0.01 max **Zn** 1.8-2.4 **Zr** 0.3-1.0 **Other** impurities 0.30 max | **ASTM** B93 (K1A) |
| **M18210** | Magnesium Alloy Casting (QH21A) | **Cu** 0.10 max **Mg** rem **Ni** 0.01 max **Ag** 2.0-3.0 **Th** 0.6-1.6 **Zn** 0.20 max **Zr** 0.40-1.0 **Other** Rare Earths 0.6-1.5 **Other** Metallic Impurities 0.30 max | **AMS** 4419 **ASTM** B80 (QH21A); B93 (QH21A) **SAE** J465 (QH21A) |
| **M18211** | Magnesium Alloy-Casting (QH21A) | **Ag** 2.0-3.0 **Cu** 0.03 max **Mg** rem **Mn** 0.15 max **Ni** 0.010 max **Re** 0.6-1.6 **Si** 0.01 max **Zn** 0.2 max **Zr** 0.3-1.0 **Other** impurities 0.30 max | **ASTM** B93 (QH21A) |
| **M18220** | Magnesium Alloy- Casting (QE22A) | **Ag** 2.0-3.0 **Cu** 0.10 max **Mg** rem **Ni** 0.01 max **Zr** 0.40-1.0 **Other** total 0.30 max, Rare Earths 1.8-2.5 | **AMS** 4418 **ASTM** B80 (QE22A); B93 (QE22A); B199 (QE22A); B403 (QE22A) **SAE** J465 (QE22A) |
| **M18221** | Magnesium Alloy-Casting (QE22A) | **Ag** 2.0-3.0 **Cu** 0.03 max **Mg** rem **Mn** 0.15 max **Ni** 0.010 max **Re** 1.9-2.4 **Si** 0.01 max **Zn** 0.2 max **Zr** 0.3-1.0 **Other** impurities 0.30 max | **ASTM** B93 (QE22A) |
| **M18330** | Magnesium Alloy-Casting (EQ21A) | **Cu** 0.05-0.10 **Mg** rem **Ni** 0.01 max **Ag** 1.3-1.7 **Zr** 0.40-1.0 **Other** Rare Earths 1.5-3.0 (Neodymium rich), Impurities 0.30 max | **ASTM** B80 (EQ21A); B93 (EQ21A); B199 (EQ21A); B403 (EQ21A) |
| **M18410** | Magnesium Alloy-Casting (WE54A) | **Cu** 0.03 max **Mg** rem **Mn** 0.15 **Ni** 0.005 max **Si** 0.01 max **Zn** 0.20 **Zr** 0.4-1.0 **Y** 4.75-5.5 **Other** Rare Earth Metals 1.5-4.0 (Neodymium rich), Impurities 0.30 max | **ASTM** B80 (WE54A); B93 (WE54A) |
| **M18430** | Magnesium Alloy Casting (WE43A) | **Cu** 0.03 max **Mg** rem **Ni** 0.01 max **Si** 0.01 max **V** 3.7-4.3 **Zr** 0.3-1.0 **Other** 0.30 max total, Rare Earths 2.4-4.4 (Neodymium rich) | **ASTM** B80 (WE43A); B93 (WE43A); B275 (WE43A) |
| **M18432** | Magnesium Alloy Casting (WE43B) | **Cu** 0.020 max **Fe** 0.010 max **Li** 0.18 max **Mg** rem **Mn** 0.030 max **Ni** 0.005 max **Si** 0.01 max **Y** 3.7-4.3 **Zr** 0.3-1.0 **Other** (Zn+Ag) 0.10 max Rare Earths 2.4-4.4 Other 0.01 max, each | **AMS** 4427 **ASTM** B80; B93 |
| **M18540** | Magnesium Alloy (TA54A) | **Cu** 0.03 max **Li** 0.20 max **Mg** rem **Ni** 0.005 max **Si** 0.01 max **Y** 4.7-5.5 **Zr** 0.40-1.0 **Other** total 0.30 max, Rare Earths 2.0-4.0 (Neodymium rich) | **ASTM** B275 (TA54A) |
| **M19001** | Magnesium Brazing Filler Metal | **Al** 8.3-9.7 **Be** 0.0002-0.0008 **Cu** 0.05 max **Fe** 0.005 max **Mg** rem **Mn** 0.15-1.5 min **Ni** 0.005 max **Si** 0.05 max **Zn** 1.7-2.3 **Other** 0.30 | **ASME** SFA5.8 (BMg-1) **AWS** A5.8 (BMg-1) |
| **M19980** | Magnesium Remelt Alloy (9980A) | **Cu** 0.02 max **Mg** 99.80 min **Mn** 0.10 max **Na** 0.006 max **Ni** 0.001 max **Pb** 0.01 max **Sn** 0.01 max **Other** each 0.05 max | **ASTM** B92 (9980A); B275 (9980A) |
| **M19990** | Magnesium Remelt Alloy (9990A) | **Al** 0.003 max **B** 0.00007 max or 0.00003 **Cd** 0.0001 max or 0.00005 **Fe** 0.04 max **Mg** 99.90 min **Mn** 0.004 max **Ni** 0.001 max **Si** 0.005 max **Other** each 0.01 max | **ASTM** B92 (9990A); B275 (9990A) |
| **M19991** | Magnesium Remelt Alloy (9990B) | **Cu** 0.02 max **Mg** 99.80 min **Ni** 0.005 max **Pb** 0.01 max **Sn** 0.01 max | **ASTM** B92 (9990B) |
| **M19995** | Magnesium Remelt Alloy (9995A) | **Al** 0.01 max **B** 0.00003 max **Cd** 0.00005 max **Fe** 0.003 max **Mg** 99.95 min **Mn** 0.004 max **Ni** 0.001 max **Si** 0.005 max **Ti** 0.01 max **Other** each 0.005 max | **ASTM** B92 (9995A); B275 (9995A) |

The chemical compositions listed are for identification purposes and should not be used in lieu of the cross referenced specifications.

| UNIFIED NUMBER | DESCRIPTION | CHEMICAL COMPOSITION | CROSS REFERENCE SPECIFICATIONS |
|---|---|---|---|
| **M19998** | Magnesium Remelt Alloy (9998A) | **Al** 0.004 max **B** 0.00003 max **Cd** 0.00005 max **Cu** 0.0005 max **Fe** 0.002 max **Mg** 99.98 min **Mn** 0.002 max **Ni** 0.0005 max **Pb** 0.001 max **Si** 0.003 max **Ti** 0.001 max **Other** each 0.005 max | **ASTM** B92 (9998A); B275 (9998A) |
| **M26800** | Manganese Brazing Alloy (U500) | **B** 0.5-1.1 **C** 0.06 max **Co** 14.0-18.0 **Mn** rem **Ni** 14.0-18.0 **Si** 1.00 max | **AMS** 4780 **MIL SPEC** MIL-B-7883 (Mn-Ni-Co-B) |
| **M27200** | Manganese Thermostat Alloy | **Cu** 18 **Mn** 72 **Ni** 10 | **ASTM** B388 |
| **M29350** | Electrolytic Manganese | **Mn** 93-94 **N** 5.5-6.5 **S** 0.035 max | **ASTM** A601 (E) |
| **M29450** | Electrolytic Manganese | **Mn** 94-95 **N** 4.0-5.4 **S** 0.035 max | **ASTM** A601 (D) |
| **M29951** | Electrolytic Manganese | **H** 0.015 max **Mn** 99.5 min **S** 0.030 max | **ASTM** A601 (A) |
| **M29952** | Electrolytic Manganese | **H** 0.005 max **Mn** 99.5 min **S** 0.030 max | **ASTM** A601 (B) |
| **M29953** | Electrolytic Manganese | **H** 0.0010 max **Mn** 99.5 min **S** 0.030 max | **ASTM** A601 (C) |
| **M29954** | Electrolytic Manganese | **H** 0.0030 max **Mn** 99.5 max **S** 0.035 max | **ASTM** A601 (F) |

The chemical compositions listed are for identification purposes and should not be used in lieu of the cross referenced specifications.

# UNS NUMBERS ASSIGNED TO DATE

## With Description of Each Material Covered and References To Documents In Which The Same or Similar Materials are Described

## Nxxxxx Number Series
## Nickel and Nickel Alloys

# NICKEL AND NICKEL ALLOYS

| UNIFIED NUMBER | DESCRIPTION | CHEMICAL COMPOSITION | CROSS REFERENCE SPECIFICATIONS |
|---|---|---|---|
| **N01001** | Ni-Al Precipitation Hardenable Thermal Spray Wire (W-NiAl-1) | **Al** 4-6 **Ni** rem | **AWS** C2.25 (W-NiAl-1) |
| **N01002** | Ni-Al Precipitation Hardenable Thermal Spray Wire (W-NiAl-2) | **Al** 17-27 **Ni** rem **Other** Impurities 1.0 max | **AWS** C2.25 (W-NiAl-2) |
| **N01555** | Ni-Ti Alloy, Shape Memory Alloy (Nitinol 55) | **C** 0.07 max **Cb** 0.025 max **Co** 0.05 max **Cr** 0.01 max **Cu** 0.01 max **Fe** 0.05 max **H** 0.005 max **Ni** 54.0-57.00 **O** 0.05 max **Ti** rem | **ASTM** F2005; F2005 |
| **N02016** | Ni-Mn-Al Thermocouple Alloy (Alumel) | **Al** 1.75-2.25 **C** 0.15 max **Fe** 0.50 max **Mn** 2.0-3.0 **Ni** rem **Si** 1.6 max | |
| **N02061** | Ni Welding Filler Metal (Nickel FM61) | **Al** 1.5 max **C** 0.15 max **Cu** 0.25 max **Fe** 1.0 max **Mn** 1.0 max **Ni** 93.0 min **P** 0.03 max **S** 0.015 max **Si** 0.75 max **Ti** 2.0-3.5 | **ASME** SFA5.14 (ERNi-1); SFA5.30 (IN61) **AWS** A5.14 (ERNi-1); A5.30 (IN61) **MIL SPEC** MIL-E-21562 (EN61, RN61); MIL-I-23413 (MIL-61) |
| **N02100** | Nickel Alloy (ACI CZ-100) | **C** 1.00 max **Cu** 1.25 max **Fe** 3.00 max **Mn** 1.50 max **Ni** rem **Si** 2.00 max | **ACI** CZ100 **ASTM** A494 (CZ-100) |
| **N02200** | Commercially Pure Ni Alloy (Nickel 200) | **C** 0.15 max **Cu** 0.25 max **Fe** 0.40 max **Mn** 0.35 max **Ni** 99.0 min **S** 0.010 max **Si** 0.35 max | **ASME** SB-160; SB-161; SB-162; SB-163 **ASTM** B160; B161; B162; B163; B366; B564; B725; B730 **AWS** C2.25 (W-Ni-2) |
| **N02201** | Commercially Pure Ni Alloy (Nickel 201) | **C** 0.02 max **Cu** 0.25 max **Fe** 0.40 max **Mn** 0.35 max **Ni** 99.0 min **S** 0.010 max **Si** 0.35 max | **ASTM** B160; B161; B162; B163; B366; B725; B730 |
| **N02202** | Ni Thermal Spray Wire (W-Ni-3) | **C** 0.15 max **Cu** 0.25 max **Fe** 0.6 max **Mn** 0.35 max **Ni** 97.0 min **Si** 0.5 max **S** 0.04 max | **AWS** C2.25 (W-Ni-3) **AMS** 5553 |
| **N02205** | Commercially Pure Ni Alloy (Nickel 205) | **C** 0.15 max **Cu** 0.15 max **Fe** 0.20 max **Mg** 0.01-0.08 **Mn** 0.35 max **Ni** 99.0 min **S** 0.008 max **Si** 0.15 max **Ti** 0.01-0.05 | **AMS** 5555 **ASTM** F1; F3 |
| **N02211** | Nickel Alloy, Solution Strengthened (Nickel 211) | **C** 0.20 max **Cu** 0.25 max **Fe** 0.75 max **Mn** 4.25-5.25 **Ni** 93.7 min **S** 0.015 max **Si** 0.15 max | **ASTM** F290 |
| **N02215** | Ni-Cr-Fe Weld Filler Metal for Cast Iron (ERNi-CI) | **C** 1.0 max **Cu** 4.0 max **Fe** 4.0 max **Mn** 2.5 max **Ni** rem **S** 0.03 max **Si** 0.75 max **Other** 1.0 max total | **ASME** SFA5.15 (ERNi-CI) **AWS** A5.15 (ERNi-CI) |
| **N02216** | Ni-Fe-Mn Weld Filler Metal for Cast Iron (ERNiFeMn-CI) | **Al** 1.0 max **C** 0.50 max **Cu** 2.5 max **Fe** rem **Mn** 10.0-14.0 **Ni** 35.0-45.0 **S** 0.03 max **Si** 1.0 max **Other** 1.0 max total | **ASME** SFA5.15 (ERNiFeMn-CI) **AWS** A5.15 (ERNiFeMn-CI) |
| **N02220** | Commercially Pure Ni Alloy (Nickel 220) | **C** 0.15 max **Cu** 0.10 max **Fe** 0.10 max **Mg** 0.01-0.08 **Mn** 0.20 max **Ni** 99.00 min **S** 0.008 max **Si** 0.01-0.05 **Ti** 0.01-0.05 | |
| **N02225** | Commercially Pure Ni Alloy (Nickel 225) | **C** 0.15 max **Cu** 0.10 max **Fe** 0.10 max **Mg** 0.01-0.08 **Mn** 0.20 max **Ni** 99.00 min **S** 0.008 max **Si** 0.15-0.25 **Ti** 0.01-0.05 | |
| **N02230** | Commercially Pure Ni Alloy (Nickel 230) | **C** 0.15 max **Cu** 0.10 max **Fe** 0.10 max **Mg** 0.04-0.08 **Mn** 0.15 max **Ni** 99.00 min **S** 0.008 max **Si** 0.010-0.035 **Ti** 0.005 max | |
| **N02233** | Commercially Pure Ni Alloy (Nickel 233) | **C** 0.15 max **Cu** 0.10 max **Fe** 0.10 max **Mg** 0.01-0.10 **Mn** 0.30 max **Ni** 99.00 min **S** 0.008 max **Si** 0.10 max **Ti** 0.005 max | **ASTM** F1; F3; F4 |

The chemical compositions listed are for identification purposes and should not be used in lieu of the cross referenced specifications.

| UNIFIED NUMBER | DESCRIPTION | CHEMICAL COMPOSITION | CROSS REFERENCE SPECIFICATIONS |
|---|---|---|---|
| **N02253** | Commercially Pure Ni Alloy | **C** 0.02 max **Co** 0.10 max **Cr** 0.005 max **Cu** 0.01 max **Fe** 0.05 max **Mg** 0.005 max **Mn** 0.003 max **Ni** 99.9 min **S** 0.003 max **Si** 0.005 max **Ti** 0.005 max | **ASTM** F1 (3); F3 (3) |
| **N02270** | Commercially Pure Ni Alloy (Nickel 270) | **C** 0.02 max **Co** 0.001 max **Cr** 0.001 max **Cu** 0.001 max **Fe** 0.005 max **Mg** 0.001 max **Mn** 0.001 max **Ni** 99.97 min **S** 0.001 max **Si** 0.001 max **Ti** 0.001 max | **ASTM** F1; F3 |
| **N02290** | Commercially Pure Nickel | **Al** 0.001 max **C** 0.006 max **Cr** 0.001 max **Cu** 0.02 max **Fe** 0.015 max **Mg** 0.001 max **Mn** 0.001 max **N** 0.001 max **Ni** rem **O** 0.025 max **S** 0.0008 max **Si** 0.001 max | |
| **N03220** | Beryllium Nickel, Precipitation Hardenable Castings (M220C) | **Be** 1.80-2.30 **C** 0.30-0.50 **Ni** rem | |
| **N03260** | Nickel - Thoria Dispersion Strengthened | **C** 0.02 max **Co** 0.20 max **Cr** 0.05 max **Cu** 0.15 max **Fe** 0.05 max **Ni** rem **S** 0.0025 max **Ti** 0.05 max **Other** $ThO_2$ 1.80-2.60 | **AMS** 5865; 5890 |
| **N03300** | Nickel Alloy, Precipitation Hardenable (Permanickel 300) | **C** 0.40 max **Cu** 0.25 max **Fe** 0.60 max **Mg** 0.20-0.50 **Mn** 0.50 max **Ni** 97.0 min **S** 0.01 max **Si** 0.35 max **Ti** 0.20-0.60 | **ASTM** F290 |
| **N03301** | Nickel Alloy, Precipitation Hardenable (Duranickel 301) | **Al** 4.00-4.75 **C** 0.30 max **Cu** 0.25 max **Fe** 0.60 max **Mn** 0.50 max **Ni** 93.0 min **S** 0.01 max **Si** 1.00 max **Ti** 0.25-1.00 | |
| **N03360** | Commercially Pure Be-Ni Alloy Precipitation Hardenable | **Be** 1.85-2.05 **Ni** rem **Ti** 0.4-0.6 | |
| **N04019** | Nickel Base Castings | **C** 0.25 max **Cu** 27.0-31.0 **Fe** 2.50 max **Mn** 1.50 max **Ni** 60.0 min **S** 0.015 max **Si** 3.50-4.50 max | **AMS** 4892; 4893 **MIL SPEC** MIL-V-17647 |
| **N04020** | Ni-Cu Casting Alloy (Weldable Grade) | **Al** 0.50 max **C** 0.35 max **Cu** 26.0-33.0 **Fe** 2.50 max **Mn** 1.5 max **Ni** rem **Si** 2.0 max | **ASTM** A494 (M-35-2) |
| **N04060** | Ni-Cu Welding Filler Metal (Monel FM60) | **Al** 1.25 max **C** 0.15 max **Cu** rem **Fe** 2.5 max **Mn** 4.0 max **Ni** 62.0-69.0 **P** 0.02 max **S** 0.015 max **Si** 1.25 max **Ti** 1.5-3.0 | **ASME** SFA5.14 (ERNiCu-7); SFA5.30 (IN60) **AWS** A5.14 (ERNiCu-7); A5.30 (IN60) **MIL SPEC** MIL-E-21562 (EN60, RN60); MIL-I-23413 (MIL-60) |
| **N04400** | Ni-Cu Alloy Solid Solution Strengthened (Monel 400) | **C** 0.3 max **Cu** rem **Fe** 2.50 max **Mn** 2.00 max **Ni** 63.00-70.00 **S** 0.024 max **Si** 0.50 max | **AMS** 4544; 4574; 4575; 4675; 4730; 4731; AS7233 **ASME** SB-127; SB-163; SB-164; SB-165; SB-564 **ASTM** B127; B163; B164; B165; B366; B564; B725; B730; F96; F467 (400); F468 (400) **FED** QQ-N-281 **MIL SPEC** MIL-N-24106; MIL-T-1368; MIL-V-17547 **Other:** NACE MR-01-75 |
| **N04401** | Ni-Cu Alloy, Solid Solution Strengthened (Monel 401) | **C** 0.10 max **Co** 0.25 max **Cu** rem **Fe** 0.75 max **Mn** 2.25 max **Ni** 40.0-45.0 **S** 0.015 max **Si** 0.25 max | |
| **N04402** | Ni-Cu Alloy Solid Solution Strengthened (Low Carbon) | **C** 0.04 max **Cu** 28.0-34.0 **Fe** 2.5 max **Mn** 2.0 max **Ni** 63.0 min **S** 0.025 max **Si** 0.5 max | **Other:** ISO 6207, 6208, 9721, 9723 |
| **N04404** | Ni-Cu Alloy Solid Solution Strengthened (Monel 404) | **Al** 0.05 max **C** 0.15 max **Cu** rem **Fe** 0.50 max **Mn** 0.10 max **Ni** 52.0-57.0 **S** 0.024 max **Si** 0.10 max | **ASTM** F96 |
| **N04405** | Ni-Cu Alloy Solid Solution Strengthened (Monel R405) | **C** 0.30 max **Cu** rem **Fe** 2.5 max **Mn** 2.0 max **Ni** 63.0-70.0 **S** 0.025-0.060 **Si** 0.50 max | **AMS** 4674; AS7234 **ASME** SB-164 **ASTM** B164; B165; F467 (405); F468 (405) **AWS** C2.25 (W-NiCu-1) **FED** QQ-N-281 **Other:** NACE MR-01-75 |

The chemical compositions listed are for identification purposes and should not be used in lieu of the cross referenced specifications.

| UNIFIED NUMBER | DESCRIPTION | CHEMICAL COMPOSITION | CROSS REFERENCE SPECIFICATIONS |
|---|---|---|---|
| **N04406** | Ni-Cu Thermal Spray Wire (W-NiCu-2) | **Al** 0.1 max **C** 0.25 max **Cu** 26-28 **Fe** 2 max **Mn** 1.5 max **Ni** rem **Si** 0.25 max | **AWS** C2.25 (W-NiCu-2) |
| **N05500** | Ni-Cu Alloy Precipitation Hardenable (Monel K500) | **Al** 2.30-3.15 **C** 0.25 max **Cu** rem **Fe** 2.00 max **Mn** 1.50 max **Ni** 63.0-70.0 **S** 0.01 max **Si** 0.50 max **Ti** 0.35-0.85 | **AMS** 4676 **ASTM** B865; F467 (500); F468 (500) **FED** QQ-N-286 **MIL SPEC** MIL-N-24549; MIL-V-17547 **Other:** NACE MR-01-75 |
| **N05502** | Ni-Cu Alloy Precipitation Hardenable (Monel 502) | **Al** 2.50-3.50 **C** 0.10 max **Cu** rem **Fe** 2.00 max **Mn** 1.50 max **Ni** 63.0-70.0 **S** 0.010 max **Si** 0.5 max **Ti** 0.50 max | **AMS** 4677 |
| **N05504** | Ni-Cu Welding Filler Metal (ERNiCu-8) | **Al** 2.0-4.0 **C** 0.25 max **Cu** rem **Fe** 2.00 max **Mn** 1.50 max **Ni** 63.0-70.0 **P** 0.030 max **Pb** 0.010 max **S** 0.015 max **Si** 1.00 max **Ti** 0.25-1.00 | **ASME** SFA5.14 (ERNiCu-8) **AWS** A5.14 (ERNiCu-8) **MIL SPEC** MIL-E-21562 (EN64, RN64) |
| * **N06001** | Ni-Cr Alloy Solid Solution Strengthened (Hastelloy F) | **C** 0.05 max **Cb** 1.8-2.5 **Co** 2.5 max **Cr** 21.0-23.0 **Fe** rem **Mn** 1.00-2.00 **Mo** 5.5-7.5 **Ni** 44.0-47.0 **P** 0.04 max **S** 0.03 max **Si** 1.00 max **W** 1.00 max | **ASME** SB-436 **ASTM** B366 **SAE** J467 (Hastelloy F) |
| **N06002** | Ni-Cr Alloy Solid Solution Strengthened (Hastelloy X) | **C** 0.05-0.15 **Co** 0.5-2.5 **Cr** 20.5-23.0 **Fe** 17.0-20.0 **Mn** 1.00 max **Mo** 8.0-10.0 **Ni** rem **P** 0.040 max **S** 0.030 max **Si** 1.00 max **W** 0.20-1.0 | **AMS** 5390; 5536; 5587; 5588; 5754; 5798; AS7237 **ASME** SB-435; SB-572; SB-619; SB-622; SB-626; SB-751; SFA5.14 (ERNiCrMo-2) **ASTM** B366; B435; B572; B619; B622; B626 **AWS** A5.14 (ERNiCrMo-2) **SAE** J467 (Hastelloy X) **Other:** NACE MR-01-75 |
| **N06003** | Ni-Cr Alloy Solid Solution Strengthened (Nichrome V) | **C** 0.15 max **Cr** 19-21 **Fe** 1.0 max **Mn** 2.5 max **Ni** rem **S** 0.01 max **Si** 0.75-1.6 | **AMS** 5676; 5677; 5682 **ASTM** B344 |
| **N06004** | Ni-Cr Alloy Solid Solution Strengthened (Nichrome) | **C** 0.15 max **Cr** 14-18 **Fe** rem **Mn** 1.0 max **Ni** 57 min **S** 0.01 max **Si** 0.75-1.6 | **ASTM** B344 **AWS** C2.25 (W-NiCrFe-3) |
| **N06005** | Ni-Cr Alloy Solid Solution Strengthened (Eatonite) | **C** 2.40 nom **Co** 10.00 nom **Cr** 29.00 nom **Fe** 6.5 max **Ni** 39.00 nom **Si** 0.70 nom **W** 15.00 nom | **SAE** J775 (VF3) |
| **N06006** | Ni-Cr Alloy Solid Solution Strengthened (ACI HX) | **C** 0.35-0.75 **Cr** 15.0-19.0 **Fe** rem **Mn** 2.00 max **Mo** 0.50 max **Ni** 64.0-68.0 **P** 0.04 max **S** 0.04 max **Si** 2.50 max | **ACI** HX **ASTM** A297 (HX) |
| **N06007** | Ni-Cr Alloy (Hastelloy G) | **C** 0.05 max **Cb** 1.75-2.5 **Co** 2.5 max **Cr** 21.0-23.5 **Cu** 1.5-2.5 **Fe** 18.0-21.0 **Mn** 1.0-2.0 **Mo** 5.5-7.5 **Ni** rem **P** 0.04 max **S** 0.03 max **Si** 1.0 max **W** 1.0 max | **ASME** SB-366; SB-619; SFA5.14 (ERNiCrMo-1) **ASTM** B366; B581; B582; B619; B622; B626; G28 **AWS** A5.14 (ERNiCrMo-1) **Other:** NACE MR-01-75 |
| **N06008** | Ni-Cr 30 Alloy Solid Solution Strengthened | **Al** 0.20 max **C** 0.15 max **Cr** 29.0-31.0 **Fe** 1.0 max **Mn** 0.10 max **Ni** rem **P** 0.030 max **S** 0.010 max **Si** 0.75-1.60 | |
| **N06009** | Ni-20 Cr, Columbium Stabilized | **C** 0.15 max **Cb** 0.75-1.50 **Cr** 19.0-21.0 **Fe** 1.00 max **Mn** 2.5 max **Ni** rem **S** 0.010 max **Si** 0.75-1.60 | |
| **N06010** | Ni-Cr 10 Thermocouple Alloy (Chromel) | **Al** 0.20 max **C** 0.15 max **Cr** 9.0-11.0 **Fe** 0.50 max **Mn** 0.10 max **Ni** rem **Si** 1.60 max | |
| **N06013** | Ni-Cr-Mo-Fe Hard Facing Alloy (Eatonite 3) | **C** 1.80-2.20 **Cr** 28.0-30.0 **Fe** 1.0-8.0 **Mn** 1.0 max **Mo** 4.00-6.00 **Ni** rem **P** 0.030 max **S** 0.030 max **Si** 0.80-1.20 | **SAE** J775 (VF 9) |
| **N06015** | Ni-Cr-Mo-Fe Hard Facing Alloy (Eatonite 5) | **C** 1.80-2.20 **Cr** 28.0-30.0 **Fe** 1.0-8.0 **Mn** 1.0 max **Mo** 7.0-9.0 **Ni** rem **P** 0.030 max **S** 0.030 max **Si** 0.80-1.20 | **SAE** J775 (VF 10) |
| **N06017** | Ni-Cr Alloy ERNiCrMo-9 replaced by N06985 | | |
| **N06021** | Ni-Cr-Mo Thermal Spray Wire (W-NiCrMo) | **Cr** 21-23 **Fe** 1-2 **Mo** 9-11 **Ni** rem **Other** 3-4 | **AWS** C2.25 (W-NiCrMo) |

*Boxed entries are no longer active and are retained for reference purposes only.

The chemical compositions listed are for identification purposes and should not be used in lieu of the cross referenced specifications.

| UNIFIED NUMBER | DESCRIPTION | CHEMICAL COMPOSITION | CROSS REFERENCE SPECIFICATIONS |
|---|---|---|---|
| **N06022** | Ni-Cr-Mo Alloy (Hastelloy C-22) | **C** 0.015 max **Co** 2.5 max **Cr** 20.0-22.5 **Fe** 2.0-6.0 **Mn** 0.50 max **Mo** 12.5-14.5 **Ni** rem **P** 0.02 max **S** 0.02 max **Si** 0.08 max **V** 0.35 max **W** 2.5-3.5 | **ASME** SB-366; SB-574; SB-575; SB-619; SB-622; SB-626; SFA5.14 (ERNiCrMo-10) **ASTM** B366; B564; B574; B575; B581; B582; B619; B622; B626; G28 **AWS** A5.14 (ERNiCrMo-10) **Other:** NACE MR-01-75 |
| **N06025** | Ni-Cr-Fe-Al Alloy (Nicrofer 6025 HT) | **Al** 1.8-2.4 **C** 0.15-0.25 **Cr** 24.0-26.0 **Cu** 0.1 max **Fe** 8.0-11.0 **Mn** 0.15 max **Ni** rem **P** 0.020 max **S** 0.010 max **Si** 0.5 max **Ti** 0.1-0.2 **Zr** 0.01-0.10 **Y** 0.05-0.12 | **ASTM** B163; B166; B167; B168; B366; B516; B517; B546; B564 |
| **N06030** | Wrought Ni-Cr-Fe-Mo-Cu Alloy Solid Solution Strengthened (Hastelloy G-30) | **C** 0.03 max **Cb** 0.30-1.50 **Co** 5.0 max **Cr** 28.0-31.5 **Cu** 1.0-2.4 **Fe** 13.0-17.0 **Ni** rem **Mn** 1.5 max **Mo** 4.0-6.0 **P** 0.04 max **S** 0.02 max **Si** 0.8 max **W** 1.5-4.0 | **ASME** SB-366; SB-581; SB-582; SB-619; SB-622; SB-626; SFA5.14 (ErNiCrMo-11) **ASTM** B366; B581; B582; B619; B622; B626; G28 **AWS** A5.14 (ERNiCrMo-11) **Other:** NACE MR-01-75 |
| **N06040** | Ni-Cr Alloy Solid Solution Strengthened (ACI CY-40) | **C** 0.40 max **Cr** 14.0-17.0 **Fe** 11.0 max **Mn** 1.50 max **Ni** rem **Si** 3.00 max | **ACI** CY40 **ASTM** A494 |
| **N06045** | Ni-Cr-Fe-Si Alloy (Nicrofer 45) | **C** 0.05-0.12 **Cr** 26.0-29.0 **Cu** 0.3 max **Fe** 21.0-25.0 **Mn** 1.00 max **N** 0.05-0.12 **Ni** 45.0 min **P** 0.020 max **S** 0.010 max **Si** 2.5-3.0 **Other** total rare earth elements 0.05-0.15, approximately 50% of which is cerium | **ASTM** B163; B166; B167; B168; B366; B516; B517; B546; B564 |
| **N06050** | Ni-Cr Alloy Solid Solution Strengthened (ACI HX-50) | **C** 0.40-0.60 **Cr** 15.0-19.0 **Fe** rem **Mn** 1.50 max **Mo** 0.50 max **Ni** 64.0-68.0 **P** 0.04 max **S** 0.04 max **Si** 0.50-2.00 | **ASTM** A608 (HX-50) |
| **N06052** | Ni-Cr Welding Filler Metal (Inconel FM 52) | **Al** 1.10 max **C** 0.04 max **Cr** 28.0-31.5 **Cb** 0.10 max **Cu** 0.30 max **Fe** 7.0-11.0 **Mn** 1.0 max **Mo** 0.50 max **Ni** rem **P** 0.020 max **S** 0.015 max **Si** 0.50 max **Ti** 1.0 max **Other** 0.50 max total, Al+Ti 1.50 max | **ASME** SFA5.14 (ERNiCrFe-7); SFA5.30 (IN52) **AWS** A5.14 (ERNiCrFe-7); A5.30 (IN52) |
| **N06058** | Low Carbon Ni-Cr-Mo-N Alloy Solid Solution Strengthened (Alloy 2120) | **Al** 0.40 max **C** 0.010 max **Co** 0.3 max **Cr** 20.0-23.0 **Cu** 0.50 max **Fe** 1.5 max **Mn** 0.50 max **Mo** 19.0-21.0 **N** 0.02-0.15 **Ni** rem **P** 0.015 max **S** 0.010 max **Si** 0.10 max **W** 0.3 max | **ASTM** B366; B564; B574; B575; B619; B622; B626 |
| **N06059** | Low Carbon Ni-Cr-Mo Alloy (Alloy 59) | **Al** 0.1-0.4 **C** 0.010 max **Co** 0.3 max **Cr** 22.0-24.0 **Cu** 0.50 max **Fe** 1.5 max **Mn** 0.5 max **Mo** 15.0-16.5 **Ni** rem **P** 0.015 max **S** 0.010 max **Si** 0.10 max | **ASME** SFA5.14 (ERNiCrMo-13) **ASTM** B564; B574; B575; B619; B622; B626; G28 **AWS** A5.14 (ERNiCrMo-13) |
| **N06060** | Ni-Cr-Mo-Fe Alloy (SM2060Mo) | **C** 0.03 max **Cr** 19.0-22.0 **Cb** 0.50-1.25 **Cu** 0.25-1.25 **Fe** rem **Mn** 1.50 max **Mo** 12.0-14.0 **Ni** 54.0-60.0 **P** 0.030 max **Si** 0.50 max **S** 0.005 max **W** 0.25-1.25 | **ASTM** B622 **Other:** NACE MR-01-75 |
| **N06062** | Ni-Cr Alloy Solid Solution Strengthened (Inconel FM62) | **C** 0.035 max **Cb** 1.0-3.0 **Cr** 14.0-17.0 **Cu** 0.5 max **Fe** 6.0-10.0 **Mn** 1.0 max **Ni** 72.0 min **P** 0.030 max **S** 0.015 max **Si** 0.5 max | **AMS** 5679 **ASME** SFA5.14 (ERNiCrFe-5); SFA5.30 (IN62) **AWS** A5.14 (ERNiCrFe-5); A5.30 (IN62) **MIL SPEC** MIL-E-21562 (EN62, RN62); MIL-I-23413 (MIL-62) |
| **N06070** | Ni-Cr-Ti Thermal Spray Wire (W-NiCrTi) | **Cr** 44-46 **Ni** rem **Ti** 3-4 | **AWS** C2.25 (W-NiCrTi) |
| **N06072** | Ni-Cr Alloy Filler Metal (Inconel Filler Metal 72) | **C** 0.01-0.10 **Cr** 42.0-46.0 **Cu** 0.50 max **Fe** 0.50 max **Mn** 0.20 max **Ni** rem **P** 0.020 max **S** 0.015 max **Si** 0.20 max **Ti** 0.30-1.00 **Other** 0.50 max total | **ASME** SFA5.14 (ERNiCr-4) **AWS** A5.14 (ERNiCr-4) |
| **N06075** | Ni-Cr Alloy Solid Solution Strengthened (Nimonic 75) | **C** 0.08-0.15 **Cr** 18.0-21.0 **Cu** 0.50 max **Fe** 5.00 max **Mn** 1.00 max **Ni** rem **Si** 1.00 max **Ti** 0.20-0.60 | **SAE** J467 (Nimonic 75) **Other:** ISO 6207; 6208; 9721; 9723 |

The chemical compositions listed are for identification purposes and should not be used in lieu of the cross referenced specifications.

| UNIFIED NUMBER | DESCRIPTION | CHEMICAL COMPOSITION | CROSS REFERENCE SPECIFICATIONS |
|---|---|---|---|
| **N06076** | Ni-Cr Welding Filler Metal (ERNiCr-6) | **Al** 0.40 max **C** 0.08-0.15 **Cr** 19.0-21.0 **Cu** 0.50 max **Fe** 2.00 max **Mn** 1.00 max **Ni** 75.0 min **P** 0.030 max **Pb** 0.010 max **S** 0.015 max **Si** 0.30 max **Ti** 0.15-0.50 | **ASME** SFA5.14 (ERNiCr-6) **AWS** A5.14 (ERNiCr-6); C2.25 (W-NiCrFe-1) **MIL SPEC** MIL-E-21562 (EN6N, RN6N) |
| **N06078** | Ni-Cr-Fe Thermal Spray Wire (W-NiCrFe-2) | **C** 0.10 max **Cr** 14.0-17.0 **Cu** 0.5 max **Fe** 6.0-10.0 **Mn** 1.0 max **Ni** 72.0 min **Si** 0.5 max **S** 0.02 max | **AWS** C2.25 (W-NiCrFe-2) |
| **N06082** | Ni-Cr Welding Filler Metal (Inconel FM82) | **C** 0.10 max **Cb** (incl Ta) 2.0-3.0 **Cr** 18.0-22.0 **Cu** 0.50 max **Fe** 3.0 max **Mn** 2.5-3.5 **Ni** 67.0 min **P** 0.03 max **S** 0.015 max **Si** 0.50 max **Ti** 0.75 max | **ASME** SFA5.14 (ERNiCr-3); SFA5.30 (IN82) **AMS** 5836 **AWS** A5.14 (ERNiCr-3); A5.30 (IN82) **MIL SPEC** MIL-E-21562 (EN82, EN82H, RN82, RN82H); MIL-I-23413 (MIL-82) |
| * **N06102** | Ni-Cr Alloy Solid Solution Strengthened (IN-102) | **Al** 0.30-0.60 **B** 0.003-0.008 **C** 0.08 max **Cb** 2.75-3.25 **Cr** 14.0-16.0 **Fe** 5.0-9.0 **Mg** 0.01-0.05 **Mn** 0.75 max **Mo** 2.75-3.25 **Ni** rem **P** 0.010 max **S** 0.010 max **Si** 0.40 max **Ti** 0.40-0.70 **W** 2.75-3.25 **Zr** 0.01-0.05 | |
| **N06110** | Austenitic Corrosion Resistant Nickel Alloy (Allcorr) | **Al** 1.50 max **C** 0.15 max **Cb** 2.00 max **Co** 12.0 max **Cr** 27.0-33.0 **Mo** 8.00-12.0 **Ni** rem **Ti** 1.50 max **W** 4.00 max | **ASTM** B564; B755; B756; B757; B758; B759 **Other:** NACE MR-01-75 |
| **N06132** | Ni-Cr Alloy Solid Solution Strengthened (Inconel WE132) replaced by W86132 | | |
| **N06200** | Ni-Cr-Mo-Cu Alloy Solid Solution Strengthened (HASTELLOY C-2000) | **Al** 0.50 max **C** 0.010 max **Co** 2.0 max **Cr** 22.0-24.0 **Cu** 1.3-1.9 **Fe** 3.0 max **Mn** 0.50 max **Mo** 15.0-17.0 **Ni** rem **P** 0.025 max **S** 0.010 max **Si** 0.08 max | **ASTM** B366; B564; B574; B575; B619; B622; B626; G28 |
| **N06210** | Ni-Cr-Mo-Ta Alloy Solid Solution Strengthened (MAT21) | **C** 0.015 max **Co** 1.0 max **Cr** 18.0-20.0 **Fe** 1.0 max **Mn** 0.50 max **Mo** 18.0-20.0 **Ni** rem **P** 0.02 max **S** 0.02 max **Si** 0.08 max **Ta** 1.5-2.2 **V** 0.35 max | **ASME** SB-564; SB-574; SB-575; SB-619; SB-622; SB-626 **ASTM** B564; B574; B575; B619; B622; B626 |
| **N06219** | Ni-Cr-Mo-Si Alloy Solid Solution Strengthened (Nicrofer 6219Si) | **Al** 0.50 max **C** 0.05 max **Co** 1.0 max **Cr** 18.0-22.0 **Cu** 0.50 max **Fe** 2.0-4.0 **Mn** 0.50 max **Mo** 7.0-9.0 **Ni** rem **P** 0.020 max **S** 0.010 max **Si** 0.70-1.10 **Ti** 0.50 max | **ASTM** B366; B443; B444; B446; B564; B704; B705 **Other:** Werkstoff Number 2.4855 |
| **N06230** | Ni-Cr-W-Mo Alloy, Solid Solution Strengthened (Alloy No. 230) | **Al** 0.20-0.50 **B** 0.015 max **C** 0.05-0.15 **Cr** 20.0-24.0 **Co** 5.0 max **Fe** 3.0 max **La** 0.005-0.05 **Mn** 0.30-1.00 **Mo** 1.0-3.0 **Ni** rem **P** 0.03 max **S** 0.015 max **Si** 0.25-0.75 **W** 13.0-15.0 | **AMS** 5878; 5891; 5896 **ASTM** B366; B564; B435; B572; B619; B622; B626 |
| **N06231** | Ni-Cr-W-Mo Welding Filler Metal (Haynes 230-W) | **Al** 0.2-0.5 **B** 0.003 max **C** 0.05-0.15 **Cr** 20.0-24.0 **Co** 3.0 max **Cu** 0.5 max **Fe** 3.0 max **La** 0.050 max **Mn** 0.3-1.0 **Mo** 1.0-3.0 **Ni** rem **P** 0.030 max **S** 0.015 max **Si** 0.25-0.75 **W** 13.0-15.0 | **AMS** 5839 **ASME** SFA5.14 (ERNiCrWMo-1) **AWS** A5.14 (ERNiCrWMo-1) |
| **N06250** | Ni-Cr-Fe-Mo Alloy (SM2050) | **C** 0.020 max **Cr** 20.0-23.0 **Cu** 0.25-1.25 **Fe** rem **Mn** 1.0 max **Mo** 10.1-12.0 **Ni** 50.0-54.0 **P** 0.030 max **S** 0.005 max **Si** 0.09 max **W** 0.25-1.25 | **ASTM** B622 |
| **N06255** | Ni-Cr-Fe-Mo-W Alloy (SM2550) | **C** 0.03 max **Cr** 23.0-26.0 **Cu** 1.20 max **Fe** rem **Mn** 1.00 max **Mo** 6.0-9.0 **Ni** 47.0-52.0 **P** 0.03 max **S** 0.03 max **Si** 1.0 max **Ti** 0.69 max **W** 3.0 max | **ASTM** B622 |

*Boxed entries are no longer active and are retained for reference purposes only.

The chemical compositions listed are for identification purposes and should not be used in lieu of the cross referenced specifications.

| UNIFIED NUMBER | DESCRIPTION | CHEMICAL COMPOSITION | CROSS REFERENCE SPECIFICATIONS |
|---|---|---|---|
| **N06333** | Ni-Cr Alloy Solid Solution Strengthened (RA333) | **C** 0.08 max **Co** 2.50-4.00 **Cr** 24.00-27.00 **Cu** 0.50 max **Fe** rem **Mn** 2.00 max **Mo** 2.50-4.00 **Ni** 44.00-47.00 **P** 0.030 max **Pb** 0.025 max **S** 0.030 max **Si** 0.75-1.50 **Sn** 0.025 max **W** 2.50-4.00 | **AMS** 5593; 5717 **ASTM** B718; B719; B722; B723; B726 **SAE** J467 (RA333) |
| **N06455** | Ni-Cr Alloy (Hastelloy C-4) | **C** 0.015 max **Co** 2.0 max **Cr** 14.0-18.0 **Fe** 3.0 max **Mn** 1.0 max **Mo** 14.0-17.0 **Ni** rem **P** 0.04 max **S** 0.03 max **Si** 0.08 max **Ti** 0.70 max | **ASME** SB-366; SB-574; SB-575; SB-619; SB-622; SB-626; SFA5.14 (ERNiCrMo-7) **ASTM** B366; B574; B575; B619; B622; B626; G28 **AWS** A5.14 (ERNiCrMo-7) |
| **N06600** | Ni-Cr Alloy Solid Solution Strengthened (Inconel 600) | **C** 0.15 max **Cr** 14.00-17.00 **Cu** 0.50 max **Fe** 6.00-10.00 **Mn** 1.00 max **Ni** 72.0 min **S** 0.015 max **Si** 0.50 max | **AMS** 5540; 5580; 5665; 5687; 5961; AS7232 **ASME** SB-163; SB-166; SB-167; SB-168; SB-564 **ASTM** B163; B166; B167; B168; B366; B516; B517; B564; B751; G28 **MIL SPEC** MIL-N-23228 **SAE** J467 (Inconel 600) **Other:** NACE MR-01-75 |
| **N06601** | Ni-Cr Alloy Solid Solution Strengthened (Inconel 601) | **Al** 1.0-1.7 **C** 0.1 max **Cr** 21.0-25.0 **Cu** 1.0 max **Fe** rem **Mn** 1.0 max **Ni** 58.0-63.0 **S** 0.015 max **Si** 0.50 max | **AMS** 5715; 5870 **ASME** SFA5.14 (ERNiCrFe-11) **ASTM** B163; B166; B167; B168 **AWS** A5.14 (ERNiCrFe-11) |
| **N06602** | 6Ni-Cr-Fe Alloy, Solid Solution Strengthened (Low Carbon) | **C** 0.02 max **Cr** 14.0-17.0 **Cu** 0.5 max **Fe** 6.0-10.0 **Mn** 1.0 max **Ni** 72.0 min **S** 0.015 max **Si** 0.5 max | **Other:** ISO 6207, 6208, 9721, 9723 |
| **N06603** | Ni-Cr-Fe-Al-C Alloy Solid Solution and Carbide Strengthened (603GT) | **Al** 2.4-3.0 **C** 0.20-0.40 **Cr** 24.0-26.0 **Cu** 0.50 max **Fe.** 8.0-11.0 **Mn** 0.15 max **Ni** rem **P** 0.020 max **S** 0.010 max **Si** 0.50 max **Ti** 0.01-0.25 **Zr** 0.01-0.10 **Other** Y 0.01-0.15 | **ASTM** B163; B166; B167; B168; B366; B516; B517; B546; B564 |
| **N06617** | Ni-Cr-Co-Mo Alloy Solid Solution Strengthened (Alloy 617) | **Al** 0.80-1.50 **B** 0.006 max **C** 0.05-0.15 **Co** 10.0-15.0 **Cr** 20.0-24.0 **Cu** 0.50 max **Fe** 3.00 max **Mn** 1.00 max **Mo** 8.00-10.0 **Ni** 44.5 min **S** 0.015 max **Si** 1.00 max **Ti** 0.60 max | **AMS** 5887; 5888; 5889 **ASME** SB-166; SFA5.14 (ERNiCrCoMo-1) **ASTM** B166; B167; B168; B546; B564 **AWS** A5.14 (ERNiCrCoMo-1) |
| **N06621** | Combined with N06075 | | |
| **N06625** | Ni-Cr-Mo-Cb(Nb) Alloy Solid Solution Strengthened (Alloy 625) | **Al** 0.40 max **C** 0.10 max **Cb** 3.15-4.15 **Cr** 20.0-23.0 **Fe** 5.0 max **Mn** 0.50 max **Mo** 8.0-10.0 **Ni** rem **P** 0.015 max **S** 0.015 max **Si** 0.50 max **Ti** 0.40 max | **AMS** 5401; 5402; 5581; 5599; 5666; 5837; 5869; 7490 **ASME** SB-443; SB-444; SB-446; SFA5.14 (ERNiCrMo-3) **ASTM** B366; B443; B444; B446; B564; B704; B705; B751; B834; G28 **AWS** A5.14 (ERNiCrMo-3) **MIL SPEC** MIL-E-21562 (RN625, EN625) **Other:** NACE MR-01-75 |
| **N06626** | Ni-Cr-Mo-Cb(Nb) Solid Solution Strengthened Alloy (Inconel 625LCF) | **Al** 0.40 max **C** 0.03 max **Cb** 3.15-4.15 **Co** 1.0 max **Cr** 20.0-23.0 **Fe** 5.0 max **Mn** 0.50 max **Mo** 8.0-10.0 **N** 0.02 max **Ni** 58.0 min **P** 0.015 max **S** 0.015 max **Si** 0.15 max **Ti** 0.40 max | **AMS** 5879 |
| **N06635** | Ni-Cr-Mo Alloy (Hastelloy S) | **Al** 0.10-0.50 **B** 0.015 max **C** 0.02 max **Co** 2.00 max **Cr** 14.5-17.0 **Cu** 0.35 max **Fe** 3.00 max **Mn** 0.30-1.00 **Mo** 14.0-16.5 **Ni** rem **P** 0.020 max **S** 0.015 max **Si** 0.20-0.75 **W** 1.00 max **Other** La 0.01-0.10 | **AMS** 5711; 5838; 5869; 5873 |
| **N06650** | Ni-Cr-Fe-Mo-W-N Alloy (ERNiCrMo-18) (Nicrofer 5020hMo) | **Al** 0.05-0.50 **C** 0.03 max **Cb** 0.05-0.60 **Cr** 18.0-21.0 **Fe** 12.0-16.0 **Mn** 0.5 max **Mo** 9.5-12.5 **N** 0.05-0.20 **Ni** rem **P** 0.020 max **S** 0.010 max **Si** 0.50 max **W** 0.5-2.5 **Other** 0.50 max, Cb = Cb+Ta | **ASME** SFA5.14 (ERNiCrMo-18) **AWS** A5.14 (ERNiCrMo-18) |
| **N06686** | Ni-Cr-Mo-W Alloy (Inconel Alloy 686) | **C** 0.010 max **Cr** 19.0-23.0 **Fe** 5.0 max **Mn** 0.75 max **Mo** 15.0-17.0 **Ni** rem **P** 0.04 max **S** 0.02 max **Si** 0.08 max **Ti** 0.02-0.25 **W** 3.0-4.4 | **ASME** SFA5.14 (ERNiCrMo-14) **ASTM** B564; B574; B575; B619; B622; B626; G28 **AWS** A5.14 (ERNiCrMo-14) |
| **N06690** | Ni-Cr Alloy Solid Solution Strengthened (Inconel 690) | **C** 0.05 max **Cr** 27.0-31.0 **Cu** 0.50 max **Fe** 7.0-11.0 **Mn** 0.50 max **Ni** 58.0 min **S** 0.015 max **Si** 0.50 max | **ASME** Code Case N-20 (1484); SB-163; SB-166; SB-167; SB-168 **ASTM** B163; B166; B167; B168; B564 |
| **N06782** | Ni-Cr Alloy Solid Solution Strengthened (X-782) | **C** 2.00 nom **Co** 0.50 nom **Cr** 26.00 nom **Fe** 4.0 max **Mn** 0.30 nom **Ni** rem **Si** 0.30 nom **W** 8.75 nom | **SAE** J775 (VF4) |

The chemical compositions listed are for identification purposes and should not be used in lieu of the cross referenced specifications.

| UNIFIED NUMBER | DESCRIPTION | CHEMICAL COMPOSITION | CROSS REFERENCE SPECIFICATIONS |
|---|---|---|---|
| **N06804** | Ni-Cr Alloy Solid Solution Strengthened (Incoloy 804) | **Al** 0.60 max **C** 0.10 max **Cr** 28.0-31.0 **Cu** 0.50 max **Fe** rem **Mn** 1.50 max **Ni** 39.0-43.0 **S** 0.015 max **Si** 0.75 max **Ti** 1.20 max | |
| **N06920** | Ni-Cr-Fe-Mo-W Alloy Solid Solution Strengthened (Hastelloy H-9M) | **C** 0.03 max **Co** 5.0 max **Cr** 20.5-23.0 **Fe** 17.0-20.0 **Mn** 1.0 max **Mo** 8.0-10.0 **Ni** rem **P** 0.040 max **S** 0.030 max **Si** 1.0 max **W** 1.0-2.0 | **ASTM** B814 **Other:** ISO 9722 |
| **N06950** | Ni-Cr-Fe-Mo Alloy, Solid Solution Strengthened (Hastelloy G50) | **C** 0.015 max **Cb** 0.5 max **Co** 2.5 max **Cr** 19.0-21.0 **Cu** 0.5 max **Fe** 15.0-20.0 **Mn** 1.0 max **Mo** 8.0-10.0 **Ni** 50.0 min **P** 0.04 max **S** 0.015 max **Si** 1.0 max **W** 1.0 max | **Other:** NACE MR-01-75 |
| **N06952** | Ni-Cr Alloy (NIC 52) | **C** 0.03 max **Cr** 23.0-27.0 **Cu** 0.5-1.0 **Fe** rem **Mn** 1.0 max **Mo** 6.0-8.0 **Ni** 48.0-56.0 **P** 0.03 max **S** 0.003 max **Ti** 0.6-1.5 | **Other:** NACE MR-01-75 |
| **N06975** | Ni-Cr Alloy (Hastelloy G-2) | **C** 0.03 max **Cr** 23.0-26.0 **Cu** 0.70-1.20 **Fe** rem **Mn** 1.0 max **Mo** 5.0-7.0 **Ni** 47.0-52.0 **P** 0.03 max **S** 0.03 max **Si** 1.0 max **Ti** 0.70-1.50 | **ASME** SB-366; SB-581; SB-582; SB-619; SFA5.14 (ERNiCrMo-8) **ASTM** B581; B582; B619; B622; B626 **AWS** A5.14 (ERNiCrMo-8) **Other:** NACE MR-01-75 |
| **N06985** | Ni-Cr Alloy ERNiCrMo-9 (Hastelloy G-3) | **C** 0.015 max **Co** 5.0 max **Cr** 21.0-23.5 **Cu** 1.5-2.5 **Fe** 18.0-21.0 **Mn** 1.0 max **Mo** 6.0-8.0 **Ni** rem **P** 0.04 max **S** 0.03 max **Si** 1.0 max **W** 1.5 max **Other** Cb+Ta 0.50 max | **ASME** SB-581; SB-582; SB-619; SB-622; SB-626; SFA5.14 (ERNiCrMo-9) **ASTM** B366; B581; B582; B619; B622; B626; G28 **AWS** A5.14 (ERNiCrMo-9) **Other:** NACE MR-01-75 |
| **N07001** | Ni-Cr Alloy Precipitation Hardenable (Waspaloy) | **Al** 1.20-1.60 **B** 0.003-0.01 **C** 0.03-0.10 **Co** 12.00-15.00 **Cr** 18.00-21.00 **Cu** 0.50 max **Fe** 2.00 max **Mn** 1.00 max **Mo** 3.50-5.00 **Ni** rem **P** 0.030 max **S** 0.030 max **Si** 0.75 max **Ti** 2.75-3.25 **Zr** 0.02-0.12 | **AMS** 5544; 5586; 5704; 5706; 5707; 5708; 5709; 5828; AS7471 **ASTM** B637 **SAE** J467 (Waspaloy); J775 (XEV-H) |
| * **N07002** | Ni-Cr Alloy Precipitation Hardenable (TPM) | **Al** 0.05 nom **C** 0.05 nom **Co** 0.50 nom **Cr** 16.00 nom **Mn** 2.30 nom **Ni** rem **Si** 0.05 nom **Ti** 3.10 nom | |
| * **N07012** | Cr-Co-Mo-W Nickel Alloy (AF2-1DA) | **Al** 4.20-4.80 **B** 0.01-0.02 **Bi** 0.00005 max **C** 0.30-0.35 **Co** 9.50-10.50 **Cr** 11.5-12.5 **Fe** 1.00 max **Mn** 0.10 max **Mo** 2.50-3.50 **N** 0.005 max **Ni** rem **O** 0.010 max **P** 0.015 max **Pb** 0.002 max **S** 0.015 max **Si** 0.10 max **Ta** 1.00-2.00 **Ti** 2.75-3.25 **W** 5.50-6.50 **Zr** 0.05-0.15 | |
| **N07013** | Cr-Co-Mo Nickel Alloy Castings | **Al** 3.20-3.60 **B** 0.010-0.020 **C** 0.07-0.20 **Cb** 0.10 max **Co** 8.50-9.50 **Cr** 12.2-13.0 **Fe** 0.50 max **Hf** 0.75-1.05 **Mn** 0.10 max **Mo** 1.70-2.10 **Ni** rem **P** 0.015 max **S** 0.015 max **Si** 0.10 max **Ta** 3.85-4.50 **Ti** 3.85-4.15 **W** 3.85-4.50 **Zr** 0.05-0.14 **Other** Al+Ti 7.30-7.70 | **AMS** 5404 |
| **N07031** | Ni-Cr Alloy Precipitation Hardenable (Pyromet 31) | **Al** 1.00-1.70 **B** 0.003-0.007 **C** 0.03-0.06 **Cb** 0.60-1.20 **Cr** 22.0-23.0 **Fe** rem **Mn** 0.20 max **Mo** 1.70-2.30 **Ni** 55.0-58.0 **P** 0.015 max **S** 0.015 max **Si** 0.20 max **Ti** 2.10-2.60 | **Other:** NACE MR-01-75 |
| **N07032** | Ni-Cr Alloy, Precipitation Hardenable (Pyromet 31V) | **Al** 1.15-1.40 **C** 0.03-0.06 **Cb** 0.75-0.95 **Co** 1.00 max **Cr** 22.3-22.9 **Fe** rem **Mo** 1.70-2.30 **Ni** 55.0-58.0 **P** 0.015 max **S** 0.015 max **Ti** 2.10-2.40 | **SAE** J775 (HEV 8) |

*Boxed entries are no longer active and are retained for reference purposes only.

The chemical compositions listed are for identification purposes and should not be used in lieu of the cross referenced specifications.

| UNIFIED NUMBER | DESCRIPTION | CHEMICAL COMPOSITION | CROSS REFERENCE SPECIFICATIONS |
|---|---|---|---|
| **N07041** | Ni-Cr Alloy Precipitation Hardenable (Rene 41) | **Al** 1.40-1.80 **B** 0.0030-0.010 **C** 0.12 max **Co** 10.00-12.00 **Cr** 18.00-20.00 **Fe** 5.00 max **Mn** 0.10 max **Mo** 9.00-10.50 **Ni** rem **S** 0.015 max **Si** 0.50 max **Ti** 3.00-3.30 | **AMS** 5399; 5545; 5712; 5713; 5800; AS7469 **SAE** J467 (REne 41) |
| **N07048** | Ni-Cr-Fe-Mo-Cu Alloy, Age Hardenable | **Al** 0.4-0.9 **C** 0.015 max **Cb** 0.5 max **Co** 2.0 max **Cr** 20.0-23.5 **Cu** 1.0-2.2 **Fe** 18.0-21.0 **Mn** 0.8 max **Mo** 5.0-7.0 **Ni** rem **P** 0.02 max **S** 0.01 max **Si** 0.1 max **Ti** 1.5-2.1 | **Other:** NACE MR-01-75 |
| **N07069** | Ni-Cr Welding Filler Metal (Inconel FM69) | **Al** 0.40-1.00 **C** 0.08 **Cb** (incl Ta) 0.70-1.20 **Cr** 14.0-17.0 **Cu** 0.50 max **Fe** 5.0-9.0 **Mn** 1.0 max **Ni** 70.0 min **P** 0.03 max **S** 0.015 max **Si** 0.50 max **Ti** 2.00-2.75 | **AMS** 5778 **ASME** SFA5.14 (ERNiCrFe-8) **AWS** A5.14 (ERNiCrFe-8) **MIL SPEC** MIL-E-21562 (RN69) |
| **N07080** | Ni-Cr Alloy Precipitation Hardenable (Nimonic 80A) | **Al** 1.0-1.8 **B** 0.008 max **C** 0.10 max **Co** 2.0 max **Cr** 18.0-21.0 **Cu** 0.2 max **Fe** 3.0 max **Mn** 1.0 max **Ni** rem **S** 0.015 max **Si** 1.00 max **Ti** 1.8-2.7 **P** 0.045 max | **ASME** SB-637 **ASTM** B637 **SAE** J467 (Nimonic 80A); J775 (HEV-5) |
| **N07090** | Ni-Cr Alloy Precipitation Hardenable (Nimonic 90) | **Al** 0.8-2.0 **C** 0.13 max **Co** 15.0-21.0 **Cr** 18.0-21.0 **Fe** 3.0 max **Mn** 1.0 max **Ni** rem **Si** 1.5 max **Ti** 1.8-3.0 | **AMS** 5829 **SAE** J467 (Nimonic 90) **Other:** NACE MR-01-75 |
| **N07092** | Ni-Cr Alloy Precipitation Hardenable (Inconel FM92) | **C** 0.08 max **Cr** 14.00-17.00 **Cu** 0.50 max **Fe** 8.0 max **Mn** 2.00-2.75 **Ni** 67.00 min **P** 0.030 max **S** 0.015 max **Si** 0.35 max **Ti** 2.50-3.50 | **AMS** 5675 **ASME** SFA5.14 (ERNiCrFe-6); SFA5.30 (IN6A) **AWS** A5.14 (ERNiCrFe-6); A5.30 (IN6A) **MIL SPEC** MIL-E-21562 (EN6A, RN6A); MIL-I-23413 (MIL-6A) |
| **N07214** | Ni-Cr Precipitation Hardenable Alloy (214) | **Al** 4.0-5.0 **B** 0.006 max **C** 0.05 max **Co** 2.0 max **Cr** 15.0-17.0 **Fe** 2.0-4.0 **Mn** 0.5 max **Mo** 0.5 max **Ni** rem **P** 0.015 max **S** 0.015 max **Si** 0.2 max **Ti** 0.5 max **W** 0.5 max **Y** 0.002-0.040 **Zr** 0.05 max | |
| **N07252** | Ni-Cr Alloy Precipitation Hardenable (M252) | **Al** 0.75-1.25 **B** 0.003-0.01 **C** 0.10-0.20 **Co** 9.00-11.00 **Cr** 18.00-20.00 **Fe** 5.00 max **Mn** 0.50 max **Mo** 9.00-10.50 **Ni** rem **P** 0.015 max **S** 0.015 max **Si** 0.50 max **Ti** 2.25-2.75 | **ASME** SB-637 **ASTM** B637 **SAE** J467 (M-252) |
| **N07263** | Ni-Cr Alloy Precipitation Hardenable (Nimonic Alloy 263) | **Al** 0.3-0.6 **C** 0.04-0.08 **Co** 19.0-21.0 **Cr** 19.0-21.0 **Cu** 0.20 max **Fe** 0.7 max **Mn** 0.60 max **Mo** 5.6-6.1 **Ni** rem **P** 0.015 max **S** 0.007 max **Si** 0.40 max **Ti** 1.9-2.4 **Other** Al+Ti 2.4-2.8 | **AMS** 5872; 5886; 5966 |
| **N07500** | Ni-Cr Alloy Precipitation Hardenable (Udimet 500) | **Al** 2.50-3.25 **B** 0.003-0.01 **C** 0.15 max **Co** 13.00-20.00 **Cr** 15.00-20.00 **Cu** 0.15 max **Fe** 4.00 max **Mn** 0.75 max **Mo** 3.00-5.00 **Ni** rem **P** 0.015 max **S** 0.015 max **Si** 0.75 max **Ti** 2.50-3.25 | **AMS** 5384; 5751 **ASME** SB-637 **ASTM** B637 **SAE** J467 (Udimet 500) |
| **N07520** | Ni-Cr-Co-Mo-Ti-Al-W Alloy, Precipitation Hardenable (Alloy 520, Allvac 520, Udimet 520) | **Al** 1.8-2.2 **B** 0.010 max **C** 0.06 max **Co** 12.0-14.0 **Cr** 18.0-20.0 **Mo** 5.0-7.0 **Ni** rem **Ti** 2.8-3.2 **W** 0.8-1.2 | **Other:** GGC-0005; GGC-1005 |
| **N07626** | Ni-Cr Precipitation Hardenable Alloy | **Al** 0.40-0.80 **C** 0.05 max **Cr** 20.0-23.0 **Co** 1.00 max **Cb** 4.50-5.50 **Cu** 0.50 max **Fe** 6.00 max **Mn** 0.50 max **Mo** 8.0-10.0 **Ni** rem **N** 0.05 max **P** 0.020 max **S** 0.015 max **Si** 0.50 max **Ti** 0.60 max | **Other:** NACE MR-01-75 |
| **N07702** | Ni-Cr Alloy Precipitation Hardenable (Inconel 702) | **Al** 2.75-3.75 **C** 0.10 max **Cr** 14.0-17.0 **Cu** 0.5 max **Fe** 2.0 max **Mn** 1.0 max **Ni** rem **S** 0.01 max **Si** 0.7 max **Ti** 0.25-1.00 | **AMS** 5550 **SAE** J467 (Inconel 702) |

The chemical compositions listed are for identification purposes and should not be used in lieu of the cross referenced specifications.

| UNIFIED NUMBER | DESCRIPTION | CHEMICAL COMPOSITION | CROSS REFERENCE SPECIFICATIONS |
|---|---|---|---|
| **N07713** | Ni-Cr Alloy Precipitation Hardenable (IN-713) | **Al** 5.5-6.5 **B** 0.005-0.015 **C** 0.08-0.20 **Cb** 1.8-2.8 **Cr** 12.00-14.00 **Fe** 2.50 max **Mn** 0.25 max **Mo** 3.8-5.2 **Ni** rem **Si** 0.50 max **Ti** 0.5-1.0 **Zr** 0.05-0.15 | **AMS** 5377; 5391 |
| **N07716** | Ni-Cr Precipitation Hardenable Alloy | **Al** 0.35 max **C** 0.03 max **Cr** 19.0-22.0 **Cb** 2.75-4.00 **Fe** rem **Mn** 0.20 max **Mo** 7.00-9.50 **Ni** 57.0-63.0 **P** 0.015 max **S** 0.010 max **Si** 0.20 max **Ti** 1.00-1.60 | **AMS** 5854 **ASTM** B805 **Other:** NACE MR-01-75 |
| **N07718** | Ni-Cr Alloy Precipitation Hardenable (Inconel 718) | **Al** 0.20-0.80 **B** 0.006 max **C** 0.08 max **Cb** 4.75-5.50 **Co** 1.00 max **Cr** 17.0-21.0 **Cu** 0.30 max **Fe** rem **Mn** 0.35 max **Mo** 2.80-3.30 **Ni** 50.0-55.0 **P** 0.015 max **S** 0.015 max **Si** 0.35 max **Ti** 0.65-1.15 | **AMS** 5383; 5589; 5590; 5596; 5597; 5662; 5663; 5664; 5832; AS7466; AS7467 **ASME** SFA5.14 (ERNiFeCr-2) **ASTM** B637; B670 **AWS** A5.14 (ERNiFeCr-2) **SAE** J467 (Inconel 718) **Other:** NACE MR-01-75 |
| **N07719** | Ni-Cr-Cb-Mo Precipitation Hardenable Alloy (Inconel 718SPF) | **Al** 0.20-0.80 **B** 0.006 max **C** 0.05 max **Cb** 4.75-5.25 **Co** 1.0 max **Cr** 17.0-21.0 **Cu** 0.30 max **Fe** rem **Mn** 0.35 max **Mo** 2.80-3.30 **N** 0.01 max **Ni** 50.0-55.0 **P** 0.015 max **S** 0.002 max **Si** 0.35 max **Ti** 0.65-1.15 | **AMS** 5914; 5950 **MAM** 5914 |
| **N07720** | Ni-Cr-Co-Ti-Mo-Al-W Alloy, Precipitation Hardenable (Alloy 720, Allvac 720, Udimet 720) | **Al** 2.0-3.0 **B** 0.02 max **C** 0.03 max **Co** 14.0-16.0 **Cr** 15.0-17.0 **Mo** 2.5-3.5 **Ni** rem **Ti** 4.5-5.5 **W** 1.0-2.0 **Zr** 0.05 max | **Other:** MSRR 7252; EMS 73105 |
| **N07721** | Ni-Cr Alloy Precipitation Hardenable (Inconel 721) | **Al** 0.10 max **C** 0.07 max **Cr** 15.0-17.0 **Cu** 0.20 max **Fe** 8.00 max **Mn** 2.00-2.50 **Ni** rem **S** 0.01 max **Si** 0.15 max **Ti** 2.75-3.35 | |
| **N07722** | Ni-Cr Alloy Precipitation Hardenable (Inconel 722) | **Al** 0.4-1.0 **C** 0.08 max **Cr** 14.0-17.0 **Cu** 0.5 max **Fe** 5.0-9.0 **Mn** 1.0 max **Ni** 70.0 min **S** 0.01 max **Si** 0.07 max **Ti** 2.00-2.75 | **AMS** 5541; 5714 **SAE** J467 (Inconel 722) |
| **N07725** | Ni-Cr-Mo Precipitation Hardenable Alloy (Inconel Alloy 725) | **Al** 0.35 max **C** 0.03 max **Cr** 19.0-22.5 **Cb** 2.75-4.00 **Fe** rem **Mn** 0.35 max **Mo** 7.00-9.50 **Ni** 55.0-59.0 **P** 0.015 max **S** 0.010 max **Si** 0.20 max **Ti** 1.00-1.70 | **ASME** SFA5.14 (ERNiCrMo-15) **ASTM** B805 **AWS** A5.14 (ERNiCrMo-15) **Other:** NACE MR-01-75 |
| **N07750** | Ni-Cr Alloy Precipitation Hardenable (Inconel X750) | **Al** 0.40-1.0 **C** 0.08 max **Cb** 0.70-1.20 **Cr** 14.0-17.0 **Cu** 0.5 max **Fe** 5.0-9.0 **Mn** 1.0 max **Ni** 70.0 min **S** 0.01 max **Si** 0.50 max **Ti** 2.25-2.75 | **AMS** 5542; 5582; 5583; 5598; 5667; 5668; 5669; 5670; 5671; 5698; 5699; 5747; 5779; AS7245; AS7246 **ASME** SB-637 **ASTM** B637 **MIL SPEC** MIL-N-7786 **SAE** J467 (Inconel X-750) **Other:** NACE MR-01-75 |
| **N07751** | Ni-Cr Alloy Precipitation Hardenable (Inconel 751) | **Al** 0.90-1.50 **C** 0.10 max **Cb** 0.70-1.20 **Cr** 14.0-17.0 **Cu** 0.50 max **Fe** 5.00-9.00 **Mn** 1.00 max **Ni** 70.0 min **S** 0.01 max **Si** 0.50 max **Ti** 2.00-2.60 | **SAE** J775 (HEV-3) |
| **N07752** | Ni-Cr-Fe-Ti-Al Alloy, Precipitation Hardenable | **Al** 0.40-1.00 **B** 0.007 max **C** 0.020-0.060 **Co** 0.050 max **Cr** 14.5-17.0 **Cu** 0.50 max **Fe** 5.00-9.00 **Mn** 1.00 max **Ni** 70.0 min **P** 0.008 max **S** 0.003 max **Si** 0.50 max **Ti** 2.25-2.75 **V** 0.10 max **Zr** 0.050 max **Other** (Cb+Ta) 0.70-1.20 | **ASTM** B637 |
| **N07754** | Ni-Cr Alloy Y Dispersion Strengthened (Inconel MA754) | **Al** 0.2-0.5 **C** 0.05 max **Cr** 19.0-23.0 **Fe** 2.5 max **Ni** rem **Ti** 0.3-0.6 **Other** $Y_2O_3$ 0.5-0.7 | |
| **N07773** | Ni-Cr-Fe-Cb-Mo Alloy Precipitation Hardenable (PH3) | **Al** 2.0 max **C** 0.03 max **Cb** 2.5-6.0 **Cr** 18.0-27.0 **Fe** rem **Mn** 1.00 max **Mo** 2.5-5.5 **Ni** 45.0-60.0 **P** 0.030 max **S** 0.010 max **Si** 0.50 max **Ti** 2.0 max **W** 0.5 max **Other** (Mo+0.5W) = 2.5-5.5 | **ASTM** B805 **Other:** NACE MR-01-75 |

The chemical compositions listed are for identification purposes and should not be used in lieu of the cross referenced specifications.

| UNIFIED NUMBER | DESCRIPTION | CHEMICAL COMPOSITION | CROSS REFERENCE SPECIFICATIONS |
|---|---|---|---|
| **N07776** | Ni-Cr-Mo-Fe-Cb-W Alloy Precipitation Hardenable (PH6) | **Al** 2.0 max **C** 0.03 max **Cb** 4.0-6.0 **Cr** 12.0-22.0 **Fe** rem **Mn** 1.00 max **Mo** 9.0-15.0 **Ni** 50.0-60.0 **P** 0.030 max **S** 0.010 max **Si** 0.50 max **Ti** 1.00 max **W** 0.5-2.5 | **ASTM** B805 **Other:** NACE MR-01-75 |
| **N07924** | Ni-Cr-Fe-Mo-Cb Alloy Precipitation Hardenable | **Al** 0.75 max **C** 0.020 max **Cb** 2.75-3.50 **Co** 3.0 max **Cr** 20.5-22.5 **Cu** 1.0-4.0 **Fe** 7.0-13.0 **Mg** 0.0050 max **Mn** 0.20 max **Mo** 5.50-7.0 **N** 0.020 max **Ni** 52.0 min **P** 0.030 max **S** 0.0050 max **Si** 0.20 max **Ti** 1.0-2.0 **W** 0.50 max | **Other:** NACE MR-01-75 |
| **N08001** | Ni-Fe-Cr Alloy Solid Solution Strengthened (ACI HW) | **C** 0.35-0.75 **Cr** 10.0-14.0 **Fe** rem **Mn** 2.00 max **Mo** 0.50 max **Ni** 58.0-62.0 **P** 0.04 max **S** 0.04 max **Si** 2.50 max | **ACI** HW **ASTM** A297 (HW) |
| **N08002** | Ni-Fe-Cr Alloy Solid Solution Strengthened (ACI HT) | **C** 0.35-0.75 **Cr** 13.0-17.0 **Fe** rem **Mn** 2.00 max **Mo** 0.50 max **Ni** 33.0-37.0 **P** 0.04 max **S** 0.04 max **Si** 2.50 max | |
| **N08004** | Ni-Fe-Cr Alloy Solid Solution Strengthened (ACI HU) | **C** 0.35-0.75 **Cr** 17.0-21.0 **Fe** rem **Mn** 2.00 max **Mo** 0.50 max **Ni** 37.0-41.0 **P** 0.04 max **S** 0.04 max **Si** 2.50 max | **ACI** HU **ASTM** A297 (HU) |
| **N08005** | Ni-Fe-Cr Alloy Solid Solution Strengthened (ACI HU-50) | **C** 0.40-0.60 **Cr** 17.0-21.0 **Fe** rem **Mn** 1.50 max **Mo** 0.50 max **Ni** 37.0-41.0 **P** 0.04 max **S** 0.04 max **Si** 0.50-2.00 | **ACI** HU-50 **ASTM** A608 (HU-50) |
| **N08006** | Ni-Fe-Cr Alloy Solid Solution Strengthened (ACI HW-50) | **C** 0.40-0.60 **Cr** 10.0-14.0 **Fe** rem **Mn** 1.50 max **Mo** 0.50 max **Ni** 58.0-62.0 **P** 0.04 max **S** 0.04 max **Si** 0.50-2.00 | **ACI** HW-50 **ASTM** A608 (HW-50) |
| **N08007** | Ni-Fe-Cr Alloy Solid Solution Strengthened (ACI CN-7M) | **C** 0.07 max **Cr** 19.0-22.0 **Cu** 3.00-4.00 **Fe** rem **Mn** 1.50 max **Mo** 2.00-3.00 **Ni** 27.5-30.5 **Si** 1.50 max | **ACI** CN-7M **ASTM** A351 (CN7M); A743 (CN7M); A744 (CN7M) |
| **N08008** | Ni-Fe-Cr Alloy Solid Solution Strengthened (ACI HT-50C) | **C** 0.40-0.60 **Cb** 0.75-1.25 **Cr** 13.0-17.0 **Fe** rem **Mo** 0.50 max **Ni** 33.0-37.0 | **ACI** HT-50C |
| **N08020** | Ni-Fe-Cr Alloy Solid Solution Strengthened (Carpenter 20Cb3) | **C** 0.07 max **Cb** 8\mtC-1.00 **Cr** 19.00-21.00 **Cu** 3.00-4.00 **Fe** rem **Mn** 2.00 max **Mo** 2.00-3.00 **Ni** 32.00-38.00 **P** 0.045 max **S** 0.035 max **Si** 1.00 max | **ASME** SB-462; SB-463; SB-464 **ASTM** A182 (F20); A240; A358; A480; A959; B366; B462; B463; B464; B468; B472; B473; B474; B475; B729; G28 **Other:** NACE MR-01-75 |
| **N08021** | Austenitic Cr-Ni-Cu-Mo Alloy Welding Filler Metal | **C** 0.07 max **Cr** 19.0-21.0 **Cu** 3.0-4.0 **Fe** rem **Mn** 2.5 max **Mo** 2.0-3.0 **Ni** 32.0-36.0 **P** 0.03 max **S** 0.03 max **Si** 0.60 max **Other** Cb+Ta 8\mtC-1.0 | **ASME** SFA5.9 (ER320) **AWS** A5.9 (ER320) |
| **N08022** | Austenitic Cr-Ni-Cu-Mo Alloy Welding Filler Metal | **C** 0.025 max **Cr** 19.0-21.0 **Cu** 3.0-4.0 **Fe** rem **Mn** 1.5-2.0 **Mo** 2.0-3.0 **Ni** 32.0-36.0 **P** 0.015 max **S** 0.020 max **Si** 0.15 max **Other** Cb+Ta 8\mtC-0.40 | **ASME** SFA5.9 (ER320LR) **AWS** A5.9 (ER320LR) |
| **N08024** | Cr-Ni-Fe-Mo-Cu Alloy Columbium Stabilized | **C** 0.03 max **Cb** 0.15-0.35 **Cr** 22.5-25.0 **Cu** 0.50-1.50 **Fe** rem **Mn** 1.00 max **Mo** 3.50-5.00 **Ni** 35.0-40.0 **P** 0.035 max **S** 0.035 max **Si** 0.50 max | **ASTM** B462; B463; B464; B468; B472; B473; B474; B475; B729 **Other:** NACE MR-01-75 |
| **N08026** | Ni-Cr Wrought Alloy (Carpenter 20Mo6) | **C** 0.03 max **Cr** 22.0-26.0 **Cu** 2.00-4.00 **Fe** rem **Mn** 1.00 max **Mo** 5.00-6.70 **N** 0.10-0.16 **Ni** 33.0-37.2 **P** 0.03 max **S** 0.03 max **Si** 0.50 max | **ASME** SB-463; SB-464; SB-468 **ASTM** B462; B463; B464; B468; B472; B473; B474; B475; B729 |
| **N08028** | Ni-Fe-Cr Alloy Solid Solution Strengthened (Sanicro 28) | **C** 0.03 max **Cr** 26.0-28.0 **Cu** 0.6-1.4 **Fe** rem **Mn** 2.50 max **Mo** 3.0-4.0 **Ni** 30.0-34.0 **P** 0.030 max **S** 0.030 max **Si** 1.00 max | **ASME** SB-668; SB-709 **ASTM** B668; B709 **Other:** NACE MR-01-75 |
| **N08030** | Ni-Fe-Cr Alloy Solid Solution Strengthened (ACI HT-30) | **C** 0.25-0.35 **Cr** 13.0-17.0 **Fe** rem **Mn** 2.00 max **Mo** 0.50 max **Ni** 33.0-37.0 **P** 0.040 max **S** 0.040 max **Si** 2.50 max | |

The chemical compositions listed are for identification purposes and should not be used in lieu of the cross referenced specifications.

| UNIFIED NUMBER | DESCRIPTION | CHEMICAL COMPOSITION | CROSS REFERENCE SPECIFICATIONS |
|---|---|---|---|
| **N08031** | Ni-Fe-Cr-Mo-Cu Alloy (Alloy 31) | **C** 0.015 max **Cr** 26.0-28.0 **Cu** 1.0-1.4 **Fe** rem **Mn** 2.0 max **Mo** 6.0-7.0 **N** 0.15-0.25 **Ni** 30.0-32.0 **P** 0.03 max **S** 0.005 max **Si** 0.05 max | **ASTM** A314; B366; B564; B581; B619; B622; B625; B626; B649 |
| **N08032** | Fe-Ni-Cr-Mo Alloy (INCO Alloy 032) | **C** 0.03 max **Cr** 20.0-23.0 **Fe** rem **Mn** 1.0 max **Mo** 4.0-5.0 **Ni** 30.0-34.0 **P** 0.03 max **S** 0.005 max **Si** 0.05 max | |
| **N08036** | Ni-Fe-Cr-Mo-Cu Alloy (20Mo-6HS) | **C** 0.06 max **Cr** 22.0-26.0 **Cu** 1.00-3.00 **Fe** rem **Mn** 1.00 max **Mo** 5.00-6.70 **Ni** 33.0-37.2 **N** 0.17-0.40 **P** 0.030 max **S** 0.030 max **Si** 0.50 max | |
| **N08042** | Ni-Fe-Cr-Mo-Cu Alloy (NIC 42M) | **C** 0.03 max **Cr** 20.0-23.0 **Cu** 1.5-3.0 **Fe** rem **Mn** 1.0 max **Mo** 5.0-7.0 **Ni** 40.0-44.0 **P** 0.03 max **S** 0.003 max **Si** 0.5 max **Ti** 0.6-1.2 | **Other:** NACE MR-01-75 |
| **N08050** | Ni-Fe-Cr Alloy Solid Solution Strengthened (ACI HT-50) | **C** 0.40-0.60 **Cr** 15.0-19.0 **Fe** rem **Mn** 1.50 max **Mo** 0.50 max **Ni** 33.0-37.0 **P** 0.04 max **S** 0.04 max **Si** 0.50-2.00 | **ACI** HT-50 **ASTM** A608 (HT-50) |
| **N08065** | Ni-Fe-Cr Welding Filler Metal (Incoloy FM65) | **Al** 0.20 max **C** 0.05 max **Cr** 19.5-23.5 **Cu** 1.50-3.0 **Fe** 22.0 min **Mn** 1.0 max **Mo** 2.5-3.5 **Ni** 38.0-46.0 **P** 0.03 max **S** 0.03 max **Si** 0.50 max **Ti** 0.60-1.2 | **ASME** SFA5.14 (ERNiFeCr-1) **AWS** A5.14 (ERNiFeCr-1) **MIL SPEC** MIL-E-21562 (RN65) |
| **N08120** | Ni-Fe-Cr Alloy (Alloy HR-120) | **Al** 0.40 max **B** 0.010 max **C** 0.02-0.1 **Cb** 0.4-0.9 **Co** 3.0 max **Cr** 23.0-27.0 **Cu** 0.50 max **Fe** rem **Mn** 1.5 max **Mo** 2.50 max **N** 0.15-0.30 **Ni** 35.0-39.0 **P** 0.040 max **S** 0.03 max **Si** 1.0 max **Ti** 0.20 max **W** 2.50 max | **ASTM** B163; B366; B407; B408; B409; B514; B515; B564 |
| **N08135** | Ni-Fe-Cr-Mo-W Alloy (SM 2035) | **C** 0.03 max **Cr** 20.5-23.5 **Cu** 0.70 max **Fe** rem **Mn** 1.00 max **Mo** 4.0-5.0 **Ni** 33.0-38.0 **P** 0.03 max **S** 0.03 max **Si** 0.75 max **W** 0.20-0.80 | **ASTM** B622 |
| **N08151** | Cast Fe-Ni-Cr-Cb Austenitic Stainless Steel Casting (CT15C) | **C** 0.05-0.15 **Cr** 19.0-21.0 **Cb** 0.50-1.50 **Fe** rem **Mn** 0.15-1.50 **Ni** 31.0-34.0 **P** 0.03 max **S** 0.03 max **Si** 0.50-1.50 | **ACI** CT15C **ASME** SA-351 **ASTM** A351 (CT15C); S351 |
| **N08221** | Ni-Fe-Cr-Cu Alloy | **Al** 0.20 max **C** 0.025 max **Cr** 20.0-22.0 **Cu** 1.50-3.00 **Fe** rem **Mn** 1.00 max **Mo** 5.00-6.50 **Ni** 36.0-46.0 **S** 0.03 max **Si** 0.50 max **Ti** 0.60-1.00 | **ASTM** B423; B424; B425 |
| * **N08245** | Ni-Fe-Cr Alloy (ARMCO 20-45-5) | **C** 0.08 max **Cb** 0.40 max **Cr** 18.0-22.0 **Fe** rem **Mn** 3.0-7.0 **Mo** 1.5-3.0 **Ni** 43.0-49.0 **P** 0.045 max **S** 0.030 max **Si** 1.00 max | |
| * **N08310** | Ni-Fe-Cr-Mo Alloy | **C** 0.020 max **Cr** 24.0-26.0 **Fe** rem **Mn** 2.00-4.00 **Mo** 2.00-4.00 **N** 0.20-0.40 **Ni** 18.0-22.0 **P** 0.035 max **S** 0.015 max **Si** 0.050 max | |
| **N08320** | Ni-Fe-Cr Alloy (Haynes No. 20 Mod) | **C** 0.05 max **Cr** 21.0-23.0 **Fe** rem **Mn** 2.5 max **Mo** 4.0-6.0 **Ni** 25.0-27.0 **P** 0.04 max **S** 0.03 max **Si** 1.0 max **Ti** 4\mtC min | **ASME** SB-366; SB-620; SB-621; SB-622 **ASTM** B619; B620; B621; B622; B626 |
| **N08321** | Ni-Fe-Cr Alloy Solution Strengthened (20 Cb3 Modified) | **C** 0.035 max **Cb** 8\mtC-0.4 max **Cr** 19.0-21.0 **Cu** 3.0-4.0 **Fe** rem **Mn** 1.5-2.5 **Mo** 2.0-3.0 **Ni** 32.0-36.0 **P** 0.02 max **S** 0.015 max **Si** 0.3 max | |

*Boxed entries are no longer active and are retained for reference purposes only.

The chemical compositions listed are for identification purposes and should not be used in lieu of the cross referenced specifications.

| UNIFIED NUMBER | DESCRIPTION | CHEMICAL COMPOSITION | CROSS REFERENCE SPECIFICATIONS |
|---|---|---|---|
| **N08330** | Ni-Fe-Cr Alloy Solid Solution Strengthened (RA-330) | **C** 0.08 max **Cr** 17.0-20.0 **Cu** 1.00 max **Fe** rem **Mn** 2.00 max **Ni** 34.0-37.0 **P** 0.03 max **Pb** 0.005 max **S** 0.03 max **Si** 0.75-1.50 **Sn** 0.025 max | **AMS** 5592; 5716 **ASME** SB-366; SB-511; SB-536; SB-710 **ASTM** A554; B366; B511; B512; B535; B536; B546; B710; B739 **SAE** J412 (30330); J467 (RA330) |
| **N08331** | Ni-Fe-Cr Welding Filler Metal (ER330) | **C** 0.18-0.25 **Cr** 15.0-17.0 **Cu** 0.75 max **Fe** rem **Mn** 1.0-2.5 **Mo** 0.75 max **Ni** 34.0-37.0 **P** 0.03 max **S** 0.03 max **Si** 0.30-0.65 | **ASME** SFA5.9 (ER330) **AWS** A5.9 (ER330) |
| **N08332** | Ni-Fe-Cr-Si Alloy (RA 330TX) | **Al** 0.10-0.50 **C** 0.05-0.10 **Cr** 17.0-20.0 **Cu** 1.00 max **Fe** rem **Mn** 2.00 max **Ni** 34.0-37.0 **P** 0.03 max **Pb** 0.005 max **S** 0.03 **Si** 0.75-1.50 **Sn** 0.025 max **Ti** 0.20-0.60 **Note:** Iron shall be determined arithmetically by difference | **ASME** SB-366; SB-511; SB-535; SB-536 **ASTM** B511; B535; B536; B546; B710; B739 |
| **N08334** | Ni-Fe-Cr-Mn Alloy, Solid Solution Strengthened (RA 330-04) | **C** 0.18-0.29 **Cr** 17.0-20.0 **Cu** 0.5 max **Fe** rem **Mn** 4.25-6.5 **Mo** 0.7 max **Ni** 33.0-37.0 **P** 0.025 max **S** 0.02 max **Si** 0.65-1.3 | |
| **N08366** | Austenitic Ni-Cr-Mo Stainless Steel (AL-6X) | **C** 0.035 max **Cr** 20.0-22.0 **Fe** rem **Mn** 2.00 max **Mo** 6.00-7.00 **Ni** 23.5-25.5 **P** 0.030 max **S** 0.030 max **Si** 1.00 max | **ASME** SB-675 **ASTM** B688; B690; B691 |
| **N08367** | Austenitic Ni-Cr-Mo-N Stainless Steel (AL-6XN) | **C** 0.030 max **Cr** 20.0-22.0 **Fe** rem **Mn** 2.00 max **Mo** 6.00-7.00 **N** 0.18-0.25 **Ni** 23.50-25.50 **P** 0.040 max **S** 0.030 max **Si** 1.00 max | **ASTM** A182; A240; A249; A269; A270; A312; A358; B366; A409; A479; A480; A688; A813; A814; A959; A988; B462; B472; B564; B675; B676; B688; B690; B691; B804; B834; G28 **SAE** J405 **Other:** NACE MR-01-75 |
| **N08421** | Ni-Fe-Cr-Mo-Cu Alloy | **Al** 0.2 max **C** 0.025 max **Cr** 20.0-22.0 **Cu** 1.5-2.0 **Fe** rem **Mn** 1.00 max **Mo** 5.0-6.5 **Ni** 39.0-41.0 **S** 0.03 max **Si** 0.5 max **Ti** 0.6-1.0 | |
| **N08535** | Ni-Cr-Mo-Fe-Cu Alloy | **C** 0.03 max **Cr** 24.0-27.0 **Cu** 1.50 max **Fe** rem **Mn** 1.00 max **Mo** 2.5-4.0 **Ni** 29.0-36.5 **P** 0.03 max **S** 0.03 max **Si** 0.50 max | **ASTM** B622 **Other:** NACE MR-01-75 |
| **N08603** | Alloy Casting (HT-30) | **C** 0.25-0.35 **Cr** 13.0-17.0 **Fe** rem **Mn** 2.00 max **Mo** 0.50 max **Ni** 33.0-37.0 **P** 0.040 max **S** 0.040 max **Si** 2.50 max | **ACI** HT-30 **ASTM** A351 (HT-30) |
| **N08604** | Alloy Casting (HL) | **C** 0.20-0.60 **Cr** 28.0-32.0 **Fe** rem **Mn** 2.00 max **Mo** 0.50 max **Ni** 16.0-22.0 **P** 0.04 max **S** 0.04 max **Si** 2.00 max | **ACI** HL **ASTM** A297 (HL) |
| **N08605** | Alloy Casting (HT) | **C** 0.35-0.75 **Cr** 15.0-19.0 **Fe** rem **Mn** 2.00 max **Ni** 33.0-37.0 **P** 0.04 max **S** 0.04 max **Si** 2.50 max | **ACI** HT **ASTM** A297 (HT) |
| **N08613** | Alloy Casting (HL-30) | **C** 0.25-0.35 **Cr** 28.0-32.0 **Fe** rem **Mn** 1.50 max **Mo** 0.50 max **Ni** 18.0-22.0 **P** 0.04 max **S** 0.04 max **Si** 0.50-2.00 | **ACI** HL-30 **ASTM** A608 (HL-30) |
| **N08614** | Alloy Casting (HL-40) | **C** 0.35-0.45 **Cr** 28.0-32.0 **Fe** rem **Mn** 1.50 max **Mo** 0.50 max **Ni** 18.0-22.0 **P** 0.04 max **S** 0.04 max **Si** 0.50-2.00 | **ASTM** A608 (HL-40) |
| **N08700** | Fe-Ni-Cr Alloy (JS 700) | **C** 0.04 max **Cb** 8 x C to 0.50 **Cr** 19.0-23.0 **Cu** 0.50 max **Fe** rem **Mn** 2.00 max **Mo** 4.3-5.0 **Ni** 24.0-26.0 **P** 0.04 max **Pb** 0.005 max **S** 0.03 max **Si** 1.00 max **Sn** 0.035 max | **ASME** SB-599; SB-672 **ASTM** B599; B672 |
| **N08705** | Alloy Casting (HP) | **C** 0.35-0.75 **Cr** 24.0-28.0 **Fe** rem **Mn** 2.00 max **Mo** 0.50 max **Ni** 35.0-37.0 **P** 0.04 max **S** 0.04 max **Si** 2.50 max | **ASTM** A297 (HP) |

The chemical compositions listed are for identification purposes and should not be used in lieu of the cross referenced specifications.

| UNIFIED NUMBER | DESCRIPTION | CHEMICAL COMPOSITION | CROSS REFERENCE SPECIFICATIONS |
|---|---|---|---|
| **N08800** | Fe-Ni-Cr Alloy Solid Solution Strengthened (alloy 800) | **Al** 0.15-0.60 **C** 0.10 max **Cr** 19.0-23.0 **Cu** 0.75 max **Fe** rem **Mn** 1.5 max **Ni** 30.0-35.0 **P** 0.045 max **S** 0.015 max **Si** 1.0 max **Ti** 0.15-0.60 | **AMS** 5766; 5871 **ASME** SB-163; SB-407; SB-408; SB-409; SB-564 **ASTM** A240; A358; A480; A959; B163; B366; B407; B408; B409; B514; B515; B564; G28 **SAE** J405 **Other:** NACE MR-01-75 |
| **N08801** | Fe-Ni-Cr Alloy Solid Solution Strengthened (Incoloy 801) | **C** 0.10 max **Cr** 19.0-22.0 **Cu** 0.5 max **Fe** rem **Mn** 1.5 max **Ni** 30.0-34.0 **S** 0.015 max **Si** 1.0 max **Ti** 0.75-1.5 | **AMS** 5552 **ASTM** B163; B407 **SAE** J467 (Incoloy 801) |
| **N08802** | Ni-Fe-Cr Alloy Solid Solution Strengthened (Incoloy 802) | **Al** 0.15-1.00 **C** 0.20-0.50 **Cr** 19.0-23.0 **Cu** 0.75 max **Fe** rem **Mn** 1.50 max **Ni** 30.0-35.0 **S** 0.015 max **Si** 0.75 max **Ti** 0.25-1.25 | |
| **N08810** | Fe-Ni-Cr Alloy (Alloy 800H) | **Al** 0.15-0.60 **C** 0.05-0.10 **Cr** 19.0-23.0 **Cu** 0.75 max **Fe** rem **Mn** 1.5 max **Ni** 30.0-35.0 **P** 0.045 max **S** 0.015 max **Si** 1.0 max **Ti** 0.15-0.60 | **ASME** SB-163; SB-407; SB-408; SB-409; SB-514; SB-564 **ASTM** A240; A358; A480; A959; B163; B366; B407; B408; B409; B514; B515; B564 **SAE** J405 |
| **N08811** | Fe-Ni-Cr Alloy (Alloy 800HT) | **Al** 0.15-0.60 **C** 0.06-0.10 **Cr** 19.0-23.0 **Cu** 0.75 max **Fe** 39.5 min **Mn** 1.5 max **Ni** 30.0-35.0 **P** 0.045 max **S** 0.015 max **Si** 1.0 max **Ti** 0.15-0.60 **Other** Al+Ti 0.85-1.20 | **ASTM** A240; A480; A959; B163; B366; B407; B408; B409; B515; B564 |
| **N08825** | Ni-Fe-Cr Alloy Solid Solution Strengthened (Incoloy 825) | **Al** 0.2 max **C** 0.05 max **Cr** 19.5-23.5 **Cu** 1.5-3.0 **Fe** rem **Mn** 1.0 max **Mo** 2.5-3.5 **Ni** 38.0-46.0 **P** 0.03 max **S** 0.03 max **Si** 0.5 max **Ti** 0.6-1.2 | **ASME** SB-163; SB-423; SB-424; SB-425 **ASTM** B163; B366; B423; B424; B425; B564; B704; B705 **Other:** NACE MR-01-75 |
| **N08826** | Ni-Fe-Cr-Mo-Cu Casting Alloy (CU5MCuC) | **C** 0.05 max **Cb** 0.6-1.2 **Cr** 19.5-23.5 **Cu** 1.5-3.5 **Fe** 22.0 min **Mn** 1.00 max **Mo** 2.5-3.5 **Ni** 38.0-44.0 **P** 0.03 max **S** 0.03 max **Si** 1.00 max | **ACI** CU5MCuC **ASME** SA-494 **ASTM** A494 |
| **N08904** | Ni-Cr-Mo Alloy | **C** 0.020 max **Cr** 19.0-23.0 **Cu** 1.00-2.00 **Fe** rem **Mn** 2.00 max **Mo** 4.00-5.00 **Ni** 23.0-28.0 **P** 0.045 max **S** 0.035 max **Si** 1.00 max | **ASME** SB-625; SB-649; SB-673; SB-674; SB-677 **ASTM** A240; A249; A269; A312; A358; A480; A959; B366; B625; B649; B673; B674; B677 **SAE** J405 (904L) |
| **N08925** | Ni-Fe-Cr-Mo-Cu Alloy, Low Carbon | **C** 0.020 max **Cr** 19.0-21.0 **Cu** 0.80-1.50 **Fe** rem **Mn** 1.00 max **Mo** 6.00-7.00 **N** 0.10-0.20 **Ni** 24.0-26.0 **P** 0.045 max **S** 0.030 max **Si** 0.50 max | **ASME** SB-625; SB-649; SB-673; SB-674 **ASTM** B366; B625; B649; B673; B674; B677 |
| **N08926** | Ni-Fe-Cr-Mo-Cu-N Alloy, Low Carbon, Nitrogen Modified | **C** 0.020 max **Cr** 19.0-21.0 **Cu** 0.5-1.5 **Fe** rem **Mn** 2.00 max **Mo** 6.0-7.0 **N** 0.15-0.25 **Ni** 24.0-26.0 **P** 0.030 max **S** 0.010 max **Si** 0.50 max | **ASME** SB-366; SB-625; SB-649; SB-673; SB-674; SB-677 **ASTM** A240; A249; A269; A270; A314; A358; A480; A688; A959; B366; B472; B625; B649; B673; B674; B677; B804 **SAE** J405 |
| **N08932** | Ni-Fe-Cr-Mo-Cu, Low Carbon Alloy (Creusot UR SB 8) | **C** 0.020 max **Cr** 24.0-26.0 **Cu** 1.0-2.0 **Fe** rem **Mn** 2.0 max **Mo** 4.7-5.7 **Ni** 24.0-26.0 **N** 0.17-0.25 **P** 0.025 max **S** 0.010 max **Si** 0.50 max | **ASTM** B625 |
| *** N09027** | Ni-Fe-Cr Alloy Precipitation Hardenable (CG27) | **Al** 1.45-1.75 **B** 0.003-0.015 **C** 0.02-0.08 **Cb** 0.60-1.10 **Cr** 12.50-14.00 **Fe** rem **Mn** 0.25 max **Mo** 5.00-6.00 **Ni** 36.50-39.50 **P** 0.015 max **S** 0.015 max **Si** 0.25 max **Ti** 2.30-2.70 | |
| **N09706** | Ni-Fe-Cr Alloy Precipitation Hardenable (Inconel 706) | **Al** 0.40 max **B** 0.006 max **C** 0.06 max **Cb** 2.5-3.3 **Cr** 14.5-17.5 **Cu** 0.30 max **Fe** rem **Mn** 0.35 max **Ni** 39.0-44.0 **P** 0.020 max **S** 0.015 max **Si** 0.35 max **Ti** 1.5-2.0 | **AMS** 5605; 5606; 5701; 5702; 5703 |
| **N09777** | Ni-Fe-Cr-Mo-Ti Alloy Precipitation Hardenable (PH7) | **Al** 0.35 max **C** 0.03 max **Cb** 0.1 max **Cr** 14.0-19.0 **Fe** rem **Mn** 1.00 max **Mo** 2.5-5.5 **Ni** 34.0-42.0 **P** 0.030 max **S** 0.010 max **Si** 0.50 max **W** 2.0-3.0 | **ASTM** B805 **Other:** NACE MR-01-75 |

*Boxed entries are no longer active and are retained for reference purposes only.

The chemical compositions listed are for identification purposes and should not be used in lieu of the cross referenced specifications.

| UNIFIED NUMBER | DESCRIPTION | CHEMICAL COMPOSITION | CROSS REFERENCE SPECIFICATIONS |
|---|---|---|---|
| **N09901** | Ni-Fe-Cr Alloy Precipitation Hardenable (Incoloy 901) | **Al** 0.35 max **B** 0.010-0.020 **C** 0.10 max **Cr** 11.00-14.00 **Cu** 0.50 max **Fe** rem **Mn** 1.00 max **Mo** 5.00-7.00 **Ni** 40.00-45.00 **S** 0.030 max **Si** 0.60 max **Ti** 2.35-3.10 | **AMS** 5660; 5661; 5830 **Other:** ISO 9721; 9723 **SAE** J467 (Incoloy 901) |
| **N09902** | Ni-Fe-Cr Alloy Precipitation Hardenable (Ni-Span-C 902) | **Al** 0.30-0.80 **C** 0.06 max **Cr** 4.90-5.75 **Fe** rem **Mn** 0.80 max **Ni** 41.0-43.5 **P** 0.04 max **S** 0.04 max **Si** 1.0 max **Ti** 2.20-2.75 | **AMS** 5221; 5223; 5225 |
| **N09908** | Ni-Fe-Cr-Cb-Ti-Al Alloy Precipitation Hardenable (Incoloy Alloy 908) | **Al** 0.75-1.25 **B** 0.012 max **C** 0.03 max **Cb** 2.7-3.3 **Co** 0.5 max **Cr** 3.75-4.5 **Cu** 0.5 max **Fe** rem **Mn** 1.0 max **Ni** 47.0-51.0 **P** 0.015 max **S** 0.005 max **Si** 0.5 max **Ti** 1.20-1.80 | **ASTM** B872 |
| **N09911** | Combined with N09901 | | |
| **N09925** | Ni-Fe-Cr Alloy Precipitation Hardenable | **Al** 0.10-0.50 **C** 0.03 max **Cb** 0.50 max **Cr** 19.5-23.5 **Cu** 1.50-3.00 **Fe** 22.0 min **Mn** 1.00 max **Mo** 2.50-3.50 **Ni** 38.0-46.0 **S** 0.03 max **Si** 0.50 max **Ti** 1.90-2.40 | **Other:** NACE MR-01-75 |
| **N09926** | Ni-Fe-Cr-Cu Alloy Precipitation Hardenable (Incoloy 926) | **Al** 0.3 max **C** 0.04 max **Cr** 14.0-18.0 **Cu** 3.5-5.5 **Fe** 39.0 min **Mn** 1.5 max **Mo** 2.5-3.5 **Ni** 26.0-30.0 **S** 0.015 max **Si** 0.75 max **Ti** 1.5-2.3 | |
| **N09979** | Ni-Fe-Cr Alloy Precipitation Hardenable (D979) | **Al** 0.75-1.30 **B** 0.008-0.016 **C** 0.08 max **Cr** 14.00-16.00 **Fe** rem **Mn** 0.75 max **Mo** 3.75-4.50 **Ni** 42.00-48.00 **P** 0.015 max **S** 0.015 max **Si** 0.75 max **Ti** 2.70-3.30 **W** 3.75-4.50 **Zr** 0.050 max | **AMS** 5746 **SAE** J467 (D-979) |
| **N10001** | Ni-Mo Alloy Solid Solution Strengthened (Hastelloy B) | **C** 0.12 max **Co** 2.50 max **Cr** 1.00 max **Fe** 6.00 max **Mn** 1.00 max **Mo** 26.0-33.0 **Ni** rem **P** 0.040 max **S** 0.030 max **Si** 1.00 max **V** 0.60 max | **AMS** 5396 **ASME** SB-333; SB-335; SB-366; SB-619; SFA5.14 (ERNiMo-1) **ASTM** B333; B335; B366; B619; B622; B626; F467 (335); F468 (335) **AWS** A5.14 (ERNiMo-1) **SAE** J467 (Hastelloy B) **Other:** NACE MR-01-75 |
| **N10002** | Ni-Mo Alloy Solid Solution Strengthened (Hastelloy C) | **C** 0.08 max **Co** 2.5 max **Cr** 14.5-16.5 **Fe** 4.0-7.0 **Mn** 1.00 max **Mo** 15.0-17.0 **Ni** rem **P** 0.040 max **S** 0.030 max **Si** 1.00 max **V** 0.35 max **W** 3.0-4.5 | **AMS** 5388; 5389; 5530; 5750 **ASTM** F467; F468 **SAE** J467 (Hastelloy C) **Other:** NACE MR0175 |
| **N10003** | Ni-Mo Alloy Solid Solution Strengthened (Hastelloy N) | **Al** 0.50 max **B** 0.010 max **C** 0.04-0.08 **Co** 0.20 max **Cr** 6.0-8.0 **Cu** 0.35 max **Fe** 5.0 max **Mn** 1.00 max **Mo** 15.0-18.0 **Ni** rem **P** 0.015 max **S** 0.020 max **Si** 1.00 max **V** 0.50 max **W** 0.50 max | **AMS** 5607; 5771 **ASME** SB-434; SFA5.14 (ERNiMo-2) **ASTM** B366; B434; B573 **AWS** A5.14 (ERNiMo-2) **SAE** J467 (Hastelloy N) |
| **N10004** | Ni-Mo Alloy Solid Solution Strengthened (Hastelloy W) | **C** 0.12 max **Cr** 4.00-6.00 **Fe** 4.00-7.00 **Mn** 1.00 max **Mo** 23.00-26.00 **Ni** rem **P** 0.050 max **S** 0.050 max **Si** 1.00 max **V** 0.60 max | **AMS** 5755; 5786 **ASME** SFA5.14 (ERNiMo-3) **AWS** A5.14 (ERNiMo-3) **SAE** J467 (Hastelloy W) |
| **N10006** | Ni-Mo-Cr-Fe-W Weld Filler Metal for Surfacing (ERNiCrMo-5A) | **C** 0.12 max **Cr** 14-18 **Fe** 4.0-7.0 **Mn** 1.0 max **Mo** 14-18 **Ni** rem **Si** 1.0 max **W** 3.0-5.0 **V** 0.40 max | **ASME** SFA5.21 (ERNiCrMo-5A) **AWS** A5.21 (ERNiCrMo-5A) |
| **N10008** | Ni-Mo Alloy Welding Filler Metal (ERNiMo-8) (TGS-709S) | **C** 0.10 max **Cr** 0.5-3.5 **Cu** 0.50 max **Mn** 1.0 max **Mo** 18.0-21.0 **Ni** 60.0 min **P** 0.015 max **Si** 0.50 max **S** 0.015 max **W** 2.0-4.0 | **ASME** SFA5.14 (ERNiMo-8) **AWS** A5.14 (ERNiMo-8) |
| **N10009** | Ni-Mo Alloy Welding Filler Metal (ERNiMo-9) (NITTETSU Filler 196) | **Al** 1.0 max **C** 0.10 max **Cu** 0.3-1.3 **Mn** 1.0 max **Mo** 19.0-22.0 **Ni** 65.0 min **P** 0.015 max **Si** 0.50 max **S** 0.015 max **W** 2.0-4.0 | **ASME** SFA5.14 (ERNiMo-9) **AWS** A5.14 (ERNiMo-9) |
| **N10242** | Ni-Mo-Cr Alloy Solid Solution Strengthened (HAYNES 242) | **Al** 0.50 max **B** 0.006 max **C** 0.03 max **Co** 1.00 max **Cr** 7.0-9.0 **Cu** 0.50 max **Fe** 2.0 max **Mn** 0.80 max **Mo** 24.0-26.0 **Ni** rem **P** 0.030 max **S** 0.015 max **Si** 0.80 max | **ASTM** B366; B434; B564; B573; B619; B622; B626 |

The chemical compositions listed are for identification purposes and should not be used in lieu of the cross referenced specifications.

| UNIFIED NUMBER | DESCRIPTION | CHEMICAL COMPOSITION | CROSS REFERENCE SPECIFICATIONS |
|---|---|---|---|
| **N10276** | Ni-Mo-Cr Alloy Solid Solution Strengthened (Alloy C276) | **C** 0.02 max **Co** 2.5 max **Cr** 14.5-16.5 **Fe** 4.0-7.0 **Mn** 1.0 max **Mo** 15.0-17.0 **Ni** rem **P** 0.030 max **S** 0.030 max **Si** 0.08 max **V** 0.35 max **W** 3.0-4.5 | **ASME** SB-366; SB-574; SB-575; SB-619; SB-622; SB-626; SFA5.14 (ERNiCrMo-4) **ASTM** B366; B564; B574; B575; B619; B622; B626; F467 (276); F468 (276) **AWS** A5.14 (ERNiCrMo-4) **Other:** NACE MR-01-75 |
| **N10624** | Low Carbon Ni-Mo-Cr-Fe Alloy Solid Solution Strengthened (Alloy B-10) | **Al** 0.5 max **C** 0.01 max **Co** 1.0 max **Cr** 6.0-10.0 **Cu** 0.5 max **Fe** 5.0-8.0 **Mn** 1.0 max **Mo** 21.0-25.0 **Ni** rem **P** 0.025 max **S** 0.01 max **Si** 0.10 max | **ASTM** B333; B335; B366; B564; B619; B622; B626 **Other:** W.Nr. 2.4710 |
| **N10629** | Ni-Mo Alloy Solid Solution Strengthened (B-4) | **Al** 0.1-0.5 **C** 0.01 max **Co** 2.5 max **Cr** 0.5-1.5 **Cu** 0.5 max **Fe** 1.0-6.0 **Mn** 1.5 max **Mo** 26.0-30.0 **Ni** rem **P** 0.04 max **S** 0.01 max **Si** 0.05 max | **ASTM** B333; B335; B366; B564; B619; B622; B626 |
| **N10665** | Ni-Mo Alloy (Hastelloy B-2) | **C** 0.02 max **Co** 1.0 max **Cr** 1.0 max **Fe** 2.0 max **Mn** 1.0 max **Mo** 26.0-30.0 **Ni** rem **P** 0.04 **S** 0.03 max **Si** 0.10 max | **ASME** SB-333; SB-335; SB-366; SB-619; SB-622; SB-626; SFA5.14 (ERNiMo-7) **ASTM** B333; B335; B366; B564; B619; B622; B626 **AWS** A5.14 (ERNiMo-7) |
| **N10675** | Ni-Mo-Cr-Fe Alloy (Hastelloy B-3) | **Al** 0.50 max **C** 0.01 max **Cb** 0.20 max **Co** 3.0 max **Cr** 1.0-3.0 **Cu** 0.20 max **Fe** 1.0-3.0 **Mn** 3.0 max **Mo** 27.0-32.0 **Ni** 65.0 min **P** 0.030 max **S** 0.010 max **Si** 0.10 max **Ta** 0.20 max **Ti** 0.20 max **V** 0.20 max **W** 3.0 max **Zr** 0.10 max **Other** Ni+Mo 94.0-98.0 | **ASME** SFA5.14 (ERNiMo-10) **ASTM** B333; B335; B366; B564; B619; B622; B626 **AWS** A5.14 (ERNiMo-10) |
| **N12160** | Ni-Co-Cr-Si Alloy Solid Solution Strengthened (HR-160) | **C** 0.15 max **Co** 27.0-33.0 **Cr** 26.0-30.0 **Fe** 3.5 max **Mn** 1.5 max **Mo** 1.0 max **Nb** 1.00 max **Ni** rem **P** 0.030 max **S** 0.015 max **Si** 2.4-3.0 **Ti** 0.20-0.80 **W** 1.00 max | **ASTM** B366; B435; B564; B572; B619; B622; B626 |
| **N13009** | Ni-W-Co-Cr Alloy Castings | **Al** 4.75-5.25 **B** 0.010-0.020 **Bi** 0.00005 max **C** 0.12-0.17 **Cb** 0.75-1.25 **Co** 9.00-11.00 **Cr** 8.00-10.00 **Cu** 0.10 max **Fe** 1.50 max **Mn** 0.20 max **Ni** rem **Pb** 0.0010 max **S** 0.015 max **Si** 0.20 max **Ti** 1.75-2.25 **W** 11.5-13.5 **Zr** 0.03-0.08 | **AMS** 5407 |
| **N13010** | Ni-Co-Cr-Mo Alloy Castings | **Al** 5.75-6.25 **B** 0.010-0.020 **Bi** 0.00005 max **C** 0.08-0.13 **Cb** 0.10 max **Co** 9.50-10.50 **Cr** 7.50-8.50 **Fe** 0.35 max **Mn** 0.20 max **Mo** 5.75-6.25 **Ni** rem **P** 0.015 max **Pb** 0.0005 max **S** 0.015 max **Si** 0.25 max **Ta** 4.00-4.50 **Ti** 0.80-1.20 **W** 0.10 max **Zr** 0.05-0.10 | **AMS** 5405; 5406 |
| **N13017** | Co-Cr-Mo-Ni Alloy (Astroloy M) | **Al** 3.85-4.15 **B** 0.020-0.030 **Bi** 0.00005 max **C** 0.02-0.06 **Co** 16.0-18.0 **Cr** 14.0-16.0 **Cu** 0.10 max **Fe** 0.50 max **Mn** 0.15 max **Mo** 4.50-5.50 **N** 0.0050 max **Ni** rem **O** 0.010 max **P** 0.015 max **Pb** 0.0002 max **S** 0.015 max **Si** 0.20 max **Ti** 3.35-3.65 **W** 0.05 max **Zr** 0.06 max | **AMS** 5851; 5852; 5882 **SAE** J467 (Astroloy) |
| **N13020** | Co-Cr-Mo-Ni Alloy | **Al** 3.75-4.75 **B** 0.025-0.035 **Bi** 0.00005 max **C** 0.03-0.10 **Co** 17.0-20.0 **Cr** 14.0-16.0 **Cu** 0.10 max **Fe** 2.00 max **Mn** 0.15 max **Mo** 4.50-5.50 **Ni** rem **Ti** 2.75-3.75 **Zr** 0.06 max | **AMS** 5846 |
| **N13021** | Ni-Co-Cr Alloy, Precipitation Hardenable | **Ag** 0.0005 max **Al** 4.5-4.9 **B** 0.003-0.010 **Bi** 0.0001 max **C** 0.12-0.17 **Cr** 14.0-15.7 **Co** 18.0-22.0 **Cu** 0.2 max **Fe** 1.0 max **Mn** 1.0 max **Mo** 4.5-5.5 **Ni** rem **Pb** 0.0015 max **S** 0.015 max **Si** 1.0 max **Ti** 0.9-1.5 | **Other:** ISO 9721, 9723 |

The chemical compositions listed are for identification purposes and should not be used in lieu of the cross referenced specifications.

| UNIFIED NUMBER | DESCRIPTION | CHEMICAL COMPOSITION | CROSS REFERENCE SPECIFICATIONS |
|---|---|---|---|
| **N13100** | Ni-Co Alloy Precipitation Hardenable (IN-100) | **Al** 5.00-6.00 **B** 0.01-0.02 **C** 0.15-0.20 **Co** 13.0-17.0 **Cr** 8.0-11.0 **Fe** 1.0 max **Mn** 0.20 max **Mo** 2.0-4.0 **Ni** rem **S** 0.015 max **Si** 0.20 max **Ti** 4.50-5.00 **V** 0.70-1.20 **Zr** 0.03-0.09 | **AMS** 5397 **SAE** J467 (IN 100) |
| **N13246** | Ni-W-Co-Cr Precipitation Hardenable (MAR-M-Alloy, Hf modified) | **Al** 5.25-5.75 **B** 0.01-0.02 **C** 0.13-0.17 **Cr** 8.0-10.0 **Co** 9.0-10.0 **Cu** 0.10 max **Hf** 1.5-2.0 **Fe** 1.0 max **Mn** 0.20 max **Mo** 2.25-2.75 **Ni** rem **S** 0.015 max **Si** 0.20 max **Ta** 1.25-1.75 **Ti** 1.25-1.75 **W** 9.0-11.0 **Zr** 0.03-0.08 | |
| **N14052** | Ni-Fe Alloy (Alloy No. 52) | **Al** 0.10 max **C** 0.05 max **Co** 0.50 max **Cr** 0.25 max **Fe** rem **Mn** 0.60 max **Ni** 50.5 nom **P** 0.025 max **S** 0.025 max **Si** 0.30 max | **ASTM** F30 (52) |
| **N14076** | Ni-Fe Ferromagnetic Alloy | **C** 0.05 max **Co** 0.50 max **Cr** 2.0-3.0 **Cu** 4.0-6.0 **Fe** rem **Mn** 1.5 max **Mo** 0.50 max **Ni** 75.0-78.0 **P** 0.02 max **S** 0.01 max **Si** 0.50 max | **ASTM** A753 (Alloy 3) |
| **N14080** | Ni-Fe Ferromagnetic Alloy | **C** 0.05 max **Co** 0.50 max **Cr** 0.30 max **Cu** 0.30 max **Fe** rem **Mn** 0.80 max **Mo** 3.5-6.0 **Ni** 79.0-82.0 **P** 0.02 max **S** 0.01 max **Si** 0.50 max | **ASTM** A753 (Alloy 4) |
| **N19903** | Ni-Fe-Co Alloy Precipitation Hardenable (Incoloy 903) | **Al** 0.30-1.15 **B** 0.012 max **C** 0.06 max **Cb** 2.40-3.50 **Co** 13.0-17.0 **Cr** 1.00 max **Cu** 0.50 max **Fe** rem **Mn** 1.00 max **Ni** 36.0-40.0 **S** 0.015 max **Si** 0.35 max **Ti** 1.00-1.25 | **AMS** 5806 |
| **N19907** | Ni-Fe-Co Alloy Precipitation Hardenable | **Al** 0.20 max **B** 0.012 max **C** 0.06 max **Cb** 4.3-5.2 **Co** 12.0-16.0 **Cr** 1.0 max **Cu** 0.5 max **Fe** rem **Mn** 1.0 max **Ni** 35.0-40.0 **P** 0.015 max **S** 0.015 max **Si** 0.35 max **Ti** 1.2-1.8 | **AMS** 5802 |
| **N19909** | Ni-Fe-Co Alloy Precipitation Hardenable | **Al** 0.15 max **B** 0.012 max **C** 0.06 max **Cb** 4.3-5.2 **Co** 12.0-16.0 **Cr** 1.0 max **Cu** 0.5 max **Fe** rem **Mn** 1.0 max **Ni** 35.0-40.0 **P** 0.015 max **S** 0.015 max **Si** 0.25-0.50 **Ti** 1.3-1.8 | **AMS** 5884; 5892; 5893 |
| **N22000** | Ni-Si-Cu Casting Alloy | **C** 0.12 max **Cr** 1.0 max **Cu** 2.0-4.0 **Mn** 1.50 max **Ni** rem **P** 0.030 max **S** 0.030 max **Si** 8.5-10.0 | **ACI** CZ12SiCu |
| **N24025** | Ni-Cu Casting Alloy | **C** 0.25 max **Cu** 27.0-33.0 **Fe** 3.50 max **Mn** 1.50 max **Ni** rem **P** 0.03 max **S** 0.03 max **Si** 3.5-4.5 | **ASTM** A494 (M-255) |
| **N24030** | Ni-Cu Casting Alloy | **C** 0.30 max **Cu** 27.0-33.0 **Fe** 3.50 max **Mn** 1.50 max **Ni** rem **P** 0.03 max **S** 0.03 max **Si** 2.7-3.7 | **ASTM** A494 (M-30H) |
| **N24130** | Ni-Cu Casting Alloy | **C** 0.30 max **Cb** 1.0-3.0 **Cu** 26.0-33.0 **Fe** 3.50 max **Mn** 1.50 max **Ni** rem **P** 0.03 max **S** 0.03 max **Si** 1.0-2.0 | **ASTM** A494 (M-30C) |
| **N24135** | Ni-Cu Casting Alloy | **C** 0.35 max **Cb** 0.5 max **Cu** 26.0-33.0 **Fe** 3.50 max **Mn** 1.50 max **Ni** rem **P** 0.03 max **S** 0.03 max **Si** 1.25 max | **ASTM** A494 (M-35-1) |
| **N26022** | Ni-Cr-Mo Casting Alloy | **C** 0.02 max **Cr** 20.0-22.5 **Fe** 2.0-6.0 **Mn** 1.00 max **Mo** 12.5-14.5 **Ni** rem **P** 0.025 max **S** 0.025 max **Si** 0.80 max **V** 0.35 max **W** 2.5-3.5 | **ACI** CX2MW **ASTM** A494 |
| **N26055** | Ni-Cr-Bi-Sn Casting Alloy | **Bi** 3.0-5.0 **C** 0.05 max **Cr** 11.0-14.0 **Fe** 2.0 max **Mn** 1.5 max **Mo** 2.0-3.5 **Ni** rem **P** 0.03 max **S** 0.03 max **Si** 0.5 max **Sn** 3.0-5.0 | **ACI** CY5SnBiM **ASTM** A494 |

The chemical compositions listed are for identification purposes and should not be used in lieu of the cross referenced specifications.

| UNIFIED NUMBER | DESCRIPTION | CHEMICAL COMPOSITION | CROSS REFERENCE SPECIFICATIONS |
|---|---|---|---|
| **N26455** | Ni-Cr-Mo Casting Alloy | **C** 0.02 max **Cr** 15.0-17.5 **Fe** 2.0 max **Mn** 1.00 max **Mo** 15.0-17.5 **Ni** rem **P** 0.03 max **S** 0.03 max **Si** 0.80 max **W** 1.0 max | **ACI** CW2M **ASTM** A494; A494 (CW2M); A990 (CW2M) |
| **N26625** | Ni-Cr-Mo Casting Alloy | **C** 0.06 max **Cb** 3.15-4.50 **Cr** 20.0-23.0 **Fe** 5.0 max **Mn** 1.00 max **Mo** 8.0-10.0 **Ni** rem **P** 0.015 max **S** 0.015 max **Si** 1.00 max | **ACI** CW6MC **ASTM** A494 |
| * **N26641** | Ni-Cr-Fe-Si Casting Alloy (Stoody 41H, Weartech 41, WT-41) | **B** 1.2-2.0 **C** 0.2-0.8 **Cr** 10-15 **Fe** 2.0-5.0 **Ni** rem **Si** 1.2-5.0 | |
| **N26985** | Ni-Cr-Fe-Mo Casting Alloy | **C** 0.02 max **Co** 5.0 max **Cr** 21.5-23.5 **Cu** 1.5-2.5 **Fe** 18.0-21.0 **Mn** 1.00 max **Mo** 6.0-8.0 **Ni** rem **P** 0.025 max **S** 0.030 max **Si** 1.00 max **W** 1.50 max **Other** (Cb+Ta) 0.5 max | **ACI** CT2MCoCu |
| * **N28820** | Replaced by N08826 | | |
| **N28825** | Ni-Fe-Cr-Mo-Cu-Si Casting Alloy | **C** 0.05 max **Cb** 0.7-1.0 **Cr** 19.5-23.5 **Cu** 1.5-3.0 **Fe** 28.0-32.0 **Mn** 1.00 max **Mo** 2.5-3.5 **Ni** rem **P** 0.030 max **S** 0.030 max **Si** 0.75-1.20 | **ACI** CU5MCuSiC |
| **N30002** | Ni-Mo-Cr Casting Alloy | **C** 0.12 max **Cr** 15.5-17.5 **Fe** 4.5-7.5 **Mn** 1.00 max **Mo** 16.0-18.0 **Ni** rem **P** 0.040 max **S** 0.030 max **Si** 1.00 max **V** 0.20-0.40 **W** 3.75-5.25 | **ACI** CW12MW **ASTM** A494 |
| **N30007** | Ni-Mo Casting Alloy | **C** 0.07 max **Cr** 1.0 max **Fe** 3.00 max **Mn** 1.00 max **Mo** 30.0-33.0 **Ni** rem **P** 0.040 max **S** 0.030 max **Si** 1.00 max | **ASTM** A494 |
| **N30012** | Ni-Mo Casting Alloy | **C** 0.12 max **Cr** 1.00 max **Fe** 4.0-6.0 **Mn** 1.00 max **Mo** 26.0-30.0 **Ni** rem **P** 0.040 max **S** 0.030 max **Si** 1.00 max **V** 0.20-0.60 | **ASTM** A494 (N-12MV) |
| **N30107** | Ni-Mo-Cr Casting Alloy | **C** 0.07 max **Cr** 17.0-20.0 **Fe** 3.0 max **Mn** 1.00 max **Mo** 17.0-20.0 **Ni** rem **P** 0.040 max **S** 0.030 max **Si** 1.00 max | **ACI** CW6M **ASTM** A494 |
| **N94620** | Fe-Ni-Co Metal-to-Ceramic Sealing Alloy | **Al** 0.01 max **C** 0.02 max **Cr** 0.03 max **Co** 25 nom **Cu** 0.20 max **Fe** rem **Mg** 0.01 max **Mn** 0.15 max **Mo** 0.06 max **Ni** 27 nom **P** 0.006 max **S** 0.006 max **Si** 0.15 max **Ti** 0.01 max **Zr** 0.01 max | **ASTM** New specification pending |
| **N94630** | Fe-Ni-Co Metal-to-Ceramic Sealing Alloy | **Al** 0.01 max **C** 0.02 max **Cr** 0.03 max **Co** 17 nom **Cu** 0.20 max **Fe** rem **Mg** 0.01 max **Mn** 0.35 max **Mo** 0.06 max **Ni** 29 nom **P** 0.006 max **S** 0.006 max **Si** 0.15 max **Ti** 0.01 max **Zr** 0.01 max | **ASTM** New specification pending |
| **N99600** | Ni-Cr-B Brazing Filler Metal (BNi-1) | **Al** 0.05 max **B** 2.75-3.50 **C** 0.6-0.9 **Co** 0.10 max **Cr** 13.0-15.0 **Fe** 4.0-5.0 **Ni** rem **P** 0.02 max **S** 0.02 max **Se** 0.005 max **Si** 4.0-5.0 **Ti** 0.05 max **Zr** 0.05 max **Other** 0.50 max | **AMS** 4775 **ASME** SFA-5.8 (BNi-1) **AWS** A5.8 (BNi-1) **MIL SPEC** MIL-B-7883 (BNi-1) |
| **N99610** | Ni-Cr-B Brazing Filler Metal (BNi-1a) | **Al** 0.05 max **B** 2.75-3.50 **C** 0.06 **Co** 0.10 max **Cr** 13.0-15.0 **Fe** 4.0-5.0 **Ni** rem **P** 0.02 max **S** 0.02 max **Se** 0.005 max **Si** 4.0-5.0 **Ti** 0.05 max **Zr** 0.05 max **Other** 0.50 max | **AMS** 4776 **ASME** SFA5.8 (BNi-1a) **AWS** A5.8 (BNi-1a) **MIL SPEC** MIL-B-7883 (BNi-1a) |

*Boxed entries are no longer active and are retained for reference purposes only.

The chemical compositions listed are for identification purposes and should not be used in lieu of the cross referenced specifications.

| UNIFIED NUMBER | DESCRIPTION | CHEMICAL COMPOSITION | CROSS REFERENCE SPECIFICATIONS |
|---|---|---|---|
| **N99612** | Ni-Cr-B Brazing Filler Metal (BNi-9) | **Al** 0.05 max **B** 3.25-4.0 **C** 0.06 max **Co** 0.10 max **Cr** 13.5-16.5 **Fe** 1.5 max **Ni** rem **P** 0.02 max **S** 0.02 max **Se** 0.005 max **Ti** 0.05 max **Zr** 0.05 max **Other** total 0.50 max | **ASME** SFA5.8 (BNi-9) **AWS** A5.8 (BNi-9) |
| **N99620** | Ni-Cr-B Brazing Filler Metal (BNi-2) | **Al** 0.05 max **B** 2.75-3.50 **C** 0.06 max **Co** 0.10 max **Cr** 6.0-8.0 **Fe** 2.5-3.5 **Ni** rem **P** 0.02 **S** 0.02 max **Se** 0.005 max **Si** 4.0-5.0 **Ti** 0.05 max **Zr** 0.05 max **Other** 0.50 max | **AMS** 4777 **ASME** SFA-5.8 (BNi-2) **AWS** A5.8 (BNi-2) **MIL SPEC** MIL-B-7883 (BNi-2) |
| **N99622** | Ni-W-Cr-B Brazing Filler Metal (BNi-10) | **Al** 0.05 max **B** 2.0-3.0 **C** 0.40-0.55 **Co** 0.10 max **Cr** 10.0-13.0 **Fe** 2.5-4.5 **Ni** rem **P** 0.02 max **S** 0.02 max **Se** 0.005 max **Si** 3.0-4.0 **Ti** 0.05 max **W** 15.0-17.0 **Zr** 0.05 max **Other** total 0.50 max | **ASME** SFA5.8 (BNi-10) **AWS** A5.8 (BNi-10) |
| **N99624** | Ni-W-Cr-B Brazing Filler Metal (BNi-11) | **Al** 0.05 max **B** 2.2-3.1 **C** 0.30-0.50 **Co** 0.10 max **Cr** 9.0-11.75 **Fe** 2.5-4.0 **Ni** rem **P** 0.02 max **S** 0.02 max **Se** 0.005 **Si** 3.25-4.25 **Ti** 0.05 max **W** 11.50-12.75 **Zr** 0.05 max **Other** total 0.50 max | **ASME** SFA5.8 (BNi-11) **AWS** A5.8 (BNi-11) |
| **N99630** | Ni-Si-B Brazing Filler Metal (BNi-3) | **Al** 0.05 max **B** 2.75-3.50 **C** 0.06 max **Co** 0.10 max **Fe** 0.5 max **Ni** rem **P** 0.02 max **S** 0.02 max **Se** 0.005 max **Si** 4.0-5.0 **Ti** 0.05 max **Zr** 0.05 max **Other** 0.50 max | **AMS** 4778 **ASME** SFA5.8 (BNi-3) **AWS** A5.8 (BNi-3) **MIL SPEC** MIL-B-7883 (BNi-3) |
| **N99640** | Ni-Si-B Brazing Filler Metal (BNi-4) | **Al** 0.05 max **B** 1.5-2.2 **C** 0.06 max **Co** 0.10 max **Fe** 1.5 max **Ni** rem **P** 0.02 max **S** 0.02 max **Se** 0.005 max **Si** 3.0-4.0 **Ti** 0.05 max **Zr** 0.05 max **Other** 0.50 max | **AMS** 4779 **ASME** SFA5.8 (BNi-4) **AWS** A5.8 (BNi-4) **MIL SPEC** MIL-B-7883 (BNi-4) **Other:** NACE MR-01-75 |
| **N99644** | NiCr-B Weld Filler Metal for Hard Surfacing (ERNiCr-A) (Colmonoy 4) | **B** 2.00-3.00 **C** 0.30-0.60 **Co** 1.50 max **Cr** 8.0-14.0 **Fe** 1.25-3.25 **Ni** rem **Se** 0.005 max **Si** 1.25-3.25 | **ASME** SFA5.21 (ERNiCr-A) **AWS** A5.21 (ERNiCr-A) **Other:** NACE MR-01-75 |
| **N99645** | Ni-Cr-B Weld Filler Metal for Hard Surfacing (ERNiCr-B) (Colmonoy 5) | **B** 2.00-4.00 **C** 0.40-0.80 **Co** 1.25 max **Cr** 10.0-16.0 **Fe** 3.00-5.00 **Ni** rem **Se** 0.005 max **Si** 3.00-5.00 | **ASME** SFA5.21 (ERNiCr-B) **AWS** A5.21 (ERNiCr-B) **MIL SPEC** MIL-R-17131 (MIL-ERNiCr-B) **Other:** NACE MR-01-75 |
| **N99646** | Ni-Cr-B Weld Filler Metal for Hard Surfacing (ERNiCr-C) (Colmonoy 6) | **B** 2.50-4.50 **C** 0.50-1.00 **Co** 1.00 max **Cr** 12.0-18.0 **Fe** 3.50-5.50 **Ni** rem **Se** 0.005 max **Si** 3.50-5.50 | **ASME** SFA5.21 (ERNiCr-C) **AWS** A5.21 (ERNiCr-C) **MIL SPEC** MIL-R-17131 (MIL-ERNiCr-C) **Other:** NACE MR-01-75 |
| **N99647** | Ni-Cr-W-B Weld Filler Metal for Hard Surfacing (ERNiCr-D) | **B** 0.35-0.60 **C** 0.6-1.1 **Cr** 8.0-12.0 **Co** 0.1 max **Fe** 1.0-5.0 **Ni** rem **Si** 4.0-6.6 **W** 1.0-3.0 | **ASME** SFA5.21 (ERNiCr-D) **AWS** A5.21 (ERNiCr-D) |
| **N99648** | Ni-Cr-W-B Weld Filler Metal for Hard Surfacing (ERNiCr-E) | **B** 0.7-1.4 **C** 0.1-0.5 **Cr** 15-20 **Co** 0.1 max **Fe** 3.5-7.5 **Si** 5.5-8.0 **Sn** 0.5-0.9 **W** 0.5-1.5 | **ASME** SFA5.21 (ERNiCr-E) **AWS** A5.21 (ERNiCr-E) |
| **N99650** | Ni-Cr-Si Brazing Filler Metal (BNi-5) | **Al** 0.05 max **B** 0.03 max **C** 0.10 **Co** 0.10 max **Cr** 18.5-19.5 **Ni** rem **P** 0.02 max **S** 0.02 max **Se** 0.005 max **Si** 9.75-10.50 **Ti** 0.05 max **Zr** 0.05 max **Other** 0.50 max | **AMS** 4782 **ASME** SFA5.8 (BNi-5) **AWS** A5.8 (BNi-5) **MIL SPEC** MIL-B-7883 (BNi-5) |
| **N99651** | Ni-Cr-Si-B Brazing Filler Metal (BNi5a) | **Al** 0.05 max **B** 1.0-1.5 **C** 0.10 max **Co** 0.10 max **Cr** 18.5-19.5 **Fe** 0.5 max **Ni** rem **P** 0.02 max **S** 0.02 max **Se** 0.005 max **Si** 7.0-7.5 **Ti** 0.05 max **Zr** 0.05 max | **ASME** SFA5.8 (BNi-5a) **AWS** A5.8 (BNi-5a) |
| **N99652** | Ni-Cr-Si-B Brazing Filler Metal (BNi-5b) | **Al** 0.05 max **B** 1.1-1.6 **C** 0.06 max **Cr** 14.5-15.5 **Co** 1.0 max **Fe** 1.0 max **Ni** rem **P** 0.02 max **Se** 0.005 max **Si** 7.0-7.5 **S** 0.02 max **Ti** 0.05 max **Zr** 0.05 max | **ASME** SFA5.8 (BNi-5b) **AWS** A5.8 (BNi-5b) |

The chemical compositions listed are for identification purposes and should not be used in lieu of the cross referenced specifications.

| UNIFIED NUMBER | DESCRIPTION | CHEMICAL COMPOSITION | CROSS REFERENCE SPECIFICATIONS |
|---|---|---|---|
| **N99700** | Ni-P Brazing Filler Metal (BNi-6) | **Al** 0.05 max **C** 0.01 max **Co** 0.10 max **Ni** rem **P** 10.0-12.0 **S** 0.02 max **Se** 0.005 max **Ti** 0.05 max **Zr** 0.05 max **Other** 0.50 max | **ASME** SFA5.8 (BNi-6) **AWS** A5.8 (BNi-6) |
| **N99710** | Ni-Cr-P Brazing Filler Metal (BNi-7) | **Al** 0.05 max **B** 0.01 max **C** 0.08 max **Co** 0.10 max **Cr** 13.0-15.0 **Fe** 0.2 max **Mn** 0.04 max **Ni** rem **P** 9.7-10.5 **S** 0.02 max **Se** 0.005 max **Si** 0.10 max **Ti** 0.05 max **Zr** 0.05 max **Other** 0.50 max | **ASME** SFA5.8 (BNi-7) **AWS** A5.8 (BNi-7) |
| **N99720** | Ni-Cr-P Brazing Filler Metal (BNi-12) | **Al** 0.05 max **B** 0.02 max **C** 0.06 max **Cr** 24.0-26.0 **Co** 0.10 max **Fe** 0.2 max **Ni** rem **P** 9.0-11.0 **Si** 0.1 max **S** 0.02 max **Ti** 0.05 max **Zr** 0.05 max | **ASME** SFA5.8 (BNi-12) **AWS** A5.8 (BNi-12) |
| **N99800** | Ni-Mn-Si-Cu Brazing Filler Metal (BNi-8) | **Al** 0.05 max **C** 0.10 max **Co** 0.10 max **Cu** 4.0-5.0 **Mn** 21.5-24.5 **Ni** rem **P** 0.02 max **S** 0.02 max **Se** 0.005 max **Si** 6.0-8.0 **Ti** 0.05 max **Zr** 0.05 max **Other** 0.50 max | **ASME** SFA5.8 (BNi-8) **AWS** A5.8 (BNi-8) |
| **N99810** | Ni-Cr-Si-B-Cu-Cb-Mo Brazing Filler Metal (BNi-13) | **Al** 0.05 max **B** 2.75-3.5 **C** 0.06 max **Cr** 7.0-9.0 **Co** 0.10 max **Cb** 1.5-2.5 **Cu** 2.0-3.0 **Fe** 0.4 max **Mo** 1.5-2.5 **Ni** rem **P** 0.02 max **Si** 3.8-4.8 **S** 0.02 max **Ti** 0.05 max **Zr** 0.05 max | **ASME** SFA5.8 (BNi-13) **AWS** A5.8 (BNi-13) |

The chemical compositions listed are for identification purposes and should not be used in lieu of the cross referenced specifications.

# UNS NUMBERS ASSIGNED TO DATE

## With Description of Each Material Covered and References To Documents In Which The Same or Similar Materials are Described

## Pxxxxx Number Series
## Precious Metals and Alloys

# PRECIOUS METALS AND ALLOYS

| UNIFIED NUMBER | DESCRIPTION | CHEMICAL COMPOSITION | CROSS REFERENCE SPECIFICATIONS |
|---|---|---|---|
| **P00010** | Refined Gold | **Ag** 0.001 max **Au** 99.995 min **Bi** 0.001 max **Cr** 0.0003 max **Cu** 0.001 max **Fe** 0.001 max **Mg** 0.001 max **Mn** 0.0003 max **Pb** 0.001 max **Pd** 0.001 max **Si** 0.001 max **Sn** 0.001 max | **ASTM** B562 (99.995) |
| **P00015** | Refined Gold | **As** 0.003 max **Au** 99.99 min **Bi** 0.002 max **Cr** 0.0003 max **Cu** 0.005 max **Fe** 0.002 max **Mg** 0.003 max **Mn** 0.0003 max **Ni** 0.0003 max **Pb** 0.002 max **Pd** 0.005 max **Si** 0.005 max **Sn** 0.001 max | **ASTM** B562 (99.99) |
| **P00016** | Gold | **Au** 99.99 min **Other** Ag+Cu 0.009 max, each, other impurity 0.003 max | **ASTM** F72 |
| **P00020** | Refined Gold | **Ag** 0.035 max **Au** 99.95 min **Cu** 0.02 max **Fe** 0.005 max **Pb** 0.005 max **Pd** 0.02 max **Other** Ag+Cu 0.04 max | **AMS** 7731 **ASTM** B562 (99.95) |
| **P00025** | Refined Gold | **Au** 99.5 min | **ASTM** B562 (99.5) |
| **P00032** | 14-Karat Yellow Gold Jewelry Alloy | **Ag** 3.5-4.5 **Au** 58.33 (-0.3) **Cu** 31.08 (+/- 1.0) rem **Fe** 0.05 (+/- 0.5) **Ni** 0.10 (+/- 0.5) **Si** 0.01 (+/- 0.5) **Zn** 6.43 (+/- 0.5) | |
| **P00100** | Gold Alloy (990) Gold | **Au** 99.0 min **Ti** 1.0 (0.8 min) | **Other:** US Code of Federal Regulations, Title 15, Sec. 295 |
| **P00105** | 10-Karat Yellow Gold Jewelry Alloy | **Ag** 5.85 (+/- 0.5) **Au** 41.817 (-0.3) **Cu** 38.5 (rem) **Ni** 1.15 (+/- 0.5) **Zn** 12.68 (+/- 1.0) | **Other:** US Code of Federal Regulations, Title 15, Sec. 295 |
| **P00110** | 10-Karat Yellow Gold Jewelry Alloy | **Ag** 11.66 (+/- 1.0) **Au** 41.70 (-0.3) **B** 0.02 **Cu** 40.81 (rem) **Si** 0.03 **Zn** 5.83 (+/- 0.5) | **Other:** US Code of Federal Regulations, Title 15, Sec. 295 |
| **P00115** | 10-Karat Yellow Gold Jewelry Alloy | **Ag** 5.50 (+/- 0.5) **Au** 41.70 (-0.3) **Cu** 43.8 (rem) **Zn** 9.00 (+/- 0.5) | **Other:** US Code of Federal Regulations, Title 15, Sec. 295 |
| **P00120** | 10-Karat Yellow Gold Jewelry Alloy | **Ag** 6.60 (+/- 0.5) **Au** 41.70 (-0.3) **Cu** 48.00 (rem) **Zn** 3.70 (+/- 0.5) | **Other:** US Code of Federal Regulations, Title 15, Sec. 295 |
| **P00125** | 10-Karat White Gold Jewelry Alloy | **Au** 41.817 (-0.3) **Cu** 46.683 (rem) **Ni** 1.15 (+/- 0.5) **Zn** 10.35 (+/- 1.0) | **Other:** US Code of Federal Regulations, Title 15, Sec. 295 |
| **P00130** | 10-Karat White Gold Jewelry Alloy | **Au** 41.70 (-0.3) **Cu** 32.82 (rem) **Ni** 17.08 (+/- 1.0) **Zn** 8.40 (+/- 0.5) | **Other:** US Code of Federal Regulations, Title 15, Sec. 295 |
| **P00135** | 10-Karat White Gold Jewelry Alloy | **Au** 41.66 (-0.3) **Co** 1.97 (+/- 0.5) **Cu** 29.20 (rem) **Ni** 15.05 (+/- 1.0) **Zn** 12.12 (+/- 0.5) | **Other:** US Code of Federal Regulations, Title 15, Sec. 295 |
| **P00140** | 10-Karat Green Gold Jewelry Alloy | **Ag** 48.90 (+/- 1.0) **Au** 41.70 (-0.3) **Cu** 9.05 (rem) **Zn** 0.35 | **Other:** US Code of Federal Regulations, Title 15, Sec. 295 |
| **P00145** | 10-Karat Red Gold Jewelry Alloy | **Ag** 2.82 (+/- 0.5) **Au** 41.70 (-0.3) **Cu** 55.48 (rem) | **Other:** US Code of Federal Regulations, Title 15, Sec. 295 |
| **P00150** | 14-Karat White Gold Jewelry Alloy | **Ag** 2.00 (+/- 0.5) **Au** 58.484 (-0.3) **Cu** 24.00 (rem) **Ni** 8.00 (+/- 0.5) **Zn** 7.52 (+/- 0.5) | **Other:** US Code of Federal Regulations, Title 15, Sec. 295 |
| **P00155** | 14-Karat White Gold Jewelry Alloy | **Au** 58.33 (-0.3) **Cu** 22.10 (rem) **Ni** 10.80 (+/- 1.0) **Zn** 8.77 (+/- 0.5) | **Other:** US Code of Federal Regulations, Title 15, Sec. 295 |
| **P00160** | 14-Karat White Gold Jewelry Alloy | **Au** 58.33 (-0.3) **Cu** 23.47 (rem) **Ni** 12.21 (+/- 1.0) **Zn** 5.99 (+/- 0.5) | **Other:** US Code of Federal Regulations, Title 15, Sec. 295 |
| **P00165** | 14-Karat White Gold Jewelry Alloy | **Au** 58.33 (-0.3) **Cu** 28.32 (rem) **Ni** 8.55 (+/- 0.5) **Zn** 4.80 (+/- 0.5) | **Other:** US Code of Federal Regulations, Title 15, Sec. 295 |
| **P00170** | 14-Karat Red Gold Jewelry Alloy | **Ag** 2.08 (+/- 0.5) **Au** 58.33 (-0.3) **Cu** 39.59 (rem) | **Other:** US Code of Federal Regulations, Title 15, Sec. 295 |
| **P00175** | 14-Karat Yellow Gold Jewelry Alloy | **Ag** 5.00 (+/- 0.5) **Au** 58.33 (-0.3) **B** 0.02 **Cu** 30.00 (rem) **Si** 0.03 **Zn** 6.63 (+/- 0.5) | **Other:** US Code of Federal Regulations, Title 15, Sec. 295 |

The chemical compositions listed are for identification purposes and should not be used in lieu of the cross referenced specifications.

| UNIFIED NUMBER | DESCRIPTION | CHEMICAL COMPOSITION | CROSS REFERENCE SPECIFICATIONS |
|---|---|---|---|
| **P00180** | 14-Karat Yellow Gold Jewelry Alloy | **Ag** 16.50 (+/- 1.0) **Au** 58.33 (-0.3) **Cu** 24.97 (rem) **Zn** 0.20 | **Other:** US Code of Federal Regulations, Title 15, Sec. 295 |
| **P00185** | 14-Karat Yellow Gold Jewelry Alloy | **Ag** 7.46 (+/- 0.5) **Au** 58.33 (-0.3) **Cu** 26.58 (rem) **Ni** 1.00 **Zn** 6.63 (+/- 0.5) | **Other:** US Code of Federal Regulations, Title 15, Sec. 295 |
| **P00190** | 14-Karat Yellow Gold Jewelry Alloy | **Ag** 10.00 (+/- 1.0) **Au** 58.33 (-0.3) **Cu** 29.67 (rem) **Zn** 2.00 (+/- 0.5) | **Other:** US Code of Federal Regulations, Title 15, Sec. 295 |
| **P00195** | 14-Karat Yellow Gold Jewelry Alloy | **Ag** 24.78 (+/- 1.0) **Au** 58.33 (-0.3) **Cu** 16.75 (rem) **Zn** 0.14 | **Other:** US Code of Federal Regulations, Title 15, Sec. 295 |
| **P00200** | 14-Karat Yellow Gold Jewelry Alloy | **Ag** 21.20 (+/- 1.0) **Au** 58.33 (-0.3) **Cu** 20.17 (rem) **Zn** 0.30 | **Other:** US Code of Federal Regulations, Title 15, Sec. 295 |
| **P00205** | 14-Karat Yellow Gold Jewelry Alloy | **Ag** 8.31 (+/- 0.5) **Au** 58.33 (-0.3) **Cu** 29.19 (rem) **Zn** 4.17 (+/- 0.5) | **Other:** US Code of Federal Regulations, Title 15, Sec. 295 |
| **P00210** | 14-Karat Yellow Gold Jewelry Alloy | **Ag** 4.00 (+/- 0.5) **Au** 58.33 (-0.3) **Cu** 31.24 (rem) **Zn** 6.43 (+/- 0.5) | **Other:** US Code of Federal Regulations, Title 15, Sec. 295 |
| **P00215** | 14-Karat Yellow Gold Jewelry Alloy | **Ag** 24.78 (+/- 1.0) **Au** 58.33 (-0.3) **Cu** 16.75 (rem) **Zn** 0.14 | **Other:** US Code of Federal Regulations, Title 15, Sec. 295 |
| **P00220** | 14-Karat Yellow Gold Jewelry Alloy | **Ag** 21.20 (+/- 1.0) **Au** 58.33 (-0.3) **Cu** 20.17 (rem) **Zn** 0.30 | **Other:** US Code of Federal Regulations, Title 15, Sec. 295 |
| **P00225** | 14-Karat Yellow Gold Jewelry Alloy | **Ag** 4.00 (+/- 0.5) **Au** 58.33 (-0.3) **Cu** 31.24 (rem) **Zn** 6.43 (+/- 0.5) | **Other:** US Code of Federal Regulations, Title 15, Sec. 295 |
| **P00230** | 14-Karat Green Gold Jewelry Alloy | **Ag** 32.50 (+/- 1.0) **Au** 58.33 (-0.3) **Cu** 8.97 (rem) **Zn** 0.20 (+/- 0.5) | **Other:** US Code of Federal Regulations, Title 15, Sec. 295 |
| **P00235** | 14-Karat Green Gold Jewelry Alloy | **Ag** 35.00 (+/- 1.0) **Au** 58.33 (-0.3) **Cu** 6.47 (rem) **Zn** 0.20 | **Other:** US Code of Federal Regulations, Title 15, Sec. 295 |
| **P00250** | 18-Karat Yellow Gold Jewelry Alloy | **Ag** 13.00 (+/- 1.0) **Au** 75.00 (-0.3) **Cu** 12.00 (rem) | **Other:** US Code of Federal Regulations, Title 15, Sec. 295 |
| **P00255** | 18-Karat Yellow Gold Jewelry Alloy | **Ag** 15.00 (+/- 1.0) **Au** 75.00 (-0.3) **Cu** 10.00 (rem) | **Other:** US Code of Federal Regulations, Title 15, Sec. 295 |
| **P00260** | 18-Karat Yellow Gold Jewelry Alloy | **Ag** 19.00 (+/- 1.0) **Au** 75.15 (-0.3) **Cu** 2.85 (rem) **Zn** 3.00 (+/- 0.5) | **Other:** US Code of Federal Regulations, Title 15, Sec. 295 |
| **P00270** | 18-Karat White Gold Jewelry Alloy | **Au** 75.15 (-0.3) **Cu** 9.85 (rem) **Ni** 12.00 (+/- 1.0) **Zn** 3.00 (+/- 0.5) | **Other:** US Code of Federal Regulations, Title 15, Sec. 295 |
| **P00275** | 18-Karat White Gold Jewelry Alloy | **Au** 75.00 (-0.3) **Cu** 1.73 (rem) **Ni** 17.80 (+/- 1.0) **Zn** 5.47 (+/- 0.5) | **Other:** US Code of Federal Regulations, Title 15, Sec. 295 |
| **P00280** | 18-Karat Green Gold Jewelry Alloy | **Ag** 22.50 (+/- 1.0) **Au** 75.00 (-0.3) **Cu** 2.50 (rem) | **Other:** US Code of Federal Regulations, Title 15, Sec. 295 |
| **P00285** | 18-Karat Red Gold Jewelry Alloy | **Ag** 5.00 (+/- 0.5) **Au** 75.00 (-0.3) **Cu** 20.00 (rem) | **Other:** US Code of Federal Regulations, Title 15, Sec. 295 |
| **P00300** | Gold-Palladium-Nickel Brazing Alloy | **Au** 29.5-30.5 **Ni** 35.5-36.5 **Pd** 33.5-34.5 **Other** 0.15 max | **AMS** 4785 **ASME** SFA5.8 (BAu-5) **AWS** A5.8 (BAu-5) |
| **P00350** | Gold Brazing Alloy | **Au** 34.5-35.5 **Cu** rem **Ni** 2.5-3.5 **Other** 0.15 max | **ASME** SFA5.8 (BAu-3) **AWS** A5.8 (BAu-3) |
| **P00375** | Gold Brazing Alloy | **Au** 37.0-38.0 **Cu** rem **Other** 0.15 max | **ASME** SFA5.8 (BAu-1) **AWS** A5.8 (BAu-1) |
| **P00500** | Gold Brazing Alloy | **Au** 49.5-50.5 **Ni** 24.5-25.5 **Pd** 24.5-25.5 | **AMS** 4784 |
| **P00507** | Gold-Palladium-Nickel Brazing Alloy (Vacuum Grade) | **Au** 49.5-50.5 **C** 0.005 max **Cd** 0.001 max **Co** 0.06 max **Ni** 24.5-25.5 **P** 0.002 max **Pb** 0.002 max **Pd** rem **Zn** 0.001 max | **ASME** SFA5.8 (BVAu-7) **AWS** A5.8 (BVAu-7) |
| **P00560** | Gold: 14 Karat Yellow Gold - Old U.S. (Prior to October 1981) | **Ag** 3.70-4.70 **Au** 56.25-56.55 **Cu** 32.30-33.30 **Fe** 0.05 max **Ni** 0.10 max **Si** 0.01 max **Zn** 6.00-7.50 **Other** all elements except Au, Ag, Cu, Zn: max 0.15 | |

The chemical compositions listed are for identification purposes and should not be used in lieu of the cross referenced specifications.

| UNIFIED NUMBER | DESCRIPTION | CHEMICAL COMPOSITION | CROSS REFERENCE SPECIFICATIONS |
|---|---|---|---|
| **P00580** | Gold: 14 Karat Yellow Gold - New U.S. (After September 1981) | **Ag** 3.0-4.0 **Au** 58.03-58.63 **Cu** 30.74-31.74 **Fe** 0.05 max **Ni** 0.10 max **Si** 0.01 max **Zn** 5.68-7.18 **Other** all elements except Au, Ag, Cu, Zn: max 0.15 | |
| **P00691** | Gold Electrical Contact Alloy | **Ag** 23.5-26.5 **Au** 68.0-70.0 **Pt** 5.0-7.0 | **ASTM** B522 (I) |
| **P00692** | Gold Electrical Contact Alloy | **Ag** 24.5-25.5 **Au** 68.5-69.5 **Pt** 5.5-6.5 **S** 0.01 max | **ASTM** B522 (II) |
| **P00700** | Gold Brazing Alloy | **Au** 69.5-70.5 **Ni** 21.5-22.5 **Pd** 7.5-8.5 **Other** 0.15 max | **AMS** 4786 **ASME** SFA5.8 (BAu-6) **AWS** A5.8 (BAu-6) |
| **P00707** | Gold Brazing Alloy replaced by P00807 | | |
| **P00710** | Gold Electrical Contact Alloy | **Ag** 4.0-5.0 **Au** 70.5-72.5 **Cu** 13.5-15.5 **Pt** 8.0-9.0 **Zn** 0.7-1.3 | **ASTM** B541 |
| **P00750** | Gold Electrical Contact Alloy | **Ag** 21.4-22.6 **Au** 74.2-75.8 **Ni** 2.6-3.4 | **ASTM** B477 |
| **P00800** | Gold Brazing Alloy | **Au** 79.5-80.5 **Cu** rem | **ASME** SFA5.8 (BAu-2) **AWS** A5.8 (BAu-2) |
| **P00807** | Gold Brazing Alloy (Vacuum Grade) | **Au** 79.5-80.5 **C** 0.005 max **Cd** 0.001 max **Cu** rem **P** 0.002 max **Pb** 0.002 max **Zn** 0.001 max **Other** 0.15 max | **ASME** SFA5.8 (BVAu-2) **AWS** A5.8 (BVAu-2) |
| **P00820** | Gold Brazing Alloy | **Au** 81.5-82.5 **Ni** rem **Other** 0.15 max | **AMS** 4787 **ASME** SFA5.8 (BAu-4) **AWS** A5.8 (BAu-4) |
| **P00827** | Gold Brazing Alloy (Vacuum Grade) | **Au** 81.5-82.5 **C** 0.005 max **Cd** 0.001 max **Ni** rem **P** 0.002 max **Pb** 0.002 max **Zn** 0.001 max | **ASME** SFA5.8 (BVAu-4) **AWS** A5.8 (BVAu-4) |
| **P00900** | Gold-Silver Alloy | **Ag** 3.0-10.0 **Au** 90.0 min | |
| **P00901** | Gold Electrical Contact Alloy | **Au** 89.0-91.0 **Cu** 9.0-11.0 | **ASTM** B596 |
| **P00927** | Gold-Palladium Brazing Alloy (Vacuum Grade) | **Au** 91.0-93.0 **C** 0.005 max **Cd** 0.001 max **P** 0.002 max **Pb** 0.002 max **Pd** rem **Zn** 0.001 max | **ASME** SFA5.8 (BVAu-8) **AWS** A5.8 (BVAu-8) |
| **P03300** | Palladium Alloy (Paliney No. 7, ""W" Alloy 3350, No. 226 Alloy) | **Ag** 29.0-31.0 **Au** 9.5-10.5 **Cu** 13.5-14.5 **Pd** 34.0-36.0 **Pt** 9.5-10.5 **Other** elements 0.01 max total | **AMS** 7735 |
| **P03350** | Palladium Electrical Contact Alloy | **Ag** 29-31 **Au** 9.5-10.5 **Cu** 13.5-14.5 **Pd** 34.0-36.0 **Pt** 9.5-10.5 **Zn** 0.6-1.2 **Other** total 0.1 max | **ASTM** B540 |
| **P03440** | Palladium Electrical Contact Alloy | **Ag** 37.0-39.0 **Cu** 15.5-16.5 **Ni** 0.8-1.2 **Pd** 43.0-45.0 **Pt** 0.8-1.2 | **ASTM** B563 |
| *** P03590** | Palladium Alloy | **Cu** 39-41 **Pd** 59 min | |
| **P03657** | Palladium-Cobalt Brazing Alloy (Vacuum Grade) | **C** 0.005 max **Cd** 0.001 max **Co** rem **Ni** 0.06 max **P** 0.002 max **Pb** 0.002 max **Pd** 64.0-66.0 **Zn** 0.001 max | **ASME** SFA5.8 (BVPd-1) **AWS** A5.8 (BVPd-1) |
| **P03980** | Refined Palladium | **Ir** 0.05 max **Pd** 99.80 min **Pt** 0.15 max **Rh** 0.10 max **Other** Ru 0.05 max | |
| *** P03990** | Palladium | **Other** Pd 99.07 min | |
| **P03995** | Refined Palladium | **Ag** 0.005 max **Al** 0.005 max **Au** 0.01 max **Ca** 0.005 max **Co** 0.001 max **Cr** 0.001 max **Cu** 0.005 max **Fe** 0.005 max **Mg** 0.005 max **Mn** 0.001 max **Ni** 0.005 max **Pb** 0.005 max **Pd** 99.95 min **Sb** 0.002 max **Si** 0.005 max **Sn** 0.005 max **Zn** 0.0025 max **Other** Pt+Rh+Ru+Ir 0.03 max | **ASTM** B589 (99.95) |

*Boxed entries are no longer active and are retained for reference purposes only.

The chemical compositions listed are for identification purposes and should not be used in lieu of the cross referenced specifications.

| UNIFIED NUMBER | DESCRIPTION | CHEMICAL COMPOSITION | CROSS REFERENCE SPECIFICATIONS |
|---|---|---|---|
| * **P04840** | Platinum Iridium Alloy | **Ir** 13-17 **Pt** 84 min | |
| **P04980** | Refined Platinum | **Ir** 0.05 max **Pt** 99.80 min **Rh** 0.10 max **Other** Pd 0.15 max, Ru 0.05 max | **ASTM** B561 |
| **P04995** | Refined Platinum | **Ag** 0.005 max **Al** 0.005 max **As** 0.005 max **Au** 0.01 max **Bi** 0.005 max **Cd** 0.005 max **Cr** 0.005 max **Cu** 0.01 max **Ir** 0.015 max **Fe** 0.01 max **Mg** 0.005 max **Mn** 0.005 max **Mo** 0.01 max **Ni** 0.005 max **Pb** 0.005 max **Pt** 99.95 min **Rh** 0.03 max **Sb** 0.005 max **Si** 0.01 max **Sn** 0.005 max **Te** 0.005 max **Zn** 0.005 max **Other** Ca 0.005 max, Ru 0.01 max, Pd 0.02 max | **ASTM** B561 (99.95) |
| **P05980** | Rhodium | **As** 0.01 max **Bi** 0.01 max **Cd** 0.01 max **Ir** 0.01 max **Fe** 0.01 max **Pb** 0.01 max **Pd** 0.05 max **Pt** 0.01 max **Rh** 99.8 min **Ru** 0.05 max **Sn** 0.01 max **Zn** 0.01 max | **ASTM** B616 (99.80) |
| **P05981** | Refined Rhodium | **Ag** 0.02 max **Al** 0.01 max **As** 0.005 max **Au** 0.01 max **B** 0.005 max **Bi** 0.005 max **Cd** 0.005 max **Co** 0.005 max **Cr** 0.01 max **Cu** 0.01 max **Fe** 0.01 max **Ir** 0.05 max **Mg** 0.01 max **Mn** 0.005 max **Ni** 0.01 max **Pb** 0.01 max **Pt** 0.05 max **Rh** 99.90 min **Sb** 0.005 max **Si** 0.01 max **Sn** 0.01 max **Te** 0.01 max **Zn** 0.01 max **Other** Pd 0.05 max, Ru 0.05 max, Ca 0.01 max | **ASTM** B616 (99.90) |
| **P05982** | Refined Rhodium | **Ag** 0.005 max **Al** 0.005 max **As** 0.003 max **Au** 0.003 max **B** 0.001 max **Bi** 0.005 max **Cd** 0.005 max **Co** 0.001 max **Cr** 0.005 max **Cu** 0.005 max **Fe** 0.003 max **Ir** 0.02 max **Mg** 0.005 max **Mn** 0.005 max **Ni** 0.003 max **Pb** 0.005 max **Pt** 0.02 max **Rh** 99.95 min **Sb** 0.003 max **Si** 0.005 max **Sn** 0.003 max **Te** 0.005 max **Zn** 0.003 max **Other** Pd 0.005 max, Ru 0.01 max, Ca 0.005 max | **ASTM** B616 (99.95) |
| **P05990** | Rhodium | **Ag** 0.02 max **Al** 0.01 max **As** 0.005 max **Au** 0.01 max **B** 0.005 max **Bi** 0.005 max **Ca** 0.01 max **Cd** 0.005 max **Co** 0.005 max **Cr** 0.01 max **Cu** 0.01 max **Fe** 0.01 max **Ir** 0.05 max **Mg** 0.01 max **Mn** 0.005 max **Ni** 0.01 max **Pb** 0.01 max **Pd** 0.05 max **Pt** 0.05 max **Rh** 99.90 min **Ru** 0.05 max **Sb** 0.005 max **Si** 0.01 max **Sn** 0.01 max **Te** 0.01 max **Zn** 0.01 max | |
| **P05995** | Refined Rhodium | **Ag** 0.005 max **Al** 0.005 max **As** 0.003 max **Au** 0.003 max **B** 0.001 max **Bi** 0.005 max **Ca** 0.005 max **Cd** 0.005 max **Co** 0.001 max **Cr** 0.005 max **Cu** 0.005 max **Fe** 0.003 max **Ir** 0.02 max **Mg** 0.005 max **Mn** 0.005 max **Ni** 0.003 max **Pb** 0.005 max **Pd** 0.005 max **Pt** 0.02 max **Rh** 99.95 min **Ru** 0.01 max **Sb** 0.003 max **Si** 0.005 max **Sn** 0.003 max **Te** 0.005 max **Zn** 0.003 max | |
| **P06100** | Refined Iridium | **As** 0.01 max **Bi** 0.01 max **Cd** 0.01 max **Fe** 0.01 max **Ir** 99.80 min **Pb** 0.02 max **Pt** 0.10 max **Rh** 0.15 max **Si** 0.02 max **Sn** 0.01 max **Other** Pd 0.05 max, Ru 0.05 max | **ASTM** B671 (99.80) |

*Boxed entries are no longer active and are retained for reference purposes only.

The chemical compositions listed are for identification purposes and should not be used in lieu of the cross referenced specifications.

| UNIFIED NUMBER | DESCRIPTION | CHEMICAL COMPOSITION | CROSS REFERENCE SPECIFICATIONS |
|---|---|---|---|
| **P07010** | Refined Silver | **Ag** 99.99 min **Bi** 0.0005 max **Cu** 0.010 max **Fe** 0.001 max **Pb** 0.001 max **Se** 0.0005 max **Te** 0.0005 max **Other** Pd 0.001 max | **ASTM** B413 (99.99) |
| **P07015** | Refined Silver | **Ag** 99.95 min **Bi** 0.001 max **Cu** 0.04 max **Fe** 0.002 max **Pb** 0.015 max | **ASTM** B413 (99.95) **MIL SPEC** MIL-S-13282 |
| **P07016** | Silver | **Ag** 99.95 min **Cu** 0.05 max **P** 0.002 max **Other** total volatile impurities 0.010 max | **ASTM** F106 (TB Ag) |
| **P07017** | Silver Brazing Alloy (Vacuum Grade) (BVAg-0) | **Ag** 99.95 min **C** 0.005 max **Cd** 0.001 max **Cu** 0.05 max **P** 0.002 max **Pb** 0.002 max **Zn** 0.001 max | **ASME** SFA5.8 (BVAg-0) **AWS** A5.8 (BVAg-0) |
| **P07020** | Refined Silver | **Ag** 99.90 min **Bi** 0.001 max **Cu** 0.08 max **Fe** 0.002 max **Pb** 0.025 max **Other** Ag+Cu 99.95 min | **ASTM** B413 (99.90) |
| * **P07200** | Silver Brazing Alloy (BAg-25) | **Ag** 19.0-21.0 **Cu** 39.0-41.0 **Mn** 4.5-5.5 **Zn** 33.0-37.0 | **ASME** SFA5.8 (BAg-25) **AWS** A5.8 (BAg-25) |
| **P07250** | Silver Brazing Alloy (BAg-26) | **Ag** 24.0-26.0 **Cu** 37.0-39.0 **Mn** 1.5-2.5 **Ni** 1.5-2.5 **Zn** 31.0-35.0 | **ASME** SFA5.8 (BAg-26) **AWS** A5.8 (BAg-26) |
| **P07251** | Silver Brazing Alloy (BAg-27) | **Ag** 24.0-26.0 **Cd** 12.5-14.5 **Cu** 34.0-36.0 **Zn** 24.5-28.5 | **ASME** SFA5.8 (BAg-27) **AWS** A5.8 (BAg-27) |
| **P07252** | Silver Brazing Alloy (BAg-33) | **Ag** 24.0-26.0 **Cd** 16.5-18.5 **Cu** 29.0-31.0 **Zn** 26.5-28.5 | **ASME** SFA5.8 (BAg-33) **AWS** A5.8 (BAg-33) |
| **P07253** | Silver Brazing Alloy | **Ag** 24.0-26.0 **Cu** 39.0-41.0 **Sn** 1.5-2.5 **Zn** 31.0-35.0 | **ASME** SFA5.8 (BAg-37) **AWS** A5.8 (BAg-37) |
| **P07300** | Silver Brazing Alloy (BAg-2a) | **Ag** 29.0-31.0 **Cd** 19.0-21.0 **Cu** 26.0-28.0 **Zn** 21.0-25.0 | **ASME** SFA5.8 (BAg-2a) **AWS** A5.8 (BAg-2a) |
| **P07301** | Silver Brazing Alloy (BAg-20) | **Ag** 29.0-31.0 **Cu** 37.0-39.0 **Zn** 30.0-34.0 | **ASME** SFA5.8 (BAg-20) **AWS** A5.8 (BAg-20) **FED** QQ-B-654 (BAg-20) |
| **P07350** | Silver Brazing Alloy (BAg-2) | **Ag** 34.0-36.0 **Cd** 17.0-19.0 **Cu** 25.0-27.0 **Zn** 19.0-23.0 | **AMS** 4768 **ASME** SFA5.8 (BAg-2) **AWS** A5.8 (BAg-2) **FED** QQ-B-654 (VIII) |
| **P07351** | Silver Brazing Alloy | **Ag** 34.0-36.0 **Cu** 31.0-33.0 **Zn** 31.0-35.0 | **ASME** SFA5.8 (BAg-35) **AWS** A5.8 (BAg-35) |
| **P07380** | Silver Brazing Alloy (BAg-34) | **Ag** 37.0-39.0 **Cu** 31.0-33.0 **Sn** 1.5-2.5 **Zn** 26.0-30.0 | **ASME** SFA5.8 (BAg-34) **AWS** A5.8 (BAg-34) |
| **P07400** | Silver Brazing Alloy (BAg-4) | **Ag** 39.0-41.0 **Cu** 29.0-31.0 **Ni** 1.5-2.5 **Zn** 26.0-30.0 | **ASME** SFA5.8 (BAg-4) **AWS** A5.8 (BAg-4) **FED** QQ-B-654 (BAg-4) |
| **P07401** | Silver Brazing Alloy (BAg-28) | **Ag** 39.0-41.0 **Cu** 29.0-31.0 **Sn** 1.5-2.5 **Zn** 26.0-30.0 | **ASME** SFA5.8 (BAg-28) **AWS** A5.8 (BAg-28) |
| **P07402** | Silver Brazing Alloy | **Ag** 39.0-41.0 **Cu** 29.0-32.0 **Zn** 28.0-32.0 **Other** 0.15 max | **AMS** 4762 |
| **P07450** | Silver Brazing Alloy (BAg-1) | **Ag** 44.0-46.0 **Cd** 23.0-25.0 **Cu** 14.0-16.0 **Zn** 14.0-18.0 | **AMS** 4769 **ASME** SFA5.8 (BAg-1) **AWS** A5.8 (BAg-1) **FED** QQ-B-654 (VII) |
| **P07453** | Silver Brazing Alloy (BAg-5) | **Ag** 44.0-46.0 **Cu** 29.0-31.0 **Zn** 23.0-27.0 | **ASME** SFA5.8 (BAg-5) **AWS** A5.8 (BAg-5) **FED** QQ-B-654 (BAg-5) |
| **P07454** | Silver Brazing Alloy | **Ag** 44.0-46.0 **Cu** 26.0-28.0 **Sn** 2.5-3.5 **Zn** 23.0-27.0 | **ASME** SFA5.8 (BAg-36) **AWS** A5.8 (BAg-36) |
| **P07490** | Silver Brazing Alloy (BAg-22) | **Ag** 48.0-50.0 **Cu** 15.0-17.0 **Mn** 7.0-8.0 **Ni** 4.0-5.0 **Zn** 21.0-25.0 | **ASME** SFA5.8 (BAg-22) **AWS** A5.8 (BAg-22) **FED** QQ-B-654 (BAg-22) |
| **P07500** | Silver Brazing Alloy (BAg-1a) | **Ag** 49.0-51.0 **Cd** 17.0-19.0 **Cu** 14.5-16.5 **Zn** 14.5-18.5 | **AMS** 4770 **ASME** SFA5.8 (BAg-1a) **AWS** A5.8 (BAg-1a) **FED** QQ-B-654 (IV) **MIL SPEC** MIL-B-7883 (BAg-1a) |
| **P07501** | Silver Brazing Alloy (BAg-3) | **Ag** 49.0-51.0 **Cd** 15.0-17.0 **Cu** 14.5-16.5 **Ni** 2.5-3.5 **Zn** 13.5-17.5 | **AMS** 4771 **ASME** SFA5.8 (BAg-3) **AWS** A5.8 (BAg-3) **FED** QQ-B-654 (V) |

*Boxed entries are no longer active and are retained for reference purposes only.

The chemical compositions listed are for identification purposes and should not be used in lieu of the cross referenced specifications.

| UNIFIED NUMBER | DESCRIPTION | CHEMICAL COMPOSITION | CROSS REFERENCE SPECIFICATIONS |
|---|---|---|---|
| **P07502** | Silver Brazing Alloy | **Ag** 49.0-51.0 **Cd** 9.0-11.0 **Cu** 17.0-19.0 **Zn** 20.0-24.0 | **FED** QQ-B-654 (VI) |
| **P07503** | Silver Brazing Alloy (BAg-6) | **Ag** 49.0-51.0 **Cu** 33.0-35.0 **Zn** 14.0-18.0 | **ASME** SFA5.8 (BAg-6) **AWS** A5.8 (BAg-6) |
| **P07505** | Silver Brazing Alloy (BAg-24) | **Ag** 49.0-51.0 **Cu** 19.0-21.0 **Ni** 1.5-2.5 **Zn** 26.0-30.0 | **AMS** 4788 **ASME** SFA5.8 (BAg-24) **AWS** A5.8 (BAg-24) |
| **P07507** | Silver Brazing Alloy (Vacuum Grade) | **Ag** 49.0-51.0 **C** 0.005 max **Cd** 0.001 max **Cu** rem **P** 0.002 max **Pb** 0.002 max **Zn** 0.001 max | **ASME** SFA5.8 (BVAg-6b) **AWS** A5.8 (BVAg-6b) |
| **P07540** | Silver Brazing Alloy (BAg-13) | **Ag** 53.0-55.0 **Cu** rem **Ni** 0.5-1.5 **Zn** 4.0-6.0 | **AMS** 4772 **ASME** SFA5.8 (BAg-13) **AWS** A5.8 (BAg-13) |
| **P07547** | Silver-Palladium Brazing Alloy (Vacuum Grade) | **Ag** 53.0-55.0 **C** 0.005 max **Cd** 0.001 max **Cu** 20.0-22.0 **P** 0.002 max **Pb** 0.002 max **Pd** rem **Zn** 0.001 max | **ASME** SFA5.8 (BVAg-32) **AWS** A5.8 (BVAg-32) |
| **P07560** | Silver Brazing Alloy | **Ag** 55.0-57.0 **Cu** rem **Ni** 1.5-2.5 | **AMS** 4765 **ASME** SFA5.8 (BAg-13a) **AWS** A5.8 (BAg-13a) **FED** QQ-B-654 (BAg-13a) |
| **P07563** | Silver Brazing Alloy | **Ag** 55.0-57.0 **Cu** 21.0-23.0 **Sn** 4.5-5.5 **Zn** 15.0-19.0 | **AMS** 4763 **ASME** SFA5.8 (BAg-7) **AWS** A5.8 (BAg-7) **FED** QQ-B-654 (BAg-7) |
| **P07587** | Silver Brazing Alloy (Vacuum Grade) | **Ag** 57.0-59.0 **C** 0.005 max **Cd** 0.001 max **Cu** 31.0-33.0 **P** 0.002 max **Pb** 0.002 max **Pd** rem **Zn** 0.001 max | **ASME** SFA5.8 (BVAg-31) **AWS** A5.8 (BVAg-31) |
| **P07600** | Silver Brazing Alloy | **Ag** 59.0-61.0 **Cu** rem **Sn** 9.5-10.5 | **AMS** 4773 **ASME** SFA5.8 (BAg-18) **AWS** A5.8 (BAg-18) **FED** QQ-B-654 (BAg-18) |
| **P07607** | Silver Brazing Alloy (Vacuum Grade) | **Ag** 59.0-61.0 **C** 0.005 max **Cd** 0.001 max **Cu** rem **P** 0.002 max **Pb** 0.002 max **Sn** 9.5-10.5 **Zn** 0.001 max | **ASME** SFA5.8 (BVAg-18) **AWS** A5.8 (BVAg-18) |
| **P07627** | Silver Brazing Alloy (Vacuum Grade) | **Ag** 60.5-62.5 **C** 0.005 max **Cd** 0.001 max **Cu** rem **P** 0.002 max **Pb** 0.002 max **In** 14.0-15.0 **Zn** 0.001 max | **ASME** SFA5.8 (BVAg-29) **AWS** A5.8 (BVAg-29) |
| **P07630** | Silver Brazing Alloy (BAg-21) | **Ag** 62.0-64.0 **Cu** 27.5-29.5 **Ni** 2.0-3.0 **Sn** 5.0-7.0 | **AMS** 4774 **ASME** SFA5.8 (BAg-21) **AWS** A5.8 (BAg-21) |
| **P07650** | Silver Brazing Alloy | **Ag** 64.0-66.0 **Cu** 19.0-21.0 **Zn** 13.0-17.0 | **ASME** SFA5.8 (BAg-9) **AWS** A5.8 (BAg-9) **FED** QQ-B-654 (BAg-9) |
| **P07687** | Silver Brazing Alloy (Vacuum Grade) | **Ag** 67.0-69.0 **C** 0.005 max **Cd** 0.001 max **Cu** rem **P** 0.002 max **Pb** 0.002 max **Pd** 4.5-5.5 **Zn** 0.001 max | **ASME** SFA5.8 (BVAg-30) **AWS** A5.8 (BVAg-30) |
| **P07700** | Silver Brazing Alloy | **Ag** 69.0-71.0 **Cu** 19.0-21.0 **Zn** 8.0-12.0 | **ASME** SFA5.8 (BAg-10) **AWS** A5.8 (BAg-10) **FED** QQ-B-654 (BAg-10) |
| **P07720** | Silver Brazing Alloy | **Ag** 71.0-73.0 **Cu** rem | **ASME** SFA5.8 (BAg-8) **AWS** A5.8 (BAg-8) |
| **P07723** | Silver Brazing Alloy | **Ag** 71.0-73.0 **Cu** rem **Li** 0.25-0.50 | **ASME** SFA5.8 (BAg-8a) **AWS** A5.8 (BAg-8a) **FED** QQ-B-654 (BAg-8a) |
| **P07727** | Silver Brazing Alloy (Vacuum Grade) | **Ag** 71.0-73.0 **C** 0.005 max **Cd** 0.001 max **Cu** rem **P** 0.002 max **Pb** 0.002 max **Zn** 0.001 max | **ASME** SFA5.8 (BVAg-8) **AWS** A5.8 (BVAg-8) |
| **P07728** | Silver Brazing Alloy (Vacuum Grade) | **Ag** 70.5-72.5 **C** 0.005 max **Cd** 0.001 max **Cu** rem **P** 0.002 max **Pb** 0.002 max **Ni** 0.3-0.7 **Zn** 0.001 max | **ASME** SFA5.8 (BVAg-8b) **AWS** A5.8 (BVAg-8b) |
| **P07850** | Silver Brazing Alloy (BAg-23) | **Ag** 84.0-86.0 **Mn** rem | **AMS** 4766 **ASME** SFA5.8 (BAg-23) **AWS** A5.8 (BAg-23) **MIL SPEC** MIL-B-7883 (BAg-23) |
| **P07900** | Silver Electrical Contact Alloy | **Ag** 89.6-91.0 **Al** 0.005 max **Cd** 0.05 max **Cu** 9.0-10.4 **Fe** 0.05 max **Ni** 0.01 max **P** 0.02 max **Pb** 0.03 max **Zn** 0.06 max | **ASTM** B617 |
| **P07925** | Silver Brazing Alloy | **Ag** 92.0-93.0 **Cu** rem **Li** 0.15-0.30 | **AMS** 4767 **ASME** SFA5.8 (BAg-19) **AWS** A5.8 (BAg-19) |
| **P07931** | Silver, Sterling Silver-Standard Grade | **Ag** 92.10-93.50 **Cd** 0.05 max **Cu** 6.50-7.90 **Fe** 0.05 max **Pb** 0.03 max **Zn** 0.06 max **Other** 0.06 max | |

The chemical compositions listed are for identification purposes and should not be used in lieu of the cross referenced specifications.

| UNIFIED NUMBER | DESCRIPTION | CHEMICAL COMPOSITION | CROSS REFERENCE SPECIFICATIONS |
|---|---|---|---|
| **P07932** | Silver, Sterling Silver, Silversmiths Grade | **Ag** 92.50-93.50 **Cd** 0.05 max **Cu** 6.50-7.50 **Fe** 0.05 max **Pb** 0.03 max **Zn** 0.06 max **Other** 0.06 max | |

The chemical compositions listed are for identification purposes and should not be used in lieu of the cross referenced specifications.

# UNS NUMBERS ASSIGNED TO DATE

## With Description of Each Material Covered and References To Documents In Which The Same or Similar Materials are Described

## Rxxxxx Number Series
## Reactive and Refractory Metals and Alloys

# REACTIVE AND REFRACTORY METALS AND ALLOYS

| UNIFIED NUMBER | DESCRIPTION | CHEMICAL COMPOSITION | CROSS REFERENCE SPECIFICATIONS |
|---|---|---|---|
| **R03600** | Molybdenum, Unalloyed | **C** 0.010-0.040 **Fe** 0.010 max **Mo** rem **N** 0.0010 max **Ni** 0.005 max **O** 0.0030 max **Si** 0.010 max | **ASTM** B386 (360); B387 (360) |
| **R03601** | Molybdenum Sealing Alloy | **C** 0.04 max **Fe** 0.01 max **Mo** 99.90 min **N** 0.001 max **Ni** 0.01 max **O** 0.005 max **Si** 0.01 max **W** 0.02 max **Other** each 0.005 max | |
| **R03602** | Molybdenum Sealing Alloy | **C** 0.005 max **Fe** 0.01 max **Mo** 99.90 min **N** 0.002 max **Ni** 0.01 max **O** 0.008 max **Si** 0.01 max **W** 0.02 max **Other** each 0.005 max | |
| **R03603** | Molybdenum | **Al** 0.015 max **C** 0.0015 max **Fe** 0.001 max **H** 0.001 max **Mo** 99.90 min **N** 0.001 max **O** 0.0175 max **Si** 0.035 max **Sn** 0.0025 max **W** 0.02 max **Other** each 0.005 max, Ca 0.005 max, K 0.015 max | **ASTM** F364 (II) |
| **R03604** | Molybdenum | **Al** 0.015 max **C** 0.005 max **Fe** 0.01 max **H** 0.001 max **Mo** 99.90 min **N** 0.002 max **O** 0.008 max **Si** 0.01 max **Sn** 0.0025 max **W** 0.02 max **Other** each 0.005 max, Ca 0.005 max, K 0.015 max | **ASTM** F364 (1) |
| * **R03605** | Molybdenum Metal | **C** 0.030 max **Fe** 0.020 max **Mo** 99.90 min **N** 0.0010 max **Ni** 0.010 max **O** 0.0030 max **Si** 0.010 max | |
| **R03606** | Molybdenum Metal | **C** 0.030 max **Fe** 0.008 max **H** 0.0005 max **Mo** 99.95 min **N** 0.002 max **Ni** 0.002 max **O** 0.0015 max **Si** 0.008 max | |
| **R03610** | Molybdenum Unalloyed | **C** 0.010 max **Fe** 0.010 max **Mo** rem **N** 0.0020 max **Ni** 0.005 max **O** 0.0070 max **Si** 0.010 max | **ASTM** B386 (361); B387 (361) |
| **R03615** | Non-Ferrous Thermal Spray Wire (W-Mo) | **Mo** 99.0 min | **AWS** C2.25 (W-Mo) |
| **R03620** | Molybdenum Alloy | **C** 0.010-0.040 **Fe** 0.010 max **Mo** rem **N** 0.0010 max **Ni** 0.005 max **O** 0.0030 max **Si** 0.010 max **Ti** 0.40-0.55 | |
| **R03630** | Molybdenum Alloy | **C** 0.010-0.040 **Fe** 0.010 max **Mo** rem **N** 0.0010 max **Ni** 0.005 max **O** 0.0030 max **Si** 0.010 max **Ti** 0.40-0.55 **Zr** 0.06-0.12 | **AMS** 7817; 7819 **ASTM** B386 (363); B387 (363) |
| **R03640** | Molybdenum Alloy | **C** 0.010-0.040 **Fe** 0.010 max **Mo** rem **N** 0.0020 max **Ni** 0.005 max **O** 0.030 max **Si** 0.005 max **Ti** 0.40-0.55 **Zr** 0.06-0.12 | **ASTM** B386 (364); B387 (364) |
| **R03650** | Molybdenum Unalloyed, Low Carbon | **C** 0.010 max **Fe** 0.010 max **Mo** rem **N** 0.0010 max **Ni** 0.005 max **O** 0.0030 max **Si** 0.010 max | **ASTM** B386 (365); B387 (365) |
| **R04200** | Niobium (Columbium), Unalloyed, Reactor Grade | **C** 0.01 max **Fe** 0.005 max **H** 0.001 max **Hf** 0.01 max **Mo** 0.005 max **N** 0.01 max **Nb** rem **Ni** 0.005 max **O** 0.015 max **Si** 0.005 max **Ta** 0.1 max **W** 0.03 max **Zr** 0.01 max | **ASTM** B391; B392; B393; B394 |
| **R04210** | Niobium (Columbium), Unalloyed, Commercial Grade | **C** 0.01 max **Fe** 0.01 max **H** 0.001 max **Hf** 0.01 max **Mo** 0.005 max **N** 0.01 max **Nb** rem **Ni** 0.005 max **O** 0.025 max **Si** 0.005 max **Ta** 0.2 max **W** 0.05 max **Zr** 0.01 max | **ASTM** B391; B392; B393; B394 |

*Boxed entries are no longer active and are retained for reference purposes only.

The chemical compositions listed are for identification purposes and should not be used in lieu of the cross referenced specifications.

| UNIFIED NUMBER | DESCRIPTION | CHEMICAL COMPOSITION | CROSS REFERENCE SPECIFICATIONS |
|---|---|---|---|
| **R04211** | Niobium (Columbium), Unalloyed, Commercial Grade | **C** 0.005 max **Cb** rem **Fe** 0.010 max **H** 0.002 max **N** 0.010 max **O** 0.030 max **Si** 0.005 max **Ta** 0.10 max **Ti** 0.005 max **Zr** 0.010 max **Other** each 0.010 max, total 0.15 max | **AMS** 7850 |
| **R04251** | Niobium (Columbium)- 1% Zirconium, Reactor Grade | **C** 0.01 max **Fe** 0.005 max **H** 0.001 max **Hf** 0.01 max **Mo** 0.005 max **N** 0.01 max **Nb** rem **Ni** 0.005 max **O** 0.015 max **Si** 0.005 max **Ta** 0.1 max **W** 0.03 max **Zr** 0.8-1.2 | **ASTM** B391; B392; B393; B394 |
| **R04261** | Niobium (Columbium)- 1% Zirconium, Commercial Grade | **C** 0.01 max **Fe** 0.01 max **H** 0.001 max **Hf** 0.01 max **Mo** 0.005 max **N** 0.01 max **Nb** rem **Ni** 0.005 max **O** 0.025 max **Si** 0.005 max **Ta** 0.2 max **W** 0.05 max **Zr** 0.8-1.2 | **ASTM** B391; B392; B393; B394 |
| **R04271** | Niobium (Columbium) Alloy, 10W-2.5Zr | **C** 0.030 max **Cb** rem **Fe** 0.02 max **H** 0.001 max **N** 0.010 max **O** 0.020 max **Si** 0.02 max **Ta** 0.15 max **Ti** 0.01 max **W** 9.00-11.00 **Zr** 2.00-3.00 | **AMS** 7851; 7855 |
| **R04295** | Niobium (Columbium) Alloy | **C** 0.015 max **Cb** rem **H** 0.0015 max **Hf** 9-11 **N** 0.010 max **O** 0.020 max **Ta** 0.500 max **Ti** 0.7-1.3 **W** 0.500 max **Zr** 0.700 max | **AMS** 7852; 7857 **ASTM** B652; B654; B655 |
| **R05200** | Tantalum, Unalloyed, Cast | **C** 0.01 max **Cb** 0.05 max **Fe** 0.01 max **H** 0.001 max **Mo** 0.01 max **N** 0.01 max **Ni** 0.01 max **O** 0.015 max **Si** 0.005 max **Ta** rem **Ti** 0.01 max **W** 0.03 max | **ASTM** B364; B365; B521; F560 |
| **R05210** | Tantalum Metal | **C** 0.0075 max **Cb** 0.100 max **Co** 0.002 max **Fe** 0.01 max **H** 0.0010 max **Mo** 0.01 max **N** 0.0075 max **Ni** 0.005 max **O** 0.015 max **Si** 0.005 max **Ta** 99.85 min **Ti** 0.005 max **W** 0.03 max **Zr** 0.01 max **Other** total 0.15 max | **AMS** 7849 |
| **R05240** | Tantalum - 40 Nb (Cb) Alloy | **C** 0.010 max **Cb** 35.0-42.0 **H** 0.0015 max **Fe** 0.010 max **Mo** 0.020 max **Ni** 0.010 max **N** 0.010 max **O** 0.020 max **Si** 0.005 max **Ta** rem **Ti** 0.010 max **W** 0.050 max **Other** trace impurities 0.00 max | **ASTM** B364; B365; B521; B708 |
| **R05252** | 97.5 Tantalum-2.5 Tungsten | **C** 0.010 max **Cb** 0.50 max **Fe** 0.010 max **H** 0.0015 max **Mo** 0.020 max **N** 0.010 max **Ni** 0.010 max **O** 0.015 max **Si** 0.005 max **Ta** rem **Ti** 0.010 max **W** 2.0-3.5 | **ASTM** B364; B365; B521; B708 |
| **R05255** | 90 Tantalum-10 Tungsten | **C** 0.01 max **Cb** 0.10 max **Fe** 0.01 max **H** 0.001 max **Mo** 0.01 max **N** 0.01 max **Ni** 0.01 max **O** 0.015 max **Si** 0.005 max **Ta** rem **Ti** 0.01 max **W** 9.0-11.0 | **AMS** 7846; 7847; 7848 **ASTM** B364; B365; B521; B708 |
| **R05400** | Tantalum, Unalloyed, Sintered | **C** 0.01 max **Cb** 0.05 max **Fe** 0.01 max **H** 0.001 max **Mo** 0.01 max **N** 0.015 max **Ni** 0.01 max **O** 0.03 max **Si** 0.005 max **Ta** rem **Ti** 0.01 max **W** 0.03 max | **ASTM** B364; B365; B521; B708; F560 |
| *** R07004** | Tungsten | **Al** 0.003 max **C** 0.010 max **Fe** 0.005 max **H** 0.002 max **Mo** 0.005 max **N** 0.005 max **Ni** 0.008 max **O** 0.006 max **Si** 0.003 max **W** 99.95 min | **ASTM** B760 |
| **R07005** | Tungsten | **W** 99.95 min **Other** 0.01 max each, 0.05 max total | **AMS** 7897 **ASTM** F288 (1A and 1B); F290 |

*Boxed entries are no longer active and are retained for reference purposes only.

The chemical compositions listed are for identification purposes and should not be used in lieu of the cross referenced specifications.

| UNIFIED NUMBER | DESCRIPTION | CHEMICAL COMPOSITION | CROSS REFERENCE SPECIFICATIONS |
|---|---|---|---|
| **R07006** | Tungsten Metal | **Al** 0.005 max **C** 0.008 max **Fe** 0.005 max **H** 0.001 max **Mo** 0.020 max **N** 0.002 max **Ni** 0.005 max **O** 0.005 max **Si** 0.005 max **W** rem | **AMS** 7898 |
| **R07030** | Tungsten | **W** 96-98 | |
| **R07031** | Tungsten-Rhenium | **Re** 2.5-3.5 **W** rem **Other** each 0.05 max, total 0.01 max | **ASTM** F73 (Electronic Grade) |
| **R07050** | Tungsten | **W** 94-96 | |
| **R07080** | Tungsten | **W** 91-94 | |
| **R07100** | Tungsten | **W** 89-91 | |
| **R07900** | Tungsten Arc Welding Electrode | **W** 99.5 min | **ASME** SFA5.12 (EWP) **AWS** A5.12 (EWP) |
| **R07911** | Tungsten-Thorium Alloy Arc Welding Electrode | **ThO** 0.8-1.2 **W** 98.3 min | **ASME** SFA5.12 (EWTh-1) **AWS** A5.12 (EWTh-1) |
| **R07912** | Tungsten-Thorium Alloy Arc Welding Electrode | **ThO** 1.7-2.2 **W** 97.3 min | **ASME** SFA5.12 (EWTh-2) **AWS** A5.12 (EWTh-2) |
| * **R07913** | Tungsten-Thorium Alloy Arc Welding Electrode | **ThO** 0.35-0.55 **W** 98.95 min | **ASME** SFA5.12 (EWTh-3) **ASTM** F288 (2B) **AWS** A5.12(EWTh-3) |
| **R07920** | Tungsten-Zirconium Alloy Arc Welding Electrode | **W** 99.1 min **ZrO** 0.15-0.40 | **ASME** SFA5.12 (EWZr-1) **AWS** A5.12 (EWZr-1) |
| **R07932** | Tungsten-Cerium Alloy (Arc Welding Electrode) | **$CeO_2$** 1.8-2.2 **W** 97.3 min | **ASME** SFA5.12 (EWCe-2) **AWS** A5.12 (EWCe-2) |
| **R07941** | Tungsten-Lanthanum Alloy (Arc Welding Electrode) | **$La_2O_3$** 0.8-1.2 **W** 98.3 min | **ASME** SFA5.12 (EWLa-1) **AWS** A5.12 (EWLa-1) |
| **R07942** | Tungsten-Lanthanum Alloy (Arc Welding Electrode | **EWLa-1.5)W** 97.8 min **Other** Lanthanum oxide ($La_2O_3$) 1.3-1.7 | **ASME** SFAA5.12 (EWLa-1.5) **AWS** A5.12 (EWLa-1.5) |
| **R07943** | Tungsten-Lanthanum Alloy (Arc Welding Electrode, EWLa-2) | **W** 97.3 min **Other** Lanthanum oxide ($La_2O_3$) 1.8-2.2 | **ASME** SFAA5.12 (EWLa-2) **AWS** A5.12 (EWLa-2) |
| **R19800** | Beryllium | **Al** 0.20 max **Be** 98.00 min **C** 0.15 max **Fe** 0.20 max **Mg** 0.08 max **Si** 0.12 max | **AMS** 7902 |
| **R19801** | Combined with R19800 | | |
| * **R19920** | Beryllium | **Al** 0.18 max **Be** 99.20 min **C** 0.15 max **Fe** 0.20 max **Mg** 0.08 max **Si** 0.10 max | |
| **R20033** | Cr-Ni-Fe-Mo Austenitic Stainless Steel (Alloy 33) | **C** 0.015 max **Cr** 31.0-35.0 **Cu** 0.3-1.2 **Fe** rem **Mn** 2.0 max **Mo** 0.5-2.0 **N** 0.35-0.6 **Ni** 30.0-33.0 **P** 0.02 max **S** 0.01 max **Si** 0.5 max | **ASTM** B366; B462; B472; B564; B619; B622; B625; B626; B649 |
| **R20500** | Chromium-Nickel Alloy | **Al** 0.25 max **C** 0.10 max **Cr** 48.0-52.0 **Fe** 1.00 max **Mn** 0.30 max **N** 0.30 max **Ni** rem **P** 0.02 max **S** 0.02 max **Si** 1.00 max **Ti** 0.50 max | **ASTM** A560 (50Cr-50Ni) |
| **R20501** | Cr-Ni-Cb Casting Alloy | **Al** 0.25 max **C** 0.10 max **Cb** 1.4-1.7 **Cr** 47.0-52.0 **Fe** 1.00 max **Mn** 0.30 max **N** 0.16 max **Ni** rem **P** 0.02 max **S** 0.02 max **Si** 0.50 max **Ti** 0.50 max **Other** (C+N) 0.20 max | **ASTM** A560 (50Cr-50Ni-Cb) |
| **R20600** | Chromium-Nickel Alloy | **Al** 0.25 max **Cr** 58.0-62.0 **Fe** 1.00 max **Mn** 0.30 max **N** 0.30 max **Ni** rem **P** 0.02 max **S** 0.02 max **Si** 1.00 max **Ti** 0.50 max | **ASTM** A560 (60Cr-40Ni) |

*Boxed entries are no longer active and are retained for reference purposes only.

The chemical compositions listed are for identification purposes and should not be used in lieu of the cross referenced specifications.

| UNIFIED NUMBER | DESCRIPTION | CHEMICAL COMPOSITION | CROSS REFERENCE SPECIFICATIONS |
|---|---|---|---|
| **R20990** | Chromium Metal | **C** 0.050 max **Cr** 99.0 min **P** 0.010 max **S** 0.030 max **Si** 0.15 max | **ASTM** A481 (A) |
| **R20994** | Chromium Metal | **C** 0.050 max **Cr** 99.4 min **P** 0.010 max **S** 0.010 max **Si** 0.10 max | **ASTM** A481 (B) |
| **R30001** | Co-Cr-W Alloy (Stellite 1, Weartech 1, WT-1) | **C** 2.5 nom **Co** rem **Cr** 30.0 nom **Fe** 3.0 min **Mn** 0.5 nom **Mo** 0.5 nom **Ni** 1.5 nom **Si** 1.3 nom **W** 13.0 nom | **ASME** SFA5.21 (ERCoCr-C) **AWS** A5.21 (ERCoCr-C) **SAE** J775 (VF6) **MIL SPEC** MIL-C-24689 (22); MIL-R-17131 (MIL-RCoCr-C) |
| **R30002** | Co-Cr-Ni-W Alloy (Stellite F) | **C** 1.5-2.00 **Co** rem **Cr** 23.00-27.00 **Fe** 6.00 max **Mn** 1.00 nom **Mo** 1.00 max **Ni** 20.50-23.50 **P** 0.030 max **S** 0.030 max **Si** 0.80-1.50 **W** 10.50-13.50 | **ASME** SFA5.21 (ERCoCr-F) **AWS** A5.21 (ERCoCr-F) **SAE** J775 (VF5) |
| **R30003** | Co-Cr-Ni-Fe-Mo (Elgiloy) | **Be** 1.0 max **C** 0.15 max **Co** 39.0-41.0 **Cr** 19.0-21.0 **Fe** rem **Mn** 1.5-2.5 **Mo** 6.0-8.0 **Ni** 15.0-16.0 | **AMS** 5833; 5834; 5875; 5876 **SAE** J467 (Elgiloy) |
| **R30004** | Co-Cr-Ni-Fe Alloy (Havar) | **Be** 0.06 max **C** 0.17-0.23 **Co** 41.0-44.0 **Cr** 19.0-21.0 **Fe** rem **Mn** 1.35-1.80 **Mo** 2.0-2.8 **Ni** 12.0-14.0 **W** 2.3-3.3 | |
| **R30005** | Fe-Co-V Soft Magnetic Alloy | **C** 0.025 max **Co** 47.50 **Cr** 0.15 max **Fe** rem **Mn** 0.15 max **Ni** 0.25 max **P** 0.015 max **S** 0.010 max **Si** 0.15max **V** 1.75-2.10 | **ASTM** A801 (Alloy 1) **MIL SPEC** MIL-A-47182 |
| **R30006** | Co-Cr-W Alloy (Stellite 6, Weartech 6, WT-6) | **C** 0.9-1.4 **Co** rem **Cr** 27.0-31.0 **Fe** 3.0 max **Mn** 1.0 max **Mo** 1.5 max **Ni** 3.0 max **Si** 1.5 max **W** 3.5-5.5 | **AMS** 5373; 5387; 5788 **ASME** SFA5.21 (ERCoCr-A) **AWS** A5.21 (ERCoCr-A) **MIL SPEC** MIL-C-24689 (20); MIL-R-17131 (MIL-RCoCr-A) **SAE** J467 (Stellite 6); J775 (VF-2) |
| **R30007** | Age Hardening Co-Cr-Fe-Ni Alloy (ARNAVAR) | **Be** 0.70 max **C** 0.10-0.20 **Co** 41.5-43.5 **Cr** 19.0-21.0 **Fe** rem **Mn** 1.20-1.60 **Mo** 1.80-2.20 **Ni** 12.0-14.0 **Si** 0.30-0.70 **W** 2.60-3.00 | |
| **R30008** | Co-Cr-Ni-Mo-Fe Alloy | **Be** 0.001 max **C** 0.15 max **Co** 39.0-42.0 **Cr** 18.5-21.5 **Fe** rem **Mn** 1.0-2.0 **Mo** 6.5-7.5 **Ni** 15.0-18.0 **P** 0.015 max **S** 0.015 max **Si** 1.20 max | **ASTM** F1058 (2) **Other:** ISO 5832-7 |
| **R30012** | Co-Cr-W Alloy (Stellite 12, WR-12, VF-7) | **B** 1.00 max **C** 1.10-1.90 **Co** rem **Cr** 26.0-33.0 **Fe** 3.0 max **Mn** 1.00 max **Mo** 1.00 max **Ni** 3.00 max **P** 0.030 max **S** 0.030 max **Si** 0.40-2.00 **W** 7.0-9.5 | **AMS** 7238 **ASME** SFA5.21 (ERCoCr-B) **AWS** A5.21 (ERCoCr-B) **SAE** J775 (VF7) |
| **R30014** | Co-Cr-W Alloy (ERCoCr-G) | **C** 3.0-4.0 **Co** rem **Cr** 24-30 **Fe** 3.0 max **Mn** 1.0 max **Mo** 1.0 max **Ni** 4.0 max **Si** 2.0 max **W** 12-16 **Other** 0.50 max | **ASME** SFA5.21 (ERCoCr-G) **AWS** A5.21 (ERCoCr-G) |
| **R30016** | Co-Cr-W Alloy (Stellite 6B) | **C** 0.90-1.40 **Co** rem **Cr** 28.0-32.0 **Fe** 3.0 max **Mn** 0.5-2.0 **Mo** 1.50 max **Ni** 3.0 max **Si** 0.2-2.0 **W** 3.50-5.50 | **AMS** 5894 |
| **R30021** | Co-Cr-Mo-Ni Alloy (Stellite 21, Weartech 21, WR-21 | **B** 0.007 max **C** 0.20-0.30 **Co** rem **Cr** 25.00-29.00 **Fe** 3.00 max **Mn** 1.00 max **Mo** 5.00-6.00 **Ni** 1.75-3.75 **Si** 1.00 max | **AMS** 5385; 5819 **ASME** SFA5.21 (ERCoCr-E) **AWS** A5.21 (ERCoCr-E) **ASTM** A732 (21) **MIL SPEC** MIL-C-24689 (23) **SAE** J467 (HS-21) |
| **R30023** | Co-Cr-W-Ni Alloy (Stellite 23) | **C** 0.40 nom **Co** rem **Cr** 24 nom **Fe** 1.0 nom **Mn** 0.3 nom **Ni** 2 nom **Si** 0.6 nom **W** 5 nom | **AMS** 5375 |
| **R30027** | Co-Ni-Cr Alloy (Stellite 27) | **C** 0.40 nom **Co** rem **Cr** 25 nom **Fe** 1.0 nom **Mn** 0.3 nom **Mo** 5.5 nom **Ni** 32 nom **Si** 0.6 nom | **AMS** 5378 |
| **R30030** | Co-Cr-Ni-Mo Alloy (Stellite 30) | **C** 0.45 nom **Co** rem **Cr** 26 nom **Fe** 1.0 nom **Mn** 0.6 nom **Mo** 6 nom **Ni** 15 nom **Si** 0.6 nom | **AMS** 5380 |
| **R30031** | Co-Cr-Ni-Mo Alloy (Stellite 31) | **C** 0.45-0.55 **Co** rem **Cr** 24.50-26.50 **Fe** 2.00 max **Mn** 1.00 max **Ni** 9.50-11.50 **Si** 1.00 max **W** 7.00-8.00 | **AMS** 5382; 5789 **ASTM** A372 (31) **SAE** J467 (HS-31, X-40) |

The chemical compositions listed are for identification purposes and should not be used in lieu of the cross referenced specifications.

| UNIFIED NUMBER | DESCRIPTION | CHEMICAL COMPOSITION | CROSS REFERENCE SPECIFICATIONS |
|---|---|---|---|
| **R30035** | Co-Ni-Cr-Mo Alloy (MP-35-N) | **C** 0.025 max **Co** rem **Cr** 19.00-21.00 **Fe** 1.00 max **Mn** 0.15 max **Mo** 9.00-10.50 **Ni** 33.00-37.00 **P** 0.015 max **S** 0.010 max **Si** 0.15 max **Ti** 1.00 max | **AMS** 5844; 5845; 5758; AS7468 **ASTM** F562; F688; F961 |
| **R30036** | Co-Cr-W-Ni Alloy Solid Solution Strengthened (Haynes 36) | **B** 0.01-0.05 **C** 0.35-0.45 **Co** rem **Cr** 17.5-19.5 **Fe** 2.00 max **Mn** 1.50 max **Ni** 9.00-11.0 **Si** 1.00 max **W** 14.0-16.0 | |
| **R30040** | Co-Cr-Ni-Si-W Brazing Filler Metal | **Al** 0.05 max **B** 0.70-0.90 **C** 0.35-0.45 **Co** rem **Cr** 18.0-20.0 **Fe** 1.00 max **Ni** 16.0-18.0 **P** 0.02 max **S** 0.02 max **Si** 7.50-8.50 **Ti** 0.05 max **W** 3.50-4.50 **Zr** 0.05 max **Other** each 0.03 max, total 0.15 max | **AMS** 4783 |
| **R30075** | Co-Cr-Mo Alloy (Biomedical Implant Applications) | **Al** 0.30 max **B** 0.01 max **C** 0.35 max **Co** rem **Cr** 27.0-30.0 **Fe** 0.75 max **Mn** 1.0 max **Mo** 5.00-7.00 **N** 0.25 max **Ni** 1.0 max **P** 0.020 max **S** 0.010 max **Si** 1.0 max **W** 0.20 max | **ASTM** F75 (Cast); F1377 **Other:** ISO 5832/IV |
| **R30100** | Fe-Cr-Ni-Co Alloy | **C** 0.10 max **Co** 18.5-21.0 **Cr** 20.0-22.5 **Fe** rem **Mn** 1.00-2.00 **Mo** 2.50-3.50 **N** 0.10-0.20 **Ni** 19.0-21.0 **P** 0.040 max **S** 0.030 max **Si** 1.00 max **W** 2.00-3.00 | **AMS** 7724 |
| **R30102** | Co-Cr-W Alloy, P/M (Stellite Star J) | **B** 0.1-1.0 **C** 2.2-2.8 **Co** rem **Cr** 31-34 **Fe** 3.0 max **Mn** 2.0 max **Ni** 2.5 max **Si** 2.0 max **W** 16-19 | |
| * **R30103** | Co-Cr-W Alloy, P/M (Stellite 3, Weartech 3, WT-3 | **B** 0.1-1.0 **C** 2.0-2.8 **Co** rem **Cr** 28-33 **Fe** 5.0 max **Mn** 2.0 max **Ni** 3.5 max **Si** 2.0 max **W** 11-14 | |
| **R30106** | Co-Cr-W Alloy, P/M (Stellite 6, Weartech 6, WT-6) | **B** 0.1-1.0 **C** 0.7-1.4 **Cr** 25-32 **Fe** 5.0 max **Mn** 2.0 max **Mo** 1.5 max **Ni** 3.0 max **Si** 2.0 max **W** 3.0-6.0 | **AMS** 5387 **MIL SPEC** MIL-C-24689 (Alloy 20) |
| **R30155** | Fe-Cr-Ni-Co Alloy (N-155) | **C** 0.08-0.16 **Cb** 0.75-1.25 **Co** 18.50-21.00 **Cr** 20.00-22.50 **Fe** rem **Mn** 1.00-2.00 **Mo** 2.50-3.50 **N** 0.20 max **Ni** 19.00-21.00 **P** 0.040 max **S** 0.030 max **Si** 1.00 max **W** 2.00-3.00 | **AMS** 5376; 5532; 5585; 5768; 5769; 5794 **ASTM** B639 (661) **SAE** J467 (N-155) |
| **R30159** | Co-Ni-Cr-Fe-Mo Alloy (MP-159) | **Al** 0.10-0.30 **B** 0.03 max **C** 0.04 max **Cb** 0.25-0.75 **Co** 34.0-38.0 **Cr** 18.0-20.0 **Fe** 8.00-10.00 **Mn** 0.20 max **Mo** 6.00-8.00 **Ni** rem **P** 0.020 max **S** 0.010 max **Si** 0.20 max **Ti** 2.50-3.25 | **AMS** 5841; 5842; 5843; AS7475 |
| **R30188** | Co-Cr-Ni-W Alloy (HS188) | **C** 0.05-0.15 **Co** rem **Cr** 20.00-24.00 **Fe** 3.00 max **La** 0.03-0.15 **Mn** 1.25 max **Ni** 20.00-24.00 **Si** 0.20-0.50 **W** 13.00-16.00 | **AMS** 5608; 5772; 5801 |
| **R30260** | Co-Ni-Cr-Mo-W Alloy Age-Hardenable (Duratherm 2602) | **Be** 0.20-0.30 **C** 0.05 max **Cb** 0.10 max **Co** 41.0-42.0 **Cr** 11.7-12.3 **Cu** 0.30 max **Fe** 9.80-10.4 **Mn** 0.40-1.10 **Mo** 3.70-4.30 **Ni** rem **Si** 0.20-0.60 **Ti** 0.80-1.20 **W** 3.60-4.20 | |
| **R30400** | Co-Mo-Cr-Si Alloy (Tribaloy T-400) | **C** 0.80 max **Co** 61.0-63.0 **Cr** 7.50-8.50 **Mo** 27.0-29.0 **P** 0.030 max **S** 0.030 **Si** 2.20-2.60 **Other** Fe+Ni 3.00 max | **SAE** J775 (VF 8) |
| **R30404** | Co-Cr-W Alloy (Stellite 4, Weartech 4, WT-4) | **C** 0.6-1.1 **Co** rem **Cr** 28-32 **Fe** 3.0 max **Mn** 1.0 max **Ni** 3.0 max **Si** 1.5 max **W** 12-16 | **Other:** Alloy Digest Co-57; DTD 900, 4782 |

*Boxed entries are no longer active and are retained for reference purposes only.

The chemical compositions listed are for identification purposes and should not be used in lieu of the cross referenced specifications.

| UNIFIED NUMBER | DESCRIPTION | CHEMICAL COMPOSITION | CROSS REFERENCE SPECIFICATIONS |
|---|---|---|---|
| **R30477** | Co-Ni-Cr-Mo-W Alloy Age-Hardenable (Duratherm 477) | **Al** 0.15 max **Be** 0.05 max **Cb** 0.10 max **Co** 39.0-46.0 **Cr** 17.7-18. **Cu** 0.30 max **Fe** 4.50-5.50 **Mn** 0.70 max **Mo** 3.80-4.20 **Ni** rem **Si** 0.50 max **Ti** 0.80-1.20 **W** 3.80-4.20 | |
| **R30556** | Fe-Cr-Ni-Co Alloy (HS 556) | **Al** 0.10-0.50 **B** 0.02 max **C** 0.05-0.15 **Cb** 0.30 max **Co** 16.0-21.0 **Cr** 21.0-23.0 **Fe** rem **La** 0.005-0.10 **Mn** 0.50-2.00 **Mo** 2.50-4.00 **N** 0.10-0.30 **Ni** 19.0-22.5 **P** 0.04 max **S** 0.015 max **Si** 0.20-0.80 **Ta** 0.30-1.25 **W** 2.00-3.50 **Zr** 0.001-0.10 | **AMS** 5831; 5874; 5877 **ASME** SFA5.9 (ER3556) **ASTM** B366 (CRH556); B435; B572; B619; B622; B626 **AWS** A5.9 (ER3556) |
| **R30563** | Co-Cr-Ni-Fe-Mo-W Alloy | **C** 0.05 max **Co** rem **Cr** 18.00-22.00 **Fe** 4.00-6.00 **Mn** 1.00 max **Mo** 3.00-4.00 **Ni** 15.00-25.00 **S** 0.010 max **Si** 0.50 max **Ti** 0.50-3.50 **W** 3.00-4.00 | **ASTM** F563 **Other:** ISO 5832-8 |
| **R30590** | Fe-Cr-Co-Ni Alloy (S-590) | **C** 0.38-0.48 **Cb** 3.50-4.50 **Co** 18.50-21.50 **Cr** 19.00-22.00 **Cu** 0.50 max **Fe** rem **Mn** 2.00 max **Mo** 3.50-4.50 **Ni** 18.50-21.50 **P** 0.040 max **S** 0.030 max **Si** 1.00 max **W** 3.50-4.50 | **SAE** J467 (S590) |
| **R30600** | Wrought Co-Ni-Cr-Mo-W Alloy Age-Hardenable (Duratherm 600) | **Al** 0.6-0.8 **Be** 0.05 max **C** 0.05 max **Cb** 0.1 max **Co** 41.0-42.0 **Cr** 11.7-12.3 **Cu** 0.3 max **Fe** 8.5-8.9 **Mn** 0.4-1.1 **Mo** 3.7-4.3 **Ni** rem **Si** 0.2-0.6 **Ti** 1.8-2.2 **W** 3.6-4.2 | |
| **R30605** | Co-Cr-Ni-W Alloy (L-605) | **C** 0.05-0.15 **Co** rem **Cr** 19.0-21.0 **Fe** 3.0 max **Mn** 2.00 max **Ni** 9.0-11.0 **Si** 1.00 max **W** 14.0-16.0 | **AMS** 5537; 5759; 5796; AS7236 **ASTM** F90; F1091 **SAE** J467 (L-605) |
| **R30700** | Co-Ni-Cr-Mo-W Alloy Age-Hardenable (Duratherm 700) | **Al** 1.00-1.40 **Be** 0.05 max **C** 0.05 max **Cb** 0.10 max **Co** 41.0-42.0 **Cr** 11.7-12.3 **Cu** 0.30 max **Fe** 6.40-7.00 **Mn** 0.40-1.10 **Mo** 3.70-4.30 **Ni** rem **Si** 0.20-0.60 **Ti** 3.30-3.70 **W** 3.60-4.20 | |
| **R30783** | Co-Ni-Fe-Al-Cb-Cr Alloy (Inconel Alloy 783) | **Al** 5.0-6.0 **B** 0.003-0.012 **C** 0.03 max **Cb** 2.5-3.5 **Co** rem **Cr** 2.5-3.5 **Cu** 0.50 max **Fe** 24.0-27.0 **Mn** 0.50 max **Ni** 26.0-30.0 **P** 0.015 max **S** 0.005 max **Si** 0.50 max **Ti** 3.40 max | **AMS** 5940 **SAE** Draft F-95AB; AMS 5940 |
| **R30816** | Co-Cr-Ni-Cb-Mo-W Alloy (S-816) | **C** 0.32-0.42 **Cb** 3.50-4.50 **Co** 40.00 min **Cr** 19.00-21.00 **Fe** 5.00 max **Mn** 1.00-2.00 **Mo** 3.50-4.50 **Ni** 19.00-21.00 **P** 0.040 max **S** 0.030 max **Si** 1.00 max **W** 3.50-4.50 | **ASTM** B639 (671) **SAE** J467 (S618) |
| **R31000** | Co-Cr-W Alloy (Stoody 1 API) | **C** 1.5-2.0 **Co** rem **Cr** 23.0-28.0 **Fe** 3.0 max **Mn** 1.0 max **Ni** 2.5 max **Si** 0.8-1.5 **W** 11.0-14.0 | **Other:** NACE MR-01-76-76 |
| **R31001** | Co-Cr-W Alloy (Stellite Star J) | **C** 2.2-2.8 **Co** rem **Cr** 31-34 **Fe** 3.0 max **Mn** 2.0 max **Ni** 2.5 max **Si** 2.0 max **W** 16-19 | |
| **R31233** | Co-Cr-Ni-Mo Alloy, Solid Solution Strengthened (ULTIMET Alloy) | **B** 0.015 max **C** 0.02-0.10 **Co** rem **Cr** 23.5-27.5 **Fe** 1.0-5.0 **Mn** 0.1-1.5 **Mo** 4.0-6.0 **N** 0.03-0.12 **Ni** 7.0-11.0 **P** 0.030 max **S** 0.020 max **Si** 0.05-1.00 **W** 1.0-3.0 | **ASTM** B815; B818 |
| **R31537** | Co-Cr-Mo Alloy | **C** 0.149 max **Co** rem **Cr** 26.0-30.0 **Fe** 0.75 max **Mn** 1.0 max **Mo** 5.0-7.0 **N** 0.25 max **Ni** 1.0 max **Si** 1.0 max | **ASTM** F799; F1537 **Other:** ISO 5832-12 |
| **R31538** | Co-Cr-Mo Alloy for Surgical Implants | **C** 0.15-0.35 **Co** rem **Cr** 26.0-30.0 **Fe** 0.75 max **Mn** 1.0 max **Mo** 5.0-7.0 **N** 0.25 max **Ni** 1.0 max **Si** 1.0 max | **ASTM** F799; F1537 |

The chemical compositions listed are for identification purposes and should not be used in lieu of the cross referenced specifications.

| UNIFIED NUMBER | DESCRIPTION | CHEMICAL COMPOSITION | CROSS REFERENCE SPECIFICATIONS |
|---|---|---|---|
| **R31539** | Co-Cr-Mo Alloy for Surgical Implants | **Al** 1.0 max **C** 0.35 max **Co** rem **Cr** 26.0-30.0 **Fe** 0.75 max **La** 0.5 max **Mn** 1.0 max **Mo** 5.0-7.0 **N** 0.25 max **Ni** 1.0 max **Si** 1.0 max | **ASTM** f799; F1537 |
| **R39001** | Cobalt Brazing Filler Metal | **Al** 0.05 max **B** 0.7-0.9 **C** 0.35-0.45 **Co** rem **Cr** 18.0-20.0 **Fe** 1.0 max **Ni** 16.0-18.0 **P** 0.02 max **S** 0.02 max **Se** 0.005 max **Si** 7.5-8.5 **Ti** 0.05 max **W** 3.5-4.5 **Zr** 0.05 max **Other** 0.50 max | **ASME** SFA5.8 (BCo-1) **AWS** A5.8 (BCo-1) |
| **R50100** | Titanium Welding Filler Metal, Unalloyed (ERTi-1) | **C** 0.03 max **Fe** 0.10 max **H** 0.005 max **N** 0.015 max **O** 0.10 max **Ti** rem | **AWS** A5.16 (ERTi-1) **ASME** SFA5.16 (ERTi-1) |
| **R50120** | Titanium Welding Filler Metal (ERTi-2, W-Ti) | **C** 0.03 max **Fe** 0.20 max **H** 0.008 max **N** 0.020 max **O** 0.10 max **Ti** rem | **ASME** SFA5.16 (ERTi-2) **AWS** A5.16 (ERTi-2); C2.25 (W-Ti) |
| **R50125** | Titanium Welding Filler Metal | **C** 0.03 max **Fe** 0.20 max **H** 0.008 max **N** 0.020 max **O** 0.10-0.15 **Ti** rem | **AMS** 4951 **ASME** SFA5.16 (ERTi-3) **AWS** A5.16 (ERTi-3) |
| **R50130** | Titanium Welding Filler Metal | **C** 0.03 max **Fe** 0.30 max **H** 0.008 max **N** 0.020 max **O** 0.15-0.25 **Ti** rem | **AWS** A5.16 (ERTi-4) |
| **R50250** | Titanium, Unalloyed | **C** 0.10 max **Fe** 0.20 max **H** 0.015 max **N** 0.03 max **O** 0.18 max **Ti** rem | **AMS** AMS-T-81915 **ASTM** F67 (1); B265 (1); B338 (1); B348 (1); B381 (F-1); B861 (1); B862 (1); B863 (1); F467 (1); F468 (1); F1341 **MIL SPEC** MIL-T-81556 |
| **R50400** | Titanium, Unalloyed | **C** 0.10 max **Fe** 0.30 max **H** 0.015 max **N** 0.03 max **O** 0.25 max **Ti** rem | **AMS** 4902; 4941; 4942; AMS-T-9046 **ASTM** F67 (2); B265 (2); B337 (2); B338 (2); B348 (2); B367 (C-2); B381 (F-2); B861 (2); B862 (2); B863 (2); F467 (2); F468 (2); F1341 **MIL SPEC** MIL-T-81556 **SAE** J467 (A40) |
| **R50550** | Titanium, Unalloyed | **C** 0.10 max **Fe** 0.30 max **H** 0.015 max **N** 0.05 max **O** 0.35 max **Ti** rem **Other** each 0.1 max, total 0.4 max | **AMS** 4900; AMS-T-9046 **ASTM** F67 (3); B265 (3); B338 (3); B348 (3); B367 (C-3); B381 (F-3); B861 (3); B862 (3); B863 (3); F467; F468; F1341 **MIL SPEC** MIL-T-81556 **SAE** J467 (A55) |
| **R50700** | Titanium, Unalloyed | **C** 0.10 max **Fe** 0.50 max **H** 0.015 max **N** 0.05 max **O** 0.40 max **Ti** rem **Other** each 0.1 max, total 0.4 max | **AMS** 4901; 4921; AMS-T-9046 **ASTM** B265 (4); B348 (4); B381 (F-4); F67 (4); F467 (4); F468 (4); F1341; F1580 **MIL SPEC** MIL-T-81556 **SAE** J467 (A70) |
| **R52250** | Titanium, Low Alloyed | **C** 0.10 max **Fe** 0.20 max **H** 0.015 max **N** 0.03 max **O** 0.18 max **Ti** rem **Other** Pd 0.12-0.25 | **ASTM** B265 (11); B338 (11); B348 (11); B381 (F-11); B861 (11); B862 (11); B863 (11) |
| **R52252** | Titanium, Low Alloyed, with Pd | **C** 0.10 max **Fe** 0.20 max **H** 0.015 max **N** 0.03 max **O** 0.18 max **Pd** 0.04-0.08 **Other** residuals each 0.1 max, total 0.4 max | **ASTM** B265 (17); B338 (17); B348 (17); B381 (17); B861 (17); B862 (17); B863 (17) |
| **R52254** | Titanium, Low Alloyed, with Ru (Grade 27) | **C** 0.08 max **Fe** 0.20 max **H** 0.015 max **N** 0.03 max, 0.18 max **Ti** rem **Other** Ru 0.08-0.14 | **ASTM** B265 (27); B338 (27); B348 (27); B363 (27); B381 (27); B861 (27); B862 (27); B863 (27) |
| **R52255** | Titanium Alloy Filler Metal with Ru (ERTi-27) | **C** 0.03 max **Fe** 0.08 max **H** 0.008 max **N** 0.015 max **O** 0.1 max **Ti** rem **Other** Ru 0.10-0.14 | **AWS** A5.16 (ERTi-27) |
| **R52400** | Titanium, Low Alloyed | **C** 0.10 max **Fe** 0.30 max **H** 0.015 max **N** 0.03 max **O** 0.25 max **Ti** rem **Other** Pd 0.12-0.25 | **ASTM** B265 (7); B338 (7); B348 (7); B381 (F-7); B861 (7); B862 (7); B863 (7); F467 (7); F468 (7) |
| **R52401** | Titanium Welding Filler Metal | **C** 0.03 max **Fe** 0.20 max **H** 0.008 max **N** 0.020 max **O** 0.10 max **Pd** 0.12-0.25 **Ti** rem | **ASTM** B265 (7, 11); B338 (7, 11); B348 (7, 11); B367 (Ti-Pd7B, Ti-Pd8A); B381 (7, 11) **AWS** A5.16 (ERTi-7) |
| **R52402** | Titanium, Low Alloyed, with Pd | **C** 0.10 max **Fe** 0.30 max **H** 0.010 max **N** 0.03 max **O** 0.25 max **Pd** 0.04-0.08 **Other** residuals each 0.1 max, total 0.4 max | **ASTM** B265 (16); B338 (16); B348 (16); B381 (16); B861 (16); B862 (16); B863 (16) |
| **R52404** | Titanium, Low Alloyed, with Ru (Grade 26) | **C** 0.08 max **Fe** 0.30 max **H** 0.015 max **N** 0.03 max **O** 0.25 max **Ti** rem **Other** Ru 0.08-0.14 | **ASTM** B265 (26); B338 (26); B348 (26); B363 (26); B381 (26); B861 (26); B862 (26); B863 (26) |
| **R52550** | Titanium, Low Alloyed | **C** 0.10 max **Fe** 0.30 max **H** 0.015 max **N** 0.05 max **O** 0.35 max **Ti** rem **Other** Pd 0.12 min | **ASTM** B367 (C-2, C-3) |

The chemical compositions listed are for identification purposes and should not be used in lieu of the cross referenced specifications.

| UNIFIED NUMBER | DESCRIPTION | CHEMICAL COMPOSITION | CROSS REFERENCE SPECIFICATIONS |
|---|---|---|---|
| **R52700** | Titanium Alloy | **C** 0.10 max **Fe** 0.50 max **H** 0.0100 max **N** 0.05 max **O** 0.40 max **Pd** 0.12 min **Ti** rem | |
| **R53400** | Titanium Alloy | **C** 0.08 max **Fe** 0.30 max **H** 0.015 max **Mo** 0.2-0.4 **N** 0.03 max **Ni** 0.6-0.9 **O** 0.25 max **Ti** rem | **ASTM** B265 (12); B338 (12); B348 (12); B381 (F-12); B861 (12); B862 (12); B863 (12) |
| **R53401** | Titanium Alloy Welding Filler Metal | **C** 0.03 max **Fe** 0.30 max **H** 0.008 max **Mo** 0.2-0.4 **N** 0.020 max **Ni** 0.6-0.9 **O** 0.25 max **Ti** rem | **ASTM** B265 (12); B338 (12); B348 (12); B381 (F-12) **AWS** A5.16 (ERTi-12) |
| **R53413** | Titanium Alloy with Ni and Ru | **C** 0.10 max **Fe** 0.20 max **H** 0.015 max **N** 0.03 max **Ni** 0.4-0.6 **O** 0.10 max **Ru** 0.04-0.06 **Other** residuals each 0.1 max, total 0.4 max | **ASTM** B265 (13); B338 (13); B348 (13); B381 (13); B861 (13); B862 (13); B863 (13) |
| **R53414** | Titanium Alloy with Ni and Ru | **C** 0.10 max **Fe** 0.30 max **H** 0.015 max **N** 0.03 max **Ni** 0.40-0.60 **O** 0.15 max **Ru** 0.04-0.06 **Other** residuals each 0.1 max, total 0.4 max | **ASTM** B265 (14); B338 (14); B348 (14); B381 (14); B861 (14); B862 (14); B863 (14) |
| **R53415** | Titanium Alloy with Ni and Ru | **C** 0.10 max **Fe** 0.30 max **H** 0.015 max **N** 0.03 max **Ni** 0.40-0.60 **O** 0.25 max **Ru** 0.04-0.06 **Other** residuals each 0.1 max, total 0.4 max | **ASTM** B265 (15); B338 (15); B348 (15); B381 (15); B861 (15); B862 (15); B863 (15) |
| **R54520** | Titanium Alloy | **Al** 4.0-6.0 **C** 0.10 max **Fe** 0.50 max **H** 0.020 max **N** 0.05 max **O** 0.20 max **Sn** 2.0-3.0 **Ti** rem | **AMS** 4910; 4926; 4966; AMS-T-9046; AMS-T-81915 **ASTM** B265 (6); B348 (6); B367 (C-6); B381 (F-6); B863 (9) **MIL SPEC** MIL-T-81556 **SAE** J437 (5Al-2.5Sn) |
| **R54521** | Titanium Alloy | **Al** 5 **Sn** 2.5 **Ti** rem | **AMS** 4909; 4924; AMS-T-9046 **MIL SPEC** MIL-T-81556 |
| **R54522** | Titanium Alloy Welding Filler Metal | **Al** 4.7-5.6 **C** 0.05 max **Fe** 0.40 max **H** 0.008 max **N** 0.030 max **O** 0.12 max **Sn** 2.0-3.0 **Ti** rem | **AMS** 4953 **ASME** SFA5.16 (ERTi-6) **AWS** A5.16 (ERTi-6) |
| **R54523** | Titanium Alloy Welding Filler Metal | **Al** 4.5-5.8 **C** 0.03 max **Fe** 0.50 max **H** 0.005 max **N** 0.012 max **O** 0.10 max **Sn** 2.0-3.0 **Ti** rem **Y** 0.005 max | **ASME** SFA5.16 (ERTi-6EL1 **AWS** A5.16 (ERTi-6ELI) |
| * **R54550** | Titanium Alloy | **Al** 5 **Sn** 5 **Ti** rem **Zr** 5 | |
| **R54560** | Titanium Alloy | **Al** 5 **Mo** 1 **Si** 0.2 **Sn** 6 **Ti** rem **Zr** 2 | **AMS** AMS-T-9046 |
| **R54620** | Titanium Alloy | **Al** 6 **Mo** 2 **Sn** 2 **Ti** rem **Zr** 4 | **AMS** 4919; 4975; 4976; AMS-T-9046; AMS-T-81915 **SAE** J467 (Ti-6-4-2) |
| **R54621** | Titanium Alloy Welding Filler Metal | **Al** 5.50-6.50 **C** 0.04 max **Cr** 0.25 max **Fe** 0.05 max **H** 0.15 max **Mo** 1.80-2.20 **N** 0.015 max **O** 0.30 max **Sn** 1.80-2.20 **Ti** rem **Zr** 3.60-4.40 | **AMS** 4575; 4952 |
| * **R54624** | Titanium 6Al-1.7Fe Alloy | **Al** 5.5-6.5 **C** 0.08 max **Fe** 1.3-2.0 **H** 0.0125 max **N** 0.04 max **O** 0.15-0.20 **Si** 0.07-0.13 **Ti** rem **Other** residual elements 0.30 max total | **ASTM** B265 |
| * **R54625** | Titanium Al-Fe-Pd Alloy | **Al** 5.5-6.5 **C** 0.08 max **Fe** 1.3-2.0 **H** 0.0125 max **N** 0.04 max **O** 0.15-0.20 **Pd** 0.04-0.08 **Si** 0.07-0.13 **Ti** rem **Other** 0.03 max total | |
| **R54790** | Titanium Alloy | **Al** 2 **Mo** 1 **Si** 0.2 **Sn** 11 **Ti** rem **Zr** 5 | **AMS** 4974 **SAE** J467 (Ti-169) |
| **R54810** | Titanium Alloy | **Al** 8 **Mo** 1 **Ti** rem **V** 1 | **AMS** 4915; 4916; 4933; 4955; 4972; 4973; AMS-T-9046 **MIL SPEC** MIL-T-81556 **SAE** J437 (8Al-1Mo-1V) |

*Boxed entries are no longer active and are retained for reference purposes only.

The chemical compositions listed are for identification purposes and should not be used in lieu of the cross referenced specifications.

| UNIFIED NUMBER | DESCRIPTION | CHEMICAL COMPOSITION | CROSS REFERENCE SPECIFICATIONS |
|---|---|---|---|
| **R55111** | Titanium Alloy 5Al-1Sn-1V-1Zr-1Mo (Grade 32) | **Al** 4.5-5.5 **C** 0.08 max **Fe** 0.25 max **H** 0.015 max **Mo** 0.6-1.2 **N** 0.03 max **O** 0.11 max **Si** 0.06-0.14 **Sn** 0.6-1.4 **Ti** rem **V** 0.6-1.4 **Zr** 0.6-1.4 | **ASTM** B265; B348; F467; F467 (32); F468; F468 (32) |
| **R56080** | Titanium Alloy | **Mn** 8 **Ti** rem | **AMS** 4908; AMS-T-9045; AMS-T-9046 **SAE** J437 (8Mn) |
| **R56210** | Titanium Alloy | **Al** 6 **Cb** 2 **Mo** 0.8 **Ta** 1 **Ti** rem | **AMS** AMS-T-9046 **AWS** A5.16 (ERTi-15) |
| **R56260** | Titanium Alloy | **Al** 6 **Mo** 6 **Sn** 2 **Ti** rem **Zr** 4 | **AMS** 4981 |
| **R56320** | Titanium Alloy | **Al** 2.5-3.5 **C** 0.05 max **Fe** 0.25 max **H** 0.013 max **N** 0.02 max **O** 0.12 max **Ti** rem **V** 2.0-3.0 | **AMS** 4943; 4944; 4945; AMS-T-9046 **ASME** SFA5.16 (ERTi-9) **ASTM** B265 (9); B338 (9); B348 (9); B381 (F-9); B861 (9); B862 (9); B863 (9) **AWS** A5.16 (ERTi-9) |
| **R56321** | Titanium Welding Filler Metal (ERTi-9EL1) | **Al** 2.5-3.5 **C** 0.03 max **Fe** 0.25 max **H** 0.005 max **N** 0.012 max **O** 0.10 max **Ti** rem **V** 2.0-3.0 **Y** 0.005 max | **ASME** SFA5.16 (ERTi-9EL1) **AWS** A5.16 (ERTi-9ELI) |
| **R56322** | Titanium Alloy 3Al-2.5V-0.05Pd | **Al** 2.5-3.5 **C** 0.10 max **Fe** 0.25 max **H** 0.015 max **N** 0.02 max **O** 0.15 max **Pd** 0.04-0.08 **V** 2.0-3.0 **Other** residuals each 0.1 max, total 0.4 max | **ASTM** B265 (18); B338 (18); B348 (18); B381 (18); B861 (18); B862 (18); B863 (18) |
| **R56323** | Titanium Alloy 3Al-2.5V-Ru (Grade 28) | **Al** 2.5-3.5 **C** 0.08 max **Fe** 0.25 max **H** 0.015 max **N** 0.03 max **O** 0.15 max **Ru** 0.08-0.14 **Ti** rem **V** 2.0-3.0 | **ASTM** B265 (28); B338 (28); B348 (28); B363; B861; B862; B863; B863 (28) |
| **R56400** | Titanium Alloy | **Al** 5.5-6.75 **C** 0.10 max **Fe** 0.40 max **H** 0.015 max **N** 0.05 max **O** 0.20 max **Ti** rem **V** 3.5-4.5 | **AMS** 4905; 4911; 4920; 4928; 4930; 4931; 4932; 4934; 4935; 4954; 4963; 4965; 4967; 4993; AMS-T-9046; AMS-T-81915; AS7460; AS7461 **ASTM** B265 (5); B348 (5); B367 (C-5); B381 (F-5); B861 (5); B862 (5); B863 (5); F1472 **AWS** A5.16 (ERTi-5) **MIL SPEC** MIL-T-81556 |
| **R56401** | Titanium Alloy 6Al-4V ELI | **Al** 5.5-6.50 **C** 0.08 max **Fe** 0.25 max **H** 0.012 max **N** 0.05 max **O** 0.13 max **Ti** rem **V** 3.5-4.5 | **AMS** 4907; 4985; 4991; 4996; 4998; AMS-T-9046 **ASTM** F136; F467 (5); F467 (23); F468 (5); F468 (23) **MIL SPEC** MIL-T-81556 **SAE** J775 (XEV-J) |
| **R56402** | Titanium Alloy Welding Filler Metal | **Al** 5.5-6.75 **C** 0.04 max **Fe** 0.15 max **H** 0.005 max **N** 0.012 max **O** 0.10 max **Ti** rem **V** 3.5-4.5 **Y** 0.005 max | **AMS** 4956 **AWS** A5.16 (ERTi-5ELI) |
| **R56403** | Titanium Alloy with V and Pd (Grade 25) | **Al** 5.5-6.75 **C** 0.10 max **Fe** 0.40 max **H** 0.0125 max **N** 0.05 max **Ni** 0.3-0.8 **O** 0.20 max **Pd** 0.04-0.08 **Ti** rem **V** 3.5-4.5 **Other** residuals each 0.1 max, total 0.4 max | **ASME** SB-265; SB-348; SB-363; SB-381 **ASTM** B265; B348 (25); B363; B381; B381 (F-25); B861 (25); B862 (25); B863 (25) **Other:** NACE MR0175 |
| **R56404** | Titanium Alloy 6Al-4V-Ru (Grade 29) | **Al** 5.5-6.5 **C** 0.08 max **Fe** 0.25 max **H** 0.015 max **N** 0.03 max **O** 0.13 max **Ru** 0.08-0.14 **Ti** rem **V** 3.5-4.5 | **ASTM** B265; B337 (29); B348 (29); B363; B861; B862; B863 |
| **R56405** | Titanium 6Al-4V-Pd Alloy (Grade 24) | **Al** 5.5-6.75 **C** 0.10 max **Fe** 0.40 max **H** 0.015 max **N** 0.05 max **O** 0.20 max **Pd** 0.04-0.08 **V** 3.5-4.5 **ASME** SB-265; SB-348; SB-363; SB-381 **Other** residual elements 0.10 max each, 0.40 max total | **ASTM** B265; B348 (24); B363; B381; B861 (240); B862 (24); B863 (24) |
| **R56406** | Titanium Alloy 6Al-4V (Cast Surgical Implants) | **Al** 5.5-6.75 **C** 0.10 max **Fe** 0.30 max **H** 0.015 max **N** 0.05max **O** 0.20 max **Ti** rem **V** 3.5-4.5 | **ASTM** F1108 **SAE** J437 (6Al-4V) |
| **R56407** | Titanium Alloy, 6Al-4V ELI (Grade 23) | **Al** 5.5-6.5 **C** 0.08 max **Fe** 0.25 max **H** 0.0125 max **N** 0.03 max **O** 0.13 max **Ti** rem **V** 3.5-4.5 | **ASME** SB-265; SB-348; SB-365; SB-381 **ASTM** B265 (23); B348 (23); B363 (23); B381 (23); B381 (F23); B861 (23); B862 (23); B863 (23); B865 (23); F620 **MIL SPEC** MIL-T-81556 (AB-2) |
| **R56410** | Titanium - 10V - 3A1 - 2 Fe Alloy | **Al** 2.6-3.4 **C** 0.05 max **H** 0.015 max **Fe** 1.6-2.2 **N** 0.05 max **O** 0.13 max **Ti** rem **V** 9.0-11.0 **Y** 0.005 max **Other** Residuals each 0.10 max, total 0.30 max | **AMS** 4984; 4986; 4987 |
| **R56430** | Titanium Alloy | **Al** 4 **Mo** 3 **Ti** rem **V** 1 | **AMS** AMS-T-9046 **SAE** J437 (4Al-3Mo-1V) |
| *** R56440** | Titanium Alloy | **Al** 4 **Mn** 4 **Ti** rem | |

*Boxed entries are no longer active and are retained for reference purposes only.

The chemical compositions listed are for identification purposes and should not be used in lieu of the cross referenced specifications.

| UNIFIED NUMBER | DESCRIPTION | CHEMICAL COMPOSITION | CROSS REFERENCE SPECIFICATIONS |
|---|---|---|---|
| **R56620** | Titanium Alloy | **Al** 5.5 **Sn** 2 **Ti** rem **V** 5.5 | **AMS** 4918; 4936; 4937; 4971; 4978; 4979; AMS-T-9046 **MIL SPEC** MIL-T-81556 |
| **R56700** | Ti-6 Al-7 Nb Alloy | **Al** 5.50-6.50 **C** 0.08 max **Cb** 6.50-7.50 **Fe** 0.25 max **H** 0.009 max **N** 0.05 max **O** 0.20 max **Ta** 0.50 max **Ti** rem | **ASTM** F1295 **Other:** ISO 5832-11 |
| **R56740** | Titanium Alloy | **Al** 7 **Mo** 4 **Ti** rem | **AMS** 4970 **SAE** J437 (7Al-4Mo) |
| **R58010** | Titanium Alloy | **Al** 3 **Cr** 11 **Ti** rem **V** 13 | **AMS** 4917; 4959; AMS-T-9046 **SAE** J437 (3Al-13V-11Cr) |
| **R58030** | Titanium Alloy | **C** 0.10 max **Fe** 0.35 max **H** 0.020 max **Mo** 10.0-13.0 **N** 0.05 max **O** 0.18 max **Sn** 3.75-5.25 **Ti** rem **Zr** 4.50-7.50 | **AMS** AMS-T-9046 |
| **R58120** | Titanium Alloy, Ti-12Mo-6Zr-2Fe (Surgical Implants) | **C** 0.05 max **Fe** 1.5-2.5 **H** 0.020 max **Mo** 10.0-13.0 **N** 0.05 max **O** 0.08-0.28 **Ti** rem **Zr** 5.0-7.0 | **ASTM** F1813 |
| **R58130** | Titanium Alloy, Ti-13Nb-13Zr (Surgical Implants) | **C** 0.08 max **Cb** 12.5-14.0 **Fe** 0.25 max **H** 0.012 max **N** 0.05 max **O** 0.15 max **Ti** rem **Zr** 12.5-14.0 | **ASTM** F1713 |
| **R58150** | Titanium-15 Mo Alloy (biomedical) | **C** 0.10 max **Fe** 0.10 max **H** 0.015 max **Mo** 14.0-16.0 **N** 0.05 max **O** 0.20 max **Ti** rem | **ASTM** F2066 |
| **R58153** | Titanium 15V-3Al-3Cr-3Sn Alloy | **Al** 2.5-3.5 **C** 0.05 max **Cr** 2.5-3.5 **Fe** 0.25 max **H** 0.015 max **N** 0.05 max **O** 0.13 max **Sn** 2.5-3.5 **V** 14.0-16.0 **Other** residual elements 0.10 max each, 0.40 max total | **AMS** 4914 |
| **R58210** | Titanium 15Mo-3Al-2.5Cb Alloy (Grade 21) | **Al** 2.5-3.5 **C** 0.05 max **Cb** 2.2-3.2 **Fe** 0.20-0.40 **H** 0.015 max **Mo** 14.0-16.0 **N** 0.05 max **O** 0.11-0.15 **Si** 0.15-0.25 **Other** residual elements 0.10 max each, 0.40 max total | **ASME** SB-265; SB-348; SB-363; SB-381 **ASTM** B265; B265 (21); B348 (21); B363; B381 (F-21); B861 (21); B862 (21); B863 (21) |
| **R58350** | Titanium 35Nb-7Zr-5Ta Alloy (Biomedical) (TiOsteum, TNZT) | **Al** 0.05 max **C** 0.02 max **H** 0.02 max **N** 0.02 max **Nb** 34.0-37.0 **O** 0.6 max **Si** 0.5 max **Ta** 4.5-6.5 **Ti** rem **Zr** 6.0-8.0 | **ASTM** pending |
| **R58450** | Titanium Alloy | **Cb** 45 **Ti** rem | **AMS** 4982 |
| **R58640** | Titanium Alloy Beta C (Grade 19) | **Al** 3.0-4.0 **C** 0.05 max **Cr** 5.5-6.5 **Fe** 0.30 max **H** 0.02 max **Mo** 3.5-4.5 **N** 0.03 max **O** 0.12 max **Ti** rem **V** 7.5-8.5 **Zr** 3.5-4.5 **Other** residuals each 0.15 max, total 0.4 max | **AMS** 4957; 4958; AMS-T-9046 **ASME** SB-265; SB-348; SB-363; SB-381 **ASTM** B265; B348; B363; B381; F467 (19); F468 (19) |
| **R58645** | Titanium Alloy Beta C plus 0.05 Pd | **Al** 3.0-4.0 **C** 0.05 max **Cr** 5.5-6.5 **Fe** 0.30 max **Mo** 3.5-4.5 **N** 0.03 max **O** 0.14 max **Pd** 0.04-0.08 **V** 7.5-8.5 **Zr** 3.5-4.5 **Other** residuals each 0.15 max, total 0.4 max | **ASTM** B265 (20); B348 (20); B381 (20) |
| **R58650** | Titanium Alloy, 5Al-2Sn-2Zr-4Cr-4Mo; B861 (20); B862 (20); B863 (20) | **Al** 4.50-5.50 **C** 0.05 max **Cr** 3.50-4.50 **Cu** 0.10 max **Fe** 0.30 max **H** 0.0125 max **Mn** 0.10 max **Mo** 3.50-4.50 **N** 0.04 max **O** 0.08-0.13 **Sn** 1.50-2.50 **Ti** rem **Y** 0.0050 max **Zr** 1.50-2.50 **Other** each 0.10 max, total 0.30 max | **AMS** 4997 |
| **R58820** | Titanium Alloy | **Al** 3 **Fe** 2 **Mo** 8 **Ti** rem **V** 8 | **AMS** AMS-T-9046 |
| **R60001** | Zirconium Unalloyed Reactor Grade | **Al** 0.0075 max **B** 0.00005 max **C** 0.027 max **Cd** 0.00005 max **Co** 0.0020 max **Cr** 0.020 max **Cu** 0.0050 max **Fe** 0.150 max **H** 0.0025 max **Hf** 0.010 max **Mn** 0.0050 max **N** 0.0065 max **Ni** 0.0070 max **Si** 0.0120 max **Ti** 0.0050 max **U** 0.00055 max **W** 0.010 max **Zr** rem | **ASTM** B349; B350; B351; B352; B353 |

*Boxed entries are no longer active and are retained for reference purposes only.

The chemical compositions listed are for identification purposes and should not be used in lieu of the cross referenced specifications.

| UNIFIED NUMBER | DESCRIPTION | CHEMICAL COMPOSITION | CROSS REFERENCE SPECIFICATIONS |
|---|---|---|---|
| * **R60701** | Zirconium, Unalloyed | **C** 0.05 max **H** 0.005 max **Hf** 4.5 max **N** 0.025 max **Other** Zr+Hf 99.5 min, Fe+Cr 0.05 max | |
| **R60702** | Zirconium, Unalloyed | **C** 0.05 max **H** 0.005 max **Hf** 4.5 max **N** 0.025 max $O_2$ 0.16 max **Other** Zr+Hf 99.2 min, Fe+Cr 0.2 max | **ASME** SB-493; SB-494; SB-495; SB-523; SB-550; SB-551; SFA5.24 (ERZr2) **ASTM** B493; B494; B495; B523; B550; B551; B653; B658; B752 (702C) **AWS** A5.24 (ERZr2) |
| **R60703** | Zirconium Alloy | **Hf** 4.5 max **Other** Zr+Hf 98.0 min | **ASTM** B494; B495 |
| **R60704** | Zirconium Alloy | **C** 0.05 max **H** 0.005 max **Hf** 4.5 max **N** 0.025 max $O_2$ 0.18 max **Sn** 1.00-2.00 **Other** Zr+Hf 97.5 min, Fe+Cr 0.20-0.40 | **ASME** SFA5.24 (ERZr3) **ASTM** B493; B495; B523; B550; B551; B653; B658; B752 (704C) **AWS** A5.24 (ERZr3) |
| **R60705** | Zirconium Alloy | **C** 0.05 max **Cb** 2.0-3.0 **H** 0.005 max **Hf** 4.5 max **N** 0.025 max **O** 0.18 max **Other** (Zr+Hf) 95.5 min, (Fe+Cr) 0.2 max | **ASME** SFA5.24 (ERZr4) **ASTM** B493; B495; B523; B550; B551; B653; B658; B752 (705C) **AWS** A5.24 (ERZr4) |
| **R60706** | Zirconium Alloy | **C** 0.050 max **Cb** 2.0-3.0 **H** 0.005 max **Hf** 4.5 max **N** 0.025 max **O** 0.16 max **Other** Zr+Hf 95.5 min Fe+Cr 0.2 max | **ASTM** B495; B551 |
| * **R60707** | Zirconium Alloy (2.5 Cb) Welding Filler Metal | **C** 0.05 max **Cb** 2.0-3.0 **Fe** 0.20 max includes Cr **H** 0.005 max **Hf** 4.5 max **N** 0.025 max **Zr+Hf** 95.5 min | **ASME** SFA5.24 (ERZr4) **AWS** A5.24 (ERZr4) |
| **R60802** | Zirconium Alloy Reactor Grade | **Al** 0.0075 max **B** 0.00005 max **C** 0.027 max **Cd** 0.00005 max **Co** 0.0020 max **Cr** 0.05-0.15 **Cu** 0.0050 max **Fe** 0.07-0.20 **H** 0.0025 max **Hf** 0.010 max **Mn** 0.0050 max **N** 0.0065 max **Ni** 0.05-0.08 **Si** 0.020 max **Sn** 1.20-1.70 **Ti** 0.0050 max **U** 0.00035 max **W** 0.010 max **Zr** rem **Other** Fe+Cr+Ni 0.18-0.38 | **ASTM** B350; B351; B352; B353; B811 |
| **R60804** | Zirconium Alloy Reactor Grade | **Al** 0.0075 max **B** 0.00005 max **C** 0.027 max **Cd** 0.00005 max **Co** 0.0020 max **Cr** 0.07-0.13 **Cu** 0.0050 max **Fe** 0.18-0.24 **H** 0.0025 max **Hf** 0.010 max **Mn** 0.0050 max **N** 0.0065 max **Ni** 0.0070 max **Si** 0.0120 max **Sn** 1.20-1.70 **Ti** 0.0050 max **U** 0.00035 max **W** 0.010 max **Zr** rem **Other** Fe+Cr 0.28-0.37 | **ASTM** B350; B351; B352; B353; B811 |
| **R60901** | Zirconium Alloy Reactor Grade | **Al** 0.0075 max **B** 0.00005 max **C** 0.027 max **Cb** 2.40-2.80 **Cd** 0.00005 max **Co** 0.0020 max **Cr** 0.020 max **Cu** 0.0050 max **Fe** 0.150 max **H** 0.0025 max **Hf** 0.010 max **Mn** 0.0050 max **N** 0.0065 max **Ni** 0.0070 max **O** 0.09-0.13 **Si** 0.0120 max **Ti** 0.0050 max **U** 0.00035 max **W** 0.010 max **Zr** rem | **ASTM** B350; B351; B352; B353 |
| **R60902** | Zirconium Alloy Reactor Grade | **Al** 0.0075 max **B** 0.00005 max **C** 0.030 max **Cb** 2.4-2.8 **Cd** 0.00005 max **Co** 0.0020 max **Cr** 0.020 max **Cu** 0.3-0.7 **Fe** 0.15 max **H** 0.0025 max **Hf** 0.010 max **Mg** 0.0020 max **Mn** 0.0050 max **N** 0.0065 max **Ni** 0.0070 max **O** 0.08-0.12 **Pb** 0.013 max **Si** 0.012 max **Sn** 0.0050 max **Ta** 0.020 max **Ti** 0.0050 max **U** 0.00035 max **V** 0.0050 max **W** 0.010 max **Zr** rem | |

*Boxed entries are no longer active and are retained for reference purposes only.

The chemical compositions listed are for identification purposes and should not be used in lieu of the cross referenced specifications.

| UNIFIED NUMBER | DESCRIPTION | CHEMICAL COMPOSITION | CROSS REFERENCE SPECIFICATIONS |
|---|---|---|---|
| **R60904** | Zirconium - 0.65 Nb | **Al** 0.0075 max **B** 0.00005 max **Cb** 2.50-2.80 **Co** 0.0020 max **Cu** 0.0050 max **Fe** 0.0650 max **H** 0.0020 max **Hf** 0.0050 max **Mg** 0.0020 max **Mn** 0.0050 max **Mo** 0.0050 max **N** 0.0065 max **Ni** 0.0035 max **Pb** 0.0050 max **Si** 0.0120 max **Sn** 0.0100 max **Ta** 0.0100 max **Ti** 0.0050 max **U** 0.00035 max **V** 0.0050 max **W** 0.0050 max **Zr** rem | **ASTM** B353 |

The chemical compositions listed are for identification purposes and should not be used in lieu of the cross referenced specifications.

## UNS NUMBERS ASSIGNED TO DATE

### With Description of Each Material Covered and References To Documents In Which The Same or Similar Materials are Described

### Sxxxxx Number Series
### Heat and Corrosion Resistant Steels (Including Stainless), Valve Steels, and Iron-Base "Superalloys"

# HEAT AND CORROSION RESISTANT STEELS (INCLUDING STAINLESS), VALVE STEELS, AND IRON-BASE "SUPERALLOYS"

| UNIFIED NUMBER | DESCRIPTION | CHEMICAL COMPOSITION | CROSS REFERENCE SPECIFICATIONS |
|---|---|---|---|
| **S13800** | Precipitation Hardenable Cr-Ni-Al-Mo Stainless Steel (PH 13-8 Mo) | **Al** 0.90-1.35 **C** 0.05 max **Cr** 12.25-13.25 **Mn** 0.20 max **Mo** 2.00-2.50 **N** 0.01 max **Ni** 7.50-8.50 **P** 0.01 max **S** 0.008 max **Si** 0.10 max | **AISI** S13800 **AMS** 5412; 5629; 5840; 5864 **ASME** SA-705 (XM-13) **ASTM** A564 (XM-13); A693 (XM-13); A705 (XM-13) |
| **S13889** | Precipitation Hardenable Cr-Ni-Al-Mo Stainless Steel Bare Filler Metal (PH 13-8 Mo) | **Al** 0.90-1.35 **C** 0.05 max **Cr** 12.25-13.25 **H** 0.0025 max **Mn** 0.10 max **Mo** 2.00-2.50 **N** 0.01 max **Ni** 7.50-8.50 **O** 0.005 max **P** 0.008 max **S** 0.010 max **Si** 0.10 max | **AMS** 5840 |
| **S14800** | Precipitation Hardenable Cr-Ni-Mo-Al Stainless Steel (PH 14-8 Mo) | **Al** 0.75-1.50 **C** 0.05 max **Cr** 13.75-15.00 **Mn** 1.00 max **Mo** 2.00-3.00 **Ni** 7.75-8.75 **P** 0.015 max **S** 0.010 max **Si** 1.00 max | **AMS** 5603 |
| **S15500** | Precipitation Hardenable Cr-Ni-Cu Stainless Steel (15-5 PH) | **C** 0.07 max **Cb** 0.15-0.45 **Cr** 14.00-15.50 **Cu** 2.50-4.50 **Mn** 1.00 max **Ni** 3.50-5.50 **P** 0.040 max **S** 0.030 max **Si** 1.0 max | **AISI** S15500 **AMS** 5659; 5826; 5862 **ASME** SA-705 (XM-12) **ASTM** A564 (XM-12); A693 (XM-12); A705 (XM-12) |
| **S15700** | Precipitation Hardenable Cr-Ni-Mo-Al Stainless Steel (PH 15-7 Mo) | **Al** 0.75-1.50 **C** 0.09 max **Cr** 14.00-16.00 **Mn** 1.00 max **Mo** 2.00-3.00 **Ni** 6.50-7.75 **P** 0.04 max **S** 0.03 max **Si** 1.00 max | **AMS** 5520; 5657 **ASTM** A564 (632); A579 (63); A693 (632); A705 (632) **SAE** J467 (PH15-7-Mo) |
| **S15780** | Precipitation Hardenable Cr-Ni-Mo-Al Stainless Steel Bare Filler Metal (PH 15-7 Mo) | **Al** 0.75-1.25 **C** 0.09 max **Cr** 14.00-15.25 **Mn** 1.00 max **Mo** 2.00-2.75 **Ni** 6.50-7.75 **P** 0.025 max **S** 0.025 max **Si** 0.50 max | **AMS** 5812 |
| **S15789** | Precipitation Hardenable Cr-Ni-Mo-Al Stainless Steel Bare Filler Metal (PH 15-7 Mo) | **Al** 0.75-1.25 **C** 0.09 max **Cr** 14.00-15.25 **H** 0.0025 max **Mn** 1.00 max **Mo** 2.00-2.75 **Ni** 6.50-7.75 **O** 0.005 max **P** 0.010 max **S** 0.010 max **Si** 0.50 max | **AMS** 5812 |
| * **S16600** | Precipitation Hardenable Cr-Ni-Ti-Al Stainless Steel (Croloy 16-6 PH) | **Al** 0.25-0.40 **C** 0.025-0.045 **Cr** 15.00-16.00 **Mn** 0.70-0.90 **Ni** 7.00-8.00 **P** 0.025 max **S** 0.025 max **Si** 0.50 max **Ti** 0.30-0.50 | |
| **S16800** | Cr-Ni-Mo Stainless Steel (16-8-2-H) | **C** 0.05-0.10 **Cr** 14.5-16.5 **Mn** 2.00 max **Mo** 1.50-2.00 **Ni** 7.50-9.50 **P** 0.040 max **S** 0.030 max **Si** 0.75 max | **ASTM** A376 (16-8-2-H) |
| **S16880** | Cr-Ni-Mo Stainless Steel Bare Filler Metal (16-8-2) | **C** 0.10 max **Cr** 14.5-16.5 **Cu** 0.75 max **Mn** 1.00-2.50 **Mo** 1.00-2.00 **Ni** 7.50-9.50 **P** 0.03 max **S** 0.03 max **Si** 0.30-0.65 | **ASME** SFA5.9 (ER16-8-2) **AWS** A5.9 (ER16-8-2) |
| **S17400** | Precipitation Hardenable Cr-Ni-Cu Stainless Steel (17-4 PH) | **C** 0.07 max **Cb** 0.15-0.45 **Cr** 15.0-17.5 **Cu** 3.00-5.00 **Mn** 1.00 max **Ni** 3.00-5.00 **P** 0.040 max **S** 0.030 max **Si** 1.00 max | **AISI** S17400 **AMS** 5604; 5622; 5643; AS7474 **ASME** SA-564 (630); SA-705 (630) **ASTM** A564 (630); A693 (630); A705 (630); F593 (630); F593 (XM7); F594 (630); F594 (XM7); F738 (630); F836 (630); F899 (630) **SAE** J467 (17-4PH) |
| **S17480** | Precipitation Hardenable Cr-Ni-Cu Stainless Steel Bare Filler Metal (17-4 PH) | **C** 0.05 max **Cr** 16.00-16.75 **Cu** 3.25-4.00 **Mn** 0.25-0.75 **Mo** 0.75 max **Ni** 4.50-5.00 **P** 0.04 max **S** 0.03 max **Si** 0.75 max **Other** Cb+Ta 0.15-0.30 | **AMS** 5803; 5825 **ASME** SFA5.9 (ER630) **AWS** A5.9 (ER630) |
| **S17600** | Precipitation Hardenable Cr-Ni-Al-Ti Stainless Steel (Stainless W) | **Al** 0.40 max **C** 0.08 max **Cr** 15.00-17.50 **Mn** 1.00 max **Ni** 6.00-7.50 **P** 0.040 max **S** 0.030 max **Si** 1.00 max **Ti** 0.40-1.20 | **ASTM** A564 (635); A693 (635); A705 (635) **SAE** J467 (Stainless W) |
| **S17700** | Precipitation Hardenable Cr-Ni-Al Stainless Steel (17-7 PH) | **Al** 0.75-1.50 **C** 0.09 max **Cr** 16.00-18.00 **Mn** 1.00 max **Ni** 6.50-7.75 **P** 0.040 max **S** 0.040 max **Si** 1.00 max | **AISI** S17700 **AMS** 5528; 5529; 5568; 5644; 5678 **ASME** SA-705 (631) **ASTM** A313 (631); A564 (631); A579 (62); A693 (631); A705 (631); F899 (631) **SAE** J467 (17-7PH) |
| **S17780** | Precipitation Hardenable Cr-Ni-Al Stainless Steel Bare Filler Metal (17-7-PH) | **Al** 0.75-1.25 **C** 0.09 max **Cr** 16.00-17.25 **Mn** 1.00 max **Ni** 6.50-7.75 **P** 0.025 max **S** 0.025 max **Si** 0.50 max | **AMS** 5824 |

*Boxed entries are no longer active and are retained for reference purposes only.

The chemical compositions listed are for identification purposes and should not be used in lieu of the cross referenced specifications.

| UNIFIED NUMBER | DESCRIPTION | CHEMICAL COMPOSITION | CROSS REFERENCE SPECIFICATIONS |
|---|---|---|---|
| **S18200** | Ferritic Cr-Mo Free-Machining Stainless Steel (18-2FM, XM-34) | **C** 0.08 max **Cr** 17.50-19.50 **Mn** 1.25-2.50 **Mo** 1.50-2.50 **P** 0.04 max **S** 0.15 min **Si** 1.00 max | **ASTM** A581(XM-34); A582(XM-34) |
| **S18235** | Ferritic Cr Stainless Steel, Low Carbon, Free Machining | **C** 0.025 max **Cr** 17.5-18.5 **Mn** 0.50 max **Mo** 2.00-2.50 **N** 0.025 max **Ni** 1.00 max **P** 0.030 max **S** 0.15-0.35 **Si** 1.00 max **Ti** 0.30-1.00 **Other** C+N 0.035 max | **ASTM** A581; A582 |
| **S20100** | Austenitic Cr-Mn-Ni Stainless Steel (201) | **C** 0.15 max **Cr** 16.00-18.00 **Mn** 5.50-7.50 **N** 0.25 max **Ni** 3.50-5.50 **P** 0.060 max **S** 0.030 max **Si** 1.00 max | **AISI** 201 **ASTM** A213 (TP201); A240 (201); A249 (TP201); A473 (201); A666 (201) **SAE** J405 (201); J412 (20201) |
| **S20103** | Cr-Mn-Ni Austenitic Stainless Steel, Low Carbon (201L) | **C** 0.03 max **Cr** 16.0-18.0 **Mn** 5.5-7.5 **N** 0.25 max **Ni** 3.5-5.5 **P** 0.045 max **S** 0.030 max **Si** 0.75 max | **ASME** SA-240 **ASTM** A240; A480; A666 **SAE** J405 |
| **S20153** | Austenitic Cr-Mn-Ni Stainless Steel, Low Carbon, High Nitrogen (T201LN) | **C** 0.03 max **Cr** 16.0-17.5 **Mn** 6.4-7.5 **N** 0.10-0.25 **Ni** 4.0-5.0 **P** 0.045 max **S** 0.015 max **Si** 0.75 max | **ASME** SA-240; SA-480 **ASTM** A240; A480; A666 **SAE** J405 |
| **S20161** | Austenitic Cr-Mn-Ni-Si Stainless Steel | **C** 0.15 max **Cr** 15.0-18.0 **Mn** 4.00-6.00 **Ni** 4.00-6.00 **N** 0.08-0.20 **P** 0.040 max **S** 0.040 **Si** 3.00-4.00 | **AMS** 5931 **ASME** SA-240 **ASTM** A240; A276; A314; A479; A580 **SAE** J405 |
| **S20200** | Austenitic Cr-Mn-Ni Stainless Steel (202) | **C** 0.15 max **Cr** 17.00-19.00 **Mn** 7.50-10.00 **N** 0.25 max **Ni** 4.00-6.00 **P** 0.060 max **S** 0.030 max **Si** 1.00 max | **AISI** 202 **AMS** AMS-QQ-S-763 **ASTM** A213 (TP202); A240 (202); A249 (TP202); A314 (202); A473 (202); A666 (202) **MIL SPEC** MIL-W-6712 **SAE** J405 (202); J412 (20202) |
| **S20280** | Thermal Spray Wire (W-FeCrNi-4) | **C** 0.06 max **Cr** 17-19 **Fe** rem **Mn** 7.0-9.0 **Ni** 4.0-6.0 **Si** 0.80 max | **AWS** C2.25 (W-FeCrNi-4) |
| **S20281** | Thermal Spray Wire (W-FeCrNi-3) | **C** 0.15 max **Cr** 17-19 **Fe** rem **Mn** 7.5-10.0 **Ni** 4.0-6.0 **P** 0.060 max **Si** 1.0 max **S** 0.03 max | **AWS** C2.25 (W-FeCrNi-3) |
| **S20300** | Austenitic Cr-Mn-Ni-Cu Free Machining Stainless Steel (203 EZ) | **C** 0.08 max **Cr** 16.00-18.00 **Cu** 1.75-2.25 **Mn** 5.00-6.50 **Mo** 0.50 max **Ni** 5.00-6.50 **P** 0.040 max **S** 0.18-0.35 **Si** 1.0 max | **AMS** 5762 **ASTM** A581 (XM-1); A582 (XM-1); F593 (XM-1); F594 (XM-1); F738 (XM-1); F836 (XM-1); F837 (XM-1); F880 (XM-1) |
| **S20400** | Austenitic Cr-Mn-Ni-N Stainless Steel (Nitronic 30) | **C** 0.03 max **Cr** 15.0-17.0 **Mn** 7.00-9.00 **N** 0.15-0.30 **Ni** 1.50-3.00 **P** 0.04 **S** 0.03 max **Si** 1.00 max | **ASTM** A240; A409; A666 |
| **S20430** | Austenitic Cr-Mn-Cu-Ni-N Stainless Steel (204 Cu) | **C** 0.15 max **Cr** 15.5-17.5 **Cu** 2.00-4.00 **Mn** 6.5-9.0 **N** 0.05-0.25 **Ni** 1.50-3.50 **P** 0.060 max **S** 0.030 max **Si** 1.00 max | **ASTM** A313 **SAE** J405 |
| **S20500** | Cr-Mn-Ni-N Stainless Steel (205) | **C** 0.12-0.25 **Cr** 16.00-18.00 **Mn** 14.00-15.50 **N** 0.32-0.40 **Ni** 1.00-1.75 **P** 0.060 max **S** 0.030 max **Si** 1.00 max | **AISI** 205 **ASTM** A276 (205); A473 (205); A666 (205) |
| **S20910** | Austenitic Cr-Ni-Mn-Mo Stainless Steel (22-13-5) | **C** 0.06 max **Cb** 0.10-0.30 **Cr** 20.50-23.50 **Mn** 4.00-6.00 **Mo** 1.50-3.00 **N** 0.20-0.40 **Ni** 11.50-13.50 **P** 0.040 max **S** 0.030 max **Si** 1.00 max **V** 0.10-0.30 | **AMS** 5764 **ASME** SA-182 (XM-19); SA-213 (XM-19); SA-240 (XM-19); SA-249 (XM-19); SA-312 (XM-19); SA-403 (XM-19); SA-479 (XM-19) **ASTM** A182 (FXM-19); A184 (TPXM-19); A193 (B8R, B8RA); A194 (8R, 8RA); A240 (XM-19); A249 (XM-19); A269 (XM-19); A312 (XM-19); A314 (XM-19); A336 (FXM-19); A358 (XM-19); A403 (XM-19); A412 (XM-19); A479 (XM-19); A580 (XM-19); A813 (TPXM-19); A814 (TPXM-19); A965 (FXM-19); F1813 **SAE** J405 (XM-19) |
| **S20980** | Austenitic Cr-Ni-Mn-Mo Stainless Steel Bare Filler Metal (22-10-5) | **C** 0.05 max **Cr** 20.5-24.0 **Cu** 0.70-1.5 **Mn** 4.00-7.00 **Mo** 1.50-3.00 **N** 0.10-0.30 **Ni** 9.50-12.00 **P** 0.03 max **S** 0.03 max **Si** 0.90 max | **ASME** SFA5.9 (ER209) **AWS** A5.9 (ER209) |
| * **S21000** | Austenitic Cr-Mn-Mo-Ni-N Strengthened Stainless Steel | **C** 0.10 max **Cr** 18.0-23.0 **Cu** 2.00 max **Mn** 4.00-7.00 **Mo** 4.00-6.00 **Ni** 16.0-20.0 **N** 0.15 min **P** 0.03 max **S** 0.03 max **Si** 0.60 max | |

*Boxed entries are no longer active and are retained for reference purposes only.

The chemical compositions listed are for identification purposes and should not be used in lieu of the cross referenced specifications.

| UNIFIED NUMBER | DESCRIPTION | CHEMICAL COMPOSITION | CROSS REFERENCE SPECIFICATIONS |
|---|---|---|---|
| * **S21300** | Austenitic Cr-Cu-Mn-N Strengthened Stainless Steel | **C** 0.25 max **Cr** 16.0-21.0 **Cu** 0.50-2.00 **Mn** 15.0-18.0 **Mo** 0.50-3.00 **Ni** 3.00 max **N** 0.20-0.80 **P** 0.05 max **S** 0.05 max **Si** 1.00 max | |
| **S21400** | Austenitic Cr-Mn Stainless Steel (Tenelon) | **C** 0.12 max **Cr** 17.00-18.50 **Mn** 14.50-16.00 **N** 0.35 max **Ni** 0.75 max **P** 0.045 max **S** 0.030 max **Si** 0.30-1.00 | **ASTM** A240 (XM-31); A580 (XM-31) |
| **S21460** | Austenitic Cr-Mn-Ni-Mo Stainless Steel (Cryogenic Tenelon) | **C** 0.12 max **Cr** 17.00-19.00 **Mn** 14.00-16.00 **N** 0.35-0.50 **Ni** 5.00-6.00 **P** 0.060 max **S** 0.030 max **Si** 1.00 max | **ASTM** A666 (XM-14) |
| **S21500** | Austenitic Cr-Mn-Ni-Mo Stainless Steel (Esshete 1250) | **B** 0.003-0.009 **C** 0.06-0.15 **Cb** 0.75-1.25 **Cr** 14.0-16.0 **Mn** 5.50-7.00 **Mo** 0.80-1.20 **Ni** 9.0-11.0 **P** 0.040 max **S** 0.030 max **Si** 0.20-1.20 **V** 0.15-0.40 | **ASTM** A213 |
| **S21600** | Austenitic Cr-Mn-Ni-Mo Stainless Steel (216) | **C** 0.08 max **Cr** 17.50-22.00 **Mn** 7.50-9.00 **Mo** 2.00-3.00 **N** 0.25-0.50 **Ni** 5.00-7.00 **P** 0.045 max **S** 0.030 max **Si** 1.00 max | **ASME** SA-240 (XM-17); SA-479 (XM-17) **ASTM** A240 (XM-17); A479 (XM-17); A492 (XM-17) **SAE** J405 (XM-17) |
| **S21603** | Austenitic Cr-Mn-Ni-Mo Stainless Steel, Low Carbon (216L) | **C** 0.03 max **Cr** 17.50-22.00 **Mn** 7.50-9.00 **Mo** 2.00-3.00 **N** 0.25-0.50 **Ni** 5.00-7.00 **P** 0.045 max **S** 0.030 max **Si** 1.00 max | **ASME** SA-240 (XM-18); SA-479 (XM-18) **ASTM** A240 (XM-18); A479 (XM-18); A492 (XM-18) **SAE** J405 (XM-18) |
| **S21800** | Austenitic Cr-Ni-Mn Stainless Steel (Nitronic 60) | **C** 0.10 max **Cr** 16.00-18.00 **Mn** 7.00-9.00 **N** 0.08-0.18 **Ni** 8.00-9.00 **P** 0.040 max **S** 0.030 max **Si** 3.50-4.50 | **AMS** 5848 **ASME** Code Case 1817 **ASTM** A193 (B8S, B8SA); A194 (8S, 8SA); A240; A276; A314; A479; A580 **SAE** J405 |
| **S21880** | Austenitic Cr-Ni-Mn Stainless Steel Bare Filler Metal (Nitronic 60W) | **C** 0.10 max **Cr** 16.0-18.0 **Cu** 0.75 max **Mn** 7.00-9.00 **Mo** 0.75 max **N** 0.08-0.18 **Ni** 8.00-9.00 **P** 0.03 max **S** 0.03 max **Si** 3.50-4.50 | **ASME** SFA5.9 (ER218) **AWS** A5.9 (ER218) |
| **S21900** | Austenitic Cr-Mn-Ni Stainless Steel (21-6-9) | **C** 0.08 max **Cr** 19.00-21.50 **Mn** 8.00-10.00 **N** 0.15-0.40 **Ni** 5.50-7.50 **P** 0.060 max **S** 0.030 max **Si** 1.00 max | **AMS** 5561 **ASTM** A269 (TPXM-10); A276 (XM-10); A312 (TPXM-10); A314 (XM-10); A473 (XM-10); A580 (XM-10); A813 (TPXM-10); A814 (TPXM-10) |
| **S21904** | Austenitic Cr-Mn-Ni Stainless Steel, Low Carbon (21-6-9 LC) | **C** 0.04 max **Cr** 19.00-21.50 **Mn** 8.00-10.00 **N** 0.15-0.40 **Ni** 5.50-7.50 **P** 0.060 max **S** 0.030 max **Si** 1.00 max | **AMS** 5562; 5595; 5656 **ASME** SA-412 (21904) **ASTM** A182 (FXM-11); A269 (XM-11); A276 (XM-11); A312 (XM-11); A314 (XM-11); A336 (FXM-11); A473 (XM-11); A479 (XM-11); A580 (XM-11); A666 (XM-11); A813 (TPXM-11); A814 (TPXM-11); A988 |
| **S21980** | Austenitic Cr-Mn-Ni Stainless Steel Bare Filler Metal (21-6-9) | **C** 0.05 max **Cr** 19.0-21.5 **Cu** 0.75 max **Mn** 8.00-10.00 **Mo** 0.75 max **N** 0.10-0.30 **Ni** 5.50-7.00 **P** 0.03 max **S** 0.03 max **Si** 1.00 max | **ASME** SFA5.9 (ER219) **ASTM** A814 (TPXM-11) **AWS** A5.9 (ER219) |
| **S23980** | Combined with S24080 | | |
| **S24000** | Austenitic Cr-Mn-Ni Stainless Steel (18-3-Mn) | **C** 0.08 max **Cr** 17.00-19.00 **Mn** 11.50-14.50 **N** 0.20-0.40 **Ni** 2.50-3.75 **P** 0.060 max **S** 0.030 max **Si** 1.00 max | **ASME** SA-240 (XM-29); SA-249 (XM-29); SA-312 (XM-29); SA-688 (XM-29) **ASTM** A240 (XM-29); A249 (XM-29); A269 (XM-29); A312 (XM-29); A314 (XM-29); A358 (XM-29); A479 (XM-29); A580 (XM-29); A688 (XM-29); A813 (TPXM-29); A814 (TPXM-29) |
| **S24080** | Austenitic Cr-Mn-Ni Stainless Steel Bare Filler Metal (18-4-Mn) | **C** 0.05 max **Cr** 17.0-19.0 **Cu** 0.75 max **Mn** 10.5-13.5 **Mo** 0.75 max **N** 0.10-0.20 **Ni** 4.00-6.00 **P** 0.03 max **S** 0.03 max **Si** 1.00 max | **ASME** SFA5.9 (ER240) **AWS** A5.9 (ER240) |
| **S24100** | Austenitic Cr-Mn-Ni Stainless Steel (18-2-Mn) | **C** 0.15 max **Cr** 16.50-19.50 **Mn** 11.00-14.00 **N** 0.20-0.45 **Ni** 0.50-2.50 **P** 0.060 max **S** 0.030 max **Si** 1.00 max | **ASTM** A313 (XM-28); A314 (XM-28); A580 (XM-28) |
| **S28200** | Austenitic Cr-Mn-Cu-Mo Stainless Steel (18-18-Plus) | **C** 0.15 max **Cr** 17.00-19.00 **Cu** 0.50-1.50 **Mn** 17.00-19.00 **Mo** 0.50-1.50 **N** 0.40-0.60 **P** 0.045 max **S** 0.030 max **Si** 1.00 max | **ASTM** A276; A314; A473; A493 (XM-29); A580; F899 |

*Boxed entries are no longer active and are retained for reference purposes only.

The chemical compositions listed are for identification purposes and should not be used in lieu of the cross referenced specifications.

| UNIFIED NUMBER | DESCRIPTION | CHEMICAL COMPOSITION | CROSS REFERENCE SPECIFICATIONS |
|---|---|---|---|
| **S30100** | Austenitic Cr-Ni Stainless Steel (301) | **C** 0.15 max **Cr** 16.00-18.00 **Mn** 2.00 max **Ni** 6.00-8.00 **P** 0.045 max **S** 0.030 max **Si** 1.00 max | **AISI** 301 **AMS** 5517; 5518; 5519; 5901; 5902 **ASTM** A554 (301); A666 (301); F899 (301) **SAE** J405 (301); J412 (30301) |
| **S30103** | Austenitic Cr-Ni Stainless Steel (301L) | **C** 0.030 max **Cr** 16.0-18.0 **Mn** 2.00 max **N** 0.20 max **Ni** 6.0-8.0 **P** 0.045 max **S** 0.030 max **Si** 1.0 max | **ASTM** A240; A666 |
| **S30115** | Austenitic Cr-Ni Stainless Steel | **C** 0.07-0.11 **Cr** 16.50-17.50 **Mn** 1.00-1.50 **Mo** 0.60-0.80 **Ni** 7.70-8.30 **P** 0.03 max **S** 0.03 max **Si** 0.90-1.40 | |
| **S30153** | Austenitic Cr-Ni-N Stainless Steel (301LN) | **C** 0.030 max **Cr** 16.0-18.0 **Mn** 2.00 max **N** 0.07-0.20 **Ni** 6.0-8.0 **P** 0.045 max **S** 0.030 max **Si** 1.0 max | **ASTM** A240; A666 |
| **S30200** | Austenitic Cr-Ni Stainless Steel (302) | **C** 0.15 max **Cr** 17.00-19.00 **Mn** 2.00 max **Ni** 8.00-10.00 **P** 0.045 max **S** 0.030 max **Si** 1.00 max | **AISI** 302 **AMS** 5500; 5515; 5516; 5600; 5636; 5637; 5688; 5693; 5866; 5903; 5904; 5905; 5906; 6693; 7241; AS7210; AS7228; AS7472; AMS-QQ-S-763 (302); AMS-S-7720 **ASME** SA-240 (302); SA-479 (302) **ASTM** A240 (302); A276 (302); A313 (302); A314 (302); A368 (302); A473 (302); A478 (302); A479 (302); A492 (302); A493 (302); A511 (302); A554 (302); A580 (302); A666 (302); F899 (302) **SAE** J405 (302); J412 (30302); J467 (302) |
| **S30210** | Austenitic Cr-Ni Stainless Steel, Low Permeability | **C** 0.26-0.33 **Cr** 17.0-19.0 **Mn** 3.00-4.00 **Ni** 8.0-10.0 **P** 0.18-0.33 **S** 0.035 max **Si** 1.00 max | |
| **S30215** | Austenitic Cr-Ni Stainless Steel (302B) | **C** 0.15 max **Cr** 17.00-19.00 **Mn** 2.00 max **Ni** 8.00-10.00 **P** 0.045 max **S** 0.030 max **Si** 2.00-3.00 | **AISI** 302B **ASTM** A167 (302B); A276 (302B); A314 (302B); A473 (302B); A580 (302B) |
| **S30260** | Austenitic Cr-Ni Stainless Steel, Low Permeability | **C** 0.15 max **Cr** 16.0-18.0 **Mn** 1.00 max **Ni** 9.5-12.0 **P** 0.20-0.40 **S** 0.040 max **Si** 1.00 max | |
| **S30280** | Thermal Spray Wire (W-FeCrNi-2) | **C** 0.12 max **Cr** 17.0-19.0 **Fe** rem **Mn** 1.5 max **Ni** 7.0-9.0 **Si** 0.50 max | **AWS** C2.25 (W-FeCrNi-2) |
| **S30300** | Austenitic Cr-Ni Free-Machining Stainless Steel (303) | **C** 0.15 max **Cr** 17.00-19.00 **Mn** 2.00 max **Mo** 0.60 max (Optional) **Ni** 8.00-10.00 **P** 0.20 max **S** 0.15 min **Si** 1.00 max | **AISI** 303 **AMS** 5640 (Type 1) **ASME** SA-194 (303); SA-320 (303) **ASTM** A194 (303, 8F, 8FA); A314 (303); A320 (303, B8F, B8FA); A473 (303); A581 (303); A582 (303); A895 (303); F593 (303); F594 (303); F738 (303); F836 (303); F837 (303); F880 (303); F899 (303) **SAE** J412 (30303) |
| **S30310** | Austenitic Cr-Ni-Mn Free Machining Stainless Steel | **C** 0.15 max **Cr** 17.00-19.00 **Mn** 2.50-4.50 **Mo** 0.75 max **Ni** 7.00-10.00 **P** 0.20 max **S** 0.25 min **Si** 1.00 max | **AMS** 5640 (Type 3) **ASTM** A581 (XM-5); A582 (XM-5) **SAE** J412 (303 plus X) |
| **S30323** | Austenitic Cr-Ni Free-Machining Stainless Steel (303Se) | **C** 0.15 max **Cr** 17.00-19.00 **Mn** 2.00 max **Ni** 8.00-10.00 **P** 0.20 max **S** 0.060 max **Se** 0.15 min **Si** 1.00 max | **AISI** 303 Se **AMS** 5640 (Type 2); 5641; 5738 **ASME** SA-194 (303 Se); SA-320 (303 Se) **ASTM** A194 (303 Se, 8F, 8FA); A314 (303 Se); A320 (303 Se, B8F, B8FA); A473 (303 Se); A511 (303Se); A581 (303 Se); A582 (303 Se); A895 (303 Se); F593 (303Se); F594 (303Se); F738 (303Se); F836 (303Se); F880 (303Se) |
| **S30330** | Austenitic Cr-Ni-Cu Free-Machining Stainless Steel (303Cu) | **C** 0.15 max **Cr** 17.00-19.00 **Cu** 2.50-4.00 **Mn** 2.00 max **Ni** 6.00-10.00 **P** 0.15 max **S** 0.10 max **Se** 0.10 max **Si** 1.00 max | |
| **S30345** | Austenitic Cr-Ni-Al Free Machining Stainless Steel (303 MA) | **Al** 0.60-1.00 **C** 0.15 max **Cr** 17.00-19.00 **Mn** 2.00 max **Mo** 0.40-0.60 **Ni** 8.00-10.00 **P** 0.05 max **S** 0.11-0.16 **Si** 1.00 max | **AMS** 5638 **ASTM** A581 (XM-2); A582 (XM-2) |
| **S30360** | Austenitic Cr-Ni-Pb Free Machining Stainless Steel (303 Pb) | **C** 0.15 max **Cr** 17.00-19.00 **Mn** 2.00 max **Mo** 0.75 max **Ni** 8.00-10.00 **P** 0.040 max **Pb** 0.12-0.30 **S** 0.12-0.30 **Si** 1.00 max | **AMS** 5635 |

The chemical compositions listed are for identification purposes and should not be used in lieu of the cross referenced specifications.

| UNIFIED NUMBER | DESCRIPTION | CHEMICAL COMPOSITION | CROSS REFERENCE SPECIFICATIONS |
|---|---|---|---|
| **S30400** | Austenitic Cr-Ni Stainless Steel (304) | **C** 0.08 max **Cr** 18.00-20.00 **Mn** 2.00 max **Ni** 8.00-10.50 **P** 0.045 max **S** 0.030 max **Si** 1.00 max | **AISI** 304 **AMS** 5501; 5513; 5560; 5563; 5564; 5565; 5566; 5567; 5639; 5697; 5857; 5910; 5911; 5912; 5913; 5868; AMS-QQ-S-763 (304); AMS-T-6845 (304); AMS-T-8506 (304); AS7245 **ASME** SA-182 (304); SA-194 (8); SA-213 (304); SA-240 (304); SA-249 (304); SA-312 (304); SA-320 (B8); SA-358 (304); SA-376 (304); SA-403 (304); SA-409 (304); SA-479 (304); SA-688 (304) **ASTM** A182 (304); A193 (304, B8, B8A); A194 (304, 8, 8A); A213 (304); A240 (304); A249 (304); A264 (304); A269 (304); A270 (304); A276 (304); A312 (304); A313 (304); A314 (304); A320 (304, B8, B8A); A336 (F304); A358 (304); A368 (304); A376 (304); A376 (TP304); A403 (304); A409 (304); A430 (304); A473 (304); A478 (304); A479 (304); A492 (304); A493 (304); A511 (304); A554 (304); A580 (304); A632 (304); A666 (304); A688 (304); A774 (304); A774 (TP304); A793 (304); A813 (TP304); A814 (TP304); A851 (TP304); A908 (304); A943 (TP304); A965 (304); A988; A988 (304); F593 (304); F738 (304); F836 (304); F837 (304); F879 (304); F880 (304); F899 (304) **MIL SPEC** MIL-S-21149/6MIL-S-24149/3; MIL-S-24249/6; MIL-T-5695; MIL-T-8504 (304) **SAE** J405 (304); J412 (30304); J467 (304) |
| **S30403** | Austenitic Cr-Ni Stainless Steel (304L) | **C** 0.03 max **Cr** 18.00-20.00 **Mn** 2.00 max **Ni** 8.00-12.00 **P** 0.045 max **S** 0.030 max **Si** 1.00 max | **AISI** 304 L **AMS** 5511; 5569; 5584; 5647; AMS-QQ-S-763 (304L); AMS-T-6845 **ASME** SA-182 (304L); SA-213 (304L); SA-240 (304L); SA-249 (304L); SA-312 (304L); SA-403 (304L); SA-479 (304L); SA-688 (304L) **ASTM** A182 (304L); A213 (TP304L); A240 (304L); A249 (304L); A269 (304L); A270 (TP304L); A276 (304L); A312 (304L); A314 (304L); A336 (F304L); A358 (304L); A368 (304L); A403 (304L); A409 (304L); A473 (304L); A478 (304L); A479 (304L); A493 (304L); A511 (304L); A554 (304L); A580 (304L); A632 (304L); A666 (304L); A688 (304L); A774 (TP304L); A778 (TP304L); A793 (304L); A813 (TP304L); A814 (TP304L); A851 (TP304L); A924 (304L); A924 (TP304L); A943 (TP304L); A965 (F304L); A988; F593 (304L); F594 (304L); F738 (304L); F836 (304L); F837 (304L); F879 (304L); F880 (304L) **MIL SPEC** MIL-S-4043 **SAE** J405 (304L); J467 (304L) |
| **S30409** | Austenitic Cr-Ni Stainless Steel, High Carbon (304 H) | **C** 0.04-0.10 **Cr** 18.00-20.00 **Mn** 2.00 max **Ni** 8.00-11.00 **P** 0.040 max **S** 0.030 max **Si** 1.00 max | **ASME** SA-182 (304H); SA-213 (304H); SA-240 (304H); SA-249 (304H); SA-312 (304H); SA-376 (304H); SA-403 (304H); SA-479 (304H) **ASTM** A182 (304H); A213 (TP304H); A240 (304H); A249 (304H); A312 (304H); A336 (F304H); A358 (304H); A376 (TP304H); A403 (304H); A479 (304H); A813 (TP304H); A814 (TP304H); A924 (TP304H); A943 (TP304H); A965 (F304H) **MIL SPEC** MIL-W-6712 **SAE** J405 (304H) |
| **S30415** | Austenitic Cr-Ni-Si Heat Resisting Steel (153 MA) | **C** 0.04-0.06 **Ce** 0.03-0.08 **Cr** 18.0-19.0 **Mn** 0.8 max **Mo** 0.5 max **Ni** 9.0-10.0 **N** 0.12-0.18 **P** 0.045 max **S** 0.030 max **Si** 1.0-2.0 | **ASTM** A240; A245; A249; A312 **DIN** 1.4891 |
| **S30430** | Austenitic Cr-Ni-Cu Low Work Hardening Stainless Steel (18-9LW) | **C** 0.10 max **Cr** 17.00-19.00 **Cu** 3.00-4.00 **Mn** 2.00 max **Ni** 8.00-10.00 **P** 0.045 max **S** 0.030 max **Si** 1.00 max | **AISI** S30430 **ASTM** A493; F593 (XM-7); F594 (XM-7); F738 (XM-7); F836 (XM-7); F837 (XM-7); F879 (XM-7); F880 (XM-7); F899 (XM-7) **MIL SPEC** MIL-S-24149/3; MIL-S-24149/6 **SAE** J775 (EV 17) |
| **S30431** | Cr-Ni-Cu Low Work Hardening Free Machining Stainless Steel | **C** 0.06 max **Cr** 16.0-19.0 **Cu** 1.30-2.40 **Mn** 2.00 max **Ni** 9.00-11.0 **P** 0.040 max **S** 0.14 max **Si** 1.00 max | |
| **S30432** | Austenitic Cr-Ni-Cu-Cb-N Stainless Steel | **Al** 0.003-0.030 **B** 0.001-0.010 **C** 0.07-0.13 **Cb** 0.20-0.60 **Cr** 17.0-19.0 **Cu** 2.5-3.5 **Mn** 0.50 max **N** 0.05-0.12 **Ni** 7.5-10.5 **P** 0.045 max **S** 0.030 max **Si** 0.30 max | **ASTM** A213 |
| **S30433** | Austenitic Cr-Ni-Cu Low Work Hardening Stainless Steel (302HQ, 302Cu, XM-7) | **C** 0.030 max **Cr** 17.0-19.0 **Cu** 3.0-4.0 **Mn** 2.00 max **Ni** 8.0-10.0 **P** 0.045 max **S** 0.030 max **Is** 1.00 max | **ASTM** A493; F593; F594; F738; F836; F837; F879; F880 |
| **S30451** | Austenitic Nitrogen Bearing Cr-Ni Stainless Steel (304 N) | **C** 0.08 max **Cr** 18.00-20.00 **Mn** 2.00 max **N** 0.10-0.16 **Ni** 8.00-10.50 **P** 0.045 max **S** 0.030 max **Si** 1.00 max | **AISI** 304 N **ASME** SA-182 (304N); SA-213 (304N); SA-240 (304N); SA-249 (304N); SA-312 (304N); SA-358 (304N); SA-376 (304N); SA-479 (304N) **ASTM** A182 (304N); A193 (B8N, B8NA); A194 (8N, 8NA); A213 (TP304N); A240 (304N); A249 (304N); A276 (304N); A312 (304N); A336 (F304N); A358 (304N); A376 (TP304N); A403 (304N); A479 (304N); A666 (304N); A688 (TP304N); A813 (TP304N); A814 (TP304N); A924 (TP304N); A943 (TP304N); A965 (F304N); A988 **SAE** J405; J405 (304N) |

The chemical compositions listed are for identification purposes and should not be used in lieu of the cross referenced specifications.

| UNIFIED NUMBER | DESCRIPTION | CHEMICAL COMPOSITION | CROSS REFERENCE SPECIFICATIONS |
|---|---|---|---|
| **S30452** | Austenitic Cr-Ni Stainless Steel, High Nitrogen | **C** 0.08 max **Cr** 18.00-20.00 **Mn** 2.00 max **N** 0.16-0.30 **Ni** 8.00-10.50 **P** 0.045 max **S** 0.030 max **Si** 1.00 max | **ASME** SA-240 (XM-21) **ASTM** A240 (XM-21); A276 (XM-21) |
| **S30453** | Austenitic Nitrogen Bearing Cr-Ni Stainless Steel (304 LN) | **C** 0.030 max **Cr** 18.00-20.00 **Mn** 2.00 max **N** 0.10-0.16 **Ni** 8.00-12.00 **P** 0.045 max **S** 0.030 max **Si** 1.00 max | **ASTM** A182 (F304LN); A193 (B8LN, B8LNA); A194 (8LN, 8LNA); A213 (TP304LN); A240 (304LN); A249 (TP304LN); A269 (TP304LN); A276 (304LN); A312 (TP304LN); A320 (B8LN, B8LNA); A336 (F304LN); A376 (TP304LN); A358 (304LN); A403 (304LN); A403 (WP304LN); A479 (304LN); A666 (304LN); A688 (304LN); A688 (TP304LN); A813 (TP304LN); A814 (TP304LN); A924 (TP304LN); A943 (TP304LN); A965 (F304LN); A988 **SAE** J405 (304LN) |
| **S30454** | Austenitic Nitrogen Bearing Cr-Ni Stainless Steel (304L (Hi)N) | **C** 0.03 max **Cr** 18.00-20.00 **Mn** 2.00 max **N** 0.16-0.30 **Ni** 8.00-12.00 **P** 0.045 max **S** 0.030 max **Si** 1.00 max | **ASTM** A276 |
| **S30460** | Austenitic Cr-Ni Stainless Steel with 0.20-0.29% Boron | **B** 0.20-0.29 **C** 0.08 max **Cr** 18.0-20.0 **Mn** 2.00 max **N** 0.10 max **Ni** 12.0-15.0 **P** 0.045 max **S** 0.030 max **Si** 0.75 max | **ASTM** A887 (304B) |
| **S30461** | Austenitic Cr-Ni Stainless Steel with 0.30-0.49% Boron | **B** 0.30-0.49 **C** 0.08 max **Cr** 18.0-20.0 **Mn** 2.00 max **N** 0.10 max **Ni** 12.0-15.0 **P** 0.045 max **S** 0.030 max **Si** 0.75 max | **ASTM** A887 (304B1) |
| **S30462** | Austenitic Cr-Ni Stainless Steel with 0.50-0.74% Boron | **B** 0.50-0.74 **C** 0.08 max **Cr** 18.0-20.0 **Mn** 2.00 max **N** 0.10 max **Ni** 12.0-15.0 **P** 0.045 max **S** 0.030 max **Si** 0.75 max | **ASTM** A887 (304B2) |
| **S30463** | Austenitic Cr-Ni Stainless Steel with 0.75-0.99% Boron | **B** 0.75-0.99 **C** 0.08 max **Cr** 18.0-20.0 **Mn** 2.00 max **N** 0.10 max **Ni** 12.0-15.0 **P** 0.045 max **S** 0.030 max **Si** 0.75 max | **ASTM** A887 (304B3) |
| **S30464** | Austenitic Cr-Ni-Stainless Steel with 1.00-1.24% Boron | **B** 1.00-1.24 **C** 0.08 max **Cr** 18.0-20.0 **Mn** 2.00 max **N** 0.10 max **Ni** 12.0-15.0 **P** 0.045 max **S** 0.030 max **Si** 0.75 max | **ASTM** A887 (304B4) |
| **S30465** | Austenitic Cr-Ni Stainless Steel with 1.25-1.49% Boron | **B** 1.25-1.49 **C** 0.08 max **Cr** 18.0-20.0 **Mn** 2.00 max **N** 0.10 max **Ni** 12.0-15.0 **P** 0.045 max **S** 0.030 max **Si** 0.75 max | **ASTM** A887 (304B5) |
| **S30466** | Austenitic Cr-Ni Stainless Steel with 1.50-1.74% Boron | **B** 1.50-1.74 **C** 0.08 max **Cr** 18.0-20.0 **Mn** 2.00 max **N** 0.10 max **Ni** 12.0-15.0 **P** 0.045 max **S** 0.030 max **Si** 0.75 max | **ASTM** A887 (304B6) |
| **S30467** | Austenitic Cr-Ni Stainless Steel with 1.75-2.25% Boron | **B** 1.75-2.25 **C** 0.08 max **Cr** 18.0-20.0 **Mn** 2.00 max **N** 0.10 max **Ni** 12.0-15.0 **P** 0.045 max **S** 0.030 max **Si** 0.75 max | **ASTM** A887 (304B7) |
| **S30480** | Austenitic Cr-Ni Stainless Steel Bare Filler Metal (ER19-10H) | **C** 0.04-0.08 **Cb** 0.05 max **Cr** 18.5-20.0 **Cu** 0.75 max **Mn** 1.0-2.0 **Mo** 0.25 max **Ni** 9.0-11.0 **P** 0.03 max **S** 0.03 max **Si** 0.30-0.65 **Ti** 0.05 max | **ASME** SFA5.9 (ER19-10H) **AWS** A5.9 (ER19-10H) |
| **S30481** | Thermal Spray Wire (W-FeCrNi-1) | **C** 0.080 max **Cr** 18-20 **Fe** rem **Mn** 2.0 max **Ni** 8.0-11.0 **P** 0.045 max **Si** 1.0 max **S** 0.045 max | **AWS** C2.25 (W-FeCrNi-1) |
| **S30483** | Austenitic Cr-Ni Stainless Steel Bare Filler Metal, Low Carbon | **C** 0.03 max **Cr** 18.0 min **Mn** 2.00 max **Ni** 8.00 min **P** 0.030 max **S** 0.030 max **Si** 1.00 max **Other** others (total) 1.00 max | |

The chemical compositions listed are for identification purposes and should not be used in lieu of the cross referenced specifications.

| UNIFIED NUMBER | DESCRIPTION | CHEMICAL COMPOSITION | CROSS REFERENCE SPECIFICATIONS |
|---|---|---|---|
| **S30500** | Austenitic Cr-Ni Stainless Steel (Low Work Hardening) (304) | **C** 0.12 max **Cr** 17.00-19.00 **Mn** 2.00 max **Ni** 10.00-13.00 **P** 0.045 max **S** 0.030 max **Si** 1.00 max | **AISI** 305 **AMS** 5514; 5685; 5686; AMS-QQ-S-763 **ASME** SA-193 (305); SA-194 (305); SA-240 (305) **ASTM** A193 (B8P, B8PA); A194 (8P, 8PA); A240 (305); A249 (305); A276 (305); A313 (305); A314 (305); A320 (B8P, B8PA); A368 (305); A473 (305); A478 (305); A492 (305); A493 (305); A511 (305); A554 (305); A580 (305); F593 (305); F594 (305); F738 (305); F836 (305); F837 (305); F879 (305); F880 (305) **MIL SPEC** MIL-S-24149/3; MIL-S-24149/6 **SAE** J405 (305); J412 (30305) |
| **S30560** | Austenitic Cr-Ni Stainless Steel Low Permeability | **C** 0.10 max **Cr** 17.0 min **Cu** 0.50 max **Mn** 0.2-2.5 **Ni** 8.5-14.0 **P** 0.035 max **S** 0.035 max **Si** 0.2-1.5 | |
| **S30600** | Austenitic Cr-Ni Stainless Steel (18-15) | **C** 0.018 max **Cr** 17.0-18.5 **Cu** 0.50 max **Mn** 2.00 max **Mo** 0.20 max **Ni** 14.0-15.5 **P** 0.02 max **S** 0.02 max **Si** 3.70-4.30 | **ASTM** A182 (F46); A240; A269; A312; A336 (F46); A358; A479; A928; A943; A965 (F46) **SAE** J405 |
| **S30601** | Austenitic Cr-Ni Stainless Steel, Low Carbon (18-17 LC) | **C** 0.015 max **Cr** 17.0-18.0 **Cu** 0.35 max **Mn** 1.0 max **N** 0.05 max **Ni** 17.0-18.0 **P** 0.030 max **S** 0.013 max **Si** 5.00-5.60 | **ASTM** A240 **SAE** J405 |
| **S30615** | Austenitic Cr-Ni-Si Heat Resisting Steel (RA85H) | **Al** 0.8-1.5 **C** 0.16-0.24 **Cr** 17.0-19.5 **Mn** 2.0 max **Ni** 13.5-16.0 **P** 0.03 max **S** 0.03 max **Si** 3.2-4.0 | **ASME** SA-240 **ASTM** A240; A249; A312 **SAE** J405 |
| **S30780** | Austenitic Cr-Ni-Mn Stainless Steel Bare Filler Metal | **C** 0.04-0.14 **Cr** 19.5-22.0 **Cu** 0.75 max **Mn** 3.30-4.75 **Mo** 0.50-1.50 **Ni** 8.00-10.70 **P** 0.03 max **S** 0.03 max **Si** 0.30-0.65 | **ASME** SFA5.9 (ER307) **AWS** A5.9 (ER307) |
| **S30800** | Austenitic Cr-Ni Heat Resisting Steel | **C** 0.08 max **Cr** 19.00-21.00 **Mn** 2.00 max **Ni** 10.00-12.00 **P** 0.045 max **S** 0.030 max **Si** 1.00 max | **AISI** 308 **ASTM** A167 (308); A276 (308); A314 (308); A473 (308); A580 (308) |
| **S30815** | Austenitic Cr-Ni-Si Heat Resisting Steel (253 MA) | **C** 0.10 max **Cr** 20.00-22.00 **Mn** 0.80 max **N** 0.14-0.20 **Ni** 10.00-12.00 **P** 0.040 max **S** 0.030 max **Si** 1.40-2.00 **Other** cerium 0.03-0.08 | **ASME** SA-182 (F45); SA-213; SA-240; SA-249; SA-312; SA-479 **ASTM** A182 (F45); A213; A240; A249; A276; A312; A358; A409; A473; A479; A813; A814; A928; A943 **SAE** J405 |
| **S30880** | Austenitic Cr-Ni Stainless Steel Bare Filler Metal (ER308) | **C** 0.08 max **Cr** 19.50-22.00 **Cu** 0.75 max **Mn** 1.00-2.50 **Mo** 0.75 max **Ni** 9.00-11.00 **P** 0.03 max **S** 0.03 max **Si** 0.30-0.65 | **ASME** SFA5.9 (ER308, ER308H); SFA5.30 (IN308) **AWS** A5.9 (ER308, ER308H); A5.30 (IN308) **MIL SPEC** MIL-E-19933 (MIL-308, MIL-308Co); MIL-I-23413 (MIL-308, MIL-308Co) |
| **S30881** | Austenitic Cr-Ni Heat Resisting Steel Bare Filler Metal | **C** 0.08 max **Cr** 19.50-22.00 **Cu** 0.5 max **Mn** 1.00-2.50 **Mo** 0.50 max **Ni** 9.00-11.00 **P** 0.03 max **S** 0.03 max **Si** 0.65-1.00 | **ASME** SFA5.9 (ER308Si) **AWS** A5.9 (ER308Si) |
| **S30882** | Austenitic Cr-Ni-Mo Heat Resisting Steel Bare Filler Metal | **C** 0.08 max **Cr** 18.00-21.00 **Cu** 0.75 max **Mn** 1.00-2.50 **Mo** 2.00-3.00 **Ni** 9.00-12.00 **P** 0.03 max **S** 0.03 max **Si** 0.30-0.65 | **ASME** SFA5.9 (ER308Mo) **AWS** A5.9 (ER308Mo) |
| **S30883** | Austenitic Cr-Ni Heat Resisting Steel Bare Filler Metal (Low Carbon) | **C** 0.03 max **Cr** 19.50-22.00 **Cu** 0.75 max **Mn** 1.00-2.50 **Mo** 0.50 max **Ni** 9.00-11.00 **P** 0.03 max **S** 0.03 max **Si** 0.30-0.65 | **ASME** SFA5.9 (ER308L); SFA5.30 (IN308L) **AWS** A5.9 (ER308L); A5.30 (IN308L) **MIL SPEC** MIL-E-19933 (MIL-308L, MIL-308CoL); MIL-I-23413 (MIL-308L, MIL-308CoL) |
| **S30884** | Austenitic Cr-Ni Heat Resisting Steel Bare Filler Metal (19-9 High Carbon) | **C** 0.08-0.15 **Cr** 18.0-20.0 **Mn** 1.00-2.50 **Ni** 9.00-11.00 **P** 0.03 max **S** 0.03 max **Si** 0.25-0.60 | **MIL SPEC** MIL-E-19933D, (MIL-308HC) |
| **S30886** | Austenitic Cr-Ni-Mo Heat Resisting Steel Bare Filler Metal (19-9-Mo Low Carbon) | **C** 0.04 max **Cr** 18.00-21.00 **Cu** 0.75 max **Mn** 1.00-2.50 **Mo** 2.00-3.00 **Ni** 9.00-12.00 **P** 0.03 max **S** 0.03 max **Si** 0.30-0.65 | **ASME** SFA5.9 (ER308LMo) **AWS** A5.9 (ER308MLMo) |
| **S30888** | Austenitic Cr-Ni Heat Resisting Steel Bare Filler Metal (308LSi) | **C** 0.03 max **Cr** 19.5-22.0 **Cu** 0.5 max **Mn** 1.00-2.50 **Mo** 0.5 max **Ni** 9.00-11.00 **P** 0.03 max **S** 0.03 max **Si** 0.65-1.00 | **ASME** SFA5.9 (ER308LSi) **AWS** A5.9 (ER308LSi) |

The chemical compositions listed are for identification purposes and should not be used in lieu of the cross referenced specifications.

| UNIFIED NUMBER | DESCRIPTION | CHEMICAL COMPOSITION | CROSS REFERENCE SPECIFICATIONS |
|---|---|---|---|
| **S30900** | Cr-Ni Heat Resisting Steel (309) | **C** 0.20 max **Cr** 22.00-24.00 **Mn** 2.00 max **Ni** 12.00-15.00 **P** 0.045 max **S** 0.030 max **Si** 1.00 max | **AISI** 309 **AMS** AMS-QQ-S-763 **ASME** SA-249 (309); SA-312 (309); SA-358 (309); SA-403 (309); SA-409 (309) **ASTM** A167 (309); A276 (309); A314 (309); A358 (309); A403 (309); A409 (309); A473 (309); A580 (309) **MIL SPEC** MIL-S-24149/3; MIL-S-24149/6 **SAE** J412 (30309) |
| **S30908** | Cr-Ni Heat Resisting Steel (309S) | **C** 0.08 max **Cr** 22.00-24.00 **Mn** 2.00 max **Ni** 12.00-15.00 **P** 0.045 max **S** 0.030 max **Si** 1.00 max | **AISI** 309 S **AMS** 5523; 5574; 5650; 7490 **ASME** SA-240 (309S) **ASTM** A213 (TP309S); A240 (309S); A249 (TP309S); A276 (309S); A312 (TP309S); A314 (309S); A358 (309S); A409 (309S); A473 (309S); A478 (309S); A479 (309S); A511 (309S); A554 (309S); A580 (309S); A813 (TP309S); A814 (TP309S); A924 (TP309S); A943 (TP309S) **SAE** J405 (309S); J467 (309S) |
| **S30909** | Austenitic Cr-Ni Heat Resisting Steel (309H) | **C** 0.04-0.10 **Cr** 22.0-24.0 **Mn** 2.00 max **Ni** 12.0-16.0 **P** 0.040 max **S** 0.030 max **Si** 0.75 max | **ASTM** A213 (TP309H); A240 (309H); A249 (TP309H); A312 (TP309H); A336 (F309H); A479 (309H); A943 (TP309H); A965 (F309H) **SAE** J405 (309H) |
| **S30940** | Austenitic Cr-Ni-Cb Heat Resisting Stainless Steel (309Cb) | **C** 0.08 max **Cb** 10\mtC-1.00 **Cr** 22.00-24.00 **Mn** 2.00 max **Ni** 12.00-15.00 **P** 0.040 max **S** 0.030 max **Si** 1.00 max | **ASTM** A213 (TP309Cb); A240 (309Cb); A249 (TP309Cb); A276 (309Cb); A312 (TP309Cb); A314 (309Cb); A409 (309Cb); A358 (309Cb); A478 (309Cb); A479 (309Cb); A554 (309SCb); A580 (309Cb); A813 (TP309Cb); A814 (TP309Cb); A924 (TP309Cb); A943 (TP309Cb) **SAE** J405 (309Cb) |
| **S30941** | Austenitic Cr-Ni Heat Resisting Steel (309HCb) | **C** 0.04-0.10 **Cr** 22.0-24.0 **Cb+Ta** 10XC min-1.10 max **Cu** 0.75 max **Mn** 2.00 max **Mo** 0.75 max **Ni** 12.0-16.0 **P** 0.040 max **S** 0.030 max **Si** 0.75 max | **ASTM** A213 (TP309HCb); A240 (309HCb); A249 (TP309HCb); A312 (TP309HCb); A924 (309HCb); A924 (TP309HCb); A943 (309HCb); A943 (TP309HCb) **SAE** J405 (309HCb) |
| **S30980** | Austenitic Cr-Ni Heat Resisting Steel Bare Filler Metal | **C** 0.12 max **Cr** 23.0-25.0 **Cu** 0.75 max **Mn** 1.00-2.50 **Mo** 0.75 max **Ni** 12.0-14.0 **P** 0.03 max **S** 0.03 max **Si** 0.30-0.65 | **ASME** SFA5.9 (ER309); SFA5.30 (IN309) **AWS** A5.9 (ER309); A5.30 (IN309) **MIL SPEC** MIL-E-19933 (MIL-309, MIL-309Co) |
| **S30981** | Austenitic Cr-Ni Heat Resisting Steel Bare Filler Metal (309Si) | **C** 0.12 max **Cr** 23.0-25.0 **Cu** 0.5 max **Mn** 1.00-2.50 **Mo** 0.5 max **Ni** 12.0-14.0 **P** 0.03 max **S** 0.03 max **Si** 0.65-1.00 | **ASME** SFA5.9 (ER309Si) **AWS** A5.9 (ER309Si) |
| **S30982** | Austenitic Cr-Ni-Mo Bare Filler Metal (ER 309Mo) | **C** 0.12 max **Cr** 23.0-25.0 **Cu** 0.75 max **Mn** 1.0-2.5 **Mo** 2.0-3.0 **Ni** 12.0-14.0 **P** 0.03 max **S** 0.03 max **Si** 0.30-0.65 | **ASME** SFA5.9 (ER 309Mo) **AWS** A5.9 (ER 309Mo) |
| **S30983** | Austenitic Cr-Ni Heat Resisting Steel Bare Filler Metal (309L) | **C** 0.03 max **Cr** 23.0-25.0 **Cu** 0.75 max **Mn** 1.00-2.50 **Mo** 0.75 max **Ni** 12.0-14.0 **P** 0.03 max **S** 0.03 max **Si** 0.30-0.65 | **ASME** SFA5.9 (ER309L); SFA5.30 (IN309L) **AWS** A5.9 (ER309L); A5.30 (IN309L) |
| **S30986** | Austenitic Cr-Ni-Mo Bare Filler Metal (ER309LMo) | **C** 0.03 max **Cr** 23.0-25.0 **Cu** 0.75 max **Mn** 1.0-2.5 **Mo** 2.0-3.0 **Ni** 12.0-14.0 **P** 0.03 max **S** 0.03 max **Si** 0.30-0.65 | **ASME** SFA5.9 (ER309LMo) **AWS** A5.9 (ER309LMo) |
| **S30988** | Austenitic Cr-Ni Stainless Steel Bare Welding Filler Metal (309 LSi) | **C** 0.03 max **Cr** 23.0-25.0 **Cu** 0.75 max **Mn** 1.0-2.5 **Mo** 0.75 max **Ni** 12.0-14.0 **P** 0.03 max **S** 0.03 max **Si** 0.65-1.00 | **ASME** SFA5.9 (ER 309LSi) **AWS** A5.9 (ER 309LSi) |
| * **S30989** | Austenitic Cr-Ni Heat Resisting Bare Filler Metal (ER 309MoSi-L) | **C** 0.03 max **Cr** 23.0-25.0 **Cu** 0.75 max **Mn** 1.0-2.5 **Mo** 2.0-3.0 **Ni** 12.0-14.0 **P** 0.03 max **S** 0.03 max **Si** 0.65-1.00 | **ASME** SFA5.9 (ER 309LMoSi) **AWS** A5.9 (ER 309LMoSi) |
| **S31000** | Austenitic Cr-Ni Heat Resisting Steel (310) | **C** 0.25 max **Cr** 24.00-26.00 **Mn** 2.00 max **Ni** 19.00-22.00 **P** 0.045 max **S** 0.030 max **Si** 1.50 max | **AISI** 310 **AMS** AMS-QQ-S-763 **ASME** SA-182 (310); SA-213 (310); SA-312 (310); SA-358 (310); SA-403 (310); SA-409 (310) **ASTM** A167 (310); A182 (310); A213 (TP310); A276 (310); A314 (310); A336 (F310); A358 (310); A403 (310); A409 (310); A473 (310); A580 (310); A632 (310); A965 (F310) **MIL SPEC** MIL-S-24149/3; MIL-S-24149/6 **SAE** J412 (30310); J467 (310) |
| **S31002** | Austenitic Cr-Ni Stainless Steel (310 low carbon, 2RE10) | **C** 0.015 max **Cr** 24.0-26.0 **Mn** 2.00 max **Mo** 0.10 max **N** 0.10 max **Ni** 19.0-22.0 **P** 0.020 max **Si** 0.15 max **S** 0.015 max | **ASTM** A213/A213M; A312/A312M **Other:** EN 1.4335 |

*Boxed entries are no longer active and are retained for reference purposes only.

The chemical compositions listed are for identification purposes and should not be used in lieu of the cross referenced specifications.

| UNIFIED NUMBER | DESCRIPTION | CHEMICAL COMPOSITION | CROSS REFERENCE SPECIFICATIONS |
|---|---|---|---|
| **S31008** | Austenitic Cr-Ni Heat Resisting Steel (310 S) | **C** 0.08 max **Cr** 24.00-26.00 **Mn** 2.00 max **Ni** 19.00-22.00 **P** 0.045 max **S** 0.030 max **Si** 1.50 max | **AISI** 310 S **AMS** 5521; 5572; 5577; 5651; 7490; AMS-QQ-S-763 **ASME** SA-240 (310S); SA-479 (310S) **ASTM** A213 (TP310S); A240 (310S); A249 (TP310S); A276 (310S); A312 (TP310S); A314 (310S); A473 (310S); A358 (310S); A409 (310S); A479 (310S); A511 (310S); A554 (310S); A580 (310S); A813 (TP310S); A814 (TP310S); A924 (TP310S); A943 (TP310S) **SAE** J405 (310S) |
| **S31009** | Austenitic Cr-Ni Heat Resisting Steel (310H) | **C** 0.04-0.10 **Cr** 24.0-26.0 **Mn** 2.00 max **Ni** 19.0-22.0 **P** 0.040 max **S** 0.030 max **Si** 0.75 max | **ASTM** A213 (TP310H); A240 (310H); A249 (310H); A312 (TP310H); A336 (F310H); A479 (310H); A943 (TP310H); A965 (F310H) **SAE** J405 (310H) |
| **S31040** | Austenitic Cr-Ni Heat Resisting Steel (310Cb) | **C** 0.08 max **Cr** 24.0-26.00 **Mn** 2.00 max **N** 0.10 max **Ni** 19.00-22.00 **P** 0.045 max **S** 0.030 max **Si** 1.50 max **Other** Cb+Ta 10\mtC min-1.10 max | **ASTM** A213 (TP310Cb); A240 (310Cb); A249 (TP310Cb); A276 (310Cb); A312 (TP310Cb); A358 (310Cb); A409 (310Cb); A478 (310Cb); A479 (310Cb); A813 (TP310Cb); A814 (TP310Cb); A924 (TP310Cb); A943 (T310Cb) **SAE** 405 (310Cb) |
| **S31041** | Austenitic Cr-Ni Heat Resisting Steel (310HCb) | **C** 0.04-0.10 **Cr** 24.0-26.0 **Cb+Ta** 10XC min-1.10 max **Cu** 0.75 max **Mn** 2.00 max **Mo** 0.75 max **Ni** 19.0-22.0 **P** 0.045 max **S** 0.030 max **Si** 0.75 max | **ASTM** A213 (TP310HCb); A240 9310HCb); A249 (TP310HCb); A312 (TP310HCb); A924 (TP310HCb); A943 (TPHCb) **SAE** J405 (310HCb) |
| **S31042** | Austenitic Cr-Ni-Cb-N Stainless Steel (HR3C) | **C** 0.04-0.10 **Cr** 24.0-26.0 **Mn** 2.00 max **N** 0.15-0.35 **Ni** 17.0-23.0 **P** 0.030 max **S** 0.030 max **Si** 0.75 max **Other** Cb+Ta 0.20-0.60 | **ASME** SA-213 **ASTM** A213 (TP310HCbN) |
| **S31050** | Austenitic Cr-Ni-Mo Stainless Steel, Low Carbon, with Nitrogen | **C** 0.030 max **Cr** 24.0-26.0 **Mn** 2.00 max **Mo** 1.6-3.0 **Ni** 20.5-23.5 **N** 0.09-0.16 **P** 0.020 max **S** 0.015 max **Si** 0.4 max | **ASTM** A213; A240; A249; A312; A943 **SAE** J405 (310MoLN) |
| **S31080** | Austenitic Cr-Ni Heat Resisting Steel Bare Filler Metal | **C** 0.08-0.15 **Cr** 25.0-28.0 **Cu** 0.75 max **Mn** 1.00-2.50 **Mo** 0.75 max **Ni** 20.0-22.5 **P** 0.03 max **S** 0.03 max **Si** 0.30-0.65 | **AMS** 5694 **ASME** SFA5.9 (ER310); SFA5.30 (IN310) **AWS** A5.9 (ER310); A5.30 (IN310) **MIL SPEC** MIL-E-19933 (MIL-310); MIL-I-23413 (MIL-310) |
| **S31100** | Duplex (Austenitic-Ferritic) Cr-Ni Stainless Steel (744X) | **C** 0.05 max **Cr** 25.00-27.00 **Mn** 1.00 max **Ni** 6.00-7.00 **P** 0.030 max **S** 0.030 max **Si** 0.60 max **Ti** 0.25 max | **ASTM** A240 (XM-26); A276; A276 (XM-26) |
| **S31200** | Ferritic-Austenitic Cr-Ni-Mo-N Stainless Steel (44LN) | **C** 0.030 max **Cr** 24.0-26.0 **Mn** 2.00 max **Mo** 1.20-2.00 **N** 0.14-0.20 **Ni** 5.50-6.50 **P** 0.045 max **S** 0.030 max **Si** 1.00 max | **ASTM** A182 (F50); A240; A789; A790; A928; A949 **SAE** J405 |
| **S31254** | Austenitic Cr-Ni-Mo-Cu-N Stainless Steel (254 SMO) | **C** 0.020 max **Cr** 19.50-20.50 **Cu** 0.50-1.00 **Mn** 1.00 max **Mo** 6.00-6.50 **N** 0.180-0.220 **Ni** 17.50-18.50 **P** 0.030 max **S** 0.010 max **Si** 0.80 max | **ASME** SA-182; SA-240; SA-249; SA-312; SA-358; SA-409; SA-479; SA-480 **ASTM** A182 (F44); A193; A240; A249; A269; A270; A276; A312; A358; A403; A409; A473; A479; A480; A813; A814; A988 **SAE** J405 |
| **S31260** | Duplex (Austenitic-Ferritic) Cr-Ni-Mo-W Stainless Steel (DP-3) | **C** 0.03 max **Cr** 24.0-26.0 **Cu** 0.20-0.80 **Mn** 1.00 max **Mo** 2.50-3.50 **N** 0.10-0.30 **Ni** 5.50-7.50 **P** 0.030 max **S** 0.030 max **Si** 0.75 max **W** 0.10-0.50 | **ASTM** A240; A789; A790; A928; A949 **SAE** J405 |
| **S31266** | Austenitic Cr-Ni-Mo-W-Cu-N Stainless Steel (Uranus B66) | **C** 0.030 max **Cr** 23.0-25.0 **Cu** 0.50-3.00 **Mn** 2.0-4.0 **Mo** 5.0-7.0 **N** 0.35-0.60 **Ni** 21.0-24.0 **P** 0.035 max **S** 0.020 max **Si** 1.00 max **W** 1.00-3.00 | **ASTM** A240; A240M; A358; A480; A480M **SAE** J405 |
| **S31272** | Austenitic Cr-Ni-Mo-Ti Stainless Steel | **B** 0.004-0.008 **C** 0.08-0.12 **Cr** 14.0-16.0 **Mn** 1.5-2.0 **Mo** 1.0-1.4 **Ni** 14.0-16.0 **P** 0.030 max **S** 0.015 max **Si** 0.3-0.7 **Ti** 0.3-0.6 | **ASTM** A213; A312; A943 |
| **S31380** | Austenitic-Ferritic Cr-Ni Stainless Steel Bare Filler Metal (ER312) | **C** 0.15 max **Cr** 28.0-32.0 **Cu** 0.75 max **Mn** 1.00-2.50 **Mo** 0.75 max **Ni** 8.00-10.5 **P** 0.03 max **S** 0.03 max **Si** 0.30-0.65 | **ASME** SFA5.9 (ER312); SFA5.30 (IN312) **AWS** A5.9 (ER312); A5.30 (IN312) **MIL SPEC** MIL-E-19933 (MIL-312, MIL-312Co) |
| **S31400** | Austenitic Cr-Ni Heat Resisting Steel | **C** 0.25 max **Cr** 23.00-26.00 **Mn** 2.00 max **Ni** 19.00-22.00 **P** 0.045 max **S** 0.030 max **Si** 1.50-3.00 | **AISI** 314 **AMS** 5652; 7490 **ASTM** A276 (314); A314 (314); A473 (314); A580 (314) **SAE** J467 (314) |

The chemical compositions listed are for identification purposes and should not be used in lieu of the cross referenced specifications.

| UNIFIED NUMBER | DESCRIPTION | CHEMICAL COMPOSITION | CROSS REFERENCE SPECIFICATIONS |
|---|---|---|---|
| **S31500** | Duplex (Austenitic-Ferritic) Cr-Ni-Mo-Si Stainless Steel (3RE60) | **C** 0.030 max **Cr** 18.0-19.0 **Mn** 1.20-2.00 **Mo** 2.50-3.00 **N** 0.05-0.10 **Ni** 4.25-5.25 **P** 0.030 max **S** 0.030 max **Si** 1.40-2.00 | **ASTM** A789; A790; A928; A949 |
| **S31600** | Austenitic Cr-Ni-Mo Stainless Steel (316) | **C** 0.08 max **Cr** 16.00-18.00 **Mn** 2.00 max **Mo** 2.00-3.00 **Ni** 10.00-14.00 **P** 0.045 max **S** 0.030 max **Si** 1.00 max | **AISI** 316 **AMS** 5524; 5573; 5690; 5691; 5692; 5696; 5907; AMS-QQ-S-763 **ASME** SA-182 (316); SA-193 (316); SA-194 (316); SA-213 (316); SA-240 (316); SA-249 (316); SA-312 (316); SA-320 (316); SA-358 (316); SA-376 (316); SA-403 (316); SA-409 (316); SA-479 (316); SA-688 (316) **ASTM** A182 (316); A193 (316, B8M, B8MA, B8M2, B8M3); A194 (316, 8M, 8MA); A213 (TP316); A240 (316); A249 (316); A269 (316); A270 (TP316); A271 (TP316); A276 (316); A312 (316); A313 (316); A314 (316); A320 (316, B8M, B8MA); A336 (F316); A358 (316); A368 (316); A376 (316); A376 (TP316) A403 (316); A409 (316); A430 (316); A580 (316); A632 (316); A632 (TP316); A666 (316); A688 (316); A688 (TP316); A793 (316); A813 (TP316); A814 (TP316); A826 (TP316); A831 (316); A943 (TP316); A965 (F316); A988; F593 (316); F594 (316); F738 (316); F836 (316); F837 (316); F879 (316); F899 (316) **MIL SPEC** MIL-S-24149/3; MIL-S-24149/6 **SAE** J405 (316); J412 (30316); J467 (316) |
| **S31603** | Austenitic Cr-Ni-Mo Stainless Steel (Low Carbon) (316L) | **C** 0.030 max **Cr** 16.00-18.00 **Mn** 2.00 max **Mo** 2.00-3.00 **Ni** 10.00-14.00 **P** 0.045 max **S** 0.030 max **Si** 1.00 max | **AISI** 316 L **AMS** 5507; 5584; 5653; AMS-QQ-S-763 **ASME** SA-182 (316L); SA-213 (316L); SA-240 (316L); SA-249 (316L); SA-312 (316L); SA-403 (316L); SA-479 (316L); SA-688 (316L) **ASTM** A182 (316L); A213 (TP316L); A240 (316L); A249 (316L); A269 (316L); A270 (TP316L); A276 (316L); A312 (316L); A314 (316L); A336 (F316L); A358 (316L); A403 (316L); A409 (316L); A473 (316L); A478 (316L); A479 (316L); A493 (316L); A511 (316L); A554 (316L); A580 (316L); A632 (316L); A666 (316L); A688 (316L); A774 (TP316L); A778 (TP316L); A793 (316L); A813 (TP316L); A814 (TP316L); A943 (TP316L); A965 (F316L); A988; F593 (316L); F594 (316L); F738 (316L); F836 (316L); F837 (316L); F879 (316L) **SAE** J405 (316L) |
| **S31609** | Austenitic Cr-Ni-Mo Stainless Steel (High Carbon) (316H) | **C** 0.04-0.10 **Cr** 16.00-18.00 **Mn** 2.00 max **Mo** 2.00-3.00 **Ni** 10.00-14.00 **P** 0.040 max **S** 0.030 max **Si** 1.00 max | **ASME** SA-182 (316H); SA-213 (316H); SA-240 (316H); SA-249 (316H); SA-312 (316H); SA-376 (316H); SA-403 (316H); SA-479 (316H) **ASTM** A182 (316H); A240 (316H); A249 (316H); A312 (316H); A336 (F316H); A358 (316H); A376 (TP316H); A403 (316H); A479 (316H); A813 (TP316H); A814 (TP316H); A814 (TP304H); A943 (TP316H); A965 (F316H); A988 (316H) **SAE** J405 (316H) |
| **S31620** | Austenitic Cr-Ni-Mo Free Machining Stainless Steel | **C** 0.08 max **Cr** 17.00-19.00 **Mn** 2.00 max **Mo** 1.75-2.50 **Ni** 12.00-14.00 **P** 0.20 max **S** 0.10 min **Si** 1.00 max | **AISI** 316 F **AMS** 5649 |
| **S31630** | Austenitic Cr-Ni-Mo-Cu-N Stainless Steel | **C** 0.08-0.12 **Cr** 16.0-17.0 **Cu** 2.5-3.5 **Mn** 2.0-3.0 **Mo** 2.50-3.00 **N** 0.12-0.19 **Ni** 9.0-10.0 **P** 0.020 max **S** 0.015 max **Si** 0.25 max | **SAE** J775 |
| **S31635** | Austenitic Cr-Ni-Mo Stainless Steel (Titanium Stabilized) (316 Ti) | **C** 0.08 max **Cr** 16.00-18.00 **Mn** 2.00 max **Mo** 2.00-3.00 **N** 0.10 max **Ni** 10.00-14.00 **P** 0.045 max **S** 0.030 max **Si** 1.00 max **Ti** 5x(C+N) min-0.70 max | **ASTM** A240 (316 Ti); A276 (316Ti); A314 (316Ti); A368 (316 Ti); A478 (316 Ti); A479 (316 Ti) **SAE** J405 (316Ti) |
| **S31640** | Austenitic Cr-Ni-Mo Stainless Steel (Columbium Stabilized) (316Cb) | **C** 0.08 max **Cr** 16.00-18.00 **Mn** 2.00 max **Mo** 2.00-3.00 **N** 0.10 max **Ni** 10.00-14.00 **P** 0.045 max **S** 0.030 max **Si** 1.50 max **Other** Cb+Ta 10\mtC min-1.10 max | **ASTM** A240 (316Cb); A276 (316Cb); A314 (316Cb); A368 (316Cb); A478 (316Cb); A479 (316Cb) **SAE** J405 (316Cb) |
| **S31651** | Austenitic Cr-Ni-Mo-N Stainless Steel (316N) | **C** 0.08 max **Cr** 16.00-18.00 **Mn** 2.00 max **Mo** 2.00-3.00 **N** 0.10-0.16 **Ni** 10.00-14.00 **P** 0.045 max **S** 0.030 max **Si** 1.00 max | **AISI** 316 N **ASME** SA-182 (316N); SA-213 (316N); SA-240 (316N); SA-249 (316N); SA-312 (316N); SA-358 (316N); SA-376 (316N); SA-403 (316N); SA-479 (316N) **ASTM** A182 (316N); A193 (B8MN, B8MNA); A194 (8MN, 8MNA); A213 (TP316N); A240 (316N); A249 (316N); A276 (316N); A312 (316N); A336 (F316N); A358 (316N); A376 (TP316N); A403 (316N); A479 (316N); A666 (316N); A688 (316N); A688 (TP316N); A813 (TP316N); A814 (TP316N); A943 (TP316N); A965 (F316N); A988 **SAE** J405 (316N) |

The chemical compositions listed are for identification purposes and should not be used in lieu of the cross referenced specifications.

| UNIFIED NUMBER | DESCRIPTION | CHEMICAL COMPOSITION | CROSS REFERENCE SPECIFICATIONS |
|---|---|---|---|
| **S31653** | Austenitic Nitrogen Bearing Cr-Ni-Mo Stainless Steel (316LN) | **C** 0.030 max **Cr** 16.00-18.00 **Mn** 2.00 max **Mo** 2.00-3.00 **N** 0.10-0.16 **Ni** 10.00-14.00 **P** 0.045 max **S** 0.030 max **Si** 1.00 max | **ASTM** A182 (F316LN); A193 (B8MLN, B8MLNA); A194 (8MLN, 8MLNA); A213 (TP316LN); A240 (316LN); A249 (TP316LN); A269 (TP316LN); A276 (316LN); A312 (TP316LN); A320 (B8MLN, B8MLNA); A336 (F316LN); A358 (316LN); A376 (TP316LN); A403 (316LN); A403 (WP316LN); A479 (316LN); A688 (316LN); A688 (TP316LN); A813 (TP316LN); A814 (TP316LN); A943 (TP316LN); A965 (F316LN); A988 **SAE** J405 (316LN) |
| **S31654** | Austenitic Nitrogen Bearing Cr-Ni-Mo Stainless Steel (316 L(Hi)N) | **C** 0.03 max **Cr** 16.00-18.00 **Mn** 2.00 max **Mo** 2.00-3.00 **N** 0.16-0.30 **Ni** 10.00-14.00 **P** 0.045 max **S** 0.030 max **Si** 1.00 max | **ASTM** A276 |
| **S31670** | Austenitic Cr-Ni-Mo Stainless Steel (Surgical Implant 316) | **C** 0.08 max **Cr** 17.00-19.00 **Cu** 0.50 max **Mn** 2.00 max **Mo** 2.00-3.00 **N** 0.10 max **Ni** 13.00-15.50 **P** 0.025 max **S** 0.010 max **Si** 0.75 max | |
| **S31673** | Austenitic Cr-Ni-Mo Stainless Steel (Surgical Implant 316L) | **C** 0.030 max **Cr** 17.00-19.00 **Cu** 0.50 max **Mn** 2.00 max **Mo** 2.00-3.00 **N** 0.10 max **Ni** 13.00-15.50 **P** 0.025 max **S** 0.010 max **Si** 0.75 max | **ASTM** F1350 |
| **S31675** | Austenitic Cr-Ni-Mn-Mo-N Stainless Steel (Surgical Implants) | **C** 0.08 max **Cr** 19.5-22.0 **Cu** 0.25 max **Mn** 2.00-4.25 **Mo** 2.0-3.0 **N** 0.25-0.5 **Nb** (Cb) 0.25-0.8 **Ni** 9.0-11.0 **P** 0.025 max **S** 0.01 max **Si** 0.75 max, | **ASTM** F1586 **Other:** ISO 5832-9 |
| **S31680** | Austenitic Cr-Ni-Mo Stainless Steel Bare Filler Metal | **C** 0.08 max **Cr** 18.0-20.0 **Cu** 0.75 max **Mn** 1.00-2.50 **Mo** 2.00-3.00 **Ni** 11.0-14.0 **P** 0.03 max **S** 0.03 max **Si** 0.30-0.65 | **AMS** 5692 **ASME** SFA5.9 (ER316, 316H); SFA5.30 (IN316) **AWS** A5.9 (ER316, 316H); A5.30 (IN316); C2.25 (W-FeCrNi-5) **MIL SPEC** MIL-E-19933 (MIL-316); MIL-I-23413 (MIL-316) |
| **S31681** | Austenitic Cr-Ni-Mo Stainless Steel Bare Filler Metal (316 Si) | **C** 0.08 max **Cr** 18.0-20.0 **Cu** 0.5 max **Mn** 1.00-2.50 **Mo** 2.00-3.00 **Ni** 11.0-14.0 **P** 0.03 max **S** 0.03 max **Si** 0.65-1.00 | **ASME** SFA5.9 (ER316Si) **AWS** A5.9 (ER316Si); C2.25 (W-FeCrNi-7) |
| **S31683** | Austenitic Cr-Ni-Mo Stainless Steel Bare Filler Metal (316L) | **C** 0.03 max **Cr** 18.0-22.0 **Cu** 0.75 max **Mn** 1.00-2.50 **Mo** 2.00-3.00 **Ni** 11.0-14.0 **P** 0.03 max **S** 0.03 max **Si** 0.30-0.65 | **ASME** SFA5.9 (ER316L); SFA5.30 (IN316L) **AWS** A5.9 (ER316L); A5.30 (IN316L); C2.25 (W-FeCrNi-6) **MIL SPEC** MIL-E-19933 (MIL-316L); MIL-I-23413 (MIL-316L) |
| **S31688** | Austenitic Cr-Ni-Mo Stainless Steel Bare Filler Metal (316LSi) | **C** 0.03 max **Cr** 18.0-20.0 **Cu** 0.5 max **Mn** 1.00-2.50 **Mo** 2.00-3.00 **Ni** 11.0-14.0 **P** 0.03 max **S** 0.03 max **Si** 0.65-1.00 | **ASME** SFA5.9 (ER316LSi) **AWS** A5.9 (ER316LSi) |
| **S31700** | Austenitic Cr-Ni-Mo Stainless Steel (317) | **C** 0.08 max **Cr** 18.00-20.00 **Mn** 2.00 max **Mo** 3.00-4.00 **Ni** 11.00-15.00 **P** 0.045 max **S** 0.030 max **Si** 1.00 max | **AISI** 317 **AMS** AMS-QQ-S-763 **ASME** SA-240 (317); SA-249 (317); SA-312 (317); SA-403 (317); SA-409 (317) **ASTM** A182 (F317); A213 (TP317); A240 (317); A249 (317); A269 (317); A276 (317); A312 (317); A314 (317); A403 (317); A409 (317); A473 (317); A478 (317); A479 (317); A511 (317); A554 (317); A580 (317); A632 (317); A813 (TP317); A814 (TP317); A943 (TP317); A988; F899 (317) **SAE** J405 (317) |
| **S31703** | Austenitic Cr-Ni-Mo Stainless Steel (Low Carbon) (317L) | **C** 0.030 max **Cr** 18.00-20.00 **Mn** 2.00 max **Mo** 3.00-4.00 **Ni** 11.00-15.00 **P** 0.045 max **S** 0.030 max **Si** 1.00 max | **AISI** 317 L **ASME** SA-240 (317L) **ASTM** A182 (F317L); A213 (TP317L); A240 (317L); A249 (TP317L); A312 (TP317L); A403 (317L); A774 (TP317L); A778 (TP317L); A813 (TP317L); A814 (TP317L); A903 (317L); A943 (TP317L); A988 **SAE** J405 (317L) |
| **S31725** | Austenitic Cr-Ni-Mo Stainless Steel (317LM) | **C** 0.03 max **Cr** 18.0-20.0 **Cu** 0.75 max **Mn** 2.00 max **Mo** 4.00-5.00 **N** 0.10 max **Ni** 13.0-17.0 **P** 0.045 max **S** 0.030 max **Si** 0.75 max | **ASTM** A182 (F47); A213; A240 (317LM); A249; A269; A276; A312; A358; A376; A376 (317LM); A403 (317LM); A409; A479; A943; A988 **SAE** J405 |
| **S31726** | Austenitic Cr-Ni-Mo-N Stainless Steel (317LMN) | **C** 0.03 max **Cr** 17.0-20.0 **Cu** 0.75 max **Mn** 2.00 max **Mo** 4.00-5.00 **N** 0.10-0.20 **Ni** 13.5-17.5 **P** 0.045 max **S** 0.030 max **Si** 0.75 max | **ASTM** A182 (F48); A213; A240 (317LMN); A249; A269; A276; A312; A358; A376; A376 (317LMN); A403 (317LMN); A409; A479; A943; A988 **SAE** J405 |
| **S31753** | Austenitic Nitrogen-Bearing Stainless Steel (317LN) | **C** 0.030 max **Cr** 18.0-20.0 **Mn** 2.00 max **Mo** 3.00-4.00 **N** 0.10-0.22 **Ni** 11.0-15.0 **P** 0.045 max **S** 0.30 max **Si** 1.00 max | **ASTM** A240 (317LN) **SAE** J405 (317LN) |

The chemical compositions listed are for identification purposes and should not be used in lieu of the cross referenced specifications.

| UNIFIED NUMBER | DESCRIPTION | CHEMICAL COMPOSITION | CROSS REFERENCE SPECIFICATIONS |
|---|---|---|---|
| **S31780** | Austenitic Cr-Ni-Mo Stainless Steel Bare Filler Metal | **C** 0.08 max **Cr** 18.5-20.5 **Cu** 0.75 max **Mn** 1.00-2.50 **Mo** 3.00-4.00 **Ni** 13.0-15.0 **P** 0.03 max **S** 0.03 max **Si** 0.30-0.65 | **ASME** SFA5.9 (ER317) **AWS** A5.9 (ER317) **MIL SPEC** MIL-E-19933 (MIL-317) |
| **S31783** | Austenitic Cr-Ni-Mo Stainless Steel Bare Filler Metal (317L) | **C** 0.03 max **Cr** 18.5-20.5 **Cu** 0.75 max **Mn** 1.00-2.50 **Mo** 3.00-4.00 **Ni** 13.0-15.0 **P** 0.03 max **S** 0.03 max **Si** 0.30-0.65 | **ASME** SFA5.9 (ER 317L) **AWS** A5.9 (ER 317L) |
| **S31803** | Duplex (Austenitic-Ferritic) Cr-Ni-Mo-N Stainless Steel | **C** 0.030 max **Cr** 21.0-23.0 **Mn** 2.00 max **Mo** 2.50-3.50 **N** 0.08-0.20 **Ni** 4.50-6.50 **P** 0.030 max **S** 0.020 max **Si** 1.00 max | **ASME** SA-182 (F51); SA-240; SA-789; SA-790; SA-815 **ASTM** A240; A276; A479; A789; A790; A815; A923; A928; A949; A988 **SAE** J405 |
| **S31905** | Austenitic Cr-Ni-Mo Stainless Steel with Boron | **B** 0.20-0.50 **C** 0.05 max **Cr** 22.00-24.00 **Mn** 2.00 max **Mo** 3.00-4.00 **N** 0.10 max **Ni** 17.00-19.00 **P** 0.03 max **S** 0.03 max **Si** 1.00 max | **ASTM** B853 |
| **S31980** | Austenitic Cr-Ni-Mo Stainless Steel (Columbium Stabilized) Bare Filler Metal | **C** 0.08 max **Cr** 18.0-20.0 **Cu** 0.75 max **Mn** 1.00-2.50 **Mo** 2.00-3.00 **Ni** 11.0-14.0 **P** 0.03 max **S** 0.03 max **Si** 0.30-0.65 **Other** Cb+Ta 8\mtC-1.0 max | **ASME** SFA5.9 (ER318) **AWS** A5.9 (ER318) **MIL SPEC** MIL-E-19933 (MIL-318) |
| **S32001** | Duplex (Austenitic-Ferritic) Cr-Mn-Ni-N Stainless Steel (19D) | **C** 0.030 max **Cr** 19.5-21.5 **Cu** 1.00 max **Mn** 4.0-6.0 **Mo** 0.60 max **Ni** 1.0-3.0 **N** 0.05-0.17 **P** 0.040 max **Si** 1.00 max **S** 0.030 max | **ASTM** A240; A480 |
| **S32050** | Austenitic Cr-Ni-Mo-N Stainless Steel (SR50A) | **C** 0.030 max **Cr** 22.0-24.0 **Cu** 0.4 max **Mn** 1.5 max **Mo** 6.0-6.8 **N** 0.21-0.32 **Ni** 20.0-23.0 **P** 0.035 max **S** 0.020 max **Si** 1.0 max | **ASTM** A240; A358; A479; A480 |
| **S32100** | Austenitic Cr-Ni-Ti Stainless Steel (Titanium Stabilized) (321) | **C** 0.08 max **Cr** 17.00-19.00 **Mn** 2.00 max **Ni** 9.00-12.00 **P** 0.045 max **S** 0.030 max **Si** 1.00 max **Ti** 5XC min | **AISI** 321 **AMS** 5510; 5557; 5559; 5570; 5576; 5645; 5689; 5896; 7490; AS7325; AMS-QQ-S-763 **ASME** SA-182 (321); SA-193 (321); SA-194 (321); SA-213 (321); SA-240 (321); SA-249 (321); SA-312 (321); SA-320 (321); SA-358 (321); SA-376 (321); SA-403 (321); SA-409 (321); SA-479 (321) **ASTM** A182 (321); A193 (321, B8T, B8TA); A194 (321, 8T, 8TA); A213 (TP321); A240 (321); A249 (321); A269 (321); A276 (321); A312 (321); A313 (321); A314 (321); A320 (321, B8T, B8TA); A336 (F321); A358 (321); A376 (321); A376 (TP321); A403 (321); A409 (321); A430 (321); A473 (321); A479 (321); A493 (TP321); A511 (321); A554 (321); A580 (321); A632 (321); A774 (TP321); A778 (TP321); A813 (TP321); A814 (TP321); A943 (TP321); A965 (F321); F593 (321); F594 (321); F738 (321); F836 (321) **MIL SPEC** MIL-T-16282 (TP321) **SAE** J405 (321); J412 (30321); J467 (321) |
| **S32109** | Austenitic Cr-Ni Stainless Steel (Titanium Stabilized), High Carbon (321H) | **C** 0.04-0.10 **Cr** 17.00-20.00 **Mn** 2.00 max **Ni** 9.00-12.00 **P** 0.040 max **S** 0.030 max **Si** 1.00 max **Ti** 4\mtC-0.60 | **ASME** SA-182 (321H); SA-213 (321H); SA-240 (321H); SA-249 (321H); SA-312 (321H); SA-376 (321H); SA-403 (321H); SA-479 (321H) **ASTM** A182 (321H); A213 (TP321H); A240 (321H); A249 (321H); A312 (321H); A336 (F321H); A376 (TP321H); A403 (321H); A479 (321H); A813 (TP321H); A814 (TP321H); A943 (TP321H); A965 (F321H) **SAE** J405 |
| **S32180** | Austenitic Cr-Ni Stainless Steel (Titanium Stabilized) Bare Filler Metal | **C** 0.08 max **Cr** 18.5-20.5 **Cu** 0.75 max **Mn** 1.00-2.50 **Mo** 0.75 max **Ni** 9.00-10.50 **P** 0.03 max **S** 0.03 max **Si** 0.30-0.65 **Ti** 9\mtC-1.0 | **ASME** SFA5.9 (ER321) **AWS** A5.9 (ER321) **MIL SPEC** MIL-E-19933 (MIL-321) |
| **S32200** | Ni-Cr-Mo Alloy (NIC 25) | **C** 0.03 max **Cr** 20.0-23.0 **Mn** 1.0 max **Mo** 2.5-3.5 **Ni** 23.0-27.0 **P** 0.03 max **S** 0.005 max **Si** 0.5 max | **Other:** NACE MR-01-75 |
| **S32205** | Duplex (Austenitic-Ferritic) Cr-Ni-Mo-N Stainless Steel (2205) | **C** 0.030 max **Cr** 22.0-23.0 **Mn** 2.00 max **Mo** 3.00-3.50 **N** 0.14-0.20 **Ni** 4.50-6.50 **P** 0.030 max **S** 0.020 max **Si** 1.00 max | **ASTM** A240; A240M; A276; A479; A480; A480M; A789; A790; A815; A923; A928; A988 **SAE** J405 |
| **S32304** | Duplex (Austenitic-Ferritic) Cr-Ni-N Stainless Steel (2304) | **C** 0.030 max **Cr** 21.5-24.5 **Cu** 0.05-0.60 **Mn** 2.50 max **Mo** 0.05-0.60 **N** 0.05-0.20 **Ni** 3.0-5.5 **P** 0.040 max **S** 0.040 max **Si** 1.0 max | **ASTM** A240; A276; A789; A790; A928; A949 **SAE** J405 |

The chemical compositions listed are for identification purposes and should not be used in lieu of the cross referenced specifications.

| UNIFIED NUMBER | DESCRIPTION | CHEMICAL COMPOSITION | CROSS REFERENCE SPECIFICATIONS |
|---|---|---|---|
| **S32404** | Duplex (Austenitic-Ferritic) Cr-Ni-Mo-Cu-N Stainless Steel (Uranus 50) | **C** 0.04 max **Cr** 20.5-22.5 **Cu** 1.0-2.0 **Mn** 2.0 max **Mo** 2.0-3.0 **N** 0.20 max **Ni** 5.5-8.5 **P** 0.030 max **S** 0.010 max **Si** 1.0 max | |
| **S32520** | Duplex (Austenitic-Ferritic) Cr-Ni-Mo-Cu-N Stainless Steel (Uranus 52N+) | **C** 0.030 max **Cr** 24.0-26.0 **Cu** 0.50-2.00 **Mn** 1.5 max **Mo** 3.0-4.0 **N** 0.20-0.35 **Ni** 5.5-8.0 **P** 0.035 max **S** 0.020 max **Si** 0.8 max | **ASTM** A240; A240M; A480; A480M; A790 **SAE** J405 |
| **S32550** | Duplex (Austenitic-Ferritic) Cr-Ni-Mo-Cu-N Stainless Steel (Ferralium 255) | **C** 0.04 max **Cr** 24.0-27.0 **Cu** 1.50-2.50 **Mn** 1.5 max **Mo** 2.9-3.9 **N** 0.10-0.25 **Ni** 4.50-6.50 **P** 0.04 max **S** 0.030 max **Si** 1.00 max | **ASTM** A240; A276; A473; A479; A789; A790; A815; A928; A949 **SAE** J405 |
| **S32615** | Austenitic Cr-Ni-Si-Cu-Mo Stainless Steel (Alloy SX) | **C** 0.07 max **Cr** 16.0-21.0 **Cu** 1.5-2.5 **Mn** 2.00 max **Mo** 0.3-1.5 **Ni** 17.5-22.5 **P** 0.045 max **S** 0.030 max **Si** 4.8-6.0 | **ASTM** A213; A240; A312; A479; A943 **SAE** J405 |
| **S32654** | Austenitic Cr-Ni-Mo-n Stainless Steel (654 SMO) | **C** 0.020 max **Cr** 24.0-25.0 **Cu** 0.30-0.60 **Mn** 2.00-4.00 **Mo** 7.00-8.00 **N** 0.45-0.55 **Ni** 21.0-23.0 **P** 0.030 max **S** 0.005 max **Si** 0.50 max | **ASTM** A240; A249; A269; A312; A358; A480; A988 **SAE** J405 |
| **S32750** | Duplex (Austenitic-Ferritic) Cr-Ni-Mo-N Stainless Steel (2507) | **C** 0.030 max **Cr** 24.0-26.0 **Mn** 1.20 max **Mo** 3.0-5.0 **N** 0.24-0.32 **Ni** 6.0-8.0 **P** 0.035 max **S** 0.020 max **Si** 0.8 max | **ASTM** A183 (F53); A240; A479; A789; A790; A815; A928; A949; A988 **SAE** J405 |
| **S32760** | Duplex (Austenitic-Ferritic) Cr-Ni-Mo-Cu-W-N Stainless Steel (Zeron 100) | **C** 0.03 max **Cr** 24.0-26.0 **Cu** 0.5-1.0 **Mn** 1.0 max **Mo** 3.0-4.0 **N** 0.2-0.3 **Ni** 6.0-8.0 **P** 0.03 max **S** 0.01 max **Si** 1.0 max **W** 0.5-1.0 **Other** %Cr+3.3x%Mo+16x%N = 40 min | **ASTM** A812 (F55); A240; A276; A314; A473, A479; A789; A790; A815; A988 **SAE** J405 |
| **S32803** | Ferritic Cr-Ni-Mo Stainless Steel with Cb (Cronifer 2803Mo) | **C** 0.010 max **Cb** 0.15-0.50 **Cr** 28.0-29.0 **Mn** 0.50 max **Mo** 1.8-2.5 **N** 0.025 max **Ni** 3.0-4.0 **P** 0.020 max **S** 0.005 max **Si** 0.50 max **Other** ratio Cb/C+N = 12 min, C+N = 0.030 max | **ASTM** A240; A268 **SAE** J405 |
| **S32900** | Duplex (Austenitic-Ferritic) Cr-Ni-Mo Stainless Steel (329) | **C** 0.08 max **Cr** 23.00-28.00 **Mn** 1.00 max **Mo** 1.00-2.00 **Ni** 2.50-5.00 **P** 0.040 max **S** 0.030 max **Si** 0.75 max | **AISI** 329 **ASME** SA-789 (329) **ASTM** A240 (329); A789 (329); A790; A928; A949 |
| **S32906** | Duplex (Austenitic-Ferritic) Cr-Ni-Mo-N Stainless Steel (Safurex) | **C** 0.030 max **Cr** 28.0-30.0 **Cu** 0.80 max **Mn** 0.80-1.50 **Mo** 1.50-2.60 **N** 0.30-0.40 **Ni** 5.8-7.5 **P** 0.030 max **S** 0.030 **Si** 0.50 max | **ASTM** A479; A789; A790; A249 |
| **S32950** | Duplex (Austenitic-Ferritic) Cr-Ni-Mo-N Stainless Steel (7-Mo Plus) | **C** 0.03 max **Cr** 26.0-29.0 **Mn** 2.00 max **Mo** 1.00-2.50 **N** 0.15-0.35 **Ni** 3.50-5.20 **P** 0.035 max **S** 0.010 max **Si** 0.60 max | **ASTM** A182 (F52); A240; A314; A473; A479; A789; A790; A815; A928; A929; A949; A988 **SAE** J405 |
| **S33100** | Ni-Cr Stainless Steel (F-10) | **C** 0.10-0.20 **Cr** 7.00-9.00 **Mn** 0.50-0.80 **Ni** 19.00-22.00 **P** 0.030 max **S** 0.030 max **Si** 1.00-1.40 | **ASTM** A182 (F-10) |
| **S33228** | Austenitic Ni-Cr Heat Resisting Stainless Steel (Nicrofer 3228 NbCe, Alloy AC66) | **Al** 0.025 max **C** 0.04-0.08 **Cb** 0.6-1.0 **Ce** 0.05-0.10 **Cr** 26.0-28.0 **Mn** 1.0 max **Ni** 31.0-33.0 **P** 0.020 max **S** 0.015 max **Si** 0.30 max | **ASTM** A182 (56); A182 (F56); A213; A240; A249; A312; A314; A403; A479; A480 **SAE** J405 |
| **S33400** | Austenitic Cr-Ni-Ti Stainless Steel (334) | **Al** 0.15-0.60 **C** 0.080 max **Cr** 18.0-20.0 **Mn** 1.00 max **Ni** 19.0-21.0 **P** 0.030 max **S** 0.015 max **Si** 1.00 max **Ti** 0.15-0.60 | **ASTM** A240 (334) |
| **S34565** | Austenitic Cr-Ni-Mn-Mo, High N, Stainless Steel (4565S) | **C** 0.03 max **Cb** 0.1 max **Cr** 23.0-25.0 **Mn** 5.0-7.0 **Mo** 4.0-5.0 **N** 0.4-0.6 **Ni** 16.0-18.0 **P** 0.030 max **S** 0.010 max **Si** 1.0 max | **ASTM** A182 (F49); A240; A249; A269; A276; A312; A358; A403; A409; A479; A480; A943 **SAE** J405 |

The chemical compositions listed are for identification purposes and should not be used in lieu of the cross referenced specifications.

| UNIFIED NUMBER | DESCRIPTION | CHEMICAL COMPOSITION | CROSS REFERENCE SPECIFICATIONS |
|---|---|---|---|
| **S34700** | Austenitic Cr-Ni-Cb Stainless Steel (Columbium/Niobium-Stabilized) (347) | **C** 0.08 max **Cb** 10XC min **Cr** 17.00-19.00 **Mn** 2.00 max **Ni** 9.00-13.00 **P** 0.045 max **S** 0.030 max **Si** 1.00 max | **AISI** 347 **AMS** 5512; 5556; 5558; 5571; 5575; 5646; 5654; 5674; 5790; 5897; 7490; AS7229; AMS-QQ-S-763 **ASME** SA-182 (347); SA-193 (347); SA-194 (347); SA-213 (347); SA-240 (347); SA-249 (347); SA-312 (347); SA-320 (347); SA-358 (347); SA-376 (347); SA-403 (347); SA-409 (347); SA-479 (347) **ASTM** A182 (347); A193 (347, B8CA); A194 (347, 8C, 8CA); A213 (TP347); A240 (347); A249 (347); A269 (347); A276 (347); A312 (347); A313 (347); A314 (347); A320 (347, B8C, B8CA); A336 (F347); A358 (347); A376 (347); A376 (TP347); A403 (347); A409 (347); A430 (347); A473 (347); A479 (347); A511 (347); A554 (347); A580 (347); A632 (347); A774 (TP347); A778 (TP347); A813 (TP347); A814 (TP347); A943 (TP347); A965 (F347); F593 (347); F594 (347); F738 (347); F836 (347) **MIL SPEC** MIL-T-16286 (TP347) **SAE** J405 (347); J412 (30347); J467 9347) |
| **S34709** | Austenitic Cr-Ni-Cb Stainless Steel (Columbium/Niobium-Stabilized, High Carbon) (347H) | **C** 0.04-0.10 **Cb** 8\mtC-1.00 **Cr** 17.00-20.00 **Mn** 2.00 max **Ni** 9.00-13.00 **P** 0.040 max **S** 0.030 max **Si** 1.00 max | **ASME** SA-182 (347H); SA-213 (347H); SA-240 (347H); SA-249 (347H); SA-312 (347H); SA-376 (347H); SA-403 (347H); SA-479 (347H) **ASTM** A182 (347H); A213 (TP347H); A240 (347H); A249 (347H); A271 (347H); A312 (347H); A336 (F347H); A376 (TP347H); A403 (347H); A403 (348H); A430 (347H); A479 (347H); A813 (TP347H); A814 (TP347H); A943 (TP347H); A965 (F347H) **SAE** J405 (347H) |
| **S34710** | Austenitic Cr-Ni Stainless Steel (Columbium Stabilized, High Carbon) (347 HFG) | **C** 0.06-0.10 **Cr** 17.0-20.0 **Cb** 8\mtC-1.0 **Mn** 2.00 max **Ni** 9.00-13.0 **P** 0.040 max **S** 0.030 max **Si** 0.75 max | **ASTM** A213 **ASME** Code Case 2159 |
| **S34720** | Austenitic Cr-Ni Free Machining Stainless Steel (Columbium Stabilized) | **C** 0.08 max **Cb** 10\mtC-1.10 **Cr** 17.00-19.00 **Mn** 2.00 max **Ni** 9.00-12.00 **P** 0.040 max **S** 0.18-0.35 **Si** 1.00 max | **AMS** 5642 (1) |
| **S34723** | Austenitic Cr-Ni Selenium Bearing Free Machining Stainless Steel (Columbium Stabilized) | **C** 0.08 max **Cb** 10\mtC-1.10 **Cr** 17.00-19.00 **Mn** 2.00 max **Ni** 9.00-12.00 **P** 0.11-0.17 **S** 0.030 max **Se** 0.15-0.35 **Si** 1.00 max | **AMS** 5642 (2) |
| * **S34740** | Austenitic Cr-Ni Free Machining Stainless Steel (Columbium Stabilized) | **C** 0.08 max **Cb** 10\mtC-1.10 **Cr** 17.00-19.00 **Mn** 2.00 max **Ni** 9.00-12.00 **P** 0.040 max **S** 0.18-0.35 **Si** 1.00 max | |
| * **S34741** | Austenitic Cr-Ni Selenium Bearing Free Machining Stainless Steel (Columbium Stabilized) | **C** 0.08 max **Cb** 10\mtC-1.10 **Cr** 17.00-19.00 **Mn** 2.00 max **Ni** 9.00-12.00 **P** 0.11-0.17 **S** 0.030 max **Se** 0.15-0.35 **Si** 1.00 max | |
| **S34751** | Austenitic Cr-Ni Stainless Steel (Columbium Stabilized, Nitrogen Strengthened) (347LN) | **C** 0.005-0.020 **Cr** 17.0-20.0 **Mn** 2.00 max **N** 0.06-0.10 **Ni** 9.0-13.0 **P** 0.040 max **S** 0.030 max **Si** 0.75 max **Other** (Cb+Ta) 0.2-0.5; (Cb+Ta)/C 15 min | **ASME** SA-213; SA-312 **ASTM** A213 (TP347LN); A312 (347LN); A312 (TP347LN) |
| **S34780** | Austenitic Cr-Ni Stainless Steel (Columbium Stabilized) Bare Filler Metal | **C** 0.08 max **Cb** 10\mtC-1.0 **Cr** 19.0-21.5 **Cu** 0.75 max **Mn** 1.00-2.50 **Mo** 0.75 max **Ni** 9.00-11.00 **P** 0.03 max **S** 0.03 max **Si** 0.30-0.65 | **AMS** 5790 **ASME** SFA5.9 (ER347); SFA5.30 (IN348) **AWS** A5.9 (ER347); A5.30 (IN348) **MIL SPEC** MIL-E-19933 (MIL-347, MIL-347Co); MIL-I-23413 (MIL-348, MIL-348Co) |
| **S34781** | Austenitic Cr-Ni Stainless Steel (Columbium Stabilized) Bare Filler Metal | **C** 0.07 max **Cr** 17.0-19.0 **Cu** 0.50 max **Mn** 2.00 max **Mo** 0.75 max **Ni** 9.00-13.00 **P** 0.040 max **S** 0.030 max **Si** 0.50-1.00 **Other** Cb+Ta, 12\mtC-0.50 | **AMS** 5680 |
| **S34788** | Austenitic Cr-Ni Stainless Steel (Columbium Stabilized) Bare Filler Metal (347Si) | **C** 0.08 max **Cr** 19.0-21.5 **Cu** 0.5 max **Mn** 1.00-2.50 **Mo** 0.5 max **Ni** 9.00-11.00 **P** 0.03 max **S** 0.03 max **Si** 0.65-1.00 **Other** Cb+Ta, 10\mtC-1.0 | **ASME** SFA5.9 (ER347Si) **AWS** A5.9 (ER347Si) |

*Boxed entries are no longer active and are retained for reference purposes only.

The chemical compositions listed are for identification purposes and should not be used in lieu of the cross referenced specifications.

| UNIFIED NUMBER | DESCRIPTION | CHEMICAL COMPOSITION | CROSS REFERENCE SPECIFICATIONS |
|---|---|---|---|
| **S34800** | Austenitic Cr-Ni Stainless Steel (Columbium Stabilized) (Ta and Co Restricted) | **C** 0.08 max **Cb** 10XC min **Co** 0.20 max **Cr** 17.00-19.00 max **Mn** 2.00 max **Ni** 9.00-13.00 **P** 0.045 max **S** 0.030 max **Si** 1.00 max **Ta** 0.10 max | **AISI** 348 **ASME** SA-182 (348); SA-213 (348); SA-240 (348); SA-249 (348); SA-312 (348); SA-358 (348); SA-376 (348); SA-403 (348); SA-409 (348); SA-479 (348) **ASTM** A182 (348); A213 (TP348); A240 (348); A249 (348); A269 (348); A276 (348); A312 (348); A314 (348); A336 (F348); A358 (348); A376 (TP348); A403 (348); A409 (348); A473; A473 (348); A479 (348); A580 (348); A632 (348); A813 (TP348); A814 (TP348); A943 (TP348H); A965 (F348); F593 (348); F594 (348) **FED** QQ-S-766 (348) **SAE** J405 (348) |
| **S34809** | Austenitic Cr-Ni Stainless Steel (Columbium Stabilized), High Carbon (Ta and Co Restricted) (348-H) | **C** 0.04-0.10 **Cb** 8\mtC-1.00 **Co** 0.20 max **Cr** 17.00-20.00 **Mn** 2.00 max **Ni** 9.00-13.00 **P** 0.045 max **S** 0.030 max **Si** 1.00 max **Ta** 0.10 max | **ASME** SA-182 (348H); SA-213 (348H); SA-240 (348H); SA-249 (348H); SA-312 (348H); SA-479 (348H) **ASTM** A182 (348H); A213 (TP348H); A240 (348H); A249 (348H); A312 (348H); A336 (F348H); A479 (348H); A813 (TP348H); A814 (TP348H); A943 (TP348H); A965 (F348H) **SAE** J405 (348H) |
| **S35000** | Precipitation Hardenable Cr-Ni-Mo Stainless Steel (AM 350) | **C** 0.07-0.11 **Cr** 16.00-17.00 **Mn** 0.50-1.25 **Mo** 2.50-3.25 **N** 0.07-0.13 **Ni** 4.00-5.00 **P** 0.040 max **S** 0.030 max **Si** 0.50 max | **AMS** 5546; 5548; 5554; 5745; AMS-S-8840 **ASTM** A579 (61); A693 (633) **SAE** J467 (AM-350) |
| **S35045** | Austenitic Cr-Ni-Al-Ti Heat Resisting Steel (alloy 803) | **Al** 0.15-0.60 **C** 0.06-0.10 **Cr** 25.0-29.0 **Cu** 0.75 max **Mn** 1.50 max **Ni** 32.0-37.0 **P** 0.045 max **S** 0.015 max **Si** 1.00 max **Ti** 0.15-0.60 | **ASTM** A240; A480 |
| **S35080** | Precipitation Hardenable Cr-Ni-Mo Stainless Steel Bare Filler Metal (AM350) | **C** 0.08-0.12 **Cr** 16.0-17.0 **Mn** 0.50-1.25 **Mo** 2.50-3.25 **N** 0.07-0.13 **Ni** 4.00-5.00 **P** 0.040 max **S** 0.030 max **Si** 0.50 max | **AMS** 5774 |
| **S35135** | Austenitic Cr-Ni-Mo-Ti Heat Resisting Steel (alloy 864) | **C** 0.08 max **Cr** 20.0-25.0 **Cu** 0.75 max **Mn** 1.00 max **Mo** 4.0-4.8 **Ni** 30.0-38.0 **P** 0.045 **S** 0.015 max **Si** 0.60-1.00 **Ti** 0.40-1.00 | **ASTM** A32; A240; A240/A240M; A480/A240M |
| **S35315** | Austenitic Ni-Cr-Si Heat Resisting Steel (353MA) | **C** 0.04-0.08 **Ce** 0.03-0.08 **Cr** 24.0-26.0 **Mn** 2.0 max **N** 0.12-0.18 **Ni** 34.0-36.0 **P** 0.040 max **S** 0.030 max **Si** 1.2-2.0 | **ASTM** A240; A249; A312; A479; A480 **SAE** J405 |
| **S35500** | Precipitation Hardenable Cr-Ni-Mo Stainless Steel (AM 355) | **C** 0.10-0.15 **Cr** 15.00-16.00 **Mn** 0.50-1.25 **Mo** 2.50-3.25 **N** 0.07-0.13 **Ni** 4.00-5.00 **P** 0.040 max **S** 0.030 max **Si** 0.50 max | **AMS** 5547; 5549; 5743; 5744; AMS-S-8840 **ASTM** A564 (634); A579 (64); A693 (634); A705 (634) **SAE** J467 (AM-355) |
| **S35580** | Precipitation Hardenable Cr-Ni-Mo Stainless Steel Bare Filler Metal (AM355) | **C** 0.10-0.15 **Cr** 15.0-16.0 **Cu** 0.50 max **Mn** 0.50-1.25 **Mo** 2.50-3.25 **N** 0.07-0.13 **Ni** 4.00-5.00 **P** 0.04 max **S** 0.03 max **Si** 0.50 max | **AMS** 5780 |
| **S36200** | Precipitation Hardenable Cr-Ni-Ti Stainless Steel (Almar 362) | **Al** 0.10 max **C** 0.05 max **Cr** 14.00-15.00 **Mn** 0.50 max **Ni** 6.00-7.00 **P** 0.030 max **S** 0.030 max **Si** 0.30 max **Ti** 0.55-0.90 | **AMS** 5739; 5740 **ASTM** A693 (XM-9) |
| * **S37000** | Austenitic Cr-Ni-Mo-Ti Stainless Steel | **Al** 0.050 max **As** 0.030 max **B** 0.0020 max **C** 0.030-0.050 **Cb** 0.050 max **Co** 0.050 max **Cr** 12.5-14.5 **Cu** 0.04 max **Mn** 1.65-2.35 **Mo** 1.50-2.50 **N** 0.005 max **Ni** 14.5-16.5 **P** 0.040 max **S** 0.010 max **Si** 0.50-1.00 **Ta** 0.020 max **Ti** 0.10-0.40 **V** 0.05 max **Other** Ti aim for 0.25 | |
| **S38100** | Austenitic Cr-Ni-Si Stainless Steel (18-18-2) | **C** 0.08 max **Cr** 17.00-19.00 **Mn** 2.00 max **Ni** 17.50-18.50 **P** 0.030 max **S** 0.030 max **Si** 1.50-2.50 | **ASME** SA-213 (XM-15); SA-240 (XM-15); SA-249 (XM-15); SA-312 (XM-15) **ASTM** A213 (XM-15); A240 (XM-15); A249 (XM-15); A269 (XM-15); A312 (XM-15); A813 (TPXM-15); A814 (TPXM-15) **SAE** J405 (XM-15) |
| **S38400** | Austenitic Cr-Ni Stainless Steel (Low Work Hardening) (384) | **C** 0.08 max **Cr** 15.00-17.00 **Mn** 2.00 max **Ni** 17.00-19.00 **P** 0.045 max **S** 0.030 max **Si** 1.00 max | **AISI** 384 **ASTM** A473 (384); A493 (384); F593 (384); F594 (384); F738 (384); F836 (384); F837 (384); F879 (384); F880 (384) |

*Boxed entries are no longer active and are retained for reference purposes only.

The chemical compositions listed are for identification purposes and should not be used in lieu of the cross referenced specifications.

| UNIFIED NUMBER | DESCRIPTION | CHEMICAL COMPOSITION | CROSS REFERENCE SPECIFICATIONS |
|---|---|---|---|
| * **S38500** | Austenitic Cr-Ni Stainless Steel (Low Work Hardening) | **C** 0.08 max **Cr** 11.50-13.50 **Mn** 2.00 max **Ni** 14.00-16.00 **P** 0.045 max **S** 0.030 max **Si** 1.00 max | **AISI** 385 |
| **S38660** | Austenitic Cr-Ni-Mo Stainless Steel-Reactor Grade | **Al** 0.050 max **As** 0.030 max **B** 0.0020 max **C** 0.03-0.05 **Cb** 0.050 max **Co** 0.050 max **Cr** 12.5-14.5 **Cu** 0.04 max **Mn** 1.65-2.35 **Mo** 1.50-2.50 **N** 0.005 max **Ni** 14.5-16.5 **P** 0.040 max **S** 0.010 max **Si** 0.50-1.00 **Ta** 0.020 max **Ti** 0.10-0.40 **V** 0.05 max | **ASTM** A771; A826; A831 |
| **S38815** | Cr-Ni-Si-Cu-Mo Austenitic Stainless Steels (Alloy 388) | **Al** 0.30 max **C** 0.030 max **Cr** 13.0-15.0 **Cu** 0.75-1.50 **Mn** 2.00 max **Mo** 0.75-1.50 **Ni** 15.0-17.0 **P** 0.040 max **Si** 5.5-6.5 **S** 0.020 max | **ASTM** A240 |
| **S39205** | Replaced by S31803 | | |
| **S39209** | Duplex Stainless Steel Welding Filler Metal (ER 2209) | **C** 0.03 max **Cr** 21.5-23.5 **Cu** 0.75 **Mn** 0.5-2.0 **Mo** 2.5-3.5 **N** 0.08-0.20 **Ni** 7.0-9.0 **P** 0.03 max **S** 0.03 max **Si** 0.90 max | **ASME** SFA5.9 (ER2209) **AWS** A5.9 (ER2209) |
| **S39215** | Replaced by S31500 | | |
| **S39226** | Replaced by S31260 | | |
| **S39230** | Replaced by S32304 | | |
| **S39240** | Replaced by S32404 | | |
| **S39253** | Replaced by S32550 | | |
| * **S39255** | Duplex Stainless Steel Welding Filler Metal (ER2553) | **C** 0.04 max **Cr** 24.0-27.0 **Cu** 1.5-2.5 **Mn** 0.50-1.5 **Mo** 2.9-3.9 **N** 0.10-0.25 **Ni** 4.5-6.5 **P** 0.03 max **S** 0.03 max **Si** 0.75 max | **ASME** SFA5.9 (ER2553) **AWS** A5.9 (ER2553) |
| **S39274** | Ferritic-Austenitic Cr-Ni-Mo-W-N Stainless Steel (DP3W) | **C** 0.030 max **Cr** 24.0-26.0 **Cu** 0.20-0.80 **Mn** 1.0 max **Mo** 2.50-3.50 **N** 0.24-0.32 **Ni** 6.0-8.0 **P** 0.030 max **S** 0.020 max **Si** 0.80 max **W** 1.50-2.50 | **ASTM** A182 (F54); A815; A988 **Other:** NACE MR-01-75 |
| **S39275** | Replaced by S32750 | | |
| **S39276** | Replaced by S32760 | | |
| **S39277** | Duplex (Austenitic-Ferritic) Cr-Ni-Mo-Cu-W-N Stainless Steel (AF918) | **C** 0.025 max **Cr** 24.0-26.0 **Cu** 1.2-2.0 **Mo** 3.0-4.0 **N** 0.23-0.33 **Ni** 6.5-8.0 **P** 0.025 max **S** 0.002 max **Si** 0.80 max **W** 0.80-1.20 | **ASTM** A182 (F57); A479; A789; A790 **Other:** NACE MR-01-75 |
| **S39295** | Replaced by S32950 | | |
| **S39553** | Duplex Stainless Steel Bare Welding Filler Metal (ER2553) | **C** 0.04 max **Cr** 24.0-27.0 **Cu** 1.5-2.5 **Mn** 1.5 max **Mo** 2.9-3.9 **N** 0.10-0.25 **Ni** 4.5-6.5 **P** 0.04 max **S** 0.03 max **Si** 1.0 max | **ASME** SFA5.9 (ER2553) **AWS** A5.9 (ER2553) |
| **S40300** | Martensitic Stainless Steel (Hardenable by Heat Treatment) (403) | **C** 0.15 max **Cr** 11.50-13.00 **Mn** 1.00 max **P** 0.040 max **S** 0.030 max **Si** 0.50 max | **AISI** 403 **AMS** AMS-QQ-S-763 **ASTM** A176 (403); A276 (403); A314 (403); A473 (403); A479 (403); A511 (403); A580 (403) |
| **S40500** | Ferritic Cr Stainless Steel (Not Hardenable by Heat Treatment) (405) | **Al** 0.10-0.30 **C** 0.08 max **Cr** 11.50-14.50 **Mn** 1.00 max **P** 0.040 max **S** 0.030 max **Si** 1.00 max | **AISI** 405 **AMS** AMS-QQ-S-763 **ASME** SA-240 (405); SA-268 (405); SA-479 (405) **ASTM** A240 (405); A268 (405); A276 (405); A473 (405); A479 (405); A511 (405); A580 (405) |
| **S40800** | Ferritic Cr Stainless Steel | **C** 0.08 max **Cr** 11.5-13.0 **Mn** 1.00 max **Ni** 0.80 max **P** 0.045 max **S** 0.045 max **Si** 1.00 max **Ti** 12\mtC-1.10 | **ASTM** A268 |

*Boxed entries are no longer active and are retained for reference purposes only.

The chemical compositions listed are for identification purposes and should not be used in lieu of the cross referenced specifications.

| UNIFIED NUMBER | DESCRIPTION | CHEMICAL COMPOSITION | CROSS REFERENCE SPECIFICATIONS |
|---|---|---|---|
| **S40900** | Ferritic Cr Stainless Steel (409) | **C** 0.08 max **Cr** 10.50-11.75 **Mn** 1.00 max **Ni** 0.50 max **P** 0.045 max **S** 0.045 max **Si** 1.00 max **Ti** 6\mtC-0.75 | **AISI** 409 **ASME** SA-268 (409) **ASTM** A268 (409); A463 (409); A803 (TP409) **SAE** J412 (51409) |
| **S40910** | Ferritic Cr Stainless Steel with Ti | **C** 0.030 max **Cr** 10.5-11.7 **Mn** 1.00 max **N** 0.030 max **Ni** 0.50 max **P** 0.040 max **S** 0.010 max **Si** 1.00 max **Ti:** 6x(C+N) min **0.50** max | **ASTM** A240 |
| **S40920** | Ferritic Cr Stainless Steel with Ti | **C** 0.030 max **Cr** 10.5-11.7 **Mn** 1.00 max **N** 0.030 max **Ni** 0.50 max **P** 0.040 max **S** 0.010 max **Si** 1.00 max **Ti:** 0.15-0.50; 8x(C+N) min | **ASTM** A240 |
| **S40930** | Ferritic Cr Stainless Steel with Cb+Ti | **C** 0.030 max **Cr** 10.5-11.7 **Mn** 1.00 max **N** 0.030 max **Ni** 0.50 max **P** 0.040 max **S** 0.010 max **Si** 1.00 max **Ti** 0.50 min **Other** (Cb+Ti): 0.08+8x(C+N) min; 0.75 max | **ASTM** A240 |
| **S40940** | Ferritic Cr Stainless Steel, with Cb | **C** 0.06 max **Cr** 10.50-11.75 **Cb** 10\mtC min-0.75 max **Mn** 1.00 max **Ni** 0.50 max **P** 0.045 max **S** 0.040 max **Si** 1.00 max | **ASME** SFA5.9 (ER409Cb) **ASTM** A493 **AWS** A5.9 (ER409Cb) **Other:** ISO 4954 |
| **S40945** | Ferritic Cr Stainless Steel with Cb | **C** 0.030 max **Cb** 0.18-0.40 **Cr** 10.50-11.75 **Mn** 1.00 max **N** 0.030 max **Ni** 0.50 max **P** 0.040 max **S** 0.030 max **Si** 1.00 max **Ti** 0.05-0.20 | **ASTM** A240; A276; A480 **SAE** J405 |
| **S40975** | Ferritic Cr Stainless Steel with Ni (409 Ni) | **C** 0.030 max **Cr** 10.50-11.75 **Mn** 1.00 max **N** 0.030 max **Ni** 0.50-1.00 **P** 0.040 max **S** 0.030 max **Si** 1.00 max **Ti** 6x(C+N) min, 0.75 max | **ASTM** A240; A240M **SAE** J405 |
| **S40976** | Ferritic Cr Stainless Steel with Ni and Cb (409Ni-Cb) | **C** 0.030 max **Cb** 10X(C+N)-0.80 **Cr** 10.5-11.7 **Mn** 1.00 max **N** 0.040 max **Ni** 0.75-1.00 **P** 0.040 max **S** 0.030 max **Si** 1.00 max | **ASTM** A276; A580 |
| **S40977** | Ferritic Cr Stainless Steel with Ni | **C** 0.03 max **Cr** 10.5-12.5 **Mn** 1.50 max **N** 0.030 max **Ni** 0.30-1.00 **P** 0.040 max **S** 0.015 max **Si** 1.00 max | **ASTM** A240 **Other:** EN 1.4003 |
| **S41000** | Martensitic Cr Stainless Steel (Hardenable by Heat Treatment) (410) | **C** 0.15 max **Cr** 11.50-13.50 **Mn** 1.00 max **P** 0.040 max **S** 0.030 max **Si** 1.00 max | **AISI** 410 **AMS** 5350; 5504; 5505; 5591; 5613; 5776; 5777; 5867; AMS-QQ-S-763 **ASME** SA-194 (6); SA-240 (410); SA-268 (410); SA-479 (410) **ASTM** A182 (F6a); A193 (410, B6, B6X); A194 (410, 6); A240 (410); A268 (410); A276 (410); A314 (410); A336 (F6); A473 (410); A479 (410); A493 (410); A511 (410); A580 (410); A815 (WP410); A837 (410); A988; F593 (410); F594 (410); F738 (410); F899 (410) **SAE** J412 (51410); J467 (410) |
| **S41001** | Cr Super Strength Steel | **Al** 0.05 max **C** 0.15 max **Cr** 11.5-13.5 **Cu** 0.50 max **Mn** 1.00 max **Mo** 0.50 max **Ni** 0.75 max **P** 0.025 max **S** 0.025 max **Si** 1.00 max **Sn** 0.05 max | **AMS** 5611; 5612 **ASTM** A579 (51) |
| **S41003** | Ferritic-Martensitic Cr Stainless Steel | **C** 0.030 max **Cr** 10.5-12.5 **Mn** 1.50 max **N** 0.030 max **Ni** 1.50 max **P** 0.040 max **S** 0.030 max **Si** 1.00 max | **ASTM** A240; A1010 **SAE** J405 |
| **S41005** | Martensitic Cr Stainless Steel | **Al** 0.025 max **C** 0.10-0.15 **Cr** 11.5-13.0 **Cb** 0.2 max **Cu** 0.5 max **Mn** 0.25-0.80 **Mo** 0.50 max **Ni** 0.75 max **N** 0.08 max **P** 0.018 max **Si** 0.5 max **S** 0.015 max **Sn** 0.05 max **Ti** 0.05 max **W** 0.1 max | **ASTM** A1xyz (B) (pending new spec) |
| **S41008** | Martensitic Cr Stainless Steel (Hardenable by Heat Treatment, Low Carbon) (410S) | **C** 0.08 max **Cr** 11.50-13.50 **Mn** 1.00 max **Ni** 0.60 max **P** 0.040 max **S** 0.030 max **Si** 1.00 max | **ASME** SA-240 (410S) **ASTM** A240 (410S); A473 (410S) |
| **S41025** | Martensitic Cr Stainless Steel with Mo | **C** 0.15 max **Cr** 11.50-13.50 **Mn** 1.00 max **Mo** 0.40-0.60 **Ni** 0.60 max **P** 0.040 max **S** 0.030 max **Si** 1.00 max | **AMS** 5614 |

The chemical compositions listed are for identification purposes and should not be used in lieu of the cross referenced specifications.

| UNIFIED NUMBER | DESCRIPTION | CHEMICAL COMPOSITION | CROSS REFERENCE SPECIFICATIONS |
|---|---|---|---|
| **S41026** | Martensitic Cr Stainless Steel with Mo | **C** 0.15 max **Cr** 11.5-13.5 **Cu** 0.50 max **Mn** 1.00 max **Mo** 0.40-0.60 **Ni** 1.0-2.0 **P** 0.02 max **S** 0.02 max **Si** 1.00 max | **ASTM** A182 (F6b); A988 |
| **S41040** | Martensitic Cr Stainless Steel with Cb | **C** 0.15 max **Cb** 0.05-0.20 **Cr** 11.50-13.50 **Mn** 1.00 max **P** 0.040 max **S** 0.030 max **Si** 1.00 max | **AMS** 5609 **ASME** SA-479 (XM-30) **ASTM** A276 (XM-30); A479 (XM-30) |
| **S41041** | Martensitic Cr Stainless Steel with Cb | **Al** 0.05 max **C** 0.13-0.18 **Cb** 0.15-0.45 **Cr** 11.50-13.00 **Mn** 0.40-0.60 **Mo** 0.20 max **Ni** 0.50 max **P** 0.030 max **S** 0.030 max **Si** 0.50 max **Sn** 0.05 max | **AMS** 5612 **ASTM** A565 |
| **S41045** | Ferritic Cr Stainless Steel with Cb | **C** 0.030 max **Cb** 9(C+N) min-0.60 max **Cr** 12.0-13.0 **Mn** 1.00 max **N** 0.030 max **Ni** 0.50 max **P** 0.040 max **S** 0.030 max **Si** 1.00 max | **ASTM** A240; A480 **SAE** J405 |
| **S41050** | Ferritic Cr Stainless Steel with Ni (E-4) | **C** 0.040 max **Cr** 10.50-12.50 **Mn** 1.00 max **N** 0.10 max **Ni** 0.60-1.10 **P** 0.045 max **S** 0.030 max **Si** 1.00 max | **ASTM** A240 **SAE** J405 |
| **S41080** | Martensitic Cr Stainless Steel Bare Filler Metal | **C** 0.12 max **Cr** 11.5-13.5 **Cu** 0.75 max **Mn** 0.6 max **Mo** 0.75 max **Ni** 0.6 max **P** 0.03 max **S** 0.03 max **Si** 0.50 max | **AMS** 5776 **ASME** SFA5.9 (ER410) **AWS** A5.9 (ER410) **MIL SPEC** MIL-E-19933 (MIL-410) |
| **S41081** | Martensitic Cr Stainless Steel Bare Filler Metal | **Al** 0.05 max **C** 0.11-0.15 **Cr** 11.5-12.5 **Cu** 0.50 max **Mn** 0.60 max **Mo** 0.20 max **N** 0.08 max **Ni** 0.75 max **P** 0.025 max **S** 0.025 max **Si** 0.50 max **Ti** 0.05 max | **AMS** 5821 |
| **S41086** | Martensitic Cr-Ni-Mo Stainless Steel Bare Filler Metal (410NiMo) | **C** 0.06 max **Cr** 11.0-12.5 **Cu** 0.75 max **Mn** 0.60 max **Mo** 0.40-0.70 **Ni** 4.00-5.00 **P** 0.03 max **S** 0.03 max **Si** 0.50 max | **ASME** SFA5.9 (ER410NiMo) **AWS** A5.9 (ER410NiMo) |
| **S41400** | Martensitic Cr Stainless Steel (Hardenable by Heat Treatment, Low Nickel) (414) | **C** 0.15 max **Cr** 11.50-13.50 **Mn** 1.00 max **Ni** 1.25-2.50 **P** 0.040 max **S** 0.030 max **Si** 1.00 max | **AISI** 414 **AMS** AMS-QQ-S-763 **ASTM** A276 (414); A314 (414); A473 (414); A479 (414); A511 (414); A580 (414) **SAE** J412 (51414) |
| **S41425** | Martensitic Cr-Ni-Mo-N Stainless Steel (AF913) | **C** 0.050 max **Cr** 12.00-15.00 **Cu** 0.30 max **Mn** 0.50-1.00 **Mo** 1.50-2.00 **N** 0.06-0.12 **Ni** 4.0-7.0 **P** 0.020 max **S** 0.005 max **Si** 0.5 max | **ASTM** A276; A314; A473; A479 **Other:** NACE MR-01-75 |
| **S41426** | Martensitic Cr-Ni-Mo-Ti Stainless Steel (SM13CRS) | **C** 0.03 max **Cr** 11.5-13.5 **Mn** 0.50 max **Mo** 1.5-3.0 **Ni** 4.5-6.5 **P** 0.02 max **Si** 0.50 max **S** 0.005 max **Ti** 0.01-0.50 **V** 0.50 max | **Other:** NACE MR0175 |
| **S41427** | Martensitic Cr-Ni-Mo-V Stainless Steel (SF13/2/5) | **C** 0.03 max **Cr** 11.5-13.5 **Mn** 1.0 max **Mo** 1.5-2.5 **Ni** 4.5-6.0 **P** 0.02 max **S** 0.005 max **Si** 0.50 max **Ti** 0.01 max **V** 0.10-0.50 | **Other:** NACE MR0175 |
| **S41428** | Martensitic Cr-Ni-Mo-V-N Stainless Steel | **Al** 0.025 **C** 0.10-0.17 **Cr** 11.25-12.75 **Cu** 0.5 max **Mn** 0.65-1.05 **Mo** 1.50-2.00 **N** 0.020-0.045 **Ni** 2.25-3.25 **P** 0.02 max **S** 0.015 max **Si** 0.35 max **Sn** 0.05 max **Ti** 0.05 max **V** 0.25-0.40 **W** 0.1 max | **ASTM** A1xyz (C) (pending new spec) |
| **S41429** | Martensitic Cr-Ni-Mo-Ti Stainless Steel | **Al** 0.05 max **B** 0.01 max **C** 0.10 max **Cb** 0.2 max **Co** 1.0 max **Cr** 10.5-14.0 **Cu** 0.5 max **Mn** 0.75 max **Mo** 0.4-1.5 **N** 0.03 max **Ni** 0.5-2.8 **P** 0.03 max **S** 0.03 max **Se** 0.01 max **Si** 1.0 max **Sn** 0.02 max **Ti** 0.25-0.85 **V** 0.25 max | **Other:** NACE MR0175 |
| **S41500** | Cr-Ni-Mo Martensitic Stainless Steel (F6NM, wrought CA-6NM) | **C** 0.05 max **Cr** 11.5-14.0 **Mn** 0.50-1.00 **Mo** 0.50-1.00 **Ni** 3.50-5.50 **P** 0.030 max **S** 0.030 max **Si** 0.60 max | **ASTM** A182 (F6NM); A240; A268; A473; A479; A815; A988 **SAE** J405 |

The chemical compositions listed are for identification purposes and should not be used in lieu of the cross referenced specifications.

| UNIFIED NUMBER | DESCRIPTION | CHEMICAL COMPOSITION | CROSS REFERENCE SPECIFICATIONS |
|---|---|---|---|
| **S41600** | Martensitic Free-Machining Stainless Steel (416) | **C** 0.15 max **Cr** 12.00-14.00 **Mn** 1.25 max **Mo** 0.60 max (Optional) **P** 0.060 max **S** 0.15 min **Si** 1.00 max | **AISI** 416 **AMS** 5610 **ASME** SA-194 (416) **ASTM** A194 (416, 6F); A314 (416); A473 (416); A581 (416); A582 (416); A895 (416); A593 (416); A594 (416); F738 (416); F836 (416); F899 (416) **SAE** J412 (51416) |
| **S41603** | Cr Stainless Steel (Free Machining, Low Hardenable) | **C** 0.08 max **Cr** 12.0-14.0 **Mn** 1.25 max **P** 0.06 max **S** 0.15 min **Si** 1.00 max | **ASTM** A581; A582 |
| **S41610** | Martensitic Cr Free Machining Stainless Steel | **C** 0.15 max **Cr** 12.00-14.00 **Mn** 1.50-2.50 **Mo** 0.60 max **P** 0.06 max **S** 0.15 min **Si** 1.00 max | **ASTM** A581 (XM-6); A582 (XM-6) |
| **S41623** | Martensitic Cr Free-Machining Stainless Steel (416Se) | **C** 0.15 max **Cr** 12.00-14.00 **Mn** 1.25 max **P** 0.060 max **S** 0.060 **Se** 0.15 min **Si** 1.00 max | **AISI** 416Se **AMS** 5610; AMS-QQ-S-763 **ASME** SA-194 (416Se) **ASTM** A194 (416Se, 6F); A314 (416Se); A473 (416Se); A511 (416Se); A581 (416Se); A582 (416Se); A895 (416Se); F593 (416Se); F594 (416Se); F738 (416Se); F836 (416Se) |
| **S41683** | Thermal Spray Wire (W-FeCr-Ni-8) | **C** 0.030 max **Cr** 12-14 **Fe** rem **Mn** 1.0 max **Ni** 1.0 max **Si** 0.080 max | **AWS** C2.25 (W-FeCrNi-8) |
| **S41780** | Ferritic Cr-Ni-Co-Mo Stainless Steel Bare Filler Metal | **C** 0.10-0.15 **Co** 1.30-2.00 **Cr** 11.0-12.5 **Cu** 0.75 max **Mn** 0.40-1.30 **Mo** 1.50-2.00 **N** 0.045 **Ni** 2.50-3.00 **P** 0.030 max **S** 0.030 max **Si** 0.40 max **V** 0.25-0.40 | **AMS** 5822; 5823 |
| **S41800** | Martensitic Cr-W-Ni Stainless and Heat Resisting Steel (Greek Ascoloy) | **C** 0.15-0.20 **Cr** 12.00-14.00 **Mn** 0.50 max **Ni** 1.80-2.20 **P** 0.040 max **S** 0.030 max **Si** 0.50 max **W** 2.50-3.50 | **AMS** 5508; 5616; AS7470 **ASTM** A565 (615) **SAE** J467 (Greek Ascoloy) |
| **S41880** | Martensitic Cr-W-Ni Stainless and Heat Resisting Steel Bare Filler Metal (Greek Ascoloy) | **Al** 0.15 max **C** 0.15-0.20 **Cr** 12.0-14.0 **Cu** 0.50 max **Mn** 0.50 max **Mo** 0.50 max **Ni** 1.80-2.20 **P** 0.040 max **S** 0.030 max **Si** 0.50 max **Sn** 0.05 max **W** 2.50-3.50 | **AMS** 5817 |
| **S42000** | Martensitic Cr Stainless Steel (Hardenable by Heat Treatment) (420) | **C** over 0.15 **Cr** 12.00-14.00 **Mn** 1.00 max **P** 0.040 max **S** 0.030 max **Si** 1.00 max | **AISI** 420 **AMS** 5506; 5621; AS7202; AMS-QQ-S-763 **ASTM** A176 (420); A276 (420); A314 (420); A473 (420); A580 (420) **SAE** J412 (51420) |
| **S42002** | Martensitic Cr Stainless Steel (High Carbon) (SF100) | **C** 0.65-0.75 **Cr** 12.0-14.0 **Mn** 1.00 max **MO** 0.50 max **Ni** 1.00 max **P** 0.040 max **Si** 1.00 max **S** 0.030 max | **Other:** ISO 7153-1, Grade 6 |
| **S42010** | Hardenable Cr Stainless Steel (Low Ni - Low Mo) | **C** 0.15-0.30 **Cr** 13.5-15.0 **Mn** 1.00 max **Mo** 0.40-1.00 **Ni** 0.25-1.00 **P** 0.040 max **S** 0.030 max **Si** 1.00 max | **ASTM** A276; A314; A493 |
| **S42020** | Martensitic Cr Stainless Steel (Hardenable by Heat Treatment) (420F) | **C** over 0.15 **Cr** 12.00-14.00 **Mn** 1.25 max **Mo** 0.60 max (Optional) **P** 0.060 max **S** 0.15 min **Si** 1.00 max | **AISI** 420F **AMS** 5620 **ASTM** A582 (420F); A895 (420F); F899 (420F) **SAE** J412 (51420F) |
| **S42023** | Martensitic Cr Stainless Steel (Hardenable by Heat Treatment) (420FSe) | **C** 0.30-0.40 **Cr** 12.00-14.00 **Mn** 1.25 max **Mo** 0.60 max **P** 0.06 max **S** 0.06 max **Se** 0.15 min **Si** 1.00 max **Other** Zr or Cu 0.60 max | **AMS** 5620 **ASTM** A582 (420FSe); A895 (420FSe) |
| **S42025** | Martensitic Cr-Mo-V-N Stainless Steel | **C** 0.37-0.45 **Cr** 15.0-16.5 **Mn** 0.60 max **Mo** 1.50-1.90 **N** 0.16-0.25 **Ni** 0.30 max **P** 0.020 max **S** 0.005 max **Si** 0.60 max **V** 0.20-0.40 | **AMS** 5925 **Other:** EN 1.4123 |
| **S42035** | Martensitic Cr-Ni-Mo Stainless Steel with Ti | **C** 0.08 max **Cr** 13.5-15.5 **Mn** 1.00 max **Mo** 0.20-1.20 **Ni** 1.00-2.50 **P** 0.045 max **S** 0.030 max **Si** 1.00 max **Ti** 0.30-0.50 | **ASTM** A240 **Other:** EN 1.4589 |
| **S42080** | Hardenable Cr Stainless Steel Bare Filler Metal | **C** 0.25-0.40 **Cr** 12.0-14.0 **Cu** 0.75 max **Mn** 0.60 max **Mo** 0.75 max **Ni** 0.60 max **P** 0.03 max **S** 0.03 max **Si** 0.50 max | **ASME** SFA5.9 (ER420) **AWS** A5.9 (ER420); C2.25 (W-FeCrNi-9) |
| **S42100** | Martensitic Cr Stainless Steel | **Al** 0.05 max **C** 0.17-0.23 **Cb** 0.050 max **Cr** 11.0-12.5 **Mn** 0.40-0.70 **Mo** 0.80-1.20 **Ni** 0.30-0.80 **P** 0.040 max **S** 0.010 max **Si** 0.20-0.30 **V** 0.25-0.35 **W** 0.40-0.60 | **ASTM** A771; A826; A831 |

The chemical compositions listed are for identification purposes and should not be used in lieu of the cross referenced specifications.

| UNIFIED NUMBER | DESCRIPTION | CHEMICAL COMPOSITION | CROSS REFERENCE SPECIFICATIONS |
|---|---|---|---|
| **S42200** | Martensitic Cr-Mo-Ni-W Stainless Steel (422, 616) | **C** 0.20-0.25 **Cr** 11.00-12.5 **Cu** 0.50 max **Mn** 1.00 max **Mo** 0.75-1.25 **Ni** 0.50-1.00 **P** 0.040 max **S** 0.030 max **Si** 0.75 max **V** 0.15-0.30 **W** 0.75-1.25 | **AISI** 422 **AMS** 5655 **ASTM** A176 (422); A565 (616) **SAE** J467 (422); J775 (HNV 8) |
| **S42201** | Martensitic Cr, Superstrength Steel | **Al** 0.05 max **C** 0.20-0.25 **Co** 0.25 max **Cr** 11.0-13.5 **Mn** 0.50-1.00 **Mo** 0.75-1.25 **Ni** 0.75-1.25 **P** 0.025 max **S** 0.025 max **Si** 0.20-0.60 **Sn** 0.04 max **Ti** 0.05 max **V** 0.20-0.30 **W** 0.75-1.25 | **ASTM** A579 (52) |
| **S42225** | Martensitic Cr-Mo-W-Ni-V Stainless Steel | **Al** 0.025 max **C** 0.20-0.25 **Cr** 11.0-12.5 **Co** 0.2 max **Cu** 0.5 max **Mn** 0.50-1.00 **Mo** 0.90-1.25 **Ni** 0.50-1.00 **P** 0.025 max **Si** 0.5 max **S** 0.01 max **Sn** 0.02 max **Ti** 0.025 max **W** 0.90-1.25 **V** 0.20-0.30 | **ASTM** A437 (B4B, B4C); A982 (D); A1xyz (D) (pending new spec) |
| **S42300** | Martensitic Cr-Mo-V Stainless Steel (Lapelloy) | **C** 0.27-0.32 **Cr** 11.00-12.00 **Mn** 0.95-1.35 **Mo** 2.50-3.00 **Ni** 0.50 max **P** 0.025 max **S** 0.025 max **Si** 0.50 max **V** 0.20-0.30 | **ASTM** A565 (619) |
| **S42390** | Martensitic Cr-Mo-V-Cb Stainless Steel (APM 2390) | **C** 0.18-0.25 **Cb** 0.08-0.15 **Cr** 11.5-12.5 **Mn** 1.00 max **Mo** 0.80-1.20 **N** 0.03-0.08 **Ni** 0.30-0.80 **P** 0.030 max **S** 0.030 max **Si** 1.00 max **V** 0.25-0.35 | **ASTM** A988 |
| **S42400** | Ferritic Cr-Ni-Mo Stainless Steel | **C** 0.06 max **Cr** 12.00-14.00 **Mn** 0.50-1.00 **Mo** 0.30-0.70 **Ni** 3.50-4.50 **P** 0.03 max **S** 0.03 max **Si** 0.30-0.60 | **Other:** NACE MR-01-75-92 |
| **S42500** | Ferritic Cr-Ni-Mo Stainless Steel | **C** 0.08-0.20 **Cr** 14.0-16.0 **Mn** 1.0 max **Mo** 0.3-0.7 **N** 0.20 max **Ni** 1.0-2.0 **P** 0.020 max **S** 0.010 max **Si** 1.0 max | **Other:** NACE MR-01-75 |
| **S42640** | Martensitic Cr-Co-Mo-Ni-V-Cb Stainless Steel | **C** 0.10-0.25 **Cb** 0.01-0.05 **Co** 11.00-14.00 **Cr** 13.00-16.00 **Cu** 0.10 max **Mn** 1.00 max **Mo** 3.00-5.00 **Ni** 1.75-2.75 **P** 0.020 max **S** 0.010 max **Si** 1.0 max **V** 0.40-0.80 **W** 0.25 max | **AMS** 5932 |
| **S42670** | Martensitic Cr-Co-Ni-Mo Stainless Steel (Pyrowear 675) | **C** 0.05-0.09 **Co** 4.00-7.00 **Cr** 12.0-14.0 **Mn** 0.50-1.00 **Mo** 1.50-2.50 **Ni** 2.00-3.00 **P** 0.015 max **S** 0.010 max **Si** 0.10-0.70 **V** 0.40-0.80 | **AMS** 5930 |
| **S42700** | Martensitic Cr-Mo-V Stainless Steel (for Bearings) | **C** 1.10-1.20 **Cr** 14.0-15.0 **Cu** 0.35 max **Mn** 0.30-0.60 **Mo** 3.75-4.25 **Ni** 0.40 max **P** 0.015 max **S** 0.010 max **Si** 0.20-0.40 **V** 1.10-1.30 | **AMS** 5749 |
| **S42800** | Martensitic Cr-Mo-V Stainless Steel (for Bearings) | **C** 1.05-1.15 **Cb** 0.25-0.35 **Cr** 13.7-14.8 **Cu** 0.35 max **Mn** 0.25-0.50 **Mo** 1.90-2.25 **Ni** 0.35 max **P** 0.015 max **S** 0.010 max **Si** 0.20-0.40 **V** 0.90-1.15 | |
| **S42900** | Ferritic Cr Stainless Steel (429) | **C** 0.12 max **Cr** 14.00-16.00 **Mn** 1.00 max **P** 0.040 max **S** 0.030 max **Si** 1.00 max | **AISI** 429 **AMS** AMS-QQ-S-763 **ASME** SA-182 (429); SA-240 (429); SA-268 (429) **ASTM** A182 (429); A240 (429); A268 (429); A276 (429); A314 (429); A473 (429); A493 (429); A511 (429); A554 (429); A815 (WP429) |
| **S43000** | Ferritic Cr Stainless Steel (430) | **C** 0.12 max **Cr** 16.00-18.00 **Mn** 1.00 max **P** 0.040 max **S** 0.030 max **Si** 1.00 max | **AISI** 430 **AMS** 5503; 5627; AMS-QQ-S-763 **ASME** SA-182 (430); SA-240 (430); SA-268 (430); SA-479 (430) **ASTM** A182 (430); A240 (430); A268 (430); A276 (430); A314 (430); A473 (430); A479 (430); A493 (430); A511 (430); A554 (430); A580 (430); A815 (WP430); F593 (430); F594 (430); F738 (430); F836 (430) **SAE** J412 (51430) |
| **S43020** | Ferritic Cr Free-Machining Stainless Steel (430FS) | **C** 0.12 max **Cr** 16.00-18.00 **Mn** 1.25 max **Mo** 0.60 max (Optional) **P** 0.060 max **S** 0.15 min **Si** 1.00 max | **AISI** 430 F **ASTM** A314 (430F); A473 (430F); A581 (430F); A582 (430F); A895 (430F); F593 (430F); F594 (430F); F738 (430F); F836 (430F); F899 (430F) **SAE** J412 (51430F) |
| **S43023** | Ferritic Cr Free-Machining Stainless Steel (430FSe) | **C** 0.12 max **Cr** 16.00-18.00 **Mn** 1.25 max **P** 0.060 max **S** 0.060 max **Se** 0.15 min **Si** 1.00 max | **AISI** 430FSe **ASTM** A314 (430FSe); A473 (430FSe); A581 (430FSe); A582 (430FSe); A895 (430FSe) |

The chemical compositions listed are for identification purposes and should not be used in lieu of the cross referenced specifications.

| UNIFIED NUMBER | DESCRIPTION | CHEMICAL COMPOSITION | CROSS REFERENCE SPECIFICATIONS |
|---|---|---|---|
| **S43035** | Ferritic Cr-Ti Stainless Steel (439) | **Al** 0.15 max **C** 0.030 max **Cr** 17.00-19.00 **Mn** 1.00 max **N** 0.030 max **Ni** 0.50 max **P** 0.040 max **S** 0.030 max **Si** 1.00 max **Ti** [0.20 + 4(C+N)] min; 1.10 max | **AISI** 439 **ASME** SA-240 (439); SA-268 (439); SA-479 (439) **ASTM** A240 (439); A268 (439); A463 (439); A479 (439); A803 (TP439) |
| **S43036** | Ferritic Cr Stainless Steel with Titanium (430 Ti) | **C** 0.10 max **Cr** 16.00-19.50 **Mn** 1.00 max **Ni** 0.75 max **P** 0.040 max **S** 0.030 max **Si** 1.00 max **Other** Ti 5\mtC-0.75 | **ASTM** A268 (430 Ti); A554 (430 Ti); A815 (430 Ti) |
| **S43080** | Ferritic Cr Stainless Steel Bare Filler Metal | **C** 0.10 max **Cr** 15.5-17.0 **Cu** 0.75 max **Mn** 0.60 max **Mo** 0.75 max **Ni** 0.60 max **P** 0.03 max **S** 0.03 max **Si** 0.50 max | **ASME** SFA5.9 (ER430) **AWS** A5.9 (ER430) |
| **S43100** | Ni Bearing Hardenable Cr Stainless Steel | **C** 0.20 max **Cr** 15.00-17.00 **Mn** 1.00 max **Ni** 1.25-2.50 **P** 0.040 max **S** 0.030 max **Si** 1.00 max | **AISI** 431 **AMS** 5628; AMS-S-18372 **ASTM** A176 (431); A276 (431); A314 (431); A473 (431); A479 (431); A493 (431); A511 (MT 431); A579 (53); A580 (431); A593 (431); A594 (431); F738 (431); F836 (431); F899 (431) **MIL SPEC** MIL-S-8967 **SAE** J412 (51431) |
| **S43400** | Ferritic Cr-Mo Stainless Steel (434) | **C** 0.12 max **Cr** 16.00-18.00 **Mn** 1.00 max **Mo** 0.75-1.25 **P** 0.040 max **S** 0.030 max **Si** 1.00 max | **AISI** 434 **ASTM** A240 (434) **SAE** J405 (51434); J412 (51434) |
| **S43600** | Ferritic Cr-Mo Stainless Steel (436) | **C** 0.12 max **Cr** 16.00-18.00 **Mn** 1.00 max **Mo** 0.75-1.25 **P** 0.040 max **S** 0.030 max **Si** 1.00 max **Other** Cb+Ta, 5XC-0.70 | **AISI** 436 **ASTM** A240 (436) **SAE** J405; J412 (51436) |
| **S43932** | Ferritic Cr Stainless Steel with Titanium and Columbium (439LT) | **Al** 0.15 max **C** 0.020 max **Cb** 0.02-0.095 **Cr** 17.0-19.0 **Mn** 1.0 max **N** 0.020 max **Ni** 0.50 max **P** 0.040 max **S** 0.030 max **Si** 1.00 max **Ti** 0.20-0.40 **Other** (Ti+Cb): [0.20+4(C+N)] min, 0.75 max | **ASTM** A240; A480 **SAE** J405 |
| **S43940** | Ferrtitic Cr Stainless Steel with Ti and Cb (Nb) | **C** 0.03 max **Cb** (3\mtC+0.30) min **Cr** 17.5-18.5 **Mn** 1.00 max **P** 0.040 max **S** 0.015 max **Si** 1.00 max **Ti** 0.10-0.60 | **ASTM** A240 **Other:** EN 1.4509 |
| **S44002** | Hardenable Cr Stainless Steel (440 A) | **C** 0.60-0.75 **Cr** 16.00-18.00 **Mn** 1.00 max **Mo** 0.75 max **P** 0.040 max **S** 0.030 max **Si** 1.00 max | **AISI** 440 A **AMS** 5631; 7445; AMS-QQ-S-763 **ASTM** A276 (440A); A314 (440A); A473 (440A); A511 (440A); A580 (440A); F899 (440A) **SAE** J412 (51440A) |
| **S44003** | Hardenable Cr Stainless Steel (440 B) | **C** 0.75-0.95 **Cr** 16.00-18.00 **Mn** 1.00 max **Mo** 0.75 max **P** 0.040 max **S** 0.030 max **Si** 1.00 max | **AISI** 440 B **AMS** 7445; AMS-QQ-S-763 **ASTM** A276 (440B); A314 (440B); A473 (440B); A580 (440B); F899 (440B) **SAE** J412 (51440B) |
| **S44004** | Hardenable Cr Stainless Steel (440 C) | **C** 0.95-1.20 **Cr** 16.00-18.00 **Mn** 1.00 max **Mo** 0.75 max **P** 0.040 max **S** 0.030 max **Si** 1.00 max | **AISI** 440 C **AMS** 5618; 5630; 5880; 5899; 7445; AMS-QQ-S-763 **ASTM** A276 (440C); A314 (440C); A473 (440C); A493 (440C); A580 (440C); F899 (440C) **SAE** J412 (51440C); J467 (440C) |
| **S44020** | Martensitic Cr Free Machining Stainless Steel (440 F) | **C** 0.95-1.20 **Cr** 16.00-18.00 **Mn** 1.25 max **Mo** 0.40-0.60 **N** 0.08 max **Ni** 0.75 max **P** 0.040 max **S** 0.10-0.35 **Si** 1.00 max | **AMS** 5632 (1) **ASTM** F899 (440F) |
| **S44023** | Martensitic CrSe Selenium Bearing Free Machining Stainless Steel (440 F Se) | **C** 0.95-1.20 **Cr** 16.00-18.00 **Mn** 1.25 max **Mo** 0.60 max **N** 0.08 max **Ni** 0.75 max **P** 0.040 max **S** 0.030 max **Se** 0.15 min **Si** 1.00 max | **AMS** 5632 (2) |
| **S44025** | Hardenable Cr Stainless Steel with Mo | **C** 0.95-1.10 **Cr** 16.00-18.00 **Cu** 0.50 max **Mn** 1.00 max **Mo** 0.40-0.65 **Ni** 0.75 max **P** 0.025 max **S** 0.025 max **Si** 1.00 max | **ASTM** A756 (440C) |
| * **S44100** | Ferritic Cr Stainless Steel (Columbium- Titanium Stabilized) | **C** 0.03 max **Cb** 0.3+(9\mtC) min-0.9 max **Cr** 17.5-19.5 **Mn** 1.00 max **N** 0.03 max **Ni** 1.00 max **P** 0.04 max **S** 0.03 max **Si** 1.00 max **Ti** 0.10-0.50 | |

*Boxed entries are no longer active and are retained for reference purposes only.

The chemical compositions listed are for identification purposes and should not be used in lieu of the cross referenced specifications.

| UNIFIED NUMBER | DESCRIPTION | CHEMICAL COMPOSITION | CROSS REFERENCE SPECIFICATIONS |
|---|---|---|---|
| **S44200** | Ferritic Cr Heat Resisting Steel (442) | **C** 0.20 max **Cr** 18.00-23.00 **Mn** 1.00 max **P** 0.040 max **S** 0.030 max **Si** 1.00 max | **AISI** 442 **ASTM** A176 (442) **SAE** J412 (51442) |
| **S44300** | Ferritic Cr-Cu Heat Resisting Steel (443) | **C** 0.20 max **Cr** 18.00-23.00 **Cu** 0.90-1.25 **Mn** 1.00 max **Ni** 0.50 max **P** 0.040 max **S** 0.030 max **Si** 1.00 max | **AMS** AMS-QQ-S-763 **ASTM** A268 (443); A511 (443) |
| **S44400** | Ferritic Cr-Mo Stainless Steel (Stabilized, Low Interstitial) (444, 18Cr-2Mo) | **C** 0.025 max **Cr** 17.5-19.5 **Mn** 1.00 max **Mo** 1.75-2.50 **N** 0.025 max **Ni** 1.00 max **P** 0.040 max **S** 0.030 max **Si** 1.00 max **Other** Cb+Ti [0.20+4(C+N)] min-0.80 max | **ASTM** A213 (18Cr-2Mo); A240; A268 (18Cr-2Mo); A276; A479; A493; A580; A791 (18-2); A803 (18-2) |
| **S44500** | High Cr-Cu Heat Resisting Cb Stabilized Steel | **C** 0.02 max **Cb** 10(C+N)-0.8 **Cr** 19.0-21.0 **Cu** 0.30-0.60 **Mn** 1.0 max **N** 0.03 max **Ni** 0.60 max **P** 0.040 max **S** 0.012 max **Si** 1.0 max | **ASTM** A240 **SAE** J405 |
| **S44600** | Ferritic Cr Heat Resisting Steel (446) | **C** 0.20 max **Cr** 23.00-27.00 **Mn** 1.50 max **N** 0.25 max **P** 0.040 max **S** 0.030 max **Si** 1.00 max | **AISI** 446 **AMS** AMS-QQ-S-763 **ASME** SA-268 (446) **ASTM** A176 (446); A268 (446-1, 446-2); A276 (446); A314 (446); A473 (446); A511 (446); A580 (446); A815 (WP446) **SAE** J412 (51446) |
| * **S44625** | Low Interstitial Ferritic Cr Stainless Steel | **C** 0.01 max **Cr** 25.00-27.50 **Cu** 0.20 max **Mn** 0.40 max **Mo** 0.75-1.50 **N** 0.015 max **Ni** 0.50 max **P** 0.020 max **S** 0.020 max **Si** 0.40 max **Other** Ni+Cu 0.50 max | **ASME** SA-182 (XM-27); SA-240 (XM-27); SA-268 (XM-27); SA-479 (XM-27) **ASTM** A182 (XM-27); A268 (XM-27); A314 (XM-27); A493 **SAE** J405 (XM-33) |
| **S44626** | Ferritic Cr-Mo-Ti Stainless Steel (Stabilized, Low Interstitial) | **C** 0.06 max **Cr** 25.00-27.00 **Cu** 0.20 max **Mn** 0.75 max **Mo** 0.75-1.50 **N** 0.04 max **Ni** 0.50 max **P** 0.040 max **S** 0.020 max **Si** 0.75 max **Ti** 0.20-1.00 **Other** Ti 7x(C+N) min | **ASME** SA-240 (XM-33); SA-268 (XM-33) **ASTM** A240 (XM-33); A268 (XM-33); A803 (TPXM-33); A815 (WP33) **SAE** J405 (XM-27) |
| **S44627** | Ferritic Cr-Mo-Cb Stainless Steel (Low Interstitial) (E-BRITE 26-1) | **C** 0.010 max **Cb** 0.05-0.20 **Cr** 25.0-27.0 **Cu** 0.20 max **Mn** 0.40 max **Mo** 0.75-1.50 **N** 0.015 max **Ni** 0.50 max **P** 0.020 max **S** 0.020 max **Si** 0.40 max | **ASME** SA-240; SA-268; SA-479; SA-731; SFA5.9 (Er26-1) **ASTM** A182 (FXM-27Cb); A240 (XM-27); A268; A276 (XM-27); A314; A336 (FXM-27Cb); A479; A803 (TPXM-27); A815 (WP27) |
| **S44635** | Ferritic Cr-Mo-Ni Stainless Steel (Stabilized, Low Interstitial) (MONIT, 25-4-4) | **C** 0.025 max **Cr** 24.5-26.0 **Mn** 1.00 max **Mo** 3.50-4.50 **N** 0.035 max **Ni** 3.50-4.50 **P** 0.040 max **S** 0.030 max **Si** 0.75 max **Other** Cb+Ti 0.20+4(C+N)-0.80 max | **ASTM** A240; A268 (25-4-4); A803 (25-4-4) **SAE** J405 |
| **S44660** | Ferritic Cr-Mo-Ni Stainless Steel (Stabilized, Low Interstitial) (SEA-CURE, SC-1) | **C** 0.030 max **Cr** 25.0-28.0 **Mn** 1.00 max **Mo** 3.00-4.00 **N** 0.040 max **Ni** 1.00-3.50 **P** 0.040 max **S** 0.030 max **Si** 1.00 max **Other** Cb+Ti [0.20+6(C+N)] min-1.00 max | **ASTM** A240; A268 (26-3-3); A803 (26-3-3) **SAE** J405 |
| **S44687** | Ferritic Cr-Mo Stainless Steel Bare Filler Metal | **C** 0.015 max **Cr** 25.0-27.5 **Cu** 0.20 max **Mn** 0.40 max **Mo** 0.75-1.50 **N** 0.015 max **P** 0.02 max **S** 0.02 max **Si** 0.40 max **Other** Ni+Cu 0.50 max | **ASME** SFA5.9 (ER446LMo) **AWS** A5.9 (ER446LMo) |
| **S44700** | Ferritic High Cr+Mo Stainless Steel (29-4) | **C** 0.010 max **Cr** 28.0-30.0 **Cu** 0.15 max **Mn** 0.30 max **Mo** 3.5-4.2 **N** 0.020 max **Ni** 0.15 max **P** 0.025 max **S** 0.020 max **Si** 0.20 max **Other** C+N 0.025 max | **ASTM** A240; A268 (29-4); A276; A479; A493; A511 (29-4); A580; A803 (29-4) **SAE** J405 |
| **S44735** | Ferritic High Cr+Mo Stainless Steel (29-4C) | **C** 0.030 max **Cr** 28.0-30.0 **Mn** 1.00 max **Mo** 3.60-4.20 **N** 0.045 max **Ni** 1.00 max **P** 0.040 max **S** 0.030 max **Si** 1.00 max **Other** (Cb+Ti) = 0.20-1.00 and 6(C+N) min | **ASTM** A240; A268; A511; A803 (29-4C) **SAE** J405 |

*Boxed entries are no longer active and are retained for reference purposes only.

The chemical compositions listed are for identification purposes and should not be used in lieu of the cross referenced specifications.

| UNIFIED NUMBER | DESCRIPTION | CHEMICAL COMPOSITION | CROSS REFERENCE SPECIFICATIONS |
|---|---|---|---|
| **S44736** | Ferritic High Cr+Mo Stainless Steel (29-4-2C) | **C** 0.030 max **Cr** 28.0-30.0 **Mn** 1.00 max **Mo** 3.60-4.20 **Ni** 2.00-4.50 **N** 0.045 max **P** 0.040 max **S** 0.030 max **Si** 1.00 max **Other** (Cb+Ti) = 0.20-1.00 and 6(C+N) min | |
| **S44800** | Ferritic High Cr Stainless Steel (29-4-2) | **C** 0.010 max **Cr** 28.0-30.0 **Cu** 0.15 max **Mn** 0.30 max **Mo** 3.5-4.2 **N** 0.020 max **Ni** 2.0-2.5 **P** 0.025 max **S** 0.020 max **Si** 0.20 max **Other** C+N 0.025 max | **ASTM** A240; A268 (29-4-2); A276; A479; A493; A511 (29-4-2); A580; A803 (29-4-2) **SAE** J405 |
| **S45000** | Stainless Steel (Custom 450) | **C** 0.05 max **Cb** 8\mtC min **Cr** 14.00-16.00 **Cu** 1.25-1.75 **Mn** 1.00 max **Mo** 0.50-1.00 **Ni** 5.00-7.00 **P** 0.030 max **S** 0.030 max **Si** 1.00 max | **AMS** 5763; 5773; 5859; 5863 **ASME** SA-564 (XM-25); SA-705 (XM-25) **ASTM** A564 (XM-25); A693 (XM-25); A705 (XM-25); F899 (XM-25) |
| **S45500** | Stainless Steel (Custom 455) | **C** 0.05 max **Cb** 0.10-0.50 **Cr** 11.00-12.50 **Cu** 1.50-2.50 **Mn** 0.50 max **Mo** 0.50 max **Ni** 7.50-9.50 **P** 0.040 max **S** 0.030 max **Si** 0.50 max **Ti** 0.80-1.40 | **AMS** 5578; 5617; 5672; 5860 **ASTM** A313 (XM-16); A564 (XM-16); A693 (XM-16); A705 (XM-16); F899 (XM-16) |
| **S45503** | Martensitic, Age Hardenable Stainless Steel (Custom 455 ELC) | **C** 0.010 max **Cr** 11.00-12.50 **Cu** 1.50-2.50 **Mn** 0.50 max **Mo** 0.50 max **Ni** 7.50-9.50 **P** 0.010 max **S** 0.010 max **Si** 0.20 max **Ti** 1.00-1.35 **Other** (Cb+Ta) 0.10-0.50 | **AMS** 5617 **ASME** SA-564; SA-705 |
| **S46500** | Precipitation Hardenable Cr-Ni-Mo-Ti Martensitic Stainless Steel (Custom 465) | **C** 0.02 max **Cr** 11.00-12.50 **Mn** 0.25 max **Mo** 0.75-1.25 **N** 0.01 max **Ni** 10.75-11.25 **P** 0.015 max **S** 0.010 max **Si** 0.25 max **Ti** 1.50-1.80 | **ASTM** A564 |
| **S46800** | Ferritic Cr Stainless Steel with Ti and Cb | **C** 0.030 max **Cb** 0.10-0.60 **Cr** 18.00-20.00 **Mn** 1.00 max **N** 0.030 max **Ni** 0.50 max **P** 0.040 max **S** 0.030 max **Si** 1.00 max **Ti** 0.07-0.30 **Other** (Ti+Cb): [0.20+4(C+N)] min, 0.80 max | **ASTM** A240; A480 **SAE** J405 |
| **S50100** | Cr Heat Resisting Steel | **C** over 0.10 **Cr** 4.00-6.00 **Mn** 1.00 max **Mo** 0.40-0.65 **P** 0.040 max **S** 0.030 max **Si** 1.00 max | **AISI** 501 **ASME** SA-194 (3); SA-387 (5) **ASTM** A182 (B5, F7); A193 (501, B5); A194 (501, 3); A314 (501); A387 (5); A473 (501) **SAE** J412 (51501) |
| **S50180** | Cr Heat Resisting Steel Bare Filler Metal | **C** 0.25-0.40 **Cr** 4.80-6.00 **Cu** 0.35 max **Mn** 0.75-1.00 **Mo** 0.45-0.65 **P** 0.025 max **S** 0.030 max **Si** 0.25-0.50 | **ASME** SFA5.23 (EB6H) **AWS** A5.23 (EB6H) |
| **S50181** | Cr Heat Resisting Steel Bare Filler Metal (Nuclear Grade) | **C** 0.25-0.40 **Cr** 4.80-6.00 **Cu** 0.08 max **Mn** 0.75-1.00 **Mo** 0.45-0.65 **P** 0.012 max **S** 0.030 max **Si** 0.25-0.50 **V** 0.05 max | **ASME** SFA5.23 (EB6HN) **AWS** A5.23 (EB6HN) |
| **S50200** | Cr Heat Resisting Steel | **C** 0.10 max **Cr** 4.00-6.00 **Mn** 1.00 max **Mo** 0.40-0.65 **P** 0.040 max **S** 0.030 max **Si** 1.00 max | **AISI** 502 **ASME** SA-387 (5) **ASTM** A314 (502); A387 (5); A473 (502) |
| **S50280** | Cr Heat Resisting Steel Base Filler Metal | **C** 0.10 max **Cr** 4.00-6.00 **Cu** 0.75 max **Mn** 0.60 max **Mo** 0.45-0.65 **Ni** 0.60 max **P** 0.03 max **S** 0.03 max **Si** 0.50 max | **AMS** 6466 **ASME** SFA5.9 (ER502); SFA5.23 (EB6); SFA5.28 (ER80S-B6); SFA5.30 (IN502) **AWS** A5.9 (ER502); A5.23 (EB6); A5.28 (ER80S-B6); A5.30 (IN502) |
| **S50281** | Cr Heat Resisting Steel Bare Filler Metal (Nuclear Grade) | **C** 0.10 max **Cr** 4.50-6.00 **Cu** 0.08 max **Mn** 0.40-0.65 **Mo** 0.45-0.65 **P** 0.012 max **S** 0.025 max **Si** 0.20-0.50 **V** 0.05 max | **ASME** SFA5.23 (EB6N) **AWS** A5.23 (EB6N) |
| **S50288** | Cr Heat Resisting Steel Bare Filler Metal (ER502Si) | **C** 0.10 max **Cr** 4.6-6.0 **Cu** 0.75 max **Mn** 0.60 max **Mo** 0.45-0.65 **Ni** 0.6 max **P** 0.03 max **S** 0.03 max **Si** 0.65-1.0 | **ASME** SFA5.9 (ER502Si) **AWS** A5.9 (ER502Si) |
| **S50300** | Cr Heat Resisting Steel | **C** 0.15 max **Cr** 6.00-8.00 **Mn** 1.00 max **Mo** 0.45-0.65 **P** 0.040 max **S** 0.040 max **Si** 1.00 max | **ASTM** A387 (7) |

*Boxed entries are no longer active and are retained for reference purposes only.

The chemical compositions listed are for identification purposes and should not be used in lieu of the cross referenced specifications.

| UNIFIED NUMBER | DESCRIPTION | CHEMICAL COMPOSITION | CROSS REFERENCE SPECIFICATIONS |
|---|---|---|---|
| * **S50400** | Replaced by K90941 | | |
| * **S50460** | Replaced by K90901 | | |
| **S50480** | Cr Heat Resisting Steel Bare Filler Metal | **C** 0.10 max **Cr** 8.00-10.50 **Cu** 0.75 max **Mn** 0.60 max **Mo** 0.80-1.20 **Ni** 0.50 max **P** 0.04 max **S** 0.03 max **Si** 0.50 max | **ASME** SFA5.9 (ER505); SFA5.23 (EB8); SFA5.28 (ER80S-B8) **AWS** A5.9 (ER505); A5.23 (EB8); A5.28 (ER80S-B8) |
| **S50481** | Cr Heat Resisting Steel Bare Filler Metal, Nuclear Grade | **C** 0.10 max **Cr** 8.00-10.5 **Cu** 0.08 max **Mn** 0.30-0.65 **Mo** 0.80-1.20 **P** 0.012 max **S** 0.030 max **Si** 0.05-0.50 **V** 0.05 max | **ASME** SFA5.23 (EB8N) **AWS** A5.23 (EB8N) |
| **S50482** | Cr Heat Resisting Bare Filler Metal (EB 9) | **Al** 0.04 max **C** 0.07-0.13 **Cr** 8.00-10.00 **Cu** 0.10 max **Mn** 1.25 max **Mo** 0.80-1.10 **N** 0.03-0.07 **Nb** 0.02-0.10 **Ni** 1.00 max **P** 0.010 max **S** 0.010 max **Si** 0.30 max **V** 0.15-0.25 | **ASME** SFA5.23 (EB 9) **AWS** A5.23 (EB 9); A5.28 (ER80S-B9) |
| **S50488** | Cr Heat Resisting Steel Bare Filler Metal (ER505Si) | **C** 0.10 max **Cr** 8.0-10.5 **Cu** 0.75 max **Mn** 0.6 max **Mo** 0.8-1.2 **Ni** 0.5 max **P** 0.03 max **S** 0.03 max **Si** 0.65-1.00 | **ASME** SFA5.9 (ER505Si) **AWS** A5.9 (ER505Si) |
| * **S59180** | Cr Heat Resisting Bare Filler Metal (B9) | **Al** 0.04 max **C** 0.07-0.13 **Cb** 0.03-0.10 **Cr** 8.0-9.5 **Cu** 0.20 max **Mn** 1.25 max **Mo** 0.80-1.10 **N** 0.03-0.07 **Ni** 1.00 max **P** 0.010 max **S** 0.010 max **Si** 0.15-0.30 **V** 0.15-0.25 | **ASME** SFA5.28 (ER90S-B9) |
| **S63005** | Valve Steel (CNS) | **C** 0.25-0.35 **Cr** 12.00-13.50 **Mn** 0.50 max **Mo** 0.50 max **Ni** 7.00-8.50 **P** 0.030 max **S** 0.030 max **Si** 2.00-3.00 | **AMS** 5705 |
| * **S63007** | Valve Steel (21-55N) | **C** 0.15-0.25 **Cr** 20.0-22.0 **Mn** 5.00-6.50 **N** 0.20-0.35 **Ni** 4.50-6.00 **P** 0.04 max **S** 0.03 max **Si** 1.00 max | |
| **S63008** | Valve Steel (21-4N) | **C** 0.48-0.58 **Cr** 20.00-22.00 **Mn** 8.00-10.00 **N** 0.35-0.50 **Ni** 3.25-4.50 **P** 0.050 max **S** 0.030 max **Si** 0.25 max | **SAE** J775 (EV 8) **Other:** ISO 683-15 (X 53 CrMnNiN 21 9) |
| * **S63011** | Valve Steel (746) | **C** 0.65-0.75 **Cr** 20.50-22.00 **Mn** 5.50-6.90 **N** 0.18-0.28 **Ni** 1.40-1.90 **P** 0.040 max **S** 0.035 max **Si** 0.45-0.85 | |
| **S63012** | Valve Steel (21-2 N) | **C** 0.50-0.60 **Cr** 19.25-21.50 **Mn** 7.00-10.00 **N** 0.20-0.40 **Ni** 1.50-2.75 **P** 0.050 max **S** 0.030 max **Si** 0.25 max | **SAE** J775 (EV 12) |
| **S63013** | Valve Steel (Gaman H) | **C** 0.47-0.57 **Cr** 20.00-22.00 **Mn** 11.00-13.00 **N** 0.40-0.50 **P** 0.030 max **S** 0.050 max **Si** 2.00-3.00 | **SAE** J775 (EV 13) |
| * **S63014** | Valve Steel (10) | **C** 0.30-0.45 **Cr** 18.0-20.0 **Mn** 0.80-1.30 **Ni** 7.75-8.25 **P** 0.030 max **S** 0.030 max **Si** 2.75-3.25 | |
| * **S63015** | Valve Steel (10 N) | **C** 0.35-0.45 **Cr** 18.0-20.0 **Mn** 0.80-1.30 **N** 0.15-0.25 **Ni** 7.75-8.15 **P** 0.030 max **S** 0.030 max **Si** 2.75-3.25 | |

*Boxed entries are no longer active and are retained for reference purposes only.

The chemical compositions listed are for identification purposes and should not be used in lieu of the cross referenced specifications.

| UNIFIED NUMBER | DESCRIPTION | CHEMICAL COMPOSITION | CROSS REFERENCE SPECIFICATIONS |
|---|---|---|---|
| * **S63016** | Valve Steel (21-12) | **C** 0.15-0.25 **Cr** 20.50-22.00 **Mn** 1.00-1.40 **Ni** 10.50-12.00 **P** 0.030 max **S** 0.030 max **Si** 0.70 max | |
| **S63017** | Valve Steel (21-12N) | **C** 0.15-0.25 **Cr** 20.00-22.00 **Mn** 1.00-1.50 **N** 0.15-0.25 **Ni** 10.50-12.50 **P** 0.045 max **S** 0.030 max **Si** 0.70-1.25 | **SAE** J775 (EV 4) |
| **S63018** | Valve Steel (23-8N) (Nitronic 20) | **C** 0.28-0.38 **Cr** 22.00-24.00 **Mn** 1.50-3.50 **Mo** 0.50 max **N** 0.25-0.35 **Ni** 7.00-9.00 **P** 0.050 max **S** 0.030 max **Si** 0.50-1.00 **W** 0.50 max | **SAE** J775 (EV 16) **Other:** ISO 683-15 (X 33 CrNiMnN 23 8) |
| **S63020** | Valve Steel, Hardfacing Powder (VMS-646) | **C** 1.70-1.85 **Cr** 19.00-20.50 **Mn** 11.00-13.00 **Mo** 9.00-10.50 **Ni** 19.00-21.00 **O** 0.030 max **Si** 0.50 max | **SAE** J775 |
| **S63021** | Valve Steel, Hardfacing Powder (VMS-637) | **C** 2.0-2.3 **Cr** 23.5-26.0 **Mo** 5.0-5.6 **N** 0.10 max **Ni** 11.0-12.0 **O** 0.020 max **Si** 1.0-1.3 | **SAE** J775 |
| **S63019** | Valve Steel (21-4N+Nb+W) | **C** 0.45-0.55 **Cb+Ta** 1.80-2.50 **Cr** 20.00-22.00 **Mn** 8.00-10.00 **N** 0.40-0.60 **Ni** 3.50-5.50 **P** 0.050 max **S** 0.030 max **Si** 0.45 max **W** 0.80-1.50 | **SAE** J775 (XEV-F) **Other:** ISO 683-15 (X 50 CrMnNiNbN 21 9) |
| **S63197** | Iron Base Superalloy, Welding Filler Metal (ER349) | **C** 0.07-0.13 **Cb+Ta** 1.00-1.40 **Cr** 19.0-21.5 **Cu** 0.75 max **Mn** 1.00-2.50 **Mo** 0.35-0.65 **Ni** 8.00-9.50 **P** 0.03 max **S** 0.03 max **Si** 0.30-0.65 **Ti** 0.10-0.30 **W** 1.25-1.75 | **AMS** 5782 **MIL SPEC** MIL-E-19933 (MIL-349) |
| **S63198** | Iron Base Superalloy (19-9-DL) | **C** 0.28-0.35 **Cb** 0.25-0.60 **Cr** 18.0-21.0 **Cu** 0.50 max **Mn** 0.75-1.50 **Mo** 1.00-1.75 **Ni** 8.00-11.00 **P** 0.040 max **S** 0.030 max **Si** 0.30-0.80 **Ti** 0.10-0.35 **W** 1.00-1.75 | **AMS** 5526; 5579 **ASTM** A453 (651) **SAE** J467 (19-9DL) |
| **S63199** | Iron Base Superalloy (19-9-DX or 19-9-W-Mo) | **C** 0.28-0.35 **Cr** 18.0-21.0 **Cu** 0.50 max **Mn** 0.75-1.50 **Mo** 1.25-2.00 **Ni** 8.00-11.00 **P** 0.040 max **S** 0.030 max **Si** 0.30-0.80 **Ti** 0.40-0.75 **W** 1.00-1.75 | **SAE** J467 (19-9DX) |
| **S64004** | Valve Steel (SUH 11M) | **C** 0.47-0.55 **Cr** 7.50-9.50 **Mn** 0.60 max **Ni** 0.60 max **P** 0.030 max **S** 0.030 max **Si** 1.00-2.00 | **SAE** J775 (SUH 11M) **Other:** JIS G 4311 SUH 11 Mod. |
| * **S64005** | Valve Steel (2) | **C** 0.50-0.60 **Cr** 7.50-8.50 **Mn** 0.20-0.60 **Mo** 0.60-0.90 **P** 0.020 max **S** 0.020 max **Si** 1.25-1.75 | |
| * **S64006** | Valve Steel (F) | **C** 0.35-0.50 **Cr** 1.50-2.50 **Mn** 0.20-0.60 **P** 0.030 max **S** 0.030 max **Si** 3.50-4.50 | |
| **S64007** | Valve Steel (GM-8440) | **C** 0.35-0.45 **Cr** 1.85-2.50 **Cu** 0.25 max **Mn** 0.20-0.40 **Ni** 0.25 max **P** 0.030 max **S** 0.040 max **Si** 3.60-4.20 | **SAE** J775 (NV 8) |
| **S64152** | High Strength Alloy Steel (M152) | **C** 0.08-0.15 **Cr** 11.00-12.50 **Mn** 0.50-0.90 **Mo** 1.50-2.00 **N** 0.01-0.05 **Ni** 2.00-3.00 **P** 0.025 max **S** 0.025 max **Si** 0.35 max **V** 0.25-0.40 | **AMS** 5718; 5719 **ASTM** A565 (XM-32) |
| **S64299** | Iron Base Superalloy (29-9) | **C** 0.08-0.15 **Cr** 27.0-31.0 **Cu** 0.50 max **Mn** 1.00-2.00 **Mo** 0.50 max **Ni** 8.50-10.50 **P** 0.040 max **S** 0.030 max **Si** 0.75 max | **AMS** 5784 |

*Boxed entries are no longer active and are retained for reference purposes only.

The chemical compositions listed are for identification purposes and should not be used in lieu of the cross referenced specifications.

| UNIFIED NUMBER | DESCRIPTION | CHEMICAL COMPOSITION | CROSS REFERENCE SPECIFICATIONS |
|---|---|---|---|
| **S65006** | Valve Steel (Silchrome XB) | **C** 0.75-0.90 **Cr** 19.00-21.00 **Mn** 0.80 max **Ni** 1.00-1.70 **P** 0.040 max **S** 0.040 max **Si** 1.75-2.60 | **AMS** 5710 **SAE** J775 (HNV 6) |
| **S65007** | Valve Steel (Silchrome 1) | **C** 0.40-0.50 **Cr** 8.00-10.00 **Mn** 0.80 max **Ni** 0.60 max **P** 0.040 max **S** 0.030 max **Si** 2.70-3.30 | **SAE** J775 (HNV 3) **Other:** ISO 683-15 (X45 CrSi 9 3) |
| * **S65150** | Iron Base Superalloy (Pyromet X-15) | **C** 0.03 max **Co** 19.0-21.0 **Cr** 14.50-16.00 **Mn** 0.10 max **Mo** 2.50-3.00 **Ni** 0.20 max **P** 0.015 max **S** 0.015 max **Si** 0.10 max | |
| * **S65770** | Iron Base Superalloy (AFC 77) | **C** 0.12-0.17 **Co** 13.0-14.0 **Cr** 13.50-14.50 **Mn** 0.30 max **Mo** 4.50-5.50 **Ni** 0.30-0.70 **P** 0.015 max **S** 0.015 max **Si** 0.25 max **V** 0.10-0.30 | |
| **S66009** | Valve Steel (TPA) | **C** 0.35-0.50 **Cr** 12.0-15.0 **Mn** 1.00 max **Mo** 0.20-0.50 **Ni** 12.0-15.0 **P** 0.045 max **S** 0.030 max **Si** 0.30-0.80 **W** 1.50-3.00 **Other** if Mo not used, W 2.00-3.00 | **AMS** 5700 |
| **S66220** | Iron Base Superalloy (Discaloy) | **Al** 0.35 max **B** 0.0010-0.010 **C** 0.08 max **Cr** 12.0-15.0 **Cu** 0.50 max **Mn** 1.50 max **Mo** 2.50-3.50 **Ni** 24.0-28.0 **P** 0.040 max **S** 0.030 max **Si** 1.00 max **Ti** 1.55-2.00 | **AMS** 5732 **ASTM** A453 (662); A638 (662) **SAE** J467 (Discaloy) |
| **S66286** | Iron Base Superalloy (A286) | **Al** 0.35 max **B** 0.0010-0.010 **C** 0.08 max **Cr** 13.50-16.00 **Mn** 2.00 max **Mo** 1.00-1.50 **Ni** 24.0-27.0 **P** 0.040 max **S** 0.030 max **Si** 1.00 max **Ti** 1.90-2.35 **V** 0.10-0.50 | **AMS** 5525; 5726; 5731; 5732; 5734; 5737; 5804; 5805; 5810; 5853; 5858; 5895; AS7235; AS7477; AS7478; AS7479; AS7481; AS7482 **ASME** SA-638 (660) **ASTM** A453 (660); A638 (660); A891; A891 (A286) **SAE** J467 (A286) |
| **S66315** | Iron Base Ni-Cr-Mo Precipitation Hardening Stainless Steel/Valve Steel (NCF 3015) | **Al** 1.6-2.2 **B** 0.010 max **C** 0.08 max **Cb** 0.4-0.9 **Cr** 13.5-15.5 **Fe** rem **Mn** 0.50 max **Mo** 0.4-1.0 **Ni** 29.5-33.5 **P** 0.015 max **S** 0.015 max **Si** 0.50 max **Ti** 2.3-2.9 | **SAE** J775 |
| **S66545** | Iron Base Superalloy (W545) | **Al** 0.25 max **B** 0.01-0.07 **C** 0.08 max **Cr** 12.0-15.0 **Cu** 0.25 max **Mn** 1.25-2.00 **Mo** 1.25-2.25 **Ni** 24.0-28.0 **P** 0.040 max **S** 0.030 max **Si** 0.10-0.80 **Ti** 2.70-3.30 | **ASTM** A453 (665) **SAE** J467 (W545) |
| * **S67956** | Iron Base Superalloy, Dispersion Strengthened (MA 956) | **Al** 3.75-5.75 **C** 0.10 max **Co** 0.30 max **Cr** 18.5-21.5 **Cu** 0.15 max **Mn** 0.30 max **Ni** 0.50 max **P** 0.02 max **S** 0.02 max **Si** 0.30 max **Ti** 0.20-0.60 **Other** Yttria ($Y_2O_3$) 0.30-0.70% by weight | |
| **S68000** | Iron, Cr-Ni-Mo Hardfacing Alloy (Eatonite 6) | **C** 1.50-2.00 **Cr** 26.00-30.00 **Mn** 1.00 max **Mo** 4.00-5.00 **Ni** 15.00-18.00 **P** 0.025 max **S** 0.020 max **Si** 1.10-1.50 | **SAE** J467 (Eatonite 6) |
| **S70003** | Fe-Ni-Cr-Si Corrosion Resistant Steel (Alloy 700Si) | **Al** 0.50 max **C** 0.02 max **Cr** 8.0-11.0 **Mn** 2.00 max **Mo** 0.50 max **Ni** 22.0-25.0 **P** 0.025 max **S** 0.010 max **Si** 6.5-8.0 | **ASTM** A213; A946; A953; A954; A968 |

*Boxed entries are no longer active and are retained for reference purposes only.

The chemical compositions listed are for identification purposes and should not be used in lieu of the cross referenced specifications.

# UNS NUMBERS ASSIGNED TO DATE

## With Description of Each Material Covered and References To Documents In Which The Same or Similar Materials are Described

## Txxxxx Number Series
## Tool Steels, Wrought and Cast

# TOOL STEELS, WROUGHT AND CAST

| UNIFIED NUMBER | DESCRIPTION | CHEMICAL COMPOSITION | CROSS REFERENCE SPECIFICATIONS |
|---|---|---|---|
| **T11301** | High-Speed Tool Steel (M-1) | **C** 0.78-0.88 **Cr** 3.50-4.00 **Mn** 0.15-0.40 **Mo** 8.20-9.20 **P** 0.03 max **S** 0.03 max **Si** 0.20-0.50 **V** 1.00-1.35 **W** 1.40-2.10 | **AISI** M-1 **ASTM** A600 (M-1) **SAE** J437 (M1); J438 (M1) |
| **T11302** | High-Speed Tool Steel (M-2) | **C** 0.78-1.05 **Cr** 3.75-4.50 **Mn** 0.15-0.40 **Mo** 4.50-5.50 **P** 0.03 max **S** 0.03 max **Si** 0.20-0.45 **V** 1.75-2.20 **W** 5.50-6.75 | **AISI** M-2 **ASTM** A597 (CM-2); A600 (M-2) **SAE** J437 (M2); J438 (M2); J467 (M2) |
| **T11304** | High-Speed Tool Steel (M-4) | **C** 1.25-1.40 **Cr** 3.75-4.75 **Mn** 0.15-0.40 **Mo** 4.25-5.50 **P** 0.03 max **S** 0.03 max **Si** 0.20-0.45 **V** 3.75-4.50 **W** 5.25-6.50 | **AISI** M-4 **ASTM** A600 (M-4) **SAE** J437 (M4); J438 (M4) |
| **T11306** | High-Speed Tool Steel (M-6) | **C** 0.75-0.85 **Co** 11.00-13.00 **Cr** 3.75-4.50 **Mn** 0.15-0.40 **Mo** 4.50-5.50 **P** 0.03 max **S** 0.03 max **Si** 0.20-0.45 **V** 1.30-1.70 **W** 3.75-4.75 | **AISI** M-6 **ASTM** A600 (M-6) |
| **T11307** | High-Speed Tool Steel (M-7) | **C** 0.97-1.05 **Cr** 3.50-4.00 **Mn** 0.15-0.40 **Mo** 8.20-9.20 **P** 0.03 max **S** 0.03 max **Si** 0.20-0.55 **V** 1.75-2.25 **W** 1.40-2.10 | **AISI** M-7 **ASTM** A600 (M-7) |
| **T11310** | High-Speed Tool Steel (M-10) | **C** 0.84-1.05 **Cr** 3.75-4.50 **Mn** 0.10-0.40 **Mo** 7.75-8.50 **P** 0.03 max **S** 0.03 max **Si** 0.20-0.45 **V** 1.80-2.20 | **AISI** M-10 **ASTM** A600 (M-10) **SAE** J467 (M10) |
| **T11313** | High-Speed Tool Steel (M-3, Class 1) | **C** 1.00-1.10 **Cr** 3.75-4.50 **Mn** 0.15-0.40 **Mo** 4.75-6.50 **P** 0.03 max **S** 0.03 max **Si** 0.20-0.45 **V** 2.25-2.75 **W** 5.00-6.75 | **AISI** M-3 (Cl.1) **ASTM** A600 (M-3, Cl.1) **SAE** J437 (M3); J438 (M3) |
| **T11323** | High-Speed Tool Steel (M-3, Class 2) | **C** 1.15-1.25 **Cr** 3.75-4.50 **Mn** 0.15-0.40 **Mo** 4.75-6.50 **P** 0.03 max **S** 0.03 max **Si** 0.20-0.45 **V** 2.75-3.25 **W** 5.00-6.75 | **AISI** M-3 (Cl.2) **ASTM** A600 (M-3, Cl.2) **SAE** J437 (M3); J438 (M3) |
| **T11330** | High-Speed Tool Steel (M-30) | **C** 0.75-0.85 **Co** 4.50-5.50 **Cr** 3.50-4.25 **Mn** 0.15-0.40 **Mo** 7.75-9.00 **P** 0.03 max **S** 0.03 max **Si** 0.20-0.45 **V** 1.00-1.40 **W** 1.30-2.30 | **AISI** M-30 **ASTM** A600 (M-30) |
| **T11333** | High-Speed Tool Steel (M-33) | **C** 0.85-0.92 **Co** 7.75-8.75 **Cr** 3.50-4.00 **Mn** 0.15-0.40 **Mo** 9.00-10.00 **P** 0.03 max **S** 0.03 max **Si** 0.25-0.55 **V** 1.00-1.35 **W** 1.30-2.10 | **AISI** M-33 **ASTM** A600 (M-33) |
| **T11334** | High-Speed Tool Steel (M-34) | **C** 0.85-0.92 **Co** 7.75-8.75 **Cr** 3.50-4.00 **Mn** 0.15-0.40 **Mo** 7.75-9.20 **P** 0.03 max **S** 0.03 max **Si** 0.20-0.45 **V** 1.90-2.30 **W** 1.40-2.10 | **AISI** M-34 **ASTM** A600 (M-34) |
| **T11336** | High-Speed Tool Steel (M-36) | **C** 0.80-0.90 **Co** 7.75-8.75 **Cr** 3.75-4.50 **Mn** 0.15-0.40 **Mo** 4.50-5.50 **P** 0.03 max **S** 0.03 max **Si** 0.20-0.45 **V** 1.75-2.25 **W** 5.50-6.50 | **AISI** M-36 **ASTM** A600 (M-36) |
| **T11341** | High-Speed Tool Steel (M-41) | **C** 1.05-1.15 **Co** 4.75-5.75 **Cr** 3.75-4.50 **Mn** 0.20-0.60 **Mo** 3.25-4.25 **P** 0.03 max **S** 0.03 max **Si** 0.15-0.50 **V** 1.75-2.25 **W** 6.25-7.00 | **AISI** M-41 **ASTM** A600 (M-41) |
| **T11342** | High-Speed Tool Steel (M-42) | **C** 1.05-1.15 **Co** 7.75-8.75 **Cr** 3.50-4.25 **Mn** 0.15-0.40 **Mo** 9.00-10.00 **P** 0.03 max **S** 0.03 max **Si** 0.15-0.65 **V** 0.95-1.35 **W** 1.15-1.85 | **AISI** M-42 **ASTM** A600 (M-42) |
| * **T11343** | High-Speed Tool Steel (M-43) | **C** 1.15-1.25 **Co** 7.75-8.75 **Cr** 3.50-4.25 **Mn** 0.20-0.40 **Mo** 7.50-8.50 **P** 0.03 max **S** 0.03 max **Si** 0.15-0.65 **V** 1.50-1.75 **W** 2.25-3.00 | **AISI** M-43 **ASTM** A600 (M-43) |

*Boxed entries are no longer active and are retained for reference purposes only.

The chemical compositions listed are for identification purposes and should not be used in lieu of the cross referenced specifications.

| UNIFIED NUMBER | DESCRIPTION | CHEMICAL COMPOSITION | CROSS REFERENCE SPECIFICATIONS |
|---|---|---|---|
| * **T11344** | High-Speed Tool Steel (M-44) | **C** 1.10-1.20 **Co** 11.00-12.25 **Cr** 4.00-4.75 **Mn** 0.20-0.40 **Mo** 6.00-7.00 **P** 0.03 max **S** 0.03 max **Si** 0.30-0.55 **V** 1.85-2.20 **W** 5.00-5.75 | **AISI** M-44 **ASTM** A600 (M-44) |
| **T11346** | High-Speed Tool Steel (M-46) | **C** 1.22-1.30 **Co** 7.80-8.80 **Cr** 3.70-4.20 **Mn** 0.20-0.40 **Mo** 8.00-8.50 **P** 0.03 max **S** 0.03 max **Si** 0.40-0.65 **V** 3.00-3.30 **W** 1.90-2.20 | **AISI** M-46 **ASTM** A600 (M-46) |
| * **T11347** | High-Speed Tool Steel (M-47) | **C** 1.05-1.15 **Co** 4.75-5.25 **Cr** 3.50-4.00 **Mn** 0.15-0.40 **Mo** 9.25-10.00 **P** 0.03 max **S** 0.03 max **Si** 0.20-0.45 **V** 1.15-1.35 **W** 1.30-1.80 | **AISI** M-47 **ASTM** A600 (M-47) |
| **T11348** | High-Speed Tool Steel, Powder Metallurgy (M-48) | **C** 1.50 **Co** 9.00 **Cr** 3.75 **Mo** 5.25 **V** 3.10 **W** 10.0 | **AISI** M-48 **ASTM** A600 (M-48) |
| **T11350** | Bearing Steel | **C** 0.77-0.85 **Co** 0.25 max **Cr** 3.75-4.25 **Cu** 0.10 max **Mn** 0.35 max **Mo** 4.00-4.50 **Ni** 0.15 max **P** 0.015 max **S** 0.015 max **Si** 0.25 max **V** 0.90-1.10 **W** 0.25 max | **AMS** 6490; 6491 **ASTM** A600 (M-50) **SAE** J467 (M-50) |
| **T11352** | High-Speed Tool Steel, Powder Metallurgy (M-52) | **C** 0.90 **Cr** 4.00 **Mo** 4.00 **V** 2.00 **W** 1.25 | **AISI** M-52 **ASTM** A600 (M-52) |
| **T11361** | High-Speed Tool Steel, Powder Metallurgy (M-61) | **C** 1.60 **Cr** 4.00 **Mo** 6.50 **V** 5.00 **W** 12.0 | **AISI** M-61 |
| **T11362** | High-Speed Tool Steel, Powder Metallurgy (M-62) | **C** 1.30 **Cr** 3.75 **Mo** 10.5 **V** 2.00 **W** 6.25 | **AISI** M-62 **ASTM** A600 (M-62) |
| **T12001** | High-Speed Tool Steel (T-1) | **C** 0.65-0.80 **Cr** 3.75-4.50 **Mn** 0.10-0.40 **P** 0.03 max **S** 0.03 max **Si** 0.20-0.40 **V** 0.90-1.30 **W** 17.25-18.75 | **AISI** T-1 **ASTM** A600 (T-1) **SAE** J437 (T1); J438 (T1) |
| * **T12002** | High-Speed Tool Steel (T-2) | **C** 0.80-0.90 **Cr** 3.75-4.50 **Mn** 0.20-0.40 **Mo** 1.00 max **P** 0.03 max **S** 0.03 max **Si** 0.20-0.40 **V** 1.80-2.40 **W** 17.50-19.00 | **AISI** T-2 **ASTM** A600 (T-2) **SAE** J437 (T2); J438 (T2) |
| **T12004** | High-Speed Tool Steel (T-4) | **C** 0.70-0.80 **Co** 4.25-5.75 **Cr** 3.75-4.50 **Mn** 0.10-0.40 **Mo** 0.40-1.00 **P** 0.03 max **S** 0.03 max **Si** 0.20-0.40 **V** 0.80-1.20 **W** 17.50-19.00 | **AISI** T-4 **ASTM** A600 (T-4) **SAE** J437 (T4); J438 (T4) |
| **T12005** | High-Speed Tool Steel (T-5) | **C** 0.75-0.85 **Co** 7.00-9.50 **Cr** 3.75-5.00 **Mn** 0.20-0.40 **Mo** 0.50-1.25 **P** 0.03 max **S** 0.03 max **Si** 0.20-0.40 **V** 1.80-2.40 **W** 17.50-19.00 | **AISI** T-5 **ASTM** A600 (T-5) **SAE** J437 (T5); J438 (T5) |
| **T12006** | High-Speed Tool Steel (T-6) | **C** 0.75-0.85 **Co** 11.00-13.00 **Cr** 4.00-4.75 **Mn** 0.20-0.40 **Mo** 0.40-1.00 **P** 0.03 max **S** 0.03 max **Si** 0.20-0.40 **V** 1.50-2.10 **W** 18.50-21.00 | **AISI** T-6 **ASTM** A600 (T-6) |
| **T12008** | High-Speed Tool Steel (T-8) | **C** 0.75-0.85 **Co** 4.25-5.75 **Cr** 3.75-4.50 **Mn** 0.20-0.40 **Mo** 0.40-1.00 **P** 0.03 max **S** 0.03 max **Si** 0.20-0.40 **V** 1.80-2.40 **W** 13.25-14.75 | **AISI** T-8 **ASTM** A600 (T-8) **SAE** J437 (T8); J438 (T8) |
| **T12015** | High-Speed Tool Steel (T-15) | **C** 1.50-1.60 **Co** 4.75-5.25 **Cr** 3.75-5.00 **Mn** 0.15-0.40 **Mo** 1.00 max **P** 0.03 max **S** 0.03 max **Si** 0.15-0.40 **V** 4.50-5.25 **W** 11.75-13.00 | **AISI** T-15 **ASTM** A600 (T-15) |
| **T20810** | Hot-Work Tool Steel (H-10) | **C** 0.35-0.45 **Cr** 3.00-3.75 **Mn** 0.25-0.70 **Mo** 2.00-3.00 **P** 0.030 max **S** 0.030 max **Si** 0.80-1.20 **V** 0.25-0.75 | **AISI** H-10 **ASTM** A681 (H-10) |

*Boxed entries are no longer active and are retained for reference purposes only.

The chemical compositions listed are for identification purposes and should not be used in lieu of the cross referenced specifications.

| UNIFIED NUMBER | DESCRIPTION | CHEMICAL COMPOSITION | CROSS REFERENCE SPECIFICATIONS |
|---|---|---|---|
| **T20811** | Hot-Work Tool Steel (H-11) | **C** 0.33-0.43 **Cr** 4.75-5.50 **Mn** 0.20-0.50 **Mo** 1.10-1.60 **P** 0.030 max **S** 0.030 max **Si** 0.80-1.20 **V** 0.30-0.60 | **AISI** H-11 **AMS** 6437; 6485; 6487; 6488 **ASTM** A646 (H-11); A681 (H-11) **SAE** J437 (H-11); J438 (H-11); J467 (H-11) |
| **T20812** | Hot-Work Tool Steel (H-12) | **C** 0.30-0.40 **Cr** 4.75-5.50 **Mn** 0.20-0.50 **Mo** 1.25-1.75 **P** 0.030 max **S** 0.030 max **Si** 0.80-1.20 **V** 0.50 max **W** 1.00-1.70 | **AISI** H-12 **ASTM** A681 (H-12) **SAE** J437 (H12); J438 (H12); J467 (H12) |
| **T20813** | Hot-Work Tool Steel (H-13) | **C** 0.32-0.45 **Cr** 4.75-5.50 **Mn** 0.20-0.50 **Mo** 1.10-1.75 **P** 0.030 max **S** 0.030 max **Si** 0.80-1.20 **V** 0.80-1.20 | **AISI** H-13 **AMS** 6408 **ASTM** A681 (H-13) **SAE** J437 (H13); J438 (H13); J467 (H13) |
| **T20814** | Hot-Work Tool Steel (H-14) | **C** 0.35-0.45 **Cr** 4.75-5.50 **Mn** 0.20-0.50 **P** 0.030 max **S** 0.030 max **Si** 0.80-1.20 **W** 4.00-5.25 | **AISI** H-14 **ASTM** A681 (H-14) |
| **T20819** | Hot-Work Tool Steel (H-19) | **C** 0.32-0.45 **Co** 4.00-4.50 **Cr** 4.00-4.75 **Mn** 0.20-0.50 **Mo** 0.30-0.55 **P** 0.030 max **S** 0.030 max **Si** 0.20-0.50 **V** 1.75-2.20 **W** 3.75-4.50 | **AISI** H-19 **ASTM** A681 (H-19) |
| **T20821** | Hot-Work Tool Steel (H-21) | **C** 0.26-0.36 **Cr** 3.00-3.75 **Mn** 0.15-0.40 **P** 0.030 max **S** 0.030 max **Si** 0.15-0.50 **V** 0.30-0.60 **W** 8.50-10.00 | **AISI** H-21 **ASTM** A681 (H-21) **SAE** J437 (H21); J438 (H21) |
| **T20822** | Hot-Work Tool Steel (H-22) | **C** 0.30-0.40 **Cr** 1.75-3.75 **Mn** 0.15-0.40 **P** 0.030 max **S** 0.030 max **Si** 0.15-0.40 **V** 0.25-0.50 **W** 10.00-11.75 | **AISI** H-22 **ASTM** A681 (H-22) |
| **T20823** | Hot-Work Tool Steel (H-23) | **C** 0.25-0.35 **Cr** 11.00-12.75 **Mn** 0.15-0.40 **P** 0.030 max **S** 0.030 max **Si** 0.15-0.60 **V** 0.75-1.25 **W** 11.00-12.75 | **AISI** H-23 **ASTM** A681 (H-23) |
| **T20824** | Hot-Work Tool Steel (H-24) | **C** 0.42-0.53 **Cr** 2.50-3.50 **Mn** 0.15-0.40 **P** 0.030 max **S** 0.030 max **Si** 0.15-0.40 **V** 0.40-0.60 **W** 14.00-16.00 | **AISI** H-24 **ASTM** A681 (H-24) |
| * **T20825** | Hot-Work Tool Steel (H-25) | **C** 0.22-0.32 **Cr** 3.75-4.50 **Mn** 0.15-0.40 **P** 0.030 max **S** 0.030 max **Si** 0.15-0.40 **V** 0.40-0.60 **W** 14.00-16.00 | **AISI** H-25 **ASTM** A681 (H-25) |
| **T20826** | Hot-Work Tool Steel (H-26) | **C** 0.45-0.55 **Cr** 3.75-4.50 **Mn** 0.15-0.40 **P** 0.030 max **S** 0.030 max **Si** 0.15-0.40 **V** 0.75-1.25 **W** 17.25-19.00 | **AISI** H-26 **ASTM** A681 (H-26) |
| * **T20841** | Hot-Work Tool Steel (H-41) | **C** 0.60-0.75 **Cr** 3.50-4.00 **Mn** 0.15-0.40 **Mo** 8.20-9.20 **P** 0.030 max **S** 0.030 max **Si** 0.20-0.45 **V** 1.00-1.30 **W** 1.40-2.10 | **AISI** H-41 **ASTM** A681 (H-41) |
| **T20842** | Hot-Work Tool Steel (H-42) | **C** 0.55-0.70 **Cr** 3.75-4.50 **Mn** 0.15-0.40 **Mo** 4.50-5.50 **P** 0.030 max **S** 0.030 max **Si** 0.20-0.45 **V** 1.75-2.20 **W** 5.50-6.75 | **AISI** H-42 **ASTM** A681 (H-42) |
| * **T20843** | Hot-Work Tool Steel (H-43) | **C** 0.50-0.65 **Cr** 3.75-4.50 **Mn** 0.15-0.40 **Mo** 7.75-8.50 **P** 0.030 max **S** 0.030 max **Si** 0.20-0.45 **V** 1.80-2.20 | **AISI** H-43 **ASTM** A681 (H-43) |
| **T30102** | Cold-Work Tool Steel (A-2) | **C** 0.95-1.05 **Cr** 4.75-5.50 **Mn** 1.00 max **Mo** 0.90-1.40 **P** 0.030 max **S** 0.030 max **Si** 0.50 max **V** 0.15-0.50 | **AISI** A-2 **ASTM** A681 (A-2) **SAE** J437 (A2); J438 (A2) |

*Boxed entries are no longer active and are retained for reference purposes only.

The chemical compositions listed are for identification purposes and should not be used in lieu of the cross referenced specifications.

| UNIFIED NUMBER | DESCRIPTION | CHEMICAL COMPOSITION | CROSS REFERENCE SPECIFICATIONS |
|---|---|---|---|
| * **T30103** | Cold-Work Tool Steel (A-3) | **C** 1.20-1.30 **Cr** 4.75-5.50 **Mn** 0.40-0.60 **Mo** 0.90-1.40 **P** 0.030 max **S** 0.030 max **Si** 0.50 max **V** 0.80-1.40 | **AISI** A-3 **ASTM** A681 (A-3) |
| **T30104** | Cold-Work Tool Steel (A-4) | **C** 0.95-1.05 **Cr** 0.90-2.20 **Mn** 1.80-2.20 **Mo** 0.90-1.40 **P** 0.030 max **S** 0.030 max **Si** 0.50 max | **AISI** A-4 **ASTM** A681 (A-4) |
| * **T30105** | Cold-Work Tool Steel (A-5) | **C** 0.95-1.05 **Cr** 0.90-1.20 **Mn** 2.80-3.20 **Mo** 0.90-1.40 **P** 0.030 max **S** 0.030 max **Si** 0.50 max | **AISI** A-5 **ASTM** A681 (A-5) |
| **T30106** | Cold-Work Tool Steel (A-6) | **C** 0.65-0.75 **Cr** 0.90-1.20 **Mn** 1.80-2.50 **Mo** 0.90-1.40 **P** 0.030 max **S** 0.030 max **Si** 0.50 max | **AISI** A-6 **ASTM** A681 (A-6) |
| **T30107** | Cold-Work Tool Steel (A-7) | **C** 2.00-2.85 **Cr** 5.00-5.75 **Mn** 0.80 max **Mo** 0.90-1.40 **P** 0.030 max **S** 0.030 max **Si** 0.50 max **V** 3.90-5.15 **W** 0.50-1.50 | **AISI** A-7 **ASTM** A681 (A-7) |
| **T30108** | Cold-Work Tool Steel (A-8) | **C** 0.50-0.60 **Cr** 4.75-5.50 **Mn** 0.50 max **Mo** 1.15-1.65 **P** 0.030 max **S** 0.030 max **Si** 0.75-1.10 **W** 1.00-1.50 | **AISI** A-8 **ASTM** A681 (A-8) |
| **T30109** | Cold-Work Tool Steel (A-9) | **C** 0.45-0.55 **Cr** 4.75-5.50 **Mn** 0.50 max **Mo** 1.30-1.80 **Ni** 1.25-1.75 **P** 0.030 max **S** 0.030 max **Si** 0.95-1.15 **V** 0.80-1.40 | **AISI** A-9 **ASTM** A681 (A-9) |
| **T30110** | Cold-Work Tool Steel (A-10) | **C** 1.25-1.50 **Mn** 1.60-2.10 **Mo** 1.25-1.75 **Ni** 1.55-2.05 **P** 0.030 max **S** 0.030 max **Si** 1.00-1.50 | **AISI** A-10 **ASTM** A681 (A-10) |
| **T30111** | Cold-Work Tool Steel (A-11) | **C** 2.45 **Cr** 5.25 **Mn** 0.50 **Mo** 1.30 **Si** 0.90 **V** 9.75 | **AISI** A-11 |
| **T30402** | Cold-Work Tool Steel (D-2) | **C** 1.40-1.60 **Co** 1.00 max **Cr** 11.00-13.00 **Mn** 0.60 max **Mo** 0.70-1.20 **P** 0.030 max **S** 0.030 max **Si** 0.60 max **V** 1.10 max | **AISI** D-2 **ASTM** A681 (D-2) **SAE** J437 (D2); J438 (D2) |
| **T30403** | Cold-Work Tool Steel (D-3) | **C** 2.00-2.35 **Cr** 11.00-13.50 **Mn** 0.60 max **P** 0.030 max **S** 0.030 max **Si** 0.60 max **V** 1.00 max **W** 1.00 max | **AISI** D-3 **ASTM** A681 (D-3) **SAE** J437 (D3); J438 (D3) |
| **T30404** | Cold-Work Tool Steel (D-4) | **C** 2.05-2.40 **Cr** 11.00-13.00 **Mn** 0.60 max **Mo** 0.70-1.20 **P** 0.030 max **S** 0.030 max **Si** 0.60 max **V** 1.00 max | **AISI** D-4 **ASTM** A681 (D-4) |
| **T30405** | Cold-Work Tool Steel (D-5) | **C** 1.40-1.60 **Co** 2.50-3.50 **Cr** 11.0-13.00 **Mn** 0.60 max **Mo** 0.70-1.20 **P** 0.030 max **S** 0.030 max **Si** 0.60 max **V** 1.00 max | **AISI** D-5 **ASTM** A681 (D-5) **SAE** J437 (D5); J438 (D5) |
| **T30407** | Cold-Work Tool Steel (D-7) | **C** 2.15-2.50 **Cr** 11.50-13.50 **Mn** 0.60 max **Mo** 0.70-1.20 **P** 0.030 max **S** 0.030 max **Si** 0.60 max **V** 3.80-4.40 | **AISI** D-7 **ASTM** A681 (D-7) **SAE** J437 (D7); J438 (D7) |
| **T31501** | Oil-Hardening Tool Steel (O-1) | **C** 0.85-1.00 **Cr** 0.40-0.60 **Mn** 1.00-1.40 **P** 0.030 max **S** 0.030 max **Si** 0.50 max **V** 0.30 max **W** 0.40-0.60 | **AISI** O-1 **ASTM** A681 (O-1) **SAE** J437 (O1); J438 (O1) |
| **T31502** | Oil-Hardening Tool Steel (O-2) | **C** 0.85-0.95 **Cr** 0.35 max **Mn** 1.40-1.80 **Mo** 0.30 max **P** 0.030 max **S** 0.030 max **Si** 0.50 max **V** 0.30 max | **AISI** O-2 **ASTM** A681 (O-2) **SAE** J437 (O2); J438 (O2) |
| **T31506** | Oil-Hardening Tool Steel (O-6) | **C** 1.25-1.55 **Cr** 0.30 max **Mn** 0.30-1.10 **Mo** 0.20-0.30 **P** 0.030 max **S** 0.030 max **Si** 0.55-1.50 | **AISI** O-6 **ASTM** A681 (O-6) **SAE** J437 (06); J438 (06) |

*Boxed entries are no longer active and are retained for reference purposes only.

The chemical compositions listed are for identification purposes and should not be used in lieu of the cross referenced specifications.

| UNIFIED NUMBER | DESCRIPTION | CHEMICAL COMPOSITION | CROSS REFERENCE SPECIFICATIONS |
|---|---|---|---|
| **T31507** | Oil Hardening Tool Steel (O-7) | **C** 1.10-1.30 **Cr** 0.35-0.85 **Mn** 1.00 max **Mo** 0.30 max **P** 0.030 max **S** 0.030 max **Si** 0.60 max **V** 0.40 max **W** 1.00-2.00 | **AISI** O-7 **ASTM** A681 (0-7) |
| **T41901** | Shock Resisting Tool Steel (S-1) | **C** 0.40-0.55 **Cr** 1.00-1.80 **Mn** 0.10-0.40 **Mo** 0.50 max **P** 0.030 max **S** 0.030 max **Si** 0.15-1.20 **V** 0.15-0.30 **W** 1.50-3.00 | **AISI** S-1 **ASTM** A681 (S-1) **SAE** J437 (S-1); J438 (S-1) |
| **T41902** | Shock-Resisting Tool Steel (S-2) | **C** 0.40-0.55 **Mn** 0.30-0.50 **Mo** 0.30-0.60 **P** 0.030 max **S** 0.030 max **Si** 0.90-1.20 **V** 0.50 max | **AISI** S-2 **ASTM** A681 (S-2) **SAE** J437 (S2); J438 (S2) |
| **T41904** | Shock-Resisting Tool Steel (S-4) | **C** 0.50-0.65 **Cr** 0.35 max **Mn** 0.60-0.95 **P** 0.030 max **S** 0.030 max **Si** 1.75-2.25 **V** 0.35 max | **AISI** S-4 **ASTM** A681 (S-4) |
| **T41905** | Shock-Resisting Tool Steel (S-5) | **C** 0.50-0.65 **Cr** 0.35 max **Mn** 0.60-1.00 **Mo** 0.20-1.35 **P** 0.030 max **S** 0.030 max **Si** 1.75-2.25 **V** 0.35 max | **AISI** S-5 **ASTM** A681 (S-5) **SAE** J437 (S5); J438 (S5) |
| **T41906** | Shock-Resisting Tool Steel (S-6) | **C** 0.40-0.50 **Cr** 1.20-1.50 **Mn** 1.20-1.50 **Mo** 0.30-0.50 **P** 0.030 max **S** 0.030 max **Si** 2.00-2.50 **V** 0.20-0.40 | **AISI** S-6 **ASTM** A681 (S-6) |
| **T41907** | Shock-Resisting Tool Steel (S-7) | **C** 0.45-0.55 **Cr** 3.00-3.50 **Mn** 0.20-0.80 **Mo** 1.30-1.80 **P** 0.030 max **S** 0.030 max **Si** 0.20-1.00 **Other** V 0.20-0.30 (Optional) | **AISI** S7 **ASTM** A681 (S-7) |
| * **T51602** | Mold Steel (P-2) | **C** 0.10 max **Cr** 0.75-1.25 **Mn** 0.10-0.40 **Mo** 0.15-0.40 **Ni** 0.10-0.50 **P** 0.030 max **S** 0.030 max **Si** 0.10-0.40 | **AISI** P-2 **ASTM** A681 (P-2) |
| * **T51603** | Mold Steel (P-3) | **C** 0.10 max **Cr** 0.40-0.75 **Mn** 0.20-0.60 **Ni** 1.00-1.50 **P** 0.030 max **S** 0.030 max **Si** 0.40 max | **AISI** P-3 **ASTM** A681 (P-3) |
| * **T51604** | Mold Steel (P-4) | **C** 0.12 max **Cr** 4.00-5.25 **Mn** 0.20-0.60 **Mo** 0.40-1.00 **P** 0.030 max **S** 0.030 max **Si** 0.10-0.40 | **AISI** P-4 **ASTM** A681 (P-4) |
| * **T51605** | Mold Steel (P-5) | **C** 0.10 max **Cr** 2.00-2.50 **Mn** 0.20-0.60 **Ni** 0.35 max **P** 0.030 max **S** 0.030 max **Si** 0.40 max | **AISI** P-5 **ASTM** A681 (P-5) |
| **T51606** | Mold Steel (P-6) | **C** 0.05-0.15 **Cr** 1.25-1.75 **Mn** 0.35-0.70 **Ni** 3.25-3.75 **P** 0.030 max **S** 0.030 max **Si** 0.10-0.40 | **AISI** P-6 **ASTM** A681 (P-6) |
| **T51620** | Mold Steel (P-20) | **C** 0.28-0.40 **Cr** 1.40-2.00 **Mn** 0.60-1.00 **Mo** 0.30-0.55 **P** 0.030 max **S** 0.030 max **Si** 0.20-0.80 | **AISI** P-20 **ASTM** A681 (P-20) |
| **T51621** | Mold Steel (P-21) | **Al** 1.05-1.25 **C** 0.18-0.22 **Cr** 0.20-0.30 **Mn** 0.20-0.40 **Ni** 4.00-4.25 **P** 0.030 max **S** 0.030 max **Si** 0.20-0.40 **V** 0.15-0.25 | **AISI** P-21 **ASTM** A681 (P-21) |
| * **T60601** | C-W Tool Steel (F-1) | **C** 0.95-1.25 **Mn** 0.50 max **P** 0.030 max **S** 0.030 max **Si** 0.50 max **W** 1.00-1.75 | **AISI** F-1 **ASTM** A681 (F-1) |

*Boxed entries are no longer active and are retained for reference purposes only.

The chemical compositions listed are for identification purposes and should not be used in lieu of the cross referenced specifications.

| UNIFIED NUMBER | DESCRIPTION | CHEMICAL COMPOSITION | CROSS REFERENCE SPECIFICATIONS |
|---|---|---|---|
| * **T60602** | C-W Tool Steel (F-2) | **C** 1.20-1.40 **Cr** 0.20-0.40 **Mn** 0.50 max **P** 0.030 max **S** 0.030 max **Si** 0.50 max **W** 3.00-4.50 | **AISI** F-2 **ASTM** A681 (F-2) |
| **T61202** | Low-Alloy Tool Steel (L-2) | **C** 0.45-1.00 **Cr** 0.70-1.20 **Mn** 0.10-0.90 **Mo** 0.25 max **P** 0.030 max **S** 0.030 max **Si** 0.50 max **V** 0.10-0.30 | **AISI** L-2 **ASTM** A681 (L-2) |
| * **T61203** | Low-Alloy Tool Steel (L-3) | **C** 0.95-1.10 **Cr** 1.30-1.70 **Mn** 0.25-0.50 **P** 0.030 max **S** 0.030 max **Si** 0.50 max **V** 0.10-0.30 | **AISI** L-3 **ASTM** A681 (L-3) |
| **T61206** | Low-Alloy Tool Steel (L-6) | **C** 0.65-0.75 **Cr** 0.60-1.20 **Mn** 0.25-0.80 **Mo** 0.50 max **Ni** 1.25-2.00 **P** 0.030 max **S** 0.030 max **Si** 0.50 max | **AISI** L-6 **ASTM** A681 (L-6) **SAE** J437 (L6); J438 (L6) |
| **T72301** | Water Hardening Tool Steel (W-1) | **C** 0.70-1.50 **Cr** 0.15 max **Cu** 0.20 max **Mn** 0.10-0.40 **Mo** 0.10 max **Ni** 0.20 max **P** 0.025 max **S** 0.025 max **Si** 0.10-0.40 **V** 0.10 max **W** 0.15 max | **AISI** W-1 **ASTM** A686 (W-1) **SAE** J437 (W108, W109, W110, W112); J438 (W108, W109, W110, W112) |
| **T72302** | Water Hardening Tool Steel (W-2) | **C** 0.85-1.50 **Cr** 0.15 max **Cu** 0.20 max **Mn** 0.10-0.40 **Mo** 0.10 max **Ni** 0.20 max **P** 0.030 max **S** 0.030 max **Si** 0.10-0.40 **V** 0.15-0.35 **W** 0.15 max | **AISI** W-2 **ASTM** A686 (W-2) **SAE** J437 (W209, W210); J438 (W209, W210) |
| **T72305** | Water-Hardening Tool Steel (W-5) | **C** 1.05-1.15 **Cr** 0.40-0.60 **Cu** 0.20 max **Mn** 0.10-0.40 **Mo** 0.10 max **Ni** 0.20 max **P** 0.030 max **S** 0.030 max **Si** 0.10-0.40 **V** 0.10 max **W** 0.15 max | **AISI** W-5 **ASTM** A686 (W-5) |
| **T74000** | Cr-Steel Solid Welding Wire for Machinable Surfaces (ERFe-1) | **C** 0.25-0.5 **Cr** 0.5-3.5 **Fe** rem **Mn** 0.5-2.0 **Mo** 1.5 max **Si** 1.0 max | **ASME** SFA5.21 (ERFe-1 **AWS** A5.21 (ERFe-1) |
| **T74001** | Cr-Steel Solid Welding Wire for Machinable Surfaces (ERFe-1A) | **C** 0.25-0.5 **Cr** 0.5-3.5 **Fe** rem **Mn** 1.7-3.5 **Si** 1.0 max | **ASME** SFA5.21 (ERFe-1A) **AWS** A5.21 (ERFe-1A) |
| **T74002** | Cr-Steel Solid Welding Wire for Machinable Surfaces (ERFe-2) | **C** 0.10-0.30 **Cr** 1.8-3.8 **Fe** rem **Mn** 0.5-2.0 **Mo** 1.0 max **Ni** 1.0 max **Si** 1.0 max **V** 0.35 max | **ASME** SFA5.21 (ERFe-2) **AWS** A5.21 (ERFe-2) |
| **T74003** | Cr-Steel Solid Welding Wire for Tool and Die Surfacing (ERFe-3) | **C** 0.50-0.80 **Cr** 4.0-8.0 **Fe** rem **Mn** 0.5-1.5 **Mo** 1.0 max **Si** 1.0 max | **ASME** SFA5.21 (ERFe-3) **AWS** A5.21 (ERFe-3) |
| **T74005** | Cr-Steel Solid Welding Wire for Tool and Die Surfacing (ERFe-5) | **C** 0.50-0.80 **Cr** 1.5-3.0 **Fe** rem **Mn** 1.5-2.5 **Si** 0.9 max | **ASME** SFA5.21 (ERFe-5) **AWS** A5.21 (ERFe-5) |
| **T75006** | High Speed Steel Solid Welding Wire for Surfacing (ERFe-6) | **C** 0.6-1.0 **Cr** 3.0-5.0 **Fe** rem **Mn** 0.4-1.0 **Mo** 7.0-9.5 **Si** 1.0 max **W** 0.5-1.5 **V** 0.5-1.5 | **ASME** SFA5.21 (ERFe-6) **AWS** A5.21 (ERFe-6) |
| **T75008** | High Speed Steel Solid Welding Wire for Surfacing (ERFe-8) | **C** 0.30-0.60 **Cr** 4.0-8.0 **Fe** rem **Mn** 1.0-2.0 **Mo** 1.0-2.0 **Si** 1.0 max **W** 1.0-2.0 **V** 0.50 max | **ASME** SFA5.21 (ERFe-8) **AWS** A5.21 (ERFe-8) |
| **T87510** | Thermal Spray Wire (W-FeCrC) | **C** 1.0 max **Cr** 1.35-1.65 **Fe** rem **Mn** 0.25-0.40 **P** 0.020 max **S** 0.02 max **Si** 0.50 max | **AWS** C2.25 (W-FeCrC) |
| **T87515** | Thermal Spray Wire (W-FeCrMn) | **C** 1.0 max **Cr** 1.6-2.0 **Fe** rem **Mn** 1.7-2.0 **Mo** 0.15-0.25 **P** 0.010 max **S** 0.040 max **Si** 0.35 max **Ti** 0.11-0.15 | **AWS** C2.25 (W-FeCrMn) |
| **T87520** | Thermal Spray Wire (W-FeNiCr) | **C** 0.10 max **Cr** 1.0-2.0 **Fe** rem **Mn** 2.5 max **Mo** 1.0-3.0 **Ni** 3.7-5.0 **P** 0.030 max **S** 0.030 max **Si** 0.35 max | **AWS** C2.25 (W-FeNiCr) |

*Boxed entries are no longer active and are retained for reference purposes only.

The chemical compositions listed are for identification purposes and should not be used in lieu of the cross referenced specifications.

| UNIFIED NUMBER | DESCRIPTION | CHEMICAL COMPOSITION | CROSS REFERENCE SPECIFICATIONS |
|---|---|---|---|
| **T90102** | Cast Cold Work Tool Steel (CA-2) | **C** 0.95-1.05 **Cr** 4.75-5.50 **Mn** 0.75 max **Mo** 0.90-1.40 **P** 0.030 max **S** 0.030 max **Si** 1.50 max **Other** V optional, if present 0.20-0.50 | **ASTM** A597 (CA-2) |
| **T90402** | Cast Cold Work Tool Steel (CD-2) | **C** 1.40-1.60 **Cr** 11.00-13.00 **Mn** 1.00 max **Mo** 0.70-1.20 **P** 0.030 max **S** 0.030 max **Si** 1.50 max **Other** Co optional, if present 0.70-1.00; V optional, if present 0.40-1.00 | **ASTM** A597 (CD-2) |
| **T90405** | Cast Cold Work Tool Steel (CD-5) | **C** 1.35-1.60 **Co** 2.50-3.50 **Cr** 11.00-13.00 **Mn** 0.75 max **Mo** 0.70-1.20 **P** 0.030 max **S** 0.030 max **Si** 1.50 max **V** 0.35-0.55 **Other** Ni optional, if present 0.40-0.60 | **ASTM** A597 (CD-5) |
| **T90812** | Cast Hot Work Tool Steel (CH-12) | **C** 0.30-0.40 **Co** 0.20-0.50 **Cr** 4.75-5.75 **Mn** 0.75 max **Mo** 1.25-1.75 **P** 0.030 max **S** 0.030 max **Si** 1.50 max **V** 0.20-0.50 **W** 1.00-1.70 | **ASTM** A597 (CH-12) |
| **T90813** | Cast Hot Work Tool Steel (CH-13) | **C** 0.30-0.42 **Cr** 4.75-5.75 **Mn** 0.75 max **Mo** 1.25-1.75 **P** 0.030 max **S** 0.030 max **Si** 1.50 max **V** 0.75-1.20 | **ASTM** A597 (CH-13) |
| **T91501** | Cast Oil Hardening Tool Steel (CO-1) | **C** 0.85-1.00 **Cr** 0.40-1.00 **Mn** 1.00-1.30 **P** 0.030 max **S** 0.030 max **Si** 1.50 max **V** 0.30 max **W** 0.40-0.60 | **ASTM** A597 (CO-1) |
| **T91905** | Cast Shock Resisting Tool Steel (CS-5) | **C** 0.50-0.65 **Cr** 0.35 max **Mn** 0.60-1.00 **Mo** 0.20-0.80 **P** 0.030 max **S** 0.030 max **Si** 1.75-2.25 **V** 0.35 max | **ASTM** A597 (CS-5) |
| **T91907** | Cast Tool Steel (CS-7) | **C** 0.45-0.55 **Cr** 3.00-3.50 **Mn** 0.40-0.80 **Mo** 1.20-1.60 **P** 0.03 max **Si** 0.60-1.00 **S** 0.03 max | **ASTM** A597 (CS-7) |

The chemical compositions listed are for identification purposes and should not be used in lieu of the cross referenced specifications.

# UNS NUMBERS ASSIGNED TO DATE

## With Description of Each Material Covered and References To Documents In Which The Same or Similar Materials are Described

## Wxxxxx Number Series
## Welding Filler Metals

# WELDING FILLER METALS

| UNIFIED NUMBER | DESCRIPTION | CHEMICAL COMPOSITION | CROSS REFERENCE SPECIFICATIONS |
|---|---|---|---|
| **W06010** | Mild Steel Arc Welding Electrode (E6010) | **C** 0.20 max **Mn** 1.20 max **Mo** 0.30 max **Ni** 0.30 max **Si** 1.00 max **V** 0.08 max | **ASME** SFA5.1 (E6010) **AWS** A5.1 (E6010) |
| **W06011** | Mild Steel Arc Welding Electrode (E6011) | **C** 0.20 max **Cr** 0.20 max **Mn** 1.20 max **Mo** 0.30 max **Ni** 0.30 max **Si** 1.00 max **V** 0.08 max | **ASME** SFA5.1 (E6011) **AWS** A5.1 (E6011) |
| **W06012** | Mild Steel Arc Welding Electrode (E6012) | **C** 0.20 max **Cr** 0.20 max **Mn** 1.20 max **Mo** 0.30 max **Ni** 0.30 max **Si** 1.00 max **V** 0.08 max | **ASME** SFA5.1 (E6012) **AWS** A5.1 (E6012) |
| **W06013** | Mild Steel Arc Welding Electrode (E6013) | **C** 0.20 max **Cr** 0.20 max **Mn** 1.20 max **Mo** 0.30 max **Ni** 0.30 max **Si** 1.00 max **V** 0.08 max | **AMS** 5031 **ASME** SFA5.1 (E6013) **AWS** A5.1 (E6013) |
| **W60617** | Aluminum Bronze Arc Welding Electrode (ECuAl-A2) | **Al** 8.5-11.0 **Cu** rem **Fe** 0.5-5.0 **Pb** 0.02 max **Si** 1.5 max | **AWS** A5.13 (ECuAl-A2) |
| **W06019** | Mild Steel Arc Welding Electrode (E6019) | **C** 0.20 max **Cr** 0.20 max **Mn** 1.20 max **Mo** 0.30 max **Ni** 0.30 max **Si** 1.00 max **V** 0.08 max | **ASME** SFA5.1 (E6019) **AWS** A5.1 (E6019) |
| **W06020** | Mild Steel Arc Welding Electrode (E6020) | **C** 0.20 max **Cr** 0.20 max **Mn** 1.20 max **Si** 1.00 max **Ni** 0.30 max **Mo** 0.30 max **V** 0.08 max | **ASME** SFA5.1 (E6020) **AWS** A5.1 (E6020) |
| **W06022** | Mild Steel Arc Welding Electrode (E6022) | **C** 0.20 max **Cr** 0.20 max **Mn** 1.20 max **Mo** 0.30 max **Ni** 0.30 max **Si** 1.00 max **V** 0.08 max | **ASME** SFA5.1 (E6022) **AWS** A5.1 (E6022) |
| **W06027** | Mild Steel Arc Welding Electrode (E6027) | **C** 0.20 max **Cr** 0.20 max **Mn** 1.20 max **Mo** 0.30 max **Ni** 0.30 max **Si** 1.00 max **V** 0.08 max | **ASME** SFA5.1 (E6027) **AWS** A5.1 (E6027) |
| **W06040** | Mild Steel Composite Wire for Electroslag Welding (EWT1) | **C** 0.13 max **Mn** 2.00 max **P** 0.03 max **S** 0.03 max **Si** 0.60 max | **ASME** SFA5.25 (EWT1) **AWS** A5.25 (EWT1) |
| **W06041** | Mild Steel Composite Wire for Submerged Arc Welding (EC 1) | **C** 0.15 max **Cu** 0.35 max **Mn** 1.80 max **P** 0.035 max **S** 0.035 max **Si** 0.90 max | **ASME** SFA5.17 (EC1) **AWS** A5.17 (EC1) |
| **W06301** | Mild Steel Flux Cored Wire for Electrogas Welding (EG6XT-1, EG43XT-1) | **Cr** 0.20 max **Cu** 0.35 max **Mn** 1.7 max **Mo** 0.35 max **Ni** 0.30 max **P** 0.03 max **S** 0.03 max **Si** 0.50 max **V** 0.08 max | **ASME** SFA5.26 (EG6XT-1, EG43XT-1) **AWS** A5.26 (EG6XT-1, EG43XT-1) |
| **W06302** | Mild Steel Flux Cored Wire for Electrogas Welding (EG6XT-2, EG43XT-2) | **Cr** 0.20 max **Cu** 0.35 max **Mn** 2.0 max **Mo** 0.35 max **Ni** 0.30 max **P** 0.03 max **S** 0.03 max **Si** 0.90 max **V** 0.08 max | **ASME** SFA5.26 (EG6XT-2, EG43XT-2) **AWS** A5.26 (EG6XT-2, EG43XT-2) |
| * **W06601** | Mild Steel Flux Cored Wire (E6XT-1) | **Al** 1.8 max **Cr** 0.20 max **Mn** 1.75 max **Mo** 0.30 max **Ni** 0.50 max **P** 0.04 max **S** 0.03 max **Si** 0.90 max **V** 0.08 max | **ASME** SFA5.20 (E6XT-1) **AWS** A5.20 (E6XT-1) |
| * **W06604** | Mild Steel Flux Cored Wire (E6XT-4) | **Al** 1.8 max **Cr** 0.20 max **Mn** 1.75 max **Mo** 0.30 max **Ni** 0.50 max **P** 0.04 max **S** 0.03 max **Si** 0.90 max **V** 0.08 max | **ASME** SFA5.20 (E6XT-4) **AWS** A5.20 (E6XT-4) |
| * **W06605** | Mild Steel Flux Cored Wire (E6XT-5) | **Al** 1.8 max **Cr** 0.20 max **Mn** 1.75 max **Mo** 0.30 max **Ni** 0.50 max **P** 0.04 max **S** 0.03 max **Si** 0.90 max **V** 0.08 max | **ASME** SFA5.20 (E6XT-5) **AWS** A5.20 (E6XT-5) |
| * **W06606** | Mild Steel Flux Cored Wire (E6XT-6) | **Al** 1.8 max **Cr** 0.20 max **Mn** 1.75 max **Mo** 0.30 max **Ni** 0.50 max **P** 0.04 max **S** 0.03 max **Si** 0.90 max **V** 0.08 max | **ASME** SFA5.20 (E6XT-6) **AWS** A5.20 (E6XT-6) |

*Boxed entries are no longer active and are retained for reference purposes only.

The chemical compositions listed are for identification purposes and should not be used in lieu of the cross referenced specifications.

| UNIFIED NUMBER | DESCRIPTION | CHEMICAL COMPOSITION | CROSS REFERENCE SPECIFICATIONS |
|---|---|---|---|
| * **W06607** | Mild Steel Flux Cored Wire (E6XT-7) | **Al** 1.8 max **Cr** 0.20 max **Mn** 1.75 max **Mo** 0.30 max **Ni** 0.50 max **P** 0.04 max **S** 0.03 max **Si** 0.90 max **V** 0.08 max | **ASME** SFA5.20 (E6XT-7) **AWS** A5.20 (E6XT-7) |
| * **W06608** | Mild Steel Flux Cored Wire (E6XT-8) | **Al** 1.8 max **Cr** 0.20 max **Mn** 1.75 max **Mo** 0.30 max **Ni** 0.50 max **P** 0.04 max **S** 0.03 max **Si** 0.90 max **V** 0.08 max | **ASME** SFA5.20 (E6XT-8) **AWS** A5.20 (E6XT-8) |
| * **W06611** | Mild Steel Flux Cored Wire (E6XT-11) | **Al** 1.8 max **Cr** 0.20 max **Mn** 1.75 max **Mo** 0.30 max **Ni** 0.50 max **P** 0.04 max **S** 0.03 max **Si** 0.90 max **V** 0.08 max | **ASME** SFA5.20 (E6XT-11) **AWS** A5.20 (E6XT-11) |
| **W06613** | Mild Steel Flux Cored Wire (E6XT-13) | **Other** not specified | **ASME** SFA5.20 (E6XT-13) **AWS** A5.20 (E6XT-13) |
| **W07014** | Mild Steel Arc Welding Electrode (E7014) | **Cr** 0.20 max **Mn** 1.25 max **Mo** 0.30 max **Ni** 0.30 max **Si** 0.90 max **V** 0.08 max | **ASME** SFA5.1 (E7014) **AWS** A5.1 (E7014) |
| **W07015** | Mild Steel Arc Welding Electrode (E7015) | **Cr** 0.20 max **Mn** 1.25 max **Mo** 0.30 max **Ni** 0.30 max **Si** 0.90 max **V** 0.08 max | **ASME** SFA5.1 (E7015) **AWS** A5.1 (E7015) |
| **W07016** | Mild Steel Arc Welding Electrode (E7016) | **Cr** 0.20 max **Mn** 1.60 max **Mo** 0.30 max **Ni** 0.30 max **Si** 0.75 max **V** 0.08 max | **ASME** SFA5.1 (E7016) **AWS** A5.1 (E7016) |
| **W07018** | Mild Steel Arc Welding Electrode (E7018) | **Cr** 0.20 max **Mn** 1.60 max **Mo** 0.30 max **Ni** 0.30 max **Si** 0.75 max **V** 0.08 max | **ASME** SFA5.1 (E7018, E7018M) **AWS** A5.1 (E7018, E7018M) **MIL SPEC** MIL-E-22200/1 (MIL-7018); MIL-E-22200/10 (MIL-7018M) |
| **W07024** | Mild Steel Arc Welding Electrode (E7024) | **Cr** 0.20 max **Mn** 1.25 max **Mo** 0.30 max **Ni** 0.30 max **Si** 0.90 max **V** 0.08 max | **ASME** SFA5.1 (E7024) **AWS** A5.1 (E7024) |
| **W07027** | Mild Steel Arc Welding Electrode (E7027) | **Cr** 0.20 max **Mn** 1.60 max **Mo** 0.30 max **Ni** 0.30 max **Si** 0.75 max **V** 0.08 max | **ASME** SFA5.1 (E7027) **AWS** A5.1 (E7027) |
| **W07028** | Mild Steel Arc Welding Electrode (E7028) | **Cr** 0.20 max **Mn** 1.60 max **Mo** 0.30 max **Ni** 0.30 max **Si** 0.90 max **V** 0.08 max | **ASME** SFA5.1 (E7028) **AWS** A5.1 (E7028) |
| **W07048** | Mild Steel Arc Welding Electrode (E7048) | **Cr** 0.20 max **Mn** 1.60 max **Mo** 0.30 max **Ni** 0.30 max **Si** 0.90 max **V** 0.08 max | **ASME** SFA5.1 (E7048) **AWS** A5.1 (E7048) |
| **W07116** | Mild Steel Arc Welding Electrode (E7016-1) | **Cr** 0.20 max **Mn** 1.60 max **Mo** 0.30 max **Ni** 0.30 max **Si** 0.75 max **V** 0.08 max | **ASME** SFA5.1 (E7016-1) **AWS** A5.1 (E7016-1) |
| **W07118** | Mild Steel Arc Welding Electrode (E7018-1) | **Cr** 0.20 max **Mn** 1.60 max **Mo** 0.30 max **Ni** 0.30 max **Si** 0.75 max **V** 0.08 max | **ASME** SFA5.1 (E7018-1) **AWS** A5.1 (E7018-1) |
| **W07124** | Mild Steel Arc Welding Electrode (E7024-1) | **Cr** 0.20 max **Mn** 1.25 max **Mo** 0.30 max **Ni** 0.30 max **Si** 0.90 max **V** 0.08 max | **ASME** SFA5.1 (E7024-1) **AWS** A5.1 (E7024-1) |
| **W07301** | Mild Steel Flux Cored Wire for Electrogas Welding (EG7XT-1, EG48XT-1) | **Cr** 0.20 max **Mn** 1.7 max **Mo** 0.35 max **Ni** 0.30 max **P** 0.03 max **S** 0.03 max **Si** 0.50 max **V** 0.08 max | **ASME** SFA5.26 (EG7XT-1, EG48XT-1) **AWS** A5.26 (EG7XT-1, EG48XT-1) |
| **W07302** | Mild Steel Flux Cored Wire for Electrogas Welding (EG7XT-2, EG48XT-2) | **Cr** 0.20 max **Cu** 0.35 max **Mn** 2.0 max **Mo** 0.35 max **Ni** 0.30 max **P** 0.03 max **S** 0.03 max **Si** 0.90 max **V** 0.08 max | **ASME** SFA5.26 (EG7XT-2, EG48XT-2) **AWS** A5.26 (EG7XT-2, EG48XT-2) |
| **W07601** | Mild Steel Flux Cored Wire (E7XT-1) | **C** 0.18 max **Cr** 0.20 max **Mn** 1.75 max **Mo** 0.30 max **Ni** 0.50 max **P** 0.03 max **S** 0.03 max **Si** 0.90 max **V** 0.08 max | **ASME** SFA5.20 (E7XT-1, E7XT-1M) **AWS** A5.20 (E7XT-1, E7XT-1M) **MIL SPEC** MIL-E-24403/1 (MIL-7XT-1) |

*Boxed entries are no longer active and are retained for reference purposes only.

The chemical compositions listed are for identification purposes and should not be used in lieu of the cross referenced specifications.

| UNIFIED NUMBER | DESCRIPTION | CHEMICAL COMPOSITION | CROSS REFERENCE SPECIFICATIONS |
|---|---|---|---|
| **W07602** | Mild Steel Flux Cored Wire (E7XT-2, E7XT-2M) | **Other** not specified | **ASME** SFA5.20 (E7XT-2, E7XT-2M) **AWS** A5.20 (E7XT-2, E7XT-2M) |
| **W07603** | Carbon Steel Bare Electrode (or Rod) for Flux Cored Arc Welding (E7XT-3) | **Other** not specified | **ASME** SFA5.20 (E7XT-3) **AWS** A5.20 (E7XT-3) |
| **W07604** | Mild Steel Flux Cored Wire (E7XT-4) | **Al** 1.8 max **Cr** 0.20 max **Mn** 1.75 max **Mo** 0.30 max **Ni** 0.50 max **P** 0.03 max **S** 0.03 max **Si** 0.60 max **V** 0.08 max | **ASME** SFA5.20 (E7XT-4) **AWS** A5.20 (E7XT-4) |
| **W07605** | Mild Steel Flux Cored Wire (E7XT-5) | **C** 0.18 max **Cr** 0.20 max **Mn** 1.75 max **Mo** 0.30 max **Ni** 0.50 max **P** 0.03 max **S** 0.03 max **Si** 0.90 max **V** 0.08 max | **ASME** SFA5.20 (E7XT-5, E7XT-5M) **AWS** A5.20 (E7XT-5, E7XT-5M) **MIL SPEC** MIL-E-24403/1 (MIL-70T-5) |
| **W07606** | Mild Steel Flux Cored Wire (E7XT-6) | **Al** 1.8 max **Cr** 0.20 max **Mn** 1.75 max **Mo** 0.30 max **Ni** 0.50 max **P** 0.04 max **S** 0.03 max **Si** 0.90 max **V** 0.08 max | **ASME** SFA5.20 (E7XT-6) **AWS** A5.20 (E7XT-6) **MIL SPEC** MIL-E-24403/1 (MIL-70T-6) |
| **W07607** | Mild Steel Flux Cored Wire (E7XT-7) | **Al** 1.8 max **Cr** 0.20 max **Mn** 1.75 max **Mo** 0.30 max **Ni** 0.50 max **P** 0.03 max **S** 0.03 max **Si** 0.60 max **V** 0.08 max | **ASME** SFA5.20 (E7XT-7) **AWS** A5.20 (E7XT-7) |
| **W07608** | Mild Steel Flux Cored Wire (E7XT-8) | **Al** 1.8 max **Cr** 0.20 max **Mn** 1.75 max **Mo** 0.30 max **Ni** 0.50 max **P** 0.04 max **S** 0.03 max **Si** 0.90 max **V** 0.08 max | **ASME** SFA5.20 (E7XT-8) **AWS** A5.20 (E7XT-8) **MIL SPEC** MIL-E-24403/1 (MIL-7XT-8) |
| **W07609** | Mild Steel, Flux Cored Wire (E7XT-9) | **C** 0.18 max **Cr** 0.20 max **Cu** 0.35 max **Mn** 1.75 max **Mo** 0.30 max **Ni** 0.50 max **P** 0.03 max **S** 0.03 max **Si** 0.90 max **V** 0.08 max | **ASME** SFA5.20 (E7XT-9, E7XT-9M) **AWS** A5.20 (E7XT-9, E7XT-9M) |
| **W07610** | Mild Steel Flux Cored Wire (E7XT-10) | **Other** not specified | **ASME** SFA5.20 (E7XT-10) **AWS** A5.20 (E7XT-10) |
| **W07611** | Mild Steel Flux Cored Wire (E7XT-11) | **Al** 1.8 max **Cr** 0.20 max **Mn** 1.75 max **Mo** 0.30 max **Ni** 0.50 max **P** 0.04 max **S** 0.03 max **Si** 0.90 max **V** 0.08 max | **ASME** SFA5.20 (E7XT-11) **AWS** A5.20 (E7XT-11) |
| **W07612** | Mild Steel, Flux Cored Wire (E7XT-12) | **C** 0.15 max **Cr** 0.20 max **Cu** 0.35 max **Mn** 1.60 max **Mo** 0.30 max **Ni** 0.50 max **P** 0.03 max **S** 0.03 max **Si** 0.90 max **V** 0.08 max | **ASME** SFA5.20 (E7XT-12, E7XT-12M) **AWS** A5.20 (E7XT-12, E7XT-12M) |
| **W07613** | Mild Steel Flux cored Wire (E7XT-13) | **Other** not specified | **ASME** SFA5.20 (E7XT-13) **AWS** A5.20 (E7XT-13) |
| **W07614** | Mild Steel, Flux Cored Wire (E7XT-14) | **Other** not specified | **ASME** SFA5.20 (E7XT-14) **AWS** A5.20 (E7XT-14) |
| **W07701** | Mild Steel Flux Cored Wire | **C** 0.12 max **Cr** 0.20 max **Cu** 0.20 max **Mn** 0.50-1.75 **Mo** 0.30 max **Ni** 0.50 max **P** 0.030 max **S** 0.030 max **Si** 0.90 max **V** 0.05 max | **MIL SPEC** MIL-E-24403/1 (MIL-71T-1A) |
| **W07703** | Mild Steel Metal Cored Wire for Multipass Arc Welding (E70C-3X) | **C** 0.12 max **Mn** 1.75 max **P** 0.03 max **S** 0.03 max **Si** 0.90 max | **ASME** SFA5.18 (E70C-3X) **AWS** A5.18 (E70C-3X) |
| **W07706** | Mild Steel Metal Cored Wire for Multipass Welding (E70C-6X) | **C** 0.12 max **Mn** 1.75 max **P** 0.03 max **S** 0.03 max **Si** 0.90 max | **ASME** SFA5.18 (E70C-6X) **AWS** A5.18 (E70C-6X) |
| **W07708** | Mild Steel Flux Cored Wire | **Al** 1.10 max **C** 0.25 max **Cr** 0.20 max **Cu** 0.20 max **Mn** 0.20-1.80 **Mo** 0.30 max **Ni** 0.30 max **P** 0.030 max **S** 0.03 max **Si** 0.60 max **V** 0.05 max | **MIL SPEC** MIL-E-24403 (MIL-71T-8A) |
| **W10013** | Low Alloy Steel Arc Welding Electrode (10013) | **C** 0.06-0.12 **Cr** 0.20 max **Cu** 0.35 max **Mn** 0.35-0.70 **Mo** 0.90-1.20 **Ni** 0.25 max **P** 0.025 max **S** 0.025 max **Si** 0.30-0.60 **V** 0.10-0.30 | **AMS** 6464 |
| **W10015** | Low Alloy Steel Arc Welding Electrode (E10015-D2) | **C** 0.15 max **Mn** 1.65-2.00 **Mo** 0.25-0.45 **P** 0.03 max **S** 0.03 max **Si** 0.60 max | **ASME** SFA5.5 (E10015-D2) **AWS** A5.5 (E10015-D2) |

The chemical compositions listed are for identification purposes and should not be used in lieu of the cross referenced specifications.

| UNIFIED NUMBER | DESCRIPTION | CHEMICAL COMPOSITION | CROSS REFERENCE SPECIFICATIONS |
|---|---|---|---|
| **W10016** | Low Alloy Steel Arc Welding Electrode (E10016-D2) | **C** 0.15 max **Mn** 1.65-2.00 **Mo** 0.25-0.45 **P** 0.03 max **S** 0.03 max **Si** 0.60 max | **ASME** SFA5.5 (E10016-D2) **AWS** A5.5 (E10016-D2) |
| **W10018** | Low Alloy Steel Arc Welding Electrode (E10018-D2) | **C** 0.15 max **Mn** 1.65-2.00 **Mo** 0.25-0.45 **P** 0.03 max **S** 0.03 max **Si** 0.80 max | **ASME** SFA5.5 (E10018-D2) **AWS** A5.5 (E10018-D2) |
| * **W10235** | Low Alloy Steel Flux Cored Wire (E100T5-D2) | **C** 0.15 max **Mn** 1.65-2.25 **Mo** 0.25-0.55 **P** 0.03 max **S** 0.03 max **Si** 0.80 max | **ASME** SFA5.29 (E100T5-D2) **AWS** A5.29 (E100T5-D2) |
| **W17010** | Low Alloy Steel Arc Welding Electrode (E7010-A1) | **C** 0.12 max **Mn** 0.60 max **Mo** 0.40-0.65 **P** 0.03 max **S** 0.03 max **Si** 0.40 max | **ASME** SFA5.5 (E7010-A1) **AWS** A5.5 (E7010-A1) |
| **W17011** | Low Alloy Steel Arc Welding Electrode (E7011-A1) | **C** 0.12 max **Mn** 0.60 max **Mo** 0.40-0.65 **P** 0.03 max **S** 0.03 max **Si** 0.40 max | **ASME** SFA5.5 (E7011-A1) **AWS** A5.5 (E7011-A1) |
| **W17015** | Low Alloy Steel Arc Welding Electrode (E7015-A1) | **C** 0.12 max **Mn** 0.90 max **Mo** 0.40-0.65 **P** 0.03 max **S** 0.03 max **Si** 0.60 max | **ASME** SFA5.5 (E7015-A1) **AWS** A5.5 (E7015-A1) |
| **W17016** | Low Alloy Steel Arc Welding Electrode (E7016-A1) | **C** 0.12 max **Mn** 0.90 max **Mo** 0.40-0.65 **P** 0.03 max **S** 0.03 max **Si** 0.60 mas | **ASME** SFA5.5 (E7016-A1) **AWS** A5.5 (E7016-A1) |
| **W17018** | Low Alloy Steel Arc Welding Electrode (E7018-A1) | **C** 0.12 max **Mn** 0.90 max **Mo** 0.40-0.65 **P** 0.03 max **S** 0.03 max **Si** 0.80 max | **ASME** SFA5.5 (E7018-A1) **AWS** A5.5 (E7018-A1) |
| **W17020** | Low Alloy Steel Arc Welding Electrode (E7020-A1) | **C** 0.12 max **Mn** 0.60 max **Mo** 0.40-0.65 **P** 0.03 max **S** 0.03 max **Si** 0.40 max | **ASME** SFA5.5 (E7020-A1) **AWS** A5.5 (E7020-A1) |
| **W17027** | Low Alloy Steel Arc Welding Electrode (E7027-A1) | **C** 0.12 max **Mn** 1.00 max **Mo** 0.40-0.65 **P** 0.03 max **S** 0.03 max **Si** 0.40 max | **ASME** SFA5.5 (E7027-A1) **AWS** A5.5 (E7027-A1) |
| * **W17030** | Low Alloy Steel Composite Wire for Gas Shielded Arc Welding (E80C-A1) | **C** 0.10 max **Cu** 0.35 max **Mn** 1.25 max **Mo** 0.40-0.70 **P** 0.03 max **S** 0.03 max **Si** 0.90 max | **ASME** SFA5.28 (E80C-A1) **AWS** A5.28 (E80C-A1) |
| **W17031** | Low Alloy Steel Flux Cored Wire (E8XT1-A1) | **C** 0.12 max **Mn** 1.25 max **Mo** 0.40-0.65 **P** 0.03 max **S** 0.03 max **Si** 0.80 max | **ASME** SFA5.29 (E8XT1-A1) **AWS** A5.29 (E8XT1-A1) |
| **W17035** | Low Alloy Steel Flux Cored Wire (E7XT5-A1) | **C** 0.12 max **Mn** 1.25 max **Mo** 0.40-0.65 **P** 0.03 max **S** 0.03 max **Si** 0.80 max | **ASME** SFA5.29 (E7XT5-A1) **AWS** A5.29 (E7XT5-A1) |
| **W17041** | Low Alloy Steel Composite Wire for Submerged Arc Welding (ECA1) | **C** 0.12 max **Cu** 0.35 max **Mn** 1.00 max **Mo** 0.40-0.65 **P** 0.030 max **S** 0.030 max **Si** 0.80 max | **ASME** SFA5.23 (ECA1) **AWS** A5.23 (ECA1) |
| **W17042** | Low Alloy Steel Composite Wire for Submerged Arc Welding (ECA2) | **C** 0.12 max **Cu** 0.35 max **Mn** 1.40 max **Mo** 0.40-0.65 **P** 0.030 max **S** 0.030 max **Si** 0.80 max | **ASME** SFA5.23 (ECA2) **AWS** A5.23 (ECA2) |
| **W17043** | Low Alloy Steel Composite Wire for Submerged Arc Welding (ECA3) | **C** 0.15 max **Cu** 0.35 max **Mn** 2.10 max **Mo** 0.40-0.65 **P** 0.030 max **S** 0.030 max **Si** 0.80 max | **ASME** SFA5.23 (ECA3) **AWS** A5.23 (ECA3) |
| **W17044** | Low Alloy Steel Composite Wire for Submerged Arc Welding (ECA4) | **C** 0.15 max **Cu** 0.35 max **Mn** 1.60 max **Mo** 0.40-0.65 **P** 0.030 max **S** 0.030 max **Si** 0.80 max | **ASME** SFA5.23 (ECA4) **AWS** A5.23 (ECA4) |
| **W17110** | Low Alloy Steel Arc Welding Electrode for Pipelines (E7010-P1) | **C** 0.20 max **Cr** 0.30 max **Mn** 1.20 max **Mo** 0.50 max **Ni** 1.00 max **P** 0.03 max **S** 0.03 max **Si** 0.60 max **V** 0.10 max | **ASME** SFA5.5 (E7010-P1) **AWS** A5.5 (E7010-P1) |

*Boxed entries are no longer active and are retained for reference purposes only.

The chemical compositions listed are for identification purposes and should not be used in lieu of the cross referenced specifications.

| UNIFIED NUMBER | DESCRIPTION | CHEMICAL COMPOSITION | CROSS REFERENCE SPECIFICATIONS |
|---|---|---|---|
| * **W17141** | Low Alloy Steel Composite Wire for Submerged Arc Welding (ECA1N) Nuclear Grade | **C** 0.12 max **Cu** 0.08 max **Mn** 1.00 max **Mo** 0.40-0.65 **P** 0.012 max **S** 0.040 max **Si** 0.80 max **V** 0.05 max | **ASME** SFA5.23 (ECA1N) **AWS** A5.23 (ECA1N) |
| * **W17142** | Low Alloy Steel Composite Wire for Submerged Arc Welding (ECA2N) Nuclear Grade | **C** 0.12 max **Cu** 0.08 max **Mn** 1.40 max **Mo** 0.40-0.65 **P** 0.012 max **S** 0.040 max **Si** 0.80 max **V** 0.05 max | **ASME** SFA5.23 (ECA2N) **AWS** A5.23 (ECA2N) |
| * **W17143** | Low Alloy Steel Composite Wire for Submerged Arc Welding (ECA3N) Nuclear Grade | **C** 0.15 max **Cu** 0.08 max **Mn** 2.10 max **Mo** 0.40-0.65 **P** 0.012 max **S** 0.040 max **Si** 0.80 max **V** 0.05 max | **ASME** SFA5.23 (ECA3N) **AWS** A5.23 (ECA3N) |
| * **W17144** | Low Alloy Steel Composite Wire for Submerged Arc Welding (ECA4N) Nuclear Grade | **C** 0.15 max **Cu** 0.08 max **Mn** 1.60 max **Mo** 0.40-0.65 **P** 0.012 max **S** 0.040 max **Si** 0.80 max **V** 0.05 max | **ASME** SFA5.23 (ECA4N) **AWS** A5.23 (ECA4N) |
| **W18016** | Low Alloy Steel Arc Welding Electrode (E8016-D3) | **C** 0.12 max **Mn** 1.00-1.80 **Mo** 0.40-0.65 **P** 0.03 max **S** 0.03 max **Si** 0.60 max | **ASME** SFA5.5 (E8016-D3) **AWS** A5.5 (E8016-D3) |
| **W18018** | Low Alloy Steel Arc Welding Electrode (E8018-D3) | **C** 0.12 max **Mn** 1.00-1.80 **Mo** 0.40-0.65 **P** 0.03 max **S** 0.03 max **Si** 0.80 max | **ASME** SFA5.5 (E8018-D3) **AWS** A5.5 (E8018-D3) |
| **W18110** | Low Alloy Steel Arc Welding Electrode for Pipelines (E8010-P1) | **C** 0.20 max **Cr** 0.30 max **Mn** 1.20 max **Mo** 0.50 max **Ni** 1.00 max **P** 0.03 max **S** 0.03 max **Si** 0.60 max **V** 0.10 max | **ASME** SFA5.5 (E8010-P1) **AWS** A5.5 (E8010-P1) |
| **W18118** | Low Alloy Steel Arc Welding Electrode (E8018-D1) | **C** 0.12 max **Mn** 1.00-1.75 **Mo** 0.25-0.45 **Ni** 0.90 max **P** 0.03 max **S** 0.03 max **Si** 0.80 max | **ASME** SFA5.5 (E8018-D1) **AWS** A5.5 (E8018-D1) |
| **W19015** | Low Alloy Steel Arc Welding Electrode (E9015-D1) | **C** 0.12 max **Mn** 1.00-1.75 **Mo** 0.25-0.45 **Ni** 0.90 max **P** 0.03 max **S** 0.03 max **Si** 0.60 max | **ASME** SFA5.5 (E9015-D1) **AWS** A5.5 (E9015-D1) |
| **W19016** | Low Alloy Steel Arc Welding Electrode (9016A) | **C** 0.12 max **Cr** 0.15 max **Mn** 1.25-1.75 **Mo** 0.30-0.60 **Ni** 0.10 max **P** 0.03 max **S** 0.04 max **Si** 0.05-0.60 **V** 0.05 max | |
| **W19018** | Low Alloy Steel Arc Welding Electrode (E9018-D1) | **C** 0.12 max **Mn** 1.00-1.75 **Mo** 0.25-0.45 **Ni** 0.90 max **P** 0.03 max **S** 0.03 max **Si** 0.80 max | **ASME** SFA5.5 (E9018-D1) **AWS** A5.5 (E9018-D1) |
| **W19118** | Low Alloy Steel Arc Welding Electrode (E9018-D3) | **C** 0.12 max **Mn** 1.00-1.80 **Mo** 0.40-0.65 **Ni** 0.90 max **P** 0.03 max **S** 0.03 max **Si** 0.80 max | **ASME** SFA5.5 (E9018-D3) **AWS** A5.5 (E9018-D3) |
| **W19131** | Low Alloy Steel Flux Cored Wire (E9XT1-D1) | **C** 0.12 max **Mn** 1.25-2.00 **Mo** 0.25-0.55 **P** 0.03 max **S** 0.03 max **Si** 0.80 max | **ASME** SFA5.29 (E9XT1-D1) **AWS** A5.29 (E9XT1-D1) |
| **W19230** | Low Alloy Steel Composite Wire for Gas Shielded Arc Welding (E90C-D2) | **C** 0.12 max **Mn** 1.00-1.90 **Mo** 0.40-0.60 **P** 0.025 max **S** 0.030 max **Si** 0.90 max | **ASME** SFA5.28 (E90C-D2) **AWS** A5.28 (E90C-D2) |
| **W19235** | Low Alloy Steel Flux Cored Wire (E9XT5-D2) | **C** 0.15 max **Mn** 1.65-2.25 **Mo** 0.25-0.55 **P** 0.03 max **S** 0.03 max **Si** 0.80 max | **ASME** SFA5.29 (E9XT5-D2, E10XT5-D2) **AWS** A5.29 (E9XT5-D2, E10XT5-D2) |
| **W19331** | Low Alloy Steel Flux Cored Wire (E9XT1-D3) | **C** 0.12 max **Mn** 1.00-1.75 **Mo** 0.40-0.65 **P** 0.03 max **S** 0.03 max **Si** 0.80 max | **ASME** SFA5.29 (E9XT1-D3) **AWS** A5.29 (E9XT1-D3) |

*Boxed entries are no longer active and are retained for reference purposes only.

The chemical compositions listed are for identification purposes and should not be used in lieu of the cross referenced specifications.

| UNIFIED NUMBER | DESCRIPTION | CHEMICAL COMPOSITION | CROSS REFERENCE SPECIFICATIONS |
|---|---|---|---|
| **W20018** | Low Alloy Steel Arc Welding Electrode (E7018-W1) | **C** 0.12 max **Cr** 0.15-0.30 **Cu** 0.30-0.60 **Mn** 0.40-0.70 **Ni** 0.20-0.40 **P** 0.025 max **S** 0.025 max **Si** 0.40-0.70 **V** 0.08 max | **ASME** SFA5.5 (E7018-W1) **AWS** A5.5 (E7018-W1) |
| **W20118** | Low Alloy Steel Arc Welding Electrode (E8018-W2) | **C** 0.12 max **Cr** 0.45-0.70 **Cu** 0.30-0.75 **Mn** 0.50-1.30 **Ni** 0.40-0.80 **P** 0.03 max **S** 0.03 max **Si** 0.35-0.80 | **ASME** SFA5.5 (E8018-W2) **AWS** A5.5 (E8018-W2) |
| **W20131** | Low Alloy Steel Flux Cored Wire (E8XT1-W2) | **C** 0.12 max **Cr** 0.45-0.70 **Cu** 0.30-0.75 **Mn** 0.50-1.3 **Ni** 0.40-0.80 **P** 0.03 max **S** 0.03 max **Si** 0.30-0.80 | **ASME** SFA5.26 (EGXXT-W); SFA5.29 (E8XT1-W2) **AWS** A5.26 (EGXXT-W); A5.29 (E8XT1-W2) |
| **W20140** | Low Alloy Steel Composite Wire for Submerged Arc and Electroslag Welding (ECW) (EWT2) | **C** 0.12 max **Cr** 0.45-0.70 **Cu** 0.30-0.75 **Mn** 0.50-1.60 **Ni** 0.40-0.80 **P** 0.035 max **S** 0.030 max **Si** 0.80 max | **ASME** SFA5.23 (ECW); SFA5.25 (EWT2) **AWS** A5.23 (ECW); A5.25 (EWT2) |
| **W20240** | Low Alloy Steel Composite Wire for Submerged Arc Welding (ECF2) | **C** 0.17 max **Cu** 0.35 max **Mn** 1.25-2.25 **Mo** 0.40-0.65 **Ni** 0.40-0.80 **P** 0.030 max **S** 0.030 max **Si** 0.80 max | **ASME** SFA5.23 (ECF2) **AWS** A5.23 (ECF2) |
| * **W20241** | Low Alloy Steel Composite Wire for Submerged Arc Welding (ECF2N) Nuclear Grade | **C** 0.17 max **Cu** 0.08 max **Mn** 1.25-2.25 **Mo** 0.40-0.65 **Ni** 0.40-0.80 **P** 0.012 max **S** 0.040 max **Si** 0.80 max **V** 0.05 max | **ASME** SFA5.23 (ECF2N) **AWS** A5.23 (ECF2N) |
| **W20440** | Low Alloy Steel Composite Wire for Submerged Arc Welding (ECF4) | **C** 0.17 max **Cr** 0.60 max **Cu** 0.35 max **Mn** 1.60 max **Mo** 0.25 max **Ni** 0.40-0.80 **P** 0.030 max **S** 0.035 max **Si** 0.80 max | **ASME** SFA5.23 (ECF4) **AWS** A5.23 (ECF4) |
| * **W20441** | Low Alloy Steel Composite Wire for Submerged Arc Welding (ECF4N) Nuclear Grade | **C** 0.17 max **Cr** 0.60 max **Cu** 0.08 max **Mn** 1.60 max **Mo** 0.25 max **Ni** 0.40-0.80 **P** 0.012 max **S** 0.040 max **Si** 0.80 max **V** 0.05 max | **ASME** SFA5.23 (ECF4N) **AWS** A5.23 (ECF4N) |
| **W20918** | Low Alloy Steel Arc Welding Electrode (E7018-C3L) | **C** 0.08 max **Cr** 0.15 max **Mn** 0.40-1.40 **Mo** 0.35 max **Ni** 0.80-1.10 **P** 0.03 max **S** 0.03 max **Si** 0.50 max **V** 0.05 max | **ASME** SFA5.5 (E7018-C3L) **AWS** A5.5 (E7018-C3L) |
| * **W21015** | Low Alloy Arc Welding Electrode (8015) | **C** 0.12 max **Cr** 0.15 max **Mn** 0.40-1.10 **Mo** 0.35 max **Ni** 0.80-1.10 **P** 0.03 max **S** 0.04 max **Si** 0.05-.60 **V** 0.05 max | |
| **W21016** | Low Alloy Arc Welding Electrode (E8016-C3) | **C** 0.12 max **Cr** 0.15 max **Mn** 0.40-1.25 **Mo** 0.35 max **Ni** 0.80-1.10 **P** 0.03 max **S** 0.03 max **Si** 0.80 max **V** 0.05 max | **ASME** SFA5.5 (E8016-C3) **AWS** A5.5 (E8016-C3) |
| **W21018** | Low Alloy Steel Arc Welding Electrode (E8018-C3) | **C** 0.12 max **Cr** 0.15 max **Mn** 0.40-1.25 **Mo** 0.35 max **Ni** 0.80-1.10 **P** 0.03 max **S** 0.03 max **Si** 0.80 max **V** 0.05 max | **ASME** SFA5.5 (E8018-C3) **AWS** A5.5 (E8018-C3) **MIL SPEC** MIL-E-22200/1 (MIL-8018-C3, MIL-8018-C3SR) |
| **W21030** | Low Alloy Steel Metal Cored Wire for Gas Shielded Arc Welding (E80C-Ni1) | **C** 0.12 max **Cu** 0.35 max **Mn** 1.50 max **Mo** 0.30 max **Ni** 0.80-1.10 **P** 0.025 max **S** 0.030 max **Si** 0.90 max **V** 0.05 max | **ASME** SFA5.28 (E80C-Ni1) **AWS** A5.28 (E80C-Ni1) |
| **W21031** | Low Alloy Steel Flux Cored Wire (E8XT1-Ni1) | **Al** 1.8 max **C** 0.12 max **Cr** 0.15 max **Mn** 1.50 max **Mo** 0.35 max **Ni** 0.80-1.10 **P** 0.03 max **Si** 0.80 max **V** 0.05 max | **ASME** SFA5.29 (EXXT1-Ni1) **AWS** A5.29 (EXXT1-Ni1) **MIL SPEC** MIL-E-24403/1 (MIL-8XT1-Ni1) |

*Boxed entries are no longer active and are retained for reference purposes only.

The chemical compositions listed are for identification purposes and should not be used in lieu of the cross referenced specifications.

| UNIFIED NUMBER | DESCRIPTION | CHEMICAL COMPOSITION | CROSS REFERENCE SPECIFICATIONS |
|---|---|---|---|
| **W21033** | Low Alloy Steel Flux Cored Wire (EGXXT-Ni1) | **C** 0.10 max **Mn** 1.0-1.8 **Mo** 0.30 max **Ni** 0.70-1.10 **P** 0.03 max **S** 0.03 max **Si** 0.50 max | **ASME** SFA5.26 (EGXXT-Ni1) **AWS** A5.26 (EGXXT-Ni1) |
| **W21035** | Low Alloy Steel Flux Cored Wire (E8XT5-Ni1) | **Al** 1.8 max **C** 0.12 max **Cr** 0.15 max **Mn** 1.50 max **Mo** 0.35 max **Ni** 0.80-1.10 **P** 0.03 max **Si** 0.80 max **V** 0.05 max | **ASME** SFA5.29 (E8XT5-Ni1) **AWS** A5.29 (E8XT5-Ni1) |
| **W21036** | Low Alloy Steel Flux Cored Wire (E7XT6-Ni1) | **Al** 1.8 max **C** 0.12 max **Cr** 0.15 max **Mn** 1.50 max **Mo** 0.35 max **Ni** 0.80-1.10 **P** 0.03 max **Si** 0.80 max **S** 0.03 max **V** 0.05 max | **ASME** SFA5.29 (E7XT6-Ni1) **AWS** A5.29 (E7XT6-Ni1) |
| **W21038** | Low Alloy Steel Flux Cored Wire (E7XT8-Ni1) | **Al** 1.8 max **C** 0.12 max **Cr** 0.15 max **Mn** 1.50 max **Mo** 0.35 max **Ni** 0.75-1.10 **P** 0.03 max **S** 0.03 max **Si** 0.80 max **V** 0.05 max | **ASME** SFA5.29 (E7XT8-Ni1) **AWS** A5.29 (E7XT8-Ni1) **MIL SPEC** MIL-E-24403/1 (MIL-71T8-Ni1) |
| ***W21040** | Low Alloy Steel Composite Wire for Submerged Arc Welding (ECNi1) | **C** 0.12 max **Cr** 0.15 max **Cu** 0.35 max **Mn** 1.60 max **Mo** 0.35 max **Ni** 0.75-1.10 **P** 0.030 max **S** 0.025 max **Si** 0.80 max | **ASME** SFA5.23 (ECNi1) **AWS** A5.23 (ECNi1) |
| ***W21041** | Low Alloy Steel Composite Wire for Submerged Arc Welding (ECNi1N) Nuclear Grade | **C** 0.12 max **Cr** 0.15 max **Cu** 0.08 max **Mn** 1.60 max **Mo** 0.35 max **Ni** 0.75-1.10 **P** 0.012 max **S** 0.025 max **Si** 0.80 max **V** 0.05 max | **ASME** SFA5.23 (ECNi1N) **AWS** A5.23 (ECNi1N) |
| **W21042** | Low Alloy Steel Composite Electrode for Submerged Arc Welding (ECNi5) | **C** 0.12 max **Cu** 0.35 max **Mn** 1.60 max **Mo** 0.10-0.35 **Ni** 0.70-1.10 **P** 0.030 max **Si** 0.80 max **S** 0.025 max | **ASME** SFA5.23 (ECNi5) **AWS** A5.23 (ECNi5) |
| **W21045** | Low Alloys Steel Flux Cored Wire (EXXT5-K6, -K6M) | **Al** 1.8 max **C** 0.15 max **Cr** 0.20 max **Mn** 0.50-1.50 **Mo** 0.15 max **Ni** 0.40-1.00 **P** 0.03 max **Si** 0.80 max **S** 0.03 max **V** 0.05 max | **ASME** SFA5.29 (EXXT5-K6, -K6M) **AWS** A5.29 (EXXT5-K6, K6M) |
| **W21048** | Low Alloy Steel Flux Cored Wire (EXXT8-K6) | **Al** 1.8 max **C** 0.15 max **Cr** 0.20 max **Mn** 0.50-1.50 **Mo** 0.15 max **Ni** 0.40-1.10 **P** 0.03 max **S** 0.03 max **Si** 0.80 max **V** 0.05 max | **ASME** SFA5.29 (EXXT8-K6) **AWS** A5.29 (EXXT8-K6) **MIL SPEC** MIL-E-24403/1 (MIL-71T8-K6) |
| **W21118** | Low Alloy Steel Arc Welding Electrode (E8018-NM1) | **Al** 0.05 max **C** 0.10 max **Cr** 0.10 max **Cu** 0.10 max **Mn** 0.80-1.25 **Mo** 0.40-0.65 **Ni** 0.80-1.10 **P** 0.02 max **S** 0.02 max **Si** 0.60 max **V** 0.02 max | **ASME** SFA5.5 (E8018-NM1) **AWS** A5.5 (E8018-NM1) |
| **W21135** | Low Alloy Steel Flux Cored Wire (E8XT5-K1) | **C** 0.15 **Cr** 0.15 **Mn** 0.80-1.40 **Mo** 0.20-0.65 **Ni** 0.80-1.10 **P** 0.03 **S** 0.03 **Si** 0.80 **V** 0.05 | **ASME** SFA5.29 (E8XT5-K1) **AWS** A5.29 (E8XT5-K1) |
| **W21140** | Low Alloy Steel Composite Wire for Submerged Arc Welding (ECF3) | **C** 0.17 max **Cu** 0.35 max **Mn** 1.25-2.25 **Mo** 0.40-0.65 **Ni** 0.70-1.10 **P** 0.030 max **S** 0.030 max **Si** 0.80 max | **ASME** SFA5.23 (ECF3) **AWS** A5.23 (ECF3) |
| ***W21141** | Low Alloy Steel Composite Wire for Submerged Arc Welding (ECF3N) Nuclear Grade | **C** 0.17 max **Cu** 0.08 max **Mn** 1.25-2.25 **Mo** 0.40-0.65 **Ni** 0.70-1.10 **P** 0.012 max **S** 0.030 max **Si** 0.80 max **V** 0.05 max | **ASME** SFA5.23 (ECF3N) **AWS** A5.23 (ECF3N) |
| **W21150** | Low Alloy Steel Composite Wire for Submerged Arc Welding (ECF1) | **C** 0.12 max **Cr** 0.15 max **Cu** 0.35 max **Mn** 0.70-1.50 **Mo** 0.55 max **Ni** 0.90-1.70 **P** 0.030 max **S** 0.030 max **Si** 0.80 max | **ASME** SFA5.23 (ECF1) **AWS** A5.23 (ECF1) |

*Boxed entries are no longer active and are retained for reference purposes only.

The chemical compositions listed are for identification purposes and should not be used in lieu of the cross referenced specifications.

| UNIFIED NUMBER | DESCRIPTION | CHEMICAL COMPOSITION | CROSS REFERENCE SPECIFICATIONS |
|---|---|---|---|
| * **W21151** | Low Alloy Steel Composite Wire for Submerged Arc Welding (ECF1N) Nuclear Grade | **C** 0.12 max **Cr** 0.15 max **Cu** 0.08 max **Mn** 0.70-1.50 **Mo** 0.55 max **Ni** 0.90-1.70 **P** 0.012 max **S** 0.040 max **Si** 0.80 max **V** 0.05 max | **ASME** SFA5.23 (ECF1N) **AWS** A5.23 (ECF1N) |
| * **W21215** | Low Alloy Steel Arc Welding Electrode (9015) (10015) | **C** 0.10 max **Cr** 0.15 max **Mn** 0.70-1.10 **Mo** 0.35 max **Ni** 1.40-1.80 **P** 0.025 max **S** 0.030 max **Si** 0.05-0.60 **V** 0.20 max | |
| * **W21216** | Low Alloy Steel Arc Welding Electrode (9015) (10016) | **C** 0.10 max **Cr** 0.15 max **Mn** 0.70-1.10 **Mo** 0.35 max **Ni** 1.40-1.80 **P** 0.025 **S** 0.030 **Si** 0.05-0.60 **V** 0.20 max | **ASME** SFA5.5 (E10016M) **AWS** A5.5 (E10016M) |
| **W21218** | Low Alloy Steel Arc Welding Electrode (E9018M) | **C** 0.10 max **Cr** 0.15 max **Mn** 0.60-1.25 **Mo** 0.35 max **Ni** 1.40-1.80 **P** 0.030 max **S** 0.030 max **Si** 0.80 max **V** 0.05 max | **ASME** SFA5.5 (E9018M) **AWS** A5.5 (E9018M) **MIL SPEC** MIL-E-22200/1 (MIL-9018-M) |
| **W21231** | Low Alloy Steel Flux Cored Wire (EXXT1-K2) | **C** 0.15 max **Cr** 0.15 max **Mn** 0.50-1.75 **Mo** 0.35 max **Ni** 1.00-2.00 **P** 0.03 max **S** 0.03 max **Si** 0.80 max **V** 0.05 max | **ASME** SFA5.29 (EXXT1-K2) **AWS** A5.29 (EXXT1-K2) |
| **W21234** | Low Alloy Steel Flux Cored Wire (E7XT4-K2) | **Al** 1.8 max **C** 0.15 max **Cr** 0.15 max **Mn** 0.50-1.75 **Mo** 0.35 max **Ni** 1.00-2.00 **P** 0.03 max **S** 0.03 max **Si** 0.80 max **V** 0.05 max | **ASME** SFA5.29 (E7XT4-K2) **AWS** A5.29 (E7XT4-K2) |
| **W21235** | Low Alloy Steel Flux Cored Wire (EXXT5-K2) | **C** 0.15 max **Cr** 0.15 max **Mn** 0.50-1.75 **Mo** 0.35 max **Ni** 1.00-2.00 **P** 0.03 max **S** 0.03 max **Si** 0.80 max **V** 0.05 max | **ASME** SFA5.29 (EXXT5-K2) **AWS** A5.29 (EXXT5-K2) |
| **W21237** | Low Alloy Steel Flux Cored Wire (E7XT7-K2) | **Al** 1.8 max **C** 0.15 max **Cr** 0.15 max **Mn** 0.50-1.75 **Mo** 0.35 max **Ni** 1.00-2.00 **P** 0.03 max **Si** 0.80 max **S** 0.03 max **V** 0.05 max | **ASME** SFA5.29 (E7XT7-K2) **AWS** A5.29 (E7XT7-K2) |
| **W21238** | Low Alloy Flux Cored Wire (E7XT8-K2) | **Al** 1.8 max **C** 0.15 max **Cr** 0.15 max **Mn** 0.50-1.75 **Mo** 0.35 max **Ni** 1.00-2.00 **P** 0.03 max **S** 0.03 max **Si** 0.80 max **V** 0.05 max | **ASME** SFA5.29 (E7XT8-K2) **AWS** A5.29 (E7XT8-K2) |
| **W21239** | Low Alloy Steel Flux Cored Wire (E7XT11-K2) | **Al** 1.8 max **C** 0.15 max **Cr** 0.15 max **Mn** 0.50-1.75 **Mo** 0.35 max **Ni** 1.00-2.00 **P** 0.03 max **Si** 0.80 max **S** 0.03 max **V** 0.05 max | **ASME** SFA5.29 (E7XT11-K2) **AWS** A5.29 (E7XT11-K2) |
| **W21240** | Low Alloy Steel Composite Wire for Submerged Arc Welding (ECM1) | **C** 0.10 max **Cr** 0.15 max **Cu** 0.30 max **Mn** 0.60-1.60 **Mo** 0.35 max **Ni** 1.25-2.00 **P** 0.030 max **S** 0.030 max **Si** 0.80 max | **ASME** SFA5.23 (ECM1) **AWS** A5.23 (ECM1) |
| * **W21241** | Low Alloy Steel Composite Wire for Submerged Arc Welding (ECM1N) Nuclear Grade | **C** 0.10 **Cr** 0.15 **Cu** 0.08 **Mn** 0.60-1.60 **Mo** 0.35 **Ni** 1.25-2.00 **P** 0.012 **S** 0.040 **Si** 0.80 **V** 0.05 | **ASME** SFA5.23 (ECM1N) **AWS** A5.23 (ECM1N) |
| **W21250** | Low Alloy Steel Composite Wire for Submerged Arc Welding (ECNi4) | **C** 0.14 max **Cu** 0.35 max **Mn** 1.60 max **Mo** 0.10-0.35 max **Ni** 1.40-2.10 **P** 0.030 max **S** 0.025 max **Si** 0.80 max | **ASME** SFA5.23 (ECNi4) **AWS** A5.23 (ECNi4) |

*Boxed entries are no longer active and are retained for reference purposes only.

The chemical compositions listed are for identification purposes and should not be used in lieu of the cross referenced specifications.

| UNIFIED NUMBER | DESCRIPTION | CHEMICAL COMPOSITION | CROSS REFERENCE SPECIFICATIONS |
|---|---|---|---|
| * **W21251** | Low Alloy Steel Composite Wire for Submerged Arc Welding (ECNi4N) Nuclear Grade | **C** 0.14 max **Cu** 0.08 max **Mn** 1.60 max **Mo** 0.35 max **Ni** 1.40-2.10 **P** 0.012 max **S** 0.030 max **Si** 0.80 max **V** 0.05 max | **ASME** SFA5.23 (ECNi4N) **AWS** A5.23 (ECNi4N) |
| **W21318** | Low Alloy Steel Arc Welding Electrode (E10018M) | **C** 0.10 max **Cr** 0.35 max **Mn** 0.75-1.70 **Mo** 0.25-0.50 **Ni** 1.40-2.10 **P** 0.030 max **S** 0.030 max **Si** 0.60 max **V** 0.05 max | **ASME** SFA5.5 (E10018M) **AWS** A5.5 (E10018M) **MIL SPEC** MIL-E-22200/1 (MIL-10018-M) |
| **W21328** | Low Alloy Steel Arc Welding Electrode (10018-M1) | **C** 0.10 max **Cr** 0.40 max **Mn** 0.80-1.85 **Mo** 0.50 max **Ni** 1.25-3.00 **P** 0.025 max **S** 0.017 max **Si** 0.65 max **V** 0.05 max | **MIL SPEC** MIL-E-22200/10 (MIL-10018-M1) |
| **W21331** | Low Alloy Steel Flux Cored Wire (EXXXT1-K3) | **C** 0.15 max **Cr** 0.15 max **Mn** 0.75-2.25 **Mo** 0.25-0.65 **Ni** 1.25-2.60 **P** 0.03 max **S** 0.03 max **Si** 0.80 max **V** 0.05 max | **ASME** SFA5.29 (EXXXT1-K3) **AWS** A5.29 (EXXXT1-K3) |
| **W21335** | Low Alloy Steel Flux Cored Wire (EXXXT5-K3) | **C** 0.15 max **Cr** 0.15 max **Mn** 0.75-2.25 **Mo** 0.25-0.65 **Ni** 1.25-2.60 **P** 0.03 max **S** 0.03 max **Si** 0.80 max **V** 0.05 max | **ASME** SFA5.29 (EXXXT5-K3) **AWS** A5.29 (EXXXT5-K3) |
| **W21340** | Low Alloy Steel Composite Wire for Submerged Arc Welding (ECM2) | **C** 0.10 max **Cr** 0.35 max **Cu** 0.30 max **Mn** 0.90-1.80 **Mo** 0.25-0.65 **Ni** 1.40-2.10 **P** 0.020 max **S** 0.020 max **Si** 0.80 max | **ASME** SFA5.23 (ECM2) **AWS** A5.23 (ECM2) |
| * **W21341** | Low Alloy Steel Composite Wire for Submerged Arc Welding (ECM2N) Nuclear Grade | **C** 0.10 max **Cr** 0.35 max **Cu** 0.08 max **Mn** 0.90-1.80 **Mo** 0.25-0.65 **Ni** 1.40-2.10 **P** 0.012 max **S** 0.040 max **Si** 0.80 max **V** 0.05 max | **ASME** SFA5.23 (ECM2N) **AWS** A5.23 (ECM2N) |
| **W21345** | Low Alloy Steel Composite Electrode for Submerged Arc Welding (ECM5) | **C** 0.12 max **Cr** 0.40 max **Cu** 0.30 max **Mn** 1.60-2.50 **Mo** 0.20-0.50 **Ni** 1.40-2.10 **P** 0.015 max **Si** 0.50 max **S** 0.015 max **Ti** 0.03 max **V** 0.02 max **Zr** 0.02 max | **ASME** SFA5.23 (ECM5) **AWS** A5.23 (ECM5) |
| **W21346** | Low Alloy Steel Composite Electrode for Submerged Arc Welding (ECM6) | **C** 0.12 max **Cr** 0.40 max **Cu** 0.30 max **Mn** 1.60-2.50 **Mo** 0.70-1.00 **Ni** 1.40-2.10 **P** 0.015 max **Si** 0.50 max **S** 0.015 max **Ti** 0.03 max **V** 0.02 max **Zr** 0.02 max | **ASME** SFA5.23 (ECM6) **AWS** A5.23 (ECM6) |
| * **W21380** | Low Alloy Steel Solid Wire for Submerged Arc Welding (MI-88) | **Al** 0.10 max **C** 0.06 max **Cr** 0.10-0.30 **Cu** 0.10-0.30 **Mn** 1.00-1.50 **Mo** 0.20-0.40 **Ni** 1.40-1.90 **P** 0.010 max **S** 0.010 max **Si** 0.50 max **Ti** 0.10 max **V** 0.05 max **Zr** 0.10 max | |
| **W21418** | Low Alloy Steel Arc Welding Electrode (E11018M) | **C** 0.10 max **Cr** 0.40 max **Mn** 1.30-1.80 **Mo** 0.25-0.50 **Ni** 1.25-2.50 **P** 0.030 max **S** 0.030 max **Si** 0.60 max **V** 0.05 max | **ASME** SFA5.5 (E11018M) **AWS** A5.5 (E11018M) **MIL SPEC** MIL-E-22200/1 (MIL-11018-M) |
| **W21438** | Low Alloy Steel Flux Cored Wire (E9XT8-K8) | **Al** 1.8 max **C** 0.15 max **Cr** 0.20 max **Mn** 1.00-2.00 **Mo** 0.20 max **Ni** 0.50-1.50 **P** 0.03 max **Si** 0.40 max **S** 0.03 max **V** 0.05 max | **ASME** SFA5.29 (E9XT8-K8) **AWS** A5.29 (E9XT8-K8) |
| **W21631** | Low Alloy Steel Flux Cored Wire (E12XT1-K5) | **C** 0.10-0.25 **Cr** 0.20-0.70 **Mn** 0.60-1.60 **Mo** 0.15-0.55 **Ni** 0.75-2.00 **P** 0.03 max **S** 0.03 max **Si** 0.80 max **V** 0.05 max | **ASME** SFA5.29 (E12XT1-K5) **AWS** A5.29 (E12XT1-K5) |

*Boxed entries are no longer active and are retained for reference purposes only.

The chemical compositions listed are for identification purposes and should not be used in lieu of the cross referenced specifications.

| UNIFIED NUMBER | DESCRIPTION | CHEMICAL COMPOSITION | CROSS REFERENCE SPECIFICATIONS |
|---|---|---|---|
| * **W21780** | Low Alloy Steel Solid Wire for Submerged Arc Welding (EB82) | **C** 0.12 max **Cr** 0.30 max **Cu** 0.40-1.10 **Mn** 0.80-1.25 **Mo** 0.15-0.60 **Ni** 0.80-1.25 **P** 0.020 max **S** 0.020 max **Si** 0.80 max **V** 0.05 max | |
| **W21916** | Low Alloy Steel Arc Welding Electrode (E8016-C4) | **C** 0.10 max **Mn** 1.25 max **Ni** 1.10-2.00 **P** 0.03 max **S** 0.03 max **Si** 0.60 max | **ASME** SFA5.5 (E8016-C4) **AWS** A5.5 (E8016-C4) |
| **W21918** | Low Alloy Steel Arc Welding Electrode (E8018-C4) | **C** 0.10 max **Mn** 1.25 max **Ni** 1.10-2.00 **P** 0.03 max **S** 0.03 max **Si** 0.80 max | **ASME** SFA5.5 (E8018-C4) **AWS** A5.5 (E8018-C4) |
| **W22016** | Low Alloy Steel Arc Welding Electrode (E8016-C1) | **C** 0.12 max **Mn** 1.25 **Ni** 2.00-2.75 **P** 0.03 max **S** 0.03 max **Si** 0.60 max | **ASME** SFA5.5 (E8016-C1) **AWS** A5.5 (E8016-C1) |
| **W22018** | Low Alloy Steel Arc Welding Electrode (E8018-C1) | **C** 0.12 max **Mn** 1.25 max **Ni** 2.00-2.75 **P** 0.03 max **S** 0.03 max **Si** 0.80 max | **ASME** SFA5.5 (E8018-C1) **AWS** A5.5 (E8018-C1) |
| **W22030** | Low Alloy Steel Metal Cored Wire for Gas Shielded Arc Welding (E70C-Ni2) (E80C-Ni2) | **C** 0.08 max **Cu** 0.35 max **Mn** 1.25 max **Ni** 1.75-2.75 **P** 0.025 max **S** 0.030 max **Si** 0.90 max | **ASME** SFA5.28 (E70C-Ni2, E80C-Ni2) **AWS** A5.28 (E70C-Ni2, E80C-Ni2) |
| **W22031** | Low Alloy Steel Flux Cored Wire (EXXT1-Ni2) | **C** 0.12 max **Mn** 1.50 max **Ni** 1.75-2.75 **P** 0.03 max **S** 0.03 max **Si** 0.80 max | **ASME** SFA5.29 (EXXT1-Ni2) **AWS** A5.29 (EXXT1-Ni2) **MIL SPEC** MIL-E-24403/1 (MIL-8XT8-Ni2) |
| **W22035** | Low Alloy Steel Flux Cored Wire (E8XT5-Ni2) | **C** 0.12 max **Mn** 1.50 max **Ni** 1.75-2.75 **P** 0.03 max **S** 0.03 max **Si** 0.80 max | **ASME** SFA5.29 (E8XT5-Ni2) **AWS** A5.29 (E8XT5-Ni2) |
| **W22038** | Low Alloy Steel Flux Cored Wire (E7XT8-Ni2, E8XT8-Ni2) | **Al** 1.8 max **C** 0.12 max **Mn** 1.50 max **Ni** 1.75-2.75 **P** 0.03 max **S** 0.03 max **Si** 0.80 max | **ASME** SFA5.29 (E7XT8-Ni2, E8XT8-Ni2) **AWS** A5.29 (E7XT8-Ni2, E8XT8-Ni2) |
| **W22040** | Low Alloy Steel Composite Wire for Submerged Arc Welding (ECNi2) | **C** 0.12 max **Cu** 0.35 max **Mn** 1.60 max **Ni** 2.00-2.90 **P** 0.030 max **S** 0.025 max **Si** 0.80 max | **ASME** SFA5.23 (ECNi2) **AWS** A5.23 (ECNi2) |
| * **W22041** | Low Alloy Steel Composite Wire for Submerged Arc Welding (ECNi2N) Nuclear Grade | **C** 0.12 max **Cu** 0.08 max **Mn** 1.60 max **Ni** 2.00-2.90 **P** 0.012 max **S** 0.030 max **Si** 0.80 max **V** 0.05 max | **ASME** SFA5.23 (ECNi2N) **AWS** A5.23 (ECNi2N) |
| **W22051** | Low Alloy Steel Flux Cored Wire (E10XT1-K7) | **C** 0.15 max **Mn** 1.00-1.75 **Ni** 2.00-2.75 **P** 0.03 max **S** 0.03 max **Si** 0.80 max | **ASME** SFA5.29 (E10XT1-K7) **AWS** A5.29 (E10XT1-K7) |
| **W22115** | Low Alloy Steel Arc Welding Electrode (E7015-C1L) | **C** 0.05 max **Mn** 1.25 max **Ni** 2.00-2.75 **P** 0.03 max **S** 0.03 max **Si** 0.50 max | **ASME** SFA5.5 (E7015-C1L) **AWS** A5.5 (E7015-C1L) |
| **W22116** | Low Alloy Steel Arc Welding Electrode (E7016-C1L) | **C** 0.05 max **Mn** 1.25 max **Ni** 2.00-2.75 **P** 0.03 max **S** 0.03 max **Si** 0.50 max | **ASME** SFA5.5 (E7016-C1L) **AWS** A5.5 (E7016-C1L) |
| **W22118** | Low Alloy Steel Arc Welding Electrode (E7018-C1L) | **C** 0.05 max **Mn** 1.25 max **Ni** 2.00-2.75 **P** 0.03 max **S** 0.03 max **Si** 0.50 max | **ASME** SFA5.5 (E7018-C1L) **AWS** A5.5 (E7018-C1L) |
| **W22218** | Low Alloy Steel Arc Welding Electrode (E12018M) | **C** 0.10 max **Cr** 0.30-1.50 **Mn** 1.30-2.25 **Mo** 0.30-0.55 **Ni** 1.75-2.50 **P** 0.030 max **S** 0.030 max **Si** 0.60 max **V** 0.05 max | **ASME** SFA5.5 (E12018M) **AWS** A5.5 (E12018M) |
| **W22230** | Low Alloy Steel Flux Cored Wire for Welding HY 100 | **C** 0.10 max **Cr** 0.60 max **Cu** 0.02 max **Mn** 1.20-1.80 **Mo** 0.25-0.50 **Ni** 2.00-2.50 **P** 0.010 max **S** 0.010 max **Si** 0.60 max **V** 0.05 max | **MIL SPEC** MIL-E-24403/2 (MIL-110 TC) |

*Boxed entries are no longer active and are retained for reference purposes only.

The chemical compositions listed are for identification purposes and should not be used in lieu of the cross referenced specifications.

| UNIFIED NUMBER | DESCRIPTION | CHEMICAL COMPOSITION | CROSS REFERENCE SPECIFICATIONS |
|---|---|---|---|
| **W22231** | Low Alloy Steel Flux Cored Wire (E11XT1-K4) | **C** 0.15 max **Cr** 0.20-0.60 **Mn** 1.20-2.25 **Mo** 0.20-0.65 **Ni** 1.75-2.60 **P** 0.03 max **S** 0.03 max **Si** 0.80 max **V** 0.03 max | **ASME** SFA5.29 (E11XT1-K4) **AWS** A5.29 (E11XT1-K4) |
| **W22235** | Low Alloy Steel Flux Cored Wire (E1XXT5-K4) | **C** 0.15 max **Cr** 0.20-0.60 **Mn** 1.20-2.25 **Mo** 0.20-0.65 **Ni** 1.75-2.60 **P** 0.03 max **S** 0.03 max **Si** 0.80 max **V** 0.03 max | **ASME** SFA5.29 (E1XXT5-K4) **AWS** A5.29 (E1XXT5-K4) |
| **W22240** | Low Alloy Steel Composite Wire for Submerged Arc Welding (ECM3) | **C** 0.10 max **Cr** 0.65 max **Cu** 0.30 max **Mn** 0.90-1.80 **Mo** 0.20-0.70 **Ni** 1.80-2.60 **P** 0.020 max **S** 0.020 max **Si** 0.80 max | **ASME** SFA5.23 (ECM3) **AWS** A5.23 (ECM3) |
| * **W22241** | Low Alloy Steel Composite Wire for Submerged Arc Welding (ECM3N) Nuclear Grade | **C** 0.10 max **Cr** 0.65 max **Cu** 0.08 max **Mn** 0.90-1.80 **Mo** 0.20-0.70 **Ni** 1.80-2.60 **P** 0.012 max **S** 0.030 max **Si** 0.80 max **V** 0.05 max | **ASME** SFA5.23 (ECM3N) **AWS** A5.23 (ECM3N) |
| **W22333** | Low Alloy Steel Flux Cored Wire (EGXXT-NM2) | **C** 0.12 max **Cr** 0.20 max **Cu** 0.035 max **Mn** 1.1-2.1 **Mo** 0.10-0.35 **Ni** 1.1-2.0 **P** 0.03 max **S** 0.03 max **Si** 0.20-0.60 **V** 0.05 max | **ASME** SFA5.26 (EGXXT-NM2) **AWS** A5.26 (EGXXT-NM2) |
| **W22334** | Low Alloy Steel Flux Cored Wire (EGXXT-NM1) | **C** 0.12 max **Cr** 0.20 max **Mn** 1.0-2.0 **Mo** 0.40-0.65 **Ni** 1.5-2.0 **P** 0.02 max **S** 0.03 max **Si** 0.15-0.50 **V** 0.05 max | **ASME** SFA5.26 (EGXXT-NM1) **AWS** A5.26 (EGXXT-NM1) |
| **W22340** | Low Alloy Steel Composite Wire for Electroslag Welding (EWT3) | **C** 0.12 max **Cr** 0.20 max **Mn** 1.00-2.00 **Mo** 0.40-0.65 **Ni** 1.50-2.50 **P** 0.02 max **S** 0.03 max **Si** 0.15-0.50 **V** 0.05 max | **ASME** SFA5.25 (EWT3) **AWS** A5.25 (EWT3) |
| **W22440** | Low Alloy Steel Tubular Wire for Submerged Arc Welding (ECM4) | **C** 0.10 max **Cr** 0.80 max **Cu** 0.30 max **Mn** 1.30-2.25 **Mo** 0.30-0.80 **Ni** 2.00-2.80 **P** 0.020 max **S** 0.020 max **Si** 0.80 max | **ASME** SFA5.23 (ECM4) **AWS** A5.23 (ECM4) |
| * **W22441** | Low Alloy Steel Composite Wire for Submerged Arc Welding (ECM4N) Nuclear Grade | **C** 0.10 max **Cr** 0.80 max **Cu** 0.08 max **Mn** 1.30-2.25 **Mo** 0.30-0.80 **Ni** 2.00-2.80 **P** 0.012 max **S** 0.030 max **Si** 0.80 max **V** 0.05 max | **ASME** SFA5.23 (ECM4N) **AWS** A5.23 (ECM4N) |
| **W22540** | Low Alloy Steel Composite Wire for Submerged Arc Welding (ECF5) | **C** 0.17 max **Cr** 0.65 max **Cu** 0.50 max **Mn** 1.20-1.80 **Mo** 0.30-0.80 **Ni** 2.00-2.80 **P** 0.020 max **S** 0.020 max **Si** 0.80 max | **ASME** SFA5.23 (ECF5) **AWS** A5.23 (ECF5) |
| **W22640** | Low Alloy Steel Composite Wire for Submerged Arc Welding (ECF6) | **C** 0.14 max **Cr** 0.65 max **Cu** 0.40 max **Mn** 0.80-1.85 **Mo** 0.60 max **Ni** 1.50-2.25 **P** 0.030 max **S** 0.020 max **Si** 0.80 max | **ASME** SFA5.23 (ECF6) **AWS** A5.23 (ECF6) |
| * **W22641** | Low Alloy Steel Composite Wire for Submerged Arc Welding (ECF6N) Nuclear Grade | **C** 0.14 max **Cr** 0.65 max **Cu** 0.08 max **Mn** 0.80-1.85 **Mo** 0.60 max **Ni** 1.50-2.25 **P** 0.012 max **S** 0.030 max **Si** 0.80 max **V** 0.05 max | **ASME** SFA5.23 (ECF6N) **AWS** A5.23 (ECF6N) |
| **W22718** | Low Alloy Steel Arc Welding Electrode (10018-N1) | **C** 0.15 max **Cr** 0.90-1.20 **Mn** 0.80-1.15 **Mo** 0.45-0.75 **Ni** 1.50-2.00 **P** 0.030 max **S** 0.030 max **Si** 0.30-0.60 **V** 0.02 max | **MIL SPEC** MIL-E-22200/5 (MIL-10018-N1) |
| **W22815** | Low Alloy Steel Arc Welding Electrode (11015) (12015) | **C** 0.10 max **Cr** 1.00-1.40 **Mn** 1.00-1.50 **Mo** 0.20-0.40 **Ni** 1.75-2.25 **P** 0.03 max **S** 0.04 max **Si** 0.05-0.60 **V** 0.20 max | |

*Boxed entries are no longer active and are retained for reference purposes only.

The chemical compositions listed are for identification purposes and should not be used in lieu of the cross referenced specifications.

| UNIFIED NUMBER | DESCRIPTION | CHEMICAL COMPOSITION | CROSS REFERENCE SPECIFICATIONS |
|---|---|---|---|
| **W22816** | Low Alloy Steel Arc Welding Electrode (11016) (12016) | **C** 0.10 max **Cr** 1.00-1.40 **Mn** 1.00-1.50 **Mo** 0.20-0.40 **Ni** 1.75-2.25 **P** 0.03 max **S** 0.04 max **Si** 0.05-0.60 **V** 0.20 max | |
| **W23016** | Low Alloy Steel Arc Welding Electrode (E8016-C2) | **C** 0.12 max **Mn** 1.25 max **Ni** 3.00-3.75 **P** 0.03 max **S** 0.03 max **Si** 0.60 max | **ASME** SFA5.5 (E8016-C2) **AWS** A5.5 (E8016-C2) |
| **W23018** | Low Alloy Steel Arc Welding Electrode (W23018) | **C** 0.12 max **Mn** 1.25 max **Ni** 3.00-3.75 **P** 0.03 max **S** 0.03 max **Si** 0.80 max | **ASME** SFA5.5 (E8018-C2) **AWS** A5.5 (E8018-C2) |
| **W23030** | Low Alloy Steel Metal Cored Wire for Gas Shielded Arc Welding (E80C-Ni3) | **C** 0.12 max **Cu** 0.35 max **Mn** 1.50 max **Ni** 2.75-3.75 **P** 0.025 max **S** 0.030 max **Si** 0.90 max | **ASME** SFA5.28 (E80C-Ni3) **AWS** A5.28 (E80C-Ni3) |
| * **W23031** | Low Alloy Steel Flux Cored Wire (EXXT1-Ni3) | **Al** 1.8 max **C** 0.12 max **Mn** 1.50 max **Ni** 2.75-3.75 **P** 0.03 max **Si** 0.30 max **S** 0.03 max | **ASME** SFA5.29 (EXXT1-Ni3) **AWS** A5.29 (EXXT1-Ni3) |
| **W23035** | Low Alloy Steel Flux Cored Wire (EXXT5-Ni3) | **C** 0.12 max **Mn** 1.50 max **Ni** 2.75-3.75 **P** 0.03 max **S** 0.03 max **Si** 0.80 max | **ASME** SFA5.29 (EXXT5-Ni3) **AWS** A5.29 (EXXT5-Ni3) |
| **W23039** | Low Alloy Steel Flux Cored Wire (EXXT11-Ni3) | **Al** 1.8 max **C** 0.12 max **Mn** 1.50 max **Ni** 2.75-3.75 **P** 0.03 max **Si** 0.80 max **S** 0.03 max | **ASME** SFA5.29 (EXXT11-Ni3) **AWS** A5.29 (EXXT11-Ni3) |
| **W23040** | Low Alloy Steel Composite Wire for Submerged Arc Welding (ECNi3) | **C** 0.12 max **Cr** 0.15 max **Cu** 0.35 max **Mn** 1.60 max **Ni** 2.80-3.80 **P** 0.030 max **S** 0.025 max **Si** 0.80 max | **ASME** SFA5.23 (ECNi3) **AWS** A5.23 (ECNi3) |
| * **W23041** | Low Alloy Steel Composite Wire for Submerged Arc Welding (ECNi3N) Nuclear Grade | **C** 0.12 max **Cr** 0.15 max **Cu** 0.08 max **Mn** 1.60 max **Ni** 2.80-3.80 **P** 0.012 max **S** 0.030 max **Si** 0.80 max **V** 0.05 max | **ASME** SFA5.23 (ECNi3N) **AWS** A5.23 (ECNi3N) |
| **W23115** | Low Alloy Arc Welding Electrode (E7015-C2L) | **C** 0.05 max **Mn** 1.25 max **Ni** 3.00-3.75 **P** 0.03 max **S** 0.03 max **Si** 0.50 max | **ASME** SFA5.5 (E7015-C2L) **AWS** A5.5 (E7015-C2L) |
| **W23116** | Low Alloy Steel Arc Welding Electrode (E7016-C2L) | **C** 0.05 max **Mn** 1.25 max **Ni** 3.00-3.75 **P** 0.03 max **S** 0.03 max **Si** 0.50 max | **ASME** SFA5.5 (E7016-C2L) **AWS** A5.5 (E7016-C2L) |
| **W23118** | Low Alloy Steel Arc Welding Electrode (E7018-C2L) | **C** 0.05 max **Mn** 1.25 max **Ni** 3.00-3.75 **P** 0.03 max **S** 0.03 max **Si** 0.50 max | **ASME** SFA5.5 (E7018-C2L) **AWS** A5.5 (E7018-C2L) |
| **W23218** | Low Alloy Steel Arc Welding Electrode (E12018M1) | **C** 0.10 max **Cr** 0.65 max **Mn** 0.80-1.60 **Mo** 0.20-0.30 **Ni** 3.00-3.80 **P** 0.015 max **S** 0.012 max **Si** 0.65 max **V** 0.05 max | **ASME** SFA5.5 (E12018M1) **AWS** A5.5 (E12018M1) |
| **W23228** | Low Alloy Steel Arc Welding Electrode (12018-M2) | **C** 0.10 max **Cr** 0.65 max **Mn** 0.80-1.85 **Mo** 0.90 max **Ni** 1.50-4.00 **P** 0.025 max **S** 0.012 max **Si** 0.65 max **V** 0.05 max | **MIL SPEC** MIL-E-22200/10 (MIL-12018-M2) |
| **W23230** | Low Alloy Steel Flux Cored Wire for Welding HY 80 (E10XT1-K9, E10XT1-K9M) | **C** 0.10 max **Cr** 0.20 max **Cu** 0.06 max **Mn** 0.50-1.50 **Mo** 0.50 max **Ni** 1.30-1.75 **P** 0.015 max **S** 0.015 max **Si** 0.60 max **V** 0.05 max | **ASME** SFA5.29 (E10XT1-K9, E10XT1-K9M) **AWS** A5.29 (E10XT1-K9, E10XT1-K9M) **MIL SPEC** MIL-E-24403/2 (MIL-10XTX) |
| **W23318** | Low Alloy Steel Arc Welding Electrode (14018-M1) | **C** 0.10 max **Cr** 0.35-1.20 **Mn** 0.75-1.35 **Mo** 0.30-1.10 **Ni** 3.10-3.90 **P** 0.013 max **S** 0.010 max **Si** 0.65 max **V** 0.09 max | **MIL SPEC** MIL-E-22200/9 (MIL-14018-M1) |

*Boxed entries are no longer active and are retained for reference purposes only.

The chemical compositions listed are for identification purposes and should not be used in lieu of the cross referenced specifications.

| UNIFIED NUMBER | DESCRIPTION | CHEMICAL COMPOSITION | CROSS REFERENCE SPECIFICATIONS |
|---|---|---|---|
| **W24230** | Low Alloy Flux Cored Wire for Welding HY 100 | **C** 0.10 max **Cr** 0.55 max **Cu** 0.06 max **Mn** 0.60-1.80 **Mo** 0.70 max **Ni** 1.75-4.00 **P** 0.015 max **S** 0.012 max **Si** 0.60 max **V** 0.05 max | **MIL SPEC** MIL-E-24403/2 (MIL-12XTX) |
| **W25018** | Nickel Alloy Steel Arc Welding Electrode (E9015-C5L) | **C** 0.05 max **Mn** 0.40-1.00 **Ni** 6.00-7.25 **P** 0.03 max **S** 0.03 max **Si** 0.50 max | **ASME** SFA5.5 (E9015-C5L) **AWS** A5.5 (E9015-C5L) |
| **W26018** | Nickel Alloy Steel Arc Welding Electrode (14018-HT) | **C** 0.10 max **Cr** 0.05-0.60 **Mn** 0.40-0.80 **Mo** 0.40-0.70 **Ni** 7.00-9.00 **P** 0.013 max **S** 0.010 max **Si** 0.50 max **V** 0.04-0.10 | **MIL SPEC** MIL-E-22200/11 (MIL-14018-HT) |
| * **W30440** | Austenitic Cr-Ni Stainless Steel Composite Welding Filler Metal (EC19-10H) | **C** 0.04-0.08 **Cb** 0.05 max **Cr** 18.5-20.0 **Cu** 0.75 max **Mn** 1.0-2.0 **Mo** 0.25 max **Ni** 9.0-11.0 **P** 0.03 max **S** 0.03 max **Si** 0.30-0.60 **Ti** 0.05 max | **ASME** SFA5.9 (EC19-10H) **AWS** A5.9 (EC19-10H) |
| **W30710** | Austenitic Cr-Ni-Mn Stainless Steel Arc Welding Electrode (E307-XX) | **C** 0.04-0.14 **Cr** 18.0-21.5 **Cu** 0.75 max **Mn** 3.30-4.75 **Mo** 0.5-1.5 **Ni** 9.0-10.7 **P** 0.04 max **S** 0.03 max **Si** 0.90 max | **ASME** SFA5.4 (E307-XX) **AWS** A5.4 (E307-XX) |
| **W30731** | Austenitic Cr-Ni-Mn Stainless Steel Flux Cored Wire (E307TX-X) | **C** 0.13 max **Cr** 18.0-20.5 **Cu** 0.5 max **Mn** 3.30-4.75 **Mo** 0.5-1.5 **Ni** 9.0-10.5 **P** 0.04 max **S** 0.03 max **Si** 1.0 max | **ASME** SFA5.22 (E307TX-X) **AWS** A5.22 (E307TX-X) |
| **W30733** | Austenitic Cr-Ni-Mn Stainless Steel Flux Cored Wire (E307T0-3) | **C** 0.13 max **Cr** 19.5-22.0 **Cu** 0.5 max **Mn** 3.30-4.75 **Mo** 0.5-1.5 **Ni** 9.0-10.5 **P** 0.04 max **S** 0.03 max **Si** 1.0 max | **ASME** SFA5.22 (E307T0-3) **AWS** A5.22 (E307T0-3) |
| * **W30740** | Austenitic Cr-Ni-Mn Stainless Steel Composite Welding Filler Metal (EC307) | **C** 0.04-0.14 **Cr** 19.5-22.0 **Cu** 0.75 max **Mn** 3.3-4.75 **Mo** 0.5-1.5 **Ni** 8.0-10.7 **P** 0.03 max **S** 0.03 max **Si** 0.30-0.65 | **ASME** SFA5.9 (EC307) **AWS** A5.9 (EC307) |
| **W30810** | Austenitic Cr-Ni Stainless Steel Arc Welding Electrode (E308-XX) | **C** 0.08 max **Cr** 18.0-21.0 **Cu** 0.75 max **Mn** 0.5-2.5 **Mo** 0.75 max **Ni** 9.0-11.0 **P** 0.04 max **S** 0.03 max **Si** 0.90 max | **ASME** SFA5.4 (E308-XX) **AWS** A5.4 (E308-XX) **MIL SPEC** MIL-E-22200/2 (MIL-308, MIL-308Co) |
| **W30813** | Austenitic Cr-Ni Stainless Steel Arc Welding Electrode (E308L) | **C** 0.04 max **Cr** 18.0-21.0 **Cu** 0.75 max **Mn** 0.5-2.5 **Mo** 0.75 max **Ni** 9.0-11.0 **P** 0.04 max **S** 0.03 max **Si** 0.90 max | **ASME** SFA5.4 (E308L) **AWS** A5.4 (E308L) **MIL SPEC** MIL-E-22200/2 (MIL-308L, MIL-308CoL) |
| **W30815** | Austenitic Cr-Ni Stainless Steel Arc Welding Electrode (308HC) High Carbon | **C** 0.08-0.15 **Cr** 18.0-21.0 **Mn** 2.5 max **Ni** 9.0-11.0 **P** 0.03 max **S** 0.03 max **Si** 5\mtC max | **MIL SPEC** MIL-E-22200/2, MIL-308HC |
| **W30816** | Austenitic Cr-Ni-Si Heat Resisting Steel, Arc Welding Electrode (253MA) | **C** 0.08 max **Ce** 0.03-0.08 **Cr** 20.5-22.5 **Cu** 0.30 max **Mn** 0.5-1.5 **Mo** 0.5 max **Ni** 9.5-11.5 **N** 0.10-0.20 **P** 0.035 max **S** 0.030 max **Si** 1.0-2.0 | |
| **W30820** | Austenitic Cr-Ni-Mo Stainless Steel Arc Welding Electrode (E308Mo) | **C** 0.08 max **Cr** 18.0-21.0 **Cu** 0.75 max **Mn** 0.5-2.5 **Mo** 2.0-3.0 **Ni** 9.0-12.0 **P** 0.04 max **S** 0.03 max **Si** 0.90 max | **ASME** SFA5.4 (E308Mo-XX) **AWS** A5.4 (E308Mo-XX) |
| * **W30821** | Austenitic Cr-Ni-Mo Stainless Steel Arc Welding Electrode - for Armor | **C** 0.07-0.17 **Cr** 18.00-21.50 **Mn** 1.25-2.25 **Mo** 1.85-2.25 **Ni** 9.00-10.7 **P** 0.04 max **S** 0.03 max **Si** 0.80 max | |
| **W30823** | Austenitic Cr-Ni-Mo Stainless Steel Arc Welding Electrode (E308MoL-XX) | **C** 0.04 max **Cr** 18.0-21.0 **Cu** 0.75 max **Mn** 0.5-2.5 **Mo** 2.0-3.0 **Ni** 9.0-12.0 **P** 0.04 max **S** 0.03 max **Si** 0.90 max | **ASME** SFA5.4 (E308MoL-XX) **AWS** A5.4 (E308MoL-XX) |

*Boxed entries are no longer active and are retained for reference purposes only.

The chemical compositions listed are for identification purposes and should not be used in lieu of the cross referenced specifications.

| UNIFIED NUMBER | DESCRIPTION | CHEMICAL COMPOSITION | CROSS REFERENCE SPECIFICATIONS |
|---|---|---|---|
| **W30830** | Austenitic Cr-Ni-Mo Stainless Steel Flux Cored Wire | **C** 0.07-0.12 **Cr** 19.0-21.5 **Mn** 1.25-2.25 **Mo** 1.8-2.4 **Ni** 9.0-10.7 **P** 0.04 max **S** 0.03 max **Si** 0.25-0.80 | **ASME** SFA5.22 (E308HMoT0-3) **AWS** A5.22 (E308HMoT0-3) |
| **W30831** | Austenitic Cr-Ni Stainless Steel Flux Cored Wire (E308TX-X, E308HT0-3) | **C** 0.08 max **Cr** 18.0-21.0 **Cu** 0.5 max **Mn** 0.5-2.5 **Mo** 0.5 max **Ni** 9.0-11.0 **P** 0.04 max **S** 0.03 max **Si** 1.0 max | **ASME** SFA5.22 (E308TX-X, E308HT0-3) **AWS** A5.22 (E308TX-X, E308HT0-3) |
| **W30832** | Austenitic Cr-Ni-Mo Stainless Steel Flux Cored Wire (E308MoTX-X) | **C** 0.08 max **Cr** 18.0-21.0 **Cu** 0.5 max **Mn** 0.5-2.5 **Mo** 2.0-3.0 **Ni** 9.0-11.0 **P** 0.04 max **S** 0.03 max **Si** 1.0 max | **ASME** SFA5.22 (E308MoTX-X) **AWS** A5.22 (E308MoTX-X) |
| **W30833** | Austenitic Cr-Ni Stainless Steel Flux Cored Wire (E308T0-3, E308HT0-3) | **C** 0.08 max **Cr** 19.5-22.0 **Cu** 0.5 max **Mn** 0.5-2.5 **Mo** 0.5 max **Ni** 9.0-11.0 **P** 0.04 max **S** 0.03 max **Si** 1.0 max | **ASME** SFA5.22 (E308T0-3, E308HT0-3) **AWS** A5.22 (E308T0-3, E308HT0-3) |
| **W30835** | Austenitic Cr-Ni Stainless Steel Flux Cored Wire (E308LTX-X, R308LT1-5) | **C** 0.04 max **Cr** 18.0-21.0 **Cu** 0.5 max **Mn** 0.5-2.5 **Mo** 0.5 max **Ni** 9.0-11.0 **P** 0.04 max **S** 0.03 max **Si** 1.0 max | **ASME** SFA5.22 (E308LTX-X, R308LT1-5) **AWS** A5.22 (E308LTX-X, R308LT1-5) |
| **W30837** | Austenitic Cr-Ni Stainless Steel Flux Cored Wire (E308LT0-3) | **C** 0.03 max **Cr** 19.5-22.0 **Cu** 0.5 max **Mn** 0.5-2.5 **Mo** 0.5 max **Ni** 9.0-11.0 **P** 0.04 max **S** 0.03 max **Si** 1.0 max | **ASME** SFA5.22 (E308LT0-3) **AWS** A5.22 (E308LT0-3) |
| **W30838** | Austenitic Cr-Ni-Mo Stainless Steel Flux Cored Wire (E308LMoT0-X, E308LMoTX-X) | **C** 0.04 max **Cr** 18.0-21.0 **Cu** 0.5 max **Mn** 0.5-2.5 **Mo** 2.0-3.0 **Ni** 9.0-12.0 **P** 0.04 max **S** 0.03 max **Si** 1.0 max | **ASME** SFA5.22 (E308LMoT0-X, E308LMoTX-X) **AWS** A5.22 (E308LMoT0-X, E308LMoTX-X) |
| **W30839** | Austenitic Cr-Ni-Mo Stainless Steel Flux Cored Wire (E308MoT0-3) | **C** 0.08 max **Cr** 18.0-21.0 **Cu** 0.5 max **Mn** 0.5-2.5 **Mo** 2.0-3.0 **Ni** 9.0-11.0 **P** 0.04 max **S** 0.03 max **Si** 1.0 max | **ASME** SFA5.22 (E308MoT0-3) **AWS** A5.22 (E308MoT0-3) |
| **W30840** | Austenitic Cr-Ni Stainless Steel Composite Welding Filler Metal EC308 | **C** 0.08 max **Cr** 19.5-22.0 **Cu** 0.75 max **Mn** 1.0-2.5 **Mo** 0.75 max **Ni** 9.0-11.0 **P** 0.03 max **S** 0.03 max **Si** 0.30-0.65 | **ASME** SFA5.9 (EC308, EC308H) **AWS** A5.9 (EC308, EC308H) |
| **W30841** | Austenitic Cr-Ni Stainless Steel Composite Welding Filler Metal EC308Si | **C** 0.08 max **Cr** 19.5-22.0 **Cu** 0.5 max **Fe** rem **Mn** 1.0-2.5 **Mo** 0.5 max **Ni** 9.0-11.0 **P** 0.03 max **S** 0.03 max **Si** 0.65-1.00 | **ASME** SFA5.9 (EC308Si) **AWS** A5.9 (EC308Si) |
| **W30843** | Austenitic Cr-Ni Stainless Steel Composite Welding Filler Metal (EC308L) | **C** 0.03 max **Cr** 19.5-22.0 **Cu** 0.75 max **Mn** 1.0-2.5 **Mo** 0.75 max **Ni** 9.0-11.0 **P** 0.03 max **S** 0.03 max **Si** 0.30-0.65 | **ASME** SFA5.9 (EC308L) **AWS** A5.9 (EC308L) |
| **W30848** | Austenitic Cr-Ni Stainless Steel Composite Welding Filler Metal EC308LSi | **C** 0.03 max **Cr** 19.5-22.0 **Cu** 0.5 max **Fe** rem **Mn** 1.0-2.5 **Mo** 0.5 max **Ni** 9.0-11.0 **P** 0.03 max **S** 0.03 max **Si** 0.65-1.00 | **ASME** SFA5.9 (EC308LSi) **AWS** A5.9 (EC308LSi) |
| **W30850** | Austenitic Cr-Ni-Mo Stainless Steel Composite Welding Filler Metal (EC308Mo) | **C** 0.08 max **Cr** 18.0-21.0 **Cu** 0.75 max **Mn** 1.0-2.5 **Mo** 2.0-3.0 **Ni** 9.0-12.0 **P** 0.03 max **S** 0.03 max **Si** 0.30-0.65 | **ASME** SFA5.9 (EC308Mo) **AWS** A5.9 (EC308Mo) |
| **W30853** | Austenitic Cr-Ni-Mo Stainless Steel Composite Welding Filler Metal (EC308LMo) | **C** 0.04 max **Cr** 18.0-21.0 **Cu** 0.75 max **Mn** 1.0-2.5 **Mo** 2.0-3.0 **Ni** 9.0-12.0 **P** 0.03 max **S** 0.03 max **Si** 0.30-0.65 | **ASME** SFA5.9 (EC308LMo) **AWS** A5.9 (EC308LMo) |
| **W30910** | Austenitic Cr-Ni Stainless Steel Arc Welding Electrode (E309) | **C** 0.15 max **Cr** 22.0-25.0 **Cu** 0.75 max **Mn** 0.5-2.5 **Mo** 0.75 max **Ni** 12.0-14.0 **P** 0.04 max **S** 0.03 max **Si** 0.90 max | **ASME** SFA5.4 (E309) **AWS** A5.4 (E309) **MIL SPEC** MIL-E-22200/2 (MIL-309, MIL-309Co) |
| **W30913** | Austenitic Cr-Ni Stainless Steel Arc Welding Electrode (E309L) | **C** 0.04 max **Cr** 22.0-25.0 **Cu** 0.75 max **Mn** 0.5-2.5 **Mo** 0.75 max **Ni** 12.0-14.0 **P** 0.04 max **S** 0.03 max **Si** 0.90 max | **ASME** SFA5.4 (E309L) **AWS** A5.4 (E309L) **MIL SPEC** MIL-E-22200/2, (MIL-309L0, MIL-309CoL) |
| **W30917** | Austenitic Cr-Ni-Cb Stainless Steel Arc Welding Electrode (E309Cb) | **C** 0.12 max **Cb** (incl Ta) 0.70-1.00 **Cr** 22.0-25.0 **Cu** 0.75 max **Mn** 0.5-2.5 **Mo** 0.75 max **Ni** 12.0-14.0 **P** 0.04 max **S** 0.03 max **Si** 0.90 max | **ASME** SFA5.4 (E309Cb) **AWS** A5.4 (E309Cb) **MIL SPEC** MIL-E-22200/2 (MIL-309Cb) |

The chemical compositions listed are for identification purposes and should not be used in lieu of the cross referenced specifications.

| UNIFIED NUMBER | DESCRIPTION | CHEMICAL COMPOSITION | CROSS REFERENCE SPECIFICATIONS |
|---|---|---|---|
| **W30920** | Austenitic Cr-Ni-Mo Stainless Steel Arc Welding Electrode (E309Mo) | **C** 0.12 max **Cr** 22.0-25.0 **Cu** 0.75 max **Mn** 0.5-2.5 **Mo** 2.0-3.0 **Ni** 12.0-14.0 **P** 0.04 max **S** 0.03 max **Si** 0.90 max | **ASME** SFA5.4 (E309Mo) **AWS** A5.4 (E309Mo) |
| **W30923** | Austenitic Cr-Ni-Mo Stainless Steel Arc Welding Electrode (E309LMo) | **C** 0.04 max **Cr** 22.0-25.0 **Cu** 0.75 max **Mn** 0.5-2.5 **Mo** 2.0-3.0 **Ni** 12.0-14.0 **P** 0.04 max **S** 0.03 max **Si** 0.90 max | **ASME** SFA5.4 (E309LMo) **AWS** A5.4 (E309LMo) |
| **W30931** | Austenitic Cr-Ni Stainless Steel Flux Cored Wire (E309TX-X) | **C** 0.10 max **Cr** 22.0-25.0 **Cu** 0.5 max **Mn** 0.5-2.5 **Mo** 0.5 max **Ni** 12.0-14.0 **P** 0.04 max **S** 0.03 max **Si** 1.0 max | **ASME** SFA5.22 (E309TX-X) **AWS** A5.22 (E309TX-X) |
| **W30932** | Austenitic Cr-Ni-Cb Stainless Steel Flux Cored Wire (E309LCbTX-X) | **C** 0.04 max **Cb** (incl Ta) 0.70-1.00 **Cr** 22.0-25.0 **Cu** 0.5 max **Mn** 0.5-2.5 **Mo** 0.5 max **Ni** 12.0-14.0 **P** 0.04 max **S** 0.03 max **Si** 1.0 max | **ASME** SFA5.22 (E309LCbTX-X) **AWS** A5.22 (E309LCbTX-X) |
| **W30933** | Austenitic Cr-Ni Stainless Steel Flux Cored Wire (E309T0-3) | **C** 0.10 max **Cr** 23.0-25.5 **Cu** 0.5 max **Mn** 0.5-2.5 **Mo** 0.5 max **Ni** 12.0-14.0 **P** 0.04 max **S** 0.03 max **Si** 1.0 max | **ASME** SFA5.22 (E309T0-3) **AWS** A5.22 (E309T0-3) |
| **W30934** | Cr-Ni-Cb Stainless Steel Flux Cored Wire (E309LCbT0-3) | **C** 0.03 max **Cb** (incl Ta) 0.70-1.00 **Cr** 23.0-25.5 **Cu** 0.5 max **Mn** 0.5-2.5 **Mo** 0.5 max **Ni** 12.0-14.0 **P** 0.04 max **S** 0.03 max **Si** 1.0 max | **ASME** SFA5.22 (E309LCbT0-3) **AWS** A5.22 (E309LCbT0-3) |
| **W30935** | Austenitic Cr-Ni Stainless Steel Flux Cored Wire (E309LTX-X, R309LT1-5) | **C** 0.04 max **Cr** 22.0-25.0 **Cu** 0.5 max **Mn** 0.5-2.5 **Mo** 0.5 max **Ni** 12.0-14.0 **P** 0.04 max **S** 0.03 max **Si** 1.0 max | **ASME** SFA5.22 (E309LTX-X, R309LT1-5) **AWS** A5.22 (E309LTX-X, R309LT1-5) |
| **W30936** | Austenitic Cr-Ni Stainless Steel Flux Cored Wire | **C** 0.04 max **Cr** 20.5-23.5 **Cu** 0.5 max **Mn** 0.5-2.5 **Mo** 2.5-3.5 **Ni** 15.0-17.0 **P** 0.04 max **S** 0.03 max **Si** 1.0 max | **ASME** SFA5.22 (E309LNiMoTX-X) **AWS** A5.22 (E309LNiMoTX-X) |
| **W30937** | Austenitic Cr-Ni Stainless Steel Flux Cored Wire (E309LT0-3) | **C** 0.03 max **Cr** 23.0-25.5 **Cu** 0.5 max **Mn** 0.5-2.5 **Mo** 0.5 max **Ni** 12.0-14.0 **P** 0.04 max **S** 0.03 max **Si** 1.0 max | **ASME** SFA5.22 (E309LT0-3) **AWS** A5.22 (E309LT0-3) |
| **W30938** | Austenitic Cr-Ni-Mo Stainless Steel Flux Cored Wire (E309LMoTX-X) | **C** 0.04 max **Cr** 21.0-25.0 **Cu** 0.5 max **Mn** 0.5-2.5 **Mo** 2.0-3.0 **Ni** 12.0-16.0 **P** 0.04 max **S** 0.03 max **Si** 1.0 max | **ASME** SFA5.22 (E309LMoT0-3, E309LMoTX-X) **AWS** A5.22 (E309LMoT0-3, E309LMoTX-X) |
| **W30939** | Austenitic Cr-Ni-Mo Stainless Steel Flux Cored Wire (E309MoTX-X) | **C** 0.12 max **Cr** 21.0-25.0 **Cu** 0.5 max **Mn** 0.5-2.5 **Mo** 2.0-3.0 **Ni** 12.0-16.0 **P** 0.04 max **S** 0.03 max **Si** 1.0 max | **ASME** SFA5.22 (E309MoTX-X) **AWS** A5.22 (E309MoTX-X) |
| **W30940** | Austenitic Cr-Ni Stainless Steel Composite Welding Filler Metal (EC309) | **C** 0.12 max **Cr** 23.0-25.0 **Cu** 0.75 max **Mn** 1.0-2.5 **Mo** 0.75 max **Ni** 12.0-14.0 **P** 0.03 max **S** 0.03 max **Si** 0.30-0.65 | **ASME** SFA5.9 (EC309) **AWS** A5.9 (EC309) |
| **W30941** | Austenitic Cr-Ni Stainless Steel Composite Welding Filler Metal (EC309Si) | **C** 0.12 max **Cr** 23.0-25.0 **Cu** 0.5 max **Fe** rem **Mn** 1.0-2.5 **Mo** 0.5 max **Ni** 12.0-14.0 **P** 0.03 max **S** 0.03 max **Si** 0.65-1.00 | **ASME** SFA5.9 (EC309Si) **AWS** A5.9 (EC309Si) |
| **W30943** | Austenitic Cr-Ni Stainless Steel Composite Welding Filler Metal (EC309L) | **C** 0.03 max **Cr** 23.0-25.0 **Cu** 0.75 max **Mn** 1.0-2.5 **Mo** 0.75 max **Ni** 12.0-14.0 **P** 0.03 max **S** 0.03 max **Si** 0.30-0.65 | **ASME** SFA5.9 (EC309L) **AWS** A5.9 (EC309L) |
| **W30950** | Austenitic Cr-Ni Stainless Steel Composite Welding Filler Metal (EC309Mo) | **C** 0.12 max **Cr** 23.0-25.0 **Cu** 0.75 max **Mn** 1.0-2.5 **Mo** 2.0-3.0 **Ni** 12.0-14.0 **P** 0.03 max **S** 0.03 max **Si** 0.30-0.65 | **ASME** SFA5.9 (EC309Mo) **AWS** A5.9 (EC309Mo) |
| **W30953** | Austenitic Cr-Ni-Mo Stainless Steel Composite Welding Filler Metal (EC309LMo) | **C** 0.03 max **Cr** 23.0-25.0 **Cu** 0.75 max **Mn** 1.0-2.5 **Mo** 2.0-3.0 **Ni** 12.0-14.0 **P** 0.03 max **S** 0.03 max **Si** 0.30-0.65 | **ASME** SFA5.9 (EC309LMo) **AWS** A5.9 (EC309LMo) |
| **W31010** | Austenitic Cr-Ni Stainless Steel Arc Welding Electrode (E310-XX) | **C** 0.08-0.20 **Cr** 25.0-28.0 **Cu** 0.75 max **Mn** 1.0-2.5 **Mo** 0.75 max **Ni** 20.0-22.5 **P** 0.03 max **S** 0.03 max **Si** 0.75 max | **ASME** SFA5.4 (E310-XX) **AWS** A5.4 (E310-XX) **MIL SPEC** MIL-E-22200/2 (MIL-310) |

The chemical compositions listed are for identification purposes and should not be used in lieu of the cross referenced specifications.

| UNIFIED NUMBER | DESCRIPTION | CHEMICAL COMPOSITION | CROSS REFERENCE SPECIFICATIONS |
|---|---|---|---|
| **W31015** | Austenitic Cr-Ni Stainless Steel Arc Welding Electrode (E310H-XX) High Carbon | **C** 0.35-0.45 **Cr** 25.0-28.0 **Cu** 0.75 max **Mn** 1.0-2.5 **Mo** 0.75 max **Ni** 20.0-22.5 **P** 0.03 max **S** 0.03 max **Si** 0.75 max | **ASME** SFA5.4 (E310H-XX) **AWS** A5.4 (E310H-XX) |
| **W31017** | Austenitic Cr-Ni-Cb Stainless Steel Arc Welding Electrode (E310Cb-XX) | **C** 0.12 max **Cb** 0.70-1.00 **Cr** 25.0-28.0 **Cu** 0.75 max **Mn** 1.0-2.5 **Mo** 0.75 max **Ni** 20.0-22.0 **P** 0.03 max **S** 0.03 max **Si** 0.75 max | **ASME** SFA5.4 (E310Cb-XX) **AWS** A5.4 (E310Cb-XX) |
| **W31020** | Austenitic Cr-Ni-Mo Stainless Steel Arc Welding Electrode (E310Mo-XX) | **C** 0.12 max **Cr** 25.0-28.0 **Cu** 0.75 max **Mn** 1.0-2.5 **Mo** 2.0-3.0 **Ni** 20.0-22.0 **P** 0.03 max **S** 0.03 max **Si** 0.75 max | **ASME** SFA5.4 (E310Mo-XX) **AWS** A5.4 (E310Mo-XX) |
| **W31031** | Austenitic Cr-Ni Stainless Steel Flux Cored Wire (E310T0-3, E310TX-X) | **C** 0.20 max **Cr** 25.0-28.0 **Cu** 0.5 max **Mn** 1.0-2.5 **Mo** 0.5 max **Ni** 20.0-22.5 **P** 0.03 max **S** 0.03 max **Si** 1.0 max | **ASME** SFA5.22 (E310T0-3, E310TX-X) **AWS** A5.22 (E310T0-3, E310TX-X) |
| **W31040** | Austenitic Cr-Ni Stainless Steel Composite Welding Filler Metal (EC310) | **C** 0.08-0.15 **Cr** 25.0-28.0 **Cu** 0.75 max **Mn** 1.0-2.5 **Mo** 0.75 max **Ni** 20.0-22.5 **P** 0.03 max **S** 0.03 max **Si** 0.30-0.65 | **ASME** SFA5.9 (EC310) **AWS** A5.9 (EC310) |
| **W31231** | Austenitic Cr-Ni Stainless Steel Flux Cored Wire (E312T0-3) | **C** 0.15 max **Cr** 28.0-32.0 max **Cu** 0.5 max **Mn** 0.5-2.5 **Mo** 0.5 max **Ni** 8.0-10.5 **P** 0.04 max **S** 0.03 max **Si** 1.0 max | **ASME** SFA5.22 (E312T0-3) **AWS** A5.22 (E312T0-3) |
| **W31310** | Austenitic-Ferritic Cr-Ni Stainless Steel Arc Welding Electrode (E312-XX) | **C** 0.15 max **Cr** 28.0-32.0 **Cu** 0.75 max **Mn** 0.5-2.5 **Mo** 0.75 max **Ni** 8.0-10.5 **P** 0.04 max **S** 0.03 max **Si** 0.90 max | **ASME** SFA5.4 (E312-XX) **AWS** A5.4 (E312) **MIL SPEC** MIL-E-22200/2 (MIL-312) |
| **W31331** | Austenitic-Ferritic Cr-Ni Stainless Steel Arc Welding Electrode (E312TX-X) | **C** 0.15 max **Cr** 28.0-32.0 **Cu** 0.5 max **Mn** 0.5-2.5 **Mo** 0.5 max **Ni** 8.0-10.5 **P** 0.04 max **S** 0.03 max **Si** 1.0 max | **ASME** SFA5.22 (E312TX-X) **AWS** A5.22 (E312TX-X) |
| **W31340** | Austenitic Cr-Ni Stainless Steel Composite Welding Filler Metal (EC312) | **C** 0.15 max **Cr** 28.0-32.0 **Mn** 1.00-2.50 **Mo** 0.75 max **Ni** 8.00-10.5 **P** 0.03 max **S** 0.03 max **Si** 0.30-0.65 | **ASME** SFA5.9 (EC312) **AWS** A5.9 (EC312) |
| **W31610** | Austenitic Cr-Ni-Mo Stainless Steel Arc Welding Electrode (E316) | **C** 0.08 max **Cr** 17.0-20.0 **Cu** 0.75 max **Mn** 0.5-2.5 **Mo** 2.0-3.0 **Ni** 11.0-14.0 **P** 0.04 max **S** 0.03 max **Si** 0.90 max | **AMS** 5691 **ASME** SFA5.4 (E316, E316H) **AWS** A5.4 (E316, E316H) **MIL SPEC** MIL-E-22200/2 (MIL-316) |
| **W31613** | Austenitic Cr-Ni-Mo Stainless Steel Arc Welding Electrode (E316L-XX) | **C** 0.04 max **Cr** 17.0-20.0 **Cu** 0.75 max **Mn** 0.5-2.5 **Mo** 2.0-3.0 **Ni** 11.0-14.0 **P** 0.04 max **S** 0.03 max **Si** 0.90 max | **ASME** SFA5.4 (E316L-XX) **AWS** A5.4 (E316L-XX) **MIL SPEC** MIL-E-22200/2 (MIL-316L) |
| **W31630** | Austenitic Cr-Ni-Mo Stainless Steel Flux Cored Wire | **C** 0.04 max **Cr** 17.0-20.0 **Cu** 0.5 max **Mn** 0.5-2.5 **Mo** 2.0-3.0 **Ni** 11.0-14.0 **P** 0.04 max **S** 0.03 max **Si** 1.0 max | **ASME** SFA5.22 (E316LKT0-3) **AWS** A5.22 (E316LKT0-3) |
| **W31631** | Austenitic Cr-Ni-Mo Stainless Steel Flux Cored Wire (E316TX-X) | **C** 0.08 max **Cr** 17.0-20.0 **Cu** 0.5 max **Mn** 0.5-2.5 **Mo** 2.0-3.0 **Ni** 11.0-14.0 **P** 0.04 max **S** 0.03 max **Si** 1.0 max | **ASME** SFA5.22 (E316TX-X) **AWS** A5.22 (E316TX-X) |
| **W31633** | Austenitic Cr-Ni-Mo Stainless Steel Flux Cored Wire (E316T0-3) | **C** 0.08 max **Cr** 18.0-20.5 **Cu** 0.5 max **Mn** 0.5-2.5 **Mo** 2.0-3.0 **Ni** 11.0-14.0 **P** 0.04 max **S** 0.03 max **Si** 1.0 max | **ASME** SFA5.22 (E316T0-3) **AWS** A5.22 (E316T0-3) |
| **W31635** | Austenitic Cr-Ni-Mo Stainless Steel Flux Cored Wire (E316LTX-X, R316LT1-5) | **C** 0.03 max **Cr** 17.0-20.0 **Cu** 0.5 max **Mn** 0.5-2.5 **Mo** 2.0-3.0 **Ni** 11.0-14.0 **P** 0.04 max **S** 0.03 max **Si** 1.0 max | **ASME** SFA5.22 (E316LTX-X, R316LT1-5) **AWS** A5.22 (E316LTX-X, R316LT1-5) |
| **W31637** | Austenitic Cr-Ni-Mo Stainless Steel Flux Cored Wire (E316LT0-3) | **C** 0.03 max **Cr** 18.0-20.5 **Cu** 0.5 max **Mn** 0.5-2.5 **Mo** 2.0-3.0 **Ni** 11.0-14.0 **P** 0.04 max **S** 0.03 max **Si** 1.0 max | **ASME** SFA5.22 (E316LT0-3) **AWS** A5.22 (E316LT0-3) |
| **W31640** | Austenitic Cr-Ni-Mo Stainless Steel Composite Welding Filler Metal (EC316) | **C** 0.08 max **Cr** 18.0-20.0 **Cu** 0.75 max **Mn** 1.0-2.5 **Mo** 2.0-3.0 **Ni** 11.0-14.0 **P** 0.03 max **S** 0.03 max **Si** 0.30-0.65 | **ASME** SFA5.9 (EC316, EC316H) **AWS** A5.9 (EC316, EC316H) |

The chemical compositions listed are for identification purposes and should not be used in lieu of the cross referenced specifications.

| UNIFIED NUMBER | DESCRIPTION | CHEMICAL COMPOSITION | CROSS REFERENCE SPECIFICATIONS |
|---|---|---|---|
| **W31641** | Austenitic Cr-Ni Stainless Steel Composite Welding Filler Metal (EC316Si) | **C** 0.08 max **Cr** 18.0-20.0 **Cu** 0.5 max **Mn** 1.0-2.5 **Mo** 2.0-3.0 **Ni** 11.0-14.0 **P** 0.03 max **S** 0.03 max **Si** 0.65-1.00 | **ASME** SFA5.9 (EC316Si) **AWS** A5.9 (EC316Si) |
| **W31643** | Austenitic Cr-Ni-Mo Stainless Steel Composite Welding Filler Metal (EC316L) | **C** 0.03 max **Cr** 18.0-22.0 **Cu** 0.75 max **Mn** 1.0-2.5 **Mo** 2.0-3.0 **Ni** 11.0-14.0 **P** 0.03 max **S** 0.03 max **Si** 0.30-0.65 | **ASME** SFA5.9 (EC316L) **AWS** A5.9 (EC316L) |
| **W31648** | Austenitic Cr-Ni Stainless Steel Composite Welding Filler Metal (EC316LSi) | **C** 0.03 max **Cr** 18.0-20.0 **Cu** 0.5 max **Mn** 1.0-2.5 **Mo** 2.0-3.0 **Ni** 11.0-14.0 **P** 0.03 max **S** 0.03 max **Si** 0.65-1.00 | **ASME** SFA5.9 (EC316LSi) **AWS** A5.9 (EC316LSi) |
| **W31710** | Austenitic Cr-Ni-Mo Stainless Steel Arc Welding Electrode (E317-XX) | **C** 0.08 max **Cr** 18.0-21.0 **Cu** 0.75 max **Mn** 0.5-2.5 **Mo** 3.0-4.0 **Ni** 12.0-14.0 **P** 0.04 max **S** 0.03 max **Si** 0.90 max | **ASME** SFA5.4 (E317-XX) **AWS** A5.4 (E317-XX) **MIL SPEC** MIL-E-22200/2 (MIL-317) |
| **W31713** | Austenitic Cr-Ni-Mo Stainless Steel Arc Welding Electrode (E317L-XX) | **C** 0.04 max **Cr** 18.0-21.0 **Cu** 0.75 max **Mn** 0.5-2.5 **Mo** 3.0-4.0 **Ni** 12.0-14.0 **P** 0.04 max **S** 0.03 max **Si** 0.90 max | **ASME** SFA5.4 (E317L-XX) **AWS** A5.4 (E317L-XX) |
| **W31735** | Austenitic Cr-Ni-Mo Stainless Steel Flux Cored Wire (E317LTX-X) | **C** 0.04 max **Cr** 18.0-21.0 **Cu** 0.5 max **Mn** 0.5-2.5 **Mo** 3.0-4.0 **Ni** 12.0-14.0 **P** 0.04 max **S** 0.03 max **Si** 1.0 max | **ASME** SFA5.22 (E317LTX-X) **AWS** A5.22 (E317LTX-X) |
| **W31737** | Austenitic Cr-Ni-Mo Stainless Steel Flux Cored Wire (E317LT0-3) | **C** 0.03 max **Cr** 18.5-21.0 **Cu** 0.5 max **Mn** 0.5-2.5 **Mo** 3.0-4.0 **Ni** 13.5-15.0 **P** 0.04 max **S** 0.03 max **Si** 1.0 max | **ASME** SFA5.22 (E317LT0-3) **AWS** A5.22 (E317LT0-3) |
| **W31740** | Austenitic Cr-Ni-Mo Stainless Steel Composite Welding Filler Metal (EC317) | **C** 0.08 max **Cr** 18.5-20.5 **Cu** 0.75 max **Mn** 1.0-2.5 **Mo** 3.0-4.0 **Ni** 13.0-15.0 **P** 0.03 max **S** 0.03 max **Si** 0.30-0.65 | **ASME** SFA5.9 (EC317) **AWS** A5.9 (EC317) |
| **W31743** | Austenitic Cr-Ni-Mo Stainless Steel Composite Welding Filler Metal (EC317L) | **C** 0.03 max **Cr** 18.5-20.5 **Cu** 0.75 max **Mn** 1.0-2.5 **Mo** 3.0-4.0 **Ni** 13.0-15.0 **P** 0.04 max **S** 0.03 max **Si** 0.30-0.65 | **ASME** SFA5.9 (EC317L) **AWS** A5.9 (EC317L) |
| **W31810** | Replaced by W39209 | | |
| **W31831** | Replaced by W39239 | | |
| **W31910** | Austenitic Cr-Ni-Mo Stainless Steel Arc Welding Electrode (E318-XX) | **C** 0.08 max **Cb** 6\mtC-1.00 includes Ta **Cr** 17.0-20.0 **Cu** 0.75 max **Mn** 0.5-2.5 **Mo** 2.0-3.0 **Ni** 11.0-14.0 **P** 0.04 max **S** 0.03 max **Si** 0.90 max | **ASME** SFA5.4 (E318-XX) **AWS** A5.4 (E318-XX) **MIL SPEC** MIL-E-22200/2 (MIL-318) |
| **W31940** | Austenitic Cr-Ni-Mo-Cb Stainless Steel Composite Welding Filler Metal (EC318) | **C** 0.08 max **Cb** 8\mtC-1.0 includes Ta **Cr** 18.0-20.0 **Cu** 0.75 max **Mn** 1.0-2.5 **Mo** 2.0-3.0 **Ni** 11.0-14.0 **P** 0.03 max **S** 0.03 max **Si** 0.30-0.65 | **ASME** SFA5.9 (EC318) **AWS** A5.9 (EC318) |
| **W32140** | Austenitic Cr-Ni-Ti Stainless Steel Composite Welding Filler Metal (EC321) | **C** 0.08 max **Cr** 18.5-20.5 **Cu** 0.75 max **Mn** 1.0-2.5 **Mo** 0.75 max **Ni** 9.0-10.5 **P** 0.03 max **S** 0.03 max **Si** 0.30-0.65 **Ti** 9\mtC-1.0 | **ASME** SFA5.9 (EC321) **AWS** A5.9 (EC321) |
| **W32210** | Nitrogen Strengthened Cr-Ni-Mn-Mo Stainless Steel Arc Welding Electrode (E209-XX) | **C** 0.06 max **Cr** 20.5-24.0 **Cu** 0.75 max **Mn** 4.0-7.0 **Mo** 1.5-3.0 **N** 0.10-0.30 **Ni** 9.5-12.0 **P** 0.04 max **S** 0.03 max **Si** 0.90 max **V** 0.10-0.30 | **ASME** SFA5.4 (E209-XX) **AWS** A5.4 (E209-XX) |
| **W32240** | Nitrogen Strengthened Austenitic Cr-Ni-Mn-Mo Stainless Steel Composite Welding Filler Metal (22-10-5) | **C** 0.05 max **Cr** 20.5-24.0 **Cu** 0.75 max **Mn** 4.0-7.0 **Mo** 1.5-3.0 **N** 0.10-0.30 **Ni** 9.5-12.0 **P** 0.03 max **S** 0.03 max **Si** 0.90 max | **ASME** SFA5.9 (EC209) **AWS** A5.9 (EC209) |
| **W32310** | Nitrogen Strengthened Austenitic Cr-Ni-Mn-Mo Stainless Steel Arc Welding Electrode (E219-XX) | **C** 0.06 max **Cr** 19.0-21.5 **Cu** 0.75 max **Mn** 8.0-10.0 **Mo** 0.75 max **N** 0.10-0.30 **Ni** 5.5-7.0 **P** 0.04 max **S** 0.03 max **Si** 1.00 max | **ASME** SFA5.4 (E219-XX) **AWS** A5.4 (E219-XX) |
| **W32340** | Nitrogen Strengthened Austenitic Cr-Mn-Ni Stainless Steel Composite Welding Filler Metal (EC219) | **C** 0.05 max **Cr** 19.0-21.5 **Cu** 0.75 max **Mn** 8.0-10.0 **Mo** 0.75 max **N** 0.10-0.30 **Ni** 5.5-7.0 **P** 0.03 max **S** 0.03 max **Si** 1.00 max | **ASME** SFA5.9 (EC219) **AWS** A5.9 (EC219) |

The chemical compositions listed are for identification purposes and should not be used in lieu of the cross referenced specifications.

| UNIFIED NUMBER | DESCRIPTION | CHEMICAL COMPOSITION | CROSS REFERENCE SPECIFICATIONS |
|---|---|---|---|
| **W32410** | Nitrogen Strengthened Cr-Ni-Mn-Mo Stainless Steel Arc Welding Electrode (E240-XX) | **C** 0.06 max **Cr** 17.0-19.0 **Cu** 0.75 max **Mn** 10.5-13.5 **Mo** 0.75 max **N** 0.10-0.30 **Ni** 4.0-6.0 **P** 0.04 max **S** 0.03 max **Si** 1.00 max | **ASME** SFA5.4 (E240-XX) **AWS** A5.4 (E240-XX) |
| **W32440** | Nitrogen Strengthened Austenitic Cr-Mn-Ni Stainless Steel Composite Welding Filler Metal (EC240) | **C** 0.05 max **Cr** 17.0-19.0 **Cu** 0.75 max **Mn** 10.5-13.5 **Mo** 0.75 max **N** 0.10-0.20 **Ni** 4.0-6.0 **P** 0.04 max **S** 0.03 max **Si** 1.00 max | **ASME** SFA5.9 (EC240) **AWS** A5.9 (EC240) |
| **W32540** | Nitrogen Strengthened Austenitic Cr-Ni-Mn Stainless Steel Composite Welding Filler Metal (Nitronic 60W) | **C** 0.10 max **Cr** 16.0-18.0 **Cu** 0.75 max **Mn** 7.0-9.0 **Mo** 0.75 max **N** 0.08-0.18 **Ni** 8.0-9.0 **P** 0.03 max **S** 0.03 max **Si** 3.5-4.5 | **ASME** SFA5.9 (EC218) **AWS** A5.9 (EC218) |
| **W32550** | Changed to W39553 | | |
| **W32561** | Replaced by W39533 | | |
| **W34710** | Austenitic Cr-Ni-Cb Stainless Steel Arc Welding Electrode (E347-XX) | **C** 0.08 max **Cb** 8\mtC-1.00 includes Ta **Cr** 18.0-21.0 **Cu** 0.75 max **Mn** 0.5-2.5 **Ni** 9.0-11.0 **P** 0.04 max **S** 0.03 max **Si** 0.90 max | **ASME** SFA5.4 (E347-XX) **AWS** A5.4 (E347-XX) **MIL SPEC** MIL-E-22200/2 (MIL-347, MIL-347Co) |
| **W34712** | Austenitic Cr-Ni-Cb Stainless Steel Arc Welding Electrode (347HC) High Carbon | **C** 0.08-0.15 **Cb** 8\mtC-1.4 includes Ta **Cr** 18.0-21.0 **Mn** 2.5 max **Ni** 9.0-11.0 **P** 0.025 max **S** 0.025 max **Si** 5\mtC max | **MIL SPEC** MIL-E-22200/2 (MIL-347HC) |
| **W34731** | Austenitic Cr-Ni-Cb Stainless Steel Flux Cored Wire (E347TX-X, R347T1-5) | **C** 0.08 max **Cb** 8\mtC-1.0 **Cr** 18.0-21.0 **Cu** 0.5 max **Mn** 0.5-2.5 **Mo** 0.5 max **Ni** 9.0-11.0 **P** 0.04 max **S** 0.03 max **Si** 1.0 max | **ASME** SFA5.22 (E347TX-X, R347T1-5) **AWS** A5.22 (E347TX-X, R347T1-5) |
| **W34733** | Austenitic Cr-Ni-Cb Stainless Steel Flux Cored Wire (E347T0-3) | **C** 0.08 max **Cb** 8\mtC-1.0 **Cr** 19.0-21.5 **Cu** 0.5 max **Mn** 0.5-2.5 **Mo** 0.5 max **Ni** 9.0-11.0 **P** 0.04 max **S** 0.03 max **Si** 1.0 max | **ASME** SFA5.22 (E347T0-3) **AWS** A5.22 (E347T0-3) |
| **W34740** | Austenitic Cr-Ni-Cb Stainless Steel Composite Welding Filler Metal (EC347) | **C** 0.08 max **Cb** 10\mtC-1.0 **Cr** 19.0-21.5 **Cu** 0.75 max **Mn** 1.0-2.5 max **Mo** 0.75 max **Ni** 9.0-11.0 **P** 0.03 max **S** 0.03 max **Si** 0.30-0.65 | **ASME** SFA5.9 (EC347) **AWS** A5.9 (EC347) |
| **W34748** | Austenitic Cr-Ni Stainless Steel Composite Welding Filler Metal (EC347Si) | **C** 0.08 max **Cb** 10\mtC-1.0 includes Ta **Cr** 19.0-21.5 **Cu** 0.5 max **Mn** 1.0-2.5 **Mo** 0.5 max **Ni** 9.0-11.0 **P** 0.03 max **S** 0.03 max **Si** 0.65-1.00 | **ASME** SFA5.9 (EC347Si) **AWS** A5.9 (EC347Si) |
| **W34910** | Austenitic Cr-Ni-W-Mo Heat Resisting Arc Welding Electrode (E349-XX) | **C** 0.13 max **Cb** 0.75-1.2 **Cr** 18.0-21.0 **Cu** 0.75 max **Mn** 0.5-2.5 **Mo** 0.35-0.65 **Ni** 8.0-10.0 **P** 0.04 max **S** 0.03 max **Si** 0.90 max **Ti** 0.15 max **W** 1.25-1.75 | **AMS** 5783 **ASME** SFA5.4 (E349-XX) **AWS** A5.4 (E349-XX) **MIL SPEC** MIL-E-22200/2 (MIL-349) |
| * **W34940** | Austenitic Cr-Ni Stainless Steel Composite Welding Filler Metal (EC349) | **C** 0.07-0.13 **Cb** 1.00-1.40 **Cr** 19.0-21.5 **Mn** 1.00-2.50 **Mo** 0.35-0.65 **Ni** 8.00-9.50 **P** 0.03 max **S** 0.03 max **Si** 0.30-0.65 | **ASME** SFA5.9 (EC349) **AWS** A5.9 (EC349) |
| **W35010** | Precipitation Hardenable Cr-Ni-Mo Stainless Steel Arc Welding Electrode (350) | **C** 0.08-0.12 **Cr** 16.00-17.00 **Mn** 0.50-1.25 **Mo** 2.50-3.25 **N** 0.07-0.13 **Ni** 4.00-5.00 **P** 0.040 max **S** 0.030 max **Si** 0.50 max | **AMS** 5775 |
| **W35510** | Precipitation Hardenable Stainless Steel Arc Welding Electrode | **C** 0.10-0.15 **Cr** 15.0-16.0 **Cu** 0.50 max **Mn** 0.50-1.25 **Mo** 2.50-3.25 **N** 0.07-0.13 **Ni** 4.00-5.00 **P** 0.040 max **S** 0.030 max **Si** 0.50 max | **AMS** 5781 |
| **W36810** | Austenitic Cr-Ni-Mo Stainless Steel Arc Welding Electrode (E16-8-2-XX) | **C** 0.10 max **Cr** 14.5-16.5 **Cu** 0.75 max **Mn** 0.5-2.5 **Mo** 1.0-2.0 **Ni** 7.5-9.5 **P** 0.03 max **S** 0.03 max **Si** 0.60 max | **ASME** SFA5.4 (E16-8-2-XX) **AWS** A5.4 (E16-8-2-XX) **MIL SPEC** MIL-E-22200/2 Type MIL-16-8-2 |

*Boxed entries are no longer active and are retained for reference purposes only.

The chemical compositions listed are for identification purposes and should not be used in lieu of the cross referenced specifications.

| UNIFIED NUMBER | DESCRIPTION | CHEMICAL COMPOSITION | CROSS REFERENCE SPECIFICATIONS |
|---|---|---|---|
| **W36840** | Austenitic Cr-Ni-Mo Stainless Steel Composite Welding Filler Metal (16-8-2) | **C** 0.10 max **Cr** 14.5-16.5 **Cu** 0.75 max **Mn** 1.0-2.5 **Mo** 1.0-2.0 **Ni** 7.5-9.5 **P** 0.03 max **S** 0.03 max **Si** 0.30-0.65 | **ASME** SFA5.9 (EC16-8-2) **AWS** A5.9 (EC16-8-2) |
| **W37410** | Precipitation Hardenable Cr-Ni-Cu Stainless Steel Arc Welding Electrode (17-4PH) (E630-XX) | **C** 0.05 max **Cb** (incl Ta) 0.15-0.30 **Cr** 16.00-16.75 **Cu** 3.25-4.00 **Mn** 0.25-0.75 **Mo** 0.75 max **Ni** 4.50-5.00 **P** 0.04 max **S** 0.03 max **Si** 0.75 max | **AMS** 5827 **ASME** SFA5.4 (E630-XX) **AWS** A5.4 (E630-XX) |
| **W37440** | Precipitation Hardenable Cr-Ni-Cu Stainless Steel Composite Welding Filler Metal (17-4PH) (EC630) | **C** 0.05 max **Cb** (incl Ta) 0.15-0.30 **Cr** 16.00-16.75 **Cu** 3.25-4.00 **Mn** 0.25-0.75 **Mo** 0.75 max **Ni** 4.50-5.00 **P** 0.04 max **S** 0.03 max **Si** 0.75 max | **ASME** SFA5.9 (EC630) **AWS** A5.9 (EC630) |
| **W39209** | Duplex Stainless Steel Electrodes (E2209-XX) | **C** 0.04 max **Cr** 21.5-23.5 **Cu** 0.75 max **Mn** 0.5-2.0 **Mo** 2.5-3.5 **N** 0.08-0.20 **Ni** 8.5-10.5 **P** 0.04 max **S** 0.03 max **Si** 0.90 max | **ASME** SFA5.4 (E2209-XX) **AWS** A5.4 (E2209-XX) |
| **W39239** | Duplex Stainless Steel Flux Cored Wire (E2209T0-X) | **C** 0.04 max **Cr** 21.0-24.0 **Cu** 0.5 max **Mn** 0.5-2.0 **Mo** 2.5-4.0 **N** 0.08-0.20 **Ni** 7.5-10.0 **P** 0.04 max **S** 0.03 max **Si** 1.0 max | **ASME** SFA5.22 (E2209T0-X) **AWS** A5.22 (E2209T0-X) |
| **W39240** | Duplex Stainless Steel Welding Composite Filler Metal (EC2209) | **C** 0.03 max **Cr** 21.5-23.5 **Cu** 0.75 max **Mn** 0.50-2.0 **Mo** 2.5-3.5 **N** 0.08-0.20 **Ni** 7.5-9.5 **P** 0.03 max **S** 0.03 max **Si** 0.90 max | **ASME** SFA5.9 (EC2209) **AWS** A5.9 (EC2209) |
| **W39533** | Duplex Stainless Steel Flux Cored Wire (E2553T0-X) | **C** 0.04 max **Cr** 24.0-27.0 **Cu** 1.5-2.5 **Mn** 0.5-1.5 **Mo** 2.9-3.9 **N** 0.10-0.20 **Ni** 8.5-10.5 **P** 0.04 max **S** 0.03 max **Si** 0.75 max | **ASME** SFA5.22 (E2553T0-X) **AWS** A5.22 (E2553T0-X) |
| **W39540** | Duplex Stainless Steel Composite Welding Filler Metal (EC2553) | **C** 0.04 max **Cr** 24.0-27.0 **Cu** 1.5-2.5 **Mn** 1.5 max **Mo** 2.9-3.9 **N** 0.10-0.25 **Ni** 4.5-6.5 **P** 0.04 max **S** 0.03 max **Si** 1.0 max | **ASME** SFA5.9 (EC2553) **AWS** A5.9 (EC2553) |
| **W39553** | Duplex Stainless Steel Electrodes (E2553-XX) | **C** 0.06 max **Cr** 24.0-27.0 **Cu** 1.5-2.5 **Mn** 0.5-1.5 **Mo** 2.9-3.9 **N** 0.10-0.25 **Ni** 6.5-8.5 **P** 0.04 max **S** 0.03 max **Si** 1.0 max | **ASME** SFA5.4 (E2553-XX) **AWS** A5.4 (E2553-XX) |
| **W40931** | Ferritic Cr-Ti Stainless Steel Flux Cored Wire (E409TX-X, E409T0-3) | **C** 0.10 max **Cr** 10.5-13.5 **Cu** 0.5 max **Mn** 0.80 max **Mo** 0.5 max **Ni** 0.60 max **P** 0.04 max **S** 0.03 max **Si** 1.0 max **Ti** 10\mtC-1.5 | **ASME** SFA5.22 (E409TX-X, E409T0-3) **AWS** A5.22 (E409TX-X, E409T0-3) |
| **W40947** | Ferritic Cr-Cb Stainless Steel Composite Welding Filler Metal (EC409Cb) | **C** 0.08 max **Cb** 10(C) min-0.75 max **Cr** 10.5-13.5 **Cu** 0.75 max **Mn** 0.8 max **Mo** 0.50 max **Ni** 0.6 max **P** 0.04 max **S** 0.03 max **Si** 1.0 max | **ASME** SFA5.9 (EC409Cb) **AWS** A5.9 (EC409Cb) |
| **W41010** | Martensitic Cr Stainless Steel Arc Welding Electrode (E410-XX) | **C** 0.12 max **Cr** 11.0-13.5 **Cu** 0.75 max **Mn** 1.0 max **Mo** 0.75 max **Ni** 0.7 max **P** 0.04 max **S** 0.03 max **Si** 0.90 max | **AMS** 5777 **ASME** SFA5.4 (E410-XX) **AWS** A5.4 (E410-XX) **MIL SPEC** MIL-E-22200/8 (MIL-410) |
| **W41016** | Martensitic Cr-Ni-Mo Stainless Steel Arc Welding Electrode (E410NiMo-XX) | **C** 0.06 max **Cr** 11.0-12.5 **Cu** 0.75 max **Mn** 1.0 max **Mo** 0.40-0.70 **Ni** 4.0-5.0 **P** 0.04 max **S** 0.03 max **Si** 0.90 max | **ASME** SFA5.4 (E410NiMo-XX) **AWS** A5.4 (E410NiMo-XX) |
| **W41031** | Martensitic Cr Stainless Steel Flux Cored Wire (E410TX-X, E410T0-3) | **C** 0.12 max **Cr** 11.0-13.5 **Cu** 0.5 max **Mn** 1.2 max **Mo** 0.5 max **Ni** 0.60 max **P** 0.04 max **S** 0.03 max **Si** 1.0 max | **ASME** SFA5.22 (E410TX-X, E410T0-3) **AWS** A5.22 (E410TX-X, E410T0-3) |
| **W41036** | Martensitic Cr-Ni-Mo Stainless Steel Flux Cored Wire (E410NiMoTX-X, E410NiMoT0-3) | **C** 0.06 max **Cr** 11.0-12.5 **Cu** 0.5 max **Mn** 1.0 max **Mo** 0.40-0.70 **Ni** 4.0-5.0 **P** 0.04 max **S** 0.03 max **Si** 1.0 max | **ASME** SFA5.22 (E410NiMoTX-X, E410NiMoT0-3) **AWS** A5.22 (E410NiMoTX-X, E410NiMoT0-3); C2.25 (CW-FeCr-1) |
| **W41038** | Martensitic Cr-Ni-Ti Stainless Steel Flux Cored Wire (E410NiTiTX-X, E410NiTiT0-3) | **C** 0.04 max **Cr** 11.0-12.0 **Cu** 0.5 max **Mn** 0.70 max **Mo** 0.05 max **Ni** 3.6-4.5 **P** 0.03 max **S** 0.03 max **Si** 0.50 max **Ti** 10\mtC-1.5 | **ASME** SFA5.22 (E410NiTiTX-X, E410NiTiT0-3) **AWS** A5.22 (E410NiTiTX-X, E410NiTiT0-3) |

The chemical compositions listed are for identification purposes and should not be used in lieu of the cross referenced specifications.

| UNIFIED NUMBER | DESCRIPTION | CHEMICAL COMPOSITION | CROSS REFERENCE SPECIFICATIONS |
|---|---|---|---|
| **W41039** | Composite Thermal Spray Wire (CW-FeCr-2) | **C** 0.06 max **Cr** 12-14 **Fe** rem **Mn** 1.0 max **Ni** 3.0-4.0 | **AWS** C2.25 (CW-FeCr-2) |
| **W41040** | Martensitic Cr Stainless Steel Composite Welding Filler Metal (EC410) | **C** 0.12 max **Cr** 11.5-13.5 **Cu** 0.75 max **Mn** 0.6 max **Mo** 0.75 max **Ni** 0.6 max **P** 0.03 max **S** 0.03 max **Si** 0.50 max | **ASME** SFA5.9 (EC410) **AWS** A5.9 (EC410) |
| **W41046** | Martensitic Cr-Ni-Mo Stainless Steel Composite Welding Filler Metal (EC410NiMo) | **C** 0.06 max **Cr** 11.0-12.5 **Cu** 0.75 max **Mn** 0.6 max **Mo** 0.4-0.7 **Ni** 4.0-5.0 **P** 0.03 max **S** 0.03 max **Si** 0.50 max | **ASME** SFA5.9 (EC410NiMo) **AWS** A5.9 (EC410NiMo) |
| **W42040** | Martensitic Cr Stainless Steel Composite Welding Filler Metal (EC420) | **C** 0.25-0.40 **Cr** 12.0-14.0 **Cu** 0.75 max **Mn** 0.6 max **Mo** 0.75 max **Ni** 0.6 max **P** 0.03 max **S** 0.03 max **Si** 0.50 max | **ASME** SFA5.9 (EC420) **AWS** A5.9 (EC420); C2.25 (CW-FeCr-3) |
| **W43010** | Ferritic Cr Stainless Steel Arc Welding Electrode (E430-XX) | **C** 0.10 max **Cr** 15.0-18.0 **Cu** 0.75 max **Mn** 1.0 max **Mo** 0.75 max **Ni** 0.6 max **P** 0.04 max **S** 0.03 max **Si** 0.90 max | **ASME** SFA5.4 (E430-XX) **AWS** A5.4 (E430-XX) |
| **W43031** | Ferritic Cr Stainless Steel Flux Cored Wire (E430TX-X, E430T0-3) | **C** 0.10 max **Cr** 15.0-18.0 **Cu** 0.5 max **Mn** 1.0 max **Mo** 0.5 max **Ni** 0.60 max **P** 0.04 max **S** 0.03 **Si** 1.0 max | **ASME** SFA5.22 (E430TX-X, E430T0-3) **AWS** A5.22 (E430TX-X, E430T0-3) |
| **W43040** | Ferritic Cr Stainless Steel Composite Welding Filler Metal (EC430) | **C** 0.10 max **Cr** 15.5-17.0 **Cu** 0.75 max **Mn** 0.6 max **Mo** 0.75 max **Ni** 0.6 max **P** 0.03 max **S** 0.03 max **Si** 0.50 max | **ASME** SFA5.9 (EC430) **AWS** A5.9 (EC430) |
| **W43070** | Composite Thermal Spray Wire (CW-FeCr-4) | **C** 0.60-0.80 **Cr** 16-18 **Fe** rem **Mn** 1.0 max **Mo** 0.75 max **Si** 1.0 max | **AWS** C2.25 (CW-FeCr-4) |
| **W44647** | Ferritic Cr-Mo Stainless Steel Composite Welding Filler Metal (EC446LMo) | **C** 0.01 max **Cr** 25.0-27.5 **Cu** 0.20 max **Mn** 0.40 max **Mo** 0.75-1.50 **N** 0.015 max **Ni** 0.5 max includes Cu **P** 0.02 max **S** 0.02 max **Si** 0.40 max | **ASME** SFA5.9 (EC446LMo) **AWS** A5.9 (EC446LMo) |
| **W46440** | Composite Thermal Spray Wire (CW-FeCrB) | **B** 4.3 max **C** 0.2 max **Cr** 32 max **Fe** rem **Mn** 1.8 max **Si** 2.1 max | **AWS** C2.25 (CW-FeCrB) |
| **W46540** | Composite Thermal Spray Wire (CW-FeCrNiB) | **B** 2.5 max **C** 0.2 max **Cr** 25 max **Cu** 2.2 max **Fe** rem **Mn** 1.25 max **Mo** 4.2 max **Ni** 10 max **Si** 1.5 max | **AWS** C2.25 (CW-FeCrNiB) |
| **W46640** | Composite Thermal Spray Wire (CW-FeCrNiMoB) | **B** 3.0 max **C** 0.2 max **Cr** 25 max **Cu** 2.7 max **Fe** rem **Mn** 1.5 max **Mo** 3.5 max **Ni** 8.0 max **Si** 2.2 max | **AWS** C2.25 (CW-FeCrNiMoB) |
| **W50140** | Martensitic 5% Cr-Mo Steel Composite Wire for Submerged Arc Welding (ECB6H) High Carbon | **C** 0.10-0.25 **Cr** 4.50-6.00 **Cu** 0.35 max **Mn** 1.20 max **Mo** 0.40-0.65 **P** 0.030 max **S** 0.030 max **Si** 0.80 max | **ASME** SFA5.23 (ECB6H) **AWS** A5.23 (ECB6H) |
| * **W50141** | Martensitic 5% Cr-Mo Steel Composite Wire for Submerged Arc Welding (ECB6HN) High Carbon-Nuclear Grade | **C** 0.10 max **Cr** 4.50-6.00 **Cu** 0.08 max **Mn** 1.20 max **Mo** 0.40-0.65 **P** 0.012 max **S** 0.040 max **Si** 0.80 max **V** 0.05 max | **ASME** SFA5.23 (ECB6HN) **AWS** A5.23 (ECB6HN) |
| **W50205** | Martensitic 5% Cr-Mo Steel Arc Welding Electrode (E8015-B6L) | **C** 0.05 max **Cr** 4.0-6.0 **Mn** 1.0 max **Mo** 0.45-0.65 **Ni** 0.40 max **P** 0.03 max **S** 0.03 max **Si** 0.90 max | **ASME** SFA5.5 (E8015-B6L) **AWS** A5.5 (E8015-B6L) **MIL SPEC** MIL-E-22200/8 (MIL-502-15L) |
| **W50206** | Martensitic 5% Cr-Mo Steel Arc Welding Electrode (E8016-B6L) | **C** 0.05 max **Cr** 4.0-6.0 **Mn** 1.0 max **Mo** 0.45-0.65 **Ni** 0.40 max **P** 0.03 max **S** 0.03 max **Si** 0.90 max | **ASME** SFA5.5 (E8016-B6L) **AWS** A5.5 (E8016-B6L) **MIL SPEC** MIL-E-22200/8 (MIL-502-16L) |
| **W50208** | Martensitic 5% Cr-Mo Steel Arc Welding Electrode (E8018-B6L) | **C** 0.05 max **Cr** 4.0-6.0 **Mn** 1.0 max **Mo** 0.45-0.65 **Ni** 0.40 max **P** 0.03 max **S** 0.03 max **Si** 0.90 max | **ASME** SFA5.5 (E8018-B6L) **AWS** A5.5 (E8018-B6L) |

*Boxed entries are no longer active and are retained for reference purposes only.

The chemical compositions listed are for identification purposes and should not be used in lieu of the cross referenced specifications.

| UNIFIED NUMBER | DESCRIPTION | CHEMICAL COMPOSITION | CROSS REFERENCE SPECIFICATIONS |
|---|---|---|---|
| **W50210** | Martensitic 5% Cr-Mo Steel Arc Welding Electrode (E502-XX) | **C** 0.05 max **Cr** 4.0-6.0 **Cu** 0.75 max **Mn** 1.0 max **Mo** 0.45-0.65 **Ni** 0.4 max **P** 0.04 max **S** 0.03 max **Si** 0.90 max | **AMS** 6467 **ASME** SFA5.4 (E502-XX) **AWS** A5.4 (E502-XX) |
| * **W50213** | Martensitic 5% Cr-Mo Steel Arc Welding Electrode (502L) | **C** 0.05 max **Cr** 4.00-6.00 **Mn** 1.00 max **Mo** 0.45-0.65 **P** 0.04 max **S** 0.03 max **Si** 0.90 max | **ASME** SFA5.5 (E801X-B6L) **AWS** A5.5 (E801X-B6L) **MIL SPEC** MIL-E-22200/8 (MIL-502-1X-L) |
| **W50215** | Martensitic 5% Cr-Mo Steel Arc Welding Electrode (E8015-B6) | **C** 0.05-0.10 **Cr** 4.0-6.0 **Mn** 1.0 max **Mo** 0.45-0.65 **Ni** 0.40 max **P** 0.03 max **S** 0.03 max **Si** 0.90 max | **ASME** SFA5.5 (E8015-B6) **AWS** A5.5 (E8015-B6) **MIL SPEC** MIL-E-22200/8 (MIL-502-15) |
| **W50216** | Martensitic 5% Cr-Mo Steel Arc Welding Electrode (E8016-B6) | **C** 0.05-0.10 **Cr** 4.0-6.0 **Mn** 1.0 max **Mo** 0.45-0.65 **Ni** 0.40 max **P** 0.03 max **S** 0.03 max **Si** 0.90 max | **ASME** SFA5.5 (E8016-B6) **AWS** A5.5 (E8016-B6) **MIL SPEC** MIL-E-22200/8 (MIL-502-16) |
| **W50218** | Martensitic 5% Cr-Mo Steel Arc Welding Electrode (E8018-B6) | **C** 0.05-0.10 **Cr** 4.0-6.0 **Mn** 1.0 max **Mo** 0.45-0.65 **Ni** 0.40 max **P** 0.03 max **S** 0.03 max **Si** 0.90 max | **ASME** SFA5.5 (E8018-B6) **AWS** A5.5 (E8018-B6) |
| **W50230** | Martensitic 5% Cr-Mo Steel, Flux Cored Welding Wire (E8XT5-B6L, -B6LM) | **C** 0.05 max **Cr** 4.0-6.0 **Cu** 0.50 max **Mn** 1.25 max **Mo** 0.45-0.65 **Ni** 0.40 max **P** 0.04 max **S** 0.03 max **Si** 1.00 max | **ASME** SFA5.29 (E8XT5-B6L, B6LM) **AWS** A5.29 (E8XT5-B6L, B6LM) |
| **W50231** | Martensitic 5% Cr-Mo Steel Flux Cored Wire (E502TX-X) | **C** 0.10 max **Cr** 4.0-6.0 **Cu** 0.5 max **Mn** 1.2 max **Mo** 0.45-0.65 **Ni** 0.40 max **P** 0.04 max **S** 0.03 max **Si** 1.0 max | **ASME** SFA5.22 (E502TX-X); SFA5.29 (E8XT5-B6) **AWS** A5.22 (E502TX-X); A5.29 (E8XT5-B6) |
| **W50240** | Martensitic 5% Cr-Mo Steel Composite Wire for Submerged Arc Welding (ECB6) | **C** 0.12 max **Cr** 4.50-6.00 **Cu** 0.35 max **Mn** 1.20 max **Mo** 0.40-0.65 **P** 0.030 max **S** 0.030 max **Si** 0.80 max | **ASME** SFA5.23 (ECB6) **AWS** A5.23 (ECB6) |
| * **W50241** | Martensitic 5% Cr-Mo Steel Composite Wire for Submerged Arc Welding (ECB6N) Nuclear Grade | **C** 0.12 max **Cr** 4.50-6.00 **Cu** 0.08 max **Mn** 1.20 max **Mo** 0.40-0.65 **P** 0.012 max **S** 0.040 max **Si** 0.80 max **V** 0.05 max | **ASME** SFA5.23 (ECB6N) **AWS** A5.23 (ECB6N) |
| **W50305** | Martensitic 5% Cr-Mo Steel Arc Welding Electrode (E8015-B7L) | **C** 0.05 max **Cr** 6.0-8.0 **Mn** 1.0 max **Mo** 0.45-0.65 **Ni** 0.40 max **P** 0.03 max **S** 0.03 max **Si** 0.90 max | **ASME** SFA5.5 (E8015-B7L) **AWS** A5.5 (E8015-B7L) |
| **W50306** | Martensitic 5% Cr-Mo Steel Arc Welding Electrode (E8016-B7L) | **C** 0.05 max **Cr** 6.0-8.0 **Mn** 1.0 max **Mo** 0.45-0.65 **Ni** 0.40 max **P** 0.03 max **S** 0.03 max **Si** 0.90 max | **ASME** SFA5.5 (E8016-B7L) **AWS** A5.5 (E8016-B7L) |
| **W50308** | Martensitic 5% Cr-Mo Steel Arc Welding Electrode (E8018-B7L) | **C** 0.05 max **Cr** 6.0-8.0 **Mn** 1.0 max **Mo** 0.45-0.65 **Ni** 0.40 max **P** 0.03 max **S** 0.03 max **Si** 0.90 max | **ASME** SFA5.5 (E8018-B7L) **AWS** A5.5 (E8018-B7L) |
| **W50310** | Martensitic 7% Cr-Mo Steel Arc Welding Electrode (E7Cr-XX) | **C** 0.10 max **Cr** 6.0-8.0 **Cu** 0.75 max **Mn** 1.0 max **Mo** 0.45-0.65 **Ni** 0.4 max **P** 0.04 max **S** 0.03 max **Si** 0.90 max | **ASME** SFA5.4 (E7Cr-XX) **AWS** A5.4 (E7Cr-XX) |
| **W50315** | Martensitic 7% Cr-Mo Steel Arc Welding Electrode (E8015-B7) | **C** 0.05-0.10 **Cr** 6.0-8.0 **Mn** 1.0 max **Mo** 0.45-0.65 **Ni** 0.40 max **P** 0.03 max **S** 0.03 max **Si** 0.90 max | **ASME** SFA5.5 (E8015-B7) **AWS** A5.5 (E8015-B7) |
| **W50316** | Martensitic 7% Cr-Mo Steel Arc Welding Electrode (E8016-B7) | **C** 0.05-0.10 **Cr** 6.0-8.0 **Mn** 1.0 max **Mo** 0.45-0.65 **Ni** 0.40 max **P** 0.03 max **S** 0.03 max **Si** 0.90 max | **ASME** SFA5.5 (E8016-B7) **AWS** A5.5 (E8016-B7) |
| **W50318** | Martensitic 7% Cr-Ni Steel Arc Welding Electrode (E8018-B7) | **C** 0.05-0.10 **Cr** 6.0-8.0 **Mn** 1.0 max **Mo** 0.45-0.65 **Ni** 0.40 max **P** 0.03 max **S** 0.03 max **Si** 0.90 max | **ASME** SFA5.5 (E8018-B7) **AWS** A5.5 (E8018-B7) |
| **W50405** | Martensitic 9% Cr-Mo Steel Arc Welding Electrode (E8015-B8L) | **C** 0.05 max **Cr** 8.0-10.5 **Mn** 1.0 max **Mo** 0.85-1.20 **Ni** 0.40 max **P** 0.03 max **S** 0.03 max **Si** 0.90 max | **ASME** SFA5.5 (E8015-B8L) **AWS** A5.5 (E8015-B8L) |

*Boxed entries are no longer active and are retained for reference purposes only.

The chemical compositions listed are for identification purposes and should not be used in lieu of the cross referenced specifications.

| UNIFIED NUMBER | DESCRIPTION | CHEMICAL COMPOSITION | CROSS REFERENCE SPECIFICATIONS |
|---|---|---|---|
| **W50406** | Martensitic 9% Cr-Ni Steel Arc Welding Electrode (E8016-B8L) | **C** 0.05 max **Cr** 8.0-10.5 **Mn** 1.0 max **Mo** 0.85-1.20 **Ni** 0.40 max **P** 0.03 max **S** 0.03 max **Si** 0.90 max | **ASME** SFA5.5 (E8016-B8L) **AWS** A5.5 (E8016-B8L) |
| **W50408** | Martensitic 9% Cr-Ni Steel Arc Welding Electrode (E8018-B8L) | **C** 0.05 max **Cr** 8.0-10.5 **Mn** 1.0 max **Mo** 0.85-1.20 **Ni** 0.40 max **P** 0.03 max **S** 0.03 max **Si** 0.90 max | **ASME** SFA5.5 (E8018-B8L) **AWS** A5.5 (E8018-B8L) |
| **W50410** | Martensitic 9% Cr-Mo Steel Arc Welding Electrode (E505-XX) | **C** 0.10 max **Cr** 8.0-10.5 **Cu** 0.75 max **Mn** 1.0 max **Mo** 0.85-1.20 **Ni** 0.4 max **P** 0.04 max **S** 0.03 max **Si** 0.90 max | **ASME** SFA5.4 (E505-XX) **AWS** A5.4 (E505-XX) |
| * **W50411** | Martensitic 9% Cr-Mo Steel Arc Welding Electrode (365) | **C** 0.10 max **Cr** 8.50-10.50 **Mo** 1.00-1.50 **Si** 0.60 max | |
| **W50415** | Martensitic 9% Cr-Mo Steel Arc Welding Electrode (E8015-B8) | **C** 0.05-0.10 **Cr** 8.0-10.5 **Mn** 1.0 max **Mo** 0.85-1.20 **Ni** 0.40 max **P** 0.03 max **S** 0.03 max **Si** 0.90 max | **ASME** SFA5.5 (E8015-B8) **AWS** A5.5 (E8015-B8) **MIL SPEC** MIL-E-22200/8 (MIL-505-15) |
| **W50416** | Martensitic 9% Cr-Mo Steel Arc Welding Electrode (E8016-B8) | **C** 0.05-0.10 **Cr** 8.0-10.5 **Mn** 1.0 max **Mo** 0.85-1.20 **Ni** 0.40 max **P** 0.03 max **S** 0.03 max **Si** 0.90 max | **ASME** SFA5.5 (E8016-B8) **AWS** A5.5 (E8016-B8) **MIL SPEC** MIL-E-22200/8 (MIL-505-16) |
| **W50418** | Martensitic 9% Cr-Mo Steel Arc Welding Electrode (E8018-B8) | **C** 0.05-0.10 **Cr** 8.0-10.5 **Mn** 1.0 max **Mo** 0.85-1.20 **Ni** 0.40 max **P** 0.03 max **S** 0.03 max **Si** 0.90 max | **ASME** SFA5.5 (E8018-B8) **AWS** A5.5 (E8018-B8) |
| **W50425** | Martensitic 9% Cr-Mo-V Steel Arc Welding Electrode (E9015-B9) | **Al** 0.04 max **C** 0.08-0.13 **Cb** 0.02-0.10 **Cr** 8.0-10.5 **Cu** 0.25 max **Mn** 1.25 max **Mo** 0.85-1.20 **N** 0.02-0.07 **Ni** 1.0 max **P** 0.01 max **S** 0.01 max **Si** 0.30 max **V** 0.15-0.30 | **ASME** SFA5.5 (E9015-B9) **AWS** A5.5 (E9015-B9) |
| **W50426** | Martensitic 9% Cr-Mo-V Steel Arc Welding Electrode (E9016-B9) | **Al** 0.04 max **C** 0.08-0.13 **Cb** 0.02-0.10 **Cr** 8.0-10.5 **Cu** 0.25 max **Mn** 1.25 max **Mo** 0.85-1.20 **N** 0.02-0.07 **Ni** 1.0 max **P** 0.01 max **S** 0.01 max **Si** 0.30 max **V** 0.15-0.30 | **ASME** SFA5.5 (E9016-B9) **AWS** A5.5 (E9016-B9) |
| **W50428** | Martensitic 9% Cr-Mo-V Steel Arc Welding Electrode (E9018-B9) | **Al** 0.04 max **C** 0.08-0.13 **Cb** 0.02-0.10 **Cr** 8.0-10.5 **Cu** 0.25 max **Mn** 1.25 max **Mo** 0.85-1.20 **N** 0.02-0.07 **Ni** 1.0 max **P** 0.01 max **S** 0.01 max **Si** 0.30 max **V** 0.15-0.30 | **ASME** SFA5.5 (E9018-B9) **AWS** A5.5 (E9018-B9) |
| **W50430** | Martensitic 9% Cr-Mo Steel, Flux Cored Wire (E8XT5-B8L, -B8LM) | **C** 0.05 max **Cr** 8.00-10.5 **Cu** 0.50 max **Mn** 1.25 max **Mo** 0.85-1.20 **Ni** 0.40 max **P** 0.03 max **S** 0.03 max **Si** 1.0 max | **ASME** SFA5.29 (E8XT5-B8L, -B8LM) **AWS** A5.29 (E8XT5-B8L, -B8LM) |
| **W50431** | Martensitic 9% Cr-Mo Steel Flux Cored Wire (E505TX-X) | **C** 0.10 max **Cr** 8.0-10.5 **Cu** 0.5 max **Mn** 1.2 max **Mo** 0.85-1.20 **Ni** 0.40 max **P** 0.04 max **S** 0.03 max **Si** 1.0 max | **ASME** SFA5.22 (E505TX-X); SFA5.29 (E6XT5-B8) **AWS** A5.22 (E505TX-X); A5.29 (E6XT5-B8) |
| **W50440** | Martensitic Cr-Mo Stainless Steel Composite Welding Filler Metal (EC-505) | **C** 0.10 max **Cr** 8.0-10.5 **Cu** 0.75 max **Mn** 0.6 max **Mo** 0.8-1.2 **Ni** 0.5 max **P** 0.04 max **S** 0.03 max **Si** 0.50 max | **ASME** SFA5.9 (EC505); SFA5.23 (ECB8) **AWS** A5.9 (EC505); A5.23 (ECB8) |
| * **W50441** | Martensitic 9% Cr-Mo Steel Composite Wire for Submerged Arc Welding (ECB8N) Nuclear Grade | **C** 0.12 max **Cr** 8.00-10.00 **Cu** 0.08 max **Mn** 1.20 max **Mo** 0.80-1.20 **P** 0.012 max **S** 0.040 max **Si** 0.80 max **V** 0.05 max | **ASME** SFA5.23 (ECB8N) **AWS** A5.23 (ECB8N) |
| **W50442** | Cr-Mo Low Alloy Steel Composite Wire for Submerged Arc Welding (ECB 9) | **Al** 0.04 max **C** 0.07-0.13 **Cr** 8.00-9.50 **Cu** 0.10 max **Mn** 1.25 max **Mo** 0.80-1.20 **N** 0.03-0.07 **Nb** 0.03-0.10 **Ni** 1.00 max **P** 0.010 max **S** 0.010 max **Si** 0.15-0.30 **V** 0.15-0.25 | **ASME** SFA5.23 (ECB 9) **AWS** A5.23 (ECB 9) |
| **W51016** | Cr-Mo Low Alloy Steel Arc Welding Electrode (E8016-B1) | **C** 0.05-0.12 max **Cr** 0.40-0.65 **Mn** 0.90 max **Mo** 0.40-0.65 **P** 0.03 max **S** 0.03 max **Si** 0.60 max | **ASME** SFA5.5 (E8016-B1) **AWS** A5.5 (E8016-B1) |

*Boxed entries are no longer active and are retained for reference purposes only.

The chemical compositions listed are for identification purposes and should not be used in lieu of the cross referenced specifications.

| UNIFIED NUMBER | DESCRIPTION | CHEMICAL COMPOSITION | CROSS REFERENCE SPECIFICATIONS |
|---|---|---|---|
| **W51018** | Cr-Mo Low Alloy Steel Arc Welding Electrode (E8018-B1) | **C** 0.05-0.12 max **Cr** 0.40-0.65 **Mn** 0.90 max **Mo** 0.40-0.65 **P** 0.03 max **S** 0.03 max **Si** 0.80 max | **ASME** SFA5.5 (E8018-B1) **AWS** A5.5 (E8018-B1) |
| **W51031** | Cr-Mo Low Alloy Steel Flux Cored Wire (E8XT1-B1) | **C** 0.05-0.12 max **Cr** 0.40-0.65 **Mn** 1.25 max **Mo** 0.40-0.65 **P** 0.03 max **S** 0.03 max **Si** 0.80 max | **ASME** SFA5.29 (E8XT1-B1) **AWS** A5.29 (E8XT1-B1) |
| **W51040** | Cr-Mo Low Alloy Steel Composite Wire for Submerged Arc Welding (ECB1) | **C** 0.12 max **Cr** 0.40-0.65 **Cu** 0.35 max **Mn** 1.60 max **Mo** 0.40-0.65 **P** 0.030 max **S** 0.030 max **Si** 0.80 max | **ASME** SFA5.23 (ECB1) **AWS** A5.23 (ECB1) |
| * **W51041** | Cr-Mo Low Alloy Steel Composite Wire for Submerged Arc Welding (ECB1N) Nuclear Grade | **C** 0.12 max **Cr** 0.40-0.65 **Cu** 0.08 max **Mn** 1.60 max **Mo** 0.40-0.65 **P** 0.012 max **S** 0.040 max **Si** 0.80 max **V** 0.05 max | **ASME** SFA5.23 (ECB1N) **AWS** A5.23 (ECB1N) |
| **W51131** | Cr-Mo Low Alloy Steel, Flux Cored Wire (B1L) | **C** 0.05 max **Cr** 0.40-0.65 **Mn** 1.25 max **Mo** 0.40-0.65 **P** 0.03 max **S** 0.03 max **Si** 0.80 max | **ASME** SFA5.29 (E8XT1-B1L) **AWS** A5.29 (E8XT1-B1L) |
| **W51316** | Cr-Mo Low Alloy Steel Arc Welding Electrode (E8016-B5) | **C** 0.07-0.15 **Cr** 0.40-0.60 **Mn** 0.40-0.70 **Mo** 1.00-1.25 **P** 0.03 max **S** 0.03 max **Si** 0.30-0.60 **V** 0.05 max | **ASME** SFA5.5 (E8016-B5) **AWS** A5.5 (E8016-B5) |
| **W51340** | Cr-Mo Low Alloy Steel Composite Wire for Submerged Arc Welding (ECB5) | **C** 0.18 max **Cr** 0.40-0.65 **Cu** 0.35 max **Mn** 1.20 max **Mo** 0.90-1.20 **P** 0.030 max **S** 0.030 max **Si** 0.80 max | **ASME** SFA5.23 (ECB5) **AWS** A5.23 (ECB5) |
| * **W51341** | Cr-Mo Low Alloy Steel Composite Wire for Submerged Arc Welding (EB5N) Nuclear Grade | **C** 0.18 max **Cr** 0.40-0.65 **Cu** 0.08 max **Mn** 1.20 max **Mo** 0.90-1.20 **P** 0.012 max **S** 0.040 max **Si** 0.80 max **V** 0.05 max | **ASME** SFA5.23 (ECB5N) **AWS** A5.23 (ECB5N) |
| **W52015** | Cr-Mo Low Alloy Steel Arc Welding Electrode (8015-B2) | **C** 0.10 max **Cr** 1.00-1.50 **Mn** 0.90 max **Mo** 0.40-0.65 **P** 0.03 max **S** 0.04 max **Si** 0.60 max | **MIL SPEC** MIL-E-22200/8 (MIL-8015-B2) |
| **W52016** | Cr-Mo Low Alloy Steel Arc Welding Electrode (E8016B-2) | **C** 0.05-0.12 max **Cr** 1.00-1.50 **Mn** 0.90 max **Mo** 0.40-0.65 **P** 0.03 max **S** 0.04 max **Si** 0.60 max | **ASME** SFA5.5 (E8016-B2) **AWS** A5.5 (E8016-B2) **MIL SPEC** MIL-E-22200/8 (MIL-8016-B2) |
| **W52018** | Cr-Mo Low Alloy Steel Arc Welding Electrode (E8018-B2) | **C** 0.05-0.12 max **Cr** 1.00-1.50 **Mn** 0.90 max **Mo** 0.40-0.65 **P** 0.03 max **S** 0.030 max **Si** 0.80 max | **ASME** SFA5.5 (E8018-B2) **AWS** A5.5 (E8018-B2) **MIL SPEC** MIL-E-22200/8 (MIL-8018-B2) |
| **W52030** | Cr-Mo Low Alloy Steel Composite Wire for Gas Shielded Arc Welding | **C** 0.05-0.12 **Cr** 1.00-1.50 **Cu** 0.35 max **Mn** 0.40-1.00 **Mo** 0.40-0.65 **Ni** 0.20 max **P** 0.025 max **S** 0.030 max **Si** 0.25-0.60 | **ASME** SFA5.28 (E80C-B2) **AWS** A5.28 (E80C-B2) |
| **W52031** | Cr-Mo Low Alloy Steel Flux Cored Wire (E8XT1-B2) | **C** 0.05-0.12 **Cr** 1.00-1.50 **Mn** 1.25 max **Mo** 0.40-0.65 **P** 0.03 max **S** 0.03 max **Si** 0.80 max | **ASME** SFA5.29 (E8XT1-B2) **AWS** A5.29 (E8XT1-B2) |
| **W52035** | Cr-Mo Low Alloy Steel Flux Cored Wire (E8XT5-B2) | **C** 0.05-0.12 max **Cr** 1.00-1.50 **Mn** 1.25 max **Mo** 0.40-0.65 **P** 0.03 max **S** 0.00 max **Si** 0.00 max | **ASME** SFA5.29 (E8XT5-B2) **AWS** A5.29 (E8XT5-B2) |
| **W52040** | Cr-Mo Low Alloy Steel Composite Wire for Submerged Arc Welding (ECB2) | **C** 0.05-0.15 **Cr** 1.00-1.50 **Cu** 0.35 max **Mn** 1.20 max **Mo** 0.40-0.65 **P** 0.030 max **S** 0.030 max **Si** 0.80 max | **ASME** SFA5.23 (ECB2) **AWS** A5.23 (ECB2) |
| * **W52041** | Cr-Mo Low Alloy Steel Composite Wire for Submerged Arc Welding (ECB2N) Nuclear Grade | **C** 0.15 max **Cr** 1.00-1.50 **Cu** 0.08 max **Mn** 1.20 max **Mo** 0.40-0.65 **P** 0.012 max **S** 0.030 max **Si** 0.80 max **V** 0.05 max | **ASME** SFA5.23 (ECB2N) **AWS** A5.23 (ECB2N) |

*Boxed entries are no longer active and are retained for reference purposes only.

The chemical compositions listed are for identification purposes and should not be used in lieu of the cross referenced specifications.

| UNIFIED NUMBER | DESCRIPTION | CHEMICAL COMPOSITION | CROSS REFERENCE SPECIFICATIONS |
|---|---|---|---|
| **W52042** | Cr-Mo Low Alloy Steel Submerged Arc Welding (B2RR) | **As** 0.005 max **C** 0.15 max **Cr** 1.00-1.50 **Cu** 0.15 max **Mn** 1.20 max **Mo** 0.40-0.65 **P** 0.010 max **S** 0.010 max **Sb** 0.005 max **Si** 0.80 max **Sn** 0.005 max | **ASME** SFA5.23 (B2LR) **AWS** A5.23 (B2LR) |
| **W52115** | Cr-Mo Low Alloy Steel Arc Welding Electrode (E7015-B2L) | **C** 0.05 max **Cr** 1.00-1.50 **Mn** 0.90 max **Mo** 0.40-0.65 **P** 0.03 max **S** 0.04 max **Si** 1.00 max | **ASME** SFA5.5 (E7015-B2L) **AWS** A5.5 (E7015-B2L) **MIL SPEC** MIL-E-22200/8 (MIL-8015-B2L) |
| **W52116** | Cr-Mo Low Alloy Steel Arc Welding Electrode (E7016-B2L) | **C** 0.05 max **Cr** 1.00-1.50 **Mn** 0.90 max **Mo** 0.45-0.65 **P** 0.03 max **S** 0.03 max **Si** 0.6 max | **ASME** SFA5.5 (E7016-B2L) **AWS** A5.5 (E7016-B2L) **MIL SPEC** MIL-E-22200/8 (MIL-7016-B2L) |
| **W52118** | Cr-Mo Low Alloy Steel Arc Welding Electrode (E7018-B2L) | **C** 0.05 max **Cr** 1.00-1.50 **Mn** 0.90 max **Mo** 0.40-0.65 **P** 0.03 max **S** 0.04 max **Si** 0.80 max | **ASME** SFA5.5 (E7018-B2L) **AWS** A5.5 (E7018-B2L) **MIL SPEC** MIL-E-22200/8 (MIL-7018-B2L) |
| **W52130** | Cr-Mo Low Alloy Steel Composite Wire for Gas Shielded Arc Welding (E70C-B2L) | **C** 0.05 max **Cr** 1.00-1.50 **Cu** 0.35 max **Mn** 0.40-1.00 **Mo** 0.40-0.65 **Ni** 0.20 max **P** 0.025 max **S** 0.030 max **Si** 0.25-0.60 | **ASME** SFA5.28 (E70C-B2L) **AWS** A5.28 (E70C-B2L) |
| **W52131** | Cr-Mo Low Alloy Steel, Flux Cored Wire (E8XT1-B2L) | **C** 0.05 max **Cr** 1.00-1.50 **Mn** 1.25 max **Mo** 0.40-0.65 **P** 0.03 max **S** 0.03 max **Si** 0.80 max | **ASME** SFA5.29 (E8XT1-B2L) **AWS** A5.29 (E8XT1-B2L) |
| **W52135** | Cr-Mo Low Alloy Steel Flux Cored Wire (E8XT5-B2L) | **C** 0.05 max **Cr** 1.00-1.50 **Mn** 1.25 max **Mo** 0.40-0.65 **P** 0.03 max **S** 0.03 max **Si** 0.80 max | **ASME** SFA5.29 (E8XT5-B2L) **AWS** A5.29 (E8XT5-B2L) |
| **W52231** | Cr-Mo Low Alloy Steel Flux Cored Wire (E8XT1-B2H) | **C** 0.10-0.15 **Cr** 1.00-1.50 **Mn** 1.25 max **Mo** 0.40-0.65 **P** 0.03 max **S** 0.03 max **Si** 0.80 max | **ASME** SFA5.29 (E8XT1-B2H) **AWS** A5.29 (E8XT1-B2H) |
| **W52240** | Cr-Mo Low Alloy Steel Composite Wire for Submerged Arc Welding (ECB2H) | **C** 0.10-0.25 **Cr** 1.00-1.50 **Cu** 0.35 max **Mn** 1.20 max **Mo** 0.40-0.65 **P** 0.020 max **S** 0.020 max **Si** 0.80 max **V** 0.30 max | **ASME** SFA5.23 (ECB2H) **AWS** A5.23 (ECB2H) |
| * **W52241** | Cr-Mo Low Alloy Steel Flux Cored Wire for Submerged Arc Welding (ECB2HN) Nuclear Grade | **C** 0.10-0.25 **Cr** 1.00-1.50 **Cu** 0.08 max **Mn** 1.20 max **Mo** 0.40-0.65 **P** 0.012 max **S** 0.040 max **Si** 0.80 max **V** 0.05 max | **ASME** SFA5.23 (ECB2HN) **AWS** A5.23 (ECB2HN) |
| **W53015** | Cr-Mo Low Alloy Steel Arc Welding Electrode (E9015-B3) | **C** 0.05-0.12 **Cr** 2.00-2.50 **Mn** 0.90 max **Mo** 0.90-1.20 **P** 0.03 max **S** 0.04 max **Si** 1.0 max | **ASME** SFA5.5 (E9015-B3) **AWS** A5.5 (E9015-B3) **MIL SPEC** MIL-E-22200/8 (MIL-9015-B3) |
| **W53016** | Cr-Mo Low Alloy Steel Arc Welding Electrode (E9016-B3) | **C** 0.05-0.12 **Cr** 2.00-2.50 **Mn** 0.90 max **Mo** 0.90-1.20 **P** 0.03 max **S** 0.04 max **Si** 0.60 max | **ASME** SFA5.5 (E9016-B3) **AWS** A5.5 (E9016-B3) **MIL SPEC** MIL-E-22200/8 (MIL-9016-B3) |
| **W53018** | Cr-Mo Low Alloy Steel Arc Welding Electrode (E9018B-3) | **C** 0.05-0.12 **Cr** 2.00-2.50 **Mn** 0.90 max **Mo** 0.90-1.20 **P** 0.03 max **S** 0.04 max **Si** 0.80 max | **ASME** SFA5.5 (E9018-B3) **AWS** A5.5 (E9018-B3) **MIL SPEC** MIL-E-22200/8 (MIL-9018-B3) |
| **W53030** | Cr-Mo Low Alloy Steel Composite Wire for Gas Shielded Arc Welding (E90C-B3) | **C** 0.07-0.12 **Cr** 2.00-2.50 **Cu** 0.35 max **Mn** 0.40-1.00 **Mo** 0.90-1.20 **Ni** 0.20 max **P** 0.025 max **S** 0.030 max **Si** 0.25-0.60 | **ASME** SFA5.28 (E90C-B3) **AWS** A5.28 (E90C-B3) |
| **W53031** | Cr-Mo Low Alloy Steel Flux Cored Wire (EXXXT1-B3) | **C** 0.05-0.12 **Cr** 2.00-2.50 **Mn** 1.25 max **Mo** 0.90-1.20 **P** 0.03 max **S** 0.03 max **Si** 0.80 max | **ASME** SFA5.29 (E9XT1-B3, E10XT1-B3) **AWS** A5.29 (E9XT1-B3, E10XT1-B3) |
| **W53035** | Cr-Mo Low Alloy Steel Flux Cored Wire (E9XT5-B3) | **C** 0.05-0.12 **Cr** 2.00-2.50 **Mn** 1.25 max **Mo** 0.90-1.20 **P** 0.03 max **S** 0.03 max **Si** 0.80 max | **ASME** SFA5.29 (E9XT5-B3) **AWS** A5.29 (E9XT5-B3) |
| **W53040** | Cr-Mo Low Alloy Steel Composite Wire for Submerged Arc Welding (ECB3) | **C** 0.05-0.15 **Cr** 2.00-2.50 **Cu** 0.35 max **Mn** 1.20 max **Mo** 0.90-1.20 **P** 0.030 max **S** 0.030 max **Si** 0.80 max | **ASME** SFA5.23 (ECB3) **AWS** A5.23 (ECB3) |

*Boxed entries are no longer active and are retained for reference purposes only.

The chemical compositions listed are for identification purposes and should not be used in lieu of the cross referenced specifications.

| UNIFIED NUMBER | DESCRIPTION | CHEMICAL COMPOSITION | CROSS REFERENCE SPECIFICATIONS |
|---|---|---|---|
| * **W53041** | Cr-Mo Low Alloy Steel Composite Wire for Submerged Arc Welding (ECB3N) Nuclear Grade | **C** 0.15 max **Cr** 2.00-2.50 **Cu** 0.08 max **Mn** 1.20 max **Mo** 0.90-1.20 **P** 0.012 max **S** 0.040 max **Si** 0.80 max **V** 0.05 max | **ASME** SFA5.23 (ECB3N) **AWS** A5.23 (ECB3N) |
| * **W53042** | Cr-Mo Low Alloy Steel Composite Wire for Submerged Arc Welding (ECB3RR) | **As** 0.005 max **C** 0.15 max **Cr** 2.00-2.50 **Cu** 0.15 max **Mn** 1.20 max **Mo** 0.90-1.20 **P** 0.010 max **S** 0.015 max **Sb** 0.005 max **Si** 0.80 max **Sn** 0.010 max | **ASME** SFA5.23 (ECB3RR) **AWS** A5.23 (ECB3RR) |
| * **W53043** | Cr-Mo Low Alloy Steel Composite Wire for Submerged Arc Welding (ECB3LRN) Nuclear Grade | **As** 0.005 max **C** 0.15 max **Cr** 2.00-2.50 **Cu** 0.08 max **Mn** 1.20 max **Mo** 0.90-1.20 **P** 0.010 max **S** 0.015 max **Sb** 0.005 max **Si** 0.80 max **V** 0.05 max | **ASME** SFA5.23 (ECB3LRN) **AWS** A5.23 (ECB3LRN) |
| **W53115** | Cr-Mo Low Alloy Steel Arc Welding Electrode (E8015-B3L) | **C** 0.05 max **Cr** 2.00-2.50 **Mn** 0.90 max **Mo** 0.90-1.20 **P** 0.03 max **S** 0.04 max **Si** 1.00 max | **ASME** SFA5.5 (E8015-B3L) **AWS** A5.5 (E8015-B3L) **MIL SPEC** MIL-E-22200/8 (MIL-8015-B3L) |
| **W53116** | Cr-Mo Low Alloy Steel Arc Welding Electrode | **C** 0.05 max **Cr** 2.0-2.50 **Mn** 0.60-0.85 **Mo** 0.90-1.20 **Si** 0.50-0.90 | |
| **W53118** | Cr-Mo Low Alloy Steel Arc Welding Electrode (E8018-B3L) | **C** 0.05 max **Cr** 2.00-2.50 **Mn** 0.90 max **Mo** 0.90-1.20 **P** 0.03 max **S** 0.03 max **Si** 0.80 max | **ASME** SFA5.5 (E8018-B3L) **AWS** A5.5 (E8018-B3L) **MIL SPEC** MIL-E-22200/8 (MIL-8018-B3L) |
| **W53130** | Cr-Mo Low Alloy Steel Composite Wire for Gas Shielded Arc Welding (E80C-B3L) | **C** 0.05 max **Cr** 2.00-2.50 **Cu** 0.35 max **Mn** 0.40-1.00 **Mo** 0.90-1.20 **Ni** 0.20 max **P** 0.025 max **S** 0.030 max **Si** 0.25-0.60 | **ASME** SFA5.28 (E80C-B3L) **AWS** A5.28 (E80C-B3L) |
| **W53131** | Cr-Mo Low Alloy Steel Flux Cored Wire (E9XT1-B3L) | **C** 0.05 max **Cr** 2.00-2.50 **Mn** 1.25 max **Mo** 0.90-1.20 **P** 0.03 max **S** 0.03 max **Si** 0.80 max | **ASME** SFA5.29 (E9XT1-B3L) **AWS** A5.29 (E9XT1-B3L) |
| **W53231** | Cr-Mo Low Alloy Steel Flux Cored Wire (E9XT1-B3H) | **C** 0.10-0.15 **Cr** 2.00-2.50 **Mn** 1.25 max **Mo** 0.90-1.20 **P** 0.03 max **S** 0.03 max **Si** 0.80 max | **ASME** SFA5.29 (E9XT1-B3H) **AWS** A5.29 (E9XT1-B3H) |
| * **W53315** | Cr-Mo Low Alloy Steel Arc Welding Electrode | **C** 0.10 max **Cr** 1.75-2.25 **Mo** 0.40-0.60 **Si** 0.60 max | |
| * **W53316** | Cr-Mo Low Alloy Steel Arc Welding Electrode | **C** 0.10 **Cr** 1.75-2.25 **Mo** 0.40-0.60 **Si** 0.60 max | |
| **W53340** | Cr-Mo Low Alloy Steel Tubular Wire for Submerged Arc Welding (ECB4) | **C** 0.12 max **Cr** 1.75-2.25 **Cu** 0.35 max **Mn** 1.20 max **Mo** 0.40-0.65 **P** 0.030 max **S** 0.030 max **Si** 0.80 max | **ASME** SFA5.23 (ECB4) **AWS** A5.23 (ECB4) |
| * **W53341** | Cr-Mo Low Alloy Composite Wire for Submerged Arc Welding (ECB4N) Nuclear Grade | **C** 0.12 max **Cr** 1.75-2.25 **Cu** 0.08 max **Mn** 1.20 max **Mo** 0.40-0.65 **P** 0.012 max **S** 0.040 max **Si** 0.80 max **V** 0.05 max | **ASME** SFA5.23 (ECB4N) **AWS** A5.23 (ECB4N) |
| **W53415** | Cr-Mo Low Alloy Steel Arc Welding Electrode (E8015-B4L) | **C** 0.05 max **Cr** 1.75-2.25 **Mn** 0.90 max **Mo** 0.40-0.65 **P** 0.03 max **S** 0.03 max **Si** 0.90 max | **ASME** SFA5.5 (E8015-B4L) **AWS** A5.5 (E8015-B4L) |

*Boxed entries are no longer active and are retained for reference purposes only.

The chemical compositions listed are for identification purposes and should not be used in lieu of the cross referenced specifications.

| UNIFIED NUMBER | DESCRIPTION | CHEMICAL COMPOSITION | CROSS REFERENCE SPECIFICATIONS |
|---|---|---|---|
| * **W53416** | Cr-Mo Low Alloy Steel Arc Welding Electrode | **C** 0.05 max **Cr** 1.75-2.25 **Mn** 0.60-0.85 **Mo** 0.45-0.65 **Si** 0.50-0.90 | |
| **W60189** | Copper Arc Welding Electrode (ECu) | **Al** 0.10 max **Cu** (incl Ag) rem **Fe** 0.2 max **Mn** 0.10 max **Si** 0.10 max | **ASME** SFA5.6 (ECu) **AWS** A5.6 (ECu) |
| **W60518** | Phosphor Bronze Composite Welding Wire (ECuSn-A) | **Al** 0.01 max **Cu** rem **Fe** 0.25 max **P** 0.05-0.35 **Pb** 0.02 max **Sn** 4.0-6.0 | **ASME** SFA5.6 (ECuSn-A); SFA5.13 (ECuSn-A); SFA5.21 (ERCCuSn-A) **AWS** A5.6 (ECuSn-A); A5.13 (ECuSn-A); A5.21 (ERCCuSn-A) |
| **W60521** | Phosphor Bronze C Arc Welding Electrode (ECuSn-C) | **Al** 0.01 max **Cu** rem **Fe** 0.25 max **P** 0.05-0.35 **Pb** 0.02 max **Sn** 7.0-9.0 | **ASME** SFA5.6 (ECuSn-C); SFA5.13 (ECuSn-C) **AWS** A5.6 (ECuSn-C); A5.13 (ECuSn-C) |
| **W60524** | Phosphor Bronze Composite Welding Wire for Hard Facing (ERCCuSn-D) | **Cu** 88.5 min **Pb** 0.05 max **P** 0.10-0.35 **Sn** 9.0-11.0 | **ASME** SFA5.21 (ERCCuSn-D **AWS** A5.21 (ERCCuSn-D) |
| **W60614** | Aluminum Bronze Arc Welding Electrode (ECuAl-A2) | **Al** 6.5-9.0 **Cu** rem **Fe** 0.50-5.0 **Pb** 0.02 max **Si** 1.5 max | **ASME** SFA5.6 (ECuAl-A2) **AWS** A5.6 (ECuAl-A2) |
| **W60618** | Aluminum Bronze Composite Welding Wire for Hard Facing (ERCCuAl-A2) | **Al** 8.5-11.0 **Cu** rem **Fe** 0.5-1.5 **Pb** 0.02 max **Si** 0.10 max **Zn** 0.02 max | **ASME** SFA5.21 (ERCCuAl-A2) **AWS** A5.21 (ERCCuAl-A2) |
| **W60619** | Aluminum Bronze Arc Welding Electrode (ECuAl-B) | **Al** 7.5-10.0 **Cu** rem **Fe** 2.5-5.0 **Pb** 0.02 max **Si** 1.5 max | **ASME** SFA5.6 (ECuAl-B); SFA5.13 (ECuAl-B) **AWS** A5.6 (ECuAl-B); A5.13 (ECuAl-B) |
| **W60624** | Aluminum Bronze Composite Welding Wire for Hard Facing (ERCCuAl-A3) | **Al** 12-13 **Cu** rem **Fe** 3.0-5.0 **Pb** 0.02 max **Si** 0.10 max **Zn** 0.10 max | **ASME** SFA5.21 (ERCCuAl-A3) **AWS** A5.21 (ERCCuAl-A3) |
| **W60625** | Aluminum Bronze Arc Welding Electrode (ECuAl-C) | **Al** 12-13 **Cu** rem **Fe** 3.0-5.0 **Pb** 0.02 max **Si** 1.0 max **Zn** 0.02 max | **ASME** SFA5.13 (ECuAl-C) **AWS** A5.13 (ECuAl-C) |
| **W60626** | Aluminum Bronze Composite Welding Wire for Hard Facing (ERCCuAl-C) | **Al** 12-13 **Cu** rem **Fe** 3.0-5.0 **Pb** 0.02 max **Si** 0.04 max **Zn** 0.02 max | **ASME** SFA5.21 (ERCCuAl-C) **AWS** A5.21 (ERCCuAl-C) |
| **W60632** | Ni-Al Bronze Arc Welding Electrode (ECuNiAl) | **Al** 6.0-8.5 **Cu** rem **Fe** 3.0-6.0 **Mn** 0.50-3.5 **Ni** 4.0-6.0 **Pb** 0.02 max **Si** 1.5 max | **ASME** SFA5.6 (ECuNiAl); SFA5.13 (ECuNiAl) **AWS** A5.6 (ECuNiAl); A5.13 (ECuNiAl) |
| **W60633** | Mn-Al Bronze Arc Welding Electrode (ECuMnNiAl) | **Al** 5.0-7.5 **Cu** rem **Fe** 2.0-6.0 **Mn** 11.0-13.0 **Ni** 1.0-2.5 **Pb** 0.02 max **Si** 1.5 max | **ASME** SFA5.6 (ECuMnNiAl); SFA5.13 (ECuMnNiAl) **AWS** A5.6 (ECuMnNiAl); A5.13 (ECuMnNiAl) |
| **W60656** | Silicon Bronze Arc Welding Electrode (ECuSi) | **Al** 0.01 max **Cu** rem **Fe** 0.5 max **Mn** 1.5 max **Pb** 0.02 max **Si** 2.4-4.0 **Sn** 1.5 max | **ASME** SFA5.6 (ECuSi); SFA5.13 (ECuSi) **AWS** A5.6 (ECuSi); A5.13 (ECuSi) |
| **W60657** | Silicon Bronze Composite Welding Wire for Hard Facing (ERCCuSi-A) | **Al** 0.01 max **Cu** rem **Fe** 0.50 max **Pb** 0.02 max **Mn** 1.5 max **Si** 2.8-4.0 **Sn** 1.0 max **Zn** 1.0 max | **ASME** SFA5.21 (ERCCuSi-A) **AWS** A5.21 (ERCCuSi-A) |
| **W60658** | Composite Thermal Spray Wire (CW-CuSnSi) | **Al** 0.5-1.5 **Cu** rem **Si** 1.5-2.5 **Sn** 5.0-6.0 | **AWS** C2.25 (CW-CuSnSi) |
| **W60715** | Copper-Nickel Arc Welding Electrode (ECuNi) | **Cu** rem **Fe** 0.4-0.75 **Mn** 1.0-2.5 **Ni** 29.0-33.0 **P** 0.020 max **Pb** 0.02 max **S** 0.015 max | **ASME** SFA5.6 (ECuNi); SFA5.13 (ECuNi) **AWS** A5.6 (ECuNi); A5.13 (ECuNi) **MIL SPEC** MIL-E-22200/4 (MIL-CuNi) |
| **W61625** | Aluminum Bronze Arc Welding Electrode (ECuAl-D) | **Al** 13-14 **Cu** rem **Fe** 3.0-5.0 **Pb** 0.02 max **Si** 1.0 max **Zn** 0.02 max | **ASME** SFA5.13 (ECuAl-D) **AWS** A5.13 (ECuAl-D) |
| **W61626** | Aluminum Bronze Composite Welding Wire for Hard Facing (ERCCuAl-D) | **Al** 13-14 **Cu** rem **Fe** 3.0-5.0 **Pb** 0.02 max **Si** 0.04 max **Zn** 0.02 max | **ASME** SFA5.21 (ERCCuAl-D) **AWS** A5.21 (ERCCuAl-D) |
| **W62625** | Aluminum Bronze Arc Welding Electrode (ECuAl-E) | **Al** 14-15 **Fe** 3.0-5.0 **Pb** 0.02 max **Si** 1.0 max **Zn** 0.02 max | **ASME** SFA5.13 (ECuAl-E) **AWS** A5.13 (ECuAl-E) |

*Boxed entries are no longer active and are retained for reference purposes only.

The chemical compositions listed are for identification purposes and should not be used in lieu of the cross referenced specifications.

| UNIFIED NUMBER | DESCRIPTION | CHEMICAL COMPOSITION | CROSS REFERENCE SPECIFICATIONS |
|---|---|---|---|
| **W62626** | Aluminum Bronze Composite Welding Wire for Hard Facing (ERCCuAl-E); Thermal Spraying Filler Metal (CW-CuAlFe) | **Al** 14-15 **Cu** rem **Fe** 3.0-5.0 **Pb** 0.02 max **Si** 0.04 max **Zn** 0.02 max | **ASME** SFA5.21 (ERCCuAl-E) **AWS** A5.21 (ERCCuAl-E) |
| **W73001** | Cobalt-Base Hard Facing Electrode (ECoCr-C) (Stellite 1) | **C** 1.7-3.0 **Co** rem **Cr** 25-33 **Fe** 5.0 max **Mn** 2.0 max **Mo** 1.0 max **Ni** 3.0 max **Si** 2.0 max **W** 11-14 | **ASME** SFA5.13 (ECoCr-C) **AWS** A5.13 (ECoCr-C) |
| **W73006** | Cobalt-Base Hard Facing Electrode (ECoCr-A) (Stellite 6) | **C** 0.7-1.4 **Co** rem **Cr** 25-32 **Fe** 5.0 max **Mn** 2.0 max **Mo** 1.0 max **Ni** 3.0 max **Si** 2.0 max **W** 3.0-6.0 | **ASME** SFA5.13 (ECoCr-A) **AWS** A5.13 (ECoCr-A) |
| **W73012** | Cobalt-Base Hard Facing Electrode (ECoCr-B) (Stellite 12) | **C** 1.0-1.7 **Co** rem **Cr** 25-32 **Fe** 5.0 max **Mn** 2.0 max **Mo** 1.0 max **Ni** 3.0 max **Si** 2.0 max **W** 7.0-9.5 | **ASME** SFA5.13 (ECoCr-B) **AWS** A5.13 (ECoCr-B) |
| **W73021** | Cobalt Base Hard Facing Electrode (ECoCr-E) | **C** 0.15-0.40 **Cr** 24-29 **Co** rem **Fe** 5.0 max **Mn** 1.5 max **Mo** 4.5-6.5 **Ni** 2.0-4.0 **Si** 2.0 max **W** 0.50 max | **ASME** SFA5.13 (ECoCr-E) **AWS** A5.13 (ECoCr-E) |
| **W73031** | Cobalt-base Hard Facing Composite Welding Wire (ERCCoCr-C) | **C** 2.0-3.0 **Cr** 25-33 **Co** rem **Fe** 5.0 max **Mn** 2.0 max **Mo** 1.0 max **Ni** 3.0 max **Si** 2.0 max **W** 11-14 | **ASME** SFA5.21 (ERCCoCr-C) **AWS** A5.21 (ERCCoCr-C) |
| **W73032** | Cobalt-base Hard Facing Composite Welding Wire (ERCCoCr-G) | **C** 3.0-4.0 **Cr** 24-30 **Co** rem **Fe** 5.0 max **Mn** 1.0 max **Mo** 1.0 max **Ni** 4.0 max **Si** 2.0 max **W** 12-16 | **ASME** SFA5.21 (ERCCoCr-G) **AWS** A5.21 (ERCCoCr-G) |
| **W73036** | Cobalt-base Hard Facing Composite Welding Wire (ERCCoCr-A) | **C** 0.7-1.4 **Cr** 25-32 **Co** rem **Fe** 5.0 max **Mn** 2.0 max **Mo** 1.0 max **Ni** 3.0 max **Si** 2.0 max **W** 3.0-6.0 | **ASME** SFA5.21 (ERCCoCr-A) **AWS** A5.21 (ERCCoCr-A) |
| **W73041** | Cobalt-base Hard Facing Composite Welding Wire (ERCCoCr-E) | **C** 0.15-0.40 **Cr** 25-30 **Co** rem **Fe** 5.0 max **Mn** 2.0 max **Mo** 4.5-7.0 **Ni** 1.5-4.0 **Si** 1.5 max **W** 0.50 max | **ASME** SFA5.21 (ERCCoCr-E) **AWS** A5.21 (ERCCoCr-E) |
| **W73042** | Cobalt-base Hard Facing Composite Welding Wire (ERCCoCr-B) | **C** 1.2-2.0 **Cr** 25-32 **Co** rem **Fe** 5.0 max **Mn** 2.0 max **Mo** 1.0 max **Ni** 3.0 max **Si** 2.0 max **W** 7-10 | **ASME** SFA5.21 (ERCCoCr-B) **AWS** A5.21 (ERCCoCr-B) |
| **W73136** | Composite Thermal Spray Wire (CW-CoCrW) | **C** 0.90-1.30 **Co** rem **Cr** 27-31 **Fe** 3 max **Mn** 1.0 max **Mo** 1.0 max **Ni** 3 max **Si** 1.5 max **W** 4.0-6.0 | **AWS** C2.25 (CW-CoCrW) |
| **W73138** | Composite Thermal Spray Wire (CW-CoMoCr) | **C** 0.08 max **Co** rem **Cr** 17-19 **Mo** 27-30 **Si** 3.2-3.8 | **AWS** C2.25 (CW-CoMoCr) |
| **W73155** | Fe-Cr-Ni-Co Superalloy Arc Welding Electrode | **C** 0.15 max **Cb** 0.75-1.25 **Co** 18.5-21.0 **Cr** 20.0-22.5 **Fe** rem **Mn** 1.00-2.50 **Mo** 2.5-3.5 **Ni** 19.0-21.0 **P** 0.040 max **S** 0.030 max **Si** 1.0 max **W** 2.0-3.0 | **AMS** 5795 **MIL SPEC** MIL-E-22200/3 (MIL-3N1N) |
| **W73605** | Cobalt-Base Superalloy Arc Welding Electrode (L605) | **C** 0.15 max **Co** rem **Cr** 19.0-21.0 **Fe** 3.0 max **Mn** 0.5-2.0 **Ni** 9.0-11.0 **P** 0.030 max **S** 0.030 max **Si** 1.0 max **W** 14.0-16.0 | **AMS** 5797 **MIL SPEC** MIL-E-22200/3 (MIL-3N1L) |
| **W74001** | Steel Arc Welding Electrodes for Surfacing (EFe-1) | **C** 0.04-0.20 **Cr** 0.5-3.5 **Mn** 0.5-2.0 **Mo** 1.5 max **P** 0.035 max **S** 0.035 max **Si** 1.0 max | **ASME** SFA5.13 (EFe-1) **AWS** A5.13 (EFe-1) |
| **W74002** | Steel Arc Welding Electrodes for Surfacing (EFe-2) | **C** 0.10-0.30 **Cr** 1.8-3.8 **Mn** 0.5-2.0 **Mo** 1.0 max **Ni** 1.0 max **P** 0.035 max **S** 0.035 max **Si** 1.0 max **V** 0.35 max | **ASME** SFA5.13 (EFe-2) **AWS** A5.13 (EFe-2) |
| **W74003** | Steel Arc Welding Electrodes for Surfacing (EFe-3) | **C** 0.50-0.80 **Cr** 4.0-8.0 **Mn** 0.5-1.5 **Mo** 1.0 max **P** 0.035 max **S** 0.035 max **Si** 1.0 max | **ASME** SFA5.13 (EFe-3) **AWS** A5.13 (EFe-3) |
| **W74004** | Steel Arc Welding Electrodes for Surfacing (EFe-4) | **C** 1.0-2.0 **Cr** 3.0-5.0 **Mn** 0.5-2.0 **P** 0.035 max **S** 0.035 max **Si** 1.0 max | **ASME** SFA5.13 (EFe-4) **AWS** A5.13 (EFe-4) |
| **W74011** | Austenitic Chromium Iron Arc Welding Electrodes for Surfacing (EFeCr-A1A) | **C** 3.5-4.5 **Cr** 20-25 **Fe** rem **Mn** 4.0-6.0 **Mo** 0.5 max **P** 0.035 max **S** 0.035 max **Si** 0.5-2.0 | **ASME** SFA5.13 (EFeCr-A1A) **AWS** A5.13 (EFeCr-A1A) |

The chemical compositions listed are for identification purposes and should not be used in lieu of the cross referenced specifications.

| UNIFIED NUMBER | DESCRIPTION | CHEMICAL COMPOSITION | CROSS REFERENCE SPECIFICATIONS |
|---|---|---|---|
| **W74012** | Austenitic Chromium Iron Arc Welding Electrodes for Surfacing (EFeCr-A2) | **C** 2.5-3.5 **Cr** 7.5-9.0 **Fe** rem **Mn** 0.5-1.5 **P** 0.035 max **S** 0.035 max **Si** 0.5-1.5 **Ti** 1.2-1.8 | **ASME** SFA5.13 (EFeCr-A2) **AWS** A5.13 (EFeCr-A2) |
| **W74013** | Austenitic Chromium Iron Arc Welding Electrodes for Surfacing (EFeCr-A3) | **C** 2.5-4.5 **Cr** 14-20 **Mn** 0.5-2.0 **Mo** 1.5 max **P** 0.035 max **S** 0.035 max **Si** 1.0-2.5 | **ASME** SFA5.13 (EFeCr-A3) **AWS** A5.13 (EFeCr-A3) |
| **W74014** | Austenitic Chromium Iron Arc Welding Electrodes for Surfacing (EFeCr-A4) | **C** 3.5-4.5 **Cr** 23.0-29.0 **Fe** rem **Mn** 1.5-3.5 **Mo** 1.0-3.0 **P** 0.035 max **S** 0.035 max **Si** 1.5 max | **ASME** SFA5.13 (EFeCr-A4) **AWS** A5.13 (EFeCr-A4) |
| **W74015** | Austenitic Chromium Iron Arc Welding Electrodes for Surfacing (EFeCr-A5) | **C** 1.5-2.5 **Cr** 24-32 **Fe** rem **Mn** 0.5-1.5 **Mo** 4.0 max **Ni** 4.0 max **P** 0.035 max **S** 0.035 max **Si** 2.0 max | **ASME** SFA5.13 (EFeCr-A5) **AWS** A5.13 (EFeCr-A5) |
| **W74016** | Austenitic Chromium Iron Arc Welding Electrodes for Surfacing (EFeCr-A6) | **C** 2.5-3.5 **Cr** 24-30 **Fe** rem **Mn** 0.5-1.5 **Mo** 0.5-2.0 **P** 0.035 max **S** 0.035 max **Si** 1.0-2.5 | **ASME** SFA5.13 (EFeCr-A6) **AWS** A5.13 (EFeCr-A6) |
| **W74017** | Austenitic Chromium Iron Arc Welding Electrodes for Surfacing (EFeCr-A7) | **C** 3.5-5.0 **Cr** 23-30 **Fe** rem **Mn** 0.5-1.5 **Mo** 2.0-4.5 **P** 0.035 max **S** 0.035 max **Si** 0.5-2.5 | **ASME** SFA5.13 (EFeCr-A7) **AWS** A5.13 (EFeCr-A7) |
| **W74018** | Austenitic Chromium Iron Arc Welding Electrodes for Surfacing (EFeCr-A8) | **C** 2.5-4.5 **Cr** 30-40 **Fe** rem **Mn** 0.5-1.5 **Mo** 2.0 max **P** 0.035 max **S** 0.035 max **Si** 1.5 max | **ASME** SFA5.13 (EFeCr-A8) **AWS** A5.13 (EFeCr-A8) |
| **W74030** | Cr-Steel Composite Welding Rod for Machinable Surfaces (ERCFe-1) | **C** 0.04-0.20 **Cr** 0.5-3.5 **Fe** rem **Mn** 0.5-2.0 **Mo** 1.5 max **Si** 1.0 max | **ASME** SFA5.21 (ERCFe-1) **AWS** A5.21 (ERCFe-1) |
| **W74031** | Cr-Steel Composite Welding Rod for Machinable Surfaces (ERCFe-1A) | **C** 0.05-0.25 **Cr** 0.5-3.5 **Fe** rem **Mn** 1.7-3.5 **Si** 1.0 max | **ASME** SFA5.21 (ERCFe-1A) **AWS** A5.21 (ERCFe-1A) |
| **W74032** | Cr-Steel Composite Welding Rod for Machinable Surfaces (ERCFe-2) | **C** 0.10-0.30 **Cr** 1.8-3.8 **Fe** rem **Mn** 0.5-2.0 **Mo** 1.0 max **Ni** 1.0 max **Si** 1.0 max **V** 0.35 max | **ASME** SFA5.21 (ERCFe-2) **AWS** A5.21 (ERCFe-2) |
| **W74033** | Cr-Steel Composite Welding Rod for Tool and Die Surfacing (ERCFe-3) | **C** 0.50-0.80 **Cr** 4.0-8.0 **Fe** rem **Mn** 0.5-1.5 **Mo** 1.0 max **Si** 1.0 max | **ASME** SFA5.21 (ERCFe-3) **AWS** A5.21 (ERCFe-3) |
| **W74035** | Cr-Steel Composite Welding Rod for Tool and Die Surfacing (ERCFe-5) | **C** 0.50-0.80 **Cr** 1.5-3.0 **Fe** rem **Mn** 1.5-2.5 **Si** 0.9 max | **ASME** SFA5.21 (ERCFe-5) **AWS** A5.21 (ERCFe-5) |
| **W74211** | Austenitic Chromium Iron Arc Welding Electrodes for Surfacing (EFeCr-E1) | **C** 5.0-6.5 **Cr** 12-16 **Fe** rem **Mn** 2.0-3.0 **P** 0.035 max **S** 0.035 max **Si** 0.8-1.5 max **Ti** 4.0-7.0 | **ASME** SFA5.13 (EFeCr-E1) **AWS** A5.13 (EFeCr-E1) |
| **W74212** | Austenitic Chromium Iron Arc Welding Electrodes for Surfacing (EFeCr-E2) | **C** 4.0-6.0 **Cr** 14-20 **Fe** rem **Mn** 0.5-1.5 **Mo** 5.0-7.0 **P** 0.035 max **S** 0.035 max **Si** 1.5 max **V** 1.5 max | **ASME** SFA5.13 (EFeCr-E2) **AWS** A5.13 (EFeCr-E2) |
| **W74213** | Austenitic Chromium Iron Arc Welding Electrodes for Surfacing (EFeCr-E3) | **C** 5.0-7.0 **Cr** 18-28 **Fe** rem **Mn** 0.5-2.0 **Mo** 5.0-7.0 **P** 0.035 max **S** 0.035 max **Si** 0.5-2.0 **W** 3.0-5.0 | **ASME** SFA5.13 (EFeCr-E3) **AWS** A5.13 (EFeCr-E3) |
| **W74214** | Austenitic Chromium Iron Arc Welding Electrodes for Surfacing (EFeCr-E4) | **C** 4.0-6.0 **Cb** 4.0-7.0 **Cr** 20-30 **Fe** rem **Mn** 0.5-1.5 **Mo** 5.0-7.0 **P** 0.035 max **S** 0.035 max **Si** 1.0 max **V** 0.5-1.5 **W** 2.0 max | **ASME** SFA5.13 (EFeCr-E4) **AWS** A5.13 (EFeCr-E4) |
| * **W74510** | Austenitic Chromium Iron Arc Welding Electrode for Surfacing (EFeCr-A1) | **C** 3.0-5.0 **Cr** 26.0-32.0 **Fe** rem **Mn** 4.0-8.0 **Mo** 2.0 **Si** 1.0-2.5 | **ASME** SFA5.13 (EFeCr-A1); SFA5.21 (EFeCr-A1) **AWS** A5.13 (EFeCr-A1); A5.21 (EFeCr-A1) |
| **W74530** | Austenitic Chromium Iron Composite Welding Rod for Surfacing (ERCFeCr-A1A) | **C** 3.5-5.5 **Cr** 20-25 **Fe** rem **Mn** 4.0-6.0 **Mo** 0.5 max **Si** 0.5-2.0 | **ASME** SFA5.21 (ERCFeCr-A1A) **AWS** A5.21 (ERCFeCr-A1A) |
| **W74531** | Austenitic Chromium Iron Composite Welding Rod for Surfacing (ERCFeCr-A) | **C** 1.5-3.5 **Cr** 8.0-14.0 **Fe** rem **Mn** 0.5-1.5 **Mo** 1.0 max **Si** 2.0 max | **ASME** SFA5.21 (ERCFeCr-A) **AWS** A5.21 (ERCFeCr-A) |

*Boxed entries are no longer active and are retained for reference purposes only.

The chemical compositions listed are for identification purposes and should not be used in lieu of the cross referenced specifications.

| UNIFIED NUMBER | DESCRIPTION | CHEMICAL COMPOSITION | CROSS REFERENCE SPECIFICATIONS |
|---|---|---|---|
| **W74532** | Composite Thermal Spray Wire (CW-FeCrC-1) | **C** 4.0-5.0 **Cr** 18-21 **Fe** rem **Mn** 2.5 max **Mo** 2.0 max **Si** 2.5 max | **AWS** C2.25 (CW-FeCrC-1) |
| **W74533** | Austenitic Chromium Iron Composite Welding Rod for Surfacing (ERCFeCr-A3A) | **C** 2.5-3.5 **Cr** 14-20 **Fe** rem **Mn** 1.5-3.5 **Si** 0.5-2.0 | **ASME** SFA5.21 (ERCFeCr-A3A) **AWS** A5.21 (ERCFeCr-A3A) |
| **W74534** | Austenitic Chromium Iron Composite Welding Rod for Surfacing (ERCFeCr-A4) | **C** 3.5-4.5 **Cr** 23-29 **Fe** rem **Mn** 1.5-3.5 **Mo** 1.0-3.0 **Si** 1.5 max | **ASME** SFA5.21 (ERCFeCr-A4) **AWS** A5.21 (ERCFeCr-A4) |
| **W74535** | Austenitic Chromium Iron Composite Welding Rod for Surfacing (ERCFeCr-A5); Thermal Spraying Filler Metal (CW-FeCrC-2) | **C** 1.5-2.5 **Cr** 24-32 **Fe** rem **Mn** 0.5-1.5 **Mo** 4.0 max **Ni** 4.0 max **Si** 2.0 max | **ASME** SFA5.21 (ERCFeCr-A5) **AWS** A5.21 (ERCFeCr-A5); C2.25 (CW-FeCrC-2) |
| **W74539** | Austenitic Chromium Iron Composite Welding Rod for Surfacing (ERCFeCr-A9) | **C** 3.5-5.0 **Cr** 24-30 **Fe** rem **Mn** 0.5-1.5 **Si** 2.5 max | **ASME** SFA5.21 (ERCFeCr-A9) **AWS** A5.21 (ERCFeCr-A9) |
| **W74540** | Austenitic Chromium Iron Composite Welding Rod for Surfacing (ERCFeCr-A10) | **C** 5.0-7.0 **Cr** 20-25 **Fe** rem **Mn** 0.5-2.5 **Si** 1.5 max | **ASME** SFA5.21 (ERCFeCr-A10) **AWS** A5.21 (ERCFeCr-A10) |
| **W74630** | Composite Thermal Spray Wire (CW-CrFeB) | **B** 7-9 **Cr** rem **Ir** 34-36 **Mo** 3.5-4.5 **Si** 2.5-3.5 | **AWS** C2.25 (CW-CrFeB) |
| **W75010** | Air Hardening Steel Arc Welding Electrode for Surfacing (A2a) | **C** 0.30 max **Cr** 0.65 max **Mn** 0.50-1.60 **Mo** 0.70 max **P** 0.04 max **S** 0.04 max **Si** 0.20-0.65 | **MIL SPEC** MIL-E-19141 (MIL-I-A2a) |
| **W75110** | Steel Arc Welding Electrodes for Surfacing (EFe-5) | **C** 0.80 max **Cr** 1.5-3.0 **Mn** 1.5-2.5 **P** 0.035 max **S** 0.035 max **Si** 0.90 max | **ASME** SFA5.13 (EFe-5) **AWS** A5.13 (EFe-5) **MIL SPEC** MIL-E-19141 (MIL-I-A2c) |
| * **W77310** | High-Speed Steel Arc Welding Electrode for Surfacing (EFe5-C) | **C** 0.3-0.5 **Cr** 3.0-5.0 **Fe** rem **Mn** 0.60 max **Mo** 5.0-9.0 **Si** 0.80 max **V** 0.8-1.2 **W** 1.0-2.5 | **ASME** SFA5.13 (EFe5-C); SFA5.21 (EFe5-C) **AWS** A5.13 (EFe5-C); A5.21 (EFe5-C) |
| * **W77410** | Austenitic Manganese Steel Arc Welding Electrodes for Surfacing (EFeMn-D) | **C** 0.50-1.0 **Cr** 4.5-7.5 **Mn** 15.0-20.0 **P** 0.035 max **Si** 1.3 max **S** 0.035 max **V** 0.4-1.2 | **ASME** SFA5.13 (EFeMn-D) **AWS** A5.13 (EFeMn-D) |
| **W77510** | Steel Arc Welding Electrodes for Surfacing (EFe-6) | **C** 0.60-1.0 **Cr** 3.0-5.0 **Mn** 0.40-1.0 **Mo** 7.0-9.0 **P** 0.035 max **S** 0.035 max **Si** 1.0 max **V** 0.50-1.5 **W** 0.50-1.5 | **ASME** SFA5.13 (EFe-6) **AWS** A5.13 (EFe-6) **MIL SPEC** MIL-E-19141 (MIL-Fe5-B) |
| **W77530** | High Speed Steel Composite Welding Rod for Surfacing (ERCFe-6) | **C** 0.6-1.0 **Cr** 3.0-5.0 **Fe** rem **Mn** 0.4-1.0 **Mo** 7.5-9.5 **Si** 1.0 max **V** 0.5-1.5 **W** 0.5-1.5 | **ASME** SFA5.21 (ERCFe-6) **AWS** A5.21 (ERCFe-6) |
| **W77538** | High Speed Steel Composite Welding Rod for Surfacing (ERCFe-8) | **C** 0.30-0.60 **Cr** 4.0-8.0 **Fe** rem **Mn** 1.0-2.0 **Mo** 1.0-2.0 **Si** 1.0 max **W** 1.0-2.0 **V** 0.50 max | **ASME** SFA5.21 (ERCFe-8) **AWS** A5.21 (ERCFe-8) |
| **W77540** | Composite Thermal Spray Wire (CW-FeVWC) | **B** 1.0 max **C** 3.5-4.5 **Fe** rem **Mn** 1.5 max **Mo** 1.0 max **Ni** 1.0 max **Si** 2.0 max **V** 15-17 **W** 5.0-7.0 | **AWS** C2.25 (CW-FeVWC) |
| **W77610** | Steel Arc Welding Electrodes for Surfacing (EFe-7) | **C** 1.5-3.0 **Cr** 4.0-8.0 **Mn** 0.5-2.0 **Mo** 1.0 max **P** 0.035 max **S** 0.035 max **Si** 1.5 max | **ASME** SFA5.13 (EFe-7) **AWS** A5.13 (EFe-7) |
| **W77640** | Composite Thermal Spray Wire (CW-FeWC) | **C** 2.0-3.0 **Fe** rem **Mn** 1.0 max **W** 56-60 | **AWS** C2.25 (CW-FeWC) |
| * **W77710** | High-Speed Steel Arc Welding Electrode for Surfacing (EFe5-A) | **C** 0.7-1.0 **Cr** 3.0-5.0 **Fe** rem **Mn** 0.60 max **Mo** 4.0-6.0 **Si** 0.80 max **V** 1.0-2.5 **W** 5.0-7.0 | **ASME** SFA5.13 (EFe5-A); SFA5.21 (EFe5-A) **AWS** A5.13 (EFe5-A); A5.21 (EFe5-A) |

*Boxed entries are no longer active and are retained for reference purposes only.

The chemical compositions listed are for identification purposes and should not be used in lieu of the cross referenced specifications.

| UNIFIED NUMBER | DESCRIPTION | CHEMICAL COMPOSITION | CROSS REFERENCE SPECIFICATIONS |
|---|---|---|---|
| * W77730 | High-Speed Steel Composite Welding Rod for Surfacing (RFe5-A) | **C** 0.7-1.0 **Cr** 3.0-5.0 **Fe** rem **Mn** 0.50 max **Mo** 4.0-6.0 **Si** 0.50 max **V** 1.0-2.5 **W** 5.0-7.0 | **ASME** SFA5.21 (RFe5-A) **AWS** A5.21 (RFe5-A) |
| W79110 | Austenitic Manganese Steel Arc Welding Electrode for Surfacing (EFe-Mn-A) | **C** 0.50-1.0 **Mn** 12-16 **Ni** 2.5-5.0 **P** 0.035 max **S** 0.035 max **Si** 1.3 max | **ASME** SFA5.13 (EFe-Mn-A) **AWS** A5.13 (EFeMn-A) **MIL SPEC** MIL-E-19141 (MIL-FeMn-A) |
| W79210 | Austenitic Manganese Steel Arc Welding Electrodes for Surfacing (EFeMn-C) | **C** 0.5-1.0 **Cr** 2.5-5.0 **Mn** 12-16 **Ni** 2.5-5.0 **P** 0.035 max **Si** 1.3 max **S** 0.035 max | **ASME** SFA5.13 (EFeMn-C) **AWS** A5.13 (EFeMn-C) |
| W79230 | Austenitic Manganese Composite Welding Rod for Surfacing (ERCFeMn-C) | **C** 0.5-1.0 **Cr** 2.5-5.0 **Fe** rem **Mn** 12-16 **Ni** 2.5-5.0 **Si** 1.3 max | **ASME** SFA5.21 (ERCFeMn-C) **AWS** A5.21 (ERCFeMn-C) |
| W79231 | Austenitic Manganese Composite Welding Rod for Surfacing (ERCFeMn-G) | **C** 0.5-1.0 **Cr** 2.5-5.0 **Fe** rem **Mn** 12-16 **Ni** 1.0 max **Si** 1.3 max | **ASME** SFA5.21 (ERCFeMn-G) **AWS** A5.21 (ERCFeMn-G) |
| W79232 | Austenitic Manganese Composite Welding Rod for Surfacing (ERCFeMn-H) | **C** 0.30-0.80 **Cr** 4.5-7.5 **Fe** rem **Mn** 12-16 **Ni** 2.0 max **Si** 1.3 max | **ASME** SFA5.21 (ERCFeMn-H) **AWS** A5.21 (ERCFeMn-H) |
| W79310 | Austenitic Manganese Steel Arc Welding Electrode for Surfacing (EFeMn-B) | **C** 0.5-1.0 **Mn** 12-16 **Mo** 0.5-1.5 **P** 0.03 max **Si** 1.3 max | **ASME** SFA5.13 (EFeMn-B) **AWS** A5.13 (EFeMn-B) |
| W79410 | Austenitic Manganese Steel Arc Welding Electrode for Surfacing (EFeMn-D) | **C** 0.5-1.0 **Cr** 4.5-7.5 **Mn** 15-20 **P** 0.035 max **S** 0.035 max **Si** 1.3 max **V** 0.4-1.2 | **ASME** SFA5.13 (EFeMn-D) **AWS** A5.13 (EFeMn-D) |
| W79510 | Austenitic Manganese Steel Arc Welding Electrodes for Surfacing (EFeMn-E) | **C** 0.5-1.0 **Cr** 3.0-6.0 **Mn** 15-20 **Ni** 1.0 max **P** 0.035 max **S** 0.035 max **Si** 1.3 max | **ASME** SFA5.13 (EFeMn-E) **AWS** A5.13 (EFeMn-E) |
| W79610 | Austenitic Manganese Steel Arc Welding Electrodes for Surfacing (EFeMn-F) | **C** 0.8-1.2 **Cr** 3.0-6.0 **Mn** 17-21 **Ni** 1.0 max **P** 0.035 max **S** 0.035 max **Si** 1.3 max | **ASME** SFA5.13 (EFeMn-F) **AWS** A5.13 (EFeMn-F) |
| W79630 | Austenitic Manganese Composite Welding Rod for Surfacing (ERCFeMn-F) | **C** 0.7-1.1 **Cr** 2.5-5.0 **Fe** rem **Mn** 16-22 **Ni** 1.0 max **Si** 1.3 max | **ASME** SFA5.21 (ERCFeMn-F) **AWS** A5.21 (ERCFeMn-F) |
| W79710 | Austenitic Manganese Steel Arc Welding Electrodes for Surfacing (EFeMnCr) | **C** 0.25-0.75 **Cr** 13-17 **Mn** 12-18 **Mo** 2.0 max **Ni** 0.5-2.0 **P** 0.035 max **S** 0.035 max **Si** 1.3 max **V** 1.0 max | **ASME** SFA5.13 (EFeMnCr) **AWS** A5.13 (EFeMnCr) |
| W79730 | Austenitic Manganese Composite Welding Rod for Surfacing (ERCFeMn-Cr) | **C** 0.25-0.75 **Cr** 11-16 **Fe** rem **Mn** 12-18 **Mo** 2.0 max **Ni** 2.0 max **Si** 1.3 max | **ASME** SFA5.21 (ERCFeMn-Cr) **AWS** A5.21 (ERCFeMn-Cr) |
| W79840 | Composite Thermal Spray Wire (CW-FeMnAl) | **Al** 4-5 **Fe** rem **Mn** 6-8 **Si** 1-2 | **AWS** C2.25 (CW-FeMnAl) |
| W80001 | Ni-Mo Arc Welding Electrode 3N1B (Hastelloy B) | **C** 0.07 max **Co** 2.5 max **Cr** 1.0 max **Cu** 0.50 max **Fe** 4.0-7.0 **Mn** 1.0 max **Mo** 26.0-30.0 **Ni** rem **P** 0.04 max **S** 0.03 max **Si** 1.0 max **V** 0.60 max **W** 1.0 max | **ASME** SFA5.11 (ENiMo-1) **AWS** A5.11 (ENiMo-1) **MIL SPEC** MIL-E-22200/3 (MIL-3N1B) |
| W80002 | Ni-Mo Arc Welding Electrode 3N1C (Hastelloy C) | **C** 0.10 max **Co** 2.5 max **Cr** 14.5-16.5 **Cu** 0.50 max **Fe** 4.0-7.0 **Mn** 1.0 max **Mo** 15.0-17.0 **Ni** rem **P** 0.04 max **S** 0.03 max **Si** 1.0 max **W** 3.0-4.5 | **ASME** SFA5.11 (ENiCrMo-5); SFA5.13 (ENiCrMo-5A) **AWS** A5.11 (ENiCrMo-5); A5.13 (ENiCrMo-5A) **MIL SPEC** MIL-E-22200/3 (MIL-3N1C) |
| W80004 | Ni-Mo Arc Welding Electrode ENiMo-3 (Hastelloy W) | **C** 0.12 max **Co** 2.5 max **Cr** 2.5-5.5 **Cu** 0.50 max **Fe** 4.0-7.0 **Mn** 1.0 max **Mo** 23.0-27.0 **Ni** rem **P** 0.04 max **S** 0.03 max **Si** 1.0 max **V** 0.60 max **W** 1.0 max | **ASME** SFA5.11 (ENiMo-3) **AWS** A5.11 (ENiMo-3) **MIL SPEC** MIL-E-22200/3 (MIL-4N1W) |
| W80008 | Ni-Mo-W Arc Welding Electrode (ENiMo-8) | **C** 0.10 max **Cr** 0.5-3.5 **Cu** 0.50 max **Fe** 10.0 max **Mn** 1.5 max **Mo** 17.0-20.0 **Ni** 60.0 min **P** 0.02 max **Si** 0.75 max **S** 0.015 max **W** 2.0-4.0 | **ASME** SFA5.11 (ENiMo-8) **AWS** A5.11 (ENiMo-8) |

*Boxed entries are no longer active and are retained for reference purposes only.

The chemical compositions listed are for identification purposes and should not be used in lieu of the cross referenced specifications.

| UNIFIED NUMBER | DESCRIPTION | CHEMICAL COMPOSITION | CROSS REFERENCE SPECIFICATIONS |
|---|---|---|---|
| **W80009** | Ni-Mo-W Arc Welding Electrode (ENiMo-9) | **C** 0.10 max **Cu** 0.3-1.3 **Fe** 7.0 max **Mn** 1.5 max **Mo** 18.0-22.0 **Ni** 62.0 min **P** 0.02 max **Si** 0.75 max **S** 0.015 max **W** 2.0-4.0 | **ASME** SFA5.11 (ENiMo-9) **AWS** A5.11 (ENiMo-9) |
| * **W80032** | Ni-Cr-Mo Composite Wire for Hard Facing (ERCNiCrMo-5A) | **C** 0.12 max **Cr** 14-18 **Fe** 4.0-7.0 **Mn** 1.0 max **Mo** 14-18 **Ni** rem **Si** 2.0 max **W** 3.0-5.0 **V** 0.40 max | **ASME** SFA5.21 (ERCNICrMo-5A) **AWS** A5.21 (ERCNiCrMo-5A) |
| **W80276** | Ni-Cr-Mo Arc Welding Electrodes (ENiCrMo-4) (Hastelloy C-276) | **C** 0.02 max **Co** 2.5 max **Cr** 14.5-16.5 **Cu** 0.50 max **Fe** 4.0-7.0 **Mn** 1.0 max **Mo** 15.0-17.0 **Ni** rem **P** 0.04 max **S** 0.03 max **Si** 0.2 max **V** 0.35 max **W** 3.0-4.5 | **ASME** SFA5.11 (ENiCrMo-4); SFA5.34 (ENiCrMo4T-X) **AWS** A5.11 (ENiCrMo-4); A5.34 (ENiCrMo4T-X) |
| **W80665** | Ni-Mo Arc Welding Electrode ENiMo-7 (Hastelloy B-2) | **C** 0.02 max **Co** 1.0 max **Cr** 1.0 max **Cu** 0.50 max **Fe** 2.25 max **Mn** 1.75 max **Mo** 26.0-30.0 **Ni** rem **P** 0.04 max **S** 0.03 max **Si** 0.2 max **W** 1.0 max | **ASME** SFA5.11 (ENiMo-7) **AWS** A5.11 (ENiMo-7) |
| **W80675** | Ni-Mo Arc Welding Electrode (ENiMo-10) | **C** 0.02 max **Cr** 1.0-3.0 **Co** 3.0 max **Cu** 0.50 max **Fe** 1.0-3.0 **Mn** 2.0 max **Mo** 27.0-32.0 **Ni** rem **P** 0.04 max **Si** 0.2 max **S** 0.03 max **W** 3.0 max | **ASME** SFA5.11 (ENiMo-10) **AWS** A5.11 (ENiMo-10) |
| **W82001** | Ni Arc Welding Electrode for Cast Iron (ENi-CI) | **Al** 1.0 max **C** 2.0 max **Cu** 2.5 max **Fe** 8.0 max **Mn** 2.5 max **Ni** 85 min **S** 0.03 max **Si** 4.0 max **Other** 1.0 max total | **AWS** A5.15 (ENi-CI) |
| **W82002** | Ni-Fe Arc Welding Electrode for Cast Iron (ENiFe-CI) | **Al** 1.0 max **C** 2.0 max **Cu** 2.5 max **Fe** rem **Mn** 2.5 max **Ni** 45-60 **S** 0.03 max **Si** 4.0 max **Other** 1.0 max total | **ASME** SFA5.15 (ENiFe-CI) **AWS** A5.15 (ENiFe-CI) |
| **W82003** | Ni Arc Welding Electrode for Cast Iron (ENi-CI-A) | **Al** 1.0-3.0 **C** 2.00 max **Cu** 2.5 max **Fe** 8.0 max **Mn** 2.5 max **Ni** 85 min **S** 0.03 max **Si** 4.0 max **Other** 1.0 max total | **ASME** SFA5.15 (ENi-CI-A) **AWS** A5.15 (ENi-CI-A) |
| **W82004** | Ni-Fe Arc Welding Electrode for Cast Iron (ENiFe-CI-A) | **Al** 1.0-3.0 **C** 2.0 max **Cu** 2.5 max **Fe** rem **Mn** 2.5 max **Ni** 45-60 **S** 0.03 max **Si** 4.0 max **Other** 1.0 max total | **ASME** SFA5.15 (ENiFe-CI-A) **AWS** A5.15 (ENiFe-CI-A) |
| **W82006** | Ni Arc Welding Electrode for Cast Iron (ENiFeMn-CI) | **Al** 1.0 max **C** 2.0 max **Cu** 2.5 max **Fe** rem **Mn** 10-14 **Ni** 35-45 **S** 0.03 max **Si** 1.0 max **Other** 1.0 max total | **ASME** SFA5.15 (ENiFeMn-CI) **AWS** A5.15 (ENiFeMn-CI) |
| * **W82032** | Ni Alloy Flux Cored Wire for Welding Cast Iron (ENiFeT3-CI) | **Al** 1.0 max **C** 2.0 max **Cu** 2.5 max **Fe** rem **Mn** 3.0-5.0 **Ni** 45.0-60.0 **S** 0.03 max **Si** 1.0 max **Other** 1.0 max total | **ASME** SFA5.15 (ENiFeT3-CI) **AWS** A5.15 (ENiFeT3-CI) |
| **W82141** | Ni Arc Welding Electrode ENi-1 (Nickel W.E.141) | **Al** 1.0 max **C** 0.10 max **Cu** 0.25 max **Fe** 0.75 max **Mn** 0.75 max **Ni** 92.0 min **P** 0.03 max **S** 0.02 max **Si** 1.25 max **Ti** 1.0-4.0 | **ASME** SFA5.11 (ENi-1) **AWS** A5.11 (ENi-1) **MIL SPEC** MIL-E-22200/3 (MIL-4N11) |
| **W83002** | Ni-Cr-Mo Arc Welding Electrodes for Surfacing (ENiCrFeCo) | **C** 2.2-3.0 **Cr** 25-30 **Co** 10-15 **Fe** 20.0-25.0 **Mn** 1.0 max **Mo** 7.0-10.0 **Ni** 10-33 **Si** 0.6-1.5 **W** 2.0-4.0 | **ASME** SFA5.13 (ENiCrFeCo) **AWS** A5.13 (ENiCrFeCo) |
| **W83032** | Ni-Cr-Fe-Co Composite Welding Wire for Hard Facing (ERCNiCrFeCo) | **C** 2.2-3.0 **Cr** 25-30 **Co** 10-15 **Fe** 20-25 **Mn1.0** max **Mo** 7-10 **Ni** 10-33 **Si** 2.0 max **W** 2.0-4.0 | **ASME** SFA5.21 (ERCNiCrFeCo) **AWS** A5.21 (ERCNiCrFeCo) |
| **W84001** | Ni-Cu Arc Welding Electrode for Cast Iron (ENiCu-A) | **C** 0.35-0.55 **Cu** 35-45 **Fe** 3.0-6.0 **Mn** 2.3 max **Ni** 50-60 **S** 0.025 max **Si** 0.75 max **Other** 1.0 max total | **AWS** A5.15 (ENiCu-A) |
| **W84002** | Ni-Cu Arc Welding Electrode for Cast Iron (ENiCu-B) | **C** 0.35-0.55 **Cu** 25-35 **Fe** 3.0-6.0 **Mn** 2.3 max **Ni** 60-70 **S** 0.025 max **Si** 0.75 max **Other** 1.0 max total | **AWS** A5.15 (ENiCu-B) |

*Boxed entries are no longer active and are retained for reference purposes only.

The chemical compositions listed are for identification purposes and should not be used in lieu of the cross referenced specifications.

| UNIFIED NUMBER | DESCRIPTION | CHEMICAL COMPOSITION | CROSS REFERENCE SPECIFICATIONS |
|---|---|---|---|
| **W84190** | Ni-Cu Arc Welding Electrode ENiCu-7 (Monel W.E.190) | **Al** 0.50 max **C** 0.15 max **Cu** rem **Fe** 2.5 max **Mn** 4.0 max **Ni** 62.0-69.0 **P** 0.02 max **S** 0.015 max **Si** 1.5 max **Ti** 1.0 max | **ASME** SFA5.11 (ENiCu-7) **AWS** A5.11 (ENiCu-7) **MIL SPEC** MIL-E-22200/3 (MIL-9N10) |
| **W86002** | Ni-Cr-Mo Arc Welding Electrodes ENiCrMo-2 (Hastelloy X) | **C** 0.05-0.15 **Co** 0.50-2.50 **Cr** 20.5-23.0 **Cu** 0.50 max **Fe** 17.0-20.0 **Mn** 1.0 max **Mo** 8.0-10.0 **Ni** rem **P** 0.04 max **S** 0.03 max **Si** 1.0 max **W** 0.2-1.0 | **AMS** 5677 **ASME** SFA5.11 (ENiCrMo-2) **AWS** A5.11 (ENiCrMo-2) |
| * **W86003** | Ni-Cr Welding Electrodes (75Ni-20Cr) | **C** 0.15 max **Cb** (incl Ta) 1.0-2.2 **Co** 1.0 max **Cr** 18.0-21.0 **Fe** 4.0 max **Mn** 1.50 max **Ni** 71.0 min **S** 0.015 max **Si** 0.75 max | |
| **W86007** | Ni-Cr-Mo Welding Electrodes (ENiCrMo-1) (Hastelloy G) | **C** 0.05 max **Cb** 1.75-2.50 **Co** 2.5 max **Cr** 21.0-23.5 **Cu** 1.5-2.5 **Fe** 18.0-21.0 **Mn** 1.0-2.0 **Mo** 5.5-7.5 **Ni** rem **P** 0.04 max **S** 0.03 max **Si** 1.0 max **W** 1.0 max | **ASME** SFA5.11 (ENiCrMo-1) **AWS** A5.11 (ENiCrMo-1) |
| **W86022** | Ni-Cr-Mo Arc Welding Electrode (ENiCrMo-10) (Hastelloy C-22) | **C** 0.02 max **Co** 2.5 max **Cr** 20.0-22.5 **Cu** 0.50 max **Fe** 2.0-6.0 **Mn** 1.0 max **Mo** 12.5-14.5 **Ni** rem **P** 0.03 max **S** 0.015 max **Si** 0.2 max **V** 0.35 **W** 2.5-3.5 | **ASME** SFA5.11 (ENiCrMo-10) **AWS** A5.11 (ENiCrMo-10) |
| **W86026** | Ni-Cr-Mo-W Arc Welding Electrode (ENiCrMo-14) | **C** 0.02 max **Cr** 19.0-23.0 **Cu** 0.50 max **Fe** 5.0 max **Mn** 1.0 max **Mo** 15.0-17.0 **Ni** rem **P** 0.02 max **Si** 0.25 max **S** 0.02 max **Ti** 0.25 max **W** 3.0-4.4 | **ASME** SFA5.11 (ENiCrMo-14) **AWS** A5.11 (ENiCrMo-14) |
| **W86030** | Ni-Cr-Mo Arc Welding Electrode (ENiCrMo-11) (Hastelloy G30) | **C** 0.03 max **Cb** 0.30-1.50 **Co** 5.0 max **Cr** 28.0-31.5 **Cu** 1.0-2.4 **Fe** 13.0-17.0 **Mn** 1.5 max **Mo** 4.0-6.0 **Ni** rem **P** 0.04 max **S** 0.02 max **Si** 1.0 max **Ti** 1.5-4.0 | **ASME** SFA5.11 (ENiCrMo-11) **AWS** A5.11 (ENiCrMo-11) |
| **W86032** | Ni-Cr-Mo Welding Electrode (Avesta P12R) | **C** 0.03 max **Cr** 20.5-22.5 **Cb** 1.0-2.8 **Cu** 0.50 max **Fe** 5.0 max **Mn** 2.2 max **Mo** 8.8-10.0 **Ni** rem **P** 0.03 max **S** 0.02 max **Si** 0.7 max **Other** 0.50 max | **AWS** A5.11 (ENiCrMo-12) |
| **W86040** | Replaced by W86032 | | |
| **W86059** | Ni-Cr-Mo Arc Welding Electrode (ENiCrMo-13) | **C** 0.02 max **Cr** 22.0-24.0 **Fe** 1.5 max **Mn** 1.0 max **Mo** 15.0-16.5 **Ni** rem **P** 0.015 max **Si** 0.2 max **S** 0.01 max | **ASME** SFA5.11 (ENiCrMo-13) **AWS** A5.11 (ENiCrMo-13) |
| **W86082** | Ni-Cr Electrode for Flux-Cored Arc Welding | **C** 0.10 max **Cr** 18.0-22.0 **Cb** 2.0-3.0 **Cu** 0.50 max **Fe** 3.0 max **Mn** 2.5-3.5 **Ni** 67.0 min **P** 0.03 max **S** 0.015 max **Si** 0.50 max **Ti** 0.75 max | **ASME** SFA5.34 (ENiCr3T-X) **AWS** A5.34 (ENiCr3T-X) |
| **W86094** | Ni-Cr-Fe-Mo Arc Welding Electrode (ENiCrFe-9) | **C** 0.15 max **Cr** 12.0-17.0 **(Cb+Ta)** 0.5-3.0 **Cu** 0.50 max **Fe** 12.0 max **Mn** 1.0-4.5 **Mo** 2.5-5.5 **Ni** 55.0 min **P.** 0.02 max **Si** 0.75 max **S** 0.015 max **W** 1.5 max | **ASME** SFA5.11 (ENiCrFe-9) **AWS** A5.11 (ENiCrFe-9) |
| **W86095** | Ni-Cr-Fe-W-Mo Arc Welding Electrode (ENiCrFe-10) | **C** 0.20 max **Cr** 13.0-17.0 **(Cb+Ta)** 1.0-3.5 **Cu** 0.50 max **Fe** 12.0 max **Mn** 1.0-3.5 **Mo** 1.0-3.5 **Ni** 55.0 min **P** 0.02 max **Si** 0.75 max **S** 0.015 max **W** 1.5-3.5 | **ASME** SFA5.11 (ENiCrFe-10) **AWS** A5.11 (ENiCrFe-10) |
| **W86112** | Ni-Cr-Mo Arc Welding Electrode ENiCrMo-3 (Inconel W.E. 112) | **C** 0.10 max **Cb** (incl Ta) 3.15-4.15 **Co** 0.12 max **Cr** 20.0-23.0 **Cu** 0.50 max **Fe** 7.0 max **Mn** 1.0 max **Mo** 8.0-10.0 **Ni** 55.0 min **P** 0.03 max **S** 0.02 max **Si** 0.75 max | **ASME** SFA5.11 (ENiCrMo3); SFA5.34 (ENiCrMo 3T-X) **AWS** A5.11 (ENiCrMo3); A5.34 (ENiCrMo3T-X) **MIL SPEC** MIL-E-22200/3 (MIL-1N12) |

*Boxed entries are no longer active and are retained for reference purposes only.

The chemical compositions listed are for identification purposes and should not be used in lieu of the cross referenced specifications.

| UNIFIED NUMBER | DESCRIPTION | CHEMICAL COMPOSITION | CROSS REFERENCE SPECIFICATIONS |
|---|---|---|---|
| **W86117** | Ni-Cr-Co-Mo Arc Welding Electrode (ENiCrCoMo-1) (Inconel W. E. 117) | **C** 0.05-0.15 **Cb+Ta** 1.0 max **Co** 9.0-15.0 **Cr** 21.0-26.0 **Cu** 0.50 max **Fe** 5.0 max **Mn** 0.3-2.5 **Mo** 8.0-10.0 **Ni** rem **P** 0.03 max **S** 0.015 max **Si** 0.75 max | **ASME** SFA5.11 (ENiCrCoMo-1) **AWS** A5.11 (ENiCrCoMo-1) |
| **W86130** | Ni-Cr Arc Welding Electrode | **Al** 0.1-1.0 **C** 0.25 max **Cb** (incl Ta) 4xSi min **Co** 1.0 max **Cr** 12.5-17.0 **Fe** 11.0 max **Mn** 1.0 max **Ni** 66.0 min **S** 0.015 max **Si** 1.0 max **Ti** 1.0-2.75 | **AMS** 5779 |
| **W86132** | Ni-Cr Arc Welding Electrode ENiCrFe-1 (Inconel W.E. 132) | **C** 0.08 max **Cb** 1.5-4.0 includes Ta **Cr** 13.0-17.0 **Cu** 0.50 max **Fe** 11.0 max **Mn** 3.5 max **Ni** 62.0 min **P** 0.03 max **S** 0.02 max **Si** 0.75 max | **ASME** SFA5.11 (ENiCrFe-1); SFA5.34 (ENiCrFe1T-X) **AWS** A5.11 (ENiCrFe-1); A5.34 (ENiCrFe1T-X) **MIL SPEC** MIL-E-22200/3 (MIL-3N12, MIL-3N12H) |
| **W86133** | Ni-Cr Arc Welding Electrode ENiCrFe-2 (Inco-Weld A) | **C** 0.10 max **Cb** 0.5-3.0 includes Ta **Cr** 13.0-17.0 **Cu** 0.50 max **Fe** 12.0 max **Mn** 1.0-3.5 **Mo** 0.50-2.50 **Ni** 62.0 min **P** 0.03 max **S** 0.02 max **Si** 0.75 max | **ASME** SFA5.11 (ENiCrFe-2); SFA5.34 (ENiCrFe2T-X) **AWS** A5.11 (ENiCrFe-2); A5.34 (ENiCrFe2T-X) **MIL SPEC** MIL-E-22200/3 (MIL-4N1A, MIL-4N1AH) |
| **W86134** | Ni-Cr Arc Welding Electrode ENiCrFe-4 (Inco-Weld B) | **C** 0.20 max **Cb** 1.0-3.5 includes Ta **Cr** 13.0-17.0 **Cu** 0.50 max **Fe** 12.0 max **Mn** 1.0-3.5 **Mo** 1.0-3.5 **Ni** 60.0 min **P** 0.03 max **S** 0.02 max **Si** 1.0 max | **ASME** SFA5.11 (ENiCrFe-4) **AWS** A5.11 (ENiCrFe-4) |
| * **W86135** | Ni-Cr-Fe-Mo Arc Welding Electrode (ENiCrFeMo-9) | **C** 0.15 max **Cr** 12.0-17.0 **(Cb+Ta)** 0.5-3.0 **Cu** 0.50 max **Fe** 12.0 max **Mn** 1.0-4.5 **Mo** 2.5-5.5 **Ni** 55.0 min **P** 0.02 max **Si** 0.75 max **S** 0.015 max **W** 1.5 max | **ASME** SFA5.11 (ENiCrMo-9) **AWS** A5.11 (ENiCrMo-9) |
| * **W86136** | Ni-Cr-Fe-Mo Arc Welding Electrode (ENiCrFeMo-10) | **C** 0.20 max **Cr** 13.0-17.0 **(Cb+Ta)** 1.0-3.5 **Cu** 0.50 max **Fe** 12.0 max **Mn** 1.0-3.5 **Mo** 1.0-3.5 **Ni** 55.0 min **P** 0.02 max **Si** 0.75 max **S** 0.015 max **W** 1.5-3.5 | **ASME** SFA5.11 (ENiCrMo-10) **AWS** A5.11 (ENiCrMo-10) |
| **W86142** | Composite Thermal Spray Wire (CW-NiCrMo) | **C** 0.10 max **Cb** 3.2-4.2 **Cr** 20-23 **Fe** 5 max **Mo** 8.0-10.0 **Ni** rem | **AWS** C2.25 (CW-NiCrMo) |
| **W86152** | Ni-Cr Arc Welding Electrode (Inconel WE 152) | **Al** 0.50 max **C** 0.05 max **Cb** 1.0-2.5 **Cr** 28.0-31.5 **Cu** 0.50 max **Fe** 7.0-12.0 **Mn** 5.0 max **Mo** 0.5 max **Ni** rem **P** 0.03 max **S** 0.015 max **Si** 0.75 max **Ti** 0.50 max **Other** elements 0.50 max total | **ASME** SFA5.11 (ENiCrFe-7) **AWS** A5.11 (ENiCrFe-7) |
| **W86182** | Ni-Cr Arc Welding Electrode ENiCrFe-3 (Inconel W.E. 182) | **C** 0.10 max **Cb** 1.0-2.5 includes Ta **Cr** 13.0-17.0 **Cu** 0.50 **Fe** 10.0 **Mn** 5.0-9.5 **Ni** 59.0 min **P** 0.03 max **S** 0.015 **Si** 1.0 max **Ti** 1.0 max | **ASME** SFA5.11 (ENiCrFe-3); SFA5.34 (ENiCrFe3T-X) **AWS** A5.11 (ENiCrFe-3); A5.34 (ENiCrFe3T-X) **MIL SPEC** MIL-E-22200/3 (MIL-8N12, MIL-8N12H) |
| **W86333** | Ni-Cr-Mo-Co Arc Welding Electrode (RA333) | **C** 0.08 max **Co** 2.50-4.00 **Cr** 24.0-27.0 **Fe** rem **Mn** 3.50 max **Mo** 2.50-4.00 **Ni** 11.0-17.0 **P** 0.030 max **S** 0.030 max **Si** 0.75-1.50 **W** 2.50-4.00 | |
| **W86455** | Ni-Mo-Cr Arc Welding Electrode (ENiCrMo-7) | **C** 0.015 max **Co** 2.0 max **Cr** 14.0-18.0 **Cu** 0.50 max **Fe** 3.0 max **Mn** 1.5 max **Mo** 14.0-17.0 **Ni** rem **P** 0.04 max **S** 0.03 max **Si** 0.2 max **Ti** 0.70 max **W** 0.5 max | **ASME** SFA5.11 (ENiCrMo-7) **AWS** A5.11 (ENiCrMo-7) |
| **W86620** | Ni-Cr-Mo Arc Welding Electrode (ENiCrMo-6) | **C** 0.10 max **Cb** 0.5-2.0 includes Ta **Cr** 12.0-17.0 **Cu** 0.50 max **Fe** 10.0 max **Mn** 2.0-4.0 **Mo** 5.0-9.0 **Ni** 55.0 min **P** 0.03 max **S** 0.02 max **Si** 1.0 max **W** 1.0-2.0 | **ASME** SFA5.11 (ENiCrMo-6) **AWS** A5.11 (ENiCrMo-6) |

*Boxed entries are no longer active and are retained for reference purposes only.

The chemical compositions listed are for identification purposes and should not be used in lieu of the cross referenced specifications.

| UNIFIED NUMBER | DESCRIPTION | CHEMICAL COMPOSITION | CROSS REFERENCE SPECIFICATIONS |
|---|---|---|---|
| * **W86685** | Ni-Cr-Mo Composite Welding Wire for Hard Facing (ERCNiCrMo-5A) | **C** 0.12 max **Cr** 14-18 **Fe** 4.0-7.0 **Mn** 1.0 max **Mo** 14-18 **Ni** rem **Si** 2.0 max **W** 3.0-5.0 **V** 0.040 max | **ASME** SFA5.21 (ERCNiCrMo-5A) **AWS** A5.21 (ERCNiCrMo-5A) |
| **W86686** | Ni-Cr-Mo-W Arc Welding Electrode (ENiCrMo-14) | **C** 0.02 max **Cr19.0-23.0Cu** 0.50 max **Fe** 5.0 max **Mn** 1.0 max **Mo** 15.0-17.0 **Ni** rem **P** 0.02 max **Si** 0.25 max **S** 0.02 max **Ti** 0.25 max **W** 3.0-4.0 | **ASME** SFA5.11 (ENiCrMo-14) **AWS** A5.11 (ENiCrMo-14) |
| **W86985** | Ni-Cr-Mo Arc Welding Electrode (ENiCrMo-9) | **C** 0.02 max **Cb** 0.5 max includes Ta **Co** 5.0 max **Cr** 21.0-23.5 **Cu** 1.5-2.5 **Fe** 18.0-21.0 **Mn** 1.0 max **Mo** 6.0-8.0 **Ni** rem **P** 0.04 max **S** 0.03 max **Si** 1.0 max **W** 1.5 max | **ASME** SFA5.11 (ENiCrMo-9) **AWS** A5.11 (ENiCrMo-9) |
| **W87718** | Ni-Fe-Cr Electrode for Flux-Cored Arc Welding | **Al** 0.20-0.80 **B** 0.006 max **C** 0.08 max **Cr** 17.0-21.0 **Co** 1.0 max **Cb** 4.75-5.5 **Cu** 0.30 max **Fe** rem **Mn** 0.50 max **Mo** 2.8-3.3 **Ni** 50.0-55.0 **P** 0.015 max **S** 0.015 max **Si** 0.75 max **Ti** 0.65-1.15 | **ASME** SFA5.34 (ENiFeCr2T-X) **AWS** A5.34 (ENiFeCr2T-X) |
| **W87740** | Composite Thermal Spray Wire (CW-NiCrFe) | **Cr** 17-19 **Fe** 18-20 **Mo** 2.5-3.5 **Ni** rem **Ta** 4.5-5.5 **Ti** 0.5-1.5 | **AWS** C2.25 (CW-NiCrFe) |
| **W88021** | Ni-Fe-Cr Arc Welding Electrode (E320-XX) (Carpenter 20Cb3) | **C** 0.07 max **Cb** 8\mtC-1.00 includes Ta **Cr** 19.0-21.0 **Cu** 3.0-4.0 **Mn** 0.5-2.5 **Mo** 2.0-3.0 **Ni** 32.0-36.0 **P** 0.04 **S** 0.03 **Si** 0.60 | **ASME** SFA5.4 (E320-XX) **AWS** A5.4 (E320-XX) |
| **W88022** | Ni-Fe-Cr Arc Welding Electrode (E320LR) Low Residuals (Carpenter 20Cb3L.R.) | **C** 0.03 max **Cb** 8\mtC-0.40 includes Ta **Cr** 19.0-21.0 **Cu** 3.0-4.0 **Mn** 1.50-2.50 **Mo** 2.0-3.0 **Ni** 32.0-36.0 **P** 0.020 max **S** 0.015 max **Si** 0.30 max | **ASME** SFA5.4 (E320LR); SFA5.9 (EC320LR) **AWS** A5.4 (E320LR); A5.9 (EC320LR) |
| **W88028** | Ni-Fe-Cr Welding Electrode (E383) | **C** 0.03 max **Cr** 26.5-29.0 **Cu** 0.6-1.5 **Mn** 0.5-2.5 **Mo** 3.2-4.2 **Ni** 30.0-33.0 **P** 0.02 max **S** 0.02 max **Si** 0.90 max | **ASME** SFA5.4 (E383) **AWS** A5.4 (E383) |
| **W88331** | Ni-Fe-Cr Arc Welding Electrode (E330) | **C** 0.18-0.25 **Cr** 14.0-17.0 **Cu** 0.75 max **Mn** 1.0-2.5 **Mo** 0.75 max **Ni** 33.0-37.0 **P** 0.04 max **S** 0.03 max **Si** 0.90 max | **ASME** SFA5.4 (E330) **AWS** A5.4 (E330) **MIL SPEC** MIL-E-22200/2 (MIL-330) |
| **W88334** | Ni-Fe-Cr-Mn Arc Welding Electrode (RA 330-04) | **C** 0.18-0.29 **Cr** 17.0-20.0 **Cu** 0.5 max **Fe** rem **Mn** 4.25-6.5 **Mo** 0.7 max **Ni** 33.0-37.0 **P** 0.03 max **S** 0.02 max **Si** 0.7-1.3 | |
| **W88335** | Ni-Fe-Cr Arc Welding Electrode (E330H) High Carbon | **C** 0.35-0.45 **Cr** 14.0-17.0 **Cu** 0.75 max **Mn** 1.0-2.5 **Mo** 0.75 max **Ni** 33.0-37.0 **P** 0.04 max **S** 0.03 max **Si** 0.90 max | **ASME** SFA5.4 (E330H) **AWS** A5.4 (E330H) |
| **W88338** | Ni-Fe-Cr Arc Welding Electrode, High Carbon (RA330-80) | **C** 0.75-0.95 **Cr** 17.0-20.0 **Fe** rem **Mn** 2.50 max **Ni** 33.0-37.0 **P** 0.03 max **S** 0.03 max **Si** 0.65-1.3 max | |
| **W88904** | Ni-Fe-Cr Arc Welding Electrode (904L) | **C** 0.03 **Cr** 19.5-21.5 **Cu** 1.2-2.0 **Mn** 1.0-2.5 **Mo** 4.2-5.2 **Ni** 24.0-26.0 **P** 0.03 **S** 0.02 **Si** 0.75 | **ASME** SFA5.4 (E385) **AWS** A5.4 (E385) |
| **W89240** | Composite Thermal Spray Wire (CW-NiFeAl) | **Al** 14-16 **Cr** 3-5 **Fe** 17-23 **Ni** rem | **AWS** C2.25 (CW-NiFeAl) |
| **W89341** | Composite Thermal Spray Wire (CW-NiCrAl) | **Al** 6-7 **Cr** 18-20 **Ni** rem **Other** all others total 6 max | **AWS** C2.25 (CW-NiCrAl) |
| **W89343** | Composite Thermal Spray Wire (CW-NiCrAlY) | **Al** 9-11 **Cr** 21-23 **Ni** rem **Other** Y 0.5-1.5 | **AWS** C2.25 (CW-NiCrAlY) |
| **W89432** | Composite Thermal Spray Wire (CW-NiAlMo) | **Al** 5.0-6.0 **Mo** 4.5-5.5 **Ni** rem | **AWS** C2.25 (CW-NiAlMo) |

*Boxed entries are no longer active and are retained for reference purposes only.

The chemical compositions listed are for identification purposes and should not be used in lieu of the cross referenced specifications.

| UNIFIED NUMBER | DESCRIPTION | CHEMICAL COMPOSITION | CROSS REFERENCE SPECIFICATIONS |
|---|---|---|---|
| * **W89604** | Ni-Cr-B Arc Welding Electrode for Hard Surfacing (ENiCr-A) | **B** 2.00-3.00 **C** 0.30-0.60 **Co** 1.50 max **Cr** 8.0-14.0 **Fe** 1.25-3.25 **Ni** rem **Si** 1.25-3.25 | **AWS** A5.13 (ENiCr-A) |
| * **W89605** | Ni-Cr-B Arc Welding Electrode for Hard Surfacing (ENiCr-B) | **B** 2.00-4.00 **C** 0.40-0.80 **Co** 1.25 max **Cr** 10.0-16.0 **Fe** 3.00-5.00 **Ni** rem **Si** 3.00-5.00 | **AWS** A5.13 (ENiCr-B) |
| **W89606** | Ni-Cr-B Arc Welding Electrode for Hard Surfaciang (ENiCr-C) | **B** 2.5-4.5 **C** 0.5-1.0 **Co** 1.0 max **Cr** 12-18 **Fe** 3.5-5.5 **Ni** rem **Si** 3.5-5.5 | **AWS** A5.13 (ENiCr-C) |
| **W89634** | Ni-Cr-B Composite Welding Wire for Hard Facing (ERCNiCr-A) | **B** 1.5-3.0 **C** 0.20-0.60 **Cr** 6.5-14.0 **Fe** 1.0-3.5 **Ni** rem **Si** 1.2-4.0 | **ASME** SFA5.21 (ERCNiCr-A) **AWS** A5.21 (ERCNiCr-A) |
| **W89635** | Ni-Cr-B Composite Welding Wire for Hard Facing (ERCNiCr-B) | **B** 2.0-4.0 **C** 0.30-0.80 **Cr** 9.5-16.0 **Fe** 2.0-5.0 **Ni** rem **Si** 3.0-5.0 | **ASME** SFA5.21 (ERCNiCr-B) **AWS** A5.21 (ERCNiCr-B) |
| **W89636** | Ni-Cr-B Composite Welding Wire for Hard Facing (ERCNiCr-C) | **B** 2.5-4.5 **C** 0.50-1.00 **Cr** 12-18 **Fe** 3.0-5.5 **Ni** rem **Si** 3.0-5.5 | **ASME** SFA5.21 (ERCNiCr-C) **AWS** A5.21 (ERCNiCr-C) |
| * **W89640** | Composite Thermal Spray Wire (CW-FeCrNiB) | **B** 2.5-3.5 **Cr** 30-32 **Cu** 2.5-3.5 **Fe** rem **Mo** 3.5-4.5 **Ni** 29-31 | **AWS** C2.25 (CW-FeCrNiB) |
| **W89740** | Composite Thermal Spray Wire (CW-NiCrSiB) | **B** 2.5-3.0 **C** 0.5-1.0 **Cr** 12-18 **Fe** 1.0-3.5 **Mn** 0.5 max **Ni** rem **Si** 3.5-5.5 | **AWS** C2.25 (CW-NiCrSiB) |

*Boxed entries are no longer active and are retained for reference purposes only.

The chemical compositions listed are for identification purposes and should not be used in lieu of the cross referenced specifications.

## UNS NUMBERS ASSIGNED TO DATE

**With Description of Each Material Covered and References To Documents In Which The Same or Similar Materials are Described**

**Zxxxxx Number Series**
**Zinc and Zinc Alloys**

# ZINC AND ZINC ALLOYS

| UNIFIED NUMBER | DESCRIPTION | CHEMICAL COMPOSITION | CROSS REFERENCE SPECIFICATIONS |
|---|---|---|---|
| * **Z12001** | 99.995 Zinc | **Al** 0.005 max **Cd** 0.003 max **Cu** 0.001 max **Fe** 0.002 max **Pb** 0.003 max **Sn** 0.001 max **Zn** 99.995 min **Other** 0.005 max total non-zinc impurities | **Other:** ISO 752 (99.995) |
| **Z12002** | 99.995 Zinc (ZN-1) | **Al** 0.001 max **Cd** 0.003 max **Cu** 0.001 max **Fe** 0.002 max **Pb** 0.003 max **Sn** 0.001 max **Zn** 99.995 min **Other** 0.005 max total non-zinc impurities | **Other:** CEN EN1179 (Z1); ISO 752 (ZN-1) |
| **Z12003** | 99.995 Zinc | **Cd** 0.002 max **Fe** 0.002 max **Pb** 0.003 max **Sn** 0.001 max **Zn** 99.995 min **Other** Fe 0.001 max for anticorrosive use. | **Other:** JIS H2107 (Highest Purity Zinc Metal) |
| **Z13000** | Zinc Anodes (Type II) | **Al** 0.005 max **Cd** 0.003 max **Cu** 0.002 max **Fe** 0.0014 max **Pb** 0.003 max | **ASTM** B418 (Zinc Anodes Type II) |
| **Z13001** | Special High Grade Zinc (ZN-2) | **Al** 0.002 max **Cd** 0.003 max **Cu** 0.002 max **Fe** 0.003 max **Pb** 0.003 max **Sn** 0.001 max **Zn** 99.990 min | **ASTM** B6 (Special High Grade) **AWS** C2.25 (W-Zn-1) |
| * **Z13002** | 99.99 Zinc | **Al** 0.005 max **Cd** 0.003 max **Cu** 0.002 max **Fe** 0.003 max **Pb** 0.003 max **Sn** 0.001 max **Zn** 99.99 min **Other** 0.010 max total non-zinc impurities, Pb max is 0.005 if not intended for pressure casting use | **Other:** ISO 752 (99.99) |
| **Z13003** | 99.99 Zinc | **Cd** 0.005 max **Cu** 0.002 max **Fe** 0.003 max **Pb** 0.005 max **Sn** 0.001 max **Zn** 99.99 min **Other** 0.01 max total non-zinc impurities | **Other:** CEN EN1179 (Z2) |
| **Z13004** | Rolled Special High Grade Zinc | **Al** 0.002 max **Cd** 0.003 max **Cu** 0.003 max **Fe** 0.003 max **Pb** 0.003 max **Sn** 0.001 max **Zn** rem | **ASTM** B69 |
| **Z13005** | 99.99 Zinc Wire for Thermal Spraying (Metallizing) | **Al** 0.002 max **Cd** 0.003 max **Cu** 0.005 max **Fe** 0.003 max **Pb** 0.003 max **Sn** 0.001 max **Zn** 99.99 min | **ASTM** B833 |
| **Z13006** | 99.99 Zinc | **Cd** 0.004 max **Fe** 0.005 max **Pb** 0.007 max **Zn** 99.99 min | **Other:** JIS H2107 (Special Zinc Metal) |
| * **Z14001** | 99.95 Zinc | **Al** 0.005 max **Cd** 0.02 max **Cu** 0.002 max **Fe** 0.02 max **Pb** 0.03 max **Sn** 0.001 max **Zn** 99.95 min **Other** 0.050 max total non-zinc impurities | **Other:** ISO 752 (99.95) |
| **Z14002** | 99.5 Zinc (ZN-3) | **Cd** 0.01 max **Cu** 0.002 max **Fe** 0.02 max **Pb** 0.03 max **Sn** 0.001 max **Zn** 99.95 min **Other** 0.05 max total non-zinc impurities; Cd 0.02 max for five (5) years from adoption of standard | **Other:** CEN EN 1179 (Z3); ISO 752 (ZN-3) |
| **Z14051** | 99.97 Zinc | **Cd** 0.005 max **Fe** 0.01 max **Pb** 0.02 max **Zn** 99.97 min | **Other:** JIS H2107 (Ordinary Zinc Metal) |
| **Z15001** | High Grade Zinc | **Al** 0.01 max **Cd** 0.02 max **Fe** 0.02 max **Pb** 0.03 max **Zn** 99.90 min | **ASTM** B6 (High Grade) |
| **Z15005** | 99.9 Zinc Wire for Thermal Spraying (Metallizing) | **Al** 0.01 max **Cd** 0.02 max **Cu** 0.02 max **Fe** 0.02 max **Pb** 0.03 max **Zn** 99.9 min **Other** total non-zinc 0.10 max | **ASTM** B833 |
| **Z15006** | Commercially Pure Rolled Zinc | **Al** 0.01 max **Cd** 0.01 max **Cu** 0.08 max **Fe** 0.02 max **Pb** 0.03 max **Sn** 0.003 max **Ti** 0.02 max **Zn** rem | **ASTM** B69 |

*Boxed entries are no longer active and are retained for reference purposes only.

The chemical compositions listed are for identification purposes and should not be used in lieu of the cross referenced specifications.

| UNIFIED NUMBER | DESCRIPTION | CHEMICAL COMPOSITION | CROSS REFERENCE SPECIFICATIONS |
|---|---|---|---|
| * **Z16001** | Zinc Metal | **Cd** 0.40 max **Fe** 0.03 max **Pb** 0.20 max **Zn** 99.5 min | |
| * **Z16002** | 99.5 Zinc | **Al** 0.01 max **Cd** 0.15 max **Fe** 0.05 max **Pb** 0.45 max **Zn** 99.5 min **Other** 0.05 max, total non-zinc impurities; Al 0.005 max for rolling quality; Sn 0.003 max for rolling quality | **Other:** ISO 752 (99.5) |
| **Z16003** | 99.5 Zinc (ZN-4) | **Cd** 0.01 max **Fe** 0.05 max **Pb** 0.45 max **Zn** 99.5 min **Other** 0.05 max total non-zinc impurities; Cd 0.050 max for five (5) years after adoption of standard | **Other:** CEN EN 1179 (Z4); ISO 752 (ZN-4) |
| **Z16004** | 99.6 Zinc | **Cd** 0.1 max **Fe** 0.02 max **Pb** 0.3 max **Zn** 99.6 min | **Other:** JIS H2107 (Distilled Zinc Metal, Special) |
| * **Z17001** | Zinc Metal | **Cd** 0.05 max **Fe** 0.03 max **Pb** 0.6 max **Zn** 99.0 min | |
| * **Z18001** | 98.5 Zinc | **Al** 0.02 max **Cd** 0.20 max **Fe** 0.05 max **Pb** 1.4 max **Zn** 98.5 min **Other** 1.50 max total non-zinc impurities; Al 0.005 max for rolling quality; Sn 0.003 max for rolling quality | **Other:** ISO 752 (98.5) |
| **Z18002** | 98.5 Zinc | **Cd** 0.01 max **Fe** 0.05 max **Pb** 1.4 max **Zn** 98.5 min **Other** 1.5 max total non-zinc impurities; Cd 0.050 max for five (5) years after adoption of standard | **Other:** CEN EN 1179 (Z5) |
| **Z18003** | 98.5 Zinc | **Cd** 0.4 max **Fe** 0.025 max **Pb** 1.3 max **Zn** 98.5 min | **Other:** JIS H2107 (Distilled Zinc Metal, Class 1) |
| **Z18004** | 98.5 Zinc Metal (ZN-5) | **Cd** 0.20 max **Fe** 0.05 max **Pb** 0.5-1.4 **Zn** 98.5 min **Other** 1.5 max total non-zinc | **Other:** ISO 752 (ZN-5) |
| **Z19001** | Prime Western Zinc | **Al** 0.01 max **Cd** 0.20 max **Cu** 0.20 max **Fe** 0.05 max **Pb** 0.5-1.4 **Zn** 98.0 min | **ASTM** B6 (Prime Western) |
| **Z19002** | 98.0 Zinc | **Cd** 0.5 max **Fe** 0.1 max **Pb** 1.8 max **Zn** 98.0 min | **Other:** JIS H2107 (Distilled Zinc Metal, Class 2) |
| **Z20301** | Zinc-Lead Rolled Zinc Alloy | **Al** 0.002 max **Cd** 0.01 max **Cu** 0.005 max **Fe** 0.01 max **Pb** 0.10 max **Ti** 0.02 max **Zn** rem | **ASTM** B69 |
| * **Z21210** | Zinc Rolled | **Al** 0.001 max **Cd** 0.005 max **Cu** 0.001 max **Fe** 0.012 max **Pb** 0.10 max **Sn** 0.001 max **Zn** rem | **ASTM** B69 (Zn-0.08Pb) |
| * **Z21220** | Zinc Alloy Rolled | **Al** 0.001 max **Cd** 0.05-0.08 **Cu** 0.005 max **Fe** 0.012 max **Pb** 0.05-0.10 **Sn** 0.001 max **Zn** rem | **ASTM** B69 (Zn-0.06Pb-0.06Cd) |
| * **Z21310** | Zinc Alloy Rolled | **Cd** 0.005 max **Cu** 0.001 max **Fe** 0.012 max **Pb** 0.05-0.12 **Zn** rem | |

*Boxed entries are no longer active and are retained for reference purposes only.

The chemical compositions listed are for identification purposes and should not be used in lieu of the cross referenced specifications.

| UNIFIED NUMBER | DESCRIPTION | CHEMICAL COMPOSITION | CROSS REFERENCE SPECIFICATIONS |
|---|---|---|---|
| * **Z21540** | Zinc Alloy Rolled | **Al** 0.001 max **Cd** 0.25-0.45 **Cu** 0.005 max **Fe** 0.002 max **Pb** 0.25-0.50 **Zn** rem | **ASTM** B69 (Zn-0.3Pb-0.3Cd) |
| **Z21721** | Zinc-Lead-Cadmium Rolled Zinc Alloy | **Al** 0.002 max **Cd** 0.07 max **Cu** 0.005 max **Fe** 0.01 max **Pb** 1.0 max **Ti** 0.02 max **Zn** rem | **ASTM** B69 |
| **Z24311** | Zinc-Lead-Manganese Rolled Zinc Alloy | **Al** 0.002 max **Cd** 0.005 max **Cu** 0.005 max **Fe** 0.01 max **Mg** 0.0015 max **Mn** 0.015 max **Pb** 0.03-0.08 **Ti** 0.02 max **Zn** rem | **ASTM** B69 |
| **Z30401** | Zn-Al Alloy Thermal Spray Wire (Metallizing) (W-ZnAl-1) | **Al** 1.5-2.5 **Zn** rem **Other** total 0.1 max | **AWS** C2.25 (W-ZnAl-1) |
| **Z30402** | 98Zn-2Al Solder Alloy (Zn98) | **Ag** 0.015 max **Al** 1.5-2.5 **As** 0.002 max **Bi** 0.02 max **Cd** 0.005 max **Cu** 0.005 max **Fe** 0.02 max **Mg** 0.02 max **Ni** 0.005 max **Pb** 0.005 max **Sb** 0.10 max **Sn** 0.003 max **Zn** rem | **ASTM** B907 |
| **Z30500** | Zinc Slush Casting Alloy B | **Al** 5.25-5.75 **Cd** 0.005 max **Cu** 0.1 max **Fe** 0.10 max **Pb** 0.007 max **Sn** 0.005 max **Zn** rem | **ASTM** B792 |
| **Z30502** | 95Zn/5Al Solder Alloy (Zn95) | **Ag** 0.015 max **Al** 4.5-5.5 **As** 0.002 max **Bi** 0.02 max **Cd** 0.005 max **Cu** 0.005 max **Fe** 0.02 max **Mg** 0.02 max **Ni** 0.005 max **Pb** 0.005 max **Sb** 0.10 max **Sn** 0.003 max **Zn** rem | **ASTM** B907 |
| **Z30503** | 95Zn/5Al Hot Dip Galvanizing Master Alloy (Type A-3) | **Al** 4.5-5.5 **Cd** 0.004 max **Cu** 0.035 max **Fe** 0.05 max **Pb** 0.005 max **Sn** 0.003 max **Zn** rem **Other** total 0.01 max | **ASTM** B860 (Type A-3, 95Zn-5Al High Purity) |
| **Z30505** | Zn 97/3 Al Solder Alloy (Zn 97) | **Ag** 0.015 max **Al** 2.5-3.5 **As** 0.002 max **Bi** 0.02 max **Cd** 0.005 max **Cu** 0.005 max **Fe** 0.02 max **Mg** 0.02 max **Ni** 0.005 max **Pb** 0.005 max **Sb** 0.10 max **Sn** 0.003 max **Zn** rem | **ASTM** B907 |
| **Z30700** | 85Zn-15Al Alloy Thermal Spray Wire (Metallizing) (W-ZnAl-2) | **Al** 14.0-16.0 **Zn** rem **Other** total non (Zn+Al) 0.05 max | **ANSI** A5.33 **ASTM** B833 **AWS** C2.25 (W-ZnAl-2) |
| **Z30702** | 85Zn-15Al Solder Alloy (Zn85) | **Ag** 0.015 max **Al** 14.0-16.0 **As** 0.002 max **Bi** 0.02 max **Cd** 0.005 max **Cu** 0.005 max **Fe** 0.02 max **Mn** 0.02 max **Ni** 0.005 max **Pb** 0.005 max **Sb** 0.10 max **Sn** 0.003 max **Zn** rem | **ASTM** B907 |
| **Z30705** | 87Zn-13Al Solder Alloy (Zn87) | **Ag** 0.015 max **Al** 12.5-13.5 **As** 0.002 max **Bi** 0.02 max **Cd** 0.005 max **Cu** 0.005 max **Fe** 0.02 max **Mg** 0.02 max **Ni** 0.005 max **Pb** 0.005 max **Sb** 0.10 max **Sn** 0.003 max **Zn** rem | **ASTM** B907 |
| **Z30750** | 90Zn-10Al Hot Dip Galvanizing Master Alloy (Type A-1) | **Al** 9.5-10.5 **Cd** 0.004 max **Cu** 0.035 max **Fe** 0.05 max **Pb** 0.005 max **Sn** 0.003 max **Zn** rem **Other** total 0.01 max | **ASTM** B860 (Type A-1 90Zn-10Al High Purity) |
| **Z30800** | 80Zn-20Al Solder Alloy (Zn80) | **Ag** 0.015 max **Al** 19.5-20.5 **As** 0.002 max **Bi** 0.02 max **Cd** 0.005 max **Cu** 0.005 max **Fe** 0.02 max **Mg** 0.02 max **Ni** 0.005 max **Pb** 0.005 max **Sb** 0.10 max **Sn** 0.003 max **Zn** rem | **ASTM** B907 |
| **Z31510** | 95Zn/5Al Hot Dip Galvanizing Master Alloy (Type A-4) | **Al** 4.5-5.5 **Cu** 0.5 max **Fe** 0.15 max **Zn** rem **Other** total 0.25 max | **ASTM** B860 (Type A-4 95Zn-5Al Low Purity) |
| **Z31710** | 90Zn-10Al Hot Dip Galvanizing Master Alloy (Type A-2) | **Al** 9.5-10.5 **Cu** 0.5 max **Fe** 0.15 max **Pb** 0.4 max **Zn** rem **Other** total 0.25 max | **ASTM** B860 (Type A-2 90Zn/10Al Low Purity) |

*Boxed entries are no longer active and are retained for reference purposes only.

The chemical compositions listed are for identification purposes and should not be used in lieu of the cross referenced specifications.

| UNIFIED NUMBER | DESCRIPTION | CHEMICAL COMPOSITION | CROSS REFERENCE SPECIFICATIONS |
|---|---|---|---|
| **Z32120** | Zinc Anodes (Type I) | **Al** 0.10-0.5 **Cd** 0.025-0.07 **Cu** 0.005 max **Fe** 0.005 max **Pb** 0.006 max **Other** 0.1 max | **ASTM** B418 (Zinc Anodes - Type I) |
| **Z32121** | Zinc Anodes (Type III) | **Al** 0.1-0.5 **Cd** 0.025-0.15 **Cu** 0.005 max **Fe** 0.005 max **Pb** 0.006 max **Si** 0.125 max | **MIL SPEC** MIL-A-18001 |
| **Z33520** | Zinc Die Casting Alloy (AG40A) | **Al** 3.5-4.3 **Cd** 0.004 max **Cu** 0.25 max **Fe** 0.100 max **Mg** 0.02-0.05 **Pb** 0.005 max **Sn** 0.003 max **Zn** rem | **AMS** 4803 **ASTM** B86 (AG40A) **SAE** J468 (903) (Castings); J469 |
| **Z33521** | Zinc Die Casting Alloy (AG40A) Ingot | **Al** 3.9-4.3 **Cd** 0.003 max **Cu** 0.10 max **Fe** 0.075 max **Pb** 0.004 max **Mg** 0.025-0.05 **Sn** 0.002 max **Zn** rem | **ASTM** B240 (AG40A) **SAE** J468 (903)(Ingot) |
| **Z33522** | Zinc Die Casting Alloy (AG40B) Ingot | **Al** 3.9-4.3 **Cd** 0.0020 max **Cr** 0.02 max **Cu** 0.10 max **Fe** 0.075 max **Mg** 0.010-0.020 **Mn** 0.05 max **Ni** 0.005-0.020 **Pb** 0.0020 max **Si** 0.035 max **Sn** 0.0010 max **Zn** rem | **ASTM** B240 (AG40B) |
| **Z33523** | Zinc Die Casting Alloy (AG40B) | **Al** 3.5-4.3 **Cd** 0.0020 max **Cr** 0.02 max **Cu** 0.25 max **Fe** 0.075 max **Mg** 0.005-0.020 **Mn** 0.5 max **Ni** 0.005-0.020 **Pb** 0.0030 max **Si** 0.035 max **Sn** 0.0010 max **Zn** rem | **ASTM** B86 (AG40B) |
| **Z33730** | Zinc Die Casting Master Alloy (V12) | **Al** 11.7-12.6 **Cd** 0.004 max **Cu** 0.25 max **Fe** 0.070 max **Mg** 0.075-0.12 **Pb** 0.005 max **Sn** 0.003 max **Zn** rem | **ASTM** B327 (V12) |
| **Z34510** | Zinc Slush Casting Alloy A | **Al** 4.5-5.0 **Cd** 0.005 max **Cu** 0.2-0.3 **Fe** 0.100 max **Pb** 0.007 max **Sn** 0.005 max **Zn** rem | **ASTM** B792 |
| **Z34550** | 90Zn-5Al-5Cu Solder Alloy (Zn90) | **Ag** 0.015 max **Al** 3.0-6.0 **Bi** 0.02 max **Cd** 0.004 max **Cu** 3.0-6.0 **Fe** 0.02 max **Mg** 0.02 max **Ni** 0.005 max **Pb** 0.005 max **Sb** 0.10 max **Sn** 0.003 max **Zn** 88.0-92.0 | **ASTM** B907 |
| **Z34961** | Zinc-Aluminum Rolled Zinc Alloy | **Al** 1.4-3.4 **Cd** 0.15 max **Cu** 5.0 max **Fe** 0.1 max **Mg** 0.10 max **Pb** 0.05 max **Sn** 0.003 max **Ti** 0.2 max **Zn** rem | **ASTM** B69 |
| **Z35530** | Zinc Die Casting Alloy (AC41A) Ingot | **Al** 3.9-4.3 **Cd** 0.003 max **Cu** 0.75-1.25 **Fe** 0.075 max **Pb** 0.004 max **Mg** 0.03-0.06 **Sn** 0.002 max **Zn** rem | **ASTM** B240 (AC41A) **SAE** J468 (925)(Ingot) |
| **Z35531** | Zinc Die Casting Alloy (AC41A) | **Al** 3.5-4.3 **Cd** 0.004 max **Cu** 0.75-1.25 **Fe** 0.100 max **Mg** 0.03-0.08 **Pb** 0.005 max **Sn** 0.003 max **Zn** rem | **ASTM** B86 **SAE** J468 (925) (Castings); J469 |
| **Z35540** | Zinc Die Casting Alloy (AC43A) Ingot | **Al** 3.9-4.3 **Cd** 0.003 max **Cu** 2.6-2.9 **Fe** 0.075 max **Pb** 0.004 max **Mg** 0.025-0.050 **Sn** 0.002 max **Zn** rem | **ASTM** B240 (AC43A) |
| **Z35541** | Zinc Die Casting Alloy (AC43A) | **Al** 3.5-4.3 **Cd** 0.004 max **Cu** 2.5-3.0 **Fe** 0.100 max **Mg** 0.020-0.050 **Pb** 0.005 max **Sn** 0.003 max **Zn** rem | **ASTM** B86 (AC43A) |
| **Z35542** | Zinc Forming Die Alloy B (Kirksite II) | **Al** 3.9-4.3 **Cd** 0.003 max **Cu** 2.5-2.9 **Fe** 0.075 max **Mg** 0.02-0.05 **Pb** 0.003 max **Sn** 0.001 max **Zn** rem | **AMS** AMS-I-7068 **ASTM** B793 **Other:** ISO 301-1981 (E) (ZnAl4Cu3) |
| **Z35543** | Zinc Forming Die Alloy (Kirksite I) | **Al** 3.5-4.5 **Cd** 0.005 max **Cu** 2.5-3.5 **Fe** 0.100 max **Mg** 0.02-0.10 **Pb** 0.007 max **Sn** 0.005 max **Zn** rem | **AMS** AMS-I-7068 **ASTM** B793 |
| **Z35630** | Zinc-Aluminum Casting Alloy ZA12 Ingot | **Al** 10.8-11.5 **Cd** 0.005 max **Cu** 0.50-1.2 **Fe** 0.065 max **Mg** 0.020-0.030 **Pb** 0.005 max **Sn** 0.002 max **Zn** rem | **ASTM** B240 (ZA-12) |

The chemical compositions listed are for identification purposes and should not be used in lieu of the cross referenced specifications.

| UNIFIED NUMBER | DESCRIPTION | CHEMICAL COMPOSITION | CROSS REFERENCE SPECIFICATIONS |
|---|---|---|---|
| **Z35631** | Zinc-Aluminum Casting Alloy ZA-12 | **Al** 10.5-11.5 **Cd** 0.004 max **Cu** 0.5-1.2 **Fe** 0.075 max **Pb** 0.006 max **Mg** 0.015-0.030 **Sn** 0.003 max **Zn** rem | **ASTM** B86 (ZA-12) **Other:** Australian AS 1881; British DD 139; ZI ZA-12 |
| **Z35635** | Zinc-Aluminum Casting Alloy ZA-8 Ingot | **Al** 8.2-8.8 **Cd** 0.005 max **Cu** 0.8-1.3 **Fe** 0.10 max **Mg** 0.020-0.030 **Pb** 0.005 max **Sn** 0.002 max **Zn** rem | **ASTM** B240 (ZA-8) |
| **Z35636** | Zinc-Aluminum Casting Alloy ZA-8 | **Al** 8.0-8.8 **Cd** 0.006 max **Cu** 0.8-1.3 **Fe** 0.075 max **Pb** 0.006 max **Mg** 0.015-0.030 **Sn** 0.003 max **Zn** rem | **ASTM** B86 (ZA-8) **Other:** Australian AS 1881; British DD 139; ZI (ZA-8) |
| **Z35840** | Zinc-Aluminum Casting Alloy (ZA-27) Ingot | **Al** 25.5-28.0 **Cd** 0.005 max **Cu** 2.0-2.5 **Fe** 0.072 max **Mg** 0.012-0.020 **Pb** 0.005 max **Sn** 0.002 max **Zn** rem | **ASTM** B240 (ZA-27) **Other:** ZI ZA-27 |
| **Z35841** | Zinc-Aluminum Casting Alloy (ZA-27) | **Al** 25.0-28.0 **Cd** 0.006 max **Cu** 2.0-2.5 **Fe** 0.075 max **Pb** 0.006 max **Mg** 0.010-0.020 **Sn** 0.003 max **Zn** rem | **Other:** Australian AS 1881; British DD 139; ZI ZA-27 |
| **Z38510** | Zinc Galvanizing Alloy Ingot | **Al** 4.2-6.2 **Cd** 0.005 max **Cu** 0.1 max **Fe** 0.075 max **Mg** 0.05 max **Pb** 0.005 max **Sb** 0.002 max **Si** 0.015 max **Sn** 0.002 max **Ti** 0.02 max **Zn** rem **Zr** 0.02 max **Other** Ce+La 0.03-0.10 **Other** each 0.02 max, total 0.04 max | **ASTM** B750 |
| **Z40101** | Zinc-Low Copper Rolled Zinc Alloy | **Al** 0.01 max **Cd** 0.005 max **Cu** 0.08-0.49 **Fe** 0.01 max **Pb** 0.01 max **Sn** 0.003 max **Ti** 0.02 max **Zn** rem | **ASTM** B69 |
| **Z40301** | Zinc-High Copper Rolled Zinc Alloy | **Al** 0.01 max **Cd** 0.005 max **Cu** 0.50-1.0 **Fe** 0.01 max **Pb** 0.01 max **Sn** 0.003 max **Ti** 0.04 max **Zn** rem | **ASTM** B69 |
| * **Z40330** | Zinc Alloy, Rolled | **Al** 0.005 max **Cd** 0.02 max **Cu** 0.70-0.90 **Fe** 0.01 max **Pb** 0.02 max **Ti** 0.02 max **Zn** rem | **ASTM** B69 (Zn-0.8Cu) |
| **Z41121** | Zinc-Low Copper-Titanium Rolled Zinc Alloy | **Al** 0.01 max **Cd** 0.005 max **Cu** 0.08-0.49 **Fe** 0.01 max **Pb** 0.01 max **Sn** 0.003 max **Ti** 0.05-0.18 **Zn** rem | **ASTM** B69 |
| * **Z41320** | Zinc Alloy, Rolled | **Al** 0.001 max **Cd** 0.05 max **Cu** 0.50-1.50 **Fe** 0.012 max **Pb** 0.10 max **Sn** 0.001 max **Ti** 0.12-0.50 **Zn** rem | **ASTM** B69 (Zn-0.8Cu-0.15Ti) |
| **Z41321** | Zinc-High Copper-Titanium Rolled Zinc Alloy | **Al** 0.01 max **Cd** 0.005 max **Cu** 0.05-1.0 **Fe** 0.01 max **Pb** 0.01 max **Sn** 0.003 max **Ti** 0.08-0.18 **Zn** rem | **ASTM** B69 |
| * **Z44330** | Zinc Alloy Rolled | **Al** 0.001 max **Cd** 0.005 max **Cu** 0.85-1.25 **Fe** 0.012 max **Pb** 0.10 max **Sn** 0.001 max **Zn** rem | **ASTM** B69 (Zn-1Cu) |
| * **Z45330** | Zinc Alloy Rolled | **Al** 0.001 max **Cd** 0.04 max **Cu** 0.85-1.25 **Fe** 0.015 max **Mg** 0.006-0.016 **Pb** 0.15 max **Sn** 0.001 max **Zn** rem | **ASTM** B69 (Zn-1Cu-0.01Mg) |
| **Z46540** | Zinc-Copper-Aluminum Casting Alloy Ingot | **Al** 2.8-3.3 **Cd** 0.003 max **Cu** 5.2-6.0 **Fe** 0.05 max **Mg** 0.035-0.050 **Pb** 0.004 max **Sn** 0.002 max **Zn** rem | **ASTM** B892 |

*Boxed entries are no longer active and are retained for reference purposes only.

The chemical compositions listed are for identification purposes and should not be used in lieu of the cross referenced specifications.

| UNIFIED NUMBER | DESCRIPTION | CHEMICAL COMPOSITION | CROSS REFERENCE SPECIFICATIONS |
|---|---|---|---|
| **Z46541** | Zinc-Copper-Aluminum Die Casting Alloy | **Al** 2.5-3.3 **Cd** 0.004 max **Cu** 5.0-6.0 **Cr** 0.075 max Mg 0.025-0.05 **Pb** 0.005 max **Sn** 0.003 max **Zn** rem | **ASTM** B894 |
| **Z50940** | 90Zn/10Cd Solder Alloy (Zn/Cd90) | **Ag** 0.015 max **Al** 0.100 max **As** 0.002 max **Bi** 0.02 max **Cd** rem **Cu** 0.05 max **Fe** 0.02 max **Mg** 0.05 max **Ni** 0.005 max **Pb** 0.05 max **Sb** 0.10 max **Sn** 0.003 max **Zn** 89.0-91.0 | **ASTM** B907 |
| **Z50980** | 60Zn/40Cd Solder Alloy (Zn/Cd60) | **Ag** 0.015 max **Al** 0.100 max **As** 0.002 max **Bi** 0.02 max **Cd** rem **Cu** 0.05 max **Fe** 0.02 max **Mg** 0.05 max **Ni** 0.005 max **Pb** 0.05 max **Sb** 0.10 max **Sn** 0.003 max **Zn** 59.0-61.0 | **ASTM** B907 |
| **Z55710** | Hot Dip Galvanizing Master Alloy (Type S-1) | **As** 0.015 max **Cd** 0.003 max **Cu** 0.003 max **Fe** 0.03 max **Pb** 0.015 max **Sb** 9.5-10.5 **Sn** 0.01 max **Zn** rem **Other** total 0.03 max | **ASTM** B860; B860 (S-1) |
| **Z56900** | 50Zn/50Sn Solder Alloy (Zn/Sn50) | **Ag** 0.015 max **Al** 0.100 max **As** 0.002 max **Bi** 0.02 max **Cd** 0.005 max **Cu** 0.005 max **Fe** 0.02 max **Mg** 0.02 max **Ni** 0.005 max **Pb** 0.05 max **Sb** 0.10 max **Sn** 49.0-51.0 **Zn** rem | **ASTM** B907 |
| **Z56930** | 50 Zn/49 Sn/1 Cu Solder Alloy (Zn/Sn 49) | **Ag** 0.015 max **Al** 0.100 max **As** 0.002 max **Bi** 0.02 max **Cd** 0.005 max **Cu** 0.8-1.3 **Fe** 0.02 max **Mg** 0.05 max **Ni** 0.005 max **Pb** 0.05 max **Sb** 0.10 max **Sn** 47.5-50.5 **Zn** rem | **ASTM** B907 |
| **Z80310** | Continuous Galvanizing Grade (CGG) Alloy | **Al** 0.22-0.28 **Cd** 0.01 max **Cu** 0.01 max **Fe** 0.0075 max **Pb** 0.007 max **Zn** rem **Other** total 0.01 max | **ASTM** B852 |
| **Z80410** | Continuous Galvanizing Grade (CGG) Alloy | **Al** 0.32-0.38 **Cd** 0.01 max **Cu** 0.02 max **Fe** 0.01 max **Pb** 0.01 max **Zn** rem **Other** total 0.01 max | **ASTM** B852 |
| **Z80411** | Continuous Galvanizing Grade (CGG) Alloy | **Al** 0.31-0.39 **Cd** 0.01 max **Cu** 0.01 max **Fe** 0.0075 max **Pb** 0.007 max **Zn** rem **Other** total 0.01 max | **ASTM** B852 |
| **Z80510** | Continuous Galvanizing Grade (CGG) Alloy | **Al** 0.42-0.48 **Cd** 0.01 max **Cu** 0.02 max **Fe** 0.01 max **Pb** 0.01 max **Zn** rem **Other** total 0.01 max | **ASTM** B852 |
| **Z80511** | Continuous Galvanizing Grade (CGG) Alloy | **Al** 0.40-0.50 **Cd** 0.01 max **Cu** 0.01 max **Fe** 0.0075 max **Pb** 0.007 max **Zn** rem **Other** total 0.01 max | **ASTM** B852 |
| **Z80531** | Continuous Galvanizing Grade (CGG) Alloy | **Al** 0.40-0.50 **Cd** 0.01 max **Cu** 0.01 max **Fe** 0.0075 max **Pb** 0.01-0.03 **Zn** rem **Other** total 0.01 max | **ASTM** B852 |
| **Z80610** | Continuous Galvanizing Grade (CGG) Alloy | **Al** 0.49-0.61 **Cd** 0.01 max **Cu** 0.01 max **Fe** 0.0075 max **Pb** 0.007 max **Zn** rem **Other** total 0.01 max | **ASTM** B852 |
| **Z80710** | Continuous Galvanizing Grade (CGG) Alloy | **Al** 0.58-0.72 **Cd** 0.01 max **Cu** 0.01 max **Fe** 0.0075 max **Pb** 0.007 max **Zn** rem **Other** total 0.01 max | **ASTM** B852 |
| **Z80810** | Continuous Galvanizing Grade (CGG) Alloy | **Al** 0.67-0.83 **Cd** 0.01 max **Cu** 0.01 max **Fe** 0.0075 max **Pb** 0.007 max **Zn** rem **Other** total 0.01 max | **ASTM** B852 |
| **Z80910** | Continuous Galvanizing Grade (CGG) Alloy | **Al** 0.90-1.10 **Cd** 0.01 max **Cu** 0.01 max **Fe** 0.0075 max **Pb** 0.007 max **Zn** rem **Other** total 0.01 max | **ASTM** B852 |
| **Z81330** | Continuous Galvanizing Grade (CGG) Alloy | **Al** 0.22-0.28 **Cd** 0.01 max **Cu** 0.02 max **Fe** 0.01 max **Pb** 0.01-0.02 **Zn** rem **Other** total 0.01 max | **ASTM** B852 |

The chemical compositions listed are for identification purposes and should not be used in lieu of the cross referenced specifications.

| UNIFIED NUMBER | DESCRIPTION | CHEMICAL COMPOSITION | CROSS REFERENCE SPECIFICATIONS |
|---|---|---|---|
| **Z81430** | Continuous Galvanizing Grade (CGG) Alloy | **Al** 0.32-0.38 **Cd** 0.01 max **Cu** 0.02 max **Fe** 0.01 max **Pb** 0.01-0.03 **Zn** rem **Other** total 0.01 max | **ASTM** B852 |
| **Z81450** | Continuous Galvanizing Grade (CGG) Alloy | **Al** 0.32-0.38 **Cd** 0.01 max **Cu** 0.02 max **Fe** 0.01 max **Pb** 0.075-0.125 **Zn** rem **Other** total 0.01 max | **ASTM** B852 |
| **Z81530** | Continuous Galvanizing Grade (CGG) Alloy | **Al** 0.42-0.48 **Cd** 0.01 max **Cu** 0.02 max **Fe** 0.01 max **Pb** 0.01-0.03 **Zn** rem **Other** total 0.01 max | **ASTM** B852 |
| **Z81540** | Continuous Galvanizing Grade (CGG) Alloy | **Al** 0.42-0.48 **Cd** 0.01 max **Cu** 0.02 max **Fe** 0.01 max **Pb** 0.025-0.075 **Zn** rem **Other** total 0.01 max | **ASTM** B852 |
| **Z81550** | Continuous Galvanizing Grade (CGG) Alloy | **Al** 0.42-0.48 **Cd** 0.01 max **Cu** 0.02 max **Fe** 0.01 max **Pb** 0.075-0.125 **Zn** rem **Other** total 0.01 max | **ASTM** B852 |

The chemical compositions listed are for identification purposes and should not be used in lieu of the cross referenced specifications.

# Cross Index of Commonly Known Documents Which Describe Materials Same as or Similar to These Covered by UNS Numbers

## AA (Aluminum Association) NUMBERS

**(The Symbol *ob* Indicates an Obsolete UNS Number.)**

# AA

| CROSS REFERENCE SPECIFICATIONS | UNIFIED NUMBER |
|---|---|
| 1-1/2% Alclad 2024 | A82024 |
| 1-1/2% Alclad One Side 2024 | A82024 |
| 2-1/2% Alclad 7075 | A87075 |
| 2-1/2% Alclad One Side 7075 | A87075 |
| 100.1 | A01001 |
| 130.1 | A01301 |
| 150.1 | A01501 |
| 160.1 | A01601 |
| 170.1 | A01701 |
| 201.0 | A02010 |
| 201.2 | A02012 |
| 202.0 | A02020 ob |
| 202.2 | A02022 ob |
| 203.0 | A02030 |
| 203.2 | A02032 |
| 204.0 | A02040 |
| 204.2 | A02042 |
| 206.0 | A02060 |
| 206.2 | A02062 |
| 208.0 | A02080 ob |
| 208.1 | A02081 ob |
| 208.2 | A02082 ob |
| 213.0 | A02130 ob |
| 213.1 | A02131 ob |
| 222.0 | A02220 ob |
| 222.1 | A02221 ob |
| 224.0 | A02240 ob |
| 224.2 | A02242 ob |
| 238.0 | A02380 ob |
| 238.1 | A02381 ob |
| 238.2 | A02382 ob |
| 240.0 | A02400 |
| 240.1 | A02401 |
| 242.0 | A02420 |
| 242.1 | A02421 |
| 242.2 | A02422 |
| 243.0 | A02430 ob |
| 243.1 | A02431 ob |
| 249.0 | A02490 ob |
| 249.2 | A02492 ob |
| 295.0 | A02950 |
| 295.1 | A02951 |
| 295.2 | A02952 |
| 296.0 | A02960 |
| 296.1 | A02961 |
| 296.2 | A02962 |
| 301.0 | A03010 |
| 301.1 | A03011 |
| 302.0 | A03020 |
| 302.1 | A03021 |
| 303.0 | A03030 |
| 303.1 | A03031 |
| 305.0 | A03050 ob |
| 305.2 | A03052 ob |
| 308.0 | A03080 |
| 308.1 | A03081 |
| 308.2 | A03082 |
| 318.0 | A03180 |
| 318.1 | A03181 |
| 319.0 | A03190 |
| 319.1 | A03191 |
| 319.2 | A03192 |
| 320.0 | A03200 |
| 320.1 | A03201 |
| 324.0 | A03240 ob |
| 324.1 | A03241 ob |
| 324.2 | A03242 ob |
| 328.0 | A03280 ob |
| 328.1 | A03281 ob |
| 332.0 | A03320 |
| 332.1 | A03321 |
| 332.2 | A03322 |
| 333.0 | A03330 |
| 333.1 | A03331 |
| 336.0 | A03360 |
| 336.1 | A03361 |
| 336.2 | A03362 |
| 337.0 | A03370 ob |
| 337.1 | A03371 ob |
| 338.0 | A03380 ob |
| 338.1 | A03381 ob |
| 339.0 | A03390 |
| 339.1 | A03391 |
| 343.0 | A03430 ob |
| 343.1 | A03431 ob |
| 354.0 | A03540 |
| 354.1 | A03541 |
| 355.0 | A03550 |
| 355.1 | A03551 |
| 355.2 | A03552 |
| 356.0 | A03560 |
| 356.1 | A03561 |
| 356.2 | A03562 |
| 357.0 | A03570 |
| 357.1 | A03571 |
| 358.0 | A03580 |
| 358.2 | A03582 |
| 359.0 | A03590 |
| 359.2 | A03592 |
| 360.0 | A03600 |
| 360.2 | A03602 |
| 361.0 | A03610 |
| 361.1 | A03611 |
| 362.0 | A03620 ob |
| 362.1 | A03621 ob |
| 363.0 | A03630 |
| 363.1 | A03631 |
| 364.0 | A03640 |
| 364.2 | A03642 |
| 365.0 | A03650 |
| 365.1 | A03651 |
| 369.0 | A03690 |
| 369.1 | A03691 |
| 380.0 | A03800 |
| 380.2 | A03802 |
| 381.0 | A03810 |
| 381.2 | A03812 |
| 383.0 | A03830 |
| 383.1 | A03831 |
| 383.2 | A03832 |
| 384.0 | A03840 |
| 384.1 | A03841 |
| 384.2 | A03842 |
| 385.0 | A03850 |
| 385.1 | A03851 |
| 390.0 | A03900 |
| 390.2 | A03902 |
| 391.0 | A03910 |
| 391.1 | A03911 |
| 392.0 | A03920 |
| 392.1 | A03921 |
| 393.0 | A03930 |
| 393.1 | A03931 |
| 393.2 | A03932 |
| 408.2 | A04082 |
| 409.2 | A04092 |
| 411.2 | A04112 |
| 413.0 | A04130 |
| 413.2 | A04132 |
| 435.2 | A04352 |
| 443.0 | A04430 |
| 443.1 | A04431 |
| 443.2 | A04432 |
| 444.0 | A04440 |
| 444.2 | A04442 |
| 445.2 | A04452 |
| 511.0 | A05110 |
| 511.1 | A05111 |
| 511.2 | A05112 |
| 512.0 | A05120 |
| 512.2 | A05122 |
| 513.0 | A05130 |
| 513.2 | A05132 |
| 514.0 | A05140 |
| 514.1 | A05141 |
| 514.2 | A05142 |
| 515.0 | A05150 |
| 515.2 | A05152 |
| 516.0 | A05160 |
| 516.1 | A05161 |
| 518.0 | A05180 |
| 518.1 | A05181 |
| 518.2 | A05182 |
| 520.0 | A05200 |
| 520.2 | A05202 |
| 535.0 | A05350 |
| 535.2 | A05352 |
| 705.0 | A07050 |
| 705.1 | A07051 |
| 707.0 | A07070 |
| 707.1 | A07071 |
| 710.0 | A07100 |
| 710.1 | A07101 |
| 711.0 | A07110 |
| 711.1 | A07111 |
| 712.0 | A07120 |
| 712.2 | A07122 |
| 713.0 | A07130 |
| 713.1 | A07131 |
| 771.0 | A07710 |
| 771.2 | A07712 |
| 772.0 | A07720 |
| 772.2 | A07722 |
| 850.0 | A08500 |
| 850.1 | A08501 |
| 851.0 | A08510 |
| 851.1 | A08511 |
| 852.0 | A08520 |
| 852.1 | A08521 |
| 853.0 | A08530 |
| 853.2 | A08532 |
| 1030 | A91030 ob |
| 1035 | A91035 |
| 1040 | A91040 |
| 1045 | A91045 |
| 1050 | A91050 |
| 1050A | A91050 |
| 1055 | A91055 ob |
| 1060 | A91060 |
| 1060/Al2O3/10p | A71001 |
| 1065 | A91065 |
| 1070 | A91070 |
| 1070A | A91070 |
| 1075 | A91075 ob |
| 1080 | A91080 |

| CROSS REFERENCE SPECIFICATIONS | UNIFIED NUMBER |
|---|---|
| 1080A | A91080 |
| 1085 | A91085 |
| 1090 | A91090 |
| 1095 | A91095 ob |
| 1098 | A91098 |
| 1100 | A91100 |
| 1110 | A91110 |
| 1120 | A91120 |
| 1135 | A91135 ob |
| 1145 | A91145 |
| 1150 | A91150 |
| 1170 | A91170 ob |
| 1175 | A91175 ob |
| 1180 | A91180 ob |
| 1185 | A91185 |
| 1188 | A91188 |
| 1190 | A91190 |
| 1193 | A91193 |
| 1198 | A91198 |
| 1199 | A91199 |
| 1200 | A91200 |
| 1200A | A91200 |
| 1230 | A91230 |
| 1235 | A91235 |
| 1250 | A91250 ob |
| 1260 | A91260 |
| 1275 | A91275 |
| 1285 | A91285 |
| 1300 | A91300 |
| 1345 | A91345 |
| 1350 | A91350 |
| 1350A | A91350 |
| 1370 | A91370 |
| 1385 | A91385 |
| 1435 | A91435 |
| 1445 | A91445 |
| 1450 | A91450 |
| 2001 | A92001 |
| 2002 | A92002 |
| 2003 | A92003 ob |
| 2004 | A92004 |
| 2005 | A92005 |
| 2006 | A92006 |
| 2007 | A92007 |
| 2007A | A92007 |
| 2008 | A92008 |
| 2009 | A92009 |
| 2010 | A92010 |
| 2011 | A92011 |
| 2011A | A92011 |
| 2012 | A92012 |
| 2014 | A92014 |
| 2014A | A92014 |
| 2017 | A92017 |
| 2017A | A92017 |
| 2018 | A92018 |
| 2020 | A92020 ob |
| 2021 | A92021 |
| 2024 | A92024 |
| 2025 | A92025 |
| 2026 | A92026 |
| 2030 | A92030 |
| 2031 | A92031 |
| 2034 | A92034 |
| 2036 | A92036 |
| 2037 | A92037 |
| 2038 | A92038 |
| 2039 | A92039 |
| 2048 | A92048 |
| 2080 | A92080 |
| 2090 | A92090 |
| 2091 | A92091 |
| 2094 | A92094 |
| 2095 | A92095 |
| 2097 | A92097 |
| 2098 | A92098 |
| 2111 | A92111 |
| 2111A | A92111 |
| 2117 | A92117 |
| 2124 | A92124 |
| 2195 | A92195 |
| 2196 | A92196 |
| 2197 | A92197 |
| 2214 | A92214 |
| 2218 | A92218 |
| 2219 | A92219 |
| 2224 | A92224 |
| 2224A | A92224 |
| 2297 | A92297 |
| 2319 | A92319 |
| 2324 | A92324 |
| 2419 | A92419 |
| 2424 | A92424 |
| 2519 | A92519 |
| 2524 | A92524 |
| 2618 | A92618 |
| 2618A | A92618 |
| 3002 | A93002 |
| 3003 | A93003 |
| 3004 | A93004 |
| 3004A | A93004 |
| 3005 | A93005 |
| 3005A | A93005 |
| 3006 | A93006 |
| 3007 | A93007 |
| 3008 | A93008 ob |
| 3009 | A93009 |
| 3010 | A93010 |
| 3011 | A93011 |
| 3012 | A93012 |
| 3013 | A93013 |
| 3014 | A93014 |
| 3015 | A93015 |
| 3016 | A93016 |
| 3017 | A93017 |
| 3018 | A93018 ob |
| 3019 | A93019 |
| 3025 | A93025 |
| 3030 | A93030 |
| 3102 | A93102 |
| 3103 | A93103 |
| 3103A | A93103 |
| 3104 | A93104 |
| 3105 | A93105 |
| 3105A | A93105 |
| 3105B | A93015 |
| 3107 | A93107 |
| 3203 | A93203 |
| 3204 | A93204 |
| 3207 | A93207 |
| 3207A | A93207 |
| 3303 | A93303 ob |
| 3307 | A93307 |
| 4002 | A94002 ob |
| 4004 | A94004 |
| 4006 | A94006 |
| 4007 | A94007 |
| 4008 | A94008 |
| 4009 | A94009 |
| 4010 | A94010 |
| 4011 | A94011 |
| 4013 | A94013 |
| 4014 | A94014 |
| 4015 | A94015 |
| 4016 | A94016 |
| 4017 | A94017 |
| 4018 | A94018 |
| 4019 | A94019 |
| 4026 | A94026 |
| 4032 | A94032 |
| 4043 | A94043 |
| 4043A | A94043 |
| 4044 | A94044 |
| 4045 | A94045 |
| 4046 | A94046 |
| 4047 | A94047 |
| 4047A | A94047 |
| 4048 | A94048 |
| 4104 | A94104 |
| 4145 | A94145 |
| 4145A | A94145 |
| 4147 | A94147 |
| 4343 | A94343 |
| 4543 | A94543 ob |
| 4643 | A94643 |
| 5005 | A95005 |
| 5005A | A95005 |
| 5006 | A95006 |
| 5010 | A95010 |
| 5013 | A95013 ob |
| 5014 | A95014 ob |
| 5016 | A95016 |
| 5017 | A95017 |
| 5018 | A95018 |
| 5018A | A95018 |
| 5019 | A95019 |
| 5019A | A95019 |
| 5021 | A95021 |
| 5022 | A95022 |
| 5023 | A95023 |
| 5025 | A95025 |
| 5034 | A95034 ob |
| 5039 | A95039 ob |
| 5040 | A95040 |
| 5042 | A95042 |
| 5043 | A95043 |
| 5049 | A95049 |
| 5050 | A95050 |
| 5050A | A95050 |
| 5051 | A95051 |
| 5051A | A95051 |
| 5052 | A95052 |
| 5056 | A95056 |
| 5058 | A95058 |
| 5059 | A95059 |
| 5082 | A95082 |
| 5083 | A95083 |
| 5086 | A95086 |
| 5087 | A95087 |
| 5091 | A95091 |
| 5110 | A95110 |
| 5119 | A95119 |
| 5149 | A95149 |
| 5150 | A95150 |
| 5151 | A95151 |
| 5154 | A95154 |
| 5154A | A95154 |

| CROSS REFERENCE SPECIFICATIONS | UNIFIED NUMBER |
|---|---|
| 5154B | A95154 |
| 5180 | A95180 |
| 5182 | A95182 |
| 5183 | A95183 |
| 5186 | A95186 |
| 5205 | A95205 |
| 5210 | A95210 |
| 5249 | A95249 |
| 5250 | A95250 |
| 5251 | A95251 |
| 5251A | A95251 |
| 5252 | A95252 |
| 5254 | A95254 |
| 5257 | A95257 |
| 5280 | A95280 ob |
| 5283 | A95283 |
| 5283A | A95283 |
| 5283B | A95283 |
| 5305 | A95305 |
| 5310 | A95310 |
| 5349 | A95349 |
| 5351 | A95351 |
| 5352 | A95352 |
| 5354 | A95354 |
| 5356 | A95356 |
| 5357 | A95357 |
| 5383 | A95383 |
| 5449 | A95449 |
| 5451 | A95451 |
| 5454 | A95454 |
| 5456 | A95456 |
| 5456A | A95456 |
| 5457 | A95457 |
| 5505 | A95505 |
| 5552 | A95552 ob |
| 5554 | A95554 |
| 5556 | A95556 |
| 5556A | A95556 |
| 5557 | A95557 |
| 5605 | A95605 |
| 5652 | A95652 |
| 5654 | A95654 |
| 5657 | A95657 |
| 5754 | A95754 |
| 5854 | A95854 ob |
| 5954 | A95954 |
| 6002 | A96002 |
| 6003 | A96003 |
| 6004 | A96004 |
| 6005 | A96005 |
| 6005A | A96005 |
| 6005B | A96005 |
| 6006 | A96006 |
| 6007 | A96007 |
| 6008 | A96008 |
| 6009 | A96009 |
| 6010 | A96010 |
| 6011 | A96011 |
| 6012 | A96012 |
| 6012A | A96012 |
| 6013 | A96013 |
| 6014 | A96014 |
| 6015 | A96015 |
| 6016 | A96016 |
| 6016A | A96016 |
| 6017 | A96017 ob |
| 6018 | A96018 |
| 6019 | A96019 |
| 6020 | A96020 |
| 6021 | A96021 |
| 6022 | A96022 |
| 6053 | A96053 |
| 6056 | A96056 |
| 6060 | A96060 |
| 6061 | A96061 |
| 6061A | A96061 |
| 6063 | A96063 |
| 6063A | A96063 |
| 6066 | A96066 |
| 6069 | A96069 |
| 6070 | A96070 |
| 6081 | A96081 |
| 6082 | A96082 |
| 6082A | A96082 |
| 6090 | A96090 ob |
| 6091 | A96091 |
| 6092 | A96092 |
| 6101 | A96101 |
| 6101A | A96101 |
| 6101B | A96101 |
| 6103 | A96103 |
| 6105 | A96105 |
| 6106 | A96106 |
| 6110 | A96110 |
| 6110A | A96110 |
| 6111 | A96111 |
| 6113 | A96113 |
| 6116 | A96116 |
| 6151 | A96151 |
| 6160 | A96160 |
| 6162 | A96162 |
| 6181 | A96181 |
| 6181A | A96181 |
| 6201 | A96201 |
| 6201A | A96201 |
| 6205 | A96205 |
| 6206 | A96206 |
| 6253 | A96253 |
| 6260 | A96260 |
| 6261 | A96261 |
| 6262 | A96262 |
| 6301 | A96301 |
| 6306 | A96306 |
| 6351 | A96351 |
| 6351A | A96351 |
| 6401 | A96401 |
| 6463 | A96463 |
| 6463A | A96463 |
| 6763 | A96763 |
| 6863 | A96863 ob |
| 6951 | A96951 |
| 6963 | A96963 |
| 7001 | A97001 ob |
| 7003 | A97003 |
| 7004 | A97004 |
| 7005 | A97005 |
| 7008 | A97008 |
| 7009 | A97009 |
| 7010 | A97010 |
| 7011 | A97011 ob |
| 7012 | A97012 |
| 7013 | A97013 ob |
| 7014 | A97014 |
| 7015 | A97015 |
| 7016 | A97016 |
| 7017 | A97017 |
| 7018 | A97018 |
| 7019 | A97019 |
| 7019A | A97019 |
| 7020 | A97020 |
| 7021 | A97021 |
| 7022 | A97022 |
| 7023 | A97023 |
| 7024 | A97024 |
| 7025 | A97025 |
| 7026 | A97026 |
| 7027 | A97027 ob |
| 7028 | A97028 |
| 7029 | A97029 |
| 7030 | A97030 |
| 7031 | A97031 |
| 7032 | A97032 |
| 7033 | A97033 |
| 7034 | A97034 |
| 7039 | A97039 |
| 7040 | A97040 |
| 7046 | A97046 |
| 7046A | A97046 |
| 7049 | A97049 |
| 7049A | A97049 |
| 7050 | A97050 |
| 7050A | A87050 |
| 7050A | A97050 |
| 7051 | A97051 ob |
| 7055 | A97055 |
| 7060 | A97060 |
| 7064 | A97064 |
| 7068 | A97068 |
| 7070 | A97070 ob |
| 7072 | A97072 |
| 7075 | A97075 |
| 7076 | A97076 |
| 7079 | A97079 ob |
| 7090 | A97090 |
| 7091 | A97091 ob |
| 7093 | A97093 |
| 7104 | A97104 ob |
| 7108 | A97108 |
| 7108A | A97108 |
| 7109 | A97109 ob |
| 7116 | A97116 |
| 7122 | A97122 |
| 7129 | A97129 |
| 7146 | A97146 ob |
| 7149 | A97149 |
| 7150 | A97150 |
| 7175 | A97175 |
| 7178 | A97178 |
| 7179 | A97179 ob |
| 7229 | A97229 |
| 7249 | A97249 |
| 7250 | A97250 |
| 7277 | A97277 ob |
| 7278 | A97278 |
| 7278A | A97278 |
| 7349 | A97349 |
| 7449 | A97449 |
| 7472 | A97472 ob |
| 7475 | A97475 |
| 8001 | A98001 ob |
| 8004 | A98004 ob |
| 8005 | A98005 |
| 8006 | A98006 |
| 8007 | A98007 |
| 8008 | A98008 |
| 8009 | A98009 ob |
| 8010 | A98010 |

## *AA (Aluminum Association) NUMBERS*

| CROSS REFERENCE SPECIFICATIONS | UNIFIED NUMBER |
|---|---|
| 8011 | A98011 |
| 8011A | A98011 |
| 8013 | A98013 ob |
| 8014 | A98014 |
| 8015 | A98015 |
| 8016 | A98016 |
| 8017 | A98017 |
| 8018 | A98018 |
| 8019 | A98019 |
| 8020 | A98020 |
| 8021 | A98021 |
| 8021A | A98021 |
| 8021B | A98021 |
| 8022 | A98022 |
| 8023 | A98023 |
| 8024 | A98024 |
| 8025 | A98025 |
| 8030 | A98030 |
| 8040 | A98040 |
| 8050 | A98050 |
| 8076 | A98076 |
| 8077 | A98077 |
| 8079 | A98079 |
| 8081 | A98081 ob |
| 8090 | A98090 |
| 8091 | A98091 |
| 8093 | A98093 |
| 8111 | A98111 |
| 8112 | A98112 |
| 8130 | A98130 |
| 8150 | A98150 |
| 8176 | A98176 |
| 8177 | A98177 ob |
| 8211 | A98211 |
| 8276 | A98276 ob |
| 8280 | A98280 |
| A201.0 | A12010 |
| A201.1 | A12011 |
| A206.0 | A12060 |
| A206.2 | A12062 |
| A242.0 | A12420 |
| A242.1 | A12421 |
| A242.2 | A12422 |
| A305.0 | A13050 ob |
| A305.1 | A13051 ob |
| A305.2 | A13052 ob |
| A319.0 | A13190 |
| A319.1 | A13191 |
| A333.0 | A13330 |
| A333.1 | A13331 |
| A355.0 | A13550 |
| A355.2 | A13552 |
| A356.0 | A13560 |
| A356.1 | A13561 |
| A356.2 | A13562 |
| A357.0 | A13570 |
| A357.0 | A23570 |
| A357.2 | A13572 |
| A359.0 | A13590 |
| A359.1 | A13591 |
| A360.0 | A13600 |
| A360.1 | A13601 |
| A360.2 | A13602 |
| A380.0 | A13800 |
| A380.1 | A13801 |
| A380.2 | A13802 |
| A383.0 | A13830 |
| A383.1 | A13831 |
| A384.0 | A13840 |
| A384.1 | A13841 |
| A390.0 | A13900 |
| A390.1 | A13901 |
| A391.0 | A13910 |
| A391.1 | A13911 |
| A413.0 | A14130 |
| A413.1 | A14131 |
| A413.2 | A14132 |
| A443.0 | A14430 |
| A443.1 | A14431 |
| A444.0 | A14440 |
| A444.1 | A14441 |
| A444.2 | A14442 |
| A535.0 | A15350 |
| A535.1 | A15351 |
| Alclad 2004 | A82004 |
| Alclad 2014 | A82014 |
| Alclad 2024 | A82024 |
| Alclad 2219 | A82219 |
| Alclad 3003 | A83003 |
| Alclad 6061 | A86061 |
| Alclad 6951 | A86951 |
| Alclad 7050 | A87050 |
| Alclad 7075 | A87075 |
| Alclad 7079 | A87079 ob |
| Alclad 7178 | A87178 |
| Alclad 7475 | A87475 |
| Alclad One Side 2024 | A82024 |
| Alclad One Side 7075 | A87075 |
| B201.0 | A22010 |
| B319.0 | A23190 |
| B319.1 | A23191 |
| B356.0 | A23560 |
| B356.2 | A23562 |
| B357.0 | A23570 |
| B357.2 | A23572 |
| B380.0 | A23800 |
| B380.1 | A23801 |
| B384.0 | A23840 |
| B384.1 | A23841 |
| B390.0 | A23900 |
| B390.1 | A23901 |
| B391.1 | A23910 |
| B391.1 | A23911 |
| B413.0 | A24130 |
| B413.1 | A24131 |
| B443.0 | A24430 |
| B443.1 | A24431 |
| B535.0 | A25350 |
| B535.2 | A25352 |
| C355.0 | A33550 |
| C355.1 | A33551 |
| C355.2 | A33552 |
| C356.0 | A33560 |
| C356.2 | A33562 |
| C357.0 | A33570 |
| C357.2 | A33572 |
| C380.0 | A33800 |
| C380.1 | A33801 |
| C384.0 | A33840 |
| C384.1 | A33841 |
| C443.0 | A34430 |
| C443.1 | A34431 |
| C443.2 | A34432 |
| D356.0 | A43560 ob |
| D356.1 | A43561 ob |
| D357.0 | A43570 |
| D380.0 | A43800 |
| D380.1 | A43801 |
| E357.0 | A53570 ob |
| E357.1 | A53571 ob |
| F356.0 | A63560 |
| F356.2 | A63562 |
| No. 11 Brazing Sheet | A83003 |
| No. 7 Brazing Sheet | A83003 |
| No. 8 Brazing Sheet | A83003 |
| No. 12 Brazing Sheet | A83003 |
| No. 13 Brazing Sheet | A86951 |
| No. 14 Brazing Sheet | A86951 |
| No. 24 Brazing Sheet | A86951 |
| No. 22 Brazing Sheet | A86951 |
| No. 34 Brazing Sheet | A86951 |
| No. 33 Brazing Sheet | A86951 |
| No. 44 Brazing Sheet | A86951 |
| No. 23 Brazing Sheet | A86951 |
| No. 21 Brazing Sheet | A86951 |
| X2096 | A92096 |
| X6030 | A96030 ob |

# Cross Index of Commonly Known Documents Which Describe Materials Same as or Similar to These Covered by UNS Numbers

## ACI (Steel Founders Society of America) NUMBERS

**(The Symbol *ob* Indicates an Obsolete UNS Number.)**

# ACI

| CROSS REFERENCE SPECIFICATIONS | UNIFIED NUMBER |
|---|---|
| CA-6N | J91650 |
| CA-6NM | J91540 |
| CA-15 | J91150 |
| CA-15M | J91151 |
| CA-40 | J91153 |
| CA28MWV | J91422 |
| CA40F | J91154 |
| CB-6 | J91804 |
| CB-30 | J91803 |
| CC-50 | J92615 |
| CD-3MN | J92205 |
| CD-4MCuN | J93372 |
| CD3MCuN | J93373 |
| CD3MVCuN | J93380 |
| CD4MCu | J93370 |
| CD6MN | J93371 |
| CE-8N | J92805 |
| CE-30 | J93423 |
| CE20N | J92802 |
| CF-3 | J92500 |
| CF-3M | J92800 |
| CF-8 | J92600 |
| CF-8C | J92710 |
| CF-8M | J92900 |
| CF-10M | J92901 |
| CF-16F | J92701 |
| CF-20 | J92602 |
| CF10MC | J92971 |
| CF10SMnN | J92972 |
| CF3MN | J92804 |
| CG-3M | J92999 |
| CG-12 | J93001 |
| CH-8 | J93400 |
| CH-10 | J93401 |
| CH-20 | J93402 |
| CK-3MCuN | J93254 |
| CK-3MN | J94653 |
| CK-20 | J94202 |
| CN-3M | J94652 |
| CN-7M | N08007 |
| CN3MN | J94651 |
| CT15C | N08151 |
| CT2MCoCu | N26985 |
| CU5MCuC | N08826 |
| CU5MCuSiC | N28825 |
| CW12MW | N30002 |
| CW2M | N26455 |
| CW6M | N30107 |
| CW6MC | N26625 |
| CX2MW | N26022 |
| CY40 | N06040 |
| CY5SnBiM | N26055 |
| CZ100 | N02100 |
| CZ12SiCu | N22000 |
| HC | J92605 |
| HC-30 | J92613 |
| HD | J93005 |
| HE | J93403 |
| HE-35 | J93413 |
| HF | J92603 |
| HH | J93503 |
| HH-30 | J93513 |
| HI | J94003 |
| HK | J94224 |
| HK-30 | J94203 |
| HK-40 | J94204 |
| HL | N08604 |

| CROSS REFERENCE SPECIFICATIONS | UNIFIED NUMBER |
|---|---|
| HL-30 | N08613 |
| HN | J94213 |
| HT | N08605 |
| HT-30 | N08603 |
| HT-50 | N08050 |
| HT-50C | N08008 |
| HU | N08004 |
| HU-50 | N08005 |
| HW | N08001 |
| HW-50 | N08006 |
| HX | N06006 |

## Cross Index of Commonly Known Documents Which Describe Materials Same as or Similar to These Covered by UNS Numbers

## AISI (American Iron and Steel Institute) NUMBERS (Carbon and Low Alloy Steels) including SAE (Society of Automotive Engineers) NUMBERS (Carbon and Low Alloy Steels)

**(The Symbol *ob* Indicates an Obsolete UNS Number.)**

# AISI & SAE

| CROSS REFERENCE SPECIFICATIONS | UNIFIED NUMBER |
|---|---|
| 15B21 H | H15211 |
| 15B35 H | H15351 |
| 15B37 H | H15371 |
| 15B41 H | H15411 |
| 15B48 H | H15481 |
| 15B62 H | H15621 |
| 50B40 H | H50401 |
| 50B44 H | H50441 |
| 50B46 H | H50461 |
| 50B50 H | H50501 |
| 50B60 H | H50601 |
| 51B60 H | H51601 |
| 81B45 H | H81451 |
| 86B30 H | H86301 |
| 86B45 H | H86451 |
| 94B15 H | H94151 |
| 94B17 H | H94171 |
| 94B30 H | H94301 |
| 201 | S20100 |
| 202 | S20200 |
| 205 | S20500 |
| 301 | S30100 |
| 302 | S30200 |
| 302B | S30215 |
| 303 | S30300 |
| 303 Se | S30323 |
| 304 | S30400 |
| 304 L | S30403 |
| 304 N | S30451 |
| 305 | S30500 |
| 308 | S30800 |
| 309 | S30900 |
| 309 S | S30908 |
| 310 | S31000 |
| 310 S | S31008 |
| 314 | S31400 |
| 316 | S31600 |
| 316 F | S31620 |
| 316 L | S31603 |
| 316 N | S31651 |
| 316N | J92804 |
| 317 | S31700 |
| 317 L | S31703 |
| 321 | S32100 |
| 329 | S32900 |
| 347 | S34700 |
| 348 | S34800 |
| 384 | S38400 |
| 385 | S38500 ob |
| 403 | S40300 |
| 405 | S40500 |
| 409 | S40900 |
| 410 | S41000 |
| 414 | S41400 |
| 416 | S41600 |
| 416Se | S41623 |
| 420 | S42000 |
| 420F | J91154 |
| 420F | S42020 |
| 422 | J91422 |
| 422 | S42200 |
| 429 | S42900 |
| 430 | S43000 |
| 430 F | S43020 |
| 430FSe | S43023 |
| 431 | S43100 |
| 434 | S43400 |
| 436 | S43600 |
| 439 | S43035 |
| 440 A | S44002 |
| 440 B | S44003 |
| 440 C | S44004 |
| 442 | S44200 |
| 446 | S44600 |
| 501 | S50100 |
| 502 | S50200 |
| 1038 H | H10380 |
| 1045 H | H10450 |
| 1330 H | H13300 |
| 1335 H | H13350 |
| 1340 H | H13400 |
| 1345 H | H13450 |
| 1522 H | H15220 |
| 1524 H | H15240 |
| 1526 H | H15260 |
| 1541 H | H15410 |
| 4027 H | H40270 |
| 4028 H | H40280 |
| 4032H | H40320 |
| 4037 H | H40370 |
| 4042 H | H40420 |
| 4047 H | H40470 |
| 4118 H | H41180 |
| 4130 H | H41300 |
| 4135 H | H41350 |
| 4137 H | H41370 |
| 4140 H | H41400 |
| 4142 H | H41420 |
| 4145 H | H41450 |
| 4147 H | H41470 |
| 4150 H | H41500 |
| 4161 H | H41610 |
| 4320 H | H43200 |
| 4340H | H43400 |
| 4419 H | H44190 |
| 4620 H | H46200 |
| 4621 H | H46210 |
| 4626 H | H46260 |
| 4718 H | H47180 |
| 4720 H | H47200 |
| 4815 H | H48150 |
| 4817 H | H48170 |
| 4820 H | H48200 |
| 5046 | G50460 |
| 5046 H | H50460 |
| 5120 H | H51200 |
| 5130 H | H51300 |
| 5132 H | H51320 |
| 5135 H | H51350 |
| 5140 H | H51400 |
| 5145 H | H51450 |
| 5147 H | H51470 |
| 5150 H | H51500 |
| 5155 H | H51550 |
| 5160 H | H51600 |
| 6118 H | H61180 |
| 6150 H | H61500 |
| 8617 H | H86170 |
| 8620 H | H86200 |
| 8622 H | H86220 |
| 8625 H | H86250 |
| 8627 H | H86270 |
| 8630 H | H86300 |
| 8637 H | H86370 |
| 8640 H | H86400 |
| 8642 H | H86420 |
| 8645 H | H86450 |
| 8650 H | H86500 |
| 8655 H | H86550 |
| 8660 H | H86600 |
| 8720 H | H87200 |
| 8740 H | H87400 |
| 8822 H | H88220 |
| 9260 H | H92600 |
| 9310 H | H93100 |
| A-2 | T30102 |
| A-3 | T30103 ob |
| A-4 | T30104 |
| A-5 | T30105 ob |
| A-6 | T30106 |
| A-7 | T30107 |
| A-8 | T30108 |
| A-9 | T30109 |
| A-10 | T30110 |
| A-11 | T30111 |
| D-2 | T30402 |
| D-3 | T30403 |
| D-4 | T30404 |
| D-5 | T30405 |
| D-7 | T30407 |
| F-1 | T60601 ob |
| F-2 | T60602 ob |
| H-10 | T20810 |
| H-11 | T20811 |
| H-12 | T20812 |
| H-13 | T20813 |
| H-14 | T20814 |
| H-19 | T20819 |
| H-21 | T20821 |
| H-22 | T20822 |
| H-23 | T20823 |
| H-24 | T20824 |
| H-25 | T20825 ob |
| H-26 | T20826 |
| H-41 | T20841 ob |
| H-42 | T20842 |
| H-43 | T20843 ob |
| L-2 | T61202 |
| L-3 | T61203 ob |
| L-6 | T61206 |
| M-1 | T11301 |
| M-2 | T11302 |
| M-3 (Cl.1) | T11313 |
| M-3 (Cl.2) | T11323 |
| M-4 | T11304 |
| M-6 | T11306 |
| M-7 | T11307 |
| M-10 | T11310 |
| M-30 | T11330 |
| M-33 | T11333 |
| M-34 | T11334 |
| M-36 | T11336 |
| M-41 | T11341 |
| M-42 | T11342 |
| M-43 | T11343 ob |
| M-44 | T11344 ob |
| M-46 | T11346 |
| M-47 | T11347 ob |
| M-48 | T11348 |
| M-52 | T11352 |
| M-61 | T11361 |
| M-62 | T11362 |
| O-1 | T31501 |
| O-2 | T31502 |
| O-6 | T31506 |

| CROSS REFERENCE SPECIFICATIONS | UNIFIED NUMBER |
|---|---|
| O-7 | T31507 |
| P-2 | T51602 ob |
| P-3 | T51603 ob |
| P-4 | T51604 ob |
| P-5 | T51605 ob |
| P-6 | T51606 |
| P-20 | T51620 |
| P-21 | T51621 |
| S-1 | T41901 |
| S-2 | T41902 |
| S-4 | T41904 |
| S-5 | T41905 |
| S-6 | T41906 |
| S7 | T41907 |
| S13800 | S13800 |
| S15500 | S15500 |
| S17400 | S17400 |
| S17700 | S17700 |
| S30430 | S30430 |
| T-1 | T12001 |
| T-2 | T12002 ob |
| T-4 | T12004 |
| T-5 | T12005 |
| T-6 | T12006 |
| T-8 | T12008 |
| T-15 | T12015 |
| W-1 | T72301 |
| W-2 | T72302 |
| W-5 | T72305 |

## Cross Index of Commonly Known Documents Which Describe Materials Same as or Similar to These Covered by UNS Numbers

## AMS (SAE/Aerospace Materials Specification) NUMBERS

**(The Symbol *ob* Indicates an Obsolete UNS Number.)**

# AMS

| CROSS REFERENCE SPECIFICATIONS | UNIFIED NUMBER |
|---|---|
| 4000 | A91060 |
| 4001 | A91100 |
| 4003 | A91100 |
| 4004 | A95052 |
| 4006 | A93003 |
| 4007 | A92024 |
| 4008 | A93003 |
| 4009 | A96061 |
| 4011 | A91145 |
| 4015 | A95052 |
| 4016 | A95052 |
| 4017 | A95052 |
| 4021 | A86061 |
| 4025 | A96061 |
| 4026 | A96061 |
| 4027 | A96061 |
| 4028 | A92014 |
| 4029 | A92014 |
| 4031 | A92219 |
| 4035 | A92024 |
| 4036 | A82024 |
| 4037 | A92024 |
| 4040 | A82024 |
| 4041 | A82024 |
| 4044 | A97075 |
| 4045 | A97075 |
| 4046 | A87075 |
| 4048 | A87075 |
| 4050 | A97050 |
| 4056 | A95083 |
| 4062 | A91100 |
| 4063 | A83003 |
| 4064 | A83003 |
| 4065 | A93003 |
| 4066 | A92219 |
| 4067 | A93003 |
| 4068 | A92219 |
| 4069 | A95052 |
| 4070 | A95052 |
| 4071 | A95052 |
| 4077 | A82024 |
| 4078 | A97075 |
| 4079 | A96061 |
| 4080 | A96061 |
| 4081 | A96061 |
| 4082 | A96061 |
| 4083 | A96061 |
| 4084 | A97475 |
| 4085 | A97475 |
| 4086 | A92024 |
| 4087 | A92024 |
| 4088 | A92024 |
| 4089 | A97475 |
| 4090 | A97475 |
| 4094 | A82219 |
| 4095 | A82219 |
| 4096 | A82219 |
| 4100 | A87475 |
| 4101 | A92124 |
| 4102 | A91100 |
| 4107 | A97050 |
| 4108 | A97050 |
| 4111 | A97049 |
| 4113 | A96061 |
| 4114 | A95052 |
| 4115 | A96061 |
| 4116 | A96061 |
| 4117 | A96061 |
| 4118 | A92017 |
| 4120 | A92024 |
| 4121 | A92014 |
| 4122 | A97075 |
| 4123 | A97075 |
| 4124 | A97075 |
| 4125 | A96151 |
| 4126 | A97075 |
| 4127 | A96061 |
| 4128 | A96061 |
| 4130 | A92025 |
| 4131 | A97075 |
| 4132 | A92618 |
| 4133 | A92014 |
| 4134 | A92014 |
| 4138 | A97079 ob |
| 4140 | A92018 |
| 4141 | A97075 |
| 4143 | A92219 |
| 4144 | A92219 |
| 4146 | A96061 |
| 4147 | A97075 |
| 4148 | A97175 |
| 4149 | A97175 |
| 4150 | A96061 |
| 4152 | A92024 |
| 4153 | A92014 |
| 4154 | A97075 |
| 4156 | A96063 |
| 4157 | A97049 |
| 4159 | A97049 |
| 4160 | A96061 |
| 4161 | A96061 |
| 4162 | A92219 |
| 4164 | A92024 |
| 4165 | A92024 |
| 4166 | A97075 |
| 4167 | A97075 |
| 4168 | A97075 |
| 4169 | A97075 |
| 4172 | A96061 |
| 4173 | A96061 |
| 4174 | A97075 |
| 4177 | A95056 |
| 4179 | A97175 |
| 4180 | A91100 |
| 4181 | A94008 |
| 4182 | A95056 |
| 4184 | A94145 |
| 4185 | A94047 |
| 4186 | A97075 |
| 4189 | A94643 |
| 4190 | A94043 |
| 4191 | A92319 |
| 4193 | A92024 |
| 4194 | A82024 |
| 4195 | A82024 |
| 4200 | A97049 |
| 4201 | A97050 |
| 4202 | A97475 |
| 4203 | A97010 |
| 4204 | A97010 |
| 4205 | A97010 |
| 4206 | A97055 |
| 4207 | A87475 |
| 4208 | A92004 |
| 4209 | A82004 |
| 4210 | A03550 |
| 4212 | A03550 |
| 4214 | A03550 |
| 4215 | A33550 |
| 4216 | A96013 |
| 4217 | A03560 |
| 4218 | A13560 |
| 4219 | A13570 |
| 4221 | A92124 |
| 4223 | A12010 |
| 4224 | A02430 ob |
| 4225 | A02030 |
| 4226 | A02240 ob |
| 4229 | A12010 |
| 4232 | A92090 |
| 4233 | A02010 |
| 4235 | A12060 |
| 4236 | A12060 |
| 4237 | A02060 |
| 4241 | A03570 |
| 4242 | A02010 |
| 4243 | A87050 |
| 4244 | A02060 |
| 4245 | A03550 |
| 4246 | A03570 |
| 4247 | A97049 |
| 4248 | A96061 |
| 4249 | A43570 |
| 4251 | A92090 |
| 4252 | A97150 |
| 4255 | A86951 |
| 4255 | A87050 |
| 4256 | A86951 |
| 4256 | A87050 |
| 4259 | A98090 |
| 4260 | A03560 |
| 4261 | A03560 |
| 4266 | A98022 |
| 4270 | A92424 |
| 4273 | A92424 |
| 4277 | A97075 |
| 4278 | A87075 |
| 4279 | A82024 |
| 4280 | A03550 |
| 4281 | A03550 |
| 4284 | A03560 |
| 4285 | A03560 |
| 4286 | A03560 |
| 4290 | A03600 |
| 4290 | A13600 |
| 4291 | A13800 |
| 4295 | A92219 |
| 4297 | A92024 |
| 4303 | A92090 |
| 4306 | A97150 |
| 4307 | A97150 |
| 4308 | A98009 ob |
| 4309 | A98009 ob |
| 4310 | A97075 |
| 4311 | A97075 |
| 4312 | A96061 |
| 4313 | A92219 |
| 4314 | A92014 |
| 4318 | A94032 |
| 4319 | A94032 |
| 4320 | A97149 |
| 4321 | A97049 |
| 4322 | A95091 |
| 4323 | A97075 |
| 4333 | A97050 |

| CROSS REFERENCE SPECIFICATIONS | UNIFIED NUMBER |
|---|---|
| 4334 | A97249 |
| 4337 | A97055 |
| 4339 | A92024 |
| 4340 | A97050 |
| 4341 | A97050 |
| 4342 | A97050 |
| 4343 | A97149 |
| 4344 | A97175 |
| 4345 | A97150 |
| 4347 | A96013 |
| 4350 | M11610 |
| 4352 | M16600 |
| 4360 | M11800 |
| 4362 | M16600 |
| 4363 | M13210 |
| 4375 | M11311 |
| 4376 | M11311 |
| 4377 | M11311 |
| 4382 | M11311 |
| 4383 | M13210 |
| 4384 | M13310 |
| 4385 | M13310 |
| 4386 | M14141 |
| 4387 | M16210 |
| 4388 | M13312 |
| 4390 | M13210 |
| 4395 | M11922 |
| 4396 | M12331 |
| 4418 | M18220 |
| 4419 | M18210 |
| 4420 | M11630 |
| 4422 | M11630 |
| 4424 | M11630 |
| 4425 | M16630 |
| 4427 | M18432 |
| 4434 | M11920 |
| 4437 | M11914 |
| 4438 | M16620 |
| 4439 | M16410 |
| 4441 | M12410 |
| 4442 | M12330 |
| 4443 | M16510 |
| 4444 | M16610 |
| 4445 | M13310 |
| 4446 | M11919 |
| 4447 | M13320 |
| 4452 | M11914 |
| 4453 | M11920 |
| 4455 | M10100 |
| 4483 | M10100 |
| 4484 | M11920 |
| 4490 | M11910 |
| 4500 | C11000 |
| 4501 | C10200 |
| 4505 | C26000 |
| 4507 | C26000 |
| 4508 | C26000 |
| 4510 | C51000 |
| 4511 | C82500 |
| 4520 | C54400 |
| 4530 | C17200 |
| 4532 | C17200 |
| 4533 | C17200 |
| 4534 | C17200 |
| 4535 | C17200 |
| 4544 | N04400 |
| 4555 | C33000 |
| 4558 | C33200 |
| 4574 | N04400 |
| 4575 | N04400 |
| 4575 | R54621 |
| 4602 | C10200 |
| 4610 | C36000 |
| 4611 | C46400 |
| 4612 | C46400 |
| 4614 | C37700 |
| 4615 | C65500 |
| 4625 | C51000 |
| 4631 | C64200 |
| 4635 | C62300 |
| 4640 | C63000 |
| 4650 | C17200 |
| 4651 | C17200 |
| 4674 | N04405 |
| 4675 | N04400 |
| 4676 | N05500 |
| 4677 | N05502 |
| 4700 | C10100 |
| 4701 | C10200 |
| 4710 | C27000 |
| 4712 | C27000 |
| 4713 | C27000 |
| 4720 | C51000 |
| 4725 | C17200 |
| 4730 | N04400 |
| 4731 | N04400 |
| 4732 | C71110 ob |
| 4750 | L54950 |
| 4751 | L13630 |
| 4762 | P07402 |
| 4763 | P07563 |
| 4764 | C69950 ob |
| 4765 | P07560 |
| 4766 | P07850 |
| 4767 | P07925 |
| 4768 | P07350 |
| 4769 | P07450 |
| 4770 | P07500 |
| 4771 | P07501 |
| 4772 | P07540 |
| 4773 | P07600 |
| 4774 | P07630 |
| 4775 | N99600 |
| 4776 | N99610 |
| 4777 | N99620 |
| 4778 | N99630 |
| 4779 | N99640 |
| 4780 | M26800 |
| 4782 | N99650 |
| 4783 | R30040 |
| 4784 | P00500 |
| 4785 | P00300 |
| 4786 | P00700 |
| 4787 | P00820 |
| 4788 | P07505 |
| 4800 | L13910 |
| 4803 | Z33520 |
| 4820 | C98400 |
| 4822 | C98400 |
| 4824 | C98200 |
| 4827 | C93700 |
| 4842 | C93700 |
| 4845 | C90500 |
| 4855 | C83600 |
| 4860 | C86500 |
| 4862 | C86300 |
| 4870 | C95420 |
| 4871 | C95420 |
| 4873 | C95420 |
| 4880 | C95510 |
| 4881 | C95520 |
| 4890 | C82500 |
| 4892 | N04019 |
| 4893 | N04019 |
| 4900 | R50550 |
| 4901 | R50700 |
| 4902 | R50400 |
| 4905 | R56400 |
| 4907 | R56401 |
| 4908 | R56080 |
| 4909 | R54521 |
| 4910 | R54520 |
| 4911 | R56400 |
| 4914 | R58153 |
| 4915 | R54810 |
| 4916 | R54810 |
| 4917 | R58010 |
| 4918 | R56620 |
| 4919 | R54620 |
| 4920 | R56400 |
| 4921 | R50700 |
| 4924 | R54521 |
| 4926 | R54520 |
| 4928 | R56400 |
| 4930 | R56400 |
| 4931 | R56400 |
| 4932 | R56400 |
| 4933 | R54810 |
| 4934 | R56400 |
| 4935 | R56400 |
| 4936 | R56620 |
| 4937 | R56620 |
| 4941 | R50400 |
| 4942 | R50400 |
| 4943 | R56320 |
| 4944 | R56320 |
| 4945 | R56320 |
| 4951 | R50125 |
| 4952 | R54621 |
| 4953 | R54522 |
| 4954 | R56400 |
| 4955 | R54810 |
| 4956 | R56402 |
| 4957 | R58640 |
| 4958 | R58640 |
| 4959 | R58010 |
| 4963 | R56400 |
| 4965 | R56400 |
| 4966 | R54520 |
| 4967 | R56400 |
| 4970 | R56740 |
| 4971 | R56620 |
| 4972 | R54810 |
| 4973 | R54810 |
| 4974 | R54790 |
| 4975 | R54620 |
| 4976 | R54620 |
| 4978 | R56620 |
| 4979 | R56620 |
| 4981 | R56260 |
| 4982 | R58450 |
| 4984 | R56410 |
| 4985 | R56401 |
| 4986 | R56410 |
| 4987 | R56410 |
| 4991 | R56401 |
| 4993 | R56400 |

| CROSS REFERENCE SPECIFICATIONS | UNIFIED NUMBER |
|---|---|
| 4996 | R56401 |
| 4997 | R58650 |
| 4998 | R56401 |
| 5010 | G12120 |
| 5010 | G12150 |
| 5020 | G11374 |
| 5022 | G11170 |
| 5024 | G11370 |
| 5027 | K22925 |
| 5028 | K23725 |
| 5029 | K33517 |
| 5030 | K00606 |
| 5031 | W06013 |
| 5032 | G10200 |
| 5040 | G10100 |
| 5042 | G10100 |
| 5044 | G10100 |
| 5045 | G10200 |
| 5046 | G10200 |
| 5046 | G10250 |
| 5047 | G10100 |
| 5050 | G10100 |
| 5053 | G10100 |
| 5060 | G10150 |
| 5061 | K00802 |
| 5062 | K02508 |
| 5069 | G10180 |
| 5070 | G10220 |
| 5075 | G10250 |
| 5077 | G10250 |
| 5080 | G10350 |
| 5082 | G10350 |
| 5085 | G10500 |
| 5110 | G10800 |
| 5112 | G10900 |
| 5112 | K08500 |
| 5112 | K08501 |
| 5115 | G10700 |
| 5120 | G10740 |
| 5121 | G10950 |
| 5122 | G10950 |
| 5132 | G10950 |
| 5221 | N09902 |
| 5223 | N09902 |
| 5225 | N09902 |
| 5310 | F23330 |
| 5315 | F33101 |
| 5316 | F34100 |
| 5328 | J23260 |
| 5329 | J23260 |
| 5330 | J24060 |
| 5331 | J24060 |
| 5333 | J11442 |
| 5334 | J13042 |
| 5335 | J13050 |
| 5336 | J13046 |
| 5337 | J93150 |
| 5338 | J14046 |
| 5339 | J93010 |
| 5340 | J92240 |
| 5341 | J92711 |
| 5342 | J92200 |
| 5343 | J92200 |
| 5344 | J92200 |
| 5346 | J92110 |
| 5347 | J92110 |
| 5348 | J92110 |
| 5349 | J91161 |
| 5350 | J91152 |
| 5350 | S41000 |
| 5351 | J91150 |
| 5352 | J91639 |
| 5353 | J91651 |
| 5354 | J91631 |
| 5355 | J92200 |
| 5356 | J92110 |
| 5357 | J92110 |
| 5358 | J92602 |
| 5359 | J92001 |
| 5360 | J92951 |
| 5361 | J93072 |
| 5362 | J92811 |
| 5363 | J92641 |
| 5364 | J92641 |
| 5365 | J94211 |
| 5366 | J94211 |
| 5368 | J92001 |
| 5369 | J92843 |
| 5370 | J92620 |
| 5371 | J92620 |
| 5372 | J91601 |
| 5373 | R30006 |
| 5375 | R30023 |
| 5376 | R30155 |
| 5377 | N07713 |
| 5378 | R30027 |
| 5380 | R30030 |
| 5382 | R30031 |
| 5383 | N07718 |
| 5384 | N07500 |
| 5385 | R30021 |
| 5387 | R30006 |
| 5387 | R30106 |
| 5388 | N10002 |
| 5389 | N10002 |
| 5390 | N06002 |
| 5391 | N07713 |
| 5392 | F47004 |
| 5393 | F47005 |
| 5394 | F47006 |
| 5395 | F43030 |
| 5396 | N10001 |
| 5397 | N13100 |
| 5398 | J92200 |
| 5399 | N07041 |
| 5400 | J92110 |
| 5401 | N06625 |
| 5402 | N06625 |
| 5404 | N07013 |
| 5405 | N13010 |
| 5406 | N13010 |
| 5407 | N13009 |
| 5412 | S13800 |
| 5500 | S30200 |
| 5501 | S30400 |
| 5503 | S43000 |
| 5504 | S41000 |
| 5505 | S41000 |
| 5506 | S42000 |
| 5507 | S31603 |
| 5508 | S41800 |
| 5510 | S32100 |
| 5511 | S30403 |
| 5512 | S34700 |
| 5513 | S30400 |
| 5514 | S30500 |
| 5515 | S30200 |
| 5516 | S30200 |
| 5517 | S30100 |
| 5518 | S30100 |
| 5519 | S30100 |
| 5520 | S15700 |
| 5521 | S31008 |
| 5523 | S30908 |
| 5524 | S31600 |
| 5525 | S66286 |
| 5526 | S63198 |
| 5528 | S17700 |
| 5529 | S17700 |
| 5530 | N10002 |
| 5532 | R30155 |
| 5536 | N06002 |
| 5537 | R30605 |
| 5540 | N06600 |
| 5541 | N07722 |
| 5542 | N07750 |
| 5544 | N07001 |
| 5545 | N07041 |
| 5546 | S35000 |
| 5547 | S35500 |
| 5548 | S35000 |
| 5549 | S35500 |
| 5550 | N07702 |
| 5552 | N08801 |
| 5553 | N02202 |
| 5554 | S35000 |
| 5555 | N02205 |
| 5556 | S34700 |
| 5557 | S32100 |
| 5558 | S34700 |
| 5559 | S32100 |
| 5560 | S30400 |
| 5561 | S21900 |
| 5562 | S21904 |
| 5563 | S30400 |
| 5564 | S30400 |
| 5565 | S30400 |
| 5566 | S30400 |
| 5567 | S30400 |
| 5568 | S17700 |
| 5569 | S30403 |
| 5570 | S32100 |
| 5571 | S34700 |
| 5572 | S31008 |
| 5573 | S31600 |
| 5574 | S30908 |
| 5575 | S34700 |
| 5576 | S32100 |
| 5577 | S31008 |
| 5578 | S45500 |
| 5579 | S63198 |
| 5580 | N06600 |
| 5581 | N06625 |
| 5582 | N07750 |
| 5583 | N07750 |
| 5584 | S30403 |
| 5584 | S31603 |
| 5585 | R30155 |
| 5586 | N07001 |
| 5587 | N06002 |
| 5588 | N06002 |
| 5589 | N07718 |
| 5590 | N07718 |
| 5591 | S41000 |
| 5592 | N08330 |
| 5593 | N06333 |
| 5595 | S21904 |

| CROSS REFERENCE SPECIFICATIONS | UNIFIED NUMBER |
|---|---|
| 5596 | N07718 |
| 5597 | N07718 |
| 5598 | N07750 |
| 5599 | N06625 |
| 5600 | S30200 |
| 5603 | S14800 |
| 5604 | S17400 |
| 5605 | N09706 |
| 5606 | N09706 |
| 5607 | N10003 |
| 5608 | R30188 |
| 5609 | S41040 |
| 5610 | S41600 |
| 5610 | S41623 |
| 5611 | S41001 |
| 5612 | S41001 |
| 5612 | S41041 |
| 5613 | S41000 |
| 5614 | S41025 |
| 5616 | S41800 |
| 5617 | S45500 |
| 5617 | S45503 |
| 5618 | S44004 |
| 5620 | S42020 |
| 5620 | S42023 |
| 5621 | S42000 |
| 5622 | S17400 |
| 5623 | K91456 |
| 5624 | K91505 |
| 5625 | K91456 |
| 5627 | S43000 |
| 5628 | S43100 |
| 5629 | S13800 |
| 5630 | S44004 |
| 5631 | S44002 |
| 5632 (1) | S44020 |
| 5632 (2) | S44023 |
| 5635 | S30360 |
| 5636 | S30200 |
| 5637 | S30200 |
| 5638 | S30345 |
| 5639 | S30400 |
| 5640 (Type 1) | S30300 |
| 5640 (Type 2) | S30323 |
| 5640 (Type 3) | S30310 |
| 5641 | S30323 |
| 5642 (1) | S34720 |
| 5642 (2) | S34723 |
| 5643 | S17400 |
| 5644 | S17700 |
| 5645 | S32100 |
| 5646 | S34700 |
| 5647 | S30403 |
| 5649 | S31620 |
| 5650 | S30908 |
| 5651 | S31008 |
| 5652 | S31400 |
| 5653 | S31603 |
| 5654 | S34700 |
| 5655 | S42200 |
| 5656 | S21904 |
| 5657 | S15700 |
| 5659 | S15500 |
| 5660 | N09901 |
| 5661 | N09901 |
| 5662 | N07718 |
| 5663 | N07718 |
| 5664 | N07718 |
| 5665 | N06600 |
| 5666 | N06625 |
| 5667 | N07750 |
| 5668 | N07750 |
| 5669 | N07750 |
| 5670 | N07750 |
| 5671 | N07750 |
| 5672 | S45500 |
| 5674 | S34700 |
| 5675 | N07092 |
| 5676 | N06003 |
| 5677 | N06003 |
| 5677 | W86002 |
| 5678 | S17700 |
| 5679 | N06062 |
| 5680 | S34781 |
| 5682 | N06003 |
| 5685 | S30500 |
| 5686 | S30500 |
| 5687 | N06600 |
| 5688 | S30200 |
| 5689 | S32100 |
| 5690 | S31600 |
| 5691 | S31600 |
| 5691 | W31610 |
| 5692 | S31600 |
| 5692 | S31680 |
| 5693 | S30200 |
| 5694 | S31080 |
| 5696 | S31600 |
| 5697 | S30400 |
| 5698 | N07750 |
| 5699 | N07750 |
| 5700 | S66009 |
| 5701 | N09706 |
| 5702 | N09706 |
| 5703 | N09706 |
| 5704 | N07001 |
| 5705 | S63005 |
| 5706 | N07001 |
| 5707 | N07001 |
| 5708 | N07001 |
| 5709 | N07001 |
| 5710 | S65006 |
| 5711 | N06635 |
| 5712 | N07041 |
| 5713 | N07041 |
| 5714 | N07722 |
| 5715 | N06601 |
| 5716 | N08330 |
| 5717 | N06333 |
| 5718 | S64152 |
| 5719 | S64152 |
| 5726 | S66286 |
| 5731 | S66286 |
| 5732 | S66220 |
| 5732 | S66286 |
| 5734 | S66286 |
| 5737 | S66286 |
| 5738 | S30323 |
| 5739 | S36200 |
| 5740 | S36200 |
| 5743 | S35500 |
| 5744 | S35500 |
| 5745 | S35000 |
| 5746 | N09979 |
| 5747 | N07750 |
| 5749 | S42700 |
| 5750 | N10002 |
| 5751 | N07500 |
| 5754 | N06002 |
| 5755 | N10004 |
| 5758 | R30035 |
| 5759 | R30605 |
| 5762 | S20300 |
| 5763 | S45000 |
| 5764 | S20910 |
| 5766 | N08800 |
| 5768 | R30155 |
| 5769 | R30155 |
| 5771 | N10003 |
| 5772 | R30188 |
| 5773 | S45000 |
| 5774 | S35080 |
| 5775 | W35010 |
| 5776 | S41000 |
| 5776 | S41080 |
| 5777 | S41000 |
| 5777 | W41010 |
| 5778 | N07069 |
| 5779 | N07750 |
| 5779 | W86130 |
| 5780 | S35580 |
| 5781 | W35510 |
| 5782 | S63197 |
| 5783 | W34910 |
| 5784 | S64299 |
| 5786 | N10004 |
| 5788 | R30006 |
| 5789 | R30031 |
| 5790 | S34700 |
| 5790 | S34780 |
| 5794 | R30155 |
| 5795 | W73155 |
| 5796 | R30605 |
| 5797 | W73605 |
| 5798 | N06002 |
| 5800 | N07041 |
| 5801 | R30188 |
| 5802 | N19907 |
| 5803 | S17480 |
| 5804 | S66286 |
| 5805 | S66286 |
| 5806 | N19903 |
| 5810 | S66286 |
| 5812 | S15780 |
| 5812 | S15789 |
| 5817 | S41880 |
| 5819 | R30021 |
| 5821 | S41081 |
| 5822 | S41780 |
| 5823 | S41780 |
| 5824 | S17780 |
| 5825 | S17480 |
| 5826 | S15500 |
| 5827 | W37410 |
| 5828 | N07001 |
| 5829 | N07090 |
| 5830 | N09901 |
| 5831 | R30556 |
| 5832 | N07718 |
| 5833 | R30003 |
| 5834 | R30003 |
| 5836 | N06082 |
| 5837 | N06625 |
| 5838 | N06635 |
| 5839 | N06231 |
| 5840 | S13800 |
| 5840 | S13889 |

| CROSS REFERENCE SPECIFICATIONS | UNIFIED NUMBER |
|---|---|
| 5841 | R30159 |
| 5842 | R30159 |
| 5843 | R30159 |
| 5844 | R30035 |
| 5845 | R30035 |
| 5846 | N13020 |
| 5848 | S21800 |
| 5851 | N13017 |
| 5852 | N13017 |
| 5853 | S66286 |
| 5854 | N07716 |
| 5857 | S30400 |
| 5858 | S66286 |
| 5859 | S45000 |
| 5860 | S45500 |
| 5862 | S15500 |
| 5863 | S45000 |
| 5864 | S13800 |
| 5865 | N03260 |
| 5866 | S30200 |
| 5867 | S41000 |
| 5868 | S30400 |
| 5869 | N06625 |
| 5869 | N06635 |
| 5870 | N06601 |
| 5871 | N08800 |
| 5872 | N07263 |
| 5873 | N06635 |
| 5874 | R30556 |
| 5875 | R30003 |
| 5876 | R30003 |
| 5877 | R30556 |
| 5878 | N06230 |
| 5879 | N06626 |
| 5880 | S44004 |
| 5882 | N13017 |
| 5884 | N19909 |
| 5886 | N07263 |
| 5887 | N06617 |
| 5888 | N06617 |
| 5889 | N06617 |
| 5890 | N03260 |
| 5891 | N06230 |
| 5892 | N19909 |
| 5893 | N19909 |
| 5894 | R30016 |
| 5895 | S66286 |
| 5896 | N06230 |
| 5896 | S32100 |
| 5897 | S34700 |
| 5899 | S44004 |
| 5901 | S30100 |
| 5902 | S30100 |
| 5903 | S30200 |
| 5904 | S30200 |
| 5905 | S30200 |
| 5906 | S30200 |
| 5907 | S31600 |
| 5910 | S30400 |
| 5911 | S30400 |
| 5912 | S30400 |
| 5913 | S30400 |
| 5914 | N07719 |
| 5925 | S42025 |
| 5930 | S42670 |
| 5931 | S20161 |
| 5932 | S42640 |
| 5940 | R30783 |
| 5950 | N07719 |
| 5961 | N06600 |
| 5966 | N07263 |
| 6250 | K44910 |
| 6255 | K21940 |
| 6256 | K71350 |
| 6257 | K44220 |
| 6260 | G93106 |
| 6263 | K44414 |
| 6264 | K44414 |
| 6265 | G93106 |
| 6266 (43BV12) | K21028 |
| 6267 | G93106 |
| 6270 | G86150 |
| 6272 | G86170 |
| 6274 | G86200 |
| 6275 | G94171 |
| 6276 | G86200 |
| 6277 | G86200 |
| 6278 | K91231 |
| 6280 | G86300 |
| 6281 | G86300 |
| 6282 | G87350 |
| 6290 | G46150 |
| 6292 | G46170 |
| 6294 | G46200 |
| 6299 | H43200 |
| 6300 | G40370 |
| 6302 | K23015 |
| 6303 | K22770 |
| 6304 | K14675 |
| 6305 | K14675 |
| 6308 | K71040 |
| 6312 | K22440 |
| 6317 | K22440 |
| 6320 | G87350 |
| 6322 | G87400 |
| 6323 | G87400 |
| 6324 | K11640 |
| 6325 | G87400 |
| 6327 | G87400 |
| 6328 | K13550 |
| 6330 | K22033 |
| 6331 | G87350 |
| 6342 | G98400 |
| 6346 | G41300 |
| 6348 | G41300 |
| 6349 | G41400 |
| 6350 | G41300 |
| 6351 | G41300 |
| 6352 | G41350 |
| 6354 | K11914 |
| 6356 | G41300 |
| 6356 | K13247 |
| 6357 | G87350 |
| 6358 | G87400 |
| 6359 | G43400 |
| 6360 | G41300 |
| 6360 | G86300 |
| 6361 | G41300 |
| 6362 | G41300 |
| 6365 | G41350 |
| 6370 | G41300 |
| 6371 | G41300 |
| 6372 | G41350 |
| 6373 | G41300 |
| 6374 | G41300 |
| 6375 | G86200 |
| 6378 | K11542 |
| 6379 | K11546 |
| 6381 | G41400 |
| 6382 | G41400 |
| 6385 | K23015 |
| 6386 (1) | K11856 |
| 6386 (2) | K11630 |
| 6386 (3) | K11511 |
| 6386 (4) | K11662 |
| 6386 (5) | K11625 |
| 6395 | G41400 |
| 6396 | K22950 |
| 6406 | K34378 |
| 6407 | K33020 |
| 6408 | T20813 |
| 6409 | G43400 |
| 6411 | K23080 |
| 6412 | G43370 |
| 6413 | G43370 |
| 6414 | G43400 |
| 6415 | G43400 |
| 6415 | G43406 |
| 6417 | K44220 |
| 6418 | K32550 |
| 6419 | K44220 |
| 6422 | K11940 |
| 6423 | K24336 |
| 6424 | K22950 |
| 6425 | K32550 |
| 6426 | K18597 |
| 6427 | K23080 |
| 6428 | K23477 |
| 6429 | K33517 |
| 6430 | K33517 |
| 6431 | K24728 |
| 6432 | K24728 |
| 6433 | K33517 |
| 6434 | K33517 |
| 6435 | K33517 |
| 6436 | K22770 |
| 6437 | T20811 |
| 6438 | K24728 |
| 6439 | K24728 |
| 6440 | G52986 |
| 6442 | G50986 |
| 6443 | G51986 |
| 6444 | G52986 |
| 6445 | K22097 |
| 6446 | G51986 |
| 6447 | G52986 |
| 6448 | G61500 |
| 6449 | G51986 |
| 6450 | G61500 |
| 6451 | G92540 |
| 6453 | K32550 |
| 6454 | G43400 |
| 6455 | G61500 |
| 6456 | K24056 |
| 6457 | K13147 |
| 6458 | K23015 |
| 6459 | K22720 |
| 6460 | K11365 |
| 6461 | K13148 |
| 6462 | G41400 |
| 6462 | K13149 |
| 6463 | K93130 |
| 6464 | W10013 |
| 6465 | K91971 |
| 6466 | S50280 |
| 6467 | W50210 |
| 6468 | K91461 |

| CROSS REFERENCE SPECIFICATIONS | UNIFIED NUMBER |
|---|---|
| 6470 | K24065 |
| 6471 | K24065 |
| 6472 | K24065 |
| 6475 | K52355 |
| 6478 | K92580 |
| 6481 | K24340 |
| 6484 | G43400 |
| 6485 | T20811 |
| 6487 | T20811 |
| 6488 | T20811 |
| 6490 | T11350 |
| 6491 | T11350 |
| 6501 | K92890 |
| 6512 | K92890 |
| 6514 | K93120 |
| 6518 | K92150 |
| 6519 | K92150 |
| 6520 | K92890 |
| 6521 | K93120 |
| 6522 | K92571 |
| 6523 | K91472 |
| 6524 | K91283 |
| 6525 | K91472 |
| 6526 | K91283 |
| 6527 | K92571 |
| 6528 | G41300 |
| 6529 | G41400 |
| 6532 | K92580 |
| 6533 | K92571 |
| 6535 | G86300 |
| 6543 | K92571 |
| 6544 | K92571 |
| 6546 | K91122 |
| 6550 | K13048 |
| 6591 | K92150 |
| 6693 | S30200 |
| 7205 | G10700 |
| 7238 | R30012 |
| 7240 | G10600 |
| 7241 | S30200 |
| 7301 | G61500 |
| 7304 | G10950 |
| 7320 | C92800 |
| 7322 | C91300 |
| 7445 | S44002 |
| 7445 | S44003 |
| 7445 | S44004 |
| 7490 | N06625 |
| 7490 | S30908 |
| 7490 | S31008 |
| 7490 | S31400 |
| 7490 | S32100 |
| 7490 | S34700 |
| 7496 | G41300 |
| 7496 | G43370 |
| 7496 | G86300 |
| 7496 | G87350 |
| 7496 | G87400 |
| 7496 | K02508 |
| 7496 | K22770 |
| 7496 | K23015 |
| 7496 | K32550 |
| 7707 | K00095 |
| 7717 | K95000 |
| 7718 | K95000 |
| 7719 | K95000 |
| 7721 | L53131 |
| 7724 | R30100 |
| 7726 | K94610 |
| 7727 | K94610 |
| 7728 | K94610 |
| 7730 | M08990 |
| 7731 | P00020 |
| 7734 | K94101 |
| 7735 | P03300 |
| 7817 | R03630 |
| 7819 | R03630 |
| 7846 | R05255 |
| 7847 | R05255 |
| 7848 | R05255 |
| 7849 | R05210 |
| 7850 | R04211 |
| 7851 | R04271 |
| 7852 | R04295 |
| 7855 | R04271 |
| 7857 | R04295 |
| 7897 | R07005 |
| 7898 | R07006 |
| 7902 | R19800 |
| 7908 | K92580 |
| 7910 | K92580 |
| AMS-A-21180 (354.0) | A03540 |
| AMS-A-21180 (359.0) | A03590 |
| AMS-A-21180 (A210.0) | A12010 |
| AMS-A-21180 (A356.0) | A13560 |
| AMS-A-21180 (A357.0) | A13570 |
| AMS-A-21180 (C355.0) | A33550 |
| AMS-A-22771 (2014) | A92014 |
| AMS-A-22771 (2219) | A92219 |
| AMS-A-22771 (6151) | A96151 |
| AMS-A-22771 (7039) | A97039 |
| AMS-A-22771 (7075) | A97075 |
| AMS-A-22771 | A92618 |
| AMS-A-22771 | A93003 |
| AMS-A-22771 | A94032 |
| AMS-A-22771 | A96061 |
| AMS-A-22771 | A97175 |
| AMS-A-81596 | A92024 |
| AMS-A-81596 | A93003 |
| AMS-A-81596 | A95052 |
| AMS-A-81596 | A95056 |
| AMS-B-20148 | A94045 |
| AMS-B-20148 | A94047 |
| AMS-B-20148 | A94048 |
| AMS-B-20148 | A94343 |
| AMS-I-7068 | Z35542 |
| AMS-I-7068 | Z35543 |
| AMS-QQ-A-200/1 | A93003 |
| AMS-QQ-A-200/10 | A96066 |
| AMS-QQ-A-200/11 | A97075 |
| AMS-QQ-A-200/13 | A97178 |
| AMS-QQ-A-200/14 | A97178 |
| AMS-QQ-A-200/15 | A97075 |
| AMS-QQ-A-200/15 | A97175 |
| AMS-QQ-A-200/16 | A96061 |
| AMS-QQ-A-200/17 | A96162 |
| AMS-QQ-A-200/2 | A92014 |
| AMS-QQ-A-200/3 | A92024 |
| AMS-QQ-A-200/4 | A92014 |
| AMS-QQ-A-200/4 | A95083 |
| AMS-QQ-A-200/5 | A95086 |
| AMS-QQ-A-200/6 | A95086 |
| AMS-QQ-A-200/6 | A95454 |
| AMS-QQ-A-200/7 | A95456 |
| AMS-QQ-A-200/8 | A96061 |
| AMS-QQ-A-200/9 | A96063 |
| AMS-QQ-A-225/1 | A91100 |
| AMS-QQ-A-225/10 | A96262 |
| AMS-QQ-A-225/2 | A93003 |
| AMS-QQ-A-225/3 | A92011 |
| AMS-QQ-A-225/4 | A92014 |
| AMS-QQ-A-225/5 | A92017 |
| AMS-QQ-A-225/6 | A92024 |
| AMS-QQ-A-225/7 | A95052 |
| AMS-QQ-A-225/8 | A96061 |
| AMS-QQ-A-225/9 | A97075 |
| AMS-QQ-A-250/1 | A91100 |
| AMS-QQ-A-250/10 | A95454 |
| AMS-QQ-A-250/11 | A96061 |
| AMS-QQ-A-250/12 | A97075 |
| AMS-QQ-A-250/13 | A87075 |
| AMS-QQ-A-250/14 | A97178 |
| AMS-QQ-A-250/15 | A87178 |
| AMS-QQ-A-250/18 | A87075 |
| AMS-QQ-A-250/19 | A95086 |
| AMS-QQ-A-250/2 | A93003 |
| AMS-QQ-A-250/20 | A95456 |
| AMS-QQ-A-250/21 | A97178 |
| AMS-QQ-A-250/22 | A87178 |
| AMS-QQ-A-250/22 | A97178 |
| AMS-QQ-A-250/24 | A97075 |
| AMS-QQ-A-250/25 | A87075 |
| AMS-QQ-A-250/26 | A87075 |
| AMS-QQ-A-250/28 | A87178 |
| AMS-QQ-A-250/29 | A92124 |
| AMS-QQ-A-250/3 | A82014 |
| AMS-QQ-A-250/30 | A92219 |
| AMS-QQ-A-250/3 | A93003 |
| AMS-QQ-A-250/4 | A92024 |
| AMS-QQ-A-250/5 | A82024 |
| AMS-QQ-A-250/6 | A95083 |
| AMS-QQ-A-250/7 | A95086 |
| AMS-QQ-A-250/8 | A95052 |
| AMS-QQ-A-250/9 | A95456 |
| AMS-QQ-S-700 (1065) | G10650 |
| AMS-QQ-S-700 (1074) | G10740 |
| AMS-QQ-S-700 (1080) | G10800 |
| AMS-QQ-S-700 (C1025) | G10250 |
| AMS-Qq-S-700 (C1030) | G10300 |
| AMS-QQ-S-700 (C1035) | G10350 |
| AMS-QQ-S-700 (C1045) | G10450 |
| AMS-QQ-S-700 (C1050) | G10500 |
| AMS-QQ-S-700 (C1055) | G10550 |
| AMS-QQ-S-700 (C1084) | G10840 |
| AMS-QQ-S-700 (C1085) | G10850 |
| AMS-QQ-S-700 (C1086) | G10860 |
| AMS-QQ-S-700 (C1095) | G10950 |
| AMS-QQ-S-763 (302) | S30200 |
| AMS-QQ-S-763 (304) | S30400 |
| AMS-QQ-S-763 (304L) | S30403 |
| AMS-QQ-S-763 | S20200 |
| AMS-QQ-S-763 | S30500 |
| AMS-QQ-S-763 | S30900 |
| AMS-QQ-S-763 | S31000 |
| AMS-QQ-S-763 | S31008 |
| AMS-QQ-S-763 | S31600 |
| AMS-QQ-S-763 | S31603 |
| AMS-QQ-S-763 | S31700 |
| AMS-QQ-S-763 | S32100 |
| AMS-QQ-S-763 | S34700 |
| AMS-QQ-S-763 | S40300 |
| AMS-QQ-S-763 | S40500 |
| AMS-QQ-S-763 | S41000 |
| AMS-QQ-S-763 | S41400 |
| AMS-QQ-S-763 | S41623 |
| AMS-QQ-S-763 | S42000 |
| AMS-QQ-S-763 | S42900 |

| CROSS REFERENCE SPECIFICATIONS | UNIFIED NUMBER |
|---|---|
| AMS-QQ-S-763 | S43000 |
| AMS-QQ-S-763 | S44002 |
| AMS-QQ-S-763 | S44003 |
| AMS-QQ-S-763 | S44004 |
| AMS-QQ-S-763 | S44300 |
| AMS-QQ-S-763 | S44600 |
| AMS-QQ-W-428 | K07001 |
| AMS-S-11310 (CS1005) | G10050 |
| AMS-S-11310 (CS1006) | G10060 |
| AMS-S-11310 (CS1008) | G10080 |
| AMS-S-11310 (CS1010) | G10100 |
| AMS-S-11310 (CS1012) | G10120 |
| AMS-S-11310 (CS1018) | G10180 |
| AMS-S-11310 (CS1020) | G10200 |
| AMS-S-11310 (CS1022) | G10220 |
| AMS-S-11310 (CS1025) | G10250 |
| AMS-S-11310 (CS1030) | G10300 |
| AMS-S-11310 (CS1040) | G10400 |
| AMS-S-1310 (CS1017) | G10170 |
| AMS-S-18372 | S43100 |
| AMS-S-18728 | G86300 |
| AMS-S-18729 | G41300 |
| AMS-S-22141 (1020) | J02001 |
| AMS-S-22141 (1040) | J04001 |
| AMS-S-22141 (1050) | J05001 ob |
| AMS-S-22141 (4130) | J13048 |
| AMS-S-22141 (4140) | J14047 |
| AMS-S-22141 (4335M) | J13432 |
| AMS-S-22141 (4340) | J24055 |
| AMS-S-22141 (4620) | J12093 |
| AMS-S-22141 (52100) | J19965 |
| AMS-S-22141 (6150) | J15047 |
| AMS-S-22141 (8620) | J12047 |
| AMS-S-22141 (8640) | J14048 |
| AMS-S-22141 (8735) | J13442 |
| AMS-S-22141 (IC-1030) | J03005 |
| AMS-S-22141 (Nitralloy) | J24056 |
| AMS-S-22141 | G52986 |
| AMS-S-5000 | G43400 |
| AMS-S-5621 | G41400 |
| AMS-S-6049 | G87400 |
| AMS-S-6050 | G86300 |
| AMS-S-6098 | G87350 |
| AMS-S-6709 | K24065 |
| AMS-S-6755 (4130) | G41300 |
| AMS-S-6758 | G41300 |
| AMS-S-700 (C1055) | G10550 |
| AMS-S-700 (C1065) | G10650 |
| AMS-S-7108 | K32550 |
| AMS-S-7420 | G52986 |
| AMS-S-7720 | S30200 |
| AMS-S-7947 (C1095) | G10950 |
| AMS-S-7952 (C1025) | G10250 |
| AMS-S-7952 (CS1020) | G10200 |
| AMS-S-8503 | G33106 |
| AMS-S-8503 | G61500 |
| AMS-S-8559 (C1095) | G10950 |
| AMS-S-8690 | G86200 |
| AMS-S-8840 | S35000 |
| AMS-S-8840 | S35500 |
| AMS-T-5066 | G10250 |
| AMS-T-6735 (8630) | G86300 |
| AMS-T-6735 | G41350 |
| AMS-T-6845 (304) | S30400 |
| AMS-T-6845 | S30403 |
| AMS-T-7081 | A96061 |
| AMS-T-81915 | R50250 |
| AMS-T-81915 | R54520 |
| AMS-T-81915 | R54620 |
| AMS-T-81915 | R56400 |
| AMS-T-8506 (304) | S30400 |
| AMS-T-9045 | R56080 |
| AMS-T-9046 | R50400 |
| AMS-T-9046 | R50550 |
| AMS-T-9046 | R50700 |
| AMS-T-9046 | R54520 |
| AMS-T-9046 | R54521 |
| AMS-T-9046 | R54560 |
| AMS-T-9046 | R54620 |
| AMS-T-9046 | R54810 |
| AMS-T-9046 | R56080 |
| AMS-T-9046 | R56210 |
| AMS-T-9046 | R56320 |
| AMS-T-9046 | R56400 |
| AMS-T-9046 | R56401 |
| AMS-T-9046 | R56430 |
| AMS-T-9046 | R56620 |
| AMS-T-9046 | R58010 |
| AMS-T-9046 | R58030 |
| AMS-T-9046 | R58640 |
| AMS-T-9046 | R58820 |
| AMS-W-21425 (II) | K06000 |
| AMS-W-2145 (I, II) | K06500 |
| AS7202 | S42000 |
| AS7210 | S30200 |
| AS7220 | A91100 |
| AS7222 | A92117 |
| AS7225 | G10100 |
| AS7228 | S30200 |
| AS7229 | S34700 |
| AS7232 | N06600 |
| AS7233 | N04400 |
| AS7234 | N04405 |
| AS7235 | S66286 |
| AS7236 | R30605 |
| AS7237 | N06002 |
| AS7245 | N07750 |
| AS7245 | S30400 |
| AS7246 | N07750 |
| AS7325 | S32100 |
| AS7452 | G87400 |
| AS7454 | K14675 |
| AS7455 | K14675 |
| AS7456 | G87400 |
| AS7458 | K14675 |
| AS7459 | K14675 |
| AS7460 | R56400 |
| AS7461 | R56400 |
| AS7466 | N07718 |
| AS7467 | N07718 |
| AS7468 | R30035 |
| AS7469 | N07041 |
| AS7470 | S41800 |
| AS7471 | N07001 |
| AS7472 | S30200 |
| AS7474 | S17400 |
| AS7475 | R30159 |
| AS7477 | S66286 |
| AS7478 | S66286 |
| AS7479 | S66286 |
| AS7481 | S66286 |
| AS7482 | S66286 |

# Cross Index of Commonly Known Documents Which Describe Materials Same as or Similar to These Covered by UNS Numbers

## ASME (American Society of Mechanical Engineers) NUMBERS

**(The Symbol *ob* Indicates an Obsolete UNS Number.)**

# ASME

| CROSS REFERENCE SPECIFICATIONS | UNIFIED NUMBER |
|---|---|
| A372 (5E, 8) | K14248 |
| B16.15 | C83600 |
| B16.15 | C83800 |
| B16.15 | C84400 |
| B16.18 | C83600 |
| B16.18 | C83800 |
| B16.18 | C84400 |
| B16.22 | C12200 |
| B16.22 | C14500 |
| B16.22 | C14700 |
| B16.22 | C23000 |
| B16.23 | C83600 |
| B16.23 | C83800 |
| B16.23 | C84400 |
| B16.24 | C83600 |
| B16.24 | C95200 |
| B16.26 | C83600 |
| B16.26 | C84400 |
| B16.29 | C12200 |
| B16.29 | C23000 |
| B16.32 | C83600 |
| B16.32 | C84400 |
| Code Case 1817 | S21800 |
| Code Case 2159 | S34710 |
| Code Case N-20 (1484) | N06690 |
| SA-182 | S31254 |
| SA-182 (304) | S30400 |
| SA-182 (304H) | S30409 |
| SA-182 (304L) | S30403 |
| SA-182 (304N) | S30451 |
| SA-182 (310) | S31000 |
| SA-182 (316) | S31600 |
| SA-182 (316H) | S31609 |
| SA-182 (316L) | S31603 |
| SA-182 (316N) | S31651 |
| SA-182 (321) | S32100 |
| SA-182 (321H) | S32109 |
| SA-182 (347) | S34700 |
| SA-182 (347H) | S34709 |
| SA-182 (348) | S34800 |
| SA-182 (348H) | S34809 |
| SA-182 (429) | S42900 |
| SA-182 (430) | S43000 |
| SA-182 (F45) | S30815 |
| SA-182 (F51) | S31803 |
| SA-182 (XM-19) | S20910 |
| SA-182 (XM-27) | S44625 ob |
| SA-188 (F91) | K90901 |
| SA-193 (305) | S30500 |
| SA-193 (316) | S31600 |
| SA-193 (321) | S32100 |
| SA-193 (347) | S34700 |
| SA-194 (3) | S50100 |
| SA-194 (6) | S41000 |
| SA-194 (8) | S30400 |
| SA-194 (303) | S30300 |
| SA-194 (303 Se) | S30323 |
| SA-194 (305) | S30500 |
| SA-194 (316) | S31600 |
| SA-194 (321) | S32100 |
| SA-194 (347) | S34700 |
| SA-194 (416) | S41600 |
| SA-194 (416Se) | S41623 |
| SA-213 | S30815 |
| SA-213 | S31042 |
| SA-213 | S34751 |
| SA-213 (304) | S30400 |
| SA-213 (304H) | S30409 |
| SA-213 (304L) | S30403 |
| SA-213 (304N) | S30451 |
| SA-213 (310) | S31000 |
| SA-213 (316) | S31600 |
| SA-213 (316H) | S31609 |
| SA-213 (316L) | S31603 |
| SA-213 (316N) | S31651 |
| SA-213 (321) | S32100 |
| SA-213 (321H) | S32109 |
| SA-213 (347) | S34700 |
| SA-213 (347H) | S34709 |
| SA-213 (348) | S34800 |
| SA-213 (348H) | S34809 |
| SA-213 (T23) | K40712 |
| SA-213 (T24) | K30736 |
| SA-213 (T91) | K90901 |
| SA-213 (T122) | K91271 |
| SA-213 (T911) | K91061 |
| SA-213 (XM-15) | S38100 |
| SA-213 (XM-19) | S20910 |
| SA-232 | G61500 |
| SA-234 (WP 11) | K11597 |
| SA-234 (WP91) | K90901 |
| SA-240 | S20103 |
| SA-240 | S20153 |
| SA-240 | S20161 |
| SA-240 | S30615 |
| SA-240 | S30815 |
| SA-240 | S31254 |
| SA-240 | S31803 |
| SA-240 | S44627 |
| SA-240 (302) | S30200 |
| SA-240 (304) | S30400 |
| SA-240 (304H) | S30409 |
| SA-240 (304L) | S30403 |
| SA-240 (304N) | S30451 |
| SA-240 (305) | S30500 |
| SA-240 (309S) | S30908 |
| SA-240 (310S) | S31008 |
| SA-240 (316) | S31600 |
| SA-240 (316H) | S31609 |
| SA-240 (316L) | S31603 |
| SA-240 (316N) | S31651 |
| SA-240 (317) | S31700 |
| SA-240 (317L) | S31703 |
| SA-240 (321) | S32100 |
| SA-240 (321H) | S32109 |
| SA-240 (347) | S34700 |
| SA-240 (347H) | S34709 |
| SA-240 (348) | S34800 |
| SA-240 (348H) | S34809 |
| SA-240 (405) | S40500 |
| SA-240 (410) | S41000 |
| SA-240 (410S) | S41008 |
| SA-240 (429) | S42900 |
| SA-240 (430) | S43000 |
| SA-240 (439) | S43035 |
| SA-240 (XM-15) | S38100 |
| SA-240 (XM-17) | S21600 |
| SA-240 (XM-18) | S21603 |
| SA-240 (XM-19) | S20910 |
| SA-240 (XM-21) | S30452 |
| SA-240 (XM-27) | S44625 ob |
| SA-240 (XM-29) | S24000 |
| SA-240 (XM-33) | S44626 |
| SA-249 | S30815 |
| SA-249 | S31254 |
| SA-249 (304) | S30400 |
| SA-249 (304H) | S30409 |
| SA-249 (304L) | S30403 |
| SA-249 (304N) | S30451 |
| SA-249 (309) | S30900 |
| SA-249 (316) | S31600 |
| SA-249 (316H) | S31609 |
| SA-249 (316L) | S31603 |
| SA-249 (316N) | S31651 |
| SA-249 (317) | S31700 |
| SA-249 (321) | S32100 |
| SA-249 (321H) | S32109 |
| SA-249 (347) | S34700 |
| SA-249 (347H) | S34709 |
| SA-249 (348) | S34800 |
| SA-249 (348H) | S34809 |
| SA-249 (XM-15) | S38100 |
| SA-249 (XM-19) | S20910 |
| SA-249 (XM-29) | S24000 |
| SA-268 | S44627 |
| SA-268 (405) | S40500 |
| SA-268 (409) | S40900 |
| SA-268 (410) | S41000 |
| SA-268 (429) | S42900 |
| SA-268 (430) | S43000 |
| SA-268 (439) | S43035 |
| SA-268 (446) | S44600 |
| SA-268 (XM-27) | S44625 ob |
| SA-268 (XM-33) | S44626 |
| SA-278 (20) | F11401 |
| SA-278 (25) | F11701 |
| SA-278 (30) | F12101 |
| SA-278 (35) | F12401 |
| SA-278 (40) | F12803 |
| SA-278 (45) | F13102 |
| SA-278 (50) | F13502 |
| SA-278 (55) | F13802 |
| SA-278 (60) | F14102 |
| SA-278 (70) | F14801 |
| SA-278 (80) | F15501 |
| SA-307 (B, C) | K03002 |
| SA-312 | S30815 |
| SA-312 | S31254 |
| SA-312 | S34751 |
| SA-312 (304) | S30400 |
| SA-312 (304H) | S30409 |
| SA-312 (304L) | S30403 |
| SA-312 (304N) | S30451 |
| SA-312 (309) | S30900 |
| SA-312 (310) | S31000 |
| SA-312 (316) | S31600 |
| SA-312 (316H) | S31609 |
| SA-312 (316L) | S31603 |
| SA-312 (316N) | S31651 |
| SA-312 (317) | S31700 |
| SA-312 (321) | S32100 |
| SA-312 (321H) | S32109 |
| SA-312 (347) | S34700 |
| SA-312 (347H) | S34709 |
| SA-312 (348) | S34800 |
| SA-312 (348H) | S34809 |
| SA-312 (XM-15) | S38100 |
| SA-312 (XM-19) | S20910 |
| SA-312 (XM-29) | S24000 |
| SA-320 (303) | S30300 |
| SA-320 (303 Se) | S30323 |
| SA-320 (316) | S31600 |
| SA-320 (321) | S32100 |
| SA-320 (347) | S34700 |
| SA-320 (B8) | S30400 |

| CROSS REFERENCE SPECIFICATIONS | UNIFIED NUMBER |
|---|---|
| SA-320 (L1) | H15211 |
| SA-335 (P91) | K90901 |
| SA-335 (P122) | K91271 |
| SA-336 (F22V) | K31835 |
| SA-336 (F91) | K90901 |
| SA-351 | J92802 |
| SA-351 | J92901 |
| SA-351 | J92972 |
| SA-351 | J92999 |
| SA-351 | N08151 |
| SA-358 | S31254 |
| SA-358 (304) | S30400 |
| SA-358 (304N) | S30451 |
| SA-358 (309) | S30900 |
| SA-358 (310) | S31000 |
| SA-358 (316) | S31600 |
| SA-358 (316N) | S31651 |
| SA-358 (321) | S32100 |
| SA-358 (347) | S34700 |
| SA-358 (348) | S34800 |
| SA-369 (FP91) | K90901 |
| SA-372 (M) | K42365 |
| SA-376 (304) | S30400 |
| SA-376 (304H) | S30409 |
| SA-376 (304N) | S30451 |
| SA-376 (316) | S31600 |
| SA-376 (316H) | S31609 |
| SA-376 (316N) | S31651 |
| SA-376 (321) | S32100 |
| SA-376 (321H) | S32109 |
| SA-376 (347) | S34700 |
| SA-376 (347H) | S34709 |
| SA-376 (348) | S34800 |
| SA-387 (5) | S50100 |
| SA-387 (5) | S50200 |
| SA-387 (91) | K90901 |
| SA-395 (60-40-18) | F32800 |
| SA-403 (304) | S30400 |
| SA-403 (304H) | S30409 |
| SA-403 (304L) | S30403 |
| SA-403 (309) | S30900 |
| SA-403 (310) | S31000 |
| SA-403 (316) | S31600 |
| SA-403 (316H) | S31609 |
| SA-403 (316L) | S31603 |
| SA-403 (316N) | S31651 |
| SA-403 (317) | S31700 |
| SA-403 (321) | S32100 |
| SA-403 (321H) | S32109 |
| SA-403 (347) | S34700 |
| SA-403 (347H) | S34709 |
| SA-403 (348) | S34800 |
| SA-403 (XM-19) | S20910 |
| SA-409 | S31254 |
| SA-409 (304) | S30400 |
| SA-409 (309) | S30900 |
| SA-409 (310) | S31000 |
| SA-409 (316) | S31600 |
| SA-409 (317) | S31700 |
| SA-409 (321) | S32100 |
| SA-409 (347) | S34700 |
| SA-409 (348) | S34800 |
| SA-412 (21904) | S21904 |
| SA-420 (WPL3) | K31918 |
| SA-420 (WPL6) | K03006 |
| SA-451 | J92802 |
| SA-451 (CF10MC) | J92971 |
| SA-479 | S30815 |
| SA-479 | S31254 |
| SA-479 | S44627 |
| SA-479 (302) | S30200 |
| SA-479 (304) | S30400 |
| SA-479 (304H) | S30409 |
| SA-479 (304L) | S30403 |
| SA-479 (304N) | S30451 |
| SA-479 (310S) | S31008 |
| SA-479 (316) | S31600 |
| SA-479 (316H) | S31609 |
| SA-479 (316L) | S31603 |
| SA-479 (316N) | S31651 |
| SA-479 (321) | S32100 |
| SA-479 (321H) | S32109 |
| SA-479 (347) | S34700 |
| SA-479 (347H) | S34709 |
| SA-479 (348) | S34800 |
| SA-479 (348H) | S34809 |
| SA-479 (405) | S40500 |
| SA-479 (410) | S41000 |
| SA-479 (430) | S43000 |
| SA-479 (439) | S43035 |
| SA-479 (XM-17) | S21600 |
| SA-479 (XM-18) | S21603 |
| SA-479 (XM-19) | S20910 |
| SA-479 (XM-27) | S44625 ob |
| SA-479 (XM-30) | S41040 |
| SA-480 | S20153 |
| SA-480 | S31254 |
| SA-494 | N08826 |
| SA-541 (22V) | K31835 |
| SA-542 (D) | K31835 |
| SA-563 (C3-N) | D61440 |
| SA-564 | S45503 |
| SA-564 (630) | S17400 |
| SA-564 (XM-25) | S45000 |
| SA-638 (660) | S66286 |
| SA-675 (45) | D40450 |
| SA-675 (50) | D40500 |
| SA-675 (55) | D40550 |
| SA-675 (60) | D40600 |
| SA-675 (65) | D40650 |
| SA-675 (70) | D40700 |
| SA-675 (75) | D40750 |
| SA-675 (80) | D40800 |
| SA-675 (90) | D40900 |
| SA-688 (304) | S30400 |
| SA-688 (304L) | S30403 |
| SA-688 (316) | S31600 |
| SA-688 (316L) | S31603 |
| SA-688 (XM-29) | S24000 |
| SA-705 | S45503 |
| SA-705 (630) | S17400 |
| SA-705 (631) | S17700 |
| SA-705 (XM-12) | S15500 |
| SA-705 (XM-13) | S13800 |
| SA-705 (XM-25) | S45000 |
| SA-731 | S44627 |
| SA-789 | S31803 |
| SA-789 (329) | S32900 |
| SA-790 | S31803 |
| SA-815 | S31803 |
| SA-832 (22V) | K31835 |
| SA-905 | K08501 |
| SA5.17 (EH12K) | K01213 |
| SB-11 | C14100 ob |
| SB-12 | C14100 ob |
| SB-42 | C10200 |
| SB-42 | C12000 |
| SB-42 | C12200 |
| SB-43 | C23000 |
| SB-61 | C92200 |
| SB-62 | C83600 |
| SB-75 | C10200 |
| SB-75 | C12000 |
| SB-75 | C12200 |
| SB-75 | C14200 |
| SB-96 | C65500 |
| SB-98 | C65100 |
| SB-98 | C65500 |
| SB-98 | C66100 |
| SB-111 | C10200 |
| SB-111 | C12000 |
| SB-111 | C12200 |
| SB-111 | C14200 |
| SB-111 | C19200 |
| SB-111 | C23000 |
| SB-111 | C28000 |
| SB-111 | C44300 |
| SB-111 | C44400 |
| SB-111 | C44500 |
| SB-111 | C60800 |
| SB-111 | C68700 |
| SB-111 | C70400 |
| SB-111 | C70600 |
| SB-111 | C71000 |
| SB-111 | C71500 |
| SB-127 | N04400 |
| SB-133 | C10200 |
| SB-148 | C95200 |
| SB-148 | C95400 |
| SB-150 | C61400 |
| SB-150 | C62300 |
| SB-150 | C63000 |
| SB-150 | C64200 |
| SB-152 | C10200 |
| SB-152 | C10400 |
| SB-152 | C10500 |
| SB-152 | C10700 |
| SB-152 | C12200 |
| SB-152 | C12300 |
| SB-160 | N02200 |
| SB-161 | N02200 |
| SB-162 | N02200 |
| SB-163 | N02200 |
| SB-163 | N04400 |
| SB-163 | N06600 |
| SB-163 | N06690 |
| SB-163 | N08800 |
| SB-163 | N08810 |
| SB-163 | N08825 |
| SB-164 | N04400 |
| SB-164 | N04405 |
| SB-165 | N04400 |
| SB-166 | N06600 |
| SB-166 | N06617 |
| SB-166 | N06690 |
| SB-167 | N06600 |
| SB-167 | N06690 |
| SB-168 | N06600 |
| SB-168 | N06690 |
| SB-169 | C61000 |
| SB-169 | C61400 |
| SB-171 | C36500 |
| SB-171 | C36600 ob |
| SB-171 | C36700 ob |
| SB-171 | C36800 ob |
| SB-171 | C44300 |
| SB-171 | C44400 |

| CROSS REFERENCE SPECIFICATIONS | UNIFIED NUMBER |
|---|---|
| SB-171 | C44500 |
| SB-171 | C46400 |
| SB-171 | C46500 |
| SB-171 | C46600 ob |
| SB-171 | C46700 ob |
| SB-171 | C61400 |
| SB-171 | C63000 |
| SB-171 | C70600 |
| SB-171 | C71500 |
| SB-209 | A83003 |
| SB-209 | A86061 |
| SB-209 | A91060 |
| SB-209 | A91100 |
| SB-209 | A93003 |
| SB-209 | A95050 |
| SB-209 | A95052 |
| SB-209 | A95083 |
| SB-209 | A95086 |
| SB-209 | A95154 |
| SB-209 | A95254 |
| SB-209 | A95454 |
| SB-209 | A95652 |
| SB-209 | A96061 |
| SB-210 | A83003 |
| SB-210 | A83003 |
| SB-210 | A91060 |
| SB-210 | A93003 |
| SB-210 | A95052 |
| SB-210 | A95154 |
| SB-210 | A96061 |
| SB-211 | A91100 |
| SB-211 | A92014 |
| SB-211 | A95086 |
| SB-211 | A95454 |
| SB-211 | A96061 |
| SB-221 | A91060 |
| SB-221 | A93003 |
| SB-221 | A93102 |
| SB-221 | A95083 |
| SB-221 | A95154 |
| SB-221 | A96061 |
| SB-221 | A96063 |
| SB-234 | A83003 |
| SB-234 | A91060 |
| SB-234 | A93003 |
| SB-234 | A95052 |
| SB-234 | A95454 |
| SB-234 | A96061 |
| SB-241 | A83003 |
| SB-241 | A91060 |
| SB-241 | A91100 |
| SB-241 | A93003 |
| SB-241 | A95052 |
| SB-241 | A95083 |
| SB-241 | A95086 |
| SB-241 | A95454 |
| SB-241 | A96061 |
| SB-241 | A96063 |
| SB-241 | A96162 |
| SB-241/SB-241M | A91100 |
| SB-247 | A93003 |
| SB-247 | A95083 |
| SB-247 | A96061 |
| SB-265 | R56403 |
| SB-265 | R56407 |
| SB-265 | R58210 |
| SB-265 | R58640 |
| SB-271 | C95200 |
| SB-271 | C95400 |

| CROSS REFERENCE SPECIFICATIONS | UNIFIED NUMBER |
|---|---|
| SB-283 | C37700 |
| SB-283 | C64200 |
| SB-308 | A96061 |
| SB-315 | C65100 |
| SB-315 | C65500 |
| SB-333 | N10001 |
| SB-333 | N10665 |
| SB-335 | N10001 |
| SB-335 | N10665 |
| SB-348 | R56403 |
| SB-348 | R56407 |
| SB-348 | R58210 |
| SB-348 | R58640 |
| SB-359 | C10200 |
| SB-359 | C12000 |
| SB-359 | C12200 |
| SB-359 | C14200 |
| SB-359 | C19200 |
| SB-359 | C23000 |
| SB-359 | C44300 |
| SB-359 | C44400 |
| SB-359 | C44500 |
| SB-359 | C60800 |
| SB-359 | C68700 |
| SB-359 | C70400 |
| SB-359 | C70600 |
| SB-359 | C71000 |
| SB-359 | C71500 |
| SB-363 | R56403 |
| SB-363 | R58210 |
| SB-363 | R58640 |
| SB-365 | R56407 |
| SB-366 | N06007 |
| SB-366 | N06022 |
| SB-366 | N06030 |
| SB-366 | N06455 |
| SB-366 | N06975 |
| SB-366 | N08320 |
| SB-366 | N08330 |
| SB-366 | N08332 |
| SB-366 | N08926 |
| SB-366 | N10001 |
| SB-366 | N10276 |
| SB-366 | N10665 |
| SB-381 | R56403 |
| SB-381 | R56407 |
| SB-381 | R58210 |
| SB-381 | R58640 |
| SB-395 | C10200 |
| SB-395 | C12000 |
| SB-395 | C12200 |
| SB-395 | C14200 |
| SB-395 | C19200 |
| SB-395 | C23000 |
| SB-395 | C44300 |
| SB-395 | C44400 |
| SB-395 | C44500 |
| SB-395 | C60800 |
| SB-395 | C68700 |
| SB-395 | C70600 |
| SB-395 | C71000 |
| SB-395 | C71500 |
| SB-407 | N08800 |
| SB-407 | N08810 |
| SB-408 | N08800 |
| SB-408 | N08810 |
| SB-409 | N08800 |
| SB-409 | N08810 |
| SB-423 | N08825 |

| CROSS REFERENCE SPECIFICATIONS | UNIFIED NUMBER |
|---|---|
| SB-424 | N08825 |
| SB-425 | N08825 |
| SB-434 | N10003 |
| SB-435 | N06002 |
| SB-436 | N06001 ob |
| SB-443 | N06625 |
| SB-444 | N06625 |
| SB-446 | N06625 |
| SB-462 | N08020 |
| SB-463 | N08020 |
| SB-463 | N08026 |
| SB-464 | N08020 |
| SB-464 | N08026 |
| SB-466 | C70400 |
| SB-466 | C70600 |
| SB-466 | C71000 |
| SB-466 | C71500 |
| SB-467 | C70600 |
| SB-467 | C71500 |
| SB-468 | N08026 |
| SB-493 | R60702 |
| SB-494 | R60702 |
| SB-495 | R60702 |
| SB-505 | C95200 |
| SB-511 | N08330 |
| SB-511 | N08332 |
| SB-514 | N08810 |
| SB-523 | R60702 |
| SB-535 | N08332 |
| SB-536 | N08330 |
| SB-536 | N08332 |
| SB-543 | C12200 |
| SB-543 | C19400 |
| SB-543 | C23000 |
| SB-543 | C44300 |
| SB-543 | C44400 |
| SB-543 | C44500 |
| SB-543 | C68700 |
| SB-543 | C70400 |
| SB-543 | C70600 |
| SB-543 | C71500 |
| SB-550 | R60702 |
| SB-551 | R60702 |
| SB-564 | N04400 |
| SB-564 | N06210 |
| SB-564 | N06600 |
| SB-564 | N08800 |
| SB-564 | N08810 |
| SB-572 | N06002 |
| SB-574 | N06022 |
| SB-574 | N06210 |
| SB-574 | N06455 |
| SB-574 | N10276 |
| SB-575 | N06022 |
| SB-575 | N06210 |
| SB-575 | N06455 |
| SB-575 | N10276 |
| SB-581 | N06030 |
| SB-581 | N06975 |
| SB-581 | N06985 |
| SB-582 | N06030 |
| SB-582 | N06975 |
| SB-582 | N06985 |
| SB-584 | C92200 |
| SB-584 | C93700 |
| SB-584 | C95820 |
| SB-584 | C97600 |
| SB-599 | N08700 |
| SB-619 | N06002 |

| CROSS REFERENCE SPECIFICATIONS | UNIFIED NUMBER |
|---|---|
| SB-619 | N06007 |
| SB-619 | N06022 |
| SB-619 | N06030 |
| SB-619 | N06210 |
| SB-619 | N06455 |
| SB-619 | N06975 |
| SB-619 | N06985 |
| SB-619 | N10001 |
| SB-619 | N10276 |
| SB-619 | N10665 |
| SB-620 | N08320 |
| SB-621 | N08320 |
| SB-622 | N06002 |
| SB-622 | N06022 |
| SB-622 | N06030 |
| SB-622 | N06210 |
| SB-622 | N06455 |
| SB-622 | N06985 |
| SB-622 | N08320 |
| SB-622 | N10276 |
| SB-622 | N10665 |
| SB-625 | N08904 |
| SB-625 | N08925 |
| SB-625 | N08926 |
| SB-626 | N06002 |
| SB-626 | N06022 |
| SB-626 | N06030 |
| SB-626 | N06210 |
| SB-626 | N06455 |
| SB-626 | N06985 |
| SB-626 | N10276 |
| SB-626 | N10665 |
| SB-637 | N07080 |
| SB-637 | N07252 |
| SB-637 | N07500 |
| SB-637 | N07750 |
| SB-649 | N08904 |
| SB-649 | N08925 |
| SB-649 | N08926 |
| SB-668 | N08028 |
| SB-672 | N08700 |
| SB-673 | N08904 |
| SB-673 | N08925 |
| SB-673 | N08926 |
| SB-674 | N08904 |
| SB-674 | N08925 |
| SB-674 | N08926 |
| SB-675 | N08366 |
| SB-677 | N08904 |
| SB-677 | N08926 |
| SB-709 | N08028 |
| SB-710 | N08330 |
| SB-751 | N06002 |
| SFA-5.8 (BNi-1) | N99600 |
| SFA-5.8 (BNi-2) | N99620 |
| SFA5.1 (E6010) | W06010 |
| SFA5.1 (E6011) | W06011 |
| SFA5.1 (E6012) | W06012 |
| SFA5.1 (E6013) | W06013 |
| SFA5.1 (E6019) | W06019 |
| SFA5.1 (E6020) | W06020 |
| SFA5.1 (E6022) | W06022 |
| SFA5.1 (E6027) | W06027 |
| SFA5.1 (E7014) | W07014 |
| SFA5.1 (E7015) | W07015 |
| SFA5.1 (E7016) | W07016 |
| SFA5.1 (E7016-1) | W07116 |
| SFA5.1 (E7018, E7018M) | W07018 |
| SFA5.1 (E7018-1) | W07118 |
| SFA5.1 (E7024) | W07024 |
| SFA5.1 (E7024-1) | W07124 |
| SFA5.1 (E7027) | W07027 |
| SFA5.1 (E7028) | W07028 |
| SFA5.1 (E7048) | W07048 |
| SFA5.2 (R100) | K12147 |
| SFA5.2 (RG45) | K00045 ob |
| SFA5.2 (RG60) | K00060 ob |
| SFA5.2 (RG65) | K00065 ob |
| SFA5.3 (E1100) | A91100 |
| SFA5.3 (E3003) | A93003 |
| SFA5.3 (E4043) | A94043 |
| SFA5.4 (E7Cr-XX) | W50310 |
| SFA5.4 (E16-8-2-XX) | W36810 |
| SFA5.4 (E209-XX) | W32210 |
| SFA5.4 (E219-XX) | W32310 |
| SFA5.4 (E240-XX) | W32410 |
| SFA5.4 (E307-XX) | W30710 |
| SFA5.4 (E308-XX) | W30810 |
| SFA5.4 (E308L) | W30813 |
| SFA5.4 (E308Mo-XX) | W30820 |
| SFA5.4 (E308MoL-XX) | W30823 |
| SFA5.4 (E309) | W30910 |
| SFA5.4 (E309Cb) | W30917 |
| SFA5.4 (E309L) | W30913 |
| SFA5.4 (E309LMo) | W30923 |
| SFA5.4 (E309Mo) | W30920 |
| SFA5.4 (E310-XX) | W31010 |
| SFA5.4 (E310Cb-XX) | W31017 |
| SFA5.4 (E310H-XX) | W31015 |
| SFA5.4 (E310Mo-XX) | W31020 |
| SFA5.4 (E312-XX) | W31310 |
| SFA5.4 (E316, E316H) | W31610 |
| SFA5.4 (E316L-XX) | W31613 |
| SFA5.4 (E317-XX) | W31710 |
| SFA5.4 (E317L-XX) | W31713 |
| SFA5.4 (E318-XX) | W31910 |
| SFA5.4 (E320-XX) | W88021 |
| SFA5.4 (E320LR) | W88022 |
| SFA5.4 (E330) | W88331 |
| SFA5.4 (E330H) | W88335 |
| SFA5.4 (E347-XX) | W34710 |
| SFA5.4 (E349-XX) | W34910 |
| SFA5.4 (E383) | W88028 |
| SFA5.4 (E385) | W88904 |
| SFA5.4 (E410-XX) | W41010 |
| SFA5.4 (E410NiMo-XX) | W41016 |
| SFA5.4 (E430-XX) | W43010 |
| SFA5.4 (E502-XX) | W50210 |
| SFA5.4 (E505-XX) | W50410 |
| SFA5.4 (E630-XX) | W37410 |
| SFA5.4 (E2209-XX) | W39209 |
| SFA5.4 (E2553-XX) | W39553 |
| SFA5.5 (E801X-B6L) | W50213 ob |
| SFA5.5 (E7010-A1) | W17010 |
| SFA5.5 (E7010-P1) | W17110 |
| SFA5.5 (E7011-A1) | W17011 |
| SFA5.5 (E7015-A1) | W17015 |
| SFA5.5 (E7015-B2L) | W52115 |
| SFA5.5 (E7015-C1L) | W22115 |
| SFA5.5 (E7015-C2L) | W23115 |
| SFA5.5 (E7016-A1) | W17016 |
| SFA5.5 (E7016-B2L) | W52116 |
| SFA5.5 (E7016-C1L) | W22116 |
| SFA5.5 (E7016-C2L) | W23116 |
| SFA5.5 (E7018-A1) | W17018 |
| SFA5.5 (E7018-B2L) | W52118 |
| SFA5.5 (E7018-C1L) | W22118 |
| SFA5.5 (E7018-C2L) | W23118 |
| SFA5.5 (E7018-C3L) | W20918 |
| SFA5.5 (E7018-W1) | W20018 |
| SFA5.5 (E7020-A1) | W17020 |
| SFA5.5 (E7027-A1) | W17027 |
| SFA5.5 (E8010-P1) | W18110 |
| SFA5.5 (E8015-B3L) | W53115 |
| SFA5.5 (E8015-B4L) | W53415 |
| SFA5.5 (E8015-B6) | W50215 |
| SFA5.5 (E8015-B6L) | W50205 |
| SFA5.5 (E8015-B7) | W50315 |
| SFA5.5 (E8015-B7L) | W50305 |
| SFA5.5 (E8015-B8) | W50415 |
| SFA5.5 (E8015-B8L) | W50405 |
| SFA5.5 (E8016-B1) | W51016 |
| SFA5.5 (E8016-B2) | W52016 |
| SFA5.5 (E8016-B5) | W51316 |
| SFA5.5 (E8016-B6) | W50216 |
| SFA5.5 (E8016-B6L) | W50206 |
| SFA5.5 (E8016-B7) | W50316 |
| SFA5.5 (E8016-B7L) | W50306 |
| SFA5.5 (E8016-B8) | W50416 |
| SFA5.5 (E8016-B8L) | W50406 |
| SFA5.5 (E8016-C1) | W22016 |
| SFA5.5 (E8016-C2) | W23016 |
| SFA5.5 (E8016-C3) | W21016 |
| SFA5.5 (E8016-C4) | W21916 |
| SFA5.5 (E8016-D3) | W18016 |
| SFA5.5 (E8018-B1) | W51018 |
| SFA5.5 (E8018-B2) | W52018 |
| SFA5.5 (E8018-B3L) | W53118 |
| SFA5.5 (E8018-B6) | W50218 |
| SFA5.5 (E8018-B6L) | W50208 |
| SFA5.5 (E8018-B7) | W50318 |
| SFA5.5 (E8018-B7L) | W50308 |
| SFA5.5 (E8018-B8) | W50418 |
| SFA5.5 (E8018-B8L) | W50408 |
| SFA5.5 (E8018-C1) | W22018 |
| SFA5.5 (E8018-C2) | W23018 |
| SFA5.5 (E8018-C3) | W21018 |
| SFA5.5 (E8018-C4) | W21918 |
| SFA5.5 (E8018-D1) | W18118 |
| SFA5.5 (E8018-D3) | W18018 |
| SFA5.5 (E8018-NM1) | W21118 |
| SFA5.5 (E8018-W2) | W20118 |
| SFA5.5 (E9015-B3) | W53015 |
| SFA5.5 (E9015-B9) | W50425 |
| SFA5.5 (E9015-C5L) | W25018 |
| SFA5.5 (E9015-D1) | W19015 |
| SFA5.5 (E9016-B3) | W53016 |
| SFA5.5 (E9016-B9) | W50426 |
| SFA5.5 (E9018-B3) | W53018 |
| SFA5.5 (E9018-B9) | W50428 |
| SFA5.5 (E9018-D1) | W19018 |
| SFA5.5 (E9018-D3) | W19118 |
| SFA5.5 (E9018M) | W21218 |
| SFA5.5 (E10015-D2) | W10015 |
| SFA5.5 (E10016-D2) | W10016 |
| SFA5.5 (E10016M) | W21216 ob |
| SFA5.5 (E10018-D2) | W10018 |
| SFA5.5 (E10018M) | W21318 |
| SFA5.5 (E11018M) | W21418 |
| SFA5.5 (E12018M) | W22218 |
| SFA5.5 (E12018M1) | W23218 |
| SFA5.6 (ECu) | W60189 |
| SFA5.6 (ECuAl-A2) | W60614 |
| SFA5.6 (ECuAl-B) | W60619 |
| SFA5.6 (ECuMnNiAl) | W60633 |
| SFA5.6 (ECuNi) | W60715 |
| SFA5.6 (ECuNiAl) | W60632 |

| CROSS REFERENCE SPECIFICATIONS | UNIFIED NUMBER |
|---|---|
| SFA5.6 (ECuSi) | W60656 |
| SFA5.6 (ECuSn-A) | W60518 |
| SFA5.6 (ECuSn-C) | W60521 |
| SFA5.7 (ERCu) | C18980 |
| SFA5.7 (ERCuAl-A2) | C61800 |
| SFA5.7 (ERCuAl-A3) | C62400 |
| SFA5.7 (ERCuAl-Al) | C61000 |
| SFA5.7 (ERCuMnNiAl) | C63380 |
| SFA5.7 (ERCuNi) | C71581 |
| SFA5.7 (ERCuNiAl) | C63280 |
| SFA5.7 (ERCuSi-A) | C65600 |
| SFA5.7 (ERCuSn-A) | C51800 |
| SFA5.7 (ERCuSn-C) | C52100 |
| SFA5.8 (BAg-1) | P07450 |
| SFA5.8 (BAg-1a) | P07500 |
| SFA5.8 (BAg-2) | P07350 |
| SFA5.8 (BAg-2a) | P07300 |
| SFA5.8 (BAg-3) | P07501 |
| SFA5.8 (BAg-4) | P07400 |
| SFA5.8 (BAg-5) | P07453 |
| SFA5.8 (BAg-6) | P07503 |
| SFA5.8 (BAg-7) | P07563 |
| SFA5.8 (BAg-8) | P07720 |
| SFA5.8 (BAg-8a) | P07723 |
| SFA5.8 (BAg-9) | P07650 |
| SFA5.8 (BAg-10) | P07700 |
| SFA5.8 (BAg-13) | P07540 |
| SFA5.8 (BAg-13a) | P07560 |
| SFA5.8 (BAg-18) | P07600 |
| SFA5.8 (BAg-19) | P07925 |
| SFA5.8 (BAg-20) | P07301 |
| SFA5.8 (BAg-21) | P07630 |
| SFA5.8 (BAg-22) | P07490 |
| SFA5.8 (BAg-23) | P07850 |
| SFA5.8 (BAg-24) | P07505 |
| SFA5.8 (BAg-25) | P07200 ob |
| SFA5.8 (BAg-26) | P07250 |
| SFA5.8 (BAg-27) | P07251 |
| SFA5.8 (BAg-28) | P07401 |
| SFA5.8 (BAg-33) | P07252 |
| SFA5.8 (BAg-34) | P07380 |
| SFA5.8 (BAg-35) | P07351 |
| SFA5.8 (BAg-36) | P07454 |
| SFA5.8 (BAg-37) | P07253 |
| SFA5.8 (BAlSi-2) | A94343 |
| SFA5.8 (BAlSi-3) | A94145 |
| SFA5.8 (BAlSi-4) | A94047 |
| SFA5.8 (BAlSi-5) | A94045 |
| SFA5.8 (BAlSi-7) | A94004 |
| SFA5.8 (BAlSi-9) | A94147 |
| SFA5.8 (BAlSi-11) | A94104 |
| SFA5.8 (BAu-1) | P00375 |
| SFA5.8 (BAu-2) | P00800 |
| SFA5.8 (BAu-3) | P00350 |
| SFA5.8 (BAu-4) | P00820 |
| SFA5.8 (BAu-5) | P00300 |
| SFA5.8 (BAu-6) | P00700 |
| SFA5.8 (BCo-1) | R39001 |
| SFA5.8 (BCu-1) | C14180 |
| SFA5.8 (BCu-1b) | C11000 |
| SFA5.8 (BCu-3) | C10100 |
| SFA5.8 (BCuP-1) | C55180 |
| SFA5.8 (BCuP-2) | C55181 |
| SFA5.8 (BCuP-3) | C55281 |
| SFA5.8 (BCuP-4) | C55283 |
| SFA5.8 (BCuP-5) | C55284 |
| SFA5.8 (BCuP-6) | C55280 |
| SFA5.8 (BCuP-7) | C55282 |
| SFA5.8 (BCuP-8) | C55285 |

| CROSS REFERENCE SPECIFICATIONS | UNIFIED NUMBER |
|---|---|
| SFA5.8 (BCuZn-G) | C26380 ob |
| SFA5.8 (BMg-1) | M19001 |
| SFA5.8 (BNi-1a) | N99610 |
| SFA5.8 (BNi-3) | N99630 |
| SFA5.8 (BNi-4) | N99640 |
| SFA5.8 (BNi-5) | N99650 |
| SFA5.8 (BNi-5a) | N99651 |
| SFA5.8 (BNi-5b) | N99652 |
| SFA5.8 (BNi-6) | N99700 |
| SFA5.8 (BNi-7) | N99710 |
| SFA5.8 (BNi-8) | N99800 |
| SFA5.8 (BNi-9) | N99612 |
| SFA5.8 (BNi-10) | N99622 |
| SFA5.8 (BNi-11) | N99624 |
| SFA5.8 (BNi-12) | N99720 |
| SFA5.8 (BNi-13) | N99810 |
| SFA5.8 (BVAg-0) | P07017 |
| SFA5.8 (BVAg-6b) | P07507 |
| SFA5.8 (BVAg-8) | P07727 |
| SFA5.8 (BVAg-8b) | P07728 |
| SFA5.8 (BVAg-18) | P07607 |
| SFA5.8 (BVAg-29) | P07627 |
| SFA5.8 (BVAg-30) | P07687 |
| SFA5.8 (BVAg-31) | P07587 |
| SFA5.8 (BVAg-32) | P07547 |
| SFA5.8 (BVAu-2) | P00807 |
| SFA5.8 (BVAu-4) | P00827 |
| SFA5.8 (BVAu-7) | P00507 |
| SFA5.8 (BVAu-8) | P00927 |
| SFA5.8 (BVCu-1X) | C14181 |
| SFA5.8 (BVPd-1) | P03657 |
| SFA5.8 (RBCuZn-A) | C47000 |
| SFA5.8 (RBCuZn-B) | C68000 |
| SFA5.8 (RBCuZn-C) | C68100 |
| SFA5.8 (RBCuZn-D) | C77300 |
| SFA5.9 (EC16-8-2) | W36840 |
| SFA5.9 (EC19-10H) | W30440 ob |
| SFA5.9 (EC209) | W32240 |
| SFA5.9 (EC218) | W32540 |
| SFA5.9 (EC219) | W32340 |
| SFA5.9 (EC240) | W32440 |
| SFA5.9 (EC307) | W30740 ob |
| SFA5.9 (EC308, EC308H) | W30840 |
| SFA5.9 (EC308L) | W30843 |
| SFA5.9 (EC308LMo) | W30853 |
| SFA5.9 (EC308LSi) | W30848 |
| SFA5.9 (EC308Mo) | W30850 |
| SFA5.9 (EC308Si) | W30841 |
| SFA5.9 (EC309) | W30940 |
| SFA5.9 (EC309L) | W30943 |
| SFA5.9 (EC309LMo) | W30953 |
| SFA5.9 (EC309Mo) | W30950 |
| SFA5.9 (EC309Si) | W30941 |
| SFA5.9 (EC310) | W31040 |
| SFA5.9 (EC312) | W31340 |
| SFA5.9 (EC316, EC316H) | W31640 |
| SFA5.9 (EC316L) | W31643 |
| SFA5.9 (EC316LSi) | W31648 |
| SFA5.9 (EC316Si) | W31641 |
| SFA5.9 (EC317) | W31740 |
| SFA5.9 (EC317L) | W31743 |
| SFA5.9 (EC318) | W31940 |
| SFA5.9 (EC320LR) | W88022 |
| SFA5.9 (EC321) | W32140 |
| SFA5.9 (EC347) | W34740 |
| SFA5.9 (EC347Si) | W34748 |
| SFA5.9 (EC349) | W34940 ob |
| SFA5.9 (EC409Cb) | W40947 |
| SFA5.9 (EC410) | W41040 |

| CROSS REFERENCE SPECIFICATIONS | UNIFIED NUMBER |
|---|---|
| SFA5.9 (EC410NiMo) | W41046 |
| SFA5.9 (EC420) | W42040 |
| SFA5.9 (EC430) | W43040 |
| SFA5.9 (EC446LMo) | W44647 |
| SFA5.9 (EC505) | W50440 |
| SFA5.9 (EC630) | W37440 |
| SFA5.9 (EC2209) | W39240 |
| SFA5.9 (EC2553) | W39540 |
| SFA5.9 (ER 309LMoSi) | S30989 ob |
| SFA5.9 (ER 309LSi) | S30988 |
| SFA5.9 (ER 309Mo) | S30982 |
| SFA5.9 (ER 317L) | S31783 |
| SFA5.9 (ER16-8-2) | S16880 |
| SFA5.9 (ER19-10H) | S30480 |
| SFA5.9 (Er26-1) | S44627 |
| SFA5.9 (ER209) | S20980 |
| SFA5.9 (ER218) | S21880 |
| SFA5.9 (ER219) | S21980 |
| SFA5.9 (ER240) | S24080 |
| SFA5.9 (ER307) | S30780 |
| SFA5.9 (ER308, ER308H) | S30880 |
| SFA5.9 (ER308L) | S30883 |
| SFA5.9 (ER308LMo) | S30886 |
| SFA5.9 (ER308LSi) | S30888 |
| SFA5.9 (ER308Mo) | S30882 |
| SFA5.9 (ER308Si) | S30881 |
| SFA5.9 (ER309) | S30980 |
| SFA5.9 (ER309L) | S30983 |
| SFA5.9 (ER309LMo) | S30986 |
| SFA5.9 (ER309Si) | S30981 |
| SFA5.9 (ER310) | S31080 |
| SFA5.9 (ER312) | S31380 |
| SFA5.9 (ER316, 316H) | S31680 |
| SFA5.9 (ER316L) | S31683 |
| SFA5.9 (ER316LSi) | S31688 |
| SFA5.9 (ER316Si) | S31681 |
| SFA5.9 (ER317) | S31780 |
| SFA5.9 (ER318) | S31980 |
| SFA5.9 (ER320) | N08021 |
| SFA5.9 (ER320LR) | N08022 |
| SFA5.9 (ER321) | S32180 |
| SFA5.9 (ER330) | N08331 |
| SFA5.9 (ER347) | S34780 |
| SFA5.9 (ER347Si) | S34788 |
| SFA5.9 (ER409Cb) | S40940 |
| SFA5.9 (ER410) | S41080 |
| SFA5.9 (ER410NiMo) | S41086 |
| SFA5.9 (ER420) | S42080 |
| SFA5.9 (ER430) | S43080 |
| SFA5.9 (ER446LMo) | S44687 |
| SFA5.9 (ER502) | S50280 |
| SFA5.9 (ER502Si) | S50288 |
| SFA5.9 (ER505) | S50480 |
| SFA5.9 (ER505Si) | S50488 |
| SFA5.9 (ER630) | S17480 |
| SFA5.9 (ER2209) | S39209 |
| SFA5.9 (ER2553) | S39255 ob |
| SFA5.9 (ER2553) | S39553 |
| SFA5.9 (ER3556) | R30556 |
| SFA5.10 (ER1100) | A91100 |
| SFA5.10 (ER1188) | A91188 |
| SFA5.10 (ER2319) | A92319 |
| SFA5.10 (ER4009) | A94009 |
| SFA5.10 (ER4010) | A94010 |
| SFA5.10 (ER4043) | A94043 |
| SFA5.10 (ER4047) | A94047 |
| SFA5.10 (ER4145) | A94145 |
| SFA5.10 (ER4643) | A94643 |
| SFA5.10 (ER5183) | A95183 |

| CROSS REFERENCE SPECIFICATIONS | UNIFIED NUMBER |
|---|---|
| SFA5.10 (ER5356) | A95356 |
| SFA5.10 (ER5554) | A95554 |
| SFA5.10 (ER5556) | A95556 |
| SFA5.10 (ER5654) | A95654 |
| SFA5.10 (R-A356.0) | A13560 |
| SFA5.10 (R-A357.0) | A13570 |
| SFA5.10 (R-C355.0) | A03550 |
| SFA5.10 (R206.0) | A02060 |
| SFA5.10 (R4011) | A94011 |
| SFA5.10 (RA-357.0) | A03570 |
| SFA5.11 (ENi-1) | W82141 |
| SFA5.11 (ENiCrCoMo-1) | W86117 |
| SFA5.11 (ENiCrFe-1) | W86132 |
| SFA5.11 (ENiCrFe-2) | W86133 |
| SFA5.11 (ENiCrFe-3) | W86182 |
| SFA5.11 (ENiCrFe-4) | W86134 |
| SFA5.11 (ENiCrFe-7) | W86152 |
| SFA5.11 (ENiCrFe-9) | W86094 |
| SFA5.11 (ENiCrFe-10) | W86095 |
| SFA5.11 (ENiCrMo-1) | W86007 |
| SFA5.11 (ENiCrMo-2) | W86002 |
| SFA5.11 (ENiCrMo3) | W86112 |
| SFA5.11 (ENiCrMo-4) | W80276 |
| SFA5.11 (ENiCrMo-5) | W80002 |
| SFA5.11 (ENiCrMo-6) | W86620 |
| SFA5.11 (ENiCrMo-7) | W86455 |
| SFA5.11 (ENiCrMo-9) | W86135 ob |
| SFA5.11 (ENiCrMo-9) | W86985 |
| SFA5.11 (ENiCrMo-10) | W86022 |
| SFA5.11 (ENiCrMo-10) | W86136 ob |
| SFA5.11 (ENiCrMo-11) | W86030 |
| SFA5.11 (ENiCrMo-13) | W86059 |
| SFA5.11 (ENiCrMo-14) | W86026 |
| SFA5.11 (ENiCrMo-14) | W86686 |
| SFA5.11 (ENiCu-7) | W84190 |
| SFA5.11 (ENiMo-1) | W80001 |
| SFA5.11 (ENiMo-3) | W80004 |
| SFA5.11 (ENiMo-7) | W80665 |
| SFA5.11 (ENiMo-8) | W80008 |
| SFA5.11 (ENiMo-9) | W80009 |
| SFA5.11 (ENiMo-10) | W80675 |
| SFA5.12 (EWCe-2) | R07932 |
| SFA5.12 (EWLa-1) | R07941 |
| SFA5.12 (EWP) | R07900 |
| SFA5.12 (EWTh-1) | R07911 |
| SFA5.12 (EWTh-2) | R07912 |
| SFA5.12 (EWTh-3) | R07913 ob |
| SFA5.12 (EWZr-1) | R07920 |
| SFA5.13 (ECoCr-A) | W73006 |
| SFA5.13 (ECoCr-B) | W73012 |
| SFA5.13 (ECoCr-C) | W73001 |
| SFA5.13 (ECoCr-E) | W73021 |
| SFA5.13 (ECuAl-B) | W60619 |
| SFA5.13 (ECuAl-C) | W60625 |
| SFA5.13 (ECuAl-D) | W61625 |
| SFA5.13 (ECuAl-E) | W62625 |
| SFA5.13 (ECuMnNiAl) | W60633 |
| SFA5.13 (ECuNi) | W60715 |
| SFA5.13 (ECuNiAl) | W60632 |
| SFA5.13 (ECuSi) | W60656 |
| SFA5.13 (ECuSn-A) | W60518 |
| SFA5.13 (ECuSn-C) | W60521 |
| SFA5.13 (EFe-Mn-A) | W79110 |
| SFA5.13 (EFe-1) | W74001 |
| SFA5.13 (EFe-2) | W74002 |
| SFA5.13 (EFe-3) | W74003 |
| SFA5.13 (EFe-4) | W74004 |
| SFA5.13 (EFe-5) | W75110 |
| SFA5.13 (EFe5-A) | W77710 ob |
| SFA5.13 (EFe5-C) | W77310 ob |
| SFA5.13 (EFe-6) | W77510 |
| SFA5.13 (EFe-7) | W77610 |
| SFA5.13 (EFeCr-A1) | W74510 ob |
| SFA5.13 (EFeCr-A1A) | W74011 |
| SFA5.13 (EFeCr-A2) | W74012 |
| SFA5.13 (EFeCr-A3) | W74013 |
| SFA5.13 (EFeCr-A4) | W74014 |
| SFA5.13 (EFeCr-A5) | W74015 |
| SFA5.13 (EFeCr-A6) | W74016 |
| SFA5.13 (EFeCr-A7) | W74017 |
| SFA5.13 (EFeCr-A8) | W74018 |
| SFA5.13 (EFeCr-E1) | W74211 |
| SFA5.13 (EFeCr-E2) | W74212 |
| SFA5.13 (EFeCr-E3) | W74213 |
| SFA5.13 (EFeCr-E4) | W74214 |
| SFA5.13 (EFeMn-B) | W79310 |
| SFA5.13 (EFeMn-C) | W79210 |
| SFA5.13 (EFeMn-D) | W77410 ob |
| SFA5.13 (EFeMn-D) | W79410 |
| SFA5.13 (EFeMn-E) | W79510 |
| SFA5.13 (EFeMn-F) | W79610 |
| SFA5.13 (EFeMnCr) | W79710 |
| SFA5.13 (ENiCrFeCo) | W83002 |
| SFA5.13 (ENiCrMo-5A) | W80002 |
| SFA5.14 (ERNi-1) | N02061 |
| SFA5.14 (ERNiCr-3) | N06082 |
| SFA5.14 (ERNiCr-4) | N06072 |
| SFA5.14 (ERNiCr-6) | N06076 |
| SFA5.14 (ERNiCrCoMo-1) | N06617 |
| SFA5.14 (ERNiCrFe-5) | N06062 |
| SFA5.14 (ERNiCrFe-6) | N07092 |
| SFA5.14 (ERNiCrFe-7) | N06052 |
| SFA5.14 (ERNiCrFe-8) | N07069 |
| SFA5.14 (ERNiCrFe-11) | N06601 |
| SFA5.14 (ERNiCrMo-1) | N06007 |
| SFA5.14 (ERNiCrMo-2) | N06002 |
| SFA5.14 (ERNiCrMo-3) | N06625 |
| SFA5.14 (ERNiCrMo-4) | N10276 |
| SFA5.14 (ERNiCrMo-7) | N06455 |
| SFA5.14 (ERNiCrMo-8) | N06975 |
| SFA5.14 (ERNiCrMo-9) | N06985 |
| SFA5.14 (ERNiCrMo-10) | N06022 |
| SFA5.14 (ErNiCrMo-11) | N06030 |
| SFA5.14 (ERNiCrMo-13) | N06059 |
| SFA5.14 (ERNiCrMo-14) | N06686 |
| SFA5.14 (ERNiCrMo-15) | N07725 |
| SFA5.14 (ERNiCrMo-18) | N06650 |
| SFA5.14 (ERNiCrWMo-1) | N06231 |
| SFA5.14 (ERNiCu-7) | N04060 |
| SFA5.14 (ERNiCu-8) | N05504 |
| SFA5.14 (ERNiFeCr-1) | N08065 |
| SFA5.14 (ERNiFeCr-2) | N07718 |
| SFA5.14 (ERNiMo-1) | N10001 |
| SFA5.14 (ERNiMo-2) | N10003 |
| SFA5.14 (ERNiMo-3) | N10004 |
| SFA5.14 (ERNiMo-7) | N10665 |
| SFA5.14 (ERNiMo-8) | N10008 |
| SFA5.14 (ERNiMo-9) | N10009 |
| SFA5.14 (ERNiMo-10) | N10675 |
| SFA5.15 (ENi-CI-A) | W82003 |
| SFA5.15 (ENiFe-CI) | W82002 |
| SFA5.15 (ENiFe-CI-A) | W82004 |
| SFA5.15 (ENiFeMn-CI) | W82006 |
| SFA5.15 (ENiFeT3-CI) | W82032 ob |
| SFA5.15 (ERNi-CI) | N02215 |
| SFA5.15 (ERNiFeMn-CI) | N02216 |
| SFA5.16 (ERTi-1) | R50100 |
| SFA5.16 (ERTi-2) | R50120 |
| SFA5.16 (ERTi-3) | R50125 |
| SFA5.16 (ERTi-6) | R54522 |
| SFA5.16 (ERTi-6EL1 | R54523 |
| SFA5.16 (ERTi-9) | R56320 |
| SFA5.16 (ERTi-9EL1) | R56321 |
| SFA5.17 (EC1) | W06041 |
| SFA5.17 (EH10K) | K01210 |
| SFA5.17 (EH11K) | K11140 |
| SFA5.17 (EH14) | K11585 |
| SFA5.17 (EL8) | K01008 |
| SFA5.17 (EL8K) | K01009 |
| SFA5.17 (EL12) | K01012 |
| SFA5.17 (EM11K) | K01111 |
| SFA5.17 (EM12) | K01112 |
| SFA5.17 (EM12K) | K01113 |
| SFA5.17 (EM13K) | K01313 |
| SFA5.17 (EM14K) | K01314 |
| SFA5.17 (EM15K) | K01515 |
| SFA5.18 (E70C-3X) | W07703 |
| SFA5.18 (E70C-6X) | W07706 |
| SFA5.18 (ER70S-2) | K10726 |
| SFA5.18 (ER70S-3) | K11022 |
| SFA5.18 (ER70S-4) | K11132 |
| SFA5.18 (ER70S-5) | K11357 |
| SFA5.18 (ER70S-6) | K11140 |
| SFA5.18 (ER70S-7) | K11125 |
| SFA5.19 (ER AZ101A) | M11101 |
| SFA5.19 (ER AZ61A) | M11611 |
| SFA5.19 (ER AZ92A) | M11922 |
| SFA5.19 (ER EZ33A) | M12331 |
| SFA5.20 (E6XT-1) | W06601 ob |
| SFA5.20 (E6XT-11) | W06611 ob |
| SFA5.20 (E6XT-13) | W06613 |
| SFA5.20 (E6XT-4) | W06604 ob |
| SFA5.20 (E6XT-5) | W06605 ob |
| SFA5.20 (E6XT-6) | W06606 ob |
| SFA5.20 (E6XT-7) | W06607 ob |
| SFA5.20 (E6XT-8) | W06608 ob |
| SFA5.20 (E7XT-1, E7XT-1M) | W07601 |
| SFA5.20 (E7XT-10) | W07610 |
| SFA5.20 (E7XT-11) | W07611 |
| SFA5.20 (E7XT-12, E7XT-12M) | W07612 |
| SFA5.20 (E7XT-13) | W07613 |
| SFA5.20 (E7XT-14) | W07614 |
| SFA5.20 (E7XT-2, E7XT-2M) | W07602 |
| SFA5.20 (E7XT-3) | W07603 |
| SFA5.20 (E7XT-4) | W07604 |
| SFA5.20 (E7XT-5, E7XT-5M) | W07605 |
| SFA5.20 (E7XT-6) | W07606 |
| SFA5.20 (E7XT-7) | W07607 |
| SFA5.20 (E7XT-8) | W07608 |
| SFA5.20 (E7XT-9, E7XT-9M) | W07609 |
| SFA5.21 (EFe5-A) | W77710 ob |
| SFA5.21 (EFe5-C) | W77310 ob |
| SFA5.21 (EFeCr-A1) | W74510 ob |
| SFA5.21 (ERCCoCr-A) | W73036 |
| SFA5.21 (ERCCoCr-B) | W73042 |
| SFA5.21 (ERCCoCr-C) | W73031 |
| SFA5.21 (ERCCoCr-E) | W73041 |
| SFA5.21 (ERCCoCr-G) | W73032 |
| SFA5.21 (ERCCuAl-A2) | W60618 |
| SFA5.21 (ERCCuAl-A3) | W60624 |
| SFA5.21 (ERCCuAl-C) | W60626 |
| SFA5.21 (ERCCuAl-D) | W61626 |
| SFA5.21 (ERCCuAl-E) | W62626 |
| SFA5.21 (ERCCuSi-A) | W60657 |
| SFA5.21 (ERCCuSn-A) | W60518 |
| SFA5.21 (ERCCuSn-D | W60524 |
| SFA5.21 (ERCFe-1) | W74030 |

| CROSS REFERENCE SPECIFICATIONS | UNIFIED NUMBER |
|---|---|
| SFA5.21 (ERCFe-1A) | W74031 |
| SFA5.21 (ERCFe-2) | W74032 |
| SFA5.21 (ERCFe-3) | W74033 |
| SFA5.21 (ERCFe-5) | W74035 |
| SFA5.21 (ERCFe-6) | W77530 |
| SFA5.21 (ERCFe-8) | W77538 |
| SFA5.21 (ERCFeCr-A) | W74531 |
| SFA5.21 (ERCFeCr-A10) | W74540 |
| SFA5.21 (ERCFeCr-A1A) | W74530 |
| SFA5.21 (ERCFeCr-A3A) | W74533 |
| SFA5.21 (ERCFeCr-A4) | W74534 |
| SFA5.21 (ERCFeCr-A5) | W74535 |
| SFA5.21 (ERCFeCr-A9) | W74539 |
| SFA5.21 (ERCFeMn-C) | W79230 |
| SFA5.21 (ERCFeMn-Cr) | W79730 |
| SFA5.21 (ERCFeMn-F) | W79630 |
| SFA5.21 (ERCFeMn-G) | W79231 |
| SFA5.21 (ERCFeMn-H) | W79232 |
| SFA5.21 (ERCNiCr-A) | W89634 |
| SFA5.21 (ERCNiCr-B) | W89635 |
| SFA5.21 (ERCNiCr-C) | W89636 |
| SFA5.21 (ERCNiCrFeCo) | W83032 |
| SFA5.21 (ERCNICrMo-5A) | W80032 ob |
| SFA5.21 (ERCNiCrMo-5A) | W86685 ob |
| SFA5.21 (ERCoCr-A) | R30006 |
| SFA5.21 (ERCoCr-B) | R30012 |
| SFA5.21 (ERCoCr-C) | R30001 |
| SFA5.21 (ERCoCr-E) | R30021 |
| SFA5.21 (ERCoCr-F) | R30002 |
| SFA5.21 (ERCoCr-G) | R30014 |
| SFA5.21 (ERCuAl-A2) | C61800 |
| SFA5.21 (ERCuAl-A3) | C62400 |
| SFA5.21 (ERCuAl-C) | C62580 |
| SFA5.21 (ERCuAl-D) | C62581 |
| SFA5.21 (ERCuAl-E) | C62582 |
| SFA5.21 (ERCuSi-A) | C65600 |
| SFA5.21 (ERCuSn-A) | C51800 |
| SFA5.21 (ERCuSn-D) | C52400 |
| SFA5.21 (ERFe-1 | T74000 |
| SFA5.21 (ERFe-1A) | T74001 |
| SFA5.21 (ERFe-2) | T74002 |
| SFA5.21 (ERFe-3) | T74003 |
| SFA5.21 (ERFe-5) | T74005 |
| SFA5.21 (ERFe-6) | T75006 |
| SFA5.21 (ERFe-8) | T75008 |
| SFA5.21 (ERNiCr-A) | N99644 |
| SFA5.21 (ERNiCr-B) | N99645 |
| SFA5.21 (ERNiCr-C) | N99646 |
| SFA5.21 (ERNiCr-D) | N99647 |
| SFA5.21 (ERNiCr-E) | N99648 |
| SFA5.21 (ERNiCrFeCo) | F46100 |
| SFA5.21 (ERNiCrMo-5A) | N10006 |
| SFA5.21 (RFe5-A) | W77730 ob |
| SFA5.22 (E307T0-3) | W30733 |
| SFA5.22 (E307TX-X) | W30731 |
| SFA5.22 (E308HMoT0-3) | W30830 |
| SFA5.22 (E308LMoT0-X, E308LMoTX-X) | W30838 |
| SFA5.22 (E308LT0-3) | W30837 |
| SFA5.22 (E308LTX-X, R308LT1-5) | W30835 |
| SFA5.22 (E308MoT0-3) | W30839 |
| SFA5.22 (E308MoTX-X) | W30832 |
| SFA5.22 (E308T0-3, E308HT0-3) | W30833 |
| SFA5.22 (E308TX-X, E308HT0-3) | W30831 |
| SFA5.22 (E309LCbT0-3) | W30934 |
| SFA5.22 (E309LCbTX-X) | W30932 |
| SFA5.22 (E309LMoT0-3, E309LMoTX-X) | W30938 |

| CROSS REFERENCE SPECIFICATIONS | UNIFIED NUMBER |
|---|---|
| SFA5.22 (E309LNiMoTX-X) | W30936 |
| SFA5.22 (E309LT0-3) | W30937 |
| SFA5.22 (E309LTX-X, R309LT1-5) | W30935 |
| SFA5.22 (E309MoTX-X) | W30939 |
| SFA5.22 (E309T0-3) | W30933 |
| SFA5.22 (E309TX-X) | W30931 |
| SFA5.22 (E310T0-3, E310TX-X) | W31031 |
| SFA5.22 (E312T0-3) | W31231 |
| SFA5.22 (E312TX-X) | W31331 |
| SFA5.22 (E316LKT0-3) | W31630 |
| SFA5.22 (E316LT0-3) | W31637 |
| SFA5.22 (E316LTX-X, R316LT1-5) | W31635 |
| SFA5.22 (E316T0-3) | W31633 |
| SFA5.22 (E316TX-X) | W31631 |
| SFA5.22 (E317LT0-3) | W31737 |
| SFA5.22 (E317LTX-X) | W31735 |
| SFA5.22 (E347T0-3) | W34733 |
| SFA5.22 (E347TX-X, R347T1-5) | W34731 |
| SFA5.22 (E409TX-X, E409T0-3) | W40931 |
| SFA5.22 (E410NiMoTX-X, E410NiMoT0-3) | W41036 |
| SFA5.22 (E410NiTiTX-X, E410NiTiT0-3) | W41038 |
| SFA5.22 (E410TX-X, E410T0-3) | W41031 |
| SFA5.22 (E430TX-X, E430T0-3) | W43031 |
| SFA5.22 (E502TX-X) | W50231 |
| SFA5.22 (E505TX-X) | W50431 |
| SFA5.22 (E2209T0-X) | W39239 |
| SFA5.22 (E2553T0-X) | W39533 |
| SFA5.23 (B2LR) | W52042 |
| SFA5.23 (EA1) | K11222 |
| SFA5.23 (EA1N) | K11122 |
| SFA5.23 (EA2) | K11223 |
| SFA5.23 (EA2N) | K11123 |
| SFA5.23 (EA3) | K11423 |
| SFA5.23 (EA3K) | K21451 |
| SFA5.23 (EA3KN) | K21351 |
| SFA5.23 (EA3N) | K11323 |
| SFA5.23 (EA4) | K11424 |
| SFA5.23 (EA4N) | K11324 |
| SFA5.23 (EB 9) | S50482 |
| SFA5.23 (EB1) | K11043 |
| SFA5.23 (EB1N) | K10943 |
| SFA5.23 (EB2) | K11172 |
| SFA5.23 (EB2H) | K23016 |
| SFA5.23 (EB2HN) | K23116 |
| SFA5.23 (EB2N) | K11072 |
| SFA5.23 (EB2RR) | K11173 |
| SFA5.23 (EB3) | K31115 |
| SFA5.23 (EB3LRN) | K31117 ob |
| SFA5.23 (EB3N) | K31015 |
| SFA5.23 (EB3RR) | K31116 ob |
| SFA5.23 (EB5) | K12187 |
| SFA5.23 (EB5N) | K12087 |
| SFA5.23 (EB6) | S50280 |
| SFA5.23 (EB6H) | S50180 |
| SFA5.23 (EB6HN) | S50181 |
| SFA5.23 (EB6N) | S50281 |
| SFA5.23 (EB8) | S50480 |
| SFA5.23 (EB8N) | S50481 |
| SFA5.23 (ECA1) | W17041 |
| SFA5.23 (ECA1N) | W17141 ob |
| SFA5.23 (ECA2) | W17042 |
| SFA5.23 (ECA2N) | W17142 ob |
| SFA5.23 (ECA3) | W17043 |
| SFA5.23 (ECA3N) | W17143 ob |
| SFA5.23 (ECA4) | W17044 |

| CROSS REFERENCE SPECIFICATIONS | UNIFIED NUMBER |
|---|---|
| SFA5.23 (ECA4N) | W17144 ob |
| SFA5.23 (ECB 9) | W50442 |
| SFA5.23 (ECB1) | W51040 |
| SFA5.23 (ECB1N) | W51041 ob |
| SFA5.23 (ECB2) | W52040 |
| SFA5.23 (ECB2H) | W52240 |
| SFA5.23 (ECB2HN) | W52241 ob |
| SFA5.23 (ECB2N) | W52041 ob |
| SFA5.23 (ECB3) | W53040 |
| SFA5.23 (ECB3LRN) | W53043 ob |
| SFA5.23 (ECB3N) | W53041 ob |
| SFA5.23 (ECB3RR) | W53042 ob |
| SFA5.23 (ECB4) | W53340 |
| SFA5.23 (ECB4N) | W53341 ob |
| SFA5.23 (ECB5) | W51340 |
| SFA5.23 (ECB5N) | W51341 ob |
| SFA5.23 (ECB6) | W50240 |
| SFA5.23 (ECB6H) | W50140 |
| SFA5.23 (ECB6HN) | W50141 ob |
| SFA5.23 (ECB6N) | W50241 ob |
| SFA5.23 (ECB8) | W50440 |
| SFA5.23 (ECB8N) | W50441 ob |
| SFA5.23 (ECF1) | W21150 |
| SFA5.23 (ECF1N) | W21151 ob |
| SFA5.23 (ECF2) | W20240 |
| SFA5.23 (ECF2N) | W20241 ob |
| SFA5.23 (ECF3) | W21140 |
| SFA5.23 (ECF3N) | W21141 ob |
| SFA5.23 (ECF4) | W20440 |
| SFA5.23 (ECF4N) | W20441 ob |
| SFA5.23 (ECF5) | W22540 |
| SFA5.23 (ECF6) | W22640 |
| SFA5.23 (ECF6N) | W22641 ob |
| SFA5.23 (ECM1) | W21240 |
| SFA5.23 (ECM1N) | W21241 ob |
| SFA5.23 (ECM2) | W21340 |
| SFA5.23 (ECM2N) | W21341 ob |
| SFA5.23 (ECM3) | W22240 |
| SFA5.23 (ECM3N) | W22241 ob |
| SFA5.23 (ECM4) | W22440 |
| SFA5.23 (ECM4N) | W22441 ob |
| SFA5.23 (ECM5) | W21345 |
| SFA5.23 (ECM6) | W21346 |
| SFA5.23 (ECNi1) | W21040 ob |
| SFA5.23 (ECNi1N) | W21041 ob |
| SFA5.23 (ECNi2) | W22040 |
| SFA5.23 (ECNi2N) | W22041 ob |
| SFA5.23 (ECNi3) | W23040 |
| SFA5.23 (ECNi3N) | W23041 ob |
| SFA5.23 (ECNi4) | W21250 |
| SFA5.23 (ECNi4N) | W21251 ob |
| SFA5.23 (ECNi5) | W21042 |
| SFA5.23 (ECW) | W20140 |
| SFA5.23 (EF1) | K11160 |
| SFA5.23 (EF1N) | K11060 |
| SFA5.23 (EF2) | K21450 |
| SFA5.23 (EF2N) | K21350 |
| SFA5.23 (EF3) | K21485 |
| SFA5.23 (EF3N) | K21385 |
| SFA5.23 (EF4) | K12048 |
| SFA5.23 (EF4N) | K11948 |
| SFA5.23 (EF5) | K41370 |
| SFA5.23 (EF6) | K21135 |
| SFA5.23 (EF6N) | K21035 |
| SFA5.23 (EL12) | K01012 |
| SFA5.23 (EL12N) | K00912 |
| SFA5.23 (EM2) | K10882 |
| SFA5.23 (EM2N) | K10982 |
| SFA5.23 (EM3) | K21015 |

## ASME (American Society of Mechanical Engineers) NUMBERS

| CROSS REFERENCE SPECIFICATIONS | UNIFIED NUMBER |
|---|---|
| SFA5.23 (EM3N) | K20915 |
| SFA5.23 (EM4) | K21030 |
| SFA5.23 (EM4N) | K20930 |
| SFA5.23 (EM12K) | K01113 |
| SFA5.23 (EM12KN) | K01013 |
| SFA5.23 (ENi1K) | K11058 |
| SFA5.23 (ENi1KN) | K10958 |
| SFA5.23 (ENi1N) | K10940 |
| SFA5.23 (ENi2) | K21010 |
| SFA5.23 (ENi2N) | K20910 |
| SFA5.23 (ENi3) | K31310 |
| SFA5.23 (ENi3N) | K31210 |
| SFA5.23 (ENi4) | K11485 |
| SFA5.23 (ENi4N) | K11385 |
| SFA5.23 (ENi5) | K11240 |
| SFA5.23 (ENi5N) | K11241 |
| SFA5.23 (ENil) | K11040 |
| SFA5.23 (EW) | K11245 |
| SFA5.24 (ERZr2) | R60702 |
| SFA5.24 (ERZr3) | R60704 |
| SFA5.24 (ERZr4) | R60705 |
| SFA5.24 (ERZr4) | R60707 ob |
| SFA5.25 (EA3K-EW) | K10945 |
| SFA5.25 (EH10K-EW) | K01010 |
| SFA5.25 (EH11K-EW) | K11140 |
| SFA5.25 (EH14-EW) | K11585 |
| SFA5.25 (EM5K-EW) | K10726 |
| SFA5.25 (EM12-EW) | K01112 |
| SFA5.25 (EM12K-EW) | K01113 |
| SFA5.25 (EM13K-EW) | K01313 |
| SFA5.25 (EM15K-EW) | K01515 |
| SFA5.25 (EWS-EW) | K11245 |
| SFA5.25 (EWT1) | W06040 |
| SFA5.25 (EWT2) | W20140 |
| SFA5.25 (EWT3) | W22340 |
| SFA5.26 (EG6XT-1, EG43XT-1) | W06301 |
| SFA5.26 (EG6XT-2, EG43XT-2) | W06302 |
| SFA5.26 (EG7XT-1, EG48XT-1) | W07301 |
| SFA5.26 (EG7XT-2, EG48XT-2) | W07302 |
| SFA5.26 (EGXXS-D2) | K10945 |
| SFA5.26 (EGXXS-1) | K01313 |
| SFA5.26 (EGXXS-2) | K10726 |
| SFA5.26 (EGXXS-3) | K11022 |
| SFA5.26 (EGXXS-5) | K11325 |
| SFA5.26 (EGXXS-5) | K11357 |
| SFA5.26 (EGXXS-6) | K11140 |
| SFA5.26 (EGXXT-Ni1) | W21033 |
| SFA5.26 (EGXXT-NM1) | W22334 |
| SFA5.26 (EGXXT-NM2) | W22333 |
| SFA5.26 (EGXXT-W) | W20131 |
| SFA5.27 (RCu) | C18980 |
| SFA5.27 (RCuNi) | C71581 |
| SFA5.28 (E70C-B2L) | W52130 |
| SFA5.28 (E70C-Ni2, E80C-Ni2) | W22030 |
| SFA5.28 (E80C-A1) | W17030 ob |
| SFA5.28 (E80C-B2) | W52030 |
| SFA5.28 (E80C-B3L) | W53130 |
| SFA5.28 (E80C-Ni1) | W21030 |
| SFA5.28 (E80C-Ni3) | W23030 |
| SFA5.28 (E90C-B3) | W53030 |
| SFA5.28 (E90C-D2) | W19230 |
| SFA5.28 (ER70S-A1) | K11235 |
| SFA5.28 (ER70S-B2L) | K20500 |
| SFA5.28 (ER80S-B2) | K20900 |
| SFA5.28 (ER80S-B3L) | K30560 |
| SFA5.28 (ER80S-B6) | S50280 |
| SFA5.28 (ER80S-B8) | S50480 |
| SFA5.28 (ER80S-D2) | K10945 |
| SFA5.28 (ER80S-Ni2) | K21240 |
| SFA5.28 (ER80S-Ni3) | K31240 |
| SFA5.28 (ER80S-Nil) | K11260 |
| SFA5.28 (ER90S-B3) | K30960 |
| SFA5.28 (ER90S-B9) | S59180 ob |
| SFA5.28 (ER100S-1) | K10882 |
| SFA5.28 (ER100S-2) | K11250 ob |
| SFA5.28 (ER110S-1) | K21015 |
| SFA5.28 (ER120S-1) | K21030 |
| SFA5.29 (E1XXT5-K4) | W22235 |
| SFA5.29 (E6XT5-B8) | W50431 |
| SFA5.29 (E7XT11-K2) | W21239 |
| SFA5.29 (E7XT4-K2) | W21234 |
| SFA5.29 (E7XT5-A1) | W17035 |
| SFA5.29 (E7XT6-Ni1) | W21036 |
| SFA5.29 (E7XT7-K2) | W21237 |
| SFA5.29 (E7XT8-K2) | W21238 |
| SFA5.29 (E7XT8-Ni1) | W21038 |
| SFA5.29 (E7XT8-Ni2, E8XT8-Ni2) | W22038 |
| SFA5.29 (E8XT1-A1) | W17031 |
| SFA5.29 (E8XT1-B1) | W51031 |
| SFA5.29 (E8XT1-B1L) | W51131 |
| SFA5.29 (E8XT1-B2) | W52031 |
| SFA5.29 (E8XT1-B2H) | W52231 |
| SFA5.29 (E8XT1-B2L) | W52131 |
| SFA5.29 (E8XT1-W2) | W20131 |
| SFA5.29 (E8XT5-B2) | W52035 |
| SFA5.29 (E8XT5-B2L) | W52135 |
| SFA5.29 (E8XT5-B6) | W50231 |
| SFA5.29 (E8XT5-B6L, B6LM) | W50230 |
| SFA5.29 (E8XT5-B8L, -B8LM) | W50430 |
| SFA5.29 (E8XT5-K1) | W21135 |
| SFA5.29 (E8XT5-Ni1) | W21035 |
| SFA5.29 (E8XT5-Ni2) | W22035 |
| SFA5.29 (E9XT1-B3, E10XT1-B3) | W53031 |
| SFA5.29 (E9XT1-B3H) | W53231 |
| SFA5.29 (E9XT1-B3L) | W53131 |
| SFA5.29 (E9XT1-D1) | W19131 |
| SFA5.29 (E9XT1-D3) | W19331 |
| SFA5.29 (E9XT5-B3) | W53035 |
| SFA5.29 (E9XT5-D2, E10XT5-D2) | W19235 |
| SFA5.29 (E9XT8-K8) | W21438 |
| SFA5.29 (E10XT1-K7) | W22051 |
| SFA5.29 (E10XT1-K9, E10XT1-K9M) | W23230 |
| SFA5.29 (E11XT1-K4) | W22231 |
| SFA5.29 (E12XT1-K5) | W21631 |
| SFA5.29 (E100T5-D2) | W10235 ob |
| SFA5.29 (EXXT1-K2) | W21231 |
| SFA5.29 (EXXT1-Ni1) | W21031 |
| SFA5.29 (EXXT1-Ni2) | W22031 |
| SFA5.29 (EXXT1-Ni3) | W23031 ob |
| SFA5.29 (EXXT5-K2) | W21235 |
| SFA5.29 (EXXT5-K6, -K6M) | W21045 |
| SFA5.29 (EXXT5-Ni3) | W23035 |
| SFA5.29 (EXXT8-K6) | W21048 |
| SFA5.29 (EXXT11-Ni3) | W23039 |
| SFA5.29 (EXXXT1-K3) | W21331 |
| SFA5.29 (EXXXT5-K3) | W21335 |
| SFA5.30 (IN6A) | N07092 |
| SFA5.30 (IN52) | N06052 |
| SFA5.30 (IN60) | N04060 |
| SFA5.30 (IN61) | N02061 |
| SFA5.30 (IN62) | N06062 |
| SFA5.30 (IN67) | C71581 |
| SFA5.30 (IN82) | N06082 |
| SFA5.30 (IN308) | S30880 |
| SFA5.30 (IN308L) | S30883 |
| SFA5.30 (IN309) | S30980 |
| SFA5.30 (IN309L) | S30983 |
| SFA5.30 (IN310) | S31080 |
| SFA5.30 (IN312) | S31380 |
| SFA5.30 (IN316) | S31680 |
| SFA5.30 (IN316L) | S31683 |
| SFA5.30 (IN348) | S34780 |
| SFA5.30 (IN502) | S50280 |
| SFA5.30 (IN515) | K20900 |
| SFA5.30 (IN521) | K30960 |
| SFA5.30 (INMs1) | K10726 |
| SFA5.30 (INMs2) | K01313 |
| SFA5.30 (INMs3) | K11140 |
| SFA5.34 (ENiCr3T-X) | W86082 |
| SFA5.34 (ENiCrFe1T-X) | W86132 |
| SFA5.34 (ENiCrFe2T-X) | W86133 |
| SFA5.34 (ENiCrFe3T-X) | W86182 |
| SFA5.34 (ENiCrMo 3T-X) | W86112 |
| SFA5.34 (ENiCrMo4T-X) | W80276 |
| SFA5.34 (ENiFeCr2T-X) | W87718 |
| SFAA5.12 (EWLa-1.5) | R07942 |
| SFAA5.12 (EWLa-2) | R07943 |

# Cross Index of Commonly Known Documents Which Describe Materials Same as or Similar to These Covered by UNS Numbers

## ASTM (American Society for Testing and Materials) NUMBERS

**(The Symbol *ob* Indicates an Obsolete UNS Number.)**

**NOTE: ASTM Standards are grouped according to the committee responsible: A=Ferrous Metals and Alloys, B=Nonferrous Metals and Alloys, F=Specialty Applications**

# ASTM

| CROSS REFERENCE SPECIFICATIONS | UNIFIED NUMBER |
|---|---|
| A1 (61-80) | K06100 |
| A1 (81-89) | K07000 |
| A1 (90-114) | K07301 |
| A1 (115 and over) | K07500 |
| A1xyz (B) (pending new spec) | S41005 |
| A1xyz (C) (pending new spec) | S41428 |
| A1xyz (D) (pending new spec) | S42225 |
| A2 (A) | K06703 |
| A2 (B) | K07700 |
| A2 (C) | K08201 |
| A21 (F, G, H) | K05200 |
| A21 (U) | K04700 |
| A27 (60-30) | J03000 |
| A27 (65-35) | J03001 |
| A27 (70-36) | J03501 |
| A27 (70-40) | J02501 |
| A27 (N1, U-60-30) | J02500 |
| A27 (N2) | J03500 |
| A29 (12L13) | G12134 |
| A29 (12L14) | G12144 |
| A29 (43B40) | G43406 |
| A29 (50B44) | G50441 |
| A29 (50B46) | G50461 |
| A29 (50B60) | G50601 |
| A29 (51B60) | G51601 |
| A29 (81B45) | G81451 |
| A29 (94B17) | G94171 |
| A29 (94B30) | G94301 |
| A29 (1005) | G10050 |
| A29 (1006) | G10060 |
| A29 (1008) | G10080 |
| A29 (1010) | G10100 |
| A29 (1011) | G10110 |
| A29 (1012) | G10120 |
| A29 (1013) | G10130 |
| A29 (1015) | G10150 |
| A29 (1016) | G10160 |
| A29 (1017) | G10170 |
| A29 (1018) | G10180 |
| A29 (1019) | G10190 |
| A29 (1020) | G10200 |
| A29 (1021) | G10210 |
| A29 (1022) | G10220 |
| A29 (1023) | G10230 |
| A29 (1025) | G10250 |
| A29 (1026) | G10260 |
| A29 (1029) | G10290 |
| A29 (1030) | G10300 |
| A29 (1034) | G10340 |
| A29 (1035) | G10350 |
| A29 (1037) | G10370 |
| A29 (1038) | G10380 |
| A29 (1039) | G10390 |
| A29 (1040) | G10400 |
| A29 (1042) | G10420 |
| A29 (1043) | G10430 |
| A29 (1044) | G10440 |
| A29 (1045) | G10450 |
| A29 (1046) | G10460 |
| A29 (1049) | G10490 |
| A29 (1050) | G10500 |
| A29 (1053) | G10530 |
| A29 (1055) | G10550 |
| A29 (1059) | G10590 |
| A29 (1060) | G10600 |
| A29 (1064) | G10640 |
| A29 (1065) | G10650 |
| A29 (1069) | G10690 |
| A29 (1070) | G10700 |
| A29 (1074) | G10740 |
| A29 (1075) | G10750 |
| A29 (1078) | G10780 |
| A29 (1080) | G10800 |
| A29 (1084) | G10840 |
| A29 (1086) | G10860 |
| A29 (1090) | G10900 |
| A29 (1095) | G10950 |
| A29 (1108) | G11080 |
| A29 (1109) | G11090 |
| A29 (1110) | G11100 |
| A29 (1116) | G11160 |
| A29 (1117) | G11170 |
| A29 (1118) | G11180 |
| A29 (1119) | G11190 |
| A29 (1132) | G11320 |
| A29 (1137) | G11370 |
| A29 (1139) | G11390 |
| A29 (1140) | G11400 |
| A29 (1141) | G11410 |
| A29 (1144) | G11440 |
| A29 (1145) | G11450 |
| A29 (1146) | G11460 |
| A29 (1151) | G11510 |
| A29 (1211) | G12110 |
| A29 (1212) | G12120 |
| A29 (1213) | G12130 |
| A29 (1215) | G12150 |
| A29 (1345) | G13450 |
| A29 (1513) | G15130 |
| A29 (1518) | G15180 |
| A29 (1522) | G15220 |
| A29 (1524) | G15240 |
| A29 (1525) | G15250 |
| A29 (1526) | G15260 |
| A29 (1527) | G15270 |
| A29 (1536) | G15360 |
| A29 (1541) | G15410 |
| A29 (1547) | G15470 |
| A29 (1548) | G15480 |
| A29 (1551) | G15510 |
| A29 (1552) | G15520 |
| A29 (1561) | G15610 |
| A29 (1566) | G15660 |
| A29 (1572) | G15720 |
| A29 (4012) | G40120 |
| A29 (4023) | G40230 |
| A29 (4024) | G40240 |
| A29 (4027) | G40270 |
| A29 (4028) | G40280 |
| A29 (4037) | G40370 |
| A29 (4042) | G40420 |
| A29 (4047) | G40470 |
| A29 (4118) | G41180 |
| A29 (4120) | G41200 |
| A29 (4121) | G41210 |
| A29 (4130) | G41300 |
| A29 (4135) | G41350 |
| A29 (4137) | G41370 |
| A29 (4140) | G41400 |
| A29 (4142) | G41420 |
| A29 (4145) | G41450 |
| A29 (4147) | G41470 |
| A29 (4150) | G41500 |
| A29 (4161) | G41610 |
| A29 (4320) | G43200 |
| A29 (4340) | G43400 |
| A29 (4419) | G44190 |
| A29 (4422) | G44220 |
| A29 (4427) | G44270 |
| A29 (4615) | G46150 |
| A29 (4620) | G46200 |
| A29 (4621) | G46210 |
| A29 (4626) | G46260 |
| A29 (4718) | G47180 |
| A29 (4720) | G47200 |
| A29 (4815) | G48150 |
| A29 (4817) | G48170 |
| A29 (4820) | G48200 |
| A29 (5015) | G50150 |
| A29 (5046) | G50460 |
| A29 (5115) | G51150 |
| A29 (5120) | G51200 |
| A29 (5130) | G51300 |
| A29 (5132) | G51320 |
| A29 (5135) | G51350 |
| A29 (5140) | G51400 |
| A29 (5145) | G51450 |
| A29 (5147) | G51470 |
| A29 (5150) | G51500 |
| A29 (5155) | G51550 |
| A29 (5160) | G51600 |
| A29 (6118) | G61180 |
| A29 (6150) | G61500 |
| A29 (8115) | G81150 |
| A29 (8615) | G86150 |
| A29 (8617) | G86170 |
| A29 (8620) | G86200 |
| A29 (8622) | G86220 |
| A29 (8625) | G86250 |
| A29 (8627) | G86270 |
| A29 (8630) | G86300 |
| A29 (8637) | G86370 |
| A29 (8640) | G86400 |
| A29 (8642) | G86420 |
| A29 (8645) | G86450 |
| A29 (8650) | G86500 |
| A29 (8655) | G86550 |
| A29 (8660) | G86600 |
| A29 (8720) | G87200 |
| A29 (8740) | G87400 |
| A29 (8822) | G88220 |
| A29 (9254) | G92540 |
| A29 (9255) | G92550 |
| A29 (9259) | G92590 |
| A29 (9259) | H92590 |
| A29 (9260) | G92600 |
| A29 (E4340) | G43406 |
| A29 (E9310) | G93106 |
| A29 (E50100) | G50986 |
| A29 (E51100) | G51986 |
| A29 (E52100) | G52986 |
| A29 (M1008) | G10080 |
| A29 (M1010) | G10100 |
| A29 (M1012) | G10120 |
| A29 (M1015) | G10150 |
| A29 (M1017) | G10170 |
| A29 (M1020) | G10200 |
| A29 (M1023) | G10230 |
| A29 (M1025) | G10250 |
| A31 (B) | K03100 |
| A32 | S35135 |
| A36 | K02595 |
| A36 | K02596 |
| A36 | K02597 |
| A36 | K02598 |

| CROSS REFERENCE SPECIFICATIONS | UNIFIED NUMBER |
|---|---|
| A36 | K02599 |
| A36 (SHAPES) | K02600 |
| A47 (32510) | F22200 |
| A47 (35018) | F22400 |
| A48 (20) | F11401 |
| A48 (25) | F11701 |
| A48 (30) | F12101 |
| A48 (35) | F12401 |
| A48 (40) | F12801 |
| A48 (45) | F13101 |
| A48 (50) | F13501 |
| A48 (55) | F13801 |
| A48 (60) | F14101 |
| A49 | K04701 |
| A53 (E-B, S-B) | K03005 |
| A65 | K11210 |
| A67 | K01505 |
| A67 (2) | K06002 |
| A105 | K03504 ob |
| A106 (B) | K03006 |
| A108 | G10100 |
| A108 (12L14) | G12144 |
| A108 (1008) | G10080 |
| A108 (1010) | G10100 |
| A108 (1015) | G10150 |
| A108 (1016) | G10160 |
| A108 (1018) | G10180 |
| A108 (1020) | G10200 |
| A108 (1022) | G10220 |
| A108 (1025) | G10250 |
| A108 (1034) | G10340 |
| A108 (1035) | G10350 |
| A108 (1040) | G10400 |
| A108 (1045) | G10450 |
| A108 (1050) | G10500 |
| A108 (1090) | G10900 |
| A108 (1095) | G10950 |
| A108 (1117) | G11170 |
| A108 (1118) | G11180 |
| A108 (1137) | G11370 |
| A108 (1141) | G11410 |
| A108 (1144) | G11440 |
| A108 (1151) | G11510 |
| A108 (1211) | G12110 |
| A108 (1212) | G12120 |
| A108 (1213) | G12130 |
| A108 (1215) | G12150 |
| A109 (1, 2, 3) | K02500 |
| A109 (4,5) | K01507 |
| A126 (A) | F11501 |
| A126 (B) | F12102 |
| A126 (C) | F12802 |
| A128 (A) | J91109 |
| A128 (B-1) | J91119 |
| A128 (B-2) | J91129 |
| A128 (B-3) | J91139 |
| A128 (B-4) | J91149 |
| A128 (C) | J91309 |
| A128 (D) | J91459 |
| A128 (E-1) | J91249 |
| A128 (E-2) | J91339 |
| A128 (F) | J91340 |
| A131 (A) | K02300 |
| A131 (AH32, DH32, EH32) | K11846 |
| A131 (AH36, DH36, EH36) | K11852 |
| A131 (B) | K02102 |
| A131 (CS, DS) | K01601 |
| A131 (D) | K02101 |
| A131 (E) | K01801 |

| CROSS REFERENCE SPECIFICATIONS | UNIFIED NUMBER |
|---|---|
| A135 (B) | K03018 |
| A139 (B) | K03003 |
| A139 (C) | K03004 |
| A139 (D) | K03010 |
| A139 (E) | K03012 |
| A148 (80-40) | D50400 |
| A148 (80-50) | D50500 |
| A148 (90-60) | D50600 |
| A148 (105-85) | D50850 |
| A148 (115-95) | D50950 |
| A148 (130-115) | D51150 |
| A148 (135-125) | D51250 |
| A148 (150-135) | D51350 |
| A148 (160-145) | D51450 |
| A148 (165-150) | D51500 |
| A148 (165-150L) | D51501 |
| A148 (210-180) | D51800 |
| A148 (210-180L) | D51801 |
| A148 (260-210) | D52100 |
| A148 (260-210L) | D52101 |
| A159 (G1800) | F10004 |
| A159 (G2500) | F10005 |
| A159 (G2500a) | F10009 |
| A159 (G3000) | F10006 |
| A159 (G3500) | F10007 |
| A159 (G3500b) | F10010 |
| A159 (G3500c) | F10011 |
| A159 (G4000) | F10008 |
| A159 (G4000d) | F10012 |
| A167 (302B) | S30215 |
| A167 (308) | S30800 |
| A167 (309) | S30900 |
| A167 (310) | S31000 |
| A176 (403) | S40300 |
| A176 (420) | S42000 |
| A176 (422) | S42200 |
| A176 (431) | S43100 |
| A176 (442) | S44200 |
| A176 (446) | S44600 |
| A178 (A) | K01200 |
| A178 (C) | K03503 |
| A178 (D) | K02709 |
| A179 | K01200 |
| A181 | K03502 |
| A182 | N08367 |
| A182 (56) | S33228 |
| A182 (304) | S30400 |
| A182 (304H) | S30409 |
| A182 (304L) | S30403 |
| A182 (304N) | S30451 |
| A182 (310) | S31000 |
| A182 (316) | S31600 |
| A182 (316H) | S31609 |
| A182 (316L) | S31603 |
| A182 (316N) | S31651 |
| A182 (321) | S32100 |
| A182 (321H) | S32109 |
| A182 (347) | S34700 |
| A182 (347H) | S34709 |
| A182 (348) | S34800 |
| A182 (348H) | S34809 |
| A182 (429) | S42900 |
| A182 (430) | S43000 |
| A182 (B5, F7) | S50100 |
| A182 (F1) | K12822 |
| A182 (F2) | K12122 |
| A182 (F3V) | K31830 |
| A182 (F3VCb) | K31390 |
| A182 (F5) | K41545 |

| CROSS REFERENCE SPECIFICATIONS | UNIFIED NUMBER |
|---|---|
| A182 (F5a) | K42544 |
| A182 (F6a) | S41000 |
| A182 (F6b) | S41026 |
| A182 (F6NM) | S41500 |
| A182 (F9) | K90941 |
| A182 (F-10) | S33100 |
| A182 (F11-1) | K11597 |
| A182 (F11-2) | K11572 |
| A182 (F12-1) | K11562 |
| A182 (F12-2) | K11564 |
| A182 (F20) | N08020 |
| A182 (F21) | K31545 |
| A182 (F22-1, F22-2, F22-3) | K21590 |
| A182 (F22V) | K31835 |
| A182 (F44) | S31254 |
| A182 (F45) | S30815 |
| A182 (F46) | S30600 |
| A182 (F47) | S31725 |
| A182 (F48) | S31726 |
| A182 (F49) | S34565 |
| A182 (F50) | S31200 |
| A182 (F52) | S32950 |
| A182 (F54) | S39274 |
| A182 (F56) | S33228 |
| A182 (F57) | S39277 |
| A182 (F91) | K90901 |
| A182 (F304LN) | S30453 |
| A182 (F316LN) | S31653 |
| A182 (F317) | S31700 |
| A182 (F317L) | S31703 |
| A182 (FR) | K22035 |
| A182 (FXM-11) | S21904 |
| A182 (FXM-19) | S20910 |
| A182 (FXM-27Cb) | S44627 |
| A182 (XM-27) | S44625 ob |
| A183 | K03015 |
| A183 (1-Bolts, 1-Nuts) | G11190 |
| A183 (2-Nuts) | G10420 |
| A183 (2-Nuts) | G10430 |
| A183 (2-Nuts) | G10440 |
| A183 (2-Nuts) | G10450 |
| A183 (F53) | S32750 |
| A184 (TPXM-19) | S20910 |
| A192 | K01201 |
| A193 | S31254 |
| A193 (304, B8, B8A) | S30400 |
| A193 (316, B8M, B8MA, B8M2, B8M3) | S31600 |
| A193 (321, B8T, B8TA) | S32100 |
| A193 (347, B8CA) | S34700 |
| A193 (410, B6, B6X) | S41000 |
| A193 (501, B5) | S50100 |
| A193 (B7, B7M) | G41400 |
| A193 (B8LN, B8LNA) | S30453 |
| A193 (B8MLN, B8MLNA) | S31653 |
| A193 (B8MN, B8MNA) | S31651 |
| A193 (B8N, B8NA) | S30451 |
| A193 (B8P, B8PA) | S30500 |
| A193 (B8R, B8RA) | S20910 |
| A193 (B8S, B8SA) | S21800 |
| A193 (B16) | K14072 |
| A194 (1) | K01503 |
| A194 (2, 2HM, 2H) | K04002 |
| A194 (4) | K14510 |
| A194 (7, 7M) | G41400 |
| A194 (8LN, 8LNA) | S30453 |
| A194 (8MLN, 8MLNA) | S31653 |
| A194 (8MN, 8MNA) | S31651 |
| A194 (8N, 8NA) | S30451 |

| CROSS REFERENCE SPECIFICATIONS | UNIFIED NUMBER |
|---|---|
| A194 (8P, 8PA) | S30500 |
| A194 (8R, 8RA) | S20910 |
| A194 (8S, 8SA) | S21800 |
| A194 (16) | K14072 |
| A194 (303, 8F, 8FA) | S30300 |
| A194 (303 Se, 8F, 8FA) | S30323 |
| A194 (304, 8, 8A) | S30400 |
| A194 (316, 8M, 8MA) | S31600 |
| A194 (321, 8T, 8TA) | S32100 |
| A194 (347, 8C, 8CA) | S34700 |
| A194 (410, 6) | S41000 |
| A194 (416, 6F) | S41600 |
| A194 (416Se, 6F) | S41623 |
| A194 (501, 3) | S50100 |
| A197 | F22000 |
| A202 (A) | K11742 |
| A202 (B) | K12542 |
| A203 (A) | K21703 |
| A203 (B) | K22103 |
| A203 (D) | K31718 |
| A203 (E) | K32018 |
| A204 (A) | K11820 |
| A204 (B) | K12020 |
| A204 (C) | K12320 |
| A209 (T1) | K11522 |
| A209 (T1a) | K12023 |
| A209 (T1b) | K11422 |
| A210 (A-1) | K02707 |
| A210 (C) | K03501 |
| A213 | S21500 |
| A213 | S30432 |
| A213 | S30815 |
| A213 | S31050 |
| A213 | S31272 |
| A213 | S31725 |
| A213 | S31726 |
| A213 | S32615 |
| A213 | S33228 |
| A213 | S34710 |
| A213 | S70003 |
| A213 (18Cr-2Mo) | S44400 |
| A213 (91) | K90901 |
| A213 (304) | S30400 |
| A213/A213M | S31002 |
| A213 (T2) | K11547 |
| A213 (T5) | K41545 |
| A213 (T5b) | K51545 |
| A213 (T5c) | K41245 |
| A213 (T11) | K11597 |
| A213 (T12) | K11562 |
| A213 (T17) | K12047 |
| A213 (T21) | K31545 |
| A213 (T22) | K21590 |
| A213 (T23) | K40712 |
| A213 (T24) | K30736 |
| A213 (T92) | K92400 |
| A213 (T122) | K91271 |
| A213 (T911) | K91061 |
| A213 (TP201) | S20100 |
| A213 (TP202) | S20200 |
| A213 (TP304H) | S30409 |
| A213 (TP304L) | S30403 |
| A213 (TP304LN) | S30453 |
| A213 (TP304N) | S30451 |
| A213 (TP309Cb) | S30940 |
| A213 (TP309H) | S30909 |
| A213 (TP309HCb) | S30941 |
| A213 (TP309S) | S30908 |
| A213 (TP310) | S31000 |
| A213 (TP310Cb) | S31040 |
| A213 (TP310H) | S31009 |
| A213 (TP310HCb) | S31041 |
| A213 (TP310HCbN) | S31042 |
| A213 (TP310S) | S31008 |
| A213 (TP316) | S31600 |
| A213 (TP316L) | S31603 |
| A213 (TP316LN) | S31653 |
| A213 (TP316N) | S31651 |
| A213 (TP317) | S31700 |
| A213 (TP317L) | S31703 |
| A213 (TP321) | S32100 |
| A213 (TP321H) | S32109 |
| A213 (TP347) | S34700 |
| A213 (TP347H) | S34709 |
| A213 (TP347LN) | S34751 |
| A213 (TP348) | S34800 |
| A213 (TP348H) | S34809 |
| A213 (XM-15) | S38100 |
| A214 | K01807 |
| A216 (WCA) | J02502 |
| A216 (WCB) | J03002 |
| A216 (WCC) | J02503 |
| A217 (C5) | J42045 |
| A217 (C12) | J82090 |
| A217 (C12A) | J80490 |
| A217 (C12A) | J84090 |
| A217 (CA-15) | J91150 |
| A217 (WC 4) | J12082 |
| A217 (WC 6) | J12072 |
| A217 (WC1) | J12524 |
| A217 (WC5) | J22000 |
| A217 (WC9) | J21890 |
| A217 (WC11) | J11872 |
| A220 (40010) | F22830 |
| A220 (45006) | F23131 |
| A220 (45008) | F23130 |
| A220 (50005) | F23530 |
| A220 (60004) | F24130 |
| A220 (70003) | F24830 |
| A220 (80002) | F25530 |
| A220 (90001) | F26230 |
| A225/A225M (C) | K12524 |
| A225 (D) | K12004 |
| A227 | K06501 |
| A228 | K08500 |
| A228 | K08501 |
| A229 | K07001 |
| A230 | K06701 |
| A231 | K15048 |
| A232 | K15047 |
| A234 (WP1) | K12821 |
| A234 (WP5) | K41545 |
| A234 (WP9) | K90941 |
| A234 (WP12) | K12062 |
| A234 (WP22) | K21590 |
| A234 (WP91) | K90901 |
| A234 (WPB) | K03006 |
| A234 (WPC) | K03501 |
| A234 (WPR) | K22035 |
| A240 | N08020 |
| A240 | N08367 |
| A240 | N08800 |
| A240 | N08810 |
| A240 | N08811 |
| A240 | N08904 |
| A240 | N08926 |
| A240 | S20103 |
| A240 | S20153 |
| A240 | S20161 |
| A240 | S20400 |
| A240 | S21800 |
| A240 | S30103 |
| A240 | S30153 |
| A240 | S30415 |
| A240 | S30600 |
| A240 | S30601 |
| A240 | S30615 |
| A240 | S30815 |
| A240 | S31050 |
| A240 | S31200 |
| A240 | S31254 |
| A240 | S31260 |
| A240 | S31266 |
| A240 | S31803 |
| A240 | S32001 |
| A240 | S32050 |
| A240 | S32205 |
| A240 | S32304 |
| A240 | S32520 |
| A240 | S32550 |
| A240 | S32615 |
| A240 | S32654 |
| A240 | S32750 |
| A240 | S32760 |
| A240 | S32803 |
| A240 | S32950 |
| A240 | S33228 |
| A240 | S34565 |
| A240 | S35045 |
| A240 | S35135 |
| A240 | S35315 |
| A240 | S38815 |
| A240 | S40910 |
| A240 | S40920 |
| A240 | S40930 |
| A240 | S40945 |
| A240 | S40975 |
| A240 | S40977 |
| A240 | S41003 |
| A240 | S41045 |
| A240 | S41050 |
| A240 | S41500 |
| A240 | S42035 |
| A240 | S43932 |
| A240 | S43940 |
| A240 | S44400 |
| A240 | S44500 |
| A240 | S44635 |
| A240 | S44660 |
| A240 | S44700 |
| A240 | S44735 |
| A240 | S44800 |
| A240 | S46800 |
| A240 (201) | S20100 |
| A240 (202) | S20200 |
| A240 (302) | S30200 |
| A240 (304) | S30400 |
| A240 (304H) | S30409 |
| A240 (304L) | S30403 |
| A240 (304LN) | S30453 |
| A240 (304N) | S30451 |
| A240 (305) | S30500 |
| A240 (309Cb) | S30940 |
| A240 (309H) | S30909 |
| A240 (309HCb) | S30941 |
| A240 (309S) | S30908 |
| A240 (310Cb) | S31040 |

| CROSS REFERENCE SPECIFICATIONS | UNIFIED NUMBER |
|---|---|
| A240 (310H) | S31009 |
| A240 (310S) | S31008 |
| A240 (316) | S31600 |
| A240 (316 Ti) | S31635 |
| A240 (316Cb) | S31640 |
| A240 (316H) | S31609 |
| A240 (316L) | S31603 |
| A240 (316LN) | S31653 |
| A240 (316N) | S31651 |
| A240 (317) | S31700 |
| A240 (317L) | S31703 |
| A240 (317LM) | S31725 |
| A240 (317LMN) | S31726 |
| A240 (317LN) | S31753 |
| A240 (321) | S32100 |
| A240 (321H) | S32109 |
| A240 (329) | S32900 |
| A240 (334) | S33400 |
| A240 (347) | S34700 |
| A240 (347H) | S34709 |
| A240 (348) | S34800 |
| A240 (348H) | S34809 |
| A240 (405) | S40500 |
| A240 (410) | S41000 |
| A240 (410S) | S41008 |
| A240 (429) | S42900 |
| A240 (430) | S43000 |
| A240 (434) | S43400 |
| A240 (436) | S43600 |
| A240 (439) | S43035 |
| A240 9310HCb) | S31041 |
| A240/A240M | S35135 |
| A240M | S31266 |
| A240M | S32205 |
| A240M | S32520 |
| A240M | S40975 |
| A240 (XM-15) | S38100 |
| A240 (XM-17) | S21600 |
| A240 (XM-18) | S21603 |
| A240 (XM-19) | S20910 |
| A240 (XM-21) | S30452 |
| A240 (XM-26) | S31100 |
| A240 (XM-27) | S44627 |
| A240 (XM-29) | S24000 |
| A240 (XM-31) | S21400 |
| A240 (XM-33) | S44626 |
| A242 (1) | K11510 |
| A245 | S30415 |
| A249 | N08367 |
| A249 | N08904 |
| A249 | N08926 |
| A249 | S30415 |
| A249 | S30615 |
| A249 | S30815 |
| A249 | S31050 |
| A249 | S31254 |
| A249 | S31725 |
| A249 | S31726 |
| A249 | S32654 |
| A249 | S32906 |
| A249 | S33228 |
| A249 | S34565 |
| A249 | S35315 |
| A249 (304) | S30400 |
| A249 (304H) | S30409 |
| A249 (304L) | S30403 |
| A249 (304N) | S30451 |
| A249 (305) | S30500 |
| A249 (310H) | S31009 |
| A249 (316) | S31600 |
| A249 (316H) | S31609 |
| A249 (316L) | S31603 |
| A249 (316N) | S31651 |
| A249 (317) | S31700 |
| A249 (321) | S32100 |
| A249 (321H) | S32109 |
| A249 (347) | S34700 |
| A249 (347H) | S34709 |
| A249 (348) | S34800 |
| A249 (348H) | S34809 |
| A249 (TP201) | S20100 |
| A249 (TP202) | S20200 |
| A249 (TP304LN) | S30453 |
| A249 (TP309Cb) | S30940 |
| A249 (TP309H) | S30909 |
| A249 (TP309HCb) | S30941 |
| A249 (TP309S) | S30908 |
| A249 (TP310Cb) | S31040 |
| A249 (TP310HCb) | S31041 |
| A249 (TP310S) | S31008 |
| A249 (TP316LN) | S31653 |
| A249 (TP317L) | S31703 |
| A249 (XM-15) | S38100 |
| A249 (XM-19) | S20910 |
| A249 (XM-29) | S24000 |
| A250 (T1) | K11522 |
| A250 (T1a) | K12023 |
| A250 (T1b) | K11422 |
| A250 (T2) | K11547 |
| A250 (T11) | K11597 |
| A250 (T12) | K11562 |
| A250 (T22) | K21590 |
| A254 | K01001 |
| A264 (304) | S30400 |
| A266 (1, 2) | K03506 |
| A266 (3) | K05001 |
| A266 (4) | K03017 |
| A268 | S32803 |
| A268 | S40800 |
| A268 | S41500 |
| A268 | S44627 |
| A268 | S44735 |
| A268 (18Cr-2Mo) | S44400 |
| A268 (25-4-4) | S44635 |
| A268 (26-3-3) | S44660 |
| A268 (29-4) | S44700 |
| A268 (29-4-2) | S44800 |
| A268 (405) | S40500 |
| A268 (409) | S40900 |
| A268 (410) | S41000 |
| A268 (429) | S42900 |
| A268 (430) | S43000 |
| A268 (430 Ti) | S43036 |
| A268 (439) | S43035 |
| A268 (443) | S44300 |
| A268 (446-1, 446-2) | S44600 |
| A268 (XM-27) | S44625 ob |
| A268 (XM-33) | S44626 |
| A269 | N08367 |
| A269 | N08904 |
| A269 | N08926 |
| A269 | S30600 |
| A269 | S31254 |
| A269 | S31725 |
| A269 | S31726 |
| A269 | S32654 |
| A269 | S34565 |
| A269 (304) | S30400 |
| A269 (304L) | S30403 |
| A269 (316) | S31600 |
| A269 (316L) | S31603 |
| A269 (317) | S31700 |
| A269 (321) | S32100 |
| A269 (347) | S34700 |
| A269 (348) | S34800 |
| A269 (TP304LN) | S30453 |
| A269 (TP316LN) | S31653 |
| A269 (TPXM-10) | S21900 |
| A269 (XM-11) | S21904 |
| A269 (XM-15) | S38100 |
| A269 (XM-19) | S20910 |
| A269 (XM-29) | S24000 |
| A270 | N08367 |
| A270 | N08926 |
| A270 | S31254 |
| A270 (304) | S30400 |
| A270 (TP304L) | S30403 |
| A270 (TP316) | S31600 |
| A270 (TP316L) | S31603 |
| A271 (347H) | S34709 |
| A271 (TP316) | S31600 |
| A276 | S20161 |
| A276 | S21800 |
| A276 | S28200 |
| A276 | S30454 |
| A276 | S30815 |
| A276 | S31100 |
| A276 | S31254 |
| A276 | S31654 |
| A276 | S31725 |
| A276 | S31726 |
| A276 | S31803 |
| A276 | S32205 |
| A276 | S32304 |
| A276 | S32550 |
| A276 | S32760 |
| A276 | S34565 |
| A276 | S40945 |
| A276 | S40976 |
| A276 | S41425 |
| A276 | S42010 |
| A276 | S44400 |
| A276 | S44700 |
| A276 | S44800 |
| A276 (205) | S20500 |
| A276 (302) | S30200 |
| A276 (302B) | S30215 |
| A276 (304) | S30400 |
| A276 (304L) | S30403 |
| A276 (304LN) | S30453 |
| A276 (304N) | S30451 |
| A276 (305) | S30500 |
| A276 (308) | S30800 |
| A276 (309) | S30900 |
| A276 (309Cb) | S30940 |
| A276 (309S) | S30908 |
| A276 (310) | S31000 |
| A276 (310Cb) | S31040 |
| A276 (310S) | S31008 |
| A276 (314) | S31400 |
| A276 (316) | S31600 |
| A276 (316Cb) | S31640 |
| A276 (316L) | S31603 |
| A276 (316LN) | S31653 |
| A276 (316N) | S31651 |
| A276 (316Ti) | S31635 |
| A276 (317) | S31700 |

| CROSS REFERENCE SPECIFICATIONS | UNIFIED NUMBER |
|---|---|
| A276 (321) | S32100 |
| A276 (347) | S34700 |
| A276 (348) | S34800 |
| A276 (403) | S40300 |
| A276 (405) | S40500 |
| A276 (410) | S41000 |
| A276 (414) | S41400 |
| A276 (420) | S42000 |
| A276 (429) | S42900 |
| A276 (430) | S43000 |
| A276 (431) | S43100 |
| A276 (440A) | S44002 |
| A276 (440B) | S44003 |
| A276 (440C) | S44004 |
| A276 (446) | S44600 |
| A276 (XM-10) | S21900 |
| A276 (XM-11) | S21904 |
| A276 (XM-21) | S30452 |
| A276 (XM-26) | S31100 |
| A276 (XM-27) | S44627 |
| A276 (XM-30) | S41040 |
| A278 (20) | F11401 |
| A278 (25) | F11701 |
| A278 (30) | F12101 |
| A278 (35) | F12401 |
| A278 (40) | F12803 |
| A278 (45) | F13102 |
| A278 (50) | F13502 |
| A278 (55) | F13802 |
| A278 (60) | F14102 |
| A278 (70) | F14801 |
| A278 (80) | F15501 |
| A283 (A) | K01400 |
| A283 (B) | K01702 |
| A283 (C) | K02401 |
| A283 (D) | K02702 |
| A285/A285M (A) | K01700 |
| A285/A285M (B) | K02200 |
| A285 (C) | K02801 |
| A288 (1) | K05002 |
| A288 (2, 3) | K14542 |
| A288 (4, 5, 6, 7, 8) | K24562 |
| A290 (A, B) | K04000 |
| A290 (C, D) | K04500 |
| A290 (E, F) | K14048 |
| A290 (G, H, I, J, K, L) | K24045 |
| A291 (1) | K05500 |
| A291 (2) | K05000 |
| A291 (3) | K14507 |
| A291 (3A) | K14557 |
| A291 (4 to 7) | K24245 |
| A295 (5160) | G51600 |
| A295 (50100) | G50986 |
| A295 (51100) | G51986 |
| A295 (52100) | G52986 |
| A297 (HC) | J92605 |
| A297 (HD) | J93005 |
| A297 (HE) | J93403 |
| A297 (HF) | J92603 |
| A297 (HH) | J93503 |
| A297 (HI) | J94003 |
| A297 (HK) | J94224 |
| A297 (HL) | N08604 |
| A297 (HN) | J94213 |
| A297 (HP) | N08705 |
| A297 (HT) | N08605 |
| A297 (HU) | N08004 |
| A297 (HW) | N08001 |
| A297 (HX) | N06006 |
| A299 | K02803 |
| A302 (A) | K12021 |
| A302 (B) | K12022 |
| A302 (C) | K12039 |
| A302 (D) | K12054 |
| A304 (15B21H) | H15211 |
| A304 (15B26H) | H15240 |
| A304 (15B35H) | H15351 |
| A304 (15B37H) | H15371 |
| A304 (15B41H) | H15411 |
| A304 (15B48H) | H15481 |
| A304 (15B62H) | H15621 |
| A304 (50B40H) | H50401 |
| A304 (50B44H) | H50441 |
| A304 (50B46H) | H50461 |
| A304 (50B50H) | H50501 |
| A304 (50B60H) | H50601 |
| A304 (51B60H) | H51601 |
| A304 (81B45H) | H81451 |
| A304 (86B30H) | H86301 |
| A304 (86B45H) | H86451 |
| A304 (94B15H) | H94151 |
| A304 (94B17H) | H94171 |
| A304 (94B30H) | H94301 |
| A304 (1038H) | H10380 |
| A304 (1045H) | H10450 |
| A304 (1330H) | H13300 |
| A304 (1335H) | H13350 |
| A304 (1340H) | H13400 |
| A304 (1345H) | H13450 |
| A304 (1522H) | H15220 |
| A304 (1524H) | H15240 |
| A304 (1526H) | H15260 |
| A304 (1541H) | H15410 |
| A304 (4027H) | H40270 |
| A304 (4028H) | H40280 |
| A304 (4032H) | H40320 |
| A304 (4037H) | H40370 |
| A304 (4042H) | H40420 |
| A304 (4047H) | H40470 |
| A304 (4118H) | H41180 |
| A304 (4130H) | H41300 |
| A304 (4135H) | H41350 |
| A304 (4137H) | H41370 |
| A304 (4140H) | H41400 |
| A304 (4142H) | H41420 |
| A304 (4145H) | H41450 |
| A304 (4147H) | H41470 |
| A304 (4150H) | H41500 |
| A304 (4161H) | H41610 |
| A304 (4320H) | H43200 |
| A304 (4340H) | H43400 |
| A304 (4419H) | H44190 |
| A304 (4620H) | H46200 |
| A304 (4621H) | H46210 |
| A304 (4626H) | H46260 |
| A304 (4718H) | H47180 |
| A304 (4720H) | H47200 |
| A304 (4815H) | H48150 |
| A304 (4817H) | H48170 |
| A304 (4820H) | H48200 |
| A304 (5046H) | H50460 |
| A304 (5120H) | H51200 |
| A304 (5130H) | H51300 |
| A304 (5132H) | H51320 |
| A304 (5135H) | H51350 |
| A304 (5140H) | H51400 |
| A304 (5145H) | H51450 |
| A304 (5147H) | H51470 |
| A304 (5150H) | H51500 |
| A304 (5155H) | H51550 |
| A304 (5160H) | H51600 |
| A304 (6118H) | H61180 |
| A304 (6150H) | H61500 |
| A304 (8617H) | H86170 |
| A304 (8620H) | H86200 |
| A304 (8622H) | H86220 |
| A304 (8625H) | H86250 |
| A304 (8627H) | H86270 |
| A304 (8630H) | H86300 |
| A304 (8637H) | H86370 |
| A304 (8640H) | H86400 |
| A304 (8642H) | H86420 |
| A304 (8645H) | H86450 |
| A304 (8650H) | H86500 |
| A304 (8655H) | H86550 |
| A304 (8660H) | H86600 |
| A304 (8720H) | H87200 |
| A304 (8740H) | H87400 |
| A304 (8822H) | H88220 |
| A304 (9260H) | H92600 |
| A304 (9310H) | H93100 |
| A304 (E4340H) | H43406 |
| A311 (1018) | G10180 |
| A311 (1035) | G10350 |
| A311 (1045) | G10450 |
| A311 (1050) | G10500 |
| A311 (1137) | G11370 |
| A311 (1141) | G11410 |
| A311 (1144) | G11440 |
| A311 (1541) | G15410 |
| A312 | N08367 |
| A312 | N08904 |
| A312 | S30415 |
| A312 | S30600 |
| A312 | S30615 |
| A312 | S30815 |
| A312 | S31050 |
| A312 | S31254 |
| A312 | S31272 |
| A312 | S31725 |
| A312 | S31726 |
| A312 | S32615 |
| A312 | S32654 |
| A312 | S33228 |
| A312 | S34565 |
| A312 | S35315 |
| A312 (304) | S30400 |
| A312 (304H) | S30409 |
| A312 (304L) | S30403 |
| A312 (304N) | S30451 |
| A312 (316) | S31600 |
| A312 (316H) | S31609 |
| A312 (316L) | S31603 |
| A312 (316N) | S31651 |
| A312 (317) | S31700 |
| A312 (321) | S32100 |
| A312 (321H) | S32109 |
| A312 (347) | S34700 |
| A312 (347H) | S34709 |
| A312 (347LN) | S34751 |
| A312 (348) | S34800 |
| A312 (348H) | S34809 |
| A312/A312M | S31002 |
| A312 (TP304LN) | S30453 |
| A312 (TP309Cb) | S30940 |
| A312 (TP309H) | S30909 |
| A312 (TP309HCb) | S30941 |

## *ASTM (American Society for Testing and Materials) NUMBERS*

| CROSS REFERENCE SPECIFICATIONS | UNIFIED NUMBER |
|---|---|
| A312 (TP309S) | S30908 |
| A312 (TP310Cb) | S31040 |
| A312 (TP310H) | S31009 |
| A312 (TP310HCb) | S31041 |
| A312 (TP310S) | S31008 |
| A312 (TP316LN) | S31653 |
| A312 (TP317L) | S31703 |
| A312 (TP347LN) | S34751 |
| A312 (TPXM-10) | S21900 |
| A312 (XM-11) | S21904 |
| A312 (XM-15) | S38100 |
| A312 (XM-19) | S20910 |
| A312 (XM-29) | S24000 |
| A313 | S20430 |
| A313 (302) | S30200 |
| A313 (304) | S30400 |
| A313 (305) | S30500 |
| A313 (316) | S31600 |
| A313 (321) | S32100 |
| A313 (347) | S34700 |
| A313 (631) | S17700 |
| A313 (XM-16) | S45500 |
| A313 (XM-28) | S24100 |
| A314 | N08031 |
| A314 | N08926 |
| A314 | S20161 |
| A314 | S21800 |
| A314 | S28200 |
| A314 | S32760 |
| A314 | S32950 |
| A314 | S33228 |
| A314 | S41425 |
| A314 | S42010 |
| A314 | S44627 |
| A314 (202) | S20200 |
| A314 (302) | S30200 |
| A314 (302B) | S30215 |
| A314 (303) | S30300 |
| A314 (303 Se) | S30323 |
| A314 (304) | S30400 |
| A314 (304L) | S30403 |
| A314 (305) | S30500 |
| A314 (308) | S30800 |
| A314 (309) | S30900 |
| A314 (309Cb) | S30940 |
| A314 (309S) | S30908 |
| A314 (310) | S31000 |
| A314 (310S) | S31008 |
| A314 (314) | S31400 |
| A314 (316) | S31600 |
| A314 (316Cb) | S31640 |
| A314 (316L) | S31603 |
| A314 (316Ti) | S31635 |
| A314 (317) | S31700 |
| A314 (321) | S32100 |
| A314 (347) | S34700 |
| A314 (348) | S34800 |
| A314 (403) | S40300 |
| A314 (410) | S41000 |
| A314 (414) | S41400 |
| A314 (416) | S41600 |
| A314 (416Se) | S41623 |
| A314 (420) | S42000 |
| A314 (429) | S42900 |
| A314 (430) | S43000 |
| A314 (430F) | S43020 |
| A314 (430FSe) | S43023 |
| A314 (431) | S43100 |
| A314 (440A) | S44002 |
| A314 (440B) | S44003 |
| A314 (440C) | S44004 |
| A314 (446) | S44600 |
| A314 (501) | S50100 |
| A314 (502) | S50200 |
| A314 (XM-10) | S21900 |
| A314 (XM-11) | S21904 |
| A314 (XM-19) | S20910 |
| A314 (XM-27) | S44625 ob |
| A314 (XM-28) | S24100 |
| A314 (XM-29) | S24000 |
| A319 (I) | F10001 |
| A319 (II) | F10002 |
| A319 (III) | F10003 |
| A320 (303, B8F, B8FA) | S30300 |
| A320 (303 Se, B8F, B8FA) | S30323 |
| A320 (304, B8, B8A) | S30400 |
| A320 (316, B8M, B8MA) | S31600 |
| A320 (321, B8T, B8TA) | S32100 |
| A320 (347, B8C, B8CA) | S34700 |
| A320 (B8LN, B8LNA) | S30453 |
| A320 (B8MLN, B8MLNA) | S31653 |
| A320 (B8P, B8PA) | S30500 |
| A320 (L7, L7M, L7C) | G41400 |
| A320 (L7A, L71) | G40370 |
| A320 (L7B, L72) | G41370 |
| A320 (L7C, L73) | G87400 |
| A320 (L43) | G43400 |
| A321 | K05501 |
| A322 (50B44) | G50441 |
| A322 (50B46) | G50461 |
| A322 (50B50) | G50501 |
| A322 (50B60) | G50601 |
| A322 (51B60) | G51601 |
| A322 (81B45) | G81451 |
| A322 (94B17) | G94171 |
| A322 (94B30) | G94301 |
| A322 (1330) | G13300 |
| A322 (1335) | G13350 |
| A322 (1340) | G13400 |
| A322 (1345) | G13450 |
| A322 (4023) | G40230 |
| A322 (4024) | G40240 |
| A322 (4027) | G40270 |
| A322 (4028) | G40280 |
| A322 (4037) | G40370 |
| A322 (4047) | G40470 |
| A322 (4118) | G41180 |
| A322 (4120) | G41200 |
| A322 (4121) | G41210 |
| A322 (4130) | G41300 |
| A322 (4137) | G41370 |
| A322 (4140) | G41400 |
| A322 (4142) | G41420 |
| A322 (4145) | G41450 |
| A322 (4147) | G41470 |
| A322 (4150) | G41500 |
| A322 (4161) | G41610 |
| A322 (4320) | G43200 |
| A322 (4340) | G43400 |
| A322 (4615) | G46150 |
| A322 (4620) | G46200 |
| A322 (4621) | G46210 |
| A322 (4626) | G46260 |
| A322 (4715) | G47150 |
| A322 (4720) | G47200 |
| A322 (4815) | G48150 |
| A322 (4817) | G48170 |
| A322 (4820) | G48200 |
| A322 (5117) | G51170 |
| A322 (5120) | G51200 |
| A322 (5130) | G51300 |
| A322 (5132) | G51320 |
| A322 (5135) | G51350 |
| A322 (5140) | G51400 |
| A322 (5150) | G51500 |
| A322 (5155) | G51550 |
| A322 (5160) | G51600 |
| A322 (6118) | G61180 |
| A322 (6150) | G61500 |
| A322 (8615) | G86150 |
| A322 (8617) | G86170 |
| A322 (8620) | G86200 |
| A322 (8622) | G86220 |
| A322 (8625) | G86250 |
| A322 (8627) | G86270 |
| A322 (8630) | G86300 |
| A322 (8637) | G86370 |
| A322 (8640) | G86400 |
| A322 (8642) | G86420 |
| A322 (8645) | G86450 |
| A322 (8655) | G86550 |
| A322 (8720) | G87200 |
| A322 (8740) | G87400 |
| A322 (8822) | G88220 |
| A322 (9259) | G92590 |
| A322 (9259) | H92590 |
| A322 (9260) | G92600 |
| A322 (E51100) | G51986 |
| A322 (E52100) | G52986 |
| A325 (1) | K02706 |
| A325 (A) | K13643 |
| A325 (B) | K14358 |
| A325 (C) | K12033 |
| A325 (D) | K12059 |
| A325 (E) | K12254 |
| A325 (F) | K12238 |
| A333 (1) | K03008 |
| A333 (3) | K31918 |
| A333 (4) | K11267 |
| A333 (6) | K03006 |
| A333 (7) | K21903 |
| A333 (8) | K81340 |
| A333 (9) | K22035 |
| A333 (10) | K02011 |
| A333 (11) | K93601 |
| A334 (1) | K03008 |
| A334 (3) | K31918 |
| A334 (6) | K03006 |
| A334 (7) | K21903 |
| A334 (8) | K81340 |
| A334 (9) | K22035 |
| A334 (11) | K93601 |
| A335 (P1) | K11522 |
| A335 (P2) | K11547 |
| A335 (P5) | K41545 |
| A335 (P5b) | K51545 |
| A335 (P5c) | K41245 |
| A335 (P11) | K11597 |
| A335 (P12) | K11562 |
| A335 (P15) | K11578 |
| A335 (P21) | K31545 |
| A335 (P22) | K21590 |
| A335 (P91) | K90901 |
| A335 (P122) | K91271 |
| A335 (P911) | K91061 |
| A335 (T92) | K92460 |
| A336 (F1) | K12520 |

| CROSS REFERENCE SPECIFICATIONS | UNIFIED NUMBER |
|---|---|
| A336 (F5) | K41545 |
| A336 (F5A) | K42544 |
| A336 (F6) | S41000 |
| A336 (F9) | K90941 |
| A336 (F11-1) | K11597 |
| A336 (F11-2) | K11597 |
| A336 (F11-2, F11-3) | K11572 |
| A336 (F12) | K11564 |
| A336 (F21-1, F21-3) | K31545 |
| A336 (F22, F22a) | K21590 |
| A336 (F46) | S30600 |
| A336 (F91) | K90901 |
| A336 (F304) | S30400 |
| A336 (F304H) | S30409 |
| A336 (F304L) | S30403 |
| A336 (F304LN) | S30453 |
| A336 (F304N) | S30451 |
| A336 (F309H) | S30909 |
| A336 (F310) | S31000 |
| A336 (F310H) | S31009 |
| A336 (F316) | S31600 |
| A336 (F316H) | S31609 |
| A336 (F316L) | S31603 |
| A336 (F316LN) | S31653 |
| A336 (F316N) | S31651 |
| A336 (F321) | S32100 |
| A336 (F321H) | S32109 |
| A336 (F347) | S34700 |
| A336 (F347H) | S34709 |
| A336 (F348) | S34800 |
| A336 (F348H) | S34809 |
| A336 (FN, F11A) | K11572 |
| A336 (FXM-11) | S21904 |
| A336 (FXM-19) | S20910 |
| A336 (FXM-27Cb) | S44627 |
| A350 (LF1) | K03009 |
| A350 (LF2) | K03011 |
| A350 (LF3) | K32025 |
| A350 (LF5) | K13050 |
| A350 (LF6) | K12202 |
| A350 (LF9) | K22036 |
| A350 (LF787) | K20747 |
| A351 | J92700 |
| A351 | J92802 |
| A351 | J92972 |
| A351 (CD3MWCuN) | J93380 |
| A351 (CD4MCu) | J93370 |
| A351 (CE20N) | J92802 |
| A351 (CF-3, CF-3A) | J92500 |
| A351 (CF-3M, CF-3MA) | J92800 |
| A351 (CF3MN) | J92804 |
| A351 (CF-8, CF-8A) | J92600 |
| A351 (CF-8C) | J92710 |
| A351 (CF-8M) | J92900 |
| A351 (CF-10) | J92590 |
| A351 (CF-10M) | J92901 |
| A351 (CG-3M) | J92999 |
| A351 (CG6MMN) | J93790 |
| A351 (CG-8M) | J93000 |
| A351 (CH-8) | J93400 |
| A351 (CH-10) | J93401 |
| A351 (CH-20) | J93402 |
| A351 (CK-3MCuN) | J93254 |
| A351 (CK-20) | J94202 |
| A351 (CN3MN) | J94651 |
| A351 (CN7M) | N08007 |
| A351 (CT15C) | N08151 |
| A351 (HK-30) | J94203 |
| A351 (HK-40) | J94204 |
| A351 (HT-30) | N08603 |
| A352 (CA-6NM) | J91540 |
| A352 (LC1) | J12522 |
| A352 (LC2) | J22500 |
| A352 (LC2-1) | J42215 |
| A352 (LC3) | J31550 |
| A352 (LC4) | J41500 |
| A352 (LC9) | J31300 |
| A352 (LCA) | J02504 |
| A352 (LCB) | J03003 |
| A352 (LCC) | J02505 |
| A353M | K81340 |
| A354 | K04100 |
| A355 (A) | K24065 |
| A355 (B) | K23745 |
| A355 (C) | K52440 |
| A355 (D) | K23510 |
| A356 (1) | J03502 |
| A356 (2) | J12523 |
| A356 (5) | J12540 |
| A356 (6) | J12073 |
| A356 (8) | J11697 |
| A356 (9) | J21610 |
| A356 (10) | J22090 |
| A356 (12) | J80490 |
| A356 (CA-6NM) | J91540 |
| A358 | N08020 |
| A358 | N08367 |
| A358 | N08800 |
| A358 | N08810 |
| A358 | N08904 |
| A358 | N08926 |
| A358 | S30600 |
| A358 | S30815 |
| A358 | S31254 |
| A358 | S31266 |
| A358 | S31725 |
| A358 | S31726 |
| A358 | S32050 |
| A358 | S32654 |
| A358 | S34565 |
| A358 (304) | S30400 |
| A358 (304H) | S30409 |
| A358 (304L) | S30403 |
| A358 (304LN) | S30453 |
| A358 (304N) | S30451 |
| A358 (309) | S30900 |
| A358 (309Cb) | S30940 |
| A358 (309S) | S30908 |
| A358 (310) | S31000 |
| A358 (310Cb) | S31040 |
| A358 (310S) | S31008 |
| A358 (316) | S31600 |
| A358 (316H) | S31609 |
| A358 (316L) | S31603 |
| A358 (316LN) | S31653 |
| A358 (316N) | S31651 |
| A358 (321) | S32100 |
| A358 (347) | S34700 |
| A358 (348) | S34800 |
| A358 (XM-19) | S20910 |
| A358 (XM-29) | S24000 |
| A368 (302) | S30200 |
| A368 (304) | S30400 |
| A368 (304L) | S30403 |
| A368 (305) | S30500 |
| A368 (316) | S31600 |
| A368 (316 Ti) | S31635 |
| A368 (316Cb) | S31640 |
| A369 (FP1) | K11522 |
| A369 (FP2) | K11547 |
| A369 (FP3b) | K21509 |
| A369 (FP5) | K41545 |
| A369 (FP9) | K90941 |
| A369 (FP11) | K11597 |
| A369 (FP12) | K11562 |
| A369 (FP21) | K31545 |
| A369 (FP22) | K21590 |
| A369 (FP91) | K90901 |
| A369 (FPA) | K02501 |
| A369 (FPB) | K03006 |
| A372 (31) | R30031 |
| A372 (A) | K03002 |
| A372 (B) | K04001 |
| A372 (C) | K04801 |
| A372 (D) | K14508 |
| A372 (E) | K13047 |
| A372 (F) | G41350 |
| A372 (G) | K13547 |
| A372 (J) | K13548 |
| A372 (K) | K31820 |
| A372 (L) | K24055 |
| A372 (M) | K42348 |
| A376 | S31725 |
| A376 | S31726 |
| A376 (16-8-2-H) | S16800 |
| A376 (304) | S30400 |
| A376 (316) | S31600 |
| A376 (317LM) | S31725 |
| A376 (317LMN) | S31726 |
| A376 (321) | S32100 |
| A376 (347) | S34700 |
| A376 (TP304) | S30400 |
| A376 (TP304H) | S30409 |
| A376 (TP304LN) | S30453 |
| A376 (TP304N) | S30451 |
| A376 (TP316) A403 (316) | S31600 |
| A376 (TP316H) | S31609 |
| A376 (TP316LN) | S31653 |
| A376 (TP316N) | S31651 |
| A376 (TP321) | S32100 |
| A376 (TP321H) | S32109 |
| A376 (TP347) | S34700 |
| A376 (TP347H) | S34709 |
| A376 (TP348) | S34800 |
| A381 | K02601 |
| A387 (2) | K12143 |
| A387 (5) | K41545 |
| A387 (5) | S50100 |
| A387 (5) | S50200 |
| A387 (7) | S50300 |
| A387 (11) | K11789 |
| A387 (12) | K11757 |
| A387 (21) | K31545 |
| A387 (22, 22L) | K21590 |
| A387 (91) | K90901 |
| A389 (C23) | J12080 |
| A389 (C24) | J12092 |
| A395 (60-40-18) | F32800 |
| A401 | G92540 |
| A403 | S31254 |
| A403 | S33228 |
| A403 | S34565 |
| A403 (304) | S30400 |
| A403 (304H) | S30409 |
| A403 (304L) | S30403 |
| A403 (304LN) | S30453 |
| A403 (304N) | S30451 |

| CROSS REFERENCE SPECIFICATIONS | UNIFIED NUMBER |
|---|---|
| A403 (309) | S30900 |
| A403 (310) | S31000 |
| A403 (316H) | S31609 |
| A403 (316L) | S31603 |
| A403 (316LN) | S31653 |
| A403 (316N) | S31651 |
| A403 (317) | S31700 |
| A403 (317L) | S31703 |
| A403 (317LM) | S31725 |
| A403 (317LMN) | S31726 |
| A403 (321) | S32100 |
| A403 (321H) | S32109 |
| A403 (347) | S34700 |
| A403 (347H) | S34709 |
| A403 (348) | S34800 |
| A403 (348H) | S34709 |
| A403 (WP304LN) | S30453 |
| A403 (WP316LN) | S31653 |
| A403 (XM-19) | S20910 |
| A409 | N08367 |
| A409 | S20400 |
| A409 | S30815 |
| A409 | S31254 |
| A409 | S31725 |
| A409 | S31726 |
| A409 | S34565 |
| A409 (304) | S30400 |
| A409 (304L) | S30403 |
| A409 (309) | S30900 |
| A409 (309Cb) | S30940 |
| A409 (309S) | S30908 |
| A409 (310) | S31000 |
| A409 (310Cb) | S31040 |
| A409 (310S) | S31008 |
| A409 (316) | S31600 |
| A409 (316L) | S31603 |
| A409 (317) | S31700 |
| A409 (321) | S32100 |
| A409 (347) | S34700 |
| A409 (348) | S34800 |
| A412 (XM-19) | S20910 |
| A413 | K03700 |
| A414 (A) | K01501 |
| A414 (B) | K02201 |
| A414 (C) | K02503 |
| A414 (D) | K02505 |
| A414 (E) | K02704 |
| A414 (F) | K03102 |
| A414 (G) | K03103 |
| A417 | K06201 |
| A420 (3) | K31918 |
| A420 (WPL3) | K31918 |
| A420 (WPL6) | K03006 |
| A420 (WPL8) | K81340 |
| A423 (1) | K11535 |
| A423 (2) | K11540 |
| A426 (CP1) | J12521 |
| A426 (CP2) | J11547 |
| A426 (CP5) | J42045 |
| A426 (CP5b) | J51545 |
| A426 (CP9) | J82090 |
| A426 (CP11) | J12072 |
| A426 (CP12) | J11562 |
| A426 (CP15) | J11522 |
| A426 (CP21) | J31545 |
| A426 (CP22) | J21890 |
| A426 (CPCA-15) | J91150 |
| A426 (CPCA-15) | J91171 |
| A430 (304) | S30400 |
| A430 (316) | S31600 |
| A430 (321) | S32100 |
| A430 (347) | S34700 |
| A430 (347H) | S34709 |
| A436 (1) | F41000 |
| A436 (1b) | F41001 |
| A436 (2) | F41002 |
| A436 (2b) | F41003 |
| A436 (3) | F41004 |
| A436 (4) | F41005 |
| A436 (5) | F41006 |
| A436 (6) | F41007 |
| A437 (B4B, B4C) | K91352 |
| A437 (B4B, B4C) | S42225 |
| A437 (B4D) | K14072 |
| A439 (D-2) | F43000 |
| A439 (D-2B) | F43001 |
| A439 (D-2C) | F43002 |
| A439 (D-3) | F43003 |
| A439 (D-3A) | F43004 |
| A439 (D-4) | F43005 |
| A439 (D-5) | F43006 |
| A439 (D-5B) | F43007 |
| A447 | J93303 |
| A447 (HH) | J93503 |
| A449 (1) | K04200 |
| A451 | J92802 |
| A451 (CF10MC) | J92971 |
| A451 (CH-2, CPH-10, CPH-20) | J93402 |
| A451 (CPF3, CPF3A) | J92500 |
| A451 (CPF3, CPF3A) | J92700 |
| A451 (CPF-3M) | J92800 |
| A451 (CPF8, CPF8A) | J92600 |
| A451 (CPF8C) | J92710 |
| A451 (CPF8M) | J92900 |
| A451 (CPH-8) | J93400 |
| A451 (CPK20) | J94202 |
| A452 (CPE20N) | J92802 |
| A453 (651) | S63198 |
| A453 (660) | S66286 |
| A453 (662) | S66220 |
| A453 (665) | S66545 |
| A455 (I) | K03300 |
| A463 (409) | S40900 |
| A463 (439) | S43035 |
| A469 (1) | K14501 |
| A469 (2) | K22573 |
| A469 (3) | K22773 |
| A469 (4) | K32723 |
| A469 (5) | K33125 |
| A469 (6, 7, 8) | K42885 |
| A470 (1) | K14501 |
| A470 (2) | K22578 |
| A470 (3, 4) | K22878 |
| A470 (5, 6, 7) | K42885 |
| A470 (8) | K23010 |
| A471 (1-6) | K32800 |
| A471 (10) | K23205 |
| A473 | S28200 |
| A473 | S30815 |
| A473 | S31254 |
| A473 | S32550 |
| A473 | S32950 |
| A473 | S34800 |
| A473 | S41425 |
| A473 | S41500 |
| A473, A479 | S32760 |
| A473 (201) | S20100 |
| A473 (202) | S20200 |
| A473 (205) | S20500 |
| A473 (302) | S30200 |
| A473 (302B) | S30215 |
| A473 (303) | S30300 |
| A473 (303 Se) | S30323 |
| A473 (304) | S30400 |
| A473 (304L) | S30403 |
| A473 (305) | S30500 |
| A473 (308) | S30800 |
| A473 (309) | S30900 |
| A473 (309S) | S30908 |
| A473 (310) | S31000 |
| A473 (310S) | S31008 |
| A473 (314) | S31400 |
| A473 (316L) | S31603 |
| A473 (317) | S31700 |
| A473 (321) | S32100 |
| A473 (347) | S34700 |
| A473 (348) | S34800 |
| A473 (384) | S38400 |
| A473 (403) | S40300 |
| A473 (405) | S40500 |
| A473 (410) | S41000 |
| A473 (410S) | S41008 |
| A473 (414) | S41400 |
| A473 (416) | S41600 |
| A473 (416Se) | S41623 |
| A473 (420) | S42000 |
| A473 (429) | S42900 |
| A473 (430) | S43000 |
| A473 (430F) | S43020 |
| A473 (430FSe) | S43023 |
| A473 (431) | S43100 |
| A473 (440A) | S44002 |
| A473 (440B) | S44003 |
| A473 (440C) | S44004 |
| A473 (446) | S44600 |
| A473 (501) | S50100 |
| A473 (502) | S50200 |
| A473 (XM-10) | S21900 |
| A473 (XM-11) | S21904 |
| A476 | F34100 |
| A478 (302) | S30200 |
| A478 (304) | S30400 |
| A478 (304L) | S30403 |
| A478 (305) | S30500 |
| A478 (309Cb) | S30940 |
| A478 (309S) | S30908 |
| A478 (310Cb) | S31040 |
| A478 (316 Ti) | S31635 |
| A478 (316Cb) | S31640 |
| A478 (316L) | S31603 |
| A478 (317) | S31700 |
| A479 | N08367 |
| A479 | S20161 |
| A479 | S21800 |
| A479 | S30600 |
| A479 | S30815 |
| A479 | S31254 |
| A479 | S31725 |
| A479 | S31726 |
| A479 | S31803 |
| A479 | S32050 |
| A479 | S32205 |
| A479 | S32550 |
| A479 | S32615 |
| A479 | S32750 |
| A479 | S32906 |
| A479 | S32950 |

| CROSS REFERENCE SPECIFICATIONS | UNIFIED NUMBER |
|---|---|
| A479 | S33228 |
| A479 | S34565 |
| A479 | S35315 |
| A479 | S39277 |
| A479 | S41425 |
| A479 | S41500 |
| A479 | S44400 |
| A479 | S44627 |
| A479 | S44700 |
| A479 | S44800 |
| A479 (302) | S30200 |
| A479 (304) | S30400 |
| A479 (304H) | S30409 |
| A479 (304L) | S30403 |
| A479 (304LN) | S30453 |
| A479 (304N) | S30451 |
| A479 (309Cb) | S30940 |
| A479 (309H) | S30909 |
| A479 (309S) | S30908 |
| A479 (310Cb) | S31040 |
| A479 (310H) | S31009 |
| A479 (310S) | S31008 |
| A479 (316 Ti) | S31635 |
| A479 (316Cb) | S31640 |
| A479 (316H) | S31609 |
| A479 (316L) | S31603 |
| A479 (316LN) | S31653 |
| A479 (316N) | S31651 |
| A479 (317) | S31700 |
| A479 (321) | S32100 |
| A479 (321H) | S32109 |
| A479 (347) | S34700 |
| A479 (347H) | S34709 |
| A479 (348) | S34800 |
| A479 (348H) | S34809 |
| A479 (403) | S40300 |
| A479 (405) | S40500 |
| A479 (410) | S41000 |
| A479 (414) | S41400 |
| A479 (430) | S43000 |
| A479 (431) | S43100 |
| A479 (439) | S43035 |
| A479 (XM-11) | S21904 |
| A479 (XM-17) | S21600 |
| A479 (XM-18) | S21603 |
| A479 (XM-19) | S20910 |
| A479 (XM-29) | S24000 |
| A479 (XM-30) | S41040 |
| A480 | N08020 |
| A480 | N08367 |
| A480 | N08800 |
| A480 | N08810 |
| A480 | N08811 |
| A480 | N08904 |
| A480 | N08926 |
| A480 | S20103 |
| A480 | S20153 |
| A480 | S31254 |
| A480 | S31266 |
| A480 | S32001 |
| A480 | S32050 |
| A480 | S32205 |
| A480 | S32520 |
| A480 | S32654 |
| A480 | S33228 |
| A480 | S34565 |
| A480 | S35045 |
| A480 | S35315 |
| A480 | S40945 |
| A480 | S41045 |
| A480 | S43932 |
| A480 | S46800 |
| A480/A240M | S35135 |
| A480M | S31266 |
| A480M | S32205 |
| A480M | S32520 |
| A481 (A) | R20990 |
| A481 (B) | R20994 |
| A485 (1) | K19667 |
| A485 (2) | K19195 |
| A485 (3) | K19965 |
| A485 (4) | K19990 |
| A487 (1) | J13002 |
| A487 (2) | J13005 |
| A487 (4) | J13047 |
| A487 (6) | J13855 |
| A487 (7) | J12084 |
| A487 (8) | J22091 |
| A487 (9) | J13345 |
| A487 (10) | J23015 |
| A487 (11) | J12082 |
| A487 (12) | J22000 |
| A487 (13) | J13080 |
| A487 (14) | J15580 |
| A487 (16) | J31200 |
| A487 (CA-15) | J91171 |
| A487 (CA-15, CA-6NM) | J91540 |
| A487 (CA-15M) | J91151 |
| A489 | K04800 |
| A490 (3) | K03900 |
| A492 (302) | S30200 |
| A492 (304) | S30400 |
| A492 (305) | S30500 |
| A492 (XM-17) | S21600 |
| A492 (XM-18) | S21603 |
| A493 | S30430 |
| A493 | S30433 |
| A493 | S40940 |
| A493 | S42010 |
| A493 | S44400 |
| A493 | S44625 ob |
| A493 | S44700 |
| A493 | S44800 |
| A493 (302) | S30200 |
| A493 (304) | S30400 |
| A493 (304L) | S30403 |
| A493 (305) | S30500 |
| A493 (316L) | S31603 |
| A493 (384) | S38400 |
| A493 (410) | S41000 |
| A493 (429) | S42900 |
| A493 (430) | S43000 |
| A493 (431) | S43100 |
| A493 (440C) | S44004 |
| A493 (TP321) | S32100 |
| A493 (XM-29) | S28200 |
| A494 | N06040 |
| A494 | N08826 |
| A494 | N26022 |
| A494 | N26055 |
| A494 | N26455 |
| A494 | N26625 |
| A494 | N30002 |
| A494 | N30007 |
| A494 | N30107 |
| A494 (CW2M) | N26455 |
| A494 (CZ-100) | N02100 |
| A494 (M-30C) | N24130 |
| A494 (M-30H) | N24030 |
| A494 (M-35-1) | N24135 |
| A494 (M-35-2) | N04020 |
| A494 (M-255) | N24025 |
| A494 (N-12MV) | N30012 |
| A500 (A, B, D) | K03000 |
| A500 (C) | K02705 |
| A501 | K03000 |
| A504 (A, L) | K05700 |
| A504 (B) | K06200 |
| A504 (C) | K07201 |
| A504 (U) | K07200 |
| A506 (4118) | G41180 |
| A506 (4130) | G41300 |
| A506 (4140) | G41400 |
| A506 (4340) | G43400 |
| A506 (5140) | G51400 |
| A506 (5150) | G51500 |
| A506 (5160) | G51600 |
| A506 (8615) | G86150 |
| A506 (8620) | G86200 |
| A506 (E9310) | G93106 |
| A507 (4118) | G41180 |
| A507 (4130) | G41300 |
| A507 (8615) | G86150 |
| A507 (8620) | G86200 |
| A508 (1) | K13502 |
| A508 (1) | K14072 |
| A508 (2) | K12766 |
| A508 (3) | K12042 |
| A508 (3V) | K31830 |
| A508 (5) | K42365 |
| A508 (22) | K21590 |
| A510 (12L13) | G12134 |
| A510 (12L14) | G12144 |
| A510 (1005) | G10050 |
| A510 (1006) | G10060 |
| A510 (1008) | G10080 |
| A510 (1010) | G10100 |
| A510 (1011) | G10110 |
| A510 (1012) | G10120 |
| A510 (1015) | G10150 |
| A510 (1016) | G10160 |
| A510 (1017) | G10170 |
| A510 (1018) | G10180 |
| A510 (1019) | G10190 |
| A510 (1020) | G10200 |
| A510 (1021) | G10210 |
| A510 (1022) | G10220 |
| A510 (1023) | G10230 |
| A510 (1025) | G10250 |
| A510 (1026) | G10260 |
| A510 (1029) | G10290 |
| A510 (1030) | G10300 |
| A510 (1034) | G10340 |
| A510 (1035) | G10350 |
| A510 (1037) | G10370 |
| A510 (1038) | G10380 |
| A510 (1039) | G10390 |
| A510 (1040) | G10400 |
| A510 (1042) | G10420 |
| A510 (1043) | G10430 |
| A510 (1044) | G10440 |
| A510 (1045) | G10450 |
| A510 (1046) | G10460 |
| A510 (1049) | G10490 |
| A510 (1050) | G10500 |
| A510 (1053) | G10530 |
| A510 (1055) | G10550 |

| CROSS REFERENCE SPECIFICATIONS | UNIFIED NUMBER |
|---|---|
| A510 (1059) | G10590 |
| A510 (1060) | G10600 |
| A510 (1064) | G10640 |
| A510 (1065) | G10650 |
| A510 (1069) | G10690 |
| A510 (1070) | G10700 |
| A510 (1074) | G10740 |
| A510 (1075) | G10750 |
| A510 (1078) | G10780 |
| A510 (1080) | G10800 |
| A510 (1084) | G10840 |
| A510 (1085) | G10850 |
| A510 (1086) | G10860 |
| A510 (1090) | G10900 |
| A510 (1095) | G10950 |
| A510 (1108) | G11080 |
| A510 (1109) | G11090 |
| A510 (1110) | G11100 |
| A510 (1116) | G11160 |
| A510 (1117) | G11170 |
| A510 (1118) | G11180 |
| A510 (1119) | G11190 |
| A510 (1132) | G11320 |
| A510 (1137) | G11370 |
| A510 (1139) | G11390 |
| A510 (1140) | G11400 |
| A510 (1141) | G11410 |
| A510 (1144) | G11440 |
| A510 (1145) | G11450 |
| A510 (1146) | G11460 |
| A510 (1151) | G11510 |
| A510 (1211) | G12110 |
| A510 (1212) | G12120 |
| A510 (1213) | G12130 |
| A510 (1215) | G12150 |
| A510 (1513) | G15130 |
| A510 (1518) | G15180 |
| A510 (1522) | G15220 |
| A510 (1524) | G15240 |
| A510 (1525) | G15250 |
| A510 (1526) | G15260 |
| A510 (1527) | G15270 |
| A510 (1536) | G15360 |
| A510 (1541) | G15410 |
| A510 (1547) | G15470 |
| A510 (1548) | G15480 |
| A510 (1551) | G15510 |
| A510 (1552) | G15520 |
| A510 (1561) | G15610 |
| A510 (1566) | G15660 |
| A510 (1572) | G15720 |
| A511 | S44735 |
| A511 (29-4) | S44700 |
| A511 (29-4-2) | S44800 |
| A511 (302) | S30200 |
| A511 (303Se) | S30323 |
| A511 (304) | S30400 |
| A511 (304L) | S30403 |
| A511 (305) | S30500 |
| A511 (309S) | S30908 |
| A511 (310S) | S31008 |
| A511 (316L) | S31603 |
| A511 (317) | S31700 |
| A511 (321) | S32100 |
| A511 (347) | S34700 |
| A511 (403) | S40300 |
| A511 (405) | S40500 |
| A511 (410) | S41000 |
| A511 (414) | S41400 |

| CROSS REFERENCE SPECIFICATIONS | UNIFIED NUMBER |
|---|---|
| A511 (416Se) | S41623 |
| A511 (429) | S42900 |
| A511 (430) | S43000 |
| A511 (440A) | S44002 |
| A511 (443) | S44300 |
| A511 (446) | S44600 |
| A511 (MT 431) | S43100 |
| A512 (1008) | G10080 |
| A512 (1010, MT 1010) | G10100 |
| A512 (1012) | G10120 |
| A512 (1015, MT 1015) | G10150 |
| A512 (1016) | G10160 |
| A512 (1018) | G10180 |
| A512 (1019) | G10190 |
| A512 (1020, MT 1020) | G10200 |
| A512 (1021) | G10210 |
| A512 (1025) | G10250 |
| A512 (1026) | G10260 |
| A512 (1030) | G10300 |
| A512 (1035) | G10350 |
| A512 (1110) | G11100 |
| A512 (1117) | G11170 |
| A513 (1008) | G10080 |
| A513 (1010, MT 1010) | G10100 |
| A513 (1015, MT 1015) | G10150 |
| A513 (1016) | G10160 |
| A513 (1017) | G10170 |
| A513 (1018) | G10180 |
| A513 (1019) | G10190 |
| A513 (1020, MT 1020) | G10200 |
| A513 (1021) | G10210 |
| A513 (1022) | G10220 |
| A513 (1023) | G10230 |
| A513 (1025) | G10250 |
| A513 (1026) | G10260 |
| A513 (1027) | G15270 |
| A513 (1030) | G10300 |
| A513 (1033) | G10330 |
| A513 (1035) | G10350 |
| A513 (1040) | G10400 |
| A513 (1050) | G10500 |
| A513 (1060) | G10600 |
| A513 (1340) | G13400 |
| A513 (1524) | G15240 |
| A513 (4118) | G41180 |
| A513 (4130) | G41300 |
| A513 (4140) | G41400 |
| A513 (5130) | G51300 |
| A513 (8620) | G86200 |
| A513 (8630) | G86300 |
| A513 91012) | G10120 |
| A514 (A) | K11856 |
| A514 (B) | K11630 |
| A514 (C) | K11511 |
| A514 (E) | K21604 |
| A514 (F) | K11576 |
| A514 (H) | K11646 |
| A514 (J) | K11625 |
| A514 (M) | K11683 |
| A514 (P) | K21650 |
| A515 (55) | K02001 |
| A515 (60) | K02401 |
| A515 (65) | K02800 |
| A515 (70) | K03101 |
| A516 (55) | K01800 |
| A516 (60) | K02100 |
| A516 (65) | K02403 |
| A516 (70) | K02700 |
| A517 (A) | K11856 |

| CROSS REFERENCE SPECIFICATIONS | UNIFIED NUMBER |
|---|---|
| A517 (B) | K11630 |
| A517 (C) | K11511 |
| A517 (E) | K21604 |
| A517 (F) | K11576 |
| A517 (H) | K11646 |
| A517 (J) | K11625 |
| A517 (M) | K11683 |
| A517 (P) | K21650 |
| A519 (12L14) | G12144 |
| A519 (50B40) | G50401 |
| A519 (50B44) | G50441 |
| A519 (50B46) | G50461 |
| A519 (50B50) | G50501 |
| A519 (50B60) | G50601 |
| A519 (51B60) | G51601 |
| A519 (81B45) | G81451 |
| A519 (86B45) | G86451 |
| A519 (94B15) | G94151 |
| A519 (94B17) | G94171 |
| A519 (94B30) | G94301 |
| A519 (94B40) | G94401 |
| A519 (1008) | G10080 |
| A519 (1010, MT 1010) | G10100 |
| A519 (1012) | G10120 |
| A519 (1015, MT 1015) | G10150 |
| A519 (1016) | G10160 |
| A519 (1017) | G10170 |
| A519 (1018) | G10180 |
| A519 (1019) | G10190 |
| A519 (1020, MT 1020) | G10200 |
| A519 (1021) | G10210 |
| A519 (1022) | G10220 |
| A519 (1025) | G10250 |
| A519 (1026) | G10260 |
| A519 (1030) | G10300 |
| A519 (1035) | G10350 |
| A519 (1040) | G10400 |
| A519 (1045) | G10450 |
| A519 (1050) | G10500 |
| A519 (1118) | G11180 |
| A519 (1132) | G11320 |
| A519 (1137) | G11370 |
| A519 (1141) | G11410 |
| A519 (1144) | G11440 |
| A519 (1213) | G12130 |
| A519 (1215) | G12150 |
| A519 (1330) | G13300 |
| A519 (1335) | G13350 |
| A519 (1340) | G13400 |
| A519 (1345) | G13450 |
| A519 (1518) | G15180 |
| A519 (1524) | G15240 |
| A519 (1541) | G15410 |
| A519 (3140) | G31400 |
| A519 (4012) | G40120 |
| A519 (4023) | G40230 |
| A519 (4024) | G40240 |
| A519 (4027) | G40270 |
| A519 (4028) | G40280 |
| A519 (4037) | G40370 |
| A519 (4042) | G40420 |
| A519 (4047) | G40470 |
| A519 (4063) | G40630 |
| A519 (4118) | G41180 |
| A519 (4130) | G41300 |
| A519 (4135) | G41350 |
| A519 (4137) | G41370 |
| A519 (4140) | G41400 |
| A519 (4142) | G41420 |

| CROSS REFERENCE SPECIFICATIONS | UNIFIED NUMBER |
|---|---|
| A519 (4145) | G41450 |
| A519 (4147) | G41470 |
| A519 (4150) | G41500 |
| A519 (4320) | G43200 |
| A519 (4337) | G43370 |
| A519 (4340) | G43400 |
| A519 (4422) | G44220 |
| A519 (4427) | G44270 |
| A519 (4520) | G45200 |
| A519 (4615) | G46150 |
| A519 (4617) | G46170 |
| A519 (4620) | G46200 |
| A519 (4621) | G46210 |
| A519 (4718) | G47180 |
| A519 (4720) | G47200 |
| A519 (4815) | G48150 |
| A519 (4817) | G48170 |
| A519 (4820) | G48200 |
| A519 (5015) | G50150 |
| A519 (5046) | G50460 |
| A519 (5115) | G51150 |
| A519 (5120) | G51200 |
| A519 (5130) | G51300 |
| A519 (5132) | G51320 |
| A519 (5135) | G51350 |
| A519 (5140) | G51400 |
| A519 (5145) | G51450 |
| A519 (5147) | G51470 |
| A519 (5150) | G51500 |
| A519 (5155) | G51550 |
| A519 (5160) | G51600 |
| A519 (6118) | G61180 |
| A519 (6120) | G61200 |
| A519 (6150) | G61500 |
| A519 (8115) | G81150 |
| A519 (8615) | G86150 |
| A519 (8617) | G86170 |
| A519 (8620) | G86200 |
| A519 (8622) | G86220 |
| A519 (8625) | G86250 |
| A519 (8627) | G86270 |
| A519 (8630) | G86300 |
| A519 (8637) | G86370 |
| A519 (8640) | G86400 |
| A519 (8642) | G86420 |
| A519 (8645) | G86450 |
| A519 (8650) | G86500 |
| A519 (8655) | G86550 |
| A519 (8660) | G86600 |
| A519 (8720) | G87200 |
| A519 (8735) | G87350 |
| A519 (8740) | G87400 |
| A519 (8742) | G87420 |
| A519 (8822) | G88220 |
| A519 (9255) | G92550 |
| A519 (9260) | G92600 |
| A519 (9262) | G92620 |
| A519 (9840) | G98400 |
| A519 (9850) | G98500 |
| A519 (E3310) | G33106 |
| A519 (E4337) | G43376 |
| A519 (E4340) | G43406 |
| A519 (E7140) | K24065 |
| A519 (E9310) | G93106 |
| A519 (E50100) | G50986 |
| A519 (E51100) | G51986 |
| A519 (E52100) | G52986 |
| A522 (I) | K81340 |
| A522 (II) | K71340 |

| CROSS REFERENCE SPECIFICATIONS | UNIFIED NUMBER |
|---|---|
| A523 (A) | K02504 |
| A523 (B) | K03005 |
| A524 (I, II) | K02104 |
| A529 | K02703 |
| A532 (IA) | F45000 |
| A532 (IB) | F45001 |
| A532 (IC) | F45002 |
| A532 (ID) | F45003 |
| A532 (IIA) | F45004 |
| A532 (IIB) | F45005 |
| A532 (IID) | F45007 |
| A533 (A) | K12521 |
| A533 (B) | K12539 |
| A533 (C) | K12554 |
| A533 (D) | K12529 |
| A534 (4118H) | H41180 |
| A534 (4320H) | H43200 |
| A534 (4620H) | H46200 |
| A534 (4620H) | H46260 |
| A534 (4720H) | H47200 |
| A534 (4817H) | H48170 |
| A534 (4820H) | H48200 |
| A534 (5120H) | H51200 |
| A534 (8617H) | H86170 |
| A534 (8620H) | H86200 |
| A534 (9310H) | H93100 |
| A536 (60-40-18) | F32800 |
| A536 (65-45-12) | F33100 |
| A536 (80-55-06) | F33800 |
| A536 (100-70-03) | F34800 |
| A536 (120-90-02) | F36200 |
| A537 | K12437 |
| A539 | K01506 |
| A540 (B21) | K14073 |
| A540 (B22) | H41420 |
| A540 (B23) | H43400 |
| A540 (B24) | K24064 |
| A540 (B24V) | K24070 |
| A541 (1) | K03506 |
| A541 (1C) | K11800 |
| A541 (2) | K12765 |
| A541 (3) | K12045 |
| A541 (3V) | K31830 |
| A541 (3VCb) | K31390 |
| A541 (4N) | K42343 |
| A541 (5) | K42348 |
| A541 (11) | K11572 |
| A541 (22B) | K21390 |
| A541 (22V) | K31835 |
| A542 (A, B) | K21590 |
| A542 (C) | K31830 |
| A542 (D) | K31835 |
| A542 (E) | K31390 |
| A543/A543M(B) | K42339 |
| A544 (1110) | G11100 |
| A551 (A, AHT) | K05701 |
| A551 (B, BHT) | K06702 |
| A551 (C, CHT, DHT) | K07701 |
| A553 (II) | K71340 |
| A553M (I) | K81340 |
| A554 | N08330 |
| A554 (301) | S30100 |
| A554 (302) | S30200 |
| A554 (304) | S30400 |
| A554 (304L) | S30403 |
| A554 (305) | S30500 |
| A554 (309S) | S30908 |
| A554 (309SCb) | S30940 |
| A554 (310S) | S31008 |

| CROSS REFERENCE SPECIFICATIONS | UNIFIED NUMBER |
|---|---|
| A554 (316L) | S31603 |
| A554 (317) | S31700 |
| A554 (321) | S32100 |
| A554 (347) | S34700 |
| A554 (429) | S42900 |
| A554 (430) | S43000 |
| A554 (430 Ti) | S43036 |
| A556 (B2) | K02707 |
| A556 (C2) | K03006 |
| A560 (50Cr-50Ni) | R20500 |
| A560 (50Cr-50Ni-Cb) | R20501 |
| A560 (60Cr-40Ni) | R20600 |
| A562 | K11224 |
| A563 (C3-N) | D61440 |
| A563 (C3A) | K13643 |
| A563 (C3B) | K14358 |
| A563 (C3C) | K12033 |
| A563 (C3D) | K12059 |
| A563 (C3E) | K12254 |
| A563 (C3F) | K12238 |
| A563 (D) | K05801 |
| A563 (DH) | K03800 |
| A563 (DH3) | K13650 |
| A563 (O, A, B, C) | K05802 |
| A564 | S46500 |
| A564 (630) | S17400 |
| A564 (631) | S17700 |
| A564 (632) | S15700 |
| A564 (634) | S35500 |
| A564 (635) | S17600 |
| A564 (XM-12) | S15500 |
| A564 (XM-13) | S13800 |
| A564 (XM-16) | S45500 |
| A564 (XM-25) | S45000 |
| A565 | S41041 |
| A565 (615) | S41800 |
| A565 (616) | S42200 |
| A565 (619) | S42300 |
| A565 (XM-32) | S64152 |
| A570 (30, 33, 36, 40) | K02502 |
| A570 (45, 50, 55) | K02507 |
| A571 (D-2M) | F43010 |
| A572 (50-1) | K02303 |
| A572 (50-2) | K02304 |
| A572 (50-3) | K02305 |
| A572 (50-4) | K02306 |
| A573 (58) | K02301 |
| A573 (65) | K02404 |
| A573 (70) | K02701 |
| A574 | K03104 |
| A574 (51B37M) | K13647 |
| A574 (5137M) | K13745 |
| A575 (M1008) | G10080 |
| A575 (M1010) | G10100 |
| A575 (M1012) | G10120 |
| A575 (M1015) | G10150 |
| A575 (M1017) | G10170 |
| A575 (M1020) | G10200 |
| A575 (M1023) | G10230 |
| A575 (M1025) | G10250 |
| A575 (M1044) | G10440 |
| A576 (12L14) | G12144 |
| A576 (1008) | G10080 |
| A576 (1012) | G10120 |
| A576 (1015) | G10150 |
| A576 (1016) | G10160 |
| A576 (1017) | G10170 |
| A576 (1018) | G10180 |
| A576 (1019) | G10190 |

| CROSS REFERENCE SPECIFICATIONS | UNIFIED NUMBER |
|---|---|
| A576 (1020) | G10200 |
| A576 (1021) | G10210 |
| A576 (1022) | G10220 |
| A576 (1023) | G10230 |
| A576 (1025) | G10250 |
| A576 (1026) | G10260 |
| A576 (1029) | G10290 |
| A576 (1030) | G10300 |
| A576 (1035) | G10350 |
| A576 (1037) | G10370 |
| A576 (1038) | G10380 |
| A576 (1039) | G10390 |
| A576 (1040) | G10400 |
| A576 (1042) | G10420 |
| A576 (1043) | G10430 |
| A576 (1044) | G10440 |
| A576 (1045) | G10450 |
| A576 (1046) | G10460 |
| A576 (1049) | G10490 |
| A576 (1050) | G10500 |
| A576 (1053) | G10530 |
| A576 (1055) | G10550 |
| A576 (1060) | G10600 |
| A576 (1070) | G10700 |
| A576 (1078) | G10780 |
| A576 (1080) | G10800 |
| A576 (1084) | G10840 |
| A576 (1090) | G10900 |
| A576 (1095) | G10950 |
| A576 (1109) | G11090 |
| A576 (1110) | G11100 |
| A576 (1116) | G11160 |
| A576 (1117) | G11170 |
| A576 (1118) | G11180 |
| A576 (1119) | G11190 |
| A576 (1132) | G11320 |
| A576 (1137) | G11370 |
| A576 (1139) | G11390 |
| A576 (1140) | G11400 |
| A576 (1141) | G11410 |
| A576 (1144) | G11440 |
| A576 (1145) | G11450 |
| A576 (1146) | G11460 |
| A576 (1151) | G11510 |
| A576 (1211) | G12110 |
| A576 (1212) | G12120 |
| A576 (1213) | G12130 |
| A576 (1215) | G12150 |
| A576 (1513) | G15130 |
| A576 (1518) | G15180 |
| A576 (1522) | G15220 |
| A576 (1524) | G15220 |
| A576 (1524) | G15240 |
| A576 (1525) | G15250 |
| A576 (1526) | G15260 |
| A576 (1527) | G15270 |
| A576 (1536) | G15360 |
| A576 (1541) | G15410 |
| A576 (1547) | G15470 |
| A576 (1548) | G15480 |
| A576 (1551) | G15510 |
| A576 (1552) | G15520 |
| A576 (1561) | G15610 |
| A576 (1566) | G15660 |
| A576 (1572) | G15720 |
| A579 (11) | K42598 |
| A579 (12) | K51255 |
| A579 (13) | K13051 |
| A579 (21) | K23477 |

| CROSS REFERENCE SPECIFICATIONS | UNIFIED NUMBER |
|---|---|
| A579 (22) | K24070 |
| A579 (23) | K24728 |
| A579 (31) | K32550 |
| A579 (32) | K44220 |
| A579 (33) | K14394 |
| A579 (51) | S41001 |
| A579 (52) | S42201 |
| A579 (53) | S43100 |
| A579 (61) | S35000 |
| A579 (62) | S17700 |
| A579 (63) | S15700 |
| A579 (64) | S35500 |
| A579 (71) | K92820 |
| A579 (72) | K92940 |
| A579 (73) | K93120 |
| A579 (74) | K91930 |
| A579 (75) | K91940 |
| A579 (81) | K91122 |
| A579 (82) | K91283 |
| A579 (83) | K91094 |
| A580 | S20161 |
| A580 | S21800 |
| A580 | S28200 |
| A580 | S40976 |
| A580 | S44400 |
| A580 | S44700 |
| A580 | S44800 |
| A580 (302) | S30200 |
| A580 (302B) | S30215 |
| A580 (304) | S30400 |
| A580 (304L) | S30403 |
| A580 (305) | S30500 |
| A580 (308) | S30800 |
| A580 (309) | S30900 |
| A580 (309Cb) | S30940 |
| A580 (309S) | S30908 |
| A580 (310) | S31000 |
| A580 (310S) | S31008 |
| A580 (314) | S31400 |
| A580 (316) | S31600 |
| A580 (316L) | S31603 |
| A580 (317) | S31700 |
| A580 (321) | S32100 |
| A580 (347) | S34700 |
| A580 (348) | S34800 |
| A580 (403) | S40300 |
| A580 (405) | S40500 |
| A580 (410) | S41000 |
| A580 (414) | S41400 |
| A580 (420) | S42000 |
| A580 (430) | S43000 |
| A580 (431) | S43100 |
| A580 (440A) | S44002 |
| A580 (440B) | S44003 |
| A580 (440C) | S44004 |
| A580 (446) | S44600 |
| A580 (XM-10) | S21900 |
| A580 (XM-11) | S21904 |
| A580 (XM-19) | S20910 |
| A580 (XM-28) | S24100 |
| A580 (XM-29) | S24000 |
| A580 (XM-31) | S21400 |
| A581 | S18235 |
| A581 | S41603 |
| A581 (303) | S30300 |
| A581 (303 Se) | S30323 |
| A581 (416) | S41600 |
| A581 (416Se) | S41623 |
| A581 (430F) | S43020 |

| CROSS REFERENCE SPECIFICATIONS | UNIFIED NUMBER |
|---|---|
| A581 (430FSe) | S43023 |
| A581 (XM-1) | S20300 |
| A581 (XM-2) | S30345 |
| A581 (XM-5) | S30310 |
| A581 (XM-6) | S41610 |
| A581(XM-34) | S18200 |
| A582 | S18235 |
| A582 | S41603 |
| A582 (303) | S30300 |
| A582 (303 Se) | S30323 |
| A582 (416) | S41600 |
| A582 (416Se) | S41623 |
| A582 (420F) | S42020 |
| A582 (420FSe) | S42023 |
| A582 (430F) | S43020 |
| A582 (430FSe) | S43023 |
| A582 (XM-1) | S20300 |
| A582 (XM-2) | S30345 |
| A582 (XM-5) | S30310 |
| A582 (XM-6) | S41610 |
| A582(XM-34) | S18200 |
| A587 | K11500 |
| A588 (A) | K11430 |
| A588 (B) | K12043 |
| A588 (C) | K11538 |
| A592 (A) | K11856 |
| A592 (E) | K11695 |
| A592 (F) | K11576 |
| A593 (416) | S41600 |
| A593 (431) | S43100 |
| A594 (416) | S41600 |
| A594 (431) | S43100 |
| A595 (A) | K02004 |
| A595 (B) | K02005 |
| A595 (C) | K11526 |
| A597 (CA-2) | T90102 |
| A597 (CD-2) | T90402 |
| A597 (CD-5) | T90405 |
| A597 (CH-12) | T90812 |
| A597 (CH-13) | T90813 |
| A597 (CM-2) | T11302 |
| A597 (CO-1) | T91501 |
| A597 (CS-5) | T91905 |
| A597 (CS-7) | T91907 |
| A600 (M-1) | T11301 |
| A600 (M-2) | T11302 |
| A600 (M-3, Cl.1) | T11313 |
| A600 (M-3, Cl.2) | T11323 |
| A600 (M-4) | T11304 |
| A600 (M-6) | T11306 |
| A600 (M-7) | T11307 |
| A600 (M-10) | T11310 |
| A600 (M-30) | T11330 |
| A600 (M-33) | T11333 |
| A600 (M-34) | T11334 |
| A600 (M-36) | T11336 |
| A600 (M-41) | T11341 |
| A600 (M-42) | T11342 |
| A600 (M-43) | T11343 ob |
| A600 (M-44) | T11344 ob |
| A600 (M-46) | T11346 |
| A600 (M-47) | T11347 ob |
| A600 (M-48) | T11348 |
| A600 (M-50) | T11350 |
| A600 (M-52) | T11352 |
| A600 (M-62) | T11362 |
| A600 (T-1) | T12001 |
| A600 (T-2) | T12002 ob |
| A600 (T-4) | T12004 |

| CROSS REFERENCE SPECIFICATIONS | UNIFIED NUMBER |
|---|---|
| A600 (T-5) | T12005 |
| A600 (T-6) | T12006 |
| A600 (T-8) | T12008 |
| A600 (T-15) | T12015 |
| A601 (A) | M29951 |
| A601 (B) | M29952 |
| A601 (C) | M29953 |
| A601 (D) | M29450 |
| A601 (E) | M29350 |
| A601 (F) | M29954 |
| A602 (M3210) | F20000 |
| A608 (HC-30) | J92613 |
| A608 (HD-50) | J93015 |
| A608 (HE-35) | J93413 |
| A608 (HF30) | J92803 |
| A608 (HH30) | J93513 |
| A608 (HH33) | J93633 |
| A608 (HI35) | J94013 |
| A608 (HK-30) | J94203 |
| A608 (HK40) | J94204 |
| A608 (HL-30) | N08613 |
| A608 (HL-40) | N08614 |
| A608 (HN40) | J94214 |
| A608 (HT-50) | N08050 |
| A608 (HU-50) | N08005 |
| A608 (HW-50) | N08006 |
| A608 (HX-50) | N06050 |
| A611 (A, B, C, E) | G10170 |
| A611 (C, D) | G10120 |
| A611 (D-1) | G10180 |
| A612 | K02900 |
| A618 (Ib) | K02601 |
| A618 (II) | K12609 |
| A618 (III) | K12700 |
| A632 (304) | S30400 |
| A632 (304L) | S30403 |
| A632 (310) | S31000 |
| A632 (316) | S31600 |
| A632 (316L) | S31603 |
| A632 (317) | S31700 |
| A632 (321) | S32100 |
| A632 (347) | S34700 |
| A632 (348) | S34800 |
| A632 (TP316) | S31600 |
| A633 | K01803 |
| A633 (A) | K01802 |
| A633 (C) | K12000 |
| A633 (D) | K12037 |
| A633 (E) | K12202 |
| A635 (1006) | G10060 |
| A635 (1008) | G10080 |
| A635 (1009) | G10090 |
| A635 (1012) | G10120 |
| A635 (1015) | G10150 |
| A635 (1016) | G10160 |
| A635 (1017) | G10170 |
| A635 (1018) | G10180 |
| A635 (1019) | G10190 |
| A635 (1020) | G10200 |
| A635 (1021) | G10210 |
| A635 (1022 | G10220 |
| A635 (1023) | G10230 |
| A635 (1524) | G15240 |
| A638 (660) | S66286 |
| A638 (662) | S66220 |
| A645 | K41583 |
| A646 (18) | K92820 |
| A646 (19) | K92890 |
| A646 (20) | K93120 |

| CROSS REFERENCE SPECIFICATIONS | UNIFIED NUMBER |
|---|---|
| A646 (98BV40) | K24336 |
| A646 (300M-8) | K44220 |
| A646 (3310-1) | G33106 |
| A646 (4130) | G41300 |
| A646 (4140) | G41400 |
| A646 (4330 Mod.-5) | K23080 |
| A646 (4335 Mod.-6) | K23477 |
| A646 (4340-7) | G43400 |
| A646 (4620-3) | G46200 |
| A646 (6150) | G61500 |
| A646 (8620-4) | G86200 |
| A646 (9310-2) | G93106 |
| A646 (D6AC) | K22925 |
| A646 (E52100) | G52986 |
| A646 (H-11) | T20811 |
| A646 (Marage 200) | K92820 |
| A646 (Marage 250) | K92890 |
| A646 (Marage 300) | K93120 |
| A646 (Nit. 135) | K24065 |
| A648 (II) | K06700 |
| A648 (III) | K07100 |
| A649 (1A) | K14247 |
| A649 (1B) | K24040 |
| A649 (2) | K05001 |
| A649 (3) | K13047 |
| A649 (Class 5) | K05502 |
| A656 (1) | K11804 |
| A656 (2) | K11509 |
| A659 (1015) | G10150 |
| A659 (1016) | G10160 |
| A659 (1017) | G10170 |
| A659 (1018) | G10180 |
| A659 (1020) | G10200 |
| A659 (1021) | G10210 |
| A659 (1023) | G10230 |
| A660 (WCA) | J02504 |
| A660 (WCB) | J03003 |
| A660 (WCC) | J02505 |
| A662 (A) | K01701 |
| A662 (B) | K02203 |
| A662 (C) | K02007 |
| A666 | S20103 |
| A666 | S20153 |
| A666 | S20400 |
| A666 | S30103 |
| A666 | S30153 |
| A666 (201) | S20100 |
| A666 (202) | S20200 |
| A666 (205) | S20500 |
| A666 (301) | S30100 |
| A666 (302) | S30200 |
| A666 (304) | S30400 |
| A666 (304L) | S30403 |
| A666 (304LN) | S30453 |
| A666 (304N) | S30451 |
| A666 (316) | S31600 |
| A666 (316L) | S31603 |
| A666 (316N) | S31651 |
| A666 (XM-11) | S21904 |
| A666 (XM-14) | S21460 |
| A675 (45) | D40450 |
| A675 (50) | D40500 |
| A675 (55) | D40550 |
| A675 (60) | D40600 |
| A675 (65) | D40650 |
| A675 (70) | D40700 |
| A675 (75) | D40750 |
| A675 (80) | D40800 |
| A675 (90) | D40900 |

| CROSS REFERENCE SPECIFICATIONS | UNIFIED NUMBER |
|---|---|
| A678 (A) | K01600 |
| A678 (B) | K02002 |
| A678 (C) | K02204 |
| A678 (D) | K12202 |
| A679 | K08200 |
| A681 (0-7) | T31507 |
| A681 (A-2) | T30102 |
| A681 (A-3) | T30103 ob |
| A681 (A-4) | T30104 |
| A681 (A-5) | T30105 ob |
| A681 (A-6) | T30106 |
| A681 (A-7) | T30107 |
| A681 (A-8) | T30108 |
| A681 (A-9) | T30109 |
| A681 (A-10) | T30110 |
| A681 (D-2) | T30402 |
| A681 (D-3) | T30403 |
| A681 (D-4) | T30404 |
| A681 (D-5) | T30405 |
| A681 (D-7) | T30407 |
| A681 (F-1) | T60601 ob |
| A681 (F-2) | T60602 ob |
| A681 (H-10) | T20810 |
| A681 (H-11) | T20811 |
| A681 (H-12) | T20812 |
| A681 (H-13) | T20813 |
| A681 (H-14) | T20814 |
| A681 (H-19) | T20819 |
| A681 (H-21) | T20821 |
| A681 (H-22) | T20822 |
| A681 (H-23) | T20823 |
| A681 (H-24) | T20824 |
| A681 (H-25) | T20825 ob |
| A681 (H-26) | T20826 |
| A681 (H-41) | T20841 ob |
| A681 (H-42) | T20842 |
| A681 (H-43) | T20843 ob |
| A681 (L-2) | T61202 |
| A681 (L-3) | T61203 ob |
| A681 (L-6) | T61206 |
| A681 (O-1) | T31501 |
| A681 (O-2) | T31502 |
| A681 (O-6) | T31506 |
| A681 (P-2) | T51602 ob |
| A681 (P-3) | T51603 ob |
| A681 (P-4) | T51604 ob |
| A681 (P-5) | T51605 ob |
| A681 (P-6) | T51606 |
| A681 (P-20) | T51620 |
| A681 (P-21) | T51621 |
| A681 (S-1) | T41901 |
| A681 (S-2) | T41902 |
| A681 (S-4) | T41904 |
| A681 (S-5) | T41905 |
| A681 (S-6) | T41906 |
| A681 (S-7) | T41907 |
| A682 (1030) | G10300 |
| A682 (1035) | G10350 |
| A682 (1040) | G10400 |
| A682 (1045) | G10450 |
| A682 (1050) | G10500 |
| A682 (1055) | G10550 |
| A682 (1060) | G10600 |
| A682 (1064) | G10640 |
| A682 (1065) | G10650 |
| A682 (1070) | G10700 |
| A682 (1074) | G10740 |
| A682 (1080) | G10800 |
| A682 (1085) | G10850 |

| CROSS REFERENCE SPECIFICATIONS | UNIFIED NUMBER |
|---|---|
| A682 (1086) | G10860 |
| A682 (1095) | G10950 |
| A686 (W-1) | T72301 |
| A686 (W-2) | T72302 |
| A686 (W-5) | T72305 |
| A688 | N08367 |
| A688 | N08926 |
| A688 (304) | S30400 |
| A688 (304L) | S30403 |
| A688 (304LN) | S30453 |
| A688 (316) | S31600 |
| A688 (316L) | S31603 |
| A688 (316LN) | S31653 |
| A688 (316N) | S31651 |
| A688 (TP304LN) | S30453 |
| A688 (TP304N) | S30451 |
| A688 (TP316) | S31600 |
| A688 (TP316LN) | S31653 |
| A688 (TP316N) | S31651 |
| A688 (XM-29) | S24000 |
| A689 | G86500 |
| A689 | G86550 |
| A689 | G86600 |
| A689 | G92600 |
| A689 (8637) | G86370 |
| A690 | K12249 |
| A693 (630) | S17400 |
| A693 (631) | S17700 |
| A693 (632) | S15700 |
| A693 (633) | S35000 |
| A693 (634) | S35500 |
| A693 (635) | S17600 |
| A693 (XM-9) | S36200 |
| A693 (XM-12) | S15500 |
| A693 (XM-13) | S13800 |
| A693 (XM-16) | S45500 |
| A693 (XM-25) | S45000 |
| A694 | K03014 |
| A695 (B) | K03504 ob |
| A696 (B, C) | K03200 |
| A705 (630) | S17400 |
| A705 (631) | S17700 |
| A705 (632) | S15700 |
| A705 (634) | S35500 |
| A705 (635) | S17600 |
| A705 (XM-12) | S15500 |
| A705 (XM-13) | S13800 |
| A705 (XM-16) | S45500 |
| A705 (XM-25) | S45000 |
| A706 | K03011 |
| A707 (L1) | K02302 |
| A707 (L2) | K03301 |
| A707 (L3) | K12510 |
| A707 (L4) | K12089 |
| A707 (L5) | K20934 |
| A707 (L6) | K20902 |
| A707 (L7) | K32218 |
| A707 (L8) | K42247 |
| A710 (A) | K20747 |
| A710 (B) | K20622 |
| A711 | G13300 |
| A711 | G41400 |
| A711 | G41420 |
| A711 | G41450 |
| A711 | G41500 |
| A711 | G43400 |
| A711 | G47200 |
| A711 | G51986 |
| A711 | G61200 |
| A711 | G86600 |
| A711 | G98400 |
| A711 (1335) | G13350 |
| A713 (1055) | G10550 |
| A713 (1059) | G10590 |
| A713 (1060) | G10600 |
| A713 (1064) | G10640 |
| A713 (1065) | G10650 |
| A713 (1069) | G10690 |
| A713 (1070) | G10700 |
| A713 (1074) | G10740 |
| A713 (1075) | G10750 |
| A713 (1078) | G10780 |
| A713 (1080) | G10800 |
| A713 (1084) | G10840 |
| A713 (1086) | G10860 |
| A713 (1090) | G10900 |
| A713 (1095) | G10950 |
| A713 (1561) | G15610 |
| A713 (1566) | G15660 |
| A713 (1572) | G15720 |
| A714 (I) | K12608 |
| A714 (II) | K12609 |
| A714 (III) | K12709 |
| A714 (IV) | K11356 |
| A714 (V) | K22035 |
| A714 (VI) | K11835 |
| A715 | K11501 |
| A716 | F32900 |
| A723 (1) | K23550 |
| A723 (2) | K34035 |
| A723 (3) | K44045 |
| A724 (A) | K11831 |
| A724 (B) | K12031 |
| A724 (C) | K12037 |
| A729 | K06001 |
| A730 (A) | K01502 |
| A730 (B) | K02000 |
| A730 (C, D, E) | K04700 |
| A730 (F) | K05200 |
| A732 (1A) | J02002 |
| A732 (2A, 2Q) | J03011 |
| A732 (3A, 3Q) | J04002 |
| A732 (5N) | J13052 |
| A732 (6N) | J13512 |
| A732 (7Q) | J13045 |
| A732 (8Q) | J14049 |
| A732 (9Q) | J23055 |
| A732 (10Q) | J24054 |
| A732 (11Q) | J12094 |
| A732 (12Q) | J15048 |
| A732 (13Q) | J12048 |
| A732 (14Q) | J13051 |
| A732 (15A) | J19966 |
| A732 (21) | R30021 |
| A734 (A) | K21205 |
| A734 (B) | K11720 |
| A735 | K10623 |
| A736 | K20747 |
| A737 (B) | K12001 |
| A737 (C) | K12202 |
| A738 (A) | K12447 |
| A738 (B) | K12001 |
| A738 (C) | K02008 |
| A739 (B11) | K11797 |
| A739 (B22) | K21390 |
| A743 | J91804 |
| A743 | J92700 |
| A743 | J94652 |
| A743 | J94653 |
| A743 (CA6N) | J91650 |
| A743 (CA-6NM) | J91540 |
| A743 (CA-15) | J91150 |
| A743 (CA-15M) | J91151 |
| A743 (CA28MWV) | J91422 |
| A743 (CA-40) | J91153 |
| A743 (CA40F) | J91154 |
| A743 (CB-30) | J91803 |
| A743 (CC-50) | J92615 |
| A743 (CD3MWCuN) | J93380 |
| A743 (CE-30) | J93423 |
| A743 (CF-3) | J92500 |
| A743 (CF-3M) | J92800 |
| A743 (CF3MN) | J92804 |
| A743 (CF-8) | J92600 |
| A743 (CF-8C) | J92710 |
| A743 (CF-8M) | J92900 |
| A743 (CF10SMnN) | J92972 |
| A743 (CF-16F) | J92701 |
| A743 (CF-20) | J92602 |
| A743 (CG-3M) | J92999 |
| A743 (CG6MMN) | J93790 |
| A743 (CG6MMN) | J94652 |
| A743 (CG-8M) | J93000 |
| A743 (CG-12) | J93001 |
| A743 (CH-10) | J93401 |
| A743 (CH-20) | J93402 |
| A743 (CK-3MCuN) | J93254 |
| A743 (CK-3MN) | J94653 |
| A743 (CK-20) | J94202 |
| A743 (CN-3M) | J94652 |
| A743 (CN3MN) | J94651 |
| A743 (CN7M) | N08007 |
| A743 (CN-7MS) | J94650 |
| A744 | J92700 |
| A744 | J94653 |
| A744 (CD4MCu) | J93370 |
| A744 (CF-3) | J92500 |
| A744 (CF-3M) | J92800 |
| A744 (CF-8) | J92600 |
| A744 (CF-8C) | J92710 |
| A744 (CF-8M) | J92900 |
| A744 (CG-3M) | J92999 |
| A744 (CG-8M) | J93000 |
| A744 (CK-3MN) | J94653 |
| A744 (CN3MN) | J94651 |
| A744 (CN7M) | N08007 |
| A744 (CN-7MS) | J94650 |
| A747 (CB7Cu-1) | J92180 |
| A747 (CB7Cu-2) | J92110 |
| A752 (50B44) | G50441 |
| A752 (50B46) | G50461 |
| A752 (50B50) | G50501 |
| A752 (50B60) | G50601 |
| A752 (51B60) | G51601 |
| A752 (81B45) | G81451 |
| A752 (94B17) | G94171 |
| A752 (94B30) | G94301 |
| A752 (1330) | G13300 |
| A752 (1335) | G13350 |
| A752 (1340) | G13400 |
| A752 (1345) | G13450 |
| A752 (4012) | G40120 |
| A752 (4023) | G40230 |
| A752 (4024) | G40240 |
| A752 (4027) | G40270 |
| A752 (4028) | G40280 |
| A752 (4037) | G40370 |

| CROSS REFERENCE SPECIFICATIONS | UNIFIED NUMBER |
|---|---|
| A752 (4047) | G40470 |
| A752 (4118) | G41180 |
| A752 (4130) | G41300 |
| A752 (4137) | G41370 |
| A752 (4140) | G41400 |
| A752 (4142) | G41420 |
| A752 (4145) | G41450 |
| A752 (4147) | G41470 |
| A752 (4150) | G41500 |
| A752 (4161) | G41610 |
| A752 (4320) | G43200 |
| A752 (4340) | G43400 |
| A752 (4419) | G44190 |
| A752 (4615) | G46150 |
| A752 (4620) | G46200 |
| A752 (4621) | G46210 |
| A752 (4626) | G46260 |
| A752 (4718) | G47180 |
| A752 (4720) | G47200 |
| A752 (4815) | G48150 |
| A752 (4817) | G48170 |
| A752 (4820) | G48200 |
| A752 (5015) | G50150 |
| A752 (5120) | G51200 |
| A752 (5130) | G51300 |
| A752 (5132) | G51320 |
| A752 (5135) | G51350 |
| A752 (5140) | G51400 |
| A752 (5145) | G51450 |
| A752 (5147) | G51470 |
| A752 (5150) | G51500 |
| A752 (5155) | G51550 |
| A752 (5160) | G51600 |
| A752 (6118) | G61180 |
| A752 (6150) | G61500 |
| A752 (8615) | G86150 |
| A752 (8617) | G86170 |
| A752 (8620) | G86200 |
| A752 (8622) | G86220 |
| A752 (8625) | G86250 |
| A752 (8627) | G86270 |
| A752 (8630) | G86300 |
| A752 (8637) | G86370 |
| A752 (8640) | G86400 |
| A752 (8642) | G86420 |
| A752 (8645) | G86450 |
| A752 (8655) | G86550 |
| A752 (8720) | G87200 |
| A752 (8740) | G87400 |
| A752 (8822) | G88220 |
| A752 (9254) | G92540 |
| A752 (9255) | G92550 |
| A752 (9260) | G92600 |
| A752 (E4340) | G43406 |
| A752 (E51100) | G51986 |
| A752 (E52100) | G52986 |
| A753 (Alloy 1) | K94490 |
| A753 (Alloy 2) | K94840 |
| A753 (Alloy 3) | N14076 |
| A753 (Alloy 4) | N14080 |
| A756 (440C) | S44025 |
| A757 (A1Q) | J03002 |
| A757 (A2Q) | J02503 |
| A757 (B2N, B2Q) | J22501 |
| A757 (B3N, B3Q) | J31500 |
| A757 (B4N, B4Q) | J41501 |
| A757 (C1Q) | J12582 |
| A757 (D1N, D1Q, D1N1, D1Q1, D1N2, D1Q2, D1N3, D1Q3) | J22092 |
| A757 (E1Q) | J42220 |
| A757 (E2N, E2Q) | J42065 |
| A757 (E3N) | J91550 |
| A758 | K02741 |
| A759 | K07500 |
| A765 (I) | K03046 |
| A765 (II) | K03047 |
| A765 (III) | K32026 |
| A765 (IV) | K02009 |
| A771 | S38660 |
| A771 | S42100 |
| A774 (304) | S30400 |
| A774 (TP304) | S30400 |
| A774 (TP304L) | S30403 |
| A774 (TP316L) | S31603 |
| A774 (TP317L) | S31703 |
| A774 (TP321) | S32100 |
| A774 (TP347) | S34700 |
| A778 (TP304L) | S30403 |
| A778 (TP316L) | S31603 |
| A778 (TP317L) | S31703 |
| A778 (TP321) | S32100 |
| A778 (TP347) | S34700 |
| A787 (1008) | G10080 |
| A787 (1010) | G10100 |
| A787 (1011) | G10110 |
| A787 (1015) | G10150 |
| A787 (1016) | G10160 |
| A787 (1017) | G10170 |
| A787 (1018) | G10180 |
| A787 (1019) | G10190 |
| A787 (1021) | G10210 |
| A787 (MT 1010) | G10100 |
| A787 (MT 1015) | G10150 |
| A787 (MT 1020) | G10200 |
| A789 | S31200 |
| A789 | S31260 |
| A789 | S31500 |
| A789 | S31803 |
| A789 | S32205 |
| A789 | S32304 |
| A789 | S32550 |
| A789 | S32750 |
| A789 | S32760 |
| A789 | S32906 |
| A789 | S32950 |
| A789 | S39277 |
| A789 (329) | S32900 |
| A790 | S31200 |
| A790 | S31260 |
| A790 | S31500 |
| A790 | S31803 |
| A790 | S32205 |
| A790 | S32304 |
| A790 | S32520 |
| A790 | S32550 |
| A790 | S32750 |
| A790 | S32760 |
| A790 | S32900 |
| A790 | S32906 |
| A790 | S32950 |
| A790 | S39277 |
| A791 (18-2) | S44400 |
| A793 (304) | S30400 |
| A793 (304L) | S30403 |
| A793 (316) | S31600 |
| A793 (316L) | S31603 |
| A794 (1015) | G10150 |
| A794 (1016) | G10160 |
| A794 (1017) | G10170 |
| A794 (1018) | G10180 |
| A794 (1020) | G10200 |
| A794 (1021) | G10210 |
| A794 (1023) | G10230 |
| A801 (Alloy 1) | R30005 |
| A801 (Alloy 2) | K92650 |
| A803 (18-2) | S44400 |
| A803 (25-4-4) | S44635 |
| A803 (26-3-3) | S44660 |
| A803 (29-4) | S44700 |
| A803 (29-4-2) | S44800 |
| A803 (29-4C) | S44735 |
| A803 (TP409) | S40900 |
| A803 (TP439) | S43035 |
| A803 (TPXM-27) | S44627 |
| A803 (TPXM-33) | S44626 |
| A808 | K11852 |
| A812 (F55) | S32760 |
| A813 | N08367 |
| A813 | S30815 |
| A813 | S31254 |
| A813 (TP304) | S30400 |
| A813 (TP304H) | S30409 |
| A813 (TP304L) | S30403 |
| A813 (TP304LN) | S30453 |
| A813 (TP304N) | S30451 |
| A813 (TP309Cb) | S30940 |
| A813 (TP309S) | S30908 |
| A813 (TP310Cb) | S31040 |
| A813 (TP310S) | S31008 |
| A813 (TP316) | S31600 |
| A813 (TP316H) | S31609 |
| A813 (TP316L) | S31603 |
| A813 (TP316LN) | S31653 |
| A813 (TP316N) | S31651 |
| A813 (TP317) | S31700 |
| A813 (TP317L) | S31703 |
| A813 (TP321) | S32100 |
| A813 (TP321H) | S32109 |
| A813 (TP347) | S34700 |
| A813 (TP347H) | S34709 |
| A813 (TP348) | S34800 |
| A813 (TP348H) | S34809 |
| A813 (TPXM-10) | S21900 |
| A813 (TPXM-11) | S21904 |
| A813 (TPXM-15) | S38100 |
| A813 (TPXM-19) | S20910 |
| A813 (TPXM-29) | S24000 |
| A814 | N08367 |
| A814 | S30815 |
| A814 | S31254 |
| A814 (TP304) | S30400 |
| A814 (TP304H) | S30409 |
| A814 (TP304H) | S31609 |
| A814 (TP304L) | S30403 |
| A814 (TP304LN) | S30453 |
| A814 (TP304N) | S30451 |
| A814 (TP309Cb) | S30940 |
| A814 (TP309S) | S30908 |
| A814 (TP310Cb) | S31040 |
| A814 (TP310S) | S31008 |
| A814 (TP316) | S31600 |
| A814 (TP316H) | S31609 |
| A814 (TP316L) | S31603 |
| A814 (TP316LN) | S31653 |
| A814 (TP316N) | S31651 |
| A814 (TP317) | S31700 |
| A814 (TP317L) | S31703 |

| CROSS REFERENCE SPECIFICATIONS | UNIFIED NUMBER |
|---|---|
| A814 (TP321) | S32100 |
| A814 (TP321H) | S32109 |
| A814 (TP347) | S34700 |
| A814 (TP347H) | S34709 |
| A814 (TP348) | S34800 |
| A814 (TP348H) | S34809 |
| A814 (TPXM-10) | S21900 |
| A814 (TPXM-11) | S21904 |
| A814 (TPXM-11) | S21980 |
| A814 (TPXM-15) | S38100 |
| A814 (TPXM-19) | S20910 |
| A814 (TPXM-29) | S24000 |
| A815 | S31803 |
| A815 | S32205 |
| A815 | S32550 |
| A815 | S32750 |
| A815 | S32760 |
| A815 | S32950 |
| A815 | S39274 |
| A815 | S41500 |
| A815 (430 Ti) | S43036 |
| A815 (WP27) | S44627 |
| A815 (WP33) | S44626 |
| A815 (WP410) | S41000 |
| A815 (WP429) | S42900 |
| A815 (WP430) | S43000 |
| A815 (WP446) | S44600 |
| A821 | K06700 |
| A822 | K01807 |
| A826 | S38660 |
| A826 | S42100 |
| A826 (TP316) | S31600 |
| A827 | G10090 |
| A827 | G10200 |
| A827 | G10350 |
| A827 | G10400 |
| A827 | G10450 |
| A827 | G10500 |
| A829 | G13300 |
| A829 | G13350 |
| A829 | G13400 |
| A829 | G13450 |
| A829 | G41180 |
| A829 | G41300 |
| A829 | G41350 |
| A829 | G41370 |
| A829 | G41400 |
| A829 | G41420 |
| A829 | G41450 |
| A829 | G43400 |
| A829 | G43406 |
| A829 | G46150 |
| A829 | G46170 |
| A829 | G46200 |
| A829 | G51600 |
| A829 | G61500 |
| A829 | G86150 |
| A829 | G86170 |
| A829 | G86200 |
| A829 | G86220 |
| A829 | G86250 |
| A829 | G86270 |
| A829 | G86300 |
| A829 | G86370 |
| A829 | G86400 |
| A829 | G86550 |
| A829 | G87420 |
| A830 | G10060 |
| A830 | G10080 |
| A830 | G10090 |
| A830 | G10100 |
| A830 | G10120 |
| A830 | G10150 |
| A830 | G10160 |
| A830 | G10170 |
| A830 | G10180 |
| A830 | G10190 |
| A830 | G10200 |
| A830 | G10210 |
| A830 | G10220 |
| A830 | G10230 |
| A830 | G10250 |
| A830 | G10260 |
| A830 | G10300 |
| A830 | G10330 |
| A830 | G10350 |
| A830 | G10370 |
| A830 | G10380 |
| A830 | G10390 |
| A830 | G10400 |
| A830 | G10420 |
| A830 | G10430 |
| A830 | G10450 |
| A830 | G10460 |
| A830 | G10490 |
| A830 | G10500 |
| A830 | G10550 |
| A830 | G10700 |
| A830 | G10740 |
| A830 | G10780 |
| A830 | G10800 |
| A830 | G10840 |
| A830 | G10850 |
| A830 | G10860 |
| A830 | G10900 |
| A830 | G10950 |
| A830 | G15240 |
| A830 | G15270 |
| A830 | G15360 |
| A830 | G15410 |
| A830 | G15480 |
| A830 | G15520 |
| A831 | S38660 |
| A831 | S42100 |
| A831 (316) | S31600 |
| A831 (T91) | K90901 |
| A832 | K31830 |
| A832 (22V) | K31835 |
| A832 (23V) | K31286 |
| A834 | G33106 |
| A837 (410) | S41000 |
| A837 (4320) | G43200 |
| A837 (4620) | G46200 |
| A837 (4815) | G48150 |
| A837 (4815) | G86200 |
| A837 (8620) | G86200 |
| A837 (E3310) | G33106 |
| A841 | K02010 |
| A844 | K81340 |
| A851 (TP304) | S30400 |
| A851 (TP304L) | S30403 |
| A852 | K12043 |
| A853 (1005) | G10050 |
| A853 (1006) | G10060 |
| A853 (1008) | G10080 |
| A853 (1010) | G10100 |
| A853 (1011) | G10110 |
| A853 (1015) | G10150 |
| A853 (1018) | G10180 |
| A853 (1020) | G10200 |
| A857 | K02507 |
| A859 | K20747 |
| A866 (1030) | G10300 |
| A866 (1040) | G10400 |
| A866 (1050) | G10500 |
| A866 (1541) | G15410 |
| A866 (1552) | G15520 |
| A866 (4130) | G41300 |
| A866 (4140) | G41400 |
| A866 (4150) | G41450 |
| A866 (4150) | G41500 |
| A866 (5140) | G51400 |
| A866 (5150) | G51500 |
| A866 (5155) | G51550 |
| A866 (6150) | G61500 |
| A872 | J93183 |
| A872 | J93550 |
| A877 | G92540 |
| A887 (304B) | S30460 |
| A887 (304B1) | S30461 |
| A887 (304B2) | S30462 |
| A887 (304B3) | S30463 |
| A887 (304B4) | S30464 |
| A887 (304B5) | S30465 |
| A887 (304B6) | S30466 |
| A887 (304B7) | S30467 |
| A890 (1A) | J93370 |
| A890 (1B) | J93372 |
| A890 (1C) | J93373 |
| A890 (2A) | J93345 |
| A890 (3A) | J93371 |
| A890 (4A) | J92205 |
| A890 (5A) | J93404 |
| A890 (6A) | J93380 |
| A891 | S66286 |
| A891 (A286) | S66286 |
| A895 (303) | S30300 |
| A895 (303 Se) | S30323 |
| A895 (416) | S41600 |
| A895 (416Se) | S41623 |
| A895 (420F) | S42020 |
| A895 (420FSe) | S42023 |
| A895 (430F) | S43020 |
| A895 (430FSe) | S43023 |
| A903 (317L) | S31703 |
| A905 | K08501 |
| A908 (304) | S30400 |
| A915 (SC 1020) | J02003 |
| A915 (SC 1025) | J02508 |
| A915 (SC 1030) | J03012 |
| A915 (SC 1040) | J04003 |
| A915 (SC 1045) | J04502 |
| A915 (SC 4130) | J13502 |
| A915 (SC 4140) | J14045 |
| A915 (SC 4330) | J23259 |
| A915 (SC 4340) | J24053 |
| A915 (SC 8620) | J12095 |
| A915 (SC 8625) | J12595 |
| A915 (SC 8630) | J13095 |
| A923 | S31803 |
| A923 | S32205 |
| A924 (304L) | S30403 |
| A924 (309HCb) | S30941 |
| A924 (TP304H) | S30409 |
| A924 (TP304L) | S30403 |
| A924 (TP304LN) | S30453 |
| A924 (TP304N) | S30451 |

| CROSS REFERENCE SPECIFICATIONS | UNIFIED NUMBER |
|---|---|
| A924 (TP309Cb) | S30940 |
| A924 (TP309HCb) | S30941 |
| A924 (TP309S) | S30908 |
| A924 (TP310Cb) | S31040 |
| A924 (TP310HCb) | S31041 |
| A924 (TP310S) | S31008 |
| A928 | S30600 |
| A928 | S30815 |
| A928 | S31200 |
| A928 | S31260 |
| A928 | S31500 |
| A928 | S31803 |
| A928 | S32205 |
| A928 | S32304 |
| A928 | S32550 |
| A928 | S32750 |
| A928 | S32900 |
| A928 | S32950 |
| A929 | S32950 |
| A943 | S30600 |
| A943 | S30815 |
| A943 | S31050 |
| A943 | S31272 |
| A943 | S31725 |
| A943 | S31726 |
| A943 | S32615 |
| A943 | S34565 |
| A943 (309HCb) | S30941 |
| A943 (T310Cb) | S31040 |
| A943 (TP304) | S30400 |
| A943 (TP304H) | S30409 |
| A943 (TP304L) | S30403 |
| A943 (TP304LN) | S30453 |
| A943 (TP304N) | S30451 |
| A943 (TP309Cb) | S30940 |
| A943 (TP309H) | S30909 |
| A943 (TP309HCb) | S30941 |
| A943 (TP309S) | S30908 |
| A943 (TP310H) | S31009 |
| A943 (TP310S) | S31008 |
| A943 (TP316) | S31600 |
| A943 (TP316H) | S31609 |
| A943 (TP316L) | S31603 |
| A943 (TP316LN) | S31653 |
| A943 (TP316N) | S31651 |
| A943 (TP317) | S31700 |
| A943 (TP317L) | S31703 |
| A943 (TP321) | S32100 |
| A943 (TP321H) | S32109 |
| A943 (TP347) | S34700 |
| A943 (TP347H) | S34709 |
| A943 (TP348H) | S34800 |
| A943 (TP348H) | S34809 |
| A943 (TPHCb) | S31041 |
| A946 | S70003 |
| A949 | S31200 |
| A949 | S31260 |
| A949 | S31500 |
| A949 | S31803 |
| A949 | S32304 |
| A949 | S32550 |
| A949 | S32750 |
| A949 | S32900 |
| A949 | S32950 |
| A953 | S70003 |
| A954 | S70003 |
| A958 (SC 1020) | J02003 |
| A958 (SC 1025) | J02508 |
| A958 (SC 1030) | J03012 |
| A958 (SC 1040) | J04003 |
| A958 (SC 1045) | J04502 |
| A958 (SC 4130) | J13502 |
| A958 (SC 4140) | J14045 |
| A958 (SC 4330) | J23259 |
| A958 (SC 4340) | J24053 |
| A958 (SC 8620) | J12095 |
| A958 (SC 8625) | J12595 |
| A958 (SC 8630) | J13095 |
| A959 | N08020 |
| A959 | N08367 |
| A959 | N08800 |
| A959 | N08810 |
| A959 | N08811 |
| A959 | N08904 |
| A959 | N08926 |
| A965 (304) | S30400 |
| A965 (F46) | S30600 |
| A965 (F304H) | S30409 |
| A965 (F304L) | S30403 |
| A965 (F304LN) | S30453 |
| A965 (F304N) | S30451 |
| A965 (F309H) | S30909 |
| A965 (F310) | S31000 |
| A965 (F310H) | S31009 |
| A965 (F316) | S31600 |
| A965 (F316H) | S31609 |
| A965 (F316L) | S31603 |
| A965 (F316LN) | S31653 |
| A965 (F316N) | S31651 |
| A965 (F321) | S32100 |
| A965 (F321H) | S32109 |
| A965 (F347) | S34700 |
| A965 (F347H) | S34709 |
| A965 (F348) | S34800 |
| A965 (F348H) | S34809 |
| A965 (FXM-19) | S20910 |
| A968 | S70003 |
| A982 (D) | S42225 |
| A988 | N08367 |
| A988 | S21904 |
| A988 | S30400 |
| A988 | S30403 |
| A988 | S30451 |
| A988 | S30453 |
| A988 | S31254 |
| A988 | S31600 |
| A988 | S31603 |
| A988 | S31651 |
| A988 | S31653 |
| A988 | S31700 |
| A988 | S31703 |
| A988 | S31725 |
| A988 | S31726 |
| A988 | S31803 |
| A988 | S32205 |
| A988 | S32654 |
| A988 | S32750 |
| A988 | S32760 |
| A988 | S32950 |
| A988 | S39274 |
| A988 | S41000 |
| A988 | S41026 |
| A988 | S41500 |
| A988 | S42390 |
| A988 (304) | S30400 |
| A988 (316H) | S31609 |
| A989 | K21590 |
| A989 | K31545 |
| A989 | K90941 |
| A990 (CW2M) | N26455 |
| A995 (1B, CD4MCuN) | J93372 |
| A995 (2A | J93345 |
| A995 (3A, CD6MN) | J93371 |
| A995 (4A | J92205 |
| A995 (5A, CE3MN) | J93404 |
| A995 (6A | J93380 |
| A1010 | S41003 |
| B1 | C10100 |
| B1 | C10200 |
| B1 | C10400 |
| B1 | C10500 |
| B1 | C10700 |
| B1 | C11000 |
| B1 | C11300 |
| B1 | C11400 |
| B1 | C11500 |
| B1 | C11600 |
| B2 | C10100 |
| B2 | C10200 |
| B2 | C10400 |
| B2 | C10500 |
| B2 | C10700 |
| B2 | C11000 |
| B2 | C11300 |
| B2 | C11400 |
| B2 | C11500 |
| B2 | C11600 |
| B3 | C10100 |
| B3 | C10200 |
| B3 | C10400 |
| B3 | C10500 |
| B3 | C10700 |
| B3 | C11000 |
| B3 | C11300 |
| B3 | C11400 |
| B3 | C11500 |
| B3 | C11600 |
| B5 (DHP) | C12200 |
| B5 (DHPS) | C12300 |
| B5 (DLP) | C12000 |
| B5 (DPA) | C14200 |
| B5 (DPTE) | C14500 |
| B5 (STP) | C11300 |
| B5 (STP) | C11400 |
| B5 (STP) | C11500 |
| B5 (STP) | C11600 |
| B6 (High Grade) | Z15001 |
| B6 (Prime Western) | Z19001 |
| B6 (Special High Grade) | Z13001 |
| B8 | C10100 |
| B8 | C10200 |
| B8 | C11000 |
| B8 | C11300 |
| B8 | C11400 |
| B8 | C11500 |
| B8 | C11600 |
| B9 | C16500 |
| B9 | C50500 |
| B16 | C36000 |
| B19 | C26000 |
| B21 | C46200 |
| B21 | C46400 |
| B21 | C47940 |
| B21 | C48200 |
| B21 | C48500 |
| B22 | C86300 |
| B22 | C90500 |

| CROSS REFERENCE SPECIFICATIONS | UNIFIED NUMBER |
|---|---|
| B22 | C91100 |
| B22 | C91300 |
| B23 (1) | L13910 |
| B23 (2) | L13890 |
| B23 (3) | L13840 |
| B23 (7) | L53585 |
| B23 (8) | L53565 |
| B23 (11) | L13870 |
| B23 (13) | L53346 |
| B23 (15) | L53620 |
| B26 (201.0) | A02010 |
| B26 (204.0) | A02040 |
| B26 (208.0) | A02080 ob |
| B26 (222.0) | A02220 ob |
| B26 (242.0) | A02420 |
| B26 (295.0) | A02950 |
| B26 (319.0) | A03190 |
| B26 (328.0) | A03280 ob |
| B26 (355.0) | A03550 |
| B26 (356.0) | A03560 |
| B26 (443.0) | A04430 |
| B26 (514.0) | A05140 |
| B26 (520.0) | A05200 |
| B26 (535.0) | A05350 |
| B26 (705.0) | A07050 |
| B26 (707.0) | A07070 |
| B26 (710.0) | A07100 |
| B26 (712.0) | A07120 |
| B26 (713.0) | A07130 |
| B26 (771.0) | A07710 |
| B26 (850.0) | A08500 |
| B26 (851.0) | A08510 |
| B26 (852.0) | A08520 |
| B26 (A356.0) | A13560 |
| B26 (B443.0) | A24430 |
| B26 (C355.0) | A33550 |
| B29 (Chemical Lead) | L51120 |
| B29 (Common Lead) | L50045 |
| B29 (Copper Bearing Lead) | L51121 |
| B29 (Corroding Lead) | L50042 |
| B30 | C81400 |
| B30 | C83450 |
| B30 | C83800 |
| B30 | C84200 |
| B30 | C84400 |
| B30 | C84800 |
| B30 | C85200 |
| B30 | C85400 |
| B30 | C85700 |
| B30 | C85800 |
| B30 | C86100 |
| B30 | C86200 |
| B30 | C86300 |
| B30 | C86400 |
| B30 | C86500 |
| B30 | C86700 |
| B30 | C87300 |
| B30 | C87400 |
| B30 | C87500 |
| B30 | C87600 |
| B30 | C87610 |
| B30 | C87800 |
| B30 | C87900 ob |
| B30 | C89320 |
| B30 | C89520 |
| B30 | C89844 |
| B30 | C90300 |
| B30 | C90500 |
| B30 | C90700 |
| B30 | C90800 |
| B30 | C91000 |
| B30 | C91100 |
| B30 | C91300 |
| B30 | C91600 |
| B30 | C91700 |
| B30 | C92200 |
| B30 | C92210 |
| B30 | C92300 |
| B30 | C92500 |
| B30 | C92600 |
| B30 | C92700 |
| B30 | C92800 |
| B30 | C92900 |
| B30 | C93200 |
| B30 | C93400 |
| B30 | C93500 |
| B30 | C93600 |
| B30 | C93700 |
| B30 | C93800 |
| B30 | C93900 |
| B30 | C94000 |
| B30 | C94100 |
| B30 | C94300 |
| B30 | C94400 |
| B30 | C94500 |
| B30 | C94700 |
| B30 | C94800 |
| B30 | C94900 |
| B30 | C95200 |
| B30 | C95300 |
| B30 | C95400 |
| B30 | C95410 |
| B30 | C95500 |
| B30 | C95520 |
| B30 | C95600 |
| B30 | C95700 |
| B30 | C95800 |
| B30 | C95820 |
| B30 | C95900 |
| B30 | C96200 |
| B30 | C96400 |
| B30 | C96800 |
| B30 | C97300 |
| B30 | C97600 |
| B30 | C97800 |
| B30 | C99400 |
| B30 | C99500 |
| B30 | C99700 |
| B30 | C99750 |
| B30 (3C, 14C) | C82200 |
| B30 (4A) | C83600 |
| B30 (10C) | C82000 |
| B30 (20C, 20CT) | C82500 |
| B30 (21C) | C82510 |
| B30 (72C) | C96700 |
| B30 (165C, 165CT) | C82400 |
| B30 (245C, 245CT) | C82600 |
| B30 (275C, 275CT) | C82800 |
| B32 (1.5S) | L50131 |
| B32 (2.5S) | L50150 |
| B32 (2A, Sn2) | L54210 |
| B32 (2B) | L54211 |
| B32 (5A) | L54320 |
| B32 (5B) | L54321 |
| B32 (10B, Sn10A) | L54520 |
| B32 (15B, Sn15) | L54560 |
| B32 (20B, Sn20A) | L54711 |
| B32 (20C, Sn20B) | L54712 |
| B32 (25A) | L54720 |
| B32 (25B, Sn25A) | L54721 |
| B32 (25C, Sn25B) | L54722 |
| B32 (30A) | L54820 ob |
| B32 (30B, Sn30A) | L54821 |
| B32 (30C, Sn30B) | L54822 |
| B32 (35A) | L54850 |
| B32 (35B, Sn35A) | L54851 |
| B32 (35C, Sn35B) | L54852 |
| B32 (40A, Sn40A) | L54915 |
| B32 (40B) | L54916 |
| B32 (40C, Sn40B) | L54918 |
| B32 (45A) | L54950 |
| B32 (45B, Sn45) | L54951 |
| B32 (50A) | L55030 |
| B32 (50B, Sn50) | L55031 |
| B32 (60A) | L13600 |
| B32 (60B) | L13601 |
| B32 (63A, Sn63) | L13630 |
| B32 (63B) | L13631 |
| B32 (70A, Sn70) | L13700 |
| B32 (70B) | L13701 |
| B32 (95TA, Sb5) | L13950 |
| B32 (96.5 TS) | L13965 |
| B32 (Ag1.5) | L50132 |
| B32 (Ag2.5) | L50151 |
| B32 (Ag5.5) | L50180 |
| B32 (Sn5) | L54322 |
| B32 (Sn10B) | L54525 |
| B33 | C10100 |
| B33 | C10200 |
| B33 | C11000 |
| B33 | C11300 |
| B33 | C11300 |
| B33 | C11500 |
| B33 | C11600 |
| B36 | C21000 |
| B36 | C22000 |
| B36 | C22600 |
| B36 | C23000 |
| B36 | C24000 |
| B36 | C26000 |
| B36 | C26800 |
| B36 | C27200 |
| B36 | C28000 |
| B42 | C10200 |
| B42 | C10300 |
| B42 | C10800 |
| B42 | C12000 |
| B42 | C12200 |
| B43 | C23000 |
| B47 | C11000 |
| B47 | C11300 |
| B47 | C11400 |
| B47 | C11500 |
| B47 | C11600 |
| B48 | C10100 |
| B48 | C10200 |
| B48 | C11000 |
| B48 | C11300 |
| B48 | C11300 |
| B48 | C11400 |
| B48 | C11600 |
| B49 | C10100 |
| B49 | C10200 |
| B49 | C10400 |
| B49 | C10500 |
| B49 | C10700 |
| B49 | C11000 |

| CROSS REFERENCE SPECIFICATIONS | UNIFIED NUMBER |
|---|---|
| B49 | C11040 |
| B49 | C11300 |
| B49 | C11400 |
| B49 | C11500 |
| B49 | C11600 |
| B61 | C92200 |
| B62 | C83600 |
| B66 | C93200 |
| B66 | C93600 |
| B66 | C93700 |
| B66 | C93800 |
| B66 | C94300 |
| B66 | C94400 |
| B66 | C94500 |
| B67 | C94100 |
| B68 | C10200 |
| B68 | C10300 |
| B68 | C10800 |
| B68 | C12000 |
| B68 | C12200 |
| B69 | Z13004 |
| B69 | Z15006 |
| B69 | Z20301 |
| B69 | Z21721 |
| B69 | Z24311 |
| B69 | Z34961 |
| B69 | Z40101 |
| B69 | Z40301 |
| B69 | Z41121 |
| B69 | Z41321 |
| B69 (Zn-0.06Pb-0.06Cd) | Z21220 ob |
| B69 (Zn-0.08Pb) | Z21210 ob |
| B69 (Zn-0.3Pb-0.3Cd) | Z21540 ob |
| B69 (Zn-0.8Cu) | Z40330 ob |
| B69 (Zn-0.8Cu-0.15Ti) | Z41320 ob |
| B69 (Zn-1Cu) | Z44330 ob |
| B69 (Zn-1Cu-0.01Mg) | Z45330 ob |
| B75 | C10100 |
| B75 | C10200 |
| B75 | C10300 |
| B75 | C10800 |
| B75 | C12000 |
| B75 | C12200 |
| B80 | M11800 |
| B80 | M18432 |
| B80 (AM100A) | M10100 |
| B80 (AZ63A) | M11630 |
| B80 (AZ81A) | M11810 |
| B80 (AZ91C) | M11914 |
| B80 (AZ91E) | M11919 |
| B80 (AZ92A) | M11920 |
| B80 (EQ21A) | M18330 |
| B80 (EZ33A) | M12330 |
| B80 (HK31A) | M13310 |
| B80 (HZ32A) | M13320 |
| B80 (K1A) | M18010 |
| B80 (QE22A) | M18220 |
| B80 (QH21A) | M18210 |
| B80 (WE43A) | M18430 |
| B80 (WE54A) | M18410 |
| B80 (ZC63A) | M16331 |
| B80 (ZE41A) | M16410 |
| B80 (ZE63A) | M16630 |
| B80 (ZH62A) | M16620 |
| B80 (ZK51A) | M16510 |
| B80 (ZK61A) | M16610 |
| B85 (360.0) | A03600 |
| B85 (380.0) | A03800 |
| B85 (383.0) | A03830 |
| B85 (384.0) | A03840 |
| B85 (390.0) | A03900 |
| B85 (392.0) | A03920 |
| B85 (413.0) | A04130 |
| B85 (518.0) | A05180 |
| B85 (A360.0) | A13600 |
| B85 (A380.0) | A13800 |
| B85 (A413.0) | A14130 |
| B85 (B390.0) | A23900 |
| B85 (C443.0) | A34430 |
| B86 | Z35531 |
| B86 (AC43A) | Z35541 |
| B86 (AG40A) | Z33520 |
| B86 (AG40B) | Z33523 |
| B86 (ZA-8) | Z35636 |
| B86 (ZA-12) | Z35631 |
| B88 | C12200 |
| B90 | M11312 |
| B90 (AZ31B) | M11311 |
| B90 (HK31A) | M13310 |
| B90 (HM21A) | M13210 |
| B90 (LA141A) | M14141 |
| B90 (ZE10A) | M16100 |
| B91 (AZ31B) | M11311 |
| B91 (AZ61A) | M11610 |
| B91 (AZ80A) | M11800 |
| B91 (ZK60A) | M16600 |
| B92 (9980A) | M19980 |
| B92 (9990A) | M19990 |
| B92 (9990B) | M19991 |
| B92 (9995A) | M19995 |
| B92 (9998A) | M19998 |
| B93 | M10410 |
| B93 | M10600 |
| B93 | M11800 |
| B93 | M11912 |
| B93 | M18432 |
| B93 (2E63A) | M11630 |
| B93 (AM60A) | M10601 |
| B93 (AM60B) | M10603 |
| B93 (AM100A) | M10100 |
| B93 (AM100A) | M10101 |
| B93 (AS41A) | M10411 |
| B93 (AS41B) | M10413 |
| B93 (AZ63A) | M11630 |
| B93 (AZ63A) | M11631 |
| B93 (AZ81A) | M11811 |
| B93 (AZ91A) | M11910 |
| B93 (AZ91A) | M11911 |
| B93 (AZ91B) | M11913 |
| B93 (AZ91C) | M11914 |
| B93 (AZ91C) | M11915 |
| B93 (AZ91D) | M11917 |
| B93 (AZ91E) | M11918 |
| B93 (AZ92A) | M11920 |
| B93 (AZ92A) | M11921 |
| B93/B93M (ZC63) | M16631 |
| B93 (EQ21A) | M18330 |
| B93 (EZ33A) | M12330 |
| B93 (EZ33A) | M12331 |
| B93 (HK31B) | M13311 |
| B93 (HZ32A) | M13320 |
| B93 (HZ32A) | M13321 |
| B93 (K1A) | M18010 |
| B93 (K1A) | M18011 |
| B93/M93M | M10501 |
| B93 (QE22A) | M18220 |
| B93 (QE22A) | M18221 |
| B93 (QH21A) | M18210 |
| B93 (QH21A) | M18211 |
| B93 (WE43A) | M18430 |
| B93 (WE54A) | M18410 |
| B93 (ZC63A) | M16331 |
| B93 (ZE41A) | M16410 |
| B93 (ZE41A) | M16411 |
| B93 (ZH62A) | M16620 |
| B93 (ZH62A) | M16621 |
| B93 (ZK51A) | M16510 |
| B93 (ZK51A) | M16511 |
| B93 (ZK61A) | M16610 |
| B93 (ZK61A) | M16611 |
| B94 (AM50A) | M10500 |
| B94 (AM60A) | M10600 |
| B94 (AM60B) | M10602 |
| B94 (AS41A) | M10410 |
| B94 (AS41B) | M10412 |
| B94 (AZ91A) | M11910 |
| B94 (AZ91B) | M11912 |
| B96 | C65100 |
| B96 | C65400 |
| B96 | C65500 |
| B96 | C65800 ob |
| B98 | C65100 |
| B98 | C65500 |
| B98 | C66100 |
| B99 | C65100 |
| B99 | C65500 |
| B100 | C51000 |
| B100 | C51100 |
| B100 | C61400 |
| B100 | C65500 |
| B101 | C11000 |
| B101 | C12200 |
| B101 | C12500 |
| B102 (CY44A) | L13913 |
| B102 (PY1815A) | L13650 |
| B102 (Y10A) | L53340 |
| B102 (YC135A) | L13820 |
| B102 (YT155A) | L53560 |
| B103 | C51000 |
| B103 | C51100 |
| B103 | C51900 |
| B103 | C52100 |
| B103 | C52400 |
| B103 | C53400 |
| B103 | C54400 |
| B105 | C16200 |
| B105 | C16500 |
| B105 | C19600 ob |
| B105 | C50500 |
| B105 | C51000 |
| B105 | C65100 |
| B107 | M16710 |
| B107 (AZ31B) | M11311 |
| B107 (AZ31C) | M11312 |
| B107 (AZ61A) | M11610 |
| B107 (AZ80A) | M11800 |
| B107 (M1A) | M15100 |
| B107 (ZK40A) | M16400 |
| B107 (ZK60A) | M16600 |
| B108 (204.0) | A02040 |
| B108 (208.0) | A02080 ob |
| B108 (213.0) | A02130 ob |
| B108 (222.0) | A02220 ob |
| B108 (242.0) | A02420 |
| B108 (296.0) | A02960 |
| B108 (308.0) | A03080 |
| B108 (319.0) | A03190 |

| CROSS REFERENCE SPECIFICATIONS | UNIFIED NUMBER |
|---|---|
| B108 (332.0) | A03320 |
| B108 (333.0) | A03330 |
| B108 (336.0) | A03360 |
| B108 (354.0) | A03540 |
| B108 (355.0) | A03550 |
| B108 (356.0) | A03560 |
| B108 (357.0) | A03570 |
| B108 (359.0) | A03590 |
| B108 (443.0) | A04430 |
| B108 (513.0) | A05130 |
| B108 (535.0) | A05350 |
| B108 (705.0) | A07050 |
| B108 (707.0) | A07070 |
| B108 (711.0) | A07110 |
| B108 (713.0) | A07130 |
| B108 (850.0) | A08500 |
| B108 (851.0) | A08510 |
| B108 (852.0) | A08520 |
| B108 (A356.0) | A13560 |
| B108 (A357.0) | A13570 |
| B108 (A444.0) | A14440 |
| B108 (B443.0) | A24430 |
| B108 (C355.0) | A33550 |
| B111 | C10200 |
| B111 | C10300 |
| B111 | C10800 |
| B111 | C12000 |
| B111 | C12200 |
| B111 | C14200 |
| B111 | C19200 |
| B111 | C23000 |
| B111 | C28000 |
| B111 | C44300 |
| B111 | C44400 |
| B111 | C44500 |
| B111 | C60800 |
| B111 | C61300 |
| B111 | C61400 |
| B111 | C68700 |
| B111 | C70400 |
| B111 | C70600 |
| B111 | C71000 |
| B111 | C71500 |
| B111 | C71640 |
| B111 | C72200 |
| B111 (OFE) | C10100 |
| B116 | C11000 |
| B116 | C11300 |
| B116 | C11400 |
| B116 | C11500 |
| B116 | C11600 |
| B121 | C33500 |
| B121 | C34000 |
| B121 | C34200 |
| B121 | C35000 |
| B121 | C35300 |
| B121 | C35340 ob |
| B121 | C35600 |
| B122 | C70600 |
| B122 | C71000 |
| B122 | C71500 |
| B122 | C72200 |
| B122 | C72500 |
| B122 | C73500 |
| B122 | C74000 |
| B122 | C74400 |
| B122 | C74500 |
| B122 | C75200 |
| B122 | C76200 |
| B122 | C77000 |
| B124 | C11000 |
| B124 | C14500 |
| B124 | C14700 |
| B124 | C36500 |
| B124 | C37700 |
| B124 | C46400 |
| B124 | C48200 |
| B124 | C48500 |
| B124 | C61900 |
| B124 | C62300 |
| B124 | C63000 |
| B124 | C63020 |
| B124 | C63200 |
| B124 | C64200 |
| B124 | C64210 |
| B124 | C65500 |
| B124 | C67500 |
| B124 | C77400 |
| B127 | N04400 |
| B129 | C26000 |
| B130 | C22000 |
| B131 | C22000 |
| B134 | C21000 |
| B134 | C22000 |
| B134 | C23000 |
| B134 | C24000 |
| B134 | C26000 |
| B134 | C27000 |
| B134 | C27400 |
| B135 | C22000 |
| B135 | C23000 |
| B135 | C26000 |
| B135 | C27000 |
| B135 | C27200 |
| B135 | C27400 |
| B135 | C28000 |
| B135 | C33000 |
| B135 | C33200 |
| B135 | C37000 |
| B135 | C44300 |
| B138 | C67000 |
| B138 | C67500 |
| B139 | C51000 |
| B139 | C52100 |
| B139 | C52400 |
| B139 | C53400 |
| B139 | C54400 |
| B140 | C31400 |
| B140 | C31600 |
| B140 | C32000 |
| B148 | C95200 |
| B148 | C95300 |
| B148 | C95400 |
| B148 | C95410 |
| B148 | C95500 |
| B148 | C95520 |
| B148 | C95600 |
| B148 | C95700 |
| B148 | C95800 |
| B148 | C95820 |
| B148 | C95900 |
| B150 | C60600 ob |
| B150 | C61300 |
| B150 | C61400 |
| B150 | C61900 |
| B150) | C62300 |
| B150 | C62400 |
| B150 | C63000 |
| B150 | C63020 |
| B150 | C63200 |
| B150 | C64200 |
| B150 | C64210 |
| B151 | C70600 |
| B151 | C70620 |
| B151 | C71500 |
| B151 | C71520 |
| B151 | C74500 |
| B151 | C75200 |
| B151 | C75700 |
| B151 | C76400 |
| B151 | C77000 |
| B151 | C79200 |
| B152 | C10100 |
| B152 | C10200 |
| B152 | C10300 |
| B152 | C10400 |
| B152 | C10500 |
| B152 | C10700 |
| B152 | C10800 |
| B152 | C11000 |
| B152 | C11300 |
| B152 | C11400 |
| B152 | C11600 |
| B152 | C12000 |
| B152 | C12200 |
| B152 | C12300 |
| B152 | C14200 |
| B152 | C14420 |
| B152 | C14530 |
| B159 | C51000 |
| B159 | C52100 |
| B159 | C52400 |
| B160 | N02200 |
| B160 | N02201 |
| B161 | N02200 |
| B161 | N02201 |
| B162 | N02200 |
| B162 | N02201 |
| B163 | N02200 |
| B163 | N02201 |
| B163 | N04400 |
| B163 | N06025 |
| B163 | N06045 |
| B163 | N06600 |
| B163 | N06601 |
| B163 | N06603 |
| B163 | N06690 |
| B163 | N08120 |
| B163 | N08800 |
| B163 | N08801 |
| B163 | N08810 |
| B163 | N08811 |
| B163 | N08825 |
| B164 | N04400 |
| B164 | N04405 |
| B165 | N04400 |
| B165 | N04405 |
| B166 | N06025 |
| B166 | N06045 |
| B166 | N06600 |
| B166 | N06601 |
| B166 | N06603 |
| B166 | N06617 |
| B166 | N06690 |
| B167 | N06025 |
| B167 | N06045 |
| B167 | N06600 |

| CROSS REFERENCE SPECIFICATIONS | UNIFIED NUMBER |
|---|---|
| B167 | N06601 |
| B167 | N06603 |
| B167 | N06617 |
| B167 | N06690 |
| B168 | N06025 |
| B168 | N06045 |
| B168 | N06600 |
| B168 | N06601 |
| B168 | N06603 |
| B168 | N06617 |
| B168 | N06690 |
| B169 | C61300 |
| B169 | C61400 |
| B171 | C36500 |
| B171 | C36600 ob |
| B171 | C36700 ob |
| B171 | C36800 ob |
| B171 | C44300 |
| B171 | C44400 |
| B171 | C44500 |
| B171 | C46400 |
| B171 | C46500 |
| B171 | C46600 ob |
| B171 | C46700 ob |
| B171 | C61300 |
| B171 | C61400 |
| B171 | C63000 |
| B171 | C63020 |
| B171 | C63200 |
| B171 | C70600 |
| B171 | C71500 |
| B171 | C72200 |
| B172 | C10100 |
| B172 | C10200 |
| B172 | C11000 |
| B172 | C11300 |
| B172 | C11400 |
| B172 | C11500 |
| B172 | C11600 |
| B173 | C10100 |
| B173 | C10200 |
| B173 | C11000 |
| B173 | C11300 |
| B173 | C11400 |
| B173 | C11500 |
| B173 | C11600 |
| B174 | C10100 |
| B174 | C10200 |
| B174 | C11000 |
| B174 | C11300 |
| B174 | C11400 |
| B174 | C11500 |
| B174 | C11600 |
| B176 | C85700 |
| B176 | C85800 |
| B176 | C86500 |
| B176 | C87800 |
| B176 | C99700 |
| B176 | C99750 |
| B179 (100.1) | A01001 |
| B179 (130.1) | A01301 |
| B179 (150.1) | A01501 |
| B179 (170.1) | A01701 |
| B179 (201.2) | A02012 |
| B179 (204.2) | A02042 |
| B179 (208.1) | A02081 ob |
| B179 (208.2) | A02082 ob |
| B179 (222.1) | A02221 ob |
| B179 (242.1) | A02421 |

| CROSS REFERENCE SPECIFICATIONS | UNIFIED NUMBER |
|---|---|
| B179 (242.2) | A02422 |
| B179 (295.1) | A02951 |
| B179 (295.2) | A02952 |
| B179 (296.1) | A02961 |
| B179 (296.2) | A02962 |
| B179 (308.1) | A03081 |
| B179 (308.2) | A03082 |
| B179 (319.1) | A03191 |
| B179 (319.2) | A03192 |
| B179 (328.1) | A03281 ob |
| B179 (332.1) | A03321 |
| B179 (332.2) | A03322 |
| B179 (333.1) | A03331 |
| B179 (336.1) | A03361 |
| B179 (336.2) | A03362 |
| B179 (354.1) | A03541 |
| B179 (355.1) | A03551 |
| B179 (355.2) | A03552 |
| B179 (356.1) | A03561 |
| B179 (356.2) | A03562 |
| B179 (357.1) | A03571 |
| B179 (359.2) | A03592 |
| B179 (360.2) | A03602 |
| B179 (380.2) | A03802 |
| B179 (383.1) | A03831 |
| B179 (383.2) | A03832 |
| B179 (384.1) | A03841 |
| B179 (384.2) | A03842 |
| B179 (390.2) | A03902 |
| B179 (392.1) | A03921 |
| B179 (392.1) | A23901 |
| B179 (413.2) | A04132 |
| B179 (443.1) | A04431 |
| B179 (443.2) | A04432 |
| B179 (513.2) | A05132 |
| B179 (514.1) | A05141 |
| B179 (514.2) | A05142 |
| B179 (518.1) | A05181 |
| B179 (518.2) | A05182 |
| B179 (520.2) | A05202 |
| B179 (535.2) | A05352 |
| B179 (705.1) | A07051 |
| B179 (707.1) | A07071 |
| B179 (710.1) | A07101 |
| B179 (711.1) | A07111 |
| B179 (712.2) | A07122 |
| B179 (713.1) | A07131 |
| B179 (771.2) | A07712 |
| B179 (850.1) | A08501 |
| B179 (851.1) | A08511 |
| B179 (852.1) | A08521 |
| B179 (A242.1) | A12421 |
| B179 (A242.2) | A12422 |
| B179 (A356.2) | A13562 |
| B179 (A357.2) | A13572 |
| B179 (A360.1) | A13601 |
| B179 (A360.2) | A13602 |
| B179 (A380.1) | A13801 |
| B179 (A380.2) | A13802 |
| B179 (A413.1) | A14131 |
| B179 (A413.2) | A14132 |
| B179 (A444.2) | A14442 |
| B179 (B443.1) | A24431 |
| B179 (C355.2) | A33552 |
| B179 (C443.1) | A34431 |
| B179 (C443.2) | A34432 |
| B187 | C10100 |
| B187 | C10200 |
| B187 | C10300 |

| CROSS REFERENCE SPECIFICATIONS | UNIFIED NUMBER |
|---|---|
| B187 | C10400 |
| B187 | C10500 |
| B187 | C10700 |
| B187 | C10800 |
| B187 | C10930 |
| B187 | C10940 |
| B187 | C11000 |
| B187 | C11300 |
| B187 | C11400 |
| B187 | C11600 |
| B187 | C12000 |
| B187 | C12200 |
| B187 | C14420 |
| B188 | C10100 |
| B188 | C10200 |
| B188 | C10300 |
| B188 | C10400 |
| B188 | C10500 |
| B188 | C10700 |
| B188 | C11000 |
| B188 | C11300 |
| B188 | C11400 |
| B188 | C11600 |
| B188 | C12000 |
| B189 | C10100 |
| B189 | C10200 |
| B189 | C10400 |
| B189 | C10500 |
| B189 | C10700 |
| B189 | C11000 |
| B189 | C11300 |
| B189 | C11400 |
| B189 | C11500 |
| B189 | C11600 |
| B194 | C17000 |
| B194 | C17200 |
| B196 | C17000 |
| B196 | C17200 |
| B196 | C17300 |
| B197 | C17200 |
| B197 | C17300 |
| B199 | M11800 |
| B199 (AM100A) | M10100 |
| B199 (AZ81A) | M11810 |
| B199 (AZ91C) | M11914 |
| B199 (AZ91E) | M11919 |
| B199 (AZ92A) | M11920 |
| B199 (EQ21A) | M18330 |
| B199 (EZ33A) | M12330 |
| B199 (QE22A) | M18220 |
| B206 | C71000 |
| B206 | C74500 |
| B206 | C75200 |
| B206 | C75700 |
| B206 | C76400 |
| B206 | C77000 |
| B206 | C79200 |
| B209 | A83003 |
| B209 (1060) | A91060 |
| B209 (1100) | A91100 |
| B209 (1230) | A91230 |
| B209 (2014) | A92014 |
| B209 (2024) | A92024 |
| B209 (2124) | A92124 |
| B209 (2219) | A92219 |
| B209 (3003) | A93003 |
| B209 (3004) | A93004 |
| B209 (3005) | A93005 |
| B209 (3105) | A93105 |

| CROSS REFERENCE SPECIFICATIONS | UNIFIED NUMBER |
|---|---|
| B209 (5005) | A95005 |
| B209 (5010) | A95010 |
| B209 (5050) | A95050 |
| B209 (5052) | A95052 |
| B209 (5083) | A95083 |
| B209 (5086) | A95086 |
| B209 (5154) | A95154 |
| B209 (5252) | A95252 |
| B209 (5254) | A95254 |
| B209 (5454) | A95454 |
| B209 (5456) | A95456 |
| B209 (5457) | A95457 |
| B209 (5652) | A95652 |
| B209 (5657) | A95657 |
| B209 (6003) | A96003 |
| B209 (6061) | A96061 |
| B209 (7008) | A97008 |
| B209 (7011) | A97001 ob |
| B209 (7011) | A97011 ob |
| B209 (7072) | A97072 |
| B209 (7075) | A97075 |
| B209 (7178) | A97178 |
| B209 (Alclad 2014) | A82014 |
| B209 (Alclad 2024) | A82024 |
| B209 (Alclad 2219) | A82219 |
| B209 (Alclad 6061) | A86061 |
| B209 (Alclad 7075) | A87075 |
| B209 (Alclad 7178) | A87178 |
| B210 | A83003 |
| B210 | A92219 |
| B210 (1060) | A91060 |
| B210 (1100) | A91100 |
| B210 (2011) | A92011 |
| B210 (2014) | A92014 |
| B210 (2024) | A92024 |
| B210 (3003) | A93003 |
| B210 (3102) | A93102 |
| B210 (3303) | A93303 ob |
| B210 (5005) | A95005 |
| B210 (5050) | A95050 |
| B210 (5052) | A95052 |
| B210 (5083) | A95083 |
| B210 (5086) | A95086 |
| B210 (5154) | A95154 |
| B210 (5456) | A95456 |
| B210 (6061) | A96061 |
| B210 (6063) | A96063 |
| B210 (6262) | A96262 |
| B210 (7072) | A97072 |
| B210 (7075) | A97075 |
| B211 (1060) | A91060 |
| B211 (1100) | A91100 |
| B211 (2011) | A92011 |
| B211 (2014) | A92014 |
| B211 (2017) | A92017 |
| B211 (2024) | A92024 |
| B211 (2219) | A92219 |
| B211 (3003) | A93003 |
| B211 (5052) | A95052 |
| B211 (5056) | A95056 |
| B211 (5154) | A95154 |
| B211 (6061) | A96061 |
| B211 (6110) | A96110 |
| B211 (6253) | A96253 |
| B211 (6262) | A96262 |
| B211 (7075) | A97075 |
| B216 | C12500 |
| B216 | C14520 |
| B221 | A83003 |
| B221 (1060) | A91060 |
| B221 (1100) | A91100 |
| B221 (2014) | A92014 |
| B221 (2024) | A92024 |
| B221 (2219) | A92219 |
| B221 (3003) | A93003 |
| B221 (3004) | A93004 |
| B221 (3102) | A93102 |
| B221 (5052) | A95052 |
| B221 (5083) | A95083 |
| B221 (5086) | A95086 |
| B221 (5154) | A95154 |
| B221 (5454) | A95454 |
| B221 (5456) | A95456 |
| B221 (6005, 6005A) | A96005 |
| B221 (6060) | A96060 |
| B221 (6061) | A96061 |
| B221 (6063) | A96063 |
| B221 (6066) | A96066 |
| B221 (6070) | A96070 |
| B221 (6105) | A96105 |
| B221 (6166) | A96162 |
| B221 (6262) | A96262 |
| B221 (6351) | A96351 |
| B221 (6463) | A96463 |
| B221 (7005) | A97005 |
| B221 (7072) | A97072 |
| B221 (7075) | A97075 |
| B221 (7116) | A97116 |
| B221 (7129) | A97129 |
| B221 (7178) | A97178 |
| B224 | C14700 |
| B224 | C15000 |
| B224 (CRTP) | C11030 |
| B224 (DHP) | C12200 |
| B224 (DHPS) | C12300 |
| B224 (DLP) | C12000 |
| B224 (DLPS) | C12100 |
| B224 (DPA) | C14200 |
| B224 (DPTE) | C14500 |
| B224 (ETP) | C11000 |
| B224 (ETP) | C11100 |
| B224 (ETP) | C11300 |
| B224 (ETP) | C11400 |
| B224 (ETP) | C11500 |
| B224 (ETP) | C11600 |
| B224 (FRHC) | C11020 |
| B224 (FRSTP) | C12700 ob |
| B224 (FRSTP) | C12800 ob |
| B224 (FRSTP) | C12900 |
| B224 (FRSTP) | C13000 ob |
| B224 (FRTP) | C12500 |
| B224 (OF) | C10200 |
| B224 (OFE) | C10100 |
| B224 (OFLP) | C10800 |
| B224 (OFS) | C10400 |
| B224 (OFS) | C10500 |
| B224 (OFS) | C10700 |
| B224 (OFXLP) | C10300 |
| B224 (RHC) | C11010 |
| B226 | C10200 |
| B226 | C11000 |
| B226 | C11300 |
| B226 | C11400 |
| B226 | C11500 |
| B226 | C11600 |
| B228 | C11000 |
| B228 | C11300 |
| B228 | C11500 |
| B229 | C11000 |
| B229 | C11300 |
| B229 | C11400 |
| B229 | C11500 |
| B229 | C11600 |
| B230 | A91350 |
| B231 | A91350 |
| B232 (1350) | A91350 |
| B233 | A91350 |
| B234 | A83003 |
| B234 (1060) | A91060 |
| B234 (3003) | A93003 |
| B234 (5052) | A95052 |
| B234 (5454) | A95454 |
| B234 (6061) | A96061 |
| B234 (7072) | A97072 |
| B236 | A91350 |
| B240 (AC41A) | Z35530 |
| B240 (AC43A) | Z35540 |
| B240 (AG40A) | Z33521 |
| B240 (AG40B) | Z33522 |
| B240 (ZA-8) | Z35635 |
| B240 (ZA-12) | Z35630 |
| B240 (ZA-27) | Z35840 |
| B241 | A83003 |
| B241 (1060) | A91060 |
| B241 (1100) | A91100 |
| B241 (2014) | A92014 |
| B241 (2024) | A92024 |
| B241 (2219) | A92219 |
| B241 (3003) | A93003 |
| B241 (5052) | A95052 |
| B241 (5083) | A95083 |
| B241 (5086) | A95086 |
| B241 (5254) | A95254 |
| B241 (5454) | A95454 |
| B241 (5456) | A95456 |
| B241 (5652) | A95652 |
| B241 (6061) | A96061 |
| B241 (6063) | A96063 |
| B241 (6162) | A96162 |
| B241 (6351) | A96351 |
| B241 (7072) | A97072 |
| B241 (7075) | A97075 |
| B241 (7178) | A97178 |
| B241/B241M (6066) | A96066 |
| B246 | C10100 |
| B246 | C10200 |
| B246 | C10400 |
| B246 | C10500 |
| B246 | C10700 |
| B246 | C11000 |
| B246 | C11300 |
| B246 | C11400 |
| B246 | C11500 |
| B246 | C11600 |
| B247 (1100) | A91100 |
| B247 (2014) | A92014 |
| B247 (2018) | A92018 |
| B247 (2025) | A92025 |
| B247 (2218) | A92218 |
| B247 (2219) | A92219 |
| B247 (2618) | A92618 |
| B247 (3003) | A93003 |
| B247 (4032) | A94032 |
| B247 (5083) | A95083 |
| B247 (6061) | A96061 |
| B247 (6066) | A96066 |
| B247 (6151) | A96151 |

| CROSS REFERENCE SPECIFICATIONS | UNIFIED NUMBER |
|---|---|
| B247 (7049) | A97049 |
| B247 (7050) | A97050 |
| B247 (7075) | A97075 |
| B247 (7076) | A97076 |
| B247 (7175) | A97175 |
| B265 | R54624 ob |
| B265 | R55111 |
| B265 | R56403 |
| B265 | R56404 |
| B265 | R56405 |
| B265 | R58210 |
| B265 | R58640 |
| B265 (1) | R50250 |
| B265 (2) | R50400 |
| B265 (3) | R50550 |
| B265 (4) | R50700 |
| B265 (5) | R56400 |
| B265 (6) | R54520 |
| B265 (7) | R52400 |
| B265 (7, 11) | R52401 |
| B265 (9) | R56320 |
| B265 (11) | R52250 |
| B265 (12) | R53400 |
| B265 (12) | R53401 |
| B265 (13) | R53413 |
| B265 (14) | R53414 |
| B265 (15) | R53415 |
| B265 (16) | R52402 |
| B265 (17) | R52252 |
| B265 (18) | R56322 |
| B265 (20) | R58645 |
| B265 (21) | R58210 |
| B265 (23) | R56407 |
| B265 (26) | R52404 |
| B265 (27) | R52254 |
| B265 (28) | R56323 |
| B271 | C83600 |
| B271 | C83800 |
| B271 | C84400 |
| B271 | C84800 |
| B271 | C85200 |
| B271 | C85400 |
| B271 | C85700 |
| B271 | C86200 |
| B271 | C86300 |
| B271 | C86400 |
| B271 | C86500 |
| B271 | C86700 |
| B271 | C87200 ob |
| B271 | C87300 |
| B271 | C87400 |
| B271 | C87500 |
| B271 | C90300 |
| B271 | C90500 |
| B271 | C92200 |
| B271 | C92300 |
| B271 | C93200 |
| B271 | C93500 |
| B271 | C93600 |
| B271 | C93700 |
| B271 | C93800 |
| B271 | C94300 |
| B271 | C95200 |
| B271 | C95300 |
| B271 | C95400 |
| B271 | C95410 |
| B271 | C95500 |
| B271 | C95520 |
| B271 | C95800 |

| CROSS REFERENCE SPECIFICATIONS | UNIFIED NUMBER |
|---|---|
| B271 | C97300 |
| B271 | C97600 |
| B271 | C97800 |
| B272 | C10100 |
| B272 | C10200 |
| B272 | C10300 |
| B272 | C10500 |
| B272 | C10700 |
| B272 | C10800 |
| B272 | C11000 |
| B272 | C12200 |
| B272 | C14200 |
| B275 (9980A) | M19980 |
| B275 (9990A) | M19990 |
| B275 (9995A) | M19995 |
| B275 (9998A) | M19998 |
| B275 (A3A) | M10030 |
| B275 (AM60A) | M10600 |
| B275 (AM80A) | M10800 |
| B275 (AM90A) | M10900 |
| B275 (AM100A) | M10100 |
| B275 (AM100B) | M10102 |
| B275 (AS41A) | M10410 |
| B275 (AZ 31D) | M11313 |
| B275 (AZ10A) | M11100 |
| B275 (AZ21A) | M11210 |
| B275 (AZ31A) | M11310 |
| B275 (AZ31B) | M11311 |
| B275 (AZ31C) | M11312 |
| B275 (AZ61A) | M11610 |
| B275 (AZ63A) | M11630 |
| B275 (AZ80A) | M11800 |
| B275 (AZ81A) | M11810 |
| B275 (AZ90A) | M11900 |
| B275 (AZ91A) | M11910 |
| B275 (AZ91B) | M11912 |
| B275 (AZ91C) | M11914 |
| B275 (AZ91D) | M11916 |
| B275 (AZ91E) | M11919 |
| B275 (AZ92A) | M11920 |
| B275 (AZ101A) | M11101 |
| B275 (AZ125A) | M11125 |
| B275 (EK30A) | M12300 |
| B275 (EK41A) | M12410 |
| B275 (EZ33A) | M12330 |
| B275 (HK31A) | M13310 |
| B275 (HM21A) | M13210 |
| B275 (HM31A) | M13312 |
| B275 (HZ32A) | M13320 |
| B275 (K1A) | M18010 |
| B275 (LA141A) | M14141 |
| B275 (LS141A) | M14142 |
| B275 (LZ145A) | M14145 |
| B275 (M1A) | M15100 |
| B275 (M1B) | M15101 |
| B275 (M1C) | M15102 |
| B275 (TA54A) | M18540 |
| B275 (WE43A) | M18430 |
| B275 (ZE10A) | M16100 |
| B275 (ZE41A) | M16410 |
| B275 (ZE63A) | M16630 |
| B275 (ZH62A) | M16620 |
| B275 (ZK21A) | M16210 |
| B275 (ZK40A) | M16400 |
| B275 (ZK51A) | M16510 |
| B275 (ZK60A) | M16600 |
| B275 (ZK60B) | M16601 |
| B275 (ZK61A) | M16610 |
| B280 | C12000 |

| CROSS REFERENCE SPECIFICATIONS | UNIFIED NUMBER |
|---|---|
| B283 | C11000 |
| B283 | C14500 |
| B283 | C14700 |
| B283 | C36500 |
| B283 | C37700 |
| B283 | C46400 |
| B283 | C48200 |
| B283 | C48500 |
| B283 | C61900 |
| B283 | C62300 |
| B283 | C63000 |
| B283 | C63020 |
| B283 | C63200 |
| B283 | C64200 |
| B283 | C64210 |
| B283 | C65500 |
| B283 | C67500 |
| B283 | C67600 |
| B283 | C77400 |
| B286 | C10100 |
| B286 | C10200 |
| B286 | C11000 |
| B286 | C11300 |
| B286 | C11400 |
| B286 | C11500 |
| B286 | C11600 |
| B298 | C10100 |
| B298 | C10200 |
| B298 | C10400 |
| B298 | C10500 |
| B298 | C10700 |
| B298 | C11000 |
| B298 | C11300 |
| B298 | C11400 |
| B298 | C11500 |
| B298 | C11600 |
| B301 | C14500 |
| B301 | C14510 |
| B301 | C14520 |
| B301 | C14700 |
| B301 | C14710 ob |
| B301 | C14720 ob |
| B301 | C18700 |
| B302 | C10300 |
| B302 | C10800 |
| B302 | C12000 |
| B302 | C12200 |
| B305 | C95520 |
| B306 | C12200 |
| B308 (6061) | A96061 |
| B313 | A93005 |
| B313 (1100) | A91100 |
| B313 (3003) | A93003 |
| B313 (3004) | A93004 |
| B313 (5050) | A95050 |
| B313 (5052) | A95052 |
| B313 (5086) | A95086 |
| B313 (5154) | A95154 |
| B313 (6061) | A96061 |
| B313 (7072) | A97072 |
| B315 | C61300 |
| B315 | C61400 |
| B315 | C65100 |
| B315 | C65500 |
| B315 | C65800 ob |
| B316 (1100) | A91100 |
| B316 (2017) | A92017 |
| B316 (2024) | A92024 |
| B316 (2117) | A92117 |

| CROSS REFERENCE SPECIFICATIONS | UNIFIED NUMBER |
|---|---|
| B316 (2219) | A92219 |
| B316 (3003) | A93003 |
| B316 (5005) | A95005 |
| B316 (5052) | A95052 |
| B316 (5056) | A95056 |
| B316 (6053) | A96053 |
| B316 (6061) | A96061 |
| B316 (7050) | A97050 |
| B316 (7075) | A97075 |
| B316 (7178) | A97178 |
| B316 (7277) | A97277 ob |
| B317 (6101) | A96101 |
| B324 | A91350 |
| B327 (V12) | Z33730 |
| B333 | N10001 |
| B333 | N10624 |
| B333 | N10629 |
| B333 | N10665 |
| B333 | N10675 |
| B335 | N10001 |
| B335 | N10624 |
| B335 | N10629 |
| B335 | N10665 |
| B335 | N10675 |
| B337 (2) | R50400 |
| B337 (29) | R56404 |
| B338 (1) | R50250 |
| B338 (2) | R50400 |
| B338 (3) | R50550 |
| B338 (7) | R52400 |
| B338 (7, 11) | R52401 |
| B338 (9) | R56320 |
| B338 (11) | R52250 |
| B338 (12) | R53400 |
| B338 (12) | R53401 |
| B338 (13) | R53413 |
| B338 (14) | R53414 |
| B338 (15) | R53415 |
| B338 (16) | R52402 |
| B338 (17) | R52252 |
| B338 (18) | R56322 |
| B338 (26) | R52404 |
| B338 (27) | R52254 |
| B338 (28) | R56323 |
| B339 (A) | L13008 |
| B344 | N06003 |
| B344 | N06004 |
| B345 | A83003 |
| B345 (1060) | A91060 |
| B345 (3003) | A93003 |
| B345 (5083) | A95083 |
| B345 (5086) | A95086 |
| B345 (6061) | A96061 |
| B345 (6063) | A96063 |
| B345 (6070) | A96070 |
| B345 (6351) | A96351 |
| B345 (7072) | A97072 |
| B348 | R55111 |
| B348 | R58640 |
| B348 (1) | R50250 |
| B348 (2) | R50400 |
| B348 (3) | R50550 |
| B348 (4) | R50700 |
| B348 (5) | R56400 |
| B348 (6) | R54520 |
| B348 (7) | R52400 |
| B348 (7, 11) | R52401 |
| B348 (9) | R56320 |
| B348 (11) | R52250 |

| CROSS REFERENCE SPECIFICATIONS | UNIFIED NUMBER |
|---|---|
| B348 (12) | R53400 |
| B348 (12) | R53401 |
| B348 (13) | R53413 |
| B348 (14) | R53414 |
| B348 (15) | R53415 |
| B348 (16) | R52402 |
| B348 (17) | R52252 |
| B348 (18) | R56322 |
| B348 (20) | R58645 |
| B348 (21) | R58210 |
| B348 (23) | R56407 |
| B348 (24) | R56405 |
| B348 (25) | R56403 |
| B348 (26) | R52404 |
| B348 (27) | R52254 |
| B348 (28) | R56323 |
| B348 (29) | R56404 |
| B349 | R60001 |
| B350 | R60001 |
| B350 | R60802 |
| B350 | R60804 |
| B350 | R60901 |
| B351 | R60001 |
| B351 | R60802 |
| B351 | R60804 |
| B351 | R60901 |
| B352 | R60001 |
| B352 | R60802 |
| B352 | R60804 |
| B352 | R60901 |
| B353 | R60001 |
| B353 | R60802 |
| B353 | R60804 |
| B353 | R60901 |
| B353 | R60904 |
| B355 | C10100 |
| B355 | C10200 |
| B355 | C10400 |
| B355 | C10500 |
| B355 | C10700 |
| B355 | C11000 |
| B355 | C11300 |
| B355 | C11400 |
| B355 | C11500 |
| B355 | C11600 |
| B359 | C10100 |
| B359 | C10200 |
| B359 | C10300 |
| B359 | C10800 |
| B359 | C12000 |
| B359 | C12200 |
| B359 | C14200 |
| B359 | C19200 |
| B359 | C23000 |
| B359 | C44300 |
| B359 | C44400 |
| B359 | C44500 |
| B359 | C60800 |
| B359 | C68700 |
| B359 | C70400 |
| B359 | C70600 |
| B359 | C71000 |
| B359 | C71500 |
| B359 | C72200 |
| B360 | C10800 |
| B360 | C12200 |
| B361 (1060) | A91060 |
| B361 (1100) | A91100 |
| B361 (3003) | A93003 |

| CROSS REFERENCE SPECIFICATIONS | UNIFIED NUMBER |
|---|---|
| B361 (5083) | A95083 |
| B361 (5086) | A95086 |
| B361 (5154) | A95154 |
| B361 (6061) | A96061 |
| B361 (6063) | A96063 |
| B361 (Alclad 3003) | A83003 |
| B361 (WP3003) | A93003 |
| B363 | R56323 |
| B363 | R56403 |
| B363 | R56404 |
| B363 | R56405 |
| B363 | R58210 |
| B363 | R58640 |
| B363 (23) | R56407 |
| B363 (26) | R52404 |
| B363 (27) | R52254 |
| B364 | R05200 |
| B364 | R05240 |
| B364 | R05252 |
| B364 | R05255 |
| B364 | R05400 |
| B365 | R05200 |
| B365 | R05240 |
| B365 | R05252 |
| B365 | R05255 |
| B365 | R05400 |
| B366 | N02200 |
| B366 | N02201 |
| B366 | N04400 |
| B366 | N06001 ob |
| B366 | N06002 |
| B366 | N06007 |
| B366 | N06022 |
| B366 | N06025 |
| B366 | N06030 |
| B366 | N06045 |
| B366 | N06058 |
| B366 | N06200 |
| B366 | N06219 |
| B366 | N06230 |
| B366 | N06455 |
| B366 | N06600 |
| B366 | N06603 |
| B366 | N06625 |
| B366 | N06985 |
| B366 | N08020 |
| B366 | N08031 |
| B366 | N08120 |
| B366 | N08330 |
| B366 | N08367 |
| B366 | N08800 |
| B366 | N08810 |
| B366 | N08811 |
| B366 | N08825 |
| B366 | N08904 |
| B366 | N08925 |
| B366 | N08926 |
| B366 | N10001 |
| B366 | N10003 |
| B366 | N10242 |
| B366 | N10276 |
| B366 | N10624 |
| B366 | N10629 |
| B366 | N10665 |
| B366 | N10675 |
| B366 | N12160 |
| B366 | R20033 |
| B366 (CRH556) | R30556 |
| B367 (C-2) | R50400 |

| CROSS REFERENCE SPECIFICATIONS | UNIFIED NUMBER |
|---|---|
| B367 (C-2, C-3) | R52550 |
| B367 (C-3) | R50550 |
| B367 (C-5) | R56400 |
| B367 (C-6) | R54520 |
| B367 (Ti-Pd7B, Ti-Pd8A) | R52401 |
| B369 | C96200 |
| B369 | C96400 |
| B370 | C11000 |
| B370 | C12200 |
| B370 | C12500 |
| B371 | C69400 |
| B371 | C69430 |
| B371 | C69440 ob |
| B371 | C69450 ob |
| B371 | C69700 |
| B371 | C69710 |
| B371 | C69720 ob |
| B371 | C69730 ob |
| B372 | C10200 |
| B372 | C10300 |
| B372 | C12000 |
| B372 | C22000 |
| B373 (1145) | A91145 |
| B373 (1235) | A91235 |
| B379 | C10800 |
| B379 (DPA) | C14200 |
| B379 (OFLP) | C12200 |
| B379 (OFXLP) | C10300 |
| B381 | R56403 |
| B381 | R56405 |
| B381 | R58640 |
| B381 (7, 11) | R52401 |
| B381 (13) | R53413 |
| B381 (14) | R53414 |
| B381 (15) | R53415 |
| B381 (16) | R52402 |
| B381 (17) | R52252 |
| B381 (18) | R56322 |
| B381 (20) | R58645 |
| B381 (23) | R56407 |
| B381 (26) | R52404 |
| B381 (27) | R52254 |
| B381 (F-1) | R50250 |
| B381 (F-2) | R50400 |
| B381 (F-3) | R50550 |
| B381 (F-4) | R50700 |
| B381 (F-5) | R56400 |
| B381 (F-6) | R54520 |
| B381 (F-7) | R52400 |
| B381 (F-9) | R56320 |
| B381 (F-11) | R52250 |
| B381 (F-12) | R53400 |
| B381 (F-12) | R53401 |
| B381 (F-21) | R58210 |
| B381 (F23) | R56407 |
| B381 (F-25) | R56403 |
| B386 (360) | R03600 |
| B386 (361) | R03610 |
| B386 (363) | R03630 |
| B386 (364) | R03640 |
| B386 (365) | R03650 |
| B387 (360) | R03600 |
| B387 (361) | R03610 |
| B387 (363) | R03630 |
| B387 (364) | R03640 |
| B387 (365) | R03650 |
| B388 | K92100 |
| B388 | K92350 |
| B388 | K92510 |
| B388 | K92850 |
| B388 | K93600 |
| B388 | K93800 |
| B388 | K94000 |
| B388 | K94200 |
| B388 | K94500 |
| B388 | K95000 |
| B388 | M27200 |
| B391 | R04200 |
| B391 | R04210 |
| B391 | R04251 |
| B391 | R04261 |
| B392 | R04200 |
| B392 | R04210 |
| B392 | R04251 |
| B392 | R04261 |
| B393 | R04200 |
| B393 | R04210 |
| B393 | R04251 |
| B393 | R04261 |
| B394 | R04200 |
| B394 | R04210 |
| B394 | R04251 |
| B394 | R04261 |
| B395 | C10200 |
| B395 | C10300 |
| B395 | C10800 |
| B395 | C12000 |
| B395 | C12200 |
| B395 | C14200 |
| B395 | C19200 |
| B395 | C23000 |
| B395 | C44300 |
| B395 | C44400 |
| B395 | C44500 |
| B395 | C60800 |
| B395 | C68700 |
| B395 | C70400 |
| B395 | C70600 |
| B395 | C71000 |
| B395 | C71500 |
| B395 | C72200 |
| B396 (5005) | A95005 |
| B397 (5005) | A95005 |
| B398 (6201) | A96201 |
| B399 (6201) | A96201 |
| B400 (1350) | A91350 |
| B401 | A91350 |
| B403 | M11800 |
| B403 (AM100A) | M10100 |
| B403 (AZ81A) | M11810 |
| B403 (AZ91C) | M11914 |
| B403 (AZ91E) | M11919 |
| B403 (AZ92A) | M11920 |
| B403 (EQ21A) | M18330 |
| B403 (EZ33A) | M12330 |
| B403 (HK31A) | M13310 |
| B403 (K1A) | M18010 |
| B403 (QE22A) | M18220 |
| B403 (ZK61A) | M16610 |
| B404 | A83003 |
| B404 (1060) | A91060 |
| B404 (3003) | A93003 |
| B404 (5052) | A95052 |
| B404 (5454) | A95454 |
| B404 (6061) | A96061 |
| B404 (7072) | A97072 |
| B407 | N08120 |
| B407 | N08800 |
| B407 | N08801 |
| B407 | N08810 |
| B407 | N08811 |
| B408 | N08120 |
| B408 | N08800 |
| B408 | N08810 |
| B408 | N08811 |
| B409 | N08120 |
| B409 | N08800 |
| B409 | N08810 |
| B409 | N08811 |
| B411 | C64700 |
| B412 | C64700 |
| B413 (99.90) | P07020 |
| B413 (99.95) | P07015 |
| B413 (99.99) | P07010 |
| B418 (Zinc Anodes Type II) | Z13000 |
| B418 (Zinc Anodes - Type I) | Z32120 |
| B422 | C19025 |
| B422 | C63800 |
| B422 | C64400 ob |
| B422 | C70250 |
| B422 | C70260 |
| B422 | C72400 ob |
| B423 | N08221 |
| B423 | N08825 |
| B424 | N08221 |
| B424 | N08825 |
| B425 | N08221 |
| B425 | N08825 |
| B427 | C90700 |
| B427 | C90800 |
| B427 | C91600 |
| B427 | C91700 |
| B427 | C92900 |
| B429 (6061) | A96061 |
| B429 (6063) | A96063 |
| B432 | C10100 |
| B432 | C10200 |
| B432 | C10300 |
| B432 | C10800 |
| B432 | C12200 |
| B432 | C36500 |
| B432 | C36600 ob |
| B432 | C36700 ob |
| B432 | C36800 ob |
| B432 | C44300 |
| B432 | C44400 |
| B432 | C44500 |
| B432 | C46400 |
| B432 | C46500 |
| B432 | C46600 ob |
| B432 | C46700 ob |
| B432 | C61300 |
| B432 | C61400 |
| B432 | C65100 |
| B432 | C65500 |
| B432 | C70600 |
| B432 | C71500 |
| B434 | N10003 |
| B434 | N10242 |
| B435 | N06002 |
| B435 | N06230 |
| B435 | N12160 |
| B435 | R30556 |
| B440 | L01900 |
| B440 | L01951 |
| B440 | L01981 |
| B441 | C17500 |

| CROSS REFERENCE SPECIFICATIONS | UNIFIED NUMBER |
|---|---|
| B441 | C17510 |
| B442 B470 | C11400 |
| B443 | N06219 |
| B443 | N06625 |
| B444 | N06219 |
| B444 | N06625 |
| B446 | N06219 |
| B446 | N06625 |
| B447 | C10100 |
| B447 | C10200 |
| B447 | C10300 |
| B447 | C10800 |
| B447 | C11000 |
| B447 | C12000 |
| B447 | C12200 |
| B447 | C14200 |
| B451 | C10200 |
| B453 | C33500 |
| B453 | C34000 |
| B453 | C34500 |
| B453 | C35000 |
| B453 | C35300 |
| B453 | C35340 ob |
| B453 | C35600 |
| B455 | C38000 |
| B455 | C38500 |
| B458 (1) | C75200 |
| B462 | N08020 |
| B462 | N08024 |
| B462 | N08026 |
| B462 | N08367 |
| B462 | R20033 |
| B463 | N08020 |
| B463 | N08024 |
| B463 | N08026 |
| B464 | N08020 |
| B464 | N08024 |
| B464 | N08026 |
| B465 | C19200 |
| B465 | C19400 |
| B465 | C19500 |
| B465 | C19700 |
| B466 | C70400 |
| B466 | C70600 |
| B466 | C71000 |
| B466 | C71500 |
| B466 | C72200 |
| B467 | C70600 |
| B467 | C71500 |
| B468 | N08020 |
| B468 | N08024 |
| B468 | N08026 |
| B470 | C10100 |
| B470 | C10200 |
| B470 | C11000 |
| B470 | C11300 |
| B470 | C11500 |
| B470 | C11600 |
| B472 | N08020 |
| B472 | N08024 |
| B472 | N08026 |
| B472 | N08367 |
| B472 | N08926 |
| B472 | R20033 |
| B473 | N08020 |
| B473 | N08024 |
| B473 | N08026 |
| B474 | N08020 |
| B474 | N08024 |
| B474 | N08026 |
| B475 | N08020 |
| B475 | N08024 |
| B475 | N08026 |
| B477 | P00750 |
| B479 (1100) | A91100 |
| B479 (1145) | A91145 |
| B479 (1235) | A91235 |
| B479 (8079) | A98079 |
| B479 (8111) | A98111 |
| B483 | M15102 |
| B483 (1060) | A91060 |
| B483 (1100) | A91100 |
| B483 (1435) | A91435 |
| B483 (3003) | A93003 |
| B483 (3102) | A93102 |
| B483 (5005) | A95005 |
| B483 (5050) | A95050 |
| B483 (5052) | A95052 |
| B483 (6061) | A96061 |
| B483 (6063) | A96063 |
| B483 (6262) | A96262 |
| B491 (1050) | A91050 |
| B491 (1100) | A91100 |
| B491 (1200) | A91200 |
| B491 (1235) | A91235 |
| B491 (3003) | A93003 |
| B491 (3102) | A93102 |
| B491 (6063) | A96063 |
| B493 | R60702 |
| B493 | R60704 |
| B493 | R60705 |
| B494 | R60702 |
| B494 | R60703 |
| B495 | R60702 |
| B495 | R60703 |
| B495 | R60704 |
| B495 | R60705 |
| B495 | R60706 |
| B496 | C10100 |
| B496 | C10200 |
| B496 | C11000 |
| B496 | C11300 |
| B496 | C11400 |
| B496 | C11500 |
| B496 | C11600 |
| B505 | C83600 |
| B505 | C83800 |
| B505 | C84200 |
| B505 | C84400 |
| B505 | C84800 |
| B505 | C86200 |
| B505 | C86300 |
| B505 | C86400 |
| B505 | C86500 |
| B505 | C90300 |
| B505 | C90500 |
| B505 | C90700 |
| B505 | C91000 |
| B505 | C91300 |
| B505 | C92200 |
| B505 | C92300 |
| B505 | C92500 |
| B505 | C92700 |
| B505 | C92800 |
| B505 | C92900 |
| B505 | C93200 |
| B505 | C93400 |
| B505 | C93500 |
| B505 | C93600 |
| B505 | C93700 |
| B505 | C93800 |
| B505 | C93900 |
| B505 | C94000 |
| B505 | C94100 |
| B505 | C94300 |
| B505 | C94700 |
| B505 | C94800 |
| B505 | C95200 |
| B505 | C95300 |
| B505 | C95400 |
| B505 | C95410 |
| B505 | C95500 |
| B505 | C95700 |
| B505 | C95800 |
| B505 | C95900 |
| B505 | C96400 |
| B505 | C97300 |
| B505 | C97600 |
| B505 | C97800 |
| B506 | C10200 |
| B511 | N08330 |
| B511 | N08332 |
| B512 | N08330 |
| B514 | N08120 |
| B514 | N08800 |
| B514 | N08810 |
| B515 | N08120 |
| B515 | N08800 |
| B515 | N08810 |
| B515 | N08811 |
| B516 | N06025 |
| B516 | N06045 |
| B516 | N06600 |
| B516 | N06603 |
| B517 | N06025 |
| B517 | N06045 |
| B517 | N06600 |
| B517 | N06603 |
| B521 | R05200 |
| B521 | R05240 |
| B521 | R05252 |
| B521 | R05255 |
| B521 | R05400 |
| B522 (I) | P00691 |
| B522 (II) | P00692 |
| B523 | R60702 |
| B523 | R60704 |
| B523 | R60705 |
| B524 | A91350 |
| B524 (6201) | A96201 |
| B531 (5005) | A95005 |
| B534 | C17500 |
| B534 | C17510 |
| B535 | N08330 |
| B535 | N08332 |
| B536 | N08330 |
| B536 | N08332 |
| B540 | P03350 |
| B541 | P00710 |
| B543 | C10800 |
| B543 | C12200 |
| B543 | C19400 |
| B543 | C23000 |
| B543 | C44300 |
| B543 | C44400 |
| B543 | C44500 |
| B543 | C68700 |

| CROSS REFERENCE SPECIFICATIONS | UNIFIED NUMBER |
|---|---|
| B543 | C70400 |
| B543 | C70600 |
| B543 | C71000 |
| B543 | C71500 |
| B543 | C71640 |
| B543 | C72200 |
| B546 | N06025 |
| B546 | N06045 |
| B546 | N06603 |
| B546 | N06617 |
| B546 | N08330 |
| B546 | N08332 |
| B547 | A83003 |
| B547 (1100) | A91100 |
| B547 (3003) | A93003 |
| B547 (3004) | A93004 |
| B547 (5050) | A95050 |
| B547 (5052) | A95052 |
| B547 (5083) | A95083 |
| B547 (5086) | A95086 |
| B547 (5154) | A95154 |
| B547 (5454) | A95454 |
| B547 (6061) | A96061 |
| B547 (7072) | A97072 |
| B548 | A96061 |
| B548 (1060) | A91060 |
| B548 (1100) | A91100 |
| B548 (3003) | A93003 |
| B548 (3004) | A93004 |
| B548 (5050) | A95050 |
| B548 (5052) | A95052 |
| B548 (5083) | A95083 |
| B548 (5086) | A95086 |
| B548 (5154) | A95154 |
| B548 (5254) | A95254 |
| B548 (5454) | A95454 |
| B548 (5456) | A95456 |
| B548 (5652) | A95652 |
| B548 (6061) | A86061 |
| B548 (Alclad 3003) | A83003 |
| B549 | A91350 |
| B550 | R60702 |
| B550 | R60704 |
| B550 | R60705 |
| B551 | R60702 |
| B551 | R60704 |
| B551 | R60705 |
| B551 | R60706 |
| B552 | C70600 |
| B552 | C71500 |
| B552 | C71640 |
| B552 | C72200 |
| B560 (1) | L13911 |
| B560 (2) | L13912 |
| B560 (3) | L13963 |
| B561 | P04980 |
| B561 (99.95) | P04995 |
| B562 (99.5) | P00025 |
| B562 (99.95) | P00020 |
| B562 (99.99) | P00015 |
| B562 (99.995) | P00010 |
| B563 | P03440 |
| B564 | N02200 |
| B564 | N04400 |
| B564 | N06022 |
| B564 | N06025 |
| B564 | N06045 |
| B564 | N06058 |
| B564 | N06059 |

| CROSS REFERENCE SPECIFICATIONS | UNIFIED NUMBER |
|---|---|
| B564 | N06110 |
| B564 | N06200 |
| B564 | N06210 |
| B564 | N06219 |
| B564 | N06230 |
| B564 | N06600 |
| B564 | N06603 |
| B564 | N06617 |
| B564 | N06625 |
| B564 | N06686 |
| B564 | N06690 |
| B564 | N08031 |
| B564 | N08120 |
| B564 | N08367 |
| B564 | N08800 |
| B564 | N08810 |
| B564 | N08811 |
| B564 | N08825 |
| B564 | N10242 |
| B564 | N10276 |
| B564 | N10624 |
| B564 | N10629 |
| B564 | N10665 |
| B564 | N10675 |
| B564 | N12160 |
| B564 | R20033 |
| B566 | C10200 |
| B566 | C11000 |
| B569 | C23000 |
| B569 | C26000 |
| B569 | C26130 |
| B570 | C17000 |
| B570 | C17200 |
| B572 | N06002 |
| B572 | N06230 |
| B572 | N12160 |
| B572 | R30556 |
| B573 | N10003 |
| B573 | N10242 |
| B574 | N06022 |
| B574 | N06058 |
| B574 | N06059 |
| B574 | N06200 |
| B574 | N06210 |
| B574 | N06455 |
| B574 | N06686 |
| B574 | N10276 |
| B575 | N06022 |
| B575 | N06058 |
| B575 | N06059 |
| B575 | N06200 |
| B575 | N06210 |
| B575 | N06455 |
| B575 | N06686 |
| B575 | N10276 |
| B581 | N06007 |
| B581 | N06022 |
| B581 | N06030 |
| B581 | N06975 |
| B581 | N06985 |
| B581 | N08031 |
| B582 | N06007 |
| B582 | N06022 |
| B582 | N06030 |
| B582 | N06975 |
| B582 | N06985 |
| B584 | C83450 |
| B584 | C85700 |
| B584 | C87200 ob |

| CROSS REFERENCE SPECIFICATIONS | UNIFIED NUMBER |
|---|---|
| B584 | C87300 |
| B584 | C87600 |
| B584 | C87610 |
| B584 | C89520 |
| B584 | C92210 |
| B584 | C92600 |
| B584 | C93800 |
| B584 | C95820 |
| B584 | C96800 |
| B584 (formerly B132) | C86700 |
| B584 (formerly B132, B147) | C86400 |
| B584 (formerly B143) | C90300 |
| B584 (formerly B143) | C90500 |
| B584 (formerly B143) | C92200 |
| B584 (formerly B143) | C92300 |
| B584 (formerly B144) | C93200 |
| B584 (formerly B144) | C93500 |
| B584 (formerly B144) | C93700 |
| B584 (formerly B144) | C94300 |
| B584 (formerly B145) | C83600 |
| B584 (formerly B145) | C83800 |
| B584 (formerly B145) | C84400 |
| B584 (formerly B145) | C84800 |
| B584 (formerly B146) | C85200 |
| B584 (formerly B146) | C85400 |
| B584 (formerly B147) | C86200 |
| B584 (formerly B147) | C86300 |
| B584 (formerly B147) | C86500 |
| B584 (formerly B149) | C97300 |
| B584 (formerly B149) | C97600 |
| B584 (formerly B149) | C97800 |
| B584 (formerly B198) | C87400 |
| B584 (formerly B198) | C87500 |
| B584 (formerly B292) | C94700 |
| B584 (formerly B292) | C94800 |
| B584 (formerly B292) | C94900 |
| B587 | C21000 |
| B587 | C22000 |
| B587 | C23000 |
| B587 | C26000 |
| B587 | C26800 |
| B587 | C27000 |
| B587 | C27200 |
| B589 (99.95) | P03995 |
| B591 | C40500 |
| B591 | C40800 ob |
| B591 | C41100 |
| B591 | C41300 |
| B591 | C41500 |
| B591 | C42200 |
| B591 | C42500 |
| B591 | C43000 |
| B591 | C43400 |
| B592 | C68800 |
| B592 | C69000 ob |
| B596 | P00901 |
| B599 | N08700 |
| B603 (I) | K92500 |
| B603 (IIA) | K91870 |
| B603 (IIB) | K92400 |
| B603 (III) | K91670 |
| B603 (IV) | K91470 |
| B608 | C61300 |
| B608 | C61400 |
| B608 | C70600 |
| B608 | C71500 |
| B616 (99.80) | P05980 |
| B616 (99.90) | P05981 |
| B616 (99.95) | P05982 |

| CROSS REFERENCE SPECIFICATIONS | UNIFIED NUMBER |
|---|---|
| B617 | P07900 |
| B618 (201.0) | A02010 |
| B618 (204.0) | A02040 |
| B618 (208.0) | A02080 ob |
| B618 (222.0) | A02220 ob |
| B618 (242.0) | A02420 |
| B618 (295.0) | A02950 |
| B618 (319.0) | A03190 |
| B618 (328.0) | A03280 ob |
| B618 (355.0) | A03550 |
| B618 (356.0) | A03560 |
| B618 (443.0) | A04430 |
| B618 (514.0) | A05140 |
| B618 (520.0) | A05200 |
| B618 (535.0) | A05350 |
| B618 (705.0) | A07050 |
| B618 (707.0) | A07070 |
| B618 (710.0) | A07100 |
| B618 (712.0) | A07120 |
| B618 (713.0) | A07130 |
| B618 (771.0) | A07710 |
| B618 (850.0) | A08500 |
| B618 (851.0) | A08510 |
| B618 (852.0) | A08520 |
| B618 (A356.0) | A13560 |
| B618 (B443.0) | A24430 |
| B618 (C355.0) | A33550 |
| B619 | N06002 |
| B619 | N06007 |
| B619 | N06022 |
| B619 | N06030 |
| B619 | N06058 |
| B619 | N06059 |
| B619 | N06200 |
| B619 | N06210 |
| B619 | N06230 |
| B619 | N06455 |
| B619 | N06686 |
| B619 | N06975 |
| B619 | N06985 |
| B619 | N08031 |
| B619 | N08320 |
| B619 | N10001 |
| B619 | N10242 |
| B619 | N10276 |
| B619 | N10624 |
| B619 | N10629 |
| B619 | N10665 |
| B619 | N10675 |
| B619 | N12160 |
| B619 | R20033 |
| B619 | R30556 |
| B620 | N08320 |
| B621 | N08320 |
| B622 | N06002 |
| B622 | N06007 |
| B622 | N06022 |
| B622 | N06030 |
| B622 | N06058 |
| B622 | N06059 |
| B622 | N06060 |
| B622 | N06200 |
| B622 | N06210 |
| B622 | N06230 |
| B622 | N06250 |
| B622 | N06255 |
| B622 | N06455 |
| B622 | N06686 |
| B622 | N06975 |
| B622 | N06985 |
| B622 | N08031 |
| B622 | N08135 |
| B622 | N08320 |
| B622 | N08535 |
| B622 | N10001 |
| B622 | N10242 |
| B622 | N10276 |
| B622 | N10624 |
| B622 | N10629 |
| B622 | N10665 |
| B622 | N10675 |
| B622 | N12160 |
| B622 | R20033 |
| B622 | R30556 |
| B624 | C16200 |
| B625 | N08031 |
| B625 | N08904 |
| B625 | N08925 |
| B625 | N08926 |
| B625 | N08932 |
| B625 | R20033 |
| B626 | N06002 |
| B626 | N06007 |
| B626 | N06022 |
| B626 | N06030 |
| B626 | N06058 |
| B626 | N06059 |
| B626 | N06200 |
| B626 | N06210 |
| B626 | N06230 |
| B626 | N06455 |
| B626 | N06686 |
| B626 | N06975 |
| B626 | N06985 |
| B626 | N08031 |
| B626 | N08320 |
| B626 | N10001 |
| B626 | N10242 |
| B626 | N10276 |
| B626 | N10624 |
| B626 | N10629 |
| B626 | N10665 |
| B626 | N10675 |
| B626 | N12160 |
| B626 | R20033 |
| B626 | R30556 |
| B632 (6061) | A96061 |
| B637 | N07001 |
| B637 | N07080 |
| B637 | N07252 |
| B637 | N07500 |
| B637 | N07718 |
| B637 | N07750 |
| B637 | N07752 |
| B638 | C11000 |
| B638 (OF) | C10200 |
| B638 (OFXLP) | C10300 |
| B639 (661) | R30155 |
| B639 (671) | R30816 |
| B640 | C10100 |
| B640 | C10200 |
| B640 | C10300 |
| B640 | C12000 |
| B640 | C12200 |
| B641 | C12200 |
| B642 | C21000 |
| B643 | C17200 |
| B649 | N08031 |
| B649 | N08904 |
| B649 | N08925 |
| B649 | N08926 |
| B649 | R20033 |
| B652 | R04295 |
| B653 | R60702 |
| B653 | R60704 |
| B653 | R60705 |
| B654 | R04295 |
| B655 | R04295 |
| B658 | R60702 |
| B658 | R60704 |
| B658 | R60705 |
| B661 | M16710 |
| B668 | N08028 |
| B670 | N07718 |
| B671 (99.80) | P06100 |
| B672 | N08700 |
| B673 | N08904 |
| B673 | N08925 |
| B673 | N08926 |
| B674 | N08904 |
| B674 | N08925 |
| B674 | N08926 |
| B675 | N08367 |
| B676 | N08367 |
| B677 | N08904 |
| B677 | N08925 |
| B677 | N08926 |
| B686 (354.0) | A03540 |
| B686 (A201.0) | A12010 |
| B686 (A356.0) | A13560 |
| B686 (A357.0) | A13570 |
| B686 (C355.0) | A33550 |
| B687 | C10200 |
| B687 | C10300 |
| B687 | C10800 |
| B687 | C12000 |
| B687 | C12200 |
| B687 | C23000 |
| B688 | N08366 |
| B688 | N08367 |
| B690 | N08366 |
| B690 | N08367 |
| B691 | N08366 |
| B691 | N08367 |
| B694 | C11000 |
| B694 | C22000 |
| B694 | C66400 |
| B694 | C66410 |
| B694 | C71000 |
| B698 | C19200 |
| B698 | C19400 |
| B698 (DHP) | C12200 |
| B698 (DLP) | C12000 |
| B698 (Gilding-95%) | C21000 |
| B698 (OF) | C10200 |
| B698 (OFE) | C10100 |
| B698 (OFLP) | C10800 |
| B698 (OFXLP) | C10300 |
| B698 (Red Brass-85%) | C23000 |
| B704 | N06219 |
| B704 | N06625 |
| B704 | N08825 |
| B705 | N06219 |
| B705 | N06625 |
| B705 | N08825 |
| B706 | C69100 |
| B708 | R05240 |

| CROSS REFERENCE SPECIFICATIONS | UNIFIED NUMBER |
|---|---|
| B708 | R05252 |
| B708 | R05255 |
| B708 | R05400 |
| B709 | N08028 |
| B710 | N08330 |
| B710 | N08332 |
| B711 (6201) | A96201 |
| B716 | C12000 |
| B716 | C12200 |
| B718 | N06333 |
| B719 | N06333 |
| B722 | N06333 |
| B723 | N06333 |
| B725 | N02200 |
| B725 | N02201 |
| B725 | N04400 |
| B726 | N06333 |
| B729 | N08020 |
| B729 | N08024 |
| B729 | N08026 |
| B730 | N02200 |
| B730 | N02201 |
| B730 | N04400 |
| B736 (1060) | A91060 |
| B736 (1100) | A91100 |
| B736 (1235) | A91235 |
| B737 (1145) | A91145 |
| B738 | C10200 |
| B738 | C11000 |
| B739 | N08330 |
| B739 | N08332 |
| B740 | C72650 |
| B740 | C72700 |
| B740 | C72900 |
| B743 | C10200 |
| B743 | C10300 |
| B743 | C10800 |
| B743 | C12000 |
| B743 | C12200 |
| B747 | C15100 |
| B749 | L50042 |
| B749 | L51120 |
| B749 | L51121 |
| B749 | L51123 |
| B749 (Common Lead) | L50045 |
| B750 | Z38510 |
| B751 | N06600 |
| B751 | N06625 |
| B752 (702C) | R60702 |
| B752 (704C) | R60704 |
| B752 (705C) | R60705 |
| B755 | N06110 |
| B756 | N06110 |
| B757 | N06110 |
| B758 | N06110 |
| B759 | N06110 |
| B760 | R07004 ob |
| B763 | C83450 |
| B763 | C83800 |
| B763 | C84400 |
| B763 | C84800 |
| B763 | C85200 |
| B763 | C85400 |
| B763 | C85700 |
| B763 | C86200 |
| B763 | C86300 |
| B763 | C86400 |
| B763 | C86500 |
| B763 | C86700 |

| CROSS REFERENCE SPECIFICATIONS | UNIFIED NUMBER |
|---|---|
| B763 | C87200 ob |
| B763 | C87300 |
| B763 | C87400 |
| B763 | C87500 |
| B763 | C87600 |
| B763 | C87610 |
| B763 | C89844 |
| B763 | C90300 |
| B763 | C90500 |
| B763 | C92300 |
| B763 | C92600 |
| B763 | C93200 |
| B763 | C93500 |
| B763 | C93700 |
| B763 | C93800 |
| B763 | C94300 |
| B763 | C94700 |
| B763 | C94800 |
| B763 | C94900 |
| B763 | C95200 |
| B763 | C95300 |
| B763 | C95400 |
| B763 | C95410 |
| B763 | C95500 |
| B763 | C95600 |
| B763 | C95800 |
| B763 | C97300 |
| B763 | C97600 |
| B763 | C97800 |
| B763 | C99400 |
| B763 | C99500 |
| B768 | C17400 ob |
| B768 | C17410 |
| B770 | C81400 |
| B770 | C82000 |
| B770 | C82200 |
| B770 | C82400 |
| B770 | C82500 |
| B770 | C82510 |
| B770 | C82600 |
| B770 | C82800 |
| B770 | C96700 |
| B774 (291-325) | L54935 |
| B778 (1350) | A91350 |
| B779 (1350) | A91350 |
| B792 | Z30500 |
| B792 | Z34510 |
| B793 | Z35542 |
| B793 | Z35543 |
| B800 (8017) | A98017 |
| B800 (8030) | A98030 |
| B800 (8076) | A98076 |
| B800 (8130) | A98130 |
| B800 (8176) | A98176 |
| B800 (8177) | A98177 ob |
| B804 | N08367 |
| B804 | N08926 |
| B805 | N07716 |
| B805 | N07725 |
| B805 | N07773 |
| B805 | N07776 |
| B805 | N09777 |
| B806 | C87500 |
| B806 | C87800 |
| B806 | C95300 |
| B806 | C95400 |
| B806 | C95410 |
| B806 | C95500 |
| B806 | C95800 |

| CROSS REFERENCE SPECIFICATIONS | UNIFIED NUMBER |
|---|---|
| B811 | R60802 |
| B811 | R60804 |
| B814 | N06920 |
| B815 | R31233 |
| B818 | R31233 |
| B833 | Z13005 |
| B833 | Z15005 |
| B833 | Z30700 |
| B834 | N06625 |
| B834 | N08367 |
| B843 | M11636 |
| B843 (AZ 31D) | M11313 |
| B843 (AZ31B) | M11311 |
| B843 (AZ63B) | M11632 |
| B843 (AZ63C) | M11634 |
| B843 (AZ63D) | M11636 |
| B843 (MIC) | M15102 |
| B852 | Z80310 |
| B852 | Z80410 |
| B852 | Z80411 |
| B852 | Z80510 |
| B852 | Z80511 |
| B852 | Z80531 |
| B852 | Z80610 |
| B852 | Z80710 |
| B852 | Z80810 |
| B852 | Z80910 |
| B852 | Z81330 |
| B852 | Z81430 |
| B852 | Z81450 |
| B852 | Z81530 |
| B852 | Z81540 |
| B852 | Z81550 |
| B853 | S31905 |
| B860 | Z55710 |
| B860 (S-1) | Z55710 |
| B860 (Type A-1 90Zn-10Al High Purity) | Z30750 |
| B860 (Type A-2 90Zn/10Al Low Purity) | Z31710 |
| B860 (Type A-3, 95Zn-5Al High Purity) | Z30503 |
| B860 (Type A-4 95Zn-5Al Low Purity) | Z31510 |
| B861 | R56323 |
| B861 | R56404 |
| B861 (1) | R50250 |
| B861 (2) | R50400 |
| B861 (3) | R50550 |
| B861 (5) | R56400 |
| B861 (7) | R52400 |
| B861 (9) | R56320 |
| B861 (11) | R52250 |
| B861 (12) | R53400 |
| B861 (13) | R53413 |
| B861 (14) | R53414 |
| B861 (15) | R53415 |
| B861 (16) | R52402 |
| B861 (17) | R52252 |
| B861 (18) | R56322 |
| B861 (21) | R58210 |
| B861 (23) | R56407 |
| B861 (25) | R56403 |
| B861 (26) | R52404 |
| B861 (27) | R52254 |
| B861 (240) | R56405 |
| B862 | R56323 |
| B862 | R56404 |
| B862 (1) | R50250 |

| CROSS REFERENCE SPECIFICATIONS | UNIFIED NUMBER | CROSS REFERENCE SPECIFICATIONS | UNIFIED NUMBER | CROSS REFERENCE SPECIFICATIONS | UNIFIED NUMBER |
|---|---|---|---|---|---|
| B862 (2) | R50400 | B888 | C76200 | F467 | C64200 |
| B862 (3) | R50550 | B892 | Z46540 | F467 | C65100 |
| B862 (5) | R56400 | B894 | Z46541 | F467 | C65500 |
| B862 (7) | R52400 | B901 (1350) | A91350 | F467 | C66100 |
| B862 (9) | R56320 | B903 | C12200 | F467 | C67500 |
| B862 (11) | R52250 | B907 | Z30402 | F467 | C71000 |
| B862 (12) | R53400 | B907 | Z30502 | F467 | C71500 |
| B862 (13) | R53413 | B907 | Z30505 | F467 | N10002 |
| B862 (14) | R53414 | B907 | Z30702 | F467 | R50550 |
| B862 (15) | R53415 | B907 | Z30705 | F467 | R55111 |
| B862 (16) | R52402 | B907 | Z30800 | F467 (1) | R50250 |
| B862 (17) | R52252 | B907 | Z34550 | F467 (2) | R50400 |
| B862 (18) | R56322 | B907 | Z50940 | F467 (4) | R50700 |
| B862 (21) | R58210 | B907 | Z50980 | F467 (5) | R56401 |
| B862 (23) | R56407 | B907 | Z56900 | F467 (7) | R52400 |
| B862 (24) | R56405 | B907 | Z56930 | F467 (19) | R58640 |
| B862 (25) | R56403 | CD3MN) | J92205 | F467 (23) | R56401 |
| B862 (26) | R52404 | CD3MWCuN) | J93380 | F467 (32) | R55111 |
| B862 (27) | R52254 | CE8MN) | J93345 | F467 (110) | C11000 |
| B863 | R56323 | F1 | N02205 | F467 (276) | N10276 |
| B863 | R56404 | F1 | N02233 | F467 (335) | N10001 |
| B863 (1) | R50250 | F1 | N02270 | F467 (400) | N04400 |
| B863 (2) | R50400 | F1 (3) | N02253 | F467 (405) | N04405 |
| B863 (3) | R50550 | F3 | N02205 | F467 (500) | N05500 |
| B863 (5) | R56400 | F3 | N02233 | F467 (510) | C51000 |
| B863 (7) | R52400 | F3 | N02270 | F467 (613) | C61300 |
| B863 (9) | R54520 | F3 (3) | N02253 | F468 | A96061 |
| B863 (9) | R56320 | F4 | N02233 | F468 | C27000 |
| B863 (11) | R52250 | F15 | K94610 | F468 | C46400 |
| B863 (12) | R53400 | F29 | K94101 | F468 | C61400 |
| B863 (13) | R53413 | F30 (42) | K94100 | F468 | C63000 |
| B863 (14) | R53414 | F30 (46) | K94600 | F468 | C64200 |
| B863 (15) | R53415 | F30 (48) | K94800 | F468 | C65100 |
| B863 (16) | R52402 | F30 (52) | N14052 | F468 | C65500 |
| B863 (17) | R52252 | F31 | K94760 | F468 | C66100 |
| B863 (18) | R56322 | F67 (1) | R50250 | F468 | C67500 |
| B863 (21) | R58210 | F67 (2) | R50400 | F468 | C71000 |
| B863 (23) | R56407 | F67 (3) | R50550 | F468 | C71500 |
| B863 (24) | R56405 | F67 (4) | R50700 | F468 | N10002 |
| B863 (25) | R56403 | F68 | C10100 | F468 | R50550 |
| B863 (26) | R52404 | F68 | C46200 | F468 | R55111 |
| B863 (27) | R52254 | F72 | P00016 | F468 (1) | R50250 |
| B863 (28) | R56323 | F73 (Electronic Grade) | R07031 | F468 (2) | R50400 |
| B865 | N05500 | F75 (Cast) | R30075 | F468 (4) | R50700 |
| B865 (23) | R56407 | F90 | R30605 | F468 (5) | R56401 |
| B870 | C17500 | F96 | C70690 | F468 (7) | R52400 |
| B870 | C17510 | F96 | C71580 | F468 (19) | R58640 |
| B872 | N09908 | F96 | C71590 | F468 (23) | R56401 |
| B876 | C15815 | F96 | N04400 | F468 (32) | R55111 |
| B888 | C14530 | F96 | N04404 | F468 (110) | C11000 |
| B888 | C15100 | F106 (TB Ag) | P07016 | F468 (276) | N10276 |
| B888 | C15500 | F136 | R56401 | F468 (335) | N10001 |
| B888 | C19400 | F256 (I) | K91800 | F468 (400) | N04400 |
| B888 | C19500 | F256 (II) | K92801 | F468 (405) | N04405 |
| B888 | C19700 | F288 (1A and 1B) | R07005 | F468 (500) | N05500 |
| B888 | C23000 | F288 (2B) | R07913 ob | F468 (510) | C51000 |
| B888 | C26000 | F290 | N02211 | F468 (613) | C61300 |
| B888 | C42200 | F290 | N03300 | F560 | R05200 |
| B888 | C42500 | F290 | R07005 | F560 | R05400 |
| B888 | C42520 | F364 (1) | R03604 | F562 | R30035 |
| B888 | C51000 | F364 (II) | R03603 | F563 | R30563 |
| B888 | C51100 | F385 | K03900 | F593 | S30433 |
| B888 | C52100 | F467 | A96061 | F593 (303) | S30300 |
| B888 | C63800 | F467 | A96262 | F593 (303Se) | S30323 |
| B888 | C65400 | F467 | C27000 | F593 (304) | S30400 |
| B888 | C68800 | F467 | C46200 | F593 (304L) | S30403 |
| B888 | C70250 | F467 | C46400 | F593 (305) | S30500 |
| B888 | C70260 | F467 | C61400 | F593 (316) | S31600 |
| B888 | C75200 | F467 | C63000 | F593 (316L) | S31603 |

| CROSS REFERENCE SPECIFICATIONS | UNIFIED NUMBER |
|---|---|
| F593 (321) | S32100 |
| F593 (347) | S34700 |
| F593 (348) | S34800 |
| F593 (384) | S38400 |
| F593 (410) | S41000 |
| F593 (416Se) | S41623 |
| F593 (430) | S43000 |
| F593 (430F) | S43020 |
| F593 (630) | S17400 |
| F593 (XM-1) | S20300 |
| F593 (XM7) | S17400 |
| F593 (XM-7) | S30430 |
| F594 | S30433 |
| F594 (303) | S30300 |
| F594 (303Se) | S30323 |
| F594 (304L) | S30403 |
| F594 (305) | S30500 |
| F594 (316) | S31600 |
| F594 (316L) | S31603 |
| F594 (321) | S32100 |
| F594 (347) | S34700 |
| F594 (348) | S34800 |
| F594 (384) | S38400 |
| F594 (410) | S41000 |
| F594 (416Se) | S41623 |
| F594 (430) | S43000 |
| F594 (430F) | S43020 |
| F594 (630) | S17400 |
| F594 (XM-1) | S20300 |
| F594 (XM7) | S17400 |
| F594 (XM-7) | S30430 |
| F620 | R56407 |
| F688 | R30035 |
| F738 | S30433 |
| F738 (303) | S30300 |
| F738 (303Se) | S30323 |
| F738 (304) | S30400 |
| F738 (304L) | S30403 |
| F738 (305) | S30500 |
| F738 (316) | S31600 |
| F738 (316L) | S31603 |
| F738 (321) | S32100 |
| F738 (347) | S34700 |
| F738 (384) | S38400 |
| F738 (410) | S41000 |
| F738 (416) | S41600 |
| F738 (416Se) | S41623 |
| F738 (430) | S43000 |
| F738 (430F) | S43020 |
| F738 (431) | S43100 |
| F738 (630) | S17400 |
| F738 (XM-1) | S20300 |
| F738 (XM-7) | S30430 |
| F745 | J31670 |
| F799 | R31537 |
| F799 | R31538 |
| f799 | R31539 |
| F836 | S30433 |
| F836 (303) | S30300 |
| F836 (303Se) | S30323 |
| F836 (304) | S30400 |
| F836 (304L) | S30403 |
| F836 (305) | S30500 |
| F836 (316) | S31600 |
| F836 (316L) | S31603 |
| F836 (321) | S32100 |
| F836 (347) | S34700 |
| F836 (384) | S38400 |
| F836 (416) | S41600 |
| F836 (416Se) | S41623 |
| F836 (430) | S43000 |
| F836 (430F) | S43020 |
| F836 (431) | S43100 |
| F836 (630) | S17400 |
| F836 (XM-1) | S20300 |
| F836 (XM-7) | S30430 |
| F837 | S30433 |
| F837 (303) | S30300 |
| F837 (304) | S30400 |
| F837 (304L) | S30403 |
| F837 (305) | S30500 |
| F837 (316) | S31600 |
| F837 (316L) | S31603 |
| F837 (384) | S38400 |
| F837 (XM-1) | S20300 |
| F837 (XM-7) | S30430 |
| F879 | S30433 |
| F879 (304) | S30400 |
| F879 (304L) | S30403 |
| F879 (305) | S30500 |
| F879 (316) | S31600 |
| F879 (316L) | S31603 |
| F879 (384) | S38400 |
| F879 (XM-7) | S30430 |
| F880 | S30433 |
| F880 (303) | S30300 |
| F880 (303Se) | S30323 |
| F880 (304) | S30400 |
| F880 (304L) | S30403 |
| F880 (305) | S30500 |
| F880 (384) | S38400 |
| F880 (XM-1) | S20300 |
| F880 (XM-7) | S30430 |
| F899 | S28200 |
| F899 (301) | S30100 |
| F899 (302) | S30200 |
| F899 (303) | S30300 |
| F899 (304) | S30400 |
| F899 (316) | S31600 |
| F899 (317) | S31700 |
| F899 (410) | S41000 |
| F899 (416) | S41600 |
| F899 (420F) | S42020 |
| F899 (430F) | S43020 |
| F899 (431) | S43100 |
| F899 (440A) | S44002 |
| F899 (440B) | S44003 |
| F899 (440C) | S44004 |
| F899 (440F) | S44020 |
| F899 (630) | S17400 |
| F899 (631) | S17700 |
| F899 (XM-7) | S30430 |
| F899 (XM-16) | S45500 |
| F899 (XM-25) | S45000 |
| F901 (2024) | A92024 |
| F901 (6061) | A96061 |
| F901 (6262) | A96262 |
| F912 | K03900 |
| F961 | R30035 |
| F1058 (2) | R30008 |
| F1091 | R30605 |
| F1108 | R56406 |
| F1295 | R56700 |
| F1341 | R50250 |
| F1341 | R50400 |
| F1341 | R50550 |
| F1341 | R50700 |
| F1350 | S31673 |
| F1377 | R30075 |
| F1466 | K94620 |
| F1466 | K94630 |
| F1472 | R56400 |
| F1537 | R31537 |
| F1537 | R31538 |
| F1537 | R31539 |
| F1580 | R50700 |
| F1586 | S31675 |
| F1684 | K93050 |
| F1684 | K93500 |
| F1684 | K93603 |
| F1713 | R58130 |
| F1813 | R58120 |
| F1813 | S20910 |
| F2005 | N01555 |
| F2005 | N01555 |
| F2066 | R58150 |
| G28 | N06007 |
| G28 | N06022 |
| G28 | N06030 |
| G28 | N06059 |
| G28 | N06200 |
| G28 | N06455 |
| G28 | N06600 |
| G28 | N06625 |
| G28 | N06686 |
| G28 | N06985 |
| G28 | N08020 |
| G28 | N08367 |
| G28 | N08800 |
| K31545 | K31545 |
| New specification pending | N94620 |
| New specification pending | N94630 |
| pending | R58350 |
| S351 | N08151 |
| S372 (m) | K42365 |

## Cross Index of Commonly Known Documents Which Describe Materials Same as or Similar to These Covered by UNS Numbers

## AWS (American Welding Society) NUMBERS

**(The Symbol *ob* Indicates an Obsolete UNS Number.)**

# AWS

| CROSS REFERENCE SPECIFICATIONS | UNIFIED NUMBER |
|---|---|
| A5.1 (E6010) | W06010 |
| A5.1 (E6011) | W06011 |
| A5.1 (E6012) | W06012 |
| A5.1 (E6013) | W06013 |
| A5.1 (E6019) | W06019 |
| A5.1 (E6020) | W06020 |
| A5.1 (E6022) | W06022 |
| A5.1 (E6027) | W06027 |
| A5.1 (E7014) | W07014 |
| A5.1 (E7015) | W07015 |
| A5.1 (E7016) | W07016 |
| A5.1 (E7016-1) | W07116 |
| A5.1 (E7018, E7018M) | W07018 |
| A5.1 (E7018-1) | W07118 |
| A5.1 (E7024) | W07024 |
| A5.1 (E7024-1) | W07124 |
| A5.1 (E7027) | W07027 |
| A5.1 (E7028) | W07028 |
| A5.1 (E7048) | W07048 |
| A5.2 (R100) | K12147 |
| A5.2 (RG45) | K00045 ob |
| A5.2 (RG60) | K00060 ob |
| A5.2 (RG65) | K00065 ob |
| A5.3 (E1100) | A91100 |
| A5.3 (E3003) | A93003 |
| A5.3 (E4043) | A94043 |
| A5.4 (E7Cr-XX) | W50310 |
| A5.4 (E16-8-2-XX) | W36810 |
| A5.4 (E209-XX) | W32210 |
| A5.4 (E219-XX) | W32310 |
| A5.4 (E240-XX) | W32410 |
| A5.4 (E307-XX) | W30710 |
| A5.4 (E308-XX) | W30810 |
| A5.4 (E308L) | W30813 |
| A5.4 (E308Mo-XX) | W30820 |
| A5.4 (E308MoL-XX) | W30823 |
| A5.4 (E309) | W30910 |
| A5.4 (E309Cb) | W30917 |
| A5.4 (E309L) | W30913 |
| A5.4 (E309LMo) | W30923 |
| A5.4 (E309Mo) | W30920 |
| A5.4 (E310-XX) | W31010 |
| A5.4 (E310Cb-XX) | W31017 |
| A5.4 (E310H-XX) | W31015 |
| A5.4 (E310Mo-XX) | W31020 |
| A5.4 (E312) | W31310 |
| A5.4 (E316, E316H) | W31610 |
| A5.4 (E316L-XX) | W31613 |
| A5.4 (E317-XX) | W31710 |
| A5.4 (E317L-XX) | W31713 |
| A5.4 (E318-XX) | W31910 |
| A5.4 (E320-XX) | W88021 |
| A5.4 (E320LR) | W88022 |
| A5.4 (E330) | W88331 |
| A5.4 (E330H) | W88335 |
| A5.4 (E347-XX) | W34710 |
| A5.4 (E349-XX) | W34910 |
| A5.4 (E383) | W88028 |
| A5.4 (E385) | W88904 |
| A5.4 (E410-XX) | W41010 |
| A5.4 (E410NiMo-XX) | W41016 |
| A5.4 (E430-XX) | W43010 |
| A5.4 (E502-XX) | W50210 |
| A5.4 (E505-XX) | W50410 |
| A5.4 (E630-XX) | W37410 |
| A5.4 (E2209-XX) | W39209 |
| A5.4 (E2553-XX) | W39553 |
| A5.5 (E801X-B6L) | W50213 ob |
| A5.5 (E7010-A1) | W17010 |
| A5.5 (E7010-P1) | W17110 |
| A5.5 (E7011-A1) | W17011 |
| A5.5 (E7015-A1) | W17015 |
| A5.5 (E7015-B2L) | W52115 |
| A5.5 (E7015-C1L) | W22115 |
| A5.5 (E7015-C2L) | W23115 |
| A5.5 (E7016-A1) | W17016 |
| A5.5 (E7016-B2L) | W52116 |
| A5.5 (E7016-C1L) | W22116 |
| A5.5 (E7016-C2L) | W23116 |
| A5.5 (E7018-A1) | W17018 |
| A5.5 (E7018-B2L) | W52118 |
| A5.5 (E7018-C1L) | W22118 |
| A5.5 (E7018-C2L) | W23118 |
| A5.5 (E7018-C3L) | W20918 |
| A5.5 (E7018-W1) | W20018 |
| A5.5 (E7020-A1) | W17020 |
| A5.5 (E7027-A1) | W17027 |
| A5.5 (E8010-P1) | W18110 |
| A5.5 (E8015-B3L) | W53115 |
| A5.5 (E8015-B4L) | W53415 |
| A5.5 (E8015-B6) | W50215 |
| A5.5 (E8015-B6L) | W50205 |
| A5.5 (E8015-B7) | W50315 |
| A5.5 (E8015-B7L) | W50305 |
| A5.5 (E8015-B8) | W50415 |
| A5.5 (E8015-B8L) | W50405 |
| A5.5 (E8016-B1) | W51016 |
| A5.5 (E8016-B2) | W52016 |
| A5.5 (E8016-B5) | W51316 |
| A5.5 (E8016-B6) | W50216 |
| A5.5 (E8016-B6L) | W50206 |
| A5.5 (E8016-B7) | W50316 |
| A5.5 (E8016-B7L) | W50306 |
| A5.5 (E8016-B8) | W50416 |
| A5.5 (E8016-B8L) | W50406 |
| A5.5 (E8016-C1) | W22016 |
| A5.5 (E8016-C2) | W23016 |
| A5.5 (E8016-C3) | W21016 |
| A5.5 (E8016-C4) | W21916 |
| A5.5 (E8016-D3) | W18016 |
| A5.5 (E8018-B1) | W51018 |
| A5.5 (E8018-B2) | W52018 |
| A5.5 (E8018-B3L) | W53118 |
| A5.5 (E8018-B6) | W50218 |
| A5.5 (E8018-B6L) | W50208 |
| A5.5 (E8018-B7) | W50318 |
| A5.5 (E8018-B7L) | W50308 |
| A5.5 (E8018-B8) | W50418 |
| A5.5 (E8018-B8L) | W50408 |
| A5.5 (E8018-C1) | W22018 |
| A5.5 (E8018-C2) | W23018 |
| A5.5 (E8018-C3) | W21018 |
| A5.5 (E8018-C4) | W21918 |
| A5.5 (E8018-D1) | W18118 |
| A5.5 (E8018-D3) | W18018 |
| A5.5 (E8018-NM1) | W21118 |
| A5.5 (E8018-W2) | W20118 |
| A5.5 (E9015-B3) | W53015 |
| A5.5 (E9015-B9) | W50425 |
| A5.5 (E9015-C5L) | W25018 |
| A5.5 (E9015-D1) | W19015 |
| A5.5 (E9016-B3) | W53016 |
| A5.5 (E9016-B9) | W50426 |
| A5.5 (E9018-B3) | W53018 |
| A5.5 (E9018-B9) | W50428 |
| A5.5 (E9018-D1) | W19018 |
| A5.5 (E9018-D3) | W19118 |
| A5.5 (E9018M) | W21218 |
| A5.5 (E10015-D2) | W10015 |
| A5.5 (E10016-D2) | W10016 |
| A5.5 (E10016M) | W21216 ob |
| A5.5 (E10018-D2) | W10018 |
| A5.5 (E10018M) | W21318 |
| A5.5 (E11018M) | W21418 |
| A5.5 (E12018M) | W22218 |
| A5.5 (E12018M1) | W23218 |
| A5.6 (ECu) | W60189 |
| A5.6 (ECuAl-A2) | W60614 |
| A5.6 (ECuAl-B) | W60619 |
| A5.6 (ECuMnNiAl) | W60633 |
| A5.6 (ECuNi) | W60715 |
| A5.6 (ECuNiAl) | W60632 |
| A5.6 (ECuSi) | W60656 |
| A5.6 (ECuSn-A) | W60518 |
| A5.6 (ECuSn-C) | W60521 |
| A5.7 (ERCu) | C18980 |
| A5.7 (ERCuAl-A2) | C61800 |
| A5.7 (ERCuAl-A3) | C62400 |
| A5.7 (ERCuAl-Al) | C61000 |
| A5.7 (ERCuMnNiAl) | C63380 |
| A5.7 (ERCuNi) | C71581 |
| A5.7 (ERCuNiAl) | C63280 |
| A5.7 (ERCuSi-A) | C65600 |
| A5.7 (ERCuSn-A) | C51800 |
| A5.7 (ERCuSn-C) | C52100 |
| A5.8 (BAg-1) | P07450 |
| A5.8 (BAg-1a) | P07500 |
| A5.8 (BAg-2) | P07350 |
| A5.8 (BAg-2a) | P07300 |
| A5.8 (BAg-3) | P07501 |
| A5.8 (BAg-4) | P07400 |
| A5.8 (BAg-5) | P07453 |
| A5.8 (BAg-6) | P07503 |
| A5.8 (BAg-7) | P07563 |
| A5.8 (BAg-8) | P07720 |
| A5.8 (BAg-8a) | P07723 |
| A5.8 (BAg-9) | P07650 |
| A5.8 (BAg-10) | P07700 |
| A5.8 (BAg-13) | P07540 |
| A5.8 (BAg-13a) | P07560 |
| A5.8 (BAg-18) | P07600 |
| A5.8 (BAg-19) | P07925 |
| A5.8 (BAg-20) | P07301 |
| A5.8 (BAg-21) | P07630 |
| A5.8 (BAg-22) | P07490 |
| A5.8 (BAg-23) | P07850 |
| A5.8 (BAg-24) | P07505 |
| A5.8 (BAg-25) | P07200 ob |
| A5.8 (BAg-26) | P07250 |
| A5.8 (BAg-27) | P07251 |
| A5.8 (BAg-28) | P07401 |
| A5.8 (BAg-33) | P07252 |
| A5.8 (BAg-34) | P07380 |
| A5.8 (BAg-35) | P07351 |
| A5.8 (BAg-36) | P07454 |
| A5.8 (BAg-37) | P07253 |
| A5.8 (BAlSi-2) | A94343 |
| A5.8 (BAlSi-3) | A94145 |
| A5.8 (BAlSi-4) | A94047 |
| A5.8 (BAlSi-5) | A94045 |
| A5.8 (BAlSi-7) | A94004 |
| A5.8 (BAlSi-9) | A94147 |
| A5.8 (BAlSi-11) | A94104 |
| A5.8 (BAu-1) | P00375 |
| A5.8 (BAu-2) | P00800 |

| CROSS REFERENCE SPECIFICATIONS | UNIFIED NUMBER |
|---|---|
| A5.8 (BAu-3) | P00350 |
| A5.8 (BAu-4) | P00820 |
| A5.8 (BAu-5) | P00300 |
| A5.8 (BAu-6) | P00700 |
| A5.8 (BCo-1) | R39001 |
| A5.8 (BCu-1) | C14180 |
| A5.8 (BCu-1b) | C11000 |
| A5.8 (BCu-3) | C10100 |
| A5.8 (BCuP-1) | C55180 |
| A5.8 (BCuP-2) | C55181 |
| A5.8 (BCuP-3) | C55281 |
| A5.8 (BCuP-4) | C55283 |
| A5.8 (BCuP-5) | C55284 |
| A5.8 (BCuP-6) | C55280 |
| A5.8 (BCuP-7) | C55282 |
| A5.8 (BCuP-8) | C55285 |
| A5.8 (BCuZn-F) | C49080 ob |
| A5.8 (BCuZn-G) | C26380 ob |
| A5.8 (BMg-1) | M19001 |
| A5.8 (BNi-1) | N99600 |
| A5.8 (BNi-1a) | N99610 |
| A5.8 (BNi-2) | N99620 |
| A5.8 (BNi-3) | N99630 |
| A5.8 (BNi-4) | N99640 |
| A5.8 (BNi-5) | N99650 |
| A5.8 (BNi-5a) | N99651 |
| A5.8 (BNi-5b) | N99652 |
| A5.8 (BNi-6) | N99700 |
| A5.8 (BNi-7) | N99710 |
| A5.8 (BNi-8) | N99800 |
| A5.8 (BNi-9) | N99612 |
| A5.8 (BNi-10) | N99622 |
| A5.8 (BNi-11) | N99624 |
| A5.8 (BNi-12) | N99720 |
| A5.8 (BNi-13) | N99810 |
| A5.8 (BVAg-0) | P07017 |
| A5.8 (BVAg-6b) | P07507 |
| A5.8 (BVAg-8) | P07727 |
| A5.8 (BVAg-8b) | P07728 |
| A5.8 (BVAg-18) | P07607 |
| A5.8 (BVAg-29) | P07627 |
| A5.8 (BVAg-30) | P07687 |
| A5.8 (BVAg-31) | P07587 |
| A5.8 (BVAg-32) | P07547 |
| A5.8 (BVAu-2) | P00807 |
| A5.8 (BVAu-4) | P00827 |
| A5.8 (BVAu-7) | P00507 |
| A5.8 (BVAu-8) | P00927 |
| A5.8 (BVCu-1X) | C14181 |
| A5.8 (BVPd-1) | P03657 |
| A5.8 (RBCuZn-A) | C47000 |
| A5.8 (RBCuZn-B) | C68000 |
| A5.8 (RBCuZn-C) | C68100 |
| A5.8 (RBCuZn-D) | C77300 |
| A5.9 (EC16-8-2) | W36840 |
| A5.9 (EC19-10H) | W30440 ob |
| A5.9 (EC209) | W32240 |
| A5.9 (EC218) | W32540 |
| A5.9 (EC219) | W32340 |
| A5.9 (EC240) | W32440 |
| A5.9 (EC307) | W30740 ob |
| A5.9 (EC308, EC308H) | W30840 |
| A5.9 (EC308L) | W30843 |
| A5.9 (EC308LMo) | W30853 |
| A5.9 (EC308LSi) | W30848 |
| A5.9 (EC308Mo) | W30850 |
| A5.9 (EC308Si) | W30841 |
| A5.9 (EC309) | W30940 |
| A5.9 (EC309L) | W30943 |
| A5.9 (EC309LMo) | W30953 |
| A5.9 (EC309Mo) | W30950 |
| A5.9 (EC309Si) | W30941 |
| A5.9 (EC310) | W31040 |
| A5.9 (EC312) | W31340 |
| A5.9 (EC316, EC316H) | W31640 |
| A5.9 (EC316L) | W31643 |
| A5.9 (EC316LSi) | W31648 |
| A5.9 (EC316Si) | W31641 |
| A5.9 (EC317) | W31740 |
| A5.9 (EC317L) | W31743 |
| A5.9 (EC318) | W31940 |
| A5.9 (EC320LR) | W88022 |
| A5.9 (EC321) | W32140 |
| A5.9 (EC347) | W34740 |
| A5.9 (EC347Si) | W34748 |
| A5.9 (EC349) | W34940 ob |
| A5.9 (EC409Cb) | W40947 |
| A5.9 (EC410) | W41040 |
| A5.9 (EC410NiMo) | W41046 |
| A5.9 (EC420) | W42040 |
| A5.9 (EC430) | W43040 |
| A5.9 (EC446LMo) | W44647 |
| A5.9 (EC505) | W50440 |
| A5.9 (EC630) | W37440 |
| A5.9 (EC2209) | W39240 |
| A5.9 (EC2553) | W39540 |
| A5.9 (ER 309LMoSi) | S30989 ob |
| A5.9 (ER 309LSi) | S30988 |
| A5.9 (ER 309Mo) | S30982 |
| A5.9 (ER 317L) | S31783 |
| A5.9 (ER16-8-2) | S16880 |
| A5.9 (ER19-10H) | S30480 |
| A5.9 (ER209) | S20980 |
| A5.9 (ER218) | S21880 |
| A5.9 (ER219) | S21980 |
| A5.9 (ER240) | S24080 |
| A5.9 (ER307) | S30780 |
| A5.9 (ER308, ER308H) | S30880 |
| A5.9 (ER308L) | S30883 |
| A5.9 (ER308LSi) | S30888 |
| A5.9 (ER308MLMo) | S30886 |
| A5.9 (ER308Mo) | S30882 |
| A5.9 (ER308Si) | S30881 |
| A5.9 (ER309) | S30980 |
| A5.9 (ER309L) | S30983 |
| A5.9 (ER309LMo) | S30986 |
| A5.9 (ER309Si) | S30981 |
| A5.9 (ER310) | S31080 |
| A5.9 (ER312) | S31380 |
| A5.9 (ER316, 316H) | S31680 |
| A5.9 (ER316L) | S31683 |
| A5.9 (ER316LSi) | S31688 |
| A5.9 (ER316Si) | S31681 |
| A5.9 (ER317) | S31780 |
| A5.9 (ER318) | S31980 |
| A5.9 (ER320) | N08021 |
| A5.9 (ER320LR) | N08022 |
| A5.9 (ER321) | S32180 |
| A5.9 (ER330) | N08331 |
| A5.9 (ER347) | S34780 |
| A5.9 (ER347Si) | S34788 |
| A5.9 (ER409Cb) | S40940 |
| A5.9 (ER410) | S41080 |
| A5.9 (ER410NiMo) | S41086 |
| A5.9 (ER420) | S42080 |
| A5.9 (ER430) | S43080 |
| A5.9 (ER446LMo) | S44687 |
| A5.9 (ER502) | S50280 |
| A5.9 (ER502Si) | S50288 |
| A5.9 (ER505) | S50480 |
| A5.9 (ER505Si) | S50488 |
| A5.9 (ER630) | S17480 |
| A5.9 (ER2209) | S39209 |
| A5.9 (ER2553) | S39255 ob |
| A5.9 (ER2553) | S39553 |
| A5.9 (ER3556) | R30556 |
| A5.10 (ER1100) | A91100 |
| A5.10 (ER1188) | A91188 |
| A5.10 (ER2319) | A92319 |
| A5.10 (ER4010) | A94010 |
| A5.10 (ER4043) | A94043 |
| A5.10 (ER4047) | A94047 |
| A5.10 (ER4145) | A94145 |
| A5.10 (ER4643) | A94643 |
| A5.10 (ER5183) | A95183 |
| A5.10 (ER5356) | A95356 |
| A5.10 (ER5554) | A95554 |
| A5.10 (ER5556) | A95556 |
| A5.10 (ER5654) | A95654 |
| A5.10 (ER9004) | A94009 |
| A5.10 (R-A356.0) | A13560 |
| A5.10 (R-A357.0) | A13570 |
| A5.10 (R-C355.0) | A03550 |
| A5.10 (R206.0) | A02060 |
| A5.10 (R4011) | A94011 |
| A5.10 (RA-357.0) | A03570 |
| A5.11 (ENi-1) | W82141 |
| A5.11 (ENiCrCoMo-1) | W86117 |
| A5.11 (ENiCrFe-1) | W86132 |
| A5.11 (ENiCrFe-2) | W86133 |
| A5.11 (ENiCrFe-3) | W86182 |
| A5.11 (ENiCrFe-4) | W86134 |
| A5.11 (ENiCrFe-7) | W86152 |
| A5.11 (ENiCrFe-9) | W86094 |
| A5.11 (ENiCrFe-10) | W86095 |
| A5.11 (ENiCrMo-1) | W86007 |
| A5.11 (ENiCrMo-2) | W86002 |
| A5.11 (ENiCrMo3) | W86112 |
| A5.11 (ENiCrMo-4) | W80276 |
| A5.11 (ENiCrMo-5) | W80002 |
| A5.11 (ENiCrMo-6) | W86620 |
| A5.11 (ENiCrMo-7) | W86455 |
| A5.11 (ENiCrMo-9) | W86135 ob |
| A5.11 (ENiCrMo-9) | W86985 |
| A5.11 (ENiCrMo-10) | W86022 |
| A5.11 (ENiCrMo-10) | W86136 ob |
| A5.11 (ENiCrMo-11) | W86030 |
| A5.11 (ENiCrMo-12) | W86032 |
| A5.11 (ENiCrMo-13) | W86059 |
| A5.11 (ENiCrMo-14) | W86026 |
| A5.11 (ENiCrMo-14) | W86686 |
| A5.11 (ENiCu-7) | W84190 |
| A5.11 (ENiMo-1) | W80001 |
| A5.11 (ENiMo-3) | W80004 |
| A5.11 (ENiMo-7) | W80665 |
| A5.11 (ENiMo-8) | W80008 |
| A5.11 (ENiMo-9) | W80009 |
| A5.11 (ENiMo-10) | W80675 |
| A5.12 (EWCe-2) | R07932 |
| A5.12 (EWLa-1) | R07941 |
| A5.12 (EWLa-1.5) | R07942 |
| A5.12 (EWLa-2) | R07943 |
| A5.12 (EWP) | R07900 |
| A5.12 (EWTh-1) | R07911 |
| A5.12 (EWTh-2) | R07912 |
| A5.12(EWTh-3) | R07913 ob |
| A5.12 (EWZr-1) | R07920 |

| CROSS REFERENCE SPECIFICATIONS | UNIFIED NUMBER |
|---|---|
| A5.13 (ECoCr-A) | W73006 |
| A5.13 (ECoCr-B) | W73012 |
| A5.13 (ECoCr-C) | W73001 |
| A5.13 (ECoCr-E) | W73021 |
| A5.13 (ECuAl-A2) | W60617 |
| A5.13 (ECuAl-B) | W60619 |
| A5.13 (ECuAl-C) | W60625 |
| A5.13 (ECuAl-D) | W61625 |
| A5.13 (ECuAl-E) | W62625 |
| A5.13 (ECuMnNiAl) | W60633 |
| A5.13 (ECuNi) | W60715 |
| A5.13 (ECuNiAl) | W60632 |
| A5.13 (ECuSi) | W60656 |
| A5.13 (ECuSn-A) | W60518 |
| A5.13 (ECuSn-C) | W60521 |
| A5.13 (EFe-1) | W74001 |
| A5.13 (EFe-2) | W74002 |
| A5.13 (EFe-3) | W74003 |
| A5.13 (EFe-4) | W74004 |
| A5.13 (EFe-5) | W75110 |
| A5.13 (EFe5-A) | W77710 ob |
| A5.13 (EFe5-C) | W77310 ob |
| A5.13 (EFe-6) | W77510 |
| A5.13 (EFe-7) | W77610 |
| A5.13 (EFeCr-A1) | W74510 ob |
| A5.13 (EFeCr-A1A) | W74011 |
| A5.13 (EFeCr-A2) | W74012 |
| A5.13 (EFeCr-A3) | W74013 |
| A5.13 (EFeCr-A4) | W74014 |
| A5.13 (EFeCr-A5) | W74015 |
| A5.13 (EFeCr-A6) | W74016 |
| A5.13 (EFeCr-A7) | W74017 |
| A5.13 (EFeCr-A8) | W74018 |
| A5.13 (EFeCr-E1) | W74211 |
| A5.13 (EFeCr-E2) | W74212 |
| A5.13 (EFeCr-E3) | W74213 |
| A5.13 (EFeCr-E4) | W74214 |
| A5.13 (EFeMn-A) | W79110 |
| A5.13 (EFeMn-B) | W79310 |
| A5.13 (EFeMn-C) | W79210 |
| A5.13 (EFeMn-D) | W77410 ob |
| A5.13 (EFeMn-D) | W79410 |
| A5.13 (EFeMn-E) | W79510 |
| A5.13 (EFeMn-F) | W79610 |
| A5.13 (EFeMnCr) | W79710 |
| A5.13 (ENiCr-A) | W89604 ob |
| A5.13 (ENiCr-B) | W89605 ob |
| A5.13 (ENiCr-C) | W89606 |
| A5.13 (ENiCrFeCo) | W83002 |
| A5.13 (ENiCrMo-5A) | W80002 |
| A5.14 (ERNi-1) | N02061 |
| A5.14 (ERNiCr-3) | N06082 |
| A5.14 (ERNiCr-4) | N06072 |
| A5.14 (ERNiCr-6) | N06076 |
| A5.14 (ERNiCrCoMo-1) | N06617 |
| A5.14 (ERNiCrFe-5) | N06062 |
| A5.14 (ERNiCrFe-6) | N07092 |
| A5.14 (ERNiCrFe-7) | N06052 |
| A5.14 (ERNiCrFe-8) | N07069 |
| A5.14 (ERNiCrFe-11) | N06601 |
| A5.14 (ERNiCrMo-1) | N06007 |
| A5.14 (ERNiCrMo-2) | N06002 |
| A5.14 (ERNiCrMo-3) | N06625 |
| A5.14 (ERNiCrMo-4) | N10276 |
| A5.14 (ERNiCrMo-7) | N06455 |
| A5.14 (ERNiCrMo-8) | N06975 |
| A5.14 (ERNiCrMo-9) | N06985 |
| A5.14 (ERNiCrMo-10) | N06022 |
| A5.14 (ERNiCrMo-11) | N06030 |
| A5.14 (ERNiCrMo-13) | N06059 |
| A5.14 (ERNiCrMo-14) | N06686 |
| A5.14 (ERNiCrMo-15) | N07725 |
| A5.14 (ERNiCrMo-18) | N06650 |
| A5.14 (ERNiCrWMo-1) | N06231 |
| A5.14 (ERNiCu-7) | N04060 |
| A5.14 (ERNiCu-8) | N05504 |
| A5.14 (ERNiFeCr-1) | N08065 |
| A5.14 (ERNiFeCr-2) | N07718 |
| A5.14 (ERNiMo-1) | N10001 |
| A5.14 (ERNiMo-2) | N10003 |
| A5.14 (ERNiMo-3) | N10004 |
| A5.14 (ERNiMo-7) | N10665 |
| A5.14 (ERNiMo-8) | N10008 |
| A5.14 (ERNiMo-9) | N10009 |
| A5.14 (ERNiMo-10) | N10675 |
| A5.15 (ENi-CI) | W82001 |
| A5.15 (ENi-CI-A) | W82003 |
| A5.15 (ENiCu-A) | W84001 |
| A5.15 (ENiCu-B) | W84002 |
| A5.15 (ENiFe-CI) | W82002 |
| A5.15 (ENiFe-CI-A) | W82004 |
| A5.15 (ENiFeMn-CI) | W82006 |
| A5.15 (ENiFeT3-CI) | W82032 ob |
| A5.15 (ERNi-CI) | N02215 |
| A5.15 (ERNiFeMn-CI) | N02216 |
| A5.15 (ESt) | K01520 |
| A5.15 (RCI) | F10090 |
| A5.15 (RCI-A) | F10091 |
| A5.15 (RCI-B) | F10092 |
| A5.16 (ERTi-1) | R50100 |
| A5.16 (ERTi-2) | R50120 |
| A5.16 (ERTi-3) | R50125 |
| A5.16 (ERTi-4) | R50130 |
| A5.16 (ERTi-5) | R56400 |
| A5.16 (ERTi-5ELI) | R56402 |
| A5.16 (ERTi-6) | R54522 |
| A5.16 (ERTi-6ELI) | R54523 |
| A5.16 (ERTi-7) | R52401 |
| A5.16 (ERTi-9) | R56320 |
| A5.16 (ERTi-9ELI) | R56321 |
| A5.16 (ERTi-12) | R53401 |
| A5.16 (ERTi-15) | R56210 |
| A5.16 (ERTi-27) | R52255 |
| A5.17 (EC1) | W06041 |
| A5.17 (EH10K) | K01210 |
| A5.17 (EH11K) | K11140 |
| A5.17 (EH12K) | K01213 |
| A5.17 (EH14) | K11585 |
| A5.17 (EL8) | K01008 |
| A5.17 (EL8K) | K01009 |
| A5.17 (EL12) | K01012 |
| A5.17 (EM11K) | K01111 |
| A5.17 (EM12) | K01112 |
| A5.17 (EM12K) | K01113 |
| A5.17 (EM13K) | K01313 |
| A5.17 (EM14K) | K01314 |
| A5.17 (EM15K) | K01515 |
| A5.18 (E70C-3X) | W07703 |
| A5.18 (E70C-6X) | W07706 |
| A5.18 (ER70S-2) | K10726 |
| A5.18 (ER70S-3) | K11022 |
| A5.18 (ER70S-4) | K11132 |
| A5.18 (ER70S-5) | K11357 |
| A5.18 (ER70S-6) | K11140 |
| A5.18 (ER70S-7) | K11125 |
| A5.19 (ER AZ101A) | M11101 |
| A5.19 (ER AZ61A) | M11611 |
| A5.19 (ER AZ92A) | M11922 |
| A5.19 (ER EZ33A) | M12331 |
| A5.20 (E6XT-1) | W06601 ob |
| A5.20 (E6XT-11) | W06611 ob |
| A5.20 (E6XT-13) | W06613 |
| A5.20 (E6XT-4) | W06604 ob |
| A5.20 (E6XT-5) | W06605 ob |
| A5.20 (E6XT-6) | W06606 ob |
| A5.20 (E6XT-7) | W06607 ob |
| A5.20 (E6XT-8) | W06608 ob |
| A5.20 (E7XT-1, E7XT-1M) | W07601 |
| A5.20 (E7XT-10) | W07610 |
| A5.20 (E7XT-11) | W07611 |
| A5.20 (E7XT-12, E7XT-12M) | W07612 |
| A5.20 (E7XT-13) | W07613 |
| A5.20 (E7XT-14) | W07614 |
| A5.20 (E7XT-2, E7XT-2M) | W07602 |
| A5.20 (E7XT-3) | W07603 |
| A5.20 (E7XT-4) | W07604 |
| A5.20 (E7XT-5, E7XT-5M) | W07605 |
| A5.20 (E7XT-6) | W07606 |
| A5.20 (E7XT-7) | W07607 |
| A5.20 (E7XT-8) | W07608 |
| A5.20 (E7XT-9, E7XT-9M) | W07609 |
| A5.21 (EFe5-A) | W77710 ob |
| A5.21 (EFe5-C) | W77310 ob |
| A5.21 (EFeCr-A1) | W74510 ob |
| A5.21 (ERCCoCr-A) | W73036 |
| A5.21 (ERCCoCr-B) | W73042 |
| A5.21 (ERCCoCr-C) | W73031 |
| A5.21 (ERCCoCr-E) | W73041 |
| A5.21 (ERCCoCr-G) | W73032 |
| A5.21 (ERCCuAl-A2) | W60618 |
| A5.21 (ERCCuAl-A3) | W60624 |
| A5.21 (ERCCuAl-C) | W60626 |
| A5.21 (ERCCuAl-D) | W61626 |
| A5.21 (ERCCuAl-E) | W62626 |
| A5.21 (ERCCuSi-A) | W60657 |
| A5.21 (ERCCuSn-A) | W60518 |
| A5.21 (ERCCuSn-D) | W60524 |
| A5.21 (ERCFe-1) | W74030 |
| A5.21 (ERCFe-1A) | W74031 |
| A5.21 (ERCFe-2) | W74032 |
| A5.21 (ERCFe-3) | W74033 |
| A5.21 (ERCFe-5) | W74035 |
| A5.21 (ERCFe-6) | W77530 |
| A5.21 (ERCFe-8) | W77538 |
| A5.21 (ERCFeCr-A) | W74531 |
| A5.21 (ERCFeCr-A10) | W74540 |
| A5.21 (ERCFeCr-A1A) | W74530 |
| A5.21 (ERCFeCr-A3A) | W74533 |
| A5.21 (ERCFeCr-A4) | W74534 |
| A5.21 (ERCFeCr-A5) | W74535 |
| A5.21 (ERCFeCr-A9) | W74539 |
| A5.21 (ERCFeMn-C) | W79230 |
| A5.21 (ERCFeMn-Cr) | W79730 |
| A5.21 (ERCFeMn-F) | W79630 |
| A5.21 (ERCFeMn-G) | W79231 |
| A5.21 (ERCFeMn-H) | W79232 |
| A5.21 (ERCNiCr-A) | W89634 |
| A5.21 (ERCNiCr-B) | W89635 |
| A5.21 (ERCNiCr-C) | W89636 |
| A5.21 (ERCNiCrFeCo) | W83032 |
| A5.21 (ERCNiCrMo-5A) | W80032 ob |
| A5.21 (ERCNiCrMo-5A) | W86685 ob |
| A5.21 (ERCoCr-A) | R30006 |
| A5.21 (ERCoCr-B) | R30012 |
| A5.21 (ERCoCr-C) | R30001 |
| A5.21 (ERCoCr-E) | R30021 |
| A5.21 (ERCoCr-F) | R30002 |

| CROSS REFERENCE SPECIFICATIONS | UNIFIED NUMBER |
|---|---|
| A5.21 (ERCoCr-G) | R30014 |
| A5.21 (ERCuAl-A2) | C61800 |
| A5.21 (ERCuAl-A3) | C62400 |
| A5.21 (ERCuAl-C) | C62580 |
| A5.21 (ERCuAl-D) | C62581 |
| A5.21 (ERCuAl-E) | C62582 |
| A5.21 (ERCuSi-A) | C65600 |
| A5.21 (ERCuSn-A) | C51800 |
| A5.21 (ERCuSn-D) | C52400 |
| A5.21 (ERFe-1) | T74000 |
| A5.21 (ERFe-1A) | T74001 |
| A5.21 (ERFe-2) | T74002 |
| A5.21 (ERFe-3) | T74003 |
| A5.21 (ERFe-5) | T74005 |
| A5.21 (ERFe-6) | T75006 |
| A5.21 (ERFe-8) | T75008 |
| A5.21 (ERNiCr-A) | N99644 |
| A5.21 (ERNiCr-B) | N99645 |
| A5.21 (ERNiCr-C) | N99646 |
| A5.21 (ERNiCr-D) | N99647 |
| A5.21 (ERNiCr-E) | N99648 |
| A5.21 (ERNiCrFeCo) | F46100 |
| A5.21 (ERNiCrMo-5A) | N10006 |
| A5.21 (RFe5-A) | W77730 ob |
| A5.22 (E307T0-3) | W30733 |
| A5.22 (E307TX-X) | W30731 |
| A5.22 (E308HMoT0-3) | W30830 |
| A5.22 (E308LMoT0-X, E308LMoTX-X) | W30838 |
| A5.22 (E308LT0-3) | W30837 |
| A5.22 (E308LTX-X, R308LT1-5) | W30835 |
| A5.22 (E308MoT0-3) | W30839 |
| A5.22 (E308MoTX-X) | W30832 |
| A5.22 (E308T0-3, E308HT0-3) | W30833 |
| A5.22 (E308TX-X, E308HT0-3) | W30831 |
| A5.22 (E309LCbT0-3) | W30934 |
| A5.22 (E309LCbTX-X) | W30932 |
| A5.22 (E309LMoT0-3, E309LMoTX-X) | W30938 |
| A5.22 (E309LNiMoTX-X) | W30936 |
| A5.22 (E309LT0-3) | W30937 |
| A5.22 (E309LTX-X, R309LT1-5) | W30935 |
| A5.22 (E309MoTX-X) | W30939 |
| A5.22 (E309T0-3) | W30933 |
| A5.22 (E309TX-X) | W30931 |
| A5.22 (E310T0-3, E310TX-X) | W31031 |
| A5.22 (E312T0-3) | W31231 |
| A5.22 (E312TX-X) | W31331 |
| A5.22 (E316LKT0-3) | W31630 |
| A5.22 (E316LT0-3) | W31637 |
| A5.22 (E316LTX-X, R316LT1-5) | W31635 |
| A5.22 (E316T0-3) | W31633 |
| A5.22 (E316TX-X) | W31631 |
| A5.22 (E317LT0-3) | W31737 |
| A5.22 (E317LTX-X) | W31735 |
| A5.22 (E347T0-3) | W34733 |
| A5.22 (E347TX-X, R347T1-5) | W34731 |
| A5.22 (E409TX-X, E409T0-3) | W40931 |
| A5.22 (E410NiMoTX-X, E410NiMoT0-3) | W41036 |
| A5.22 (E410NiTiTX-X, E410NiTiT0-3) | W41038 |
| A5.22 (E410TX-X, E410T0-3) | W41031 |
| A5.22 (E430TX-X, E430T0-3) | W43031 |
| A5.22 (E502TX-X) | W50231 |
| A5.22 (E505TX-X) | W50431 |
| A5.22 (E2209T0-X) | W39239 |
| A5.22 (E2553T0-X) | W39533 |
| A5.23 (B2LR) | W52042 |
| A5.23 (EA1) | K11222 |
| A5.23 (EA1N) | K11122 |
| A5.23 (EA2) | K11223 |
| A5.23 (EA2N) | K11123 |
| A5.23 (EA3) | K11423 |
| A5.23 (EA3K) | K21451 |
| A5.23 (EA3KN) | K21351 |
| A5.23 (EA3N) | K11323 |
| A5.23 (EA4) | K11424 |
| A5.23 (EA4N) | K11324 |
| A5.23 (EB 9) | S50482 |
| A5.23 (EB1) | K11043 |
| A5.23 (EB1N) | K10943 |
| A5.23 (EB2) | K11172 |
| A5.23 (EB2H) | K23016 |
| A5.23 (EB2HN) | K23116 |
| A5.23 (EB2N) | K11072 |
| A5.23 (EB2RR) | K11173 |
| A5.23 (EB3) | K31115 |
| A5.23 (EB3LRN) | K31117 ob |
| A5.23 (EB3N) | K31015 |
| A5.23 (EB3RR) | K31116 ob |
| A5.23 (EB5) | K12187 |
| A5.23 (EB5N) | K12087 |
| A5.23 (EB6) | S50280 |
| A5.23 (EB6H) | S50180 |
| A5.23 (EB6HN) | S50181 |
| A5.23 (EB6N) | S50281 |
| A5.23 (EB8) | S50480 |
| A5.23 (EB8N) | S50481 |
| A5.23 (ECA1) | W17041 |
| A5.23 (ECA1N) | W17141 ob |
| A5.23 (ECA2) | W17042 |
| A5.23 (ECA2N) | W17142 ob |
| A5.23 (ECA3) | W17043 |
| A5.23 (ECA3N) | W17143 ob |
| A5.23 (ECA4) | W17044 |
| A5.23 (ECA4N) | W17144 ob |
| A5.23 (ECB 9) | W50442 |
| A5.23 (ECB1) | W51040 |
| A5.23 (ECB1N) | W51041 ob |
| A5.23 (ECB2) | W52040 |
| A5.23 (ECB2H) | W52240 |
| A5.23 (ECB2HN) | W52241 ob |
| A5.23 (ECB2N) | W52041 ob |
| A5.23 (ECB3) | W53040 |
| A5.23 (ECB3LRN) | W53043 ob |
| A5.23 (ECB3N) | W53041 ob |
| A5.23 (ECB3RR) | W53042 ob |
| A5.23 (ECB4) | W53340 |
| A5.23 (ECB4N) | W53341 ob |
| A5.23 (ECB5) | W51340 |
| A5.23 (ECB5N) | W51341 ob |
| A5.23 (ECB6) | W50240 |
| A5.23 (ECB6H) | W50140 |
| A5.23 (ECB6HN) | W50141 ob |
| A5.23 (ECB6N) | W50241 ob |
| A5.23 (ECB8) | W50440 |
| A5.23 (ECB8N) | W50441 ob |
| A5.23 (ECF1) | W21150 |
| A5.23 (ECF1N) | W21151 ob |
| A5.23 (ECF2) | W20240 |
| A5.23 (ECF2N) | W20241 ob |
| A5.23 (ECF3) | W21140 |
| A5.23 (ECF3N) | W21141 ob |
| A5.23 (ECF4) | W20440 |
| A5.23 (ECF4N) | W20441 ob |
| A5.23 (ECF5) | W22540 |
| A5.23 (ECF6) | W22640 |
| A5.23 (ECF6N) | W22641 ob |
| A5.23 (ECM1) | W21240 |
| A5.23 (ECM1N) | W21241 ob |
| A5.23 (ECM2) | W21340 |
| A5.23 (ECM2N) | W21341 ob |
| A5.23 (ECM3) | W22240 |
| A5.23 (ECM3N) | W22241 ob |
| A5.23 (ECM4) | W22440 |
| A5.23 (ECM4N) | W22441 ob |
| A5.23 (ECM5) | W21345 |
| A5.23 (ECM6) | W21346 |
| A5.23 (ECNi1) | W21040 ob |
| A5.23 (ECNi1N) | W21041 ob |
| A5.23 (ECNi2) | W22040 |
| A5.23 (ECNi2N) | W22041 ob |
| A5.23 (ECNi3) | W23040 |
| A5.23 (ECNi3N) | W23041 ob |
| A5.23 (ECNi4) | W21250 |
| A5.23 (ECNi4N) | W21251 ob |
| A5.23 (ECNi5) | W21042 |
| A5.23 (ECW) | W20140 |
| A5.23 (EF1) | K11160 |
| A5.23 (EF1N) | K11060 |
| A5.23 (EF2) | K21450 |
| A5.23 (EF2N) | K21350 |
| A5.23 (EF3) | K21485 |
| A5.23 (EF3N) | K21385 |
| A5.23 (EF4) | K12048 |
| A5.23 (EF4N) | K11948 |
| A5.23 (EF5) | K41370 |
| A5.23 (EF6) | K21135 |
| A5.23 (EF6N) | K21035 |
| A5.23 (EGXXS-5) | K11357 |
| A5.23 (EL12) | K01012 |
| A5.23 (EL12N) | K00912 |
| A5.23 (EM2) | K10882 |
| A5.23 (EM3) | K21015 |
| A5.23 (EM4) | K21030 |
| A5.23 (EM4N) | K20930 |
| A5.23 (EM12K) | K01113 |
| A5.23 (EM12KN) | K01013 |
| A5.23 (ENi1K) | K11058 |
| A5.23 (ENi1KN) | K10958 |
| A5.23 (ENi1N) | K10940 |
| A5.23 (ENi2) | K21010 |
| A5.23 (ENi2N) | K20910 |
| A5.23 (ENi3) | K31310 |
| A5.23 (ENi3N) | K31210 |
| A5.23 (ENi4) | K11485 |
| A5.23 (ENi4N) | K11385 |
| A5.23 (ENi5) | K11240 |
| A5.23 (ENi5N) | K11241 |
| A5.23 (ENil) | K11040 |
| A5.23 (EW) | K11245 |
| A5.24 (ERZr2) | R60702 |
| A5.24 (ERZr3) | R60704 |
| A5.24 (ERZr4) | R60705 |
| A5.24 (ERZr4) | R60707 ob |
| A5.25 (EA3K-EW) | K10945 |
| A5.25 (EH10K-EW) | K01010 |
| A5.25 (EH11K-EW) | K11140 |
| A5.25 (EH14-EW) | K11585 |
| A5.25 (EM5K-EW) | K10726 |
| A5.25 (EM12-EW) | K01112 |
| A5.25 (EM12K-EW) | K01113 |
| A5.25 (EM13K-EW) | K01313 |
| A5.25 (EM15K-EW) | K01515 |
| A5.25 (EWS-EW) | K11245 |
| A5.25 (EWT1) | W06040 |

| CROSS REFERENCE SPECIFICATIONS | UNIFIED NUMBER |
|---|---|
| A5.25 (EWT2) | W20140 |
| A5.25 (EWT3) | W22340 |
| A5.26 (EG6XT-1, EG43XT-1) | W06301 |
| A5.26 (EG6XT-2, EG43XT-2) | W06302 |
| A5.26 (EG7XT-1, EG48XT-1) | W07301 |
| A5.26 (EG7XT-2, EG48XT-2) | W07302 |
| A5.26 (EGXXS-D2) | K10945 |
| A5.26 (EGXXS-1) | K01313 |
| A5.26 (EGXXS-2) | K10726 |
| A5.26 (EGXXS-3) | K11022 |
| A5.26 (EGXXS-5) | K11325 |
| A5.26 (EGXXS-5) | K11357 |
| A5.26 (EGXXS-6) | K11140 |
| A5.26 (EGXXT-Ni1) | W21033 |
| A5.26 (EGXXT-NM1) | W22334 |
| A5.26 (EGXXT-NM2) | W22333 |
| A5.26 (EGXXT-W) | W20131 |
| A5.27 (RCu) | C18980 |
| A5.28 (E70C-B2L) | W52130 |
| A5.28 (E70C-Ni2, E80C-Ni2) | W22030 |
| A5.28 (E80C-A1) | W17030 ob |
| A5.28 (E80C-B2) | W52030 |
| A5.28 (E80C-B3L) | W53130 |
| A5.28 (E80C-Ni1) | W21030 |
| A5.28 (E80C-Ni3) | W23030 |
| A5.28 (E90C-B3) | W53030 |
| A5.28 (E90C-D2) | W19230 |
| A5.28 (ER70S-A1) | K11235 |
| A5.28 (ER70S-B2L) | K20500 |
| A5.28 (ER80S-B2) | K20900 |
| A5.28 (ER80S-B3L) | K30560 |
| A5.28 (ER80S-B6) | S50280 |
| A5.28 (ER80S-B8) | S50480 |
| A5.28 (ER80S-B9) | S50482 |
| A5.28 (ER80S-D2, ER90S-D2) | K10945 |
| A5.28 (ER80S-Ni2) | K21240 |
| A5.28 (ER80S-Ni3) | K31240 |
| A5.28 (ER80S-Nil) | K11260 |
| A5.28 (ER90S-B3) | K30960 |
| A5.28 (ER100S-1) | K10882 |
| A5.28 (ER110S-1) | K21015 |
| A5.28 (ER120S-1) | K21030 |
| A5.29 (E1XXT5-K4) | W22235 |
| A5.29 (E6XT5-B8) | W50431 |
| A5.29 (E7XT11-K2) | W21239 |
| A5.29 (E7XT4-K2) | W21234 |
| A5.29 (E7XT5-A1) | W17035 |
| A5.29 (E7XT6-Ni1) | W21036 |
| A5.29 (E7XT7-K2) | W21237 |
| A5.29 (E7XT8-K2) | W21238 |
| A5.29 (E7XT8-Ni1) | W21038 |
| A5.29 (E7XT8-Ni2, E8XT8-Ni2) | W22038 |
| A5.29 (E8XT1-A1) | W17031 |
| A5.29 (E8XT1-B1) | W51031 |
| A5.29 (E8XT1-B1L) | W51131 |
| A5.29 (E8XT1-B2) | W52031 |
| A5.29 (E8XT1-B2H) | W52231 |
| A5.29 (E8XT1-B2L) | W52131 |
| A5.29 (E8XT1-W2) | W20131 |
| A5.29 (E8XT5-B2) | W52035 |
| A5.29 (E8XT5-B2L) | W52135 |
| A5.29 (E8XT5-B6) | W50231 |
| A5.29 (E8XT5-B6L, B6LM) | W50230 |
| A5.29 (E8XT5-B8L, -B8LM) | W50430 |
| A5.29 (E8XT5-K1) | W21135 |
| A5.29 (E8XT5-Ni1) | W21035 |
| A5.29 (E8XT5-Ni2) | W22035 |
| A5.29 (E9XT1-B3, E10XT1-B3) | W53031 |
| A5.29 (E9XT1-B3H) | W53231 |
| A5.29 (E9XT1-B3L) | W53131 |
| A5.29 (E9XT1-D1) | W19131 |
| A5.29 (E9XT1-D3) | W19331 |
| A5.29 (E9XT5-B3) | W53035 |
| A5.29 (E9XT5-D2, E10XT5-D2) | W19235 |
| A5.29 (E9XT8-K8) | W21438 |
| A5.29 (E10XT1-K7) | W22051 |
| A5.29 (E10XT1-K9, E10XT1-K9M) | W23230 |
| A5.29 (E11XT1-K4) | W22231 |
| A5.29 (E12XT1-K5) | W21631 |
| A5.29 (E100T5-D2) | W10235 ob |
| A5.29 (EXXT1-K2) | W21231 |
| A5.29 (EXXT1-Ni1) | W21031 |
| A5.29 (EXXT1-Ni2) | W22031 |
| A5.29 (EXXT1-Ni3) | W23031 ob |
| A5.29 (EXXT5-K2) | W21235 |
| A5.29 (EXXT5-K6, K6M) | W21045 |
| A5.29 (EXXT5-Ni3) | W23035 |
| A5.29 (EXXT8-K6) | W21048 |
| A5.29 (EXXT11-Ni3) | W23039 |
| A5.29 (EXXXT1-K3) | W21331 |
| A5.29 (EXXXT5-K3) | W21335 |
| A5.30 (IN6A) | N07092 |
| A5.30 (IN52) | N06052 |
| A5.30 (IN60) | N04060 |
| A5.30 (IN61) | N02061 |
| A5.30 (IN62) | N06062 |
| A5.30 (IN67) | C71581 |
| A5.30 (IN82) | N06082 |
| A5.30 (IN308) | S30880 |
| A5.30 (IN308L) | S30883 |
| A5.30 (IN309) | S30980 |
| A5.30 (IN309L) | S30983 |
| A5.30 (IN310) | S31080 |
| A5.30 (IN312) | S31380 |
| A5.30 (IN316) | S31680 |
| A5.30 (IN316L) | S31683 |
| A5.30 (IN348) | S34780 |
| A5.30 (IN502) | S50280 |
| A5.30 (IN515) | K20900 |
| A5.30 (IN521) | K30960 |
| A5.30 (INMs1) | K10726 |
| A5.30 (INMs2) | K01313 |
| A5.30 (INMs3) | K11140 |
| A5.33 (W-Cu) | C13100 |
| A5.34 (ENiCr3T-X) | W86082 |
| A5.34 (ENiCrFe1T-X) | W86132 |
| A5.34 (ENiCrFe2T-X) | W86133 |
| A5.34 (ENiCrFe3T-X) | W86182 |
| A5.34 (ENiCrMo3T-X) | W86112 |
| A5.34 (ENiCrMo4T-X) | W80276 |
| A5.34 (ENiFeCr2T-X) | W87718 |
| C2.25 (CW-CoCrW) | W73136 |
| C2.25 (CW-CoMoCr) | W73138 |
| C2.25 (CW-CrFeB) | W74630 |
| C2.25 (CW-CuSnSi) | W60658 |
| C2.25 (CW-FeCr-1) | W41036 |
| C2.25 (CW-FeCr-2) | W41039 |
| C2.25 (CW-FeCr-3) | W42040 |
| C2.25 (CW-FeCr-4) | W43070 |
| C2.25 (CW-FeCrB) | W46440 |
| C2.25 (CW-FeCrC-1) | W74532 |
| C2.25 (CW-FeCrC-2) | W74535 |
| C2.25 (CW-FeCrNiB) | W46540 |
| C2.25 (CW-FeCrNiB) | W89640 ob |
| C2.25 (CW-FeCrNiMoB) | W46640 |
| C2.25 (CW-FeMnAl) | W79840 |
| C2.25 (CW-FeVWC) | W77540 |
| C2.25 (CW-FeWC) | W77640 |
| C2.25 (CW-NiAlMo) | W89432 |
| C2.25 (CW-NiCrAl) | W89341 |
| C2.25 (CW-NiCrAlY) | W89343 |
| C2.25 (CW-NiCrFe) | W87740 |
| C2.25 (CW-NiCrMo) | W86142 |
| C2.25 (CW-NiCrSiB) | W89740 |
| C2.25 (CW-NiFeAl) | W89240 |
| C2.25 (W-Al-1100) | A91100 |
| C2.25 (W-Al-1350) | A91350 |
| C2.25 (W-Al-4043) | A94043 |
| C2.25 (W-Al-4047) | A94047 |
| C2.25 (W-Al-5356) | A95356 |
| C2.25 (W-Al-Al2O3) | A71001 |
| C2.25 (W-Cd) | L01950 |
| C2.25 (W-CuAl) | C61800 |
| C2.25 (W-CuSn) | C51800 |
| C2.25 (W-CuZn-1) | C22000 |
| C2.25 (W-CuZn-2) | C26800 |
| C2.25 (W-CuZn-3) | C47000 |
| C2.25 (W-CuZnAg) | C56000 |
| C2.25 (W-FeC-1) | G10090 |
| C2.25 (W-FeC-2) | H15220 |
| C2.25 (W-FeC-3) | G10200 |
| C2.25 (W-FeC-4) | G10750 |
| C2.25 (W-FeC-5) | G10800 |
| C2.25 (W-FeCrC) | T87510 |
| C2.25 (W-FeCrMn) | T87515 |
| C2.25 (W-FeCrNi-1) | S30481 |
| C2.25 (W-FeCrNi-2) | S30280 |
| C2.25 (W-FeCrNi-3) | S20281 |
| C2.25 (W-FeCrNi-4) | S20280 |
| C2.25 (W-FeCrNi-5) | S31680 |
| C2.25 (W-FeCrNi-6) | S31683 |
| C2.25 (W-FeCrNi-7) | S31681 |
| C2.25 (W-FeCrNi-8) | S41683 |
| C2.25 (W-FeCrNi-9) | S42080 |
| C2.25 (W-FeNiCr) | T87520 |
| C2.25 (W-Mo) | R03615 |
| C2.25 (W-Ni-2) | N02200 |
| C2.25 (W-Ni-3) | N02202 |
| C2.25 (W-NiAl-1) | N01001 |
| C2.25 (W-NiAl-2) | N01002 |
| C2.25 (W-NiCrFe-1) | N06076 |
| C2.25 (W-NiCrFe-2) | N06078 |
| C2.25 (W-NiCrFe-3) | N06004 |
| C2.25 (W-NiCrMo) | N06021 |
| C2.25 (W-NiCrTi) | N06070 |
| C2.25 (W-NiCu-1) | N04405 |
| C2.25 (W-NiCu-2) | N04406 |
| C2.25 (W-Pb) | L50045 |
| C2.25 (W-Pb-Sb) | L53105 |
| C2.25 (W-PbSbSn) | L53465 |
| C2.25 (W-Sn) | L13012 |
| C2.25 (W-SnCuSb) | L13910 |
| C2.25 (W-SnSbCu-1) | L13890 |
| C2.25 (W-SnSbCu-2) | L13892 |
| C2.25 (W-SnSbCu-3) | L13893 |
| C2.25 (W-SnZn-1) | L13201 |
| C2.25 (W-SnZn-2) | L13202 |
| C2.25 (W-Ti) | R50120 |
| C2.25 (W-Zn-1) | Z13001 |
| C2.25 (W-ZnAl-1) | Z30401 |
| C2.25 (W-ZnAl-2) | Z30700 |

# Cross Index of Commonly Known Documents Which Describe Materials Same as or Similar to These Covered by UNS Numbers

## FEDERAL SPECIFICATION NUMBERS

**(The Symbol *ob* Indicates an Obsolete UNS Number.)**

**NOTE: Following U.S. Government procedure, these are indexed first on the letter between the dashes and then by the number.**

| CROSS REFERENCE SPECIFICATIONS | UNIFIED NUMBER |
|---|---|

# FEDERAL SPECIFICATIONS

| | |
|---|---|
| QQ-A-430 | A91100 |
| QQ-A-430 | A92017 |
| QQ-A-430 | A92024 |
| QQ-A-430 | A92117 |
| QQ-A-430 | A92219 |
| QQ-A-430 | A93003 |
| QQ-A-430 | A95052 |
| QQ-A-430 | A95056 |
| QQ-A-430 | A96053 |
| QQ-A-430 | A96061 |
| QQ-A-430 | A97050 |
| QQ-A-430 | A97075 |
| QQ-A-430 | A97178 |
| QQ-A-430 (5005) | A95005 |
| QQ-A-1876 | A91100 |
| QQ-A-1876 | A91145 |
| QQ-A-1876 | A91235 |
| QQ-B-575 | C10100 |
| QQ-B-575 | C10200 |
| QQ-B-575 | C11000 |
| QQ-B-575 | C11300 |
| QQ-B-575 | C11400 |
| QQ-B-575 | C11500 |
| QQ-B-575 | C11600 |
| QQ-B-639 | C46200 |
| QQ-B-639 | C46400 |
| QQ-B-639 | C48200 |
| QQ-B-639 | C48500 |
| QQ-B-650 | C14400 ob |
| QQ-B-654 (BAg-4) | P07400 |
| QQ-B-654 (BAg-5) | P07453 |
| QQ-B-654 (BAg-7) | P07563 |
| QQ-B-654 (BAg-8a) | P07723 |
| QQ-B-654 (BAg-9) | P07650 |
| QQ-B-654 (BAg-10) | P07700 |
| QQ-B-654 (BAg-13a) | P07560 |
| QQ-B-654 (BAg-18) | P07600 |
| QQ-B-654 (BAg-20) | P07301 |
| QQ-B-654 (BAg-22) | P07490 |
| QQ-B-654 (BCuP-5) | C55284 |
| QQ-B-654 (IV) | P07500 |
| QQ-B-654 (V) | P07501 |
| QQ-B-654 (VI) | P07502 |
| QQ-B-654 (VII) | P07450 |
| QQ-B-654 (VIII) | P07350 |
| QQ-C-40 (AA, C) | L50065 |
| QQ-C-40 (D) | L51124 |
| QQ-C-450 | C60600 ob |
| QQ-C-450 | C61000 |
| QQ-C-450 | C61300 |
| QQ-C-450 | C61400 |
| QQ-C-450 | C63000 |
| QQ-L-171 (A) | L50050 |
| QQ-L-171 (B) | L50080 |
| QQ-L-171 (C) | L51121 |
| QQ-L-201 | L50070 |
| QQ-L-201 (C) | L51121 |
| QQ-L-201 (D) | L51123 |
| QQ-N-281 | N04400 |
| QQ-N-281 | N04405 |
| QQ-N-286 | N05500 |
| QQ-S-766 (348) | S34800 |
| QQ-T-371 | L13007 |
| QQ-T-390 | L05120 |
| QQ-W-343 | C10100 |

| CROSS REFERENCE SPECIFICATIONS | UNIFIED NUMBER |
|---|---|
| QQ-W-343 | C11500 |
| WW-P-377 | C12200 |
| WW-P-460 | C83600 |
| WW-P-460 | C84200 |
| WW-P-460 | C92200 |
| WW-T-700/1 | A91100 |
| WW-T-700/2 | A93003 |
| WW-T-700/3 | A92024 |
| WW-T-700/4 | A95052 |
| WW-T-700/5 | A95086 |
| WW-T-700/6 | A96061 |
| WW-T-700/7 | A97075 |
| WW-T-725 | C83600 |
| WW-T-725 | C83800 |
| WW-T-725 | C84400 |
| WW-T-725 | C92200 |
| WW-T-799 | C10200 |
| WW-T-799 | C12000 |
| WW-T-799 | C12200 |
| WW-U-516 | C83600 |
| WW-U-516 | C83800 |
| WW-U-516 | C84400 |
| WW-V-1967 | C82600 |
| WW-V-1967 | C82700 |
| WW-V-1967 | C82800 |
| WW-V-1967 | C83800 |
| WW-V-1967 | C85200 |
| WW-V-1967 | C85400 |
| WW-V-1967 | C85500 |
| WW-V-1967 | C86100 |
| WW-V-1967 | C86200 |
| WW-V-1967 | C86300 |
| WW-V-1967 | C86400 |
| WW-V-1967 | C86500 |
| WW-V-1967 | C86800 |
| WW-V-1967 | C87300 |
| WW-V-1967 | C87400 |
| WW-V-1967 | C90300 |
| WW-V-1967 | C92200 |
| WW-V-1967 | C92500 |
| WW-V-1967 | C94700 |
| WW-V-1967 | C94800 |
| WW-V-1967 | C95200 |
| WW-V-1967 | C95300 |
| WW-V-1967 | C95400 |
| WW-V-1967 | C95500 |
| WW-V-1967 | C95800 |
| WW-V-1967 | C96400 |

# Cross Index of Commonly Known Documents Which Describe Materials Same as or Similar to These Covered by UNS Numbers

## MIL (Military Specification) NUMBERS

(The Symbol *ob* Indicates an Obsolete UNS Number.)

NOTE: Following U.S. Government procedure, these are indexed by the numerical portion of the designation only. The letter between the dashes is the first letter of the title.

# MIL S

| CROSS REFERENCE SPECIFICATIONS | UNIFIED NUMBER |
|---|---|
| MIL-W-85 | C10100 |
| MIL-W-85 | C10200 |
| MIL-W-85 | C12000 |
| MIL-W-85 | C22000 |
| MIL-S-645 | K01602 |
| MIL-S-980 | K19964 ob |
| MIL-T-1368 | N04400 |
| MIL-S-3039 | K04600 |
| MIL-T-3235 | C10300 |
| MIL-S-3289 | G10350 |
| MIL-S-3289 | K02901 |
| MIL-C-3383 | C22000 |
| MIL-T-3520 | G10250 |
| MIL-T-3595 | C51000 |
| MIL-F-3922 | A02080 ob |
| MIL-F-3922 | A02950 |
| MIL-F-3922 | A03550 |
| MIL-F-3922 | A03560 |
| MIL-F-3922 | A05180 |
| MIL-F-3922 | A13600 |
| MIl-F-3922 (C712.0) | A07110 |
| MIL-F-3922 (D712.0) | A07120 |
| MIL-S-4043 | S30403 |
| MIL-T-5695 | S30400 |
| MIL-W-5712 | G10120 |
| MIL-W-6712 | A91100 |
| MIL-W-6712 | A94043 |
| MIL-W-6712 | C10200 |
| MIL-W-6712 | C11000 |
| MIL-W-6712 | C22000 |
| MIL-W-6712 | C26800 |
| MIL-W-6712 | C46400 |
| MIL-W-6712 | C51000 |
| MIL-W-6712 | C61800 |
| MIL-W-6712 | G10080 |
| MIL-W-6712 | S20200 |
| MIL-W-6712 | S30409 |
| MIL-S-7393 | G33106 |
| MIL-S-7393 (9310) | G93106 |
| MIL-N-7786 | N07750 |
| MIL-B-7883 (BAg-1a) | P07500 |
| MIL-B-7883 (BAg-23) | P07850 |
| MIL-B-7883 (BNi-1) | N99600 |
| MIL-B-7883 (BNi-1a) | N99610 |
| MIL-B-7883 (BNi-2) | N99620 |
| MIL-B-7883 (BNi-3) | N99630 |
| MIL-B-7883 (BNi-4) | N99640 |
| MIL-B-7883 (BNi-5) | N99650 |
| MIL-B-7883 (Mn-Ni-Co-B) | M26800 |
| MIL-B-7883 (RBCuZn-A) | C47000 |
| MIL-T-8504 (304) | S30400 |
| MIL-W-8957 | K08700 |
| MIL-S-8967 | S43100 |
| MIL-C-10375 | C26100 ob |
| MIL-A-10841 | M00995 |
| MIL-A-10841 | M00998 |
| MIL-A-11356 | J13025 |
| MIL-S-11713 | G10690 |
| MIL-S-11713 (2) | G10700 |
| MIL-A-12560 | K11918 ob |
| MIL-S-13282 | P07015 |
| MIL-B-13501 | C54600 ob |
| MIL-T-15005 | C70600 |
| MIL-T-15005 | C71500 |
| MIL-C-15726 | C70600 |
| MIL-C-15726 | C71500 |
| MIL-B-16166 | C62300 |
| MIL-B-16166 | C63000 |
| MIL-S-16216 | K31820 |
| MIL-S-16216 | K32045 |
| MIL-S-16216 (HY80) | J42015 |
| MIL-S-16216 (HY100) | J42240 |
| MIL-T-16282 (TP321) | S32100 |
| MIL-T-16286 (TP347) | S34700 |
| MIL-T-16420 | C70600 |
| MIL-T-16420 | C71500 |
| MIL-R-17131 (MIL-ERNiCr-B) | N99645 |
| MIL-R-17131 (MIL-ERNiCr-C) | N99646 |
| MIL-R-17131 (MIL-RCoCr-A) | R30006 |
| MIL-R-17131 (MIL-RCoCr-C) | R30001 |
| MIL-V-17547 | C65100 |
| MIL-V-17547 | C92200 |
| MIL-V-17547 | C95800 |
| MIL-V-17547 | N04400 |
| MIL-V-17547 | N05500 |
| MIL-V-17647 | N04019 |
| MIL-A-18001 | Z32121 |
| MIL-G-18014 | A95052 |
| MIL-G-18014 | A95454 |
| MIL-V-18436 | C31400 |
| MIL-V-18436 | C31600 |
| MIL-V-18436 | C36000 |
| MIL-V-18436 | C83600 |
| MIL-V-18436 | C96200 |
| MIL-V-18436 | C97600 |
| MIL-V-18436 | C97800 |
| MIL-B-18907 | C10100 |
| MIL-B-18907 | C10200 |
| MIL-B-18907 | C12000 |
| MIL-B-18907 | C12200 |
| MIL-B-18907 | C22000 |
| MIL-B-18907 | C70200 |
| MIL-E-19141 (MIL-Fe5-B) | W77510 |
| MIL-E-19141 (MIL-FeMn-A) | W79110 |
| MIL-E-19141 (MIL-I-A2a) | W75010 |
| MIL-E-19141 (MIL-I-A2c) | W75110 |
| MIL-C-19311 | C18500 ob |
| MIL-S-19434 | G10350 |
| MIL-S-19434 | G98500 |
| MIL-E-19933D, (MIL-308HC) | S30884 |
| MIL-E-19933 (MIL-308, MIL-308Co) | S30880 |
| MIL-E-19933 (MIL-308L, MIL-308CoL) | S30883 |
| MIL-E-19933 (MIL-309, MIL-309Co) | S30980 |
| MIL-E-19933 (MIL-310) | S31080 |
| MIL-E-19933 (MIL-312, MIL-312Co) | S31380 |
| MIL-E-19933 (MIL-316) | S31680 |
| MIL-E-19933 (MIL-316L) | S31683 |
| MIL-E-19933 (MIL-317) | S31700 |
| MIL-E-19933 (MIL-318) | S31980 |
| MIL-E-19933 (MIL-321) | S32180 |
| MIL-E-19933 (MIL-347, MIL-347Co) | S34780 |
| MIL-E-19933 (MIL-349) | S63197 |
| MIL-E-19933 (MIL-410) | S41080 |
| MIL-B-20292 | C12700 ob |
| MIL-B-20292 | C12800 ob |
| MIL-B-20292 | C13000 ob |
| MIL-F-20670 | G10260 |
| MIL-S-21149/6MIL-S-24149/3 | S30400 |
| MIL-E-21562 (EN6A, RN6A) | N07092 |
| MIL-E-21562 (EN6N, RN6N) | N06076 |
| MIL-E-21562 (EN60, RN60) | N04060 |
| MIL-E-21562 (EN61, RN61) | N02061 |
| MIL-E-21562 (EN62, RN62) | N06062 |
| MIL-E-21562 (EN64, RN64) | N05504 |
| MIL-E-21562 (EN82, EN82H, RN82, RN82H) | N06082 |
| MIL-E-21562 (RN7, EN67, RN67) | C71581 |
| MIL-E-21562 (RN65) | N08065 |
| MIL-E-21562 (RN69) | N07069 |
| MIL-E-21562 (RN625, EN625) | N06625 |
| MIL-C-21657 | C17200 |
| MIL-S-21952 | K31820 |
| MIL-S-21952 | K32045 |
| MIL-C-22087 (10) | C82500 |
| MIL-E-22200/1 (MIL-7018) | W07018 |
| MIL-E-22200/1 (MIL-8018-C3, MIL-8018-C3SR) | W21018 |
| MIL-E-22200/1 (MIL-9018-M) | W21218 |
| MIL-E-22200/1 (MIL-10018-M) | W21318 |
| MIL-E-22200/1 (MIL-11018-M) | W21418 |
| MIL-E-22200/2 (MIL-308, MIL-308Co) | W30810 |
| MIL-E-22200/2, MIL-308HC | W30815 |
| MIL-E-22200/2 (MIL-308L, MIL-308CoL) | W30813 |
| MIL-E-22200/2 (MIL-309, MIL-309Co) | W30910 |
| MIL-E-22200/2 (MIL-309Cb) | W30917 |
| MIL-E-22200/2, (MIL-309L0, MIL-309CoL) | W30913 |
| MIL-E-22200/2 (MIL-310) | W31010 |
| MIL-E-22200/2 (MIL-312) | W31310 |
| MIL-E-22200/2 (MIL-316) | W31610 |
| MIL-E-22200/2 (MIL-316L) | W31613 |
| MIL-E-22200/2 (MIL-317) | W31710 |
| MIL-E-22200/2 (MIL-318) | W31910 |
| MIL-E-22200/2 (MIL-330) | W88331 |
| MIL-E-22200/2 (MIL-347, MIL-347Co) | W34710 |
| MIL-E-22200/2 (MIL-347HC) | W34712 |
| MIL-E-22200/2 (MIL-349) | W34910 |
| MIL-E-22200/2 Type MIL-16-8-2 | W36810 |
| MIL-E-22200/3 (MIL-1N12) | W86112 |
| MIL-E-22200/3 (MIL-3N12, MIL-3N12H) | W86132 |
| MIL-E-22200/3 (MIL-3N1B) | W80001 |
| MIL-E-22200/3 (MIL-3N1C) | W80002 |
| MIL-E-22200/3 (MIL-3N1L) | W73605 |
| MIL-E-22200/3 (MIL-3N1N) | W73155 |
| MIL-E-22200/3 (MIL-4N11) | W82141 |
| MIL-E-22200/3 (MIL-4N1A, MIL-4N1AH) | W86133 |
| MIL-E-22200/3 (MIL-4N1W) | W80004 |
| MIL-E-22200/3 (MIL 8N12, MIL-8N12H) | W86182 |
| MIL-E-22200/3 (MIL-9N10) | W84190 |
| MIL-E-22200/4 (MIL-CuNi) | W60715 |
| MIL-E-22200/5 (MIL-10018-N1) | W22718 |
| MIL-E-22200/8 (MIL-410) | W41010 |
| MIL-E-22200/8 (MIL-502-15) | W50215 |
| MIL-E-22200/8 (MIL-502-15L) | W50205 |
| MIL-E-22200/8 (MIL-502-16) | W50216 |
| MIL-E-22200/8 (MIL-502-16L) | W50206 |
| MIL-E-22200/8 (MIL-502-1X-L) | W50213 ob |
| MIL-E-22200/8 (MIL-505-15) | W50415 |
| MIL-E-22200/8 (MIL-505-16) | W50416 |
| MIL-E-22200/8 (MIL-7016-B2L) | W52116 |
| MIL-E-22200/8 (MIL-7018-B2L) | W52118 |
| MIL-E-22200/8 (MIL-8015-B2) | W52015 |
| MIL-E-22200/8 (MIL-8015-B2L) | W52115 |

*MIL (Military Specification) NUMBERS*

| CROSS REFERENCE SPECIFICATIONS | UNIFIED NUMBER |
|---|---|
| MIL-E-22200/8 (MIL-8015-B3L) | W53115 |
| MIL-E-22200/8 (MIL-8016-B2) | W52016 |
| MIL-E-22200/8 (MIL-8018-B2) | W52018 |
| MIL-E-22200/8 (MIL-8018-B3L) | W53118 |
| MIL-E-22200/8 (MIL-9015-B3) | W53015 |
| MIL-E-22200/8 (MIL-9016-B3) | W53016 |
| MIL-E-22200/8 (MIL-9018-B3) | W53018 |
| MIL-E-22200/9 (MIL-14018-M1) | W23318 |
| MIL-E-22200/10 (MIL-7018M) | W07018 |
| MIL-E-22200/10 (MIL-10018-M1) | W21328 |
| MIL-E-22200/10 (MIL-12018-M2) | W23228 |
| MIL-E-22200/11 (MIL-14018-HT) | W26018 |
| MIL-T-22214 | C12200 |
| MIL-T-22214 | C17500 |
| MIL-T-22214 | C70600 |
| MIL-S-22499 | C26100 ob |
| MIL-S-22664 | K31820 |
| MIL-S-22664 | K32045 |
| MIL-A-22771 | A97049 |
| MIL-A-22771 | A97050 |
| MIL-S-23008 (HY80) | J42015 |
| MIL-S-23008 (HY100) | J42240 |
| MIL-S-23009 | K31820 |
| MIL-S-23009 | K32045 |
| MIL-S-23009 (HY80) | J42015 |
| MIL-S-23009 (HY100) | J42240 |
| MIL-N-23228 | N06600 |
| MIL-I-23413 (MIL-MS-1) | K10726 |
| MIL-I-23413 (MIL-6A) | N07092 |
| MIL-I-23413 (MIL-60) | N04060 |
| MIL-I-23413 (MIL-61) | N02061 |
| MIL-I-23413 (MIL-62) | N06062 |
| MIL-I-23413 (MIL-67) | C71581 |
| MIL-I-23413 (MIL-82) | N06082 |
| MIL-I-23413 (MIL-308, MIL-308Co) | S30880 |
| MIL-I-23413 (MIL-308L, MIL-308CoL) | S30883 |
| MIL-I-23413 (MIL-310) | S31080 |
| MIL-I-23413 (MIL-316) | S31680 |
| MIL-I-23413 (MIL-316L) | S31683 |
| MIL-I-23413 (MIL-348, MIL-348Co) | S34780 |
| MIL-I-23413 (MIL-515) | K20900 |
| MIL-I-23413 (MIL-521) | K30960 |
| MIL-E-23765/1 (MIL-70S-1) | K01515 |
| MIL-E-23765/1 (MIL-70S-2) | K10726 |
| MIL-E-23765/1 (MIL-70S-3) | K11022 |
| MIL-E-23765/1 (MIL-70S-4) | K11132 |
| MIL-E-23765/1 (MIL-70S-5) | K11357 |
| MIL-E-23765/1 (MIL-70S-6) | K11140 |
| MIL-E-23765/1 (MIL-70S-7) | K11585 |
| MIL-E-23765/1 (MIL-70S-8) | K11123 |
| MIL-E-23765/1 (MIL-70S-8) | K11223 |
| MIL-E-23765/1 (MIL-706-9) | K11423 |
| MIL-E-23765/2 (MIL-80S-1) | K21485 |
| MIL-E-23765/2 (MIL-80S-2) | K11423 |
| MIL-23765/2 (MIL-80S-3) | K21451 |
| MIL-E-23765/2 (MIL-100S-1) | K10882 |
| MIL-E-23765/2 (MIL-110S-1) | K21015 |
| MIL-E-23765/2 (MIL-120S-1, MIL-120S-2) | K21030 |
| MIL-E-23765/3 (MIL-CuAl-A2) | C61800 |
| MIL-E-23765/3 (MIL-CuMnNiAl) | C63380 |
| MIL-E-23765/3 (MIL-CuNiAl) | C63280 |
| MIL-E-23765/3 (MIL-CuSi) | C65600 |
| MIL-E-23765/3 (MIL-CuSn-C) | C52100 |
| MIL-E-23765/4 (MIL-EA1) | K11222 |
| MIL-E-23765/4 (MIL-EL12) | K01012 |
| MIL-E-23765/4 (MIL-EL8) | K01008 |
| MIL-E-23765/4 (MIL-EM12) | K01112 |
| MIL-E-23765/4 (MIL-EM12K) | K01113 |
| MIL-N-24106 | N04400 |
| MIL-T-24107 | C10100 |
| MIL-T-24107 | C10200 |
| MIL-T-24107 | C10300 |
| MIL-T-24107 | C10800 |
| MIL-T-24107 | C12000 |
| MIL-T-24107 | C12200 |
| MIL-T-24107 | C14200 |
| MIL-S-24149/2 | A95083 |
| MIL-S-24149/3 | S30430 |
| MIL-S-24149/3 | S30500 |
| MIL-S-24149/3 | S30900 |
| MIL-S-24149/3 | S31000 |
| MIL-S-24149/3 | S31600 |
| MIL-S-24149/5 | A95086 |
| MIL-S-24149/5 | A95356 |
| MIL-S-24149/6 | S30430 |
| MIL-S-24149/6 | S30500 |
| MIL-S-24149/6 | S30900 |
| MIL-S-24149/6 | S31000 |
| MIL-S-24149/6 | S31600 |
| MIL-S-24238 (A) | K12042 |
| MIL-S-24249/6 | S30400 |
| MIL-E-24355 (MIL-140S-1) | K31200 |
| MIL-S-24371 | K51255 |
| MIL-E-24403 (MIL-71T-8A) | W07708 |
| MIL-E-24403/1 (MIL-7XT-1) | W07601 |
| MIL-E-24403/1 (MIL-7XT-8) | W07608 |
| MIL-E-24403/1 (MIL-8XT1-Ni1) | W21031 |
| MIL-E-24403/1 (MIL-8XT8-Ni2) | W22031 |
| MIL-E-24403/1 (MIL-70T-5) | W07605 |
| MIL-E-24403/1 (MIL-70T-6) | W07606 |
| MIL-E-24403/1 (MIL-71T-1A) | W07701 |
| MIL-E-24403/1 (MIL-71T8-K6) | W21048 |
| MIL-E-24403/1 (MIL-71T8-Ni1) | W21038 |
| MIL-E-24403/2 (MIL-10XTX) | W23230 |
| MIL-E-24403/2 (MIL-12XTX) | W24230 |
| MIL-E-24403/2 (MIL-110 TC) | W22230 |
| MIL-S-24451 | K31820 |
| MIL-S-24451 | K32045 |
| MIL-B-24480 | C95800 |
| MIL-S-24502 | K14675 |
| MIL-N-24549 | N05500 |
| MIL-C-24679 | C70600 |
| MIL-C-24689 (Alloy 20) | R30106 |
| MIL-C-24689 (20) | R30006 |
| MIL-C-24689 (22) | R30001 |
| MIL-C-24689 (23) | R30021 |
| MIL-C-24707/2 | J12520 ob |
| MIL-C-24707/3 (CF-8) | J92600 |
| MIL-C-24707/3 (CF-8C) | J92710 |
| MIL-C-24707/3 (CF-8M) | J92900 |
| MIL-P-25995 | A93003 |
| MIL-P-25995 | A96061 |
| MIL-P-25995 | A96063 |
| MIL-F-39000 | A02080 ob |
| MIL-DTL-45225 | A95083 |
| MIL-DTL-45225 | A95456 |
| MIL-DTL-45225 | A97039 |
| MIL-DTL-46027 | A95083 |
| MIL-DTL-46027 | A95456 |
| MIL-S-46047 | K14185 |
| MIL-S-46049 (1065) | G10650 |
| MIL-S-46049 (1074) | G10740 |
| MIL-DTL-46063 | A97039 |
| MIL-T-46072 | C33100 ob |
| MIL-DTL-46083 | A95083 |
| MIL-DTL-46083 | A95456 |
| MIL-DTL-46083 | A97039 |
| MIL-DTL-46118 | A92219 |
| MIL-M-46143 | M14142 |
| MIL-M-46143 | M14145 |
| MIL-A-46173 | K33370 |
| MIL-DTL-46192 | A92519 |
| MIL-S-46850 | K92710 |
| MIL-S-46850 | K93540 |
| MIL-A-47182 | R30005 |
| MIL-S-47319 | K92150 |
| MIL-C-81021 | C17500 |
| MIL-C-81021 | C17510 |
| MIL-T-81556 | R50250 |
| MIL-T-81556 | R50400 |
| MIL-T-81556 | R50550 |
| MIL-T-81556 | R50700 |
| MIL-T-81556 | R54520 |
| MIL-T-81556 | R54521 |
| MIL-T-81556 | R54810 |
| MIL-T-81556 | R56400 |
| MIL-T-81556 | R56401 |
| MIL-T-81556 | R56620 |
| MIL-T-81556 (AB-2) | R56407 |

# Cross Index of Commonly Known Documents Which Describe Materials Same as or Similar to These Covered by UNS Numbers

## SAE (Society of Automotive Engineers) "J" NUMBERS

**(The Symbol *ob* Indicates an Obsolete UNS Number.)**

# SAE

| CROSS REFERENCE SPECIFICATIONS | UNIFIED NUMBER |
|---|---|
| 405 (310Cb) | S31040 |
| AMS 5940 | R30783 |
| AS7220 (1100) | A91100 |
| Draft F-95AB | R30783 |
| J 452 | A02960 |
| J125 (G4000) | F10008 |
| J158 (M3210) | F20000 |
| J158 (M4504) | F20001 |
| J158 (M5003) | F20002 |
| J158 (M5503) | F20003 |
| J158 (M7002) | F20004 |
| J158 (M8501) | F20005 |
| J403 | G10130 |
| J403 (12L14) | G12144 |
| J403 (1005) | G10050 |
| J403 (1006) | G10060 |
| J403 (1008) | G10080 |
| J403 (1010) | G10100 |
| J403 (1012) | G10120 |
| J403 (1013) | G10130 |
| J403 (1015) | G10150 |
| J403 (1016) | G10160 |
| J403 (1017) | G10170 |
| J403 (1018) | G10180 |
| J403 (1020) | G10200 |
| J403 (1021) | G10210 |
| J403 (1022) | G10220 |
| J403 (1023) | G10230 |
| J403 (1025) | G10250 |
| J403 (1026) | G10260 |
| J403 (1029) | G10290 |
| J403 (1030) | G10300 |
| J403 (1033) | G10230 |
| J403 (1035) | G10350 |
| J403 (1037) | G10370 |
| J403 (1038) | G10380 |
| J403 (1039) | G10390 |
| J403 (1040) | G10400 |
| J403 (1042) | G10420 |
| J403 (1043) | G10430 |
| J403 (1044) | G10440 |
| J403 (1045) | G10450 |
| J403 (1046) | G10460 |
| J403 (1049) | G10490 |
| J403 (1050) | G10500 |
| J403 (1053) | G10530 |
| J403 (1055) | G10550 |
| J403 (1060) | G10600 |
| J403 (1070) | G10700 |
| J403 (1078) | G10780 |
| J403 (1080) | G10800 |
| J403 (1085) | G10850 |
| J403 (1086) | G10860 |
| J403 (1090) | G10900 |
| J403 (1095) | G10950 |
| J403 (1117) | G11170 |
| J403 (1118) | G11180 |
| J403 (1137) | G11370 |
| J403 (1140) | G11400 |
| J403 (1141) | G11410 |
| J403 (1144) | G11440 |
| J403 (1212) | G12120 |
| J403 (1213) | G12130 |
| J403 (1215) | G12150 |
| J403 (1522) | G15220 |
| J403 (1524) | G15240 |
| J403 (1526) | G15260 |
| J403 (1527) | G15270 |
| J403 (1541) | G15410 |
| J403 (1548) | G15480 |
| J403 (1552) | G15520 |
| J403 (1566) | G15660 |
| J403 (M1008) | G10080 |
| J403 (M1010) | G10100 |
| J403 (M1012) | G10120 |
| J403 (M1015) | G10150 |
| J403 (M1017) | G10170 |
| J403 (M1020) | G10200 |
| J403 (M1023) | G10230 |
| J403 (M1025) | G10250 |
| J403 (M1044) | G10440 |
| J404 (50B46) | G50461 |
| J404 (51B60) | G51601 |
| J404 (1330) | G13300 |
| J404 (1335) | G13350 |
| J404 (1340) | G13400 |
| J404 (1345) | G13450 |
| J404 (4023) | G40230 |
| J404 (4027) | G40270 |
| J404 (4037) | G40370 |
| J404 (4047) | G40470 |
| J404 (4118) | G41180 |
| J404 (4120) | G41200 |
| J404 (4130) | G41300 |
| J404 (4135) | G41350 |
| J404 (4137) | G41370 |
| J404 (4140) | G41400 |
| J404 (4142) | G41420 |
| J404 (4145) | G41450 |
| J404 (4150) | G41500 |
| J404 (4320) | G43200 |
| J404 (4340) | G43400 |
| J404 (4615) | G46150 |
| J404 (4617) | G46170 |
| J404 (4620) | G46200 |
| J404 (4820) | G48200 |
| J404 (5120) | G51200 |
| J404 (5130) | G51300 |
| J404 (5132) | G51320 |
| J404 (5140) | G51400 |
| J404 (5150) | G51500 |
| J404 (5160) | G51600 |
| J404 (6150) | G61500 |
| J404 (8615) | G86150 |
| J404 (8617) | G86170 |
| J404 (8620) | G86200 |
| J404 (8622) | G86220 |
| J404 (8625) | G86250 |
| J404 (8627) | G86270 |
| J404 (8630) | G86300 |
| J404 (8637) | G86370 |
| J404 (8640) | G86400 |
| J404 (8645) | G86450 |
| J404 (8655) | G86550 |
| J404 (8720) | G87200 |
| J404 (8822) | G88220 |
| J404 (9259) | G92590 |
| J404 (9259) | H92590 |
| J404 (9260) | G92600 |
| J404 (E4340) | G43406 |
| J404 (E52100) | G52986 |
| J405 | N08367 |
| J405 | N08800 |
| J405 | N08810 |
| J405 | N08926 |
| J405 | S20103 |
| J405 | S20153 |
| J405 | S20161 |
| J405 | S20430 |
| J405 | S21800 |
| J405 | S30451 |
| J405 | S30600 |
| J405 | S30601 |
| J405 | S30615 |
| J405 | S30815 |
| J405 | S31200 |
| J405 | S31254 |
| J405 | S31260 |
| J405 | S31266 |
| J405 | S31725 |
| J405 | S31726 |
| J405 | S31803 |
| J405 | S32109 |
| J405 | S32205 |
| J405 | S32304 |
| J405 | S32520 |
| J405 | S32550 |
| J405 | S32615 |
| J405 | S32654 |
| J405 | S32750 |
| J405 | S32760 |
| J405 | S32803 |
| J405 | S32950 |
| J405 | S33228 |
| J405 | S34565 |
| J405 | S35315 |
| J405 | S40945 |
| J405 | S40975 |
| J405 | S41003 |
| J405 | S41045 |
| J405 | S41050 |
| J405 | S41500 |
| J405 | S43600 |
| J405 | S43932 |
| J405 | S44500 |
| J405 | S44635 |
| J405 | S44660 |
| J405 | S44700 |
| J405 | S44735 |
| J405 | S44800 |
| J405 | S46800 |
| J405 (201) | S20100 |
| J405 (202) | S20200 |
| J405 (301) | S30100 |
| J405 (302) | S30200 |
| J405 (304) | S30400 |
| J405 (304H) | S30409 |
| J405 (304L) | S30403 |
| J405 (304LN) | S30453 |
| J405 (304N) | S30451 |
| J405 (305) | S30500 |
| J405 (309Cb) | S30940 |
| J405 (309H) | S30909 |
| J405 (309HCb) | S30941 |
| J405 (309S) | S30908 |
| J405 (310H) | S31009 |
| J405 (310HCb) | S31041 |
| J405 (310MoLN) | S31050 |
| J405 (310S) | S31008 |
| J405 (316) | S31600 |
| J405 (316Cb) | S31640 |
| J405 (316H) | S31609 |
| J405 (316L) | S31603 |
| J405 (316LN) | S31653 |
| J405 (316N) | S31651 |

| CROSS REFERENCE SPECIFICATIONS | UNIFIED NUMBER |
|---|---|
| J405 (316Ti) | S31635 |
| J405 (317) | S31700 |
| J405 (317L) | S31703 |
| J405 (317LN) | S31753 |
| J405 (321) | S32100 |
| J405 (347) | S34700 |
| J405 (347H) | S34709 |
| J405 (348) | S34800 |
| J405 (348H) | S34809 |
| J405 (904L) | N08904 |
| J405 (51434) | S43400 |
| J405 (XM-15) | S38100 |
| J405 (XM-17) | S21600 |
| J405 (XM-18) | S21603 |
| J405 (XM-19) | S20910 |
| J405 (XM-27) | S44626 |
| J405 (XM-33) | S44625 ob |
| J412 (50B40) | G50401 |
| J412 (50B44) | G50441 |
| J412 (50B46) | G50461 |
| J412 (50B50) | G50501 |
| J412 (50B60) | G50601 |
| J412 (51B60) | G51601 |
| J412 (81B45) | G81451 |
| J412 (86B45) | G86451 |
| J412 (94B15) | G94151 |
| J412 (94B17) | G94171 |
| J412 (303 plus X) | S30310 |
| J412 (1005) | G10050 |
| J412 (1006) | G10060 |
| J412 (1008) | G10080 |
| J412 (1010) | G10100 |
| J412 (1012) | G10120 |
| J412 (1013) | G10130 |
| J412 (1015) | G10150 |
| J412 (1016) | G10160 |
| J412 (1017) | G10170 |
| J412 (1018) | G10180 |
| J412 (1019) | G10190 |
| J412 (1020) | G10200 |
| J412 (1021) | G10210 |
| J412 (1022) | G10220 |
| J412 (1023) | G10230 |
| J412 (1025) | G10250 |
| J412 (1026) | G10260 |
| J412 (1029) | G10290 |
| J412 (1030) | G10300 |
| J412 (1035) | G10350 |
| J412 (1038) | G10380 |
| J412 (1039) | G10390 |
| J412 (1040) | G10400 |
| J412 (1042) | G10420 |
| J412 (1043) | G10430 |
| J412 (1044) | G10440 |
| J412 (1045) | G10450 |
| J412 (1046) | G10460 |
| J412 (1049) | G10490 |
| J412 (1050) | G10500 |
| J412 (1053) | G10530 |
| J412 (1055) | G10550 |
| J412 (1059) | G10590 |
| J412 (1060) | G10600 |
| J412 (1064) | G10640 |
| J412 (1065) | G10650 |
| J412 (1069) | G10690 |
| J412 (1070) | G10700 |
| J412 (1074) | G10740 |
| J412 (1075) | G10750 |
| J412 (1078) | G10780 |

| CROSS REFERENCE SPECIFICATIONS | UNIFIED NUMBER |
|---|---|
| J412 (1080) | G10800 |
| J412 (1085) | G10850 |
| J412 (1086) | G10860 |
| J412 (1090) | G10900 |
| J412 (1095) | G10950 |
| J412 (1117) | G11170 |
| J412 (1118) | G11180 |
| J412 (1137) | G11370 |
| J412 (1141) | G11410 |
| J412 (1144) | G11440 |
| J412 (1145) | G11450 |
| J412 (1146) | G11460 |
| J412 (1330) | G13300 |
| J412 (1335) | G13350 |
| J412 (1340) | G13400 |
| J412 (1513) | G15130 |
| J412 (1522) | G15220 |
| J412 (1524) | G15240 |
| J412 (1526) | G15260 |
| J412 (1527) | G15270 |
| J412 (1536) | G15360 |
| J412 (1541) | G15410 |
| J412 (1548) | G15480 |
| J412 (1551) | G15510 |
| J412 (1552) | G15520 |
| J412 (1561) | G15610 |
| J412 (1566) | G15660 |
| J412 (4012) | G40120 |
| J412 (4023) | G40230 |
| J412 (4024) | G40240 |
| J412 (4027) | G40270 |
| J412 (4028) | G40280 |
| J412 (4032) | G40320 |
| J412 (4037) | G40370 |
| J412 (4047) | G40470 |
| J412 (4118) | G41180 |
| J412 (4130) | G41300 |
| J412 (4135) | G41350 |
| J412 (4137) | G41370 |
| J412 (4140) | G41400 |
| J412 (4142) | G41420 |
| J412 (4145) | G41450 |
| J412 (4147) | G41470 |
| J412 (4161) | G41610 |
| J412 (4320) | G43200 |
| J412 (4340) | G43400 |
| J412 (4422) | G44220 |
| J412 (4427) | G44270 |
| J412 (4620) | G46200 |
| J412 (4720) | G47200 |
| J412 (4815) | G48150 |
| J412 (4820) | G48200 |
| J412 (5046) | G50460 |
| J412 (5060) | G50600 |
| J412 (5130) | G51300 |
| J412 (5132) | G51320 |
| J412 (5135) | G51350 |
| J412 (5140) | G51400 |
| J412 (5147) | G51470 |
| J412 (5150) | G51500 |
| J412 (5155) | G51550 |
| J412 (5160) | G51600 |
| J412 (6150) | G61500 |
| J412 (8615) | G86150 |
| J412 (8617) | G86170 |
| J412 (8620) | G86200 |
| J412 (8622) | G86220 |
| J412 (8625) | G86250 |
| J412 (8627) | G86270 |

| CROSS REFERENCE SPECIFICATIONS | UNIFIED NUMBER |
|---|---|
| J412 (8630) | G86300 |
| J412 (8637) | G86370 |
| J412 (8640) | G86400 |
| J412 (8642) | G86420 |
| J412 (8645) | G86450 |
| J412 (8650) | G86500 |
| J412 (8655) | G86550 |
| J412 (8660) | G86600 |
| J412 (8720) | G87200 |
| J412 (8740) | G87400 |
| J412 (8822) | G88220 |
| J412 (9254) | G92540 |
| J412 (9260) | G92600 |
| J412 (20201) | S20100 |
| J412 (20202) | S20200 |
| J412 (30301) | S30100 |
| J412 (30302) | S30200 |
| J412 (30303) | S30300 |
| J412 (30304) | S30400 |
| J412 (30305) | S30500 |
| J412 (30309) | S30900 |
| J412 (30310) | S31000 |
| J412 (30316) | S31600 |
| J412 (30321) | S32100 |
| J412 (30330) | N08330 |
| J412 (30347) | S34700 |
| J412 (50100) | G50986 |
| J412 (51100) | G51986 |
| J412 (51409) | S40900 |
| J412 (51410) | S41000 |
| J412 (51414) | S41400 |
| J412 (51416) | S41600 |
| J412 (51420) | S42000 |
| J412 (51420F) | S42020 |
| J412 (51430) | S43000 |
| J412 (51430F) | S43020 |
| J412 (51431) | S43100 |
| J412 (51434) | S43400 |
| J412 (51436) | S43600 |
| J412 (51440A) | S44002 |
| J412 (51440B) | S44003 |
| J412 (51440C) | S44004 |
| J412 (51442) | S44200 |
| J412 (51446) | S44600 |
| J412 (51501) | S50100 |
| J412 (52100) | G52986 |
| J431 (G1800) | F10004 |
| J431 (G2500) | F10005 |
| J431 (G2500a) | F10009 |
| J431 (G3000) | F10006 |
| J431 (G3500) | F10007 |
| J431 (G3500b) | F10010 |
| J431 (G3500c) | F10011 |
| J431 (G4000) | F10008 |
| J431 (G4000d) | F10012 |
| J434 (D4018) | F32800 |
| J434 (D4512) | F33100 |
| J434 (D5506) | F33800 |
| J434 (D7003) | F34800 |
| J434 (DQ and T) | F30000 |
| J435 (0022) | J01700 |
| J435 (0025) | J02507 |
| J435 (0030) | J03010 |
| J435 (0050A, 0050B) | J04501 |
| J437 (3Al-13V-11Cr) | R58010 |
| J437 (4Al-3Mo-1V) | R56430 |
| J437 (5Al-2.5Sn) | R54520 |
| J437 (06) | T31506 |
| J437 (6Al-4V) | R56406 |

| CROSS REFERENCE SPECIFICATIONS | UNIFIED NUMBER |
|---|---|
| J437 (7Al-4Mo) | R56740 |
| J437 (8Al-1Mo-1V) | R54810 |
| J437 (8Mn) | R56080 |
| J437 (A2) | T30102 |
| J437 (D2) | T30402 |
| J437 (D3) | T30403 |
| J437 (D5) | T30405 |
| J437 (D7) | T30407 |
| J437 (H-11) | T20811 |
| J437 (H12) | T20812 |
| J437 (H13) | T20813 |
| J437 (H21) | T20821 |
| J437 (L6) | T61206 |
| J437 (M1) | T11301 |
| J437 (M2) | T11302 |
| J437 (M3) | T11313 |
| J437 (M3) | T11323 |
| J437 (M4) | T11304 |
| J437 (O1) | T31501 |
| J437 (O2) | T31502 |
| J437 (S-1) | T41901 |
| J437 (S2) | T41902 |
| J437 (S5) | T41905 |
| J437 (T1) | T12001 |
| J437 (T2) | T12002 ob |
| J437 (T4) | T12004 |
| J437 (T5) | T12005 |
| J437 (T8) | T12008 |
| J437 (W108, W109, W110, W112) | T72301 |
| J437 (W209, W210) | T72302 |
| J438 (06) | T31506 |
| J438 (A2) | T30102 |
| J438 (D2) | T30402 |
| J438 (D3) | T30403 |
| J438 (D5) | T30405 |
| J438 (D7) | T30407 |
| J438 (H-11) | T20811 |
| J438 (H12) | T20812 |
| J438 (H13) | T20813 |
| J438 (H21) | T20821 |
| J438 (L6) | T61206 |
| J438 (M1) | T11301 |
| J438 (M2) | T11302 |
| J438 (M3) | T11313 |
| J438 (M3) | T11323 |
| J438 (M4) | T11304 |
| J438 (O1) | T31501 |
| J438 (O2) | T31502 |
| J438 (S-1) | T41901 |
| J438 (S2) | T41902 |
| J438 (S5) | T41905 |
| J438 (T1) | T12001 |
| J438 (T2) | T12002 ob |
| J438 (T4) | T12004 |
| J438 (T5) | T12005 |
| J438 (T8) | T12008 |
| J438 (W108, W109, W110, W112) | T72301 |
| J438 (W209, W210) | T72302 |
| J452 | A02080 ob |
| J452 | A03540 |
| J452 | A03570 |
| J452 | A03590 |
| J452 | A03600 |
| J452 | A03900 |
| J452 | A04130 |
| J452 | A05350 |
| J452 | A13570 |
| J452 | A13900 |
| J452 | A14440 |

| CROSS REFERENCE SPECIFICATIONS | UNIFIED NUMBER |
|---|---|
| J452 | A23900 |
| J452 (34) | A02220 ob |
| J452 (34) | A02221 ob |
| J452 (35) | A04430 |
| J452 (38) | A02950 |
| J452 (39) | A02420 |
| J452 (303) | A03840 |
| J452 (304) | A34430 |
| J452 (305) | A14130 |
| J452 (306) | A13800 |
| J452 (308) | A03800 |
| J452 (309) | A13600 |
| J452 (310) | A07120 |
| J452 (311) | A07050 |
| J452 (312) | A07070 |
| J452 (313) | A07100 |
| J452 (315) | A07130 |
| J452 (320) | A05140 |
| J452 (321) | A03360 |
| J452 (322) | A03550 |
| J452 (323) | A03560 |
| J452 (324) | A05200 |
| J452 (326) | A03190 |
| J452 (327) | A03280 ob |
| J452 (327) | A03281 ob |
| J452 (329) | A23190 |
| J452 (329) | A23191 |
| J452 (331) | A03330 |
| J452 (332) | A03320 |
| J452 (334) | A03391 |
| J452 (335) | A33550 |
| J452 (336) | A13560 |
| J452 (380) | A02960 |
| J452 (382) | A02010 |
| J452 (382) | A02012 |
| J452 (383) | A03830 |
| J454 (1100) | A91100 |
| J454 (2011) | A92011 |
| J454 (2024) | A92024 |
| J454 (2036) | A92036 |
| J454 (2038) | A92038 |
| J454 (2117) | A92117 |
| J454 (3002) | A93002 |
| J454 (3003) | A93003 |
| J454 (3004) | A93004 |
| J454 (5005) | A95005 |
| J454 (5052) | A95052 |
| J454 (5083) | A95083 |
| J454 (5086) | A95086 |
| J454 (5182) | A95182 |
| J454 (5252) | A95252 |
| J454 (5454) | A95454 |
| J454 (5457) | A95457 |
| J454 (5657) | A95657 |
| J454 (6009) | A96009 |
| J454 (6010) | A96010 |
| J454 (6053) | A96053 |
| J454 (6061) | A96061 |
| J454 (6063) | A96063 |
| J454 (6111) | A96111 |
| J454 (6463) | A96463 |
| J454 (7021) | A97021 |
| J454 (7029) | A97029 |
| J454 (7116) | A97116 |
| J454 (7129) | A97129 |
| J460 | C93700 |
| J460 | C93720 |
| J460 | C94330 |
| J460 | C98200 |

| CROSS REFERENCE SPECIFICATIONS | UNIFIED NUMBER |
|---|---|
| J460 | C98400 |
| J460 (13) | L53345 |
| J460 (791) | C83520 ob |
| J460 (795) | C83420 ob |
| J461 | C10200 |
| J461 | C11000 |
| J461 | C11300 |
| J461 | C11400 |
| J461 | C11500 |
| J461 | C11600 |
| J461 | C12000 |
| J461 | C12200 |
| J461 | C14700 |
| J461 | C15000 |
| J461 | C16200 |
| J461 | C17000 |
| J461 | C17200 |
| J461 | C17500 |
| J461 | C17600 ob |
| J461 | C18400 |
| J461 | C18700 |
| J461 | C19200 |
| J461 | C21000 |
| J461 | C22000 |
| J461 | C23000 |
| J461 | C24000 |
| J461 | C26000 |
| J461 | C26800 |
| J461 | C27000 |
| J461 | C33000 |
| J461 | C33100 ob |
| J461 | C34200 |
| J461 | C34500 |
| J461 | C35000 |
| J461 | C36000 |
| J461 | C37700 |
| J461 | C46400 |
| J461 | C46500 |
| J461 | C46600 ob |
| J461 | C46700 ob |
| J461 | C51000 |
| J461 | C51100 |
| J461 | C52100 |
| J461 | C52400 |
| J461 | C54400 |
| J461 | C61400 |
| J461 | C62300 |
| J461 | C62400 |
| J461 | C63000 |
| J461 | C64200 |
| J461 | C65500 |
| J461 | C67000 |
| J461 | C67300 |
| J461 | C67400 |
| J461 | C67500 |
| J461 | C70600 |
| J461 | C71000 |
| J461 | C71500 |
| J461 | C75200 |
| J461 | C77000 |
| J461 | C83600 |
| J461 | C83800 |
| J461 | C85200 |
| J461 | C85400 |
| J461 | C85800 |
| J461 | C86200 |
| J461 | C86300 |
| J461 | C86500 |
| J461 | C87200 ob |

| CROSS REFERENCE SPECIFICATIONS | UNIFIED NUMBER |
|---|---|
| J461 | C87500 |
| J461 | C87800 |
| J461 | C87900 ob |
| J461 | C90300 |
| J461 | C90500 |
| J461 | C90700 |
| J461 | C92200 |
| J461 | C92300 |
| J461 | C92500 |
| J461 | C92700 |
| J461 | C92900 |
| J461 | C93200 |
| J461 | C93500 |
| J461 | C93700 |
| J461 | C93800 |
| J461 | C94300 |
| J461 | C94700 |
| J461 | C94800 |
| J461 | C95200 |
| J461 | C95300 |
| J461 | C95400 |
| J461 | C95500 |
| J461 | C96200 |
| J461 | C97600 |
| J461 (DPTE) | C14500 |
| J461, J463 | C60800 |
| J461, J463 | C61800 |
| J462 | C83600 |
| J462 | C83800 |
| J462 | C85200 |
| J462 | C85400 |
| J462 | C85800 |
| J462 | C86200 |
| J462 | C86300 |
| J462 | C86500 |
| J462 | C87200 ob |
| J462 | C87500 |
| J462 | C87800 |
| J462 | C87900 ob |
| J462 | C90300 |
| J462 | C90500 |
| J462 | C90700 |
| J462 | C92200 |
| J462 | C92300 |
| J462 | C92500 |
| J462 | C92700 |
| J462 | C92900 |
| J462 | C93200 |
| J462 | C93500 |
| J462 | C93700 |
| J462 | C93800 |
| J462 | C94300 |
| J462 | C94700 |
| J462 | C94800 |
| J462 | C95200 |
| J462 | C95300 |
| J462 | C95400 |
| J462 | C95500 |
| J462 | C95800 |
| J462 | C96200 |
| J463 | C10200 |
| J463 | C11000 |
| J463 | C11100 |
| J463 | C11300 |
| J463 | C11400 |
| J463 | C11500 |
| J463 | C11600 |
| J463 | C12000 |
| J463 | C12200 |
| J463 | C14500 |
| J463 | C14700 |
| J463 | C15000 |
| J463 | C16200 |
| J463 | C17000 |
| J463 | C17200 |
| J463 | C17500 |
| J463 | C17600 ob |
| J463 | C18400 |
| J463 | C18700 |
| J463 | C19200 |
| J463 | C21000 |
| J463 | C22000 |
| J463 | C23000 |
| J463 | C24000 |
| J463 | C26000 |
| J463 | C26800 |
| J463 | C27000 |
| J463 | C33000 |
| J463 | C33100 ob |
| J463 | C34200 |
| J463 | C34500 |
| J463 | C35000 |
| J463 | C36000 |
| J463 | C37700 |
| J463 | C46400 |
| J463 | C46500 |
| J463 | C46600 ob |
| J463 | C46700 ob |
| J463 | C51000 |
| J463 | C52100 |
| J463 | C52400 |
| J463 | C54400 |
| J463 | C61300 |
| J463 | C61400 |
| J463 | C62300 |
| J463 | C62400 |
| J463 | C63000 |
| J463 | C64200 |
| J463 | C65500 |
| J463 | C67000 |
| J463 | C67300 |
| J463 | C67400 |
| J463 | C67500 |
| J463 | C70600 |
| J463 | C71000 |
| J463 | C71500 |
| J463 | C75200 |
| J463 | C77000 |
| J465 (AM60-A) | M10600 |
| J465 (AM100-A) | M10100 |
| J465 (AS41-A) | M10410 |
| J465 (AZ63A) | M11630 |
| J465 (AZ81A) | M11810 |
| J465 (AZ91A) | M11910 |
| J465 (AZ91B) | M11912 |
| J465 (AZ91C) | M11914 |
| J465 (AZ92A) | M11920 |
| J465 (EZ33A) | M12330 |
| J465 (HK31A) | M13310 |
| J465 (HZ32A) | M13320 |
| J465 (K1A) | M18010 |
| J465 (QE22A) | M18220 |
| J465 (QH21A) | M18210 |
| J465 (ZE41A) | M16410 |
| J465 (ZE63A) | M16630 |
| J465 (ZH62A) | M16620 |
| J465 (ZK51A) | M16510 |
| J465 (ZK61A) | M16610 |
| J466 (AZ31B) | M11311 |
| J466 (AZ61A) | M11610 |
| J466 (AZ80A) | M11800 |
| J466 (HK31A) | M13310 |
| J466 (HM21A) | M13210 |
| J466 (HM31A) | M13312 |
| J466 (LA141A) | M14141 |
| J466 (M1A) | M15100 |
| J466 (ZE10A) | M16100 |
| J466 (ZK40A) | M16400 |
| J466 (ZK60A) | M16600 |
| J467 (14-4PH) | J92240 |
| J467 (17-22-A) | K14675 |
| J467 (17-22-AS) | K23015 |
| J467 (17-22-AV) | K22770 |
| J467 (17-4PH) | S17400 |
| J467 (17-7PH) | S17700 |
| J467 (19-9DL) | S63198 |
| J467 (19-9DX) | S63199 |
| J467 (300M) | K44220 |
| J467 (302) | S30200 |
| J467 (304) | S30400 |
| J467 (304L) | S30403 |
| J467 (309S) | S30908 |
| J467 (310) | S31000 |
| J467 (314) | S31400 |
| J467 (316) | S31600 |
| J467 (321) | S32100 |
| J467 (410) | S41000 |
| J467 (422) | S42200 |
| J467 (440C) | S44004 |
| J467 9347) | S34700 |
| J467 (A40) | R50400 |
| J467 (A55) | R50550 |
| J467 (A70) | R50700 |
| J467 (A286) | S66286 |
| J467 (AM-350) | S35000 |
| J467 (AM-355) | S35500 |
| J467 (Astroloy) | N13017 |
| J467 (D-979) | N09979 |
| J467 (D6A) | K24728 |
| J467 (Discaloy) | S66220 |
| J467 (Eatonite 6) | S68000 |
| J467 (Elgiloy) | R30003 |
| J467 (Greek Ascoloy) | S41800 |
| J467 (H-11) | T20811 |
| J467 (H12) | T20812 |
| J467 (H13) | T20813 |
| J467 (Hastelloy B) | N10001 |
| J467 (Hastelloy C) | N10002 |
| J467 (Hastelloy F) | N06001 ob |
| J467 (Hastelloy N) | N10003 |
| J467 (Hastelloy W) | N10004 |
| J467 (Hastelloy X) | N06002 |
| J467 (HS-21) | R30021 |
| J467 (HS-31, X-40) | R30031 |
| J467 (IN 100) | N13100 |
| J467 (Incoloy 801) | N08801 |
| J467 (Incoloy 901) | N09901 |
| J467 (Inconel 600) | N06600 |
| J467 (Inconel 702) | N07702 |
| J467 (Inconel 718) | N07718 |
| J467 (Inconel 722) | N07722 |
| J467 (Inconel X-750) | N07750 |
| J467 (L-605) | R30605 |
| J467 (M-252) | N07252 |
| J467 (M-50) | T11350 |
| J467 (M2) | T11302 |
| J467 (M10) | T11310 |

| CROSS REFERENCE SPECIFICATIONS | UNIFIED NUMBER |
|---|---|
| J467 (N-155) | R30155 |
| J467 (Nimonic 75) | N06075 |
| J467 (Nimonic 80A) | N07080 |
| J467 (Nimonic 90) | N07090 |
| J467 (PH15-7-Mo) | S15700 |
| J467 (RA330) | N08330 |
| J467 (RA333) | N06333 |
| J467 (REne 41) | N07041 |
| J467 (S590) | R30590 |
| J467 (S618) | R30816 |
| J467 (Stainless W) | S17600 |
| J467 (Stellite 6) | R30006 |
| J467 (Ti-169) | R54790 |
| J467 (Ti-6-4-2) | R54620 |
| J467 (Udimet 500) | N07500 |
| J467 (W545) | S66545 |
| J467 (Waspaloy) | N07001 |
| J468 (903) (Castings) | Z33520 |
| J468 (903)(Ingot) | Z33521 |
| J468 (925)(Ingot) | Z35530 |
| J468 (925) (Castings) | Z35531 |
| J469 | Z33520 |
| J469 | Z35531 |
| J473 (No. 1B) | L54940 |
| J473 (No. 2B) | L54905 |
| J473 (No. 3B) | L54815 |
| J473 (No. 5B) | L54610 |
| J473 (No. 6B) | L54555 |
| J473 (No. 9B) | L54250 |
| J775 | S31630 |
| J775 | S63020 |
| J775 | S63021 |
| J775 | S66315 |
| J775 (EV 12) | S63012 |
| J775 (EV 13) | S63013 |
| J775 (EV 16) | S63018 |
| J775 (EV 17) | S30430 |
| J775 (EV 4) | S63017 |
| J775 (EV 8) | S63008 |
| J775 (HEV 8) | N07032 |
| J775 (HEV-3) | N07751 |
| J775 (HEV-5) | N07080 |
| J775 (HNV 3) | S65007 |
| J775 (HNV 6) | S65006 |
| J775 (HNV 8) | S42200 |
| J775 (K14072) | K14072 |
| J775 (NV 1) | H15410 |
| J775 (NV 4) | G31400 |
| J775 (NV 5) | G86450 |
| J775 (NV 6) | H51500 |
| J775 (NV 7) | H41400 |
| J775 (NV 8) | S64007 |
| J775 (NV-2) | G15470 |
| J775 (SUH 11M) | S64004 |
| J775 (VF 10) | N06015 |
| J775 (VF 8) | R30400 |
| J775 (VF 9) | N06013 |
| J775 (VF-2) | R30006 |
| J775 (VF3) | N06005 |
| J775 (VF4) | N06782 |
| J775 (VF5) | R30002 |
| J775 (VF6) | R30001 |
| J775 (VF7) | R30012 |
| J775 (XEV-F) | S63019 |
| J775 (XEV-H) | N07001 |
| J775 (XEV-J) | R56401 |
| J1081 (PS 59) | G51210 |
| J1081 (PS 64) | G51190 |
| J1081 (PS 65) | G51240 |
| J1099 (A538A) | K92820 |
| J1099 (A538B) | K92890 |
| J1099 (A538C) | K93120 |
| J1249 | G11230 ob |
| J1249 | G15330 ob |
| J1249 | G15340 ob |
| J1249 | G15530 ob |
| J1249 | G15700 ob |
| J1249 | G15800 ob |
| J1249 | G15900 ob |
| J1249 (50B40) | G50401 |
| J1249 (50B50) | G50501 |
| J1249 (50B60) | G50601 |
| J1249 (81B45) | G81451 |
| J1249 (86B45) | G86451 |
| J1249 (94B15) | G94151 |
| J1249 (94B17) | G94171 |
| J1249 (94B30) | G94301 |
| J1249 (94B40) | G94401 |
| J1249 (1572) | G15720 |
| J1249 (4012) | G40120 |
| J1249 (4020) | G40280 |
| J1249 (4024) | G40240 |
| J1249 (4032) | G40320 |
| J1249 (4042) | G40420 |
| J1249 (4063) | G40630 |
| J1249 (4121) | G41210 |
| J1249 (4147) | G41470 |
| J1249 (4161) | G41610 |
| J1249 (4337) | G43370 |
| J1249 (4419) | G44190 |
| J1249 (4422) | G44220 |
| J1249 (4427) | G44270 |
| J1249 (4621) | G46210 |
| J1249 (4626) | G46260 |
| J1249 (4715) | G47150 |
| J1249 (4718) | G47180 |
| J1249 (4720) | G47200 |
| J1249 (4815) | G48150 |
| J1249 (4817) | G48170 |
| J1249 (5015) | G50150 |
| J1249 (5046) | G50441 |
| J1249 (5060) | G50600 |
| J1249 (5115) | G51150 |
| J1249 (5117) | G51170 |
| J1249 (5135) | G51350 |
| J1249 (5145) | G51450 |
| J1249 (5147) | G51470 |
| J1249 (5155) | G51550 |
| J1249 (6118) | G61180 |
| J1249 (6120) | G61200 |
| J1249 (8115) | G81150 |
| J1249 (8642) | G86420 |
| J1249 (8650) | G86500 |
| J1249 (8660) | G86600 |
| J1249 (8735 | G87350 |
| J1249 (8740) | G87400 |
| J1249 (8742) | G87420 |
| J1249 (9254) | G92540 |
| J1249 (9255) | G92550 |
| J1249 (9262) | G92620 |
| J1249 (9310) | G93106 |
| J1249 (9840) | G98400 |
| J1249 (9850) | G98500 |
| J1249 (50100) | G50986 |
| J1249 (E51100) | G51986 |
| J1268 (15B21H) | H15211 |
| J1268 (15B28H) | H15281 |
| J1268 (15B30H) | H15301 |
| J1268 (15B35H) | H15351 |
| J1268 (15B37H) | H15371 |
| J1268 (15B41H) | H15411 |
| J1268 (15B48H) | H15481 |
| J1268 (15B62H) | H15621 |
| J1268 (50B40H) | H50401 |
| J1268 (50B44H) | H50441 |
| J1268 (50B46H) | H50461 |
| J1268 (50B50H) | H50501 |
| J1268 (50B60H) | H50601 |
| J1268 (51B60H) | H51601 |
| J1268 (81B45H) | H81451 |
| J1268 (86B30H) | H86301 |
| J1268 (86B45H) | H86451 |
| J1268 (94B15H) | H94151 |
| J1268 (94B17H) | H94171 |
| J1268 (94B30H) | H94301 |
| J1268 (1038H) | H10380 |
| J1268 (1045H) | H10450 |
| J1268 (1330H) | H13300 |
| J1268 (1335H) | H13350 |
| J1268 (1340H) | H13400 |
| J1268 (1345H) | H13450 |
| J1268 (1522H) | H15220 |
| J1268 (1524H) | H15240 |
| J1268 (1526H) | H15260 |
| J1268 (1541H) | H15410 |
| J1268 (4027H) | H40270 |
| J1268 (4028H) | H40280 |
| J1268 (4032H) | H40320 |
| J1268 (4037H) | H40370 |
| J1268 (4042H) | H40420 |
| J1268 (4047H) | H40470 |
| J1268 (4118H) | H41180 |
| J1268 (4120H) | H41200 |
| J1268 (4130H) | H41300 |
| J1268 (4135H) | H41350 |
| J1268 (4137H) | H41370 |
| J1268 (4140H) | H41400 |
| J1268 (4142H) | H41420 |
| J1268 (4145H) | H41450 |
| J1268 (4147H) | H41470 |
| J1268 (4150H) | H41500 |
| J1268 (4161H) | H41610 |
| J1268 (4320H) | H43200 |
| J1268 (4340H) | H43400 |
| J1268 (4620H) | H46200 |
| J1268 (4718H) | H47180 |
| J1268 (4720H) | H47200 |
| J1268 (4815H) | H48150 |
| J1268 (4817H) | H48170 |
| J1268 (4820H) | H48200 |
| J1268 (5046H) | H50460 |
| J1268 (5120H) | H51200 |
| J1268 (5130H) | H51300 |
| J1268 (5132H) | H51320 |
| J1268 (5135H) | H51350 |
| J1268 (5140H) | H51400 |
| J1268 (5147H) | H51470 |
| J1268 (5150H) | H51500 |
| J1268 (5155H) | H51550 |
| J1268 (5160H) | H51600 |
| J1268 (6118H) | H61180 |
| J1268 (6150H) | H61500 |
| J1268 (8617H) | H86170 |
| J1268 (8620H) | H86200 |
| J1268 (8622H) | H86220 |
| J1268 (8625H) | H86250 |
| J1268 (8627H) | H86270 |

| CROSS REFERENCE SPECIFICATIONS | UNIFIED NUMBER |
|---|---|
| J1268 (8630H) | H86300 |
| J1268 (8637H) | H86370 |
| J1268 (8640H) | H86400 |
| J1268 (8642H) | H86420 |
| J1268 (8645H) | H86450 |
| J1268 (8650H) | H86500 |
| J1268 (8655H) | H86550 |
| J1268 (8660H) | H86600 |
| J1268 (8720H) | H87200 |
| J1268 (8740H) | H87400 |
| J1268 (8822H) | H88220 |
| J1268 (9259H) | H92590 |
| J1268 (9260H) | H92600 |
| J1268 (9310H) | H93100 |
| J1268 (E4340H) | H43406 |
| J1268 (E9310H) | H93106 |
| J1397 (12L14) | G12144 |
| J1397 (50B40) | G50401 |
| J1397 (50B44) | G50441 |
| J1397 (50B46) | G50461 |
| J1397 (50B50) | G50501 |
| J1397 (50B60) | G50601 |
| J1397 (51B60) | G51601 |
| J1397 (81B45) | G81451 |
| J1397 (86B45) | G86451 |
| J1397 (94B15) | G94151 |
| J1397 (94B17) | G94171 |
| J1397 (94B30) | G94301 |
| J1397 (1006) | G10060 |
| J1397 (1008) | G10080 |
| J1397 (1010) | G10100 |
| J1397 (1012) | G10120 |
| J1397 (1015) | G10150 |
| J1397 (1016) | G10160 |
| J1397 (1017) | G10170 |
| J1397 (1018) | G10180 |
| J1397 (1019) | G10190 |
| J1397 (1020) | G10200 |
| J1397 (1021) | G10210 |
| J1397 (1022) | G10220 |
| J1397 (1023) | G10230 |
| J1397 (1025) | G10250 |
| J1397 (1026) | G10260 |
| J1397 (1030) | G10300 |
| J1397 (1035) | G10350 |
| J1397 (1037) | G10370 |
| J1397 (1038) | G10380 |
| J1397 (1039) | G10390 |
| J1397 (1040) | G10400 |
| J1397 (1042) | G10420 |
| J1397 (1043) | G10430 |
| J1397 (1044) | G10440 |
| J1397 (1045) | G10450 |
| J1397 (1046) | G10460 |
| J1397 (1049) | G10490 |
| J1397 (1050) | G10500 |
| J1397 (1055) | G10550 |
| J1397 (1060) | G10600 |
| J1397 (1064) | G10640 |
| J1397 (1065) | G10650 |
| J1397 (1070) | G10700 |
| J1397 (1074) | G10740 |
| J1397 (1078) | G10780 |
| J1397 (1080) | G10800 |
| J1397 (1084) | G10840 |
| J1397 (1085) | G10850 |
| J1397 (1086) | G10860 |
| J1397 (1090) | G10900 |
| J1397 (1095) | G10950 |
| J1397 (1108) | G11080 |
| J1397 (1117) | G11170 |
| J1397 (1132) | G11320 |
| J1397 (1137) | G11370 |
| J1397 (1140) | G11400 |
| J1397 (1141) | G11410 |
| J1397 (1144) | G11440 |
| J1397 (1146) | G11460 |
| J1397 (1151) | G11510 |
| J1397 (1211) | G12110 |
| J1397 (1213) | G12120 |
| J1397 (1213) | G12130 |
| J1397 (1330) | G13300 |
| J1397 (1335) | G13350 |
| J1397 (1340) | G13400 |
| J1397 (1345) | G13450 |
| J1397 (1524) | G15240 |
| J1397 (1536) | G15360 |
| J1397 (1541) | G15410 |
| J1397 (1548) | G15480 |
| J1397 (1552) | G15520 |
| J1397 (4023) | G40230 |
| J1397 (4024) | G40240 |
| J1397 (4027) | G40270 |
| J1397 (4028) | G40280 |
| J1397 (4032) | G40320 |
| J1397 (4037) | G40370 |
| J1397 (4042) | G40420 |
| J1397 (4047) | G40470 |
| J1397 (4118) | G41180 |
| J1397 (4130) | G41300 |
| J1397 (4135) | G41350 |
| J1397 (4137) | G41370 |
| J1397 (4140) | G41400 |
| J1397 (4142) | G41420 |
| J1397 (4145) | G41450 |
| J1397 (4147) | G41470 |
| J1397 (4150) | G41500 |
| J1397 (4161) | G41610 |
| J1397 (4320) | G43200 |
| J1397 (4340) | G43400 |
| J1397 (4422) | G44220 |
| J1397 (4427) | G44270 |
| J1397 (4615) | G46150 |
| J1397 (4617) | G46170 |
| J1397 (4620) | G46200 |
| J1397 (4626) | G46260 |
| J1397 (4718) | G47180 |
| J1397 (4720) | G47200 |
| J1397 (4815) | G48150 |
| J1397 (4817) | G48170 |
| J1397 (4820) | G48200 |
| J1397 (5046) | G50460 |
| J1397 (5060) | G50600 |
| J1397 (5115) | G51150 |
| J1397 (5120) | G51200 |
| J1397 (5130) | G51300 |
| J1397 (5132) | G51320 |
| J1397 (5135) | G51350 |
| J1397 (5140) | G51400 |
| J1397 (5147) | G51470 |
| J1397 (5150) | G51500 |
| J1397 (5155) | G51550 |
| J1397 (5160) | G51600 |
| J1397 (6118) | G61180 |
| J1397 (6150) | G61500 |
| J1397 (8115) | G81150 |
| J1397 (8615) | G86150 |
| J1397 (8617) | G86170 |
| J1397 (8620) | G86200 |
| J1397 (8622) | G86220 |
| J1397 (8625) | G86250 |
| J1397 (8627) | G86270 |
| J1397 (8630) | G86300 |
| J1397 (8637) | G86370 |
| J1397 (8640) | G86400 |
| J1397 (8642) | G86420 |
| J1397 (8645) | G86450 |
| J1397 (8650) | G86500 |
| J1397 (8655) | G86550 |
| J1397 (8660) | G86600 |
| J1397 (8720) | G87200 |
| J1397 (8740) | G87400 |
| J1397 (8822) | G88220 |
| J1397 (9254) | G92540 |
| J1397 (9260) | G92600 |
| J1397 (9310) | G93106 |
| J1397 (50100) | G50986 |
| J1397 (51100) | G51986 |
| J1397 (52100) | G52986 |
| J1397 (E4340) | G43406 |

# Index of Common Trade Names

| | |
|---|---|
| 18-4Mn | W32440 |
| 18-4-1 | T12001 |
| 18-5-11 | S32440 |
| 18-5-11 | W32410 |
| 18-8Mo | S31600 |
| 18-8Ti | S32100 |
| 18-9-LW | S30430 |
| 18-15 | S30600 |
| 18-15LCSi | S30600 |
| 18-17LC | S30601 |
| 18-18 Plus | S28200 |
| 18-18-2 | S38100 |
| 19 | C87200 |
| 19 | C87300 |
| 19 | K92890 |
| 19 | L54510 |
| 19 | R58640 |
| 19D | S32001 |
| 19-9DL | J92843 |
| 19-9DL | S63198 |
| 19-9DX | S63199 |
| 19-9 High Carbon | S30884 |
| 19-9Mo Low Carbon | S30886 |
| 19-9 W-Mo | S63199 |
| 19-9 W-Mo | W34910 |
| 20 | F11401 |
| 20 | K93120 |
| 20 | N08020 |
| 20 | R30006 |
| 20 | R30106 |
| 20 | R58645 |
| 20B Solder | L54711 |
| 20C | C82500 |
| 20C | G10950 |
| 20Cb3 | N08020 |
| 20Cb3 | W88021 |
| 20Cb3 Mod | N08321 |
| 20C Solder | L54712 |
| 20CT | C82500 |
| 20Mo4 | N08024 |
| 20Mo6 | N08026 |
| 20Mo-6HS | N08036 |
| 20 Mod | N08320 |
| 20 Solder | L54712 |
| 20% Tin Solder | L54710 |
| 20-45-5 | N08245 |
| 21 | A86951 |
| 21 | C86300 |
| 21 | K20477 |
| 21 | K31545 |
| 21 | R58210 |
| 21 Brazing Sheet | A86951 |
| 21C | C82510 |
| 21Cr-6Ni-9Mn | S21904 |
| 21-2N | S63012 |
| 21-4N | S63008 |
| 21-4N+Nb+W | S63019 |
| 21-4NB | S63019 |
| 21-6-9 | S21900 |
| 21-6-9 | S21904 |
| 21-6-9 | S21980 |
| 21-6-9 | W32310 |
| 21-6-9 | W32340 |
| 21-6-9LC | S21904 |
| 21-12 | S63016 |
| 21-12N | S63017 |
| 21-55N | S63007 |
| 22 | A86951 |
| 22 | C95300 |
| 22 | K21590 |
| 22 | K24070 |
| 22 | N06022 |
| 22 | R30103 |
| 22B | K21390 |
| 22B | K21590 |
| 22 Brazing Sheet | A86951 |
| 22C | K21390 |
| 22Cr-13Ni-5Mn | S20910 |
| 22D | K21390 |
| 22L | K21590 |
| 22/5 | S31803 |
| 22-10-5 | S20980 |
| 22-10-5 | W32210 |
| 22-10-5 | W32240 |
| 22-13-5 | S20910 |
| 23 | A86951 |
| 23 | C95400 |
| 23 | K24728 |
| 23 | N02270 |
| 23 | N08020 |
| 23 | R30021 |
| 23 | R56407 |
| 23-8N | S63018 |
| 24 | A86951 |
| 24 | R56405 |
| 25 | C17200 |
| 25 | C83600 |
| 25 | F11701 |
| 25 | R30605 |
| 25 | R56403 |
| 25-4-4 | S44635 |
| 25/75 Solder | L54720 |
| 25A Solder | L54720 |
| 25B Solder | L54721 |
| 25C Solder | L54722 |
| 25-6Mo | N08926 |
| 25-6-1LN | S31200 |
| 26 | C90300 |
| 26 | R52404 |
| 26-1 | S44626 |
| 26-1S | S44426 |
| 26-1Ti | S44626 |
| 26-3-3 | S44660 |
| 27 | R52254 |
| 28 | N08028 |
| 28 | R56323 |
| 28-2-3.5 | S32803 |
| 28-4Mn | S23980 |
| 29 | K94610 |
| 29 | R56404 |
| 29Cr-4Mo | S44700 |
| 29-4 | S44700 |
| 29-4C | S44735 |
| 29-4-2 | S44800 |
| 29-4-2C | S44736 |
| 29-9 | S64299 |
| 29-17 | K94610 |
| 30 | F12101 |
| 30 | K02502 |
| 30 | N08800 |
| 30 | R05250 |
| 30A Solder | L54820 |
| 30B Solder | L54821 |
| 30C Solder | L54822 |
| 30/70 Solder | L54820 |
| 30/70 Solder | L54822 |
| 31 | K32550 |
| 31 | N08031 |
| 31H | N08810 |
| 31HT | N08811 |
| 32 | K44220 |
| 32 | R55111 |
| 33 | K02502 |
| 33 | K14394 |
| 33 | R20033 |
| 33-25 | C17200 |
| 33-25 | C17300 |
| 34 | A02220 |
| 34 | A02221 |
| 35 | A04430 |
| 35 | A04431 |
| 35 | F12401 |
| 35A Solder | L54850 |
| 35B | C82200 |
| 35B Solder | L54851 |
| 35C | C82200 |
| 35C Solder | L54852 |
| 35N | R30035 |
| 35/65 Solder | L54850 |
| 36 | K02502 |
| 36 | K93600 |
| 36%Ni Steel | K93601 |
| 36%Ni Steel | K93602 |
| 38 | A02950 |
| 38 | A02952 |
| 39 | A02420 |
| 40 | F12803 |
| 40 | K02502 |
| 40 | R50400 |
| 40A Solder | L54915 |
| 40B Solder | L54916 |
| 40C Solder | L54918 |
| 40E | A07120 |
| 40/60 Solder | L54915 |
| 41 | N07041 |
| 42 | K94100 |
| 42 | K94600 |
| 42Ni-56.5Fe | K94101 |
| 43 | A04430 |
| 43 | A14430 |
| 43 | A24430 |
| 43BV12 | K21028 |
| 43.5-13 | C77600 |

# Appendix

SAE The Engineering Society For Advancing Mobility Land Sea Air and Space®
INTERNATIONAL
400 Commonwealth Drive, Warrendale, PA 15096-0001

SURFACE VEHICLE RECOMMENDED PRACTICE

SAE J1086 and ASTM E 527

REV. JUL95

Issued 1974-08
Revised 1995-07

Superseding J1086 DEC92

# NUMBERING METALS AND ALLOYS

***Foreword***—This Document has not changed other than to put it into the new SAE Technical Standards Board Format.

UNS designations shall not be used for metals and alloys which are not registered under the system described herein, or for any metal or alloy whose composition differs from those registered.

## *1. Scope*

**1.1** This SAE Recommended Practice describes a unified numbering system (UNS) for metals and alloys which have a "commercial standing" (see 6.1), and covers the procedure by which such numbers are assigned.

Section 2 describes the system of alphanumeric designations or "numbers" established for each family of metals and alloys.

Section 3 outlines the organization established for administering the system.

Section 4 describes the procedure for requesting number assignment to metals and alloys for which UNS numbers have not previously been assigned.

**1.2** The UNS provides a means of correlating many nationally used numbering systems currently administered by societies, trade associations, and individual users and producers of metals and alloys, thereby avoiding confusion caused by use of more than one identification number for the same material; and by the opposite situation of having the same number assigned to two or more entirely different materials. It provides, also, the uniformity necessary for efficient indexing, record keeping, data storage and retrieval, and cross referencing.

**1.3** A UNS number is not in itself a specification, since it establishes no requirements for form, condition, quality, etc. It is a unified identification of metals and alloys for which controlling limits have been established in specifications published elsewhere. (See 6.2.)

## 2. References

**2.1 Related Publications**—The following publications are provided for information purposes only and are not a required part of this document.

2.1.1 SAE PUBLICATIONS—Available from SAE, 400 Commonwealth Drive, Warrendale, PA 15096-0001.

SAE HS-1086—Metals and Alloys in the Unified Numbering System

2.1.2 ASTM PUBLICATIONS—Available from ASTM, 100 Barr Harbor Drive, West Conshohocken, PA 19428-2959.

ASTM E 527—Practice for Numbering Metals and Alloys (UNS)
ASTM Publication No. DS-56—Metals and Alloys and the Unified Numbering System

## 3. Description of Numbers (or Codes) Established for Metals and Alloys

**3.1** The unified numbering system (UNS) establishes 18 series of numbers for metals and alloys, as shown in Table 1. Each UNS number consists of a single letter-prefix followed by five digits. In most cases the letter is suggestive of the family of metals identified, for example, A for aluminum, P for precious metals, S for stainless steels. Table 2 shows the secondary division of some primary series of numbers.

**3.2** Whereas some of the digits in certain of the UNS number groups have special assigned meaning, each series is independent of the others in such significance; this practice permits greater flexibility and avoids complicated and lengthy UNS numbers. (See 6.3.)

**TABLE 1—PRIMARY SERIES OF NUMBERS**

| UNS Series | Metal |
|---|---|
| Nonferrous metals and alloys | |
| A00001-A99999 | Aluminum and aluminum alloys |
| C00001-C99999 | Copper and copper alloys |
| E00001-E99999 | Rare earth and rare earth-like metals and alloys (18 Items, see Table 2) |
| L00001-L99999 | Low melting metals and alloys (14 Items, see Table 2) |
| M00001-M99999 | Miscellaneous nonferrous metals and alloys (12 Items, see Table 2) |
| N00001-N99999 | Nickel and nickel alloys |
| P00001-P99999 | Precious metals and alloys (8 Items, see Table 2) |
| R00001-R99999 | Reactive and refractory metals and alloys (14 Items, see Table 2) |
| Z00001-Z99999 | Zinc and zinc alloys |
| Ferrous metals and alloys | |
| D00001-D99999 | Specified mechanical properties steels |
| F00001-F99999 | Cast irons |
| G00001-G99999 | AISI and SAE carbon and alloy steels (except tool steels) |
| H00001-H99999 | AISI H-steels |
| J00001-J99999 | Cast steels (except tool steels) |
| K00001-K99999 | Miscellaneous steels and ferrous alloys |
| S00001-S99999 | Heat and corrosion resistant (stainless) steels |
| T00001-T99999 | Tool steels |
| Welding filler metals | |
| W00001-W99999 | Welding filler metals, covered and tubular electrodes, classified by weld deposit composition (see Table 2) |

## TABLE 2—SECONDARY DIVISION OF SOME SERIES OF NUMBERS

| UNS Series | Metal | UNS Series | Metal |
|---|---|---|---|
| E00001-E99999 Rare earth and rare earthlike metals and alloys | | P00001-P99999 Precious metals and alloys | |
| E00000-E00999 | Actinium | P00001-P00999 | Gold |
| E01000-E20999 | Cerium | P01001-P01999 | Iridium |
| E21000-E45999 | Mixed rare earths[1] | P02001-P02999 | Osmium |
| E46000-E47999 | Dysprosium | P03001-P03999 | Palladium |
| E48000-E49999 | Erbium | P04001-P04999 | Platinum |
| E50000-E51999 | Europium | P05001-P05999 | Rhodium |
| E52000-E55999 | Gadolinium | P06001-P06999 | Ruthenium |
| E56000-E57999 | Holmium | P07001-P07999 | Silver |
| E58000-E67999 | Lanthanum | | |
| E68000-E68999 | Lutetium | R00001-R99999 Reactive and refractory metals and alloys | |
| E69000-E73999 | Neodymium | | |
| E74000-E77999 | Praseodymium | | |
| E78000-E78999 | Promethium | R01001-R01999 | Boron |
| E79000-E82999 | Samarium | R02001-R02999 | Hafnium |
| E83000-E84999 | Scandium | R03001-R03999 | Molybdenum |
| E85000-E86999 | Terbium | R04001-R04999 | Niobium (Columbium) |
| E87000-E87999 | Thulium | R05001-R05999 | Tantalum |
| E88000-E89999 | Ytterbium | R06001-R06999 | Thorium |
| E90000-E99999 | Yttrium | R07001-R07999 | Tungsten |
| | | R08001-R08999 | Vanadium |
| F00001-F99999 Cast irons | Gray, malleable, pearlitic malleable, and ductile (nodular) cast irons | R10001-R19999 | Beryllium |
| | | R20001-R29999 | Chromium |
| | | R30001-R39999 | Cobalt |
| | | R40001-R49999 | Rhenium |
| K00001-K99999 Miscellaneous steels and ferrous alloys | | R50001-R59999 | Titanium |
| | | R60001-R69999 | Zirconium |
| L00001-L99999 Low-melting metals and alloys | | W00001-W99999 Welding filler metals, classified by weld deposit composition | |
| L00001-L00999 | Bismuth | | |
| L01001-L01999 | Cadmium | W00001-W09999 | Carbon steel with no significant alloying elements |
| L02001-L02999 | Cesium | | |
| L03001-L03999 | Gallium | W10000-W19999 | Manganese-molybdenum low alloy steels |
| L04001-L04999 | Indium | | |
| L06001-L06999 | Lithium | W20000-W29999 | Nickel low alloy steels |
| L07001-L07999 | Mercury | W30000-W39999 | Austenitic stainless steels |
| L08001-L08999 | Potassium | W40000-W49999 | Ferritic stainless steels |
| L09001-L09999 | Rubidium | W50000-W59999 | Chromium low alloy steels |
| L10001-L10999 | Selenium | W60000-W69999 | Copper base alloys |
| L11001-L11999 | Sodium | W70000-W79999 | Surfacing alloys |
| L13001-L13999 | Tin | W80000-W89999 | Nickel base alloys |
| L50001-L59999 | Lead | | |
| | | Z00001-Z99999 Zinc and zinc alloys | Zinc |
| M00001-M99999 Miscellaneous nonferrous metals and alloys | | | |
| M00001-M00999 | Antimony | | |
| M01001-M01999 | Arsenic | | |
| M02001-M02999 | Barium | | |
| M03001-M03999 | Calcium | | |
| M04001-M04999 | Germanium | | |
| M05001-M05999 | Plutonium | | |
| M06001-M06999 | Strontium | | |
| M07001-M07999 | Tellurium | | |
| M08001-M08999 | Uranium | | |
| M10001-M19999 | Magnesium | | |
| M20001-M29999 | Manganese | | |
| M30001-M39999 | Silicon | | |

1. Alloys in which the rare earths are used in the ratio of their natural occurrence (that is, unseparated rare earths). In this mixture, cerium is the most abundant of the rare earth elements.

**3.3** Wherever feasible, identification "numbers" from existing systems are incorporated into the UNS numbers. For example: The carbon steel which is presently identified by "AISI 1020" (American Iron & Steel Institute) is covered by "UNS G10200" and the nickel alloy presently identified by "M252" is covered by "UNS N07252."

**3.4** Welding filler metals fall into two general categories: those whose compositions are determined by the filler metal analysis (e.g., solid bare wire or rods and cast rods), and those whose composition is determined by the weld deposit analysis (e.g., covered electrodes, flux-cored and other composite wire electrodes). The latter are assigned to a new primary series with the letter W as shown in Table 1. The solid bare wire and rods continue to be assigned in the established number series according to their composition.

(Readers are cautioned not to make their own "assignments" of numbers from such listings, as this can result in unintended and unexpected duplication and conflict.)

**3.5** The ASTM and the SAE periodically publish up-to-date listings of all UNS numbers assigned to specific metals and alloys, with appropriate reference information on each. (See 6.6.) Many trade associations also publish similar listings related to materials of primary interest to their organizations.

## *4. Organization for Administering Unified Numbering System for Metals and Alloys*

**4.1** The organization for administering the UNS consists of: (1) an advisory board, (2) several number-assigning offices, (3) a corps of volunteer consultants, and (4) staffs at ASTM and SAE. In addition, SAE and ASTM committees dealing with various groups of materials may be consulted.

4.1.1 The Advisory Board has approximately 20 volunteer members who are affiliated with major producing and using industries, trade associations, government agencies, and standards societies, and who have extensive experience with identification, classification, and specification of materials. The Board is the administrative arm of SAE and ASTM on all matters pertaining to the UNS. It coordinates thinking on the format of each series of numbers and the administration of each by selected experts. It sets up ground rules for determining eligibility of any material for a UNS number, for requesting such numbers, and for appealing unfavorable rulings. It is the final referee on matters of disagreement between requesters and assigners.

4.1.2 UNS number assigners for certain materials are set up at trade associations which have successfully administered their own numbering systems; for other materials, assigners are located at the offices of SAE and ASTM. Each of these assigners has the responsibility for administering a specific series of numbers, as shown in Table 3. Each considers requests for assignment of new UNS numbers, and informs applicants of the action taken. Trade association UNS number assigners also report immediately to both SAE and ASTM details of each number assignment. ASTM and SAE assigners collaborate with designated consultants when considering requests for assignment of new numbers.

4.1.3 Consultants are selected by the Advisory Board to provide expert knowledge of a specific field of materials. Since they are utilized primarily by the Board and the SAE and ASTM number assigners, they are not listed in this document. At the request of the ASTM or SAE number assigner, a consultant considers a request for a new number in light of the ground rules established for the material involved, decides whether a new number is justified, and informs the ASTM or SAE number assigner accordingly.

This utilization of experts (consultants and number assigners) is intended to insure prompt and fair consideration of all requests. It permits each decision to be based on current knowledge of the needs of a specific industry of producers and users.

## TABLE 3—NUMBER ASSIGNERS AND AREAS OF RESPONSIBILITY

| | |
|---|---|
| The Aluminum Association<br>900 19th Street, NW, Suite 30<br>Washington, DC 20006<br>Telephone: (202) 862-5100 | Aluminum and aluminum alloys<br>UNS Number Series: A00001-A99999 |
| American Society for Testing and Materials<br>100 Barr Harbor Drive<br>West Conshohocken, PA 19428-2959<br>Attention: Office for Unified Numbering System for Metals<br>Telephone: (215) 299-5400 | Rare earth and rare earth-like metals and alloys<br>UNS Number Series: E00001-E99999<br><br>Cast irons<br>UNS Number Series: F00001-F99999<br><br>Cast steels<br>UNS Number Series: J00001-J99999<br><br>Miscellaneous steels and ferrous alloys<br>UNS Number Series: K00001-K99999<br><br>Low melting metals and alloys<br>UNS Number Series: L00001-L99999<br><br>Miscellaneous nonferrous metals and alloys<br>UNS Number Series: M00001-M99999<br><br>Precious metals and alloys<br>UNS Number Series: P00001-P99999 |
| American Welding Society<br>550 N.W. LeJeune Road<br>P.O. Box 351040<br>Miami, FL 33135<br>Attention: Office for Unified Numbering System<br>Telephone: (800) 443-9353<br>Fax: 303-443-7559 | Welding filler metals<br>UNS Number Series: W00001-W99999 |
| Copper Development Association<br>260 Madison Avenue<br>New York, NY 10016-2401<br>Attention: Office for Unified Numbering System for Metals<br>Telephone: (212) 251-7200 | Copper and copper alloys<br>UNS Number Series: C00001-C99999 |
| SAE (Society of Automotive Engineers)<br>400 Commonwealth Drive<br>Warrendale, PA 15096-0001<br>Attention: Office for Unified Numbering System for Metals<br>Telephone: (412) 776-4841 | Carbon and alloy steels<br>UNS Number Series: G00001 - G99999<br><br>H-steels<br>UNS Number Series: H00001-H99999<br><br>Nickel and nickel alloys<br>UNS Number Series: N00001-N99999<br><br>Heat and corrosion resistant (stainless) steels<br>UNS Number Series: S00001 - S99999<br><br>Tool steels<br>UNS Number Series: T00001 - T99999 |
| Zinc Institute, Inc.<br>292 Madison Avenue<br>New York, NY 10017<br>Attention: Office for Unified Numbering System for Metals<br>Telephone: (212) 578-4750 | Zinc and zinc alloys<br>UNS Number Series: Z00001 - Z99999 |

4.1.4 Staff members at SAE and ASTM maintain duplicate master listings of all UNS numbers assigned.

4.1.5 Established SAE and ASTM committees which normally deal with standards and specifications for the materials covered by the UNS, and other knowledgeable persons, are called upon by the Advisory Board for advice when considering appeals from unfavorable rulings in the matter of UNS number assignments.

## *5. Procedure for Requesting Number Assignment to Metals and Alloys Not Already Covered by UNS Numbers (or Codes)*

**5.1** UNS numbers are assigned only to metals and alloys which have a commercial standing (as defined in 6.1).

**5.2** The need for a new number should always be verified by determining from the latest complete listing of already assigned UNS numbers that a usable number is not available. (See 6.4.)

**5.3** For a new UNS number to be assigned, the composition (or other properties, as applicable) must be significantly different from those of any metal or alloy which has already been assigned a UNS number.

5.3.1 In the case of metals or alloys that are normally identified or specified by chemical composition, the chemical composition limits must be reported.

5.3.2 In the case of metals or alloys which are normally identified or specified by mechanical (or other) properties, such properties and limits thereof must be reported. Only those chemical elements and limits, if any, which are significant in defining such materials need be reported.

**5.4** Requests for new numbers shall be submitted on "Application for UNS Number Assignment" forms (Figure 1). Copies of these are available from any UNS number assigning office (Table 3) or facsimiles may be made of the one herein.

**5.5** All instructions on the printed application form should be read carefully and all information provided as indicated. (See 6.5.)

**5.6** To further assist in assigning UNS numbers, the requester is encouraged to suggest a possible UNS number in each request, giving appropriate consideration to any existing number presently used by a trade association, standards society, producer, or user.

**5.7** Each completed application form shall be sent to the UNS number assigning office having responsibility for the series of numbers which appears to most closely relate to the material described on the form (Table 3).

## 6. Notes

**6.1** The terms "commercial standing," "production usage," and others, are intended to portray a material in active industrial use, although the actual amount of such use will depend, among other things, upon the type of materials. (Obviously gold will not be used in the same "tonnages" as hot rolled steel.)

Different standardizing groups use different criteria to define the status that a material has to attain before a standard number will be assigned to it. For instance, the American Iron and Steel Institute requires for stainless steels "two or more producers with combined production of 200 tons per year for at least two years"; the Copper Development Association requires that the material be "in commercial use (without tonnage limits)"; the Aluminum Association requires that the alloy must be "offered for sale (not necessarily in commercial use)"; the SAE Aerospace Materials Division calls for "repetitive procurement by at least two users."

While it is apparent that no hard and fast usage definition can be set up for an all-encompassing system, the UNS numbers are intended to identify metals and alloys that are in more or less regular production and use.

A UNS number will not ordinarily be issued for a material which has just been conceived or which is still in only experimental trial.

**6.2** Organizations that issue specifications should report to appropriate UNS number assigning offices (see 4.1.2) any specification changes which affect descriptions shown in published UNS listings.

**6.3** This arrangement of alphanumeric six character numbers is a compromise of the thinking that identification numbers should indicate many characteristics of the material, and the thinking that numbers should be short and uncomplicated to be widely accepted and used.

**6.4** In assigning UNS numbers, and consequently in searching complete listings of numbers, the predominant element of the metal or alloy usually determines the prefix letter of the series to which it is assigned. In certain instances where no one element predominates, arbitrary decisions are made as to what prefix letter to use, depending upon the producing industry and other factors.

**6.5** The application form is designed to serve also as a data input sheet to facilitate processing each request through to final printout of the data on electronic data processing equipment and to minimize transcription errors at number-assigning offices and data processing centers.

**6.6** One such listing is ASTM Publication No. DS-56 and SAE Handbook Supplement HS-1086 (a joint ASTM-SAE publication).

**6.7** **Marginal Indicia**—The change bar (l) located in the left margin is for the convenience of the user in locating areas where technical revisions have been made to the previous issue of the report. An (R) symbol to the left of the document title indicates a complete revision of the report.

PREPARED BY THE SAE UNIFIED NUMBERING SYSTEM ADVISORY BOARD

**APPLICATION FOR UNS NUMBER ASSIGNMENT**
**and**
**Data Input Sheet for Entering a Specific Material in the**
**SAE-ASTM Unified Numbering System for Metals and Alloys**
(See Reverse Side for Instructions for Completing This Form)

Material Description ______________________________

______________________________

______________________________

______________________________ Suggested UNS No. ______________

*UNS Assigned Description ______________________________

______________________________

______________________________ *UNS Assigned No. ______________

*Chemical Composition (percent by wt.)

| | | | | | | | | |
|---|---|---|---|---|---|---|---|---|
| Silver | Ag | ______ | Hafnium | Hf | ______ | Sulfur | S | ______ |
| Aluminum | Al | ______ | Mercury | Hg | ______ | Antimony | Sb | ______ |
| Arsenic | As | ______ | Indium | In | ______ | Selenium | Se | ______ |
| Gold | Au | ______ | Iridium | Ir | ______ | Silicon | Si | ______ |
| Boron | B | ______ | Lithium | Li | ______ | Tin | Sn | ______ |
| Beryllium | Be | ______ | Magnesium | Mg | ______ | Tantalum | Ta | ______ |
| Bismuth | Bi | ______ | Manganese | Mn | ______ | Tellurium | Te | ______ |
| Carbon | C | ______ | Molybdenum | Mo | ______ | Thorium | Th | ______ |
| Columbium | Cb | ______ | Nitrogen | N | ______ | Titanium | Ti | ______ |
| Cadmium | Cd | ______ | Nickel | Ni | ______ | Uranium | U | ______ |
| Cobalt | Co | ______ | Oxygen | O | ______ | Vanadium | V | ______ |
| Chromium | Cr | ______ | Phosphorus | P | ______ | Tungsten | W | ______ |
| Copper | Cu | ______ | Lead | Pb | ______ | Zinc | Zn | ______ |
| Iron | Fe | ______ | Platinum | Pt | ______ | Zirconium | Zr | ______ |
| Germanium | Ge | ______ | Rhenium | Re | ______ | | | |
| Hydrogen | H | ______ | Rhodium | Rh | ______ | | | |

Other ______________________________

______________________________

______________________________

*Cross References

AA ______________________________
ACI ______________________________
AISI ______________________________
AMS ______________________________
ANSI ______________________________
ASME ______________________________
ASTM ______________________________
AWS ______________________________
CDA ______________________________
FED ______________________________
MIL SPEC ______________________________
SAE ______________________________
OTHER ______________________________

Requesting Person and Organization (full address) ______________________________

______________________________ Date of Request ______________

*Assigning Org. ______________________________ *Date of UNS Assignment ______________
Assigner's Name and Office ______________________________

Applicant: DO NOT write in shaded areas.

* These items for Computer Operator

RULES, NOMENCLATURE AND UNS
APPLICATION FORM FOR UNS NUMBER ASSIGNMENT

## GENERAL

Before attempting to complete this form, the applicant should be thoroughly familiar with the objectives of the UNS and the ground rules for assigning numbers, as stated in Section 4 of SAE J1086 and ASTM E 527.

## MATERIAL DESCRIPTION

Identify: the base element; the single alloying element that constitutes 50% or more of the total alloy content; other distinguishing predominant characteristics (such as "casting"); and common or generic names if any (such as "ounce metal" or "Waspalloy"). When no single element makes up 50% or more of the total alloy content, list in decreasing order of abundance the two alloying elements which together constitute the largest portion of the total alloy content; except that if no two elements make up at least 50% of the total alloy content, list the three most abundant, and so on. Instead of "iron," use "steel" to identify the base element of those iron-low-carbon alloys commonly known as steels.

When mechanical properties or physical characteristics are the primary defining criteria and chemical composition is secondary or nonsignificant, enter such properties and characteristics with the appropriate values or limits for each.

## SUGGESTED UNS NO.

While applicant's suggestion may or may not be the one finally assigned, it will assist proper identification of the material by the UNS Number Assigner.

## CHEMICAL COMPOSITION

Enter limits such as 0.13-0.18 (not .13-.18 or 0.13 to 0.18), 1.5 max, 0.040 min, and balance. In space designated "other," enter information such as "0.05 max each, 0.15 max total" and "Sn+Pb 2.0 min." Additional specific elements not included in the list on this form may be entered in the spaces provided at the end of the list.

## CROSS REFERENCES

Letter symbols listed indicate widely known trade associations and standards issuing organizations. Enter after appropriate symbols any known specification numbers or identification numbers issued by such groups to cover material equivalent to, similar to, or closely resembling the subject material.

Examples: SAE J404 (50B44), AISI 415, ASTM A638 (660)

In space designated "other" enter any pertinent numbers issued by groups not listed above. In these instances, the full name and address of the issuing group shall be included.

**SUBMIT COMPLETED FORM TO**
**SAE**
**PRODUCT DEVELOPMENT**
**400 COMMONWEALTH DRIVE**
**WARRENDALE, PA 15096-0001**
**ATTENTION: UNS**

## ADOPTION NOTICE

SAE-HS1086, "Metals and Alloys in the Unified Numbering System," was adopted on 02-MAR-93 for use by the Department of Defense (DoD). Proposed changes by DoD activities must be submitted to the DoD Adopting Activity: Department of the Army, US Army Research Laboratory Materials Directorate, ATTN: AMSRL-MA-S, Watertown, MA 02172-0001. DoD activities may obtain copies of this standard from the Standardization Document Order Desk, 700 Robbins Avenue, Building 4D, Philadelphia, PA 19111-5094. The private sector and other Government agencies may purchase copies from the Society of Automotive Engineers, Inc. (SAE), 400 Commonwealth Drive, Warrendale, PA 15096.

**Custodians:**
Army - MR
Air Force - 99

**Adopting Activities:**
Army - MR
(Project 95GP-0400)

**Reviewer Activities:**
Army - AR
Nave - OS, SH
Air Force - 84
DLA - IS

**FSC 95GP**

**DISTRIBUTION STATEMENT A**: Approved for public release; distribution is unlimited.